Astronomy Today

ABOUT THE AUTHORS

Eric Chaisson Eric holds a doctorate in Astrophysics from Harvard University, where he spent ten years on the faculty of Arts and Sciences. For five years, Eric was Senior Scientist and Director of Educational Programs at the Space Telescope Science Institute and Adjunct Professor of Physics at Johns Hopkins University. He recently joined Tufts University, where he is now Professor of Physics, Professor of Education, and Director of the Dudley Wright Center for Innovative Science Education. He has written eight books on astronomy, which have received such literary awards as the Phi Beta Kappa Prize, the American Institute of Physics Award, and Harvard's Smith Prize for Literary Merit. He has published more than 100 scientific papers in professional journals, and has also received Harvard's Bok Prize for original contributions to astrophysics.

Steve McMillan Steve holds a bachelor's and master's degree in Mathematics from Cambridge University and a doctorate in Astronomy from Harvard University. He held post-doctoral positions at the University of Illinois and Northwestern University, where he continued his research in theoretical astrophysics, star clusters, and numerical modeling. He serves regularly as a visiting scholar at Princeton's Institute for Advanced Study, Cambridge's Institute of Astronomy, and the University of Tokyo. He is currently an Associate Professor of Physics at Drexel University. He has published over 30 scientific papers in professional journals. This is Steve's first book.

Astronomy Today

Eric Chaisson
Tufts University

Steve McMillan
Drexel University

Prentice Hall
Englewood Cliffs, NJ 07632

Library of Congress Cataloging-in-Publication Data

Chaisson, Eric.
Astronomy today/Eric Chaisson, Steve McMillan.
p. cm.
Includes index.
ISBN 0-13-050824-1
1. Astronomy. I. McMillan, S. (Stephen). II. Title.
QB43.2.C44 1993
520—dc20 92-37424
CIP

Acquisition Editor: Ray Henderson
Editor in Chief: Tim Bozik
Development Editor: Stephen Deitmer
Production Text Editor: Ed Thomas
Production Art Editor: Jennifer Fischer
Marketing Manager: Gary June
Copy Editor: Margo Quinto
Designer: Judith A. Matz-Coniglio
Cover Designer: Bill McClosky
Prepress Buyer: Paula Massenaro
Manufacturing Buyer: Lori Bulwin
Editorial Assistant: Joan Dellostritto
Design Director: Florence Dara Silverman
Text Compositor: York Graphics
Art Studio: Network Graphics
Page Make-up: Judith A. Matz-Coniglio
Photo Editor: Lorinda Morris-Nantz
Photo Researcher: Anita Dickhuth

Printed in the United States of America
10 9 8 7 6 5 4 3

ISBN 0-13-050824-1

Prentice-Hall International (UK) Limited, *London*
Prentice-Hall of Australia Pty. Limited, *Sydney*
Prentice-Hall Canada Inc., *Toronto*
Prentice-Hall Hispanoamericana, S.A., *Mexico*
Prentice-Hall of India Private Limited, *New Delhi*
Prentice-Hall of Japan, Inc., *Tokyo*
Simon & Schuster Asia Pte. Ltd., *Singapore*
Editora Prentice-Hall do Brasil, Ltda., *Rio de Janeiro*

Brief Contents

Contents

5

Spectroscopy: *The Inner Workings of Atoms* 84

6

Telescopes: *Observational Methods of Astronomy* 106

7

The Earth: *Our Home in Space* 138

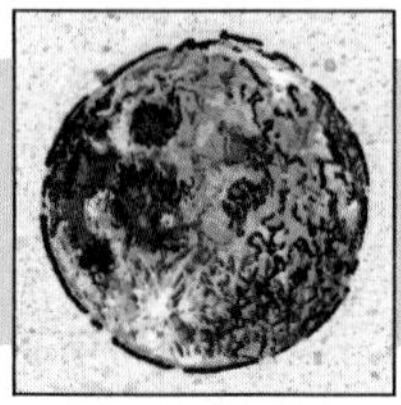

8

The Moon: *Our Nearest Neighbor* 169

9

The Solar System: *An Introduction to Comparitive Planetology* 189

10

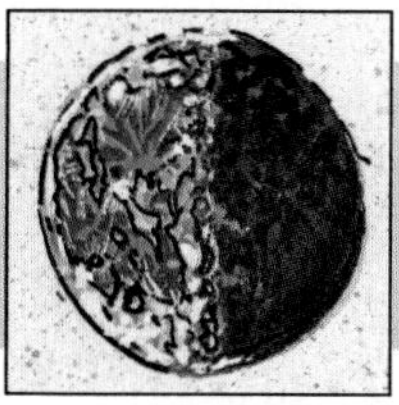

Mercury: *A Scorched and Battered World* 202

11

Venus: *Earth's Sister Planet* 217

12

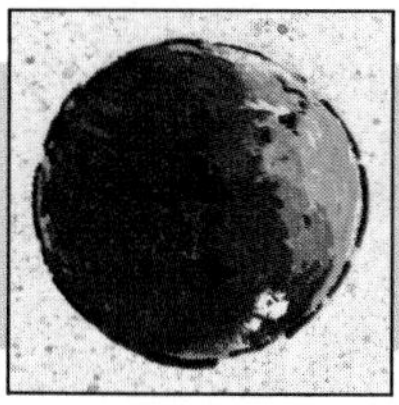

Mars: *A Near Miss for Life?* 241

13

Jupiter: *The Giant of the Solar System* 260

14

Saturn: *Spectacular Rings and Mysterious Moons* 282

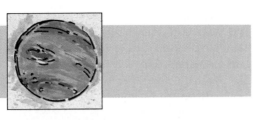

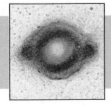

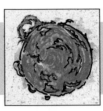

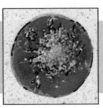

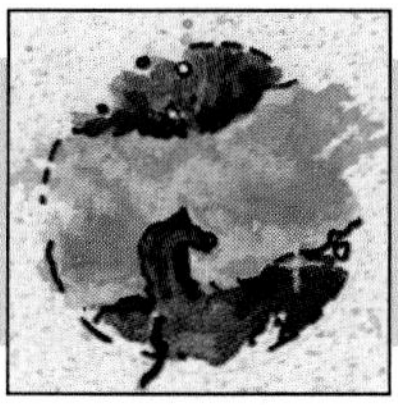

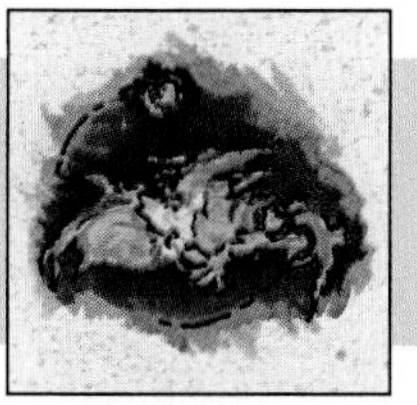

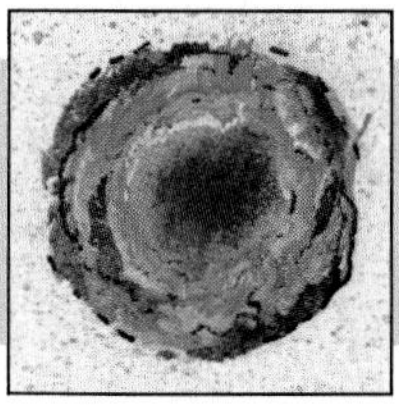

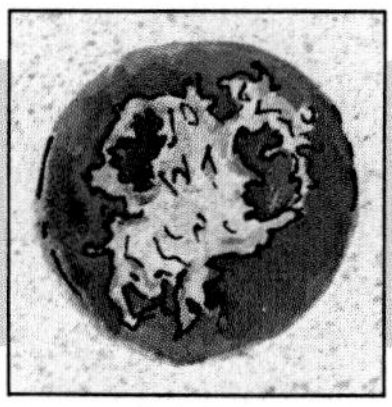

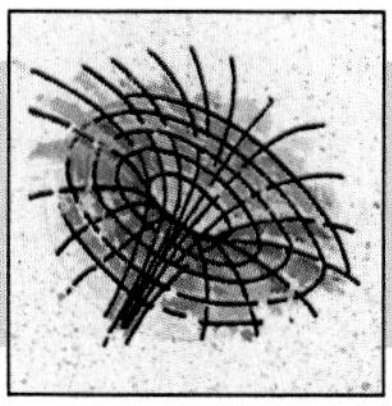

25

The Milky Way Galaxy: *A Grand Design* 520

26

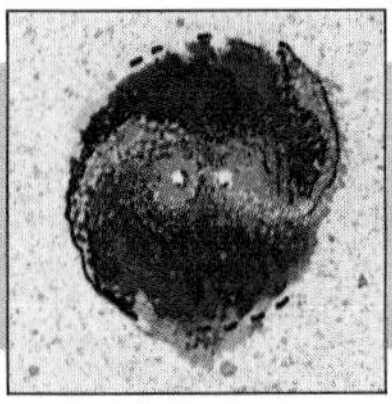

Normal Galaxies: *The Local Group and Beyond* 543

27

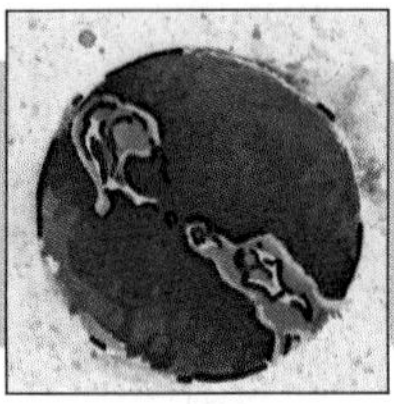

Active Galaxies and Quasars: *Limits of the Observable Universe* 566

28

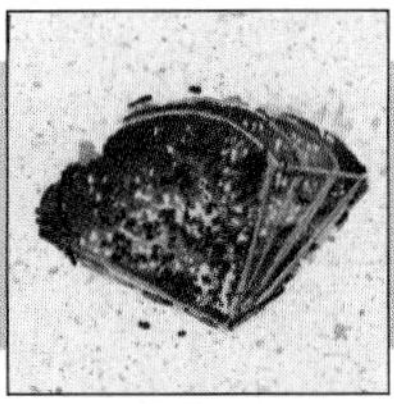

Cosmology: *The Big Bang and the Fate of the Universe* 591

29

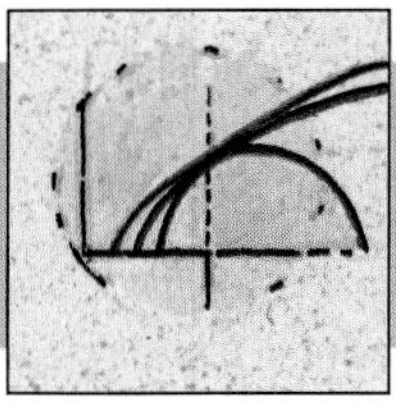

30

To the Student

We hope that you will enjoy learning about astronomy. As you go through this course, you will encounter new terms, as you would in any other subject. Important astronomy terms appear in **boldface** type when they are introduced. At the end of the book you will find a glossary, in which all key terms are defined. Use the glossary as a reference if you need to refamiliarize yourself with astronomy terms as you go through the course.

At the start of each chapter we present learning goals, which summarize the essential ideas we would like you to understand as you read the chapter. When we reach the first point in the text that touches on a given goal, a check mark appears. The check mark will help you locate each goal quickly as you review material.

In astronomy, not all photographs are taken in visible light (as we will discuss in Chapters 4 and 6). We have attached a marker to most images we present, indicating what kind of radiation was used in making the images. For example, a highlighted V indicates that an image was taken with visible light. A highlighted X indicates that X-rays were used. The different kinds of radiation that exist, along with their use and significance, will become clear once you have read the introductory chapters.)

At the end of each chapter we present Review Questions, which we strongly urge you to try to answer. Some of these questions require some calculations (although most do not) and are marked with a ''calculator'' symbol. We also offer Discussion Questions, Projects, and Suggested Readings at the end of each chapter.

Learning astronomy also means learning the metric system. Astronomers and nearly every other scientist across the globe use the metric system of measurement. You may wish to use the information that follows to ''translate'' from the English system of measurement—which uses inches, feet, miles, pounds, and other terms generally familiar to people in the United States—to the metric system. We hope that as you read this book you will grow familiar enough with the metric system that conversion of measurement units becomes unnecessary.

English	Metric
1 inch	= 2.54 centimeters (cm)
1 foot (ft)	= 0.3048 meters (m)
1 mile	= 1.609 kilometers (km)

To convert a measurement of, say, 250 kilometers into miles, we would consult the conversion table and perform this calculation:

$$250 \text{ km} = 250 \text{ km} \times (1 \text{ mile}/1.609 \text{ km}) = 155.38 \text{ mile}.$$

To convert a measurement of 300 feet into meters, we would carry through this operation:

$$300 \text{ ft} = 300 \text{ ft} \times (0.3048 \text{ m}/1 \text{ ft}) = 91.44 \text{ m}.$$

You will often read measurements of speed, or velocity, given in terms of kilometers per second (km/s). You may be more familiar with measurements given in miles per hour (mph). The following equality links these two measures:

$$1 \text{ km/s} = 2237 \text{ mph}.$$

Thus a speed of 150 km/s is equivalent to

$$150 \text{ km/s} = 150 \text{ km/s} \times (2237 \text{ mph}/1 \text{ km/s}) = 335{,}500 \text{ mph}.$$

In science it is necessary to draw a very clear distinction between mass, which is the amount of matter contained within a body, and weight, which is the force with which the Earth (or some other object) attracts it. A body's mass is the same no matter where we measure it. Its weight, however, varies. The weight of any body would be lower on the Moon (because the Moon's gravity is weaker than the Earth's) and greater near Jupiter (because that planet's gravity is stronger). In the metric system, the unit of mass is the gram or often the kilogram. (The unit of weight is the same as the unit of force, called a newton, as we discuss in Chapter 3.) In the English system the pound is actually a unit of weight, although most people use it more or less as both a unit of weight and a unit of mass. As long as we stay on Earth, weight and mass are equivalent.

1 pound (lb)
= 453.6 grams (g) or .4536 kilograms (kg) (on Earth)

So how much does a 55-kg object weigh, in pounds, on Earth?

$$55 \text{ kg} = 55 \text{ kg} \times (1 \text{ lb}/.4536 \text{ kg}) = 121.25 \text{ lb}$$

At the end of the book we provide appendices that summarize information on planets, the Sun, and selected stars and galaxies. We also present star charts, which you can use to find your way around the sky, which we now turn to describe.

To the Instructor

Astronomy Today has been written for students who have taken no previous college science courses and who will likely not major in physics or astronomy. This book is suitable for one-semester and two-semester courses. We present a broad view of astronomy, straightforwardly descriptive and without complex mathematics. The absence of sophisticated mathematics in no way prevents discussion of important conceptual issues. We rely on analogies to processes and objects familiar to the student to explain the complexities of our field without gross oversimplification. Whenever possible, we emphasize the process of science, focusing on the actual experiments and the observations that have expanded our knowledge of the field. Generally, we have tried to impart the excitement we feel about astronomy and to awaken students to the marvelous universe that surrounds us.

We feel that it is important for authors of introductory books to be active in research and in teaching. Between us we have published over 100 papers in scientific journals and have introduced thousands of students to astronomy. In our classrooms we have used headline events and new developments in the field to motivate students, but we have also grounded our teaching in solid coverage of the fundamentals. An important goal that we have pursued throughout this project has been to prepare students for their future participation in the policy-making process relating to science. Often our discussions focus on points of general scientific literacy that students must grasp to understand national initiatives and the public debate over scientific funding.

Outline. Our overall organization follows the traditional but effective ''Earth-out'' progression. We have found that most students, especially those with little or no scientific background, are much more comfortable learning first about the more familiar solar system objects before tackling stars and galaxies. Thus, the Earth is the first object we discuss in detail. With the Earth providing the planetary blueprint, we move through the solar system, drawing on comparative planetology to provide rich explanations. We conclude coverage of the solar system with a discussion of its formation, which leads us to the Sun. With the Sun as our model star, we broaden the scope of our discussion to include stars in general. This journey naturally leads us to coverage of the Milky Way Galaxy, which then serves as a benchmark for our treatment of other galaxies. We then reach the large-scale structure of the universe.

We have placed most of the physics in the first few chapters, although additional physics material is introduced as needed later, both in the text narrative and in the boxed features we call Interludes. We realize that these introductory chapters may not capture students' imaginations in the way that later chapters do, so we have attempted to keep the material modular. You can insert it at later stages in your course as you desire.

Illustrations. Even the least sophisticated students in our classes have come to expect dazzling imagery from an astronomy textbook, and we have aimed to produce an aesthetically beautiful art program. Still, our most important criteria in developing the illustration program were pedagogical soundness, scientific accuracy, and tight coordination with the text. The wizardry of Network Graphics, with help from the Prentice Hall developmental and production staffs, has produced the most visually compelling text we have ever seen, and each piece is carefully crafted to enhance student learning.

Our photograph program is similarly visually compelling. It would not have been possible without the contributions of images and ideas from hundreds of colleagues. We have frequently—and intentionally—blurred the distinction between art and photograph. You will find many composite illustrations that combine these visual media, and photographs are often enhanced with labels. Our aim has been to use the most creative means possible to capture what can sometimes be elusive points of science. We hope that you find these pieces a particular advantage in your classroom. Color acetates are included in our supplements package.

We have found that students often review a chapter by ''looking at the pictures.'' In fact, we encourage this in our own classrooms. This in mind, you will find that our captions are often a bit longer and more detailed than those in other texts.

Chapter objectives ✓1 . Studies indicate that beginning students often have difficulty prioritizing textual material. We address this concern by providing a few well-defined learning objectives at the start of each chapter. If students have met these objectives after reading the chapter, they have learned at least the most important material. Learning objectives are numbered. Within the chapter, each learning objective is checked off in red next to the text that covers that objective. This in-text highlighting of the most important aspects of the chapter also helps students review.

Interlude boxes. Throughout *Astronomy Today* numerous Interlude boxes appear. These boxes provide expanded coverage of selected topics within the text or interesting asides. To some extent, they may be considered optional material, although we very much hope that the instructor will cover them in lectures or assign them as reading and that students will find them interesting. The material they present adds depth to the book's coverage.

Spectrum icons [**R** I V U X G]. A unique feature of our photographic program is the inclusion of icons that identify the radiation used to capture the images. The majority of our images are made in visible light, but it is sometimes quite hard (even for a professional) to tell at a glance which images are visible-light photographs and which are false-color images made in some other region of the electromagnetic spectrum. The spectrum icon is a simple version of the electromagnetic spectrum. The icon shows a boldfaced **R** to signal an image made in radio, an **I** to identify an image made in infrared, and so on with **V**isible, **U**ltraviolet, **X**-rays, and **G**amma rays. This feature is intrinsically interesting, and seeing that images are captured in different wavelengths emphasizes to students how astronomy relies on the full range of the spectrum to gather information from the cosmos.

End-of-chapter pedagogy. Each chapter ends with a comprehensive Chapter Summary, a very useful review tool for students. A list of Key Words follows. This list contains all the chapter's important terms, which are boldfaced on their first use. A full glossary appears at the end of the book. There each term is defined, and the page number identifying where the term first appears is also provided.

Review Questions, about twenty per chapter, may be used in-class or for assignment. The last few Review Questions involve some actual numerical calculation; they are identified by a calculator icon. Discussion Questions explore particular topics more deeply than do Review Questions. They often extend student investigation into important topics by asking for opinion, not just fact, and they have no single "correct" answer. Next, most Projects aim to get the student out of the classroom and under the sky, but some entail research, in libraries or through magazine articles. The chapter concludes with an annotated selection of Suggested Readings.

Supplementary material. *Astronomy Today* is accompanied by a truly remarkable set of instructional aids. Our COMETS supplement is a unique media subscription for adopting instructors. Twice annually, adopting professors will receive a coordinated kit with updates for their course materials, including timely software simulations from Dance of the Planets, video highlights from ABC News, a special edition of *The New York Times* on Astronomy (in the quantity needed for your classroom), audio highlights keyed to the text from the nationally syndicated program "Earth and Sky Radio," and 25 additional slides to supplement your course materials with late-breaking imagery. Also, the COMETS newsletter will provide instructional comments on integrating these items into your course.

In addition, an extensive set of color acetates and a comprehensive 35mm slide set are available at the time of adoption. Testing materials in both hardcopy and electronic form are offered free upon adoption (in both Macintosh and IBM-compatible formats). More information about these materials is available from the publisher.

Acknowledgments. Throughout the many drafts that have led to this book, we have relied on the critical analysis of many people. Their recommendations ranged from the macroscopic issue of the book's overall organization to the micro—but no less important—points of each statement's technical accuracy. We offer them all our sincerest thanks.

Robert Allen, *University of Wisconsin-LaCrosse*
Jeff E. Arrington, *Abilene Christian University*
Dr. Harry Ashworth, *Seton Hall University*
Edgar A. Bering III, *University of Houston*
Ronald J. Bieniek, *University of Missouri-Rolla*
Patricia Boyd, *Drexel University*
G. O. Brink, *SUNY-Buffalo*
David Burstein, *Arizona State University*
Bruce Carney, *University of North Carolina-Chapel Hill*
John Cowan, *Columbia University*
James N. Douglas, *University of Texas-Austin*
Robert A. Egler, *North Carolina State University*
Roger Freedman, *University of California-Santa Barbara*
Michael Friedlander, *Washington University*
Solomon Gartenhaus, *Purdue University*
Chris Godfrey, *Missouri Western State College*
Paul Goldsmith, *University of Massachusetts-Amherst*
Austin Gulliver, *Brandon University*
D. Bruce Hanna, *Old Dominion University*
John Hoessel, *University of Wisconsin-Madison*
Darrel Hoff, *Harvard College Observatory*
John K. Lawrence, *California State University-Northridge*
Michael Lieber, *University of Arkansas*
James LoPresto, *Edinboro University of Pennsylvania*
Marie E. Machacek, *Northeastern University*
Ellis W. Miller, *Ricks College*
Dr. Tom Mote, *St. Mary's University*
Charles L. Perry, *Louisiana State University*

Charles J. Peterson, *University of Missouri-Columbia*
James Roberts, *University of North Texas*
Edward G. Schmidt, *University of Nebraska-Lincoln*
Richard L. Sears, *University of Michigan*
Marc Sher, *College of William & Mary*
Stephen R. Walton, *California State University-Northridge*
Louis Winkler, *Pennsylvania State University*

The team at Prentice Hall assisted us at every step along the way. Our thanks to Steve Deitmer, our Developmental Editor, who read and revised numerous drafts of the manuscript. He was a constant source of knowledge and inspiration in this effort. Our production staff performed admirably under the strain of producing a book with this level of complexity. Ed Thomas handled the text manuscript efficiently and with tremendous attention to detail. Jennifer Fischer skillfully managed the illustrations and shepherded our crude sketches through the iterations required to produce the pleasing finished work. Judy Matz-Coniglio and Florence Silverman designed (and frequently redesigned) the book to accommodate our evolving needs. Gary June and Jeanne Ambrosio have shown their proficiency and class in marketing and advertising efforts. We are indebted to Ray Henderson, our Acquisitions Editor, for securing the substantial investment and human resources required to produce a book of this quality and for managing the other parties that participated along the way. We look forward to working with them all on future editions of the text.

For more information about the text or supplements, please contact your local Prentice Hall representative, or write to:

Ray Henderson
Astronomy Editor
Prentice Hall
Englewood Cliffs, NJ 07632

CHAPTER

1

Introduction

The Sky at Night

Lunar Earthrise. The Earth is here majestically pictured, nearly 400,000 kilometers away from the Moon, a portion of which appears at the bottom. This stunning image was captured by an astronaut during one of the Apollo *missions to the Moon during the early 1970s. This one picture, perhaps more than any other, well illustrates the fragile nature of our home in space. (Courtesy NASA)*

R I **V** U X G

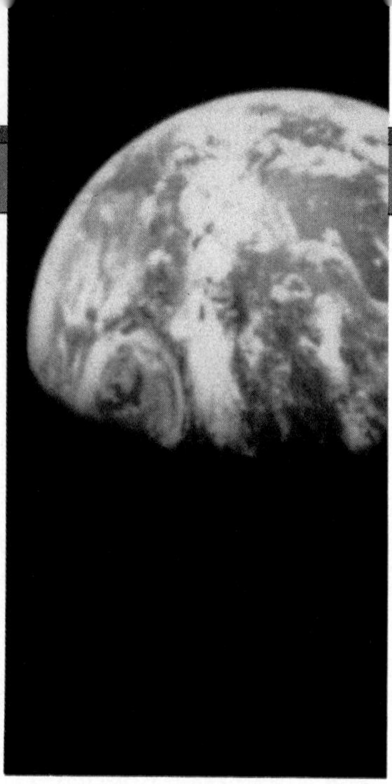

LEARNING GOALS

1 to appreciate the age-old quest to understand our world and the universe we live in

2 to understand the use of scientific notation, astronomical measurements and terminology, and angular measure

3 to understand the notion of the celestial sphere and the system of celestial coordinates on the sky

4 to understand how the Sun, the Moon, and the stars appear to change their positions from night to night and from month to month

5 to appreciate how our notions of timekeeping and seasons are tied to the changing nighttime sky

6 to recognize how the relative motions of the Earth, the Sun, and the Moon lead to eclipses

Nature offers no greater splendor than the starry sky on a clear, dark night. Silent, jeweled with the constellations of ancient myth and legend, the night sky has inspired wonder throughout the ages—a wonder that leads our imaginations far from the confines of Earth and the pace of the present day and out into the distant reaches of space and cosmic time itself.

Astronomy, born in response to that wonder, is built on two of the most basic traits of human nature: the need to explore and the need to understand. Through the interplay of curiosity, discovery, and analysis—the keys to exploration and understanding—people have sought answers to questions about the universe since the earliest times. Astronomy is the oldest of all the sciences, and never has it been more exciting than it is today.

In all of history, there have been only two periods in which our understanding of the universe has been revolutionized within a single human lifetime. The first occurred nearly four centuries ago in the time of Galileo—the second is now under way. Our generation has broken away from planet Earth and in doing so has achieved a whole new perspective on the universe in which we live.

OUR PLACE IN SPACE

✓1 Of all the science accomplished to date, one result stands out most boldly: Earth is neither central nor special. We inhabit no unique place in the universe. Astronomical research, especially within the past few decades, strongly suggests that we live on what seems to be an ordinary rocky *planet* called Earth, one planet orbiting an average *star* called the Sun, one star near the edge of a huge collection of stars called the Milky Way *galaxy,* one galaxy among countless billions of others spread throughout the observable *universe*.

We are connected to these distant realms of space and time not only by our imaginations, but also through a common cosmic heritage: Most of the chemical elements in our bodies were created billions of years ago in the hot centers of long-vanished stars. Their fuel supply spent, these giant stars died in huge explosions, scattering afar the elements cooked deep within their cores. Eventually, this matter collected into clouds of gas that slowly collapsed to give birth to a new generation of stars. In this way, the Sun and its family of planets were formed nearly five billion years ago. Everything on Earth embodies atoms from other parts of the universe and from a past far more remote than the beginning of human evolution.

Although ours is the only planetary system we know of, others may orbit many of the billions upon billions of stars in the universe. Elsewhere, other beings, perhaps with an intelligence much greater than our own, might at this very moment be gazing in wonder at their own nighttime sky. Our own Sun might be nothing more than an insignificant point of light to them, if it is visible at all. Nevertheless, if such beings exist, they too must share our cosmic origin.

We are at the dawn of a new age in space science. Astronomy has evolved a long way from the popular notion of eccentric intellectuals spending long, cold nights on distant mountaintops, peering through telescope tubes to count the stars and catalog their properties. Astronomers still make great use of ground-based telescopes to collect information, but today the range of their efforts is greatly expanded by the Hubble Space Telescope and other sophisticated orbiting instruments from many countries. Together, they gather increasingly more comprehensive data on a vibrant, changing universe.

OUR UNDISTINGUISHED HOME

Modern science has arrived at a view diametrically opposite to that held by the ancient philosophers who believed that the Earth in general, and humankind in particular, were absolutely central to the workings of the universe. By contrast, our present-day outlook is that the Earth, the solar system—and, some would argue, humanity—are ordinary in every way. This idea is often (and only half-jokingly) called the "principle of mediocrity," and it is deeply embedded in modern scientific thought. Nowadays, any theory or observation that even *appears* to single out the Earth, the solar system, or the Milky Way Galaxy as in some way special is immediately regarded with great suspicion, and sometimes even outright hostility, in scientific circles.

The principle of mediocrity extends far beyond a mere statement of philosophical preference, however. Simply put, if we do not make this hypothesis, then we cannot make much headway in science, and we cannot do astronomy at all. Virtually every statement made in this text (and all other astronomy books, for that matter) rests squarely on the premise that the laws of physics, as we know them here on Earth, apply everywhere else too, without modification and without exception. If the principle of mediocrity were not valid, then we could not use our knowledge of optics and atomic structure to interpret the light we receive from distant stars, we could not use the familiar laws of mechanics and thermodynamics to understand what we see, and we could not hope to apply our familiar terrestrial science to the conditions found elsewhere in the universe.

Happily, there is no reason to believe that we are the victims of such a cruel cosmic conspiracy. The universe appears to be knowable in terms of terrestrial physics. This does *not* mean, however, that we do not check our assumptions whenever and wherever possible. The **scientific method,** which we will discuss in more detail in Chapter 2, demands that the principle of mediocrity, like all other scientific principles, be continually *tested* and replaced if found inadequate. At present, there is no evidence that it needs modification.

THE UNIVERSE AROUND US

Before going any further, we should perhaps clarify just what we mean by "the universe." We might define it poetically as the vast tracts of space and enormous stretches of time populated sparsely by stars and galaxies glowing in the dark. More scientifically, the **universe** is the totality of all space, time, matter, and energy. Consult Figures 1.1 through 1.4, and put some of these objects in perspective by studying Figure 1.5.

R I **V** U X G

Figure 1.1 *Earth is a planet, a mostly solid and molten object, though it has some liquid in its oceans and gas in its atmosphere. (You can see here the African continent and parts of the Middle East.) (NASA)*

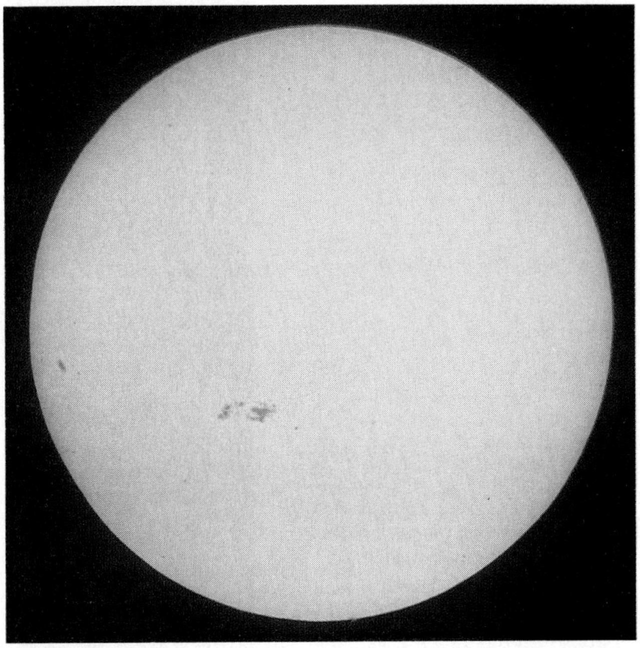

R I **V** U X G

Figure 1.2 *The Sun is a star, a very hot ball of gas, much bigger than Earth. It is held together by its own gravity. (AURA)*

Figure 1.3 A typical galaxy is a collection of a hundred billion stars, each separated by vast regions of nearly empty space. This galaxy is the one-thousand five-hundred sixty-sixth entry in the New General Catalog, called NGC 1566 for short. Our Sun is a rather undistinguished star near the edge of another such galaxy called the Milky Way. (AURA)

Figure 1.4 This photograph shows a cluster of several galaxies far away in the universe. Each contains hundreds of billions of stars, probably planets, and possibly living creatures. (AURA)

Figure 1.5 This artist's conception puts each of the previous figures in perspective. The bottom of this figure shows spacecraft (and astronauts) in Earth orbit, a view that widens progressively in each of the next five cubes drawn from bottom to top—Earth, the planetary system, the local neighborhood of stars, the Milky Way Galaxy, and the closest cluster of galaxies. (D. Berry)

THE SCALE OF THINGS

☐2 Take another look at the galaxy in Figure 1.3. This galaxy, whose catalog name is NGC 1566, is a swarm of about a hundred billion stars—more stars than people who have ever lived on Earth. The entire assemblage is spread across some 100,000 light-years. A **light-year** is the *distance* traveled by light, at a velocity of about 300,000 kilometers per second, in a year. It equals about 10 trillion kilometers (or around 6 trillion miles). Typical galactic systems are truly "astronomical" in size.

A thousand (1000), a million (1,000,000), a billion (1,000,000,000), and even a trillion (1,000,000,000,000)—these words occur regularly in everyday speech. (Note that we use *billion* in the American sense to mean a "thousand million.") But let's take a moment to understand the magnitude of these numbers and to appreciate the differences among them. One thousand is easy enough to understand; at the rate of one number per second, you could count to a thousand in about 16 minutes. However, if you wanted to count to a million, you would need more than 2 weeks of counting at the rate of one number per second, 16 hours per day (allowing 8 hours per day for sleep). To count from 1 to a billion at the same rate of one number per second and 16 hours per day would take nearly 50 years.

Throughout our text we will consider spatial domains spanning not just billions of kilometers but also billions of light-years. We will discuss objects containing not just trillions of atoms but also trillions of stars. We will contemplate time intervals of not just billions of seconds or hours but billions of years. Become accustomed to gargantuan numbers of things, enormous intervals of space, and extremely long durations of time. Learn to recognize just how much larger than a thousand is a million, and how much larger still is a billion.

THE OBVIOUS VIEW

How have we come to know the universe around us? How do we know the proper perspective sketched in Figure 1.5? Our study of the universe, the subject of **astronomy,** begins by examining the sky.

Over the course of a clear night, we can see some 3000 points of light. Include the view from the opposite side of Earth and nearly 6000 stars are visible to the naked eye. A natural human tendency is to see patterns, and so people connected the brightest stars into configurations called **constellations,** which ancient astronomers named after mythological beings, heroes, and animals—whatever was important to them. Figure 1.6 shows a constellation especially prominent in the nighttime sky from October through March: the "hunter" named Orion. Orion was a mythical

INTERLUDE 1-1 *Scientific Notation*

The work of astronomers ranges from the smallest particles to the largest possible object—the universe itself. Subatomic particles have sizes of about 0.0000000000001 centimeter, and galaxies typically measure approximately 100,000,000,000,000,000,000,000 centimeters. The most distant known objects in the universe lie on the order of 10,000,000,000,000,000,000,000,000,000 centimeters from Earth. To avoid writing so many zeros, scientists use a shorthand notation in which the number of zeros is denoted by a superscript power, or *exponent*, of 10. This superscript is negative if the number is smaller than unity (1) and the zeros lie to the right of the decimal point; the superscript is positive if the number exceeds unity and the zeros lie to the left of the decimal.

For example, we can shorten the number describing subatomic particles to 10^{-13} centimeter and write the number describing galaxies as 10^{23} centimeters. In this way we avoid carrying around the excess baggage of all those zeros. We also lessen the chance of making an error by writing one too many or one too few. More complicated numbers are expressed as a combination of a power of 10 and a multiplying factor: for example, 0.000000025 centimeters can be more concisely written as 2.5×10^{-8} centimeters.

Some other examples of scientific notation are:

- the approximate distance to the Andromeda Galaxy = 2,000,000 light-years = 2×10^6 light-years
- the size of a hydrogen atom = 0.000000005 centimeters = 5×10^{-9} centimeters
- the diameter of the Sun = 1,392,000 kilometers = 1.392×10^9 meters (recall that 1 kilometer = 10^3 meters)
- 2 billion, 546 million, and 2 hundred thousand dollars = 2.5462×10^9 dollars.

The rule for multiplication (or division) of numbers expressed in this way is simple: Just multiply (or divide) the factors and add (or subtract) the exponents. Thus, 200,000 centimeters is $2 \times 10^5 \times 10^{-2}$ meters, or 2×10^3 meters = 2 kilometers. Similarly, 5×10^6 divided by 2×10^4 is simply $(5/2) \times 10^{6-4}$, or 2.5×10^2. Verify these rules yourself with a few more examples of your own. The advantages of this notation when considering astronomical objects will soon become obvious.

Greek hero famed for his great beauty and stature, his hunting prowess, and his amorous pursuit of the Pleiades, the seven daughters of the giant Atlas. In order to save the Pleiades from Orion's unwanted earthly attentions, the gods placed them among the stars, where Orion nightly stalks them across the sky, but never catches them. Many constellations have similarly fabulous connections with the culture of ancient Greece.

For the most part, the stars that make up a particular constellation are not actually close to one another, even by astronomical standards. These stars merely are bright enough to observe with the naked eye and happen to lie in

INTERLUDE 1-2 *Astronomical Measurement*

Astronomers are notorious for their inconsistent and often rather arbitrary use of units. Unfortunately, we will perpetuate this reputation in this book simply because no single system of units will do. Rather than the meter-kilogram-second (MKS) metric system used in most high-school and college science classes, many professional astronomers still prefer the older centimeter-gram-second (CGS) system. However, astronomers commonly introduce new units when convenient. For example, when discussing stars, the mass and radius of the Sun are often used as reference points. The solar mass, written as $M_\odot$ is equal to 2.0×10^{33} g, or 2.0×10^{30} kg. The solar radius, $R_\odot$, is equal to 7.0×10^{10} cm. The subscript $\odot$ always stands for Sun.

Of particular importance are the units of length astronomers use. On small scales, the *angstrom* (1 Å = 10^{-8} cm) and the *micron* (1 μm = 10^{-6} m = 10^{-4} cm) are used. Distances within the solar system are usually expressed in terms of the *astronomical unit* (A.U.), the mean distance between the Earth and the Sun. One A.U. is approximately equal to 150,000,000 km, or 1.5×10^{13} cm. On larger scales, the *light-year* (1 ly = 9.5×10^{17} cm) and the *parsec* (1 pc = 3.1×10^{18} cm = 3.3 ly) are commonly used. Still larger distances use the regular prefixes of the metric system: *kilo* for one thousand and *mega* for one million. Thus, 1 *kiloparsec* (kpc) = 3.1×10^{21} cm, 20 megaparsecs (Mpc) = 6.2×10^{25} cm, and so on.

Astronomers use units that make sense within a context, but as contexts change, so do the units. For example, we might measure densities in grams per cubic centimeter (g/cm³), in atoms per cubic meter (atoms/m³), or even in solar masses per cubic megaparsec ($M_\odot$/Mpc³), depending on the circumstances. Some of the more common units used in astronomy and the contexts in which they are most likely to be encountered follow.

Length:	1 angstrom (Å)	= 10^{-10} m	atomic physics, spectroscopy, interstellar dust and gas
	1 nanometer (nm)	= 10^{-9} m	
	1 micron (μm)	= 10^{-6} m	
	1 centimeter (cm)	= 0.01 m	in widespread use throughout all of astronomy
	1 meter (m)	= 100 cm	
	1 kilometer (km)	= 1000 m = 10^5 cm	
	Earth radius ($R_\oplus$)	= 6378 km	planetary astronomy
	solar radius ($R_\odot$)	= 6.96×10^8 m	solar system, stellar evolution
	1 astronomical Unit (A.U.)	= 1.496×10^{11} m	
	1 light-year (ly)	= 9.46×10^{15} m	galactic astronomy, stars and star clusters
		= 63,200 A.U.	
	1 parsec (pc)	= 3.09×10^{16} m	
		= 3.26 ly	
	1 kiloparsec (kpc)	= 1000 pc	galaxies, galaxy clusters, cosmology
	1 megaparsec (Mpc)	= 1000 kpc	
Mass:	1 gram (g)		in widespread use in many different areas
	1 kilogram (kg)	= 1000 g	
	Earth mass ($M_\oplus$)	= 5.98×10^{27} g	planetary astronomy
	solar mass ($M_\odot$)	= 1.99×10^{33} g	"standard" unit for all mass scales larger than the Earth
Time:	1 second (s)		in widespread use throughout astronomy
	1 hour (h)	= 3600 s	planetary and stellar scales
	1 day (d)	= 86400 s	
	1 year (yr)	= 3.16×10^7 s	virtually all processes occurring on scales larger than a star

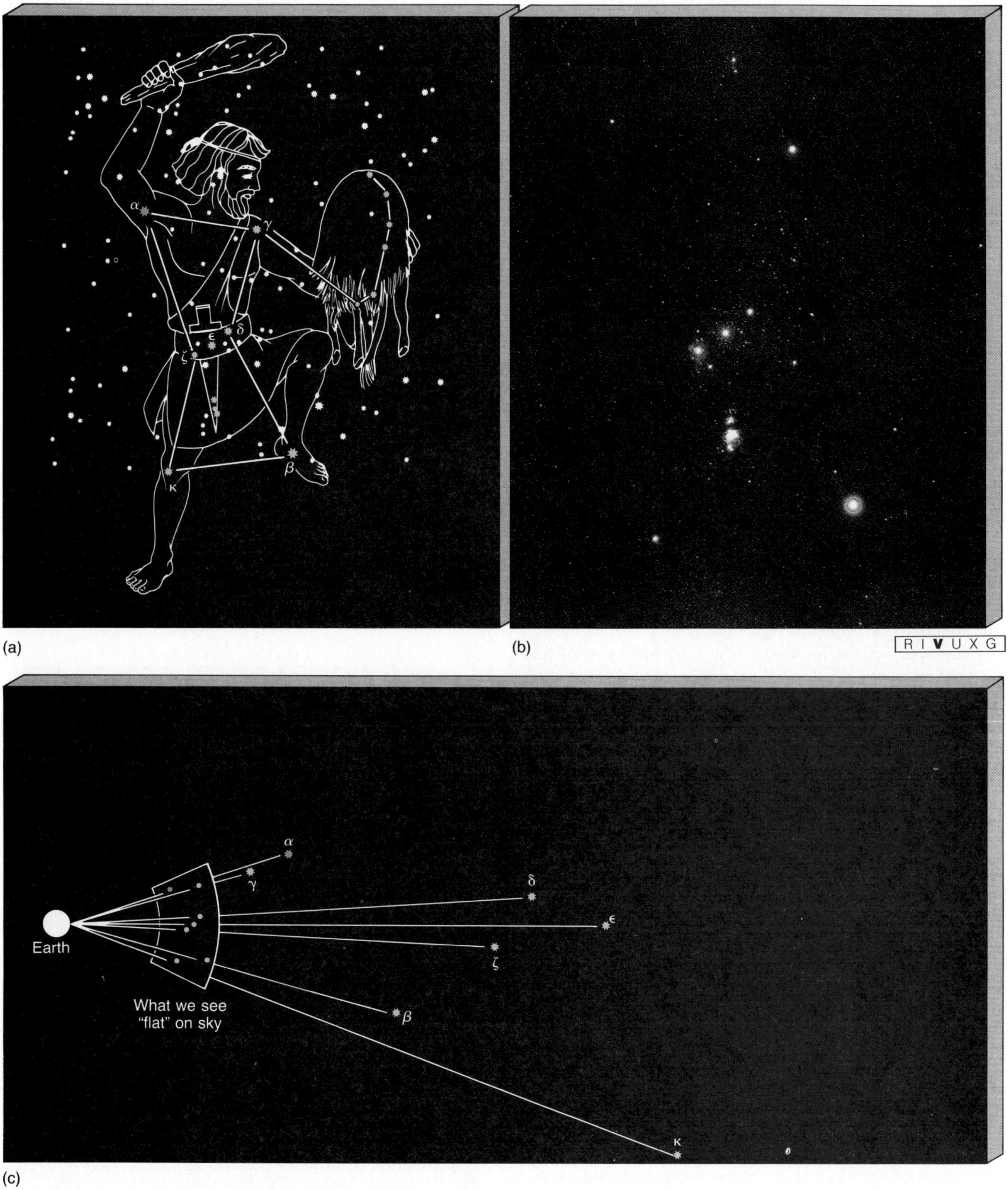

Figure 1.6 *(a) The groupings of stars often gave rise to constellations like this one, called Orion. Here, the brightest stars in this region of the sky are connected to form the outlines of a hunter. (b) This is an actual photograph of most of this same region. You can orient yourself by identifying the line of three bright stars in the hunter's ''belt.'' (Harvard College Observatory) (c) The true relationships between the stars, in three dimensions.*

roughly the same direction in the sky as seen from Earth. That these stars have become associated with one another over the centuries is a tribute to the power of the human brain, which is extremely good at recognizing patterns and relationships between objects even when no true connection exists.

Not surprisingly, the patterns seen have a strong cultural bias. The astronomers of ancient China saw mythical figures different from those seen by the ancient Greeks, the Babylonians, and the people of other cultures, even though they were all looking at the same stars. For example, the group of seven stars usually known in North America as "the Big Dipper" is known as "the Wagon" or "the Plough" in Western Europe. The ancient Greeks regarded these same stars as the tail of "the Great Bear," the Egyptians saw them as the leg of an ox, and the Siberians discerned a stag. Some Native Americans saw seven mythical brothers, others an ermine, and others still a funeral procession. The Chinese apparently saw the pattern as a minor government official, dealing with the day-to-day concerns of the Emperor. It is interesting to note that, despite all these different interpretations, these cultures made the same *grouping* of stars, which is the case for many constellations.

The origins of most constellations, and of their names, date back to the dawn of recorded history. Some constellations served as navigational guides. For example, the constancy of the location of the Pole Star, part of the Little Dipper, has aided travelers since ancient times. Other constellations served as primitive calendars to predict planting and harvesting seasons. For example, many cultures knew well that the appearance of certain stars on the horizon just before daybreak signaled the beginning of spring and the end of winter. Astronomy came into being because people believed that there was a practical benefit in being able to predict the positions of stars. The roots of both astronomy and astrology—originally indistinguishable from each other—are in the patterns of stars that sweep nightly and yearly across the sky.

Today we recognize that the constellations have no occult power to determine our destinies and no special astrophysical significance. Most people recognize that astrology is nothing more than an amusing diversion (although millions still study their horoscopes in the newspaper every morning!). Nevertheless, the ancient astrological terminology—the names of the constellations and some terms used to describe the locations and motions of the planets, for example—is still used throughout the astronomical world. The constellations still help astronomers to specify bulk areas of the sky, much as geologists use land continents or politicians use voting precincts to identify certain localities on planet Earth. In all, there are 88 constellations, most of them visible from North America at some time during the year. See Appendix A for sky charts that can be used to locate prominent stars and constellations at different times of the year.

THE CELESTIAL SPHERE

✓3 Following the constellations nightly, ancient skywatchers noted that the star patterns seemed unchanging. The stars of Orion moved across the sky as a unit thousands of years ago much as they do today. It was natural, perhaps even patently obvious, to conclude that the stars must be firmly attached to a **celestial sphere** surrounding the Earth—a canopy of stars resembling an astronomical painting on a heavenly ceiling. Figure 1.7 shows how the stars appear to move with this celestial sphere as it turns around a fixed, unmoving Earth once every 24 hours.

When talking about the stars, it is necessary to devise a means of describing how to locate them in the sky. The oldest method, still in use today, simply specifies the constellation and then ranks the stars in it in order of brightness. The brightest star is denoted by the Greek letter α (alpha), the second brightest by β (beta), and so on. Thus, the two brightest stars in the constellation Orion—Betelgeuse and Rigel—are also known as α Orionis and β Orionis, respectively. (Improved observations show that Rigel is actually the brighter of the two, but the names are now permanent.) Because there are many more stars in any given constellation than there are letters in the Greek alphabet, this method is of limited utility. However for naked-eye astronomy, where only bright stars are involved, it is quite satisfactory.

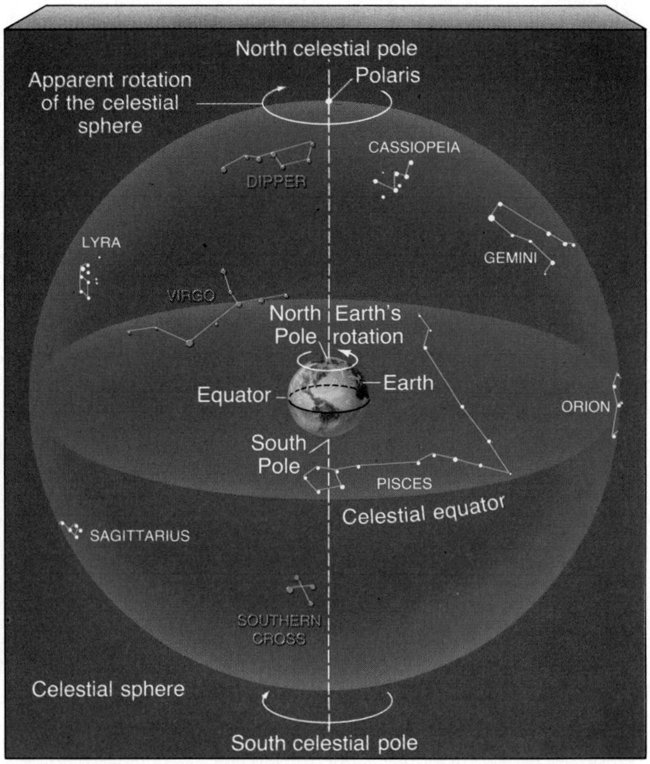

Figure 1.7 *Planet Earth sits fixed at the hub of the celestial sphere, which contains all the stars. This is one of the simplest possible models of the universe, but it doesn't agree with all the facts that astronomers know about the universe.*

While this system has a certain immediacy, it is not very precise. In order to quantify measurements of the stars, astronomers find it helpful to lay down a system of **celestial coordinates** on the sky. If we think of the stars as being attached to a giant sphere centered on the Earth, then the familiar system of latitude and longitude on the Earth's surface extends quite naturally to cover the sky.

For the most part, stars rise in the east, move across the sky, and set in the west each night. Some sweep out a large arc high above the horizon, but others appear to move very little. In fact, closer scrutiny (or time-lapse photography—see Figure 1.8) shows that all stars move in circles around a point in the sky very close to the star Polaris (better known as the Pole Star, or the North Star). From the point of view of the ancients, the Pole Star represented the point around which the entire celestial sphere rotated.

From our modern standpoint, Polaris indicates the direction—due north—of the Earth's axis of **rotation,** or spin. That point in the sky is known as the **north celestial pole,** and it is directly above the Earth's North Pole. In the Southern Hemisphere, the extension of the Earth's axis in the opposite direction defines the **south celestial pole,** directly above the Earth's South Pole. Midway between the north and south celestial poles lies the **celestial equator,** representing the intersection of the Earth's equatorial plane with the celestial sphere.

Figure 1.8 *Time-lapse photograph of the sky looking north one evening. Each trail is the path of a single star as it sweeps across the nighttime sky. The duration of the exposure is about 5 hours. The center of the concentric circles is near the star Polaris. (AURA)*

The analogs of latitude and longitude on the Earth's surface are known as **declination** and **right ascension,** respectively. Declination (dec) is measured in **degrees** north or south of the celestial equator, just as latitude is measured in degrees north or south of the Earth's equator. (See Interlude 1-3 for a discussion of angular measure.) Thus, the celestial equator is at a declination of 0 degrees (°), the

INTERLUDE 1-3 *Angular Measure*

The size and scale of things are often specified by measuring lengths and angles. The concept of *length measurement* is fairly intuitive. We've already noted the dimensions of some objects found in the universe, ranging from subatomic particles with sizes of 10^{-13} centimeter to whole galaxies with sizes of 10^{23} centimeters.

The concept of *angular measurement* may seem less intuitive, but it too can become second nature if you remember a few simple facts. One of these facts is obvious—namely, that a full circle contains 360 **arc degrees** (or just 360°). Therefore, the half-circle that stretches from horizon to horizon, passing directly overhead, and spans the portion of the sky visible to one person at any one time contains 180°.

Each 1° increment can be further subdivided into fractions of an arc degree, called **arc minutes;** there are 60 arc minutes in 1 arc degree. For example, both the Sun and the Moon project an angular size of 30 arc minutes on the sky. Your little finger, held at arm's length, does just about the same, covering a small (actually, about 40 arc minute) slice of the entire 180 arc degrees from horizon to horizon.

Finally, an arc minute can be further sliced into 60 equal **arc seconds.** Put another way, an arc minute is $\frac{1}{60}$ of an arc degree, and an arc second is $\frac{1}{60} \times \frac{1}{60}$—or $\frac{1}{3600}$—of an arc degree. An arc second is an extremely small unit of angular measure. It is, in fact, the angle projected by a centimeter-sized object at a distance of about 2 kilometers. Expressed in more familiar terms, an arc second is the angle subtended by an American dime when viewed from a distance of nearly 3 miles away.

Don't be confused by the units used to measure angles. Arc minutes and arc seconds have nothing to do with the measurement of time, and arc degrees have nothing to do with temperature. Arc degrees, arc minutes, and arc seconds are simply ways to measure the size, shape, and position of objects in the universe. Just as 1 arc degree is usually written as 1°, 1 arc minute is denoted 1′, and 1 arc second as 1″.

The angular size of an object depends both on its actual size and on its distance from us. For example, the Moon, at its present distance from Earth, has an angular diameter of 30′. If the Moon were twice as far away, it would appear half as big—15′ across—even though its actual size would be the same. Similarly, if the Moon were three times closer, its angular diameter would be three times as great—1.5°, and so on. Thus, angular size by itself is not enough to determine the actual diameter of an object—the distance must also be known.

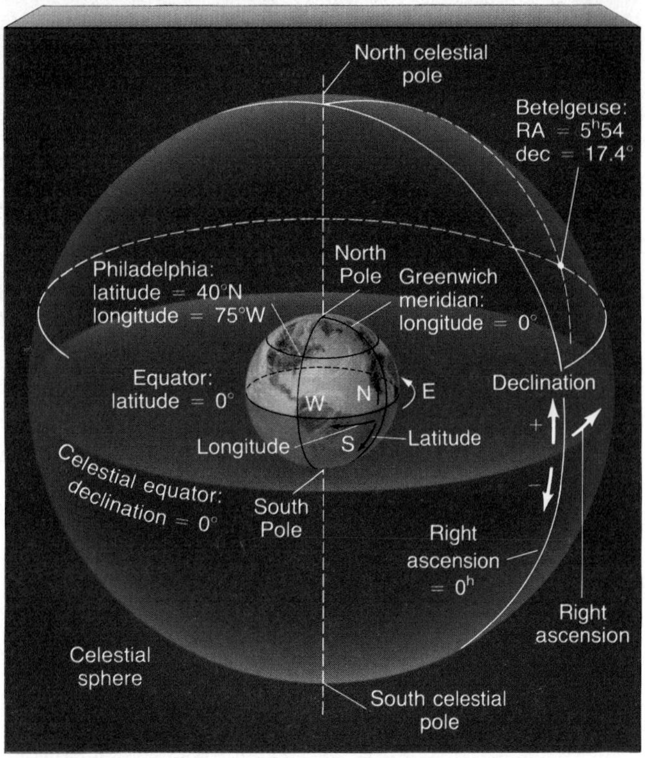

Figure 1.9 *The right ascension and declination of a star on the celestial sphere are defined similarly to longitude and latitude on the surface of the Earth. Just as longitude and latitude allow us to locate a point on the surface of the Earth, right ascension and declination specify locations on the sky. For example, to find Philadelphia on Earth, look 75° west of the Greenwich Meridian and 40° north of the Equator. Similarly, to locate the star Betelgeuse on the celestial sphere, look 5^h54^m east of the vernal equinox (the line on the sky with a right ascension of zero) and 7°24′ north of the celestial equator.*

north celestial pole is at +90°, and the south celestial pole is at −90° (the minus sign here just means "south of the celestial equator"). Right ascension (RA) is measured in units called **hours, minutes,** and **seconds,** and it increases to the east. Notice that, although there are strong connections with the measurement of time, these are *angular* units. One hour (1^h) is 15° (so that 24 hours make up one complete revolution), 1 minute (1^m) is 1/60 of an hour, and 1 second (1^s) is 1/60 of a minute.

The angular units used to measure right ascension are constructed to parallel the units of time; the two sets of units are connected by the rotation of the Earth (or of the celestial sphere). In one day, the Earth rotates once on its axis, or through 360°. Thus, in a time period of 1 hour, the Earth rotates through 360/24 = 15°, or 1^h. In 1 minute of time, the Earth rotates through 1^m; in 1 second, the Earth rotates through 1^s. These units are used *only* for right ascension and are defined in this way to assist astronomical observation. The use of the names minutes and seconds is rather unfortunate, however, as they are *not* the same units as defined in Interlude 1-3. From the above definition, it can be seen that $1^m = 15°/60 = 0.25°$, or 15 arc minutes (15′). Similarly, 1^s = 15 arc seconds (15″). Just remember that all angular measurements except right ascension use arc minutes (′) and arc seconds (″), and you should avoid undue confusion.

The choice of zero right ascension is quite arbitrary, just as the choice of terrestrial longitude is. The right ascension equivalent of the Greenwich Meridian (the line on the Earth with a longitude of zero) is a semicircle through the two celestial poles, passing through a point in the sky midway between the constellations Pisces and Aquarius. We will return to the reason for this choice in a moment. Figure 1.9 illustrates the meanings of right ascension and declination on the celestial sphere and compares them with longitude and latitude on Earth.

Just as latitude and longitude are tied to the Earth, right ascension and declination are fixed in the sky. Although the stars appear to move across the sky because of the Earth's rotation, their celestial coordinates remain *constant*. A system of coordinates that changed from minute to minute and differed from place to place on the Earth would obviously be very inconvenient and quite impractical. Thus, we have a quantitative alternative to the use of constellations in specifying the positions of stars in the sky. For example, Rigel and Betelgeuse can be precisely located by looking in the directions $5^h\,13^m\,36^s$ (RA), −8°13′ (dec) and $5^h\,54^m\,0^s$ (RA), 07°24′ (dec), respectively. The use of right ascension and declination is much more accurate than simply referring to "the *n*th brightest star in constellation X." Nevertheless, many amateur and professional astronomers have a much better "feel" for the whereabouts of regions in the sky that are identified by constellations rather than by their coordinates.

THE APPARENT MOTION OF THE SUN AND THE STARS

Day-to-Day Changes

✓4 We measure time by the Sun. The rhythm of day and night is central to our lives, so it is hardly surprising that the period of time from one sunrise (or noon, or sunset) to the next is our basic social time unit. The day measured relative to the Sun is called a **solar day.** The daily progress of the Sun and the stars across the sky is known as **diurnal** motion. It is a consequence of the rotation of the Earth (or of the celestial sphere, depending on your point of view), and it repeats itself every 24 hours. Actually, the repetition is not exact. Each night, the whole celestial sphere appears to be shifted a little relative to the horizon, compared with the night before. The easiest way to confirm this difference is by noticing the stars visible just after sunset or just before dawn. You will find that they are in slightly different locations from the previous night. Because of this shift, a day measured by the stars—called a **sidereal day**—differs from a solar day. Evidently, there is more to the apparent motion of the heavens than just simple rotation. In fact, the motion of the Earth relative to the Sun—the Earth's **revolution**—is of great importance.

The reason for the difference between a solar day and a sidereal day is sketched in Figure 1.10. Every 24 hours, not only does Earth rotate once on its axis, it also moves a small distance along its orbit about the Sun. Thus, the interval of time between noon one day and noon the next (a solar day) is slightly greater than one true rotation period (one sidereal day). The Earth has to rotate through slightly more than 360° for the Sun to return to the same apparent location in the sky. The Earth takes 365 days to orbit the Sun, so the additional angle is only 360°/365 = 0.986°. Because the Earth takes about 3.9 minutes to rotate through this angle, the solar day is 3.9 minutes longer than the sidereal day. This is the modern explanation for the difference between a solar day and a sidereal day. From the point of view of the ancients, the "fact" that the Sun moved relative to the stars made it necessary to envisage not just one sphere for all heavenly objects, but *two* spheres: one for the stars and one for the Sun, each rotating about the Earth but at slightly different rates and with different orientations.

After an entire year, the Sun returns to its starting point on the celestial sphere, and the cycle begins anew. The apparent motion of the Sun in the sky, expressed relative to the stars, follows a path known as the **ecliptic.** It forms a great circle on the celestial sphere, inclined at an angle of about 23½° to the celestial equator. This tilt is just a consequence of the inclination of the Earth's rotation axis to the plane of its orbit, as illustrated in Figure 1.11.

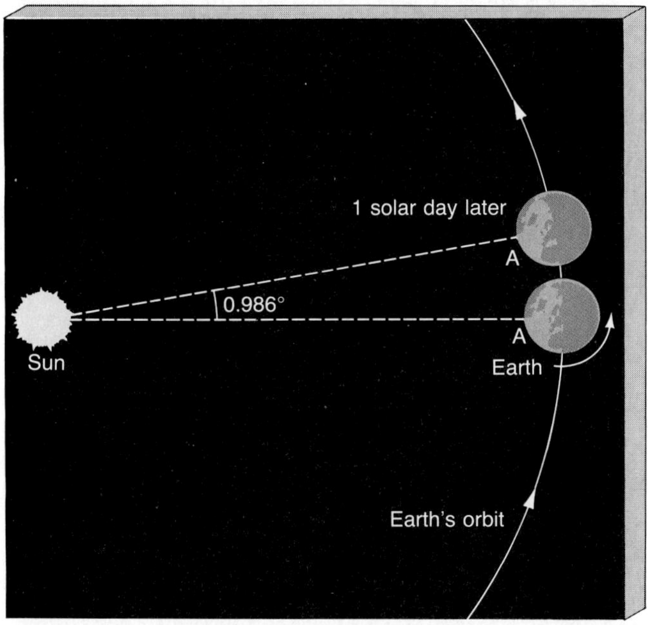

Figure 1.10 *The difference between a solar and a sidereal day can be easily explained once we understand that the Earth revolves around the Sun at the same time as it rotates on its axis. A solar day is the time from one noon to the next. In that time, the Earth moves a little in its orbit. Because the Earth completes one circuit (360°) around the Sun in 1 year (365 days), it moves through about 1° (actually, 360°/365 = 0.986°) in 1 day. Thus, from noon at point A to noon the next day, Earth actually rotates through about 361°, and the solar day exceeds the sidereal day (360° rotation) by about 24/360 hours, or about 4 minutes.*

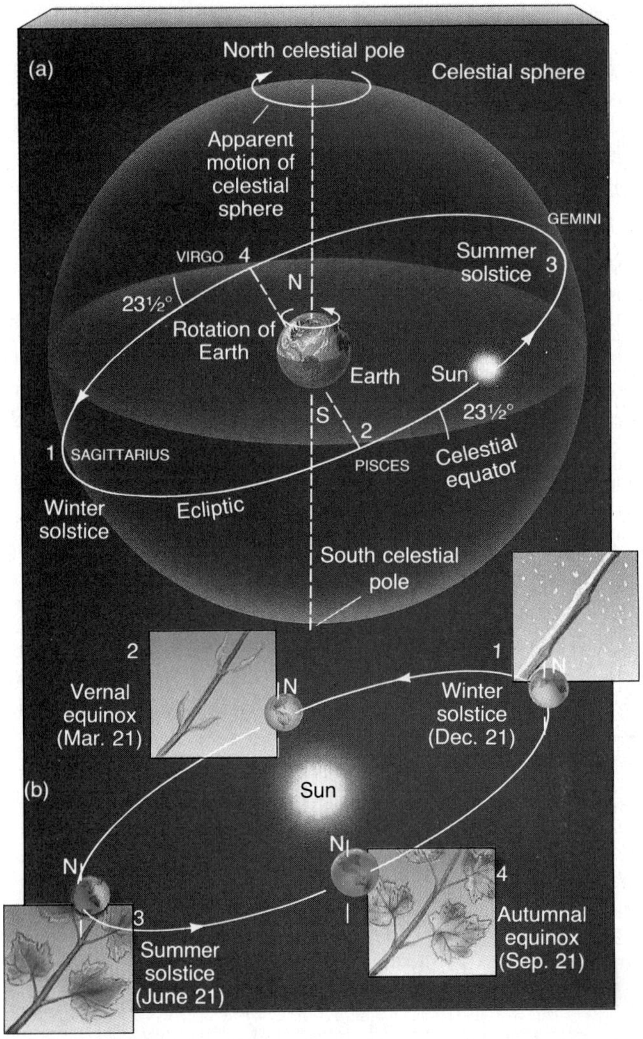

Figure 1.11 *(a) The apparent path of the Sun on the celestial sphere and (b) its relation to the Earth's rotation and revolution. The seasons result from the changing height of the sun above the horizon.*

Seasonal Changes

The Sun completes one trip around the ecliptic in a year (in fact, in one **tropical year,** or 365.242 solar days). The point on this path where the Sun is at its northernmost point above the celestial equator (declination +23½°) is known as the **summer solstice.** This occurs on or near June 21—the exact date varies slightly from year to year because the actual length of a year is not a whole number of days. As the Earth rotates, points north of the equator spend the greatest fraction of their time in sunlight on that date, so the summer solstice corresponds to the longest day of the year in the Northern Hemisphere and the shortest day in the Southern Hemisphere. Six months later, the Sun is at its southernmost point (declination −23½°), and we have reached the

winter solstice (December 21)—the shortest day in the Northern Hemisphere and the longest in the Southern Hemisphere. These two effects—the height of the Sun above the horizon and the length of the day—combine to account for the **seasons** we experience. In summer, the Sun is high in the sky and the days are long, so that temperatures are generally much higher than in winter, when the Sun is low and the days are short.

The two points where the ecliptic intersects the celestial equator are known as **equinoxes.** On those dates, day and night are of equal duration. In the fall, as the Sun crosses from the Northern into the Southern Hemisphere, we have the **autumnal equinox** (on September 22). The **vernal equinox** occurs in the spring, on or near March 21, as the Sun crosses the celestial equator moving north. Because of its association with the end of winter and the start of a new growing season, the vernal equinox was particularly important to early astronomers and astrologers. Today, it still figures in astronomical measurements because the position of the Sun in the sky *at the instant of the vernal equinox* is taken to have a right ascension of zero. This position in the sky happens to lie between the constellations Pisces and Aquarius, mentioned earlier in discussing how astronomers chose the zero-point for right ascension.

Summer and Winter Constellations

✓5 Figure 1.12(a) illustrates the major stars visible from most locations in the United States on clear summer evenings. The brightest stars, Vega, Deneb, and Altair, form a conspicuous triangle high above the constellations Sagittarius and Capricornus, low on the southern horizon. From the same location in the winter, however, these stars have been replaced, as shown in Figure 1.12(b), by several well-known constellations including Orion, Leo, and Gemini, a group dominated by Sirius (the Dog Star), the brightest star in the sky. Year after year, the same stars and constellations return, each in its proper season. Every winter evening, Orion is high overhead; every summer, it is gone.

The reason for these regular seasonal changes is, once again, the revolution of the Earth around the Sun. Earth's darkened hemisphere faces in a slightly different direction each evening. Since our planet's yearly orbit takes 365 days to complete and since a full circle has 360°, this change in direction is only about 1° per night—too small to be easily noticed with the naked eye from one evening to the next, but clearly noticeable over the course of weeks and months, as illustrated in Figure 1.13. After 6 months, Earth has reached the opposite side of its orbit, and we see an entirely different group of stars and constellations. The 12 constellations through which the Sun passes as it moves along the ecliptic had special significance for astrologers of old. These constellations are collectively known as the **zodiac.** In the alternative view, the Sun has moved to the opposite side of the celestial sphere, so that a different set of stars is visible at night. The time required for the constellations to complete one cycle around the sky and to return to their starting points as seen from a given point on Earth is one

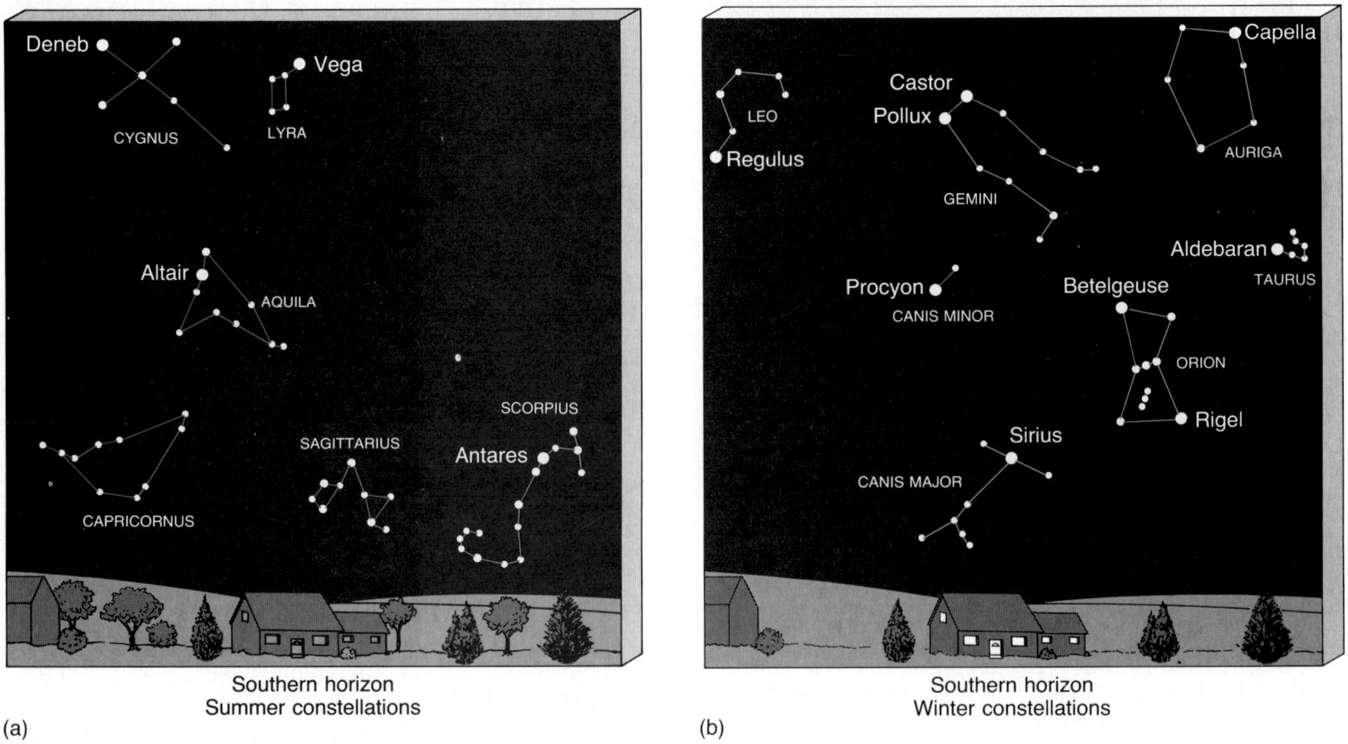

Figure 1.12 *(a) A typical summer sky above the United States. Some prominent stars (labeled in larger print) and constellations (labeled in small capital letters) are shown. (b) A typical winter sky above the United States.*

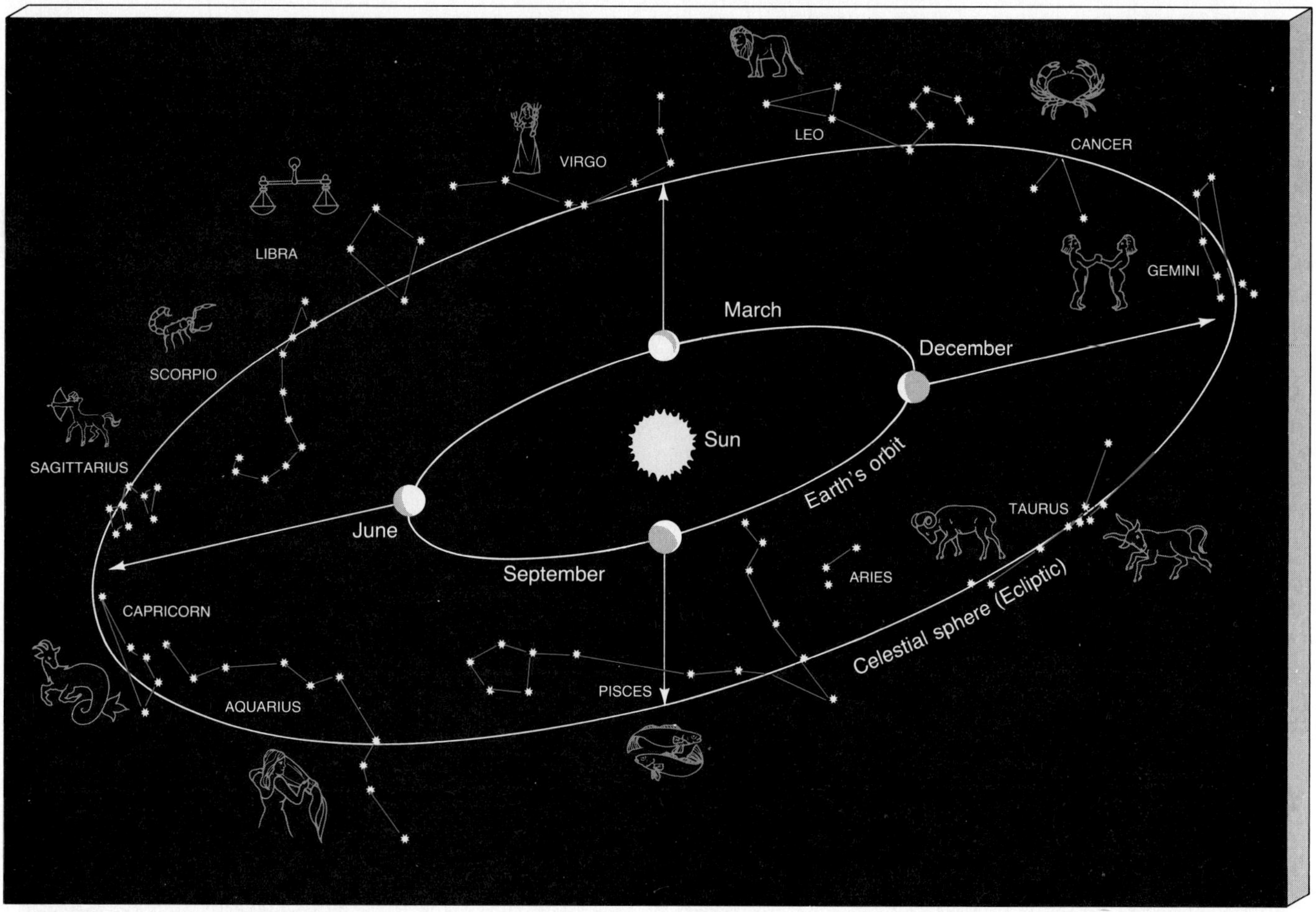

Figure 1.13 *The view of the night sky changes as Earth moves in its orbit about the Sun. As drawn here, the night side of Earth faces a different set of constellations at different times of the year.*

sidereal year. The Earth completes exactly one orbit around the Sun in this time. One sidereal year is 365.256 solar days long.

THE MOTION OF THE MOON

The Moon is our nearest neighbor in space. Apart from the Sun, it is by far the brightest object in the sky. Unlike the Sun and the stars, however, it emits no light of its own. Instead, it shines by reflecting sunlight. Again unlike the Sun and the stars, its appearance changes from night to night—in fact, on some nights it cannot be seen at all. Also, the Moon's daily rising and setting and its nightly motion through the sky differ from the motion of the celestial sphere. Like the Sun, the Moon appears to move relative to the stars—it crosses the sky at a rate of about 12° per day (so that it moves an angular distance equal to its own diameter in about an hour). Today, we explain these observations in terms of the Moon's revolution around the Earth. For ancient astronomers, however, the Moon necessitated the introduction of yet another sphere in the sky, with a motion separate from both that of the stars and that carrying the Sun.

The Moon's appearance undergoes a cycle of changes, or **phases,** taking a little more than 29 days to complete. (The word *month* is derived from the word *Moon*.) Figure 1.14 illustrates the appearance of the Moon at different times in this monthly cycle. Starting from the so-called **full moon,** visible as a complete circular disk in the sky, the Moon appears to **wane** (or shrink) a little each night, passing through the **gibbous** phase until, 1 week later, only half of the disk can be seen. This phase is known as a **quarter Moon.** During the next week, the moon is visible as a shrinking **crescent.** Two weeks after full moon, the crescent has shrunk to nothing, and the Moon is all but invisible. This phase is called a **new Moon.** During the next 2 weeks, the Moon **waxes** (or grows) again, passing through the crescent, quarter, and gibbous phases and eventually becoming full. The waning and waxing phases are not merely time reversals of each other, however. The waning Moon shrinks toward the eastern edge of the disk, and the waxing Moon grows from the west.

The Moon doesn't actually change its size and shape on a monthly basis, of course. The full circular disk of the Moon is present at all times. Why then don't we always see

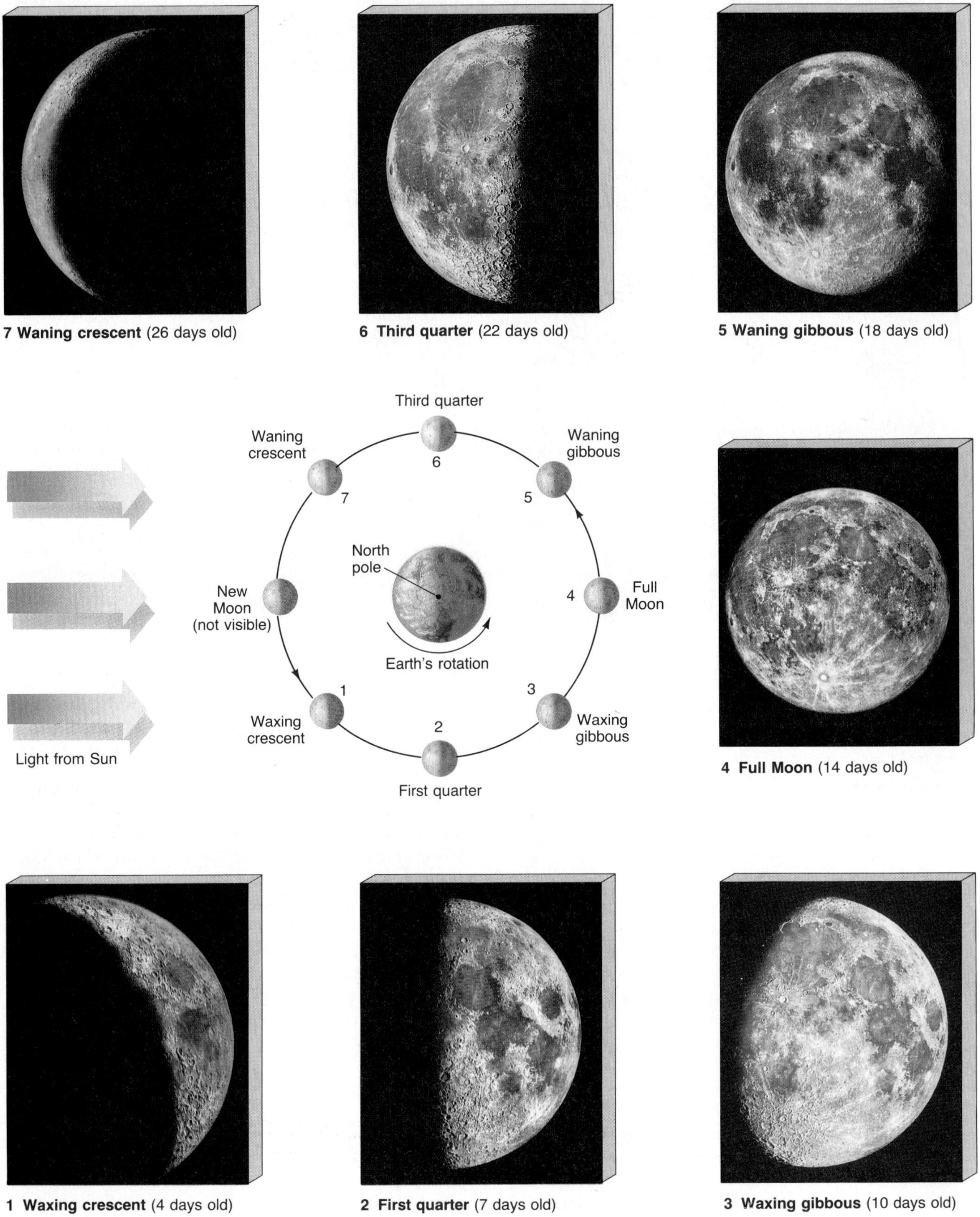

Figure 1.14 *Because the Moon orbits the Earth as it does, the visible fraction of the sunlit face differs from night to night. (Lick Observatory)*

R I **V** U X G

a full Moon? Because usually only a portion of the disk is actually visible from Earth. As illustrated in Figure 1.14, half of the Moon's surface is illuminated by the Sun at any instant (except during eclipses, as we will see in a moment). However, not all of the Moon's sunlit face can be seen because of the Moon's position with respect to the Earth and the Sun. When the Moon is full, we see the entire "daylit" face because the Sun and the Moon are in opposite directions from the Earth in the sky. In the case of a new moon, the Moon and the Sun are in almost the same part of the sky, and the sunlit side of the Moon is oriented away from us. A moment's thought tells us that this must mean the Moon is closer to the Earth than is the Sun—during the new Moon, the Sun must be almost behind the Moon, as we view it.

As the Moon revolves around the Earth, its position in the sky changes with respect to the stars. In one **sidereal month** (27.3 days), the Moon returns to its starting point on the celestial sphere, having traced out a great circle in the sky. The time required for the Moon to complete a full cycle of phases, one **synodic month,** is a little longer—about 29.5 days. The synodic month is a little longer than the sidereal month for the same reason that a solar day is slightly longer than a sidereal day: Because of the motion of the Earth around the Sun, the Moon must complete slightly more than one full revolution to return to the same phase in its orbit (see Figure 1.15).

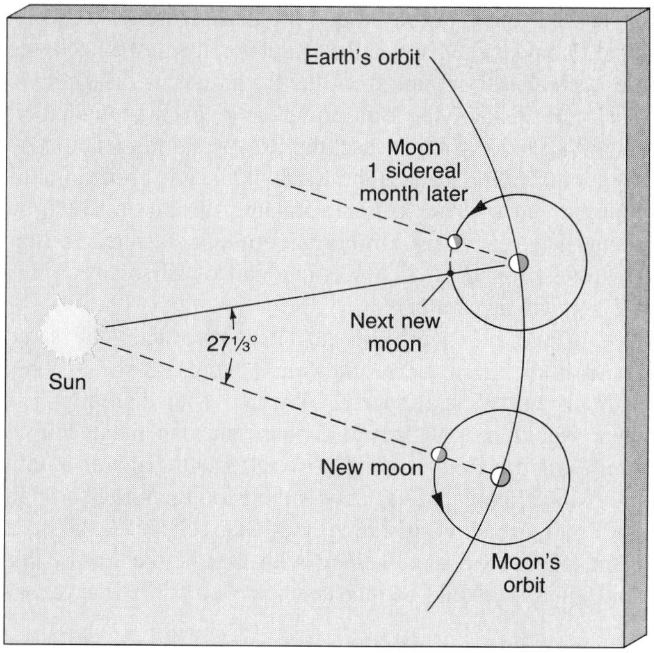

Figure 1.15 *The difference between a synodic and a sidereal month stems from the motion of the Earth relative to the Sun. Because the Earth orbits the Sun (in 365 days), in the 29.5 days from one full Moon to the next (one synodic month), Earth moves through an angle of 360° × (29.5/365), or approximately 29°. Thus the Moon must revolve more than 360° between full Moons. The sidereal month, which is the time taken for the Moon to revolve through exactly 360°, is about 2 days shorter.*

ECLIPSES

✓6 From time to time—at new or full Moon—the Sun and the Moon line up precisely as seen from Earth, and we observe the spectacular phenomenon known as an **eclipse.** Eclipses fall into two categories. When the Sun and the Moon are in exactly *opposite* directions, as seen from Earth, the Earth's shadow sweeps across the Moon, temporarily blotting it out in a **lunar eclipse.** From Earth, we see the curved edge of the Earth's shadow begin to cut across the face of the full moon and slowly eat its way into the lunar disk. More often than not, the alignment of the Sun, Earth, and Moon is imperfect, and the shadow never quite covers the Moon completely. This is known as a **partial eclipse.** Occasionally, however, the entire lunar surface is obscured in a **total eclipse** (Figure 1.16). Lunar eclipses last only as long as is needed for the Moon to pass through the Earth's shadow—no more than about 100 minutes. During that time, the Moon often acquires an eerie, deep red coloration—the result of a small amount of sunlight being refracted (bent) by the Earth's atmosphere onto the lunar surface, preventing the shadow from being completely black. It is perhaps understandable that many primitive cultures interpreted lunar eclipses as harbingers of disaster.

When the Moon and the Sun are in exactly the *same* direction, an even more awe-inspiring event occurs. The Moon passes directly in front of the Sun, briefly turning day into night in a **solar eclipse.** In a **total eclipse,** when the alignment is perfect, the stars become visible in the daylight as the Sun's light is reduced to nearly nothing. We can also

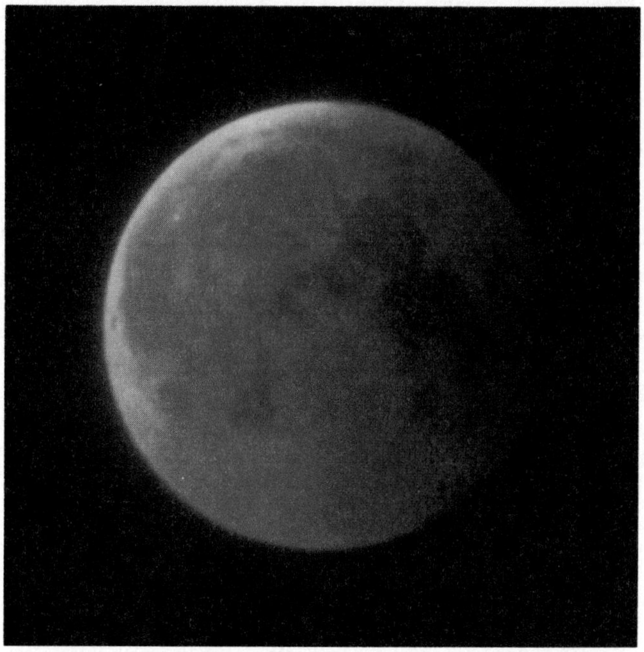

R I V U X G

Figure 1.16 *A total lunar eclipse. The dark red coloration is caused by sunlight deflected by the Earth's atmosphere onto the Moon's surface. (G. Schneider)*

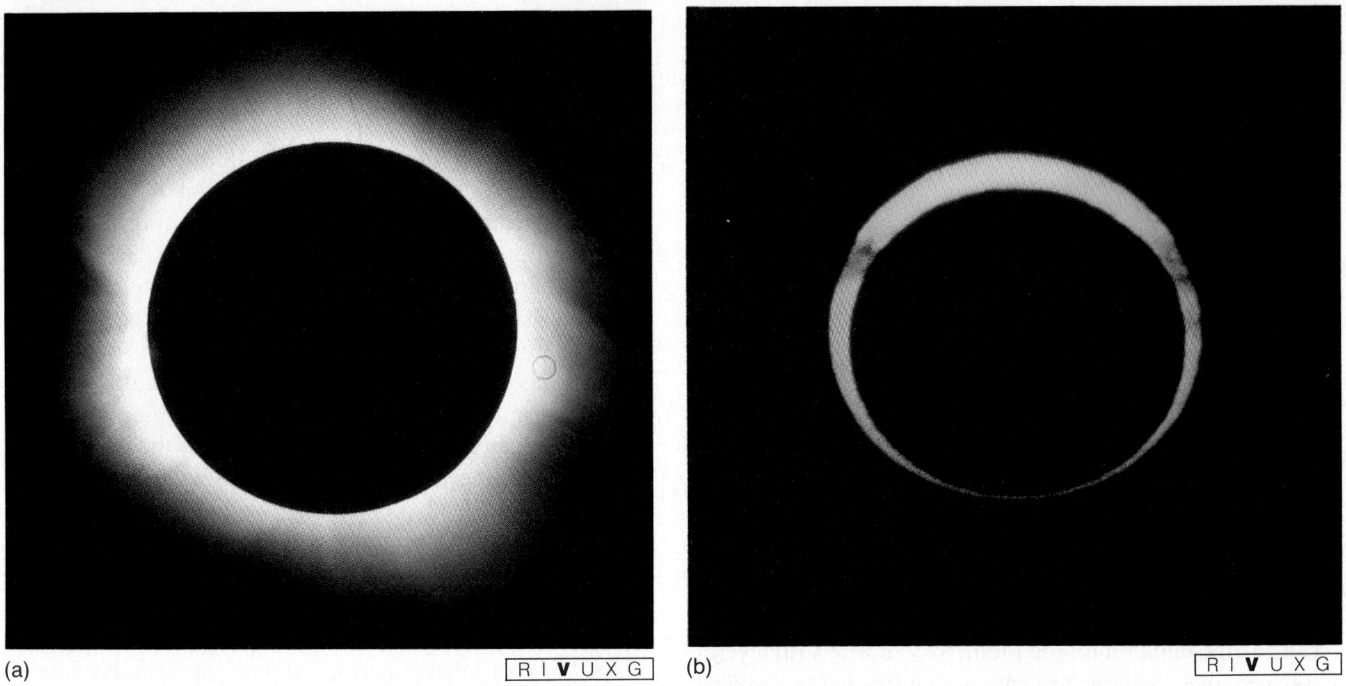

Figure 1.17 *(a) During a total solar eclipse, the Sun's corona becomes visible as an irregularly shaped halo surrounding the blotted-out disk of the Sun. This was the July, 1991 eclipse, as seen from the Baja Peninsula. (b) During an annular eclipse, the Moon fails to completely hide the Sun, so a thin ring of light remains. No corona is seen in this case because even the small amount of the Sun still visible completely overwhelms the corona's faint glow. This was the December, 1973 eclipse, as seen from Algiers. (The gray fuzziness at top left and right are clouds in Earth's atmosphere.) (G. Schneider)*

see the Sun's ghostly outer atmosphere, or corona (Figure 1.17[a]). (Actually, although a total solar eclipse is undeniably a spectacular occurrence, the visibility of the corona is probably the most important astronomical aspect of such an event today. It enables us to study this otherwise hard-to-see part of our Sun.) In a partial eclipse, the Moon's path is slightly off-center and only a portion of the Sun's face is covered. In either case, the sight of the Sun apparently being swallowed up by the black disk of the Moon is disconcerting even today. It must surely have inspired fear in early observers. Small wonder, then, that the ability to predict such events was a highly prized skill.

Unlike a lunar eclipse, which is simultaneously visible from all locations on the nighttime side of the Earth, only a small portion of the daytime side is privileged to witness a total solar eclipse. The Moon's shadow on the Earth's surface is about 3500 kilometers wide—roughly the diameter of the Moon. Outside of that shadow, no eclipse is seen. However, only within the central region of the shadow, the **umbra,** is the eclipse total. Outside the umbra, in the **penumbra,** the eclipse is partial, with less and less of the Sun being obscured the farther one travels from the shadow's center. The connections between the umbra, the penumbra, and the relative locations of the Earth, Sun, and Moon are illustrated in Figure 1.18. One of the reasons that total solar eclipses are rare is that, although the penumbra is some 3500 kilometers across, the umbra is always very small—even under the most favorable circumstances, its diameter never exceeds 270 kilometers. Moreover, the shadow sweeps across the Earth's surface at some 1700 kilometers per hour, so the duration of a total eclipse at any point can never exceed 7.5 minutes.

The Moon's orbit around the Earth is not exactly circular. Thus, it can happen that, at the moment of an eclipse, the Moon is far enough from the Earth that its disk fails to cover the disk of the Sun completely, even though their centers coincide. In that case, there is no region of totality—the umbra never reaches the Earth at all, and a thin ring of sunlight can still be seen surrounding the Moon. Such an occurrence, called an **annular eclipse,** is depicted in Figures 1.17(b) and 1.18. More than half of all solar eclipses are annular in nature.

If the Moon orbited the Earth in exactly the same plane as the Earth orbits the Sun—that is, if the Moon's path on the celestial sphere coincided with the ecliptic—there would be precisely one solar and one lunar eclipse every synodic month, with the two alternating at half-month intervals. Figure 1.19(a) shows the Moon passing through the Earth's shadow, resulting in a lunar eclipse. If the orbit plane of the Moon coincided with that of the Earth, this configuration would be repeated every month. However, as

Figure 1.18 *(a) The Moon's shadow on the Earth during a solar eclipse consists of the umbra, where the eclipse is total, and the penumbra, where the Sun is only partially obscured. If the Moon is too far from the Earth at the moment of the eclipse, there is no region of totality; instead, an annular eclipse is seen. (b) Actual photograph taken by an Earth-orbiting weather satellite of the Moon's shadow projected onto Earth's surface (near the Baja Peninsula) during the total solar eclipse of July 11, 1991. (G. Schneider; NOAA)*

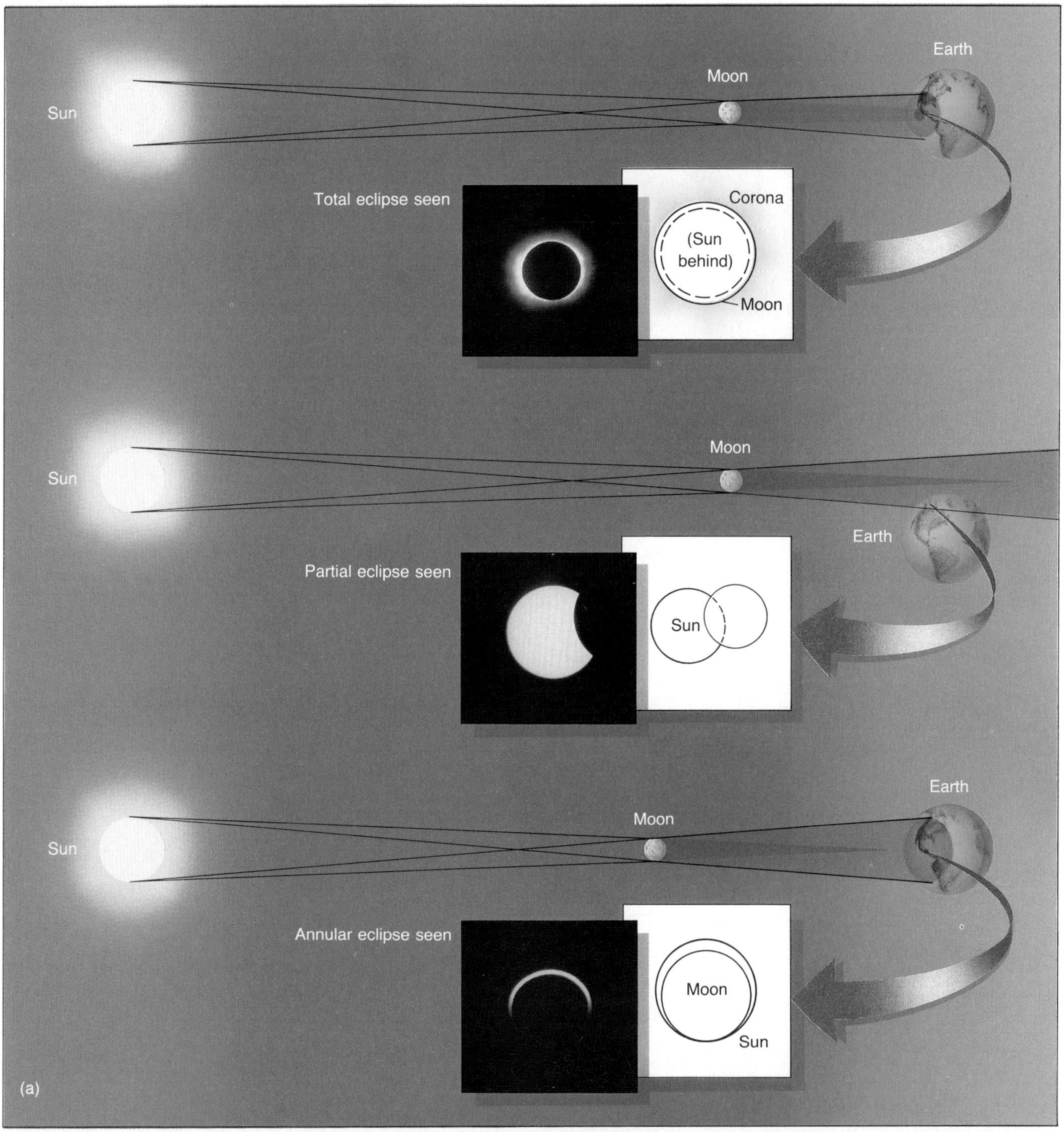
Earth
Moon
Sun
Total eclipse seen
Corona
(Sun
behind)
Moon
Moon
Sun
Earth
Partial eclipse seen
Sun
Earth
Moon
Sun
Annular eclipse seen
Moon
Sun
(a)

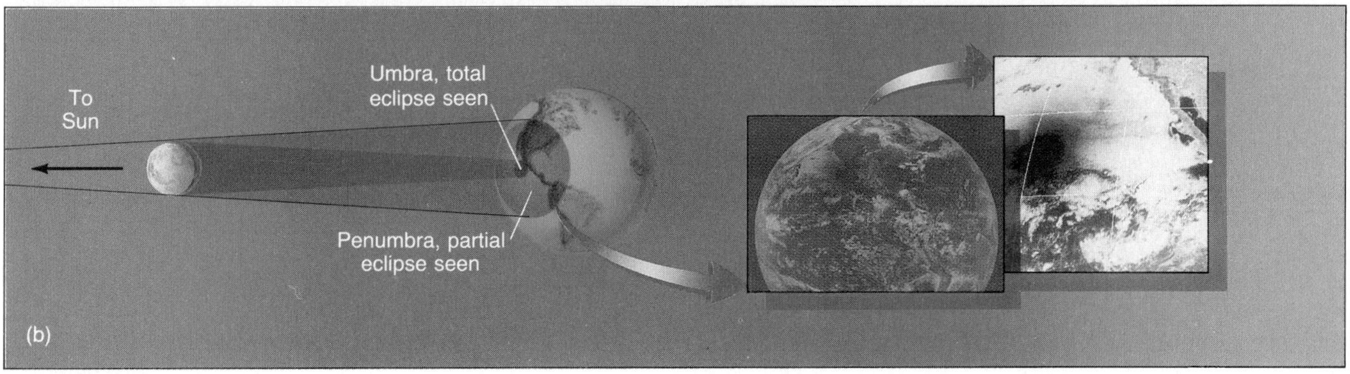
To
Sun
Umbra, total
eclipse seen
Penumbra, partial
eclipse seen
(b)

R I **V** U X G

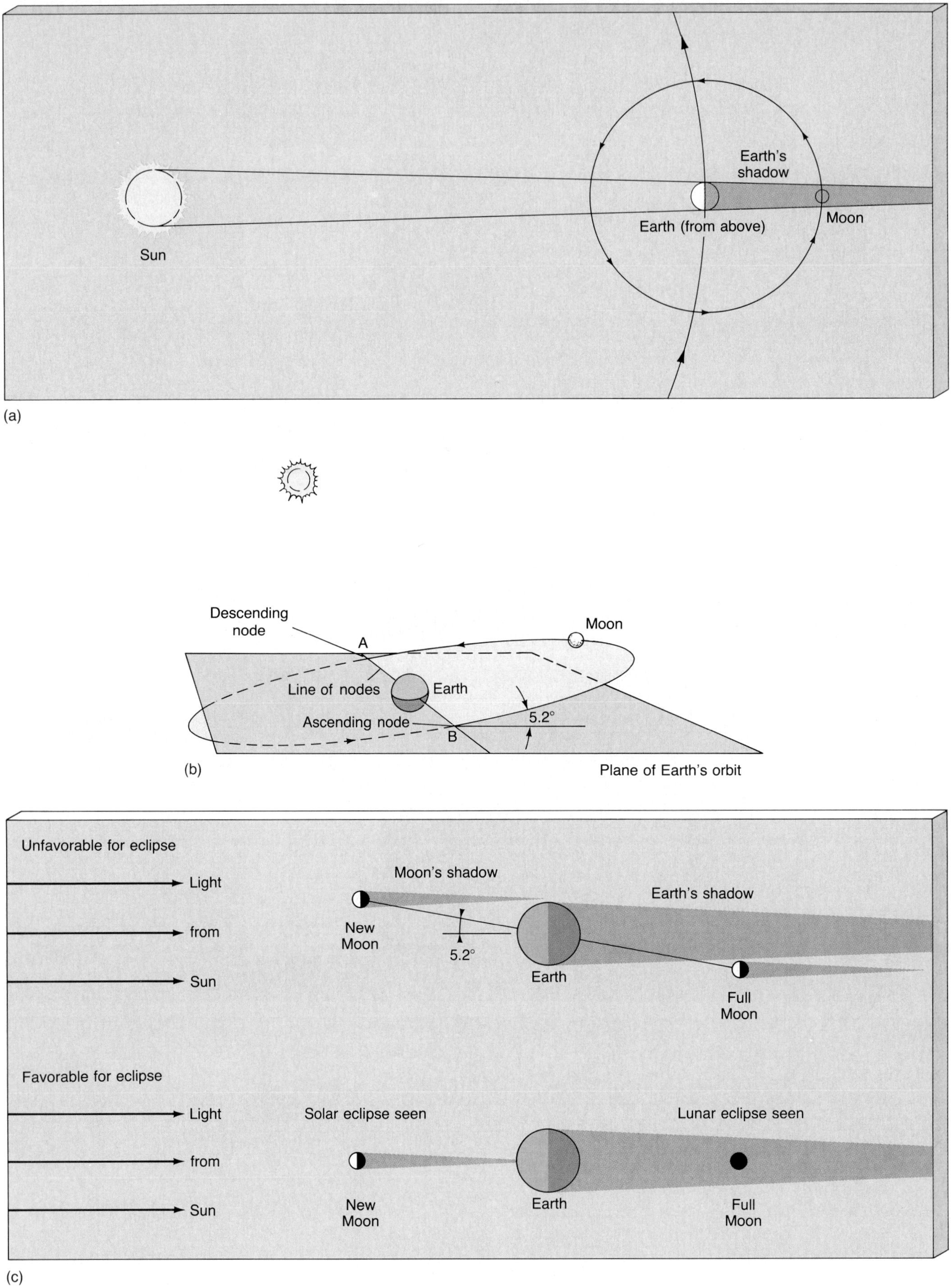
Earth's shadow
Moon
Earth (from above)
Sun
(a)
Descending node
A
Moon
Line of nodes
Earth
5.2°
Ascending node
B
(b)
Plane of Earth's orbit
Unfavorable for eclipse
Light
from
Sun
Moon's shadow
Earth's shadow
New Moon
5.2°
Earth
Full Moon
Favorable for eclipse
Light
from
Sun
Solar eclipse seen
Lunar eclipse seen
New Moon
Earth
Full Moon
(c)

(d)

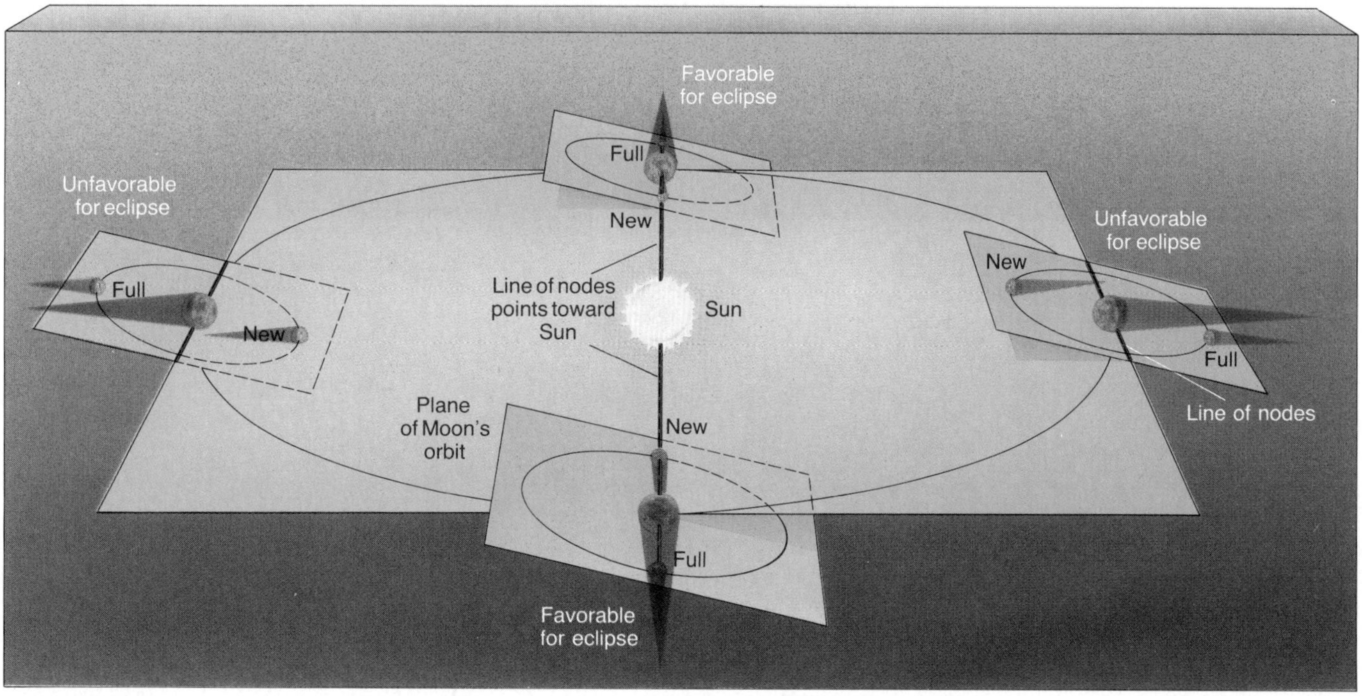

Figure 1.19 *(a) An eclipse occurs when the Earth, Moon, and Sun are precisely aligned. Here, the Moon passes through the Earth's shadow and a lunar eclipse occurs. If the Moon's orbital plane lay in exactly the plane of the ecliptic, this alignment would occur once a month. (b) The Moon's orbit is actually inclined at about 5° to the ecliptic. The intersection of the Moon's and the Earth's orbital planes is known as the line of nodes of the Moon's orbit. (c) Unfavorable and favorable orbital configurations for producing an eclipse. (d) For an eclipse to occur, the line of nodes of the Moon's orbit must lie along the Earth-Sun line. Thus, eclipses can occur only at specific times of the year.*

shown in Figure 1.19(b), the Moon's orbit is actually slightly inclined to the ecliptic, at an angle of 5.2°, so that the chance of an Earth-Moon-Sun alignment occurring just as the Moon crosses the ecliptic plane is greatly reduced. Figure 1.19(c) illustrates two configurations—one termed *unfavorable,* the other *favorable*—for producing an eclipse. In the unfavorable configuration, the Moon is above the plane of the ecliptic when new and below it when full, so that an eclipse cannot occur. In the favorable configuration, the Moon happens to be new (or full) just as it crosses the ecliptic plane and eclipses are seen. Unfavorable configurations are much more common than favorable ones.

Thus, eclipses are relatively rare events. Moreover, they can occur only at certain times of the year. In Figure 1.19(b), the two points on the Moon's orbit where it crosses the ecliptic plane are known as the *nodes* of the orbit. The line joining them, which is also the line of intersection of the Earth's and the Moon's orbital planes, is known as the **line of nodes.** As the Earth-Moon system orbits the Sun, the orientation of the Moon's orbit, and hence of the line of nodes, remains the same, as shown in Figure 1.19(d). Times when the line of nodes is not directed toward the Sun are unfavorable for eclipses. However, at the two times when the line of nodes briefly lies along the Earth-Sun line, eclipses are possible. These two times of year are known as **eclipse seasons.** They are the only times at which an eclipse can occur. Notice that there is no guarantee that an eclipse *will* occur—the lunar phase also has to be right. Thus, the condition for a solar eclipse is that we have a new Moon during an eclipse season.

In fact, the gravitational tug of the Sun causes the Moon's orbital orientation, and hence the line of nodes, to change with time. The result is that the eclipse seasons gradually progress backward through the calendar, occurring about 20 days earlier each year and taking 18.6 years to make one complete circuit. This backward progression of the Moon's orbit is known as the **regression of the nodes.** In 1991, the eclipse seasons were in January and July; on July 11, a total eclipse actually occurred, and was visible in Hawaii, Mexico, and parts of Central and South America. Because we know the orbits of the Earth and the Moon to great accuracy, we can predict eclipses far into the future. Figure 1.20 shows the location and duration of all total eclipses of the Sun until the year 2017.

The solar eclipses that we do see highlight a remarkable cosmic coincidence. Although the Sun is many times farther away from the Earth than is the Moon, it is also much larger. In fact, the ratio of distances is almost exactly the same as the ratio of sizes, so that, as seen from the Earth, the Sun and the Moon both have roughly the *same* angular size—about half a degree in diameter. Thus, the Moon covers the face of the Sun almost exactly. If the Moon were larger, we would never see annular eclipses, and total eclipses would be much more common. If the Moon were a little smaller, we would see only annular eclipses.

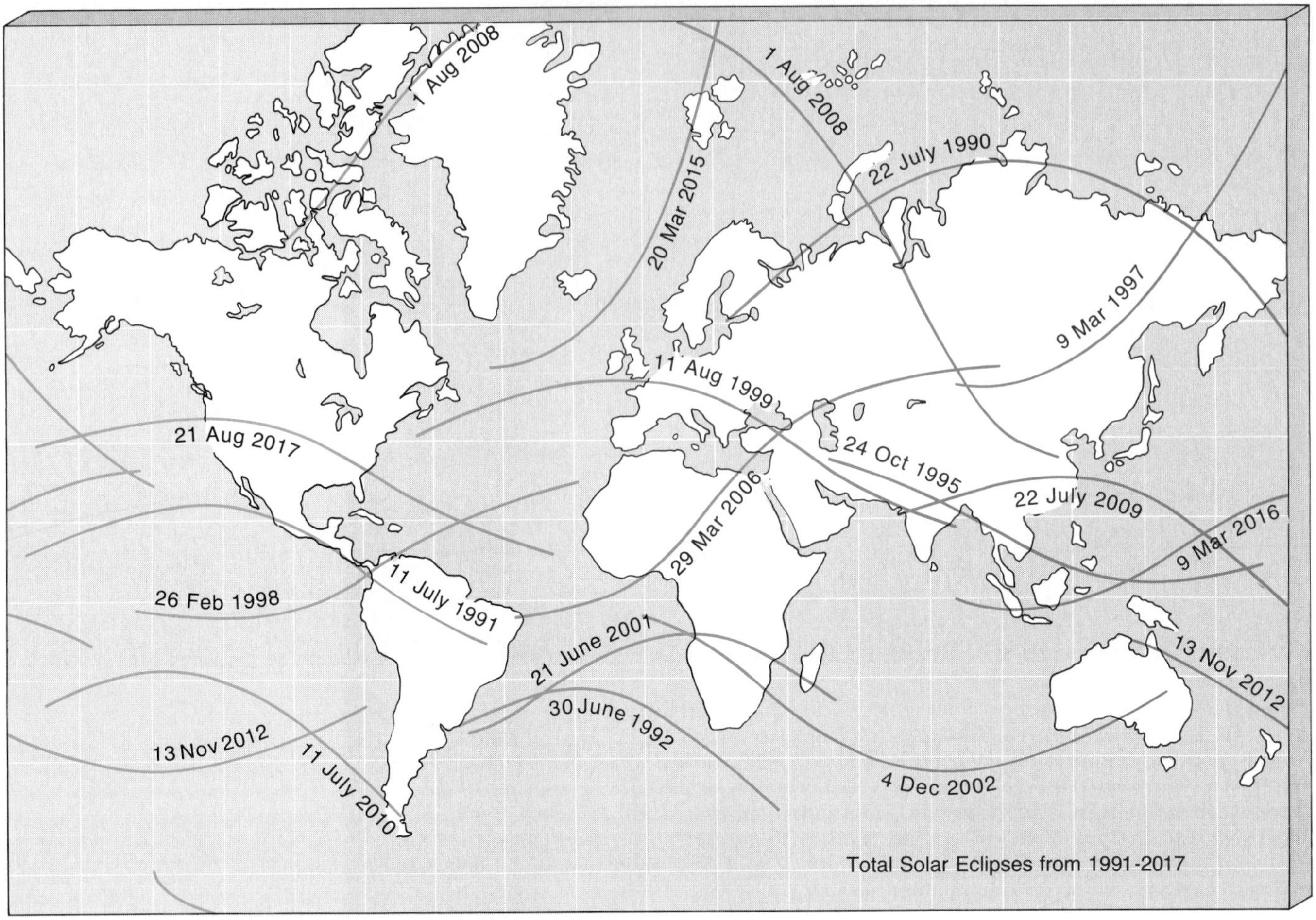

Figure 1.20 *Regions of the Earth that will see total solar eclipses between the years 1991 and 2017. Each dark line represents the path of the Moon's umbra across the Earth's surface during an eclipse. The eclipse tracks are marked with the date when they will occur.*

That *both* annular and total eclipses are observed tells us an important fact about the solar system: Assuming that the sizes of the Moon and the Sun are constant, we know their *distances* from the Earth cannot be. Otherwise these objects would always have the same angular size and so would always produce the same type of eclipse. Thus, even without detailed measurement we can conclude that the Earth's orbit around the Sun, the Moon's orbit around the Earth, or both orbits deviate from a perfect circle. We take this for granted today, but ignorance of this fact played a crucial role in the development of many early theories of the solar system, as we will see in Chapter 2.

THE EARTH MOVES

Up to now, we have been fairly liberal with our description of the night sky, sometimes picturing it in terms of a rotating sky and a fixed, central Earth, and sometimes picturing it in precisely the opposite manner. For some purposes, the two views have similar descriptive power. However, the purpose of science is understanding and prediction, and we will find in the next two chapters that one of these pictures allows us to construct a simple and powerful model of the solar system and that the other fails miserably because the Earth is not fixed, nor is it at the center of anything.

Earth has many motions—it spins on its axis, it travels around the Sun, and it moves with the Sun through the Galaxy. We have just seen how some of these motions can account for the changing nighttime sky and the changing seasons. In fact, Earth's true motion is even more complicated.

Currently, Earth's axis is tilted by about 23.5° to the plane of its orbit. This was shown in Figure 1.11(b) as the angle between Earth's axis and a perpendicular to the ecliptic plane. However, owing to the gravitational influence of the Moon and of the planet Jupiter, this tilt varies back and forth between 22° and 24°. This variation is very slow, with one cycle completed only every 41,000 years.

In addition, like a spinning top that rotates rapidly on its own axis while slowly turning about the vertical, Earth's axis changes its *direction* over the course of time (although the angle between the axis and the ecliptic remains close to 23.5°). This slow change in the direction of the axis of a spinning object is called **precession.** Figure 1.21 illustrates this precession which, in the case of Earth, is caused mostly by the gravitational tugs of the Moon and the Sun. A complete cycle of precession traces out a cone and takes about 26,000 years. One consequence of this is that the vernal equinox drifts slowly around the zodiac as the Earth's axis changes its orientation. Another term for this drift is the **precession of the equinoxes.**

Because of the Earth's precession, the length of time from one vernal equinox to the next—one tropical year, or 365.242 solar days—is not quite the same as one sidereal year—the time required for the Earth to complete one orbit, 365.256 solar days. Because of the slight shift in the Earth's axis over the course of a year, one tropical year is about 20 minutes less than one sidereal year. The tropical year is the year that our calendars measure. If our timekeeping were tied to the sidereal year, the seasons would slowly march around the calendar as the Earth precessed—13,000 years from now, summer in the northern hemisphere would be at its height in late February! By using the tropical year instead, we ensure that July and August will always be summer months. However, in 13,000 years' time, Orion will be a summer constellation.

There are still more complexities to the Earth's motion. For example, the cone traced out by Earth's precession is not as clean as that drawn in Figure 1.21. As mentioned previously, the gravitational effect of the Sun causes the Moon's orbit around the Earth to rotate slowly. This regression of the line of nodes of the Moon's orbit in turn causes the Earth's rotation axis to wobble ever so slightly, changing the angle between Earth's axis and the ecliptic by plus or minus 9 arc seconds ($\pm 9''$) every 18.6 years. This additional motion, superimposed on the Earth's precession, is known as **nutation.**

Finally, the Sun itself travels through space. Currently, the Sun is moving at a speed of about 20 kilometers per second (relative to our neighboring stars) toward Vega, a bright star almost directly overhead in the early evening autumn sky (see the sky charts at the end of the book).

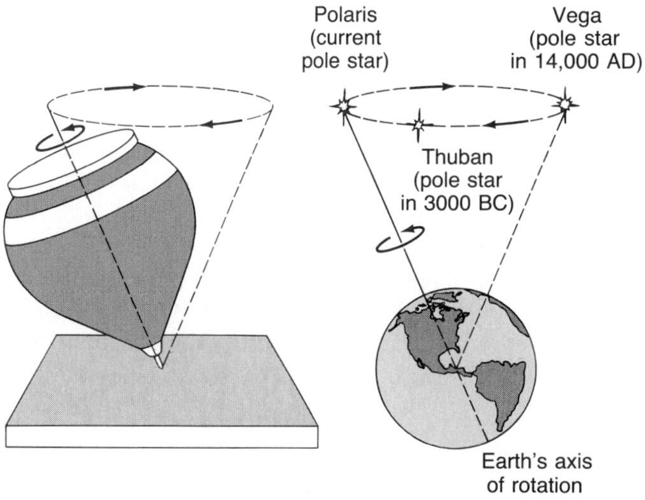

Figure 1.21 *Earth's axis currently points toward the North Star. Some twelve millennia from now—nearly half-way through one cycle of precession—Earth's axis will point toward a star called Vega. Five millennia ago, it was pointed at the star named Thuban in the constellation Draco.*

CHAPTER SUMMARY

Astronomical research suggests that we live on a planet orbiting a star on the edge of one galaxy, among billions of galaxies in the universe. Astronomical science relies on the premise that the laws of physics apply in distant realms of space and time, just as they do at home. To an astronomer, the universe is the totality of all space, time, matter, and energy.

Astronomy deals in gigantic numbers, enormous intervals of space, and extremely long durations of time. Scientific notation lets astronomers avoid writing out long numbers by hand. It is a shorthand notation in which the number of zeros is denoted by a superscript power, or exponent, of 10. Astronomers use units that make sense within a context, but as contexts change, so do the units.

Ancient skywatchers imagined connections between the brightest stars in the sky. The patterns they saw and named are called constellations. The names generally originated as names of mythical gods, heroes, and animals. For the most part, the stars that make up a particular constellation are not gravitationally bound or even physically close to one another.

The stars appear to be attached to a celestial sphere—a canopy of stars entirely surrounding Earth. This imaginary sphere is useful today because it lets astronomers devise a coordinate system for precisely locating stars. Celestial coordinates are a natural extension from the system of longitude and latitude used to locate points on Earth's surface. Instead of longitude and latitude, astronomers generally use right ascension and declination. Astronomers measure the width of objects on the dome of the sky. These are called angular measurements. The portion of the sky visible to one person at any one time spans 180 degrees. Each degree can be subdivided into 60 arc minutes. Each arc minute can be subdivided into 60 arc seconds. The angular size of an object depends both on its actual size and on its distance from us.

Earth both rotates on its axis and revolves around the Sun. Earth's rotation creates day and night and thereby establishes the basic rhythm of our lives. As Earth revolves around the sun, there is a change in both the height of the Sun above the horizon and the length of the day. These changes combine to account for seasonal changes on the planet. We see different stars in summer than in winter because, over a period of 6 months, our night sky shifts to reveal an opposite region of the galaxy.

The Moon shines with reflected light. It moves across the sky at a rate of about 12° per day. This movement stems from the Moon's revolution around Earth. One hemisphere of the Moon is always fully lighted by the Sun, but from Earth we usually see only a fraction of this lighted hemisphere. We describe the change in our view of the Moon's lighted hemisphere as a change in the Moon's "phase."

A lunar eclipse takes place when the Sun and the Moon are in exactly opposite directions as seen from Earth, and the Earth's shadow sweeps across the Moon's face. A solar eclipse takes place when the Moon passes directly in front of the Sun, and day briefly turns to night.

Earth spins on its axis, travels around the Sun, and moves with the Sun through the Galaxy. These motions account for the changing nighttime sky and the changing seasons. Earth's 23.5° axial tilt also changes, and the axis wobbles and precesses like a top.

KEY WORDS

annular eclipse
arc degree
arc minute
arc second
astronomy
autumnal equinox
celestial coordinates
celestial equator
celestial sphere
constellation
crescent
declination
degree
diurnal
eclipse
eclipse season
ecliptic
equinox
full moon
gibbous
hour
light-year
line of nodes
lunar eclipse
lunar phase
minute
new moon
north celestial pole
nutation
partial eclipse
penumbra
precession
precession of the equinoxes
quarter moon
regression of the nodes
revolution
right ascension
rotation
scientific method
seasons
second
sidereal day
sidereal month
sidereal year
solar day
solar eclipse
south celestial pole
summer solstice
synodic month
total eclipse
tropical year
umbra
vernal equinox
universe
wane
winter solstice
wax
zodiac

REVIEW QUESTIONS

1. What truth is there to the poet's declaration that people are made of stardust?
2. Why is it important to astronomical science that the laws of physics as we know them apply equally well throughout the whole universe?
3. What does an astronomer mean by "the universe?"
4. What is a light-year?
5. Write these numbers in scientific notation: 100; 1000; 1,000,000; 1,000,000,000,000,000; 0.01; 0.001; 123,000; 0.000456.
6. How many grams are there in a kilogram? How many centimeters in a meter? How many meters in a kilometer?
7. What is a constellation?
8. What old method, still in use today, can be used to describe the location of stars in the sky?
9. Why does the Sun rise in the east and set in the west each day? Does the Moon also rise in the east and set in the west? Why? Do stars do the same? Why?
10. If you look at a group of stars at the same time every evening or in the hour before dawn, you'll notice that they shift progressively westward as the weeks and months pass. Why?
11. How many times in your life have you traveled around the sun?
12. What is a solar day? A sidereal day?
13. Why do we have seasons on Earth?
14. Why do we see different stars in summer than in winter?
15. At what phase of the Moon would you expect there to be the most burglaries? Why?
16. If one complete hemisphere of the Moon is always lighted by the sun, why do we see different phases of the Moon?
17. What causes a lunar eclipse? A solar eclipse?
18. What does the fact that there are both annular and total eclipses tell us about the Moon's distance from Earth?
19. Why aren't there lunar and solar eclipses every month?
20. What is an eclipse season?

21. From what parts of Earth is the Pole Star (Polaris) never visible?

22. In 1 second, light leaving Los Angeles will reach approximately as far as (a) the Moon (400,000 km), (b) Venus (0.3 AU from Earth at closest approach), (c) the nearest star (about 1 pc from Earth), (d) London (roughly 10,000 km), or (e) San Francisco (about 500 km). Which is correct?

23. Through how many degrees, arc minutes, or arc seconds does the Moon move in (a) one hour of time, (b) one minute, (c) one second? How long does it take for the Moon to move a distance equal to its own diameter?

DISCUSSION QUESTIONS

1. Do you think your view of yourself would be different if you thought you lived in a universe bounded by encircling crystal spheres?
2. Can you imagine any other cosmology—any other way of interpreting the universe around us—besides those described in this chapter?
3. What do stars in constellations have in common, when viewed from our earthly vantage point? If you traveled to the outermost planet in our solar system, do you think the constellations would appear to change their shapes? What would happen if you traveled to the next-nearest star? If you traveled to the center of the Galaxy, could you still see the familiar constellations found in Earth's night sky?

PROJECTS

1. Before learning any constellations, go to a country location on a clear dark night. Imagine patterns among the stars, and name the patterns yourself. Note (or better yet, draw) these stars' location with respect to trees or buildings in the foreground. Do this every week or so for a couple of months. Be sure to look at the same time every night. What happens?
2. Now learn some constellations! Monthly maps of the night sky can be found in both *Astronomy* and *Sky and Telescope* magazines.
3. Find the star Polaris, also known as the North Star, in the evening sky. Identify any separate pattern of stars in the same general vicinity of the sky. Wait several hours, at least until after midnight, and then locate Polaris again. Has Polaris moved? What has happened to the nearby pattern of stars? Why?
4. Hold your little finger out at arm's length. Can you cover the disk of the Moon? The Moon projects an angular size of 30′ (half a degree), and your finger should more than cover it. How can you apply this fact in making sky measurements?
5. Advanced idea: Can you see why the sun rises and sets due east and due west at the equinoxes? You'll probably need a sky chart or map that shows the celestial equator to see why.

SUGGESTED READINGS

Burnham, Robert, Jr., *Burnham's Celestial Handbook*. 3 vol. New York: Dover, 1978. Comprehensive work on the contents of outer space, as we knew them a few years ago—plenty on the lore of the sky, too.

Ferris, T. *Coming of Age in the Milky Way*. New York: Morrow, 1988. Another book about how our perception of outer space has changed, by a popular author.

Koestler, A. *The Sleepwalkers*. New York: Grosset and Dunlap, 1963. This is a wonderful but dense book tracing the history of our understanding of the cosmos.

Peltier, L. *Starlight Nights*. Cambridge, MA: Sky Publishing, 1965. Lyrical account of a man's love affair with the night sky.

There are two excellent magazines for beginning astronomers. One discusses astronomy mainly from the amateur, or hobbyist, point of view. Nice pictures. Monthly sky columns and charts. Twelve issues per year for $24.

Astronomy
Kalmbach Publishing
P.O. Box 1612
Waukesha, WI 53187-9950

The other astronomy magazine is read by many professionals as well as amateurs. It presents a good view of the world of astronomical research, but some articles are a little dense for most beginners. Monthly sky columns and charts. Twelve issues per year for $24.

Sky and Telescope
Sky Publishing Corporation
P.O. Box 9111
Belmont, MA 02178-9918

CHAPTER

2

Measurement and Modeling
The Foundations of Astronomy

These San Francisco buildings seem to be rising up to meet the Moon. To be sure, these structures are built with steel, concrete, and glass, but just as surely they are built on the principles of geometry. These principles allow us to measure the length of a girder, to project the height of a skyscraper, and so to construct our buildings. Humans have extended geometry to measure the distance to the Moon itself and beyond, and we have used what we have learned to build our models of the universe. (C. Trost)

R I **V** U X G

LEARNING GOALS

1 to appreciate the importance of the scientific method in astronomy

2 to realize that some ancient people knew the size and shape of the Earth

3 to appreciate the straightforward geometric principles that form the modern foundation of the cosmic distance scale

4 to see how some ancient civilizations attempted to explain the heavens in terms of Earth-centered models of the universe

Living in the Space Age, we may find it hard to imagine that Earth is anything but spherical. Images of the Earth taken from space leave no doubt that our planet is round. Yet there was a time, not too long ago, when our ancestors maintained that the Earth was flat. Indeed, until the beginning of the Renaissance, in the fifteenth century, most people believed Earth to be flat. There were some exceptions, however. These were people schooled in geometry, the study of the size, shape, and scale of things.

Geometers of old, at least as far back as the ancient Greeks and probably the Sumerians and Babylonians before them, realized that Earth and many other cosmic objects are spherical. Paintings on cave walls imply that even some people of the Stone Age may have had some idea of the true shape of our planet thousands of years before then. We can't be sure that they recognized Earth to be a near-perfect sphere, but they surely had a strong interest in the phenomena of the skies.

Our view of the universe has undergone a radical transformation since those ancient times. Humankind has been taken from its throne at the center of the cosmos and relegated to a rather unremarkable position on the periphery of the Galaxy. However, we have been amply compensated for our loss of prominence by the wealth of scientific knowledge we have gained in the process. The story of how all this came about is the story of the rise of the scientific method and the genesis of modern astronomy.

THE SCIENTIFIC METHOD

✓1 The idea that Earth is at the center of the universe was overthrown only a few centuries ago. During the fifteenth and sixteenth centuries, people began to inquire more critically about themselves and the universe. They realized that thinking about Nature was no longer sufficient; *looking* at it was also necessary. Experiments and observations became a central part of the process of inquiry. To be effective, theories—the framework of ideas and assumptions used to explain observations of the real world—had to be tested and either refined, if experiment favored them, or rejected, if it did not. This new approach to investigation, combining thinking and doing—that is, theory and experiment—was the **scientific method,** and it ushered in the age of modern science.

The scientific method flowered during the Renaissance, but its seeds were planted long before. The earliest steps toward substituting myth with critical thinking were largely based on the school of thought championed by the philosopher Thales (625–547 B.C.). Especially intrigued by astronomy and the origin of things, Thales is perhaps best known for having predicted an eclipse of the Sun. Using high-speed computers to trace back the positions of the Sun, Earth, and Moon, we know that a total solar eclipse did in fact occur in Asia Minor in 585 B.C.

Thales' predictive success might not have been entirely his own, however. His hometown of Miletus, Greece, had close ties with Babylonia, whose "astropriests" apparently knew how to predict eclipses as early as about 1000 B.C. Doubtless, trading and commercial contacts with Babylonia (and perhaps with Egypt as well) also provided the Greeks with rich cultural interactions that helped soften the superstitions of earlier, mythological times. As knowledge from all sources was sought after and embraced for its own sake, the influence of logic and reasoned argument grew, and the power of myth diminished.

One of the greatest achievements of the ancient Greeks was that they conceived of the universe as something to be studied, questioned, and even explained—in rational terms. Moreover, their inquiries resulted not from some practical need but from a passion for knowing. An intellectualism had seized humankind more than 20 centuries ago.

The Greek attitude toward nature considerably influenced the philosophies and historical systems developed by the classical thinkers. Some of the earliest known models of the universe were based largely on imagination and pure thought, with little attempt to explain the workings of the heavens in terms of known earthly experience. The early Greeks rarely paused to test their theories by further observation and experiment. Their disdain for manual labor may have had much to do with their attitude that thoughts and observations need not be exhaustively checked by tests and more tests. However, some did come to realize the importance of prediction, experimentation (or observation), and correction to the formulation of their theories. The success of their approach slowly but surely led to the modern scientific method and opened the door to a fuller understanding of nature.

Scientists throughout the world today use an approach that relies heavily on testing ideas. In what is usually a three-step process, scientists gather some data, form a theory, and test that theory. This is a rational, methodical approach used to investigate all natural phenomena. Experiment and observation are integral parts of the process of scientific inquiry. Theories unsupported by such evidence rarely gain any measure of acceptance in scientific circles. Used properly over a period of time, the scientific method enables us to arrive at conclusions that are mostly free of the personal bias and human values of any one scientist. The scientific method is designed to yield an objective view of the universe we inhabit.

✓2 Aristotle (384–322 B.C.) performed one of the first documented uses of the scientific method nearly 25 centuries ago. Aristotle noticed that during a lunar eclipse, when Earth is positioned between the Sun and the Moon, it casts a *curved* shadow onto the surface of the Moon. Figure 2.1 shows a series of photographs taken during a recent lunar eclipse. The Earth's shadow, projected onto the Moon's surface, is curved. This is what Aristotle must have seen and recorded so long ago.

On the basis of this observation, Aristotle theorized that all lunar eclipses would show that Earth's shadow was curved, regardless of the orientation of our planet. Furthermore, because the observed shadow is always an arc of the same circle, he concluded that the Earth must be round. Aristotle was not the first person to argue that Earth is round, but he was apparently the first to offer a proof of it using the lunar-eclipse method.

In concluding that the Earth was round, Aristotle made a basic assumption. He assumed that the geometrical relationships of everyday terrestrial objects—balls, sticks,

R I **V** U X G

Figure 2.1 *In this series of photographs taken of the Moon during a lunar eclipse, Earth's shadow, projected onto the Moon, is curved. Not only is it curved, but that curvature seems to be part of a circle or a sphere, implying that the Earth is round. (G. Schneider)*

planes, triangles—also held true for much larger objects. That is, he assumed that the terrestrial geometry he knew, namely Euclidean geometry, could be extended into the astronomical domain. He used the same simple, straightforward geometry now taught in high school—the geometry of flat space, where the circumference of a circle is π (= 3.14159...) times the diameter, the angles of a triangle add up to 180°, and parallel lines never meet—and applied it to the Earth and the Moon. In particular, he took the known fact that a sphere is the only object whose shadow is always circular, regardless of its orientation, and used it to infer the shape of our planet.

Aristotle argued further that if Earth were really round, distant stars in the nighttime sky would appear at different positions to observers at different latitudes. He quoted the experience of travelers as confirmation of his theory, especially the then well-known fact that a bright star named Canopus was visible in Egypt but not farther north.

The reasoning procedure used by Aristotle forms the basis of all scientific inquiry today. He first made an observation. He then formulated a theory. And finally, he tested the validity of the theory by making predictions that could be confirmed or refuted by further observation. *Observation, theory, and testing*—these are the cornerstones of the scientific method, a technique whose power will be demonstated again and again throughout our text.

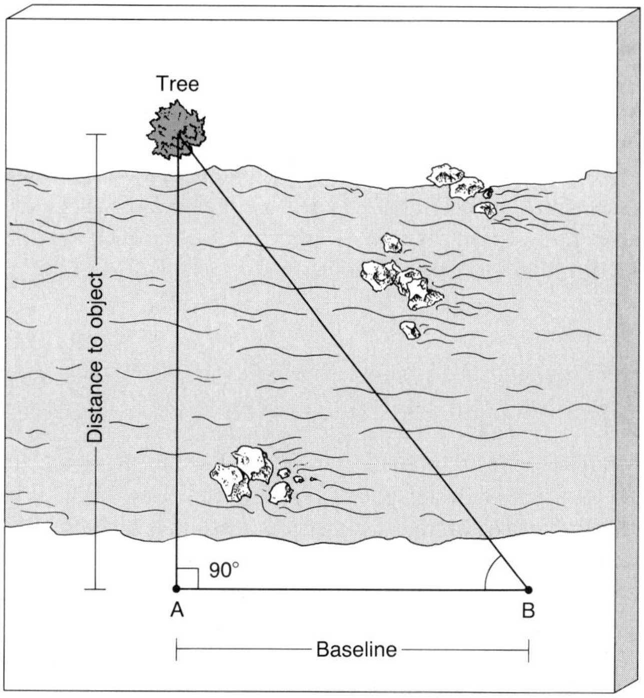

Figure 2.2 *Surveyors often use simple geometry and trigonometry to estimate the distance to a faraway object.*

THE MEASUREMENT OF DISTANCE

Triangulation

✓3 In Chapter 1, we concentrated on the changing positions of the Sun, the Moon, and the stars in the sky. Of course, knowing just the direction to something is only part of the information needed to locate it in space. Before we can make a systematic study of the heavens, we must find a way of measuring *distances,* too. One such method is known as **triangulation.** It is based on the principles of Euclidean geometry and finds widespread application today in both terrestrial and astronomical settings. The method Eratosthenes used to measure the size of the Earth (see Interlude 2-1) relied on the same principles. Today's engineers, especially surveyors, use these age-old geometrical ideas to measure indirectly the distance to faraway objects. In astronomical contexts, triangulation forms the foundation of the family of distance-measurement techniques that together make up the **cosmic distance scale.**

For example, imagine trying to measure the distance to a tree on the other side of a river. The most direct method is to lay a tape across the river, but that's not the simplest way. A smart surveyor would make the measurement by visualizing an *imaginary* triangle, sighting the tree on the far side of the river from two positions on the near side, as illustrated in Figure 2.2. The simplest possible triangle is a right triangle, in which one of the angles is exactly 90°, so it is usually convenient to set up one observation position directly opposite the object, as at point A. The surveyor then moves to another observation position at point B, noting the distance covered between points A and B. This distance is the **baseline** of the imaginary triangle. Finally, the surveyor sights toward the tree whose distance is to be measured and notes the angle at point B. No further observations are required. The rest of the problem is a matter of calculation. Knowing the value of one side (AB) and two angles (the right angle itself, at point A, and the angle at point B) of the right triangle, the surveyor can geometrically construct the remaining sides and angles and so establish the distance to the tree.

To use triangulation to measure distances, a surveyor must be familiar with trigonometry, the mathematics of geometrical angles. However, even if we knew no trigonometry at all, we could still solve the problem by graphical means, as shown in Figure 2.3. Pacing off the baseline, letting it equal so many intervals on a piece of graph paper, and setting the angle at point B, we can draw the complete triangle. The number of graphed intervals from point A to the tree then approximates the desired distance. We have solved the real problem by **modeling** it on paper.

One further point: The right triangle is not essential to estimate distance in this way. We could use any triangle, although an additional measurement is needed when one of the angles of the triangle is not equal to 90°. All the sides and all the angles of a triangle can be indirectly determined by simple geometry, provided that we know either two sides and the included angle, or two angles and the included side.

The point to remember is this: Nothing more complex than simple geometry is needed to infer the distance, the size, and even the shape of an object too far away or too inconvenient to measure directly.

We stress again that this type of modeling assumes that the geometry of *real* terrestrial objects—rivers and trees, for example—mimics the geometry of much smaller *scale* models, like those drawn on paper in Figures 2.2 and 2.3. That is, we assume that the principles of geometry are independent of an object's size. This assumption, used centuries ago by the Greek geometers, is the assumption we will continue to make now as we extend the imaginary triangle to measure the distances to extraterrestrial objects.

INTERLUDE 2-1 *Earth Dimensions*

About 200 B.C. another Greek, Eratosthenes (276–194 B.C.), used Aristotle's ideas to measure the physical size of the Earth. The simple geometrical reasoning he employed now provides a basis for all measurements of distance outside of our own solar system.

Eratosthenes knew that although observers at a certain location saw the Sun directly overhead on a certain day of the year, observers at other locations saw the Sun at some angle displaced from directly overhead. In particular, he knew that on the first day of summer the Sun passed *directly* overhead at Syene, Egypt, as shown in the figure below. At midday of the same day in Alexandria, a city 5000 stadia to the north, he measured the angular displacement of the Sun from the overhead to be 7.2 arc degrees. (The stadium was a Greek unit of length believed to have been about 0.16 km, although the exact value is uncertain—not all Greek stadia were the same size.)

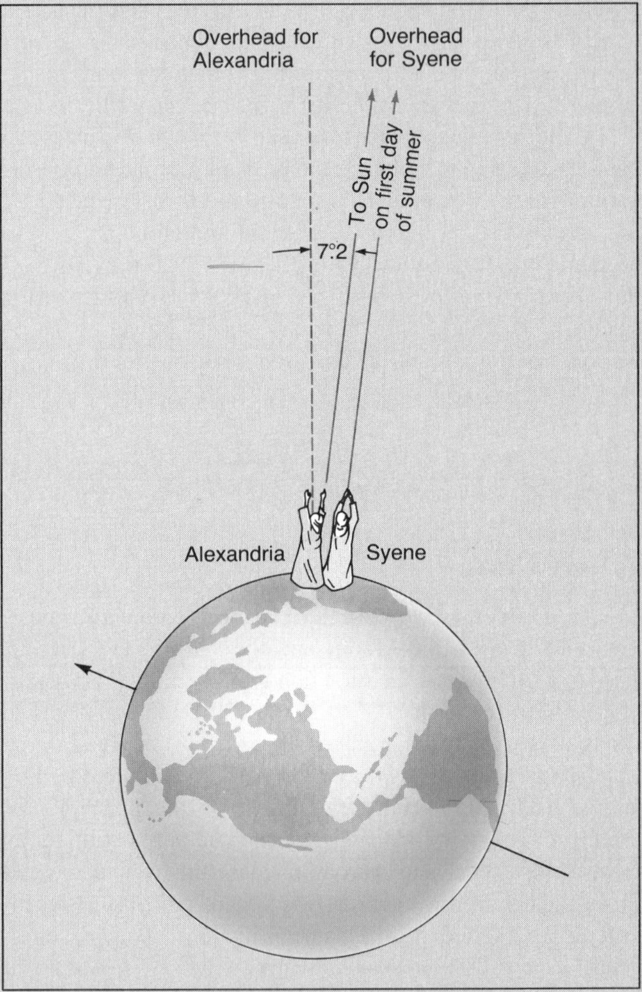

Rays of light reaching us from a very distant object, such as the Sun, travel almost parallel to one another. Consequently, as shown in the figure, the angle measured at Alexandria between the Sun's rays and the overhead is also proportional to the small portion of Earth's total circumference between Syene and Alexandria. Since there are 360 degrees in a full circle, we see that 7.2° is $^1/_{50}$ of a full circle. The entire Earth's circumference can then be estimated by multiplying the distance between the two cities by a factor of 50. We can express this reasoning as follows:

7.2° is to 360° as 5000 stadia is to ?

The answer is 50 × 5000, or 250,000 stadia. If we take the stadium unit to be 0.16 km, we find that Eratosthenes estimated the circumference of Earth at about 40,000 km. In fact, he went one step further. Knowing that the circumference C of a circle is related to its radius R by the relation $C = 2\pi R$, he was also able to infer that the Earth's radius is $250{,}000/2\pi$ stadia, or 6366 km. The correct values for the Earth's circumference and radius, now measured accurately by orbiting spacecraft, are 40,070 and 6378 km, respectively.

Eratosthenes' reasoning was a remarkable accomplishment. More than 20 centuries ago, he estimated the circumference of the Earth to within 1 percent accuracy. Even more remarkable, he used only simple Euclidean geometry—the geometry of terrestrial familiarity. Even if our modern value for the size of one stadium turns out to be incorrect, the real achievement—that a person making measurements on only a small portion of the Earth's surface was able to compute the size of the entire planet on the basis of observation and logic—is undiminished.

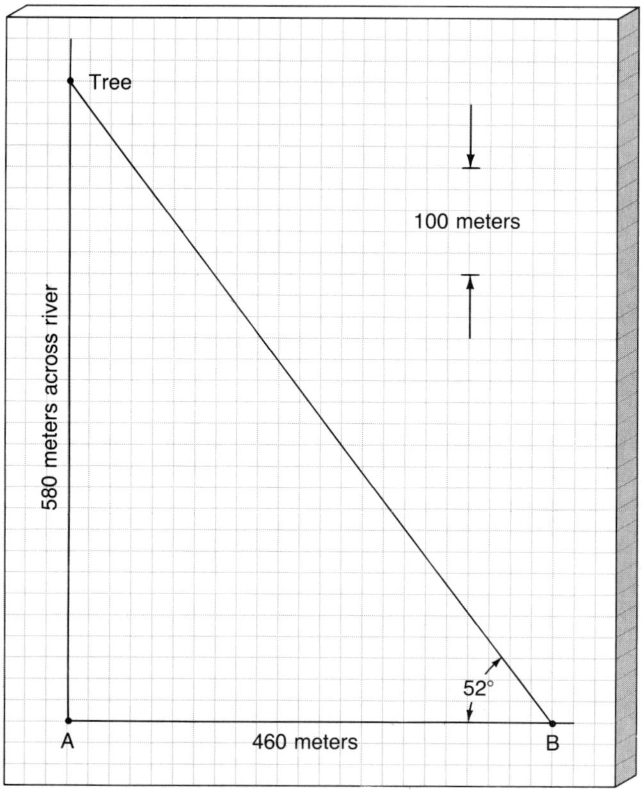

Figure 2.3 Knowledge of trigonometry is not even needed to estimate distances indirectly. Scaled estimates, like this one on a piece of paper, often suffice.

Extraterrestrial Applications

Triangles with larger baselines are needed if we are to measure greater distances. Figure 2.4 shows a triangle having a fixed baseline between two observation positions at points A and B. Note how the triangle becomes narrower as an object's distance becomes progressively greater. Narrow triangles cause problems because the angles at points A and B are hard to measure accurately. The measurements can be made easier by "fattening" the triangle—in other words, by lengthening the baseline.

Consider an imaginary triangle extending from Earth to a nearby object in space, perhaps the Moon or a neighboring planet. The imaginary triangle is extremely long and narrow even for the nearest cosmic objects. Figure 2.5(a) illustrates case where Earth's diameter, measured from point A to point B, is the baseline. Two observers could, in principle, sight the object from opposite sides of the Earth and thus measure the triangle's angles at points A and B, but in practice, these angles cannot be accurately measured.

It is actually easier to measure the third angle of the imaginary triangle, namely the very small one near the object. This may seem odd, because the angle of interest is so far removed from our point of view on Earth, yet a special observational technique enables astronomers to measure this narrow angle very accurately.

Observers on either side of the Earth sight toward the object, taking note of its position *relative to some distant stars* seen on the plane of the sky. The observer at point A sees the object projected against a field of very distant stars. Call its apparent location A′, as indicated in Figure 2.5(a). Similarly, the object appears projected at point B′ to an observer at point B on Earth. If each observer takes a photograph of the appropriate region of the sky, the object will appear at slightly different places in the two images. In other words, the object's photographic image is slightly displaced, or shifted, relative to the field of distant background stars, as shown in Figure 2.5(b). The background stars themselves appear undisplaced because of their much greater distance from the observer. This apparent displacement of a foreground object relative to the background as the observer's location changes is known as **parallax.** The size of the shift in Figure 2.5(b) equals the very small angle shown in Figure 2.5(a). For historical reasons, one-*half* of this angle is called the **parallactic angle.**

Parallax is the *apparent* shift of an object as seen from two different points of observation. The closer an object is to the observer, the fatter is the imaginary triangle and the larger the parallax. To understand this concept, hold a pencil vertically in front of your nose, as sketched in Figure

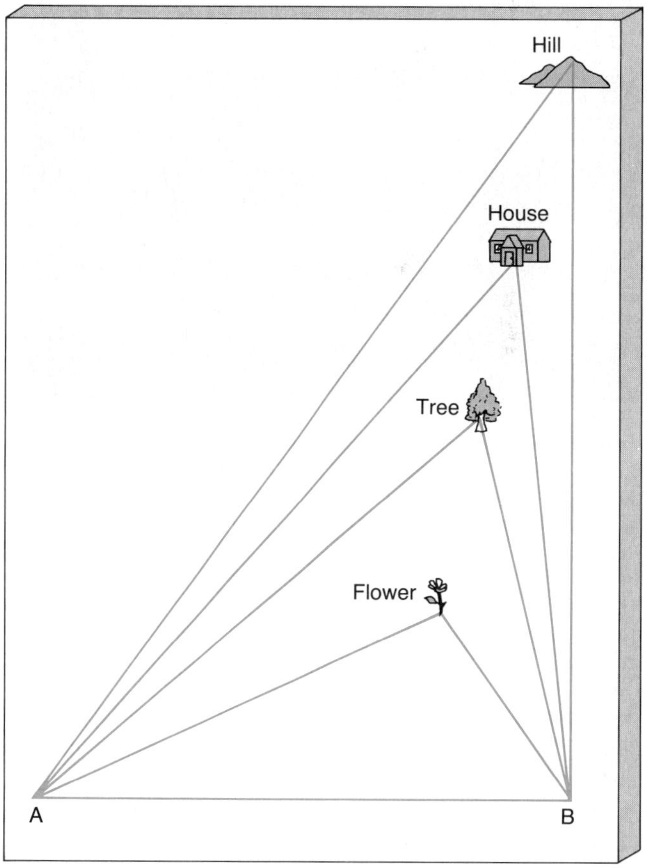

Figure 2.4 A triangle of fixed baseline (distance between points A and B) is narrower the farther away the object. As shown here, the imaginary triangle is much thinner when estimating the distance to a remote hill than it is when estimating the distance to a nearby flower.

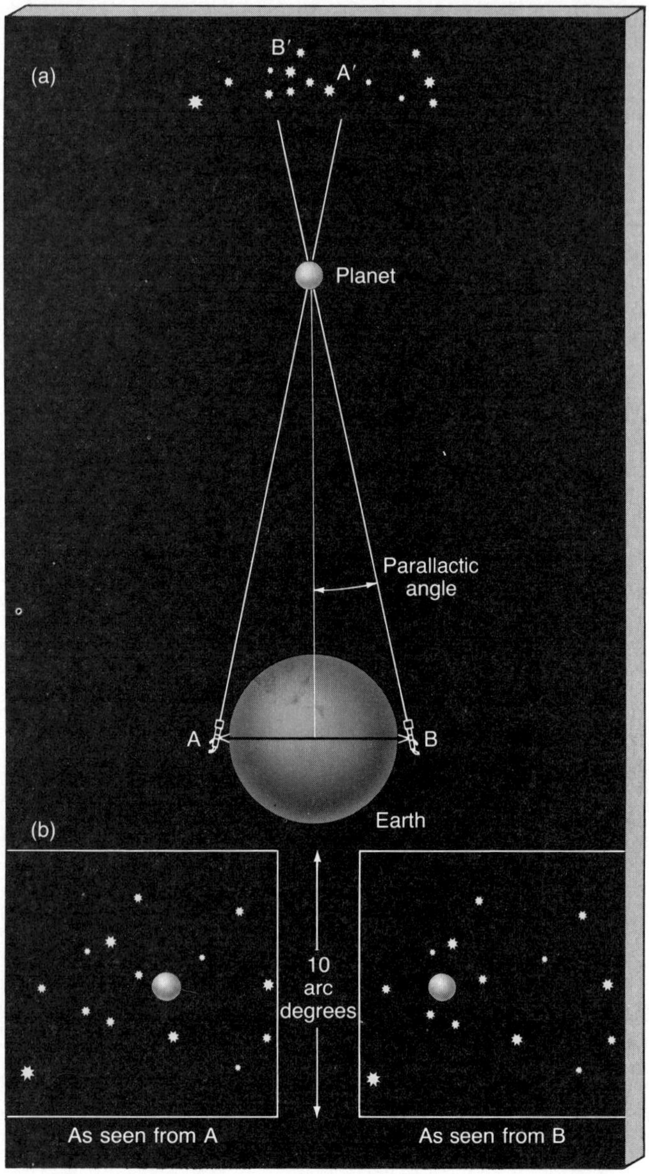

Figure 2.5 (a) This imaginary triangle extends from Earth to a nearby object in space (such as a planet). The group of stars at the top represents a background field of very distant stars. (b) Hypothetical photographs of the same star field showing the nearby object's apparent displacement, or shift, relative to the more distant, undisplaced stars.

2.6. Concentrate on some remote view, say a distant wall. Now close one eye. Then open it while closing the other. By blinking in this way, you should be able to see a large shift of the apparent position of the pencil projected onto the distant wall. In this example, one eye corresponds to point A, the other eye to point B, the distance between your eyeballs to the baseline, the pencil to the nearby object, and the distant wall to a remote field of stars. If you now hold the pencil at arm's length, corresponding to the case of a more distant object but one still not as far away as the distant stars, the apparent shift of the pencil will be less. By moving the pencil farther away, we are narrowing the triangle and decreasing the parallax (and, in the process, making its accurate measurement more difficult). If you were to paste the pencil to the wall, corresponding to the case where the object of interest is as far away as the background star field, blinking would produce no apparent shift of the pencil at all.

The amount of parallax is thus inversely proportional to an object's distance. Small parallax implies large distance. Conversely, large parallax implies small distance. Knowing the amount of parallax and the length of the baseline, we can derive that distance through triangulation. As surveyors of the sky, we use the same information as the surveyor of the land—the angles and one side of a triangle. The calculation is basically the same. Only the means used to obtain the angles are different.

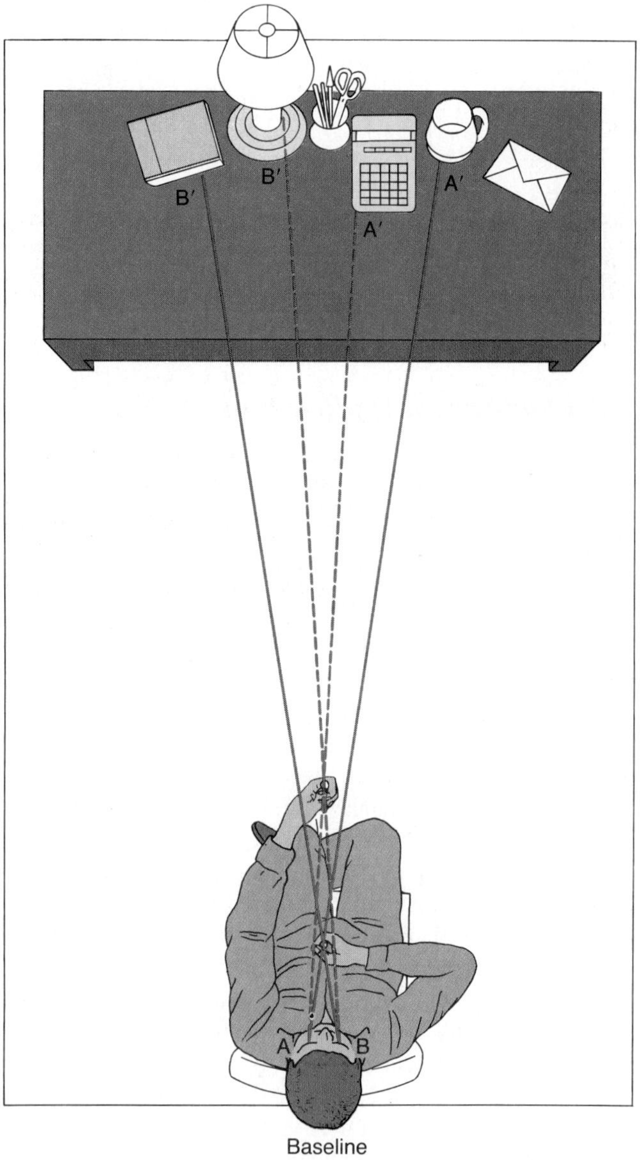

Figure 2.6 Parallax is inversely proportional to an object's distance. An object near your nose has a much larger parallax than an object held at arm's length.

THE MOTION OF THE PLANETS

The preceding sections presented two vital parts of any attempt to understand the universe—the construction of a theory and the refinement of that theory through observation, modeling, and measurement. Let us now look at the roles these key ideas played in answering the central questions facing early astronomers—the description, prediction, and explanation of the ever-changing night sky. We will see how the growth of the scientific method, together with continually improving observations and models, eventually led to a comprehensive understanding of the solar system. In the process, we will become acquainted with some basic astronomical concepts that will serve us well throughout the remainder of the text.

The Greeks of antiquity, and perhaps civilizations before them, built models of the universe. The study of the workings of the universe on the very largest scales is called **cosmology.** Today, cosmology entails looking at the universe on scales so large that even entire galaxies can be regarded as mere points of light scattered throughout space. To the Greeks, however, the universe comprised essentially the **solar system**—namely, the Sun, Earth, Moon, and the known planets. The stars beyond were surely part of the universe, but they were considered to be fixed, unchanging beacons on a mammoth celestial dome. The Greeks did not consider the Sun, Moon, and planets to be part of the celestial sphere. There was something different about those objects.

Chapter 1 presented a broad sketch of our home in space, of some of the changing motions of planet Earth and how these changes in turn produce apparent changes in the nighttime sky. We have already seen how ancient astronomers had to introduce the idea of additional spheres, separate from that bearing the stars, to account for the monthly and yearly motions of the Moon and the Sun across the sky. With the inclusion of some small corrections to allow for the precession of the equinoxes and the regression of the Moon's orbit, this model actually did a fairly good job of describing and predicting the positions of the Sun and the Moon. However, the Greeks were also aware of five other bodies in the sky—the planets Mercury, Venus, Mars, Jupiter, and Saturn—whose behavior was not so easily explained. Their motions ultimately led to the downfall of an entire theory of the solar system and to a fundamental change in humankind's view of the universe.

Over the course of a night, the stars slide smoothly across the sky. Over the course of a month, the Moon moves smoothly and steadily along its path on the sky relative to the stars. Over the course of a year, the Sun progresses along the ecliptic at an almost constant rate. Allowing for weather and the seasons, the Sun and Moon have roughly the same brightness every time we see them. In short, their locations and appearance are fairly easy to predict.

Planets do not behave in such a regular and predictable fashion. They vary in brightness, and they don't maintain a fixed position in the sky. Unlike the Sun and the Moon, they seem to wander against the backdrop of distant stars—indeed, the word "planet" derives from the Greek word *planetes,* meaning "wanderer." Planets never stray far from the ecliptic and generally traverse the celestial sphere from west to east, but they seem to speed up and slow down as they go. Three planets—Mars, Jupiter, and Saturn—even appear to loop back and forth relative to the stars, as shown in Figure 2.7. In other words, there are periods when the eastward motion (relative to the stars)

Figure 2.7 *The movements of several planets over the course of many years are reproduced here on the inside dome of a planetarium. Their seemingly erratic motions differ markedly from the background stars, also shown here. The background stars are represented as points. The motion of the planets relative to the stars produces continuous streaks on the planetarium "sky." For most of the time, the planets move west to east, but occasionally they change direction and undergo retrograde motion before looping back. (Boston Museum of Science)*

of each of those planets stops, and the planet appears to move westward in the sky for a month or two before reversing direction again and continuing on its eastward journey.

Motion in the eastward sense is usually referred to as **direct motion;** the backward (westward) loops are known as **retrograde motion.** Mars, Jupiter, and Saturn are always brightest—which means closest to the Earth—during the retrograde portions of their orbits. This planetary behavior requires an explanation more complex than the relatively simple motions of the Moon and the Sun. The occasional retrograde loops of some planets, and the brightness variations of all of them, necessitated major modifications to the basic cosmological model describing the Sun, the Moon, and the stars.

THE GEOCENTRIC UNIVERSE

✓4 The earliest models of the solar system followed the teachings of Aristotle and were **geocentric** in nature. They held that the Earth lay at the *center* of the universe, and that all other bodies moved around it. The Aristotelian school presented some simple and (at the time) compelling arguments is favor of their views. First, of course, the Earth doesn't *feel* like it's moving. Second, if it were, wouldn't there be a strong wind as we moved at high speed around the Sun? Third, why don't we see stellar parallax as the vantage point from which we view the stars' changes over the course of a year? Nowadays, we might be inclined to dismiss the first two points as simply naive, but the third is a valid argument and the reasoning is essentially sound. We now know that there *is* stellar parallax as the Earth orbits the Sun. However, because the stars are so distant, it amounts to less than 1″, even for the closest stars. Early astronomers would not have noticed it. We will encounter other instances in astronomy where correct reasoning has led to the wrong conclusions because it was based on inadequate data.

These early models employed what Aristotle had taught was the perfect curve: the circle. Aristotle's influence was so strong that his teachings carried great weight, even centuries after his death. The geocentric view went largely unchallenged until the sixteenth century A.D. The logic of their construction demanded that they be built from "perfect" Aristotelian materials, namely, uniform circular motion around the Earth. The simplest configuration satisfying these criteria—motion around a circle centered on the Earth—was a good description of the orbits of the Sun and the Moon, but it could not possibly account for variations in planetary brightness or retrograde motion. A more complex model was needed to describe the planets.

The first step toward this new model modified the idea that the planets moved on circles centered on the Earth. Instead, each planet was taken to move uniformly around a small circle, called an **epicycle,** whose *center* moved uniformly around the Earth, on a second, larger, circle known as the **deferent,** as depicted in Figure 2.8. The motion was now composed of two separate circular orbits, which created the possibility that, at some times, the planet's apparent motion in the sky could be retrograde. Also, the distance from the planet to the Earth would vary, accounting for changes in brightness. By tinkering with the relative sizes of the epicycle and the deferent, and with the planet's speed on the epicycle, and the epicycle's speed along the deferent, this "epicyclic" motion could be brought into fairly good agreement with the observed paths of the planets in the sky. This model even had good predictive power, at least to the accuracy of observations at the time.

However, as time went by and the number and the quality of observations increased, it became clear that the epicyclic model was not perfect. Small corrections had to be introduced to bring it into line with new observations. The center of the deferent had to be offset from the Earth, and the motion of each epicycle's center had to be uniform with respect not to the Earth, but to yet another point in space.

In the second century (A.D.), a Greek astronomer named Ptolemy (about A.D. 140) constructed perhaps the best geocentric model of all time. It explained surprisingly well the paths of the five planets then known, as well as the paths of the Sun and the Moon. Figure 2.9 illustrates this

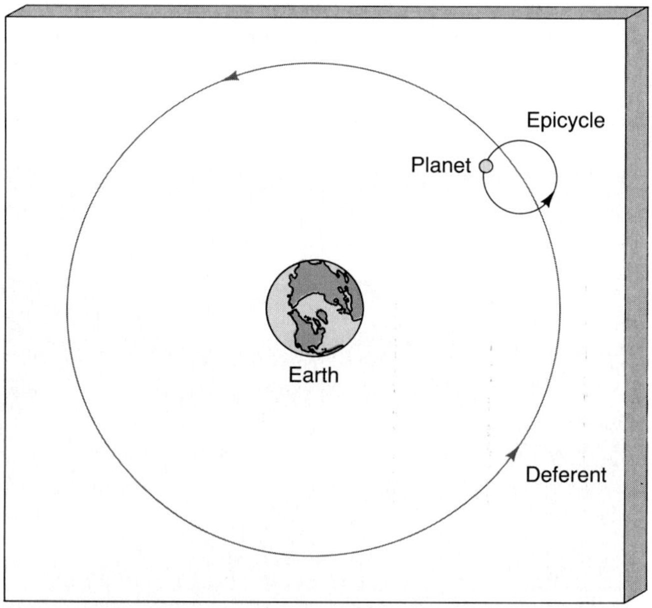

Figure 2.8 *In the geocentric model of the solar system, the observed motions of the planets made it impossible to assume that they moved on simple circular paths around the Earth. Instead, each planet was thought to follow a small circular orbit, or epicycle, about an imaginary point that itself traveled in a large, circular orbit (the deferent) about the Earth.*

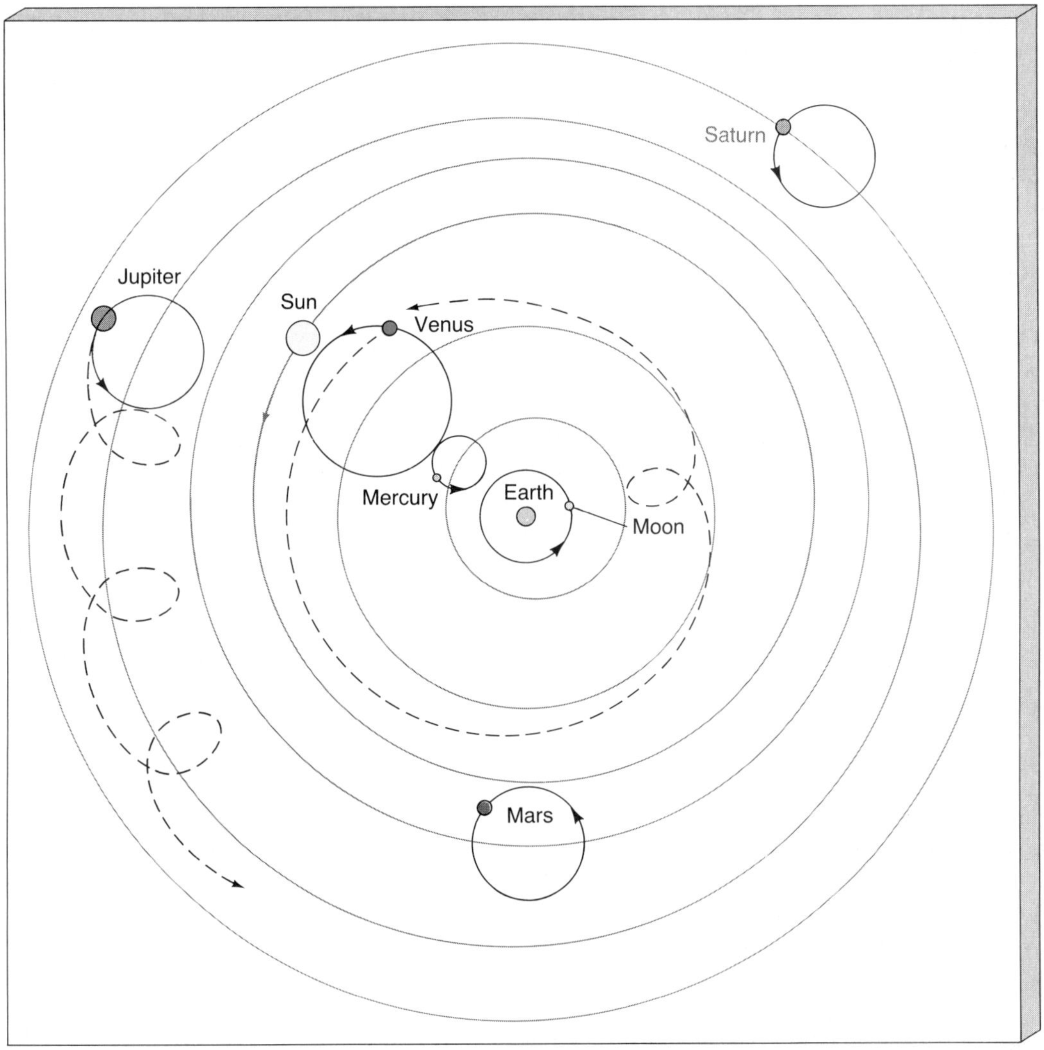

Figure 2.9 *The basic features, drawn roughly to scale, of the geocentric model of the inner solar system that enjoyed widespread popularity prior to the Renaissance. To avoid confusion, we have drawn (dashed) partial paths of only two planets, Venus and Jupiter.*

geocentric view, showing some of the planets' epicycles. However, to achieve its explanatory and predictive power, the **Ptolemaic model** was based on a series of no fewer than 80 circles, each offset in its own peculiar way. To account for the paths of the Sun, the Moon, and all the nine planets (and their moons) that we know today would require a much vaster array of circles within circles within circles. Nevertheless, Ptolemy's text *Syntaxis* provided the intellectual framework for all discussion of the universe for well over a thousand years.

Today, our scientific training leads us to seek simplicity, because simplicity is so often an indicator of truth. We would regard the intricacy of a model as complicated as the Ptolemaic system as a clear sign of a fundamentally flawed theory. With the benefit of hindsight, we now recognize that the major error lay in the assumption of a geocentric universe. This was compounded by the insistence on uniform circular motion, whose basis was largely aesthetic, rather than scientific, in nature. Perhaps to early astronomers the very complexity of the Ptolemaic model argued in its favor—how could something so intricate and precise possibly be wrong?

Actually, history records that some ancient Greek astronomers reasoned differently about the motions of heavenly bodies. Foremost among them was Aristarchus of Samos (310–230 B.C.), who proposed that all the planets, including Earth, revolve around the Sun and, furthermore, that the Earth rotates on its axis once each day. This, he argued, would create an *apparent* motion of the sky, much as the landscape seems to move past a person on a merry-go-round. However, Aristotle's influence proved too widespread, his followers too numerous, his writings too comprehensive. The idea of the fixed Earth at the center of the universe prevailed for two millennia.

ANCIENT ASTRONOMY

Our discussion of the early history of astronomy has so far focused on the Greeks. Their beliefs, methods, and knowledge had a profound influence on the scientific thinkers of the Renaissance. We will see in the next chapter how the reawakening of scientific—and astronomical—thought during the Renaissance had a profound impact on the later development of European science. We are heirs to this scientific legacy, so it is perhaps fitting that we single out the contributions of the Greeks for special study. However, we should also appreciate that other ancient cultures built their own astronomical systems, based on their own beliefs, methods, knowledge, and needs.

The major driving force behind the development of astronomy in early cultures (including that of the Greeks) was most likely practical rather than religious. Seafarers needed to navigate their vessels, and farmers had to know when to plant their crops. The ability to predict the arrival of the seasons, as well as other astronomical events, was undoubtedly a highly-prized, and perhaps also jealously guarded, skill. In some cases the keepers of the secrets of the sky eventually enshrined their knowledge in myth and ritual, so that many ancient astronomical sites were also used for religious rites.

In Chapter 1 we noted that the human brain's ability to perceive patterns in the stars led to the "invention" of constellations as a convenient means of labeling regions of the celestial sphere. The realization that these patterns returned to the night sky at the same time each year met the need for a practical means of tracking the seasons. Many separate cultures, all over the world, built large and elaborate structures to serve, at least in part, as primitive calendars.

Perhaps the best known ancient site is *Stonehenge,* located on Salisbury Plain, in England, and shown in Figure 2.10. This ancient stone circle, which today is one of the most popular tourist attractions in Britain, dates back to the Stone Age. Researchers believe it was an early astronomical observatory. Its construction apparently spanned almost 2000 years, beginning around 2800 B.C. Additions and modifications continued up to about 1100 B.C., indicating its ongoing importance to the Stone Age and later Bronze Age people who built, maintained, and used Stonehenge. The largest stones shown in Figure 2.10 weigh up to 50 tons and were transported from quarries many miles away.

Over the years, some of the stones making up the circle have fallen over or have been carried away by local residents more interested in keeping a roof over their heads than in studying the sky. Figure 2.11 shows a reconstruction of what Stonehenge may have looked like when intact. Many of the stones are aligned with important astronomical events. For example, the line joining the center of the inner circle to the so-called *heel stone,* set off at some distance from the rest of the structure, points in the direction of the rising Sun on the summer solstice, as illustrated in Figure 2.12. Other alignments are related to the rising and setting of the Sun and the Moon at various other times of the year.

Although the accurate alignments (within a degree or so) of the stones of Stonehenge were first noted in the eighteenth century, it is only relatively recently—in the second half of the twentieth century, in fact—that the scientific community has credited Stone Age technology with the ability to carry out such a precise feat of engineering. While some of Stonehenge's purposes remain uncertain and controversial, the site's function as an astronomical almanac seems well established. In fact, many ancient cultures are now known to have been capable of similarly precise accomplishments. Although Stonehenge is the most impressive and the best preserved, other stone circles, found all over Europe, are believed to have performed similar astronomical functions.

People in many other parts of the world built astronomical structures. Figure 2.13 shows the Caracol temple, built by the Mayans around 1000 A.D. in Mexico's Yucatán peninsula. This temple is much more sophisticated than

Figure 2.10 *Stonehenge, on Salisbury Plain, was probably constructed as a primitive calendar and almanac. The fact that the largest stones were carried to the site from many miles away attests to the importance of this structure to its Stone Age builders. (O. Gingerich)*

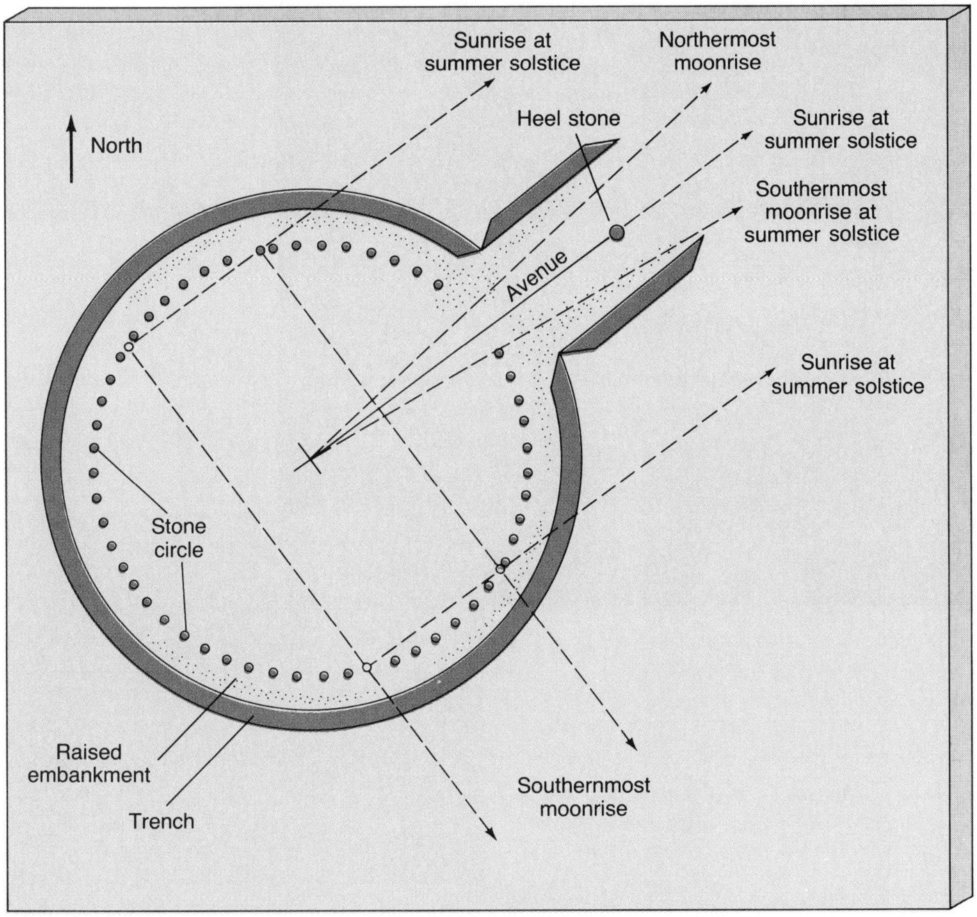

Figure 2.11 Artist's conception of how Stonehenge may have been laid out when it was first built. Notice the many alignments between the larger stones and key astronomical events.

Figure 2.12 Sunrise at Stonehenge on the summer solstice. As seen from the center of the stone circle, the Sun rises directly over the "heel stone" on the longest day of the year. (c. English Heritage)

Figure 2.13 Caracol temple in Mexico. The many windows of this Mayan construct are aligned with astronomical events, indicating that at least part of Caracol's function was to keep track of the seasons and the heavens. (J. Cornell)

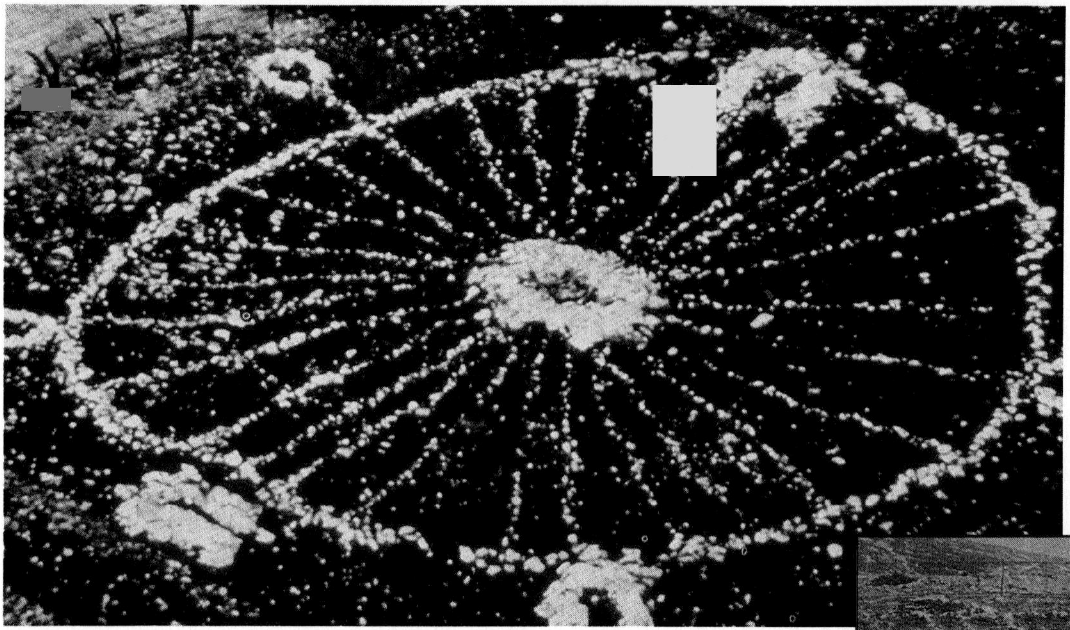

Figure 2.14 *The Big Horn Medicine Wheel, in Wyoming, was built by the Plains Indians. Its spokes and other features are aligned with risings and settings of the Sun and other stars. Above: aerial view. Right: the view from ground level. (J. Eddy)*

Stonehenge, but it played a similar role as an astronomical observatory. Its many windows are accurately aligned with astronomical events, such as sunrise and sunset at the solstices and equinoxes and the risings and settings of the planet Venus. Astronomy was of more than mere academic interest to the Mayans, however. Caracol was also the site of countless human sacrifices, prompted by Venus's appearance in the morning or evening sky.

The Big Horn Medicine Wheel in Wyoming (Figure 2.14) is similar to Stonehenge in design—and, presumably, intent—although it is somewhat simpler in execution. As marked on the figure, the Medicine Wheel's alignments with the rising and setting Sun and with some bright stars indicate that its builders—the plains Indians—had much more than a passing familiarity with the changing nighttime sky.

The ancient Chinese too observed the heavens. Their astrology attached great importance to ''omens'' such as comets and ''guest stars''—stars which appeared suddenly in the sky and then slowly faded away—and they kept careful and extensive records of such events. Twentieth-century astronomers still turn to the Chinese records to obtain observational data from the time of the Dark Ages (roughly from the fifth to the tenth century A.D.), when turmoil in Europe largely halted the progress of Western science. Perhaps the best known guest star was one that appeared in 1054 A.D. and was visible in the daytime sky for many months. We now know that the event was actually a supernova, where a giant star exploded, scattering most of its mass into space. It left behind a remnant that is still detectable today, nine centuries later. The Chinese data are a prime source of historical information for supernova research.

The Chinese theorized about the motions of the heavens, asking many of the same questions as the Greeks and coming up with many of the same answers. For example, Chinese astronomers independently conceived of epicycles and deferents to explain the behavior of the planets and undoubtedly encountered many of the same difficulties with the geocentric model as did their European counterparts. However, the Chinese apparently never developed the notion that Earth and the planets might orbit the Sun. That idea was imported from the West.

A vital link between the astronomy of ancient Greece and that of medieval Europe was provided by Arab astronomers. For six centuries, from the depths of the Dark Ages to the beginning of the Renaissance, Islamic astronomy flourished and grew, preserving and augmenting the knowledge of the Greeks. The Arab influence on modern astronomy is subtle but quite pervasive. Many of the mathematical techniques involved in trigonometry were developed by Muslim astronomers in response to very practical problems, such as determining the precise dates of holy days or the direction of Mecca from any given location on Earth. Astronomical terms like ''zenith'' and ''azimuth'' and the names of many stars, such as Rigel, Betelgeuse, and Vega, all bear witness to this extended period of Muslim scholarship. Even Ptolemy's *Syntaxis,* the blueprint for the medieval universe, returned to Europe after centuries of Arab stewardship as the *Almagest* (which is Arabic for "the greatest").

Islamic astrology was concerned primarily with planetary motion and attached little or no significance to omens, so Islamic records contain virtually no mention of the individual events, such as the supernova of 1054, that were of such importance to the Chinese. While they never strayed outside the confines of the geocentric cosmology, Muslim astronomers refined and extended the work of Ptolemy, finding ways to eliminate some of the more awkward, or aesthetically unsatisfying, constructs in his model. Their work, when reimported into pre-Renaissance Europe, played a major part in shaping the future course of European astronomy.

Astronomy, we see, is not the property of any one culture, civilization, or era. The same ideas, the same tools, and even the same misconceptions have been invented and reinvented by human societies all over the world, in response to the same basic driving forces. The link between these early attempts to understand the universe and the pursuit of science as we know it today will be the subject of the next chapter.

CHAPTER SUMMARY

Humankind once viewed itself as central to the cosmos. Now we have been relegated to an unremarkable position on the periphery of the galaxy. In the process, we have gained a wealth of scientific knowledge.

The idea that Earth lay at the center of the universe persisted until only a few centuries ago. The scientific method, a new approach to investigation that combined theory and experimentation, ushered in the age of modern science. The seeds of the scientific method were planted by the Greeks. They conceived of the universe as something to be studied, questioned, and explained in rational terms. The scientific method is usually a three-step process: scientists gather data, form a theory, and test the theory. Aristotle used the scientific method in the 4th century B.C. when he concluded, based on observations of the Moon, that the Earth is round.

Triangulation lets astronomers find distances to objects in outer space. It is based on the principles of Euclidean geometry. It forms the foundation of the family of distance-measurement techniques that together make up the cosmic distance scale. The distance to an object can be found by constructing an imaginary triangle, with the object at one point of the triangle. Knowing the value of the baseline between the other two points, plus the values of at least two angles used in the triangle, it is possible to calculate the distance to the object.

Astronomers use parallax, the apparent shift in an object's position as seen from two different vantage points, to find distances to objects in space. For example, a star's position in front of background objects is noted and then recorded again 6 months later, when Earth is on the opposite side of its orbit around the Sun. The amount of shift in the star's apparent position is its parallax. Large parallax implies small distance. Small parallax implies large distance. Most objects in space show no parallax at all, because they are so far away.

The universe of the Greeks consisted mainly of the solar system—Sun, Earth, Moon and the then-known planets. The stars were also part of the universe, but they were considered to be fixed, unchanging beacons on a mammoth celestial dome. The stars remain fixed relative to each other, but planets wander among the stars. The more complicated motion of the planets ultimately brought an end to the idea of an Earth-centered universe.

Geocentric, or Earth-centered, models of the universe devised by the Greeks employed what Aristotle had taught was the perfect curve, the circle. Each planet was thought to move uniformly around a small circle called an epicycle, whose center moved uniformly around the Earth in a second larger circle known as the deferent.

The driving force behind the development of astronomy in early cultures was most likely practical rather than religious. Many separate cultures built large and elaborate structures whose function was at least in part to serve as a primitive calendar. The best known of these is Stonehenge, located in England. American Indians, Chinese, and Islamic cultures, among others, also studied the sky.

KEY WORDS

baseline
cosmic distance scale
cosmology
deferent
direct motion
epicycle
geocentric
modeling
parallactic angle
parallax
Ptolemaic model
retrograde motion
solar system
triangulation

REVIEW QUESTIONS

1. What perception of the ancient Greeks helped lead to the development of modern science?
2. What is the scientific method?
3. Explain how Aristotle used the scientific method to conclude that the Earth is round.
4. What assumption did Aristotle have to make in order to reach the conclusion that the Earth is round?
5. What further argument did Aristotle make about the Earth based on his conclusion that it is round?
6. How did Eratosthenes measure the physical size of the Earth in about 200 B.C.?
7. What is triangulation?
8. Why is it necessary to have a long baseline when using triangulation to measure the distances to objects in space?
9. What is parallax? Give an example.
10. What are epicycles and deferents?
11. Give a brief description of the Ptolemaic model of the universe.
12. Why is the Ptolemaic model perhaps ''the best geocentric model of all time''?
13. The benefit of our current knowledge makes it possible to see flaws in the Ptolemaic model of the universe. What is the basic flaw?
14. What arguments from the Aristotelian school of thought were used to counter the model of Aristarchus of Samos?
15. Give an example of how Stonehenge may have been used to track important events in the heavens. What is the ''heel stone?''
16. What other early cultures observed the skies? How do we know?
17. A surveyor wishes to measure the distance between two points on either side of a river, as illustrated in Figure 2.2. She measures the distance AB to be 250 m and the angle at B to be 30°. What is the distance between the two points?
18. At what distance is an object if its parallax, as measured from either end of a 1000 km baseline, is (a) 1°, (b) 1′, (c) 1″?
19. Given that the distance to the Moon is 384,000 km and its angular size is 0.5°, calculate the Moon's diameter.
20. Given that the distance to the Moon is 384,000 km and the distance to the Sun is 150,000,000 km, calculate how many times larger than the Moon is the Sun. (Recall that the Moon and the Sun have the same angular diameter.)
21. What angle would Eratosthenes have measured at Alexandria (see Interlude 2-1) had the Earth been flat?

DISCUSSION QUESTIONS

1. Modern scientists believe that simplicity is often an indicator of truth. The Greeks, though, might have prized the most intricate theories most highly. Which way of thinking ''feels'' right to you, and why?
2. Explain why geometry has been so important in astronomy.
3. Would Earth exhibit retrograde motion as seen from the planet Venus? Draw a picture to explain your answer.
4. Consider the Aristotelian arguments against the idea that Earth moves around the Sun. Why doesn't Earth feel as though it's moving? Why isn't there a strong wind as we move at high speed around the Sun?

PROJECTS

1. For thousands of years, some Northern Hemisphere skywatchers have thrilled to the sight of Canopus, the sky's second-brightest star. It is a particularly special sight for those who live at northern latitudes because it is located far to the south on the celestial sphere. If you live in the southern United States, watch in the evening during the month of February for Canopus to make its low arc across across the southern sky. Why isn't this star visible from the northern states? Where would Canopus appear in your sky if you were in the Southern Hemisphere?
2. Look in an almanac for the date of opposition of one or all of these bright planets: Mars, Jupiter, and Saturn. Observe these planets. How long before opposition does each planet's retrograde motion begin? How long after opposition does it end? (Hint: your almanac might use the word *stationary* to indicate the beginning and end of retrograde motion. Can you explain why?)
3. Some months before opposition, note the location of Mars, Jupiter, or Saturn in front of the stars. Is the planet moving toward the east or toward the west with respect to these stars? Track its changing location in front of the star background, drawing its picture perhaps once a week for several months. How long before opposition does the planet begin moving in a retrograde manner? How long after opposition does it resume eastward motion?

SUGGESTED READINGS

Aveni, A. F. (ed.), *Native American Astronomy*. Austin: University of Texas Press, 1975. Specialists in archaeology, astronomy, architecture, art history, mathematics, solar physics, and anthropology explore the astronomical knowledge and beliefs of pre-Columbian peoples in the Americas.

Carlson, J. B. "America's Ancient Skywatchers." *National Geographic* (March 1990). Descriptions of the cosmologies of the Maya, Navajo, Inca and other American civilizations.

Chapman, A. "Gauging Angles in the 17th Century." *Sky and Telescope* (April 1987). The author built copies of early astronomical instruments: a modified camera obscura, a long refracting telescope, and a visual micrometer. He used these instruments to make 17th century sky measurements.

Frazier, K. "Stars, Sky and Culture." *Science News* 116 (August 4, 1979): 90. From the Pawnee of Nebraska to the Quechua of Peru, a look at how archeoastronomy provides insight into early American cultures, plus how these early people integrated astronomical knowledge into their lives and social systems.

Hadingham, E. *Early Man and the Cosmos*. New York: Walker Publishing, 1984. Reliable introduction to the controversial field of archeoastronomy: Stonehenge, Nazca lines, Native American observatories, solstice alignments, and more. Available from Sky Publishing.

Kovalesvsky, J. "Astrometry from Earth and Space." *Sky and Telescope* (May 1990). Astrometry: its usefulness to astronomy and some prospects for the near future.

Ley, W. *Watchers of the Skies*. New York: The Viking Press, 1969. Willie Ley was a famous rocket scientist, but he also had a wonderful ability to tell a story. Good reading!

Mohlenbrock, R. H. "Medicine Mountain, Wyoming." *Natural History* (January 1990). Interesting story about Wyoming's Big Horn Medicine Wheel, thought to have been used centuries ago as an astronomical observatory.

Robbins R. R., and M. K. Hemenway. *Modern Astronomy: An Activities Approach*. Austin: University of Texas Press, 1982. Chapters include "The Principles of Measurement: Using a Cross-Staff and a Quadrant," "Mapping the Night Sky and Its Motions," and "The Motions of the Planets."

Thomsen, D. E. "Star Tracks." *Science News* (February 28, 1987). Descriptions of proposed astronometric endeavors, such as TAU—the Thousand Astronomical Unit Project—a suggested spaceflight to establish an astronometric baseline of 1,000 A.U.

Van Helden, A. *Measuring the Universe—Cosmic Dimensions from Aristarchus to Halley*. Chicago: University of Chicago Press, 1985. Contains information on Aristarchus, Ptolemy, Copernicus, Kepler, Galileo, Eratosthenes.

CHAPTER

3

The Copernican Revolution

The Birth of Modern Science

On August 20, 1977, the United States launched a Titan/Centaur-7 rocket that propelled the first of two Voyager *craft into space. As the* Voyagers *toured our solar system, they opened our eyes with their many discoveries. In a real sense, their journey was launched much longer ago, back in the sixteenth century, when Copernicus opened many eyes with a new—and accurate—model of the solar system. Today's astronomers owe a great deal to Copernicus, Galileo, Kepler, and Newton as the builders of modern astronomy. (Courtesy A. Tannenbaum-Sygma)*

LEARNING GOALS

1 to recognize the importance of Renaissance science in the history of astronomy

2 to know how the observed motions of the planets led to our modern view of a Sun-centered solar system

3 to understand Kepler's laws of planetary motion

4 to know the experimental method used to determine the distance from the Earth to the Sun

5 to understand how Newton's law of gravity explains Kepler's laws

6 to understand the concept of gravity as well as the role that gravity plays in the planets' movements around the Sun and moons' movements around planets

The geocentric view of the solar system, as embodied in the writings of Ptolemy, dominated all reflections upon the nature of the universe for many centuries. With its wheels within wheels and its seemingly inexhaustible complexity, this view managed to provide a satisfactory description of the motions of the Sun, the Moon, and the planets, and it even achieved considerable predictive power. It was a coherent model of the universe. It was entirely in accord with established scientific doctrine. And it was completely wrong.

The death of the geocentric universe resulted from several factors. First, the Ptolemaic model simply became too elaborate, as new details were added to accommodate new observations. Even with its great intricacy, the model was ultimately unable to account precisely for the motions in the heavens, and astronomers began to suspect that it was fundamentally flawed. Second, technological developments improved observational astronomy to the point where it came into direct and irreconcilable contradiction with Ptolemaic theory. Finally, and perhaps most important, the scientific method became the established way of looking at Nature, and the stage was set for the overthrow of an entire world view.

The chain of discoveries that led to the replacement of the complex and imprecise geocentric theory by a much simpler and vastly more accurate heliocentric (Sun-centered) model began with a sixteenth-century Polish astronomer named Copernicus. The result of the intellectual revolution that now bears his name was the birth of science as we know it today.

THE HELIOCENTRIC MODEL OF THE SOLAR SYSTEM

The Ptolemaic picture of the universe survived, more or less intact, for almost 13 centuries. Given the scope of the cultural, scientific, and technological changes that have occurred in even the last 50 years, it may be difficult for us to grasp how *any* theory, and especially one so wrong, could have persisted for such a long time. There is no easy explanation. Certainly, during the Dark Ages (from the fifth to the tenth century), intellectual development in Europe withered. It is also true that the Church, one of the few stable repositories of knowledge for almost a millennium, accepted the teachings of Aristotle as entirely consistent with its own doctrine. At the same time, there seems to have been a genuine lack of scientific curiosity during much of this period and a willingness to accept dogma rather than to question and to probe. We will not delve deeper here into the causes of this extended intellectual drought. Suffice it to say that the Ptolemaic model of the solar system became deeply embedded in European culture at all levels until the fifteenth century, when the **Renaissance** marked a rebirth of artistic, philosophical, and scientific expression.

THE COPERNICAN REVOLUTION

✓1 The Renaissance began in Italy in the early fifteenth century. Almost instantaneously, Western scholarship evolved from passive religious dogma and static beliefs into

Figure 3.1 *Nicholas Copernicus (1473–1543). (Erich Lessing/Art Resource)*

critical thinking and observational testing. In astronomy, thinking (theory) and looking (observation) merged to produce a model of the solar system simpler than the geocentric one embraced by the ancients. A sixteenth-century Polish cleric, Nicholas Copernicus (1473–1543; see Figure 3.1), rediscovered Aristarchus's **heliocentric** model—one centered on the Sun—and showed how, in its harmony and organization, it better fit the observed facts than did the tangled geocentric cosmology.

In his most important work, entitled *De Revolutionibus Orbium Coelestium* (*On the Revolution of the Celestial Spheres*), Copernicus asserted that the Earth moves around the Sun every year and spins on its axis every day. The seven crucial statements from that work that form the basis for what is now known as the "Copernican Revolution" are presented in Summary 3-1. They stand in stark contrast to the conventional beliefs of the preceding two millennia. The Copernican picture presented a more ordered and natural explanation of the observed facts than any geocentric model could provide. Figure 3.2 shows how the Copernican system accounts for both the varying brightness of the planets and their observed looping motions. Note that the looping motions are now only apparent, not real. If we suppose that the Earth moves faster than the planet Mars, then every so often we will "overtake" Mars. Mars will appear to move backwards in the sky, in much the same way as a car we overtake on the highway seems to slip backwards relative to us.

Copernicus's major motivation for introducing the heliocentric model was simplicity. Even so, he was still influenced by Greek thinking and clung to the idea of circles to model the planets' motions. In order to bring his theory into agreement with observations, he was forced to retain the idea of epicyclic motion, only now the deferent was centered on the Sun rather than on the Earth and the epicycles were smaller than in the geocentric picture. Thus, he retained unnecessary complexity and actually gained only a

SUMMARY 3-1 *The Foundations of the Copernican Revolution*

The following seven points are essentially Copernicus's own words. The italicized material is additional explanation.

1. The celestial spheres do not have just one common center. *Specifically, the Earth is not at the center of everything.*
2. The center of the Earth is not the center of the universe, but is instead only the center of gravity and of the lunar orbit. *Only the Moon orbits the Earth.*
3. All the spheres revolve around the Sun. *By spheres here, Copernicus was referring to the planets.*
4. The ratio of the Earth's distance from the Sun to the height of the firmament is so much smaller than the ratio of the Earth's radius to the distance to the Sun that the distance to the Sun is imperceptible when compared with the height of the firmament. *By firmament, Copernicus meant the distant stars. The point he was making is that the stars are very much farther away than the Sun.*
5. The motions appearing in the firmament are not its motions, but those of the Earth. The Earth performs a daily rotation around its poles, while the firmament remains immobile as the highest heaven. *Because the stars are so far away, any apparent daily motion we see in them is the result of the Earth's rotation.*
6. The motions of the Sun are not its motions, but the motion of the Earth. *Similarly, the Sun's apparent daily and yearly motion are actually due to the various motions of the Earth.*
7. What appears to us as retrograde and forward motion of the planets is not their own, but that of the Earth. *The heliocentric picture provides a natural explanation for retrograde planetary motion, again as a consequence of the Earth's motion.*

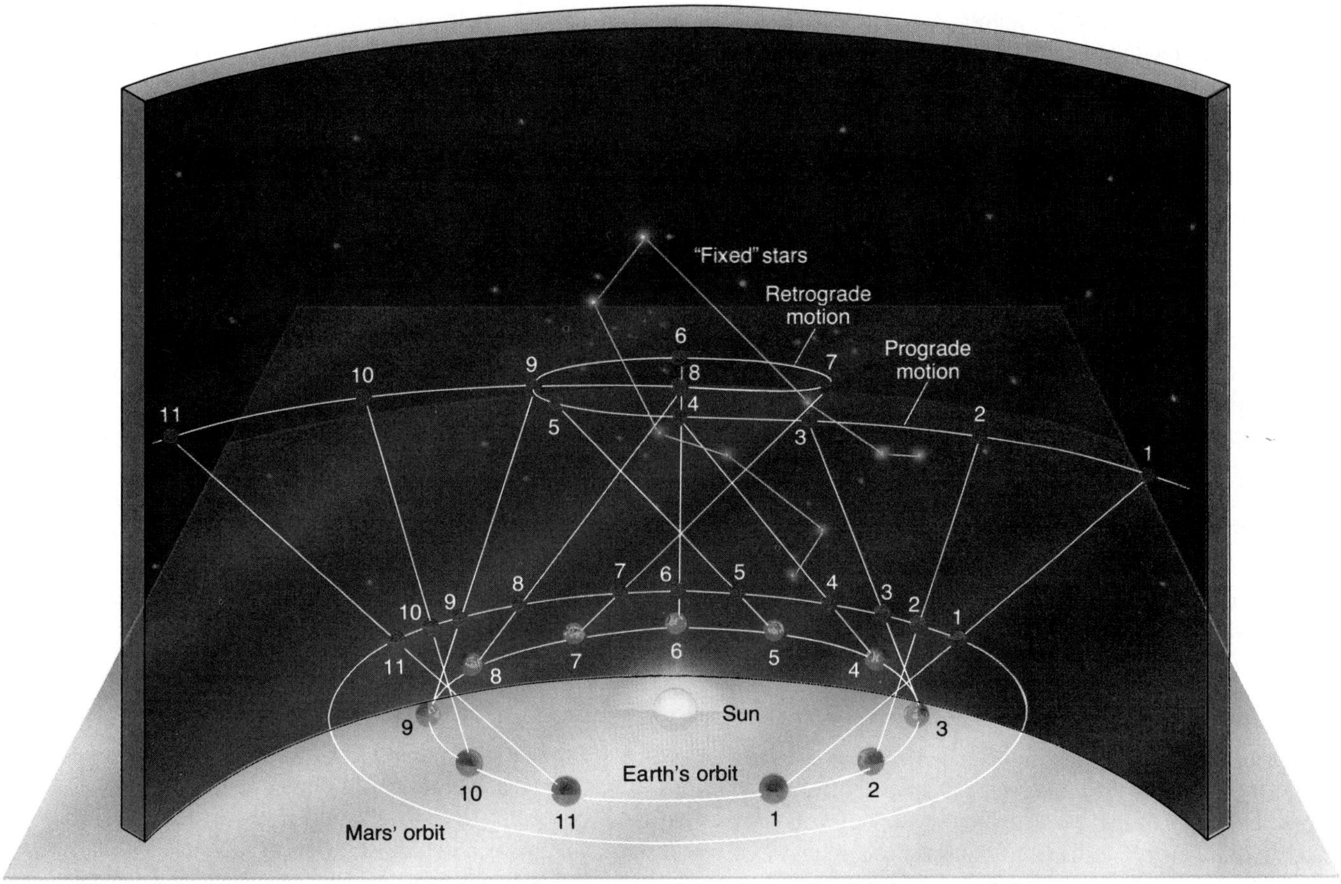

Figure 3.2 *The Copernican model of the solar system explains the varying brightnesses of the planets, something the Ptolemaic system largely ignores. Here, for example, when Earth and Mars are relatively close to one another in their respective orbits (as at position 6), Mars seems brighter; when farther away (as at position 1), Mars seems dimmer. Also, because the line of sight from Earth to Mars changes as the two planets smoothly orbit the Sun, Mars would appear to loop back and forth, and hence undergo retrograde motion. The line of sight changes because Earth, on the inside track, moves faster in its orbit than does Mars.*

little in accuracy over the Ptolemaic model. Although there was indeed an accumulation of small discrepancies and inconsistencies in the Ptolemaic system, which the heliocentric model rectified, to Copernicus heliocentricity was primarily "more pleasing to the mind," something he felt more than he could prove. Even today, scientists are guided by simplicity, symmetry, and beauty in modeling all aspects of the universe.

Copernicus's model was not easily accepted by his fellow scholars, nor by the public, despite some supporting observational data. Heliocentricity rubbed against the grain of much previous thinking and violated many of the religious teachings of the time, largely because it relegated Earth to a noncentral and undistinguished place within the solar system and the universe. Earth became just one of several planets. In fact, Copernicus's work had little impact on the general populace, in part because it was published in Latin, which most people could not read. Only when others—notably Galileo Galilei—popularized Copernicus's ideas, long after Copernicus's death, did the Roman Catholic church take them seriously enough to bother to ban them. *De Revolutionibus* was placed on the *Index of Prohibited Books* in 1616, 73 years after its publication, and remained there until the end of the 18th century.

Copernicus was an unlikely revolutionary. He was a quiet, even timid man—certainly not one to challenge the status quo if he could help it. *De Revolutionibus* was not published until the year he died, 1543. Even though he had completed much of the work decades earlier, Copernicus finished the manuscript only in 1542. He was apparently unwilling to publish because he feared the scorn and ridicule of his colleagues. He was certainly aware of the significance of his work, however. He had sought to "tuck in the tails and tighten the weakened seams" of the traditional model, but instead he discovered to his horror that he had to replace the whole thing. The original version of Copernicus's work contained an anonymous preface relegating the whole heliocentric picture to the status of a mere aid to calculation and suggesting that the author did not believe the Earth actually orbited the Sun. This preface was actually inserted by a Lutheran theologian by the name of Osiander, who happened to be in charge of printing the book. Whether he sought to dilute the impact of the work or to save Copernicus from possible retribution by the Church is unclear, but the preface was certainly not a reflection of Copernicus's attitude toward his new cosmology.

THE BIRTH OF MODERN ASTRONOMY

✓2 In the century following the death of Copernicus and the publication of his theory of the solar system, two scientists made indelible imprints on the study of astronomy. Contemporaries, they were aware of each other's work and corresponded from time to time about their theories. Each achieved fame for his discoveries and made great strides in popularizing the Copernican viewpoint. Yet, in their approach to astronomy—and in personality—they were as different as day and night.

Johannes Kepler was a German mathematician and astronomer. Through his painstaking and exhaustive efforts to explain the behavior of the five known planets, he eventually succeeded in describing their motion in a simple and completely deterministic way. We will return to Kepler's contributions to astronomy in the next section. Galileo Galilei (see Figure 3.3) was an Italian mathematician and philosopher. By his willingness to perform experiments to test his ideas—then a rather radical idea—and by embracing the brand-new technology of the telescope, he revolutionized the way science was done, so much so that he is now widely regarded as the father of experimental science. Together, Kepler and Galileo exemplify the scientific method and symbolize the beginnings of astronomy as we know it today.

Figure 3.3 *Galileo Galilei (1564–1642). (Art Resource)*

Galileo's Greatness

The telescope, invented in Holland in the early seventeenth century, quickly provided much new data in support of the ideas of Copernicus. From having heard of its invention, yet never having seen one, Galileo built a telescope for himself in 1609 and aimed it at the sky. What he saw conflicted greatly with the philosophy of Aristotle (in fact, Galileo had already abandoned Aristotle in favor of Copernicus, although he had never published his beliefs) and helped propel a revolution in astronomy.

Through his telescope, Galileo discovered that the Moon had mountains, valleys, and craters—a surface reminiscent more of the Earth than of some perfect celestial orb, as the Moon was conventionally held to be. Looking at the Sun—something that should *not* be done directly and which eventually blinded Galileo—he saw dark blemishes. These blemishes, which we now call sunspots and which we will study in Chapter 18, directly contradicted the ancient Greek notion that the Sun was a perfect, jewel-like body. Furthermore, he could see that the blemishes moved across the face of the Sun. From this he inferred that the Sun *rotates* on an axis roughly perpendicular to the ecliptic about once a month.

In studying the planet Jupiter, Galileo saw four small points of light, invisible to the naked eye, orbiting it. He recognized that Jupiter and its natural satellites—now called the Galilean moons, which we will study in Chapter 13—formed a miniature solar system, much like the Sun and its family of planets. Clearly then, Earth was not the center of all things. The Galilean moons circulated about another planet, an idea again in direct conflict with Aristotelianism. To Galileo, the moons of Jupiter provided the strongest support for the Copernican model.

Galileo also discovered that Venus showed phases like those of our Moon (see Figure 3.4[a]). Figure 3.4(b) shows how Galileo reasoned that the full or crescent views of Venus must be caused by the planet's motion around the Sun. These observations too were strong evidence that Earth is not the center of things and that at least one planet orbited the Sun. Although they didn't *prove* Copernicus correct, the phases of Venus lent further support to Copernicus's idea that the Sun is at the center, with the planets, including Earth, revolving about it.

Galileo immediately published his findings and his controversial conclusions supporting the Copernican theory in a book called *Sidereus Nuncius* (*The Starry Messenger*)

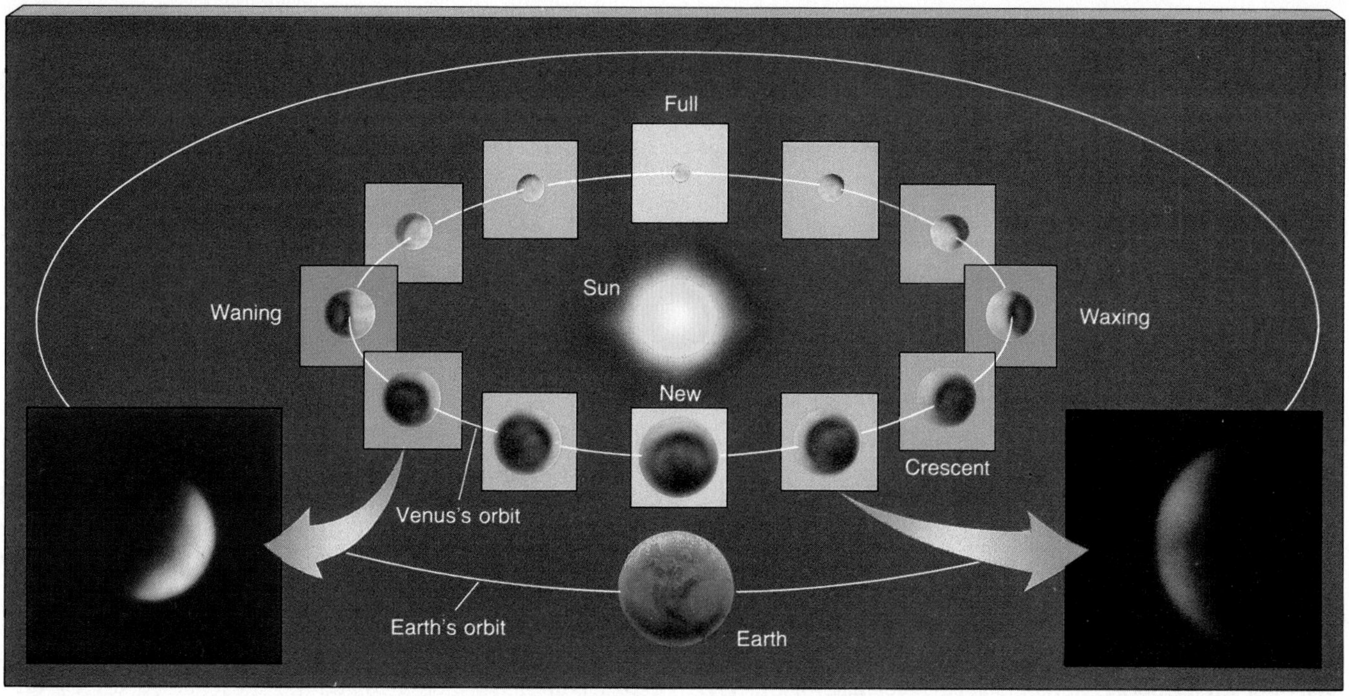

Figure 3.4 *The phases of Venus, drawn for different points in the planet's orbit, and photographed within the two inserts. If Venus orbits the Sun and is closer to the Sun than is Earth, as Copernicus maintained, then Venus should display phases much like our Moon does. As shown here, when directly between Earth and the Sun, Venus's unlit side faces us and is invisible to us. As Venus moves in its orbit (at a faster speed than Earth in its orbit), progressively more of its illuminated face is visible from Earth. Note also the connection between orbital phase and the apparent size of the planet. Venus seems much larger in its crescent phase than when it is full because it is much closer to us during its crescent phase. (Lick Observatory)*

in 1610. In reporting these and other wondrous observations made with his new telescope, Galileo was challenging the scientific establishment and religious dogma of the time and aggressively urging people to change their basic view of the universe. He was (literally) playing with fire. He certainly was aware that only a few years earlier, in 1600, the astronomer Giordano Bruno had been burned at the stake in Rome for his heretical teaching that the Earth orbited the Sun. By all accounts, however, Galileo delighted in publicly ridiculing, and no doubt irritating, his Aristotelian colleagues. In 1616, his ideas were judged heretical, and Copernicus's works were banned by the Roman Church. Galileo was instructed to abandon his cosmological pursuits.

But Galileo would not desist. In 1632, he raised the stakes. He published *Dialogue Concerning the Two Chief World Systems,* a comparison of the Ptolemaic and Copernican models. The book presented a discussion among three people: one of them a dull-witted Aristotelian, whose views were roundly defeated by the arguments of one of his companions, an erudite proponent of the heliocentric system. To make the book accessible to a wide popular audience, Galileo wrote it in Italian rather than Latin. Perhaps not surprisingly, the Church could not tolerate Galileo's actions. The Inquisition, under threat of torture, forced Galileo to retract his claim that Earth orbits the Sun. He was placed under house arrest in 1633 and remained imprisoned for the rest of his life. Not until 1992 were Galileo's "crimes" publicly forgiven by the Church. However, the damage to the orthodox view of the universe was done, and the Copernican genie was out of the bottle once and for all.

The Ascendency of the Copernican System

Though Renaissance scholars were correct, none of them could prove that our planetary system is centered on the Sun or even that Earth moves through space. (Unambiguous proof of Earth's movement was obtained only in the mid-nineteenth century with the first successful measurement of stellar parallax.) Verification of the heliocentricity of the solar system came gradually, with innumerable observational tests that culminated with the expeditions of our unmanned space probes of the 1960s, 1970s, and 1980s. The Copernican world view is fully confirmed today.

The Copernican episode is a good example of how the scientific method, though affected at any given time by the subjective whims and human biases of researchers, does lead to a definite degree of objectivity. Over time, many groups of scientists checking, confirming, and refining experimental tests will neutralize the subjective attitudes of individuals. Usually one generation of scientists can bring sufficient objectivity to bear on a problem, though some especially revolutionary concepts are so swamped by tradition, religion, and politics that more time is necessary. In the case of heliocentricity, objective confirmation was not obtained until about three centuries after Copernicus published his work and more than 2000 years after Aristarchus had proposed the concept. Nonetheless, that objectivity *did in fact* eventually reveal reality.

The development and eventual acceptance of the heliocentric model were milestones in our thinking. Understanding our planetary system freed us from an Earth-centered view of the universe and eventually enabled us to realize that Earth orbits only one of myriad similar stars in the Milky Way Galaxy, which is itself one of myriad galaxies. This removal of Earth from any position of great cosmic significance is generally known, even today, by the term *Copernican principle*.

The Picture Today

Figure 3.5 shows the modern model of the solar system. As currently explored, the solar system's main objects include one star (the Sun), nine planets, 60 moons (at last count), and more than 5000 ''minor planets'' (or asteroids). The planets, in order of increasing average distance from the Sun, are Mercury, Venus, Earth, Mars, Jupiter, Saturn, Uranus, Neptune, and Pluto. All the planets orbit the Sun in nearly the same plane as Earth's orbit (the ecliptic), and they all orbit in the same direction (counterclockwise, as seen from terrestrial north).

KEPLER'S LAWS OF PLANETARY MOTION

✓3 At about the same time that Galileo was becoming famous for his telescopic observations, Johannes Kepler (1571–1630; see Figure 3.6) announced his discovery of a set of simple empirical ''laws'' that accurately describe the motions of the planets. Although Galileo was the first ''modern'' observer, Kepler was a pure theorist. Kepler based his work almost entirely on the observations of another (in part because of his own poor eyesight). Those observations, which predated the telescope by several decades, were made by Kepler's employer, Tycho Brahe (1546–1601), arguably one of the greatest observational astronomers who ever lived.

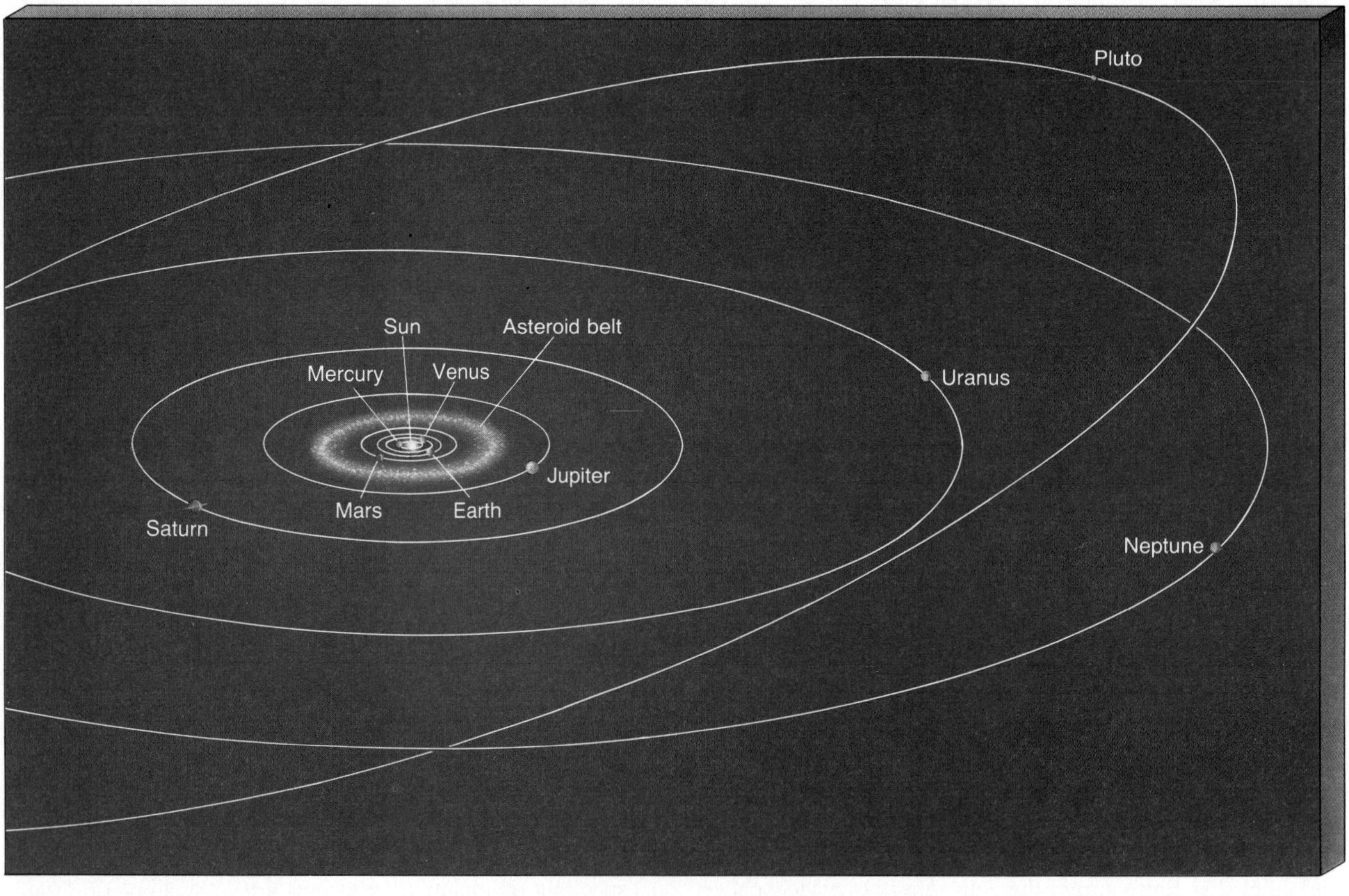

Figure 3.5 *Schematic diagram of the planetary orbits drawn to scale. The entire solar system extends end to end for about 80 times the distance between the Sun and the Earth. That may sound large, but it's only about a thousandth of a light-year.*

Figure 3.6 Johannes Kepler (1571–1630). (Erich Lessing/Art Resource)

Brahe's Complex Observations

Tycho, as he is often known, was both an eccentric aristocrat and a talented observer. He was born in Denmark and educated at some of the best universities in Europe, where he studied astrology, alchemy, and medicine. He was impossibly rude and insulting to everyone he met, an attitude that cost him dearly, both personally and professionally, throughout his life. Part of that price was the loss of his nose in a duel while still a student. Thereafter, he sported a gold and silver replacement held in place by glue.

Despite his great skill as an astronomer, Tycho's personality ultimately resulted in his leaving Denmark in 1597. Having alienated virtually everyone of importance in the entire country, he moved to Prague, which happens to be fairly close to Graz, in Austria, where Kepler lived and worked. Kepler joined Tycho in Prague in 1600 and was put to work trying to find a theory explaining Brahe's planetary data. When Tycho died a year later, Kepler inherited not only Brahe's position as Imperial Mathematician of the Holy Roman Empire (then actually located in Eastern Europe), but also his most priceless possession—the accumulated observations of the planets, spanning several decades, made mostly from Brahe's private observatory, Uraniborg, back in Denmark. Tycho's observations, though made with the naked eye, were nevertheless of very high quality—in most cases, his measured positions of stars and planets were accurate to within about 1′. Thus, Kepler set to work seeking a unifying principle to explain in detail the motions of the planets, without the need for epicycles. The effort was to occupy much of the remaining 29 years of his life.

Kepler and Tycho held different theories of the solar system. Brahe never fully accepted the Copernican view. Instead, he had his own pet "hybrid" cosmology, in which the Sun orbited the Earth, but the planets orbited the Sun—a picture which, from a purely observational viewpoint, was just as consistent with observations as Copernicus's, and was philosophically more satisfying to Tycho. Kepler accepted the fully heliocentric picture, yet he was careful not to say so in such a way as to antagonize his contemporaries. He was concerned about the relationship of his work to established Church doctrine, not so much because he feared retribution, but simply because he was a religious man. Kepler's goal was to find a simple, elegant description of the solar system that fit Tycho's complex mass of detailed observations. In the end, he found that it was necessary to abandon Copernicus's original idea of circular planetary orbits. However, as a result, an even greater simplicity emerged. After long years studying Brahe's planetary data, and after many false starts and blind alleys, Kepler developed the laws of planetary motion that now bear his name.

Kepler determined the shape of each planet's orbit by triangulation—not from different points on Earth, but by using Brahe's detailed observations made from many different points along the Earth's orbit at many different times of the year. By using a portion of the Earth's orbit as a baseline, he was able to measure the sizes of the other planetary orbits in terms of the orbit of the Earth. By noting where the planets were on successive nights, he also found the speeds at which the planets moved. We do not know how many geometrical shapes Kepler tried for the orbits before he hit upon the correct one. His task was complicated by the fact that the Earth's orbit, too, had to be determined. Suffice it to say that he eventually triumphed and that he succeeded in summarizing the motions of all the known planets, including the Earth, in just three simple laws. **Kepler's Laws of Planetary Motion** are summarized in Summary 3-2.

SUMMARY 3-2 *Kepler's Laws of Planetary Motion*

1. The orbital paths of the planets are elliptical, with the Sun at one focus.
2. An imaginary line connecting the Sun to any planet sweeps out equal areas of the ellipse in equal intervals of time.
3. The square of a planet's orbital period is proportional to the cube of its orbital semi-major axis:

$$P(\text{years})^2 = a(\text{astronomical units})^3$$

Kepler's Simple Laws

Kepler's first law has to do with the *shapes* of the planetary orbits: *The orbital paths of the planets are elliptical (not circular) with the Sun at one focus*. Figure 3.7 illustrates a means of constructing an **ellipse,** which is simply an elongated circle. Almost any ellipse can be drawn by varying the length of a string attached by tacks at each of two points, called **focus.** The long axis of the ellipse, containing the two foci, is called the **major axis.** We conventionally refer to half the length of this long axis—the **semi–major axis**—as a measure of the ellipse's "size." The **eccentricity** of the ellipse is the ratio of the distance between the foci to the length of the major axis. Notice that a circle is a special kind of ellipse in which the two foci coincide, and so its eccentricity is zero. The semi–major axis of a circle is just its radius.

These two numbers—the semi–major axis and the eccentricity—are all that we need to describe the size and shape of the orbital path. From them, we can derive other useful quantities. For example, if a planet's orbit has semi–major axis a and eccentricity e, we can compute that its **perihelion** (closest approach to the Sun, at one focus) is at a distance $a(1 - e)$ and that its **aphelion** (greatest distance) is $a(1 + e)$. Thus, a (hypothetical) planet with a semi–major axis of 400 million km and an eccentricity of 0.5 would range between $400 \times (1 - 0.5) = 200$ million km and $400 \times (1 + 0.5) = 600$ million km from the Sun over the course of one complete orbit.

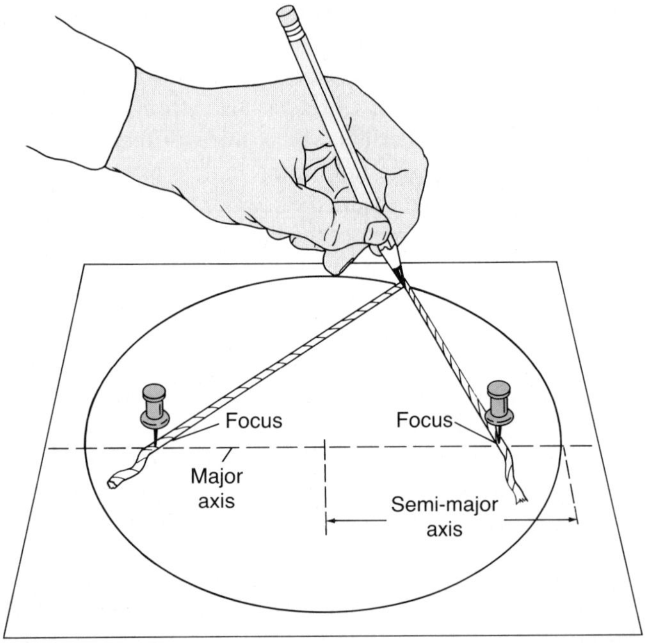

Figure 3.7 *Any ellipse can be drawn with the aid of a string, a pencil, and two tacks. The wider the foci are separated, the more elongated, or eccentric, is the ellipse. In the special case where the two foci are at one and the same place, then the drawn curve is "perfect"—a circle. Note that the semi-major axis of a circle is simply its radius.*

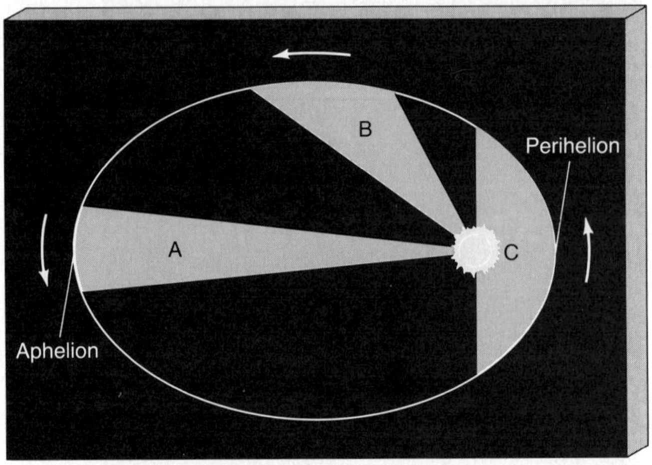

Figure 3.8 *A schematic diagram illustrating Kepler's Second Law: Equal areas are swept out in equal intervals of time. The three shaded areas are equal. Note that an object would travel the length of each of the three arrows in the same amount of time. Therefore, planets move faster when closer to the Sun. (Most planetary orbits are not nearly this eccentric.)*

None of the planets' elliptical orbits are as pronounced as that shown in Figure 3.7. Most, in fact, are quite close to circular. With two exceptions, Mercury and Pluto, our eyes would have trouble distinguishing a real planetary orbit from a true circle. Only because the planets' orbits are so close to being circles were the Ptolemaic and Copernican models able to come as close as they did to describing reality.

The substitution of ellipses for circles was no small advance. It amounted to abandoning an aesthetic bias—Aristotle's belief in the perfection of the circle—that had governed astronomy since Greek antiquity. And it was another heavy blow to Aristotelian philosophy. Even Galileo, not known for his conservatism in these scholarly matters, clung to the idea of circular motion and never accepted that the planets move on elliptical paths. Nevertheless, throughout the Renaissance, beliefs unsupported by hard evidence continued to give way to new beliefs based on observation and experimental tests.

Kepler's Second Law is sketched schematically in Figure 3.8. This law addresses the *speed* at which a planet traverses different portions of its orbit: *An imaginary line connecting the Sun to any planet sweeps out equal areas of the ellipse in equal intervals of time*. In other words, while orbiting the Sun, a planet traces the arcs labeled A, B, and C in Figure 3.8 in equal times. Consequently, when a planet is close to the Sun, as in sector C, it moves much faster than when farther away, as in sector A. This law is not restricted to planets. It applies to any orbiting object. Spy satellites, for example, move very rapidly as they swoop close to

Earth's surface—not (necessarily) because they are propelled with powerful on-board rockets, but because any object in a highly elliptical orbit moves much faster as it approaches the focus.

By taking into account the relative speeds and positions of the planets in their elliptical orbits about the Sun, Kepler's first two laws explained the variations in brightness and the peculiar nonuniform motions observed for some of the planets that could not be accommodated within the assumption of circular motion, even with the inclusion of epicycles. Gone at last were the circles within circles that rolled across the sky. Kepler's modification of Copernicus's theory to allow the possibility of elliptical orbits both greatly simplified the model of the solar system and at the same time provided much greater predictive accuracy than had hitherto been possible.

Kepler published these two laws in 1609 in *Astronomia Nova* (*The New Astronomy*), stating that he had proven them only for the orbit of Mars. Ten years later he wrote *Harmonice Mundi* (*The Harmony of the Worlds*), in which he extended the first and second laws to all the known planets and added a third law relating the size of a planet's orbit to its orbital period. *Kepler's Third Law* states that *the square of a planet's orbital period is proportional to the cube of its semi–major axis.* In other words, the time taken for a planet to complete one circuit around the Sun—the planet's ''year,'' or, more technically, its (sidereal) orbital period P—increases more rapidly than the size of its orbit a, according to the rule $P^2 \propto a^3$ (the symbol $\propto$ just means ''is proportional to''). Because we know by definition that the Earth, at a mean distance of one **astronomical unit** (see Interlude 1-2), orbits the Sun in exactly 1 year, we can determine the constant of proportionality and simply say that

$$P(\text{in years})^2 = a(\text{in astronomical units})^3.$$

By the definition of our units, the Earth, with an orbital semi-major axis of 1 A.U., orbits the Sun in exactly 1 year. A planet closer to the Sun than 1 A.U. (for example Venus, with $a \approx 0.7$ A.U.) takes less than 1 year to complete one circuit (about 0.6 Earth years, or 225 days, in the case of Venus). The planet Saturn, almost 10 A.U. out, takes nearly 30 Earth years to orbit the Sun just once. Because of their importance, Kepler's three laws are collected and summarized in Summary 3-2.

Comparison with Modern Observations

Table 3-1 presents basic data describing the orbits of the nine known planets. Renaissance astronomers knew these properties for the innermost six planets and used them to construct the currently accepted heliocentric model of the solar system.

The second column presents each planet's orbital semi–major axis—essentially its average distance from the Sun. The third column gives the orbital period. The fourth column lists the planets' orbital eccentricities. For purposes of verifying Kepler's Third Law, the rightmost column lists the ratio P^2/a^3.

TABLE 3-1 ***Some Solar System Dimensions***

Planet	Semi–major Axis, a (astronomical units)	Orbital Period, P (Earth years)	Orbital Eccentricity	P^2/a^3
Mercury	0.387	0.241	0.206	1.002
Venus	0.723	0.615	0.007	1.001
Earth	1.000	1.000	0.017	1.000
Mars	1.524	1.881	0.093	1.000
Jupiter	5.203	11.86	0.048	0.999
Saturn	9.539	29.46	0.056	1.000
Uranus	19.19	84.07	0.046	1.000
Neptune	30.06	164.82	0.010	1.000
Pluto	39.53	248.6	0.248	1.001

The tabulated distances and times are expressed in terms of Earth values. For example, we write the period of our own planet as 1.0 Earth year, and we express Venus's orbital period as 0.615 of an Earth year, or roughly 225 days. Compare this relatively brief ''year'' with that of Pluto, which equals nearly 2.5 Earth centuries. Similarly, the second column shows, for example, that the distance of Jupiter from the Sun averages about 5.2 times the distance of Earth from the Sun.

The main points to be grasped from Table 3-1 are these: (1) With the exception of Mercury and Pluto, the planets' orbits really are very nearly circular (that is, their eccentricities are close to zero), and (2) the farther a planet is from the Sun, the greater is its orbital period. Kepler's Third Law is exactly obeyed by *all* the planets and not just by the six on which he based his conclusions. For example, although Pluto is ''only'' about 40 times the distance of Earth from the Sun, its orbital period is a good deal more than 40 years: In fact, it is almost 250 years. Plugging in the exact numbers, we see that $39.53^3 = 248.6^2$, in precise accordance with Kepler's law, to within the observational errors in the numbers.

SOLAR SYSTEM DIMENSIONS

✓4 Kepler's laws allow us to construct a scale model of the solar system, with the correct shapes, locations, and relative sizes of all the planetary orbits. However, they do not tell us the *actual* size of any of the orbits. In our above statement of the third law, we expressed distances in terms of the astronomical unit, the mean distance from the Earth to the Sun. If we could somehow measure its value—in kilometers, say—then we would be able to compute the exact distances to all the planets and the size of the solar system as a whole.

We might propose using triangulation to measure the distance to the Sun. However, we would find it virtually impossible to measure the Sun's parallax using the Earth's diameter as a baseline. The Sun is too bright, too big, too fuzzy, and too distant for us to distinguish any apparent displacement relative to the field of distant stars. To measure the Sun's distance from Earth, we must resort to some indirect method.

As noted earlier, we *can* measure the Moon's parallax using Earth's diameter as a baseline. Accordingly, we can express the lunar distance in a familiar unit—for example, kilometers—or in whatever unit we choose for Earth's diameter. But the distance to the Moon is not related to the astronomical unit, because the Moon follows right along with Earth's orbital motion around the Sun. Thus there is no easy way to determine how many Earth-Moon distance units make up one Earth-Sun distance unit. We need another method.

We can measure parallax for some of the neighboring planets, especially when we use the larger baseline of Earth's orbit. Of course, individual planetary motions, including that of Earth, must be taken into account. But even here, we can express the distance to each planet only in terms of the astronomical unit. Why? Because all the measurements are based on the astronomical unit, the absolute value of which we have not yet determined. In this way we can construct a perfectly good model of our solar system's geometry (in fact, this is what Kepler did), but the *absolute* distance—that is, the real distance—to each planet remains unspecified. The result would be analogous to a road map showing the *relative* positions of cities and towns but without the all-important scale marker indicating distances in kilometers or miles.

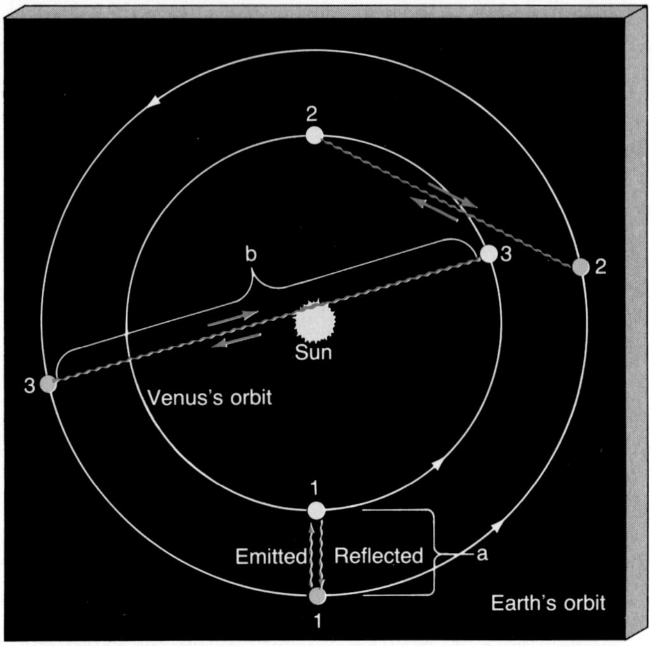

Figure 3.9 *Simplified geometry of the orbits of Earth and Venus as they move around the Sun. The wavy lines represent some paths along which radar signals might be transmitted toward Venus and received back at Earth. The distances* a *and* b *are, respectively, the minimum and maximum distances between Earth and Venus. Because the radius of the Earth's orbit is 1 A.U. and that of Venus is about 0.7 A.U., we know that* a $\approx$ *0.3 A.U. and* b $\approx$ *1.7 A.U. Thus, radar measurement of* a *or* b *(or both, to provide a check) in kilometers allows us to determine the astronomical unit.*

Measuring the Astronomical Unit

The modern method for deriving the absolute scale of the solar system uses radar. The word **radar** is an acronym for **ra**dio **d**etection **a**nd **r**anging. In this technique, radio waves are transmitted toward an object. Their returning echo indicates the object's direction and range, or distance, in absolute terms (that is, in kilometers rather than in astronomical units)*.

Unfortunately, the Sun is not a good target for radar observation. It has no solid surface from which radio waves can be bounced to yield a clear echo. (Echos do return, but from variable heights in the Sun's atmosphere.) Instead, the target most often used in this experiment is the planet Venus. This planet is the closest to Earth and therefore returns the most easily detectable echo, although the strength of the received echo is much weaker than the transmitted signal.

The calculations involved in using radar to measure astronomical distances resemble those used to derive the distance between two cities if our car's speed and travel time are known. Twice the distance to Venus (back and forth) equals the speed of light (300,000 km/s, which is also the speed of radio waves) multiplied by the time elapsed between transmission of the signal and reception of the echo. The round-trip travel time can be measured with high precision—in fact, good enough to determine the dimensions of Venus's orbit to an accuracy of 1 km.

Figure 3.9 is an idealized diagram of the Sun-Earth-Venus orbital geometry. The planetary orbits are drawn as circles here, but in reality they are slight ellipses. This is a subtle difference, and we can correct for it using detailed knowledge of orbital motions. The important point is that transmissions of radio waves at various times of the year toward Venus yield accurate distances between Venus and Earth as they orbit the Sun.

Let's not lose track of the objective here. We are not now concerned with the details of Venus's orbit as much as we are with determining the absolute value of the astronomical unit. Assuming that the orbits are perfect circles, we see

*In fact, the astronomical unit was measured long before the invention of radar. The earliest attempts involved the use of triangulation, with the Earth's diameter as a baseline, to measure the distances of nearby astronomical bodies whose orbits were known. Because of the distances involved, however, these determinations were generally rather inaccurate. Some of the later techniques used—namely, measuring the time taken for light to traverse a portion of the solar system—are conceptually similar to the modern radar method, but they are much more complicated in detail.

from Figure 3.9 that twice the astronomical unit equals the sum of two terms: the Earth-Venus distance when Venus is farthest from Earth (1.7 A.U.), labeled *b,* plus the Earth-Venus distance when Venus is closest to Earth (0.3 A.U.), labeled *a*. In other words, the astronomical unit equals $(a + b)/2$, and both a and b can be directly measured. Through precise radar ranging, the astronomical unit is now known to be 149,597,870 km. In this text, we round off this number to a value of 1.5×10^8 km.

We can express the Earth-Sun distance in many ways. Here is another one. Because the velocity of light is 3×10^5 km/s—the metric equivalent of 186,000 miles per second—a calculation shows that the Sun is about 8 *light-minutes* from Earth.

Size and Scale

Having determined the value of the astronomical unit, we can reexpress the sizes of the other planetary orbits in terms of a more familiar unit, such as kilometers or even light-years. The entire scale of the solar system can then be calibrated, and the absolute distances among the nine planets and their parent Sun can be determined to high precision.

As we probe matter in progressively more distant realms of the universe (in Parts II–IV of our text), we will study other methods astronomers use to chart truly remote objects. With each new distance technique, we will encompass larger volumes of space. A succession of many techniques will take us from Earth to the limits of the observable universe. But each new technique is based on the previous one. We use objects whose distances we have already measured by previous methods to calibrate in turn each newer technique, all the way up to the largest scales in the universe. Thus, measurement of the scale of the solar system is the first small step in our cosmic journey through the universe.

NEWTON'S LAWS

✓5 Kepler's three laws, which so simplified the solar system, were discovered empirically. In other words, they resulted solely from the analysis of observational data and were not derived from a theory or mathematical model. Indeed, Kepler did not have any appreciation for the physics underlying his laws. Nor did Copernicus understand the basic reasons *why* his heliocentric model of the solar system worked. Even Galileo, often called the father of modern physics, failed to understand why the planets orbit the Sun.

What prevents the planets from flying off into space or from falling into the Sun? What causes the planets to revolve about the Sun, apparently endlessly? To be sure, the motions of the planets obey Kepler's three laws, but only by considering something more fundamental than those laws can we understand these motions. The heliocentric system was secured when, in the seventeenth century, the British mathematician Isaac Newton (see Figure 3.10) developed a deeper understanding of the way *all* objects move and interact with one another.

Isaac Newton was born in Lincolnshire, England, on Christmas Day in 1642, the year that Galileo died. Newton studied at Trinity College, Cambridge, but when the bubonic plague reached Cambridge in 1665, he returned to the relative safety of his home for two years. During that time he made probably the most famous of his discoveries, the Law of Gravity (although it is but one of the many major scientific advances Newton made during his lifetime). However, either because he regarded the theory as incomplete or possibly because he was afraid that he would be attacked or plagiarized by his colleagues (he was an extremely secretive and suspicious man), he did not tell anyone of his monumental achievement for almost 20 years. It was not until 1684, when Newton was discussing with Edmund Halley (of Halley's comet fame) the leading astronomical problem of the day—*Why* do the planets move according to Kepler's laws?—that he astounded his companion by off-handedly remarking that he had solved the problem in its entirety nearly two decades before.

Prompted by Halley, Newton published his theories in his most famous work, the *Philosophiae Naturalis Principia Mathematica,* known simply as Newton's *Principia.* The ideas expressed in the *Principia* form the basis for what today is known as **Newtonian mechanics.** Three basic Laws of Motion, the Law of Gravity, and the calculus (which Newton also invented) are sufficient to explain and quantify virtually all of the complex dynamical behavior we

Figure 3.10 *Isaac Newton (1642–1727). (The Granger Collection)*

SUMMARY 3-3 *Newton's Laws of Motion and Gravitation*

The Three Laws of Motion

1. *Every body continues in a state of rest or in a state of uniform motion in a straight line unless it is compelled to change that state by forces acting upon it.*

It requires *no* force to maintain motion in a straight line with constant velocity. The tendency of a body to remain in a state of uniform motion is usually referred to as inertia. When velocity does change, its rate of change is known as acceleration. The relation of acceleration to any forces acting on a body is the subject of the second law of motion:

2. *When a force F acts on a body of mass m, it produces in it an acceleration a equal to the force divided by the mass. Thus, $a = F/m$, or $F = ma$.*

In honor of Newton, the MKS unit of force is named after him. By definition, 1 newton (N) is the force required to cause a mass of 1 kilogram to accelerate at a rate of 1 meter per second every second.

Newton's third law relates the forces acting between separate bodies:

3. *To every action, there is an equal and opposite reaction.*

This law means, for example, that you attract the Earth with exactly the same force as it attracts you—that force is called your weight. It is governed by one final law:

The Law of Universal Gravitation ("Newton's Law of Gravity")

Every particle of matter in the universe attracts every other particle with a force that is directly proportional to the product of the masses of the particles and inversely proportional to the square of the distance between them.

In other words, two bodies of masses m_1 and m_2, separated by a distance R, attract each other with a force F that is proportional to $(m_1 \times m_2)/R^2$. The constant of proportionality is known as the **gravitational constant,** or simply as Newton's constant, and is always denoted by the letter G. We can express the Law of Gravity as

$$F = G\frac{m_1 m_2}{R^2}.$$

The value of G has been measured in extremely delicate laboratory experiments. In CGS units, its value is 6.67×10^{-8} centimeter^3/gram/second^2 (cm^3/g/s^2).

see on Earth and throughout the universe. Newton's laws are listed in Summary 3-3. Only in extreme circumstances do Newton's laws break down, and this fact was not realized until the twentieth century, when Albert Einstein's Theories of Relativity once again revolutionized our view of the universe (see Chapter 24). Most of the time, however, Newtonian mechanics provides an excellent description of the motion of planets, stars, and galaxies through the cosmos.

Figure 3.11 illustrates the essence of Newton's first law of motion and also serves as an operational definition of the phenomenon of **force** (the subject of the second law). As shown, a moving object will, in principle, move forever in a straight line unless a force changes its direction of motion. For example, the object might glance off a brick wall or be hit with a baseball bat. These are examples of *momentary,* or instantaneous, forces that can change the original path of the object. This tendency of an object to keep moving at the same speed unless acted upon by a force is known as **inertia.**

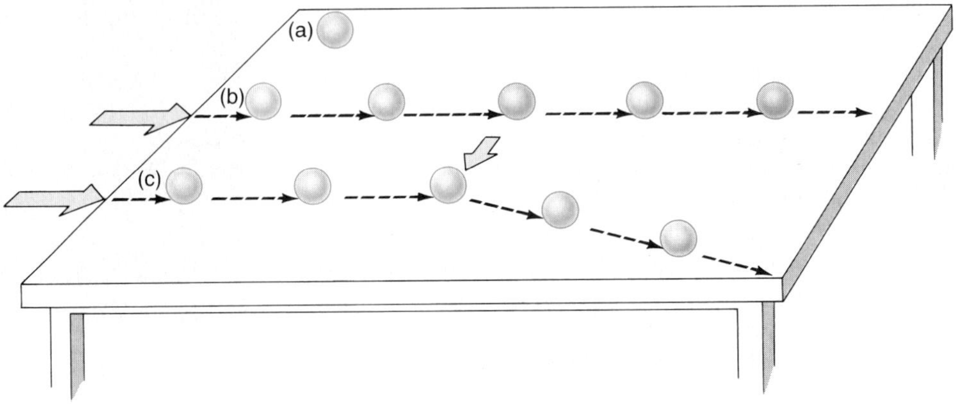

Figure 3.11 An object at rest will remain at rest (a) until some force momentarily acts on it (b). It will then remain in that state of uniform motion until another force acts on it. The arrow in (c) shows a second force acting at a direction different from the first, which causes the object to change direction.

Newton's first law contrasts sharply with the view of Aristotle, who incorrectly maintained that the natural state of an object was to be *at rest*—most probably an opinion based on Aristotle's observations of the effect of friction. In our discussion, we will neglect the familiar concept of friction—the force that slows balls skirting along the ground or pencils rolling across a tabletop. In any case, it is not an issue for the planets, because there is no appreciable friction in outer space. The fallacy in Aristotle's argument was first realized and exposed by Galileo, who conceived of the notion of inertia long before Newton formalized it into a law.

Gravity

✓6 In addition to momentary forces, forces can also act *continuously*. A continuous force produces an **acceleration,** rather than a sudden change in velocity (see Summary 3-3). A good example of a continuous force is the one that started Newton on the path to the discovery of his laws: gravity. Any object having mass always exerts an inward, attractive **gravitational force** on all other massive objects. The more massive an object, the stronger its gravitational pull. To Newton, gravity was a force that acted at a distance, with no obvious way in which it was actually transmitted from place to place. Newton was not satisfied with this explanation, but he had none better.

To appreciate the modern view of gravity, consider a piece of matter having some mass. This matter can be smaller than an atom or larger than a galaxy. Extending outward in all directions from it is a **gravitational field.** We now regard such a field as a property of space itself—a property that determines the influence of one massive object on another. The piece of matter produces the field. Any other matter feels the field as a gravitational force. Thus, the force does not act at a distance, without affecting the intervening space—the force is transmitted through the gravitational field. Although the difference between a force acting at a distance and a force mediated by a field may seem slight, the distinction is in actuality a very important part of modern physics.

The continuous pull of Earth's gravity can be visualized if we consider a baseball thrown upward. Figure 3.12 illustrates how the baseball's path changes continuously. The baseball, having some mass of its own, also exerts a gravitational pull on the Earth. By Newton's third law (see again Summary 3-3), this force is equal and opposite to the weight of the ball. But, by Newton's second law, the more massive Earth has a much greater effect on the baseball than the baseball has on the Earth. The ball and the Earth each feel the same gravitational force, but the Earth's *acceleration* is much smaller.

Now consider the trajectory of a baseball batted from the surface of the Moon, which has much less mass than the Earth. The pull of gravity is about six times weaker on the Moon than on Earth, so a baseball's path changes more slowly near the Moon. A typical home run in a ballpark on Earth would travel nearly a kilometer on the Moon. The Moon, less massive than Earth, has less gravitational influence on the baseball.

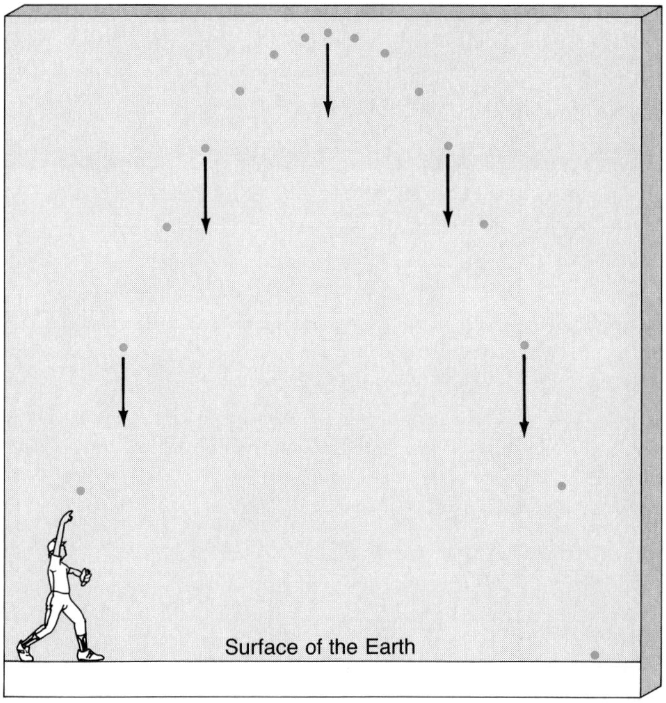

Figure 3.12 A ball thrown up from the surface of a massive object such as a planet is pulled continuously by the gravitational field of that planet (and, conversely, the field of the ball continuously pulls the planet).

The magnitude of the gravitational force, then, depends on the mass of the attracting bodies. Theoretical insight, as well as detailed experiments, tells us that the force is in fact directly proportional to the product of the two masses.

Studying the motions of the planets around the Sun reveals a second aspect of the gravitational force. At all locations equidistant from the Sun's center, the gravitational force has the same strength. Furthermore, it is always directed toward the Sun. What's more, in much the same way that temperature decreases with distance from a fire, gravity weakens with distance from any object that has mass.

Force fields that decrease with distance from their source are encountered throughout all of science. Many of them, including gravity, decrease as the *square* of the distance. They are said to obey an **inverse-square law.** As shown in Figure 3.13, inverse-square fields decrease rapidly with distance from their source, becoming, for example, 9 (3^2) times weaker at a distance 3 times greater and 25 (5^2) times weaker at a distance 5 times greater. Despite this rapid decrease, the force never quite reaches zero. Accord-

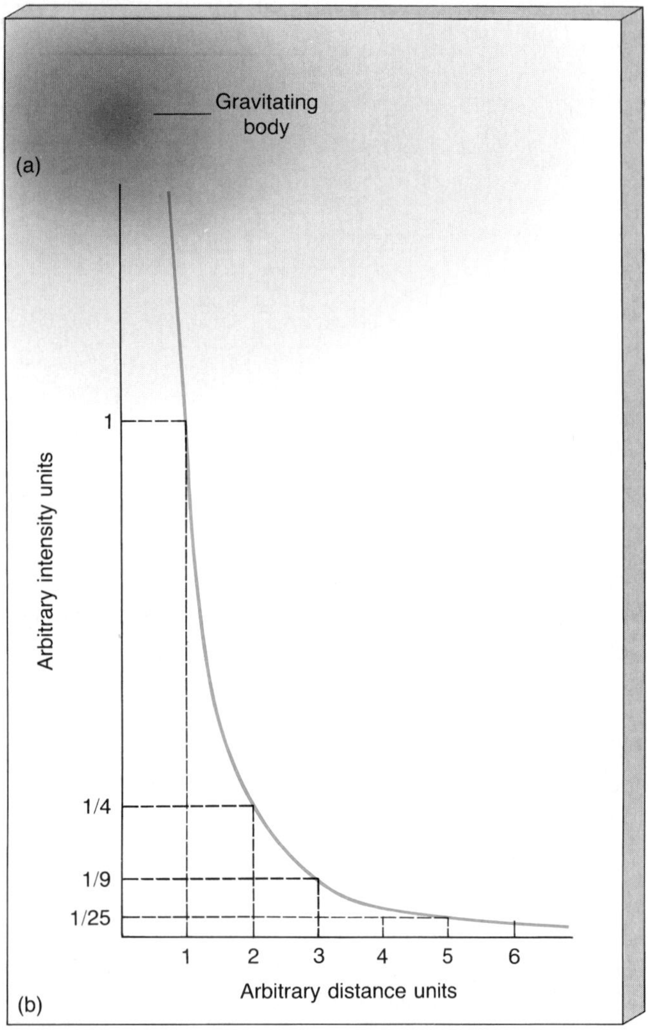

Figure 3.13 *Inverse-square fields rapidly weaken with distance from the source of that field. (a) The field is stronger near the massive particle where the shading is heavier. The strength of the gravitational force decreases with the square of the distance from the Sun. The force never quite diminishes to zero, however, no matter how far away from the Sun we go. (b) The graph illustrates just how fast such forces decrease with distance.*

ingly, the gravitational fields of objects having some mass can never be completely extinguished.

We can combine statements about mass and distance to form a law of gravity that dictates the way that *all* material objects attract each other. As a proportionality, Newton's Law of Gravity is:

$$\textit{gravitational force} \propto \frac{\textit{mass of object } 1 \times \textit{mass of object } 2}{\textit{distance}^2}.$$

This relationship is a compact way of stating that the gravitational pull between two objects is directly proportional to the product of their masses and inversely proportional to the square of the distance separating them. See Summary 3-3 for a more precise statement of this law.

Planetary Motion

It is the mutual gravitational attraction of the Sun and the planets, as expressed in Newton's Law of Gravity, that produces the observed planetary orbits. Because the Sun is much more massive than any of the planets, it dominates the interaction. We might say the Sun "controls" the planets, not the other way around. As the Sun pulls, it tries to draw the planets directly toward itself. But this gravitational pull is counteracted somewhat by each planet's forward motion, or **momentum,** which would, in the absence of gravity, cause them to escape into deep space. Figure 3.14 depicts this competition between gravity and inertia.

The Sun-planet interaction sketched here is analogous to whirling a rock at the end of a string above your head. The Sun's gravitational field is your hand and the string, and the planet is the rock at the end of that string. The tension in the string provides the necessary **centripetal force** (literally, "center-seeking") for the rock to move in a circular path. If you suddenly release the string—which would be like eliminating the Sun's gravity—the rock would fly away along a tangent to the circle, in accordance with Newton's first law.

Returning to the solar system, we can calculate the relationship between the distance (R) and the velocity (V) of

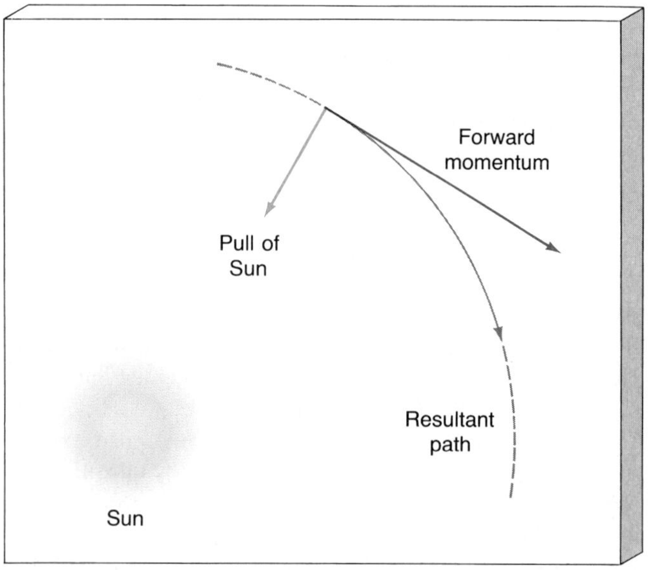

Figure 3.14 *The Sun's inward pull of gravity on a planet is counteracted continuously by the forward motion of that planet. These two effects combine, causing the planet to move smoothly along an intermediate path, which continually "falls over the edge" of the Sun. This continual tug-of-war between the Sun's gravity and the planet's inertia results in a stable orbit.*

a planet (of mass m) moving in a circular orbit around the Sun (of mass M). By calculating the centripetal force required to keep the planet moving in a circle and comparing it with the gravitational force due to the Sun, it is found that the circular velocity is simply

$$V = \sqrt{\frac{GM}{R}}$$

(the constant G is defined in Summary 3-3). At this very moment, Earth is moving under the influence of these two effects: the competition between gravity and inertia. The net result is orbital stability—an Earth that feels ''rock solid'' despite its continuous and rapid motion through space. In fact, the Earth orbits the Sun with a velocity of about 30 km/s, or some 70,000 miles per hour.

Kepler's Laws Reconsidered

Newton's Laws of Motion and his Law of Universal Gravitation provided a theoretical explanation for Kepler's empirical laws of planetary motion. However, just as Kepler modified Copernicus's model with the introduction of ellipses rather than circles, so too did Newton make slight corrections to Kepler's first and third laws. It turns out that a planet does not orbit the exact center of the Sun. Instead, both the planet and the Sun orbit their common **center of mass.** Since the Sun and the planet feel equal and opposite gravitational forces (by Newton's third law), the Sun must move (by Newton's first law). Because the Sun is so much more massive than any planet, the center of mass is very close to the center of the Sun, which is why Kepler's laws are so accurate. Thus, Kepler's First Law becomes: *The orbit of a planet around the Sun is an ellipse, with the common center of mass of the planet and the Sun at one focus.*

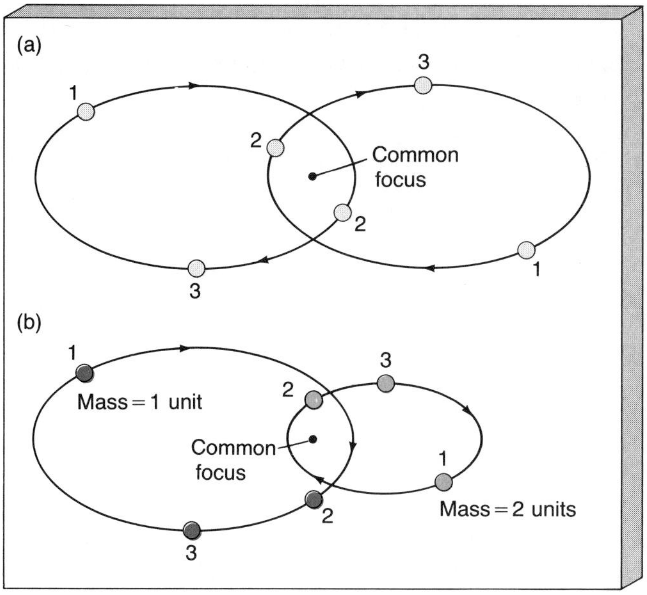

Figure 3.15 *(a) The orbits of two bodies (stars, for example) with equal masses, under the influence of their mutual gravity, are identical ellipses with a common focus. That focus is not at the center of either star, but instead is located at the center of mass of the pair. (b) The orbits of two bodies, one of which is twice as massive as the other. Again, the elliptical orbits have a common focus, and the two ellipses have the same eccentricity. However, the more massive body moves in a smaller orbit, staying closer to the center of mass (at the common focus). In this particular case, the larger ellipse is twice the size of the smaller one.*

As depicted in Figure 3.15, however, the center of mass for two objects of comparable mass does not lie within either object. For identical masses (Figure 3.15[a]), the orbits are identical ellipses, with a common focus located midway between the two objects. For unequal masses (as in Figure 3.15[b]), the elliptical orbits still share a focus and both have the same eccentricity, but the more massive object moves on a tighter orbit.

The change to Kepler's Third Law is also small in the case of a planet orbiting the Sun but again is very important in other circumstances, such as the orbital motion of two stars that happen to be gravitationally bound to one another. Following through the mathematics of Newton's theory, we find that the true relationship between the semi-major axis a of the planet's orbit relative to the Sun and its orbital period P is:

$$P^2 = \left(\frac{4\pi^2}{GM_{\text{total}}}\right)a^3.$$

Notice that Newton's restatement of Kepler's Third Law (see Summary 3-2) preserves the proportionality between P^2 and a^3. The important difference from Kepler's original is that now we have an explicit formula for the constant of proportionality. Because that constant includes M_{total}, the *sum* of the masses of the planet and the Sun, the proportionality is not the same for all the planets. The Sun's mass is so great, however, that the differences are almost unnoticeable, and Kepler's Third Law, as originally stated, is a very good approximation.

In the spirit of our previous discussion, we can once again choose units to make unwanted constants go away. Expressing distance in astronomical units, time in years, and mass in units of the mass of the Sun, we can now say:

$$P(\text{in years})^2 = a(\text{in astronomical units})^3/M_{\text{total}}\ (\text{in solar masses})$$

Before we leave this topic, let's make one more important point. Because we have measured G in the laboratory on Earth and because we know the length of a year and the size of the astronomical unit, we can use our modified version of Kepler's Third Law to *weigh* the Sun. Plugging in the known values for P, a, and G, into the above equation

INTERLUDE 3-1 *Interplanetary Navigation*

''Celestial mechanics'' is the study of the motions of gravitationally interacting objects. These days, the people studying the subject bring powerful computers to bear on Newton's laws in order to understand the intricate movements of astronomical bodies. Computerized celestial mechanics enables astronomers to calculate the orbits of the planets with high precision, taking their (small) gravitational influences on one another into account. Even before the computer age, the discovery of one of the outermost planets, Neptune, came about almost entirely through studies of the distortions of Uranus's orbit that were caused by Neptune's gravity.

Celestial mechanics has become an essential tool as scientists and engineers navigate manned and unmanned spacecraft throughout the solar system. Robot probes can now be sent on stunningly accurate trajectories, expressed in the trade with such slang phrases as ''sinking a corner shot on a billion-kilometer pool table.'' In fact, near-flawless rocket launches, aided by occasional midcourse changes in flight paths, now enable interplanetary navigators to steer spacecraft remotely controlled into an imaginary aperture just a few kilometers wide and a billion kilometers away.

Sophisticated knowledge of celestial mechanics can also facilitate navigation of a single space probe toward several planets. For example, in 1974, the *Mariner 10* spacecraft was guided into the part of Venus's gravitational field that would swing it around to precisely the right path for an additional trek toward Mercury. In other words, Venus itself propelled the probe in a new direction, a course alteration that required no fuel.

The figure at right illustrates how a **gravitational slingshot** works. From the point of view of the planet, the spacecraft arrives along a hyperbolic (unbound) trajectory, passes close by, and then escapes along the same trajectory, in a new direction, with the same speed relative to the planet. However, if the planet itself is moving, some of the planet's momentum is transferred to the spacecraft as it passes by. If the orbit of the spacecraft is chosen correctly, the craft can gain energy, and speed up, as a result of the encounter. Of course, there is no ''free lunch''—the extra energy acquired by the spacecraft comes from the planet's motion, causing its orbit to change ever so slightly. However, since planets are so much more massive than spacecraft, the effect is tiny—*Mariner 10* had no measurable effect on Venus's orbit.

Such a slingshot maneuver has been used several times in subsequent years. For example, the *Voyager 2* spacecraft (launched in late 1977; see Chapter 13) closely bypassed Jupiter (in 1979), Saturn (in 1981), Uranus (in 1986), and Neptune (in 1989). It is now moving through the outermost reaches of the solar system. The gravitational fields of the giant planets whipped the craft around at each visitation, in turn enabling flight controllers to get considerable extra ''mileage'' out of *Voyager 2*. Instruments on board then radioed back information about each of those planets, as we will discuss in later chapters.

The *Galileo* mission to Jupiter, launched in 1990, and

(the one still containing G), we learn that the mass of the Sun is 2×10^{33} g—an enormous mass, by terrestrial standards. Similarly, knowing the distance to the Moon and the length of the (sidereal) month, we can measure the mass of the Earth to be 6×10^{27} g. In fact, this is how *all* masses are measured in astronomy. Because we can't just go out and weigh an astronomical object when we need to know its mass, we must look for its gravitational influence on something else. This principle applies to planets, stars, galaxies, and even clusters of galaxies—very different objects, but all subject to the same physical laws.

The law of gravity that describes the orbits of planets around the Sun applies equally well to natural moons and artificial satellites orbiting any planet. All of our Earth-orbiting, humanmade satellites move along paths governed by a combination of the inward pull of Earth's gravity and the forward motion gained during the rocket launch. If the rocket initially imparts enough velocity to the satellite, it counteracts gravity enough for the satellite to go into orbit. Satellites not given enough velocity at launch (such as intercontinental ballistic missiles [ICBMs]) fail to achieve orbit and fall back to Earth (see Figure 3.16). (Technically, ICBMs actually do orbit the Earth's attracting center, but their orbits intersect Earth's surface.)

Some space vehicles—such as the robot probes that visit the other planets—attain enough velocity to escape our planet's gravitational field and move away from Earth forever. This velocity, known as the **escape velocity,** is about 41 percent greater (actually, $\sqrt{2} = 1.41421 \ldots$ times greater) than the velocity of a circular orbit at the Earth's surface.* Below escape velocity, the old adage ''what goes up must come down'' (or at least stay in orbit) still applies.

*In terms of our earlier formula, the escape velocity is just

$$V_{\text{escape}} = \sqrt{\frac{2GM}{R}}.$$

due to arrive at its target in 1995, is scheduled to receive *three* gravitational assists en route—one from Venus, and two from Earth itself (see Chapter 13 for more details). Once in the Jupiter system, one of its probes will use the gravity of Jupiter and its moons to propel it through a complex series of maneuvers designed to bring it close to all the major moons, as well as the planet itself. Every encounter with every moon will have a slingshot effect—sometimes accelerating, sometimes slowing the probe, and moving it onto a different orbit—and every one of them was carefully calculated long before *Galileo* ever left Earth.

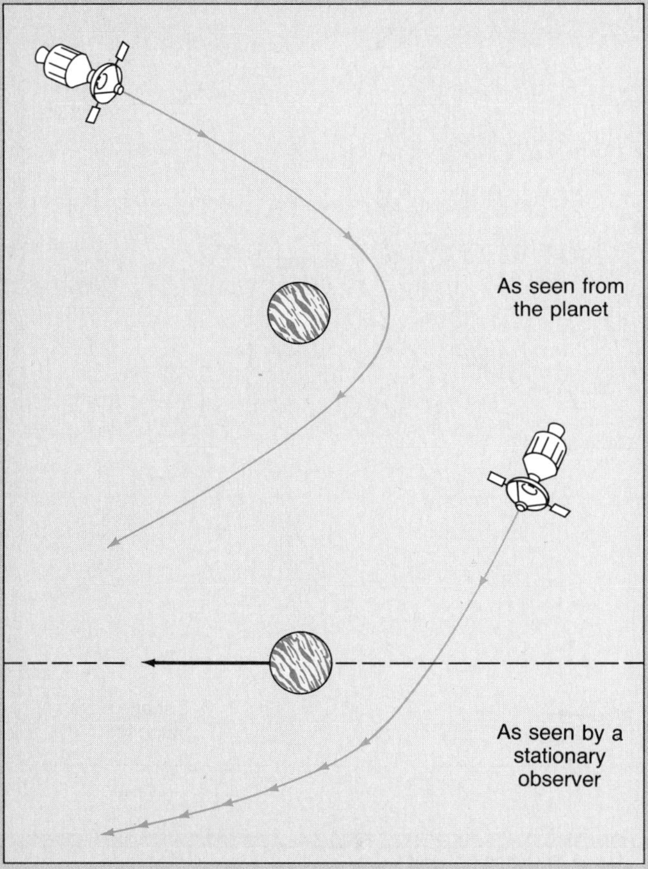

Gravity-assisted flights such as these may well increase in the future. No additional rocket fuel (or money) is needed for a spacecraft redirected in this way from one planet to another. The crux is knowing enough celestial mechanics to decide when and where to aim the space probe into the strong gravitational field of the appropriate planet. A small mistake in calculation can lead to an enormous positional error in space—especially when compounded over the large distances between planets. For example, a 0.1″ mistake at launch from Earth can produce nearly a 1000-km error in the position of a spacecraft by the time it reaches Saturn. Of course, even with perfect calculations, the planets still have to be in the right place—the *Voyager 1* probe was able to visit only Jupiter and Saturn because the alignment of planets did not allow for the "Grand Tour" of the planets undertaken by *Voyager 2*.

With better knowledge of planetary masses, accurate radio tracking of space vehicles, and fortuitous positioning of the planets, we might someday be able to send a single robot probe to every planet in our solar system. We would then enter into a new era of planetary exploration, acquiring a large amount of data about our cosmic neighbors for a small investment of money.

Above escape velocity, our spacecraft has left the Earth for good (unless we turn it around using an on-board rocket motor). Planets, stars, galaxies—all gravitating bodies—have escape velocities. No matter how massive the body, gravity decreases with distance. As a result, the escape velocity diminishes with increasing separation. The farther we go from Earth (or any gravitating body), the easier it becomes to escape.

The velocity of a satellite in a circular orbit just above the Earth's atmosphere is 7.9 km/s (roughly 18,000 mph). The satellite would have to travel at 11.2 km/s (about 25,000 mph) to escape from the Earth altogether. Newton's laws still accurately describe its motion, however. In the case of an object exceeding the escape velocity, the motion is said to be **unbound,** and the orbit is no longer an ellipse. In fact, the path of the spacecraft relative to the Earth is a related geometrical figure called a **hyperbola.** In the intermediate case, where the spacecraft has exactly the escape velocity and so has just enough energy to get away, the orbital trajectory is an intermediate geometrical shape—a **parabola.** If we simply change the word *ellipse* to *hyperbola,* the modified version of Kepler's First Law still applies, as does Kepler's Second Law. (Kepler's Third Law does not extend to unbound orbits because it doesn't make sense to talk about a period in those cases.) Figure 3.17 illustrates the idea of hyperbolic motion for the same bodies as depicted in Figure 3.15.

Newton's laws explain the paths of objects moving at any point in space near a gravitating body. These laws provide a firm physical and mathematical foundation for Copernicus's heliocentric model of the solar system and for Kepler's laws of planetary motions. But they do much more than that. Newtonian gravitation governs not only the planets, moons, and satellites in their elliptical orbits, but also the stars and galaxies in their motion throughout our universe—as well as apples falling to the ground.

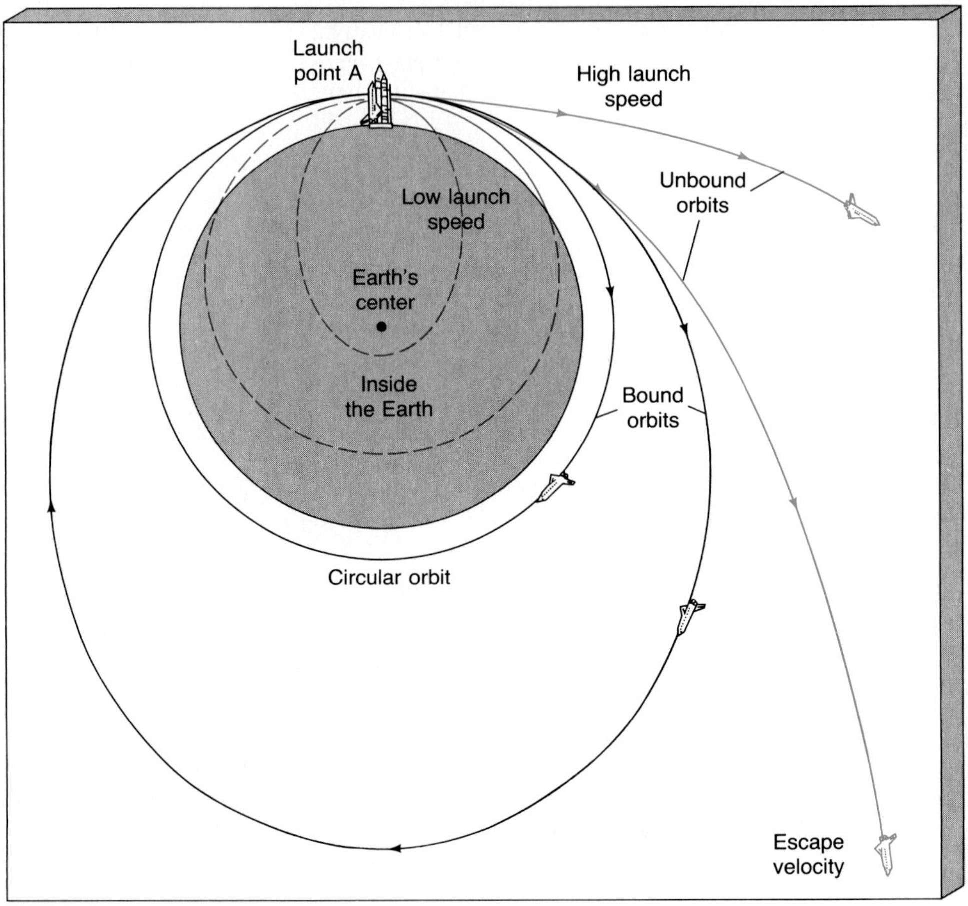

Figure 3.16 The effect of launch velocity on the trajectory of a satellite. With too low a velocity at point A the satellite will simply fall back to Earth. Given enough speed, however, the satellite will go into orbit—it "falls around the Earth." As the initial speed at point A is increased, the orbit will become more and more elongated. Eventually, as the initial speed exceeds the escape velocity, the satellite will become unbound from the Earth and will escape along a hyperbolic trajectory.

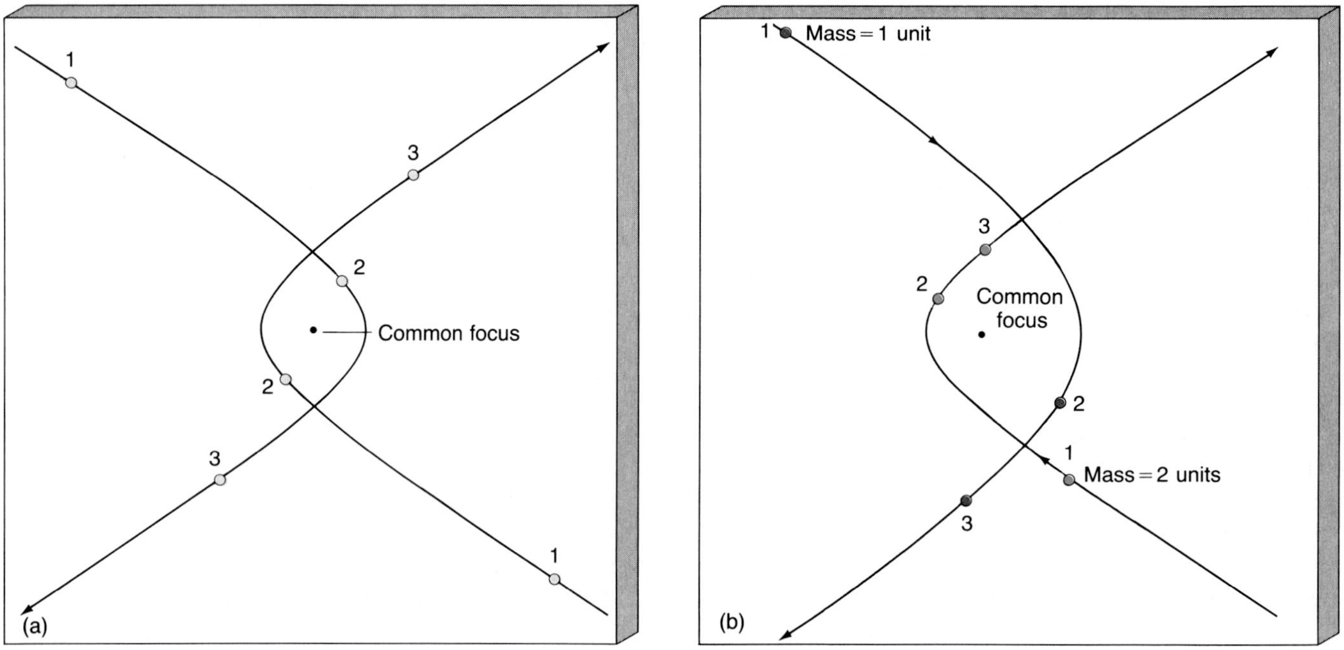

Figure 3.17 Hyperbolic, or unbound, motion of two fast-moving bodies (a) of equal mass, and (b) of unequal mass, with one twice as massive as the other. As in the elliptical case (Figure 3.15), the orbits have a common focus, at the center of the mass of the pair. The heavier body moves in a tighter orbit around the focus.

CHAPTER SUMMARY

The Ptolemaic model ultimately could not account precisely for the motions of the planets. As technological developments improved observational astronomy and as the scientific method came into widespread use, the stage was set for a new world view.

The Ptolemaic model probably survived for so long (almost thirteen centuries) for a variety of reasons. One was that the teachings of Aristotle were consistent with those of the Church. Another was that there seems to have been a genuine lack of scientific curiosity during much of this time.

In the sixteenth century, Nicholas Copernicus started a revolution in thought by asserting that the Earth moves around the Sun every year and spins on its axis every day.

In the century following Copernicus, Johannes Kepler and Galileo Galilei made great strides in popularizing the Copernican viewpoint. Galileo used one of the first telescopes to discover mountains, valleys, and craters on the Moon and dark spots on the face of the Sun. He also discovered four moons in orbit around Jupiter, and he saw that Venus exhibits phases. His discoveries bolstered the idea that Earth does not occupy a central position in the cosmos. Verification that the solar system is Sun-centered came gradually, with innumerable observational tests climaxing with the unmanned space expeditions of the past several decades. The removal of Earth from a position of great cosmic significance is generally known as the ''Copernican principle.'' Today the solar system is known to include one star, nine planets, at least 60 moons, and thousands of asteroids.

Johannes Kepler discovered a set of simple empirical laws that accurately describe the motions of the planets around the Sun. His first law describes the shapes of planetary orbits as ellipses rather than as circles. His second law addresses the speed at which a planet moves at different portions of its orbit, showing that planets travel fastest when closest to the Sun and slowest when farthest away. The third law describes the way in which a planet's orbital period increases more rapidly than the size of its orbit. His work was based largely on the observations of Tycho Brahe.

Kepler's laws let us construct a scale model of the solar system, with the correct shapes, locations, and relative sizes of all the planetary orbits. However, they don't reveal the absolute distance to each planet. Precise radar measurements of the distance to Venus give the Earth-Sun distance. Knowing the Earth-Sun distance and applying Kepler's laws, we can express the sizes of the other planetary orbits in whatever units we like. Distance estimates to increasingly remote objects are made by basing each new estimate on a previous one. Thus the measurement of the scale of the solar system is the first step in the distance ''ladder'' to the cosmos.

Kepler's laws describe the planet's motions, but they do not explain why the planets move as they do. The heliocentric system was secured when, in the seventeenth century, the British mathematician Isaac Newton developed a deeper understanding of the way in which all objects interact with one another. His three basic Laws of Motion, the Law of Gravity, and the calculus are sufficient to explain and quantify virtually all of the complex dynamical behavior we see on Earth and throughout the universe. Newton discovered that any object having mass always exerts an inward, attractive gravitational force on all other massive objects. Gravity weakens rapidly with distance, but it never disappears entirely.

According to Newton, the Sun dominates the motions of the planets because it is so much more massive than the planets. As the Sun pulls, it tries to draw the planets directly toward itself, but this inward pull is counteracted somewhat by each planet's forward motion, or momentum.

Newton's laws show that small objects moving around larger ones actually are in orbit around a common center of mass shared with the larger objects. Planets, moons, and artificial satellites—as well as stars and galaxies in their motions through space—all are governed by Newtonian gravitation.

KEY WORDS

acceleration
aphelion
astronomical unit (A.U.)
center of mass
centripetal force
eccentricity
ellipse
escape velocity
focus
force
gravitational constant
gravitational field
gravitational force
gravitational slingshot
heliocentric
hyperbola
inertia
inverse-square law
Kepler's laws of Planetary Motion
major axis
momentum
Newtonian mechanics
parabola
perihelion
radar
Renaissance
semi-major axis
unbound
weight

REVIEW QUESTIONS

1. Why did the Ptolemaic picture of the universe survive for so long?
2. What was the great contribution of Copernicus?
3. What was Copernicus's major motivation for introducing the heliocentric model of the universe?
4. When were Copernicus's ideas finally accepted?
5. What is the Copernican principle?
6. What discoveries of Galileo helped confirm the views of Copernicus?
7. What was Kepler's contribution to astronomy?
8. What did Kepler use as the basis for his ideas?
9. Do Kepler's laws let us specify the actual distances between orbits of the planets and thereby the scale of the solar system?
10. How can radar be used to find the distance between Earth and Venus?
11. How can radar be used to find the distance between the Earth and the Sun?
12. Scientists sometimes say that Kepler's laws aren't true natural laws, while Newton's laws are. What's the difference?
13. Do Newton's laws hold true in all circumstances?
14. What is inertia? Give an example.
15. Consider the gravitational interaction between Earth and a baseball thrown into the air. If the force of gravity is acting on both of them, why does the baseball move toward Earth and not Earth toward the baseball?
16. Why would a baseball go higher if it were thrown upward from the surface of the Moon?
17. What can you deduce about gravity from the answers to questions 15 and 16?
18. Why can the motion of a planet around the Sun be described as a tug-of-war?
19. Do planets orbit the center of mass of the Sun?
20. What is a gravitational slingshot?
21. How long does an Earth-Venus radar signal take to complete its round trip when Earth and Venus are at their closest (0.3 AU)?
22. A hypothetical planet has an orbit that just grazes Earth's orbit at perihelion and that of Mars at aphelion. What is its orbital eccentricity and semi-major axis? What is its orbital period?
23. Jupiter's moon Callisto orbits it at a distance of 1.9 million km. Its orbital period about the planet is 16.7 days. What is the mass of Jupiter? (Assume that Callisto's mass is negligible compared with that of Jupiter.) Use the modified version of Kepler's Third Law (see p. 47).
24. The escape velocity from Earth's surface is 11.2 km/s. What is the escape velocity (a) from Mars, with mass 0.11 Earth masses and radius 3400 km, and (b) from Saturn, with mass 95 Earth masses and radius 60,000 km?

DISCUSSION QUESTIONS

1. Discuss the role of empirical knowledge in studies of nature. Why is this sort of knowledge particularly important in exploring the natural world? What pitfalls might be associated with the acquisition of empirical knowledge?
2. Is the climate for new ideas better today than it was during the time of Copernicus? Why or why not?
3. Employ the concepts of inertia, momentum, acceleration, and gravitational force to explain why Earth orbits the Sun. What would happen to Earth if the Sun's gravity suddenly stopped?
4. Does Earth's gravity affect you as powerfully when you are floating in a swimming pool as when you are standing on the side of the pool? Does it affect you as powerfully when you are flying high above Earth's surface in an airplane or orbiting Earth in the Space Shuttle? If Earth's gravity is powerful enough to keep the Moon in orbit, are astronauts on the Moon more subject to the Earth's gravity than to the Moon's?
5. Given that the gravitational slingshot method is an efficient technique for propelling spacecraft from one planet to another, what arrangement of planets (or their present orbits) in our solar system would be most convenient for an interplanetary voyager? Consider the fact that the spacecraft requires many months or years to travel from one planet to another! Draw a picture representing your ideas. Show the locations of all the planets at the time of *each* planetary encounter.

PROJECTS

1. Draw an ellipse. (See Figure 3.7, p. 48) You'll need two thumbtacks, a piece of string, and a pencil. Tie the string to the thumbtacks, push them into a surface (a piece of cardboard works well), and run the pencil around the

inside of the string. The two tacks represent the two focal points of the ellipse. If you imagine the Sun at one focus, can you see why a planet's distance from the Sun varies throughout the year? What is the eccentricity of the ellipse you have drawn?

2. Use a small telescope to replicate Galileo's observations of Jupiter's four largest moons. Note the moons' brightnesses and their locations with respect to Jupiter. If you watch over a period of several nights, draw what you see; you'll notice that these moons change their positions as they orbit the giant planet. It's also likely that as you watch you'll see one or more of the moons going into eclipse behind Jupiter.

3. Use a small telescope to replicate Galileo's observation of the planet Venus as a crescent. You'll need to consult an almanac to find out when Venus next passes between the Earth and the Sun. In what phase does the planet appear?

SUGGESTED READINGS

Blumenberg, H., *The Genesis of the Copernican World*. Cambridge, MA: Massachusetts Institute of Technology, 1987. Major work by a German philosopher on the significance of the Copernican revolution for our understanding of modern thought. Presents a new account of the history of philosophical interpretations of the significance of the heavens for human beings.

Byard, M. M., ''Poetic Responses to the Copernican Revolution.'' *Scientific American* (June 1977). How English poetry of the 17th century reflected a new awareness of a vaster universe.

Christianson, G. E., ''Newton's Principia: A Retrospective.'' *Sky and Telescope* (July 1987). Written in honor of the 300th anniversary of one of the most influential books in the history of science.

Kinoshita, J., ''Weighty Matters: A Test of Gravity in Greenland Casts Doubt on Newton's Law.'' *Scientific American* (October 1988). A measurement of the Earth's gravity in a mile-deep borehole in Greenland's ice sheet.

Laeser, R. et al, ''Engineering *Voyager 2*'s Encounter with Uranus.'' *Scientific American* (November 1986). Contains information on the technique of using gravitational assists to send a spacecraft from one planet to another.

Lerner and Gosselin, ''Galileo and the Specter of Bruno.'' *Scientific American* (November 1986). The authors suggest that political tensions and Bruno's earlier heresy—not astronomy—put Galileo before the Inquisition.

MacRobert, A., ''Astronomy with a $5. Telescope.'' *Sky and Telescope* (April 1990). How to build a simple telescope, as outlined by Project Star, an astronomy curriculum development program.

Monastersky, R., ''Evidence for New Force—May Be No. 6.'' *Science News* (December 19 & 26 1987). Report on gravity experiments.

O'Neil, W. M., *Early Astronomy—From Babylonia to Copernicus*. Sydney University Press, 1986. The early background of modern astronomy, spanning three-and-a-half millenia, ending in the 16th century.

Pannekoek, A., *A History of Astronomy*. New York: Dover, 1961. Classic, scholarly history of astronomy from ancient times to the middle of the 20th century. Translated into English in 1961, it was hailed as the best history of astronomy written in over half a century. 521 pages.

Ronan, C. A., ''Galileo Galilei—1564–1642,'' *Sky and Telescope* (February 1964). Biographical article about the first person to peer at the heavens through a telescope, on the 400th anniversary of his birth.

Rosen, E., ''Copernicus' Place in the History of Astronomy.'' *Sky and Telescope* (February 1973). Short biographical article, written on the 500th anniversary of the birth of Copernicus.

Stephenson, B., *Kepler's Physical Astronomy*. New York; Springer-Verlag, 1987. (Subtitled ''Studies in the History of Mathematics and the Physical Sciences.'') An explanation of how physical astronomy works.

Waters, T., ''Gravity Under Siege.'' *Discover* (April 1989). Tests of Newton's theory of gravity performed in Greenland and elsewhere. Lively discussion about the possibility that gravity may be more than Newton or Einstein knew.

CHAPTER

4

Radiation
Information from the Cosmos

What does color tell us about temperature? At its highest temperature, molten steel burns white-hot. As the steel flows away from the source of the heat and cools, it radiates blue light. Farther still, the steel shines yellow, then red. Following the same principle, stars in the sky, too, reveal their temperature by the color in which they shine. Check the nighttime sky, and you will see blue stars, yellow stars, and red stars. But temperature is only a part of the information that radiation, speeding at nearly 300,000 kilometers per second, carries to us from these faraway objects. (Courtesy A. Tannenbaum-Sygma)

LEARNING GOALS

1 to realize that astronomy is primarily an observational, rather than experimental, science

2 to understand how radiation moves, and thus transfers energy and information, through the near-void of outer space

3 to realize that radiation behaves sometimes as a wave and at other times as a particle

4 to know that radiation consists of much more than just visible light

5 to understand how the strength or intensity of radiation can help us determine the temperature of an object

6 to appreciate how motion can change our perception of the basic properties of radiation

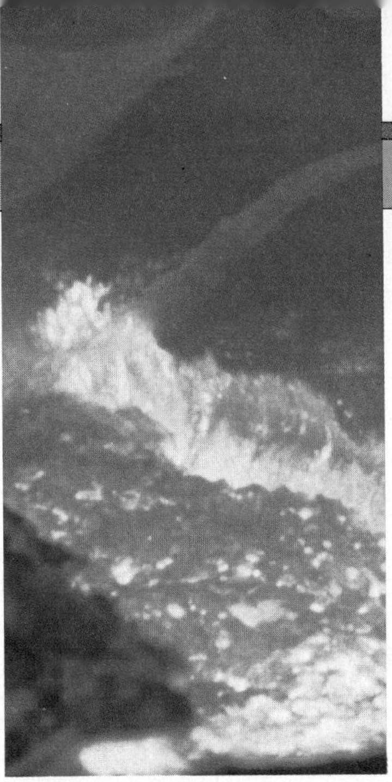

Astronomical objects are more than just things of beauty in the night sky. Planets, stars, and galaxies are of vital significance if we are to understand fully our place in the big picture—the "grand design" of the universe. Each object is a source of information about the material aspects of our universe—its state of motion, its temperature, its chemical composition, and its history. When we look at the stars, the light we see actually began its journey to Earth decades, centuries—even millennia—ago. The faint rays from the most distant galaxies have taken billions of years to reach us. The stars and galaxies in the night sky light up the far away and the long ago.

In this chapter, we study how astronomers gather information from the light that astronomical objects emit. These basic concepts of radiation underlie much of modern astronomy.

INFORMATION

✓1 How do astronomers know anything about stars or galaxies far from Earth? How do they obtain detailed information about any planet, star, or galaxy when these objects (with the exception, so far, of the Moon, Mars, and Venus) are all far too distant for a personal visit or any kind of controlled experiment? The answer is that we use the laws of physics, as we know them on Earth, to interpret the radiation emitted by these objects. Unlike researchers in all other branches of science, astronomers cannot perform laboratory experiments on the objects they wish to study. To gather information about the universe, then, astronomers must examine the radiation that happens to arrive at our planet from afar.

Figure 4.1 shows a neighboring spiral galaxy, in the constellation Andromeda. On a dark, clear night, far from cities or other sources of light, this galaxy can be seen with the naked eye as a faint, fuzzy patch on the sky comparable in diameter to the Moon. The Andromeda Galaxy, as it is generally called, is roughly 2 million light-years from Earth. We can begin to appreciate the enormous distances involved in astronomy when we realize that a light-year is the *distance* traveled by light in a full year and that light moves at the fastest velocity known, around 300,000 kilometers per *second*. A single light-year equals about 10 trillion kilometers (or 6 trillion miles), and Andromeda is 2 million times farther away than that.

Apart from marveling at the immensity of this distance, there are two important points to note. First, the stars that make up the constellation Andromeda are much closer than the Andromeda Galaxy—in fact, they all lie within our own Galaxy and generally less than a few hundred light-years from Earth. The Andromeda Galaxy happens to lie in the same direction as the constellation, but it is 10 thousand times more distant. Second, an object at such a distance is truly inaccessible, in any realistic human sense. Even if a space probe could miraculously travel at the speed of light, 2 million years would be needed for it to reach its destination and 2 million more for it to return with its findings. Considering that civilization has existed on Earth for less than 10 thousand years (and its prospects for the next 10 thousand are far from certain), even this unattainable technological feat would not provide us with a practical means of exploring other galaxies—or even the farthest reaches of

R I **V** U X G

Figure 4.1 *The pancake-shaped Andromeda Galaxy is about 2 million light-years away and contains a few hundred billion stars. (AURA)*

our own, several tens of thousands of light years away. Nearly all that we know about Andromeda, and other galaxies like it, results from studies of the light that they emit.

The velocity of light moving through the vacuum of space is the ultimate speed limit in the universe. As far as we know, nothing—no particle, no signal, no information—can travel faster than 300,000 km/s. While this is a very large velocity, it is still *finite*. Light does not travel instantaneously from place to place. This fact has some interesting consequences for our study of distant objects. It takes time—often lots of time—for light to travel through space. The light we see from Andromeda must have left that galaxy 2 million years ago—about the time our human ancestors first appeared on planet Earth. We can know nothing about this galaxy as it exists today. For all we know, Andromeda might no longer even exist! Only our descendants 2 million years into the future will know for sure if it really exists now.

So, as we study cosmic objects, remember that the light now seen left those objects long ago. We can never observe the universe as it is—only as it was.

THE WAVE NATURE OF RADIATION

Wave Motion

✓2 Light is a special kind of **radiation.** It is also a way in which *information* is transmitted—virtually all of our knowledge about the universe beyond Earth's atmosphere has come to us in the form of light radiation. Visible light is just the kind of radiation to which our human eyes are sensitive. As light enters our eye, the cornea and lens focus it onto the retina, whereupon small chemical reactions send electrical impulses to the brain to record the sensation of light.

Exactly what is radiation? How does it travel? Especially relevant for us, how can it travel through the emptiness of outer space? Radiation travels in the form of a **wave.** In order to understand the behavior of light, we thus must know a little about wave motion.

Wave Properties

A wave is a way in which energy is transferred from place to place. Figure 4.2 shows how waves are quantified and

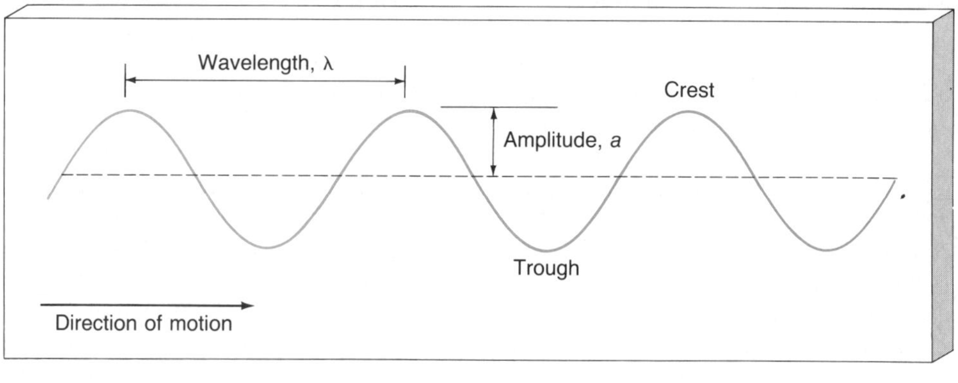

Figure 4.2 *A typical wave, showing its direction of motion, wavelength, and amplitude.*

establishes some standard terminology. The wave pattern repeats itself cyclically in both time and in space. We characterize waves not only by the velocity with which they move, but also by the length of their cycle. How many seconds does it take for a wave to repeat itself at some point in space? This is the **wave period.** How many centimeters does it take for the wave to repeat itself at a given moment in time? This is the **wavelength.** The wavelength (denoted by the Greek letter λ, lambda, in the figure) is defined as the length of an *individual* wave cycle. This can be the distance between two adjacent **wave crests,** two adjacent **wave troughs,** or any two similar points on adjacent wave cycles. The maximum height (or depth) of the wave above (or below) the undisturbed state is called its **amplitude,** marked on Figure 4.2 by the symbol a.

If a wave moves at high velocity, then the number of crests or cycles passing any given point per unit time, defined as the **frequency,** will be reasonably high. Conversely, if a wave moves slowly, with only a few crests passing per unit time, we say that it has a low frequency. The frequency, f, of a wave is just 1 divided by the wave's period, T: $f = 1/T$.

Simple experiments demonstrate that wavelength and wave frequency are *inversely* related—$\lambda \propto 1/f$. In other words, doubling the frequency halves the wavelength, halving the frequency doubles the wavelength, and so on. Together, the product of wavelength and frequency equals the wave velocity, V:

$$\textit{wavelength} \times \textit{frequency} = \textit{velocity}$$

or

$$\lambda f = V.$$

Thus, a wave has a high frequency when its wavelength is small and a low frequency when its wavelength is large. This inverse relationship is easily understood: For a given wave velocity, if the wave crests are close together, more of them pass by a given point each second; when the crests are far apart, few of them pass by per unit time.

Not surprisingly, wavelength has units of length, often expressed in centimeters or meters. Units of velocity amount to a length divided by a time, for example, centimeters per second (cm/s). Frequency has units of inverse time, or cycles per second, termed hertz (Hz), in honor of a nineteenth-century German scientist Heinrich Hertz, who studied radio waves (which, as we will see, are actually a long-wavelength form of light radiation).

Diffraction and Interference

√3 Until the early nineteenth century, a great debate raged in scientific circles regarding the true nature of light. On the one hand, the particle, or *corpuscular*, theory, first expounded in detail by Isaac Newton in the seventeenth century, held that light consisted of tiny particles, or corpuscles, moving in straight lines with constant velocity. Different colors were presumed to correspond to different types of particles. On the other hand, the *wave* theory, championed by the seventeenth-century Dutch astronomer Christian Huygens, sought to explain light in terms of wave motion, where the color was determined by the wavelength. In the nineteenth century, the discovery of two key properties of light, *diffraction* and *interference*, argued strongly in favor of the wave theory.

Diffraction is the ability of waves to "bend around corners." It occurs in any type of wave motion, but it is not to be expected if light is composed of particles moving in straight lines. Thus, a sharp-edged gap in a wall, as depicted in Figure 4.3, would produce a sharp shadow according to the particle theory but a "fuzzy" shadow if light behaves as a wave. The reason that this experiment did not settle the wave-particle argument long before the nineteenth century is that diffraction, while present, is very small for visible light—only fairly sophisticated equipment can measure it. (The effect is much more noticeable for sound waves, however. No one thinks twice about our ability to

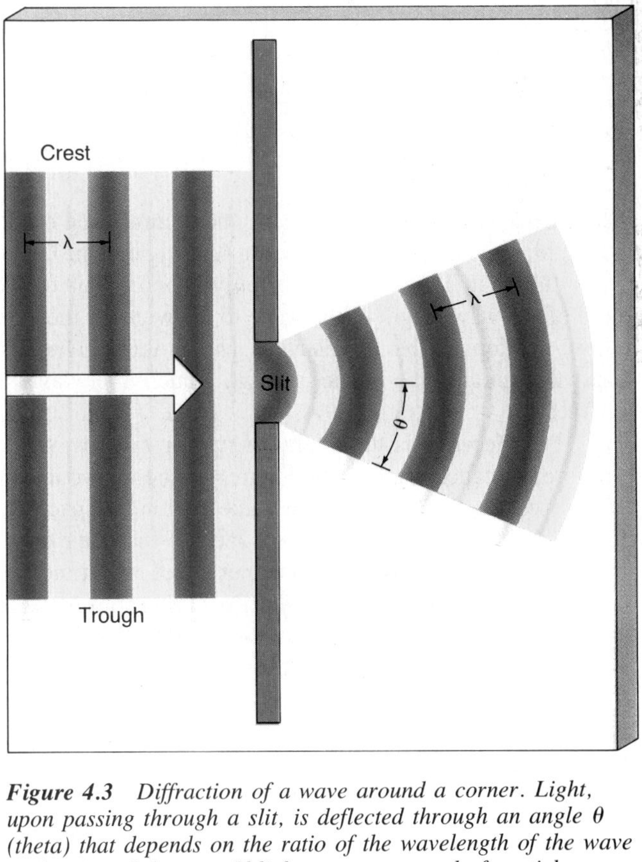

Figure 4.3 Diffraction of a wave around a corner. Light, upon passing through a slit, is deflected through an angle θ (theta) that depends on the ratio of the wavelength of the wave to the size of the gap. If light were composed of particles moving in straight lines, this bending would not occur—thus, the existence of diffraction is experimental evidence in favor of the wave theory of light.

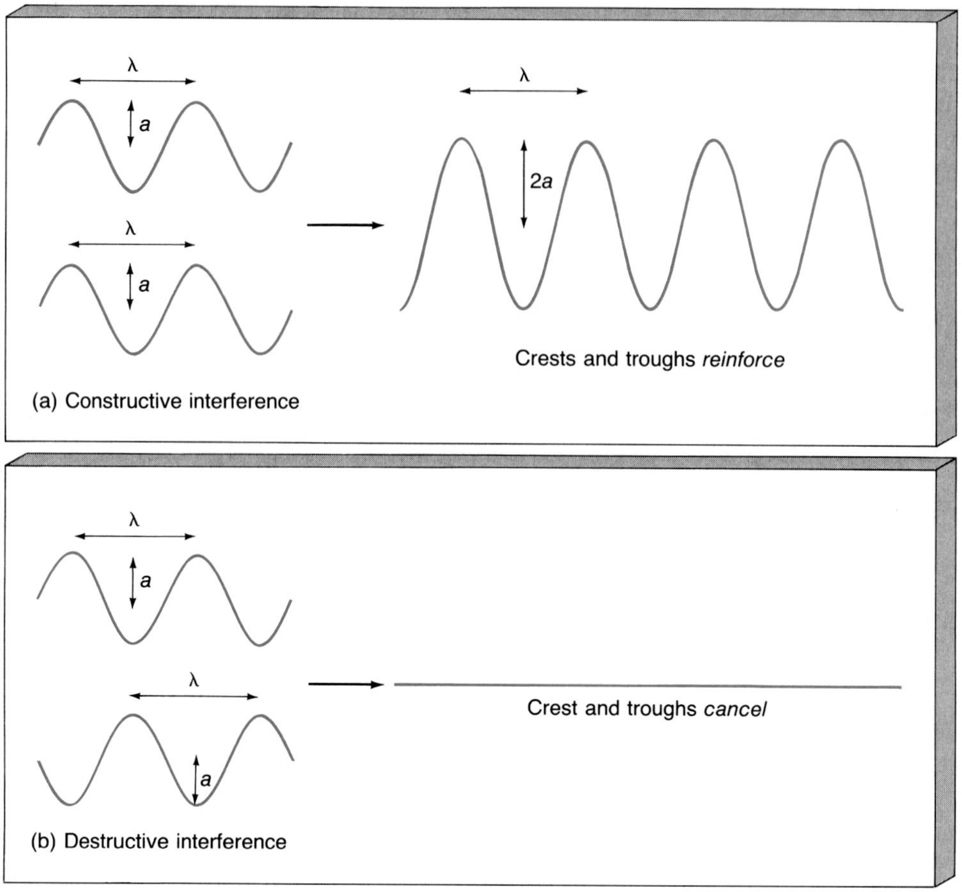

Figure 4.4 *Interference of two identical waves: (a) constructive and (b) destructive. In constructive interference, the two waves reinforce each other to produce a larger-amplitude wave. However, in destructive interference, the two waves exactly cancel out.*

hear people even when they are around a corner and out of our sight.) The amount of diffraction (the angle θ, the Greek letter theta, in Figure 4.3) is simply proportional to the ratio of the wavelength of the wave to the width of the gap. Thus, the longer the wavelength, and/or the smaller the gap, the greater the angle through which the wave is diffracted.

Interference is the ability of two or more waves to reinforce or cancel each other. Figure 4.4 shows two identical waves moving through the same region of space. In Figure 4.4(a), the waves are positioned so that their crests and troughs exactly coincide. The net effect is that the two wave motions reinforce each other, resulting in a wave of greater amplitude. This is known as **constructive interference.** In Figure 4.4(b), the two waves exactly cancel, so that no net motion remains. This is **destructive interference.** Interference is a property common to all wave motions. Like diffraction, interference supports the wave theory of light. The particle theory cannot account for it.

In the early nineteenth century, an English physicist named Thomas Young demonstrated that light can undergo interference. His famous double-slit experiment (Figure 4.5) was regarded by the scientific community as conclusive proof that the wave theory of light was correct. This theory reigned supreme for almost a century.

WAVES IN WHAT?

By means of wave motion, energy and information can be transmitted from one place to another. Imagine a twig floating in a pond of water. A pebble, thrown at some distance from the twig, will generate waves from the point of impact, causing the twig to move up and down when the wave reaches it. In this way, information—the fact that the pebble entered the water—is transferred from the place where the pebble landed to the location of the twig. At the same time, some of the pebble's energy is imparted to the twig, resulting in its eventual motion. By studying various aspects of the wave motion, we can learn about the pebble, the twig, and even the pond itself.

However, waves of radiation differ fundamentally from water waves, sound waves, or any other waves that travel through a material medium. Radiation needs no such medium. When light radiation travels from the Andromeda Galaxy or from any other cosmic object, it moves through the virtual vacuum of empty space. Sound waves, by contrast, cannot do this. If we were to remove all the air from a room, oral conversation would be impossible, but communication through flashlight or radio intercom would be entirely feasible.

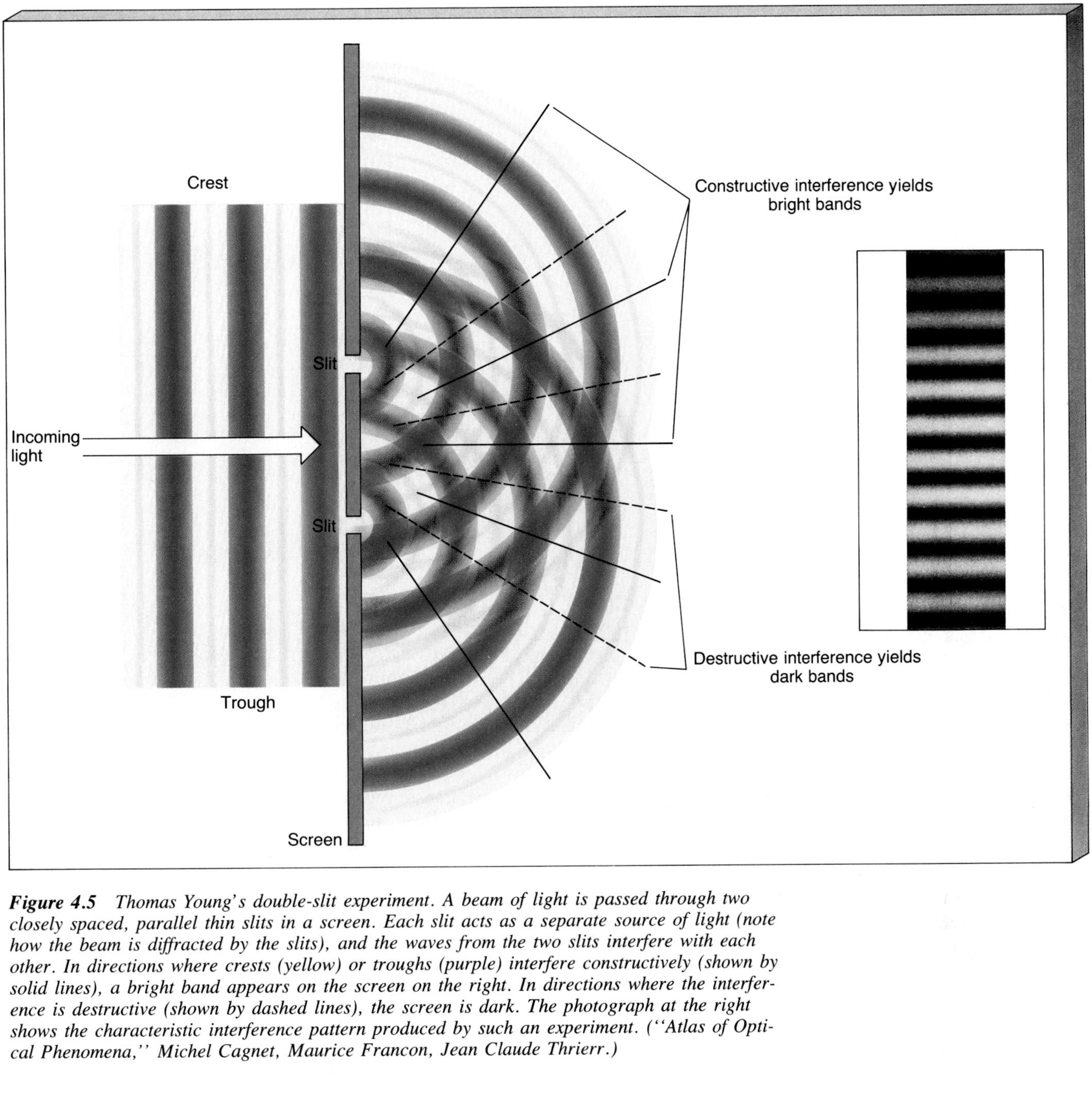

Figure 4.5 *Thomas Young's double-slit experiment. A beam of light is passed through two closely spaced, parallel thin slits in a screen. Each slit acts as a separate source of light (note how the beam is diffracted by the slits), and the waves from the two slits interfere with each other. In directions where crests (yellow) or troughs (purple) interfere constructively (shown by solid lines), a bright band appears on the screen on the right. In directions where the interference is destructive (shown by dashed lines), the screen is dark. The photograph at the right shows the characteristic interference pattern produced by such an experiment. (''Atlas of Optical Phenomena,'' Michel Cagnet, Maurice Francon, Jean Claude Thrierr.)*

The ability of light to travel through empty space was once a great mystery. The idea that light, or any other kind of radiation, could move as a wave through nothing at all seemed to violate common sense, yet it is now a cornerstone of modern physics. Let's discuss it in a little more detail.

Interactions Between Charged Particles

To understand more about the nature of radiation, consider for a moment a charged particle, such as an electron or a proton. Extending outward in all directions from the particle is an **electric field,** which determines the electric force exerted by the particle on all other particles in the universe. The strength of the electric field, like the strength of the gravitational field, decreases with increasing distance from the charge, according to an inverse-square law. By means of the electric field, the particle's presence is ''felt'' by other charged particles, near and far. Now suppose that our particle begins to move, perhaps because it becomes heated or collides with some other object. Its changing position causes its associated electric field to change, and this changing field in turn causes the force exerted on other charges to vary. Thus, *information about the particle's state of motion is transmitted through space via its changing electric field.*

Of prime importance is this question: How *quickly*

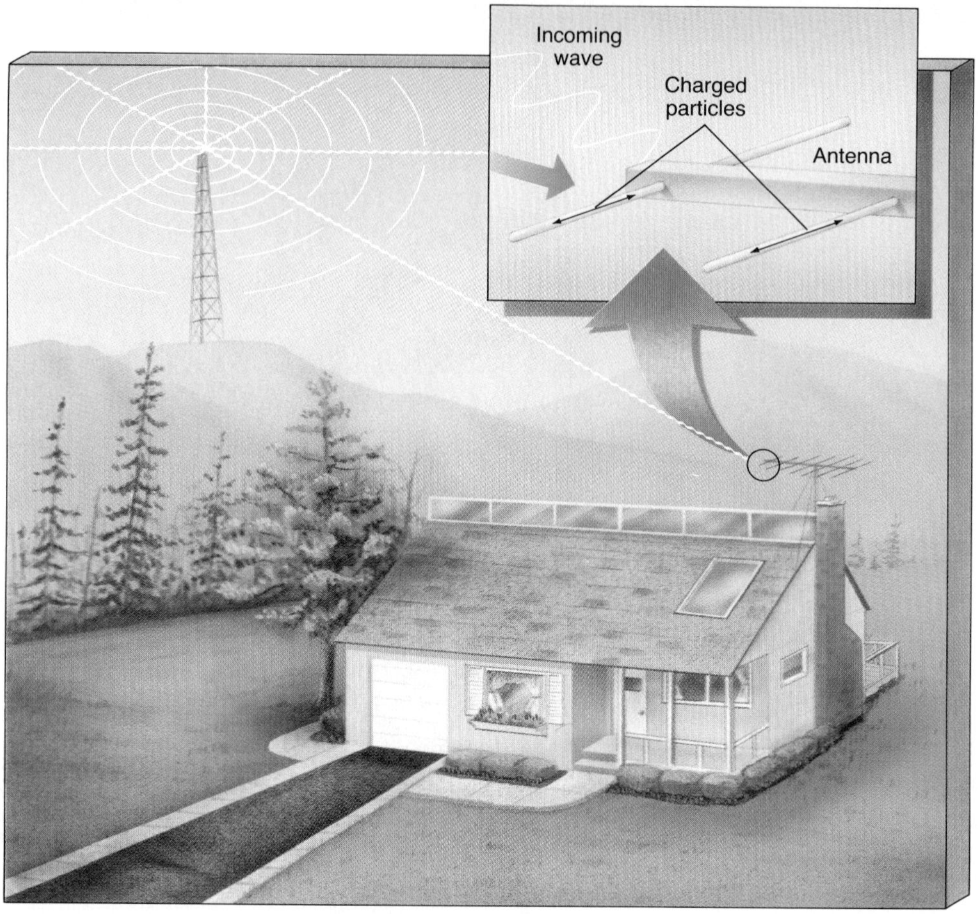

Figure 4.6 Charged particles in an ordinary household television antenna vibrate in response to the charged particle movements in the transmitter tower that sends out waves of radiation.

does one charge feel the change in the electric field when another begins to oscillate (vibrate)? Is the information transmitted at some measurable velocity, or is it an instantaneous flash? Both theory and experiment tell us that the information moves as a wave—a *light wave*—at a very specific velocity—the **velocity of light.** We can imagine the wave as a series of vibrations in the electric field, moving at about 3×10^5 kilometers (186,000 miles) per second.

The velocity of light is always denoted by the letter c. Its exact value, measured in the laboratory, is 299,792 ± 1 km/s in a vacuum (and somewhat less in material substances such as air or water). However, it is usually rounded off for convenience, and we can say $c = 3.00 \times 10^5$ km/s to good accuracy. This is an extremely high velocity. In the time needed to snap a finger—about a tenth of a second—light radiation can travel once around the entire Earth. If the currently known laws of physics are correct, the velocity of light is the fastest speed possible.

Now consider a real cosmic object—a star, say. When some of its charged contents move around, the electric fields change, and we can detect that change. The resulting ripples in the electric field travel outward in waves. They are waves of light radiation, and they require no material medium to propagate. Small charged particles, either in our eyes or in our experimental equipment, eventually respond to the field changes by vibrating in tune with the received radiation. This response is how we detect the radiation—this is how we see.

Figure 4.6 shows a familiar example of information being transferred by radiation. A television transmitter forces electric charges to oscillate up and down a metal rod near the tower's top, thereby radiating waves of information into the air. This radiation can be detected by rooftop antennas having metal booms in which electric charges respond by vibrating in tune with the transmitted wave frequency.

Electromagnetism

One further critical aspect of electric fields must be mentioned. We now know that a **magnetic field** must accompany every changing electric field. Magnetic fields govern the influence of *magnetized* objects on one another, much as electric fields govern interactions between charged particles. The fact that a compass needle always points north is the result of the interaction between the magnetized needle and the Earth's magnetic field.

Here, then, is the way that radiative waves emanate from any terrestrial or extraterrestrial source. As illustrated in Figure 4.7, they consist of both electric and magnetic field vibrations, linked to one another and moving at the speed of light. The waves of electricity and magnetism do not exist as independent quantities; instead, they are just two different aspects of a single physical phenomenon—**electromagnetism.** For this reason, radiative waves, or electric waves, or magnetic waves, are collectively termed "electromagnetic waves." It is this **electromagnetic radi-**

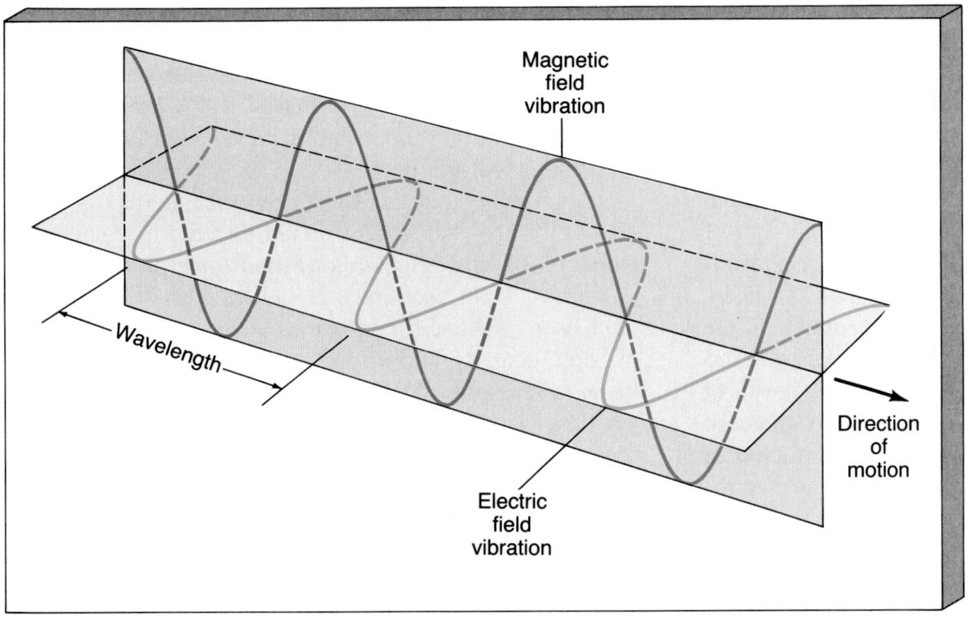

Figure 4.7 Electric and magnetic fields vibrate perpendicular to each other. Together they constitute electromagnetic radiation, which always moves at the velocity of light.

ation that transfers energy and information from one place to another. We will use the terms *light* and *electromagnetic radiation* more or less interchangeably.

COMPONENTS OF LIGHT

White light is a mixture of six colors—red, orange, yellow, green, blue, and violet. We can identify each of these colors, as shown in Figure 4.8, by passing light through a prism, an experiment first performed by Isaac Newton over 300 years ago. Each color is one component of white light, and each is perceived differently by our eyes because of the colors' different wavelengths. The original beam of white light could be produced once again by passing the entire red-to-violet color spectrum through a second, oppositely oriented prism.

Red light has a wavelength of about 7×10^{-5} cm, nearly twice as long as violet light, whose wavelength measures about 4×10^{-5} cm. Wavelengths have been measured for the other colors, too, covering the whole **visible**

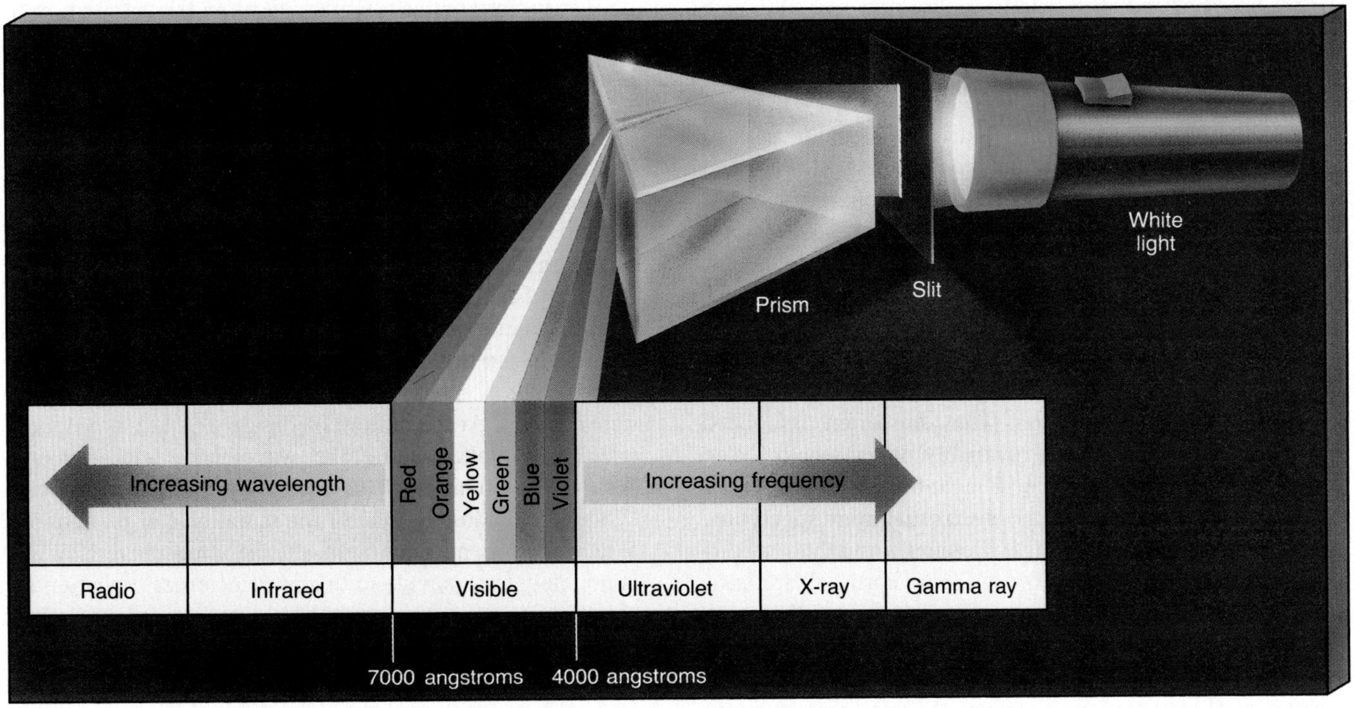

Figure 4.8 While passing through a prism, white light splits into its component colors, spanning red to violet in the visible part of the electromagnetic spectrum.

spectrum shown in Figure 4.8. Human eyes are insensitive to radiation of wavelength shorter than 4×10^{-5} cm or longer than 7×10^{-5} cm. Thus, radiation outside this range is invisible.

For convenience, astronomers often use the angstrom (Å) when describing the wavelength of light. This unit of length is named after the nineteenth-century Swedish physicist Anders Ångström (pronounced "Ongstrem"). There are 10^8 Å in 1 cm (see Interlude 1-2). Accordingly, the visible spectrum spans the wavelength range from 4000 to 7000 Å. The kind of radiation to which our eyes are most sensitive has a wavelength near the middle of this range, at about 5500 Å. It is no coincidence that this is also the range of wavelengths at which the Sun emits most of its electromagnetic energy.

THE ELECTROMAGNETIC SPECTRUM

The Full Range of Radiation

√4 Visible light is only one kind of electromagnetic radiation—the kind to which human eyes are sensitive. There is also **invisible radiation,** to which our eyes are completely insensitive. This radiation includes **radio, infrared,** and **ultraviolet** waves, as well as **x-rays** and **gamma rays.** Despite the different names, the terms *rays*, *radiation*, and *waves* all refer to the same thing. The various names are historical accidents, reflecting the fact that it took many years for scientists to realize that these apparently very different types of radiation are in reality one and the same phenomenon.

Figure 4.9 plots the entire range of all electromagnetic radiation. To the long-wavelength side of light lie radio and infrared radiation. Radio wavelengths include radar, microwave radiation, and the familiar AM, FM, and TV bands. We perceive infrared radiation as heat. At shorter wavelengths are the domains of ultraviolet, x-ray, and gamma-ray radiation. Ultraviolet radiation, lying just shortward of visible blue light, is responsible for suntans and sunburns. X-rays are perhaps best known for their ability to penetrate human tissue and reveal the state of our insides without the need for surgery. Gamma rays are the shortest-wavelength radiation. They are often associated with radioactivity, and are invariably damaging to living cells. All these spectral regions, including the visible spectrum, collectively make up the **electromagnetic spectrum.** Despite their greatly differing wavelengths and the very different roles they play in everyday life on Earth, all are basically the same phenomenon, and all move at the same speed—the speed of light, c.

Figure 4.9 is worth studying carefully. It contains a great deal of information. Note how the wave frequency (in hertz) is plotted increasing toward the right, and the wavelength (in centimeters) increases toward the left. These wave properties behave in opposite ways because, as noted earlier, they are inversely related: $\lambda \propto 1/f$. We will adhere to this convention for presenting wavelengths and frequencies throughout the book.

Note also that the wavelength and frequency scales are not linear—that is, they do not increase by equal increments of 10. Instead, the horizontal scale plots additional *factors* of 10 for each interval shown. This type of plot is often used in science in order to condense a very large range of some quantity into a manageable size. Had we used a linear scale for the same wavelength range shown in Figure 4.9, the figure would have been many light-years long! Throughout the text we will often find it convenient to compress a wide range of some quantity onto a single easy-to-view plot by using such a scale.

Figure 4.9 shows that wavelengths extend from the size of mountains for radio radiation to the size of an atomic nucleus for gamma-ray radiation. We can immediately see how small is the visible portion of the electromagnetic spectrum. Most objects in the universe emit large amounts of invisible radiation. Indeed, many of them emit only a tiny fraction of their total energy in the visible range. Thus, there is a wealth of extra knowledge to be gained by studying the invisible regions of the electromagnetic spectrum.

The part of the electromagnetic spectrum to the high-frequency (short-wavelength) side of the visible spectrum is often called the "high-energy" domain. Scientists who study high-frequency radiation (especially x-rays and gamma rays) from space are therefore sometimes known as high-energy astronomers. There is no corresponding collective term for those who study wavelengths longer than visible light—infrared astronomers work in the infrared, radio astronomers in the radio.

Atmospheric Blockage

Our eyes are sensitive to only a minute portion of the many different kinds of radiation known to exist. In addition, only a small fraction of the radiation produced by astronomical objects actually reaches our eyes, in part because of the **opacity** of the Earth's atmosphere. Opacity is the extent to which radiation is blocked by the material through which it is passing—in this case, air. The more opaque an object is, the less radiation gets through it. Opacity is the opposite of transparency. At the bottom of Figure 4.9, the atmospheric opacity is plotted along the wavelength and frequency scales. The extent of shading is proportional to the opacity. Where the shading is greatest, no radiation can get in or out. Where there is no shading at all, the atmosphere is almost completely transparent, so that extraterrestrial radiation can reach the Earth's surface and terrestrial radiation from transmissions made by humans can pass virtually unhindered into space.

What causes opacity to vary along the spectrum? Why is it that visible radiation can penetrate our atmosphere but most ultraviolet cannot? How can relatively long-wave-

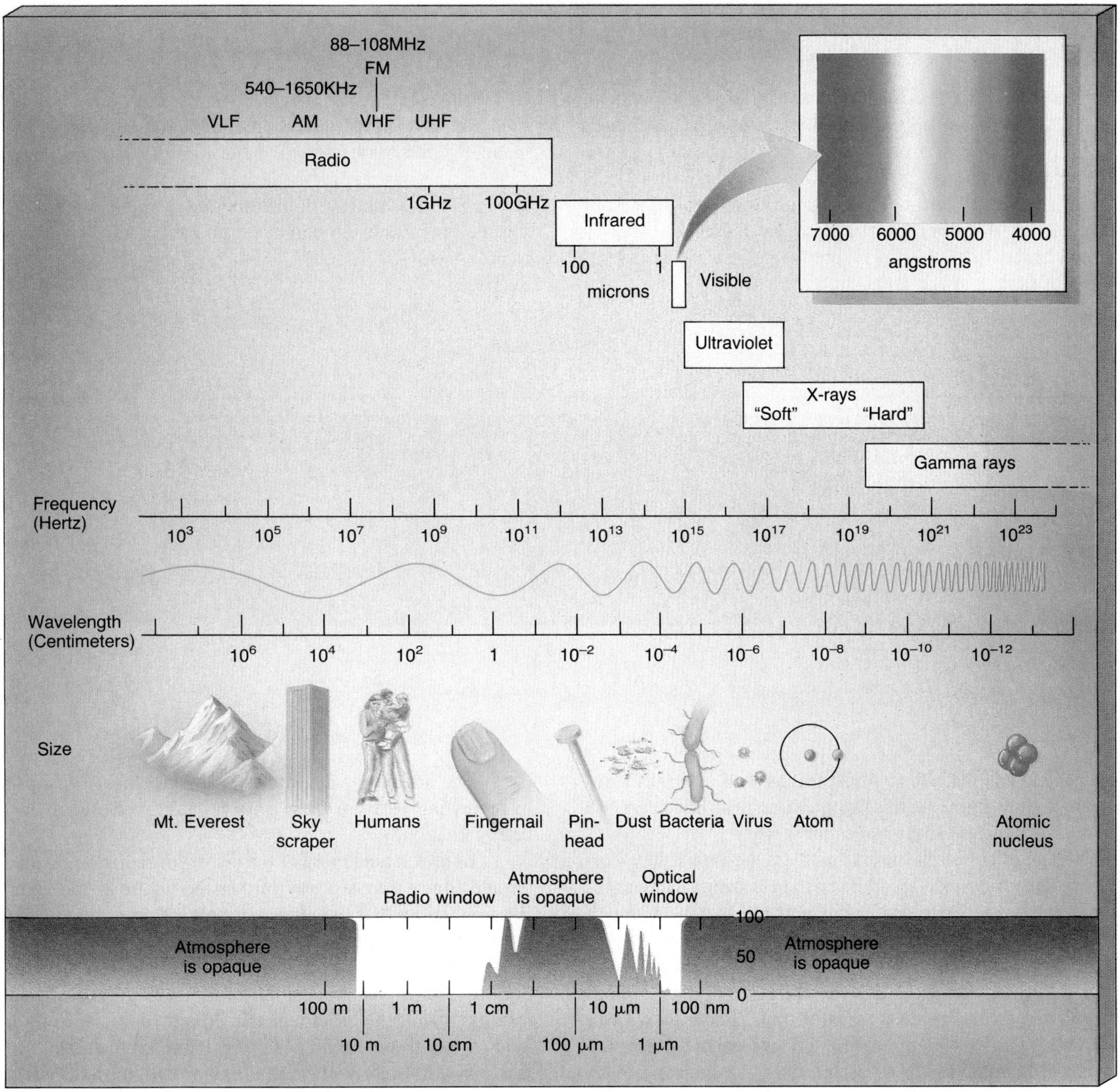

Figure 4.9 *The entire electromagnetic spectrum.*

length radio waves (those of about 1 to 1000 cm wavelength) pass unhindered through the atmosphere but short-wavelength radio waves (those of about 0.1 to 1 cm wavelength) are blocked? Some answers can be found by studying the gaseous content of the Earth's atmosphere.

Certain atmospheric gases are known to absorb radiation very efficiently at some wavelengths. For example, water vapor (H_2O) and oxygen (O_2) absorb radio waves having wavelengths less than about a centimeter, while water vapor and carbon dioxide (CO_2) are strong absorbers of infrared radiation. Ultraviolet, x-ray, and gamma-ray radiation are completely blocked by the **ozone layer** high in Earth's atmosphere. A transient but unpredictable source of atmospheric opacity in the visible part of the spectrum is the blockage of light by atmospheric clouds.

Absorption of radiation by molecular gas is not the only cause of the partial opacity of Earth's atmosphere. The interaction between the Sun's ultraviolet radiation and the upper atmosphere produces a thin, electrically conducting layer at high altitudes (above about 80 km). From this layer, called the ionosphere, long-wavelength radio waves are reflected. In fact, the ionosphere reflects long radio waves

INTERLUDE 4-1 *More on Size and Scaling*

Figure 4.9 includes a general scale of objects found on Earth. We can usefully extend this scale to include much larger objects encountered in space. For example, the sketch below is a continuation of the size scale for objects larger than Mt. Everest. These extend from Earth-sized objects about 10 billion, or 10^{10}, cm in diameter to galaxy superclusters that, at 10^{26} cm, are the largest known assemblies of matter in the universe.

This scale is not meant to imply that some kinds of radiation have wavelengths comparable to these cosmic systems. Scientists are currently unaware of naturally occurring electromagnetic radiation with wavelength longer than about 1000 kilometers. In this Interlude, we are merely taking the opportunity to complete our sketch of the size and scale—from 10^{-14} to 10^{26} cm—of all things known in the universe.

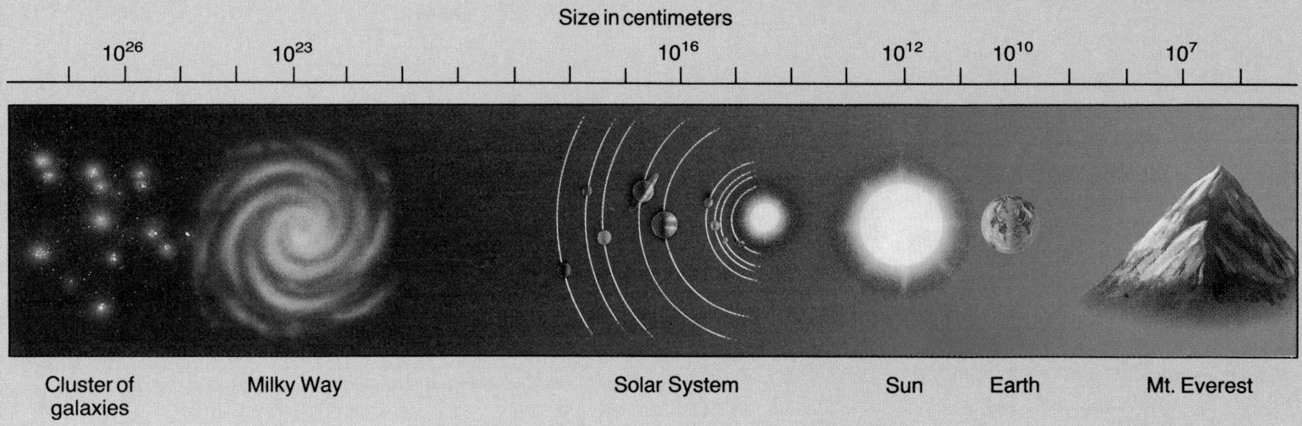

(wavelengths greater than 1000 cm) as well as a mirror reflects visible light. In this way, extraterrestrial waves are kept out, and terrestrial waves are kept in.

The effect of all this is that there are only a few *windows*, at well-defined locations in the electromagnetic spectrum, where the Earth's atmosphere is transparent. In much of the radio and in the visible portions of the spectrum, the opacity is low, so we can observe the universe at those wavelengths from ground level. In parts of the infrared range, the atmosphere is partially transparent, so we can make infrared observations from the ground. Moving to the tops of mountains, above as much of the atmosphere as possible, improves observations. In the rest of the spectrum, however, the atmosphere is opaque. Ultraviolet, x-ray, and gamma-ray observations can be made *only* from above the atmosphere, from orbiting satellites.

THE PARTICLE NATURE OF RADIATION

Radiation surrounds us at all times. Radio, infrared, and visible light radiation pose no significant threats to our health. However, short-wavelength radiation is less benign. As a rule of thumb, ultraviolet radiation will burn us, x-ray radiation will sterilize us, and gamma-ray radiation will kill us. As we have just seen, our atmosphere plays a vital role in protecting us from these harmful forms of electromagnetic radiation.

To understand more of the effects of radiation on matter and life, we must study further the nature of radiation. We now know that radiation not only behaves as a wave, but it sometimes also acts as a *particle*.

Albert Einstein deduced the particle nature of radiation in 1905 in his explanation of a puzzling experimental result known as the **photoelectric effect.** Figure 4.10 illustrates this experiment, in which light shines on a metal surface. When high-frequency ultraviolet light is used, a nearby detector picks up bursts of minute particles, called electrons (see Chapter 5). The metal does not merely reflect the incident radiation. It also absorbs some of it, dislodging electrons from the surface, rather like one billiard ball hitting another and knocking it off the table. For lower-frequency light—blue, say—the detector still records bursts of particles, but now their velocities, and hence their *energies*, are less. For even lower frequencies—red or infrared light—*no* electrons are kicked out of the metal surface at all.

Further experimentation shows that the speed with which the particles are ejected from the metal depends only on the *color* of the light. The shorter the wavelength, the higher the speed. Einstein realized that the only way to explain the presence of an abrupt cutoff at the detector—that

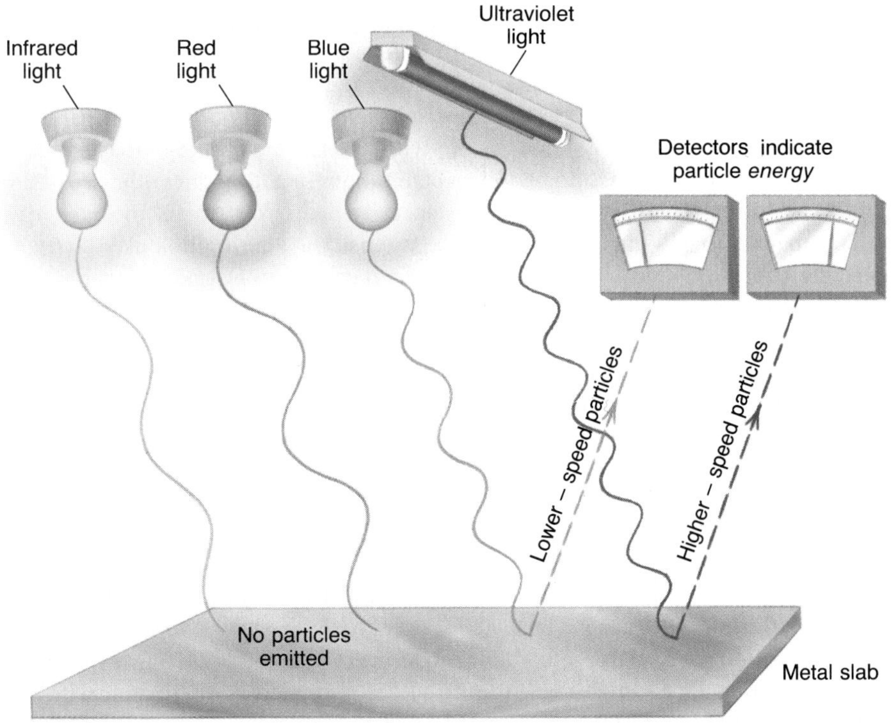

Figure 4.10 *Schematic diagram of the equipment used to study the photoelectric effect. When energy (upper left) from different light bulbs is directed toward metal, electrons are sometimes ejected from the metal and can be recorded by a detector (at upper right). Only the concept of discrete particles (photons) of radiation can explain the results of this experiment. A smooth wave of radiation cannot.*

is, that the detector registered nothing when the frequency of the incoming radiation dropped below a certain level—and the increase in electron speed with light frequency was to envision radiation as traveling as "bullets," or particles. We call these particles **photons.** They can be thought of as little packets of electromagnetic energy.

To explain the experimental results, the energy contained within a photon must be proportional to the *frequency* of the radiation. Thus, for example, a "red" photon with a wavelength of 7000 Å carries roughly half (actually 4/7) the energy of a "blue" photon with wavelength 4000 Å. High-frequency radiation (consisting of high-*energy* photons) ejects particles from the metal at high speed. As the frequency decreases, so too does the energy, and hence the speed of the liberated particles. If we suppose that some minimum amount of energy is needed just to "unglue" the electrons from the metal, then we can also see why no electrons are emitted below some critical frequency—red photons just don't carry enough energy.

Einstein found that:

$$\textit{photon energy} \propto \textit{radiation frequency}.$$

What is the precise connection between the energy of a photon and the frequency of the radiation it represents? The German physicist Max Planck determined the numerical value of the constant of proportionality in this relation. It is now known as **Planck's constant** in his honor, and is denoted by the symbol h. The equation relating photon energy E to radiation frequency f is thus:

$$E = hf.$$

The CGS unit of energy is called the **erg.** In CGS units, the value of Planck's constant is 6.63×10^{-27} erg seconds. How "big" is an erg? Very small, at least by everyday human standards—more than 40 million ergs of energy would be required to raise the temperature of a single gram of water by only 1 degree Celsius. Alternatively, a 100-watt light bulb emits energy (mostly in the form of infrared and visible light) at a rate of 10^9 ergs per second. By plugging in 10^{22} Hz as the radiation frequency for a gamma ray, we see that a single photon's energy is tiny: only 7×10^{-5} erg, even for the most energetic type of electromagnetic radiation.

The above equation explains why, for example, gamma radiation is more harmful than radio radiation. High-frequency gamma rays carry much more energy in every photon, enough energy to harm living tissue—the energy carried by a gamma ray isn't enough to heat a gram of water significantly, but it is more than enough to cause serious damage to a human cell.

This explanation for the photoelectric effect won Einstein the 1921 Nobel prize. The phenomenon occurs whenever radiation interacts with matter, and the wave theory is totally unable to account for it. In addition to bringing about the birth of a whole new branch of physics (now known as *quantum mechanics*), the photoelectric effect, and its explanation, radically changed the way physicists view light and all other forms of radiation. As well as exhibiting smooth, continuous wavelike behavior, radiation can also act in a jerky, discontinuous way. The environmental conditions ultimately determine which way—as a wave or as particles—the radiation behaves. As a general rule of thumb, in the macroscopic realm of large bulk objects, radiation is more usefully described as a wave, while in the microscopic domain of atoms, radiation is best characterized as a series of particles.

The discovery and explanation of the photoelectric effect is another example of the scientific method at work. Despite the enormous success of the wave theory of radiation, the experimental evidence led scientists inexorably to the conclusion that the theory was incomplete and that the picture had to be modified to allow for the fact that light exhibits particle characteristics. Many people find confusing the idea that radiation can behave in two such different ways. To be truthful, modern physicists don't yet fully understand *why* nature displays this wave-particle duality. Nonetheless, there is irrefutable experimental evidence for both of these aspects of radiation.

INTENSITY

✓5 All macroscopic objects emit radiation at all times, regardless of their size, shape, or chemical composition. They radiate mainly because heat causes the microscopic charged particles they contain to move around, and whenever charges change their positions, electromagnetic radiation is emitted. The amount of microscopic motion in an object is determined by its **temperature** (see Interlude 4-2). The hotter the object is, the faster its constituent particles move—and, as we will see, the more energy they radiate.

Some objects emit more radiation than others. A 100-watt bulb, for instance, radiates four times more energy than a 25-watt bulb. **Intensity** is a term often used to specify the amount or strength of radiation. Like frequency and wavelength, intensity is a basic property of radiation. In the wave picture, intensity is simply related to the wave amplitude (in fact, it is just proportional to the square of the amplitude). From the particle standpoint, intensity is proportional to the number of photons received per unit area per second.

Distribution of Radiation

No natural object emits all of its radiation at just one frequency. Instead, its energy is spread out over some portion of the electromagnetic spectrum. By studying the way in which the intensity of this radiation is distributed with respect to frequency (that is, to color), we can gain direct

INTERLUDE 4-2 *The Kelvin Temperature Scale*

Temperature is a measure of the heat of an object. Like the archaic English system used to measure length (in feet) and mass (in pounds), the familiar Fahrenheit temperature scale is of somewhat dubious value. In fact, the "degree Fahrenheit" is now a peculiarity of American society. Most of the world uses the Celsius scale (which is also called the centigrade scale) of temperature measurement. In this system, water freezes at 0 degrees Celsius and boils at 100 degrees Celsius (°C), as illustrated in the accompanying figure.

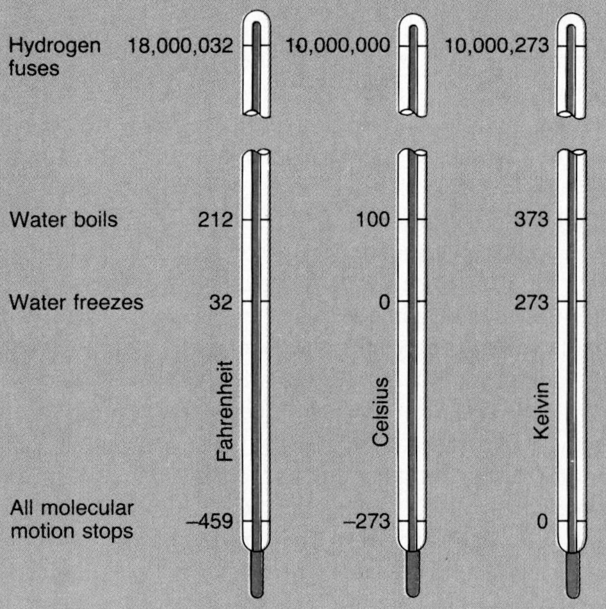

Temperature can go below the freezing point of water, reaching as low as −273°C. Although we know of no matter anywhere in the universe to be this cold, all atomic and molecular motion is expected to stop at this lowest possible temperature. It is natural to choose a system based on this temperature, often called *absolute zero*. Its precise value is −273.15°C.

For convenience, scientists use the *Kelvin scale*, named in honor of the nineteenth-century British physicist Lord Kelvin. It differs from the Celsius scale by 273.15°. In this book, we round off the decimal places and simply use:

$$\text{kelvins} = \text{degrees Celsius} + 273.$$

Consequently:

- motion ceases at 0 kelvins (0 K),
- water freezes at 273 kelvins (273 K),
- water boils at 373 kelvins (373 K).

Note that the unit is "kelvins," not "degrees kelvin." Occasionally, the term "degrees absolute" is used instead.

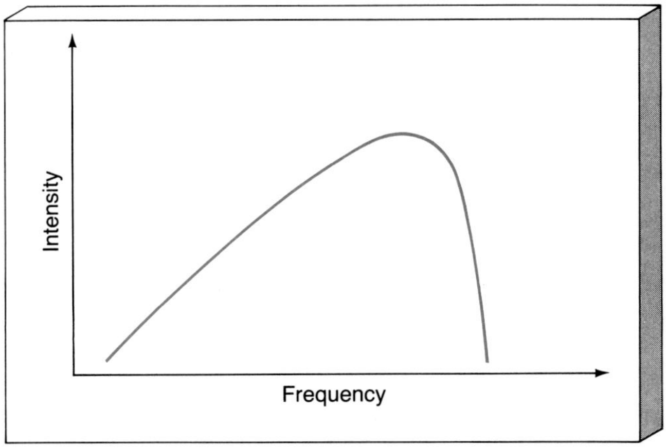

Figure 4.11 *The Planck curve represents the distribution of the intensity of radiation emitted by any heated object.*

information about the object's temperature and often obtain a lot of other more indirect information about its nature.

The intensity of radiation is generally maximized at some particular frequency and falls off to lesser values on either side of the peak. Figure 4.11 shows the standard distribution of radiation emitted by any heated object. We call it the "curve of a perfect thermal emitter," or sometimes a **black-body curve,** or simply the **Planck curve,** after Max Planck, who first quantified these effects in 1900. We will use the term *Planck curve* in this text. However, the term *black-body curve* is also widely used, and the two can be used interchangeably. (The expression *black-body* refers to an idealized concept of an object that absorbs all radiation falling upon it. In a steady state, the object must reemit exactly the same amount of energy as it absorbs—the black-body curve describes the distribution of that reemitted radiation.) Note that the curve is not shaped like a symmetrical bell that declines evenly on either side of the peak. Instead, the intensity falls off more slowly as we move to lower frequencies than it does as we go to high frequencies. This peculiar shape is characteristic of the emission of radiation from *all* heated objects, regardless of their temperature.

The frequency at which the intensity peaks depends on the temperature of the emitting object. In fact, as illustrated in Figure 4.12, the entire Planck curve shifts toward higher frequencies and greater intensities as an object's temperature increases. Even so, the shape of the curve remains the same. This shifting of radiation's peak frequency with temperature is familiar to us all: Glowing objects, such as fire or stars, emit visible radiation. Cooler objects, such as warm rocks or smoldering wood, produce invisible radiation. The latter, warm to the touch but not glowing hot to the eye, emit their radiation most intensely as low-frequency infrared and radio waves.

We can further illustrate the shift of the Planck curve with temperature. A piece of metal in a hot furnace first becomes warm, though it doesn't change its visible appear-

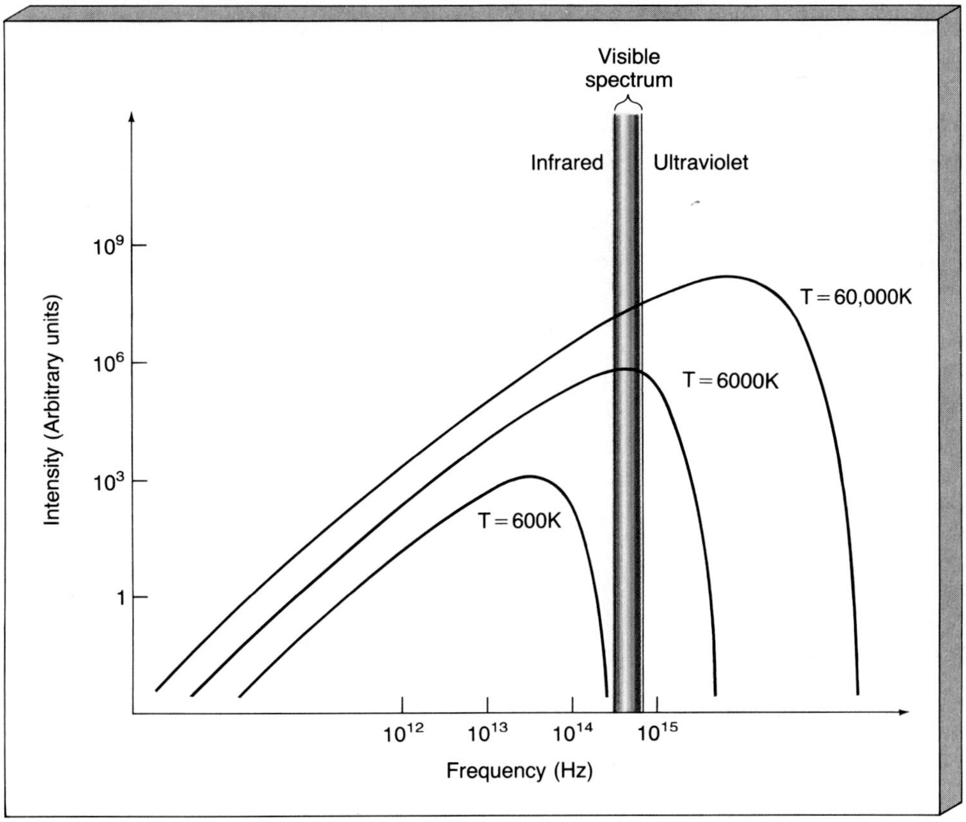

Figure 4.12 *As an object is heated, the radiation it emits is still described by the Planck curve, but the curve shifts to peak at higher and higher frequency as the temperature rises. Shown here are the curves corresponding to temperatures of 600 K, 6000 K, and 60,000 K, which peak, respectively, in the infrared, the visible, and the ultraviolet regions of the electromagnetic spectrum.*

ance. As it heats up, it begins to glow dull red, then orange, brilliant yellow, and finally white. How do we explain this? At first, when the metal was warm, it emitted invisible radio and infrared radiation. As the metal became hotter, the peak of the metal's emitted radiation shifted toward higher frequencies, producing more radiation in the visible domain. The Planck curve for an object of several thousand **kelvins**—a unit generally used by scientists to measure temperature, as noted in Interlude 4-2—peaks in the visible part of the spectrum regardless of its chemical composition spectrum. As it becomes even hotter, its color gradually changes from red to orange to yellow to blue-white.

From more detailed study of the precise form of the Planck curve, we can obtain a very simple connection between the wavelength λ_{max} at which most light is emitted and the temperature of the emitter—they are inversely related:

$$\lambda_{max} \propto \frac{1}{temperature}.$$

If we measure the temperature T in kelvins and λ_{max} in centimeters, we find

$$\lambda_{max} = \frac{0.29}{T}.$$

This relation is usually called **Wien's Law,** after Wilhelm Wien, the German scientist who formulated the law in 1897. In words, Wien's Law tells us that the hotter the

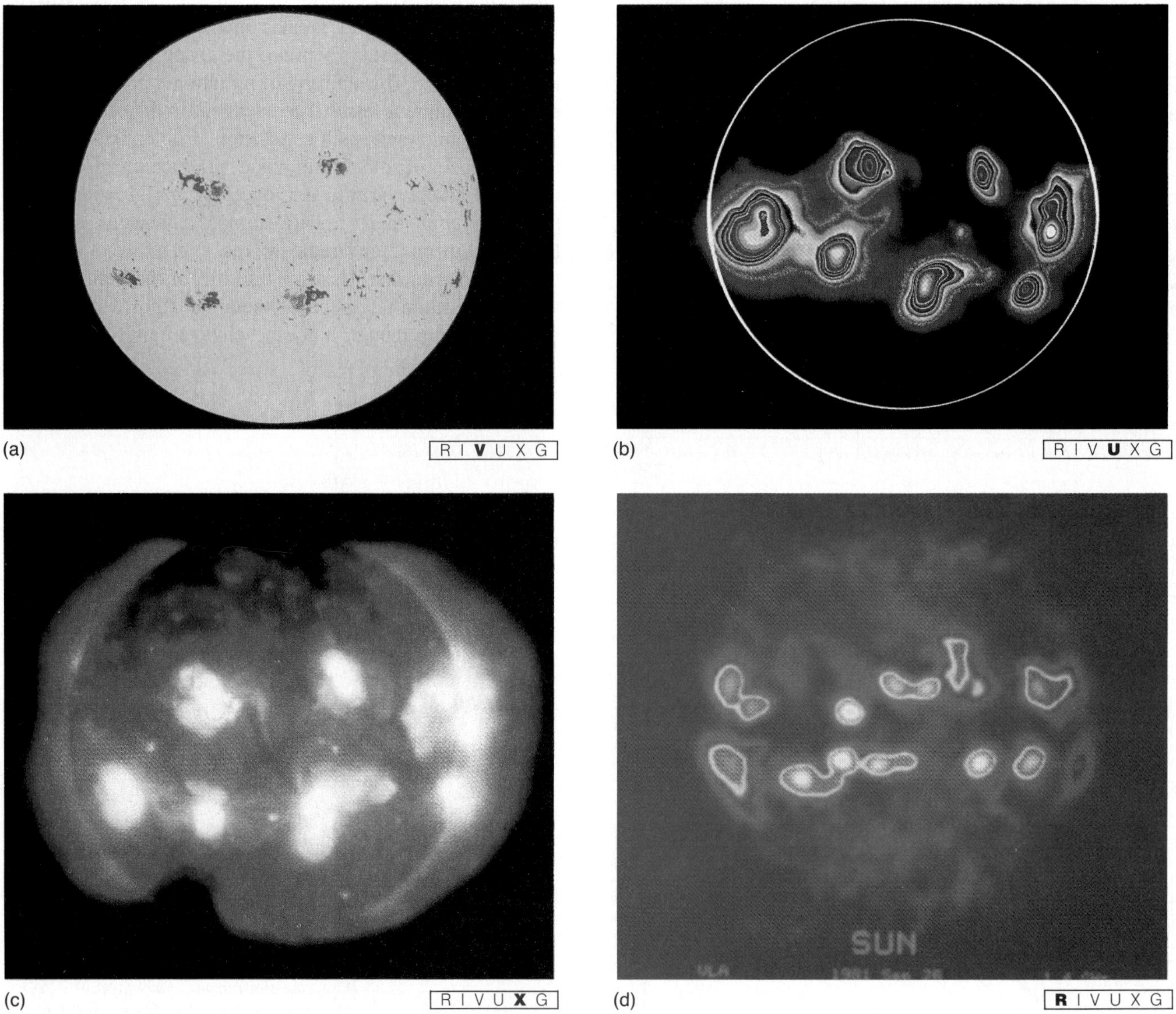

Figure 4.13 *Four images of the Sun, made using (a) visible light, (b) ultraviolet light, (c) x-rays, and (d) radio waves. By studying the similarities and differences among these views of the same object, important clues to its structure and composition can be found. (AURA; NASA; NASA; NRAO)*

object, the bluer its radiation. Thus we can see, for example, that an object with a temperature of 1000 K emits most energy at a wavelength of 0.29/1000 cm, or 2.9 μm, which is a wavelength well into the infrared portion of the spectrum (recall that 1 micron—1 μm—is equal to 10^{-6} m, or 10^{-4} cm, or 10^4 Å). At a temperature of 10,000 K, the peak has moved all the way through the visible spectrum to a wavelength of 0.29 μm, or 2900 Å, in the ultraviolet range. At 100 K, the peak would be in the far infrared range, at 29 μm, and so on.

Objects eventually become white hot (provided that they are not destroyed at high temperatures) because Planck curves that peak in the green or blue part of the spectrum still have a low-frequency "tail" extending through the longer-wavelength portion of the visible spectrum. This means that all colors are emitted, and together they combine to produce the neutral color white. This is the reverse of the dispersion of white light into its component colors by a prism, shown in Figure 4.8.

It is also a matter of everyday experience that, as the temperature of an object increases, the *total* amount of energy it radiates at all frequencies (which is, in fact, just proportional to the area under the Planck curve) increases rapidly. Careful experimentation (or a thorough understanding of the Planck curve) leads to the conclusion that the total amount of energy radiated per unit time is actually proportional to the *fourth power* of the object's temperature:

$$\textit{energy radiated} \propto \textit{temperature}^4.$$

If we measure temperature T in kelvins, then the total energy emitted per square centimeter of its surface per second (a quantity known as the energy flux, F) is given by

$$F = \sigma T^4.$$

This equation is called **Stefan's Law.** The constant σ, the Greek letter sigma, has the value 5.67×10^{-5} erg/s/cm^2/K^4 and is called the **Stefan-Boltzmann constant,** or often simply Stefan's constant, after Josef Stefan, the Austrian scientist who formulated the equation. Notice just *how* rapid is the increase in the energy flux with increasing temperature. A red-hot poker, with a temperature of, say, 3500 K, radiates energy at a rate of 8.5×10^9 ergs per second for every square centimeter of its surface area. Doubling its temperature to 7000 K (so that the poker becomes yellow to white-hot) increases the energy emitted by a factor of 16, to 1.36×10^{11} erg/s/cm^2.

No known natural *terrestrial* objects can become hot enough to emit high-energy radiation. Only human-made thermonuclear explosions are hot enough for their Planck spectrum to peak in the x-ray or gamma-ray range. Most other human inventions that emit short-wavelength radiation, such as x-ray machines, are designed to emit only a specific range of wavelengths and do not operate at high temperatures. (They are said to produce a *non-thermal* spectrum of radiation.) However, many extraterrestrial objects do emit copious quantities of ultraviolet, x-ray, and even gamma-ray radiation. Figure 4.13 shows the appearance of a familiar object—our Sun—when viewed in different regions of the electromagnetic spectrum. While most sunlight is visible, a great deal of information about our parent star can be obtained by studying it at other wavelengths.

Applications

Astronomers use Planck curves to determine the temperatures of distant objects without the necessity of making on-site measurements with, for example, a thermometer. In fact, studying the Planck curve is the main method used by astronomers to measure the temperature of the Sun's surface. Observations of the radiation from our Sun at many different frequencies yield a curve shaped somewhat like the one shown in Figure 4.11. The Sun's curve peaks in the visible part of the electromagnetic spectrum; it also emits much infrared and a little ultraviolet radiation (see Figures 4.13 and 4.14). The overall curve suggests that the temperature of the *surface* of the Sun is approximately 6000 K.

Many stars, including all those seen on a clear night, obviously emit visible radiation. Most of these stars have surface temperatures comparable to the Sun's (within a factor of 2 or 3 one way or the other). Other cosmic objects, however, have surfaces very much cooler or hotter than our Sun's. These objects emit radiation mostly in invisible parts of the spectrum. For example, much cooler regions, such as the surfaces of very young stars, measure about 600 K and emit mostly infrared radiation. The brightest stars have surface temperatures as high as 60,000 K, and hence emit mostly ultraviolet radiation.

Figure 4.14 compares the Planck curves for a variety of objects. The temperatures of these cosmic objects pertain only to the part of the object actually emitting the observed radiation. In the case of stars, for example, such Planck curves tell us that the surface temperatures generally range from about 2000 to 60,000 K.

These surface temperatures, however, fall far short of the extremely hot, millions-of-kelvin temperatures found in the cores of normal stars. Fortunately for us, the lethal x-ray and gamma-ray radiation from the extremely hot interior of any star doesn't reach Earth directly. Instead, the stars' internal structure acts to "downgrade" the radiation, ultimately converting it from high-energy gamma rays into the visible light we see.

By measuring the frequency distribution of emitted radiation, we can infer the surface temperature of *any* object—terrestrial or extraterrestrial. Provided that nothing interferes with the emitted radiation, the Planck curve yields a reasonably accurate temperature for whatever matter emits the radiation. In effect, the Planck curve acts as a thermometer to help us estimate the temperatures of many remote objects in the universe.

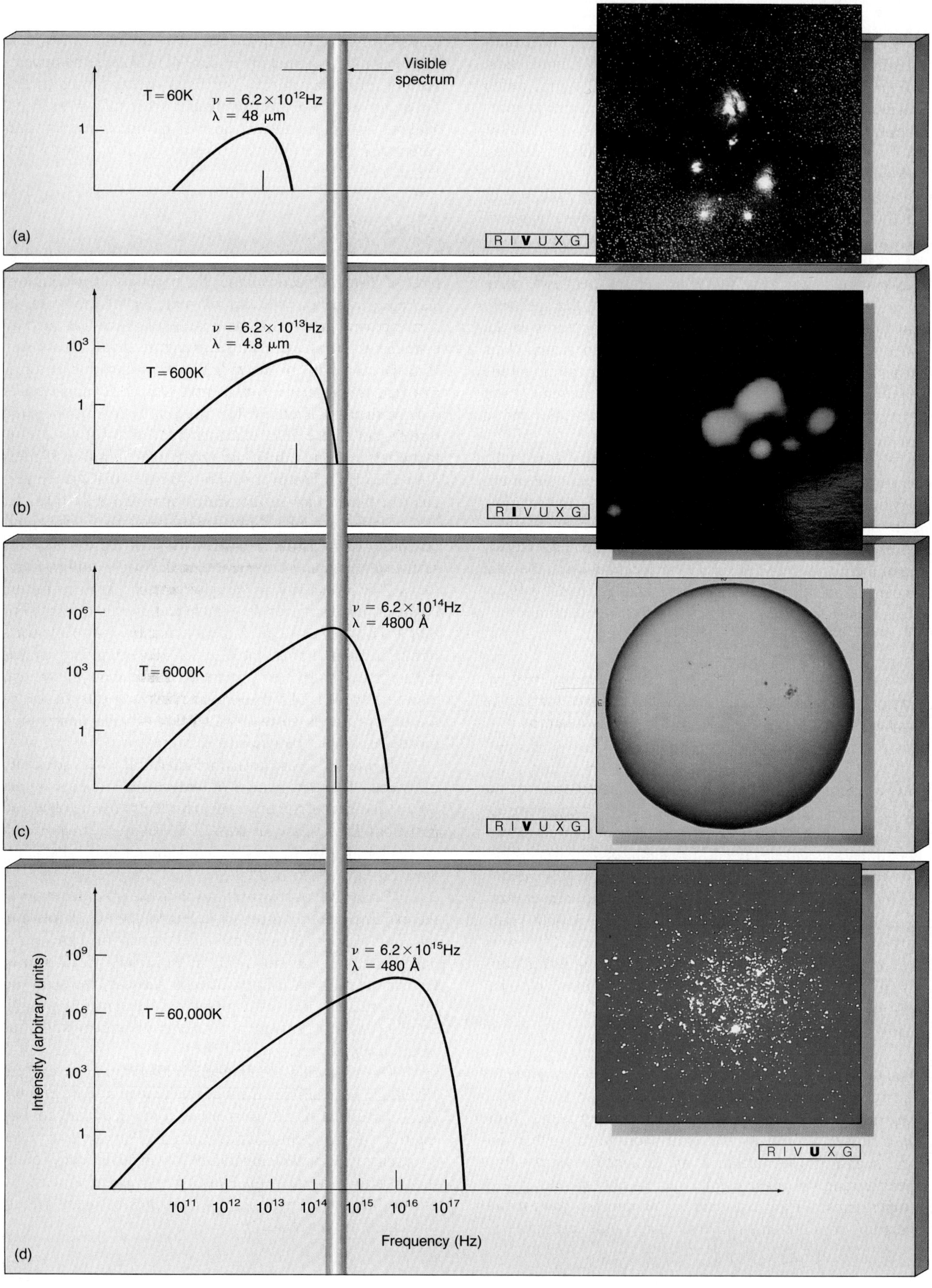

Visible spectrum
T=60K
ν = 6.2×10¹²Hz
λ = 48 μm
1
(a)
R I V U X G
ν = 6.2×10¹³Hz
λ = 4.8 μm
10³
T=600K
1
(b)
R I V U X G
ν = 6.2×10¹⁴Hz
λ = 4800 Å
10⁶
10³
T=6000K
1
(c)
R I V U X G
ν = 6.2×10¹⁵Hz
λ = 480 Å
10⁹
10⁶
T=60,000K
10³
1
Intensity (arbitrary units)
R I V U X G
10¹¹ 10¹² 10¹³ 10¹⁴ 10¹⁵ 10¹⁶ 10¹⁷
Frequency (Hz)
(d)

Figure 4.14 *Comparison of Planck curves for four cosmic objects having different temperatures. (a) A cool, invisible galactic gas cloud (called Rho Ophiuchi) at 60K, which emits mostly long-wavelength radio radiation. (b) A young star's dim atmosphere at 600K, emitting primarily in the infrared (and shown here in red near the center of the Orion Nebula). (c) The Sun's surface, at 6000K. (d) A cluster of very bright stars at 60,000K, a strong region of ultraviolet radiation, called Omega Centauri, as observed by a telescope aboard the Space Shuttle. (Harvard-Smithsonian Center for Astrophysics; Anglo-Australian Telescope Board; AURA; NASA)*

THE DOPPLER EFFECT

√6 Radiation can change in many ways. Wavelength, frequency, and intensity can each vary in response to environmental circumstances. However, the *speed* of radiation never changes. It always equals the velocity of light, *c*. Interestingly, our *perception* of radiation can sometimes change, depending on our own state of motion. Let's use a hypothetical example as an illustration.

A Description

Suppose that a rocket ship of the future is launched from Earth with enough fuel to allow it to accelerate constantly. Eventually, the ship would reach a very high velocity. Travelers looking out from the front of the spacecraft would note that the light emitted from the star system toward which they were moving seemed to be getting bluer. They might think that the star's surface temperature was increasing, emitting more radiation in the blue part of the spectrum, but this is not the reason for the change in color. As the travelers would soon realize, *all* stars seem to become bluer than normal when approached at high speed, and the greater the velocity of approach, the greater the color change. If the spacecraft slowed down and came to rest, the stars would seemingly emit normally once again, just as they had appeared through Earth-based telescopes prior to the voyage.

Furthermore, if the travelers looked back as the spacecraft surged forward, they would notice the opposite effect: the stars behind them would seem *redder* than normal, and the greater the velocity of recession, the greater the color change. Again, when the spacecraft stopped, the stars would regain their usual appearance.

Because all the stars in the spacecraft's forward direction would seem bluer than normal and all those behind redder than normal, the travelers would be forced to conclude that the stars changed their colors because of the spacecraft's motion and not because of a real change in any physical properties of the stars.

Motion is the key here. Motion toward or away from a source of radiation can change the way we perceive that radiation. Such motion along any particular line of sight is known as **radial motion.** It should be distinguished from motion perpendicular to the line of sight, known as **transverse motion.** Radial (but not transverse) motion induces apparent changes in the properties of a wave. Specifically, as we saw in our illustration, it changes our measurement of the wavelength (or frequency) of radiation, because blue light differs from red light only by its shorter wavelength (or higher frequency). Radial motion doesn't change our perception of any other property of a radiative wave. In particular, radiation still travels at the speed of light.

Change due to motion is not restricted to radiation or to fast-moving spacecraft. Waiting at a railroad crossing for a train to pass, we hear the pitch of the engine's horn change from high shrill to low blare as the train approaches and then recedes. Because high-pitched sound (treble) has shorter wavelength than low-pitched sound (bass), this well-known sound change is similar to the light change our space travelers experienced. If the train had a white light (a mixture of all colors) atop its engine, we could, in principle, witness the color change of this light, from bluish to reddish, as the train passed by. In practice, however, the light's color change would be impossible to perceive because trains travel far too slowly for us to perceive the effect. Notice, however, that in this case it is the *source* of the wave that is in motion. In the spaceship example, it was the *observer* that was moving. In fact, it is only the *relative* motion between the source and the observer that matters.

This motion-induced change in the observed wavelength (or frequency) of a wave—whether it is a radiative light wave or an acoustical sound wave—is known as the **Doppler effect,** in honor of Christian Doppler, the nineteenth-century Austrian physicist who first explained it. Because the Doppler effect plays such a crucial role in much of astronomy, we will study it in a little more detail.

An Explanation

Imagine a wave moving from the place where it is generated toward an unmoving rocket ship, like that sketched in Figure 4.15(a). By counting the number of wave crests passing us per unit time, we could determine the frequency and the wavelength.

Suppose now that the ship is moving away from the source of the wave, as in Figure 4.15(b). This case is analogous to the space travelers receding from the distant star system. The number of wave crests now counted per unit time would be smaller than for a fixed observer because the travelers are moving away from the advancing waves. The *observed* wave will thus seem to have a lower frequency or, equivalently, a longer wavelength. We say that the wave has become **redshifted,** because red radiation has longer wavelength than blue radiation. (This convention holds even for invisible radiation for which red and blue color have no meaning. Any shift toward longer wavelengths is called a red shift, and any shift toward shorter wavelengths is called a blue shift. Ultraviolet light would be redshifted to blue, and red light would be redshifted into the infrared.) The greater the observer's velocity away from the source of the waves, the greater the increase in wavelength and the

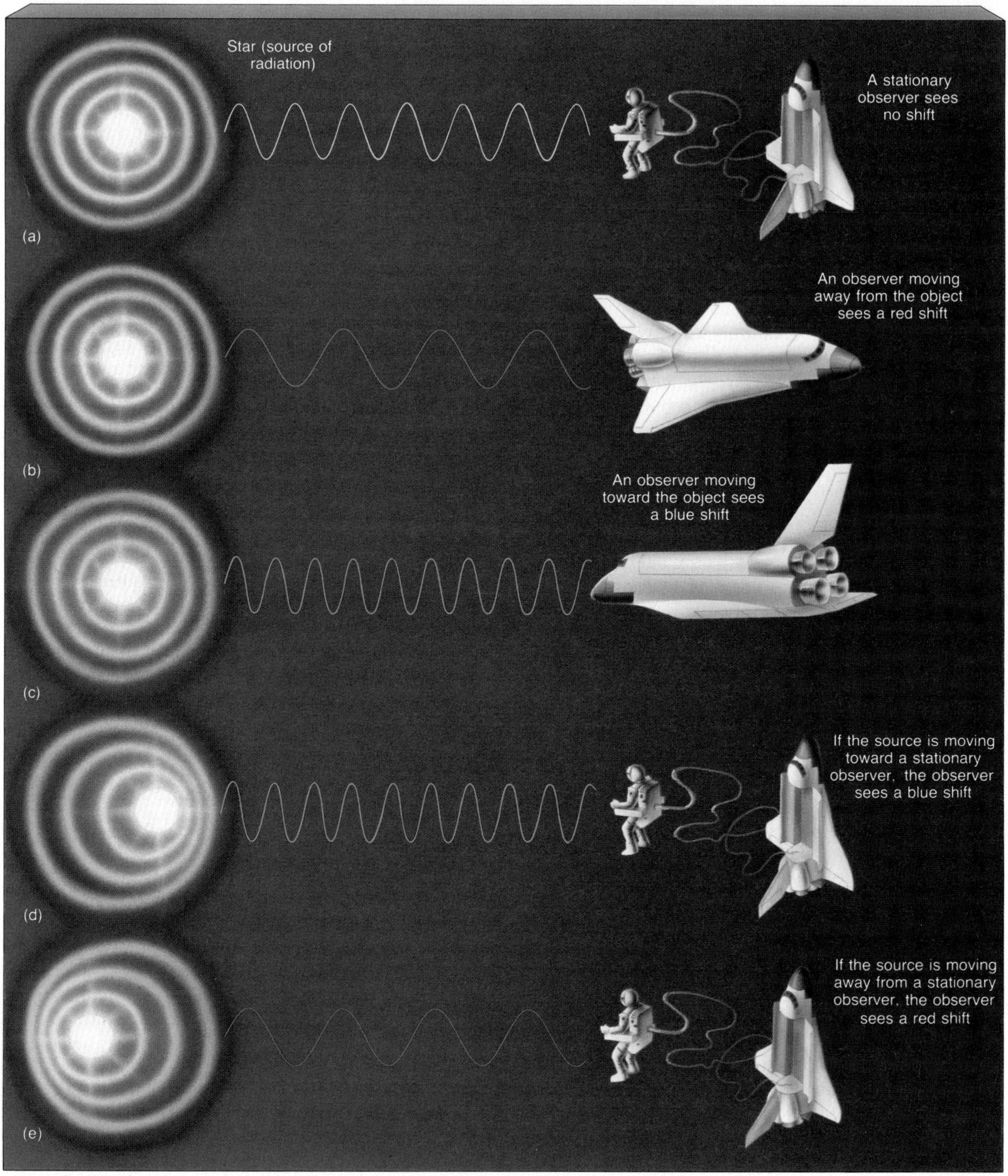
Star (source of radiation)
A stationary observer sees no shift
(a)
An observer moving away from the object sees a red shift
(b)
An observer moving toward the object sees a blue shift
(c)
If the source is moving toward a stationary observer, the observer sees a blue shift
(d)
If the source is moving away from a stationary observer, the observer sees a red shift
(e)

Figure 4.15 *(a) Wave motion from a fixed source toward a fixed observer. (b) Wave motion from a fixed source toward an observer moving away from the source. (c) Wave motion from a fixed source toward an observer moving toward the source. (d) Wave motion from a source moving toward a fixed observer. (e) Wave motion from a source moving away from a fixed observer.*

greater the extent of the red shift:

$$\frac{\textit{apparent increase in wavelength}}{\textit{true wavelength}} \propto \textit{velocity of recession}.$$

We can quantify this statement quite easily. For recession with velocity V, the true (that is, as measured by an observer moving with the source) and the apparent (as measured by our moving observer) wavelengths are related by

$$\frac{\lambda(apparent)}{\lambda(true)} = 1 + \frac{V}{c}$$

where c is the speed of the wave (in this case, the velocity of light). Notice that the effect is extremely small for everyday terrestrial velocities, which are generally much smaller than that of light. Even with its source receding with the Earth's orbital velocity of 30 km/s, a beam of blue light would be redshifted only from 4000 to 4000.4 Å—a very small change indeed, one that the human eye could not discern (but one that is easily detectable with modern instruments).

The red shift would also be noted if the source of the waves were moving away from a fixed observer, as in Figure 4.15(e). In fact, a red shift occurs whenever the source and the observer have *any* net motion of *recession* between them. If V denotes this net recessional velocity, the above equation still applies.

The opposite is true for approaching velocities. If the source of the waves is moving toward us (as shown in Figure 4.15[d]) or if we are moving toward a fixed source, (as shown in Figure 4.15[c])—that is, if there is *any net velocity of approach*—we would count more wave crests per unit time. The observed wave is then said to have a **blue shift** (because its apparent wavelength is reduced). If we adopt the convention that $V < 0$ corresponds to approaching motion (that is we would use a negative number for approach), the above equation gives the correct Doppler shift in all cases.

In summary, we conclude that the extent of Doppler shift is directly proportional to the net radial velocity. We emphasize the word *net*: If the net velocity were zero (either for a source and an observer both at rest or for both source and an observer moving in the same direction with the same velocity), no Doppler shift would be observed. Notice that, while the Doppler effect is a wave phenomenon, the wave-particle nature of light means that light particles—photons—are also affected by relative motion. Photons from an approaching source are blueshifted to higher frequency, and hence higher energy. Similarly, photons from a receding object are detected at lower energy than would be measured by an observer moving with the source. Finally, realize that the Doppler effect does not depend on *distance* in any way. It depends strictly on the net radial velocity.

Of what value is the Doppler effect? In principle, we could use the extent of blue shift or red shift to measure the motion of some cosmic object along the line of sight. In practice, however, it is hard to measure Doppler shifts of the entire Planck curve, simply because it is widely spread over many frequencies, and small shifts are extremely hard to measure with any accuracy. However, if the radiation were more narrowly defined and took up just a ''sliver'' of the spectrum, precise measurements of Doppler effect *could* be made, as discussed in Chapter 5.

CHAPTER SUMMARY

Every planet, star, and galaxy is a source of information about the material aspects of our universe—its state of motion, its temperature, its composition, and its history.

Astronomers use the laws of physics, as we know them on Earth, to interpret the light emitted by objects in space. Because light does not travel instantaneously from place to place, we can know nothing about these objects as they exist at the present moment. We can know them only as they were when the light we now see left them—minutes, centuries, or billions of years ago.

It is possible to picture light traveling through empty space in the form of a wave. Repeating itself cyclically in both time and space, a wave can be characterized by its velocity and by the length of its cycle. Until the early nineteenth century, a debate raged in scientific circles regarding the question of whether light consisted of waves or particles. The discovery of two key properties of light, diffraction and interference, argued strongly in favor of the wave theory. Diffraction is the ability of waves to bend around corners. Interference is the ability of two or more waves to reinforce or cancel each other.

Waves of radiation differ fundamentally from other sorts of waves because they move through empty space. Waves of light can be imagined as a series of vibrations in an electric field, moving at about 300,000 km/s. According to the known laws of physics, this is the fastest speed possible. A magnetic field accompanies every changing electric field. Thus radiative waves from any source consist of both electric and magnetic field vibrations, linked to one another and moving at the speed of light. It is this electromagnetic radiation that transfers energy and information from one place to another.

White light is a mixture of red, orange, yellow, green, blue, and violet. We perceive individual colors because of their different wavelengths. Visible light is just one form of electromagnetic radiation. Our eyes cannot detect radio, infrared, and ultraviolet waves and x-rays and gamma rays. Most objects in the universe emit large amounts of invisible radiation. Many emit only a tiny fraction of their total energy in the visible range.

Earth's atmosphere blocks a large fraction of the radiation produced by astronomical objects. We can observe the universe from Earth in much of the radio, in the visible, and in parts of the infrared. Ultraviolet, x-ray, and gamma-ray observations must be made from above the atmosphere.

Radiation behaves not only as waves, but also sometimes as particles. We call these particles photons. They can be thought of as little packets of electromagnetic energy. As explained by Albert Einstein, the photoelectric effect shows that the energy contained within a photon is proportional to the frequency of the radiation.

The hotter the object, the faster its constituent particles move and the more energy they radiate. Intensity is a term often used to specify the amount or strength of radiation. The energy of natural objects is spread out over some portion of the electromagnetic spectrum. The expression "black-body" refers to an idealized concept of an object that absorbs all radiation falling upon it. In a steady state, the object must reemit exactly the same amount of energy as it absorbs. A black-body curve, or Planck curve, shows the standard distribution of radiation emitted by any heated object. For all objects, the frequency at which the intensity peaks depends on the temperature of the emitting object. Wien's Law tells us that the hotter the object, the bluer its radiation. Stefan's Law describes the rate of increase in the energy flux with increasing temperature. Astronomers use Planck curves to gauge the temperatures of far away objects.

Motion toward or away from a source of radiation changes our measurement of the wavelength (or frequency) of radiation. This motion-induced change in the observed wavelength (or frequency) of a wave is known as the Doppler effect. If the direction of motion is toward the observer, more wave crests will be counted per unit time and thus the radiation will appear bluer. This radiation is blueshifted. If the direction of motion is away from the observer, fewer wave crests will be counted per unit time and thus the radiation will appear redder. It is redshifted.

KEY WORDS

absolute zero
amplitude
black-body curve
blue shift
constructive interference
destructive interference
diffraction
Doppler effect
electric field
electromagnetic radiation
electromagnetic spectrum
electromagnetism
erg
frequency
gamma rays
infrared
intensity
interference
invisible radiation
Kelvin scale
magnetic field
opacity
ozone layer
photoelectric effect
photon
Planck's constant
Planck curve
radial motion
radiation
radio
red shift
Stefan-Boltzmann constant
Stefan's law
temperature
transverse motion
trough
ultraviolet
velocity of light
visible spectrum
wave
wave crest
wavelength
wave period
wave trough
Wien's Law
x-rays

REVIEW QUESTIONS

1. Astronomers generally cannot perform laboratory experiments on the objects they wish to study. How do they find out about the universe?
2. Define the following: wave period, wavelength, amplitude, frequency.
3. What is the relationship between wavelength and wave frequency?
4. What is diffraction, and how does it relate to the behavior of light as a wave?
5. Describe Thomas Young's double-slit experiment. What did it prove?
6. What's so special about *c*?
7. Describe the way in which light radiation leaves a star, travels through the vacuum of space, and finally is seen by someone on Earth.
8. What do magnetic fields have to do with the phenomenon of radiation traveling in outer space?
9. Name the colors that combine to make white light. What is it about the various colors that cause us to perceive them differently?
10. What do radio waves, infrared radiation, visible light, ultraviolet radiation, x-rays, and gamma rays have in common? How do they differ?
11. Why don't cosmic x-rays reach Earth's surface?
12. In what regions of the electromagnetic spectrum is the atmosphere transparent enough to make observations from the ground?

13. What is the photoelectric effect, and why is it significant?
14. Why are gamma rays generally harmful to life forms but radio waves generally harmless?
15. What is a black body? Describe a typical black-body, or Planck, curve.
16. Why do astronomers care about black body, or Planck, curves?
17. What does Wien's Law reveal about stars in the sky?
18. What is the Doppler effect, and how does it alter the way in which we perceive radiation?
19. What is the wavelength of a 100 megahertz ("FM 100") radio signal?
20. What is the frequency of a 6000 Å red photon?
21. How many times more energy has a 1 Å gamma ray than a 10 MHz radio photon?
22. According to Wien's law, how many times hotter is an object whose Planck spectrum peaks in the ultraviolet, at a wavelength of 2000 Å, than an object whose spectrum peaks in the red, at 6500 Å? According to Stefan's law, how much more energy does it radiate per area per second?
23. At what speed, and in what direction, would a spacecraft have to be moving for a radio station transmitting at 100 MHz to be picked up by a radio tuned to 99.9 MHz?

DISCUSSION QUESTIONS

1. If Earth were completely blanketed with clouds and we couldn't see the sky, would light still play a role in what we could surmise about the realm beyond the clouds? In your answer, consider the daily rotation of the Earth and the light of the Sun and Moon.
2. What does the finite speed of electromagnetic radiation mean with respect to future communication among societies that may be scattered throughout the Galaxy?
3. Describe what happens when a red-hot glowing coal cools off, in terms of its Planck curve.
4. What if, like Superman, we could see in other regions of the electromagnetic spectrum? What if we had x-ray vision, or ultraviolet vision, or radio vision? Would this improved vision have allowed us to advance our knowledge of the universe more quickly?

PROJECTS

1. Locate the constellation Orion. Its two brightest stars are Betelgeuse and Rigel. Which of these is the hotter star? Which is the cooler star? Which of the other stars scattered in the night sky are hot, and which are cool? How can you tell?
2. Stand near (but not too near!) a train track and wait for a train to pass by. Can you notice the Doppler effect in the sound of its whistle? How does the sound frequency depend on (a) the train's speed, (b) the train's motion (toward or away from you)?

SUGGESTED READINGS

Achinstein, P. *Particles and Waves: Historical Essays in the Philosophy of Science*. New York, Oxford University Press, 1991. Contains historical background and more on the wave-particle debate about the nature of light.

Cowen, R. "Heavenly Bodies Make Their UV Film Debut." *Science News* (January 26, 1991). Results from NASA's Ultraviolet Imaging Telescope.

Davies, J. K. "The Extreme Ultraviolet: A Promising New Window on the Universe." *Astronomy* (July 1987). The latest on using the extreme ultraviolet, or EUV, to understand the universe.

Field, G. B. and Chaisson, E. J. *The Invisible Universe*. New York: Vintage Books, 1987. The universe as revealed by modern studies across the electromagnetic spectrum.

Griffin, R. "The Radial-Velocity Revolution." *Sky and Telescope* (September 1989). How increasingly sensitive radial velocity measurements are used in studies of possible binary systems and in the search for distant solar systems.

Minnaert, M. *The Nature of Light and Color in the Open Air*. New York: Dover, 1954. A classic for half a century, this book explains the simple physics behind hundreds of everyday light phenomena.

Morris, R. *Light*. Indianapolis, New York: The Bobbs, Merrill Company, 1979. Off-beat but interesting history of man's thinking about light. Chapters include "Lasers" and "Light in Painting."

Schorn, R. A. "Listening to the Universe." *Sky and Telescope* (November 1988). An update on amateur radio astronomy, in a special issue devoted to amateur astronomy.

Vershuur, G. *The Invisible Universe Revealed*. New York: Springer Verlag, 1987. The story of modern radio astronomy.

CHAPTER

5

Spectroscopy
The Inner Workings of Atoms

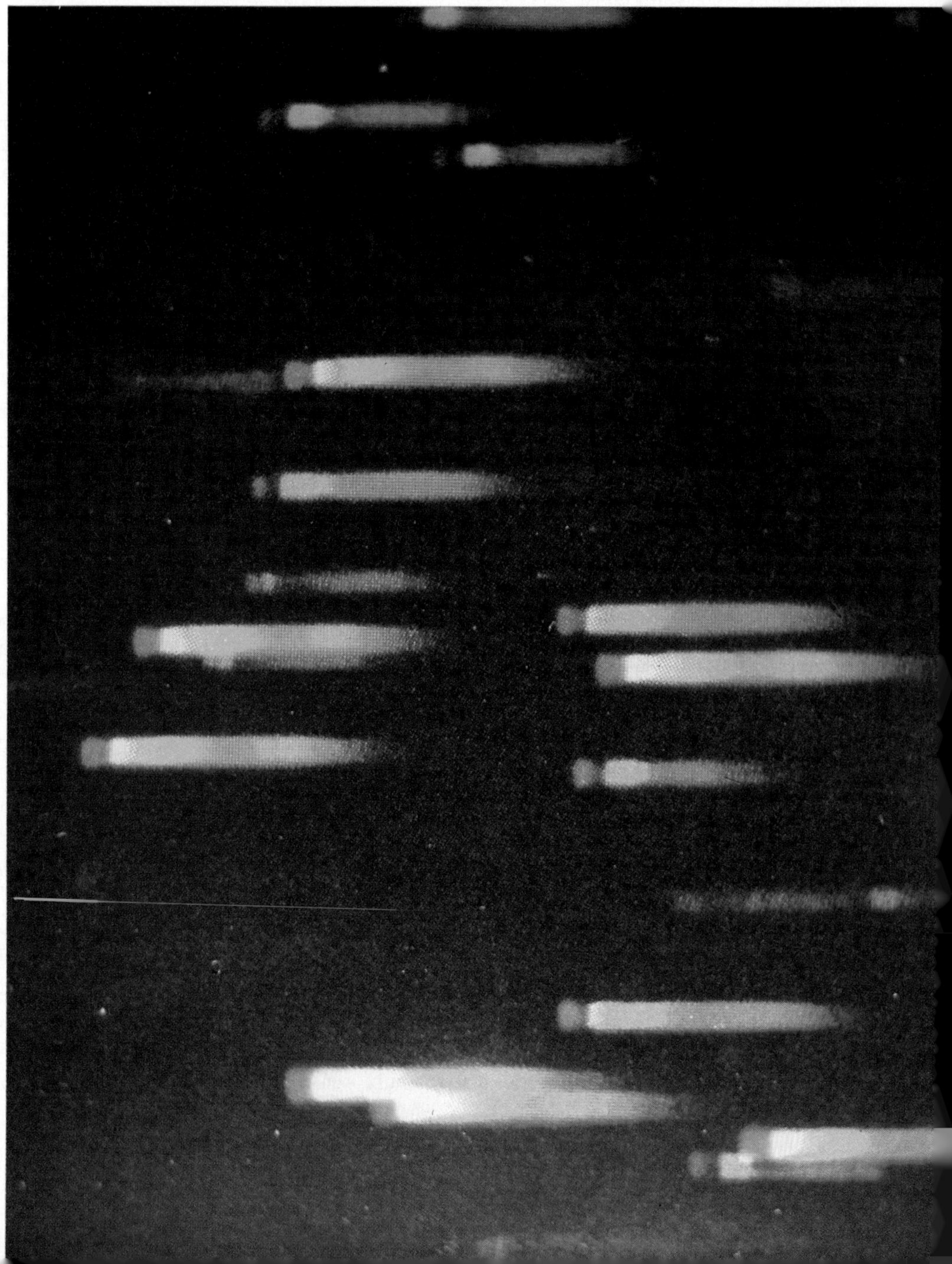

These rectangles of rainbows across the sky communicate a wealth of information to the people who study them. Each rectangle in this image is a spectrum, a ''snapshot'' of the radiation that an object emits in visible light. How valuable is the information that this radiation carries? Consider this fundamental astronomical question: What are stars made of? We cannot go out and sample stars, even with robot probes, but an investigation of a star's spectrum reveals much about its composition. (Courtesy Harvard College Observatory)

R I **V** U X G

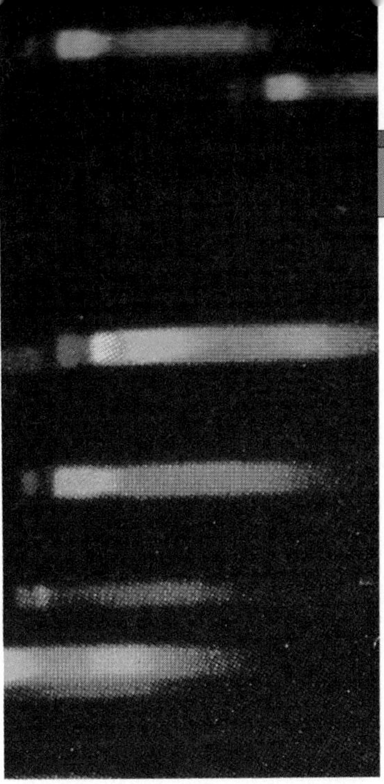

LEARNING GOALS

1 to realize that every element produces a distinctive and recognizable pattern of spectral lines
2 to understand the basic concepts of modern atomic theory
3 to understand how electron changes within atoms produce emission and absorption features in the spectra of those atoms
4 to examine the kinds of information obtained by analyzing spectral features
5 to recognize that molecules also have characteristic spectral features

If matter on microscopic scales can emit and absorb radiation in a discontinuous way—as we saw in our discussion of photons and the photoelectric effect in the last chapter—the inescapable conclusion is that matter itself is not continuous, but instead is made up of individual building blocks. Those building blocks are called atoms. The observational technique that enables scientists to analyze the nature of atoms by the way they emit and absorb radiation is called spectroscopy.

Coupling spectroscopy as a technique with atomic physics as a theory, we now know how each type of atom emits and absorbs radiation in its own distinct way. We also understand how an atom's surroundings affect its emission and absorption of light. Spectroscopy is an indispensable tool in modern astronomy. By analyzing the fine details of the radiation emitted by a cosmic object, we can deduce its physical and chemical properties without having to pay a personal visit to make direct measurements. Ultimately, this knowledge yields a wealth of information about the birth, evolution, and death of myriad objects throughout the universe.

SPECTRAL LINES

The analysis of the way in which **atoms** absorb and emit light is called **spectroscopy.** Simple spectroscopic studies can be performed with an instrument known as a **spectroscope.** In its most basic form, this might consist of an opaque mask with a slit in it (to define a beam of light), a prism (to split the beam into its component colors), and an eyepiece, or even just a screen (to allow the user to view the resulting spectrum). Figure 5.1 shows such an arrangement.

Although we could use this instrument to illustrate most of the basic points about atomic spectra discussed in this chapter, it is quite inadequate for astronomical observations. Professional astronomers require much more detail than can be obtained by eye with such a simple setup. Instead of looking directly at the spectrum, they use a more complex device—a **spectrograph** (also known as a spectrometer)—to record the spectrum on a photographic plate (or, more commonly nowadays, in electronic form on a computer) for later analysis. The resulting photograph (or digitized image) is known as a **spectrogram.** Essentially, the modern research spectrograph consists of a telescope (to capture the radiation), a dispersing device (to spread it out into a spectrum), and a detector (to record the result). Despite its greater complexity, however, the basic operation of the spectrograph is conceptually similar to the simple spectroscope shown on the next page.

Emission Lines

✓1 The spectra we encountered in Chapter 4 are examples of **continuous spectra.** A white-hot metal bar emits radiation of all frequencies (mostly in the visible range), and the distribution of intensity with respect to frequency is described by the Planck curve corresponding to the bar's temperature. Viewed through a spectroscope, the spectrum of the light from the bar would show the familiar rainbow of colors, from red to blue, without interruption, as shown in Figure 5.2(a). The light from an ordinary incandescent light bulb is another example of a continuous spectrum.

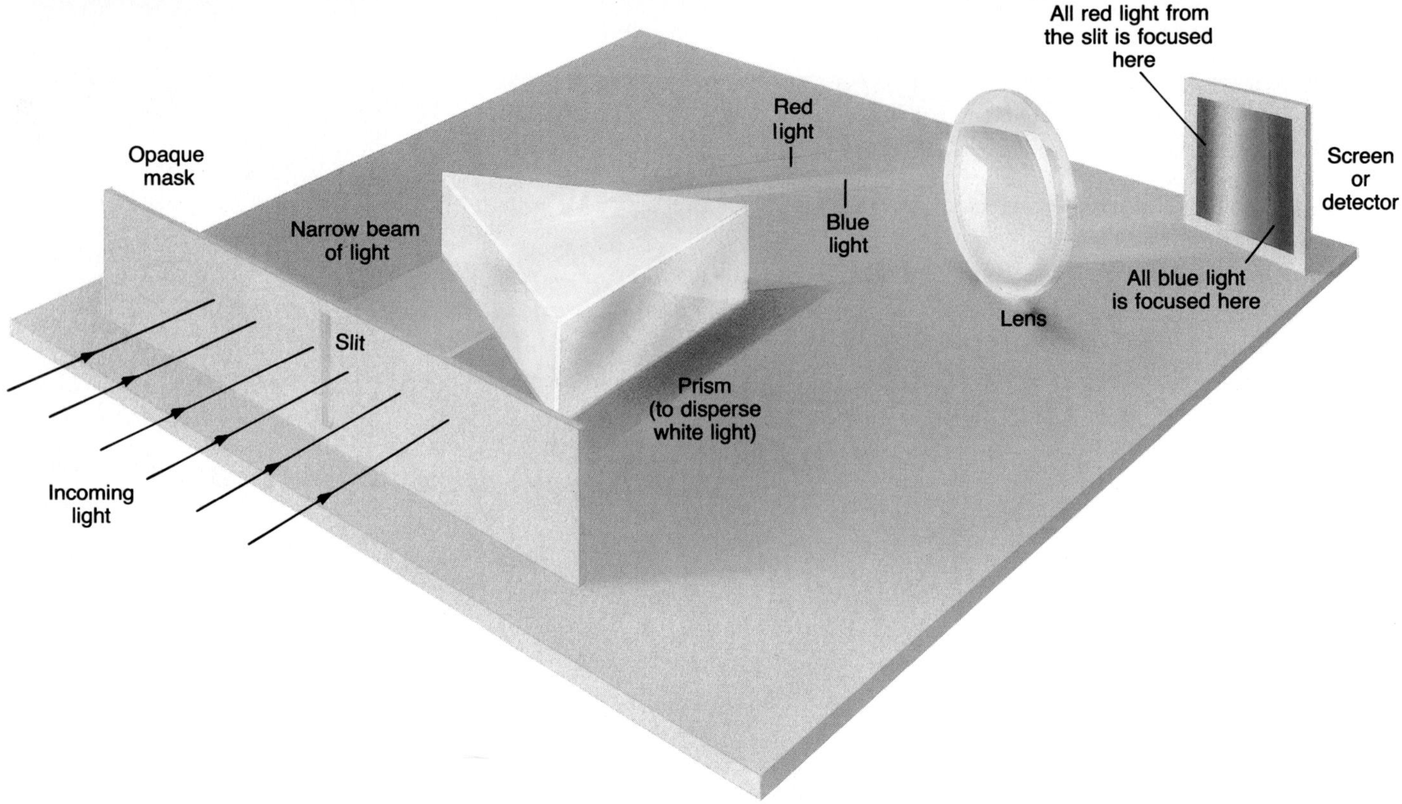

Figure 5.1 *Diagram of a simple spectroscope. A small slit in the mask on the left allows a narrow beam of light to pass. The light passes through a prism and is split up into its component colors. The resulting spectrum can be viewed through an eyepiece or simply cast onto a screen.*

Not all spectra are of this type. For instance, if we took a glass jar containing pure hydrogen gas and passed an electrical discharge through it (a little like a lightning bolt arcing through the Earth's atmosphere), the gas would begin to glow—that is, it would emit radiation. If we were to examine that radiation with our spectroscope, we would find that its spectrum consisted of only a few bright lines on an otherwise dark background, quite unlike the continuous spectrum described for the incandescent light bulb. Figure 5.2(b) shows this schematically. An actual photograph of the spectrum of hydrogen appears in Figure 5.3. The light produced by the hydrogen in this experiment does *not* consist of all possible colors, but instead includes only a few well-defined **emission lines.**

After further experimentation, we would also find that although we could alter the *intensity* of the lines (for example, by changing the amount of hydrogen in the jar or the strength of the electrical discharge), we could not alter their *color* (in other words, their frequency or wavelength). The pattern of spectral emission lines is a property of the element hydrogen. Whenever we perform this experiment, the same characteristic colors always result.

By the early nineteenth century, scientists had carried out on many different gases experiments similar to the one just described. By vaporizing solids and liquids in a flame, they extended their inquiries to include materials that are not normally found in the gaseous state. Sometimes the pattern of lines was fairly simple, sometimes it was very complex. Always it was *unique*. Thus, even though the origin of the lines was a mystery, scientists quickly realized that the lines provided a unique "fingerprint" of the substance under investigation. Scientists could deduce the presence of a particular type of atom solely through the study of the light it emitted.

Scientists have by now accumulated huge catalogs of the specific frequencies at which all known atoms emit radiation. For any given atom, the particular pattern of the light it emits is known as its **emission spectrum.** Examples of the emission spectra of a number of common elements are shown in Figure 5.3.

Absorption Lines

When sunlight is split up by a spectroscope, at first glance it appears to produce a continuous spectrum. This is what Isaac Newton must have seen over three centuries ago. However, closer scrutiny shows that the solar spectrum is actually interrupted by a large number of narrow dark lines, as shown in Figure 5.4. We now know that these lines are formed by the removal of light from the Sun's otherwise continuous spectrum by intervening gases. These gases are present in the outer layers of the Sun or in the Earth's atmosphere, and they "absorb" certain wavelengths, which then do not appear in the spectrum. The lines are called **absorption lines.**

The English astronomer William Wollaston first no-

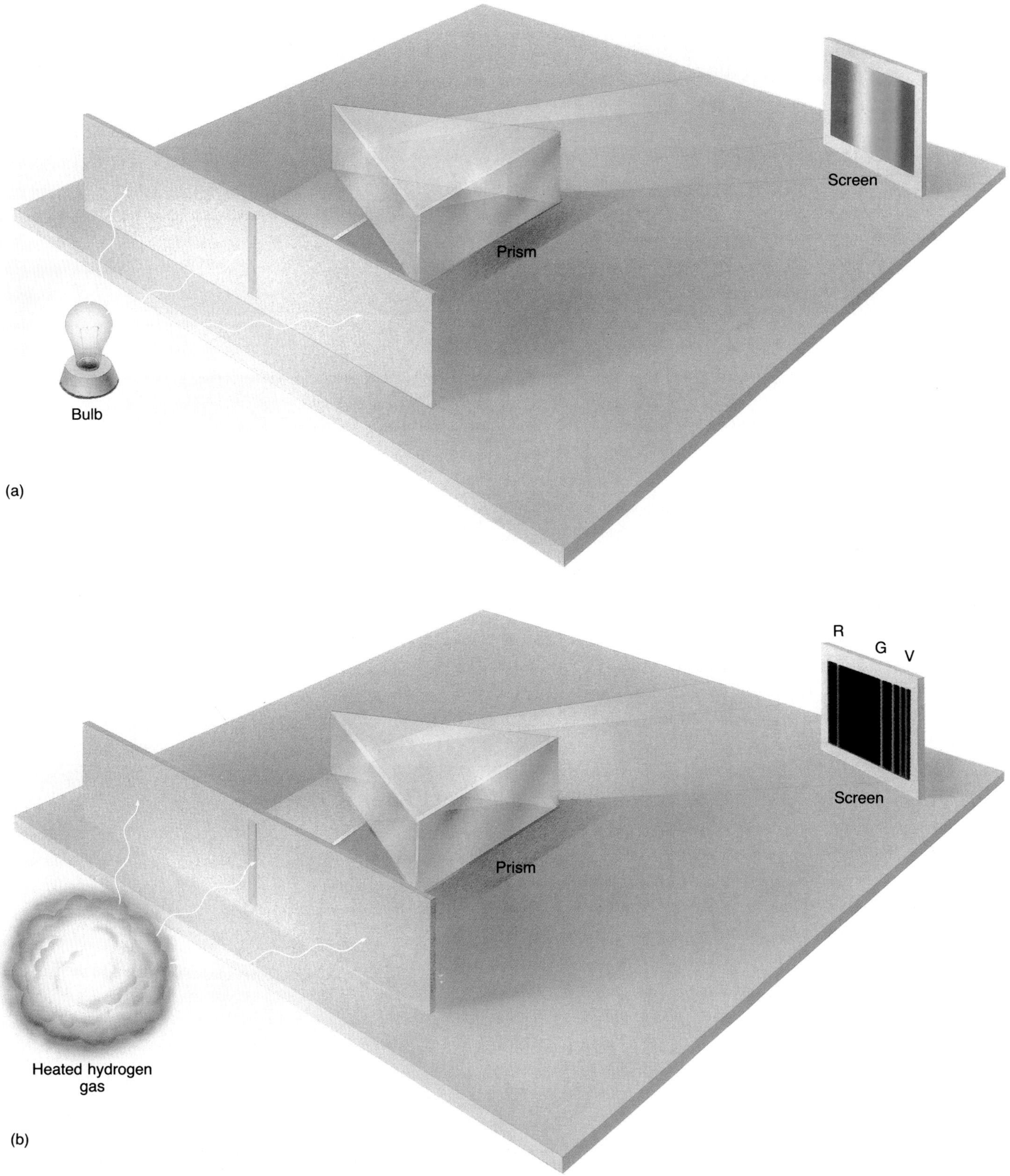

Figure 5.2 *When passed through a slit and split up by a prism, light from a source of continuous radiation (a) gives rise to the familiar rainbow of colors. By contrast, the light from hydrogen gas (b) consists of a series of distinct spectral lines.*

Figure 5.3 *The emission spectra of some well-known elements. From top to bottom, the photographs show spectra of hydrogen, sodium, helium, neon, mercury. (Bausch & Lomb)*

Figure 5.4 *This visible spectrum of the Sun shows hundreds of dark absorption lines superposed on a bright continuous spectrum. Here, the scale extends from long wavelengths (red) at the upper left to short wavelengths (blue) at the lower right. (AURA)*

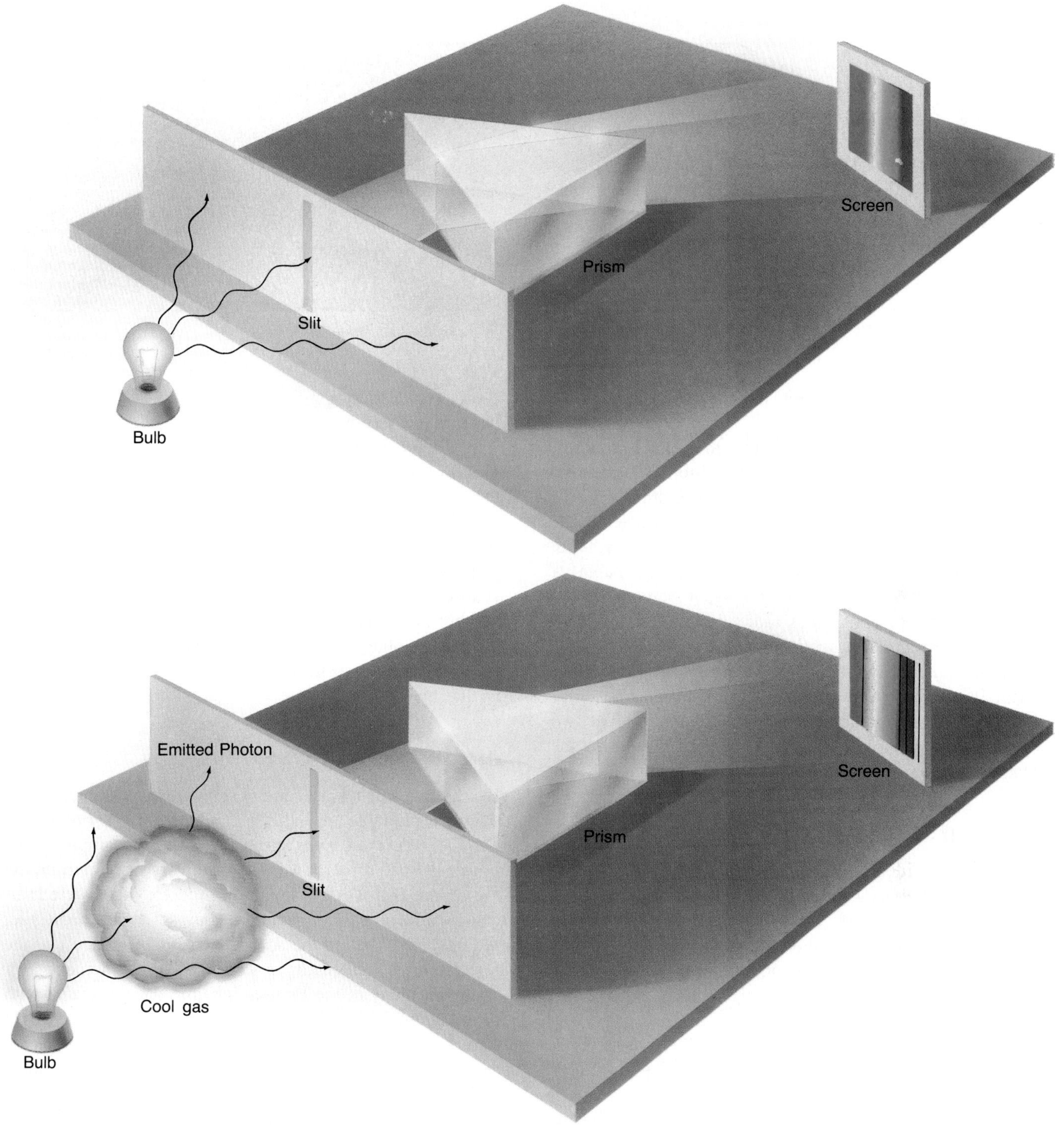

Figure 5.5 *(a) A hot light bulb produces a continuous emission spectrum. (b) When some cool gas is placed between the bulb and the detector, the resulting emission spectrum is crossed by a series of dark absorption lines.*

ticed the solar absorption lines in 1802. They were studied in greater detail about 10 years later by the German physicist Joseph Fraunhofer, who measured and cataloged over 600 of them. They are now referred to collectively as **Fraunhofer lines.** Although the Sun is by far the easiest star to study, and so has the most extensive set of observed absorption lines, similar lines exist in the spectra of all stars.

After the discovery of the Fraunhofer lines, scientists found that absorption lines could also be produced in the laboratory by passing a beam of light from a continuous source through a cool gas (as shown in Figure 5.5), and a vital connection between emission and absorption lines was uncovered. The absorption lines associated with a given gas occur at precisely the *same* frequencies as the emission lines produced when the gas is heated. Thus, if the emission

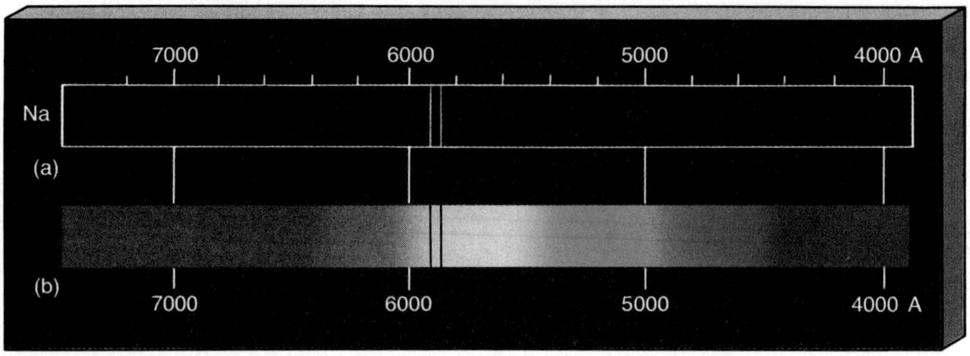

Figure 5.6 (a) *The characteristic emission lines of sodium. The two bright lines in the center appear in the yellow part of the spectrum. (b) The absorption spectrum of sodium. The two dark lines appear at exactly the same wavelengths as the bright lines in the sodium emission spectrum.*

lines form a unique fingerprint, the absorption lines do so also.

To illustrate this point, consider the element sodium, whose emission spectrum appears in Figure 5.3. When sodium vapor is heated, or energized in some other way, its atoms emit photons at just two particular frequencies in the visible range: The two characteristic yellow lines, with wavelengths of 5890 Å and 5896 Å, are clearly visible. If instead, some relatively *cool* sodium vapor were placed between a source of continuous radiation and our detector, then two sharp, dark *absorption* lines would be noted in the spectrum at precisely the same wavelengths as the two lines normally emitted by hot sodium. The emission and absorption spectra of sodium are compared in Figure 5.6. The two spectra are shown to the same scale, clearly illustrating the relation between emission and absorption features.

Kirchhoff's Laws

The observed relationships between the three types of spectra—continuous, emission-line, and absorption-line—were summarized by the German physicist Gustav Kirchhoff in 1859. He listed three spectroscopic rules, now known as **Kirchhoff's Laws,** governing the formation of spectra.

1. A luminous solid or liquid (or, in fact, a sufficiently dense gas) emits light of all wavelengths and so produces a *continuous spectrum* of radiation.
2. A low-density, hot gas emits light whose spectrum consists of a series of bright *emission lines*. These lines are characteristic of the chemical composition of the gas.
3. A cool, rarefied gas absorbs certain wavelengths from a continuous spectrum, leaving dark *absorption lines* in their place, superposed on the continuous spectrum. Once again, these lines are characteristic of the composition of the intervening gas, and occur at precisely the same wavelengths as the emission lines produced by the gas at higher temperatures.

We see that hot gas situated in front of a cool background produces bright emission lines, and cool intervening gas causes dark absorption lines to appear in the spectrum of a background source.

Figure 5.7 combines Kirchhoff's three laws into a single illustration. When viewed directly, the light source on the left has a continuous Planck spectrum. When viewed through a cloud of cool hydrogen gas, a series of dark absorption lines appear at wavelengths characteristic of hydrogen. The lines appear because the light at those wavelengths is absorbed by the hydrogen and subsequently reradiated in another direction. When the cloud is viewed from the side against an otherwise dark background, a series of bright emission lines is seen. These lines contain the energy lost by the forward beam.

Astronomical Applications

By the late nineteenth century, spectroscopists had developed a formidable arsenal of techniques for interpreting the radiation received from space. A striking example of the power of spectroscopy was the discovery of helium on the Sun. Once astronomers knew that spectral lines were indicators of chemical composition, they set about identifying the observed lines in the solar spectrum. Almost all of the lines were accounted for in terms of known elements (for example, many of the Fraunhofer lines are associated with the element iron), giving astronomers immediate insight into the Sun's composition. However, some lines were not familiar from terrestrial experiments. In 1868, astronomers realized that those lines must correspond to a previously unknown element. It was given the name helium, after the Greek word *helios*, meaning "Sun." Helium was discovered on Earth only in 1895, almost three decades after its detection on the Sun. (A laboratory spectrum of helium is shown in Figure 5.3.)

For all the information that nineteenth-century astronomers could glean from observations of stellar spectra, they still lacked a theory explaining how the spectra themselves arose. Despite their sophisticated spectroscopic equipment, they knew hardly more about the physics of stars than did Galileo or Newton. To understand how spectroscopy can be used to extract detailed information about astronomical objects from the light they emit, we must delve a little

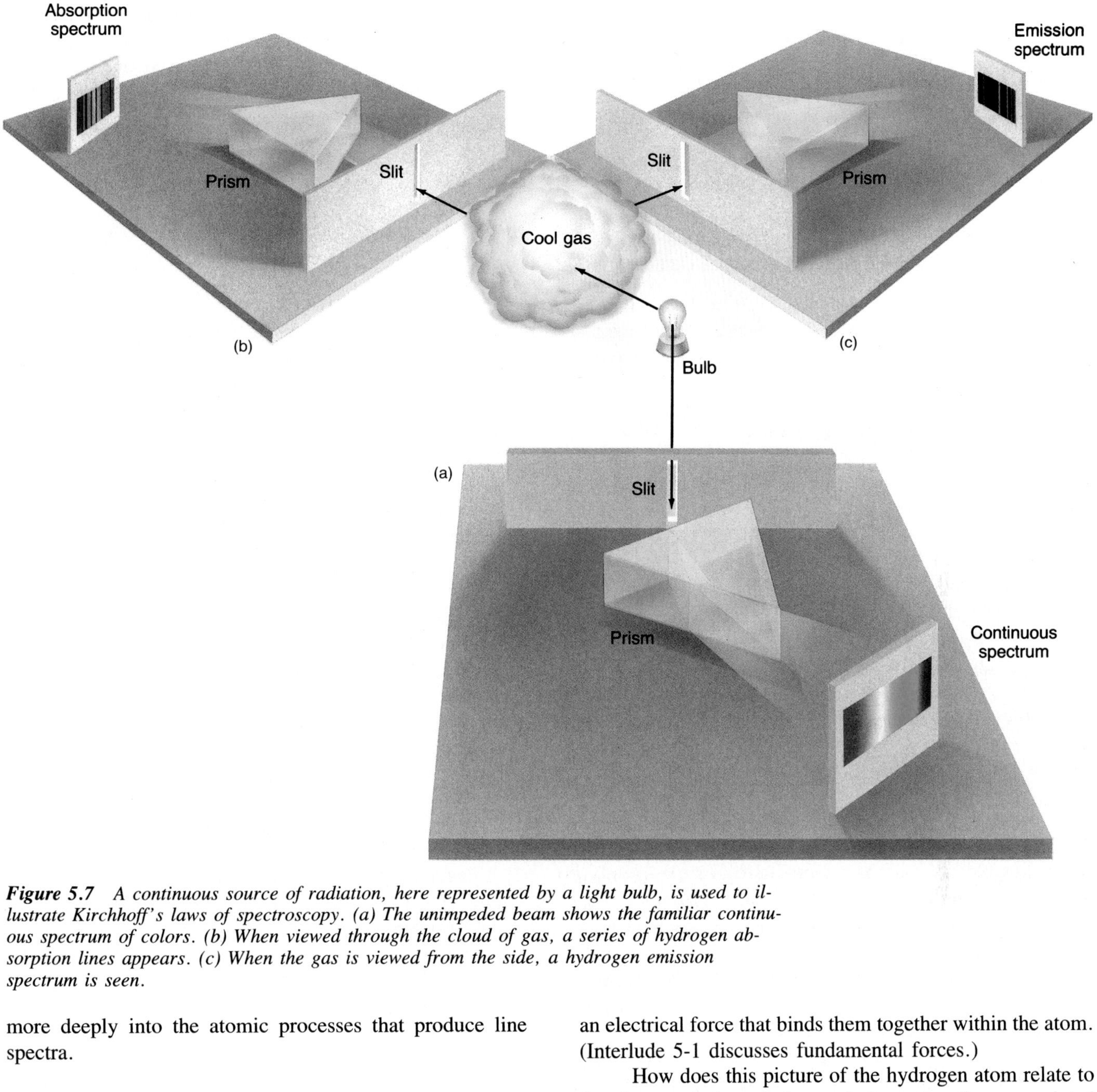

Figure 5.7 *A continuous source of radiation, here represented by a light bulb, is used to illustrate Kirchhoff's laws of spectroscopy. (a) The unimpeded beam shows the familiar continuous spectrum of colors. (b) When viewed through the cloud of gas, a series of hydrogen absorption lines appears. (c) When the gas is viewed from the side, a hydrogen emission spectrum is seen.*

more deeply into the atomic processes that produce line spectra.

ATOMIC MODELS

The Bohr Atom

✓2 To begin to appreciate the structure and behavior of atoms, let us start with the simplest atom of all—hydrogen. A hydrogen atom consists of an **electron,** with a negative electrical charge, orbiting a **proton,** which carries a positive charge and forms the central **nucleus** (plural: nuclei) of the atom. The charges of the electron and proton are equal in magnitude but opposite in kind, making the hydrogen atom as a whole electrically neutral. The equal and opposite charges of the proton nucleus and the orbiting electron exert an electrical force that binds them together within the atom. (Interlude 5-1 discusses fundamental forces.)

How does this picture of the hydrogen atom relate to the characteristic emission and absorption lines associated with hydrogen gas? The law of conservation of energy demands that if an atom emits some energy in the form of radiation, that energy has to come from somewhere within the atom. Similarly, if energy is absorbed, it must cause some internal change. It is reasonable (and correct) to suppose that the energy emitted or absorbed by the atom is associated with changes in the motion of the orbiting electron.

The first theory of the hydrogen atom to provide an explanation of hydrogen's observed spectral lines was propounded by the Danish physicist Niels Bohr. This theory is now known simply as the **Bohr model** of the atom. The essential features of the Bohr model are as follows. First, there is a state of lowest energy—the **ground state**—which represents the ''normal'' condition of the electron as it orbits the nucleus. Second, there is a maximum energy that the electron can have and still be part of the atom. Beyond that energy, the electron is no longer bound to the nucleus, and the atom is said to be **ionized.** Third, between those two energy levels, the electron can exist only in certain sharply defined energy states, often referred to as **orbitals.**

Originally, each electron orbital was pictured as having a specific radius, much like a planetary orbit in the solar

INTERLUDE 5-1 *Fundamental Forces*

As best we can tell, the behavior of all matter in the universe—from elementary particles to clusters of galaxies—is ruled by just four (or fewer) basic forces, which are *fundamental* to *everything* in the universe. In a sense, the search to understand the nature of the universe is the quest to understand the nature of these forces.

The *gravitational force* is perhaps the best known. Gravity binds galaxies, stars, and planets together and holds humans on the surface of the Earth. As we have seen, its magnitude decreases with distance, according to the inverse-square law. Its strength is also proportional to the masses of each of the two objects involved. Thus, the gravitational field of an atom is extremely weak, but that of a galaxy, consisting of huge numbers of atoms, is very powerful. An important aspect of gravity is that nothing can cancel its attractive pull. There is no such thing as antigravity that repels objects, and no material can shield us from gravity's effects: The result is a gravitational force that is always attractive. Although gravity is by far the weakest of the forces of nature, its effect accumulates as we move to larger and larger volumes of space. Gravity is the dominant force in the universe on all scales larger than, say, a planet.

The *electromagnetic force* is another of nature's basic agents. Its strength also decreases with distance and, like gravity, it obeys an inverse-square law. Any particle having a net electric charge, such as an electron or a proton in an atom, exerts an electromagnetic force. The everyday things we see around us are held together by this force. Electromagnetism is much stronger than gravity. For example, the electromagnetic force between two protons exceeds their gravitational attraction by a factor of about 10^{36}. However, unlike gravity, electromagnetic forces can repel (between like charges) as well as attract (between opposite charges). Such forces can then sometimes cancel one another. Similar numbers of positive and negative charges tend to neutralize each other, greatly diminishing their net electromagnetic influence. Above the microscopic level, most objects are in fact very close to being electrically neutral. Thus, except in unusual circumstances, the electromagnetic force is relatively unimportant on macroscopic scales.

A third fundamental force of nature is simply termed the *weak force*. It is much weaker than electromagnetism, and its influence is somewhat more subtle. The weak force governs the emission of radiation from some radioactive atoms. It is now known that the weak force is not really a separate force at all but just a form of the electromagnetic force, acting under peculiar circumstances. As such, physicists often speak of the ''electroweak force.'' However, when acting in its ''weak'' mode, the electroweak force does not obey the inverse-square law. Its effective range is less than the size of an atomic nucleus, about 10^{-13} cm.

The strongest force of all is the *strong* (or *nuclear*) *force*. It binds atomic nuclei together and governs the generation of energy in the Sun and all other stars. Like the weak force, and unlike the forces of gravity and electromagnetism, which obey the inverse-square law, the strong force operates only at very close range. It is unimportant outside a distance of a millionth of a millionth (10^{-12}) of a centimeter. However, within this range (for example, in atomic nuclei), particles are bound with enormous strength. In fact, it is the range of the strong force that determines the typical sizes of atomic nuclei. Only when two protons are brought within about 10^{-13} cm of one another can the (attractive) strong force overcome their electromagnetic repulsion.

Notice that not all particles are subject to all types of force. All particles interact through gravity because all carry mass. However, only *charged* particles interact electromagnetically. Protons and neutrons are affected by the strong force—electrons are not. Under the right circumstances, the weak force can affect any type of subatomic particle, regardless of its charge.

Scientists now believe that protons, neutrons, and many other elementary particles are actually made of subparticles called *quarks*. (The name derives from a meaningless word coined by novelist James Joyce in his book *Finnegans Wake*.) It is thought that the strong force is itself just a manifestation of an even more basic *color* force that binds quarks to one another. Note that the term *color* here has *nothing* whatever to do with electromagnetic radiation—it is just a term introduced by theorists to categorize the many possible interactions between quarks. The color force binds quarks on a scale less than 10^{-13} cm. Many physicists now maintain that ultimately all the basic forces of nature will be shown to be different aspects of a single ''superforce,'' just as the weak force is linked to the electromagnetic force by the electroweak interaction.

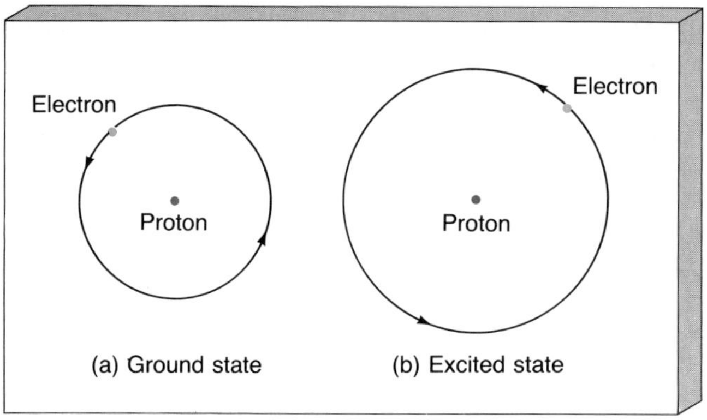

Figure 5.8 *An early conception of the hydrogen atom pictured its electron orbiting the central proton in a well-defined orbit, rather like a planet orbiting the Sun. Two electron orbits of different energies are shown. The left-hand figure represents the ground state, the right-hand figure an excited state.*

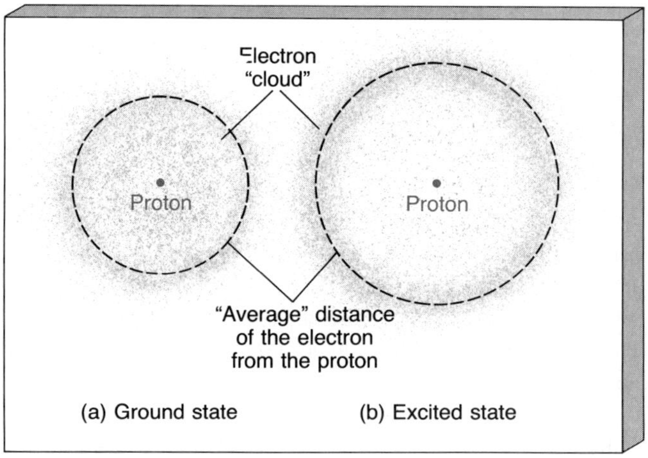

Figure 5.9 *The modern view of the hydrogen atom sees the electron as a "cloud" surrounding the nucleus. The same two energy states are shown as in Figure 5.8.*

system, as shown in Figure 5.8. However, the modern view is not so simple. Although each orbital *does* have a precise energy, the electron is now envisioned as being smeared out in an "electron cloud" surrounding the nucleus, as illustrated in Figure 5.9. It is common to speak of the *mean* (average) distance from the cloud to the nucleus as the "radius" of the electron's orbit. When a hydrogen atom is in its ground state, the radius of the orbit is about 0.5 Å. As the orbital energy increases, the radius increases, too.

Excitation

Atoms do not always remain in their ground states. They sometimes become "excited." An **excited state** is a condition that occurs when an electron temporarily resides at a greater-than-normal distance from its parent nucleus. An atom in such an excited state has a greater-than-normal amount of energy. How can atoms become excited? Generally, in one of two ways: They can become *radiatively excited* by absorbing a photon of light energy from some source of electromagnetic radiation, or they can become *collisionally excited* by colliding with a nearby atom or elementary particle.

Because electrons may exist only in orbitals with specific energies, atoms can absorb only specific amounts of energy as their electrons are boosted into excited states. When an atom becomes excited, the amount of energy absorbed must correspond precisely to the energy difference between two orbitals. In the case of radiative excitation, that corresponds to the absorption of a photon with a definite wavelength, or color.

However, the electron cannot stay "out of place" in this higher orbital forever—the ground state is the only level where it can remain indefinitely. After about 10^{-8} s, the electron returns to its normal ground state orbital. In doing so, it emits a photon with energy precisely equal to the energy difference between the two orbitals involved. The emitted photon thus carries exactly the same energy as was required to excite the electron in the first place—an atom therefore absorbs and emits radiation at the same characteristic frequencies. The processes of absorption and emission of photons by a hydrogen atom are illustrated schematically in Figure 5.10.

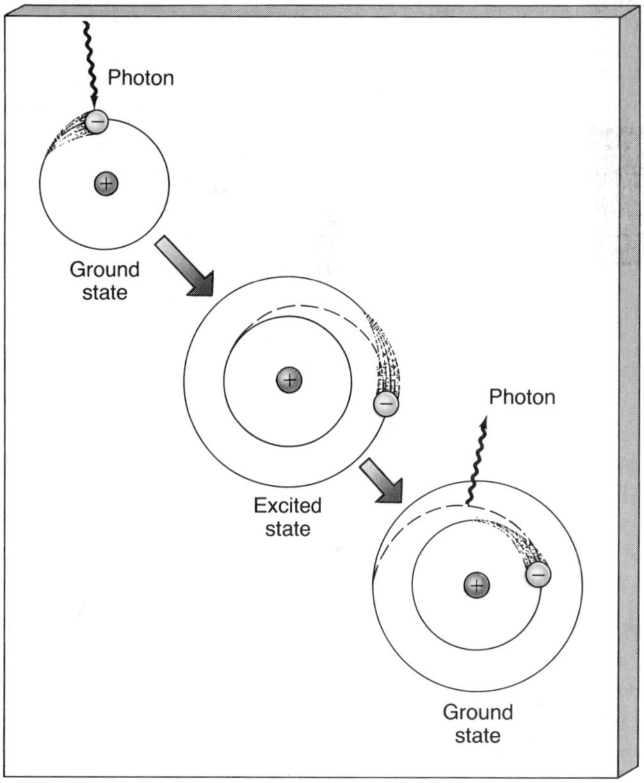

Figure 5.10 *Diagram of a photon being absorbed by a hydrogen atom (top), causing the momentary excitation of that atom (center). Eventually, the atom returns to its ground state, accompanied by the emission of a photon of the same energy as the original photon (bottom).*

The Formation of Spectral Lines

✓3 Let's reconsider our earlier discussion of emission and absorption lines in the light of the Bohr model, to see how spectral lines are produced.

In Figure 5.7, a beam of continuous radiation shines through a cloud of hydrogen gas. The gas is cool, so collisions between atoms do not have enough energy to boost hydrogen into its first excited state, and nearly all of the atoms in the gas are in their ground state until the beam enters the cloud. The beam contains photons of all energies, but most of them cannot interact with the gas. The gas absorbs only those photons having precisely the right energy to cause an electronic transition (that is, a change in an electron's orbit from one state to another). All other photons in the beam—with energies that cannot produce a transition—do not interact with the gas at all and pass through it unhindered. In this way, external radiation excites the gas. In the process, some photons are removed from the beam.

The atoms rapidly return to their ground states by emitting a photon—a photon having precisely the same energy as was absorbed in the first place. We might reason, then, that although some photons from the bulb are absorbed by the intervening cool gas, their reemission would cancel that effect, and we would never observe the effects of absorption. This is not the case, however. Although the photons not absorbed by the intervening gas follow a clear path from the bulb to the detector, the reemitted photons do not all follow that path. The reemitted photons can leave in *any* direction. Thus, many of the photons viewable in the absence of the intervening cool gas are absorbed and reemitted at other angles. These photons are effectively lost from the original beam.

A detector looking through the cloud at the source of the radiation (a star, say) records a continuous spectrum, except at those precise wavelengths where photons have been subtracted from the beam. The dark absorption lines thus produced are characteristic of the intervening gas. They are direct indicators of the energy differences between atomic orbitals. The bright emission surrounding the dark lines is unchanged by the intervening cloud. It results from those photons in the beam that had no effect on the gas.

A detector looking at the cloud from the side records the photons that are reemitted from the cloud after absorption within the gas. Every absorbed photon ultimately results in the emission of an identical photon in some direction. The emission spectrum seen from the side contains photons of exactly the same energies as those removed from the forward beam.

Absorption and emission spectra are created by the same atomic processes. They correspond to the same atomic transitions. They contain the same information about the composition of the cloud. In the lab, we can move our detector and can measure both. In astronomy, we are not able to change our vantage point—the type of spectrum we see depends on our chance location with respect to both the source and the cloud.

More Complex Atoms

All hydrogen atoms have basically the same structure, but there are many other kinds of atoms, with very different structures. Different kinds of atoms are called **elements.** The number of protons in the nucleus determines the element that an atom represents.

The next simplest element after hydrogen is helium. The central nucleus of helium is made up of two protons and two **neutrons** (another kind of elementary particle having a mass slightly larger than that of a proton but having no electrical charge at all), about which orbit two electrons. As with hydrogen and all other atoms, the "normal" condition for helium is to be electrically neutral, with the negative charge of the orbiting electrons exactly canceling the positive charge of the nucleus. The ground state of helium is illustrated schematically in Figure 5.11.

More complex atoms contain more protons (and neutrons) in the nucleus and have correspondingly more orbiting electrons. For example, an atom of carbon, shown in Figure 5.12, consists of six electrons orbiting a nucleus containing six protons and six neutrons. As we progress to heavier and heavier elements, the number of orbiting electrons increases, and the number of possible electronic transitions rises rapidly. The result is that very complicated spectra can be produced. The complexity of atomic spectra generally reflects the complexity of the atoms themselves.

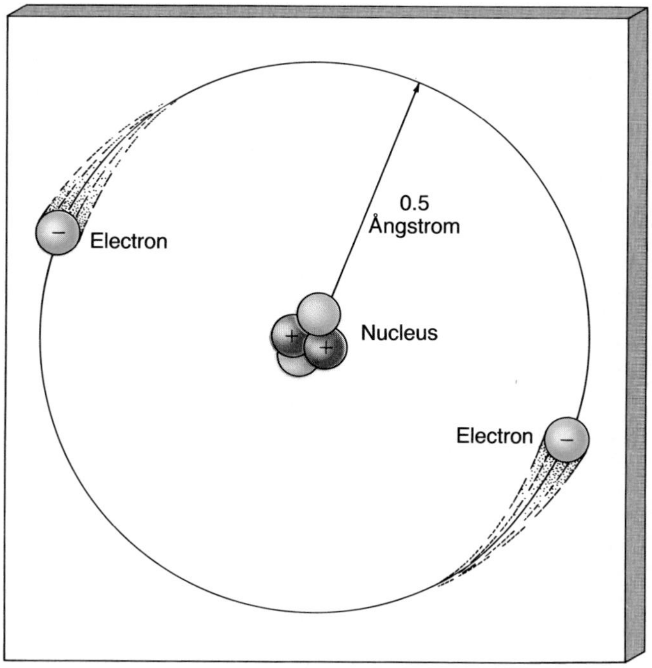

Figure 5.11 Diagram of a helium atom in its normal, ground state. Two electrons occupy the lowest-energy orbital around a nucleus containing two protons and two neutrons.

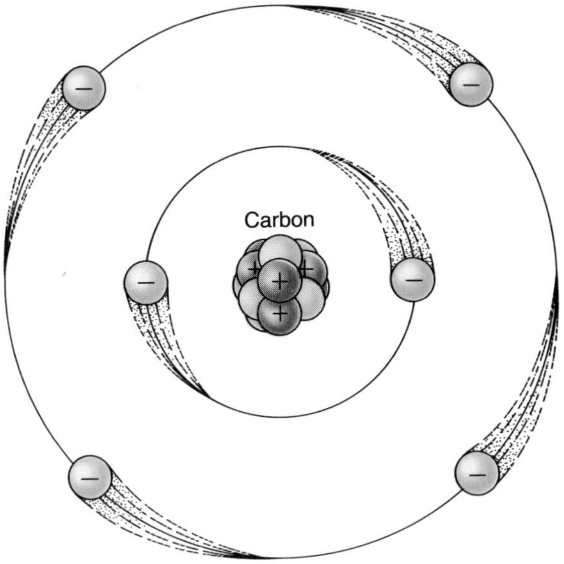

Figure 5.12 Diagram of a carbon atom in its normal, ground state. Six electrons orbit a six-proton, six-neutron nucleus, two in an inner orbital, the other four at a greater distance from the center.

ATOMIC SPECTRA

A Characteristic Fingerprint

Transitions among the various orbitals are unique for each element. Consequently, as electrons return to lower orbitals from excited states, the emitted photons carry energies characteristic of only one kind of atom. In effect, individual atoms emit tiny flashes of radiation having energies characteristic of that element—*and only that element*. This phenomenon is the powerful essence of spectroscopy. By focusing on specific frequencies, spectroscopy enables us to study one kind of atom to the exclusion of all others, even though many different kinds of atoms might be mixed together in a gas.

The Emission Spectrum of Hydrogen

In developing his model of the hydrogen atom, Niels Bohr derived a useful formula for the energy of any given orbital of hydrogen's lone electron. That formula is presented in Interlude 5-2. We will use it here to illustrate the quantitative relation between electron transitions in hydrogen and hydrogen's observed emission spectrum. Notice that Bohr's formula, and the discussion in this section, is expressed in terms of a convenient energy unit called an *electron volt* (eV), defined in Interlude 5-2. For reference, a photon of red light has an energy of about 2 eV.

Consider once again a hydrogen atom in its ground state. If left alone, the atom will not emit any energy. But if we provide some energy (by heating the gas, for instance, or by shining ultraviolet light on it), the electron may reach a higher orbital. The atom will then return almost immediately to its normal state, in the process radiating a photon, as depicted in Figure 5.13(a).

INTERLUDE 5-2 *The Energy Levels of the Hydrogen Atom*

By observing the emission spectrum of hydrogen, Niels Bohr determined early in the twentieth century what the energy differences between the various energy levels must be. Using that information, he was then able to take the next step and infer the actual energies of the excited states of hydrogen.

A unit of energy often used in atomic physics is the *electron volt* (eV). Its name derives from the amount of energy imparted to an electron by accelerating it through an electric potential of 1 volt. For our purposes, however, it is just a convenient quantity of energy, numerically equal to 1.6×10^{-12} erg, roughly half the energy carried by a single photon of red light. The minimum amount of energy needed to ionize hydrogen from its ground state is 13.6 eV.

Bohr numbered the energy levels of hydrogen, with level 1 the ground state, level 2 the first excited state, and so on. He found that by assigning zero energy to the ground state, the energy of any state (the nth) could be calculated as follows:

$$E_n = 13.6\left(1 - \frac{1}{n^2}\right)\ eV$$

Thus, the ground state has energy $E_1 = 0$ (as it must, by our definition), the first excited state has an energy of $E_2 =$ 10.2 eV, the second excited state has an energy of $E_3 =$ 12.1 eV, and so on. Notice that there is an *infinite* number of excited states between the ground state and the energy at which the atom is ionized, crowding closer and closer together as n becomes large and E_n approaches 13.6 eV.

Knowing the energy of each electron orbital, we can reverse Bohr's original reasoning and calculate the energy associated with a transition between any two given states. For example, to boost an electron from the second state to the third, an atom must be supplied with $E_3 - E_2 = 1.9$ eV of energy. Using the formulas given in Chapter 4, we find that this quantity corresponds to a photon with a wavelength 6563 Å, lying in the red portion of the spectrum. Similarly, the jump from level 3 to level 4 requires $E_4 - E_3 = 0.66$ eV of energy, corresponding to an infrared photon with a wavelength 18,900 Å, or 1.89 μm, and so on.

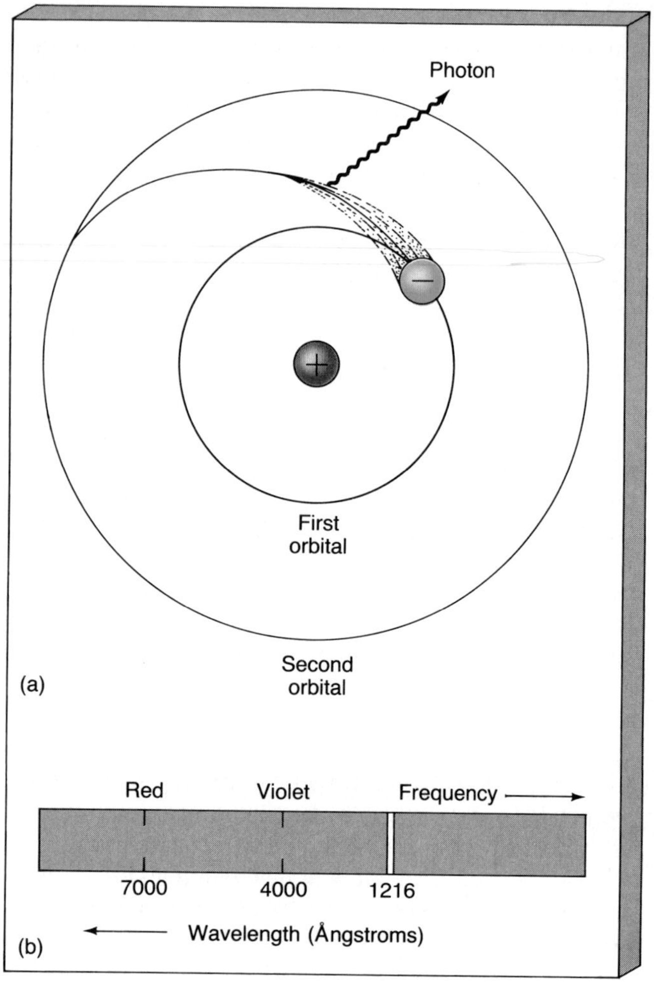

Figure 5.13 *(a) Orbital arrangement for a slightly excited hydrogen atom in the process of returning to its ground state. (b) The resulting spectrum of the emitted radiation. The visible part of the electromagnetic spectrum covers a range of wavelengths from about 7000 Å (red) down to 4000 Å (blue). The photon emitted by this rearrangement of the hydrogen atom has a wavelength of 1216 Å and so lies shortward of blue, in the invisible ultraviolet portion of the spectrum.*

Let's imagine that the atom in our example is temporarily pushed into its first excited state ($n = 2$ in Interlude 5-2). The amount of energy required is 10.2 eV. This is also the energy of the photon emitted when the atom returns to its ground state. Knowing the energy, we can calculate the photon's wavelength from the formulas given in Chapter 4: It is 1216 Å. Recall that the visible spectrum extends from 7000 Å for red light to 4000 Å for violet light, so this flash occurs in the invisible ultraviolet part of the spectrum, as shown in Figure 5.13(b).

Now, imagine exciting a hydrogen atom enough to boost the electron from the ground orbital to a higher orbital. Shown schematically in Figure 5.14(a), the electron has been raised to the third orbital ($E_3 = 12.1$ eV), far removed from the nucleus. As usual, the electron will return very rapidly to the ground state, but this time it can do so in two possible ways. It could proceed directly from the third orbital down to the first, in the process emitting an ultraviolet photon with an energy of 12.1 eV and a wavelength of 1026 Å. Alternatively, it could *cascade* down one orbital at a time, emitting two photons, one with an energy equal to the difference between the third and second orbitals, $E_3 - E_2 = 1.9$ eV, and the other with the energy difference between the second and first orbitals, or 10.2 eV.

As illustrated in Figure 5.14, the second step of the cascade process emits a 1216-Å ultraviolet photon, just as in Figure 5.13. The direct transition from the third orbital to the ground state emits a photon with wavelength 1026 Å, even farther into the ultraviolet portion of the spectrum. However, the first step of the cascade process—the transition from the third to the second orbital—produces a photon with a wavelength of 6563 Å, which is in the visible part of the spectrum. This photon is seen as red light. An individual atom—if one could be isolated—would emit a momentary red flash.

Additional heating (or absorption of any type of en-

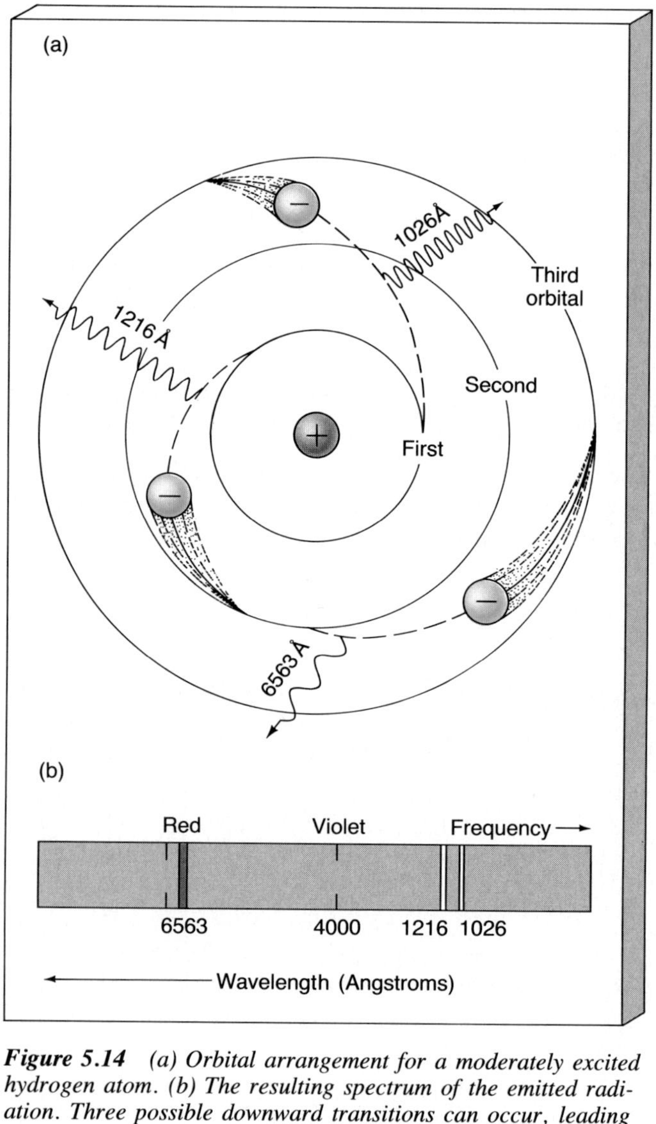

Figure 5.14 *(a) Orbital arrangement for a moderately excited hydrogen atom. (b) The resulting spectrum of the emitted radiation. Three possible downward transitions can occur, leading to three spectral lines—two in the ultraviolet and one in the visible.*

ergy) could boost the electron to even higher orbitals within the atom. Upon cascading back to the ground state, the atom could emit many photons, each with a different energy and hence a different wavelength, and the resulting spectrum would show many spectral lines. In a sample of heated hydrogen gas, at any instant, atoms are found in many different excited states. The complete emission spectrum therefore consists of frequencies corresponding to all possible transitions between those states and states of lower energy. The visible hydrogen spectrum shown earlier in Figure 5.3 is the result.

More Complex Spectra

Atoms more complex than hydrogen have a much larger number of possible electronic changes, and the resultant spectrum may contain many visible emission lines. A good example is the element iron (which contributes several hundred of the Fraunhofer absorption lines seen in the solar spectrum). The many possible transitions of its 26 orbiting electrons yield an extremely rich line spectrum, as shown in Figure 5.3.

Even more complex emission spectra can be produced by heating a *mixture* of different kinds of atoms—that is, a group of different elements. Similarly, a cool intervening gas cloud of many elements will produce a complex absorption spectrum in the light received from a continuous source. Usually, absorption lines are never numerous enough to blend together and entirely obliterate the background spectrum. They can, however, significantly alter its appearance. Figure 5.15 shows an actual spectrum observed toward a cosmic object, with some spectral lines labeled.

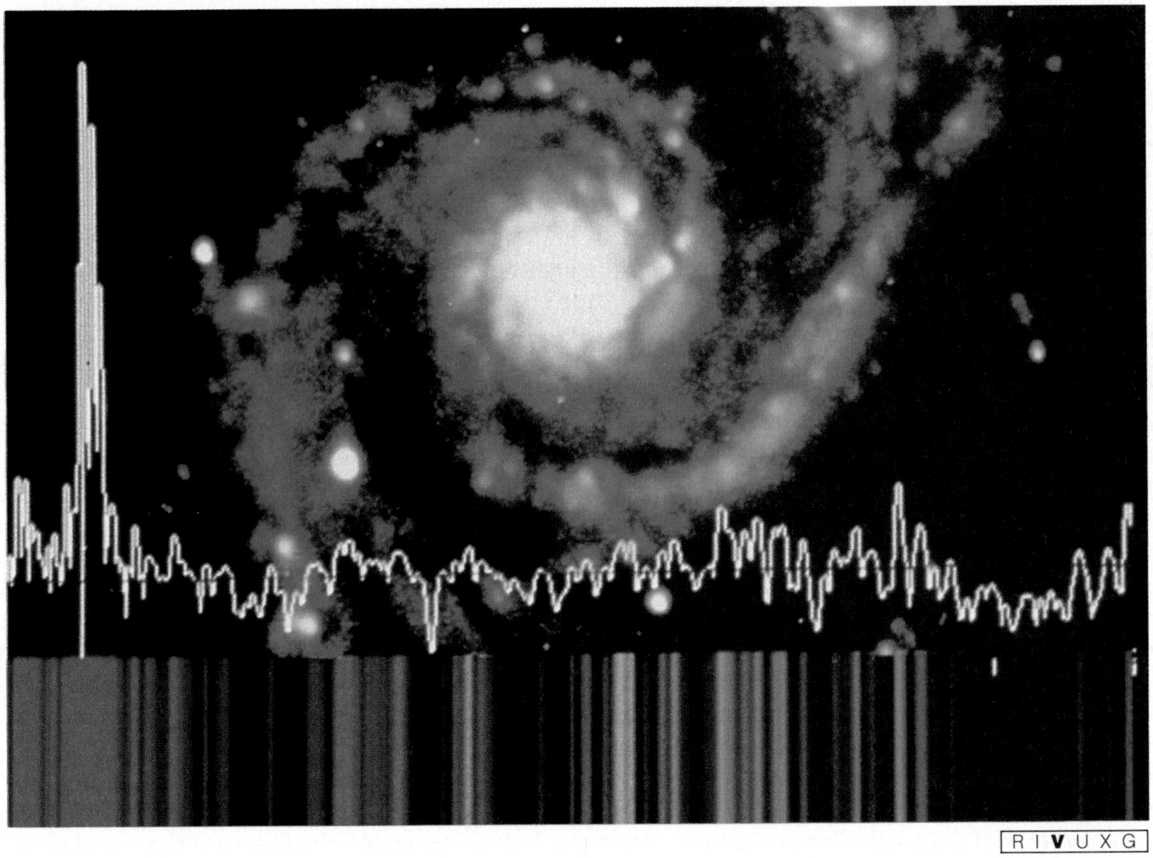

Figure 5.15 *The visible spectrum of the hot gases in a nearby spiral galaxy (called M51, the Whirlpool galaxy). Shining by the light of billions of stars, the galaxy produces a complex spectrum of bright and dark lines (bottom), also shown here as an intensity trace from red to blue (center). (Harvard-Smithsonian Center for Astrophysics)*

Beyond the Visible Spectrum

Spectral lines occur throughout the whole electromagnetic spectrum. Usually, electron changes among the lowest orbitals of the lighter elements produce visible and ultraviolet spectral lines. Changes among the lowest-lying orbitals in heavier, more complex elements produce x-ray spectral lines. These spectral lines have been observed in the laboratory. Some of them are also observed in stars and other cosmic objects.

Electron changes among very highly excited states of hydrogen and other elements can produce spectral lines in the infrared and radio parts of the electromagnetic spectrum. For example, a transition from the 20th to the 19th excited state of hydrogen produces an emission line in the far infrared region of the spectrum, with a wavelength of about 390 μm (consult Interlude 5-2). A transition from the 200th excited state to the 199th would result in a radio line, at a wavelength of 37 cm. Conditions on the Earth make it all but impossible to detect these radio and infrared features in the laboratory, but they *can* be routinely observed coming from space. Our discussion is cast in terms of emission lines, but bear in mind that it applies equally well to absorption lines, simply by reversing the directions of the transitions.

SPECTRAL-LINE ANALYSIS

✓4 Astronomers apply the laws of spectroscopy in analyzing radiation from beyond the Earth. A nearby star or a distant galaxy takes the place of the bulb of our previous examples. A galactic cloud or a stellar (or even planetary) atmosphere plays the role of the intervening cool gas. And a spectrograph attached to a telescope replaces our simple prism and detector.

Real Spectra

Stars are very hot, especially deep down in their cores, where the temperature is measured in millions of kelvins. Because of the heat, the atoms are ionized and the spectrum of radiation is continuous. However, right at the relatively cool surface of a star, not all atoms are completely ionized. Some retain a few, or even most, of their orbital electrons—that is, they are only partially ionized. Because some electrons remain bound to the nucleus, transitions between different energy states can occur as the ions interact with each other and the radiation from within the star. The ions near the surface modify the continuous spectrum from the interior by emitting and absorbing radiation at many different wavelengths.

Thus, when we observe a star through a spectrograph attached to a telescope, we see a nearly continuous spectrum with numerous dark absorption lines superposed on it. These lines are produced in the star's atmosphere—the cooler gas that absorbs some of the photons emitted from deeper in the star.

The spectrum of the nearest star, the Sun, appears in Figure 5.4. As we have already discussed, literally thousands of dark absorption lines cover the Sun's visible spectrum, nearly 800 of them produced by variously excited atoms and ions of just one element: iron. Atoms of a single element, such as iron, can yield many lines for two reasons. First, iron's 26 electrons can move among the numerous orbitals of the iron atoms in the Sun's gas, leading to a large number of possible transitions and hence spectral lines. Second, some of the iron is ionized, with some of the 26 electrons stripped away. Because the removal of electrons alters an atom's electromagnetic structure, the energy levels of ionized iron are quite *different* from those of neutral iron. Each new level of ionization introduces a whole new set of spectral lines. Besides iron, many other elements, also in different stages of excitation and ionization, absorb photons at visible wavelengths. When we observe the entire Sun, all these atoms and ions absorb simultaneously to yield the rich spectrum we see.

Line Intensity

The same basic properties that characterize continuous radiation (see Chapter 4) characterize spectral-line radiation too. A line's wavelength (or frequency) is measured directly from its location in the spectrum. The line's *intensity* is the source of a great deal of valuable astronomical information.

The intensity of a particular line depends in part on the number of atoms giving rise to the line, because the intensity is proportional to the number of photons emitted or absorbed by those atoms. The more atoms present to emit or absorb the photons corresponding to a given line, the stronger (brighter or darker, depending on whether it is an emission or absorption feature) that line is. However, intensity also depends on the *temperature* of the atoms—that is, the temperature of the entire gas of which the atoms are members—because temperature determines what fraction of the atoms at any instant are in the right orbital to undergo any particular transition.

As an example of this second very important point, consider the absorption of radiation by hydrogen atoms in the outer atmospheres of stars. If all the hydrogen were in its ground state, the only transitions that could occur would be the Lyman series (see Interlude 5-3), resulting in absorption lines in the ultraviolet portion of the spectrum. In that case, astronomers would observe *no* visible hydrogen absorption lines (for example, the Balmer series), not because there is no hydrogen, but because there would be no hydrogen atoms in any excited state (as is required to produce visible absorption features).

As the temperature rises, more and more energy becomes available in the form of collisions, and more and more electrons are boosted into an excited state. It takes a little time—about 10 nanoseconds (10^{-8} s), typically—for

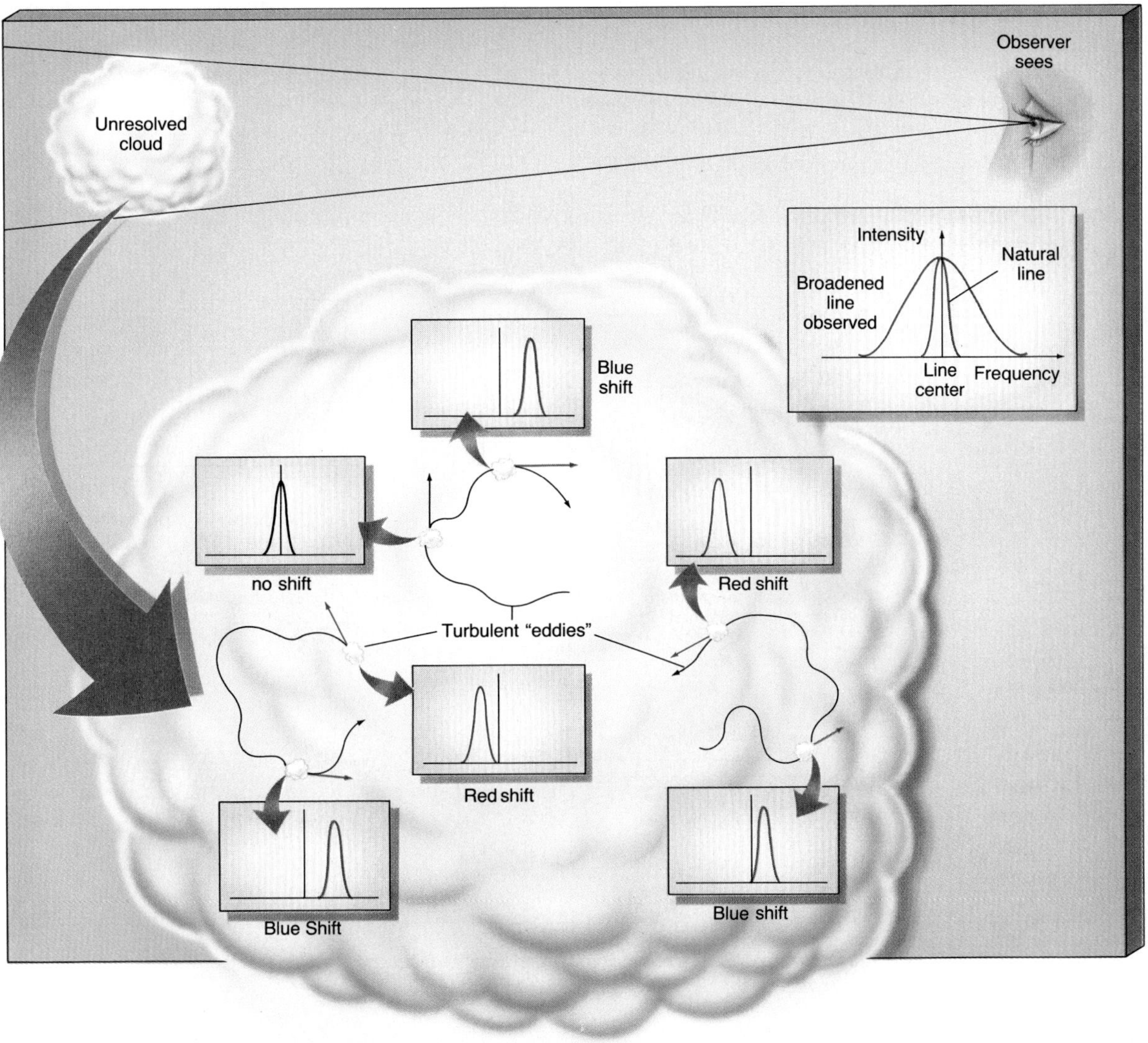

Figure 5.18 *The swirling motion of gas in a cloud can produce an overall broadening of a spectral line if the cloud is too small for a telescope to resolve.*

tor. Similarly, if the atom is moving toward us, its light is received at a shorter wavelength and so is blueshifted.

In a cloud of gas, atoms are in constant thermal motion. Some atoms have motion toward us, some away from us. Still others are moving transverse to our line of sight and are unaffected by the Doppler effect (at least, from our perspective). Throughout the whole cloud, atoms move in every possible direction. The result is that many atoms emit or absorb photons at slightly different wavelengths than would normally be the case if all the atoms were motionless.

Most atoms in a typical cloud have very small thermal velocities. As a result, most atoms emit or absorb radiation that is Doppler-shifted only a little, and very few atoms have large shifts. So, the center of any spectral line is much more pronounced than either of its "wings." The result is a bell-shaped spectral feature like that in Figure 5.17(b).

Thus, even if all atoms emitted and absorbed at only one specific frequency, the effect of this **thermal broadening** would be to smear the line out over a small range in wavelength. The hotter the gas, the larger the spread of Doppler motions and the greater the width of the line. Thus, by measuring a line's width, astronomers can estimate the temperature of a gas.

In principle, then, the width of just one spectral line can yield the temperature of a group of atoms. However, in practice the analysis is not so simple. Temperature is not the only environmental factor that broadens spectral lines. Several other physical mechanisms can also produce a similar effect.

One such mechanism is gas **turbulence,** where the gas in the cloud is not at rest or flowing smoothly, but instead is seething and churning in eddies and vortices of many sizes. Motion of this type causes Doppler-shifting of spectral lines, but lines from different parts of the cloud are shifted more or less randomly. Very often, the cloud is too small or too far away for our equipment to be able to distinguish, or *resolve*, different parts from one another, so the light from the entire cloud is blended together in our detector. As shown in Figure 5.18, when averaged over the

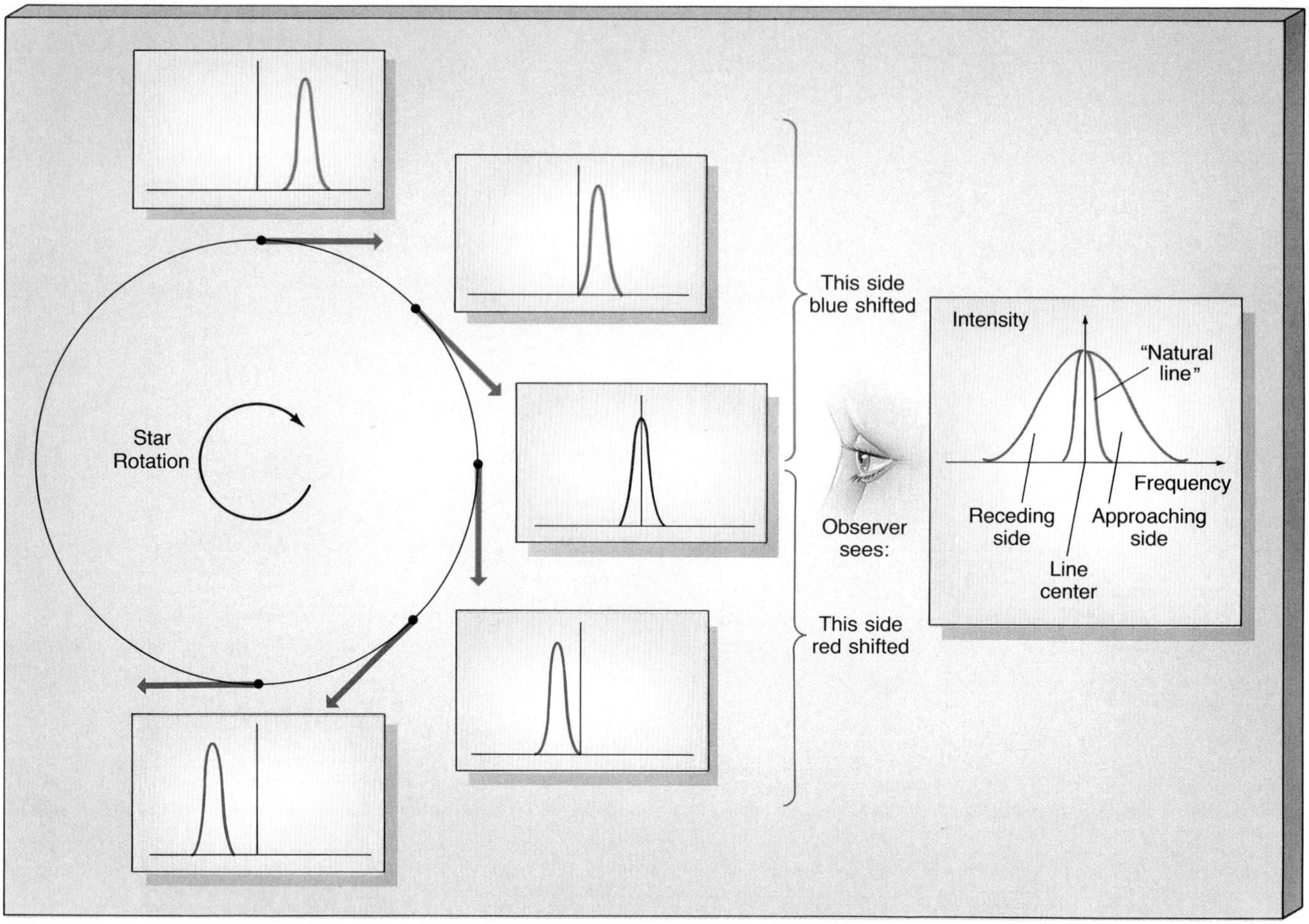

Figure 5.19 *The rotation of a star can cause spectral line broadening. Since most stars are unresolved, light rays from all parts of the star merge together to produce wide lines.*

whole cloud the net effect is rather similar to the thermal broadening just discussed. An important difference is that this broadening mechanism has *nothing* to do with the temperature of the gas.

A third influence on spectral-line width is rotation. Consider a star or a gas cloud oriented so that we see it spinning. Photons emitted from the side spinning toward us are blueshifted by the Doppler effect. Photons emitted from the side spinning away from us are redshifted. As with turbulence, if our equipment is unable to resolve the object, a net broadening of its observed spectral lines results, as illustrated in Figure 5.19. Thus, rotation is really another Doppler-broadening mechanism, but it involves neither heat nor turbulence.

A fourth broadening mechanism does not depend on the Doppler effect at all. If electrons are moving between orbitals while their parent atom is colliding with another atom, the energy of the emitted or absorbed photons changes slightly, thus ''blurring'' the spectral lines. This mechanism occurs most often in dense gases, when collisions are most frequent. It is usually called **pressure broadening.**

Magnetism is a fifth cause of spectral-line broadening. The electrons and nuclei within atoms behave as tiny, spinning magnets. As a result, the basic emission and absorption rules of atomic physics change slightly whenever atoms are immersed in a magnetic field, as is found in many stars to a greater or lesser degree. Generally, the greater the magnetic field, the more pronounced the spectral-line broadening.

Deciphering the extent to which each of these five broadening mechanisms affects spectral lines is often a very difficult task, as many of the mechanisms occur simultaneously. Any observed spectral line is likely to be affected by a combination of several of them. Sometimes, all five mechanisms act simultaneously to broaden a given line. The challenge facing astronomers is to unravel the extent to which each mechanism contributes to spectral-line profiles and so obtain meaningful information about the gas producing the lines.

MOLECULES

✓5 A **molecule** is a tightly bound group of atoms held together by the electromagnetic fields of those atoms. Like atoms, molecules can emit or absorb photons and so produce characteristic emission and absorption spectral lines. Molecular states are restricted much like atomic states. However, because molecules are more complex than individual atoms, the rules of molecular physics are also much more complex.

Painstaking experimental work over the past few decades has determined the precise wavelengths (or frequencies) at which millions of molecules emit and absorb radiation. Molecules, like atoms, produce emission or absorption spectral lines because of electron changes among the orbitals of the clustered atoms. In addition, molecules yield lines because of two other kinds of changes not possible in atoms. Molecules can rotate, and they can vibrate. Figure 5.20 illustrates these basic molecular motions.

Molecules rotate and vibrate in specific ways. Only certain spins and vibrations are allowed by the rules of molecular physics. When a molecule *changes* its rotational state or its vibrational state, a photon is emitted or absorbed. Spectral lines characteristic of the specific kind of molecule result. These lines are molecular fingerprints, just like their atomic counterparts. And like atomic lines, they enable researchers to identify and study one kind of molecule to the exclusion of all others. Finally, like atomic lines, molecular spectral lines are often broadened by some or all of the five mechanisms we noted in the preceding section.

As a rule of thumb, we can say that electron changes within molecules produce visible and ultraviolet spectral-line features. Changes in molecular vibration produce infrared spectral features. And changes in molecular rotation produce spectral lines in the radio part of the electromagnetic spectrum. Figure 5.21(a) shows the emission spectrum of molecular hydrogen. Notice how different it is from the spectrum of atomic hydrogen shown in part (b) of the figure.

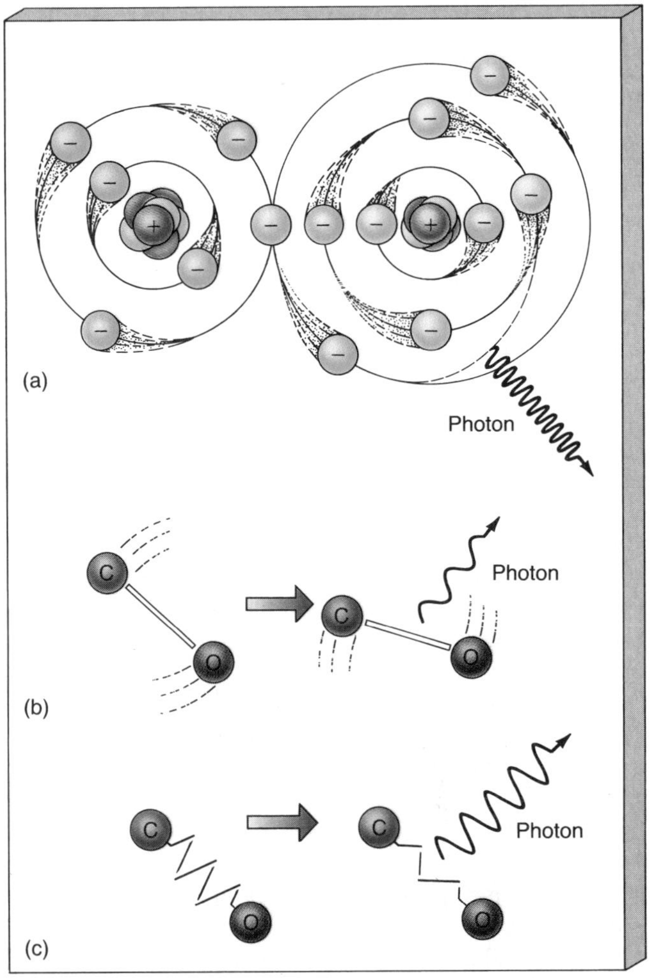

Figure 5.20 *Molecules can change in three ways while emitting or absorbing electromagnetic radiation. Sketched here is the molecule carbon monoxide (CO) experiencing (a) a change in electron arrangement, (b) a change in rotation, and (c) a change in vibration.*

Figure 5.21 *(a) The spectrum of molecular hydrogen. Notice how it differs from the spectrum of the simpler atomic hydrogen (b). (Bausch & Lomb)*

CHAPTER SUMMARY

Objects in space emit radiation, which can be analyzed in fine detail. In this way, astronomers can deduce an object's physical and chemical properties from afar.

The observational technique enabling scientists to analyze the nature of atoms by the way they emit and absorb radiation is called spectroscopy. Because every element absorbs and emits light in a characteristic way, a spectrum provides a unique "fingerprint" of the substance under investigation. A continuous spectrum consists of a rainbow of colors with a wide range of frequencies. Some spectra show bright emission lines. These lines occur when a gas, such as hydrogen, is heated enough to glow, or produce radiation. Spectra can also show dark absorption lines. These lines are formed by the removal of light from an otherwise continuous spectrum by an intervening gas. The absorption lines associated with a given gas occur at precisely the same frequencies as the emission lines produced when the gas is heated. Kirchhoff's Laws summarize the relationship between these three types of spectra.

An understanding of spectroscopy comes with an understanding of the structure of atoms. There is a state of lowest energy—the ground state—representing the "normal" condition of the electron as it orbits the nucleus. The electron can go to higher energy levels, but it must move in sharply defined steps to predetermined orbitals. Atoms are said to be excited when one or more of their electrons reside temporarily at greater-than-normal distances from the parent nucleus. Atoms can become radiatively excited by absorbing a photon of light energy from some source of electromagnetic radiation, or they can become collisionally excited by colliding with a nearby atom or elementary particle. When the electron returns to its normal ground-state orbital, it emits a photon with energy precisely equal to the energy difference between the two orbitals involved. Both absorption and emission lines are produced when a beam of starlight shines through a cloud of hydrogen gas. A detector looking through the cloud at the star records a bright stellar emission spectrum with dark absorption lines, characteristic of the intervening gas. A detector looking at the cloud from the side records the photons that are reemitted from the gas as an emission spectrum.

Spectroscopy is a powerful tool, in part because it enables astronomers to study objects in exacting detail. In a sample of heated gas, at any instant different atoms would be found in many different excited states. The complete spectrum therefore consists of frequencies corresponding to all possible transitions between those states. More complex atoms have more possible electronic changes, and more visible emission lines. Very complex emission spectra can be produced by heating a mixture of different kinds of atoms. Similarly, a cool intervening gas cloud of many elements will produce a complex absorption spectrum in the light received from a continuous source.

Astronomers apply the laws of spectroscopy in analyzing radiation from beyond the Earth. Because stars are so hot, their atoms are ionized and their spectra are continuous. Ions near the surface modify the continuous spectrum from the star's interior by emitting and absorbing radiation at many different wavelengths. Several physical mechanisms can broaden spectral lines. The most important mechanism is the Doppler effect. It occurs because stars are hot and their atoms are in motion. Broadening can also be caused by gas turbulence, or the star's rotation. Further mechanisms for line broadening involve collisions between atoms in the star and magnetic fields.

A molecule is a tightly bound group of atoms held together by the electromagnetic fields of those atoms. Molecules can produce emission or absorption lines because of electron changes among the orbitals of their component atoms. Molecules can also yield lines because of two other kinds of changes not possible in atoms. Molecules can rotate and they can vibrate, and changes in their rotational or vibrational states are accompanied by the emission or absorption of photons.

KEY WORDS

absorption line
atom
Bohr model
continuous spectrum
electron
element
emission line
emission spectrum
excited state
Fraunhofer lines
ground state
ionized
Kirchhoff's Laws
molecule
neutron
nucleus
orbital
pressure broadening
proton
spectrogram
spectrograph
spectroscope
spectroscopy
thermal broadening
turbulence

REVIEW QUESTIONS

1. What is spectroscopy? Why is it so important to astronomy?
2. Describe the basic components of a simple spectroscope.
3. Give a brief description of a hydrogen atom.
4. What is the normal condition for atoms? What is an excited atom? What are orbitals?
5. What is a continuous spectrum?
6. How are emission lines produced in a spectrum?
7. How are absorption lines produced?
8. How would absorption lines in the spectrum of a star reveal information about a cloud of cool gas lying between us and the star?

9. Explain how a cool hydrogen cloud can produce emission lines.
10. Give an example of an early discovery made possible by spectroscopy.
11. Name and briefly describe the fundamental forces of nature.
12. In the context of elementary particles, what is color?
13. Why do excited atoms absorb and reemit radiation at characteristic frequencies?
14. Why are there both emission and absorption lines in the spectrum of an ordinary star?
15. How does the intensity of a spectral line yield information?
16. Why is the H-α absorption line of hydrogen in the Sun so weak, even though the Sun has abundant hydrogen?
17. What is line broadening, and what mechanisms cause it?
18. What attributes of molecules enable them to produce spectral lines?
19. How does the Doppler effect cause line broadening?
20. Calculate the wavelength of radiation emitted by the electronic transition from the third to the second excited state in hydrogen (the Paschen-α line).
21. Calculate the wavelength and frequency of the radiation emitted by the electronic transition from the 100th to the 99th excited state in hydrogen.
22. How many different photons (that is, photons of different frequencies) can be emitted as a hydrogen atom in the second excited state falls back, directly or indirectly, to the ground state? What about a hydrogen atom in the third excited state?
23. The H-α line of a certain star is received on Earth at a wavelength of 6550 Å. What is the star's velocity with respect to the Earth?

DISCUSSION QUESTIONS

1. Why do you think scientists are so concerned with identifying an underlying "superforce"?
2. Just a few centuries ago, philosophers and scientists believed that we would never answer the question, "What are stars made of?" Now the answer has become possible, thanks to spectroscopy. Discuss some questions in the modern world that most people think will never be answered. Do you think these questions will ever be answered? What role will science play?
3. Why is spectroscopy so important to astronomers?

PROJECTS

Projects in spectroscopy generally require fairly lengthy explanations. We refer the reader to the following clearly written pieces.

Sky Publishing has a variety of "Laboratory Exercises in Astronomy," prepared or approved by Owen Gingerich. Notes for Teachers accompany the exercises. In one, called "Spectral Classification," "the student compares 30 numbered spectra and a dozen detailed reproductions of standard spectra, covering the entire spectral sequence." Available from Sky Publishing.

Colin A. Ronan has a couple of chapters on making a simple spectroscope, analyzing sunlight, and learning about Fraunhofer lines in his book *The Practical Astronomer,* New York: Macmillan, 1981.

Michael K. Gainer has a chapter called "Objective Prism Spectra" and one on "The Analysis of Stellar Spectra" in his book *Astronomy Laboratory and Observation Manual,* from Prentice Hall.

R. Robert Robbins and Mary Kay Hemenway have chapters on "Introduction to Spectroscopy," "A Spectral Comparison of the Sun and Beta Draconis," and Astronomical Spectroscopy at the Telescope" in their book *Modern Astronomy: An Activities Approach,* Austin: University of Texas Press, 1982.

SUGGESTED READINGS

Davies, P., *Superforce.* New York: Simon and Schuster, 1984. (Subtitled "The Search for a Grand Unified Theory of Nature.")

Devorkin, D. H., "The Dawn of Balloon Astronomy." *Sky and Telescope* (December 1986). The first airborne astronomical spectroscope, launched by balloon in the nineteenth century.

Goldberg, L., "Atomic Spectroscopy and Astrophysics." *Physics Today* (August 1988). A personal retrospective on the importance of atomic physics to astronomy.

Hearnshaw, J. B., *The Analysis of Starlight.* Cambridge: Cambridge University Press, 1986. Subtitled "One Hundred and Fifty Years of Astronomical Spectroscopy."

Kaler, J. B., "Origins of the Spectral Sequence." *Sky and Telescope* (February 1986). How astronomers classify and interpret spectra.

Kaler, J. B., "Extraordinary Spectral Types." *Sky and Telescope* (February 1988). Objects that defy classification under the standard systems.

CHAPTER

6

Telescopes

Observational Methods of Astronomy

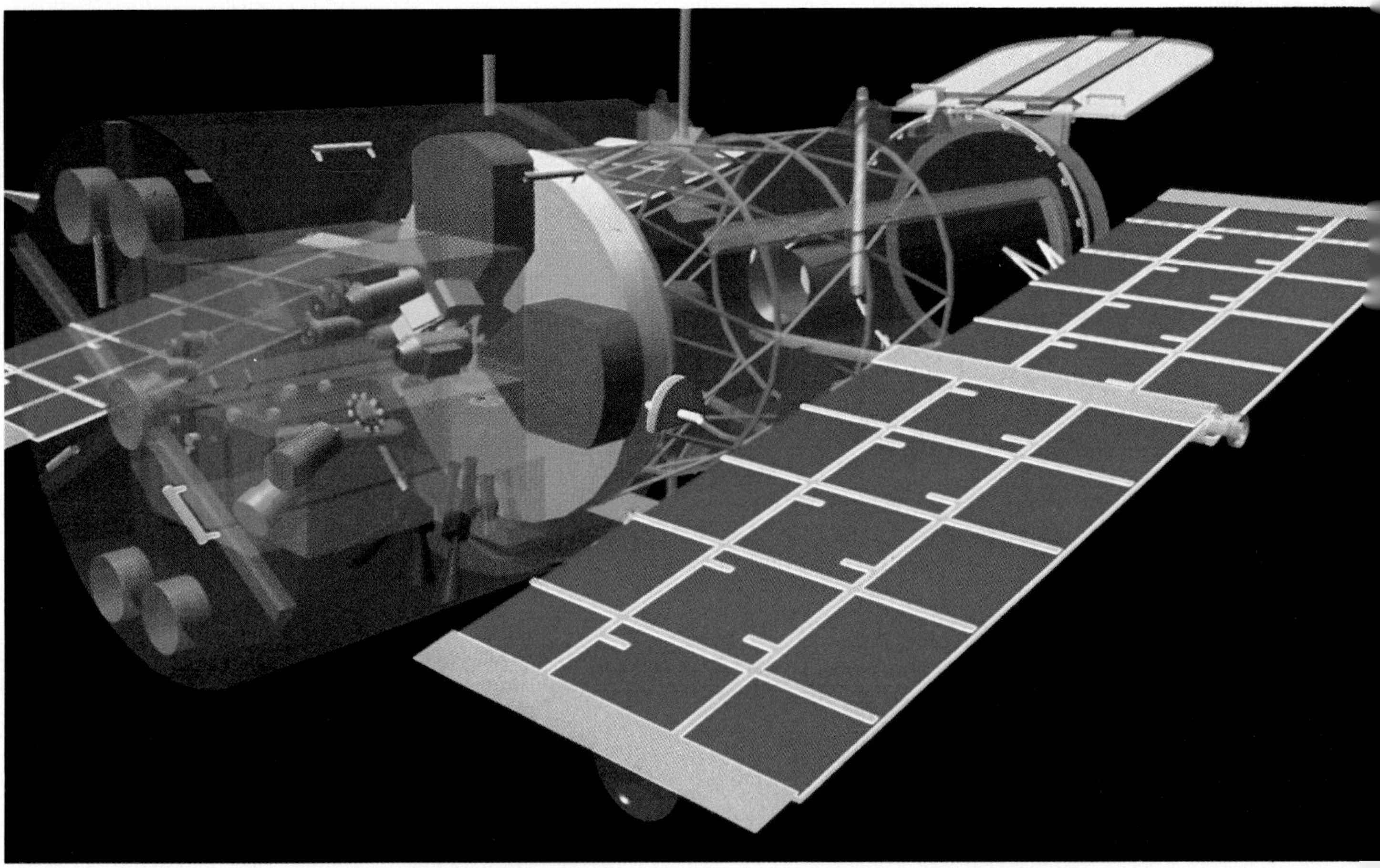

Called the "family portrait" of all the many devices aboard the Hubble Space Telescope, *this "see-through" illustration shows some of the principle features of this spaceborne observatory. The large blue disk at the center of the spacecraft is the main mirror, and the red gadgets to its rear are the sensors that guide the pointing of the telescope. The open aperture door is at upper right. The huge solar panels are shown in blue at right and partly obscured at left. Looking inside the aft bay of the vehicle, we can see key components of each of the science instruments. For example, the detectors for* Hubble's *two spectrometers are shown in copper, and blue (in the foreground). Camera components are shown in pink, green, and lavender (mostly in the background). The entire vehicle is the size of a city bus. (Courtesy D. Berry)*

LEARNING GOALS

1 to understand the basic modes of operation of optical telescopes
2 to appreciate the need for large telescopes
3 to appreciate the current efforts to improve ground-based astronomy
4 to recognize the advantages of radio astronomy
5 to recognize that radiation collected in each part of the electromagnetic spectrum contributes to our understanding of cosmic objects

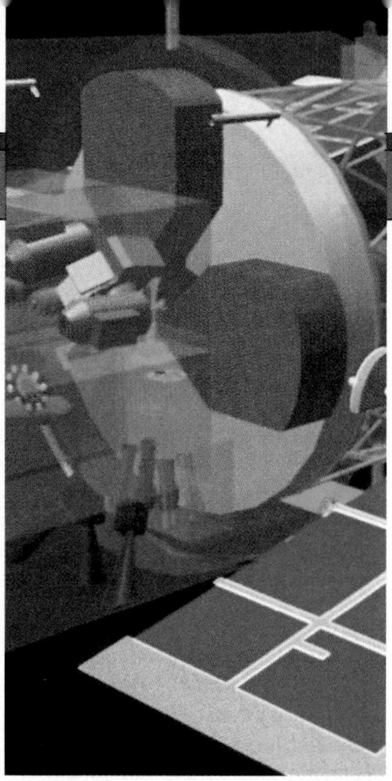

Observational knowledge of the cosmos normally advances in three phases. First, radiation from space is collected and measured, using a device known as a telescope. Then the resulting data are stored for future use, usually on photographic film or magnetic tape. Finally, the data are analyzed and interpreted. The laws of physics are applied, and a model, or theory, that explains the data is developed and tested. Theoretical work plays a vital role in this process and often suggests what new data need to be collected. At its heart, however, astronomy is an observational science. More often than not, observations of cosmic phenomena precede any clear theoretical understanding of their nature.

This chapter discusses some of the equipment astronomers use to extract information from the radiation emitted by cosmic objects. Although all types of electromagnetic radiation, from radio waves to gamma rays, are really just different forms of the same basic phenomenon, they have widely differing properties. Our detecting instruments—our telescopes—have evolved to observe as broad a range of these properties as possible. Whatever the details of its design, a telescope is a machine whose purpose is to collect light and deliver it to a detector for detailed study.

OPTICAL TELESCOPES

✓1 In essence, a **telescope** is a "light bucket," whose primary function is to capture as many photons as possible from a given region of the sky and then to concentrate them into a focused beam for analysis. An *optical* telescope is one designed specifically to collect wavelengths visible to the human eye. Optical telescopes have a long history, reaching back to the days of Galileo in the early seventeenth century. They are probably also the best-known type of telescope, so it is fitting that we begin our study of astronomical hardware with these devices.

Modern astronomical telescopes have evolved a long way from Galileo's simple apparatus. Their development over the years has seen a steady increase in *size* for one simple, but very important, reason: Large telescopes can gather and focus more radiation than can their smaller counterparts, allowing astronomers to study fainter objects and to obtain more detailed information about bright ones. This fact has played a central role in determining the design of contemporary instruments.

Types of Telescope

Figure 6.1 illustrates the two basic ways in which optical telescopes operate. Most of the world's large telescopes are **reflectors,** shown schematically in Figure 6.1(a). A reflecting telescope uses a carefully shaped *mirror* to focus light to a point known as the **prime focus,** where the light is viewed by an eye, a camera, or a computer. Light from objects as distant as stars reaches us as parallel or very nearly parallel rays. Any ray of light that enters parallel to the telescope's axis is reflected through the prime focus. The distance from the mirror to this point is known as the **focal length** of the telescope.

A fundamentally different type of telescope is known as a **refractor,** illustrated in Figure 6.1(b). Refraction is simply the bending of a beam of light as it passes from one transparent medium (for example, air) into another (such as

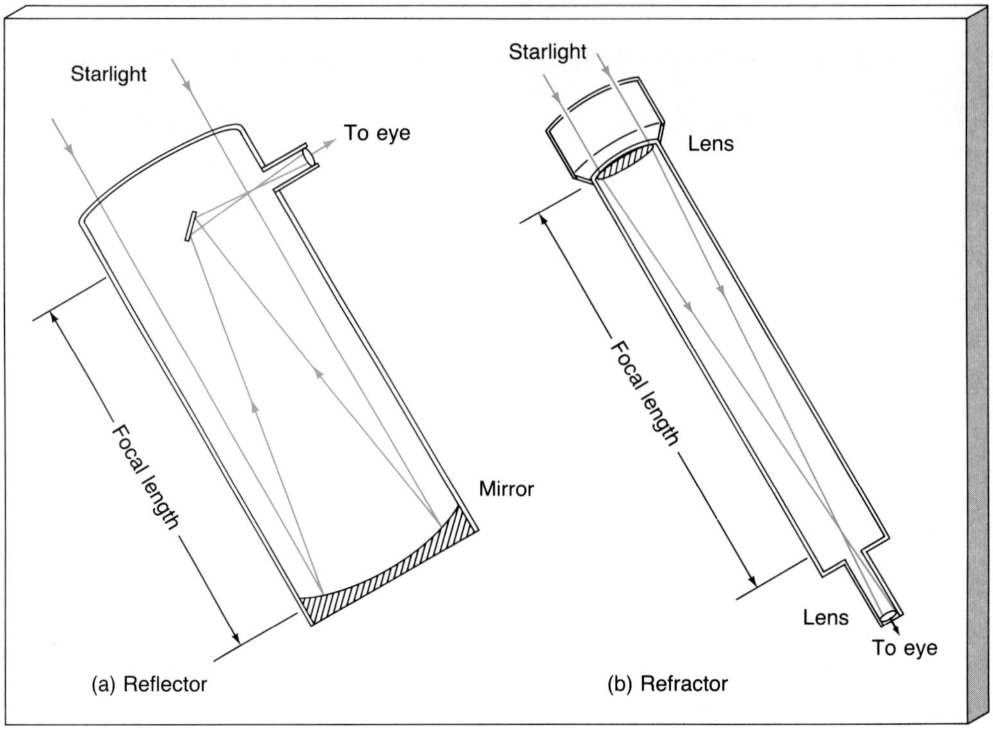

Figure 6.1 *Comparison of (a) reflecting and (b) refracting telescope systems. Both types are used to gather and focus cosmic radiation—to be observed by human eyes or recorded on photographs or in computers.*

glass). Consider how a stick in water looks bent. The stick itself remains straight, of course, but the light by which we see it is bent—refracted—as it leaves the water and enters the air. A prism splits sunlight into its component colors through refraction. The amount of refraction of any light depends on the wavelength of the light involved. Sunlight is made up of many different colors, and so the sunlight is spread out by wavelength. The blue component of sunlight is bent more than the green component, the green component more than the yellow component, and so on.

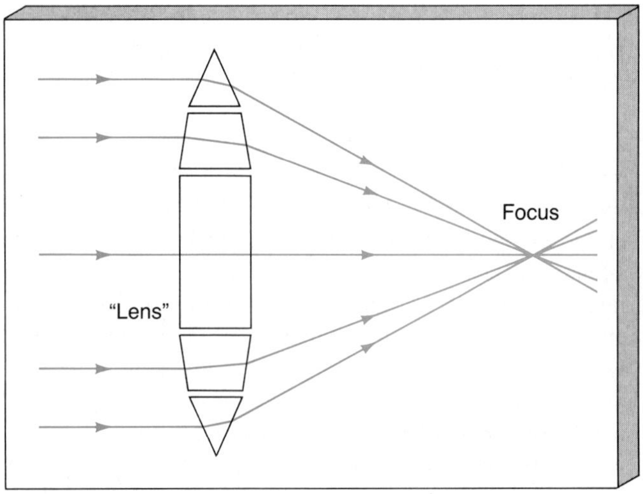

Figure 6.2 *A lens can be thought of as a series of prisms. A light ray traveling along the axis of the lens is undeflected as it passes through the lens. Rays arriving at progressively greater distances from the axis are deflected by increasing amounts. Thus all rays parallel to the axis are ultimately focused to a point.*

A refracting telescope uses a *lens* as its light-gathering device. As shown in Figure 6.2, we can think of a lens as a series of prisms. These prisms are chosen in such a way that a ray of light at any given distance from the axis is refracted so that it passes through a single point—the *focus* of the lens. The focal length for a refracting telescope is the distance from the center of the lens to the focus.

These two telescope designs—reflecting and refracting—achieve the same basic result: A beam of light initially parallel to the axis of the instrument is focused to a point. On the face of it, then, it might appear that there is little reason to prefer either mirrors or lenses in telescope construction. However, there are other factors to consider when deciding which type to buy or build. In large telescopes, these other factors all argue against refractors and in favor of reflectors.

The fact that light must pass through the lens is a major disadvantage of refracting telescopes. Large lenses cannot be constructed in such a way that light passes through them uniformly. Just as a prism disperses light, the lens focuses red and blue light differently. This deficiency is known as **chromatic aberration.** Figure 6.3 shows schematically how chromatic aberration occurs and indicates how it affects the image of a star. Careful design and choice of materials can largely correct chromatic aberration, but it is very difficult to eliminate entirely. And as the lens diameter increases, so do the aberration and the problems necessary to correct it. In short, large lenses cannot be focused well. Mirrors do not suffer from this defect.

Lenses suffer from other drawbacks, too. As light passes through them, some of it is absorbed by the glass. This absorption is a particular problem for nonoptical radiation, because glass is opaque to (that is, it blocks com-

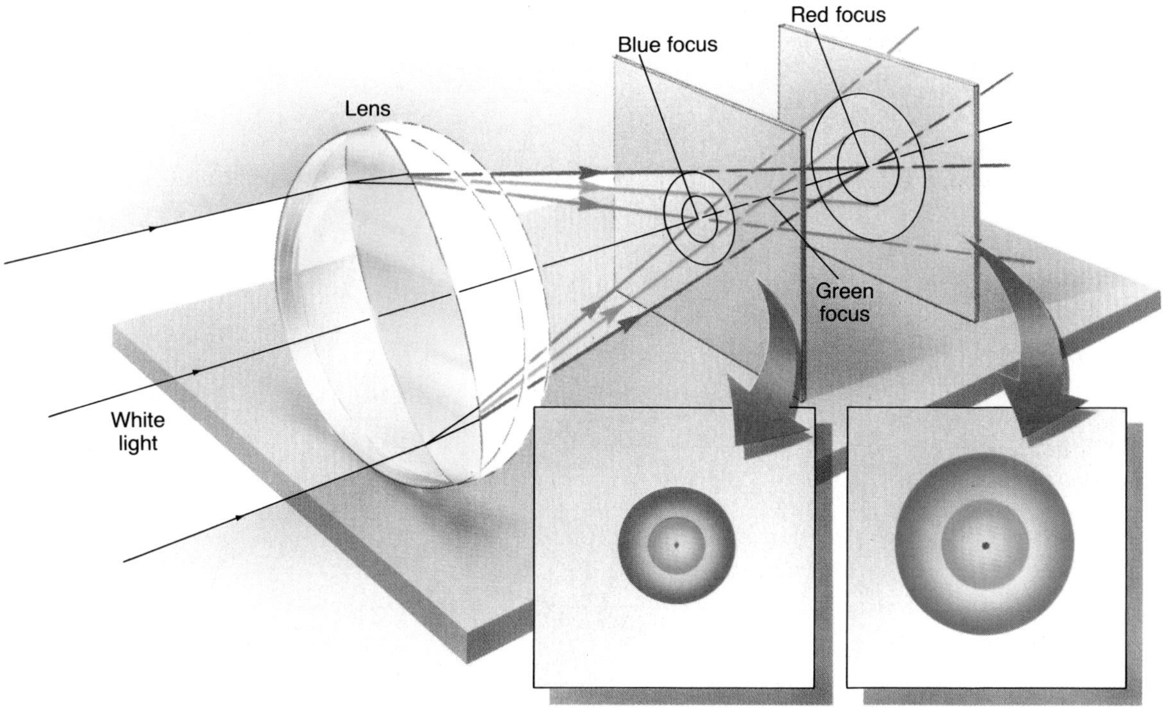

Figure 6.3 Chromatic aberration. A prism bends blue light more than it bends red light, so the blue component of light passing through a lens is focused slightly closer to the lens than is the red component. As a result, the image of an object acquires a colored "halo," no matter where we place our detector.

pletely) much of the infrared and ultraviolet regions of the spectrum. The problem of opacity obviously does not affect mirrors. Another serious difficulty is that a large lens can be quite heavy. Because it can be supported only around its edge (so as not to block the incoming radiation), the lens tends to deform under its own weight. A mirror, conversely, can be supported over its entire back surface. Finally, a lens has two surfaces that must be accurately machined and polished—which can be quite difficult jobs—but a mirror has only one.

For these reasons, *all* large modern telescopes use mirrors as their primary light gatherers. Figure 6.4 shows the world's largest refractor, installed in 1897 at the Yerkes Observatory in Wisconsin and still in use today. It has a lens diameter of 1 m (about 40 inches). By contrast, as we will see, the largest reflecting telescopes currently have mirror diameters in the 4- to 6-m range, and much larger instruments are on the way.

Telescope Design

Figure 6.5 presents a schematic diagram of some basic reflecting telescope designs. Radiation from a star enters the instrument, passes down through the main telescope tube,

Figure 6.4 Photograph of the Yerkes Observatory's 1-m diameter refracting telescope. (University of Chicago)

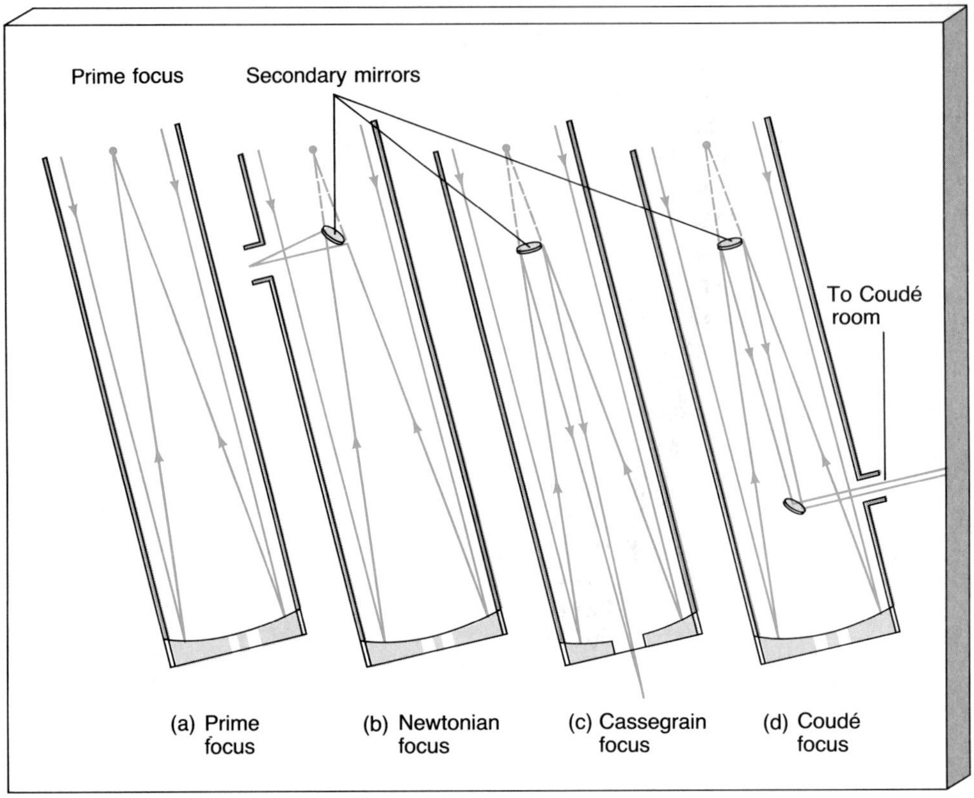

Figure 6.5 *Diagrams of the essentials of any apparatus used to collect, focus, and record information from cosmic objects. Shown here are four different telescope designs: (a) Prime focus, (b) Newtonian focus, (c) Cassegrain focus, and (d) Coudé focus. Each uses a primary mirror at the bottom of the telescope to capture radiation, which is then directed along different paths for analysis.*

hits the curved surface of the *primary mirror*, and is reflected back toward the prime focus—near the top of the tube. Sometimes astronomers simply place their instruments right at the prime focus, as in Figure 6.5(a). However, it can be very inconvenient, or even impossible, to suspend bulky pieces of equipment there. More often, the light is intercepted on its path to the focus by a smaller *secondary mirror* and is redirected to a more convenient location. Three such arrangements are shown in Figure 6.5(b)–(d).

Notice, by the way, that, in all of these cases, the detector or the secondary mirror actually *blocks* some of the incoming light. This is generally not a problem because the shadow produced covers only a small fraction of the mirror surface and so does not seriously impair the light-gathering capability of the telescope. For example, with a 4-m telescope (that is, a telescope with a mirror 4 m in diameter), even a relatively large 1-m wide detector has an area only about 6 percent that of the mirror, so that 94 percent of the incoming starlight is unaffected. Still, when very large detectors are needed, it is obviously preferable to move them elsewhere and bring the light to them rather than to keep them at the prime focus.

In a **Newtonian telescope,** the light is intercepted before it reaches the prime focus and is deflected by 90°, usually to an eyepiece at the side of the instrument. This is a particularly popular design for smaller reflecting telescopes, such as those used by amateur astronomers. Alternatively, astronomers may choose to work on a rear platform where they can use equipment too heavy to hoist to the prime focus. In this case, light proceeds from a star toward the primary mirror and reflects up toward the prime focus, where a smaller secondary mirror reflects it back down through a small hole at the center of the main mirror. This arrangement is known as a **Cassegrain telescope,** and the point behind the primary mirror where the light from the star finally converges is called the Cassegrain focus.

Another, more complex observational configuration requires starlight to reflect from several mirrors. The radiation first reflects normally from the primary mirror toward the focus, after which a secondary mirror reflects it back down the tube. A third, much smaller, mirror then deflects the light into an environmentally controlled laboratory. Known as the coudé room (after the French word for "bent"), this laboratory is separate from the telescope itself, enabling astronomers to use very heavy and finely tuned equipment that could not possibly be lifted to either the prime focus or the elevated platform at the rear of the telescope.

In order to illustrate some of the points we have discussed, let us now briefly consider an instrument that has been at or near the forefront of astronomical research for much of the last half-century. Figure 6.6 depicts the Hale 5-m (200-inch) diameter optical telescope on California's Palomar Mountain, dedicated in 1948. As the size of the figure in the observer's cage at the prime focus indicates, this is indeed a very large telescope. In fact, for almost three decades, the Hale telescope was the largest in the world. As illustrated, observations can be made at the prime, the Cassegrain, or the coudé focus, depending on the needs of the user. Only the auxiliary coudé mirror is shown, and the coudé room itself is out of the picture, to the lower right.

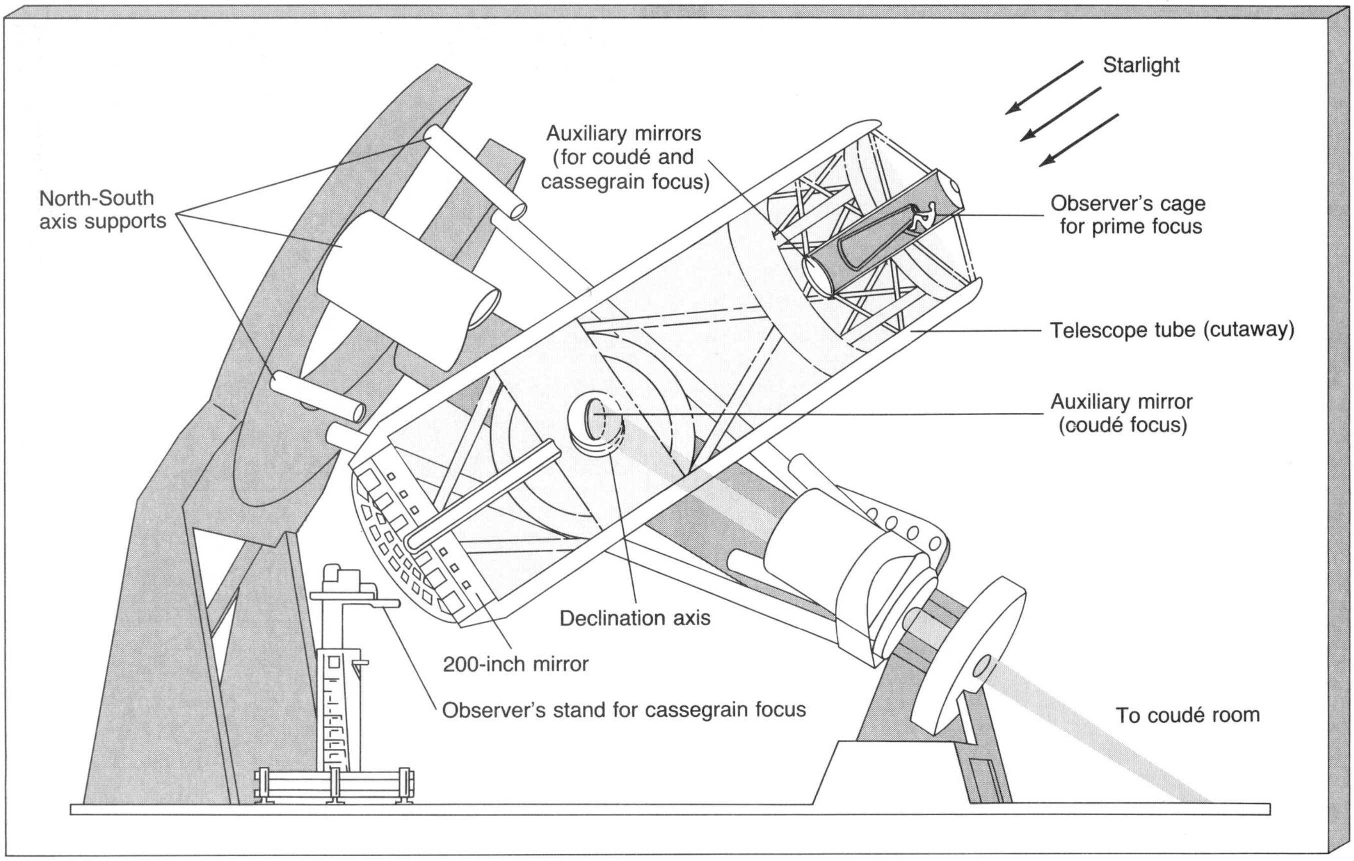

(a)

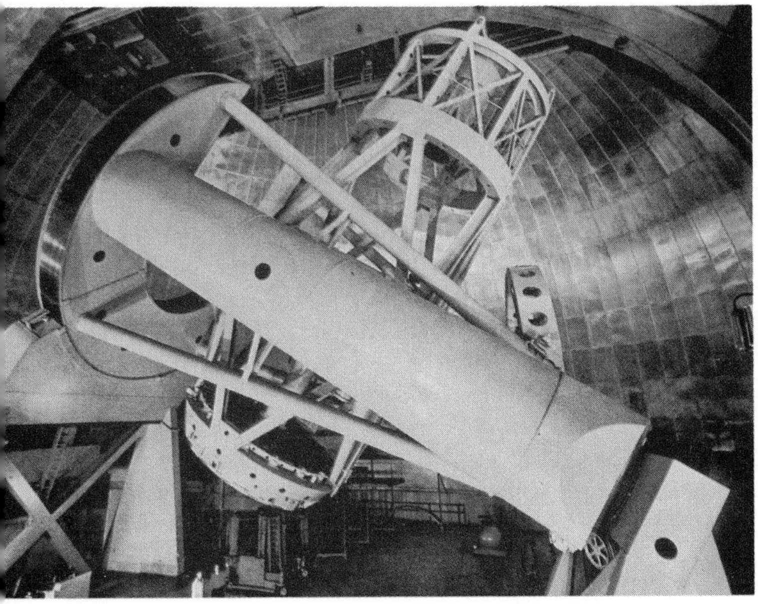

(b)

(c)

Figure 6.6 *(a) An artist's illustration of the 5-m diameter Hale optical telescope on Palomar Mountain in California. (b) A photograph of the telescope. (c) Astronomer Edwin Hubble in the observer's cage at the Hale prime focus. (Palomar Observatory)*

Images and Detectors

Light entering a telescope along the axis of the mirror is reflected to the prime focus. As shown in Figure 6.7, light coming from a slightly different direction is focused onto a slightly different point, so that an **image** is formed near the prime focus. Each point on the image corresponds to a different point on the sky. Most large telescopes actually produce quite small prime-focus images. For example, the image of the Hale 5-m telescope's 2′ (arc minute) field of view is only about 1 cm across. Usually, the image is magnified with a lens before being observed by eye, or, more likely, recorded as a photograph. In principle, there is no limit to the magnification that can be achieved just by using more and more powerful lenses. However, as we will see in the next section, there are important practical restrictions on how much detail can be extracted in this way.

Large reflectors are good at forming images of narrow fields of view (typically, only a few arc minutes), where all the light that strikes the mirror surface moves almost parallel to the axis of the instrument. However, if the light enters at an appreciable angle, it cannot be accurately focused. This inability to focus leads to an effect known as **coma,** where off-center star images acquire "tails," degrading the overall quality of the image, as illustrated in Figure 6.8. Coma worsens as we move farther from the center of the field of view. Eventually, the image quality is reduced to the point where it is no longer usable. The distance from the center to where the image becomes unacceptable defines the useful field of view of the telescope.

An alternative design that overcomes this problem is the *Schmidt telescope*, named after its inventor, Bernhard Schmidt, who built the first such instrument in the 1930s. The telescope uses a correcting lens, which sharpens the final image of the entire field of view. Consequently, a Schmidt telescope is well suited to producing wide-angle photographs, covering several degrees of the sky.

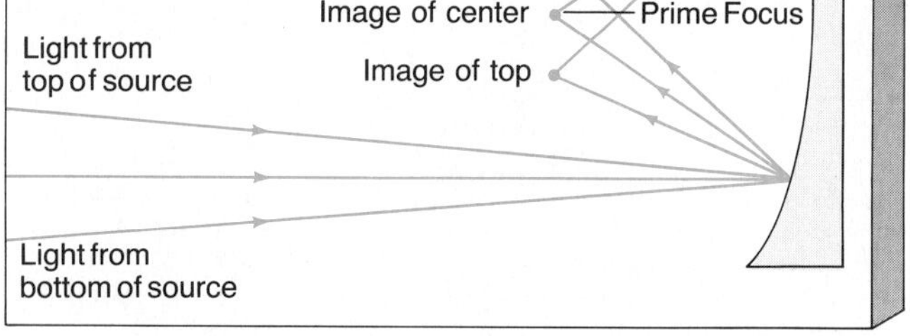

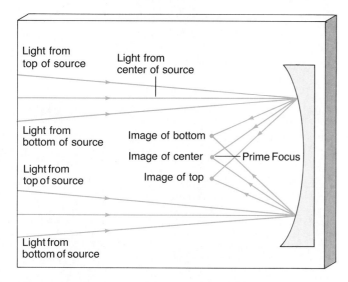

Figure 6.7 Formation of an image. Rays of light coming from different points on a distant object are focused to slightly different locations. The result is that an image of the object is formed around the prime focus. Notice that the image is inverted (that is, upside down).

Figure 6.8 The effects of coma can be seen in this image, as stellar images farther from the center acquire more and more pronounced tails. Notice the slightly egg shaped stars near the left and right edges. (Palomar Observatory)

The design of the Schmidt telescope results in a curved image that is not suitable for viewing with an eyepiece. Instead, the image is recorded on a specially shaped piece of photographic film. For this reason, the instrument is often called a *Schmidt camera*. The Palomar Observatory Schmidt camera, one of the largest in the world (with a 1.8-m mirror and a 1.2-m lens), performed a survey of the entire northern sky in the 1950s. The Palomar Observatory Sky Survey, as it is now known, remains an invaluable research tool for all professional observers.

When a light-sensitive device, such as a photographic plate, is placed at the focus to record an image of the field of view, the telescope is acting in effect as a high-powered camera lens. However, this is by no means the only instrument that can be placed at the focus to analyze the radiation received from space. When very accurate and rapid measurements of light intensity are required, a device known as a **photometer** is used instead. A photometer measures the total amount of light received in all or part of the image. The region of interest is selected simply by masking out the rest of the field of view. Thus, use of a photometer often "throws away" spatial detail, but in return allows more information to be obtained about the intensity and time-variability of a source.

Often, astronomers want to study the *spectrum* of the incoming light. Large spectrometers frequently work in tandem with optical telescopes. Light radiation collected by the primary mirror may be redirected to the underground coudé room, defined by a narrow slit, passed through a prism, and projected onto a screen—a process not so different from the operation of the simple spectroscope described in Chapter 4. The spectrum can be studied in real time (that is, as it happens) or stored on a photographic plate (or, more commonly nowadays, in a computer) for later analysis.

Some Modern Telescopes

The world's largest conventional reflector is the 6-m (240-inch) diameter telescope in the Caucasus Mountains in Russia, shown in Figure 6.9. It is similar in design to the Palomar 5-m instrument and took more than a decade to build. It was finally completed in 1976.

The U.S. National Observatory for Optical Astronomy in the Northern Hemisphere is located high on Kitt Peak near Tucson, Arizona. The site was chosen because of its many dry, clear nights, and construction was completed in 1973. Several separate telescopes are sited there. The largest has a mirror diameter of 4 m (160 inches).

Figure 6.10 is an aerial view of the major U.S. observatory in the Southern Hemisphere, completed in 1974 and located at Cerro Tololo, in the Chilean Andes. Numerous domes house optical telescopes of different sizes, each with varied support equipment, making this the most versatile observatory south of the equator. Economic difficulties in the United States and political difficulties in Chile have postponed further construction of several major telescopes at this exceptionally clear site.

A radically different type of telescope sits on Mt. Hopkins, Arizona, a few miles from Kitt Peak. Figure 6.11 shows the Multiple Mirror Telescope, usually known as the MMT, which was completed in 1979. The primary optical device in this reflector is not a single huge, heavy mirror but rather six smaller, lighter mirrors. These mirrors, each 1.8 m (72 inches) in diameter, work together, under the control of a central computer, to approximate the capabilities of a single mirror nearly 5 m in diameter. The smaller mirrors are much easier and cheaper to construct, although the complex paths the incoming light waves must follow to reach a common focus make the telescope tricky to operate and time-consuming to align. Even so, the success of the MMT has demonstrated that multiple-mirror telescopes can work well.

Figure 6.10 *Located high in the Andes Mountains, the Cerro Tololo Inter-American Observatory, a major Southern Hemisphere operation, is run by the same group of U.S. universities that oversees the Kitt Peak National Observatory. (AURA)*

Figure 6.9 *One of the world's largest (6-m diameter) reflecting telescopes in the Caucasus, in Russia. Notice the people standing on top of the dome. (Russian Space Agency)*

Figure 6.11 *Photograph of a large optical telescope of radical design—the Multiple Mirror Telescope atop Mt. Hopkins, Arizona. Notice its six separate mirrors. (Harvard-Smithsonian Center for Astrophysics)*

TELESCOPE SIZE

Light-gathering Power

✓2 As we have already noted, astronomers generally prefer large telescopes over small ones. For optical work, the main reason for this preference is simply that a larger telescope has a greater **collecting area,** which is the total area of a telescope capable of capturing radiation. The larger the telescope (either reflecting mirror or refracting lens), the more light it collects, and the easier it is to measure and study an object's radiative properties. For this reason, astronomers can observe weak cosmic sources only with large telescopes. Figure 6.12 illustrates the effect of increasing telescope size by comparing images of the Andromeda Galaxy taken with different instruments.

The observed brightness of any astronomical object is directly proportional to the area of our telescope's mirror—in other words, to the *square* of the mirror diameter. Thus, a 5-m telescope will produce an image 25 times as bright as a 1-m instrument because a 5-m mirror has 25 ($=5^2$) times the collecting area of a 1-m mirror.

We can also think of this relationship in terms of the length of *time* required for a telescope to collect a given amount of light energy—for example, the energy needed to create a recognizable image on a photographic plate. Our 5-m telescope will produce an image 25 times faster than the 1-m device because it gathers energy at a rate 25 times greater. Expressed in another way, a 1-hour time exposure with a 1-m telescope is roughly equivalent to a 2.4-minute time exposure with a 5-m instrument.

Figure 6.12 *Effect of increasing telescope size on an image of the Andromeda Galaxy. Both photographs had the same exposure time. Fainter detail can be seen as the diameter of the telescope mirror increases because larger telescopes are able to collect more photons per unit time. (AURA)*

Resolving Power

A second advantage of large telescopes is their resolution. **Angular resolution** is the ability to distinguish between two adjacent objects in the sky. The finer the angular resolution, the better we can make such a distinction, and the better we can see the details of any given object. Figure 6.13 illustrates how the appearance of two nearby objects might change as the angular resolution varies. When the angular resolution is much greater than the separation of the objects, the objects cannot be distinguished from one another and appear as a single fuzzy "blob." As the resolution improves, the two sources eventually become discernable as separate objects.

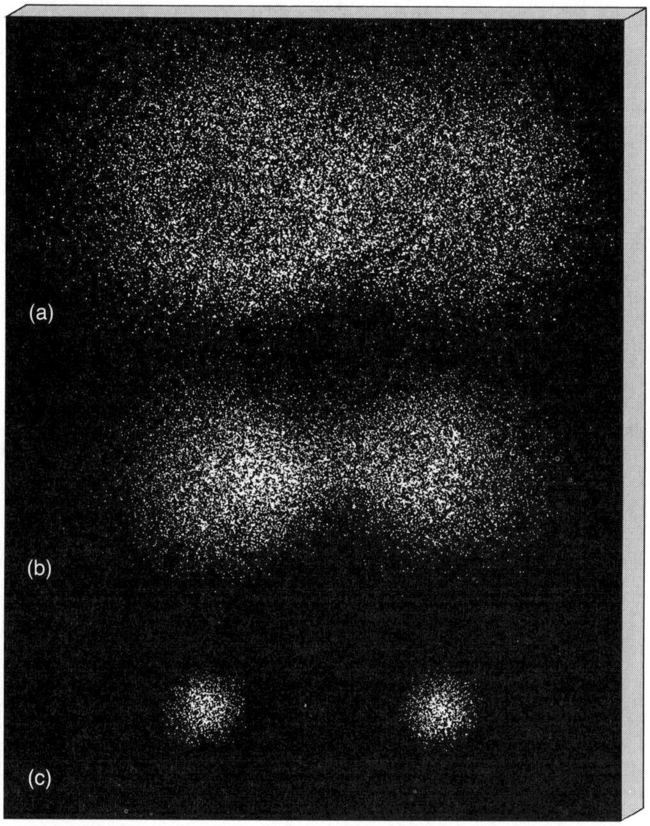

Figure 6.13 *Two comparably bright light sources become progressively clearer when viewed at finer and finer angular resolution.*

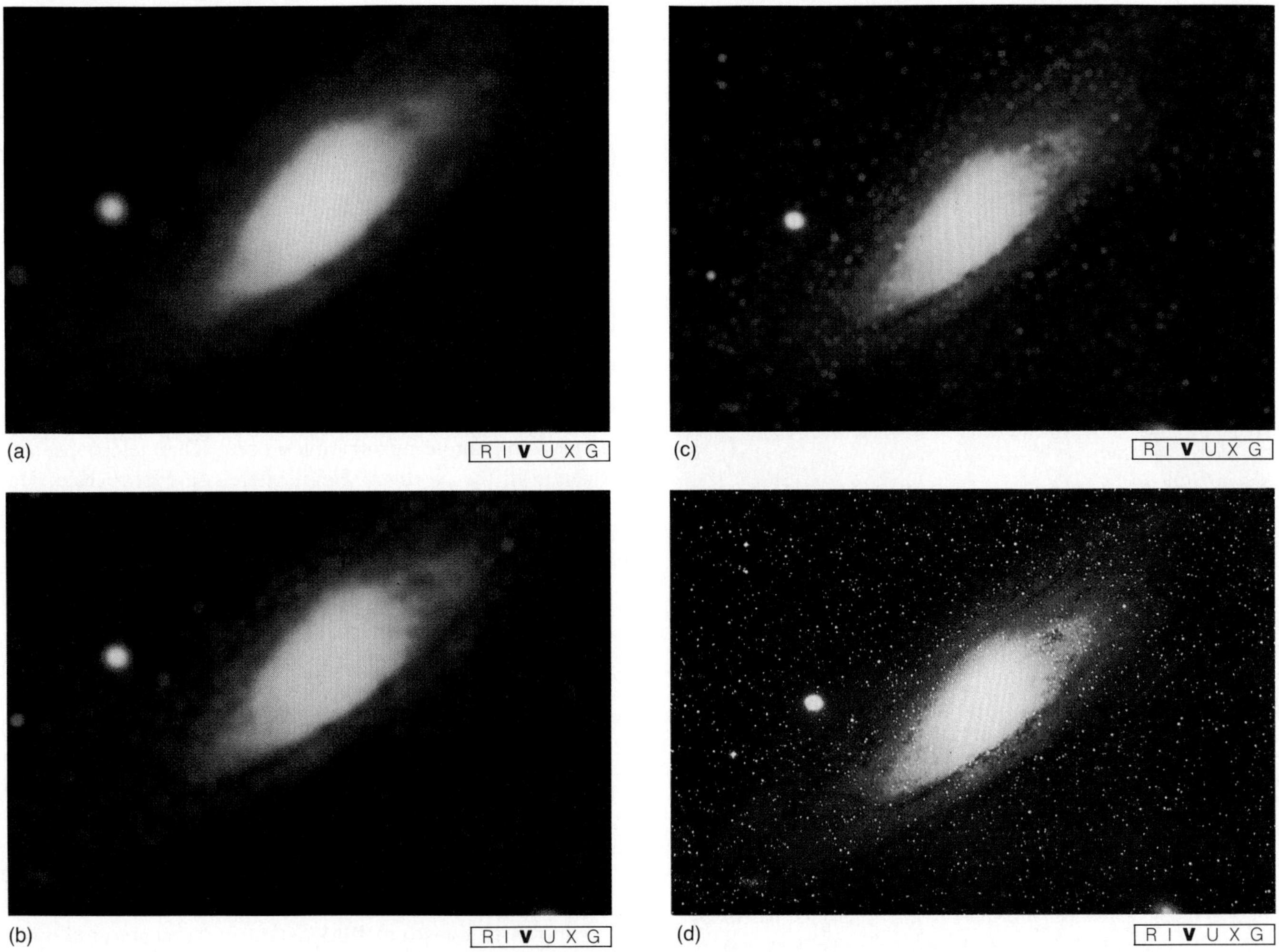

Figure 6.14 *Detail can be more readily discerned in the Andromeda Galaxy as the angular resolution is improved some 600 times, from (a) 10′, to (b) 1′, (c) 5″ and (d) 1″. (AURA)*

What limits a telescope's resolution? One factor is diffraction, which we discussed in Chapter 4. As light enters the telescope, the rays are bent slightly, and this bending makes it impossible to focus the light to a perfect point, even with a perfectly constructed mirror. The situation is similar to that shown in Figure 4.3, where a beam of light is shown spreading out slightly after passing through a narrow opening. The angle through which the beam spreads is proportional to the wavelength of the radiation divided by the width of the opening (in this case, the diameter of the mirror), and this angle sets the angular resolution of the telescope.

Thus, for light of a given wavelength, large telescopes produce less diffraction than small ones. For example, in an otherwise perfect observing environment, the best possible angular resolution that could be obtained using a 1-m telescope observing blue light (with a wavelength of 4000 Å) would be about 0.1″ (arc second). This quantity is known as the *diffraction-limited* resolution of the telescope. A 5-m telescope would have a diffraction-limited resolution five times finer, or 0.02″, and a 10-cm telescope would have a diffraction limit of 1″. Figure 6.14 shows how the Andromeda Galaxy would appear in greater detail with progressively higher resolution, when viewed in visible light through a hypothetical series of telescopes. For reference, the angular resolution of the human eye is about 0.5′ (arc minute).

We will see in a moment that no large ground-based optical telescope actually comes close to its diffraction limit because of the blurring effects of Earth's atmosphere. However, for a given telescope size, the amount of diffraction increases in proportion to the wavelength used, so that observations in the infrared or radio range are often limited by its effects. For example, if we were to use our 1-m telescope to make observations in the near infrared range, at a wavelength of 10 microns (10 μm = 10^{-3} cm = 10^5 Å), the best resolution we could obtain would be only 2.5″. A 1-m radio telescope, operating at a wavelength of 1 cm, would have an angular resolution of slightly under 1°.

The Construction of Very Large Telescopes

We have seen that, in the push toward larger and larger telescopes, refractors were abandoned long ago, in large part because of the difficulty in constructing them. While large refracting telescopes are certainly hard to build, large

Figure 6.15 *Photograph of part of the segmented mirror of the Keck telescope, taken in 1990, when the telescope was first pointed at the sky. (W. W. Keck Observatory)*

reflectors are not exactly easy. Conventional telescope mirrors are made from large blocks of quartz, glass, or some other type of polishable material capable of withstanding large temperature changes with little expansion or contraction. The construction process begins by pouring the molten material into a large cast, then cooling it slowly over the course of several years to keep it from cracking or developing internal stresses while it transforms from liquid to solid. It takes years more to grind and polish the surface to the required curvature. Finally, the surface is coated with a thin film of aluminum to provide a reflecting surface. All of these stages are slow, painstaking, and, unfortunately, not always successful. Engineers encounter severe difficulties in building very large telescopes by these means. Indeed, since the construction of the 5-m Palomar instrument in 1948, only one larger telescope (the Russian 6-m) has been completed, and this telescope, unfortunately, does not focus light well, resulting in poor-quality images.

Until quite recently, the conventional wisdom was that telescopes with mirrors larger than 5- or 6-m in diameter were simply too expensive and impractical to build. However, new manufacturing techniques, coupled with radically new mirror designs, now make the construction of telescopes in the 8- to 12-m range almost a routine matter. Experts can now make large mirrors much lighter for their size than had hitherto been believed feasible, and, pioneered by the MMT, the technology now exists to combine many smaller mirrors into the equivalent of a much larger single-mirror telescope. Several large-diameter instruments are now under construction, and many more are planned. This new generation of telescopes owes its existence in large part to the revolution that has taken place in mirror technology in the last decade.

The National Optical Astronomy Observatories (NOAO), the U.S. consortium that presently operates the Kitt Peak and Cerro Tololo observatories, plans to construct twin 8-m (single-mirror) telescopes. One telescope will operate in the Northern Hemisphere, on Mauna Kea, Hawaii, and one in the Southern Hemisphere, near Cerro Tololo. The California Institute of Technology and the University of California are jointly building two 10-m telescopes, also to be sited on Mauna Kea. Unlike the NOAO 8-m instruments, these telescopes, known as the Keck telescopes (Figure 6.15) employ a segmented design, combining 36 1.8-m six-sided mirrors into the equivalent of a single reflector. The first Keck telescope saw "first light" in 1990, at which time nine of its mirror segments were operational, and its collecting area already exceeded that of the Russian 6-m instrument. It became fully operational in 1992.

Even grander proposals exist. Both NOAO and the European Southern Observatory (ESO) are actively pursuing plans to construct MMT-style instruments employing 8-m mirrors as the "smaller" components. NOAO's hoped-for National New Technology Telescope (NNTT) would combine four 7.5-m mirrors into the equivalent of a single 15-m mirror. ESO's Very Large Telescope (the VLT, to be sited near Cerro Tololo, at Cerro Paranal) will have an effective diameter of 16-m. It now appears that the current major limitation on telescope size is the availability of funding rather than the laws of physics.

HIGH-RESOLUTION ASTRONOMY

Atmospheric Blurring

✓3 Even large telescopes have their limitations. For example, consider again the 5-m optical telescope on Palomar Mountain. According to our earlier discussion of diffraction, this telescope should achieve an angular resolution of around 0.02″. In practice, it cannot exceed 1″. In fact, apart from some special techniques developed to examine some bright stars, *no* ground-based optical telescope built before 1990 can presently resolve astronomical objects to much better than an arc second. Why?

As we observe a star, atmospheric turbulence produces continual small changes in the optical properties of the air between the star and our telescope. The light from

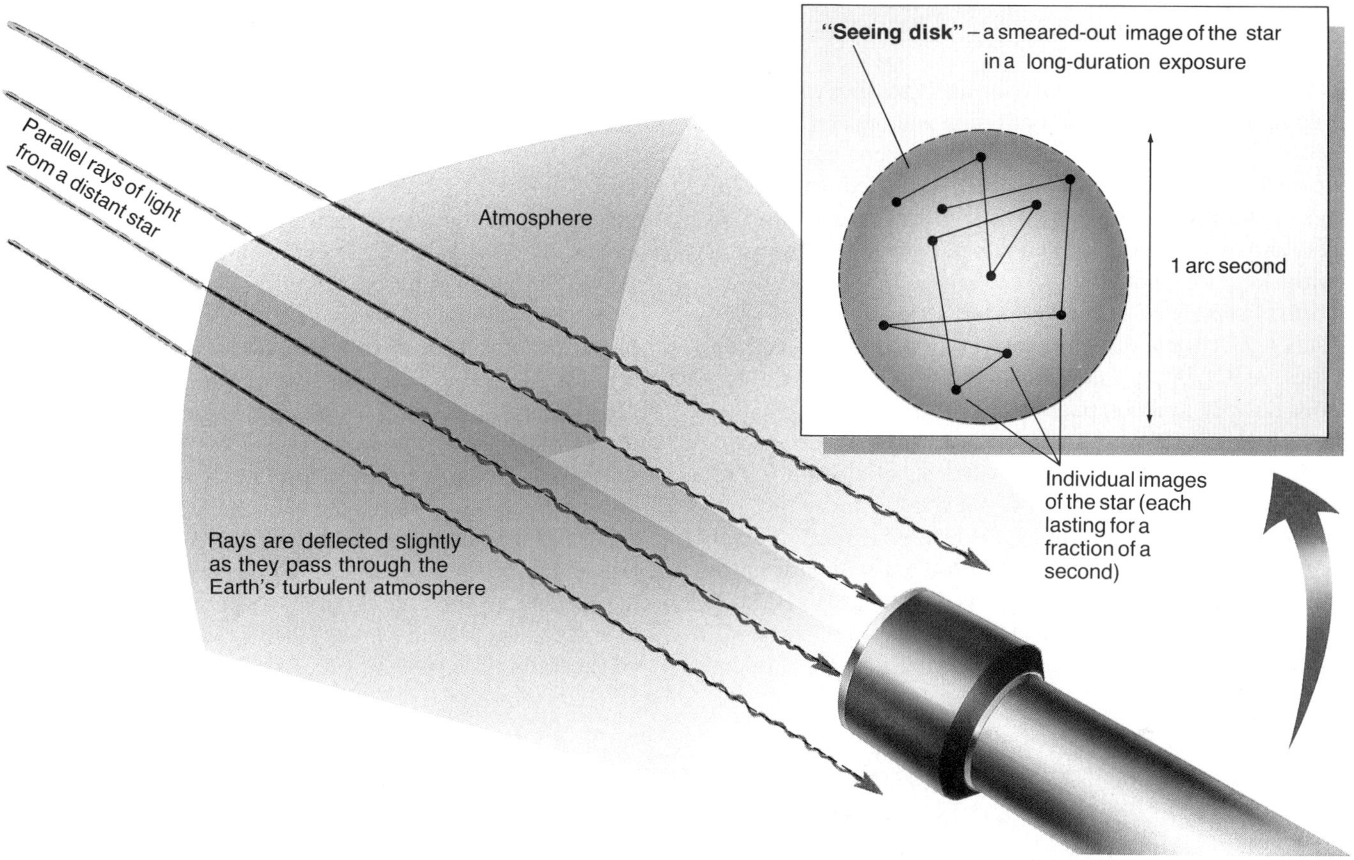

Figure 6.16 *Individual photons from a distant star strike the detector in a telescope at slightly different locations because of turbulence in the Earth's atmosphere. Over time, the individual strikes cover a roughly circular region on the detector, so that even the pointlike image of a star is recorded as a small disk, called the seeing disk.*

the star is deflected slightly, so that the stellar image dances around on the detector (or on our retina). This continual deflection is the cause of the well-known ''twinkling'' of stars. It occurs for the same reason that objects appear to shimmer when viewed across a hot roadway on a summer day.

On a good night, at the best observing sites, the maximum amount of deflection produced by the atmosphere is slightly less than 1″. After a few minutes' exposure time (long enough for the intervening atmosphere to have undergone many small, random changes), the original point of light from the star has been smeared out over a roughly circular region an arc second or so in diameter. Astronomers use the term **seeing** to describe the effects of atmospheric turbulence. The circle over which a star's light (or the light from any other astronomical source) is spread is called the **seeing disk.** Figure 6.16 illustrates the formation of the seeing disk for a small telescope. (For a large instrument—more than about 1 m in diameter—the situation is more complicated because rays striking different parts of the mirror have actually passed through different turbulent atmospheric regions. The end result is still a seeing disk.)

In order to achieve the best possible seeing, telescopes are sited on mountaintops (to get above as much of the atmosphere as possible), in regions of the world where the atmosphere is known to be fairly stable and relatively free of dust, moisture, and light pollution from cities. In the continental United States, these sites tend to be in the desert Southwest. Seeing of 1″ from such a location is regarded as good, and seeing of a few arc seconds is tolerable for many purposes. Even better conditions are found on Mauna Kea, Hawaii, and at Cerro Tololo, Chile—that is why many current large telescopes are located in one of those two sites.

An optical telescope in orbit about the Earth or placed on the Moon could obviously overcome the limitations imposed by the atmosphere on ground-based instruments. Without atmospheric blurring, extremely fine resolution—close to the diffraction limit—can be achieved, subject only to the engineering restrictions of building large structures in space. The *Hubble Space Telescope (HST)*, named after one of America's most notable astronomers, Edwin Hubble, was launched into Earth orbit by NASA's space shuttle *Discovery* in 1990. This telescope has a 2.4-m mirror, with a diffraction limit of only 0.05″. Thus, this orbiting observatory can give us a view of the universe as much as 20 times sharper than is normally available from even the much larger ground-based instruments. For more information on *HST*, read Interlude 6-1.

Image Processing

Nowadays, computers are playing an increasingly important role in the process of light collection and image formation. Most large telescopes are controlled by computers or by operators who rely heavily on their assistance, and the images themselves are recorded in a form that can be easily read and manipulated by computer programs. It is becoming fairly rare for photographic equipment to be used as the primary means of data acquisition at large observatories. Rather, electronic detectors known as **charge-coupled devices,** or **CCDs,** are now in widespread use. Their output goes directly to a computer and is stored on magnetic disk or tape.

A CCD (see Figure 6.17) consists of a wafer of silicon divided into a two-dimensional array of many tiny picture elements, known as **pixels.** When light strikes a pixel, an electron is released, and an electric charge builds up on the device. The amount of charge is directly proportional to the number of photons striking each pixel—in other words, to the intensity of the light at that point. The charge buildup is monitored electronically, so that a two-dimensional image can be obtained. The entire device is typically only 1 cm^2 in area and may contain as many as 1 million pixels, generally arranged in a square grid. As the technology improves, both the areas of CCDs and the number of pixels they contain are steadily increasing. Incidentally, the technology is not limited to astronomy—many home video cameras are based on CCD chips quite similar to those in use at the great astronomical observatories of the world.

Charge-coupled devices have many advantages over photographic plates, the staple of astronomers for over a century. CCDs are much more efficient than photographs. They record as many as 75 percent of the photons striking them, while photographic methods record less than 5 percent. This fact alone means that a CCD image can show objects 10 to 20 times fainter than can a photograph taken using the same telescope and the same exposure time. Alternatively, CCDs can record the same level of detail in less than a tenth of the time required by photographs, or it can record that detail with a much smaller telescope—important considerations, given the fierce competition for access to

INTERLUDE 6-1 *The Hubble Space Telescope*

The *Hubble Space Telescope (HST)* is the largest, most complex, and most sensitive observatory ever deployed in space. At $1.5 billion, it is also the most expensive scientific instrument ever constructed. Built jointly by NASA and the European Space Agency, *HST* is designed to allow astronomers to probe the universe with at least 10 times finer resolution and with some 50 times greater sensitivity to light than existing devices on the ground. It is operated remotely from the ground. There are no astronauts aboard the telescope, which orbits around the Earth about once every 95 minutes.

The telescope's overall dimensions approximate those of a city bus or railroad tank car—13 m (43 feet) long, 12 m (39 feet) across with solar arrays extended, and 11,000 kg (12.5 tons when weighed on the ground). The heart of *HST* is a 2.4-m (94.5-inch) diameter mirror designed to capture optical, ultraviolet, and infrared radiation before it reaches Earth's murky atmosphere. The accompanying figure shows the telescope being lifted out of the cargo bay of the space shuttle *Discovery* in the spring of 1990.

The optical system and scientific instruments aboard *HST* are compact and pioneering. The telescope reflects light from its large mirror back to a smaller, 0.3-m (12-inch) secondary mirror, which in turn sends the light through a hole in the doughnut-shaped main mirror and on into the aft bay of the spacecraft. There, any of six major scientific instruments wait to analyze the incoming radiation. Most of these instruments are about the size of a telephone booth. They include two cameras to image (or electronically photograph) various regions of the sky, two spectrographs to split the radiation into its component colors, a photometer to study the intensity of light, and a group of fine guidance sensors to measure the positions of stars in the sky.

Although not the largest ever built, *HST*'s mirror is assuredly the most finely polished mirror of its size. "Gee-whiz" statistics abound to describe its optical characteristics, one of which is this: If *HST*'s mirror were scaled up to equal the width of the continental United States, the highest hill or lowest valley would be less than 2 inches from the average surface. By contrast, skyscraper-sized imperfec-

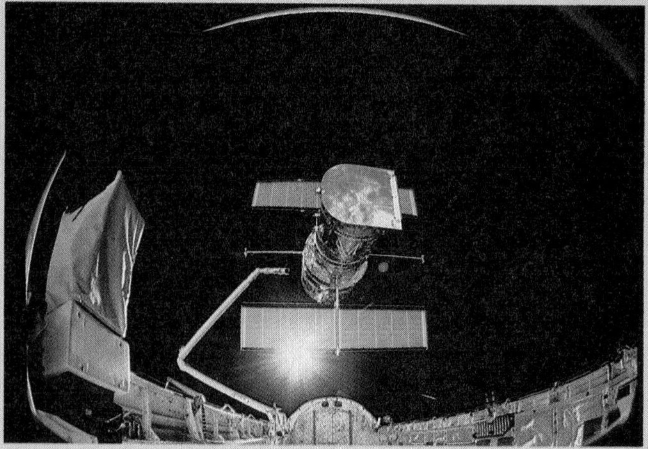

(NASA)

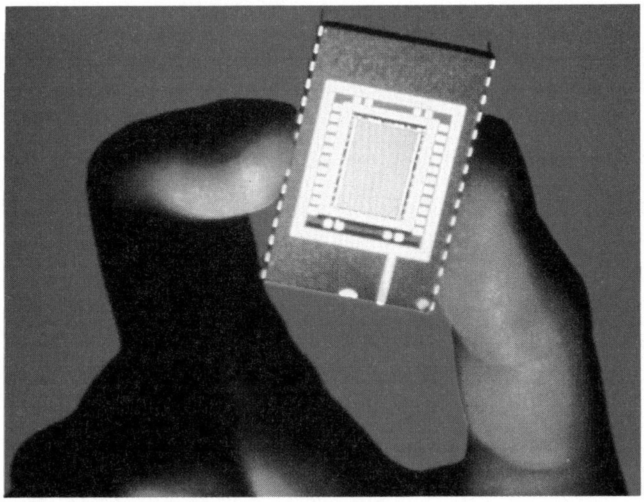

Figure 6.17 *A CCD chip. A charge-coupled device consists of hundreds of thousands, or even millions, of tiny light-sensitive cells, or pixels, usually arranged in a square array. Light striking a cell causes electrical charge to build up on it. By electronically reading out the charge on each pixel, a computer can reconstruct the pattern of light—the image—falling on the chip. (Harvard-Smithsonian Center for Astrophysics)*

large telescopes. CCDs also produce a faithful representation of an image in a digital format that can be stored on magnetic tape, stored on optical disk, or even sent directly across a computer network to an observer's home institution for detailed study.

With the aid of high-speed computers, much of the background noise found in the ''raw'' image from a telescope can be eliminated, allowing astronomers to see features that would otherwise remain hidden. Noise is anything that corrupts the integrity of a message, such as static on an AM radio or ''snow'' on a television screen. It has many causes. In part, it results from faint, unresolved sources in the telescope field of view and from light scattered into the line of sight by the Earth's atmosphere. It can also be caused by electronic ''hiss'' within the detector itself. Whatever its origin, its characteristics can be determined (for example, by observing a part of the sky where there are no known sources of radiation), and, once known, the noise can be removed.

It is also possible to compensate for known instrumental defects and even partially to compensate for the ef-

tions would result if ordinary eyeglass lenses were scaled up to reach from coast to coast. Unfortunately, soon after launch, astronomers discovered that the mirror had been polished to the wrong shape. The mirror is too flat by 2 μm, or about 1/50 the width of a human hair. Even though the above statements remain true—it *is* the smoothest mirror ever made—its imperfect shape makes it impossible to focus all of the captured light as well as expected. This is an optical flaw known as *spherical aberration* and is not easy to fix, given that *HST* orbits 610 km (380 miles) above us. The net result is that *HST* is not as sensitive as designed, but it can still see many objects in the universe with unprecedented resolution.

Despite the telescope's optical impairment, it is now sending us some remarkable new data, many examples of which appear in this book. Even the very first image taken about a month after launch illustrates how much better *HST* can resolve stars than can any telescope on the ground. The double-framed illustration at right shows on the left the kind of excellent (1″) seeing obtainable at the best ground-based observatories; the few stars shown are part of the star cluster NGC 3532, a region about 1300 light-years away in the Southern Hemisphere. On the right is *HST*'s (0.5″) image of the same group of stars, clearly showing better resolution; the two stars at the upper right are clearly distinguishable in the *HST* image, though those same stars in the ground-based image are merged together into an oval blob. Even sharper images (0.1″) are possible with *HST* by using sophisticated computer analysis, as shown in the small insert (see also Figure 6.18[b]).

NASA intends to mount a repair mission to visit *HST*

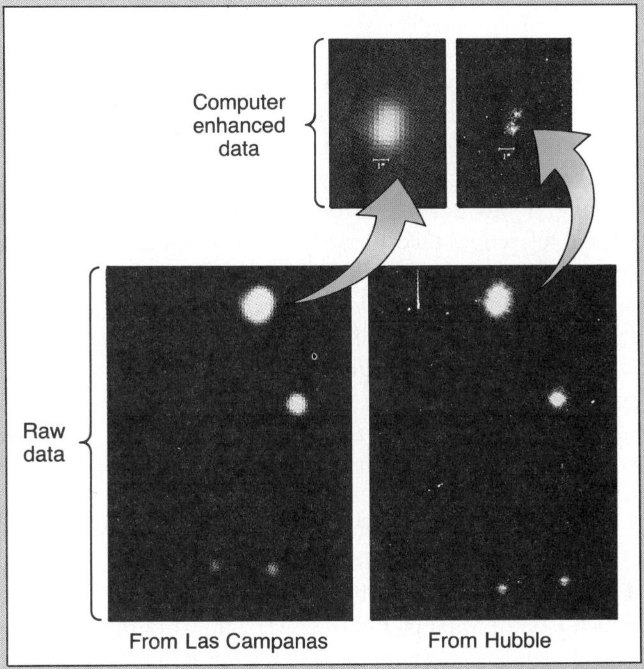

(NASA) R I V U X G

sometime in the mid-1990s. Space-suited astronauts should be able to fix several devices that are now not working properly—such as its gyroscopes and solar arrays—and perhaps install a kind of ''monocle,'' an instrument that should compensate for the telescope's aberration. Such a repair mission, if successful, should enable astronomers to see the faintest objects in the universe with the finest clarity ever.

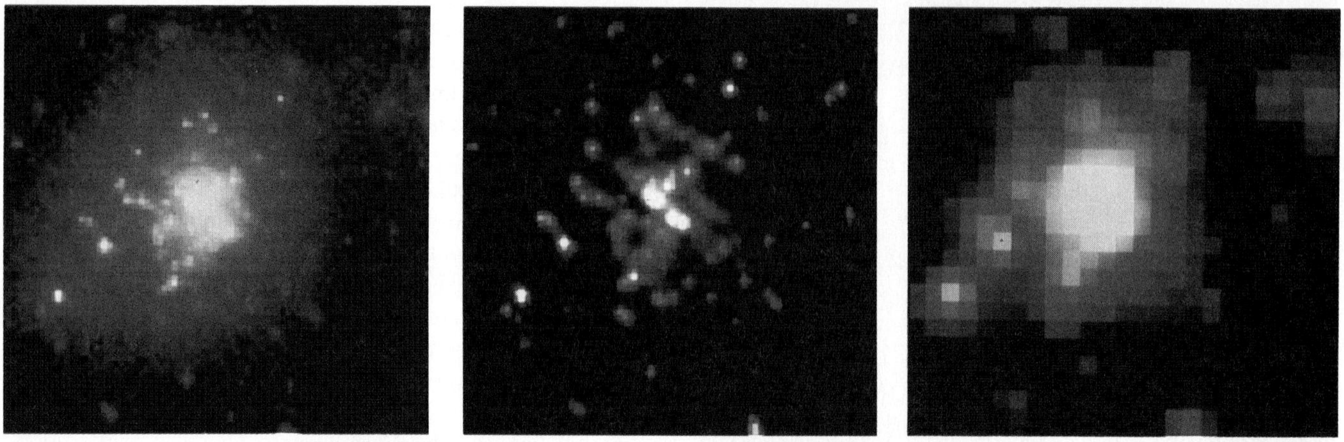

(a) R I V U X G (b) R I V U X G (c) R I V U X G

Figure 6.18 *Three images of the star cluster R136 that lies in the Large Magellanic Cloud, a nearby galaxy. (a) The "raw" image of a group of stars as seen by the space-based Hubble Space Telescope. (b) The same image seen in (a) after computerized compensation for imperfections in the mirror. (c) The best view of the same area ever obtained by a ground-based telescope equipped with a CCD camera. (NASA, ESO)*

fects of bad seeing. In addition, the computer can often carry out many of the relatively simple, but tedious and time-consuming, chores that must be performed before an image (or spectrum) reaches its final "clean" form. Figure 6.18 (a) and (b) illustrates how computerized image-processing techniques have been used to correct for known instrumental problems in the *Hubble Space Telescope,* allowing most of the planned resolution of the telescope to be recovered. For comparison, Figure 6.18(c) shows the best image obtainable from the ground.

An exciting development that promises to bring about striking improvements in the resolution of ground-based optical telescopes takes these ideas of computer control and image processing one stage further. If an image could be analyzed while the light was still being collected (a process that can take many minutes, or even hours, in some cases), it might be possible to adjust the telescope from moment to moment to correct for the effects of mirror distortion, temperature changes, and bad seeing. Perhaps the telescope could come close to the theoretical (diffraction-limited) resolution.

Some of these techniques, collectively known as **active optics,** are already in use in ESO's New Technology Telescope (NTT—not to be confused with NOAO's NNTT, which is still in the planning stage), shown in Figure 6.19. This 3.5-m telescope, employing the latest in real-time telescope controls, achieves resolution of about 0.5″ by making minute modifications to the overall configuration of the instrument as its temperature and orientation change, in order to maintain the best possible focus at all times. From its very first observing run, NTT became the highest-resolution optical telescope on Earth. The Keck 10-m instruments and the 8-m telescopes being built by NOAO will also employ these methods and may achieve resolution as fine as 0.25″.

An even more ambitious undertaking is known as **adaptive optics.** This technique would actually deform the shape of the mirror's surface, under computer control, while the image is being exposed, with the intent of undoing the effects of atmospheric turbulence. Lasers would probe the atmosphere above the telescope, and computers would continuously monitor the appearance of known reference stars, returning information that would allow another com-

Figure 6.19 *The New Technology Telescope, sited on La Silla, in Chile. The use of state-of-the-art active optics in this 3.5-m instrument has allowed it to achieve angular resolution of about 0.5″. (European Southern Observatory)*

puter to modify the mirror thousands of times per second to compensate for the air's swirling motion. Adaptive optics presents formidable theoretical and technological problems, but the rewards are so great that they are presently the subject of intense research. Recently declassified SDI ("Star Wars") technology has provided an enormous boost to this effort. In the next decade, it may well be possible to have the "best of both worlds" and combine the large size of a ground-based telescope with resolution presently achievable only from space.

RADIO ASTRONOMY

✓4 In addition to the visible radiation that normally penetrates Earth's atmosphere on a clear day, radio radiation also reaches the ground. In fact, the radio window in the electromagnetic spectrum is much wider than the optical window, as we noted in Chapter 4. The atmosphere is no hindrance to long-wavelength radiation, and radio astronomers have built many ground-based **radio telescopes** capable of detecting cosmic radio waves. These devices have all been constructed since the 1950s—radio astronomy is a much younger subject than optical astronomy.

The field originated with the work of Karl Jansky at Bell Labs in 1931, but only after the technological push of World War II did it grow into a distinct branch of astronomy. Jansky was engaged in a study of shortwave radio interference when he discovered a faint static "hiss" that had no apparent terrestrial source. He noticed that the strength of the hiss varied in time and that its peak occurred about 4 minutes earlier each day. He soon realized that the peaks were coming exactly one *sidereal day* apart, and correctly inferred that the hiss was not of terrestrial origin but came from a definite direction in space. That direction is now known to correspond to the center of our Galaxy. It took over a decade, and the realization by astronomers that interstellar gas could actually be observed in the radio, for the full importance of his work to be appreciated, but today Jansky is widely regarded as the father of radio astronomy.

Essentials of Radio Telescopes

Figure 6.20(a) shows a fairly typical radio telescope, the large 43-m (140-foot) diameter telescope located at the National Radio Astronomy Observatory in West Virginia. We will use it to illustrate some of the key elements of radio astronomy. Figure 6.20(b) diagrams its basic components.

Although much larger than any reflecting optical telescope, most radio telescopes are built in basically the same way. They have a large horseshoe-shaped mount that supports the large metal curved "dish," or mirror. The collecting area captures cosmic radio waves and reflects them to the focus, where a receiver detects the signals and channels them to the computer. The signals may be partially analyzed before being stored on magnetic tape for later processing and full analysis.

Unlike optical telescopes, which can simultaneously detect all visible frequencies, radio telescopes normally register radiation only within a narrow band of frequencies. To detect radiation at another radio frequency, we must retune the equipment, much as we tune a television set to a different channel.

Large radio telescopes are very sensitive instruments and can detect even very faint radio sources. Indeed, the *total* amount of radio energy detected by *all* the radio telescopes on Earth since Jansky built his first receiver amounts to little more than 1 erg! However, their angular resolution is generally poor compared to their optical counterparts. Indeed, *the* major disadvantage of all radio telescopes is their relatively low resolving power, despite the enormous size of many radio dishes. It is not our atmosphere that is to blame—the radio wavelengths normally studied pass through air without any significant distortion. The problem is that the typical wavelength of radio waves is roughly six orders of magnitude (that is, about a million times) larger than those of light, and these longer wavelengths impose a corresponding crudeness in angular resolution because of the effects of diffraction.

The best angular resolution obtainable with a single radio telescope is about 10″ (for the largest instruments operating at millimeter wavelengths), at least an order of magnitude coarser than the sub arc-second capabilities of the largest optical mirrors. Unlike optical systems, however, astronomers can often use radio telescopes at many different wavelengths. Their resolution varies widely, depending on the wavelength being observed. The 43-m radio telescope shown in Figure 6.20 can achieve resolution of about 1′ when observing radio waves with wavelengths of around 1 cm. However, it was designed to operate most efficiently at wavelengths closer to 5 cm, where the resolution is only about 6′, or 0.1°. Despite these limitations, the study of cosmic radio radiation provides us with a wealth of useful information about the universe, especially about the "invisible" universe, which is how astronomers refer to that part of the universe that emits radiation at nonvisible frequencies.

Radio telescopes are large not only because that is the only way they can achieve good resolution, but also because the total amount of energy arriving at the Earth in the form of radio radiation is extremely small—less than a trillionth of a watt spread over the entire surface of our planet. Radio telescopes can be built so much larger than their optical counterparts because their reflecting surface need not be as smooth as is needed for shorter-wavelength light waves. Provided that surface irregularities (dents, bumps, and the like) are much smaller than the wavelength of the waves to be detected, the surface will reflect them without distortion. Because the wavelength of visible radiation is small (approximately 10^{-4} cm), very smooth mirrors are needed to reflect the waves properly, and it is difficult to construct

(a)

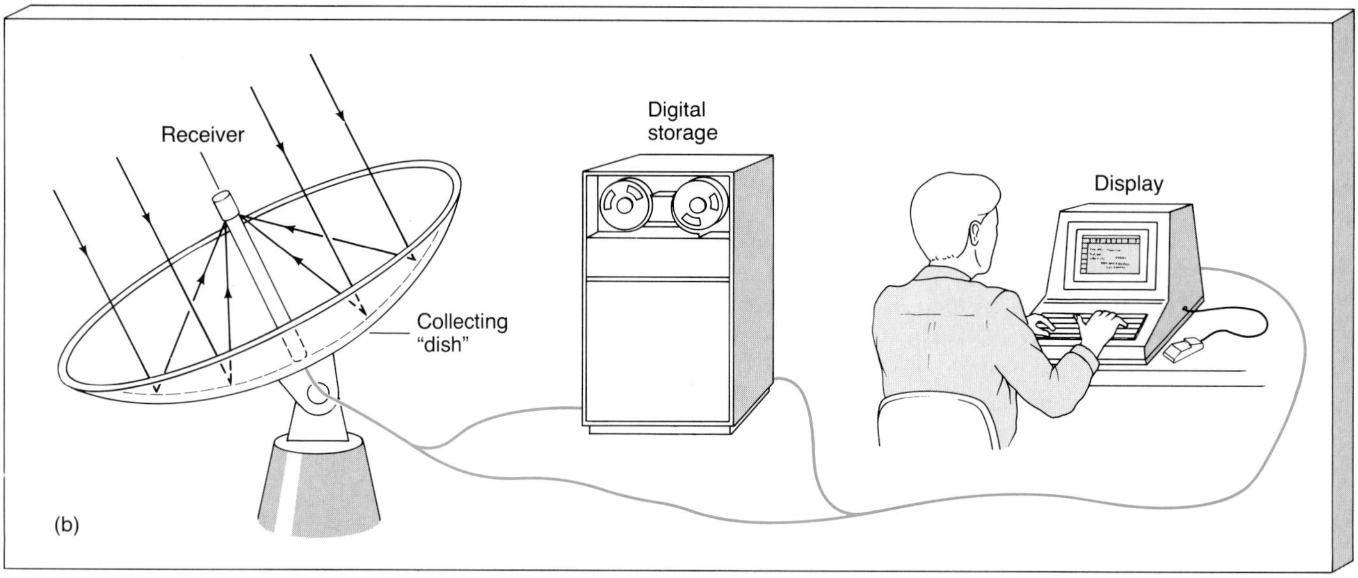

(b)

Figure 6.20 *(a) Photograph of the 43-m diameter radio telescope at the National Radio Astronomy Observatory in Green Bank, West Virginia. (b) Diagram of the basic radio receiver and data processing devices used at all radio telescopes. (NRAO)*

very large mirrors to such exacting tolerances. Radio waves are much longer, however, and can often be reflected well even from quite rough surfaces.

For example, when visible light shines on a curved surface having irregularities of about a millimeter, the light scatters and does not focus well; the image of the light source is severely distorted. You could test this statement by looking at your own blurred reflection in a piece of unpolished metal. But radio waves of a centimeter or longer wavelength are not scattered at all by slightly rough surfaces. Instead, they are reflected to an accurate focus. Very long radio waves of, say, 1-m wavelength can reflect perfectly well from surfaces having irregularities even as large as your fist.

The situation is somewhat analogous to trying to bounce ("reflect") a ball ("photon") off an irregular surface ("mirror"). If the surface irregularities are much smaller than the radius of the ball (its "wavelength," in this analogy), the bounce will be true, and the ball will travel in the intended direction (that is, toward a focal point). However, if the irregularities are comparable in size to the radius of the ball, the bounce will be unpredictable and erratic (the focus will be blurred).

Figure 6.21 shows the world's largest radio telescope, located in Arecibo, Puerto Rico. Approximately 300-m (1000 feet) in diameter, the surface of the Arecibo telescope spans nearly 20 acres. Constructed in 1963 in a natural depression in the hillside, the dish was originally surfaced

Figure 6.21 An aerial photograph of the 300-m diameter dish at the National Astronomy and Ionospheric Center near Arecibo, Puerto Rico. The receivers that detect the focused radiation are suspended nearly 300 m (about 80 stories) above the center of the dish. (NAIC)

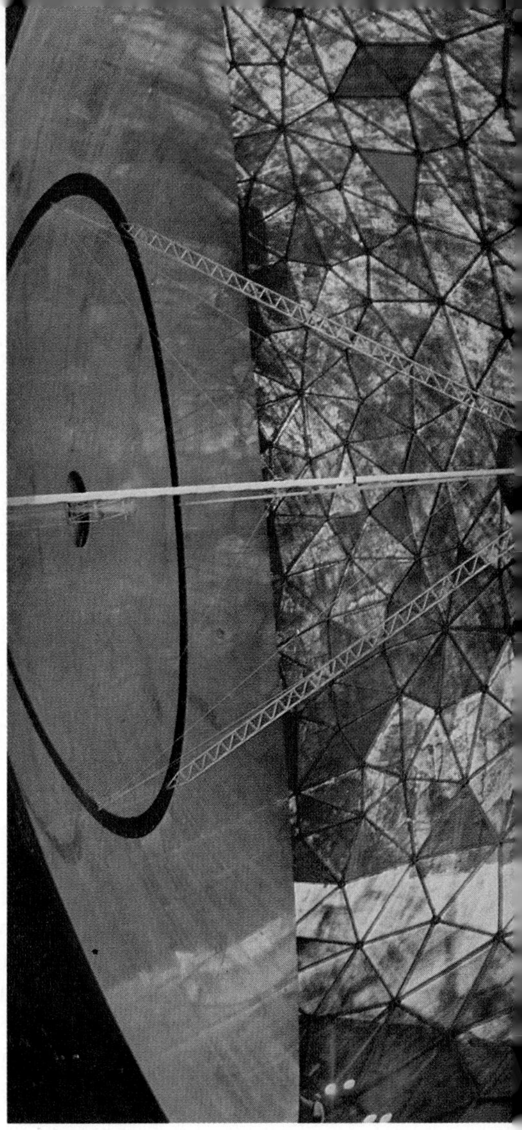

Figure 6.22 Photograph of the Haystack dish, taken from inside the radome. For scale, note the engineer at the bottom. Also note the dull shine on the telescope surface, which indicates its smooth construction. Haystack is a poor optical mirror but a superb radio telescope. Accordingly, it can be used to reflect and accurately focus radiation having short radio wavelengths, even as small as a fraction of a centimeter. (MIT)

with chicken wire, which was lightweight and cheap. Although fairly rough, the chicken wire was adequate for proper reflection because the openings between adjacent strands of wire were much smaller than the long-wavelength radio waves to be detected.

The entire Arecibo dish was resurfaced in 1974 with thin metal plates, so it can now be used to study shorter-wavelength radio radiation. Even so, useful observations are still restricted to radiation of wavelength greater than about 10 cm, making the telescope's angular resolution no better than that of some smaller radio telescopes, despite its enormous size. The huge size of the dish, in fact, creates one distinct disadvantage: The Arecibo telescope cannot be pointed very well to follow cosmic objects across the sky. The dish is literally strung among several limestone hills, restricting its observations to those objects passing nearly overhead.

Arecibo is an example of a roughly surfaced telescope capable of detecting long-wavelength radio radiation. At the other extreme, Figure 6.22 shows the 36-m diameter Haystack dish in northeastern Massachusetts. It is constructed of polished aluminum and maintains a parabolic curve to an accuracy of about a millimeter all the way across its solid surface. It can reflect and accurately focus radio radiation with a wavelength as short as a few millimeters.

The Haystack telescope is contained within the protective shell, or radome, which protects the surface from the harsh wind and weather of New England. It acts much like the protective dome of an optical telescope, except that there is no slit through which the telescope "sees." Incoming cosmic radio signals pass virtually unimpeded through the radome's fiberglass construction.

The Value of Radio Astronomy

Despite the inherent disadvantage of relatively poor angular resolution, radio astronomy enjoys many advantages. Radio telescopes can observe 24 hours a day—darkness is not needed for receiving radio signals. The reason for this is simply that the Sun is a relatively weak source of radio energy, so its emission does not swamp radio signals from elsewhere. Astronomers can often make radio observations through cloudy skies, and radio telescopes can detect the longest-wavelength radio waves even during rain or snow storms. Poor weather causes few problems because the wavelength of most radio waves is much larger than the

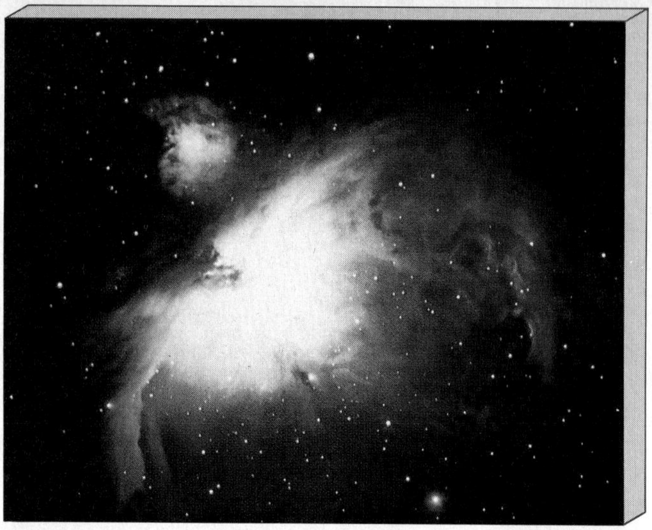

Figure 6.23 *Optical photograph of the Orion Nebula, a star-forming region some 1500 light years distant. The bright regions in this photograph are stars and clouds of glowing gas. The dark regions are not empty but are simply obscured by interstellar matter. The nebula can be seen in Figure 1.6 of the hunter's sword in the constellation Orion. (AURA)*

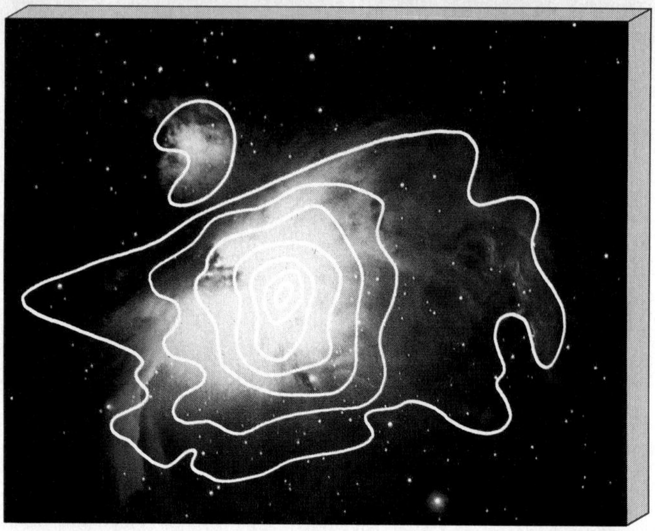

Figure 6.24 *Superposition of a radio contour map onto the optical photograph of Orion in Figure 6.23. The resolution of the optical image is about 1 arc second, that of the radio map 1 arc minute (AURA, MIT)*

typical size of atmospheric raindrops or snowflakes. Optical astronomy cannot be done under these conditions because the wavelength of visible light is smaller than a raindrop, a snowflake, and even a minute water droplet in a cloud.

However, perhaps the greatest value of radio astronomy (and, in fact, all other invisible astronomies) is that it opens up a whole new window on the universe. Objects that are bright in the optical region of the spectrum are not necessarily strong radio emitters, and very often, strong radio sources are completely undetectable in visible wavelengths. Thus, radio observations do not just afford us the opportunity of studying the same objects at different wavelengths. They have also allowed us to see whole new classes of objects that were hitherto completely unknown.

For example, Figure 6.23 shows an optical photograph of the Orion Nebula taken with the 4-m telescope on Kitt Peak. Figure 6.24 shows a radio contour map of the same region superimposed on the optical image. By aiming a radio telescope at the nebula, radio astronomers determine the strength of its radio emission. Scanning back and forth across the nebula and taking many measurements, the astronomers construct a radio map of the entire region. The map is drawn as a series of contour lines connecting locations of equal radio brightness. These radio contours are similar to pressure contours drawn by meteorologists on weather maps and height contours drawn by cartographers on geographic maps. The inner contours usually represent relatively strong radio signals, the outside contours weak signals.

The radio map shown in Figure 6.24 has many similarities to the visible image of the nebula—the radio emission is clearly strongest near the center of the optical image and declines toward the nebular edge. But there are also some subtle differences between the radio and optical images. The two maps differ mainly toward the upper left of the main cloud, where visible light seems to be absent, despite the existence of radio waves. How can radio waves be detected from locations not showing any light emission? In this particular case, the nebular region is known to be especially dusty in its top-left quadrant. The dust obscures the short-wavelength visible radiation, but not the long-wavelength radio radiation. We will study the nature of this dust in more detail in Chapter 20.

Thus, our radio map allows us to see the true extent of this cosmic source. Optical images are often distorted by intervening dust somewhere along our line of sight. In fact, many important objects and regions of the universe cannot be seen at all by optical astronomy; the very center of our Milky Way Galaxy is a prime example of a totally invisible region. Our knowledge of such regions results almost entirely from analyses of their longer-wavelength radio and infrared emissions.

We should mention one more point. Some people have the mistaken impression that radio telescopes transmit signals toward the stars, like radar. They don't. Radio telescopes normally do *not* transmit radio waves toward astronomical objects, any more than optical telescopes observe the stars with searchlight beams. The only exception to this statement is the use of radar and laser ranging to study the Moon and some of the inner planets of the solar system.

INTERFEROMETRY

The main disadvantage of radio astronomy compared with optical work is its lack of good angular resolution. Not to be outdone, however, radio astronomers have invented ways to overcome this problem. By using a technique known as **interferometry,** the angular resolution of some radio maps can be enormously improved. In fact, using interferometry,

Figure 6.25 *This large interferometer is made up of 27 separate dishes spread along a* Y*-shaped pattern about 30 km across on the Plain of San Augustin in New Mexico. The most sensitive radio device in the world, it is called the Very Large Array, or VLA for short. (NRAO)*

it is actually possible to produce radio images of much higher angular resolution than can be achieved with even the best optical telescopes.

In interferometry, two or more radio telescopes are used in tandem to observe the *same* object at the *same* wavelength and at the *same* time. The combined apparatus is called an **interferometer.** Figure 6.25 shows such an instrument—several separate radio telescopes working together as a team. By means of electronic cables or radio links, the signals each antenna receives are sent to a central computer that analyzes how the waves interfere with each other when added together. If the detected waves are in step when added, they combine positively to form a strong radio signal. If the signals are not in step, they destructively interfere and cancel each other. As the antennas track their target, a pattern of peaks and troughs emerges, which, after extensive computer processing, translates into an image of the observed object.

An interferometer is essentially a substitute for a single huge antenna. The effective telescope diameter of an interferometer equals the distance between its outermost dishes. In other words, two small dishes can act as opposite ends of an imaginary but huge single radio telescope, dramatically improving the angular resolution. For example, resolution of a few arc seconds can be achieved at typical radio wavelengths (say, 10 cm) either by using a single radio telescope 5 km in diameter (which is quite impossible to build) or by using two or more much smaller dishes *separated* by 5 km and connected electronically.

Large interferometers made up of many dishes, like the instrument shown in Figure 6.25, now routinely attain radio resolution comparable to that of optical images. Figure 6.26 compares an interferometric radio map of a nearby galaxy with a photograph of that same galaxy made using a large optical telescope. The radio clarity is superb and has little of the "fuzziness" or lack of detail evident in the radio map of Figure 6.24.

The larger the distance separating the telescopes—known as the **baseline** of the interferometer—the better the resolution attainable. Astronomers have created radio interferometers spanning very great distances, first across North America and later between continents. A typical very-long-baseline-interferometry experiment (usually known by the acronym VLBI) might use radio telescopes in North America, Europe, Australia, and Russia to achieve angular resolution on the order of 0.001″, about 1000 times better than images produced by most current optical telescopes.

Unlike visible light, radio waves are not deflected appreciably by the Earth's turbulent atmosphere. Put another way, radio "seeing" is always very good, and angular resolution at radio wavelengths is limited only by our apparatus, not by atmospheric turbulence. Radio interferometers could conceivably be constructed using the largest possible baseline on Earth, namely the diameter of the entire planet—something that the American-European-Australian VLBI linkup has almost done already. In fact, in a recent test, radio astronomers successfully used an antenna in orbit, together with several antennas on the ground, to construct an even longer baseline and achieve still better resolution. It seems that even Earth's diameter is no limit. Perhaps not surprisingly, proposals now exist to place interferometers entirely in Earth orbit, and even on the Moon.

Interferometry has its drawbacks, however. For example, interferometers can observe spectral lines only when radiation is rather intense. Their usefulness is restricted to certain parts of the radio domain, and they are less sensitive than single radio dishes when observing smoothly distributed, low-density gas. Furthermore, the politics of arranging international experiments can be quite complex. Accordingly, single radio telescopes are still highly valued instruments.

Before leaving the topic of interferometry, it is important to note that the technique is no longer restricted to the radio domain. Radio interferometry became feasible when electronic equipment and computers achieved speeds great enough to combine and analyze radio signals from separate radio detectors without loss of data. As the technology has improved, it has become possible to apply the same methods to higher-frequency radiation. Millimeter-wave interferometry has already become an established and important observational technique, and it is very likely that infrared interferometry will become commonplace in the coming

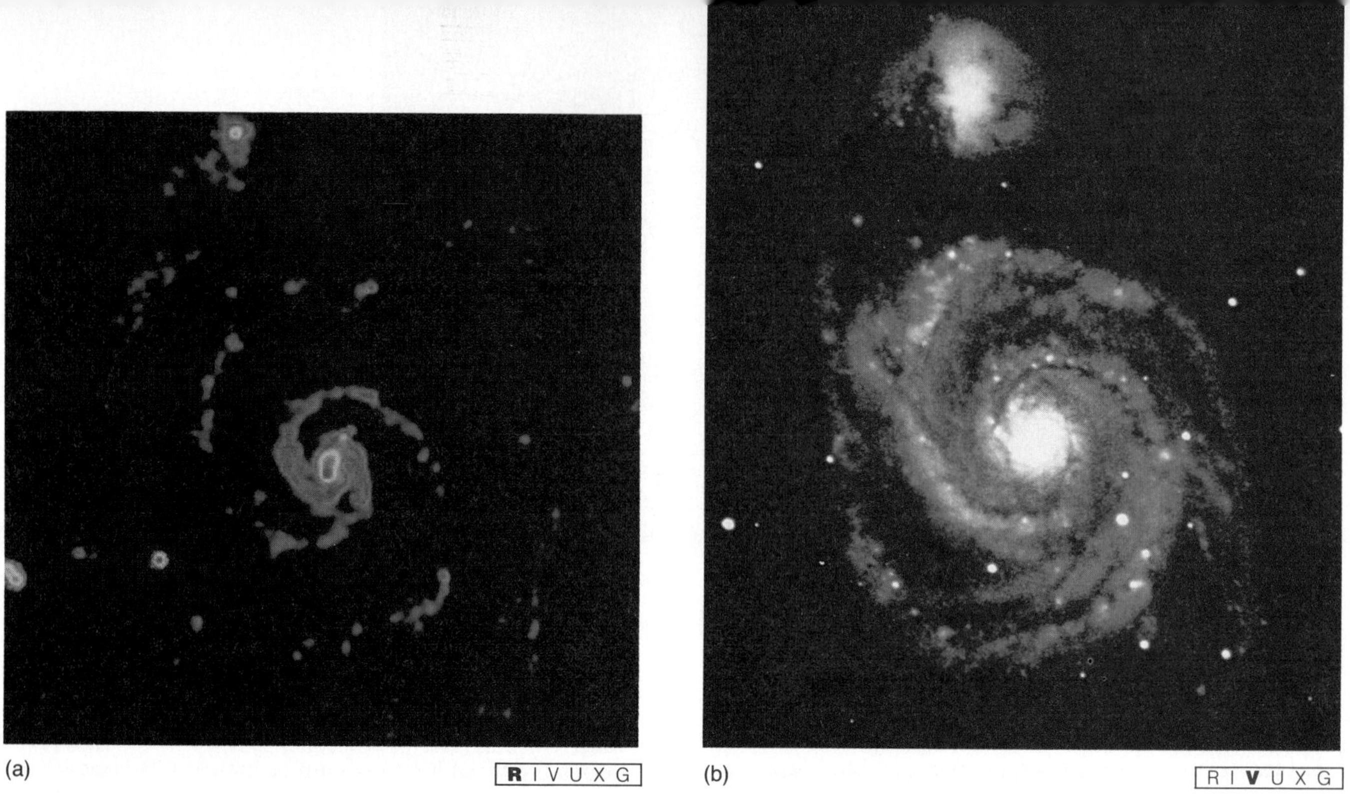

Figure 6.26 *VLA radio "photograph" (or radiograph) of a spiral galaxy observed at radio frequencies with an angular resolution of a few arc seconds (a) shows nearly as much detail as an actual (light) photograph of that same galaxy (b) made with the 4-m Kitt Peak optical telescope. (NRAO; AURA)*

few years. Interferometry is not yet widely used in optical work, because of the technical difficulties involved, but that too is the subject of intensive research.

OTHER ASTRONOMIES

✓5 The electromagnetic spectrum consists of far more than just visible light and radio waves. Although optical and radio astronomy are the oldest and best-established branches of astronomy, since the 1970s there has been a virtual explosion of observational techniques covering the many other types of electromagnetic radiation. Today, all portions of the spectrum are studied, from radio waves to gamma rays, in order to maximize the amount of information available about astronomical objects. Because of the transmission characteristics of the Earth's atmosphere, astronomers must study most wavelengths other than optical and radio from space. The rise of these "other astronomies" has therefore been closely tied to the development of the space program.

Infrared Astronomy

Infrared studies are an important component of modern observational astronomy. Although most infrared radiation is absorbed by the atmosphere (primarily by water vapor), there are still a few windows in the high-frequency part of the infrared spectrum (see Figure 4.9), where the opacity is low enough to allow ground-based observations. Indeed, some of the most useful infrared observing is done from the ground, using conventional telescopes, even though the radiation is somewhat diminished in intensity by our atmosphere.

Figure 6.27 is a photograph of the world's highest ground-based observatory, perched more than 4 km (about 14,000 feet) above sea level on top of an extinct volcano at Mauna Kea, Hawaii. Despite its remoteness, this site draws a full schedule of astronomers throughout the year. The thin air at this high altitude guarantees less atmospheric absorption of incoming radiation, and hence a clearer view, than is possible from sea level. In fact, Mauna Kea is one of the finest locations for ground-based infrared astronomy, but the air is so thin that astronomers must occasionally wear oxygen masks while performing their observations.

Astronomers can make still better infrared observations if they can place their instruments above most or all of the Earth's atmosphere. Improvements in balloon, rocket, and satellite technologies are rapidly making infrared research a powerful tool with which to study the universe. Observations are also made from high-flying aircraft. NASA's *Kuiper Airborne Observatory*, named after one of the founding fathers of the field of infrared astronomy, consists of a 0.9-m telescope carried by a Lockheed C-141 at a height of 12 km. Since the 1970s, its flights have become a matter of routine—perhaps one or two a week—and have provided a wealth of observational data on the far-infrared

Figure 6.27 Photograph of the world's highest-altitude ground-based observatory at Mauna Kea, Hawaii. The domes house the Canada-France-Hawaii 3.6-m telescope (left), the 2.2-m telescope of the University of Hawaii (center), and Britain's 3.8-m infrared facility (right). The foreground device is a 3-m infrared telescope owned and operated by NASA. (Mauna Kea Observatory)

portion (wavelengths from 40 to 100 μm) of the electromagnetic spectrum.

Generally, **infrared telescopes** resemble optical telescopes (indeed, many optical telescopes are also used for infrared work), but their detectors are sensitive to longer-wavelength radiation. As might be expected, the size of the infrared telescopes that can be carried above the atmosphere is necessarily considerably smaller than that of massive ground-based optical instruments.

Figure 6.28 shows how a typical infrared telescope is used. This telescope is part of an unmanned balloon payload usually lofted to an altitude of about 30 km (or 100,000 feet). The 1-m diameter mirror, which can be seen at the base of the payload, is one of the largest infrared telescopes

Figure 6.28 A balloon hoists a payload containing an infrared telescope high above most of Earth's atmosphere. (a) The gondola containing the instrument package; (b) filling the balloon; (c) release of the balloon and gondola; (d) the rise through the Earth's atmosphere; (e) a common end result, even after a successful mission. (Harvard-Smithsonian Center for Astrophysics)

Figure 6.29 *An infrared photograph taken near San Jose, California (left), and an optical photo of the same area taken at the same time (right). Longer wavelength infrared radiation can penetrate smog much better than short-wavelength visible light. (NASA)*

yet to get above *most* of Earth's atmosphere. At typical infrared wavelengths (around 100 μm), such a mirror provides an angular resolution of approximately 1′. Contour maps showing the spread of invisible infrared radiation across cosmic sources often resemble those made with large radio telescopes.

As with radio observations, the longer wavelength of infrared radiation often enables us to perceive objects partially hidden from optical view. As an example of the penetrating properties of infrared radiation, Figure 6.29 shows a dusty and hazy region in California, hardly viewable optically, but easily seen using infrared radiation. Note again, however, that the advantages of using different wavelengths extend far beyond the ability of the radiation to penetrate cosmic (or terrestrial) haze. As we have already noted in the context of radio astronomy, the *types* of astronomical objects that can be observed may differ quite markedly from one wavelength range to another. Thus, full-spectrum coverage is essential, not only to see things more clearly, but even to see some things at all.

The most advanced facility to function in this part of the spectrum is the *Infrared Astronomy Satellite*, called *IRAS* for short and shown in Figure 6.30. Launched into Earth orbit in 1983 but now inoperative, this British-Dutch-U.S. satellite housed a 0.6-m mirror with an angular resolution as fine as 30″ (as usual, depending on the wavelength observed). Its sensitivity was greatest for radiation in the 10 to 100 μm range. During its 10-month lifetime, *IRAS* contributed greatly to our knowledge of clouds of galactic matter that seem destined to become stars (and possibly planets)—regions composed of warm gas that can be neither seen with optical telescopes nor adequately studied

Figure 6.30 *An artist's conception of the* Infrared Astronomy Satellite, *placed in orbit in 1983. This 0.6-m telescope surveyed the infrared sky at wavelengths ranging from 10 to 100 μm. During its 10-month lifetime, it greatly increased astronomers' understanding of many different aspects of the universe, from the formation of stars and planets to the evolution of galaxies. (NASA)*

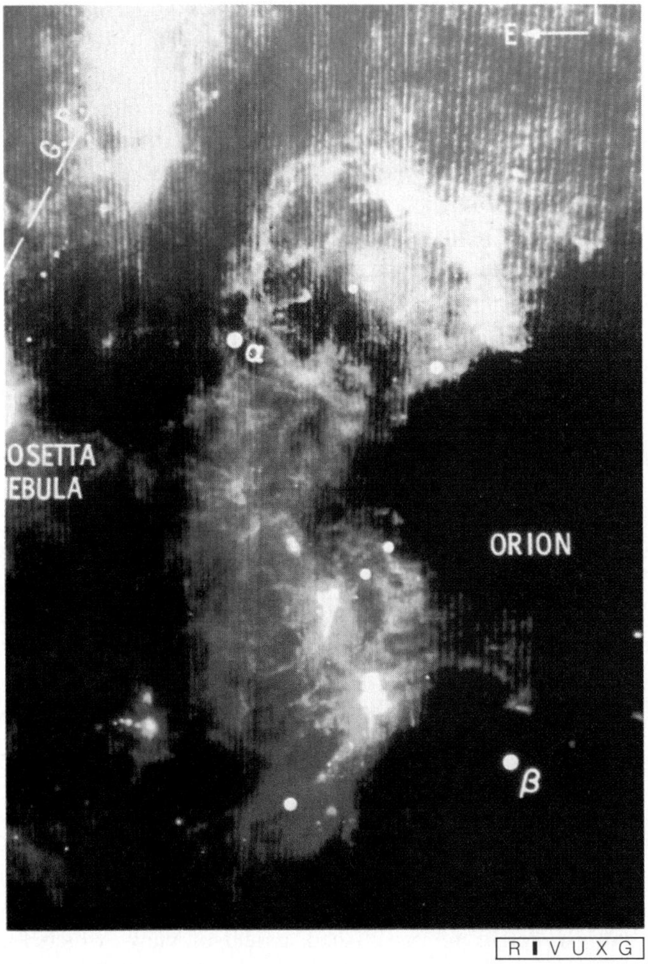

Figure 6.31 This infrared image of the Orion Nebula and its surrounding environment was made by the Infrared Astronomy Satellite. *Here, the brighter regions denote greater strength of infrared radiation; the false colors denote different temperatures. (The scale of this figure is larger than that of Figure 6.24; it is comparable to Figure 1.6.) (NASA)*

with radio telescopes. Throughout the text we will encounter many findings this satellite made about comets, stars, galaxies, and the scattered dust and rocky debris found between the stars. All such objects have some temperature. Consequently, they are sure to ''glow'' in the infrared. Because much of the material between the stars has a temperature between a few tens and a few hundred kelvins, Wien's law (see Chapter 4) tells us that the infrared domain is the natural portion of the spectrum in which to study it.

Figure 6.31 shows an *IRAS* image of the Orion Nebula. At nearly 1′ angular resolution, the fine details of Orion visible in the earlier optical image (Figure 6.23) cannot be perceived. Nonetheless, astronomers can extract useful information about this object and others like it from such observations. For example, clouds of warm dust and gas, believed to play a critical role in the process of star formation, and extensive groups of bright young stars, completely obscured at visible wavelengths, have been seen.

Unfortunately, by Wien's Law, telescopes themselves also radiate strongly in the infrared, unless they are cooled to nearly absolute zero. The end of *IRAS*'s mission came not because of any equipment malfunction or unexpected mishap, but simply because its supply of liquid helium coolant ran out. Its own thermal emission overwhelmed the radiation it was built to detect.

In the 1990s, the European Space Agency (ESA) and NASA both plan to launch successors to *IRAS* into Earth orbit. The first of these, ESA's *Infrared Space Observatory (ISO)*, should be in place by the middle of the decade and will refine and extend the groundbreaking work begun by *IRAS*. NASA's *Space Infrared Telescope Facility (SIRTF)*, a 1-m telescope (which may be deployed in conjunction with the U.S. Space Station *Freedom*), is also expected to become operational later in the decade. It may also have the distinct advantage of being serviceable by astronauts, so that its coolant can be replenished and its lifetime extended far beyond that of either *IRAS* or *ISO*.

Ultraviolet Astronomy

To the short-wavelength side of the visible spectrum lies the ultraviolet domain. This region of the spectrum, extending in wavelength from 4000 Å (blue light) down to a few tens of angstroms (''soft'' x-rays), has only relatively recently begun to be explored. Because Earth's atmosphere is partially opaque shortward of 4000 Å and is totally opaque below about 3000 Å (in part because of the ozone layer), astronomers cannot conduct any useful ultraviolet observations from the ground, not even from the highest mountaintop. Rockets, balloons, or satellites are therefore essential to any **ultraviolet telescope.**

One of the most successful ultraviolet space missions is the *International Ultraviolet Explorer*, called *IUE* for short. This satellite was placed in Earth orbit in 1978 and is still functioning as designed. Like infrared telescopes, its basic appearance and construction are quite similar to optical telescopes. In a truly collaborative effort, several hundred astronomers from all over the world have used *IUE* to explore a variety of phenomena in planets, stars, and galaxies, especially the surprisingly hot gases in stellar atmospheres. In subsequent chapters, we will learn what this relatively new window on the universe has shown us about the activity and even the violence that seems to pervade the cosmos. (For more information on *IUE,* consult Interlude 20-1. The *Hubble Space Telescope*, described in Interlude 6-1, is also a superb ultraviolet instrument.)

An alternative means of placing astronomical payloads into (temporary) Earth orbit is provided by NASA's space shuttle. In December 1990, the shuttle *Columbia* carried aloft the *Astro-1* package of ultraviolet and x-ray telescopes (see Figure 6.32). Such astronomical shuttle missions offer a new and potentially very flexible way for astronomers to obtain access to space, without the long lead times and great expense of permanent satellite missions like *HST* and *IUE*.

(a)

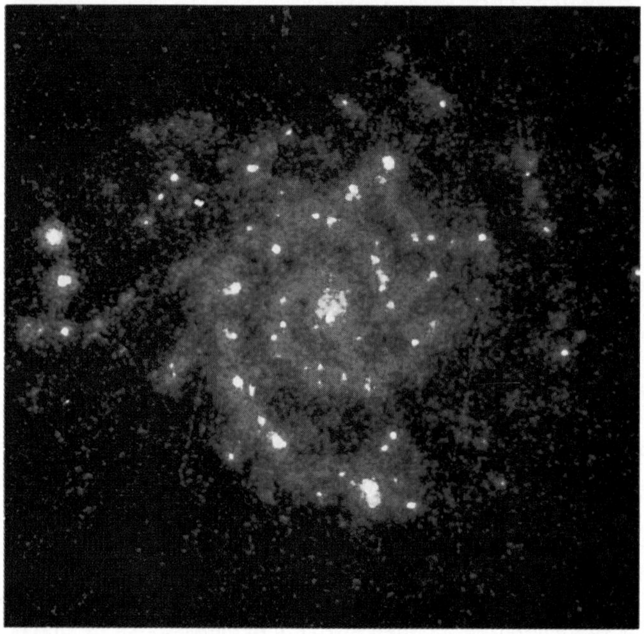

(b) R I V **U** X G

Figure 6.32 *(a) The* Astro-1 *mission was carried by space shuttle* Columbia *in 1990 and performed ultraviolet and x-ray observations from orbit for the 1-week duration of the mission. (b) This image of the spiral galaxy M74 was made by an ultraviolet telescope aboard* Astro-1. *(NASA)*

High-Energy Astronomy

High-energy astronomy studies the universe as it presents itself to us in x-ray and gamma-ray radiation. How do we detect radiation of such short wavelengths? First, it must be captured high above the Earth's atmosphere because none of it reaches the ground. Second, its detection requires the use of equipment that is basically different in design from that used to capture the relatively low-energy radiation studied up to this point.

The basic difference in the design of **high-energy telescopes** comes about because x- and gamma-rays cannot be reflected easily by any kind of surface. Rather, these rays tend to pass through or be absorbed by any material they strike. When x-rays barely graze a surface, however, they can be reflected from it in a way that can be made to form an image, although the mirror design is fairly complex. For gamma rays (with wavelengths less than 0.1 Å), no such method of producing an image has yet been devised—present-day gamma-ray telescopes simply point in a specified direction and count photons received.

In addition, detection methods using photographic plates or CCD devices do not work well. Instead, individual x-ray and gamma-ray photons are counted by on-board electronic detectors, and the results are then transmitted to the ground for further processing and analysis. Furthermore, the number of photons in the universe seems to be inversely related to frequency. Billions of visible (starlight) photons arrive at Earth each second, but hours or even days are often needed for a single gamma-ray photon to be recorded. Not only are these photons hard to focus and measure, they are also few and far between.

Figure 6.33 shows an early instrument designed to detect x-ray radiation. Still in Earth orbit but now inactive owing to lack of battery power, this high-energy telescope is on board the *Small Astronomy Satellite No. 1 (SAS-1)*. The honeycomb structure on the front (looking rather like a car radiator), called a **collimator,** rejects all x-ray photons except those coming nearly straight at the detector. This is how the telescope established a field of view, which for *SAS-1* was a few arc degrees across. No image was formed—the detectors simply registered all photons that reached them within the degree or so defined by the collimator, regardless of their precise direction. However, the satellite's position could be accurately controlled by ground commands, and cosmic sources of x-ray emission were scanned and studied to an accuracy of a few arc minutes.

Toward the end of the 1970s, a new generation of x-ray and gamma-ray telescopes was launched into Earth orbit. Called the *High-Energy Astronomy Observatories*, or *HEAO* for short, these spacecraft have made major advances in our understanding of high-energy phenomena throughout the universe. Having greater accuracy and sensitivity than all earlier high-energy satellites, these spacecraft did for x-ray astronomy what the first large optical and radio telescopes did for longer-wavelength radiation.

Figure 6.34 is a photograph of the *HEAO-2* spacecraft. In 1979, the year when it first came on-line, the satellite was renamed the *Einstein Observatory*, in honor of the birth centenary of the great scientist. This instrument featured a major design breakthrough. Unlike its x-ray–collecting predecessors, *Einstein* could *image* its 1° field of view, instead of simply recording the total number of photons intercepted. The x-rays struck a complicated set of cylindrical mirrors at grazing angles and were reflected to a focus, as depicted in Figure 6.35. Various detectors posi-

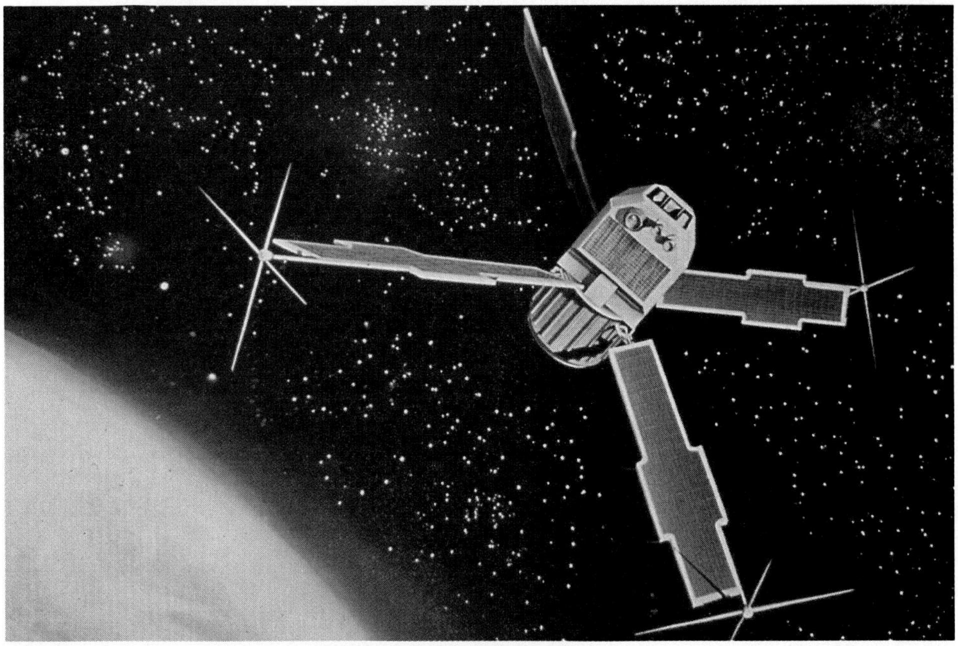

Figure 6.33 *An artist's conception of the* Small Astronomy Satellite No. 1, *placed into orbit by NASA in the early 1970s to study cosmic objects that emit high-energy radiation. Because the rocket launch occurred from a sea-going platform off the coast of Kenya, which was then celebrating its independence, astronomers nicknamed this first x-ray satellite Uhuru—which means "freedom" in Swahili. (NASA)*

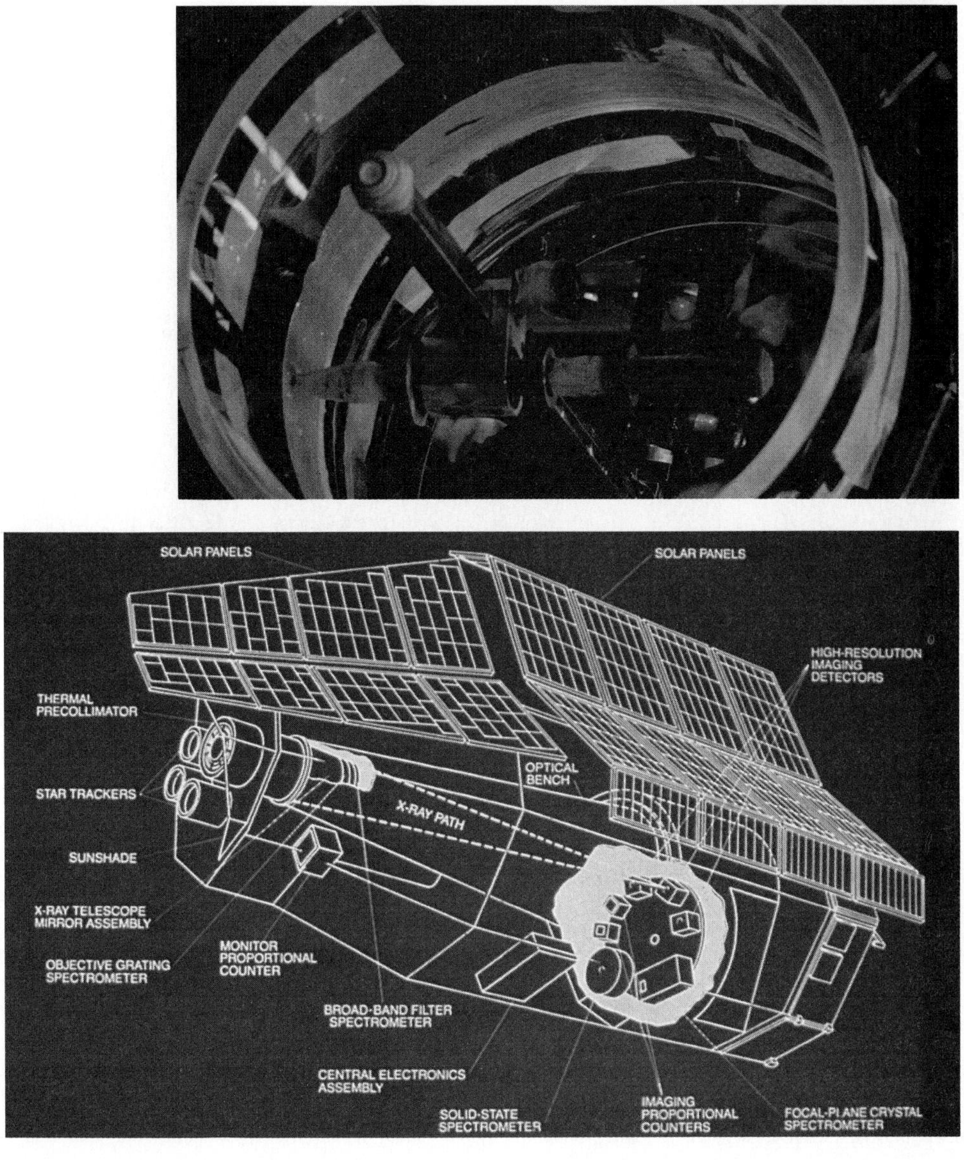

(a)

(b)

Figure 6.34 *(a) The x-ray telescope on board the* Einstein Observatory. *The basic features of the telescope's construction are sketched in (b). Note how a revolving turret can position a variety of detectors and spectrometers at the prime focus of the instrument. (NASA)*

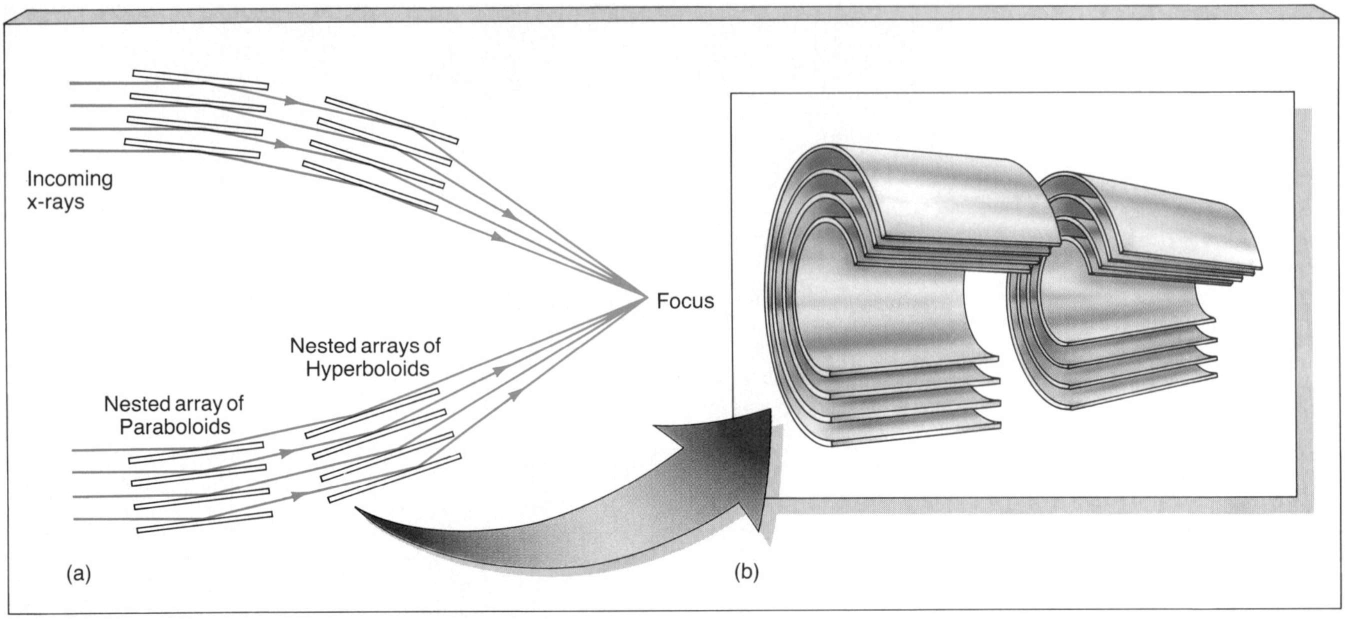

Figure 6.35 *The arrangement of mirrors in the* Einstein Observatory *allowed x-rays to be reflected at grazing angles and focused into an image. (a) and (b) depict the nested array of concentric mirrors.*

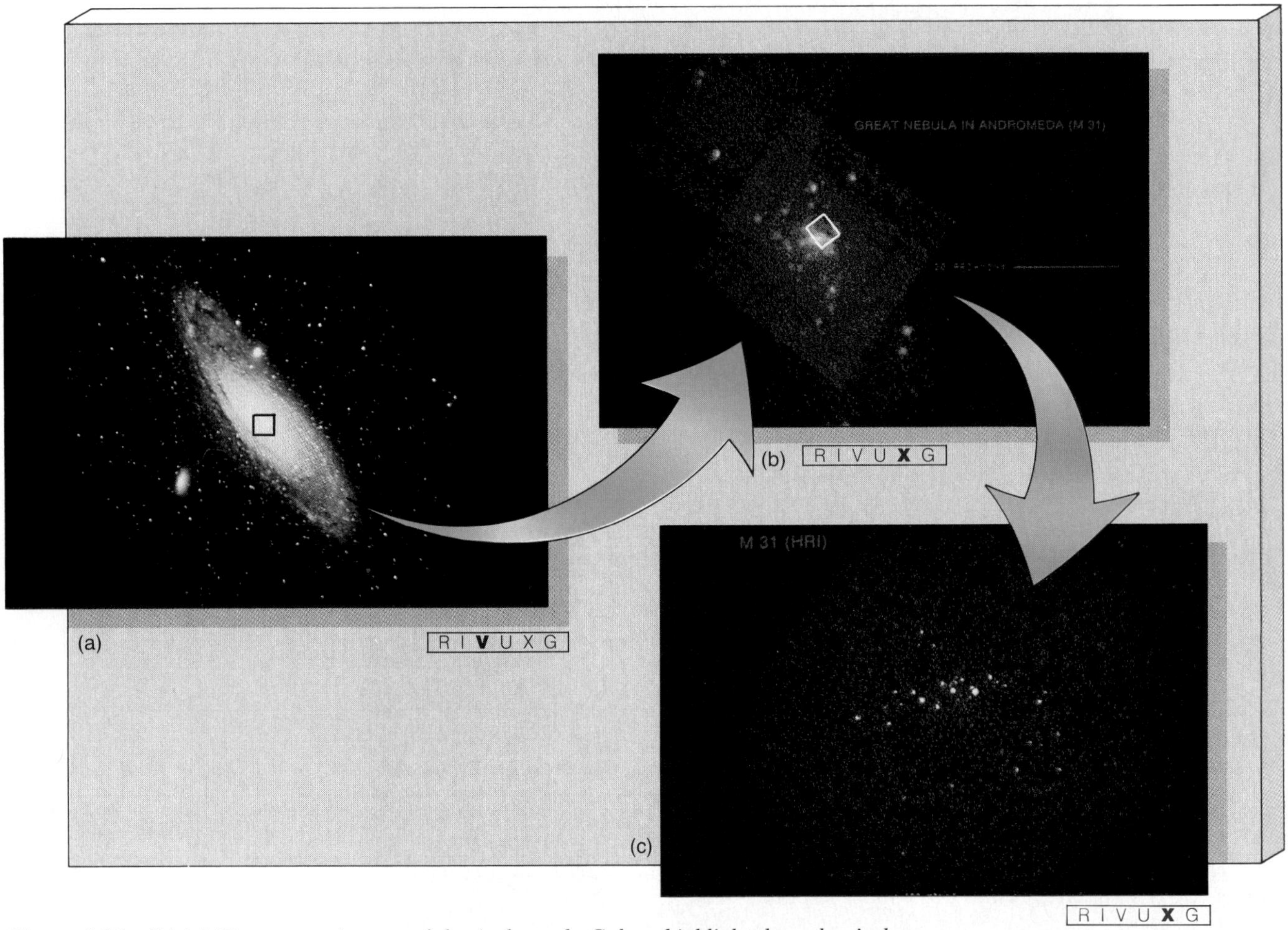

Figure 6.36 *(b),(c) These x-ray images of the Andromeda Galaxy highlight the galaxy's hottest regions. By contrast, few of the galaxy's stars and gas clouds, detectable at optical, radio, or infrared wavelengths, are hot enough to emit x-rays. For comparison, the corresponding visible-light image of the galaxy is shown in (a). (AURA; NASA)*

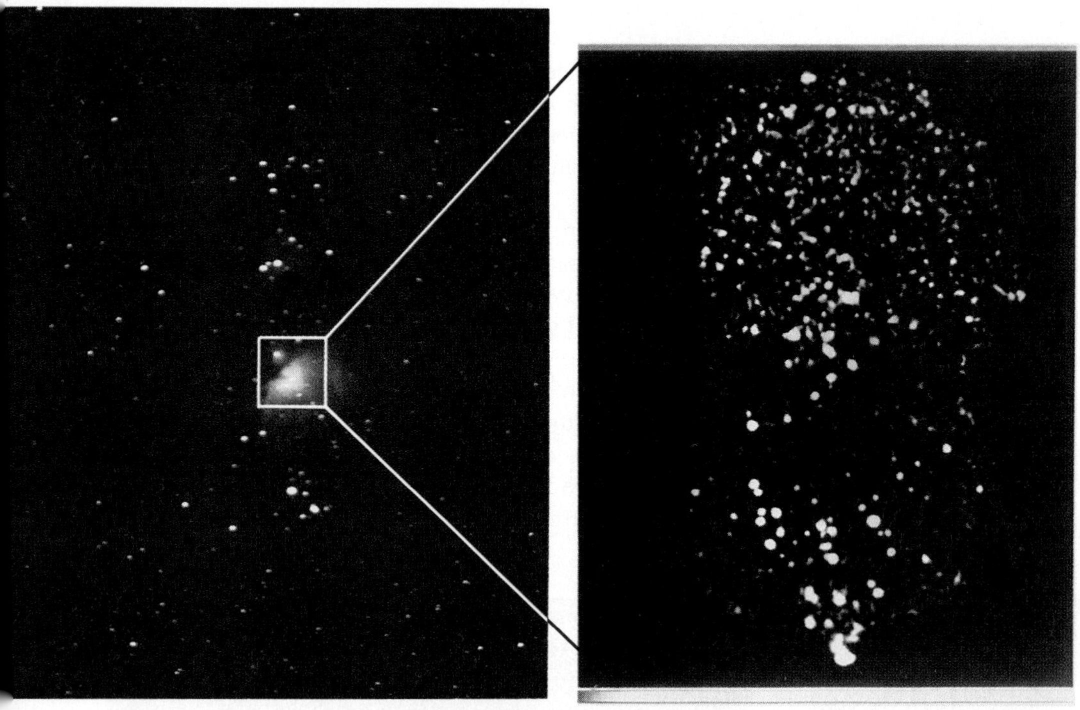

Figure 6.37 *An x-ray image of the Orion region (right), one of the first objects to be studied by the* ROSAT *x-ray satellite. In the optical view on the left, the three stars of Orion's belt can clearly be seen, with the nebula below them. (Compare with Figures 1.6, 6.24, and 6.31.) (AURA, NASA)*

tioned at the focus determined the approximate energies of the incoming photons and so allowed astronomers to perform simple spectral analysis of the radiation.

Although the collecting diameter of *Einstein* was only 0.6 m, the short wavelength of the x-rays made its angular resolution a mere 3″. Accordingly, this spacecraft could produce images of quality comparable to optical photographs. Figure 6.36 is an *Einstein* x-ray image, showing some of the many hot regions in and around the center of the Andromeda Galaxy. In this image, the hottest regions stand out most clearly because the Planck curve of their emission peaks well into the high-energy domain. Thus, they shine brightly in x-rays compared to the much cooler surrounding material, which emits primarily in the infrared and visible regions of the spectrum and hardly at all at higher energies.

The latest major x-ray satellite is the German *ROSAT* (short for *Röntgen Satellite*, after Wilhelm Röntgen, the discoverer of x-rays). Launched in 1990 by a European *Ariane* rocket, it began its mission with a detailed survey of the x-ray sky and is now making detailed observations of specific astronomical objects (see Figure 6.37). With more sensitivity, a wider field of view, and better resolution than *Einstein*, *ROSAT* is providing high-energy astronomers with new levels of observational detail. Even more powerful (if funded) will be NASA's *Advanced X-Ray Astrophysics Facility (AXAF)*, another long-duration orbiting observatory (in the spirit of *IUE*, *HST*, and *SIRTF*) that may become operational by the end of the 1990s.

The youngest entrant into the observational arena is *gamma-ray astronomy*. As mentioned above, true imaging gamma-ray telescopes do not exist, so that only fairly coarse (1° resolution) observations can be made. Nevertheless, even at that resolution, there is much to be learned. Cosmic gamma-rays were originally detected in the 1960s by the U.S. *Vela* series of satellites, whose primary mission was to monitor illegal nuclear detonations on Earth. Since then, several x-ray telescopes have also been equipped with gamma-ray detectors. By far the most advanced instrument is the *Gamma Ray Observatory (GRO)*, launched by the space shuttle in 1991. This satellite can scan the sky and study individual objects in much greater detail than has yet been attempted. As with all brand-new observational windows, the likelihood of astronomers gaining radically new insight into the cosmos through its use is very high. Figure 6.38 shows *GRO* on-station in low Earth orbit.

Figure 6.38 *This photograph of the* Gamma-Ray Observatory *(also called the* Compton Observatory, *after an American gamma-ray pioneer) was taken by an astronaut during the satellite's deployment from the space shuttle* Atlantis *over the Pacific coast of the U.S. (NASA)*

FULL-SPECTRUM COVERAGE

In this chapter we have studied some of the basic techniques and equipment used by astronomers to study the universe. Besides the familiar optical telescopes that collect light from cosmic objects, other tools are needed to capture the invisible radiation emitted by a variety of celestial sources. Often, these "invisible astronomies" are crucial in our study of objects that are totally obscured from view in the visible or simply do not emit any visible light. Radio astronomy is the oldest of the nonvisible subjects, high-energy astronomy the newest. In the end, they all supplement one another, helping us accumulate a growing store of astronomical knowledge.

As we proceed through the text, we will discuss more fully the wealth of information that high-precision astronomical instruments can provide us. It is reasonable to suppose that the future holds many further improvements in both the quality and the availability of astronomical data and that many new discoveries will be made. The current and proposed pace of technological progress presents us with the following very exciting prospect: By the mid- to late 1990s, if all goes according to plan, it will be possible, for the first time ever, to make *simultaneous* high-quality measurements of any astronomical object at *all* wavelengths, from radio to gamma-ray. The consequences of this development for our understanding of the workings of the universe may be little short of revolutionary.

As a foretaste of the sort of comparison that full-

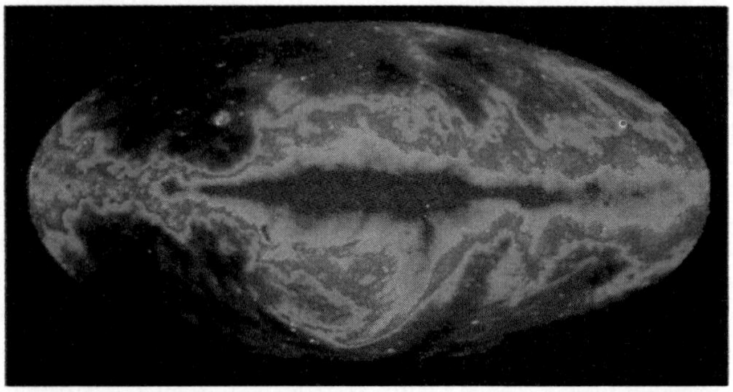

(a) **R** I V U X G

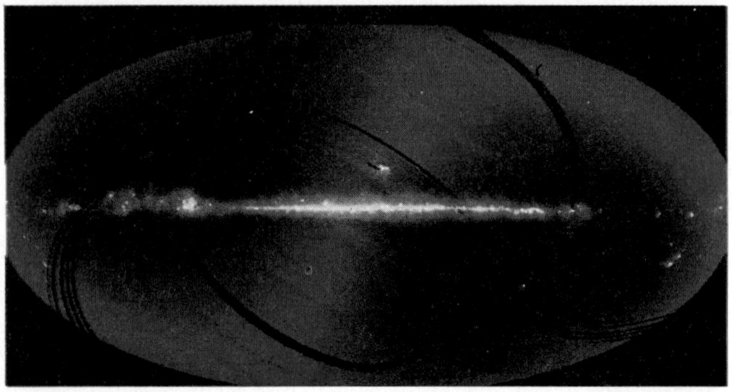

(b) R **I** V U X G

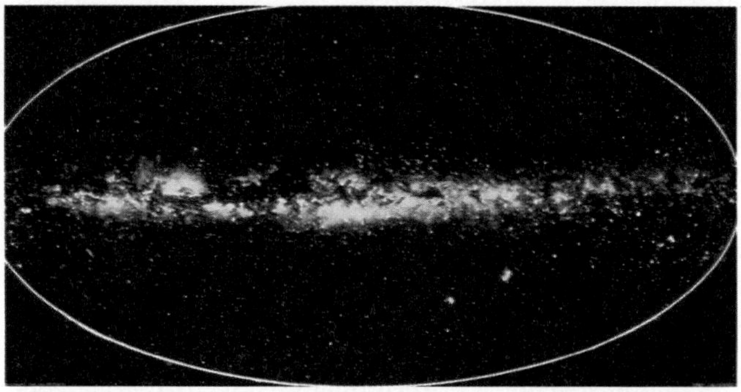

(c) R I **V** U X G

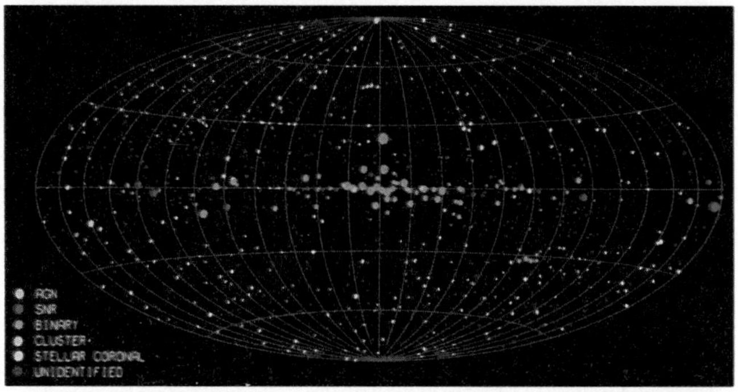

(d) R I V U **X** G

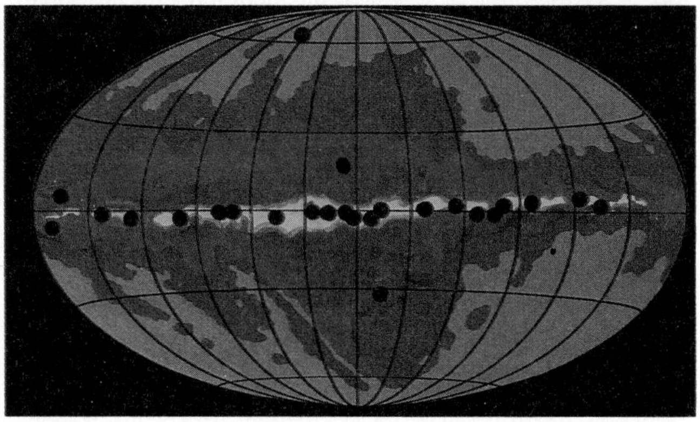

(e) R I V U X **G**

Figure 6.39 *The Milky Way Galaxy, as it appears (from top to bottom) at (a) radio, (b) infrared, (c) visible, (d) x-ray, and (e) gamma-ray wavelengths. (NRAO; Lund Observatory; NASA)* spectrum coverage allows, Figure 6.39 shows a series of images of our own Milky Way Galaxy, made by several different instruments, at wavelengths ranging from radio to gamma-ray, over a period of about five years. By comparing the features visible in each, we immediately see how multiwavelength observations can complement each other, greatly extending our perception of the universe around us.

CHAPTER SUMMARY

Telescopes are designed primarily to gather light. The bigger the telescope, the more light it can collect. Refracting telescopes use lenses to bend and focus light. Reflecting telescopes use mirrors. All large telescopes are reflectors. After it enters the telescope and bounces off the primary mirror, the light can be directed along a variety of paths to different instruments. Starlight entering a telescope can be modified and recorded in a number of ways. A photometer records changing light intensity. An optical spectrometer disperses light into its component colors. A Schmidt telescope is used for wide-angle photography. Large telescopes collect more photons faster than do small telescopes. Also, large telescopes give finer angular resolution than small ones. In the last decade, advances in mirror technology have made it possible to build enormous mirrors and to use many smaller mirrors or mirror segments in tandem.

Atmospheric turbulence hinders ground-based astronomers, limiting resolution to about an arc second. To avoid atmospheric turbulence, astronomers typically locate telescopes at sites where the atmosphere is stable and relatively free of dust, moisture, and light pollution. The orbiting *Hubble Space Telescope* gives 10 times finer resolution and has 50 times greater sensitivity to light than do existing telescopes on Earth. Computers play a critical role in both the acquisition and the subsequent processing of images in modern telescopes. In the next decade, computers will probably make it possible for large ground-based telescopes to achieve the resolution now possible only from space.

Radio wavelengths can be very long, and the total amount of radio radiation arriving at Earth is extremely small. Thus radio dishes must be enormous in order to detect and resolve radio sources. The human eye is more efficient than any telescope, but it can see only visible light. Radio telescopes and other nonoptical detectors enable us to perceive an invisible universe. Radio telescopes can observe 24 hours a day and in different kinds of weather. Radio maps show the "true" extent of cosmic sources because radio waves pass easily through intervening material.

Interferometry is used to improve the angular resolution of radio telescopes. Two or more radio telescopes are used in tandem to observe the same object. The larger the distance separating the telescopes, the better the resolution. Very-long-baseline interferometry can achieve resolution of a milli-arc second or less.

Many techniques for observing the whole electromagnetic spectrum—not just radio and visible light—have been developed since the 1970s. Infrared telescopes resemble optical telescopes, but they are sensitive to longer wavelengths of radiation. UV observations are made from telescopes housed in rockets, balloons, and satellites, because most UV radiation is absorbed by Earth's atmosphere. X-rays and gamma-rays must also be captured high above Earth's atmosphere.

In the future, the quality and availability of astronomical data are bound to improve dramatically. The result may be a revolution in our understanding of the workings of the universe.

KEY WORDS

active optics
adaptive optics
angular resolution
baseline
Cassegrain telescope
charge-coupled device (CCD)
chromatic aberration
collecting area
collimator
coma
focal length
high-energy telescope
image
infrared telescope
interferometer
interferometry
Newtonian telescope
photometer
pixel
prime focus
radio telescope
reflector
refractor
seeing
seeing disk
telescope
ultraviolet telescope

REVIEW QUESTIONS

1. Cite two technical reasons why astronomers are constantly campaigning for larger telescopes.
2. How does a refracting telescope focus light? How does a reflecting telescope focus light?
3. Why do all the world's largest optical telescopes use the reflector design?
4. Name some ways in which images from optical telescopes are recorded.
5. How does the Multiple Mirror Telescope work?
6. What is "seeing"?
7. What advantages does the *Hubble Space Telescope* have over ground-based telescopes? What disadvantages?
8. What role do computers play in image processing?
9. Why do radio telescopes have to be extremely large?
10. How does a radio telescope work?
11. What kind of information do radio waves give us?
12. What is the value of interferometry to radio astronomy?
13. What is high-energy astronomy, and how are high-energy wavelengths studied?
14. What is the main advantage of studying different wavelengths of radiation?
15. A 2-m telescope can collect a given amount of light in 1 hour. Under the same observing conditions, how much time would be required for a 6-m telescope to perform the same task? For 0.5-m telescope?
16. What effect would replacing the 2-m telescope in question 15 with the 6-m or the 0.5-m instrument have on angular resolution if the telescopes are sited (a) on the ground, (b) in space?
17. A certain space-based telescope can achieve (diffraction-limited) angular resolution of $0.05''$ for red light (wavelength = 7000 Å). What would its resolution be in (a) the infrared, at 3.5 μm; and (b) in the ultraviolet, at 1400 Å? (c) What would be the resolution of a radio telescope of the same size at a wavelength of 21 cm?
18. The photographic equipment on a telescope is replaced by a CCD. If the photographic plate records 5 percent of the light reaching it, while the CCD records 75 percent, how much time would the new system take to collect as much information as the old detector recorded in a 1-hour exposure?
19. The prime-focus image of the $2'$ field of view of the *Hale* 200-inch telescope is 1 cm across. If a photographic plate were placed there, what would be the distance on the plate between two stars whose angular separation on the sky is $10''$?
20. What would be the equivalent single-mirror area of a telescope constructed from six separate 4-m mirrors?

DISCUSSION QUESTIONS

1. Our eyes can see light with an angular resolution of $1'$. What would it be like if our eyes detected only infrared radiation with $1°$ angular resolution? Would we be able to make our way around on Earth's surface? To read? To sculpt? To create technology?
2. Compare and contrast the capture of radio waves and x-rays by modern telescopes.
3. Explain how astronomical instruments—such as photometers and spectrometers—work in tandem with telescopes. Which is more important—the telescope or the instrument?
4. Go to the library and find recent articles on discoveries made by the *Hubble Space Telescope*. Do you think this telescope was a good investment of resources? Why or why not?
5. Look up some articles on adaptive optics. What are the new developments in this field of research?
6. Explain why simultaneous high-quality measurements of astronomical objects at all wavelengths would be valuable to astronomers.
7. How do you think astronomers of the future will study outer space and its contents?

PROJECTS

1. Anyone can grind a mirror for a small telescope, and many people go the whole way in building very beautiful and serviceable small reflecting telescopes. If you are considering building a telescope, you may want to look at these books:

   How to Make a Telescope by Jean Texereau. This is the classic book on telescope making. A thorough guide, it has been used by amateur astronomers for over 30 years. It is still available through Kalmbach Publishing or Sky Publishing.

   A more recent book is *Build Your Own Telescope* by Richard Berry, former editor of *Astronomy* magazine. It contains instructions for building five amateur telescopes. This book is very well regarded. It is available through Kalmbach Publishing or Sky Publishing.
2. Many people with simple cameras take good pictures of

the night sky. Here are some reference books useful for the astrophotographer:

Exposure Guides for the Sun, Moon and Planets by Ginger LeGendre. This is a concise guide containing data and techniques that can help you photograph solar-system and some deep sky objects. Contains tables giving precise exposure times. Available from Sky Publishing.

Astrophotography by Barry Gordon. A detailed photo guide that presents the data and principles needed to choose the right telescopes, lenses, exposures, and so on for different objects. Available from Sky Publishing.

Sky Shooting—Photography for Amateur Astronomers by R. Mayall and M. Mayall. Issued in 1949 and updated in 1968, this is a proven introduction to the hobby of astrophotography. Available from Sky Publishing.

SUGGESTED READINGS

Bahcall, J., and L. Spitzer. ''The Space Telescope.'' *Scientific American* (July 1982). Excellent prelaunch article on design and capabilities of the *Hubble Space Telescope*.

Balick, B. ''Astrophysics at Apache Point.'' *Sky and Telescope* (August 1988). A fully automatic 3.5-meter telescope in New Mexico. Accompanying article describes Roger's Angel's technique for making ''spin-cast'' telescope mirrors.

Burbidge, G., and A. Hewitt, eds. *Telescopes for the 1980s*. Palo Alto, CA: Annual Reviews, 1981. Discussion of proposals to build large telescopes.

Fienberg, R. T. ''*HST:* Astronomy's Discovery Machine.'' *Sky and Telescope* (July 1990). Deployment into space of the *Hubble Space Telescope*. Includes general information about the telescope.

Chaisson, Eric J. ''Early Results from the Hubble Space Telescope.'' *Scientific American* (June 1992)

Kristian, J., and M. Blanke, ''Charge-coupled Devices in Astronomy.'' *Scientific American* (October 1982). Tiny television sensors that have enabled astronomers to obtain a new, vastly more detailed look at the heavens.

Learner, R. ''The Legacy of the 200-Inch.'' *Sky and Telescope* (April 1986). How the successful design of the Palomar 200-inch telescope suppressed fresh ideas about ways to build large telescopes for four decades.

Maran, S. P. ''Beyond Galileo's 'Vast Crowd of Stars.' '' *Smithsonian* (June 1987). Observing the sky at wavelengths other than visible light.

MacRobert, A. ''Stunning Planet Images from Earth.'' *Sky and Telescope* (June 1991). High-resolution images of solar system objects, obtained using CCDs at the Pic du Midi Observatory.

Strom, S. ''New Frontiers in Ground-based Optical Astronomy.'' *Sky and Telescope* (July 1991). What some proposed and existing ground-based telescopes are expected to reveal about the universe.

Tucker, W. *The Star Splitters*. NASA SP-466, Washington, D.C.: U.S. Government Printing Office, 1984.

Tucker, W., and K. Tucker. *The Cosmic Inquirers*. Cambridge, MA: Harvard University Press, 1986. Based on a series of interviews, this book reveals how and why prominent astronomers accomplished their goals of building and operating large observatories.

CHAPTER

7

The Earth

Our Home in Space

Among the casualties in the Persian Gulf War was the Earth's environment. Oil wells burned day and night for months, spewing ash into the atmosphere and polluting the sky. How much abuse can Earth's environment withstand? Will there come a time when the skies are so polluted that we can no longer gaze out at the stars? Although we know much about our planet, we still have a lot to learn. (Courtesy A. Tannenbaum-Sygma)

LEARNING GOALS

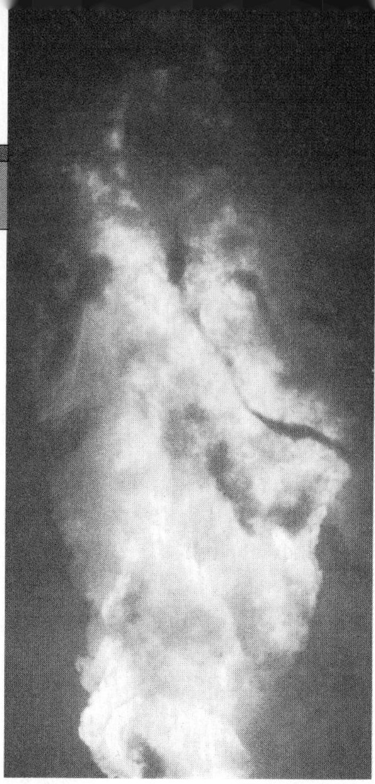

1 to know some of the average physical and chemical properties of our planet

2 to understand how Earth's atmosphere helps to heat us as well as protect us

3 to examine the nature and origin of the Earth's magnetosphere

4 to appreciate some of the experimental techniques used to infer the structure of Earth's interior

5 to understand that Earth's surface activity—such as earthquakes and volcanos—is a remnant of a hotter, more violent era

6 to examine the arguments favoring the idea known popularly as "continental drift"

Looking up at the nighttime sky, we see an almost bewildering array of stars. Thousands are visible to the naked eye, and we can observe millions with even a small telescope. Because astronomers can see and study so many stars, they have come to recognize patterns in stellar properties and have classified the stars accordingly. Ultimately, this classification has led to an understanding of the birth, maturity, and death of stars everywhere.

In studying the planets, we are not so fortunate. Unlike the legions of stars, we are aware of only a single planetary system—our solar system—and each of its nine planets differs significantly from all the others. We don't have tens of thousands of Earthlike planets to compare, so we cannot perform the statistical analyses that have taught us so much about the stars. Although we inhabit Earth, we know less about the evolution of our own planet than we do about the evolution of a typical star. Every piece of information we can glean about the structure and history of the other worlds orbiting our Sun plays a potentially vital role in helping us understand our own. We will return to this growing subject of comparative planetology in later chapters.

To fathom what Earth might have been like long ago is not easy. Imagining what it might be like in the future is even more difficult. Yet from the matter of our planet sprang life, intelligence, culture, and the technology that we now use to explore the cosmos. We ourselves are made of "earthstuff" as much as are rocks, trees, and air. If we are to appreciate the universe, we must first come to know our planet. Our study of astronomy begins at home.

THE EARTH IN BULK

Dimensions

✓1 The circumference of our planet has been known reasonably well since the time of the ancient Greeks (see Interlude 2-1). As we saw in Interlude 2-1, the Earth measures approximately 40,000 km all the way around. Knowing that the circumference of a circle is 2π times its radius, we find the (rounded-off) radius is $R_{\oplus} = 6500$ km.[1] More precise measurements are now routinely made by spacecraft, leading to the result $R_{\oplus} = 6378$ km. The volume of the Earth then follows directly. It amounts to about 10^{27} cubic centimeters.

Throughout the text, we will use the cubic centimeter (cm^3) as a unit of volume. A cubic centimeter is simply the volume of a cube each of whose sides is 1 cm long (see also Interlude 1-2). You might think of a cubic centimeter as roughly the volume of an acorn or a sewing thimble. Thus we would need approximately a billion, billion, billion acorns to fill the volume of the Earth! This calculation applies only to the solid and liquid parts of Earth. It does not include the gaseous atmosphere or the magnetosphere lying high above our planet's surface.

[1]Recall that the symbol ⊕ is the ancient astrological symbol for the Earth (see Interlude 1-2). Apart from the naming of the constellations, one of the few holdouts of astrology in modern astronomy is the use of such symbols for the Sun, the Moon, and the planets. Thus, any symbol with a ⊕ attached refers specifically to the Earth. Similarly, when the Sun sign, ⊙, is used, the symbol relates to the Sun, and so on.

Mass

Broadly defined, **mass** is the total amount of matter contained within an object. We saw in Chapter 3 how we can measure the mass of the Earth, or any other astronomical body, by observing its gravitational influence on some other nearby object and applying Newton's Law of Gravity. By studying the dynamical behavior of objects near our planet—be they baseballs, rockets, or the Moon—we can compute the Earth's mass, $M_{\oplus}$. In effect, we can weigh the Earth. We find, approximately, $M_{\oplus} = 6.0 \times 10^{27}$ g, or about 6000 billion billion tons. Virtually all of the Earth's mass is contained within the surface and interior. The atmosphere and the magnetosphere contribute hardly anything at all—less than 0.1 percent.

Density

A very important property of any object is its **density.** Density is a measure of the "compactness" of the matter within an object. It is computed by dividing the mass of an object by its volume. Dividing Earth's mass by its volume, we obtain 6×10^{27} g/10^{27} cm^3, or an average density of approximately 6 g/cm^3. A more accurate calculation gives 5.5 g/cm^3. Again, this density pertains to the mostly rocky planet, excluding the atmosphere and magnetosphere. *On average,* then, there are about 6 grams of Earth matter in every cubic centimeter of Earth volume.

This simple measurement of the Earth's average density allows us to make a very important deduction about the interior of our planet. The water that makes up much of the

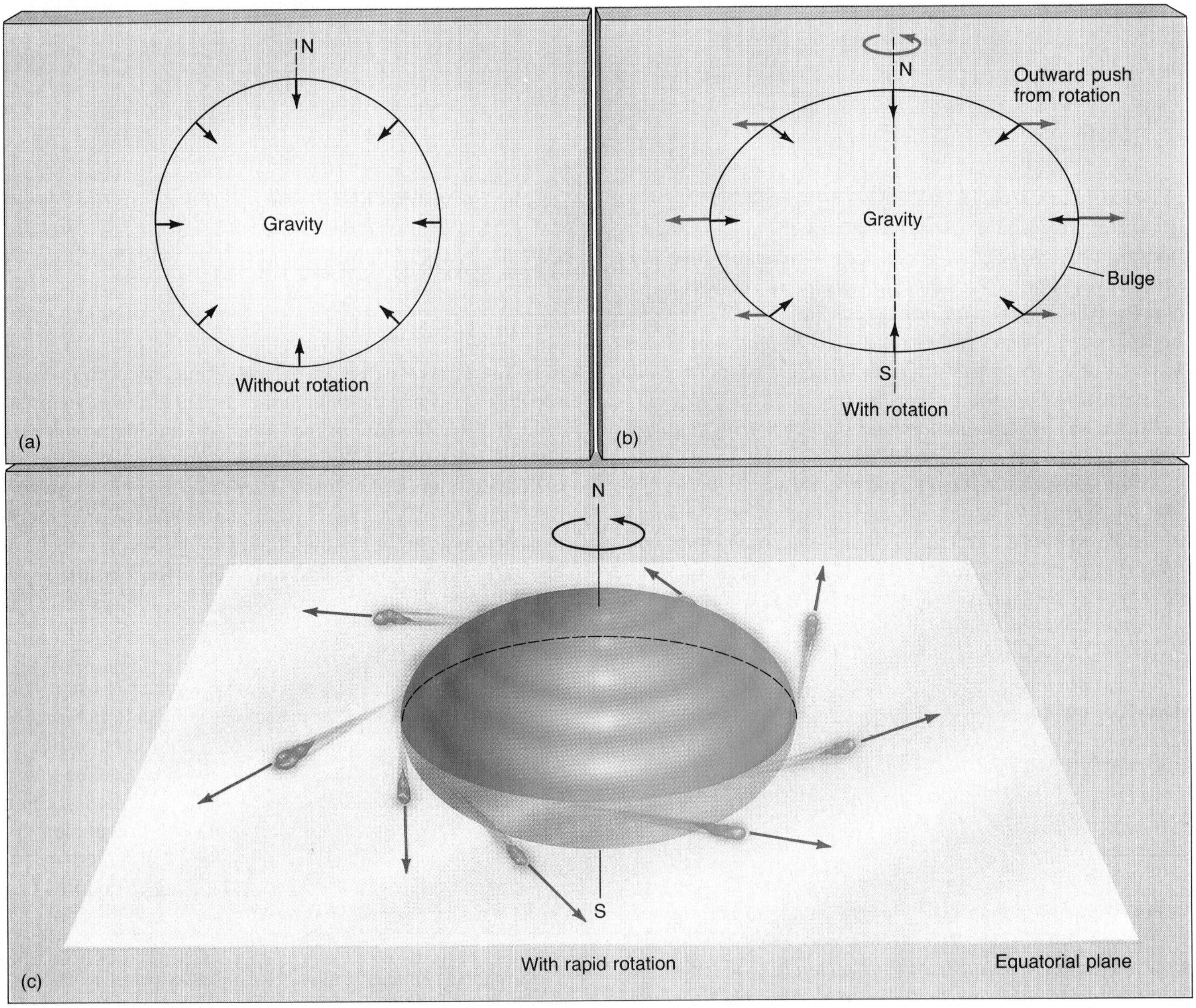

Figure 7.1 *All spinning objects tend to develop an equatorial bulge because rotation causes matter to push outward against the inward-pulling gravity. The size of the bulge depends on the mechanical strength of the matter and the rate of rotation. The inward-pointing arrows denote gravity, the outward arrows the push due to rotation; if large enough, the outward push can break the object apart.*

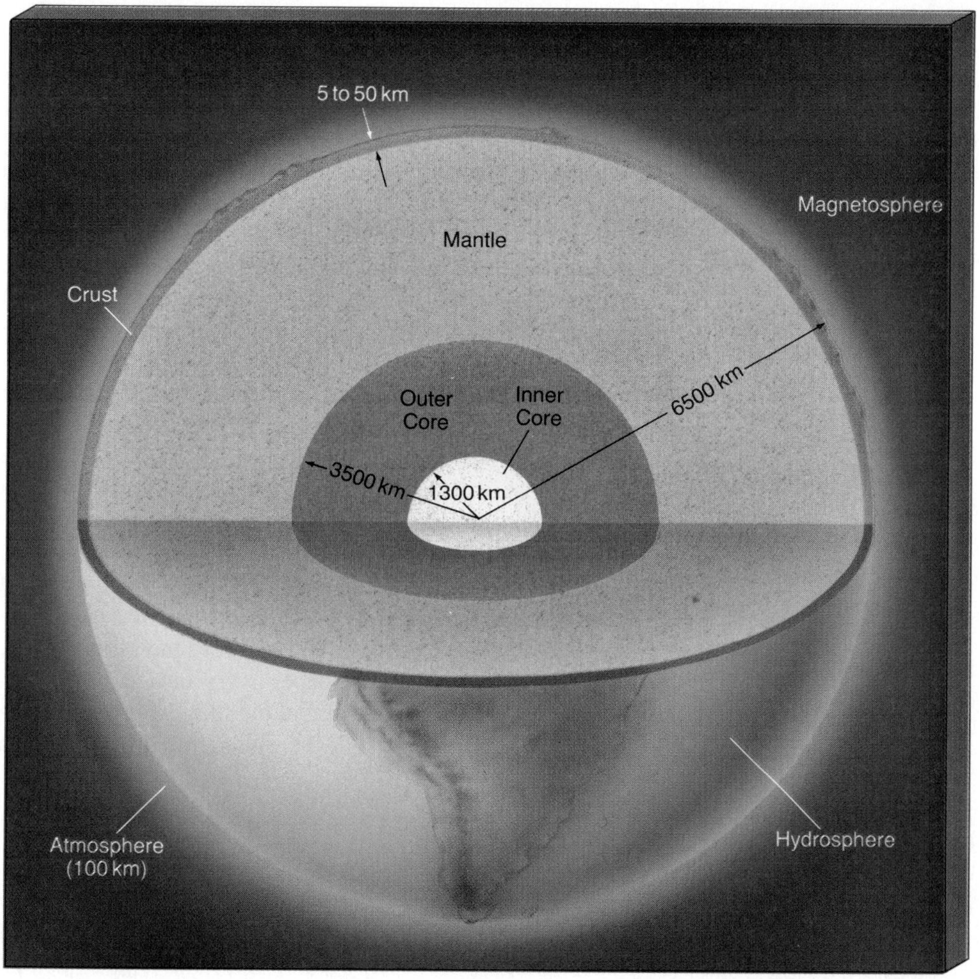

Figure 7.2 The main regions of planet Earth. At the Earth's center lies our planet's inner core, about 2600 km in diameter. Surrounding the inner core is an outer core, some 7000 km across. Most of the rest of the Earth's 13,000 km diameter is taken up by the mantle, which is topped by a thin crust only a few tens of kilometers thick. The liquid portions of the Earth's surface make up the hydrosphere. Above the hydrosphere and solid crust lies the atmosphere, most of it within 50 km of the surface. Earth's outermost region is the magnetosphere, extending thousands of kilometers out into space.

Earth's surface has a density of 1 g/cm^3, and the rock beneath us on the continents, as well as on the seafloor, has a density in the range of 2 to 4 g/cm^3. We can immediately conclude that because the surface layers have densities much less than the average, much denser material must lie deeper in. Hence we should expect that much of Earth's interior is made up of very dense matter, even more compact than the densest continental rocks on the surface.

Shape

The Earth is not a perfect sphere. It bulges slightly at the equator (or, equivalently, it is slightly flattened at the poles), so that the equatorial diameter is about 40 km larger than the polar diameter. The bulge is not very pronounced for an object of this size—40 km is small compared with the Earth's full diameter of nearly 13,000 km. In fact, relative to its overall dimensions, Earth is smoother and more spherical than a billiard ball.

To understand this slight flattening, recall from Chapter 1 that our planet spins around a north-south axis. A spinning object develops a bulge around its midsection, as illustrated in Figure 7.1. The more loosely the object's matter is bound together or the faster the object spins, the larger the bulge becomes. In objects made up of gas or loosely packed matter, high spin rates produce a pronounced bulge. Even faster rates can completely tear matter from the object. The present spin rate of Earth (one rotation every 24 hours) is great enough to cause some departure from sphericity, but it is not nearly fast enough to dislodge solid or liquid matter from the surface.

Earth's Structure: An Overview

In probing our planet, we have the obvious advantage that we live on it. We can analyze the soil, sample the water, and perform direct experiments to try to decipher its workings. Earth is now examined and monitored in nearly every conceivable way—with aircraft in the atmosphere, satellites in orbit, gauges on the land, submarines in the ocean, and drilling gear below the rocky crust.

Our planet can be divided into six main regions. As shown in Figure 7.2, a zone of charged particles trapped by the Earth's magnetic field forms the **magnetosphere.** This lies high above the **atmosphere** of air. At the surface, we have a relatively thin **crust,** comprising the solid continents and the seafloor, and the **hydrosphere,** which contains the liquid oceans and accounts for some 70 percent of our planet's total surface area. In the interior, a large **mantle** surrounds a smaller **core** (which itself is divided into an **outer core** and an even smaller **inner core**). Let's examine each of Earth's major regions in greater detail.

THE HYDROSPHERE

Earth is unique among the planets in that it has large quantities of liquid water on its surface. Approximately three-quarters of the Earth's surface is covered by water, to an average depth of about 3.6 km. Only 2 percent of the water is contained within lakes, rivers, clouds, and glaciers. The remaining 98 percent makes up the oceans. The extent of the hydrosphere's role in purifying our planet's atmosphere and regulating its temperature has become apparent since the 1970s, and it remains the subject of intensive research today. We now know that the oceans were crucial to the processes that led to the emergence of life on Earth and are essential to its continued well-being.

Tides

A familiar hydrospheric phenomenon is the daily fluctuation in ocean level known as the **tides.** At most coastal locations on Earth, there are two low tides and two high tides each day. The "height" of the tides—the magnitude of the variation in sea level—can range from a few centimeters to many meters, depending on the location on Earth and time of year. The height of a typical tide on the open ocean is about a meter, but if this tide is funneled into a narrow opening such as the mouth of a river, it can become much higher. At the Bay of Fundy, on the Maine-Canada border, the high tide can reach nearly 20 m (approximately 60 feet, or the height of a six-story building!), above the low-tide level. An enormous amount of energy is contained in the daily motion of the oceans. This energy is constantly eroding and reshaping our planet's coastlines. In some locations, it is a source of electrical power for human activities.

What causes the tides? A clue comes from the fact tides exhibit daily, monthly, and yearly cycles. The tides are a direct result of the gravitational influence of the Moon and the Sun on the Earth. We have already seen how gravity keeps the Earth and Moon in orbit about one another, and both in orbit around the Sun. For simplicity, let us first consider just the interaction between the Earth and the Moon.

Recall that the strength of gravity depends on the distance separating any two objects. Thus, different parts of the Earth feel slightly different pulls due to the Moon's gravity, depending on their distance from the Moon. For example, the Moon's gravitational attraction is greater on the side of Earth that faces the Moon than on the opposite side, some 13,000 km farther away. This difference in the gravitational force is small—only about 3 percent—but it produces a noticeable effect, a **tidal bulge.** The Earth becomes slightly elongated, with the long axis of the distortion pointing toward the Moon, as illustrated in Figure 7.3.

The liquid portions of the Earth undergo the greatest deformation, because they can most easily move around on the surface. (A bulge *is* actually raised in the solid material of the Earth, but it is about a hundred times smaller than the oceanic bulge.) Thus, the ocean becomes a little deeper in some places (along the line joining the Earth to the Moon) and shallower in others (perpendicular to this line). The daily tides result as the Earth rotates beneath this deformation. As noted in Figure 7.3, the side of Earth *opposite* the Moon also experiences a tidal bulge. The different gravitational pulls—greatest on that part of Earth closest to the Moon, weaker at the Earth's center, and weakest of all on Earth's opposite side—cause average tides on either side of our planet to be approximately equal in height. On the side nearer the Moon, the ocean water is pulled slightly toward the Moon. On the opposite side, the ocean water is literally left behind as the Earth is pulled closer to the Moon. Thus, high tide occurs twice, not once, every day.

In fact, both the Moon and the Sun exert tidal forces on our planet. Even though the Sun is roughly 375 times

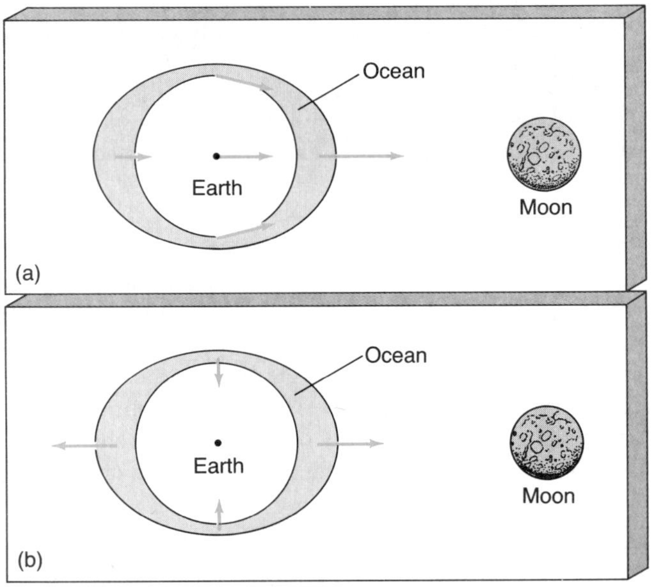

Figure 7.3 *This exaggerated illustration shows how the Moon induces tides on both the near and the far sides of the Earth. The lengths of the straight arrows indicate the relative strengths of the Moon's gravitational pull on various parts of Earth. (a) The lunar gravitational forces acting on several different locations on and in the Earth. The force is greatest on the side nearest the Moon and smallest on the opposite side. (b) The* difference *between the lunar forces experienced at those same locations and the force acting on the Earth's center. The arrows represent the force with which the Moon tends to pull matter away from, or squeeze it toward, the center of our planet. Material on the side of Earth nearest the Moon tends to be pulled away from the center, and material on the far side is "left behind," so that a bulge is formed. High and low tides result, twice per day, as the Earth rotates beneath the bulges in Earth's oceans.*

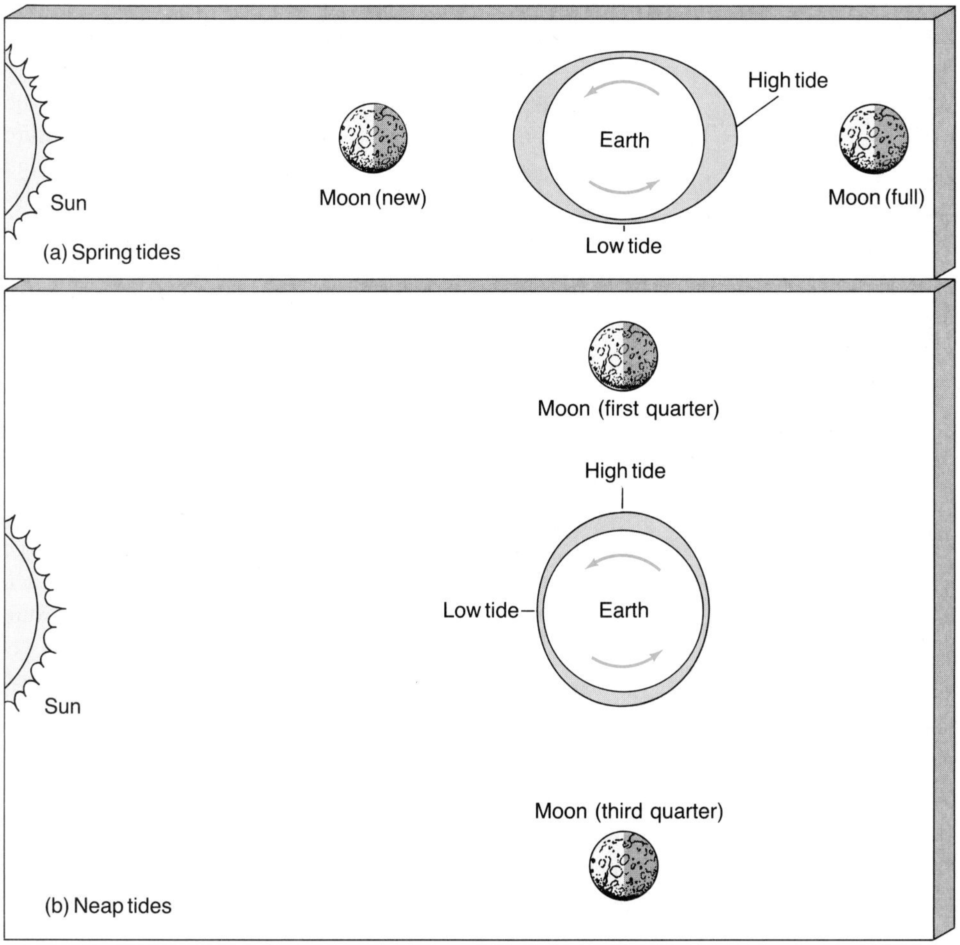

Figure 7.4 *The combined effects of the Sun and the Moon produce variations in the high and low tides. (a) When the Moon is full (or new), the Earth, Moon, and Sun are approximately aligned, and the tidal bulges raised in Earth's oceans by the Moon and the Sun reinforce one another. (b) At first (or third) quarter Moon, the tidal effects of the Moon and the Sun partially cancel, and the tides are smallest. The Moon's tidal effect is greater than that of the Sun, so the net bulge points toward the Moon.*

farther away from Earth than is the Moon, its mass is so much greater (by about a factor of 27 million) that its tidal influence is still significant—about half that of the Moon. Thus, instead of one tidal bulge, there are actually two—one pointing toward the Moon, the other toward the Sun—and the interaction between them accounts for the changes in the height of the tides over the course of a month or a year. When the Earth, Moon, and Sun are roughly lined up, the gravitational effects reinforce one another, and so the highest tides are generally found at times of new and full moon. These tides are known as **spring tides.** When the Earth-Moon line is perpendicular to the Earth-Sun line (at the first and third quarters), the daily tides are smallest. These are termed **neap tides.** The relative orientations of the Earth, Sun, and Moon at times of spring and neap tides are illustrated in Figure 7.4.

The variation of the Moon's gravity across the Earth is an example of a *differential,* or **tidal, force.** The *average* gravitational interaction between two bodies determines their orbit around one another. However, the *tidal* force, superimposed on that average, tends to deform the bodies themselves. The tidal influence of one body on another diminishes very rapidly with increasing distance (in fact, as the inverse *cube* of the separation). For example, if the distance from the Earth to the Moon were to double, the tides resulting from the Moon's gravity would decrease by a factor of eight. We will see many examples in this book of situations where tidal forces are critically important in understanding astronomical phenomena. Notice that we still use the word *tidal* in these other contexts, even though we are not discussing oceanic tides and possibly not even planets at all.

Effect of Tides on Earth's Rotation

Earth's rotation on its axis is gradually slowing down. As the spin slows, the length of the day increases. We now have direct evidence that the day is lengthening by about 0.002 seconds every century. That's a small time increase on the scale of a human lifetime, but over millions of years, this steady slowing of Earth's rotation adds up.

How do we know that the Earth's spin rate is decreasing? A number of natural biological clocks allow us to make that conclusion. For example, every time Earth spins on its axis, a growth mark is deposited on a certain type of coral in the reefs off the Bahamas. These growth marks are similar to the annual rings found in tree trunks, except that in the case of coral the marks are made daily. However, they also show yearly variations, as the coral's growth responds to the Earth's seasonal changes, allowing us to perceive annual cycles. Coral growing at the present time shows 365 marks per year, but ancient coral shows many more growth deposits per year. In fact, fossilized reefs dated to be several hundred million years old contain coral with nearly 400 deposits per year of growth. At that time, there must have been 35 more days in each Earth year, implying that Earth must have spun more rapidly in the past. Nearly 500 million years ago, the day was only 22 hours long and the year contained 400 days.

Why is the Earth's spin slowing? The main reason is the tidal effect of the Moon. In reality, the tidal bulge raised by the Moon does *not* point directly at it, as shown in Figure 7.3. Because of the effects of friction, both between the crust and the oceans and within the Earth itself, the bulge actually points slightly *ahead* of our satellite, as the rotating Earth tries to drag the tidal bulge around with it. The net effect of the Moon's pull on this slightly offset bulge is that our planet's rotation rate decreases. At the same time the Moon spirals slowly away from us, increasing its average distance from the Earth by about 4 cm per century. This process will continue until the Earth rotates on its axis at exactly the same rate as the Moon orbits the Earth. At that time the Moon will always be above the same point on Earth and will no longer lag behind the bulge it raises. The Earth's rotation period will be 47 of our present days, and the distance to the Moon will be 550,000 km (about 43 percent greater than at present). However, this will take a very long time to occur—at least a trillion years!

THE ATMOSPHERE

✓2 From a human perspective, probably the most obvious aspect of Earth's atmosphere is that we can breathe it. Air is a mixture of gases, the most common of which are nitrogen (N_2: 78 percent by volume), oxygen (O_2: 21 percent), argon (Ar: 0.9 percent), and carbon dioxide (CO_2: 0.03 percent). Water vapor (H_2O) is a variable constituent of the atmosphere, making up anywhere from 0.1 percent to 3 percent, depending on location and climatic conditions. The presence of a large amount of oxygen makes our atmosphere unique in the solar system, and the presence of even trace amounts of water and carbon dioxide play a vital role in the workings of our planet.

Our atmosphere is much more than simply the air we breathe. It protects us from most of the harsh radiation emitted by the Sun and other cosmic objects—particularly the high-frequency and often harmful ultraviolet and x-ray radiation. It also guards us from most rocky debris, or "meteoroids," falling in from space—air friction causes all but the largest to burn up long before they reach the ground. Furthermore, it helps to keep us warm. In short, the atmosphere also acts as a protective blanket, making the surface of our planet a relatively comfortable place to live.

Atmospheric Structure

Figure 7.5 shows a cross section of Earth's atmosphere. Compared with the overall dimensions of our planet, the extent of our atmosphere is not large. Half of it lies within 5 km of the surface, and all but 1 percent of it is found below 30 km. The temperature and the density both decrease with altitude in the lowest-lying atmospheric zones, where virtually all weather occurs. Climbing even a modest mountain—4 or 5 km high, say—clearly demonstrates this cooling and thinning of the air. The portion of the atmosphere below about 15 km is called the **troposphere.** Above it, extending up to an altitude of 40 to 50 km, lies the **stratosphere.** At even greater altitudes we encounter, in succession, the **mesosphere** (from 50 to 90 km), and the **thermosphere** (above 90 km). These regions are distinguished from each other by the behavior of the temperature (decreasing or increasing with altitude) in each. The outermost layers of the atmosphere (above about 250 km) are known as the **exosphere.** Both atmospheric density and pressure decrease steadily with altitude throughout all of these regions.

The troposphere is the region of Earth's (or any other planet's) atmosphere where **convection** occurs, driven by the heat of Earth's warm surface. Convection, described in more detail below, is the constant upwelling of warm air and the concurrent downward flow of cooler air to take its place. This constant churning motion is responsible for all the weather we experience. Above the **tropopause** (the top of the troposphere), in the stratosphere, the atmosphere is stable, and the air is calm. Within the stratosphere lies the **ozone layer.** There, at an altitude of 20 to 50 km, the air temperature increases as incoming solar ultraviolet radiation is absorbed by atmospheric oxygen (O_2), ozone (O_3), and nitrogen (N_2). Note that the term *stratosphere* is sometimes used to refer to *both* the stratosphere and the mesosphere.

In the thermosphere, the temperature of what little air is found increases steadily with altitude, as the high-energy portion of the Sun's radiation spectrum splits molecules into atoms and atoms into ions. Above about 100 km, the atmosphere is significantly ionized, and the degree of ionization increases with altitude. This electrically conducting portion of the upper atmosphere is also known as the **ionosphere.** Its conductivity makes it highly reflective to certain radio wavelengths. The reason that AM radio stations can be heard beyond the horizon is that their signal actually bounces off the ionosphere before reaching the receiver.

The ozone layer is one of the insulating spheres that serve to protect life on Earth from the harsh realities of outer space. Not long ago, scientists judged space to be hostile to advanced life forms because of what is missing out there—breathable air and a warm environment. Now, most scientists regard outer space harsh because of what *is* present out there—fierce radiation and energetic particles, both of which are injurious to human health. The ozone layer is one

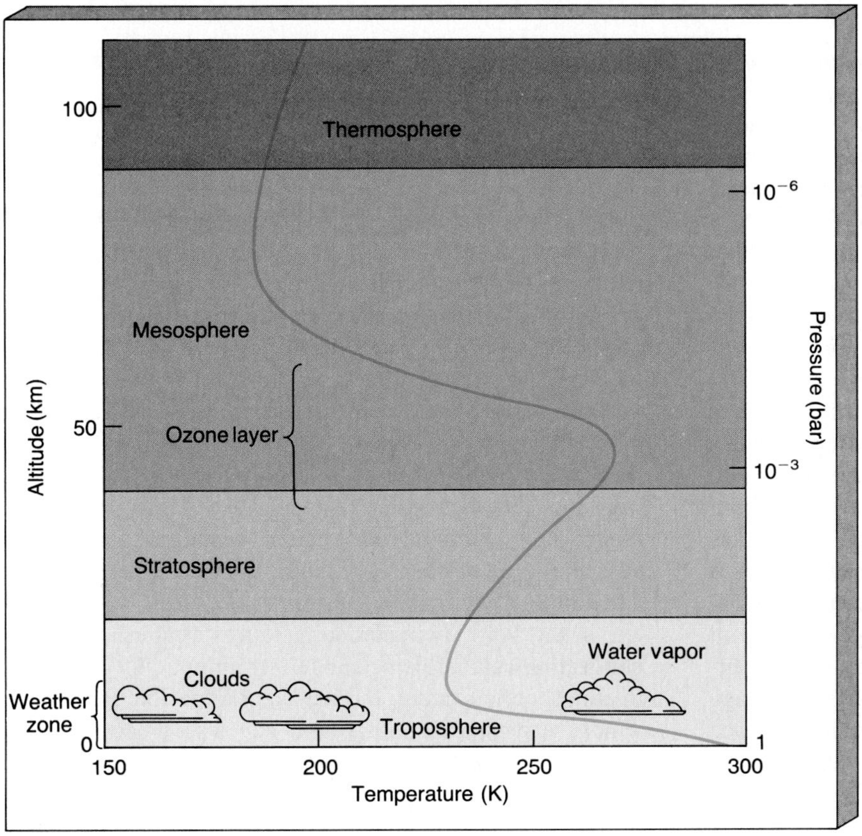

Figure 7.5 Diagram of Earth's atmosphere, showing the changes of temperature and pressure from the surface to the bottom of the ionosphere. (Note that pressure equals 1 bar at sea level.)

of our planet's umbrellas. Without it, advanced life (at least on the Earth's surface) would be at best unlikely and at worst impossible.

Gases released as waste into our atmosphere because of the daily activities of our technological civilization can damage and deplete the ozone layer surrounding and protecting our planet. In the opinion of many experts, that damage is already at a fairly advanced stage. Humanmade pollutants (such as the Freon in our refrigerators and other chlorofluorocarbons, which propel chemicals from some spray guns) thus become a socially relevant problem. A depleted ozone layer could allow potentially harmful levels of ultraviolet and x-ray radiation to reach Earth's surface. Only in the last decade has the true fragility of Earth's atmosphere become apparent.

Surface Heating

Much of the Sun's radiation manages to penetrate the Earth's atmosphere, eventually reaching the ground. Most of this energy is in the form of visible radiation—ordinary sunlight. (Solar radio waves, in the other portion of the electromagnetic spectrum to which the Earth's atmosphere is transparent, also reach Earth's surface, but the hot Sun emits comparatively little of this radiation.) All the solar radiation not absorbed or reflected from clouds in the upper atmosphere shines directly onto Earth's surface. The result is that our planet's surface and most objects on it heat up considerably during the hours of daylight. But the Earth can't possibly absorb this solar energy indefinitely. If it did, surface objects would soon become hot enough to melt, and all life forms would be cooked.

As it heats up, the surface reradiates much of its absorbed energy. The reradiated radiation follows the usual Planck curve, studied earlier in Chapter 4. As the surface temperature rises, the amount of energy radiated increases rapidly, according to Stefan's Law. Eventually the Earth radiates as much energy back into space as it receives from the Sun, and a stable balance is struck. In the absence of any complicating effects, this balance would be achieved at an average surface temperature of about 250K (−23°C). At that temperature, Wien's Law tells us that most of the reemitted energy is in the form of infrared radiation.

But there are complications. Infrared radiation is partially blocked by Earth's atmosphere. The primary reason for this is the presence of molecules of water vapor and carbon dioxide, which absorb very efficiently in the infrared portion of the spectrum. Even though these two gases are only trace constituents of our atmosphere, they manage to absorb a large fraction of all the infrared radiation emitted from the surface. Consequently, only some of that radiation escapes back into space. The remainder is trapped within our atmosphere, and the temperature rises.

This partial trapping of solar radiation is known as the **greenhouse effect.** The name comes from the fact that a very similar process operates in a greenhouse—sunlight passes relatively unhindered through glass panes, but much of the infrared (heat) radiation reemitted by the plants is blocked by the glass and cannot get out. Consequently, the interior of the greenhouse heats up, and flowers, fruits, and vegetables can grow even on cold wintery days. The radiative processes that determine the temperature of the Earth's atmosphere are shown in Figure 7.6. Earth's greenhouse effect makes our planet about 40K hotter than would otherwise be the case.

The magnitude of the effect is very sensitive to the concentration of the so-called greenhouse gases (that is, those that absorb infrared radiation efficiently) in the atmosphere. Of greatest importance among these is carbon dioxide, although water vapor too plays a significant role. The amount of carbon dioxide in Earth's atmosphere is increasing, largely as a result of the burning of fossil fuels (principally oil and coal) in the industrialized world. Carbon dioxide levels have increased by over 20 percent in the last century, and they are continuing to rise at a present rate of 4 percent per decade. In Chapter 11, we will see how a runaway increase in carbon dioxide levels in the atmosphere of the planet Venus has radically altered conditions on its surface, causing its temperature to rise to over 700K. Although no one is predicting that Earth's temperature will ever reach that of Venus, many scientists now believe that this increase—if left unchecked—may result in global temperature increases of several kelvins over the next half-century, enough to cause dramatic, and possibly catastrophic, changes in the Earth's climate.

Atmospheric Convection

Earth's air is heated not only by solar radiation. In the troposphere, convection is another important mechanism by which heat is transported from place to place. As sketched in Figure 7.7, convection occurs when hot matter rises, thus physically transferring heat from a lower (hotter) to a higher (cooler) level.

In Figure 7.7(a), part of Earth's surface is heated by the Sun. The air immediately above the warmed surface is heated, expands a little, and becomes less dense. As a result, the hot air becomes buoyant and starts to rise. At higher altitudes, the opposite effect occurs: The air gradually cools, grows denser, and sinks back to the ground. Cool air at the surface rushes in to replace the hot buoyant air. In this way, a circulation pattern is established. These **convection cells** of rising and falling air not only contribute to atmospheric heating, but are also responsible for surface winds. Atmospheric convection can also create clear-air turbulence—the bumpiness we may experience on aircraft flights. Ascending and descending parcels of air, especially below fluffy clouds (themselves the result of convective

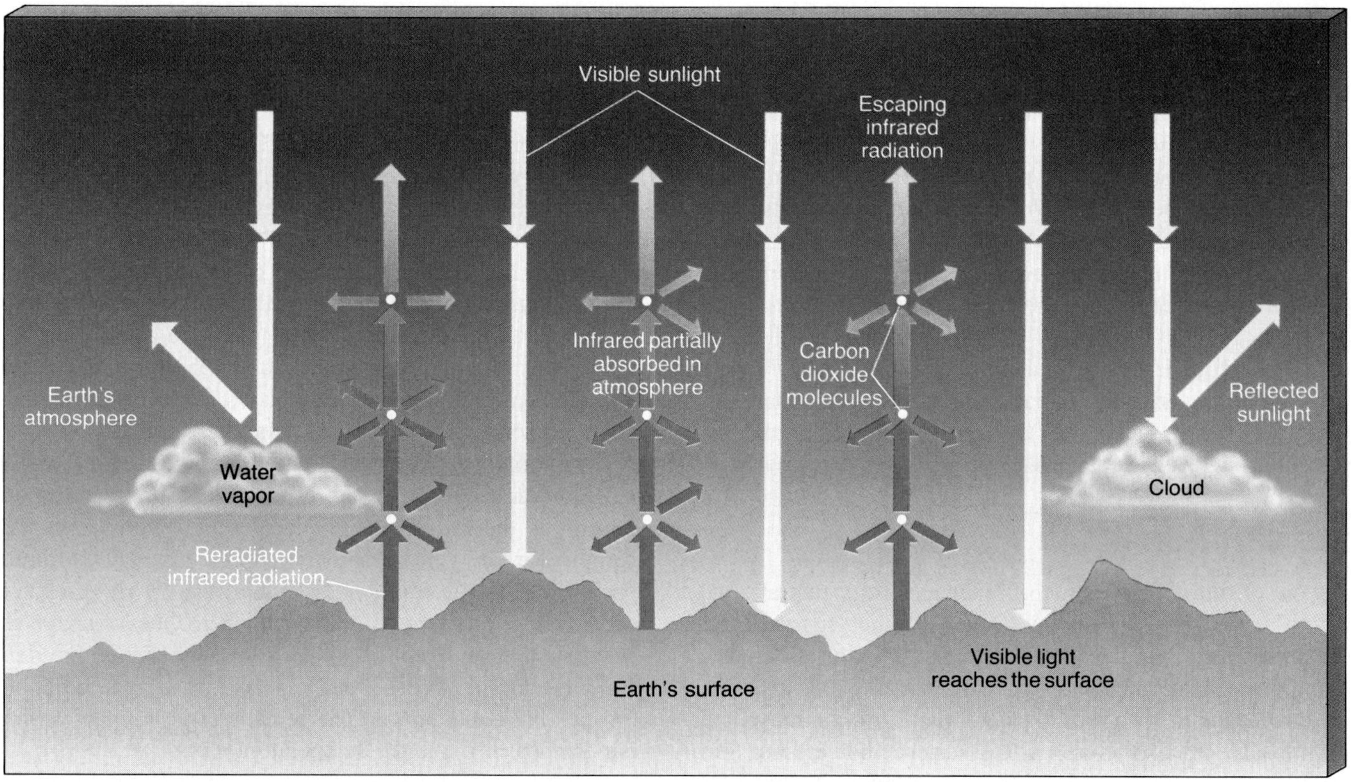

Figure 7.6 *The greenhouse effect. Sunlight that is not reflected by clouds reaches the Earth's surface, warming it up. Infrared radiation reradiated from the surface is partially absorbed by water vapor and carbon dioxide in the atmosphere, causing the overall surface temperature to rise.*

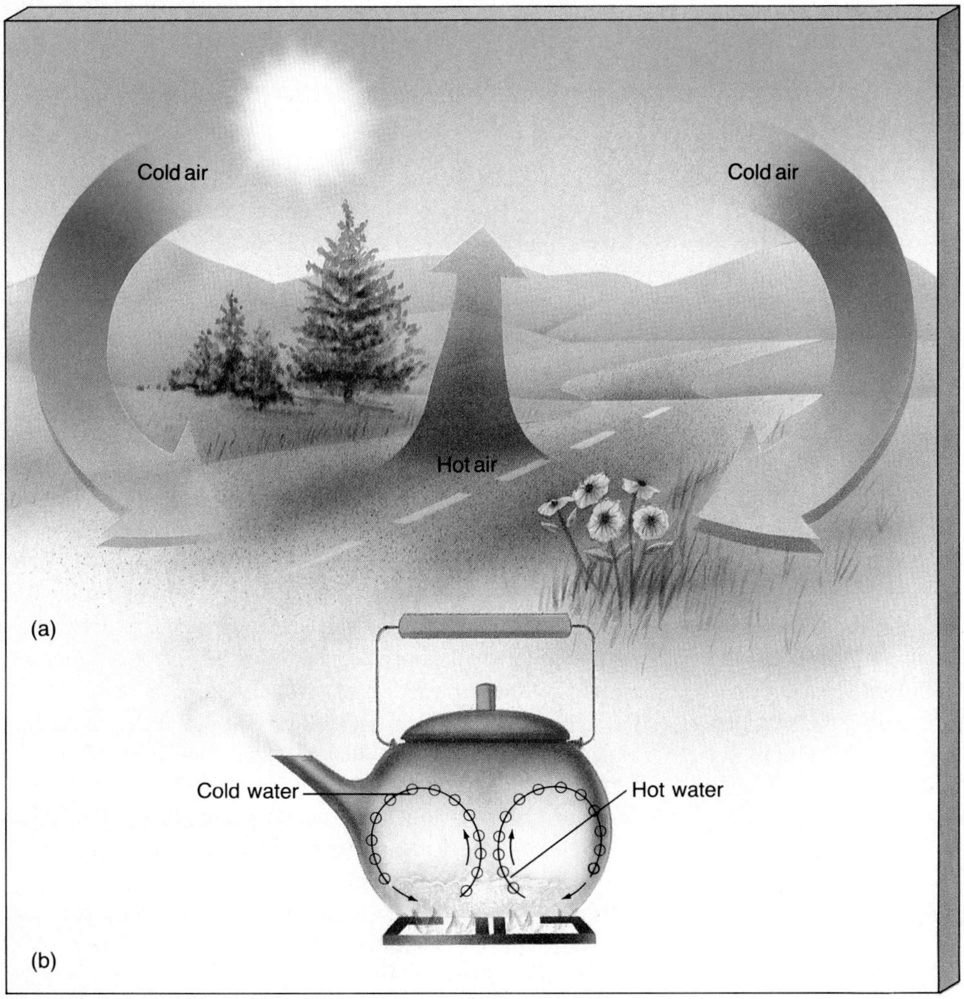

Figure 7.7 *Convection occurs whenever cool matter overlies warm matter. The resulting circulation currents are familiar to us (a) as the winds in the Earth's atmosphere caused by the solar-heated ground and (b) as the upwelling water motions in a pot of water caused by the heat of a stove. Hot air or hot fluid rises, cools, and falls repeatedly. Eventually, steady circulation patterns with rising and falling currents are established and maintained, provided that the source of heat (the Sun in the case of the atmosphere, the stove in the case of the water) remains intact.*

processes), frequently cause a choppy ride. For this reason, passenger aircraft tend to fly above the turbulence, in the lower stratosphere.

Origin of Earth's Atmosphere

Why is our atmosphere made up of its present constituents? Why is it not composed entirely of nitrogen, say, or of carbon dioxide, like the atmospheres of Venus and Mars? The origin and development of Earth's atmosphere was a fairly complex and lengthy process. The main evolutionary stages are outlined below.

When the Earth first formed, any **primary atmosphere** it might have had would have consisted of the gases most common in the early solar system—light gases, such as hydrogen, helium, methane, ammonia, and water vapor—a far cry from the atmosphere we enjoy today. Almost all of this light material, and especially any hydrogen or helium, escaped into space during the first half-billion or so years after the Earth was formed. (For more information on how planets retain or lose their atmospheres, consult Interlude 9-1.)

Subsequently, Earth developed a **secondary atmosphere,** which was *outgassed* from the planet's interior as the result of volcanic activity. Volcanic gases are rich in water vapor, methane, carbon dioxide, sulfur dioxide, and compounds containing nitrogen (such as nitrogen gas, ammonia, and nitric oxide). Solar ultraviolet radiation decomposed the lighter, hydrogen-rich gases, allowing the hydrogen to escape, and liberated much of the nitrogen from its bonds with other elements. As the Earth's surface temperature fell and the water vapor condensed, oceans formed. Much of the carbon dioxide and sulfur dioxide became dissolved in the oceans or combined with surface rocks. Oxygen is such a reactive gas that any free oxygen that appeared at early times was removed as quickly as it formed. An atmosphere consisting largely of nitrogen slowly appeared.

The final major development in the story of our planet's atmosphere is (so far) only known to have occurred on Earth. The appearance of *life* in the oceans 3.5 billion years ago eventually began to produce atmospheric oxygen. The ozone layer formed, shielding the surface from the Sun's harmful radiation. Eventually, life spread to the land and flourished. The fact that oxygen is a major constituent of the present-day atmosphere is a direct consequence of the evolution of life on Earth. Read Interlude 7-1 to learn why our sky is blue.

INTERLUDE 7-1 *Why Is the Sky Blue?*

Is the sky blue because it reflects the color of the ocean, or is the ocean blue because it reflects the color of the surrounding sky? The answer is the latter, and the reason has to do with the way that light is *scattered* by air molecules and minute dust particles. By scattering we mean the process by which radiation is absorbed and then reradiated by the material through which it passes.

As sunlight passes through our atmosphere, it is scattered by gas molecules in the air. This process turns out to be very sensitive to the wavelength of the light involved. The British physicist Lord Rayleigh first investigated this phenomenon about a century ago, and today it bears his name—it is known as Rayleigh scattering.

Rayleigh found that blue light is much more easily scattered than red light, in essence because the wavelength of blue light (4000 Å) is closer to the size of air molecules than is the wavelength of red light (7000 Å). He went on to prove that the amount of scattering is actually inversely proportional to the *fourth* power of the wavelength, so that blue light is scattered about nine ($[7000/4000]^4$) times more efficiently than red light. Dust particles also preferentially scatter blue light, but the amount of scattering by dust particles is inversely proportional only to the wavelength, so the contrast between red and blue scattering is only a factor of 1.75 (7000/4000).

Consequently, with the Sun at a reasonably high elevation, the blue component of incoming sunlight will scatter much more than any other color. Thus, some of the blue photons in the light are removed from the line of sight between us and the Sun and may scatter many times in the atmosphere before eventually entering our eyes, as shown in the figure below. Red or yellow photons are scattered relatively little and arrive at our eyes predominantly along the line of sight to the Sun. The net effect of all this is that the Sun is "reddened" slightly, because of the removal of blue photons, while the sky away from the Sun appears blue. In outer space, where there is no atmosphere, there is no Rayleigh scattering of sunlight, and the sky is black.

At dawn or dusk, with the Sun near the horizon, sunlight must pass through much more atmosphere before reaching our eyes—so much so, in fact, that the blue component of the Sun's light is almost entirely scattered out of the line of sight, and even the red component is diminished in intensity. Accordingly, the Sun itself appears orange—a combination of its normal yellow color and a reddishness caused by the subtraction of virtually all of the blue end of the spectrum—and dimmer than at noon. At the end of a particularly dusty day, when weather conditions or human activities during the daytime hours have raised excess particles into the air, short-wavelength Rayleigh scattering can be so heavy that the Sun appears brilliantly red. Reddening is often especially evident when we look at the westerly "sinking" summer Sun over the ocean, where seawater molecules have evaporated into the air, or during the weeks and months after an active volcano has released huge quantities of gas and dust particles into the air—as was the case in North America when the Philippine volcano Mt. Pinatubo erupted in 1991.

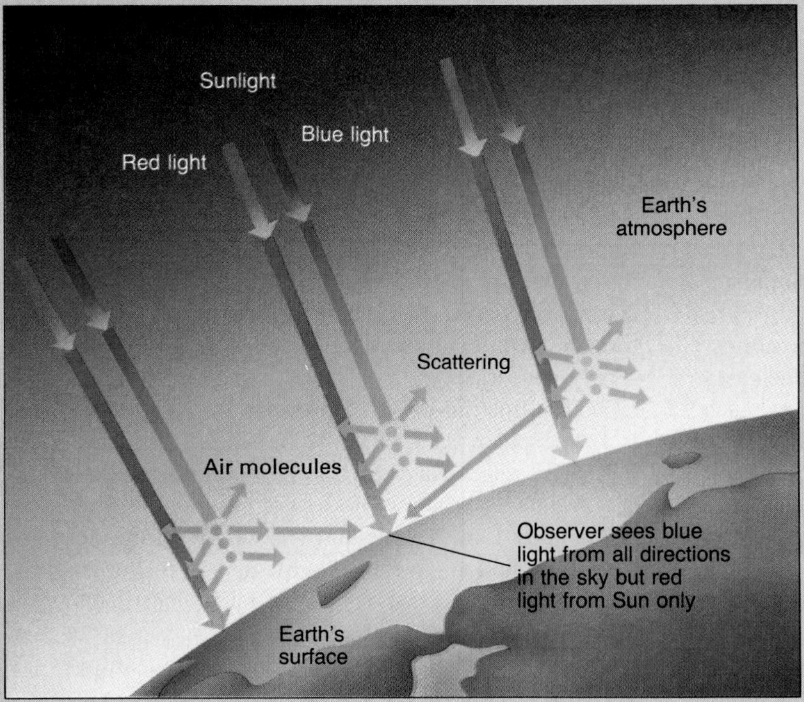

(NCAR)

THE MAGNETOSPHERE

☑3 Another part of Earth also helps protect us from the harsh realities of outer space. Discovered by artificial satellites launched in the late 1950s, the magnetosphere lies far above the atmosphere. Simply put, the magnetosphere is the region around the Earth that is influenced by our planet's magnetic field. Sketched in Figure 7.8, Earth's field is similar to that of a bar magnet. Having a north and a south pole, the magnetic field surrounds our planet three-dimensionally. The magnetic field lines, which indicate the strength and direction of the field at any point in space, run from south to north, as indicated. The north and south *magnetic poles,* where the axis of the imaginary bar magnet intersects the Earth's surface, are roughly aligned with the spin axis of the Earth.[2] The magnetic north pole is currently in northern Canada, at a latitude of about 80°N (80 degrees north), almost due north of the center of North America. The magnetic south pole lies at a latitude of about 60°S, just off the coast of Antarctica south of Adelaide, Australia.

Instrumented spacecraft have found that the magnetosphere contains two doughnut-shaped zones of high-energy particles, one about 3000 and the other 20,000 km above Earth's surface. Called the **Van Allen belts,** these zones are named for the American physicist whose instruments on board one of the first artificial satellites detected them. We call them "belts" because they are most pronounced near the Earth's equator or midsection and because they completely surround the planet. Figure 7.9 shows how these invisible regions envelop our planet, except near the north and south poles.

We could never survive unprotected in the Van Allen belts. Unlike the lower atmosphere, on which humans and other life forms rely for warmth and protection, much of the magnetosphere is subject to intense bombardment by large numbers of high-velocity, and potentially very harmful, charged particles. Colliding violently with an unprotected human body, these particles would deposit large amounts of energy wherever they made contact, causing severe damage to living organisms. Without sufficient shielding on the *Apollo* spacecraft, for example, the astronauts might not have survived the passage through the magnetosphere on their journey to the Moon.

What is the origin of the magnetosphere and the Van Allen belts within it? Recognize first of all that the Earth's magnetic field is not a permanent part of the planet, but instead is thought to be generated within the Earth's core. Like the dynamos that run industrial machines, the spin of electrically conducting metal deep inside the Earth induces our planet's magnetism. The theory that explains planetary (and other) magnetic fields in terms of rotating, conducting material flowing in the planet's interior is known as **dynamo theory.**

Why do the particles that make up the Van Allen belts stay in the Earth's magnetosphere? Traveling through space, neutral particles and electromagnetic radiation are unaffected by Earth's magnetism, but electrically *charged* particles are strongly influenced. A magnetic field exerts a

[2] When we talk about everyday bar magnets, such as a compass needle, the end that is attracted toward the North Pole is conventionally designated the "north-seeking," or simply "north," pole of the magnet. Similarly, the "south-seeking" pole points south. Because unlike magnetic poles attract, this convention has an unfortunate side effect: If we wish to imagine the Earth's field as an enormous bar magnet, then Earth's north magnetic pole, which attracts the north-seeking pole of a compass needle, must properly be labeled *S,* and the south magnetic pole must be labeled *N!*

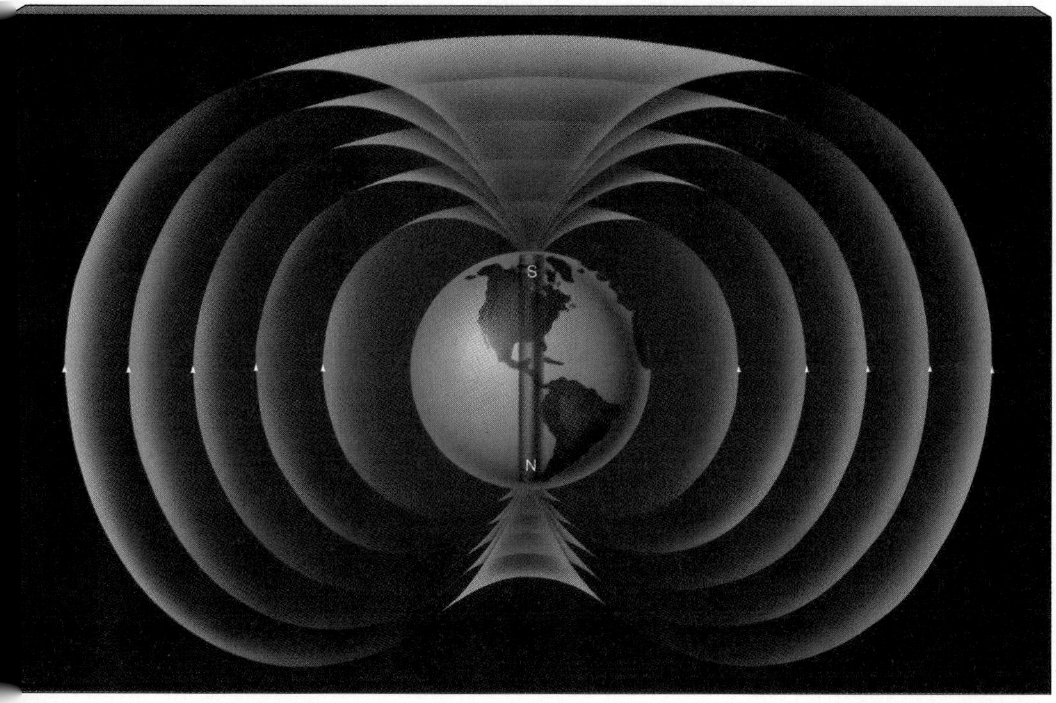

Figure 7.8 *Earth's magnetic field resembles that of an enormous bar magnet situated inside our planet. Because of the convention used in naming magnetic poles, the Earth's north magnetic pole, which attracts the "north-seeking" pole of a bar magnet, must be labeled S! The arrows on the field lines indicate the direction in which a compass needle would point.*

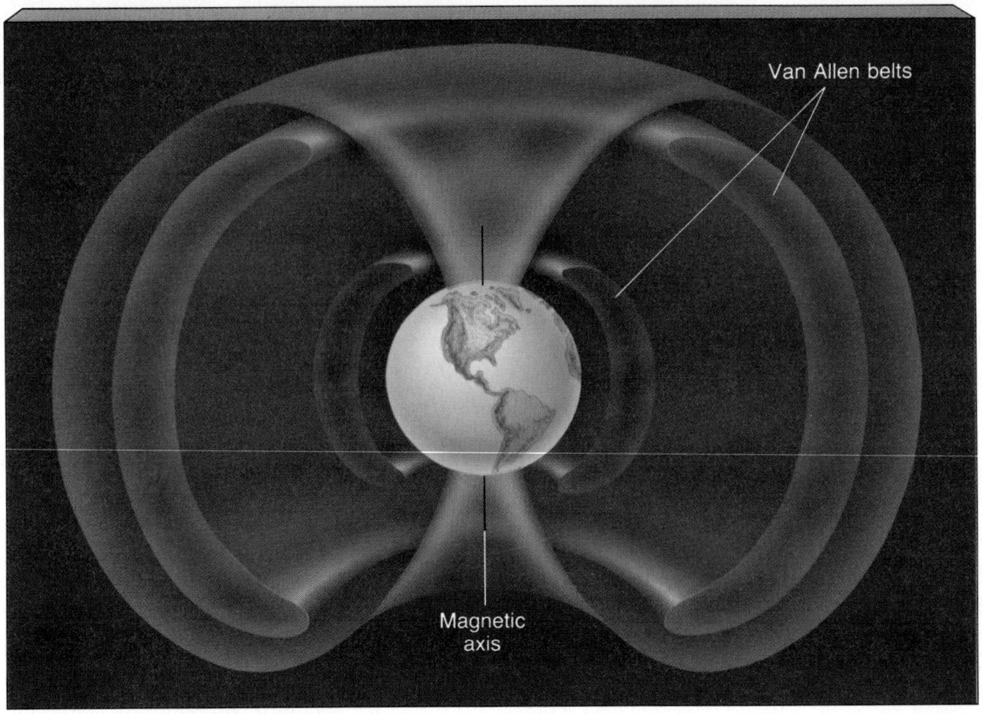

Figure 7.9 High above Earth's atmosphere, the magnetosphere (lightly shaded area) contains at least two doughnut-shaped regions (heavily shaded areas) of magnetically trapped charged particles. These are the Van Allen belts.

force on a moving charged particle, causing it to spiral around the magnetic field lines, as illustrated in Figure 7.10. Charged particles headed toward Earth, especially protons and electrons from the Sun—the so-called **solar wind**—can become trapped by the Earth's magnetism. In this way, Earth's magnetic field, sketched in Figure 7.8, exerts electromagnetic control over the particles, herding them into the shape of the Van Allen belts shown in Figure 7.9.

The charged particles often escape from the magneto-

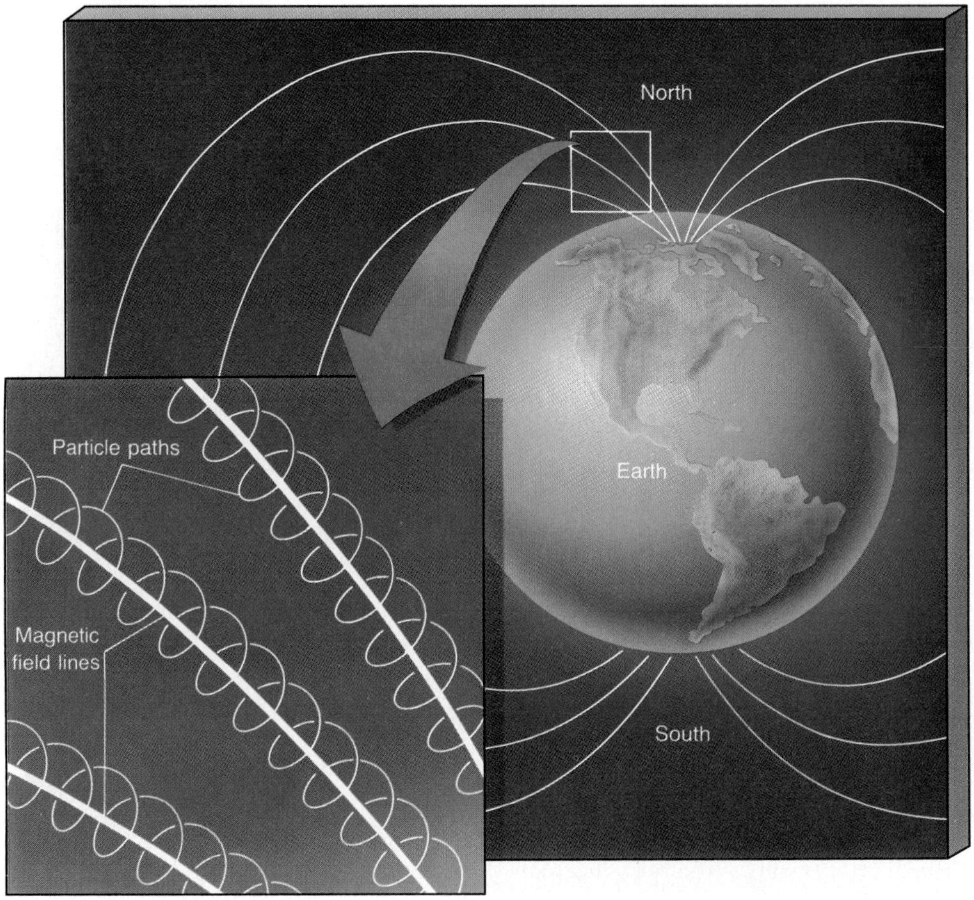

Figure 7.10 A charged particle in a magnetic field spirals around the field lines. Thus, charged particles tend to become "trapped" by strong magnetic fields.

sphere near the north and south poles, where the field lines intersect the atmosphere. Their collisions with the air rip apart some atmospheric molecules, creating a spectacular light show. Called an **aurora,** shown in Figure 7.11, this colorful display results when atmospheric molecules, excited upon collision with the charged particles, fall back to their ground states and emit visible light. Aurorae are most brilliant at high latitudes, especially within the Arctic and Antarctic circles. In the north, the spectacle is called the **aurora borealis,** or *northern lights*. In the south, it is called the **aurora australis,** or *southern lights*.

Occasionally, particularly after a storm on the Sun, the Van Allen belts can become distorted by the solar wind and overloaded with many more particles than normal, allowing some particles to escape prematurely and at lower latitudes. For example, in North America, the aurora borealis is normally seen with any regularity only in northern Canada. However, at times of greatest solar activity, the display has been seen as far south as the southern United States.

Actually, Earth's magnetosphere is not nearly as symmetrical as depicted in Figure 7.8. Satellites have mapped its true shape. As shown in Figure 7.12, the entire region of trapped particles is quite distorted, forming a teardrop-shaped cavity. On the sunlit (daytime) side of Earth, the magnetosphere is compressed by the flow of high-energy particles in the solar wind. The boundary between the magnetosphere and this flow is known as the **magnetopause.** It is found at about 10 Earth radii from our planet. On the side opposite the Sun, the belts are extended, with a long tail often reaching beyond the orbit of the Moon.

Earth's magnetic field plays an important role in controlling many of the potentially destructive charged particles

Figure 7.11 *A colorful aurora results from the emission of light radiation after the collision of magnetospheric particles with atmospheric molecules. The aurora rapidly flashes across the sky, like huge wind-blown curtains glowing in the dark. (NCAR)*

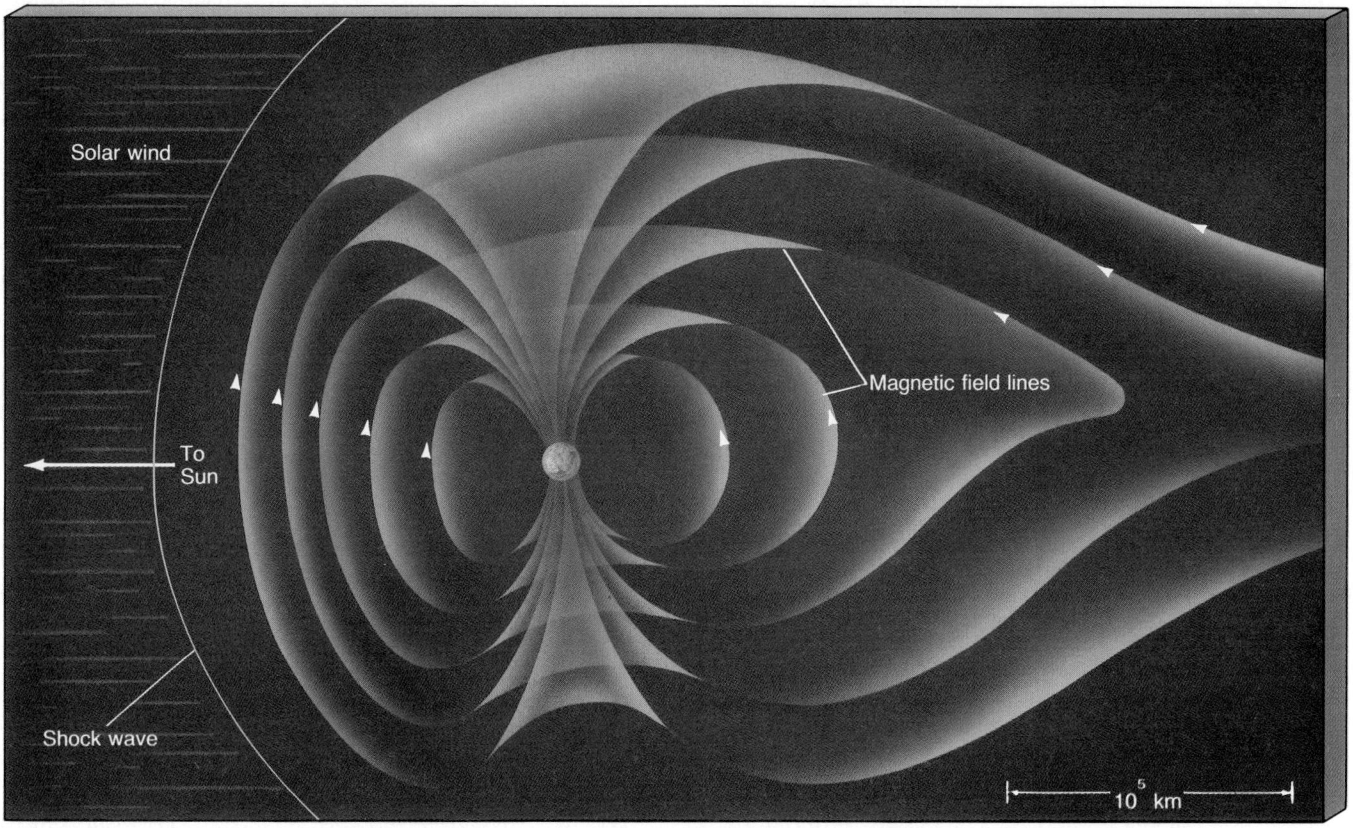

Figure 7.12 *Earth's real magnetosphere is greatly distorted by the solar wind, with a long tail extending from the nighttime side of Earth well into space.*

that venture near our planet. Without the magnetosphere, Earth's atmosphere—and perhaps the surface, too—would be bombarded by harmful particles, possibly damaging many forms of life on our planet. Some researchers have even suggested that had the magnetosphere not existed in the first place, life might never have arisen at all on planet Earth (see also Interlude 7-4).

THE INTERIOR

✓4 Although we reside on Earth, we cannot easily probe our planet's interior. Drilling gear can penetrate rock only so far before breaking. No substance—even diamond, the hardest known material—can withstand the pressure below a depth of about 10 km. That's rather shallow compared with Earth's 6500 km (approximate) radius. Fortunately, geologists have developed other techniques that indirectly probe the deep recesses of our planet.

Seismic Waves

A sudden dislocation of rocky material near Earth's surface—an **earthquake**—causes the entire planet to vibrate a little. It literally rings like a bell. These vibrations are not random. They are systematic waves, called **seismic waves,** that move outward from the site of the quake. Like all waves, they carry information. This information can be detected and recorded using sensitive equipment—a seismograph—designed to monitor Earth tremors.

Decades of earthquake research have demonstrated many kinds of seismic wave. Two types are of particular importance to the study of the Earth's internal structure. First to arrive at a monitoring site after a distant earthquake are the primary waves, or **P-waves.** These are *pressure* waves, a little like ordinary sound waves in air, that alternately expand and compress the material medium through which they move. Seismic P-waves usually travel at velocities ranging from 5 to 6 km/s and can travel through both liquids and solids. Some time later (the actual delay depending on the distance to the earthquake site), secondary waves, or **S-waves,** arrive. These are *shear* waves. Unlike P-waves, which vibrate the material through which they pass back and forth along the direction of travel of the wave, S-waves cause side-to-side motion, more like waves in a guitar string. The two types of wave are illustrated in Figure 7.13. S-waves normally travel through Earth's interior at 3 to 4 km/s. Seismic S-waves cannot travel through liquid, which absorbs them.

The velocity of both types of wave depends on the density of the matter through which they are traveling. Consequently, if we can measure the time taken for the waves to move from the site of an earthquake to one or more monitoring stations on Earth's surface, we can determine the density of matter in the interior. Figure 7.14 illustrates some

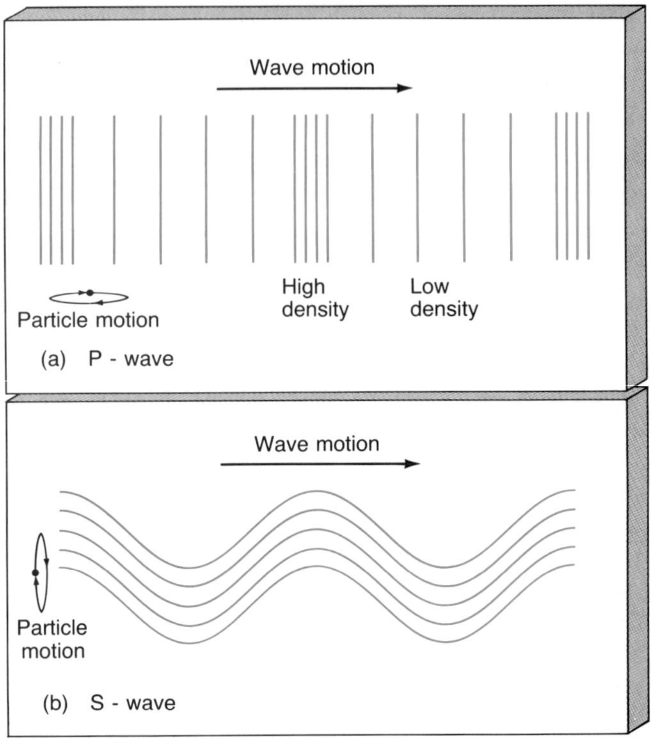

Figure 7.13 (a) A pressure (P-) wave traveling through the Earth's interior causes material to vibrate in a direction parallel to the direction of motion of the wave. Material is alternately compressed and expanded. (b) A shear (S-) wave produces motion perpendicular to the direction in which the wave travels, pushing material from side to side. Both waves are shown as if seen from above. Also shown is the motion of one typical particle. In case (a), the particle oscillates forward and backward about its initial position. In (b), the particle moves from side to side.

paths followed by P- and S-waves from the site of an earthquake. Seismographs located around the world measure the times of arrival as well as the strengths of the seismic waves. Both observations contain much useful information—both about the earthquake itself and about Earth's interior through which the waves pass. Notice that the waves do not travel in straight lines through the planet. Because the wave velocity varies with depth, the waves bend as they move through the interior.

A particularly important result emerged after numerous quakes were monitored several decades ago: Seismic stations on the side of the world opposite a quake never detect S-waves—some shear waves seem to be blocked by material within Earth's interior. Further, while P-waves never fail to arrive at stations diametrically opposite the quake, there are parts of the Earth's surface that they cannot reach (see Figure 7.14). Most geologists believe that S-waves are absorbed by a liquid core at the center of the Earth and that P-waves are "refracted" at the core boundary, much as light is refracted by a lens. The result is the S- and P-wave "shadow zones" we observe. The fact that every earthquake exhibits these shadow zones is the best

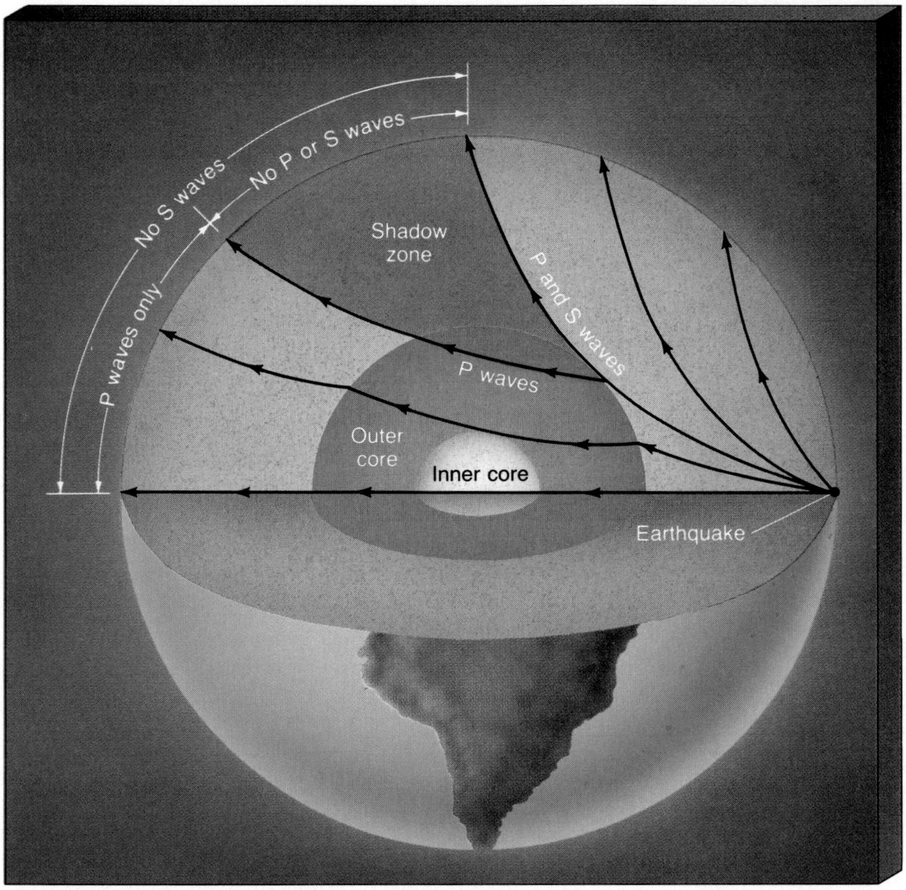

Figure 7.14 *Earthquakes generate pressure (P-) and shear (S-) waves that can be detected at seismographic stations around the world. S-waves are not detected by stations "shadowed" by the liquid core of the Earth. P-waves do reach the side of the Earth opposite the earthquake, but their interaction with Earth's core produces another shadow zone, where no P-waves are seen.*

evidence that the core of our planet is hot enough to be in the liquid state.

The radius of the core, as determined from seismic data, is about 3500 km. In fact, very faint P-waves *are* observed in the P-wave shadow zone indicated in Figure 7.14. These are believed to be reflected off the surface of a solid inner core, of radius 1300 km, lying at the center of the liquid outer core.

Modeling the Interior

Because earthquakes occur often and at widespread places across the globe, geologists have accumulated a large amount of data about shadow zones and seismic-wave properties. They have used these data, along with direct knowledge of surface rocks, to build mathematical models of the Earth's interior.

Figure 7.15 presents a recent model that most scientists accept. As the graphs show, the density and temperature both increase with depth. Specifically, from Earth's surface to its very center, the density increases from about 3 g/cm^3 to a little more than 12 g/cm^3. These densities suggest to geologists that the inner parts of the Earth must be rich in nickel and iron. Under the heavy pressure of the overlying layers, these metals (whose densities under "normal" surface conditions are around 8 g/cm^3) can be compressed to the abnormally high densities predicted by the model. The sharp density increase at the mantle-core boundary results from the difference in composition between the two regions. The mantle is composed of dense but *rocky* material. The core consists primarily of even denser *metallic* elements. There is no similar jump in density or temperature at the inner core boundary—the material there simply changes from the solid to the liquid state.

The model indicates that the core must be a mixture of nickel, iron, and some other lighter element, possibly sulfur. Without direct observations, it is difficult to be absolutely certain of the light component's identity. All geologists agree that much of the core must be in the liquid state. The existence of the shadow zone demands that, and our current explanation of the Earth's magnetic field relies on it. The outer parts of the core are liquid, because of the high temperature. However, the pressure near the center, about 4 million times the pressure of the atmosphere at the Earth's surface, is high enough to force the material there into the solid state.

The outer core is surrounded by a thick mantle and topped with a thin crust. The mantle is about 3000 km thick and accounts for the bulk (80 percent) of our planet's volume. Models of the Earth's interior suggest that much of the mantle has a density midway between the densities of the core and crust—namely, about 5 g/cm^3. The crust has an average thickness of only 15 km—a little less (around 8 km) under the oceans and somewhat more (20 to 50 km) under the continents. The average density of the material making up Earth's crust is around 3 g/cm^3.

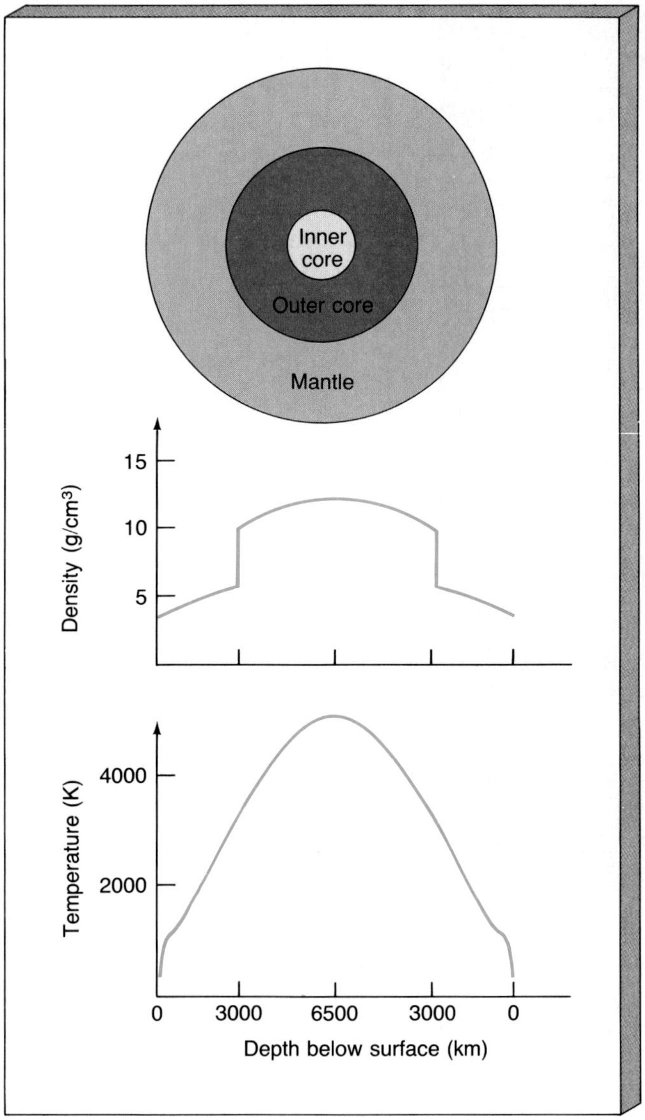

Figure 7.15 *Computer models of Earth's interior imply that the density and temperature vary considerably through the mantle and the core. Note the sharp density discontinuity between Earth's core and mantle.*

Because geologists have been unable to drill deeper than about 10 km, we recognize a startling fact: No experiment has yet succeeded in piercing Earth's crust to recover a sample of the mantle. Fortunately, we are not entirely ignorant of the mantle's properties. In a **volcano** (Figure 7.16), hot lava upwells from below the crust, bringing a little of the mantle to us and providing some inkling of the Earth's interior. The chemical makeup and physical state of the newly emerged lava are generally consistent with predictions based on the model sketched in Figure 7.15.

The chemical composition of the upper mantle is probably similar to the iron-magnesium-silicate mixtures known as **basalt.** You may have seen some dark gray basaltic rocks scattered across the Earth's surface, especially near volcanos. Basalt is formed as mantle material upwells from Earth's interior as lava, then cools and solidifies. With a density between 3.0 and 3.3 g/cm^3, basalt contrasts with the lighter **granite** (density 2.7–3.0 g/cm^3) that constitutes much of the rest of the Earth's crust. Granite is richer than basalt in the light elements silicon and aluminum, which explains why the surface continents do not sink into the interior. Their low-density composition lets the crust "float" atop the denser matter of the mantle and core below.

Earth, then, is not a homogeneous ball of rock. Instead, it has a layered structure, with a low-density crust at the surface, intermediate-density material in the mantle, and a high-density core. Such variation in density and composition is known as **differentiation.** Why isn't our planet just one big, rocky ball of uniform density? The answer appears to be that much of the Earth was *molten* at some time in the past. As a result, the higher-density matter sank to the core, and the lower-density material was displaced toward the surface. The remnant of this ancient heating can still be seen in our models of Earth's interior as it exists today. Figure 7.15 shows a sharp increase in temperature from 300K at the surface to an extremely hot 5000K in the core. Earth's central temperature is nearly equal to that of the Sun's surface.

The Cause of Differentiation

Only the outer core of the Earth is molten now. The temperature of today's mantle is not hot enough to melt its rocky matter. Yet for Earth to have become differentiated, the *entire* planet must have been molten, or at least hot enough that it became plastic (semi-solid), at some ancient time. Only then could heavy material sink toward the center, increasing the density in the core.

What processes were responsible for heating the entire planet to this extent? To answer this question, we must try to visualize the past. When the Earth formed, it did so by capturing material from its surroundings (see Chapter 17). As the young planet grew, its gravitational field strengthened, and the speed with which newly captured matter struck its surface increased. This process generated a lot of heat—so much, in fact, that the Earth may already have been partially or wholly molten by the time it reached its present size. As the Earth began to differentiate, and heavy material sank to the center, even more gravitational energy was released, and the interior temperature must have increased still further. Later, Earth continued to be bombarded with debris left over from the formation process. At its peak some 4 billion years ago, this secondary bombardment was probably intense enough to keep the surface molten, but only down to a depth of a few tens of kilometers. Erosion by wind and water has long since removed all trace of this early period from the surface of the Earth, but our Moon still bears visible scars of the onslaught.

Another important process for heating the early Earth after its formation was **radioactivity**—the release of energy by certain rare, heavy elements, such as uranium, thorium, and plutonium (see Interlude 7-2). These elements emit

INTERLUDE 7-2 *Radioactivity*

In Chapter 5 we saw that atoms are made up of electrons and nuclei and that nuclei are composed of protons and neutrons. The number of protons in a nucleus determines which element it represents. The elements we have studied so far—hydrogen, helium, carbon, iron—are all *stable*. For example, left alone, a carbon-12 nucleus, consisting of six protons and six neutrons, will remain unchanged forever. It will not break up into smaller pieces, nor will it turn into anything else.

Not all nuclei are stable, however. Many heavy nuclei—for example, uranium-235 (containing 92 protons and 143 neutrons), uranium-238 (92 protons, 146 neutrons), plutonium-239 (94 protons, 145 neutrons), and plutonium-241 (94 protons, 147 neutrons)—are inherently *unstable*. Left alone, they will eventually break up into lighter "daughter" nuclei, in the process emitting some elementary particles and releasing some energy. The change happens spontaneously, without any external influence. This instability is known as *radioactivity*. The energy released by the disintegration of the radioactive elements we just listed is the basis for nuclear fission reactors (and nuclear bombs).

Unstable heavy nuclei achieve greater stability by disintegrating into lighter nuclei, but they do not break up immediately. Each type of "parent" nucleus needs a characteristic amount of time before it decays. The *half-life* is the time required for half of a sample of parent nuclei to disintegrate. Notice that this is really a statement of probability. We cannot say *which* nuclei will decay in any given half-life interval, only that half of them are expected to do so. Thus, if we start with a billion radioactive nuclei embedded in a sample of rock, a half-billion nuclei would remain after one half-life, a quarter-billion after two half-lives, and so on.

The figure below illustrates the half-lives and decay reactions for four unstable heavy nuclei. The decay products in this case are stable nuclei of lead or bismuth, electrons, and helium nuclei. For some radioactive elements, the half-life is short, even as brief as a few seconds. Other nuclei have decay times of a million years or more. Two of the elements noted in the figure have half-lives of billions of years. Every kind of radioactive element has its own characteristic half-life, and most of them are now well known from studies conducted since the 1950s.

Lighter elements can be radioactive, too. In fact, adding neutrons to (or removing them from) any stable nucleus will eventually result in an unstable nucleus of that element. For example, carbon-12 (by far the most common form of carbon found in the universe) is stable, but carbon-14, with six protons and eight neutrons, is unstable. Eventually one of carbon-14's neutrons turns into a proton, changing the nucleus into stable nitrogen-14 (seven protons, seven neutrons) and emitting an electron. The half-life of carbon-14 is about 6000 years.

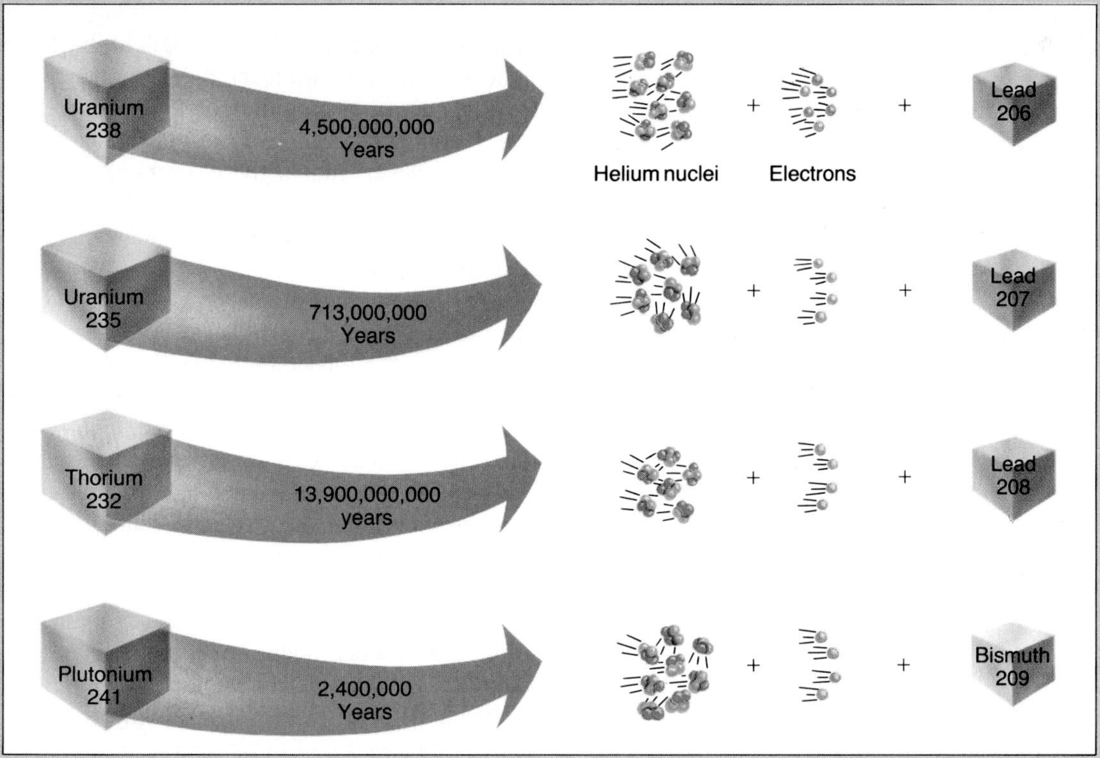

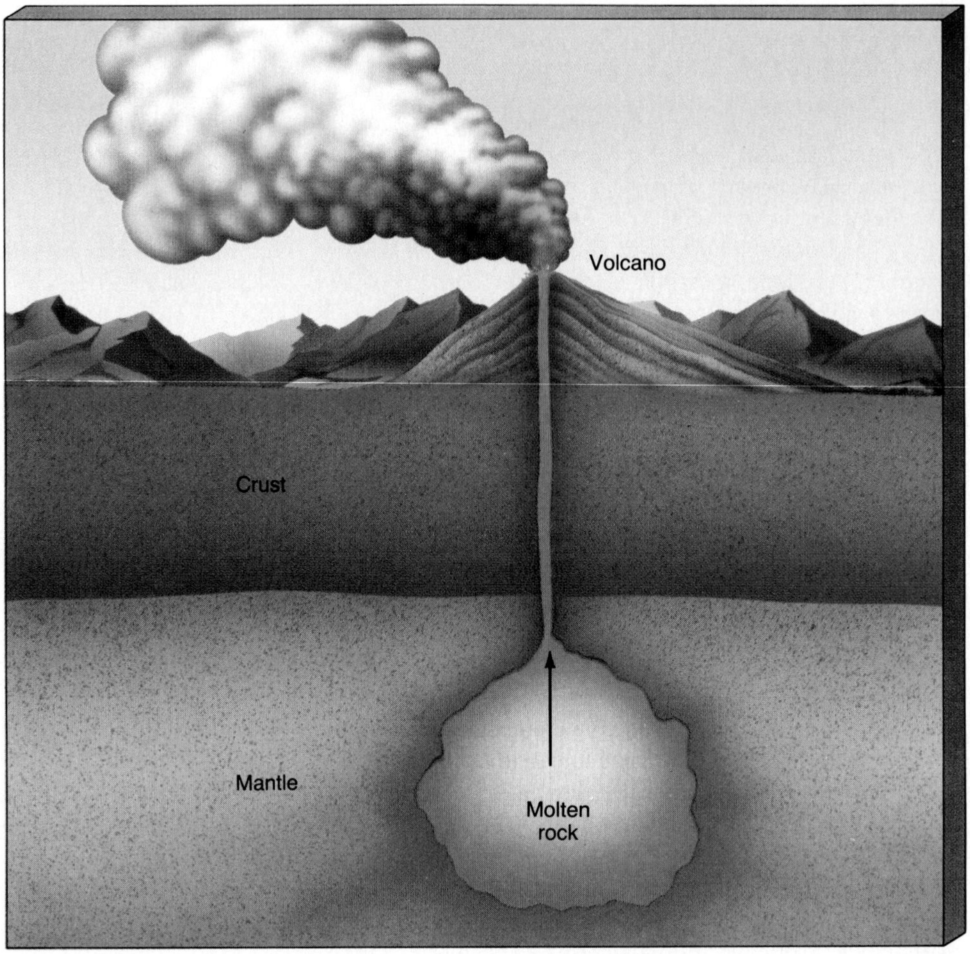

Figure 7.16 *The material that ultimately erupts from a volcano actually originates deep below the Earth's surface, in the upper mantle of our planet.*

energy as their complex, heavy nuclei decay into simpler, lighter nuclei. While the energy produced by the decay of a single radioactive atom is tiny, the Earth contained a lot of radioactive atoms, and a lot of time was available. Rock is such a poor conductor of heat that the energy would have taken a very long time to reach the surface and leak away into space, so the heat built up in the interior, adding to the energy left there by the Earth's formation. Provided that enough radioactive elements were originally spread throughout the primitive Earth, rather like raisins in a cake, the entire planet—from crust to core—could have melted and remained molten for about a billion years. That is a long time by human standards, but not so long in the cosmic scheme of things. Measurements of the ages of some surface rocks indicate that Earth finally began to solidify roughly a billion years after it originally formed (see Interlude 7-3).

Earth's Geological History

The arguments above imply that our entire planet was plastic or molten during most of its first billion years of existence. It differentiated during that time. Much of the denser matter sank toward the core, while the lighter matter rose to the surface. Radioactive heating did not stop after the first billion years. It continued even after Earth's surface cooled and solidified. However, radioactivity works in only one direction, always producing lighter elements from heavier ones. Once decayed, the heavy and rare radioactive elements cannot be replenished.

The early source of heat diminished with time, allowing the planet to cool over the past 3.5 billion years. In so doing, it cooled from the outside in, much like a hot potato. The reason for this is that it cooled by radiating its energy into space, and regions closest to the surface could most easily unload their excess heat. In this way, the surface developed a crust, and the differentiated interior attained the layered structure now implied by seismic studies. Today, radioactive heating continues throughout the Earth. However, there is probably not enough of it to melt any part of our planet. The high temperatures in the core are mainly the trapped remnant of a much hotter Earth that existed eons ago. The surface has completely cooled and solidified, making it a habitable place from which intelligent life can begin to unravel the history of our planet.

INTERLUDE 7-3 *Radioactive Dating*

The decay of unstable radioactive nuclei into more stable nuclei is an important natural phenomenon. Not only does it heat, and thus lead to differentiation on, planets such as Earth, but it also provides us with a useful tool to date any rocks we can get our hands on.

How do geologists determine the age of a rock? The first step is to measure the amount of stable nuclei of a given kind (for example, lead-206). This amount is then compared with the amount of remaining unstable "parent" nuclei (in this case, uranium-238) from which the stable "daughter" nuclei descended (see Interlude 7-2). The third step involves knowing the rate (or half-life) at which the disintegration occurs. In the simplest possible calculation, we might then assume that all of the daughter nuclei resulted from the radioactive decay of a parent. (In other words, in the above example we ignore the possibility that any lead-206 was initially present in the rock.) The age of the rock then follows directly. For example, if half of the parent nuclei of some element have decayed, so that the number of daughter nuclei equals the number of parents, the age of the rock must be equal to the half-life of the radioactive nucleus studied. Similarly, if only a quarter of the parent nuclei remain (three times as many daughters as parents), the rock's age is twice the half-life of that nucleus, and so on.

Typically, several sets of parent-daughter measurements are used in calibrating the age of a given sample. Combining these measurements results in an age determination accurate to within a few percent. In this way, the most ancient rocks on Earth are dated to be nearly 4 billion years old (3.9 billion years old, to be more precise). These rare specimens have been found in Greenland and Labrador.

The radioactive-dating technique rests on the assumption that the rock has remained *solid* while the radioactive decays have been going on. If the rock melts, there is no reason to expect the daughter nuclei to remain in the same locations their parents had occupied, and the whole method fails. Thus, radioactive dating indicates the time that has elapsed since the *last* time the rock in question solidified. Hence this 4-billion-year value represents only a portion of the true age of our planet. It does not measure the duration of its molten existence.

A variety of indirect arguments (discussed at various places in the text) suggests that 4.5 billion years ago is a special date. For example, all meteorites (undifferentiated rocky debris that once traveled through space before colliding with Earth) are radioactively dated to be 4.5 billion years old. Furthermore, the oldest Moon rocks have ages close to 4.5 billion years. Clearly, something important happened in the vicinity of Earth, and perhaps throughout the entire solar system, 4.5 billion years ago. That something, astronomers reason, was the formation of the Sun and planets.

SURFACE CHANGE

Surface Activity

✓5 Earth is geologically alive today. Its interior seethes and its surface changes. Figure 7.17 shows two indicators of surface geological activity: volcanos, where molten rock and hot ash upwell through fissures or cracks in the surface, and earthquakes, which occur when the crust suddenly dislodges under great pressure. Catastrophic volcanos and earthquakes are relatively rare events these days, but geological studies of rocks, lava, and other surface features imply that surface activity must have been more frequent, and probably more violent, long ago.

Many cases of past geological events are scattered across our globe. Erosion by wind and water has wiped away much of the evidence for ancient activity, but modern exploration has documented the sites of most of the recent activity, such as earthquakes and volcanic eruptions. Figure 7.18 is a map of the currently active areas of our planet. The hatched areas represent sites of volcanism or earthquakes. Nearly all these sites have experienced surface activity within this century, some of them suffering much damage and loss of many lives. These sites are especially abundant along the western coast of the United States, throughout the Aleutian Islands off the coast of Alaska, down along the Andes Mountains, across the Japanese Islands, up through India, and throughout much of Turkey, Greece, and the Aegean Sea.

The intriguing aspect of Figure 7.18 is that the active sites are not spread evenly across our planet. Instead, they trace well-defined lines of activity, where crustal rocks dislodge (as in earthquakes) or mantle material upwells (as in volcanos). In the mid-1960s, it became clear that these lines are really the outlines of gigantic "plates," or slabs of the Earth's surface.

Most startling of all, these plates are slowly moving—literally drifting around the surface of our planet. These plate motions have created the surface mountains, the oceanic trenches, and the other large-scale features across the face of planet Earth. In fact, plate motions have shaped the continents themselves. The process is popularly known as "continental drift." The technical term for the study of plate movement and the reasons for it is **plate tectonics.**

(a) (b) (c)

Figure 7.17 Active volcanos on (a) Mount Kilauea in Hawaii and (b) Mount St. Helens in the state of Washington. Kilauea seems to be a virtually ongoing eruption. St. Helens was a rare catastrophic eruption (equivalent to about 1000 Hiroshima-type atomic bombs), shown here blowing its top on May 18, 1980. (c) The aftermath of the earthquake that claimed dozens of lives and caused billions of dollars' worth of damage in northern California in October 1989. (NCAR; Bettmann Archive)

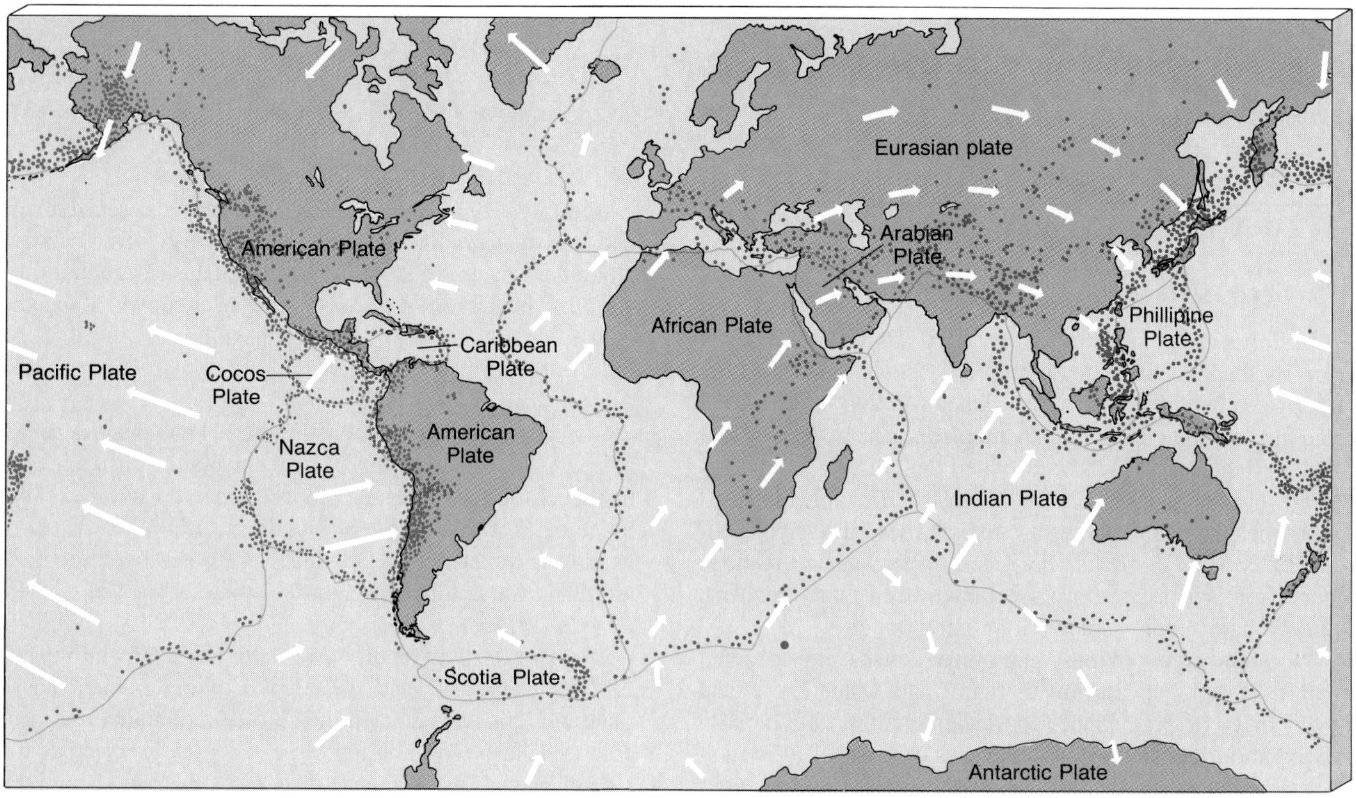

Figure 7.18 Hatched areas represent active sites where major volcanos or earthquakes have occurred in the twentieth century. Taken together, the sites outline vast "plates" that drift around on the surface of our planet. The arrows show the general directions of the plate motions.

The theory of plate tectonics suggests that the plates are not simply slowing to a stop after some initial and ancient movements. Rather, they are still drifting today, although at an extremely slow rate. Typical velocities of the plates amount to only a few centimeters per year. However, this is well within the measuring capabilities of modern equipment. Curiously enough, one of the best ways of monitoring plate motion on a global scale is by making accurate observations of very distant astronomical objects. Quasars (see Chapter 27), lying many hundreds of millions of light-years from Earth, will never show any measurable apparent motion on the sky stemming from their own motion in space. Thus any apparent change in their position (after correction for the Earth's motion, of course) can be interpreted as arising from the motion of the telescope or of the plate on which it is located.

On smaller scales, laser-ranging and other techniques now routinely track the relative motion of plates in many populated areas, such as California. During the course of Earth history, each plate has had plenty of time to move large distances, even at their sluggish pace. For example, a drift rate of only 2 cm per year can cause two continents (for example, Europe and North America) to separate by some 4000 km over the course of 200 million years. That may be a long time by human standards, but it represents only about 5 percent of the age of the Earth.

A common misconception is that the plates are the continents themselves. Some plates are indeed made mostly of continental landmasses, but other plates are made of a continent plus a large part of an ocean. For example, Figure 7.19 is a photograph of the Indian Plate, which includes all of India, much of the Indian Ocean, and all of Australia and its surrounding south seas. Still other plates are mostly ocean. The seafloor itself is the slowly drifting plate, and the oceanic water merely fills in the depressions between continents. The southeastern portion of the Pacific Ocean, called the Nazca Plate (see Figure 7.18), contains no landmass at all. For the most part, the continents are just passengers riding on much larger plates.

One might wonder why these large drifting regions are called plates. The reason is that their vertical thickness is slight compared with their surface area (like a plate). The thickness of a typical plate is approximately 50 km, usually less than 1/20 of its full horizontal size. Indeed, the plates do resemble upside-down dishes.

Taken together, the plates make up the Earth's **lithosphere,** which contains both the crust and a small part of the upper mantle. The lithosphere is the portion of the Earth that undergoes tectonic activity. The semisolid part of the mantle over which the lithosphere slides is known as the **aesthenosphere.** The relationships between these regions of the Earth are shown in Figure 7.20.

All the currently known plates of the world are marked on Figure 7.18. Note that the major boundary separating the North American Plate from the Eurasian Plate is a thin strip of activity in the middle of the Atlantic Ocean.

Figure 7.19 *Spacecraft photograph of the subcontinent of India and much of the Indian Ocean. All this (and more) is included in the Indian Plate, which is outlined in Figure 7-18. The Himalaya mountains appear near the top. (NASA)*

Discovered after World War II by oceanographic ships studying the geography of the seafloor, this giant fault is called the Mid-Atlantic Ridge. It extends, like a seam on a giant baseball, all the way from Scandinavia in the North Atlantic to the latitude of Cape Horn at the southern tip of South America. This ridge, then, also separates the South American Plate from the African Plate. The only major part of the entire ridge that rises above sea level is the island of Iceland.

As the plates drift around, we might expect collisions to be routine. Indeed, plates do collide. But unlike two automobiles that collide and then stop, the surface plates are driven by enormous forces. They do not stop easily. Instead, they just keep crunching into one another. Figure 7.21 shows a collision currently occurring between two continental landmasses. The resulting folds of rocky crust, shown in the figure, create mountains—in this case, Mount Everest, in the Himalayan mountain range. This entire mountain system results from the Indian Plate thrusting northward into the Eurasian Plate. The collision is still going on today.

Not all colliding plates produce mountain ranges. At other collision locations, called **subduction zones,** one plate slides under the other, ultimately to be destroyed. Subduction zones are responsible for most of the deep trenches in the world's oceans.

Nor do all plates experience head-on collisions. As

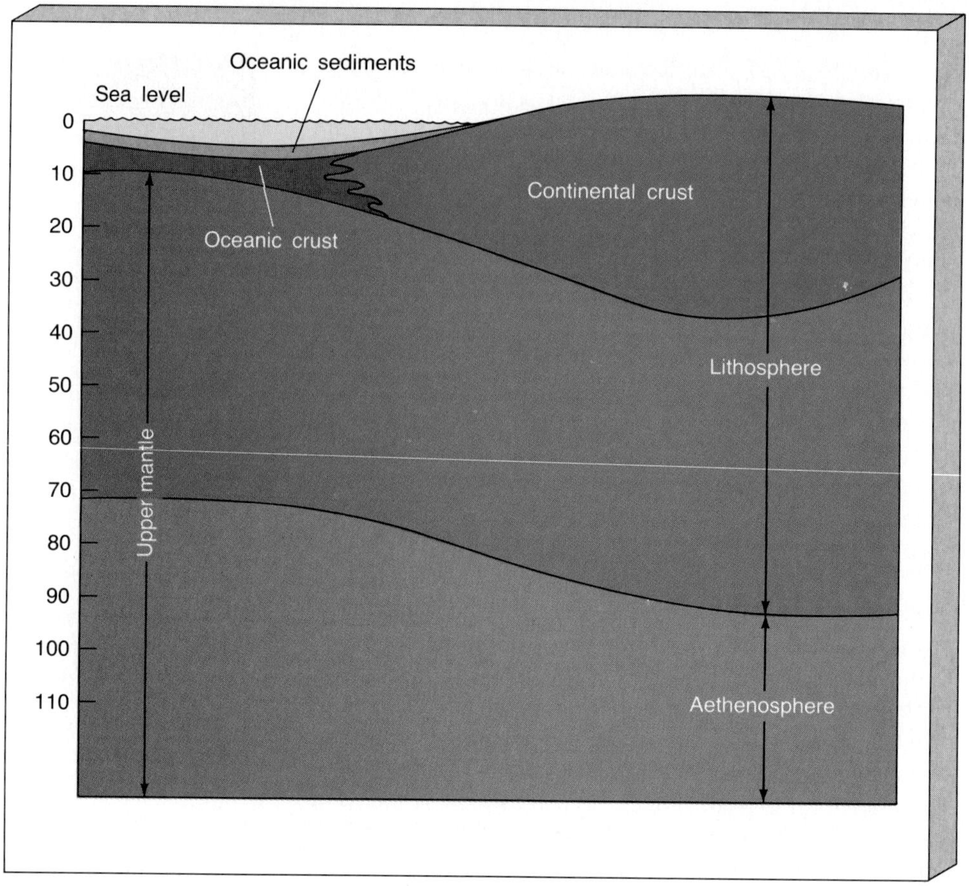

Figure 7.20 The outer layers of the Earth's interior. The rocky lithosphere comprises both the crust and part of the Earth's upper mantle. It is typically between 50 and 100 km thick. Below it lies the aesthenosphere, a relatively soft part of the mantle over which the lithosphere slips.

noted by the arrows of Figure 7.18, many plates slide or shear past one another. A good example is the most famous active region in North America—the San Andreas Fault in California. Illustrated in Figure 7.22, this fault causes much earthquake activity because the Pacific and North American plates are rubbing past one another. They are not moving in quite the same direction or at quite the same speed. Like moving parts in a poorly oiled machine, the motion of these two plates is neither steady nor smooth. The sudden, jerky movements that occur when they do move against each other are often strong enough to cause major earthquakes.

At still other locations, such as down the center of the Atlantic Ocean, the plates are moving apart. As they recede, new mantle material wells up between them, forming midocean ridges. Today, hot mantle material is rising through a crack all along the Mid-Atlantic Ridge, and radioactive dating techniques indicate that material has been upwelling along the ridge more or less steadily for the past 200 million years. The Atlantic seafloor is slowly growing, as the North and South American plates move away from their Eurasian and African counterparts. Similar events are occurring elsewhere on our planet.

What Drives the Plates?

The question is: What really drives the plates? What process is responsible for the enormous forces that drag plates apart in some locations and ram them together in others? The

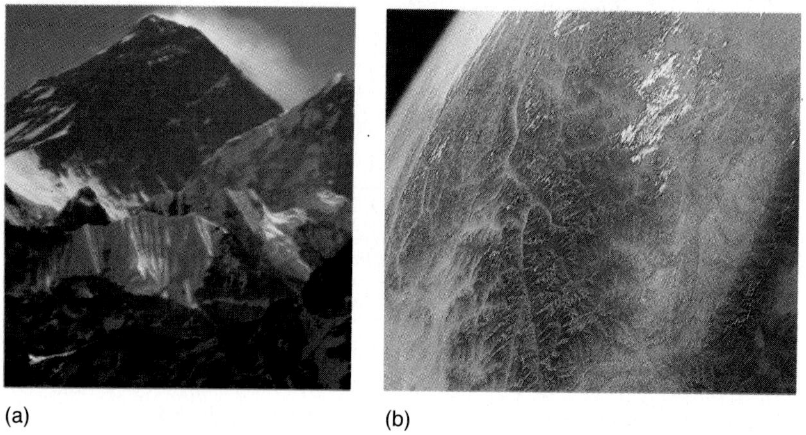

Figure 7.21 Mountain building results largely from plate collisions. Here, the folding of rock (a) is clearly seen on one side of Mount Everest, which is only a small part of the entire Himalayan mountain range (b). (Lee Day/ Black Star; NASA)

(a) (b)

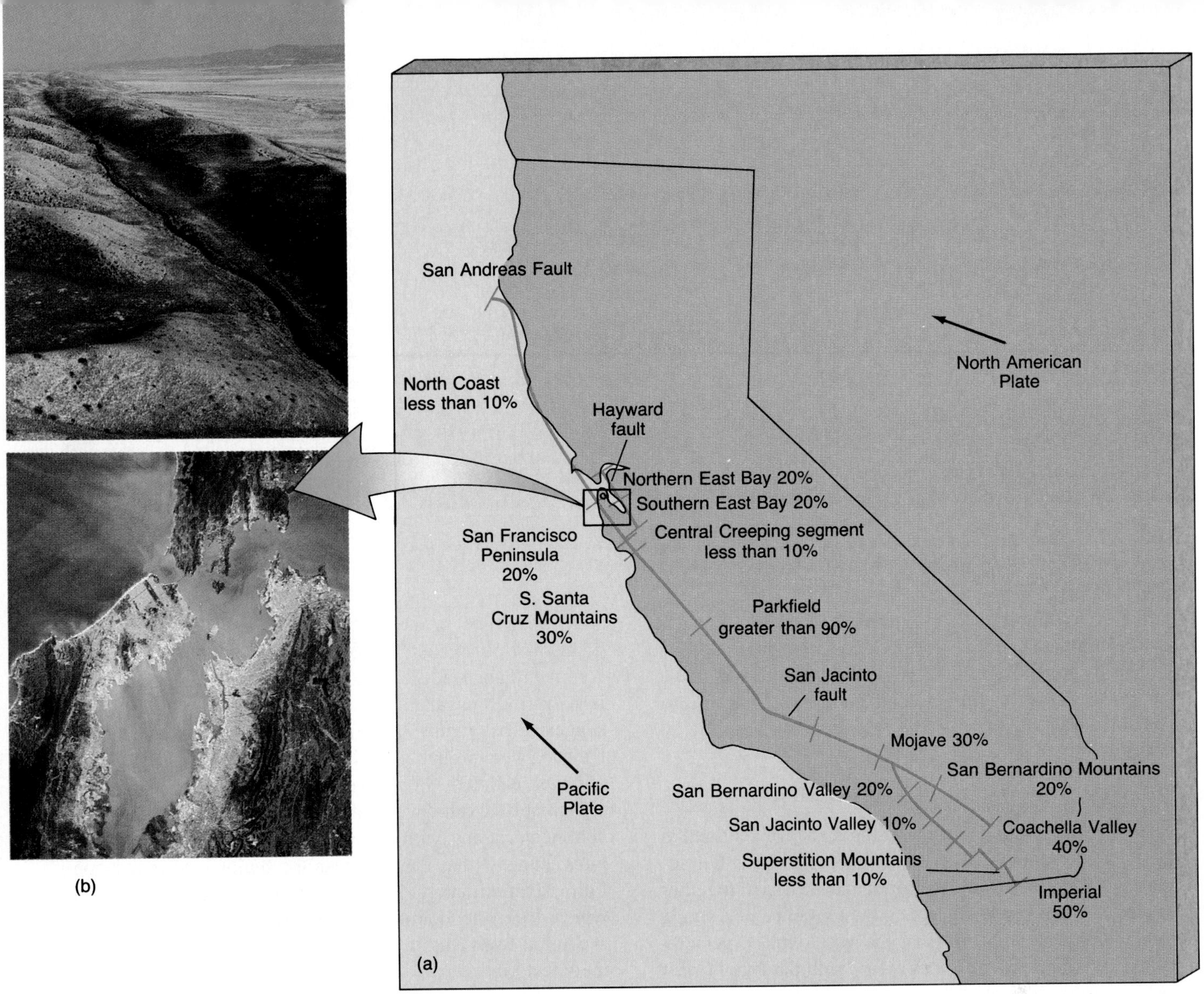

Figure 7.22 *(a) The San Andreas and associated faults in California result from the North American and Pacific plates sliding roughly past one another. The numbers represent estimates of the probability of a major earthquake occurring at various locations along the faults. (b) A small part of the fault line separating the two plates shown in (a) can be seen clearly in the top photograph and running up the peninsula in the bottom photograph, which shows the San Francisco Bay area. (P. Menzel)*

answer is probably convection—the same physical process we encountered earlier in our study of the atmosphere.

Figure 7.23 is a cross-sectional diagram of the top few hundred kilometers of our planet's interior. It depicts roughly the region in and around the Mid-Atlantic Ridge. There the ocean floor is covered with a layer of sediment—dirt, sand, and dead sea organisms that have fallen through the seawater for millions of years. Below the sediment lies about 10 km of granite, the low-density rock that makes up the crust. Deeper still lies the upper mantle, whose temperature increases with depth. Below the base of the lithosphere, at a depth of perhaps 50 km, the temperature is sufficiently high that the mantle is soft enough to flow, very slowly, although it is not molten. This region is the aesthenosphere.

This is a perfect setting for convection—warm matter underlying cool matter. The warm mantle rock rises, just as hot air rises in our atmosphere. Sometimes, the rock squeezes up through cracks in the granite crust. Every so often, such a fissure may open in the midst of a continental landmass, producing a volcano such as Mount St. Helens or possibly a geyser like those at Yellowstone National Park. However, most such cracks are under water. The Mid-Atlantic Ridge is a prime example.

Not all the rising warm rock in the upper mantle can squeeze through cracks and fissures. Some warm rock cools and falls back down to lower levels. In this way, large circulation patterns become established within the upper mantle, as depicted in Figure 7.23. Riding atop these convection patterns are the plates. The circulation is extraordinarily sluggish. Semisolid rock takes millions of years to complete one convection cycle. Although the details are far from certain and remain controversial, many researchers believe that

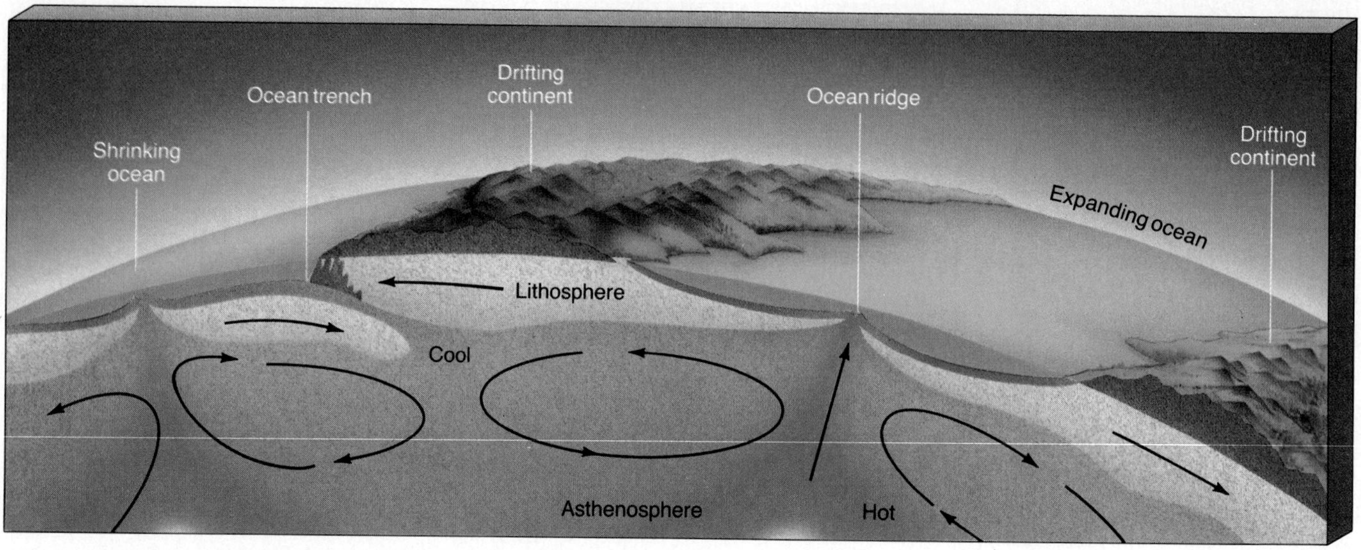

Figure 7.23 *Plate drift is probably caused by convection—in this case giant circulation patterns in the upper mantle that drag the plates across the surface.*

it is the large-scale circulation patterns near plate boundaries that ultimately drive the motion of the plates. Some scientists also conjecture that the aesthenosphere acts like a lubricant, enabling the thin plates to slide across the surface of our planet.

Evidence for Past Continental Drift

✓6 Several pieces of data support the theory of continental drift, or plate tectonics. The first is geographical in nature. Figure 7.24 shows how all the continents nearly fit together like pieces of a puzzle, suggesting the existence of a single huge landmass at some time in the past. Note especially how the Brazilian coast meshes nicely with the Ivory Coast of Africa. In fact, most of the continental landmasses in the Southern Hemisphere fit together remarkably well. Following the arrows on Figure 7.18 backwards, we can see that the fits are roughly consistent with the present motions of the plates involved. The fit is not so good in the Northern Hemisphere, but we shouldn't be terribly surprised, given all the geological "debris" in the North Atlantic—Iceland, Greenland, and the British Isles. However, if we were to consider the entire continental shelves (the continental borders, which are under water) instead of just the portions that happen to stick up above sea level, the fit would improve markedly.

The idea of continental drift was first suggested in 1912 by a German meteorologist named Alfred Wegener, and part of the evidence he cited was this geographical fit of the continents. No one took him seriously, however, in part because there was then no known mechanism that could drive the plates' motions. Nearly all scientists thought it preposterous that large segments of rocky crust could be drifting across the surface of our planet. These skeptical views persisted until the mid-1960s, when the accumulation of data in support of continental drift became overwhelming.

Sometime in the past a single gargantuan landmass dominated our planet. Geologists call this ancestral supercontinent Pangaea, meaning "all lands." The southern part is named Gondwanaland, the northern part Laurasia. As suggested by Figure 7.24(a), these two major segments of Pangaea were probably partly separated by a V-shaped body of water, called the Sea of Tethys. The rest of the planet was presumably covered with water. The current locations of the continents, along with measurements of their current drift rates, suggest that Pangaea was the major land feature on Earth approximately 200 million years ago. Dinosaurs, which were the dominant form of life then, could have sauntered from Russia to Texas via Boston without getting their feet wet.

The existence of this ancient supercontinent and its subsequent breakup explain several peculiar discoveries. For example, when the first climbers reached the summit of Mount Everest in the 1950s, they found fossils of fish. How could marine fossils possibly get to the highest point on Earth? Plate tectonics provides the answer. Whatever caused Pangaea to break apart also sent its continental fragments drifting. At some point, India began its slow motion northward across the Sea of Tethys. In the process, fossils of marine life deposited at the bottom of Tethys were apparently pushed up alongside parts of the Eurasian landmass to form the Himalayas. About 180 million years ago, Gondwanaland and Laurasia began to separate, probably near what we now call the Gulf of Mexico. About 150 million years ago, Gondwanaland itself broke into various pieces—South America, Africa, and Australia, as we now know them. Shortly thereafter, Laurasia split, producing North America and Europe.

A second piece of evidence favoring the theory of plate tectonics came when fossils of a reptile extinct for nearly 200 million years were uncovered at only two locations on Earth, one on the Brazilian coast and the other on the west coast of Africa. These two places are precisely

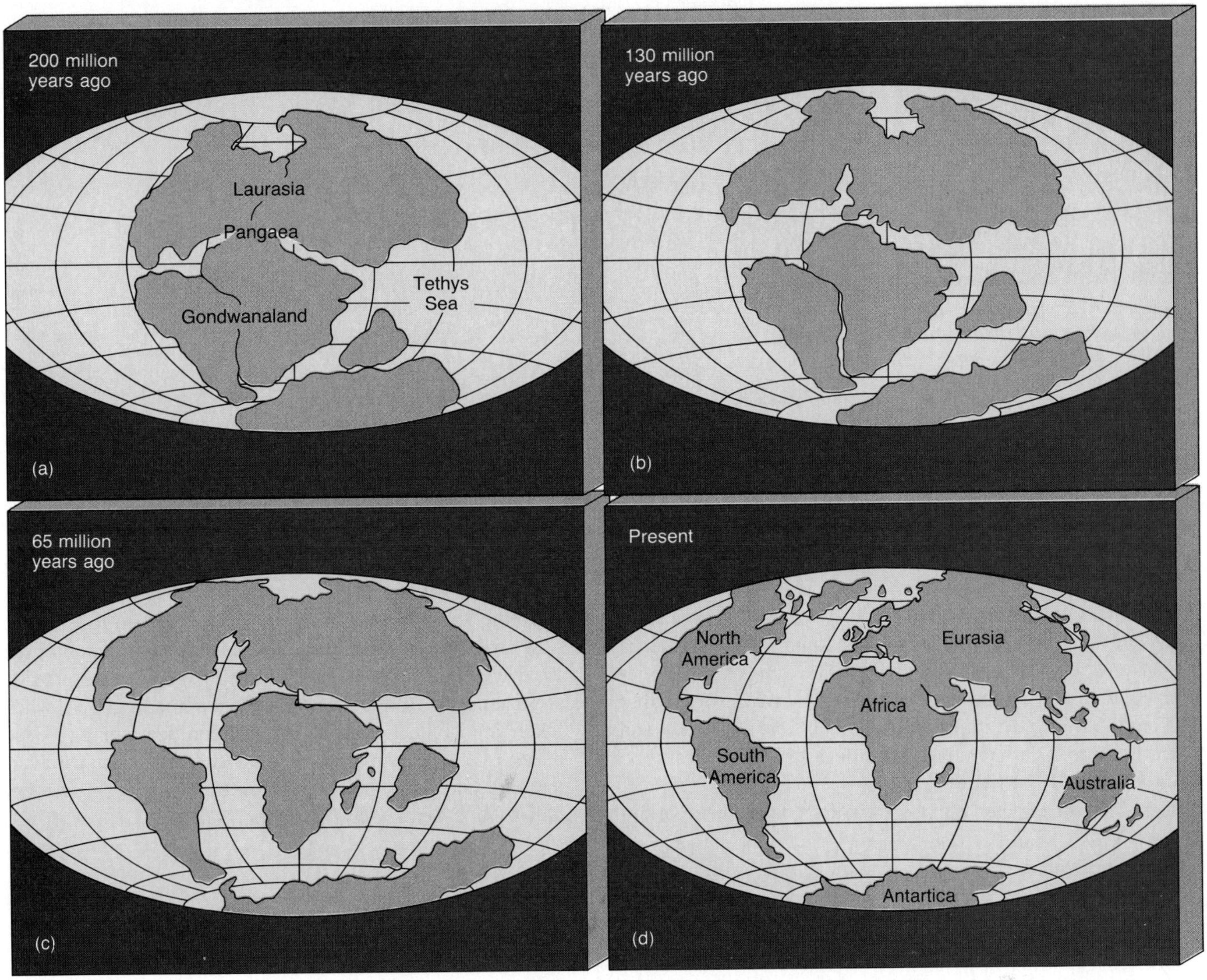

Figure 7.24 *Given the currently estimated drift rates and directions of the plates, we can trace their movements back into the past. About 200 million years ago, they would have been at the approximate positions shown in (a). The continents' current positions are shown in (d).*

where the continents apparently once meshed as part of the ancestral supercontinent of Pangaea. If Africa and South America had always been separated, these creatures could hardly have survived the swim between the coasts. Even if they had, the chances are slim that they would have departed and landed at exactly those parts of the continents that geographically mesh. A much more reasonable conclusion is that Africa and South America were once joined and that this reptile lived in the middle of Gondwanaland.

A third piece of evidence comes from active sites submerged beneath the ocean, forming a giant system of undersea cracks. One example is the Mid-Atlantic Ridge, mentioned earlier. During the 1970s, robot submarines retrieved samples of the ocean floor at a variety of locations on either side of this mountain range. As depicted in Figure 7.25, the ocean floor closest to the underwater ridge is relatively young, while material farther away, on either side, is noticeably older. This is exactly what would be expected if hot molten matter had been upwelling and solidifying as the Eurasian and North American plates were drifting apart. The plates on either side of the Atlantic Ocean must have been drifting apart for the past 200 million years, the oldest age found for any part of the seafloor.

Finally, a fourth piece of evidence comes from **paleomagnetism**—the study of ancient, or fossilized, magnetism. As hot mantle material (carrying traces of iron) upwells from cracks in the oceanic ridges and solidifies, it becomes slightly magnetized, retaining an imprint of the Earth's magnetic field *at the time of cooling*. Thus, the ocean-floor matter has preserved within it a record of Earth's magnetism during past times, rather like a tape recording.

Figure 7.26 is a schematic diagram of a small portion of the ocean floor around the Mid-Atlantic Ridge. As shown, the current magnetism of Earth is oriented in the familiar north-south fashion. When samples of ocean floor

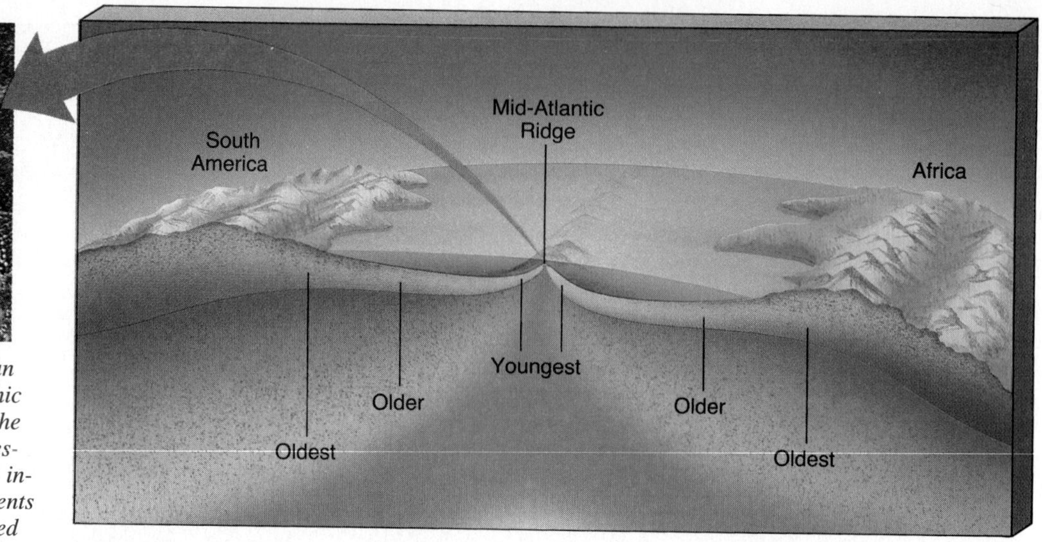

Figure 7.25 Samples of ocean floor retrieved by oceanographic vessels are youngest close to the Mid-Atlantic Ridge and progressively older farther away. The insert shows some underwater vents near the ridge, as photographed by an undersea robot. (Woods Hole Oceanographic Institute)

are examined close to the ridge, the iron deposits are oriented just as expected—north-south. This is the "young" basalt that upwelled and cooled fairly recently. Samples retrieved far from the ridge, corresponding to older material that upwelled at earlier times, however, are often magnetized with the opposite orientation. As we move farther from the ridge, the imprinted magnetic field flips back and forth, more or less regularly, and does so symmetrically on either side of the ridge.

Scientists believe that these different magnetic orientations were caused by reversals in the Earth's magnetic field as the plates drifted away from the central ridge. Working backward, we can use the fossil field to infer the past positions of the plates, as well as the orientation of the Earth's magnetic field. These paleomagnetic data provide strong support for the idea of seafloor spreading, and, when taken in conjunction with the data on seafloor age, also allow us to time our planet's magnetic reversals. On average, the Earth's magnetic field reverses itself roughly every half-million years (see Interlude 7-4).

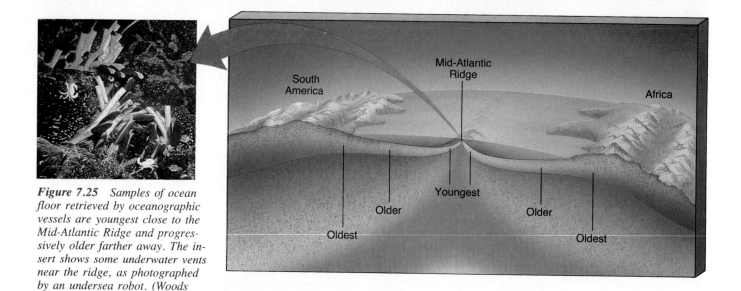

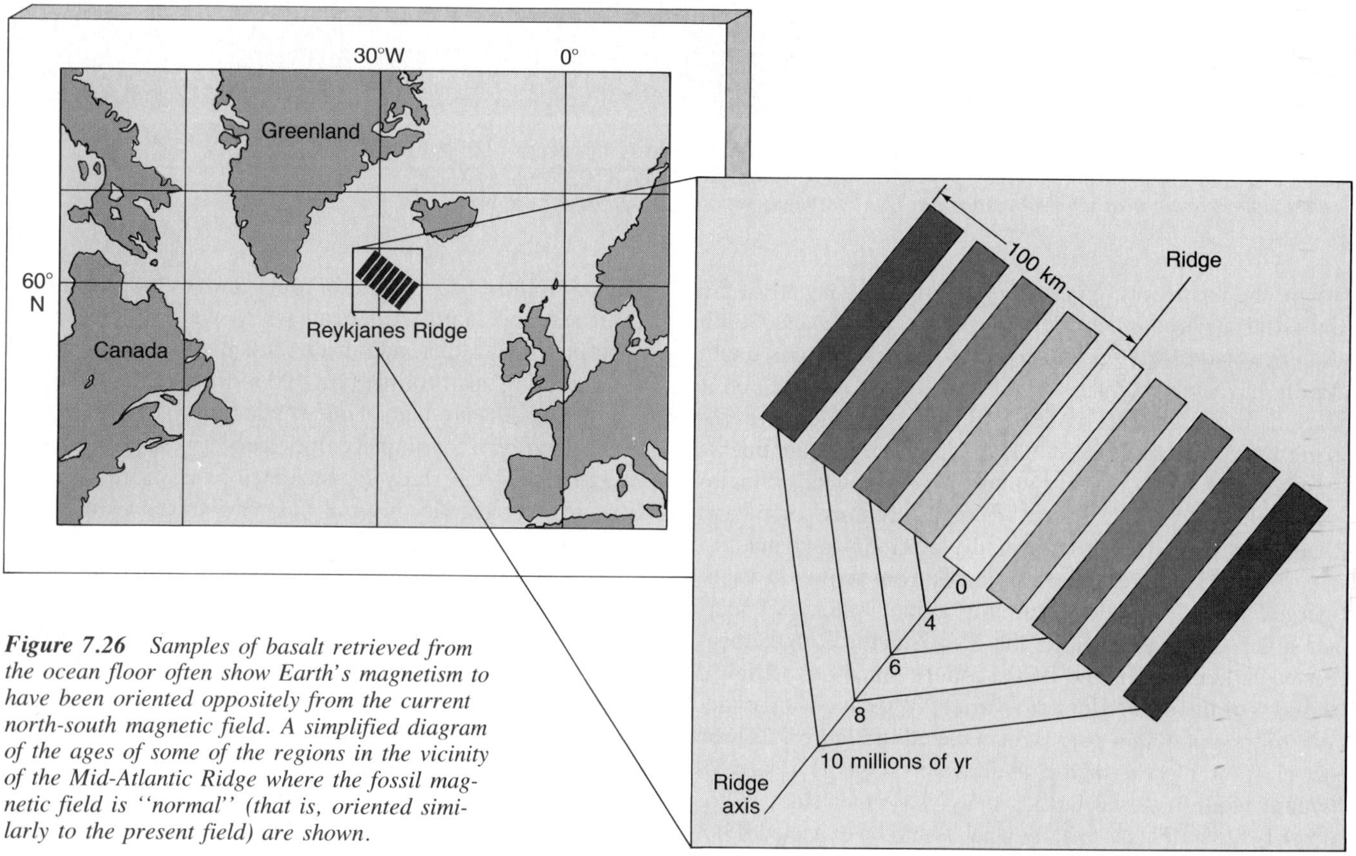

Figure 7.26 Samples of basalt retrieved from the ocean floor often show Earth's magnetism to have been oriented oppositely from the current north-south magnetic field. A simplified diagram of the ages of some of the regions in the vicinity of the Mid-Atlantic Ridge where the fossil magnetic field is "normal" (that is, oriented similarly to the present field) are shown.

INTERLUDE 7-4 *Earth's Magnetic Reversals*

The study of fossilized magnetism—paleomagnetism for short—has become a useful tool with which to study the magnetic properties of Earth. Within the past few decades, researchers have used paleomagnetic measurements of numerous seafloor samples to support the theory of plate tectonics.

As discussed in the text, paleomagnetic studies have led to the remarkable discovery that the north and south magnetic poles of our planet have flip-flopped back and forth many times over the years. The north magnetic pole, now in the Arctic, has at times been located in the Antarctic. The south magnetic pole sometimes resides near geographic north. These flip-flops have occurred many times in the past, although by human standards they are not frequent events. Seafloor samples indicate that the North Pole has been in the Arctic region for about the past 700,000 years. The oceanic data suggest that the Earth's field has reversed itself at least a dozen times in the past 10 million years and perhaps as many as a hundred times in the past 50 million years.

What could have caused such reversals? They are almost certainly related to the motion of the Earth's liquid metal outer core, because, according to the dynamo theory, that is the probable source of our planet's magnetism. Some researchers have speculated that the culprit might be the collision of Earth and a cosmic object such as a comet or an asteroid. Such catastrophic events have occurred in the past and could conceivably change the core's spin by causing the liquid there to slosh around. However, it is very unlikely that such events would occur with either the regularity or the frequency observed for magnetic reversals. Also, the huge amounts of energy released by each collision (equivalent to many millions of megatons of TNT) would surely have left many other easily recognizable imprints on the fossil record—for example, in the form of mass extinctions. These imprints are simply not seen.

The answer seems to be more complex. Because the overall motion in the Earth's core is tied to the planet's overall rotation, the magnetic axis is oriented roughly parallel to the Earth's rotation axis. However, the dynamo is fluid, not rigid, and the field it generates is not permanent. Instead, it can apparently persist stably for long periods of time, but eventually it must decay away, to be regenerated again but with the opposite orientation. The dynamo theory of the Earth's magnetism is far from fully understood and is still controversial in some quarters. But it provides a natural, and even plausible, explanation for the observed field reversals and a roughly correct estimate of the time scales involved. Proponents of the theory are buoyed by observations of the well-known 22-year solar cycle (see Chapter 18), which involves similar field reversals and whose connection with dynamo theory is somewhat better established.

Whatever the cause, a magnetic reversal is not likely to be an instantaneous event. Some time is probably needed for the magnetism to weaken and finally disappear. Some additional time—perhaps on the order of a few hundred years—might be needed for the magnetic field to become reestablished. If so, then each time a flip occurs, the magnetosphere (including the Van Allen belts) far above Earth's surface disappears for a rather long time.

No one really knows what effect a magnetic field reversal might have on the surface of our planet. Scientists in general believe that the absence of the magnetosphere would cause an increase in the high-energy particles and radioactive atoms that reach our atmosphere—and possibly the surface. These atoms would be absorbed by plants and in turn be eaten by animals, including humans. Although this higher level of radiation in plants and animals would probably not be fatal, the normal course of biological evolution would be disrupted. Reproductive errors from generation to generation—mutations—would probably increase, changing the genetic structure in some living systems.

Yet not all mutations are bad. Some are beneficial, enabling living systems to adapt better to a changing environment—in short, to evolve. Mutations act as the motor of evolution. And when the magnetosphere has temporarily collapsed, that motor apparently runs faster than normal. The species that dominates life at any given time does so largely because it enjoys a nearly optimum equilibrium with its environment. It is best suited for its natural surroundings. A mutation, then, is more likely to hurt the dominant species. Some less-than-dominant species, though, might actually profit from increased mutations.

Humans are now the dominant species on Earth.

Future Drifts

There is nothing particularly special about a time 200 million years in the past. We do not suppose that Pangaea remained intact for some 4 billion years after the crust first formed, only to break up so suddenly and so (relatively) recently. Much more likely, Pangaea itself came about after an earlier period during which other plates, carrying widely separated continental masses, were driven together by tectonic forces, merging their landmasses to form a single supercontinent. It is quite possible that there has been a long series of "Pangaeas" stretching back in time over much of the Earth's history. By the same token, there will probably be many more.

Finally, we can inquire about the future positions of Earth's landmasses. We can use the current magnitudes and

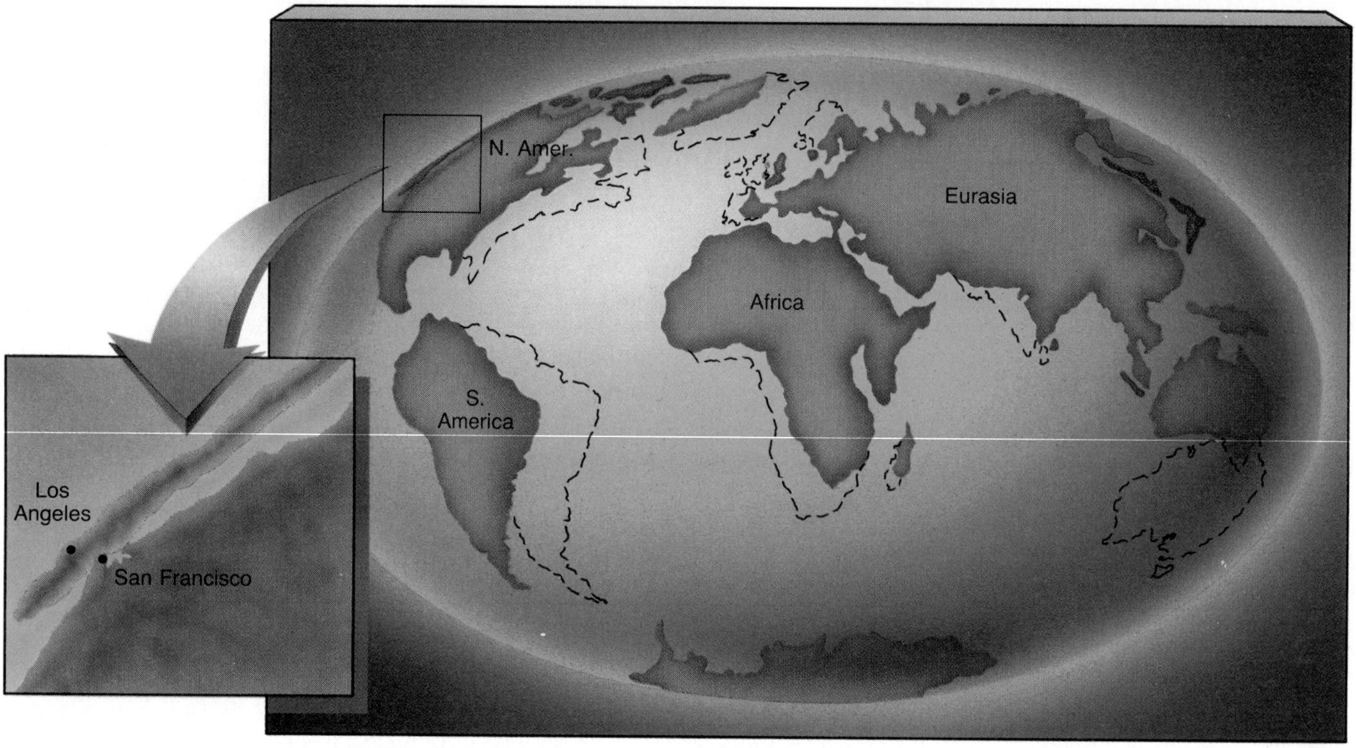

Figure 7.27 *A prediction of where the plates will have carried the continents of our planet about 50 million years from now. The dashed contours outline the continents' current positions.*

directions of plate drift to predict where the continents will be in the years ahead. Figure 7.27 charts the expected locations of the continents 50 million years from now. This prediction assumes that the Atlantic Ocean will continue to widen at the present rate. The Pacific Ocean will contract considerably. Because there is no large landmass on it, the Pacific Plate will presumably continue to be overridden by other continental plates, such as the South American Plate and the Eurasian Plate.

Australia (part of the Indian Plate) will continue its northerly movement toward the Eurasian landmass, destined for massive collisions. India will continue to thrust northward, building the Himalayas to possibly greater heights. The Mediterranean Sea is doomed because of the African Plate's northerly motion. This loss may dampen some tourists' enthusiasm, but, on the bright side, this same drift guarantees great skiing in the Alps for millions of years (provided that the climate doesn't change). And southern California, as part of the Pacific Plate, will be torn away from the North American Plate. Los Angeles will become a suburb of San Francisco in about 15 million years, before being dumped into the Aleutian Trench some 40 million years later.

CHAPTER SUMMARY

Earth is our home in the only planetary system known to us, our solar system. From the outside in, the six main regions of Earth are the magnetosphere, the atmosphere, the crust and the hydrosphere, the mantle, and the core.

Earth's oceans account for three-quarters of the planet's surface. Coastal locations have two high tides and two low tides per day, a direct result of the gravitational influence of both the Sun and the Moon. As the Moon pulls on Earth's tidal bulges, causing Earth's spin on its axis to slow, the Moon spirals away from Earth, at a rate of about 4 cm per century.

Earth's atmosphere is breatheable and acts as a protective blanket, making life on Earth possible. From the surface up, the atmosphere consists of layers called the troposphere, the stratosphere, the mesosphere, the thermosphere, and the exosphere. Atmospheric density and pressure decrease steadily with altitude. The ozone layer, a region within the stratosphere, protects us from harmful ultraviolet radiation from the Sun. The planet's surface reradiates much of the energy it absorbs from the sun in the form of infrared radiation. The outgoing radiation is partially blocked by water vapor and carbon dioxide in our atmosphere, in what is known as the greenhouse effect. This effect keeps the planet warmer than it would be otherwise. Convection in the atmosphere takes place as warm air rises and cool air sinks. This process contributes to surface heating and creates Earth's winds. The air we breathe is Earth's second atmosphere. It originated in material outgassed from the planet's interior by volcanos, then altered by solar radiation and the emergence of life.

The magnetosphere, an invisible region surrounding Earth, is produced by Earth's magnetic field. The field itself is thought to be generated by rotating, conducting material flowing in the planet's interior. The magnetosphere is dominated by two doughnut-shaped zones of high-energy particles from the Sun, called the Van Allen belts. If the magnetosphere becomes overloaded, auroral displays are seen.

Even the strongest drills known can penetrate only a few kilometers into Earth's interior. Geologists have learned to probe the interior indirectly. Earthquakes produce seismic waves, which carry useful information. Additional information comes from volcanos, which bring material from Earth's mantle to the surface. Mathematical models of the interior indicate that Earth has a layered structure, with increasing density and temperature toward the planet's center. This layered structure is called differentiation. To have become differentiated, the entire Earth must once have been molten. Perhaps meteoritic bombardment and the radioactive decay of unstable nuclei generated enough heat to melt the planet during its first billion years.

Planet Earth is a geologically active world. Erosion by wind and water has wiped away much evidence for past geologic activity, but there are many active sites on Earth today. Much of the planet's geologic activity takes place at the boundaries of tectonic plates. These are great slabs of Earth's surface, known to be moving across the planet in a process sometimes called continental drift. Tectonic plates move at a rate of a few centimeters per year.

KEY WORDS

aesthenosphere
atmosphere
aurora
aurora australis
aurora borealis
basalt
convection
convection cell
core
crust
density
differentiation
dynamo theory
earthquake
exosphere
granite
greenhouse effect
hydrosphere
inner core
ionosphere
lithosphere
magnetopause
magnetosphere
mantle
mass
mesosphere
neap tide
outer core
ozone layer
paleomagnetism
plate tectonics
primary atmosphere
P-wave
radioactivity
secondary atmosphere
seismic wave
solar wind
spring tides
stratosphere
subduction zone
S-wave
thermosphere
tidal bulge
tidal force
tides
tropopause
troposphere
Van Allen belts
volcano

REVIEW QUESTIONS

1. By comparison with Earth's average density, what do the densities of water and rocks in Earth's crust tell us about Earth's interior?
2. What is Rayleigh scattering? What is its most noticeable effect for us on Earth?
3. Give a brief description of the magnetosphere, and tell how it was discovered.
4. Compare and contrast P-waves and S-waves, and tell how they are useful.
5. How would our knowledge of Earth's interior change if the planet were geologically dead, like the Moon?
6. What conditions are needed to create a dynamo in Earth's interior? What effect is this hypothetical dynamo believed to have?
7. Give two reasons why geologists believe that Earth's core is in a liquid state.
8. What clue does Earth's differentiation provide concerning our planet's history?
9. What is a half-life? How did radioactive decay heat the Earth early in its history?
10. When did the radioactive heating of Earth end?
11. What process has created the surface mountains, oceanic trenches, and other large-scale features on Earth's surface?
12. Discuss the way in which distant quasars, which are objects lying hundreds of millions of light years from Earth, are used to monitor the motion of Earth's tectonic plates.
13. How do we know that Earth's magnetic field has undergone reversals in the past?
14. How might Earth's magnetic field reversals affect the evolution of life on our planet?

15. What is convection? What effect does it have on (a) Earth's atmosphere, and (b) Earth's interior?

16. Approximating the Earth's atmosphere as a layer of gas 10 km thick, with uniform density 1.2×10^{-3} g/cm^3, calculate the total mass of the atmosphere. Compare it with the mass of the Earth.

17. What is the ratio of the gravitational force of the Moon on the part of the Earth closest to the Moon to the force on the part of the Earth farthest away from the Moon? (Assume that the radius of the Earth is 6500 km and that the distance from the center of the Earth to the Moon is 400,000 km.)

18. At 2 cm/yr, how long would it take a typical plate to traverse the present width of the Atlantic Ocean, about 6000 km?

19. Following an earthquake, how long would it take a P-wave moving in a straight line with speed 5 km/s to reach the opposite side of the Earth?

20. Given that the Moon lies 384,000 km from Earth's center and orbits once every 27.3 days, calculate the orbit period of the *Hubble Space Telescope,* 400 km above Earth's surface.

DISCUSSION QUESTIONS

1. Tell why and how the Moon's gravity exerts a differential force across Earth, resulting in two high tides and two low tides per day. Draw a picture of what you mean. What would happen to the Earth if the strength of the Moon's gravity were to increase indefinitely?
2. Is the greenhouse effect operating in Earth's atmosphere helpful or harmful? Give examples. Go to the library and look up recent articles about the greenhouse effect. Do scientists agree about whether the global temperature in the twenty-first century will rise dramatically?
3. In 1991, NASA proposed that a Mission to Planet Earth be included in the federal budget. The program is designed to provide comprehensive observations of Earth from space. Find out the current status of this proposal. What are some of its major strategies? Do you think a Mission to Planet Earth is a good investment of resources? Explain.

PROJECTS

1. Pick out and draw a simple pattern of stars in the night sky. It doesn't have to be one of the official constellations. Draw the pattern regularly throughout the semester, with respect to foreground objects and/or the horizon. Be sure to make your observation at the same time every night. Explain what you observe with respect to Earth's yearly motion in orbit around the Sun.
2. Write or draw an explanation of spring tides and neap tides. Look in an almanac to find a month during which the Moon's perigee and the time of new or full moon more or less coincide. Sometimes predictions of especially high "spring" tides will appear in local newspapers or on television. Does a high tide actually occur on that day?

SUGGESTED READINGS

Badash, L. "The Age-of-the-Earth Debate." *Scientific American* (August 1989). The controversy that has "aged" the Earth 4.5 billion years in the past three centuries. It embroiled Archbishop Ussher, James Hutton, Lord Kelvin, Ernest Rutherford, and others.

Macdonald, K., and P. Fox. "The Mid-Ocean Ridge." *Scientific American* (June 1990). How a planet-girdling undersea mountain range is created by magma welling up from below as tectonic plates pull apart.

"Managing Planet Earth." *Scientific American* (Special issue, September 1989).

Monastersky, R. "Clouds without a Silver Lining." *Science News* (October 15, 1988). The role of clouds in global warming.

"Planet Earth." *The Planetary Report* (Special issue, January/February 1990).

The editors of *Astronomy* magazine are now publishing a magazine called *Earth.* You can get six issues per year for $14.95.

Kalmbach Publishing Co.
21027 Crossroads Circle
P.O. Box 1612
Waukesha, WI 53187
(414) 796-8776

The American Geophysical Union publishes a wonderful magazine called *Earth in Space*. There are nine issues per year for $10.

American Geophysical Union
2000 Florida Ave. N.W.
Washington, D.C. 20009
1 (800) 966-2481
1 (202) 462-6900

CHAPTER

The Moon
Our Nearest Neighbor

8

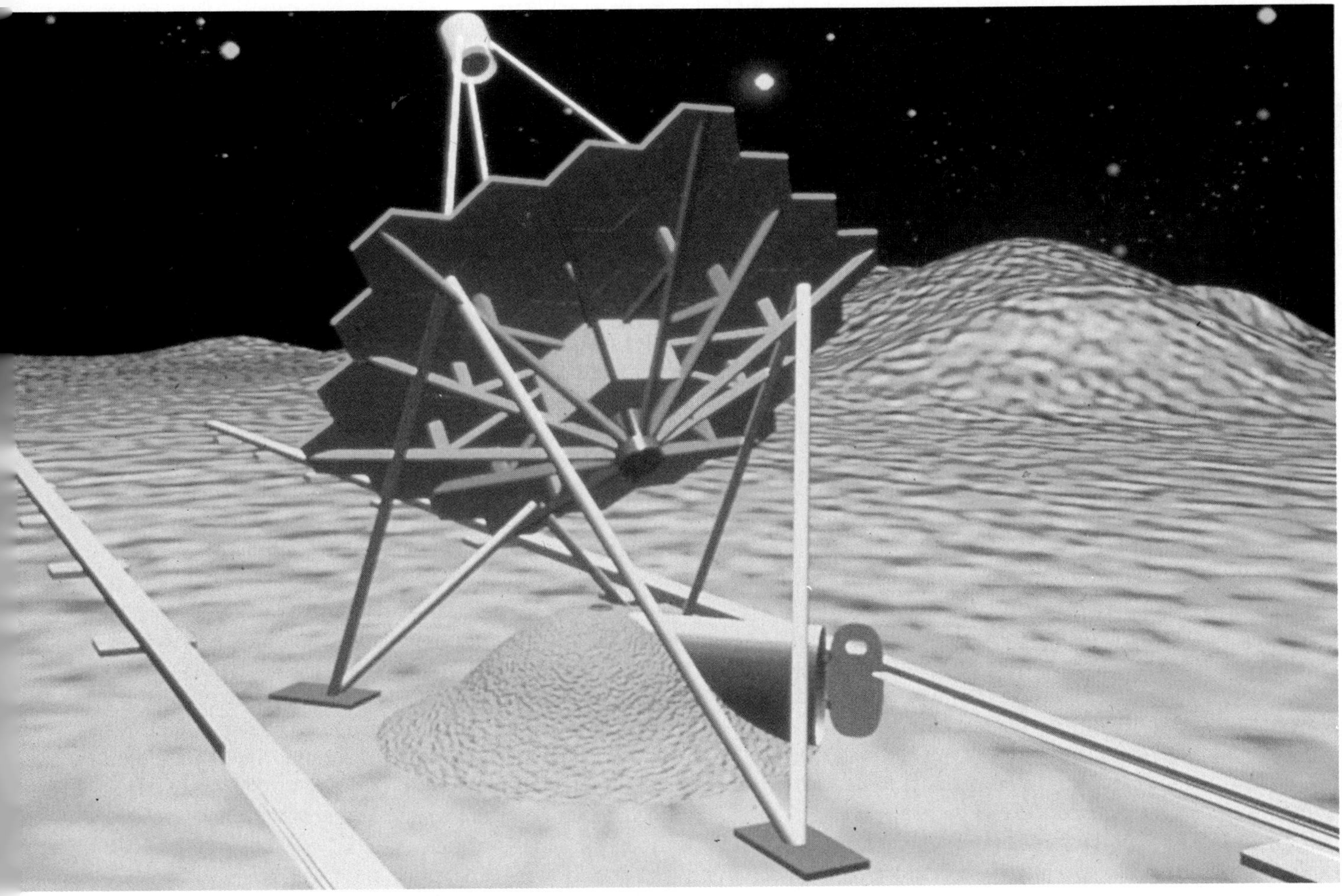

This is an artist's conception for a lunar-based telescope. It would probably be about 16 meters in diameter, or about 50% larger than the world's largest telescope, the Keck. Once humans have mastered the use of telescopes in space, the Moon would seem the next likely place to site a telescope for improved viewing of the universe. This telescope would be assembled on the Moon from segmented mirrors and would have instruments buried in the ground in the mound directly beneath the telescope. Astronauts would have access to these instruments through a little hatch door, seen open in this view. (Courtesy D. Berry)

LEARNING GOALS

1 to examine the properties of the Moon and to compare them with those of the Earth

2 to study the Moon's various surface features and to appreciate how dynamic events early in the Moon's history formed them

3 to understand how studies of lunar cratering can provide us with a means of dating other planets

4 to understand how we can piece together the Moon's origin and history and learn about its interior largely from studies of lunar rocks returned by the *Apollo* astronauts

5 to examine the various theories for the formation of the Moon

The Moon is the only natural satellite of planet Earth. Yet, despite its nearness, it is a very different world from our own. It has no air, no sound, no water. Weather is nonexistent; clouds, rainfall, and blue sky are all absent. Boulders and pulverized dust litter the landscape. The Moon is magnificent in its desolation.

For the past 4 billion years, meteorites have slowly eroded and altered the Moon's surface. Two decades ago human explorers and their machines paid a brief visit. But otherwise, the Moon shows strikingly little activity. There are no detectable movements of the crust and probably no volcanos at the present time. For all practical purposes, the Moon is a dead object in space. It is dead geologically, and it is dead biologically.

At local noontime on the Moon, surface temperatures rise as high as 400K, above the boiling point of water. But at night or in the shade, the temperature plummets to as low as 100K, well below water's freezing point. Warmth does not spread on the moon's surface, for lunar soil is made of pulverized dust that does not conduct heat well. Nor does heat move through lunar air, for the Moon has no air. Without an artificial Earth habitat such as a spacecraft or spacesuit, humans could not survive on the Moon.

If the Moon is so hostile and desolate, why bother to explore and study it? In part, simply because it is our nearest neighbor and dominates our night sky. But beyond that, because the Moon holds important clues about the origin of our solar system. The very fact that it hasn't changed much since its formation means the Moon is a truly primitive object. As such—and in contrast to the Earth and the other nearby planets—the Moon is a vital key to unlock the secrets of the solar system.

THE MOON'S ORBIT

Let's begin our study of the Moon by examining some of its general properties. We will start with its orbit, which will in turn help us determine many of its other characteristics.

We can determine the distance to the Moon rather easily. As noted in Chapter 2, parallax methods can provide us with quite accurate measurements if we use Earth's diameter as the baseline. Radar yields a more accurate lunar distance. The Moon is much closer than any of the planets, and the radar echo bounced off the Moon's surface is strong. A radio telescope receives the echo after about a 2.56-second wait. Dividing this time by 2 (to account for the round-trip the signal took) and multiplying it by the speed of light (300,000 km/s) gives us a distance of 384,000 km. This distance is really an average. The actual distance at any specific time depends on the Moon's location in its slightly elliptical orbit around the Earth. Current technology allows astronomers to measure the radar's round-trip time to submicrosecond accuracy. Repeated radar measurements allow us to determine the Moon's orbit to within a few meters (see Interlude 8-1). This precision is needed to program unmanned spacecraft to land successfully on the lunar surface.

The Moon's orbit is by now well known. It is depicted schematically in Figure 8.1. The orbital semi-major axis (relative to the Earth) is 384,000 km. The orbital eccentricity is 0.055, so that the Moon's **perigee** (closest approach to the Earth) occurs at a distance of 363,000 km and its **apogee** (greatest separation) occurs at 405,000 km. The Moon's sidereal period (relative to the stars) is 27.3 days. Its synodic period (relative to the Earth—new moon to new

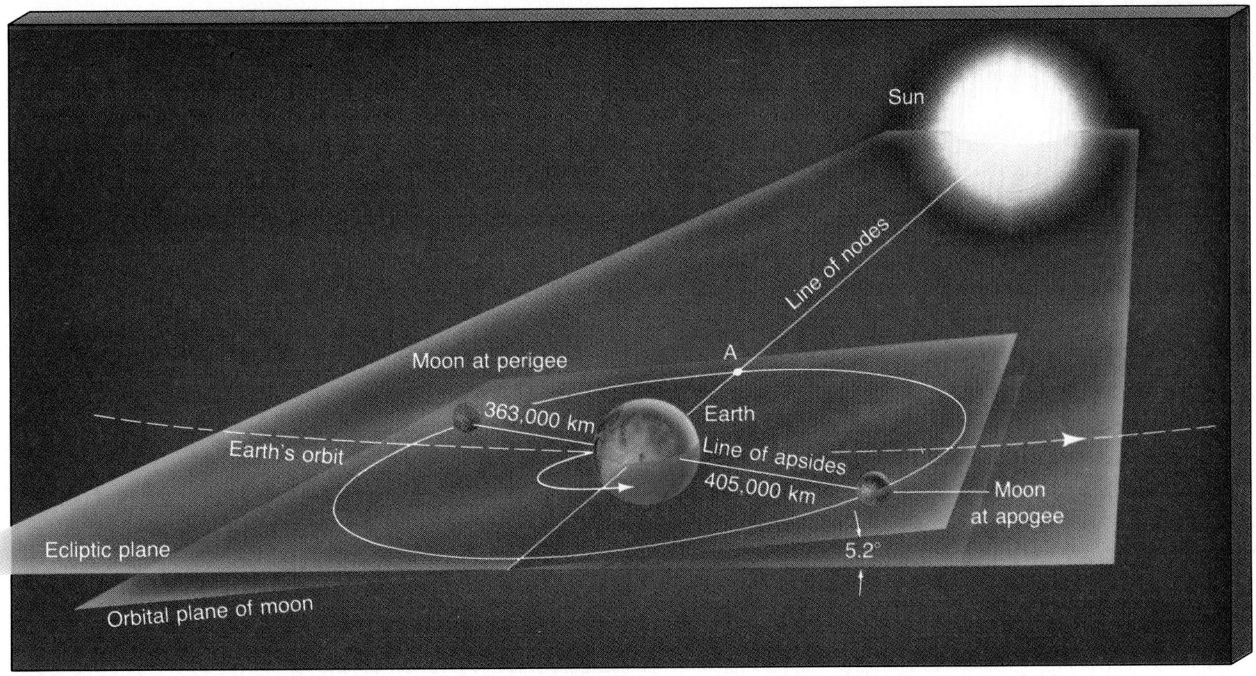

Figure 8.1 *The orbit of the Moon. As shown, the line of nodes happens to intersect the Sun, so a solar eclipse will occur when the Moon crosses point A in its orbit. The extent of the eclipse depends on the orientation of the long axis of the Moon's orbit (the line of apsides) with respect to the line of nodes. The deepest total eclipses occur when point A corresponds to lunar perigee; annular eclipses occur when A is near apogee.*

INTERLUDE 8-1 *Lunar Laser Ranging*

Several manned *Apollo* (U.S.) and unmanned *Luna* (USSR) missions left equipment on the Moon. One of the most interesting devices the Americans left is an array of mirrors designed to intercept light pulses launched from Earth and to reflect them back toward Earth. Each small mirror resembles those used on highway posts and signs to reflect automobile headlight beams. The whole array of reflectors is not much larger than this book.

Astronomers can use this equipment to measure the Moon's distance by a method similar to the radar technique described in the text but different in two important ways. First, the lunar-laser-ranging method does not use radio signals. Instead, it uses a *laser,* which transmits highly concentrated *light* pulses. (The word *laser* is an acronym for **l**ight **a**mplification by **s**timulated **e**mission of **r**adiation.) Second, the reflectors are positioned at well-known locations on the Moon, so researchers know precisely which point on the surface is being measured. Radar echoes bounce from much larger areas on the Moon's surface and are thus a little less accurate.

Aiming the laser at the reflector on the Moon is comparable to hitting a dime with a rifle at a distance of about a kilometer. And like the radar technique, the return echo of light is a lot weaker than the transmitted light. Only about one photon is detected upon return for every billion billion (10^{18}) photons in the burst of light sent toward the target.

Astronomers can currently determine the distance to the Moon—and so gather data to model its orbit—to an accuracy of about 6 cm. Further refinements in equipment during the next few years should enable researchers to decrease the error to within a centimeter or two.

(NASA)

moon, say) is 29.5 days. The orbit is prograde (meaning that it revolves around the Earth in the same sense as the Earth revolves around the Sun) and lies close to, but not exactly in, the plane of the ecliptic. The inclination of the orbital plane to the ecliptic plane is about 5.2° (degrees). Because the distance from the Earth to the Moon is not constant, the angular size of the Moon, as seen from Earth, varies noticeably. When the Moon is at apogee, it is about 29.4′ (arc minutes) in diameter, compared with 33.5′ at perigee—a difference of about 14 percent.

This variation in the Earth-Moon distance is the reason why not all solar eclipses look the same or last the same length of time. Recall from Chapter 1 that because the Moon's orbital plane is inclined to the plane of the ecliptic, eclipses can occur only at those times of year when the line of intersection of the two planes—the line of nodes—happens to point directly toward the Sun. Figure 8.1 shows just such an alignment. When the Moon crosses point A on its orbit, a solar eclipse will occur. Recall also from Chapter 1 that because of the gravitational effect of the Sun, the whole lunar orbit precesses, with a period of about 19 years. In fact, the Moon's motion is even more complicated than that. The direction of the long axis of the ellipse (from perigee to apogee)—sometimes known as the line of apsides—also rotates around the Moon's orbital plane, taking about 9 years to complete one circuit.

The effect of all these simultaneous motions is this: When an eclipse occurs, the Moon can be at essentially any point on its orbit, from apogee to perigee, so the angular size of the Moon during an eclipse can be anywhere between the minimum and maximum possible values. In Figure 8.1, the eclipse will occur with the Moon at some intermediate angular size. If the Moon happened to be near perigee at the instant of an eclipse, its angular size would be greatest, and it would blot out the Sun for the longest time. It is at these times that the longest-lasting eclipses occur. If instead the Moon were at apogee, its angular size would actually be *less* than that of the Sun, and an annular eclipse would result.

THE MOON IN BULK

Size

☑1 The Moon has an angular diameter of about 0.5°. Knowing the Moon's distance, we can then easily calculate its true size. The argument is essentially the same that Eratosthenes used to measure the size of the Earth (see Interlude 2-1): The ratio of the Moon's diameter to the circumference of its orbit ($2\pi \times 384{,}000$ km) is equal to the ratio of its angular diameter to 360°. The Moon's radius is therefore ($2\pi \times 384{,}000$ km $\times$ 0.25°/360°), or about 1700 km, roughly one-fourth the radius of Earth. Using more precise calculations, astronomers find the Moon's radius to be 1738 km.

Mass

As for any astronomical object, we determine the mass of the Moon by studying its gravitational pull on objects in its vicinity—the Earth and humanmade spacecraft, for example. This measurement requires a knowledge of the Law of Gravity, which we studied in Chapter 3. If we know the period of a spacecraft's orbit around the Moon and its distance from the Moon, we can find the Moon's mass from Kepler's Third Law (as modified by Newton), which relates any satellite's orbital period and semi-major axis to the mass of the body it orbits. Even before the space age, the Moon's mass was already well known, from studies of the motion of the Earth. The result is 7.4×10^{25} g, approximately 1/80 the mass of the Earth.

Because the Moon is so much lighter than the Earth, its gravitational field is also weaker—the force of gravity on the lunar surface is only about 1/6 that on Earth. Thus, for example, an astronaut weighing 180 pounds on Earth would weigh a mere 30 pounds on the Moon. The bulky spacesuits and backpacks used by the *Apollo* astronauts on the Moon were not nearly as heavy as they appeared!

Density

Knowing the Moon's size enables us to find its volume, and given its mass, we can determine the average lunar density. The result, 3.3 g/cm^3, contrasts with the average Earth value of about 5.5 g/cm^3, suggesting that the Moon contains fewer heavy elements (such as iron) than the Earth.

Rotation

To an astronaut standing on the Moon's surface, Earth would appear almost stationary in the sky. The Moon's rotation period about its axis equals its revolution period about the Earth (27.3 days), so that the Moon keeps the same side facing Earth at all times. Sunrise and sunset on the Moon happen roughly 2 weeks apart.

That the Moon's rotation and revolution periods are equal is no accident. Just as the Moon raises tides on the Earth, Earth produces a tidal bulge in the Moon, too. In fact, because the Earth is so much more massive, the tidal force on the Moon is about 20 times greater than that on the Earth. In Chapter 7, we saw how tidal forces are causing the Earth's spin to slow. As a result, Earth will eventually rotate on its axis at the same rate as the Moon revolves around the Earth. This condition—when the spin of one body is precisely equal to (or *synchronized* with) the period of its orbit around another—is known as a **synchronous orbit.** The Earth's rotation will not become synchronous with the Earth-Moon orbital period for trillions of years. However, the Moon's much larger tidal deformation caused it to evolve into a synchronous orbit long ago—the Moon's spin is said to have become *tidally locked* to the Earth. As a consequence, the Moon has a ''near'' side, which is always

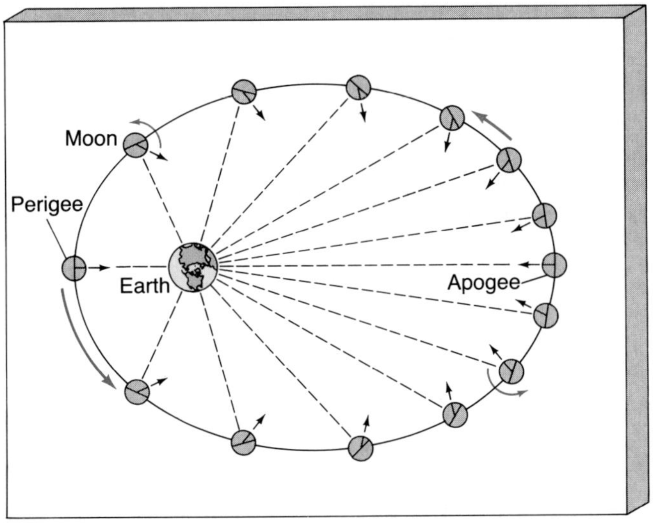

Figure 8.2 *The libration of the Moon. In this highly exaggerated diagram, the Moon is shown at several points in its orbit, separated by equal intervals of time. The areas of the "slices" formed by the lines joining the Moon to the Earth are all equal, by Kepler's Second Law. The arrows represent the point on the surface that faces directly toward Earth at perigee. As the Moon rotates, the arrow moves counterclockwise at a constant rate. When the Moon moves fastest on its orbit (near perigee), the rotation is unable to keep up with its orbital motion, so that the marked point drifts toward the east and a little more of the western side of the Moon becomes visible. Later, when the Moon moves more slowly in its orbit, the rotation rate catches up again until, by the time apogee is reached, the arrow points directly at Earth once more. On the second half of the orbit, the opposite occurs—the arrow first drifts to the west, then returns to center by the time perigee is reached.*

visible from Earth, and a "far" side, which never is. Thus, until very recently, no one on Earth had any idea what half of our satellite looked like. It was only when spacecraft actually flew around the Moon that we finally saw the far side.

Actually, as seen from Earth, the Moon appears to wobble slightly as it moves around its orbit. Although the Moon's rotation rate is virtually constant, its orbit is not perfectly circular (the eccentricity is 0.055), so its distance from the Earth varies from day to day. By Kepler's Second Law (see Chapter 3), the closer the Moon comes to the Earth, the faster it moves along its orbital path. Thus, the Moon's orbital speed also varies—the Moon moves fastest at perigee and slowest at apogee, about 2 weeks later. The Moon's rotation rate is synchronized with the *average* orbital speed. Thus, as illustrated in Figure 8.2, the Moon seems to rock back and forth slightly about a north-south axis. This rocking, or **libration,** permits about 59 percent of the Moon's surface to be seen from Earth over the course of a month. Without the effect of libration, we'd see only 50 percent.

Shape

Like the Earth, the Moon is not perfectly spherical. Its equatorial diameter exceeds its polar diameter, as might be expected for a massive, rotating object. But, as in many areas of astronomy, we find a puzzle here. The lunar bulge, measured by spacecraft orbiting the Moon, nearly 4 km, far too large to be produced by the Moon's rather sluggish rotation: We would theoretically expect the bulge to be much smaller—a mere 0.1 km.

The easiest explanation of the extra bulge is that the Moon rotated much more rapidly long ago. We have seen that the spin rates of both the Earth and the Moon are steadily slowing because of tidal forces. Shortly after the Earth-Moon system came into being, both the Earth and the Moon may have had rotation periods of only a few hours. If the Moon, like the Earth, was molten at that time, the combination of its rapid spin and fluid composition could have resulted in a bulge even larger than observed today. If the Moon became solid at that stage, it might still retain enough of its early bulge to explain its present shape.

The Earth's tidal effect may also have contributed to the Moon's bulge. Long ago, the distance from the Earth to the Moon may have been as little as two-thirds of its current value, or about 250,000 km. Earth's tidal force on the Moon would then have been more than three times greater than it is today. Again, the resulting distortion would have "set" when the Moon solidified. The surprisingly large lunar bulge would also have accelerated the synchronization of the Moon's orbit. The bulge now points directly toward Earth.

Lunar Air?

What about the Moon's atmosphere? That's easy—there is none! All of it apparently escaped. Massive objects have a better chance of retaining their atmospheres, because the more massive the object, the larger the velocities that are needed for atoms and molecules to escape. Although the Moon is only 4 times smaller than Earth, it is 80 times less massive, and its escape velocity is only 2.4 km/s, compared with 11 km/s for Earth. The Moon has a lot less pulling power. Any primary atmosphere it had initially, or secondary atmosphere that appeared later, has gone forever.

SURFACE FEATURES

✓2 The first observers to point their telescopes at the Moon, among them Galileo Galilei, noted large dark areas, resembling (they thought) Earth's oceans. They also saw light-colored areas resembling the continents. Both types of region are clear in Figure 8.3(a), a photograph of the full Moon, made possible by sunlight reflected toward the Earth from the lunar surface. Such light and dark surface features are also evident to the naked eye, creating the face of the familiar "Man-in-the-Moon."

Today we know that the dark areas are not oceans but extensive flat areas that probably resulted from lava flows

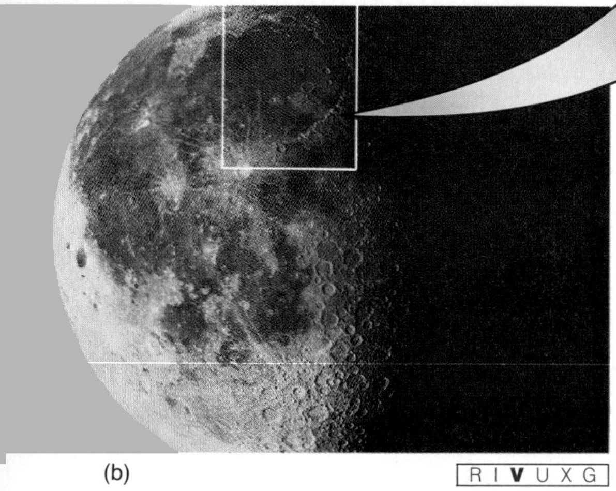

Figure 8.3 *(a) A mosaic of the full Moon, north pole at the top. Because the Moon does not emit its own visible radiation, we can see it only by the reflected light of the Sun. (b) The Moon near third quarter. Notice that surface features are much more visible near the terminator, separating light from dark, where sunlight strikes at a sharp angle and shadows highlight the topography. (c) Magnified view of a region near the terminator, as seen from Earth through a large telescope. Crater Copernicus is at bottom left, crater Eratosthenes is at bottom center; the central dark area is Mare Imbrium, ringed at bottom right by the Apennine Range mountains which includes Mt. Hadley. (d) Even greater magnification. (Lick Observatory; Palomar Observatory)*

during a much earlier period of the Moon's evolution. Nevertheless, they are still called **maria,** a Latin word meaning ''seas'' (singular: *mare*). There are 14 maria, all roughly circular. The largest of them (Mare Imbrium) is about 1100 km in diameter. The lighter areas, originally dubbed **terrae,** from the Latin word for ''land,'' are now known to be elevated several kilometers above the maria and are accordingly called the lunar **highlands.** However, neither region is chemically or structurally similar to Earth's plains or mountains.

The smallest lunar features we can distinguish with the naked eye are about 200 km across. Telescopic observations further resolve the surface into numerous bowl-shaped depressions, or **craters.** Most craters apparently formed eons ago primarily as the result of meteoritic impact. Craters are particularly clear in Figure 8.3(b), which shows a half Moon. Near the **terminator,** which separates day from night on the surface, the Sun is low in the sky, casting long shadows that enable us to distinguish quite small surface details. Because of the effects of the our atmosphere, the smallest lunar objects that telescopes on the Earth's surface can resolve are about 1 km across (see Figure 8.3[c]).

Much more detailed photographs have been taken by orbiting spacecraft and, of course, by visiting humans. Figure 8.4 shows a view of the Moon taken from an orbiting spacecraft. The smallest features that can be distinguished in this image are about 500 m across. Lunar craters are now known to come in all sizes—the largest are hundreds of kilometers in diameter, the smallest are microscopic. Craters are found everywhere on the Moon's surface, although they are much more prevalent in the highlands.

Figure 8.4 *The Moon, as seen from the* Apollo 8 *orbiter, during the first human circumnavigation of the moon in 1968. Craters ranging in size from 50 km to 500 m (also the width of the long fault lines) can be seen. (NASA)*

R I V U X G

Figure 8.5 The western hemisphere of the moon, as seen by the *Galileo* probe en route to Jupiter. The large, dark region is part of the face visible from Earth. This far side image shows one or two small maria.

To the surprise of most astronomers, when the far side of the Moon was mapped first by Soviet and later by American spacecraft, no major maria were found there. The far side (Figure 8.5) is composed almost entirely of highlands. As we will see below, this fact has great bearing on our theory of how the Moon's surface terrain came into being, for it suggests that the processes involved could *not* have been entirely internal in nature. The location of the Earth must somehow have played a role.

All of the Moon's significant surface features have names. The 14 maria bear fanciful Latin names—Mare Imbrium (''Sea of Showers''), Mare Nubium (''Sea of Clouds''), Mare Nectaris (''Sea of Nectar''), and so on. Mountain ranges in the highlands tend to bear the names of terrestrial mountain ranges—the Alps, the Carpathians, the Apennines, the Pyrenees, and so on. Craters tend to be named after great scientists or philosophers, such as Aristotle, Eratosthenes, Copernicus, and Plato.

CRATERING

Impact Craters

✓3 Despite the lack of activity on the Moon, the lunar surface is not entirely changeless. Even in the absence of wind and water, there is evidence of **erosion**—a slow but ceaseless wearing away of surface material. Figure 8.6 shows examples of this surface wear and tear, especially the well-rounded mountaintops. In the absence of erosion, these features would still be as jagged and angular today as they were when they formed. From their rounded appearance, we infer that something must have worn them down to their present condition. Note also that the smaller craters seem partially filled with either lunar or foreign material, and the larger craters have rims that are clearly smoothed to some extent.

On the Moon, the primary source of erosion is interplanetary debris that collides with the lunar surface. This meteoritic material, much of it having rocky or metallic composition, is strewn throughout the solar system (see Chapter 15). It wanders around in interplanetary space, perhaps for billions of years, until it collides with some planet or moon. On Earth, most meteoroids burn up in the atmosphere, producing the streaks of light known as *meteors,* or ''shooting stars.'' But the Moon, without an atmosphere, has no protection against this onslaught. Large and small meteoroids just zoom in and collide with the surface, sometimes producing huge craters. Over billions of years, these collisions have scarred, cratered, and generally reshaped the lunar landscape.

Meteorites generally strike the Moon at velocities of several kilometers per second. At these speeds, even a small piece of matter carries an enormous amount of energy—for example, a 1-kg meteorite hitting the Moon's surface at 10 km/s would release as much energy as the detonation of 10 kg of TNT. As illustrated in Figure 8.7, impact of a meteorite with the surface causes sudden and tremendous pressures to build up, heating the normally brittle rock and

Figure 8.6 Despite the complete lack of wind and water on the airless Moon, the surface has still eroded a little under the constant ''rain'' of impacting meteorites, especially micrometeorites. The twin tracks were made by the Apollo lunar rover. (NASA)

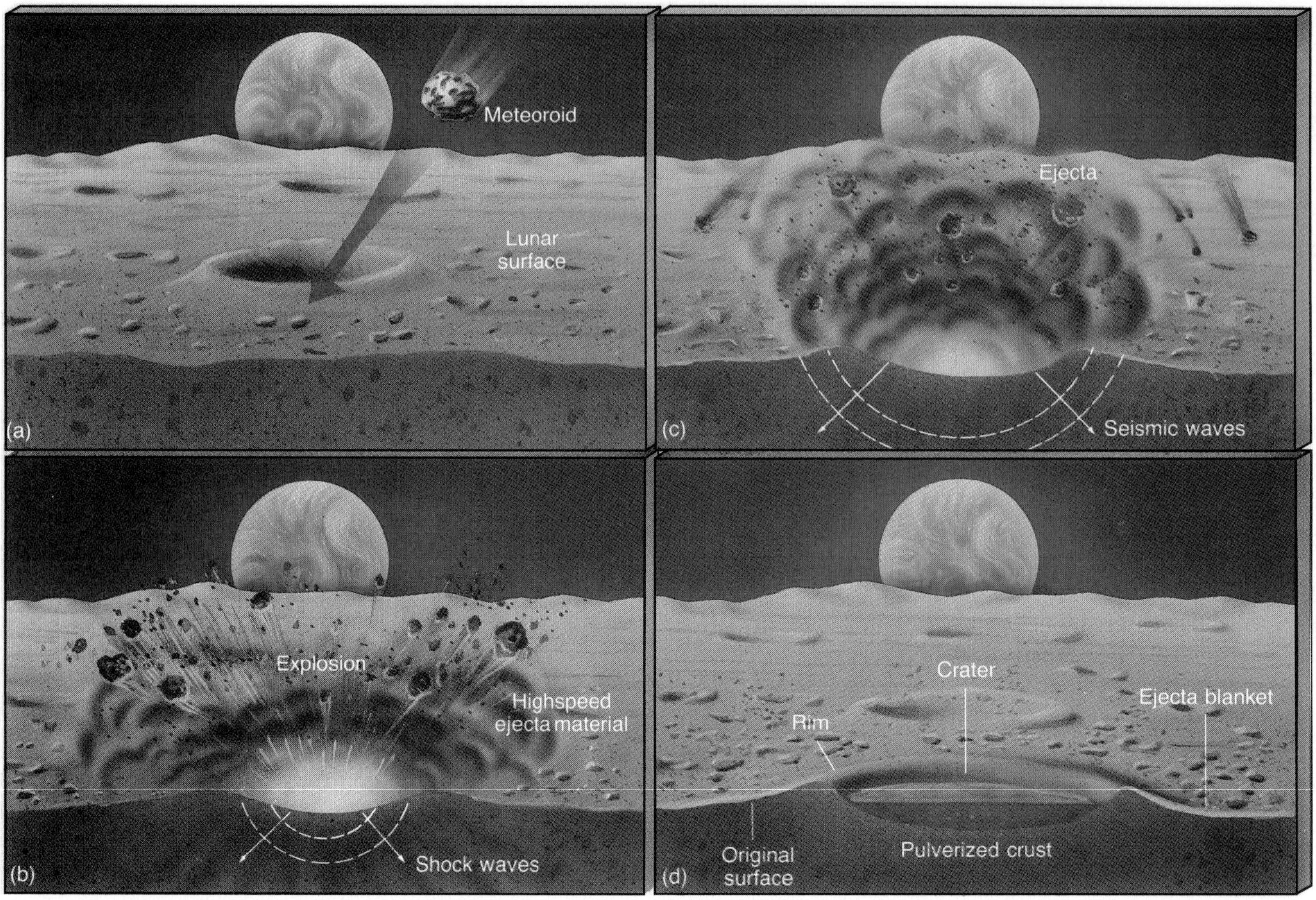

Figure 8.7 *Several stages in the formation of a crater by meteoritic impact. (a) The meteoroid strikes the surface, releasing a large amount of energy. (b, c) The resulting explosion ejects material from the impact site and sends shock waves through the underlying surface. (d) Eventually, a characteristic crater surrounded by a blanket of ejected material results.*

deforming the ground like heated plastic. The instant of impact probably mimics the opening of the petals of a flower, with previously flat layers of rock pushed up and out as a result of the explosion.

The diameter of the eventual crater is typically 10 times that of the incoming meteorite; the crater depth is about twice the meteorite's diameter. Thus, our 1-kg meteorite, measuring perhaps 10 cm across, would produce a crater about 1 m in diameter and 20 cm deep. Shock waves from the impact pulverize the lunar surface to a depth many times that of the crater itself. The material thrown out by the explosion surrounds the crater in an **ejecta blanket.** The ejected debris ranges in size from fine dust to large boulders. The larger pieces may themselves form secondary craters. Figure 8.8 shows the result of two large meteoritic impacts on the Moon.

Erosion Rates

Craters are constantly being created—even as you read this—all across the lunar surface. In addition to bombardment by meteorites with masses of a gram or more, a constant ''rain'' of micrometeorites (debris with masses ranging from micrograms up to about 1 g) also eats away at the structure of the lunar surface, contributing to the overall erosion process. Figure 8.9 shows a microscope photograph of some glassy ''beads'' brought back to Earth by the *Apollo* astronauts. The beads themselves were formed during the explosion following a meteorite impact, when surface rock was melted, ejected, and rapidly cooled. However, several of them also display fresh, miniature craters caused by micrometeorites that struck them after they cooled and solidified. Even the submicroscopic particles escaping the Sun in the solar wind (mostly high-velocity protons, with masses of about 10^{-24} g) nibble at the lunar surface. Each proton chips off a tiny particle of dust from the rock, eroding the shapes of older craters, smoothing their edges, and helping to fill in the smaller craters.

The *rate* of cratering decreases rapidly with crater size—fresh large craters are very few and far between, but small craters are very common. The reason for this is simple: There just aren't very many large chunks among the interplanetary debris, so their collisions with the Moon are rare. There are about 100 times more 10-km craters than

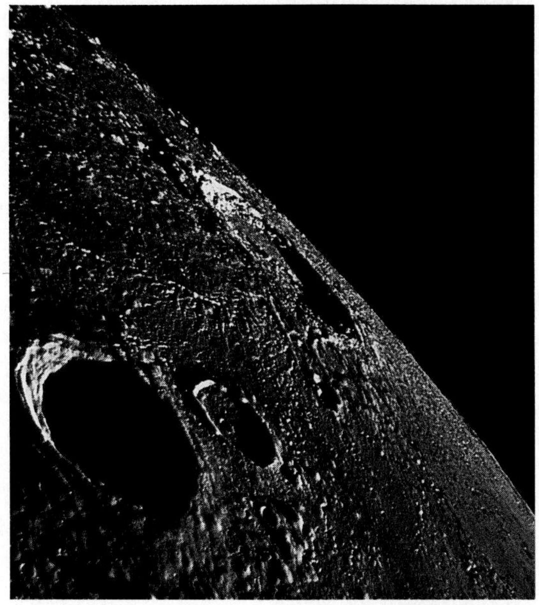

Figure 8.8 *(a) A large lunar crater, called the Orientale basin. The meteorite that produced this crater upthrust much surrounding matter, which can be seen as concentric rings of cliffs called the Cordillera Mountains. The outermost ring is nearly 1000 km in diameter. (b) Two smaller craters called Reinhold and Eddington sit amidst the secondary cratering resulting from the impact that created the 90 km wide Copernicus crater (near the horizon) about a billion years ago. The ejecta blanket from crater Reinhold, 40 km across, and in the foreground, can be seen clearly. View is looking northeast from the Lunar Module during the* Apollo 12 *mission. (NASA)*

100-km ones, about 100 times more 1-km craters than 10-km ones, and so on. At the present average rates, one new 10-km (diameter) lunar crater is formed every 10 million years, a new meter-sized crater is created about once a month, and centimeter-sized craters are formed every few minutes.

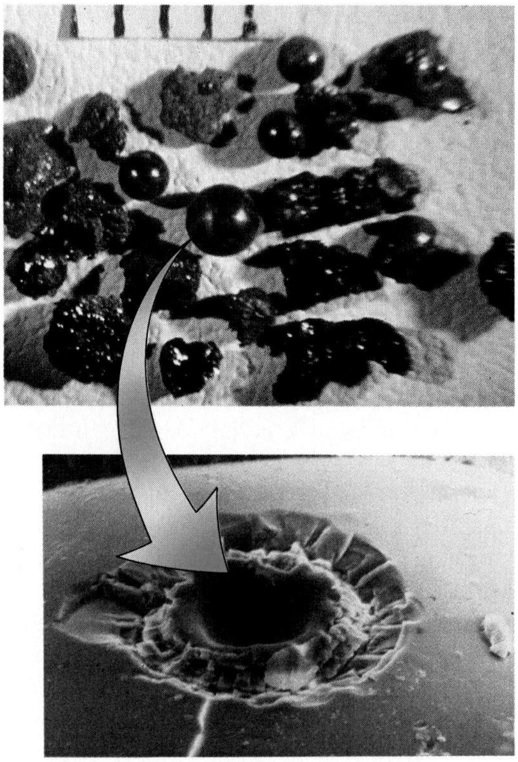

Figure 8.9 *Craters of all sizes litter the lunar landscape. Some shown here, embedded in glassy beads retrieved by American astronauts, measure only 0.01 mm across. (The scale at top is in millimeters.) (NASA)*

We can examine lunar erosion a little more quantitatively. Here's how it is done. Photographs taken from spacecraft show fewer "small" craters with depths less than about 20 m than would be expected on the basis of numbers of larger craters. This relative lack of small craters is significant. Planetary scientists now believe that most of the larger craters we see today were actually formed long ago, during the period of intense meteoritic bombardment that followed the formation of the Earth-Moon system. Since that time, the cratering rate has been much lower, and meteoritic erosion has slowly erased these old craters. All original craters less than 20 m deep have been filled in by lunar and meteoritic debris. The small craters that we see today are relatively young. In fact, they are part of the erosive process that has erased their older counterparts.

Thus, we can estimate the rate of lunar erosion. Radioactive dating of samples returned from the *Apollo* missions suggests that the large lunar craters formed about 4 billion years ago. Since then, erosion has filled in all craters less than 20 m deep. Therefore, the lunar erosion rate is about 20 m divided by 4 billion years, or 5 meters per billion years. Of course, this is only an *average* rate, over the entire lunar surface and over the Moon's lifetime. Nevertheless, it is a useful "ballpark" figure.

The Moon's average erosion rate is about 10,000 times less than the Earth's. This large difference results from an abundance of prime erosion agents on our planet—especially running water and atmospheric wind. For example, the Barringer Meteor Crater (shown in Figure 8.10) in the Arizona desert, one of the largest meteor craters on Earth, is only 25,000 years old, but it is already decaying. It will probably disappear completely in a mere million years, quite a short time geologically. If a crater that size had formed on the Moon even 4 billion years ago, it would still be plainly visible today.

Figure 8.10 *The Barringer Meteor Crater, near Winslow, Arizona, is 1.2 km in diameter and 0.2 km deep. Geologists think an interplanetary meteorite made it about 25,000 years ago. The meteorite was probably about 50 m across and likely weighed around 300,000 tons. (U.S. Geological Survey)*

Dating Lunar Terrain

We can use observations of lunar cratering to estimate the *ages* of different locations on the Moon. The basic argument is simple. Older regions have been exposed to meteoritic impact longer and so have more craters on them. This technique is most useful in determining *relative* ages rather than *absolute* ones. Because we don't know the rate of meteoritic bombardment—we don't even know if the bombardment was constant in the past—we cannot assign a definite age to the lunar terrain by this method. Note too that (like radioactive dating) this technique measures age since the rock in question last *solidified*. Obviously, all cratering is erased and the clock is "reset" if the rock melts.

Using this technique, we can infer that the highlands are older than the maria. The **crater density**—the average number of 10-km (or larger) craters per million square kilometers (about the area of the largest mare)—in the highlands is about 1000. In the maria, a more typical number is 20 to 50 such craters per million square kilometers. Some further indication of individual crater ages can be obtained by studying the degree of erosion that has occurred since they formed. The fact that the highlands are so much more heavily cratered than the maria at first led many astronomers to believe that the maria are significantly younger than the highlands and quite young geologically—perhaps only a few hundred million years. This view was disproved when another means of dating the lunar surface became available.

When the *Apollo* astronauts visited several lunar sites and brought back rock samples, it became possible to measure the ages of the highlands and the maria using radioactive dating techniques and to turn the whole problem around. Now, instead of trying to use cratering to date the Moon, we use the known ages of Moon rocks to estimate the rate of cratering in the past. Although the highlands *are* older than the maria, it is now clear that the difference in age is much less than early estimates. The highlands are typically more than 4 billion years old, and the maria have ages ranging from 3.2 to 3.9 billion years. Thus, the very different crater densities are not simply the result of different exposure times. Astronomers now generally believe that the Moon, and presumably the entire inner solar system, experienced a sudden sharp drop in meteoritic bombardment about 3.9 billion years ago. The highlands solidified and received most of their craters before that time, and the maria solidified afterward. The rate of cratering (and erosion) has been roughly constant since then. We will return to some possible causes for this sharp decline in cratering in Chapter 17.

SURFACE COMPOSITION

Surface Rocks

✓4 Generally, two kinds of surface rock exist on the Moon. The *Apollo* program demonstrated a clear contrast between the less-dense highlands (2.9 g/cm^3) and the denser maria (3.3 g/cm^3). The highlands are made largely of rocks rich in aluminum, giving it its lighter color and lower density. The maria's basaltic matter contains more iron, giving it its darker color and greater density. Samples of the basic types of rock are shown in Figure 8.11(a), (b). The highlands represent the Moon's crust, while the maria are made of mantle material. Many of the rock samples brought back by the *Apollo* astronauts are of a type known as **impact breccias** (Figure 8.11[c]). A breccia is a piece of rock that consists of many smaller rock fragments stuck together. The lunar impact breccias owe their existence to the shock waves and high temperatures produced in meteoritic impacts, which have repeatedly shattered, recemented, and redistributed the rocks of the Moon's surface.

The maria rock is quite similar to terrestrial basalt, and geologists believe that is arose on the Moon much as basalt did on Earth, through the upwelling of molten material through the lunar crust. The great basins that formed the maria are thought to have been created during the final stages of the heavy meteoritic bombardment just described, between about 4.1 and 3.9 billion years ago. Subsequent volcanic activity filled the craters with lava, creating the formations we see today. In a sense, then, the maria *are* oceans—ancient seas of molten lava, now solidified.

Not all of these great craters became flooded with lava. One of the youngest craters is the Orientale basin (Figure 8.8[a]), which formed about 3.9 billion years ago. It did not undergo much subsequent volcanism, and so we can recognize its structure as an impact crater rather than as

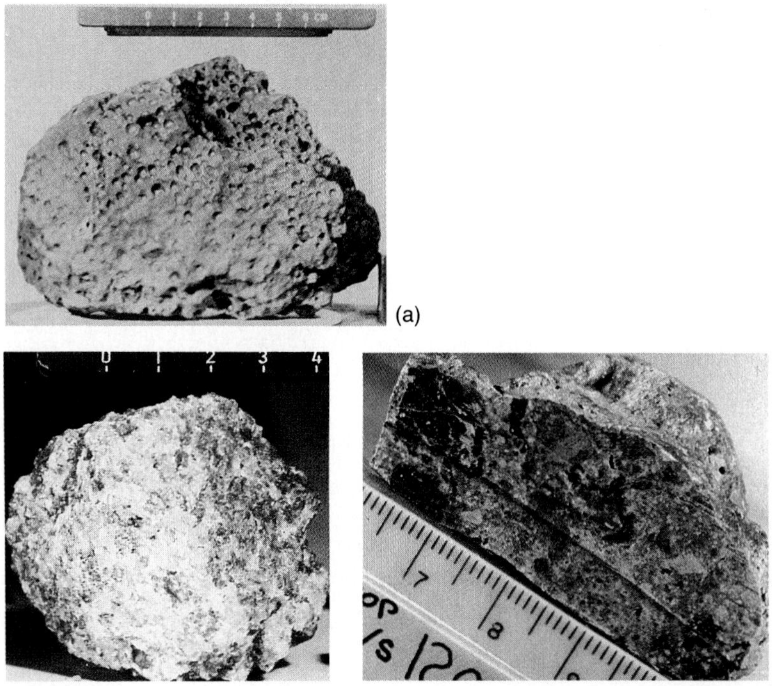

(a) (b) (c)

Figure 8.11 Rock samples brought back to Earth by the Apollo *astronauts clearly show the differences in color and texture between mare and highland material. (a) This photograph of material from Mare Imbrium, obtained by* Apollo 15, *includes a chunk of darker lunar lava having long ago upwelled from beneath the surface and cooled. (b) This sample of highland rock, from near the edge of Mare Serenitatis, returned by* Apollo 17, *shows an example of one of the first, lighter rocks to solidify when the Moon formed. (c) A lunar breccia, containing the crushed fragments of many different rocks, was collected by the* Apollo 16 *astronauts. All scales are in centimeters. (NASA)*

Figure 8.12 Photograph of an Apollo *astronaut's bootprint in the lunar dust. The astronaut's weight has compacted the regolith to a depth of a few centimeters. (NASA)*

another mare. On the far side of the Moon, similar ''unflooded'' basins can be seen.

Lunar Dust

Meteoroid collisions with the Moon are the primary cause of the layer of pulverized ejecta—also called lunar dust or **regolith** (meaning ''fine rocky layer'')—that covers the lunar landscape to an average depth of about 20 m. But that does not mean that the regolith is composed of old meteorites. Because each meteorite dislodges several hundred times its own mass of lunar matter, the regolith is mostly made up of Moon rock and contains only a small fraction of meteoritic material. *Apollo* samples show that meteoritic matter accounts for only about 1 percent of all lunar soil. The regolith is thinnest on the maria (10 m) and thickest on the highlands (over 100 m in places).

Figure 8.12 shows an astronaut's bootprint in the regolith. This microscopic dust has a typical particle size of about 0.01 mm. The regolith has some definite cohesiveness, though it lacks strength. It is much like baby powder or ready-mix dry mortar. Owing to the very low rate of lunar erosion, even those shallow bootprints will remain intact for millions of years.

In contrast to Earth's soil, lunar regolith contains no organic matter like that ordinarily produced by biological organisms. No life whatsoever exists on the Moon. Nor were any fossils found in *Apollo* samples. Lunar rocks are barren of life and apparently always have been. NASA was so confident of this that the astronauts were not even quarantined on their return from the last few *Apollo* landings. Furthermore, all the lunar samples returned by the American and Soviet Moon programs were absolutely bone dry. Lunar rock doesn't even contain minerals with water molecules locked within their crystal structure. Terrestrial rocks, conversely, are almost always 1 or 2 percent water.

VOLCANISM

Only a few decades ago, debate raged in scientific circles as to the origin of lunar craters. Most scientists believed that the craters were the result of volcanic activity. We now know that almost all lunar craters are meteoritic in origin. However, a few of them apparently are not. For example, Figure 8.13 shows an intriguing alignment of several craters in a *crater-chain* pattern so straight that it could not possibly have been produced by meteoroids colliding with the surface. The chance that interplanetary rocks could impact with such straight-line regularity is vanishingly small. Instead, the crater chain marks the location of a subsurface fault—a place where cracking or shearing of the surface

Figure 8.13 *This "chain" of well-ordered craters was photographed by an* Apollo 14 *astronaut. The largest crater is called Davy, located on the western edge of Mare Nubium. Field of view measures about 100 km across. (NASA)*

R I **V** U X G

once allowed molten matter to upwell from below. As the lava cooled, it formed a solid "dome" above each fissure. Subsequently, the underlying lava receded and the centers of the domes collapsed, forming the craters we see today. Similar features have been observed on Venus by the orbiting *Magellan* probe (see Chapter 11).

We can locate many other examples of lunar volcanism, both in telescopic observations from Earth and in the close-up photographs taken during the *Apollo* missions. Figure 8.14 shows what appear to be lava flows and the Marius Hills, which may be long-extinct volcanos. Figure 8.15(a) shows a volcanic **rille,** a ditch where molten lava once flowed. These examples make a strong case for surface volcanism early in the Moon's history, and the volcanism in turn explains the presence of the lava that formed the maria.

Apollo 15 astronauts discovered one of the best examples of lunar volcanism along the inside walls of one of the rilles. Figure 8.15(b) shows this surface feature, called Hadley Rille. Deep down inside the rille, as shown in Figure 8.15(c), we can see very definite layering of lunar deposits. The most reasonable explanation for the layering is that a series of volcanos repeatedly flooded the area with lava at various times in the past, eventually forming the Mare Imbrium. In the case of the Earth, we know that volcanos result from hot interior matter upwelling through fissures, cracks, and other geological faults. We must conclude that much of the lunar volcanism—especially that in the maria—must similarly have originated in an upper mantle that was wholly or partially molten at some time in the past. The lunar lava flows have been radioactively dated to be between 3.2 and 3.9 billion years old. Recall from Inter-

Figure 8.14 *The lunar lava beds seen here form the bulk of the mare known as Oceanus Procellarum. Rising above the lava plain are unusual volcanic features called the Marius Hills. (NASA)*

R I **V** U X G

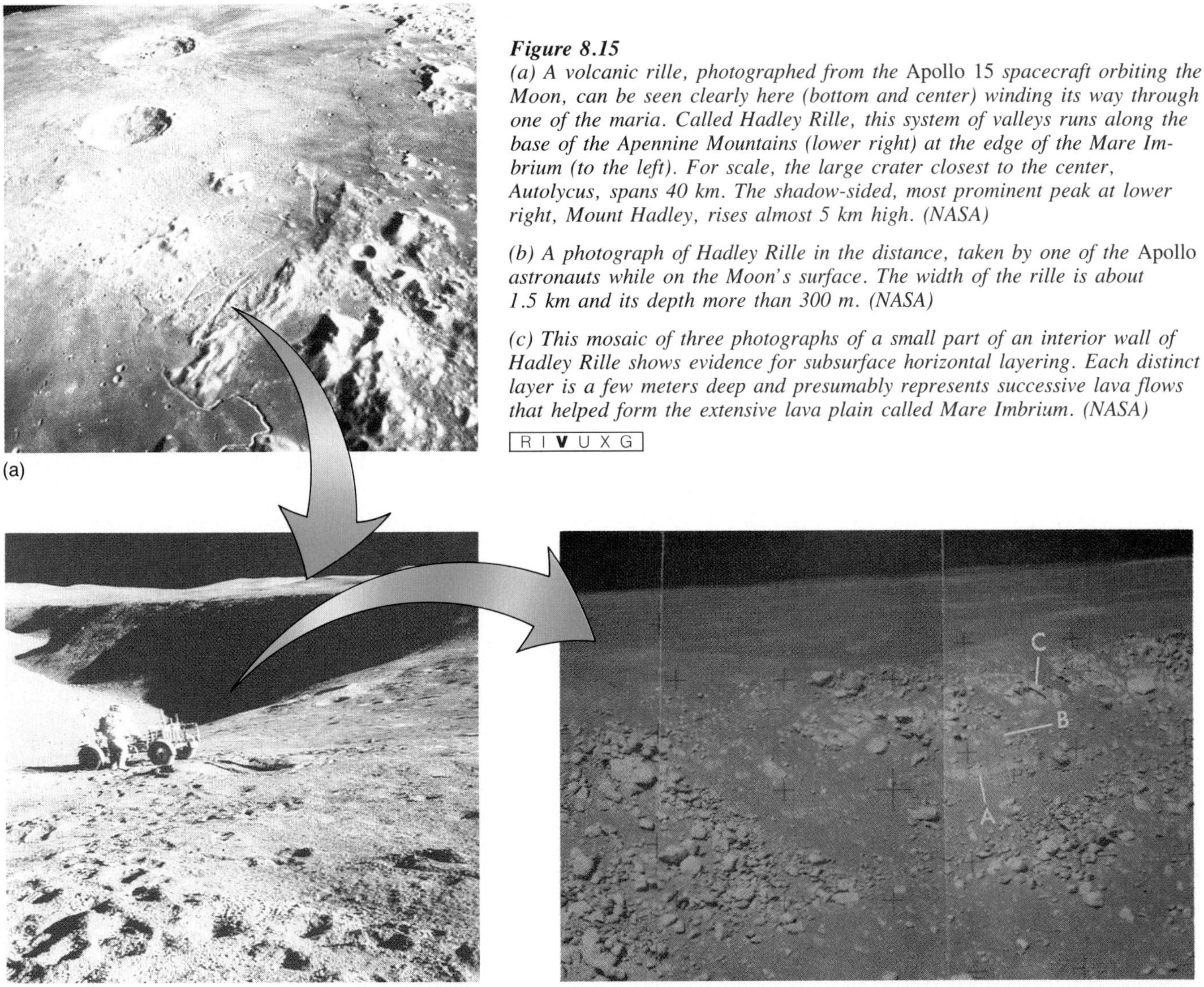

(a)

Figure 8.15
(a) A volcanic rille, photographed from the Apollo 15 *spacecraft orbiting the Moon, can be seen clearly here (bottom and center) winding its way through one of the maria. Called Hadley Rille, this system of valleys runs along the base of the Apennine Mountains (lower right) at the edge of the Mare Imbrium (to the left). For scale, the large crater closest to the center, Autolycus, spans 40 km. The shadow-sided, most prominent peak at lower right, Mount Hadley, rises almost 5 km high. (NASA)*

(b) A photograph of Hadley Rille in the distance, taken by one of the Apollo *astronauts while on the Moon's surface. The width of the rille is about 1.5 km and its depth more than 300 m. (NASA)*

(c) This mosaic of three photographs of a small part of an interior wall of Hadley Rille shows evidence for subsurface horizontal layering. Each distinct layer is a few meters deep and presumably represents successive lava flows that helped form the extensive lava plain called Mare Imbrium. (NASA)

R I **V** U X G

(b) R I **V** U X G

(c) R I **V** U X G

lude 7-3 that the radioactivity clock doesn't begin ticking until the rock solidifies, so the ages inferred for the maria represent the time of the *last* eruption.

Whatever volcanic activity existed on the Moon must have ended long ago. The low-density lunar highlands are dated to be *at least* 4 billion years old (and some are as old as 4.4 billion years). The high-density maria are in all cases found to be only a little younger. Nowhere on the Moon are rocks known to be younger than 3 billion years. Apparently, the maria solidified over 3 billion years ago, and the Moon has been dormant ever since.

THE LUNAR INTERIOR

The Moon's average density, about 3 g/cm^3, is similar to that of lunar surface rock, virtually eliminating any chance that the Moon has a large, massive, and very dense nickel-iron core like that within the Earth. In fact, the low density implies that the entire Moon is actually deficient in iron and other heavy metals relative to our planet. As discussed in Interlude 8-2, much of our information about the Moon's interior comes from seismic data, using equipment left on the surface by astronauts. There is no evidence for a lunar magnetic field at the present time. Researchers believe that planetary magnetism requires a rapidly rotating liquid metal core. Thus, the absence of a lunar magnetic field could result from the Moon's slow rotation, from the absence of a liquid core, or from both.

Modeling of all the available data indicates that the Moon's interior is of rather uniform density, although it is chemically differentiated. As depicted schematically in Figure 8.16, the models suggest a central core of perhaps 200 km radius surrounded by a region of semisolid rock with properties similar to the Earth's aesthenosphere and 500 km or so thick. The core is probably somewhat more iron-rich than the rest of the Moon, as a result of differentiation (although it is still iron-poor relative to the Earth's core). Near the center, the *current* temperature may be as low as 1500K, too cool to melt rock. However, some of the seismic data suggest that the inner parts of the core may be at least partially molten, implying a somewhat higher temperature. Our knowledge of the details of the Moon's central regions is very limited.

Above these regions lies a 1000-km thick mantle of solid rock, topped by a 60 to 150 km crust. Together, these layers constitute the Moon's lithosphere. The crust material, which forms the lunar highlands, is lighter than the mantle, which is similar in chemical composition to the lunar maria.

The Moon's crust is thicker than that of the Earth, and the crust on the far side is thicker than the crust on the side nearer Earth. If we assume that lava takes the line of least resistance in getting to the surface, we can understand why the far side of the Moon has no large maria. The lava had a shorter route to the surface on the Earth side. Volcanic activity did not occur on the far side simply because the crust was too thick to allow it to occur there.

Why is the far-side crust thicker? The answer is probably connected to the gravitational pull of the Earth. Just as heavier material tries to sink to the center of the Earth, the denser lunar mantle tended to sink below the lighter crust in the Earth's gravitational field. The effect of this was that the crust and the mantle became slightly off-center with respect to one another. The mantle was pulled a little closer to the Earth, while the crust moved slightly away. Thus, the crust became thinner on the near side and thicker on the far side.

We will see below how all these pieces of information on the Moon's internal structure constrain our theories of the Moon's origin and evolution.

INTERLUDE 8-2 *Moonquakes*

In addition to reflecting mirrors and other instruments left on the Moon, the American astronauts deployed several seismographs to measure lunar vibrations. The figure below shows one of these drum-shaped instruments, powered by panels of solar cells that capture sunlight and convert it into electricity.

Since their deployment in the early 1970s, however, the seismographs have been rather inactive. Some slight vibrations—moonquakes—have been recorded and radioed back to Earth, but these indicate only very weak activity. Even if you stood directly above one of these quakes, you would hardly feel the vibrations. The total energy released in a year of moonquaking is nearly a billion times less than in a year of earthquaking. Expressed another way, the average moonquake releases about as much energy as a firecracker.

This hardly perceptible seismic action confirms the idea that the Moon is geologically dead. There are no large quakes, volcanos, or lava flows at the present time. However, that *some* seismic waves are detected once in a while implies that the Moon is still settling a little. We shouldn't be surprised at this, because the Moon's interior is still somewhat hotter than its surface. Furthermore, the Earth exerts a large tidal strain on the Moon.

Although seismic activity is slight, researchers can use the weak lunar vibrations to learn about the Moon's interior. Signals transmitted back to Earth indicate:

1. Moonquakes occur at surprisingly great depths—roughly 1000 km below the surface, compared with about 100-km depths for earthquakes. This fact implies that the Moon is securely capped with a 1000-km-thick shell of strong and rigid rock—the lunar lithosphere. As with the Earth, the quakes occur at the plastic base of this rigid shell, in the aesthenosphere.
2. The seismic waves change speed while moving through several different layers of rock about 60 km beneath the surface (on the near side). This fact, along with direct chemical analysis of surface rocks, is evidence that the lithosphere is chemically differentiated into a crust and a mantle.
3. The Moon *might* have a small, partially molten core. As discussed in Chapter 7, seismic S-waves (shear waves) cannot move through liquid. *Apollo* seismographs operating on the near side do not detect shear waves when small meteoroids or discarded rocket stages hit the far side of the Moon. However, additional seismographs will probably be needed, especially on the far side, before space scientists fully accept this suggestion.

(NASA)

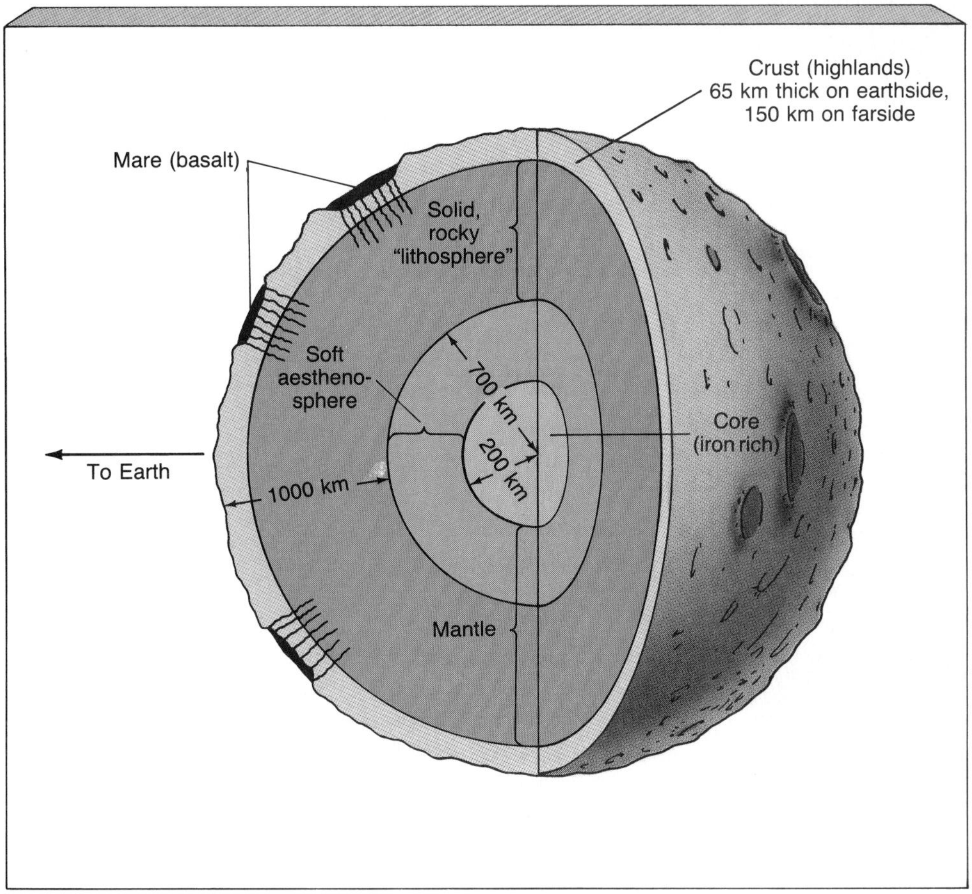

Figure 8.16 *Cross-sectional diagram of the Moon. Unlike the Earth, the Moon's rocky lithosphere is very thick—about 1000 km. Below the lithosphere, in the inner mantle, lies the lunar aesthenosphere, similar in properties to that of the Earth. At the center lies the core, which may be partly molten.*

THE ORIGIN OF THE MOON

☑5 The origin of the Moon is uncertain, although several theories have been advanced to account for it. As we will see, both the similarities between the Moon and the Earth *and* their differences conspire to confound many promising attempts to explain the Moon's existence.

One theory (the *sister,* or *coformation,* theory) suggests that the Moon formed as a separate object near the Earth and in much the same way as our own planet—the ''blob'' of material that eventually coalesced into the Earth also gave rise to the Moon at about the same time. The two objects thus formed as a double-planet system, each revolving about the common center of mass. Although once favored by many astronomers, this idea suffers from a major flaw: The Moon differs in both density and composition from the Earth, which makes it hard to understand how both could have originated from the same preplanetary blob.

A second theory (the *capture* theory) maintains that the Moon formed far from the Earth and was then later captured by it. In this way the density and composition of the two objects need not be similar, for the Moon presumably materialized in a quite different region of the early solar system. The objection to this theory is that the Moon's capture would be an extraordinarily difficult event; it might even be an impossible one. Why? Because the mass of our Moon is so large relative to that of the Earth. It's not that our Moon is the largest natural satellite in the solar system, but it is unusually large compared with its parent planet. Mathematical modeling suggests that it is unreasonable to expect Earth's gravity to have attracted our Moon in exactly the right way to capture it during a close encounter sometime in the past. Furthermore, there are too many similarities between the composition of the Earth's mantle and that of the Moon to admit the possibility that the two bodies formed entirely independently from one another.

A third theory (the *daughter,* or *fission,* theory) speculates that the Moon originated out of the Earth itself. The Pacific Ocean basin has often been mentioned as the place from which protolunar matter may have been torn—the result of the young Earth's rapid spin or even of tidal effects (from the Sun) on the young, mostly molten Earth. As absurd as this idea may seem, the early results of the *Apollo* missions seemed to favor it. Both the lunar composition and density mimic those of Earth's mantle, just below the crust. The matter constituting both the Moon and the Pacific basin is basalt largely devoid of iron. However, there remains the fundamental mystery of how Earth could have possibly have been spinning so fast that it ejected an object as large as our Moon. Also, computer simulations indicate that the ejection of the Moon into a stable orbit would not occur.

Today, many astronomers favor a hybrid of the capture and fission themes. This idea—often called the *impact* theory—postulates a collision by a large, Mars-sized object with a youthful and molten Earth. Such collisions may have

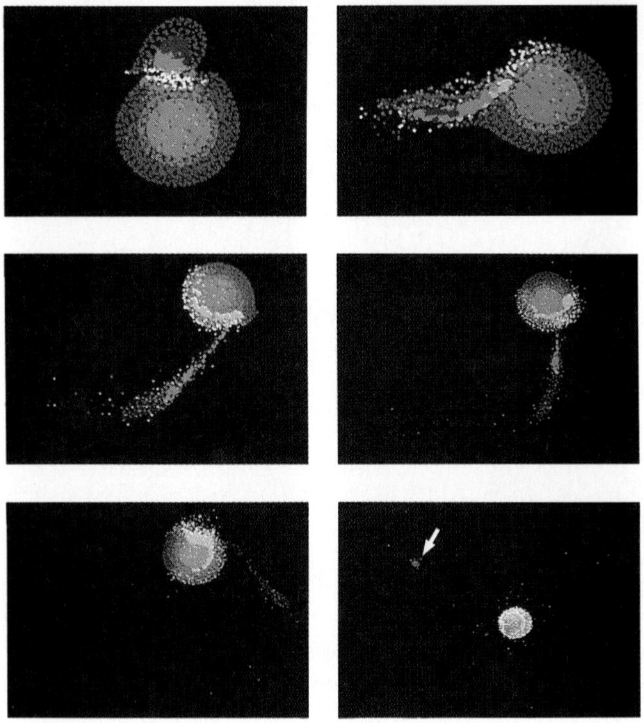

Figure 8.17 *This sequence shows a simulated collision between the Earth and an object the size of Mars. (The sequence proceeds from left to right, top to bottom, and zooms out dramatically.) The arrow in the final frame shows the newly formed moon. (W. Benz, W. L. Slatery, A. G. W. Cameron)*

been quite frequent in the early solar system (see Chapter 17). The collision presumed by the impact theory would have been more a glancing blow than a direct impact. The dislodged matter from our planet then assembled into the Moon. Computer simulations of such a catastrophic event show that most of the bits and pieces of splattered Earth could have coalesced into a stable orbit. Figure 8.17 shows some of the stages of one such simulation. If the Earth had already formed an iron core by the time the collision occurred, the Moon would indeed have ended up with a composition similar to the Earth's mantle. During the collision, any iron core in the impacting object itself would have been left behind in the Earth, eventually to become part of Earth's core. Thus, both the Moon's overall similarity to Earth's mantle and its lack of a dense central core are naturally explained.

LUNAR HISTORY

Given all the data, can we construct a reasonably consistent history of the Moon after its formation? The answer seems to be yes. Many specifics are still debated, but a consensus exists. Refer to Figure 8.18 while studying the following details.

The Moon apparently formed about 4.5 billion years ago. The approximate date of the oldest rocks discovered in the lunar highlands is 4.4 billion years, so we know that at least part of the crust must already have solidified by that time. At formation, the Moon was already depleted in heavy metals relative to the Earth.

During the earliest phases of the Moon's existence—roughly the first half billion years or so—meteoritic bombardment must have been frequent enough to heat, and remelt, most of the *surface* layers of the Moon, perhaps to a depth of 400 km. The early solar system was surely populated with lots of interplanetary matter, much of it in the form of boulder-sized fragments, capable of generating large amounts of energy on collision with planets and their moons. But the intense heat derived from such collisions could not have penetrated very deeply into the lunar interior. Rock simply does not conduct heat well.

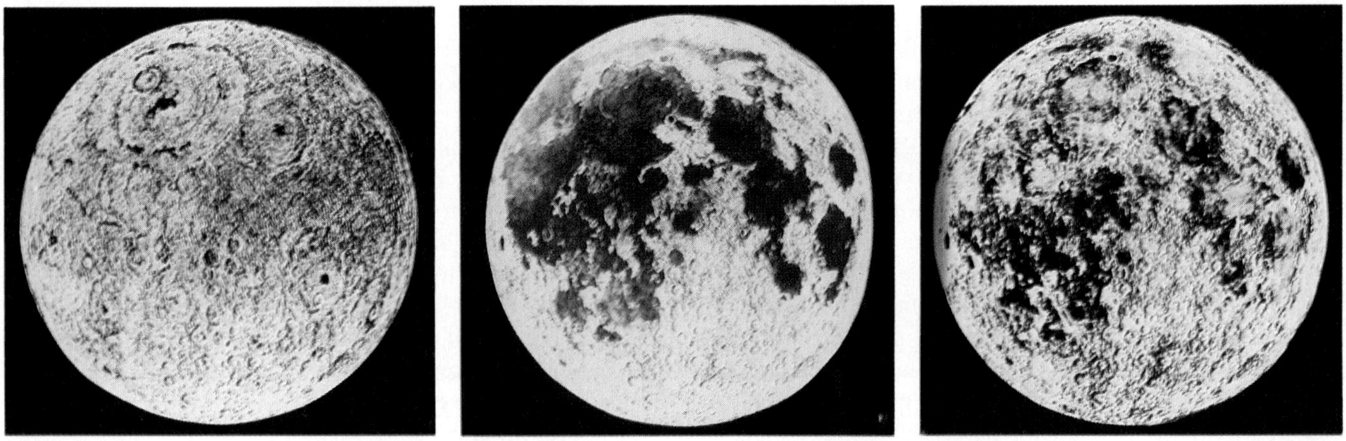

Figure 8.18 *Paintings of the Moon (a) about 4 billion years ago, after much of the meteoritic bombardment had subsided and the surface had somewhat solidified; (b) about 3 billion years ago after molten lava had made its way up through surface fissures to fill the low-lying impact basins and thus create the smooth maria; and (c) at present, with much of the originally smooth maria now heavily pitted with craters formed at various times within the past 3 billion years. (United States Geological Survey)*

This situation resembles the surface melting we suspect occurred on Earth from meteoritic impacts during the first billion years or so. But the Moon is much less massive than the Earth, and it does not contain enough radioactive elements to heat it much further. Radioactivity probably heated the Moon a little, but not sufficiently to transform it from a warm, plastic object to a completely liquid one. The Moon must have differentiated during this period. If the Moon has a small iron core, it also formed at this time.

About 3.9 billion years ago, around the time that the Earth's crust solidified, the heaviest phase of the meteoritic bombardment ceased. The Moon was left with a solid crust, ultimately to become the highlands, dented with numerous large basins, soon to flood with lava and become the maria. Between 3.9 and 3.2 billion years ago, lunar volcanism filled the maria with the basaltic material we see today. The age of the youngest maria—3.2 billion years—apparently indicates the time when the volcanic activity subsided. The mare basins are the sites of the last extensive lava flows on the Moon, over 3 billion years ago. Their smoothness, compared with the older, more rugged highlands, disguise their great age.

Small objects cool more rapidly than large ones—their interior is closer to the surface, on average. Being so small, the Moon rapidly lost its internal heat to space. As a consequence, it cooled much faster than did the Earth. As the Moon cooled, the volcanic activity ended as the thickness of the solid surface layer increased. With the exception of a few meters of surface erosion from eons of meteoritic bombardment, the lunar landscape has remained more or less structurally frozen for the past 3 billion years. The Moon is dead now, and it has been dead for a long time.

LUNAR EXPLORATION

Early Missions

The Space Age began in earnest on October 4, 1957, with the launch of the Soviet satellite *Sputnik 1*. Thirteen months later, the Soviet *Luna 1*, the first human craft to escape the Earth's gravity, passed the Moon on January 4, 1959. *Luna 2* crash-landed on the surface in September of that year, and *Luna 3* returned the first pictures of the far side a month later. The long-running *Luna* series established a clear Soviet lead in the early "space race" and returned volumes of detailed information about the lunar surface. *Luna 9* successfully soft-landed in Oceanus Procellarum (Ocean of Storms) in February 1969. Several of the *Luna* missions, ending with *Luna 24* in August 1976, subsequently returned lunar surface material to Earth.

The U.S. lunar exploration program got off to a rocky start. The first six attempts in the *Ranger* series, between 1961 and 1964, failed to accomplish their objective of just hitting the Moon. The last three were successful, however. *Ranger 7* collided with the lunar surface (as intended) on June 28, 1964. Five U.S. *Lunar Orbiter* spacecraft, launched in 1966 and 1967, were successfully placed in orbit around the Moon, and they relayed back to Earth high-resolution images of much of the lunar surface. For regions not visited by later landers, those images are often the best available today. Between 1966 and 1968, seven *Surveyor* missions soft-landed on the Moon and performed detailed analyses of the surface.

The Apollo Program

Many of the above unmanned U.S. missions were performed in support of the manned ***Apollo* program.** On May 25, 1961, at a time when the U.S. space program was in great disarray, President John F. Kennedy declared that the U.S. would "send a man to the Moon and return him safely to Earth" before the end of the decade, and the *Apollo* program was born. On July 20, 1969, less than 12 years after *Sputnik* and only 8 years after the statement of the program's goal, *Apollo 11* commander Neil Armstrong became the first human to set foot on the Moon, in Mare Tranquilitatis (Sea of Tranquility). Three and a half years later, on December 14, 1972, scientist-astronaut Harrison Schmitt, of *Apollo 17*, was the last.

The two astronauts who traveled to the lunar surface in each lunar lander (Figure 8.19) performed numerous geological and other scientific studies on the surface. The later

Figure 8.19 *Each* Apollo *lander—the Lunar Excursion Module—carried two astronauts down to the lunar surface from lunar orbit. The upper half separated from the lower portions to carry the astronauts back to the orbiting Command Module at the end of the surface expedition. The entire spacecraft was covered with reflective material to prevent it from overheating in the intense lunar sunlight. (NASA)*

landings brought with them a "lunar rover"—a small golf-cart-sized vehicle that greatly expanded the area they could cover. Probably the most important single aspect of the *Apollo* program was the collection of samples of surface rock from various locations on the Moon. In all, some 382 kg of material was returned to Earth. Chemical analysis and radioactive dating of these samples revolutionized our understanding of the Moon's surface history—no amount of Earth-based observations could have achieved the same results. (It is also true, however, that a manned mission was not essential for returning samples, as the Soviet *Luna* program demonstrated.)

Each *Apollo* lander also left behind a nuclear-powered package of scientific instruments called **ALSEP** (**A**pollo **L**unar **S**urface **E**xperiments **P**ackage), whose function was to monitor the solar wind, measure heat flow in the Moon's interior, and, perhaps most important, to record lunar seismic activity. With several ALSEPs on the surface, scientists could determine the location of moonquakes by triangulation and could map out the Moon's inner structure, obtaining information critical to our understanding of the Moon's evolution. NASA turned off the ALSEPs in 1978 to save money.

Present and Future Exploration

By any standards, the *Apollo* program was a spectacular success. It represents one of the most towering achievements of the human race. The project goals were met on schedule, within budget, and on live television, and our knowledge of the Moon, the Earth, and the solar system increased enormously. But the "Age of Apollo" was short-lived. Public interest quickly waned. Over half a billion people breathlessly watched television as Neil Armstrong set foot on the Moon, yet barely 3 years later, when the program was abruptly cancelled for largely political (rather than technological, scientific, or economic) reasons, the landings had become so routine that they no longer excited the interest of the American public. Unmanned space science moved away from the Moon and toward the other planets, and the manned space program floundered. One of the most amazing and saddest aspects of *Apollo* is that today, two decades later, *no* nation on Earth (including the United States) has the desire, the capability, or the money to replicate the feat.

Currently there are no ongoing, long-term lunar exploration programs. The only spacecraft presently studying the Moon is the small Japanese *Hiten* satellite, placed in lunar orbit in March 1990. Plans do exist to establish permanent human colonies on the Moon, either for commercial ventures, such as mining, or for scientific research. Proposals have also been made to site large optical, radio, and other telescopes on the lunar surface. Such instruments, which could be constructed larger than Earth-based devices, would enjoy perfect seeing and no light pollution. However, none of these projects is scheduled to become reality in the foreseeable future. After a brief encounter with humanity, the Moon is once again a lifeless, unchanging world.

CHAPTER SUMMARY

Earth's only natural satellite is an airless, virtually unchanging world that experiences extremes in temperature. After centuries of measurements, scientists are well acquainted with the orbit of the Moon. Lunar laser ranging lets the distance to the Moon at any instant be measured to an accuracy of a few centimeters. The Moon has roughly one-fourth the Earth's radius. Its mass is approximately 1/80 the mass of Earth. Its density is 3.3 g/cm^3, in contrast to 6 g/cm^3 for the Earth. The Moon rotates in 27.3 days, the same time it takes to orbit the Earth, so that the same lunar hemisphere always faces Earth. The large lunar equatorial bulge probably indicates that the Moon used to rotate more rapidly and that it once orbited closer to Earth. The Moon is too small to have a large enough gravitational pull to have maintained an atmosphere.

The Moon has mountains called highlands and lowlands called maria. The maria are ancient lava beds. These regions are neither chemically nor structurally similar to Earth's plains or mountains.

The Moon's surface is scarred with numerous bowl-shaped depressions called craters. Most craters on the Moon resulted from collisions with interplanetary debris. The rate of erosion on the Moon is about 10,000 times less than on Earth because the Moon lacks wind and running water. The lunar highlands are typically more than 4 billion years old, while the maria have ages ranging from 3.2 to 3.9 billion years.

The Moon's surface consists of both rocky and dusty material. Highland rocks, which are less dense than rocks from the maria, are believed to represent the Moon's crust. Maria rocks are believed to have originated in the mantle of the Moon. Lunar dust, called regolith, is made mostly of pulverized lunar rock, mixed with a small amount of material from impacting meteorites.

The lunar interior is thought to be of rather uniform density, but it is chemically differentiated, which implies a partially molten Moon in the distant past. Slight vibrations called moonquakes are due to warmer conditions inside the

Moon and to the tidal pull of Earth. They can be used to study the lunar interior.

The impact theory is the currently favored theory of lunar origin. It states that a glancing collision between Earth and a Mars-sized object sent earthly matter into orbit, where it assembled into the Moon. The Moon apparently formed 4.5 billion years ago. Bombarding meteorites must have melted its surface, but it is likely that the Moon was never entirely melted by radioactive heating. Volcanos erupted on the Moon between 3.9 and 3.2 billion years ago. The small size of the Moon caused it to lose its heat rapidly to space, and it is now a dead world.

The Moon, Earth's nearest neighbor, has been a target of numerous space missions. Early missions involved crash landings on the Moon, and orbiting spacecraft. The manned *Apollo* landings on the Moon and the Soviet *Luna* program returned lunar rocks to Earth. Moon rocks revolutionized our understanding of the Moon's surface history.

KEY WORDS

ALSEP
apogee
Apollo program
crater
crater density
ejecta blanket
erosion
highlands
impact breccia
libration
maria
perigee
regolith
rille
synchronous orbit
terrae
terminator

REVIEW QUESTIONS

1. How far away is the Moon? How do we know?
2. Explain why, if the Moon always keeps a single face toward Earth, we can see 59 percent of the lunar surface in a month.
3. How big is the Moon's equatorial bulge, and how does its size compare with what one would expect on theoretical grounds? What does the Moon's large equatorial bulge suggest about lunar history?
4. Employ the concept of escape velocity to explain why the Moon has no air.
5. Where is the best place on the Moon to aim a telescope? Why? Describe what you might see.
6. What is the primary source of erosion on the Moon? Explain why the average rate of lunar erosion is so much less than on Earth.
7. Name two pieces of evidence showing that the lunar highlands are older than the maria.
8. In what sense were the lunar maria "seas"?
9. Tell how lunar soil is different from earthly soil.
10. What is a possible explanation for the lunar crust being thicker on the far side than on the near side?
11. Give evidence for a molten core for the Moon.
12. Explain the theory of the Moon's origin favored by many of today's astronomers. Tell why they favor it.
13. What are "moonquakes"? Why, despite moonquakes, do scientists say the Moon is a "dead" world?
14. What was the most important result of the *Apollo* missions to the Moon with respect to our knowledge about it? Why?
15. What is the "synodic" period of the Earth as seen from the Moon? (That is, what is the time from earthrise to earthrise at any given point on the lunar surface)?
16. Given the average rate of lunar erosion quoted in the text, how much erosion occurs in a day? In a century?
17. The Moon's mass is $1/80$ the mass of the Earth, and the lunar radius is $1/4$ Earth's radius. Calculate the total weight on the Moon of a 100-kg astronaut with a 50-kg spacesuit and backpack, relative to the weight on Earth.
18. What was the orbit period of the Apollo command module orbiting just above the lunar surface?

DISCUSSION QUESTIONS

1. Because the Moon always keeps one face toward Earth, an observer on the moon's near side would see Earth appear almost stationary in the lunar sky. Still, Earth would change its appearance as the Moon orbited Earth. How would Earth change? Why?
2. The best place to aim a telescope or binoculars on the Moon is along the terminator line, the line between the Moon's light and dark hemispheres. Why? If you were standing on the lunar terminator, where would the sun be in your sky? What time of day is it when you're standing on the Earth's terminator line?
3. Do you think that the Moon would be a good place from which to make astronomical observations? Explain your answer.

4. Should Moon rocks brought back by Apollo astronauts and in Soviet unmanned missions be shared equally by scientists from all nations on Earth?
5. The Moon once was considered to be remote, mysterious, and forever unreachable by humanity. Do you think that the Moon's age-old mystery was sacrificed when the first human beings landed on its surface? Should humans attempt to return to the moon? Why or why not?
6. The lunar maria have fanciful Latin names, such as Mare Imbrium, or "Sea of Showers." Craters are named for dead great scientists or philosophers. Should these names be changed to make them more up-to-date?

PROJECTS

1. Consult an almanac to find out the date of the next fourth-quarter Moon. Around 10:00 or 11:00 P.M. on that date, go to a country location where many stars can be seen sprinkled across the heavens. Around midnight, the Moon will rise. Can you see as many stars?
2. If you have binoculars, turn them on the Moon when it appears at twilight and when it appears overhead. Draw pictures of what you see. What differences do you notice in your two drawings? What color is the Moon seen near the horizon? What color is the Moon seen overhead? Why is there a difference?

SUGGESTED READINGS

Beatty, J. K. "The Making of a Better Moon." *Sky and Telescope* (December 1986). An explanation of what is currently the best theory on the origin of the Moon.

Burnham, R. "How *Apollo* Changed the Moon." *Astronomy* (July 1989). On the twentieth anniversary of the first manned lunar landing.

———. "Watching the Lunar Phases." *Astronomy* (June 1980). Observing the Moon for an entire month as it goes through its various phases.

Burns, J. O., N. Duric, G. J. Taylor, and S. W. Johnson. "Observatories on the Moon." *Scientific American* (March 1990). The authors propose plans for establishing high-resolution optical, radio, infrared, gamma-ray, and x-ray observatories on the Moon.

Cameron, W. S. "Lunar Transient Phenomena." *Sky and Telescope* (March 1991). On the claim that something unusual, perhaps associated with volcanic or other activity, is still happening on the Moon.

Coles, R. R. "Moon Tales." *Sky and Telescope* (July 1948). Some ways in which the Moon has influenced civilization.

Kitt, M. T. "Eight Lunar Wonders." *Astronomy* (March 1989). Using a small telescope to investigate eight geological oddities on the Moon.

———. "One Day at Copernicus Crater." *Astronomy* (September 1988). Using a small telescope to explore sights of a large lunar crater in the course of the long lunar day.

Levitt, I. M. "Moon Illusion." *Sky and Telescope* (April 1952). Tells why the full Moon seen near the horizon looks larger than from higher in the sky.

Thomsen, D. E. "Man in the Moon." *Science News* (March 8, 1986). Discussion of the future possibility of a lunar colony.

Robertson, D. F. "Reaching for the Moon." *Astronomy* (August 1988). Plans for an observatory on the farside of the Moon.

CHAPTER 9

The Solar System
An Introduction to Comparative Planetology

Against the blackness of space, the Galileo *craft has just detached from a cradlelike device aboard the Earth-orbiting Space Shuttle* Atlantis. *The five-member crew deployed the satellite on October 18, 1989.* Galileo *is on a six-year journey to Jupiter. Much of our knowledge—including many startling discoveries—about our solar-system neighborhood has come from the satellites that have ventured beyond our Earth-Moon system. In this image, the background white is our planet Earth, its thin bluish blanket of atmosphere insulating it from the coldness of space. (Courtesy NASA)*

LEARNING GOALS

1. to gain an overall perspective on the solar system
2. to understand the basic differences between the terrestrial and the Jovian planets
3. to appreciate the essence of the science of comparative planetology

Moving away from our own home planet and its barren moon, we encounter many new objects in our solar system. Each has a unique story to tell. Our three nearest planetary neighbors are all rocky worlds, comparable to Earth in size, composition, and distance from the Sun. The similarities and differences among those other planets and between them and the Earth-Moon system offer clues to the events that formed the inner solar system and help us piece together the early history of our own world.

Farther out from the Sun, we come across four giant gaseous planets, all much larger than Earth and quite different from it in both composition and structure. Each has many moons, no two of them alike and none of them like our own. Farther still, beyond the outermost gas giant, lies one more small world, frozen and mysterious. Among the nine known planets move countless chunks of rock and ice, all orbiting the Sun, many on highly eccentric paths. All of these bodies carry information about the processes responsible for forming the planets we see today. The solar system is filled with clues to its own origin and evolution.

In less than a single generation, we have learned more about our solar system than in all the centuries that went before. Instruments aboard unmanned robots have taken close-up photographs and in some cases made on-site measurements. They have revolutionized our understanding of our cosmic neighborhood. The most recent era of solar system exploration has revealed the planets and their moons to be worlds unto themselves—alien worlds, to be sure, with unearthly conditions and histories, forbidding and hostile to human life, yet beautiful to behold and fascinating to explore.

OUR PLANETARY SYSTEM IN BRIEF

The last two chapters dealt with our own backyard—the Earth-Moon system. In the next seven chapters, we will examine the other members of our solar system. By studying the individual **planets**—the major bodies that orbit the Sun and reflect its light—and their **moons**—which orbit them—we will gain a richer outlook on our own home in space. We will use the powerful and newly emerging perspective of **comparative planetology**—comparing and contrasting the properties of these diverse worlds—to understand better the conditions under which planets form and develop. Having made the most important stops along the way, we will conclude our tour in Chapter 17 with a look at the modern theory of how our solar system formed and developed.

But before embarking on this planet-by-planet study, let us consider the solar system as a whole and try to place the planets in perspective and see what patterns we can discern in their orbits and bulk properties. We will begin by making a short historical survey of our cosmic neighborhood.

The Greeks and other astronomers of old were aware of five planets in the nighttime sky—Mercury, Venus, Mars, Jupiter, and Saturn. We have already seen in Chapters 2 and 3 how observations of the apparently erratic motions of those wanderers across the celestial sphere ultimately led to the Copernican revolution and the birth of our modern view of the cosmos. In addition to the Moon and the Sun, the ancients also knew of two other types of heavenly phenomena—comets and meteors, which we will study in Chapter 16—that are neither stars nor planets. Their role in the "big picture" of the solar system was not understood until much later.

With the invention of the telescope, more detailed observations of the known planets could be made. Galileo Galilei was the first to capitalize on this new technology (his simple telescope is shown in Figure 9.1), and his discovery of the phases of Venus and four moons around Jupiter early in the seventeenth century played a large part in changing forever humankind's vision of the universe. With continuing technological advances, knowledge of the solar system continued to improve rapidly. Astronomers began discovering objects invisible to the unaided human eye. By the end of the nineteenth century, astronomers had found Uranus, Neptune, many planetary moons, Saturn's rings, and hundreds of "minor planets," or **asteroids,** mostly orbiting in a broad band—the **asteroid belt**—between Mars and Jupiter. The seventh planet, Uranus, was discovered late in the eighteenth century; the eighth, Neptune, was found in the mid-nineteenth. The largest asteroid, Ceres, also the first to be sighted, was discovered early in the nineteenth century. A large telescope of mid nineteenth-century vintage is shown in Figure 9.2.

The twentieth century has brought continued improvements in optical telescopes. One more planet (Pluto) has been discovered, along with three more ring systems, dozens of moons, and thousands of asteroids. The century has also seen the rise of both nonoptical astronomy—especially radio and infrared—and spacecraft exploration as making vitally important contributions to the field of planetary science. Astronauts have carried out experiments on the Moon (see Figure 9.3), and numerous unmanned probes have left Earth and traveled to all but one of the other planets. Figure 9.4 shows the 1989 launch of *Galileo*, one of NASA's most recent planetary missions, carried in the cargo bay of the Space Shuttle *Atlantis*. As currently explored, we know that our solar system contains 1 star (the Sun), 9 planets, 61 moons (at last count), 6 asteroids larger than 300 km in diameter, more than 4000 smaller (but still well-studied) asteroids, myriad comets a few kilometers in diameter, and countless meteoroids less than a meter across. The list may grow as we continue to explore our neighborhood. The near-void between all these objects is termed **interplanetary space.**

Some selected planetary properties are listed in Table 9-1. As we proceed through the solar system, we will seek to understand how each planet compares with our Earth-Moon system, and we will learn what each planet contributes to our knowledge of the solar system as a whole. On the basis of our studies of the Earth and Moon, we already know some of the questions to ask. For example, does the planet have a magnetic field, an atmosphere, or geological activity? Does it have a rocky surface? Is its surface molten? Does it have a liquid core? As we will see, each planet will also present us with new questions and insights of its own. Whatever the answers, the comparison enriches our knowledge of the ways planets work.

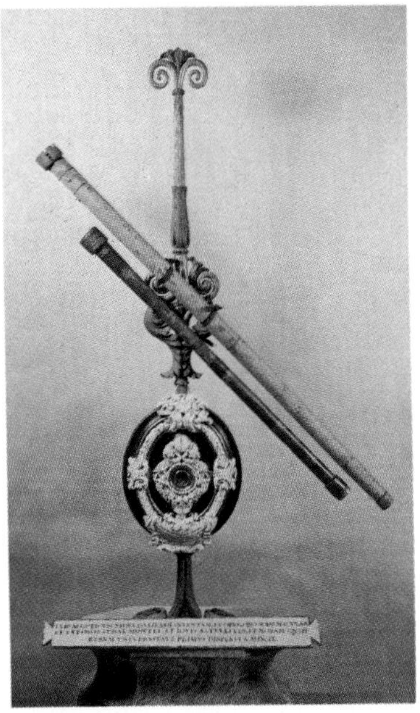

Figure 9.1 *The telescope with which Galileo made his first observations was simple, but its influence on astronomy was immeasurable. (Art Resource)*

Figure 9.2 *By the mid nineteenth century, telescopes had improved enormously in both size and quality. Shown here is the telescope built and used by Irish nobleman and amateur astronomer the Earl of Rosse. (The Birr Scientific & Heritage Foundation)*

Figure 9.3 *An* Apollo *astronaut doing some basic geology prospecting near a huge boulder near the Moon's Mare Serenitatus. (NASA)*

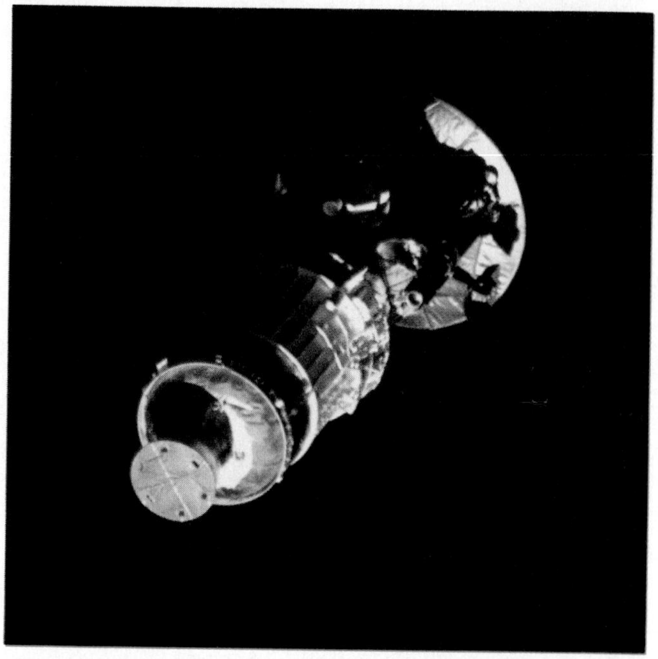

Figure 9.4 *The launch of the space probe* Galileo *on a mission to explore in detail the moons and the atmosphere of Jupiter. (NASA)*

TABLE 9-1 ***Some Planetary Properties***

Planet	Average Distance from Sun (A.U.)	Mass (Earth masses)	Number of Known Satellites	Radius (Earth radii)	Rotation Period (Earth days)	Density (Earth densities)
Mercury	0.4	0.06	0	0.4	59	1.0
Venus	0.7	0.8	0	0.9	243	0.9
Earth	1.0	1.0	1	1.0	1.0	1.0
Mars	1.5	0.1	2	0.5	1.03	0.7
Ceres (asteroid)	2.8	0.0002	0	0.07	<1	0.5
Jupiter	5.2	318	16	11.2	0.4	0.2
Saturn	9.5	95	18	9.4	0.4	0.1
Uranus	19.2	15	15	4.0	0.7	0.2
Neptune	30.1	17	8	3.9	0.6	0.3
Pluto	39.5	0.003	1	0.2	6.4	0.3

THE OVERALL LAYOUT OF THE SOLAR SYSTEM

☑1 The planet closest to the Sun is Mercury. Moving outward, we encounter in turn Venus, Earth, Mars, Jupiter, Saturn, Uranus, Neptune, and Pluto. We have already seen (in Chapter 3) the basic properties of the planets' orbits. Their paths are all ellipses, with the Sun at one focus. Most orbits have low eccentricities (the exceptions being the innermost and the outermost worlds, Mercury and Pluto), so we can reasonably think of the planets' orbits as circles centered on the Sun. Figure 9.5 is an artist's rendition of the planetary system as future generations of space voyagers might perceive it from a distant vantage point. Figure 9.6 shows some of the best photographs of our solar system as a whole—a view of the Sun and several of the planets taken by *Voyager 2* as it sped toward interstellar space. Figure 9.7 is a photograph taken during the July 1991 eclipse. Many solar system objects are visible in this one photograph.

All the planets orbit the Sun in nearly the same plane as Earth (the ecliptic plane, as we learned in Chapter 3). Again, Mercury and Pluto deviate somewhat from this rule—their orbital planes lie at 7° (degrees) and 17° to the ecliptic, respectively. Still, we can think of the solar system as being quite flat. Its "thickness" perpendicular to the ecliptic is at least 50 times smaller than the diameter of Pluto's orbit. All planets orbit the Sun in the same direction—counterclockwise, seen from above Earth's North Pole.

By terrestrial standards, the solar system is immense. The distance from the Sun to Pluto is about 40 A.U., almost a million times the radius of the Earth and roughly 15,000 times the distance from Earth to the Moon. Yet the diameter of Pluto's orbit is less than 1/1000 of a light-year, and the next nearest star is several light-years distant from the Sun. Despite its vast extent, the entire solar system lies very close to the parent Sun, astronomically speaking.

The planetary orbits are not evenly spaced. As Table 9-1 shows, they get farther and farther apart as we move farther out from the Sun. Yet there is a certain regularity in this behavior. In the eighteenth century a fairly simple rule, now known as the Titius-Bode Law (see Interlude 9-1), seemed to "predict" the planetary orbits remarkably well. Even the asteroid belt between Mars and Jupiter had a place in the scheme, which excited great interest among astronomers and numerologists alike. There is apparently no simple explanation for this empirical "law." Today, it is regarded more as a curiosity than as a fundamental property of the solar system. Still, the law is at least a convenient device for remembering the planets' orbital radii.

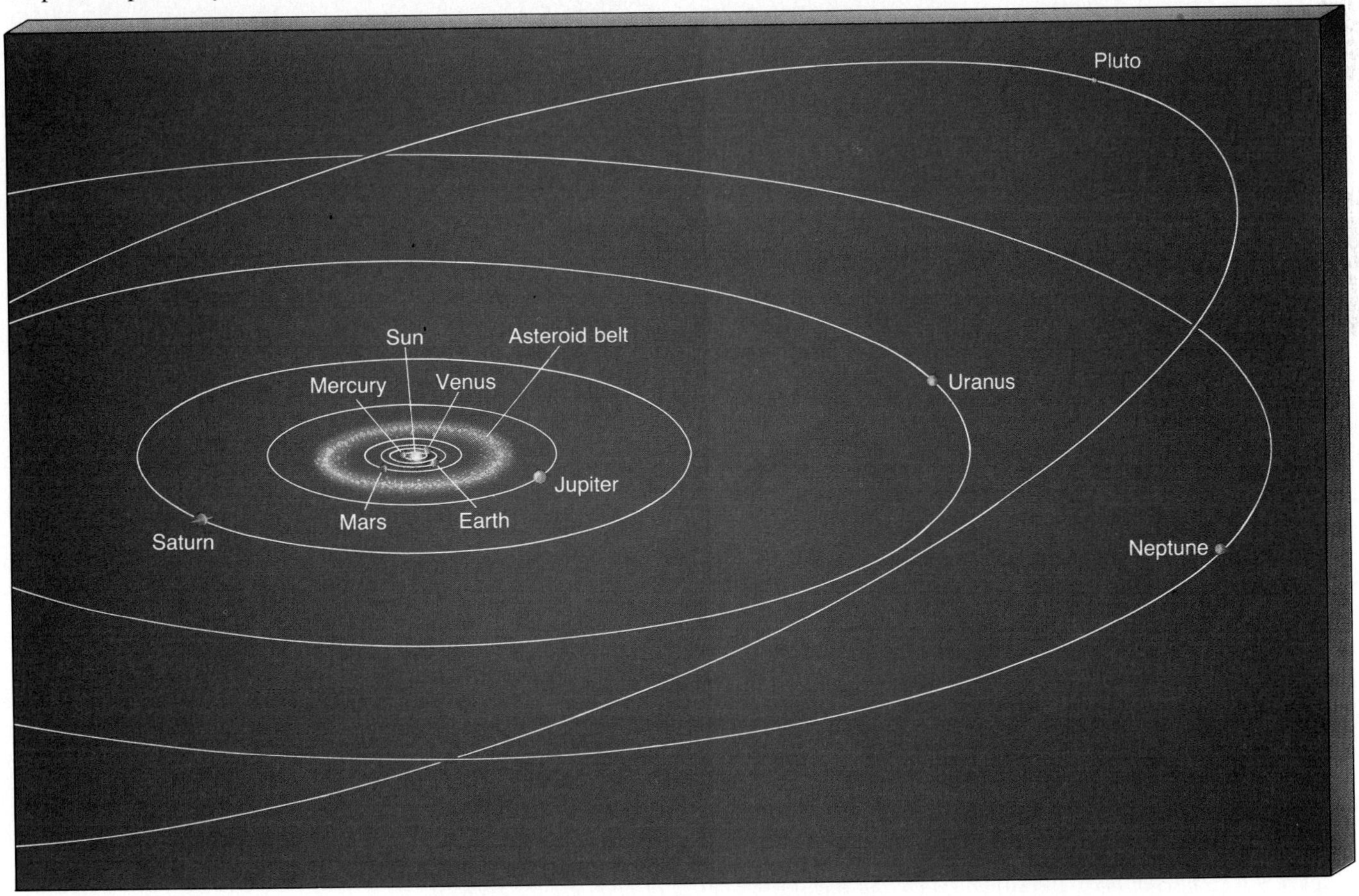

Figure 9.5 *Might future space voyagers travel far enough from Earth to gain this perspective on our solar system? Except for Mercury and Pluto, the orbits of the planets lie nearly in the same plane, so the solar system is quite flat. As we move out from the Sun, the distance between the orbits of the planets increases. The entire solar system spans nearly 80 A.U.*

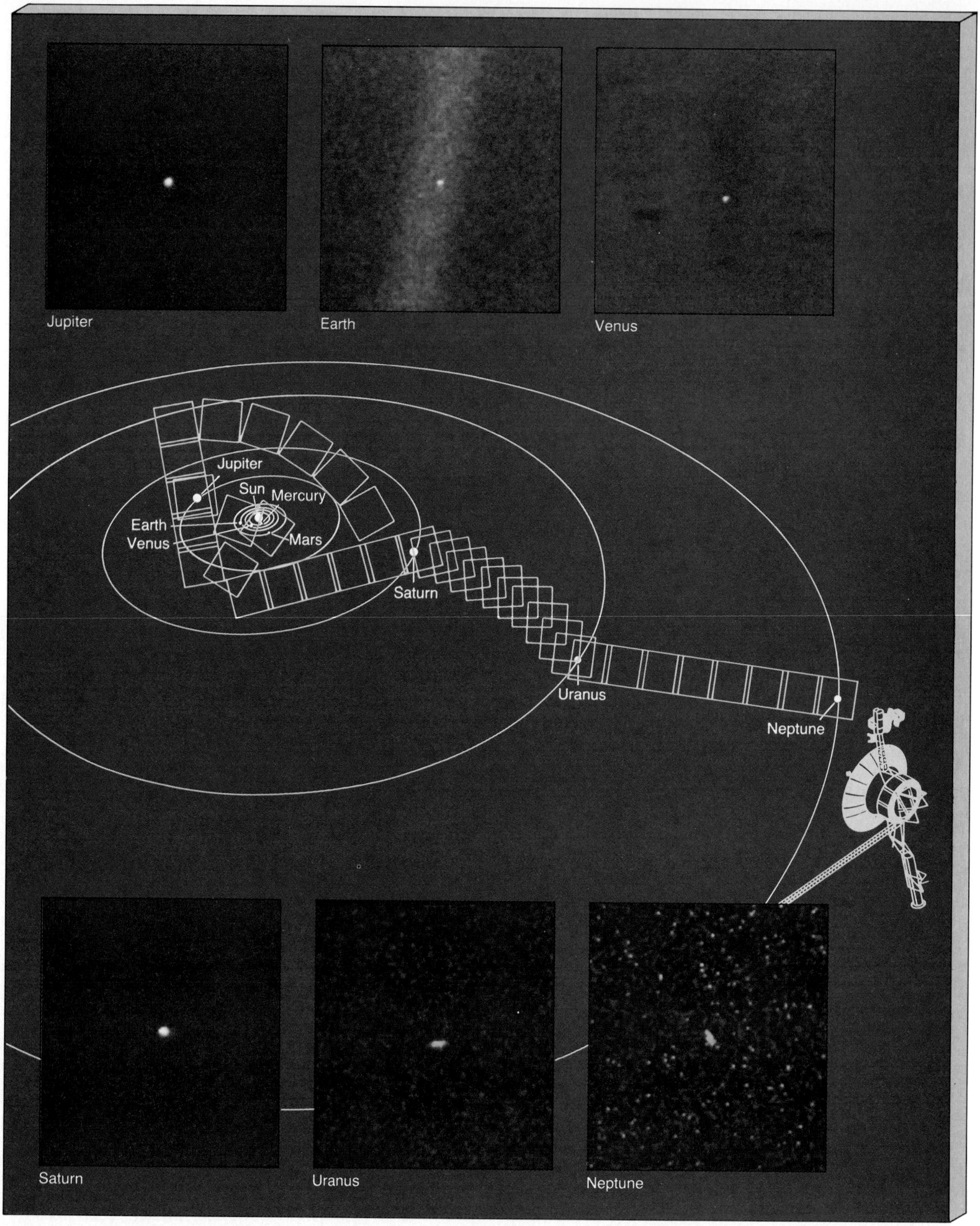

Figure 9.6 *The Sun and several planets can be seen in this series of photographs radioed back to Earth by* Voyager 2 *a year after its encounter with Neptune. Called the "family portrait," it is a diagram of numerous picture frames against a sketch of the solar system, and its inserts show some of the actual images radioed back to Earth. (NASA)*

Figure 9.7 *Taken during the July 1991 eclipse, this photograph shows Mercury, Venus, Mars, and Jupiter. We can see many solar system objects in this one photograph, in part because their orbits lie nearly in the same plane. (S. Westphal)*

TERRESTRIAL AND JOVIAN PLANETS

✓2 The solar system presents us with a sense of orderly motion. The planets move nearly in a plane, on almost concentric circular paths, in the same direction around the Sun, at steadily increasing orbital intervals. The individual properties of the planets themselves are not so regular, however. Still, definite trends can be discerned, and a clear division can be made between the inner and the outer members of our planetary system. In short, the inner planets are small and dense, while the outer worlds (with the exception of Pluto) are large and of low density. These differences are plainly evident in the data presented in Table 9-1.

The four innermost planets—Mercury, Venus, Earth, and Mars—are often called the **terrestrial planets** because their physical and chemical properties are somewhat similar to Earth's. (The word *terrestrial* derives from the Latin word *terra*, meaning "land," or "earth.") The larger, outer planets—Jupiter, Saturn, Uranus, and Neptune—are often labeled the **Jovian planets** because of their physical and chemical resemblance to Jupiter. (The word *Jovian* comes from *Jove*, another name for the Roman god Jupiter.) Pluto doesn't fit well into either category. It might once

INTERLUDE 9-1 *The Titius-Bode "Law"*

A close look at the list of semi-major axes of the known planets (see Table 9-1) reveals considerable regularity in their orbits. The *spacing* of the orbits increases more or less geometrically as we move out from the Sun: at any point in the list, the distance to the next planet out is about twice that to the next one in.

Is there some underlying structure to the solar system? In search of an answer to that question, the German astronomer Johann Titius in 1766 came up with a simple formula that "predicted" quite well the orbits of the then-known planets, Mercury through Saturn. Johann Bode, a better-known astronomer of the day, later popularized this relationship among the planets' orbits, so that it now is usually known as the Titius-Bode Law, or even just Bode's Law. We must emphasize that it is not a law at all, in the scientific sense, but rather just a rule for determining the approximate orbital semi-major axes of the planets. Nevertheless, the *law* part of the name has stuck.

The rule for determining the planet's orbits is as follows. Start with 0.4, the distance (in A.U.) from the Sun to Mercury. Then add to it 0.0 (Mercury: 0.4 A.U.), 0.3 (Venus: 0.7 A.U.), 0.6 (Earth: 1.0 A.U.), 1.2 (Mars: 1.6 A.U.), 2.4 (?: 2.8 A.U.), 4.8 (Jupiter: 5.2 A.U.), and 9.6 (Saturn: 10.0 A.U.) to arrive at the orbital distances of the known planets, plus one extra "planet" between Mars and Jupiter. The relation between successive orbits is easily seen—after Venus, the added term simply doubles at each step. Even the fictitious extra "planet" is acceptable—it corresponds to the middle of the asteroid belt! Thus it was that this rule appeared to contain some deep insight into our planetary system.

According to the rule, the next planets beyond Saturn should lie at 19.6 and 39.2 A.U. When Uranus was discovered in 1781, it fell almost exactly where the Titius-Bode Law predicted, at 19.2 A.U. Speculation was rife that the next planet would also lie where the "law" decreed it should. Sadly, the rule fails for Neptune, which lies only 30 A.U. from the Sun. However, if we ignore Neptune, Pluto is in just the right place. No one today takes the rule sufficiently seriously to bother to look for a scenario in which Neptune started out in the "right" orbit and somehow got moved farther in toward the Sun, perhaps leaving Pluto behind. Still, the regularity of the "law" was appealing to astronomers of the eighteenth and nineteenth centuries, as it suggested some fundamental harmony in the structure of the solar system, and, to some extent, that appeal remains today.

The Titius-Bode rule is not the result of any known planetary interaction, nor is it even very accurate, yet it remains a curious coincidence that many astronomers feel may be telling us *something* about the formation of the solar system—they just don't know quite *what*.

have been a moon of Neptune or perhaps even a large cometlike body and not originally a planet at all.

The four terrestrial planets exist close to the Sun, all within 1.5 A.U. of their parent star. All are small and of low mass—Earth is the largest and most massive of the four—and all have generally rocky composition and solid surfaces. Beyond that, however, the similarities end. The average density of the terrestrial worlds decreases steadily as we move farther from the Sun, indicating that their compositions differ, and the planets' present-day surface condi-

INTERLUDE 9-2 *Why Air Sticks Around*

Some planets and moons in the solar system—for example Venus and Titan (the largest moon of Saturn)—have atmospheres thicker than Earth's. Other objects, such as Mars, have thinner atmospheres. Still others—for example our own Moon and the planet Mercury—have virtually no atmosphere at all. The Jovian planets have atmospheres rich in hydrogen, while gaseous hydrogen is rare on the terrestrial worlds.

Why do planets have atmospheres? Why does a layer of air lie just above the surface of Earth? After all, experience shows that most gas naturally expands to fill all the volume available. Perfume in a room, tear gas in a riot, and steam from a teakettle all rapidly disperse until we can hardly sense them. Why, then, doesn't our atmosphere similarly disperse by just floating away into outer space? The answer is that gravity holds it down.

The gravitational field of the Earth exerts a pull on every atom and molecule in our atmosphere, preventing it from escaping. Gravity, however, can't be the only influence on the air. Other agents must compete with Earth's gravity to keep the atmosphere buoyant. Otherwise, gravity would eventually force all the air onto the ground.

Heat is the influence that competes with gravity. Solar heating of our planet gives the atmospheric molecules some random motion, much like the gas in a hot-air balloon or the molecules in a pot of boiling water. The constant movement of the heated molecules results in pressure, which exerts an upward force, preventing our atmosphere from collapsing under its own weight. If we are to understand planetary atmospheres, we must study in a little more detail this competition between gravity and heat.

An important measure of the strength of an object's gravity is its *escape velocity*, which is the speed needed for a smaller object to escape completely the larger body's gravitational pull. As we saw in Chapter 3, this speed increases as the mass of the parent body increases, or as its radius decreases. Mathematically, the escape velocity depends on the mass and radius of the parent (often a moon or a planet) in the following way:

$$\textit{escape velocity} \propto \sqrt{\frac{\textit{mass of parent object}}{\textit{radius of parent object}}}$$

Thus, if the *mass* of the parent were to quadruple, the escape velocity would double. If the parent's *radius* were to quadruple, then the escape velocity would be halved.

Earth's escape velocity is about 11 km/s. This is the minimum velocity needed for any object—molecules, baseballs, or rockets—to escape from Earth's surface. Civilian rockets designed to propel robot space probes toward the outer planets must attain larger velocities because they must overcome the gravitational pull of the Sun in addition to escaping from our planet. Military intercontinental ballistic missiles are designed to achieve less than escape velocity (actually, about 6 km/s), so that they fall back to Earth. A speeding bullet attains a maximum velocity of only about 1 km/s—some 10 times less than escape velocity—which explains why bullets don't leave the Earth when fired from guns.

Notice that the above relationship predicts that the escape velocity depends only on the physical properties of the *parent* object from which the escape is made. Generally speaking, parent objects of relatively small mass tend to have small escape velocities. For example, the Moon, which is much less massive than Earth, has an escape velocity of only about 2.4 km/s. Very massive objects usually have a large gravitational pull, making it harder to escape from their surfaces. The table below lists the escape velocities for some prominent members of the solar system.

To determine whether or not a planet will retain an atmosphere, we must compare the planet's escape velocity with the *molecular velocity*, which is the average speed of the gas particles making up the atmosphere. This molecular motion depends on the temperature of the gas and the mass

Object	Mass (Earth masses)	Radius (Earth radii)	Escape Velocity (km/s)
Mercury	0.055	0.38	4.3
Venus	0.82	0.95	10.4
Earth	1.00	1.00	11.2
Mars	0.11	0.53	5.0
Jupiter	318	11.2	60
Saturn	95.2	9.41	36
Uranus	14.6	4.01	21
Neptune	17.2	3.89	24
Pluto	0.0025	0.18	1.3
Moon	0.012	0.27	2.4
Sun	332,000	109	617
Ceres	0.00024	0.07	0.66

tions are all quite distinct. All the terrestrials but Mercury have atmospheres (consult Interlude 9-2 for a discussion of the factors determining whether or not a given planet can retain an atmosphere), but the atmospheres are about as dissimilar as we could imagine. Earth alone has oxygen in its atmosphere, and Earth alone has liquid water on its surface. Earth and Mars spin at roughly the same rate—one rotation every 24 (Earth) hours—but Mercury and Venus both take months to rotate just once. Earth and Mars have Moons, but Mercury and Venus do not. Finding the com-

of the individual molecules—the hotter the gas, the larger the average velocity of the molecules, and more massive molecules have smaller molecular velocities:

$$\textit{average molecular velocity} \propto \sqrt{\frac{\textit{temperature of gas}}{\textit{mass of molecule}}}$$

Thus, heating a sample of gas from 100 K to 400 K doubles the average speed of its constituent molecules. At any given temperature, molecules of oxygen in air move, on average, four times more slowly than molecules of hydrogen, which are 16 times lighter. Larger molecules move even more sluggishly.

Atmospheric molecules can gain or lose energy (and therefore velocity) by bumping into one another or by colliding with objects near the ground. Thus, although we can characterize a gas by an average molecular velocity, the molecules do *not* all move at the same speed. The spread in velocity is illustrated in the following figure, which plots numbers of molecules against the velocity of those molecules.

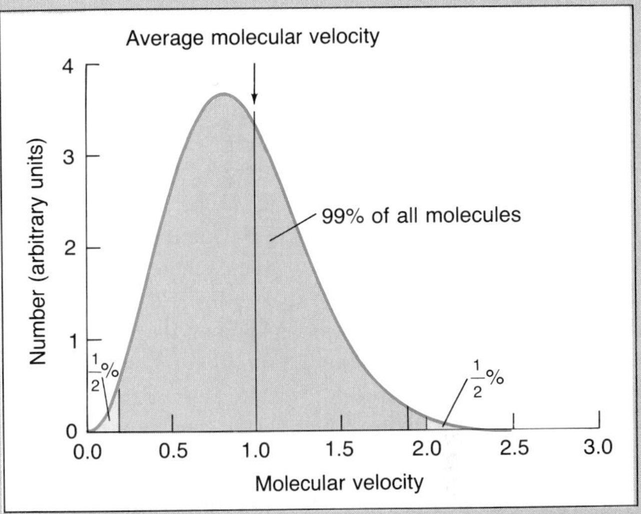

The velocity corresponding to the peak of the curve lies close to the average molecular velocity, which, as we have just seen, depends on both the temperature and the mass of the molecules. For oxygen or nitrogen in Earth's atmosphere, where the temperature near the surface is nearly 300K, the curve peaks at about 0.5 km/s. This typical molecular velocity is far smaller than the 11 km/s needed for a molecule to escape into space, so Earth is able to retain its nitrogen-oxygen atmosphere. On the whole, the gravity of our planet simply has more influence than the heat of our atmosphere.

In fact, only a tiny fraction of the molecules in any gas has velocity much greater than average. Only one molecule in 2 million has a velocity more than three times the average, and only one in 10^{16} exceeds the average by more than a factor of 5. As a rule of thumb, if the escape velocity from a planet exceeds the average velocity of a given type of molecule by a factor of 6 or more, then molecules of that type will not have escaped from the planet's atmosphere in significant quantities since the solar system formed. If, conversely, the escape velocity is less than six times the average molecular velocity, then most of the molecules will have escaped by now, and we should not expect to find them in the atmosphere.

Hydrogen molecules move (on average) four times faster than oxygen molecules—at about 2 km/s in Earth's atmosphere at sea level—so they have had time to escape since our planet formed (6×2 km/s = 12 km/s, which is greater than Earth's 11 km/s escape velocity). Consequently, we find very little hydrogen in Earth's atmosphere today. On Jupiter, where the temperature is lower (about 100K, instead of 300K), the velocity of the molecules is correspondingly lower—about 1.2 km/s. At the same time, the escape velocity is 60 km/s, over 5 times higher than on Earth. For those reasons, Jupiter has retained its hydrogen. In fact, hydrogen is the dominant constituent of Jupiter's atmosphere.

As we will see in Chapter 10, this line of reasoning almost certainly explains the lack of any significant atmosphere on the planet Mercury. That planet's escape velocity is 4.3 km/s, and its peak surface temperature is around 700 K, corresponding to an average molecular velocity for nitrogen or oxygen of about 0.8 km/s. Thus, there has been ample time for those gases to escape because the escape velocity is less than six times the average molecular velocity. Similarly, if our own Moon originally had an Earthlike atmosphere, it would have been heated by the Sun to much the same temperature as Earth's air today. The average molecular velocity would have been about 0.5 km/s. However, the Moon's escape velocity is only 2.4 km/s. Thus, any original lunar atmosphere has long ago dispersed into interplanetary space. In the coming chapters we will see many other examples of how the competition between gravity and heat determines a planet's atmospheric environment.

mon threads in the evolution of four such worlds is no simple task! Comparative planetology will be our indispensable guide as we proceed through the coming chapters.

For all the differences among the terrestrial worlds, they still seem very similar when compared with the Jovian planets. Perhaps the simplest way to express the major differences between the terrestrial and Jovian worlds is to say that the Jovian planets are everything the terrestrial planets are not. The terrestrial worlds lie close together, near the Sun; the Jovian worlds are widely spaced through the outer solar system. The terrestrial worlds, as we have just described, are small, dense, and rocky; the Jovian worlds are large and gaseous, made up predominantly of hydrogen and helium (the lightest elements, quite rare on the inner planets). The terrestrial worlds have solid surfaces; the Jovian worlds have none (their dense atmospheres thicken with depth, eventually merging with their liquid interiors). The terrestrial worlds have weak magnetospheres, if any; the Jovian worlds all have strong magnetic fields. The terrestrial worlds have only three moons among them; the Jovian worlds each have many moons and all have rings, a feature unknown on the inner planets. Despite their greater size, the Jovian worlds all rotate much faster than any terrestrial planet.

Figure 9.8 presents a diagram of the sizes of the planets relative to the Sun. The Sun is clearly the largest object in our solar system. It has over 1000 times the mass of the next largest object, the planet Jupiter. The Sun in fact contains about 99.9 percent of all solar system material. The terrestrial planets—including our home planet—are insignificant in comparison. The major bodies in our solar system are cataloged in Figure 9.9.

INTERPLANETARY DEBRIS

The final component of the solar system is the collection of interplanetary matter that orbits between the major planets. This debris ranges in size from the relatively large asteroids, through the smaller comets and even smaller meteoroids, down to the smallest grains of interplanetary dust that litter our cosmic environment. Larger bodies collide and break apart into smaller bodies. In turn, smaller bodies collide and are ground into dust, which eventually settles into the Sun or is swept away by the solar wind. The dust is generally quite difficult to detect in visible light, but studies in the infrared range reveal that interplanetary space contains sur-

Figure 9.8 Diagram, drawn to scale, of the relative sizes of the planets and our Sun. Notice how much larger the Jovian planets are than the Earth and the other terrestrials and how much larger still is the Sun.

Figure 9.9 *Spacecraft images of the planets and major moons of the solar system, shown to scale. (NASA)*

prisingly large amounts of it. Our solar system is an extremely good vacuum by terrestrial standards but positively dirty by the standards of interstellar or intergalactic space.

Asteroids and meteoroids are generally rocky, a little like the outer layers of the terrestrial planets. Their total mass is much less than Earth's Moon. They are important because they are made of material that is nearly unchanged since the early days of the solar system. They often conveniently deliver themselves right to our doorstep, in the form of meteorites, allowing us to study them in detail without the necessity of fetching them from space.

The comets are quite distinct from the other small bodies in the solar system. Comets are icy rather than rocky. In fact, comets are quite similar in composition to some of the icy moons of the outer planets, but they too represent truly ancient material dating back to the formation of the planets. Most comets are probably unchanged and probably haven't interacted with anything at all since they formed long ago along with the rest of the solar system. Comets striking Earth's atmosphere do not reach the surface intact, so we do not have actual samples of cometary material. However, they do vaporize and emit radiation as their highly elongated orbits take them near the Sun. We can determine their makeup by spectroscopic means and so gather information on what the early solar system was like 4.5 billion years ago.

COMPARATIVE PLANETOLOGY

✓3 As we catalog the similarities and differences among the planets, terrestrial and Jovian, what can we learn? And what does the makeup of the debris that orbits among the planets have to tell us? The answers may well hold the key to unraveling the origin of our solar system—and so, perhaps, the origin of other planetary systems beyond our own. The overall scale, shape, and composition of the solar system allow us to paint a fairly clear picture of the processes that led to the formation of the Sun and the planets. As we acquire more detailed information with each new space mission, our understanding of the birth of stars and planets grows.

To understand the solar system, we must try to answer basic questions such as, "Why are the terrestrial planets so different from the Jovian worlds?" and "Why did planet X evolve in one way, while planet Y turned out completely different?" and "Why are the planets' orbits so orderly, when their individual properties are not?" Our main problem in this task is that we don't have many objects to work with, even if we include the properties of all known planetary moons and of the larger asteroids as well. We thus must study these bodies in great detail to determine their common features and their differences and to find the reasons for both. Our goal is to develop a comprehensive theory of the *origin* of the solar system that explains all, or at least most, of its observed properties. We will return to this endeavor after a more thorough study of the planets and the other constituents of our solar system.

CHAPTER SUMMARY

Each object in the solar system is unique. Each holds clues about the solar system's origin and evolution. Comparative planetology is the process of comparing and contrasting the properties of the planets. It helps scientists understand better the conditions under which planets form and develop. The nine known planets orbit the Sun in nearly the same plane as the Earth. Their paths are all ellipses, with the Sun at one focus. The solar system is immense, yet the diameter of Pluto's orbit is less than $\frac{1}{1000}$ of a light-year, and the next nearest star is several light-years away. The spacing of the planets' orbits increases more-or-less geometrically as we move out from the Sun. The Titius-Bode Law describes the planets' orbits remarkably well.

The inner terrestrial worlds are small and rocky, while the outer Jovian worlds (except Pluto) are large and gaseous. The composition of a planet's atmosphere is governed by a competition between gravity and heat. Gravity keeps a planet's atmosphere from floating off into space, while the Sun's heat keeps air molecules in motion, preventing the atmosphere from collapsing under its own weight.

The solar system contains interplanetary matter apparently left over from its formation. This "debris" ranges in size from asteroids to comets and meteoroids, down to grains of cosmic dust.

KEY WORDS

asteroid belt	comparative planetology	Jovian planet	planet
asteroid	interplanetary space	moon	terrestrial planet

REVIEW QUESTIONS

1. What is comparative planetology? Why is it useful?
2. Why has our knowledge of the solar system greatly increased in recent years?
3. Give a brief description of the contents of our solar system. Include a mention of space!
4. Tell which planets are called the terrestrial planets, and explain how they came to have that name. Do the same for the planets that are labeled Jovian. Is Pluto a Jovian or a terrestrial planet? Why?
5. What is the Titius-Bode "Law"? Why isn't it considered to be a true law, in the scientific sense?
6. Name some differences between the terrestrial planets and the Jovian planets.
7. Discuss two mechanisms that operate to keep a planet's

atmosphere in place. What would happen to the atmosphere if a planet's surface temperature increased or if its mass decreased?

8. Why are asteroids and meteoroids important to scientists?
9. Because comets generally vaporize upon striking Earth's atmosphere, how do we know they are icy bodies?
10. What is the ultimate goal of comparative planetology?
11. According to the Titius-Bode "Law," where should the ninth, tenth, and eleventh planets in the solar system lie?
12. Assume that a planet will lose its atmosphere if the molecular velocity exceeds one-sixth the escape velocity. What would Jupiter's temperature have to be for its hydrogen to escape?
13. Suppose the average mass of each of the 4000 asteroids in the solar system is about 10^{20} g. Compare the total mass of all asteroids to the mass of a typical terrestrial planet.

DISCUSSION QUESTIONS

1. Go to the library and look for books about the planets from around the turn of the century. What did people then perceive to be the main differences among the planets? What did their ideas about the planets lead them to believe about the origin of the solar system?
2. Compare and contrast Kepler's Laws with the Titius-Bode "Law". Why are Kepler's Laws considered to be true natural laws, while the Titius-Bode "Law" is not?
3. With what you now know about planetary atmospheres, try to predict what you'll learn in the coming chapters about the abundance of atmosphere on Mercury, Venus, Mars, and the outer planets.
4. Why do you suppose astronomers care about the origin of the solar system? Do you think it's important to know how the solar system formed?

PROJECTS

1. You can begin to visualize the ecliptic—the plane of the planets' orbits—just by noticing the path of the Sun throughout the day and of the full Moon in the course of a single night. It helps if you watch from one spot, such as your backyard or a rooftop. It's also good to have a general notion of direction (west is where the Sun sets). You will see that the movements of the Sun and Moon are confined to a narrow pathway across our sky. The planets also travel along this path. The motion of the Sun, Moon, and planets is a two-dimensional reflection of the three-dimensional plane of our solar system.
2. Once you get a feeling for the whereabouts of the ecliptic, try locating the North Star. Knowing the direction to celestial north makes it easier to imagine the motion of the planets in the plane of the solar system. Don't worry about being too precise. Just get a sense of the ecliptic as a kind of merry-go-round of planets—that we on Earth also ride!

SUGGESTED READINGS

Beatty, J. K., and A. Chaikin, eds. *The New Solar System.* 3d ed. Cambridge, MA: Cambridge University Press and Sky Publishing, 1990. An authoritative synthesis of planetary exploration, by 26 scientists in the field.

Hartley, K. "Solar System Chaos." *Astronomy* (May 1990). The dynamics of the solar system in light of modern chaos theory.

Littmann, M. *Planets Beyond: Discovering the Outer Solar System.* New York: John Wiley & Sons, 1988. Tales of Uranus, Neptune, and Pluto. Includes descriptions of what modern spacecraft have found.

Lovi, G. "The Solar System Seen from Outside." *Sky and Telescope* (August 1986). Excellent for showing how our viewpoint on a moving planet determines the way we see other members of our solar system.

Miller, R., and W. K. Hartmann, *The Grand Tour: A Traveler's Guide to the Solar System.* New York: Workman Publishing, 1981. Imaginative guide to the solar system, with many photos and paintings by the authors.

Murray, B., M. Malin, and R. Greeley, *Earthlike Planets.* San Francisco: W. H. Freeman, 1981. Mercury, Venus, Earth, Mars—how they formed and their characteristics today.

Sheehan, W. *Planets and Perception: Telescopic Views and Interpretations 1609–1909.* Tucson: University of Arizona Press, 1988.

"Were Titius and Bode Right?" *Sky and Telescope* (News Notes, April 1987). A mechanism by modern theorists that explains how the orbits of the planets came to be where they are.

CHAPTER

10

Mercury
A Scorched and Battered World

The spacecraft Mariner 10 *sent back this image of Mercury, the closest planet to the Sun. Mercury has a desolate, cratered surface devoid of permanent atmosphere and constantly bombarded by intense solar radiation. Despite the intensity of sunlight, Mercury's polar regions, where the Sun never rises far above the horizon, are quite cool, and they may even have ice on them. (Courtesy NASA)*

R I **V** U X G

LEARNING GOALS

1 to understand how Mercury's rotation is influenced by its orbit around the Sun
2 to compare Mercury with the Moon and the Earth in terms of surface and interior structure
3 to understand some of the reasons why Mercury and the Moon evolved differently

The planet closest to the Sun is Mercury, the smallest terrestrial world. At first glance, you might mistake it for Earth's Moon—Mercury is only a little larger in diameter than our Moon, and both Mercury and the Moon exhibit heavily cratered surfaces. In this chapter, as we explore the similarities and differences between these two worlds, and seek to understand the reasons for them, we begin our comparative study of the planets and the moons that make up our solar system.

Mercury has the greatest temperature range of any body in the solar system. Because it lacks a significant atmosphere to moderate day-to-night variations in solar heating, its nearness to the Sun drives temperatures to a sizzling 675K on its solar side. The temperature on the opposite side falls to a frigid 100K. Alternately scorched and frozen as it turns, the planet is airless and waterless. Mercury is clearly unfit for humans. Most likely, it could not harbor anything resembling life as we know it. Nevertheless, as a world in some ways intermediate between Earth and the Moon in its properties, it is perhaps a fitting place to begin our tour of the planets.

MERCURY'S APPEARANCE FROM EARTH

Seen from Earth, Mercury never strays too far from the Sun. As shown in Figure 10.1, its orbit always keeps it within about half an astronomical unit of the Sun. When viewed from the Earth's distance of 1 A.U., the angular distance between the Sun and Mercury—known as Mercury's **elongation**—is never greater than 28° (degrees). Mercury is visible to the naked eye only when the Sun's light is blotted out—just before dawn, just after sunset, or, much less frequently, during a total solar eclipse. Figure 10.2 is a photograph of Mercury taken just after sunset.

Because of its proximity to the Sun, we see Mercury only low in the sky (except during an eclipse). It can lie at

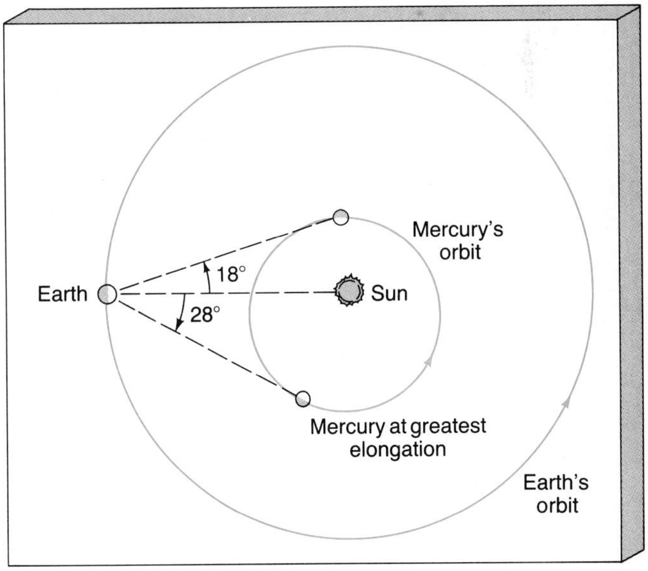

Figure 10.1 *Mercury is never found far from the Sun, and its elongation is never greater than about 28°. This maximum (western) elongation occurs when Mercury is at aphelion, 70 million km (0.47 A.U.) from the Sun. When Mercury is at perihelion, only 46 million km (0.31 A.U.) from the Sun, its greatest (eastern) elongation is at most 18°, as indicated.*

Figure 10.2 *Mercury just after sunset, to right of the crescent moon. (C. Trost)*

most 28° above the horizon. Because the Earth rotates at a rate of 15° per hour, Mercury is visible for at most 2 hours on any given night, even under the most favorable circumstances. The dashed lines on Figure 10.2 represent the location of the horizon 1 hour and 2 hours after sunset. In fact, for most observers at most times of the year, Mercury is considerably less than 28° above the horizon, so it is generally visible for a much shorter period (see Figure 10.3).

Mercury was well known to ancient astronomers. However, because the planet was visible only just before dawn and just after sunset, it was not possible to follow Mercury through a full cycle of phases. The ancients originally believed that this companion to the Sun was two different objects. The connection between the planet's morning and evening appearances took some time to figure out. The early Greek astronomers, for example, had two separate names for Mercury—Hermes in the evening and Apollo in the morning. (Mercury is the Roman name for the Greek god Hermes.) However, later Greek astronomers were certainly aware that the "two planets" were really different alignments of a single body.

Like all the planets, Mercury shines by the reflected light of the Sun. The surface of the planet is fairly unreflective. Its **albedo**—the fraction of incident sunlight it reflects into space—is only about 0.1, quite similar to the albedo of the Moon. By contrast, the Earth's albedo is much greater (about 0.4), in large part because of the Earth's highly reflective partial cloud cover. Venus, which is completely shrouded in clouds, has an albedo of almost 0.7. In spite of its low reflectivity, Mercury's nearness to the Sun makes it one of the brightest objects in the nighttime sky. The difficulty in observing Mercury is simply the result of its closeness to the horizon.

Large telescopes can actually observe Mercury during the daytime, when the planet is higher in the sky and the atmosphere's effects are reduced (the amount of air that the light from the planet has to traverse before reaching our telescope decreases as the height above the horizon increases). By filtering out the unwanted sunlight, the planet

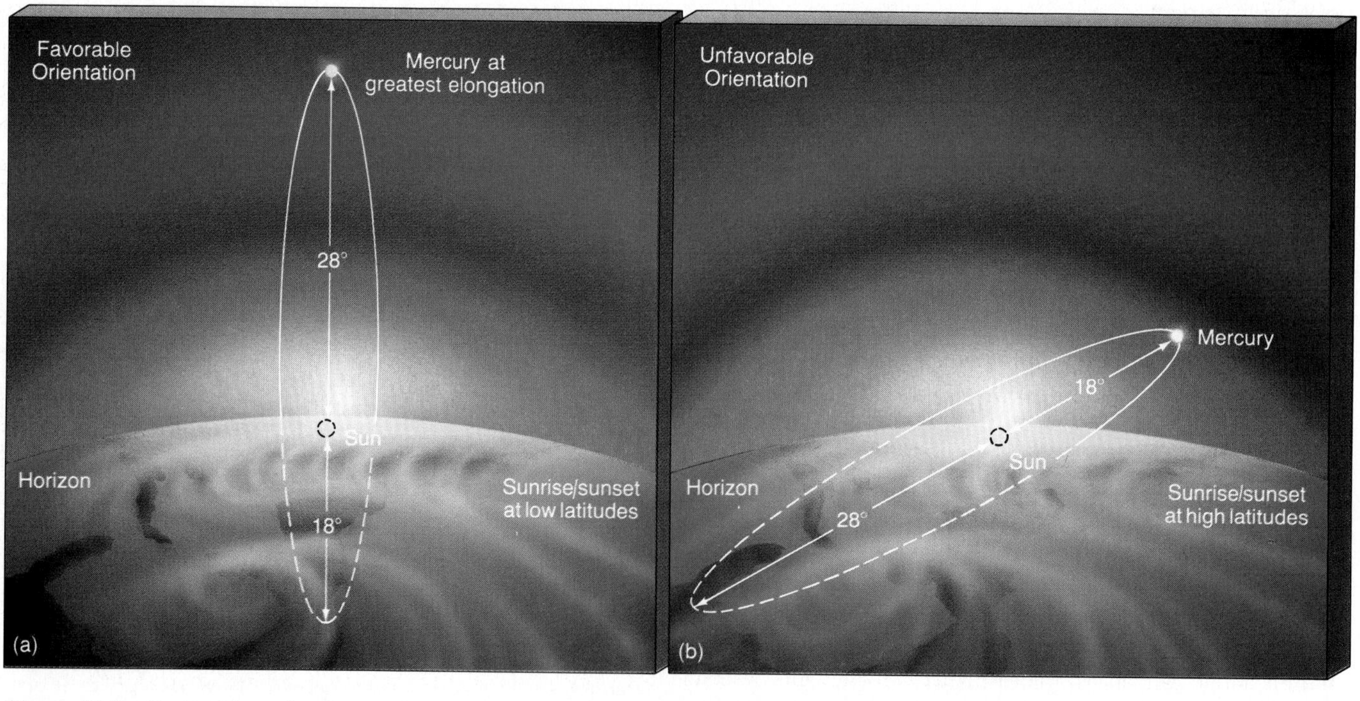

Figure 10.3 *Favorable and unfavorable orientations of Mercury's orbit result from different Earth orientations and observer locations.*

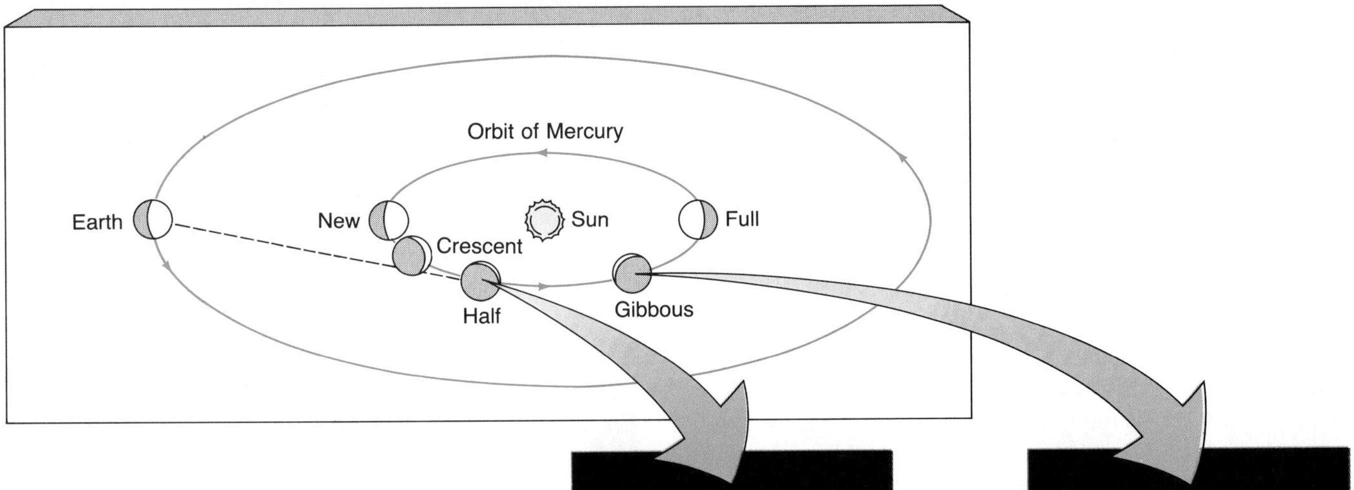

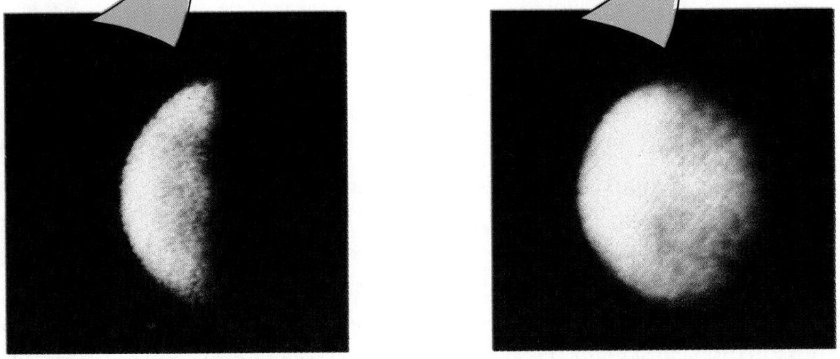

Figure 10.4 *Some views of Mercury at different points along its orbit. The best views of the planet are taken at maximum elongation from the Sun and show a "half Mercury." (New Mexico State University)*

can be imaged. The best views of Mercury are generally obtained in this way. The naked-eye or amateur astronomer, however, is generally limited to nighttime observations. In all cases, it becomes progressively more difficult to view Mercury the closer (in the sky) its orbit takes it to the Sun. The best views of the planet therefore show a "half Mercury," close to maximum elongation, in Figure 10.4.

MERCURY'S ORBIT

Mercury's orbital semi-major axis is only 0.39 A.U. (58 million km), and its orbital eccentricity is 0.21. Its distance from the Sun ranges between 0.31 A.U. (46 million km) and 0.47 A.U. (70 million km). In accordance with Kepler's Third Law, Mercury's orbital period is 88 days. (Whenever we use *day*, we mean an Earth day unless otherwise noted.) Mercury's orbit is much more eccentric than that of any other planet, with the exception of Pluto.

Mercury orbits the Sun in the same sense as the Earth—counterclockwise, as seen from above the ecliptic looking down on the Earth's North Pole. Its orbital plane is inclined at approximately 7° to the plane of the ecliptic. Because its orbit lies within our own and in roughly the same plane, it is possible from our perspective on Earth to see Mercury pass across the face of the Sun. Such an event, where a smaller, darker object passes in front of a larger, brighter one, is known as a **transit.** Because of the 7° angle between the orbits of the Earth and Mercury, the precise alignment of the Sun, Mercury, and Earth required for a transit to occur is a fairly rare event. Solar transits of Mercury happen only a dozen or so times per century. Mercury's orbital plane is oriented such that solar transits of Mercury, when they do occur, are always seen in November or May. A solar transit of Mercury is shown in Figure 10.5.

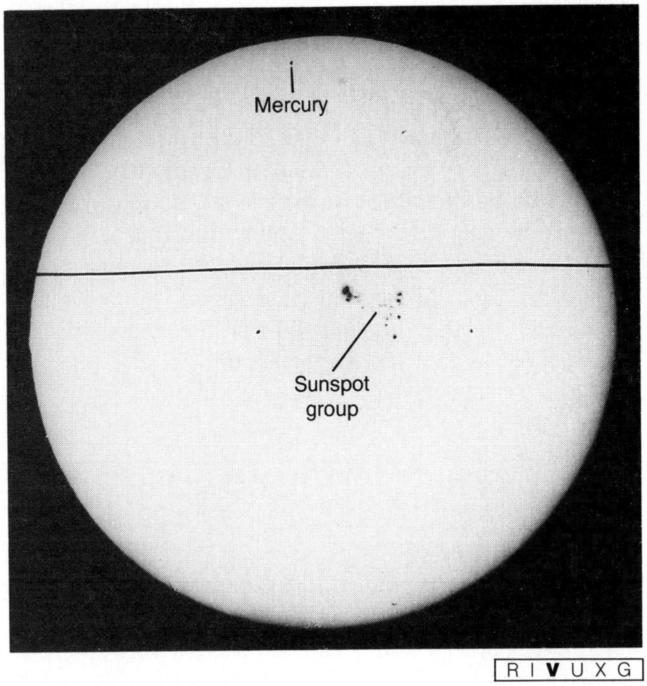

Figure 10.5 *A solar transit of Mercury. Such transits are quite rare because of the orientation of Mercury's orbit relative to the ecliptic, happening only about once per decade. (The horizontal line is an artifact.) (Yerkes Observatory)*

MERCURY IN BULK

Radius

Once we measure the astronomical unit (for example, by radar ranging on Venus), we can determine the distance to Mercury, as discussed in Chapter 3. Once we know Mercury's distance, we can determine the planet's radius from simple geometry, just as we can for the Moon and the Sun. At its closest approach to the Earth, a distance of about 0.52 A.U., Mercury's angular diameter is measured to be 26″ (arc seconds), so its radius is about 2450 km. More accurate measurements by unmanned space probes yield a result of 2439 km, or 0.38 Earth radii.

Mass

Measuring Mercury's mass is a little more difficult. Mercury has no moon, so, before the Space Age, astronomers determined its mass by studying its effects on the motion of the other inner planets. Because the orbits of the planets are so well known, even Mercury's small gravitational effect is easy to measure, and the mass of Mercury was quite well known even before spacecraft were sent there. Now that spacecraft have interacted with its gravitational field, the mass is much better known. It is measured at 3.3×10^{26} g, or 0.055 times the mass of the Earth.

Density

Knowing the mass and the radius, we can calculate the average density of Mercury. The result is 5.4 g/cm^3. Thus, although Mercury's mass (and, as we will see, its surface appearance) is like that of the Moon, Mercury's density is only slightly less than that of the Earth. Assuming that the surface rock on Mercury is of similar density to the surface rock on the Earth and the Moon, we are led to the conclusion that the interior of Mercury must contain a lot of high-density material, most probably iron. To account for the high density, we must assume that Mercury's iron core represents a much larger fraction of the planet's mass (about 60 percent) than is the case for the Earth.

MERCURY'S CURIOUS ROTATION RATE

✓1 Mercury is the second smallest planet. It is hardly larger than Earth's Moon and is actually smaller than three other moons in the solar system. As we have seen, Mercury is difficult to observe from Earth because of its closeness to the Sun. Even with a fairly large telescope, we can see it only as a slightly pinkish disk. Figure 10.6 is one of the few photographs of Mercury taken from Earth that shows some evidence of surface markings. Astronomers could only speculate about the faint, dark markings in the days before the robot spacecraft *Mariner 10* provided clearer images. We now know that these markings are much like the markings we see when gazing casually at Earth's Moon. The largest telescopes can resolve features on the surface of Mercury no better than we can perceive features on our Moon with our unaided eyes.

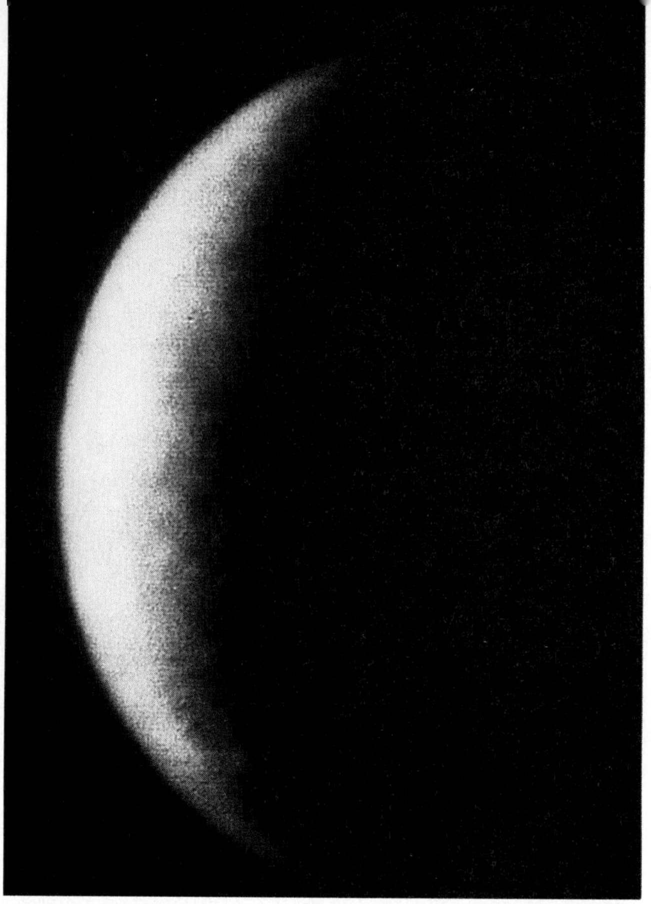

Figure 10.6 *Photograph of Mercury taken from Earth with one of the largest ground-based optical telescopes. Only a few surface features are discernable. (Palomar Observatory)*

Measurement of Mercury's Spin

In principle, the ability to discern surface features should allow us to measure Mercury's rotation rate simply by watching a particular region move around the planet. In the mid-nineteenth century, an Italian astronomer named Giovanni Schiaparelli did just that and concluded that Mercury always keeps one side facing the Sun, much as our Moon perpetually presents only one face to the Earth. The explanation suggested for this synchronous rotation was the same as for the Moon—the tidal bulge raised in Mercury by the Sun had modified the planet's rotation rate until the bulge always pointed directly at the Sun. While the surface features could not be seen clearly, the combination of Schiaparelli's observations and a plausible physical explanation was enough to convince most astronomers, and the belief that Mercury rotates synchronously with its revolution about the Sun persisted for almost half a century.

In 1965 two astronomers, Rolf Dyce and Gordon Pettengill, were making radar observations of Mercury from the Arecibo radio telescope in Puerto Rico (see Figure 6.26) when they discovered that this long-held view was in error. The technique they used is illustrated in Figure 10.7, which shows a radar signal reflecting from the surface of a hypothetical planet. Let us imagine, for the purpose of this discussion, that the pulse of outgoing radiation is of a single frequency.

The returning pulse bounced off the planet is very much weaker than the outgoing signal. Beyond this change, the reflected signal can be modified in two important ways. First, the signal as a whole might be redshifted or blueshifted, depending on the overall radial velocity of the planet with respect to the Earth. Let's assume for simplicity that this velocity is zero, so, on average, the frequency of the reflected signal is the same as the outgoing beam. Second, if the planet is *rotating*, the radiation reflected from the side of the planet moving toward us returns at a slightly higher frequency than the radiation reflected from the receding side, simply as a consequence of the Doppler effect. (Think of the two hemispheres as being separate sources of radiation and moving at slightly different velocities, one toward us and one away.) Only the radiation reflected from the portion of the surface moving transverse to our line of sight (that is, neither toward nor away from us) is unchanged in frequency. The effect is very similar to the rotational line broadening we discussed in Chapter 5, except that, in this case, the radiation we are measuring was not emitted by the planet, only reflected from its surface. What we see in the reflected signal is a spread of frequencies on either side of the original frequency. By measuring the extent of that spread, we can determine the rotational speed of the planet.

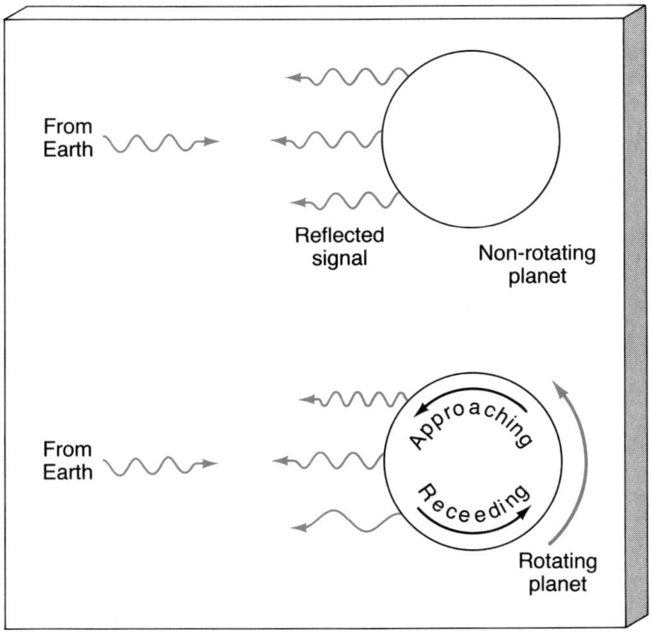

Figure 10.7 *A radar beam reflected from a rotating planet yields information about both the planet's overall motion and its rotation rate.*

Dyce and Pettengill found that the rotation period of Mercury is not 88 days, as astronomers had previously believed, but actually 59 days. The Italian astronomer Giuseppe Colombo realized that the 59-day rotation period is very close to two-thirds of the 88-day orbital period. More detailed measurements confirmed that the rotation period is in fact *exactly* two-thirds of a Mercury year. Because there are exactly three rotations for every two revolutions, we say that there is a 3:2 **spin-orbit resonance** in Mercury's motion. In this context, the term *resonance* just means that two characteristic times—here Mercury's day and year—are related to one another in a simple way. An even simpler example of a spin-orbit resonance is the Moon's orbit around the Earth. In that case, the rotation is synchronous with the revolution, so the resonance is said to be 1:1. In our study of the solar system, we will encounter many examples of resonant motion.

A Mercury Day

Figure 10.8 illustrates the implications of Mercury's resonant motion for an inhabitant of Mercury. The "person" represents his or her location on the surface. Let's imagine starting at Day 0 in the orbit, where, for our observer, the Sun is directly overhead. In other words, it is noon at that location on the planet. Roughly 15 (Earth) days later, the planet has rotated through 90° (one-quarter of a complete rotation) on its axis and moved about one-sixth of the way around the Sun on its orbit, to Day 15. (For the sake of simplifying this discussion, let's temporarily consider Mercury's orbit to be perfectly circular.) For our observer, it is now early afternoon, with the Sun about 30° away from the zenith (the position in the sky directly overhead). After another 15 days, at Day 30, it is still daylight, but now the Sun is only 30° above the horizon. Finally, 44 days after noon and half-way around the orbit, the Sun sets. Continuing to track the orbit of the planet, we see that after a complete revolution, at Day 88, Mercury has rotated one and a half times on its axis, and it is now midnight. After one more orbit around the Sun, it will finally be noon again for our observer, at Day 176.

This exercise clearly illustrates the difference between sidereal and solar days. In the case of the Earth, which rotates very rapidly compared with its orbital period, the difference is small—only a few minutes. However, for Mercury, whose sidereal day (59 Earth days) is comparable to its year (88 days), the Mercury solar day—the time from noon to noon, say—is actually two Mercury years long! The Sun stays "up" in the black Mercury sky for almost 3 Earth months at a time, after which follows nearly 3 months of darkness. At any given point in its orbit (at perihelion, say), Mercury presents the same face to the Sun not *every* time around, but *every other* time.

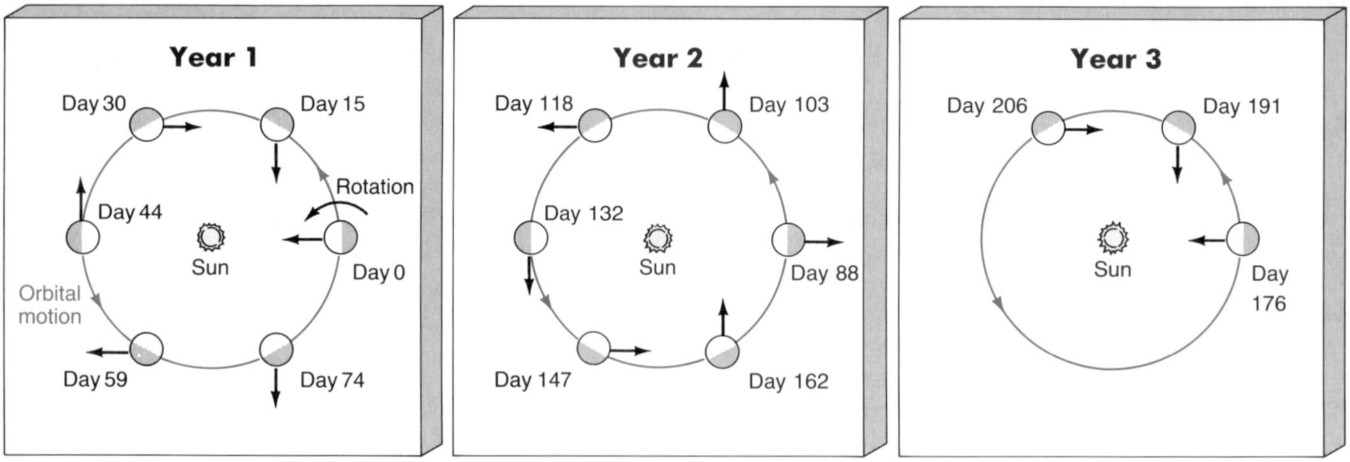

Figure 10.8 *Mercury's orbital and rotational motions combine to produce a day that is 2 years long. The arrow represents an observer standing on the surface of the planet. At Day 0, it is noon for our observer, and the Sun is directly overhead. By the time Mercury has completed one full orbit around the Sun and moved from Day 0 to Day 88, it has rotated on its axis exactly 1.5 times, so that it is now midnight at the observer's location. After another complete orbit, it is noon once again.*

Explanation of Mercury's Spin-Orbit Resonance

Mercury's 3:2 spin-orbit resonance did not occur by chance. What mechanism establishes and maintains it? In the case of the Moon orbiting the Earth, we explained in Chapter 8 the 1:1 resonance as the result of tidal forces acting on the Moon. In essence, the lunar rotation period, which probably started off much shorter than its present value, has lengthened so that the tidal bulge created by the Earth is fixed relative to the body of the Moon. Indeed, similar reasoning (with the Sun replacing the Earth) was partly responsible for the acceptance of Schiaparelli's claim that Mercury's rotation was synchronous with its revolution. We now know that tidal forces *are* also responsible for Mercury's 3:2 resonance, but in a more subtle way.

Mercury did not settle into a 1:1 resonance because its orbit around the Sun is quite eccentric ($e = 0.21$). By Kepler's Second Law, Mercury's orbital speed is greatest at perihelion and least at aphelion. A moment's thought shows that there is no way for the planet (rotating at a constant rate) to remain in a synchronous orbit—if its rotation were synchronous near perihelion, it would be too rapid at aphelion, while synchronism at aphelion would be too slow at perihelion. We saw this phenomenon earlier, in Chapter 8, in our discussion of the libration of the Moon (see especially Figure 8.2).

Tidal forces always act to try to synchronize the rotation rate with the instantaneous orbital speed. But we have just seen that such synchronization cannot be maintained over Mercury's entire orbit. However, recall from the discussion in Chapter 7 that the tidal effect decreases very rapidly with increasing distance. As a result, the tidal forces acting on Mercury at perihelion are much greater than those at aphelion. The effect of this is that perihelion wins the struggle to determine the rotation rate. As long as the tidal bulge induced in Mercury by the Sun is in the same orientation relative to the planet at each perihelion passage, the tidal forces acting elsewhere in the orbit simply don't matter. (In Figure 10.8, note that even though the planet has rotated through 180° between Day 0 and 88, the tidal bulge would look just the same in either case.)

Thus, Mercury presents the same face to the Sun at every perihelion or at every other perihelion. But this could be accomplished in many different ways. Why three rotations every two orbits? Why not synchronous rotation, like the Moon orbiting the Earth, or two rotations per orbit, or five rotations every two orbits? The answer is that in the 3:2 resonance, Mercury's orbital and rotational motion are almost exactly synchronous *at perihelion*, so that particular rotation rate is naturally "picked out" by the tidal forces on the planet.

These arguments may seem complicated, but the motion of Mercury is actually one of the simplest nonsynchronous resonances found in the solar system. Astronomers now believe that these intricate dynamical interactions are responsible for much of the fine detail observed in the motion of the solar system. Examples of resonances can be found in the orbits of many of the planets, their moons, and their rings and in the asteroid belt.

There is one further strange aspect of Mercury's rotation rate. As the planet revolves around the Sun and rotates on its axis, most of the time the rotation is faster than synchronous, and the Sun appears to move across the sky in one direction—east to west. Near perihelion, however, the rotation is almost, but slightly slower than, synchronous. For a few Earth days, at Mercury's perihelion, the Sun appears to stop in the sky and move backwards a little. Imagine the likelihood of the Copernican picture gaining acceptance if the Sun exhibited retrograde motion to us on Earth!

The influence of the Sun also determines the tilt of Mercury's spin axis. Because of the Sun's tides, Mercury's rotation axis is exactly perpendicular to its orbit plane. Thus, the Sun is always directly overhead at noon on the equator and is always on the horizon as seen from the poles.

THE SURFACE OF MERCURY

The Mariner 10 *Flybys*

✓2 In 1974, the U.S. spacecraft *Mariner 10* approached within 10,000 km of the surface of Mercury, sending back high-resolution images of the planet. These photographs, which showed surface features as small as 150 m across, revolutionized our knowledge of the planet. For the first time, we saw Mercury as a heavily cratered world, in many ways reminiscent of our own Moon.

Mariner 10 was launched from Earth in November 1973. Its planned trajectory shown in Figure 10.9, included a gravitational assist from the planet Venus (see Interlude 3-2) in February 1974, which placed the spacecraft in an elliptical 176-day orbit about the Sun. Its perihelion is close to Mercury's path, and its aphelion lies between the orbits of Venus and Earth. The 176-day period is exactly two Mercury years, so that the spacecraft revisits Mercury roughly every 6 months. However, only on the first three encounters—in March 1974, September 1974, and March 1975—did the spacecraft return data. After that, the craft's supply of maneuvering fuel was exhausted. In total, over 4000 photographs, covering about 45 percent of the planet's surface, were radioed back to Earth during the mission's active lifetime. The other 55 percent of Mercury is still unexplored. Because of Mercury's resonant rotation rate, the spacecraft saw the same face of the planet on each pass.

Surface Features

Figure 10.10 shows a picture of Mercury radioed back to Earth from a distance of 200,000 km, while the *Mariner 10* spacecraft was still approaching the planet. Figure 10.11 shows a view from a similar distance, taken as the craft receded from Mercury, after passing it on the nighttime side. Together, these images cover the known surface of Mercury. No similar photographs exist of the hemisphere that happened to be in shadow during the encounters. Figure 10.12 shows a higher-resolution photograph of the planet from a distance of 20,000 km. The similarities to the Moon are striking. We see no sign of clouds, rivers, dust storms, or other aspects of weather. Much of the cratered surface bears a strong resemblance to the Moon's highlands. The

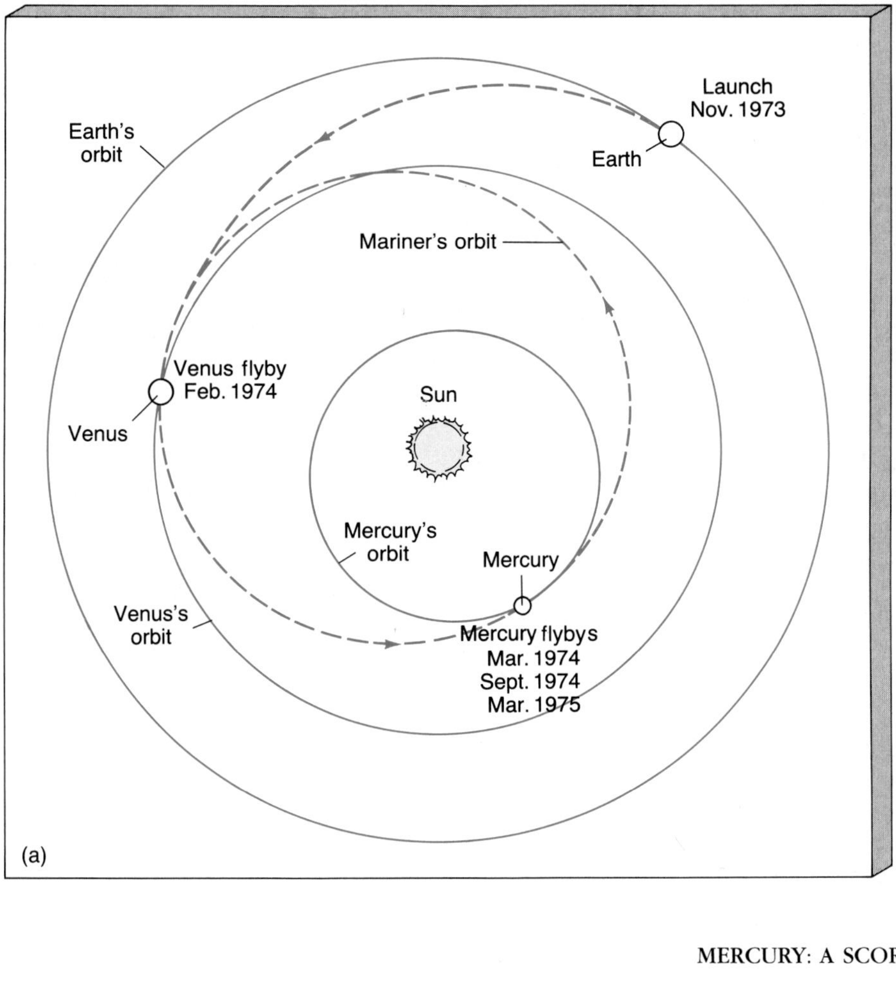

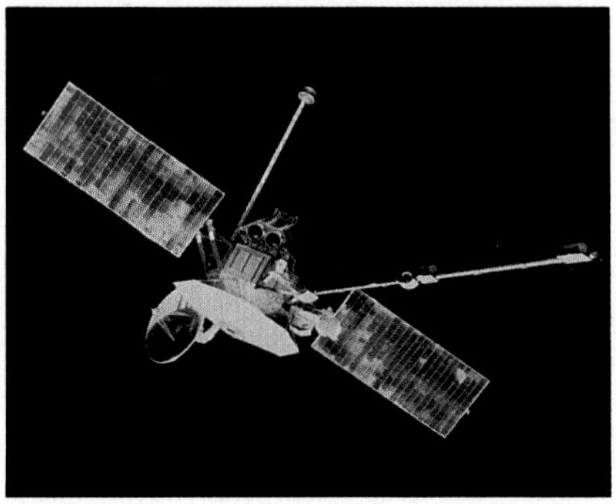

(b)

Figure 10.9 *(a) The* Mariner 10 *mission. The spacecraft's orbit included a gravitational boost from Venus. (b)* Mariner 10 *returned data until March 1975. (NASA)*

Figure 10.10 *Mercury is imaged here as a mosaic of photographs taken by the* Mariner 10 *spacecraft in the mid-1970s during its approach to the planet. At the time, the spacecraft was some 200,000 km away. (NASA)*

crater walls are generally not as high as on the Moon, and the ejected material landed closer to the impact site, exactly as we would expect on the basis of the greater surface gravity on Mercury. (Mercury's gravity is about two-fifths of the Earth's gravity. The Moon's gravity is only one-sixth that of the Earth.) Mercury, however, shows few extensive lava-flow regions akin to the lunar maria.

Like the Moon's, Mercury's craters are the result of meteoritic bombardment. But the craters are not so densely packed as their lunar counterparts, and there are extensive **intercrater plains.** One likely explanation for Mercury's relative lack of craters is that the older craters have been filled in by volcanic activity, in much the same way as the Moon's maria filled in craters as they formed. However, the plains do not look much like mare material—they are much lighter in color and not as flat. Still, most geologists believe that volcanism did occur in Mercury's past, obscuring the old craters, but the details of how Mercury's landscape came to look the way it does remain unexplained. The apparent absence of rilles or other obvious features associated with very-large-scale lava flows along with the light color of the lava-flooded regions suggest that Mercury's volcanic past was different from the Moon's. Figure 10.13 shows comparable surface areas of Mercury and the Moon.

The overall color and reflectivity of most of Mercury's surface are consistent with a composition like that of the lunar highlands. Scientists believe that the surface regolith is similar to that of the Moon, both in composition and in origin. Without actual samples of Mercury surface rock to study, though, we cannot know for certain to what extent the chemical composition of the surface mimics that of the Moon, nor can we determine with any certainty the age of the planet's surface layers.

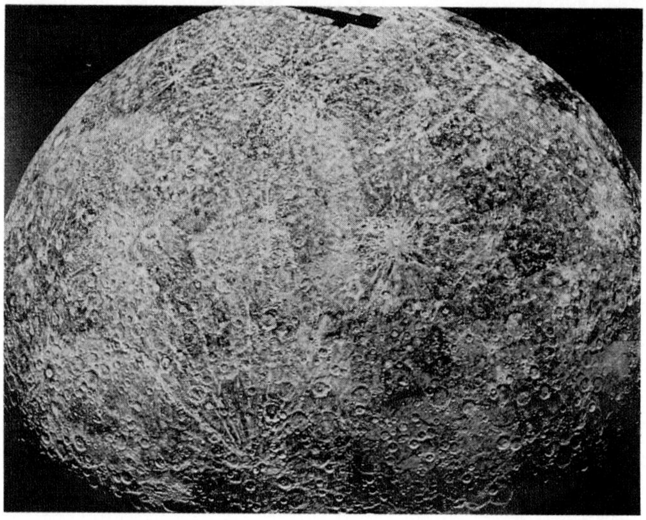

Figure 10.11 Mariner 10*'s view of Mercury as it sped away from the planet after each encounter. Again, the spacecraft was about 200,000 km away when the photographs making up this mosaic were taken. (NASA)*

Mercury has at least one type of surface feature not found on the Moon. Figure 10.14 shows a **scarp,** or cliff, on the surface that does not appear to be the result of volcanic or other familiar geological activity. The scarp cuts across several craters, which indicates that whatever produced it occurred *after* most of the meteoritic bombardment

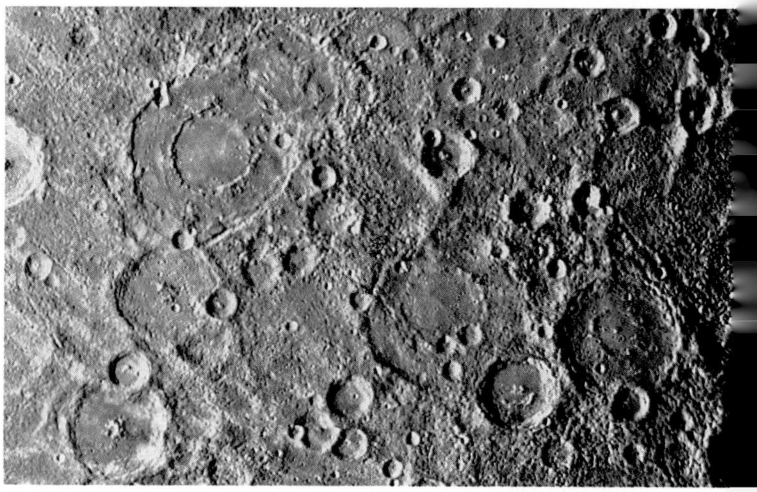

Figure 10.12 *Another photograph of Mercury by* Mariner 10, *this time from about 20,000 km above the planet's surface. The "double-ringed crater" at the upper left, named C. Bach, and spanning about 100 km across, exemplifies how many of the large craters on Mercury tend to form double, rather than single, rings. The reason is not yet understood. (NASA)*

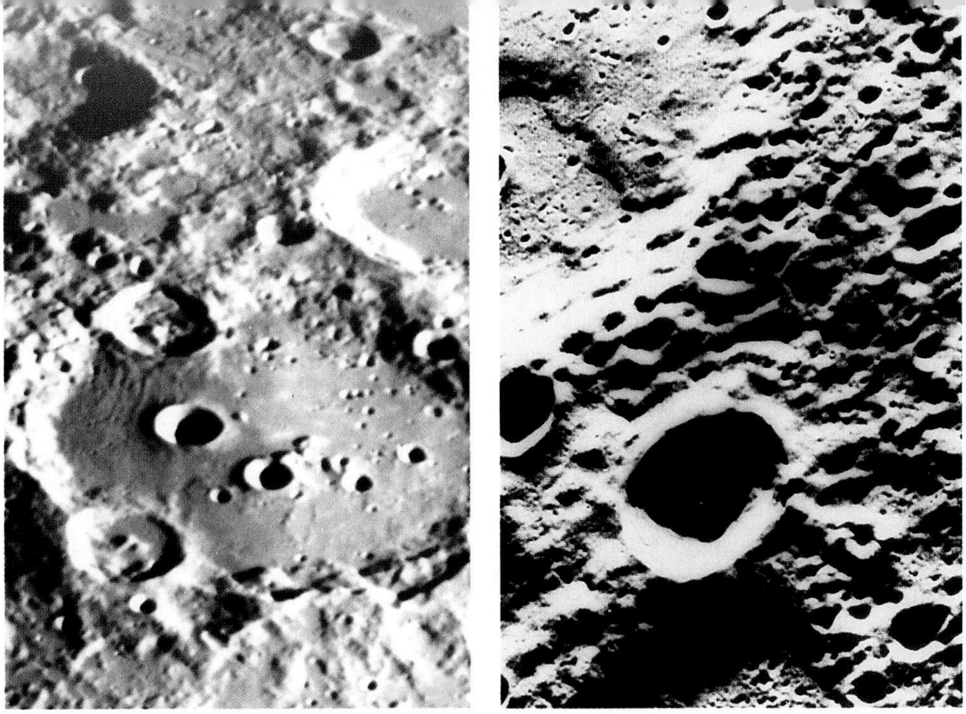

Figure 10.13 *Comparison of the surfaces of the Moon and Mercury. (a) shows the large 200-km diameter crater Clavius on the Moon. (b) is a* Mariner 10 *photo of Mercury, showing lava flow front at upper left. (Palomar Observatory; NASA)*

was over. Mercury shows no evidence for crustal motions. The scarps, of which several are known from the *Mariner* images, probably formed when the crust cooled and shrank long ago, much like wrinkles form on the skin of an old shrunken apple. If we can apply to Mercury the cratering age estimates we use for the Moon, the scarps probably formed about 4 billion years ago.

Much of the discussion in Chapter 8 about the *surface* of Earth's Moon applies equally well to Mercury (although some significant differences do exist). Figure 10.15 shows what was probably the last great event in the geological history of Mercury—an immense bull's-eye crater called the Caloris basin, formed eons ago by the impact of a large asteroid. (Because of the orientation of the planet during *Mariner 10*'s flybys, only half of the basin was visible. The

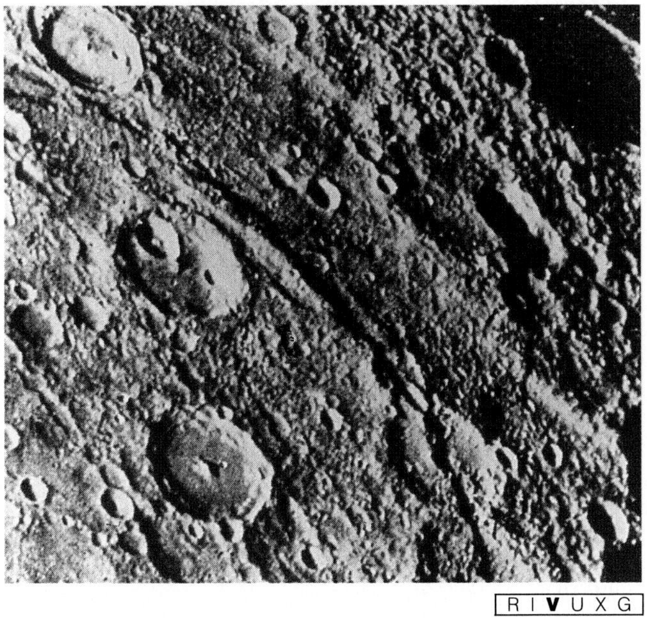

Figure 10.14 *Discovery Scarp on Mercury's surface. This appears to be a compressional feature that formed when the planet's crust cooled and contracted early in its history, causing a crease in the surface. This scarp, running diagonally across the center of the frame, is several hundred kilometers long and up to 3 km high in places. (NASA)*

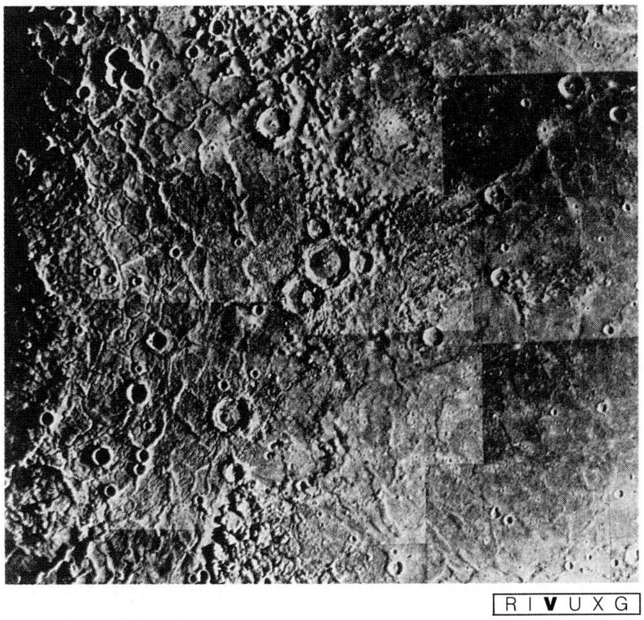

Figure 10.15 *Mercury's most prominent geological feature—the Caloris Basin—measures about 1400 km across and is ringed by concentric mountain ranges that reach more than 3 km high in places. This huge circular basin, only half of which shows (at left) in this* Mariner 10 *photo, is similar in size to the Moon's Mare Imbrium and spans more than half of Mercury's radius. (NASA)*

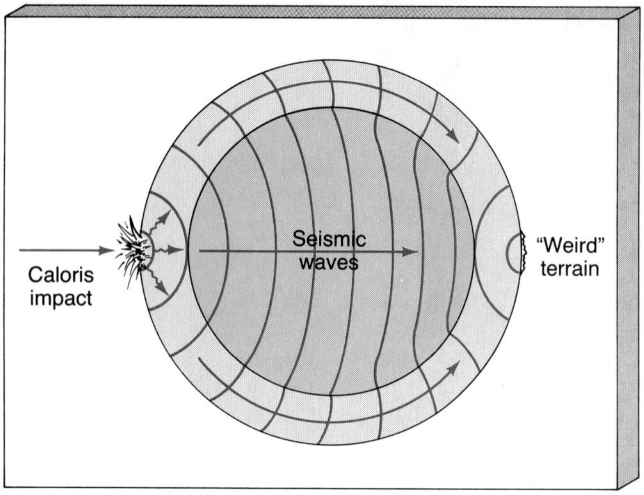

Figure 10.16 The refocusing of seismic waves after the Caloris basin impact may have created the weird terrain on the opposite side of the planet.

center of the crater is off the left-hand side of the photograph.) Compare this basin with the Orientale basin on the Moon (Figure 8.8). The impact-crater structures are quite similar, but even here there is a mystery. The patterns visible on the crater floor are unlike any seen on the Moon. Their origin, like the composition of the floor itself, is unknown.

So large was the impact that created the Caloris basin that it apparently sent strong seismic waves reverberating throughout the entire planet. On the opposite side of Mercury from Caloris there is a region of oddly rippled and wavy surface features that is usually referred to as **weird** (or *jumbled*) **terrain.** Scientists believe that this terrain was produced by seismic waves from the Caloris impact traveling around the planet and converging on the diametrically opposite point, causing large-scale disruption of the surface there, as illustrated in Figure 10.16.

MERCURY'S SURFACE TEMPERATURE AND ATMOSPHERE

Recall from the discussion in Chapter 7 that the surface temperature of a planet is determined by a balance between the (visible) energy it receives from the Sun and the (infrared) energy it radiates back into space. The closer a planet is to the Sun, the more energy it receives per unit area, and the greater is the temperature required for it to reradiate that energy. Because Mercury is so much closer to the Sun than is the Earth, its daytime temperature is considerably higher. At night, however, in the absence of any significant heat conduction from warmer parts of the planet, or atmospheric blanketing, the temperature drops sharply. Radio observations of Mercury's thermal emission indicate that the daytime temperatures can reach as high as 700K and that the nighttime hemisphere has a fairly uniform temperature of about 100K. This 600K temperature range is the largest of any planet or moon in the solar system.

Because of Mercury's eccentric orbit and the spin-orbit resonance, some points on the surface get much hotter than others. In particular, the two (diametrically opposite) points on the surface where the Sun is directly overhead at perihelion get hottest of all. They are called the **hot longitudes.** The peak temperature of 700K occurs at noon at these two locations. The Caloris basin happens to lie close to one of the hot longitudes (and so the region of weird terrain lies close to the other). In fact, the name *Caloris* is derived from the Latin word *calor*, meaning "heat." At the *warm longitudes*, where the Sun is directly overhead at aphelion, the peak temperature is about 150K cooler—a mere 550K! Near the poles, where the Sun's light strikes the surface obliquely, the peak surface temperature can be much lower. Recent Earth-based radar studies suggest that Mercury's polar temperature may be as low as 125K and that, despite the planet's scorched equator, the poles may be covered with extensive sheets of water ice.

Mercury's small mass and high temperature virtually guarantee that any gas would escape quite quickly. Astronomers have never observed an appreciable atmosphere, either spectroscopically from the Earth (in reflected sunlight or during Mercury's transits across the face of the Sun, when atmospheric absorption lines might be seen) or during the *Mariner 10* rendezvous. Although *Mariner 10* found a trace of what was at first thought to be an atmosphere, this gas is now known to be temporarily trapped hydrogen and helium from the solar wind. Mercury captures this gas and holds it for just a few weeks. Ground-based observations have also found an extremely tenuous envelope of sodium and potassium around the planet. Scientists believe that these atoms are torn out of the surface rocks by impacts with high-energy particles in the solar wind. Mercury has, then, no real atmosphere, and thus no protection from the harsh environment of interplanetary space. In addition to meteoroids, x-rays and ultraviolet radiation must constantly rain down on Mercury's surface.

MERCURY'S MAGNETIC FIELD AND INTERNAL STRUCTURE

Mercury's magnetic field, discovered by *Mariner 10*, is about 100 times weaker than Earth's. The discovery that Mercury had any magnetic field at all came as a surprise to planetary scientists who, having detected no magnetic field in the Moon (and, in fact, none in Venus or Mars, either), expected Mercury to have no measurable magnetism. In Chapters 7 and 8, we stated that a combination of a liquid metal core *and* rapid rotation is believed necessary for the production of a planetary magnetic field through dynamo action. Mercury certainly does not rotate rapidly, and it may possibly lack a liquid metal core. Yet a magnetic field unde-

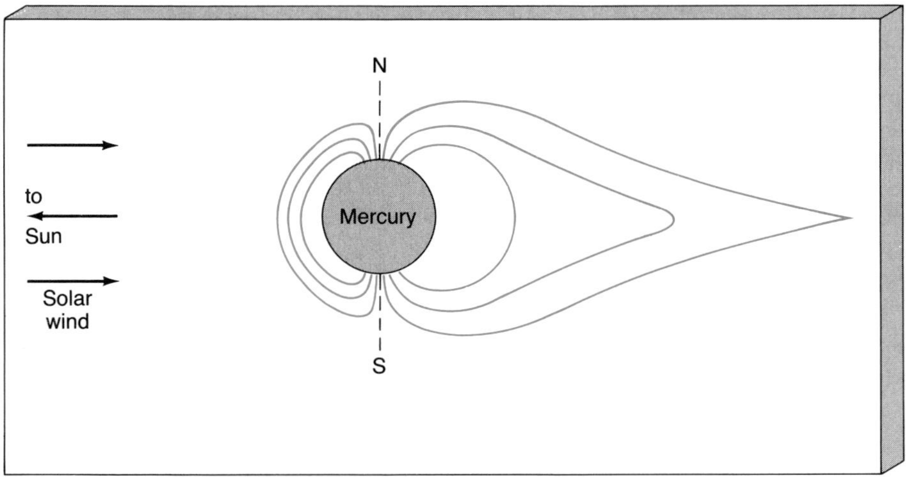

Figure 10.17 *Mercury's magnetosphere. Scientists cannot easily explain the presence of the planet's magnetic field.*

niably surrounds it. Although weak, the field is strong enough to deflect the solar wind and create a small magnetosphere around the planet, as shown in Figure 10.17.

To be honest, scientists have no clear understanding of the origin of Mercury's magnetic field. If it is produced by ongoing dynamo action, as in the Earth, Mercury's core must be at least partially molten. Yet the absence of any recent surface geological activity suggests that the outer layers are solid to a considerable depth, as on the Moon. It is difficult to reconcile these two considerations in a single theoretical model of Mercury's interior. If the field is being generated dynamically, Mercury's slow rotation might at least account for the field's weakness. Alternatively, Mercury's current weak magnetism might be a mere remnant of an extinct dynamo. Given Mercury's rather small size, its metallic core might well have long since solidified. The models are inconclusive on this issue, and no spacecraft is scheduled to revisit Mercury in the foreseeable future.

Mercury's magnetic field, as well as its large average density (roughly 5 g/cm^3), together imply that the planet is chemically differentiated. Even without the luxury of seismographs on the surface, we can infer that most of its interior is dominated by a large, heavy, iron-rich core with a radius of perhaps 1800 km. Whether that core is solid or liquid remains to be determined. A less dense, lunarlike mantle probably lies atop this to a depth of about 500 to 600 km. Thus, about 40 percent of the volume of Mercury, or 60 percent of its mass, is contained in its iron core. The ratio of core volume to total planet volume is greater for Mercury than for any other object in the solar system.

The large fraction of iron indicates that we should perhaps expect the chemical composition of the mantle to differ from that of the Moon. It is probably more correct to say that Mercury is deficient in the lighter rocky material that forms the Moon's mantle rather than to say that Mercury has an excess of metals. Figure 10.18 illustrates the relative sizes and internal structures of the Earth, the Moon, and Mercury.

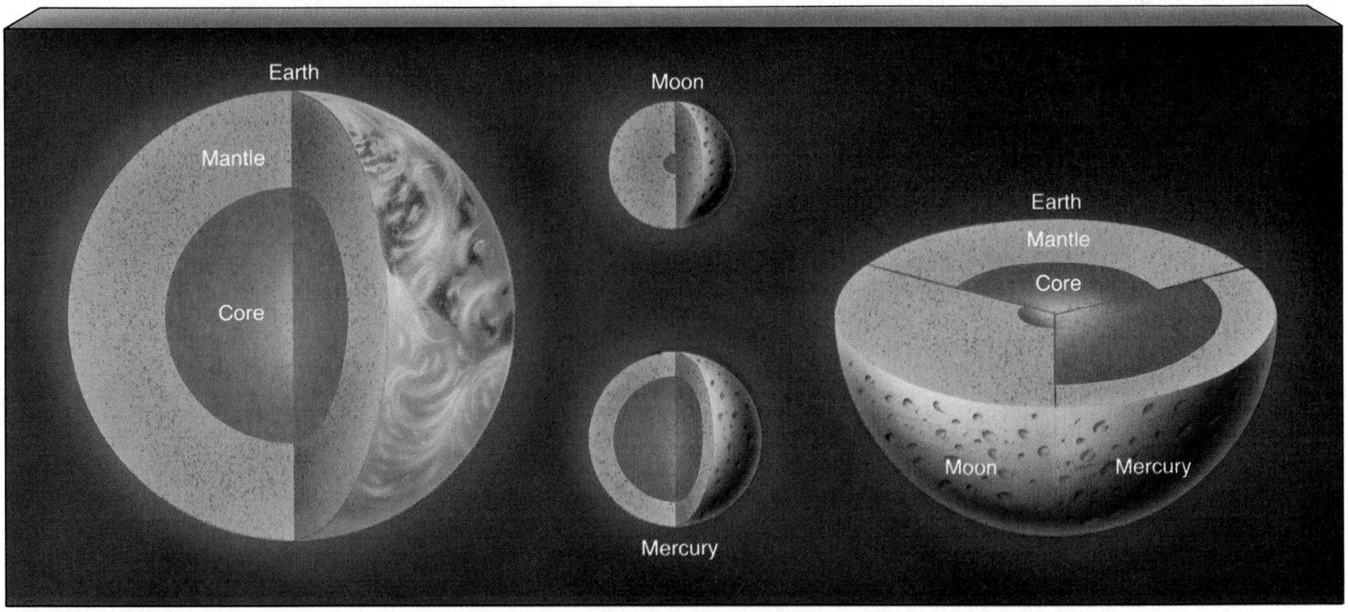

Figure 10.18 *The internal structures of the Earth, the Moon, and Mercury, drawn to the same scale. Note how large a fraction of Mercury's interior is core.*

MERCURY'S EVOLUTIONARY HISTORY

✓3 Like the Moon, Mercury seems to have been a geologically dead world for roughly the past 4 billion years. On both the Moon and Mercury, the lack of ongoing geological activity results from a thick solid mantle that prevents volcanism or tectonic motion. Because of the *Apollo* program, the Moon's early history is much better understood than Mercury's, which remains somewhat speculative. What we do know about Mercury's history is learned mostly through comparison with the Moon.

When Mercury formed some 4.5 billion years ago, it was already depleted in lighter, rocky material. We will see later that this was largely a consequence of its location in the hot inner regions of the early solar system, although possibly a collision stripped away some of its light mantle. During the next half-billion years, it melted and differentiated, like the other terrestrial worlds. It suffered the same intense meteoritic bombardment as the Moon. Mercury, being more massive than the Moon, cooled more slowly, so Mercury's crust was thinner and volcanic activity more common. More craters were erased, leading to the intercrater plains *Mariner 10* found.

As the planet's large iron core formed and then cooled, the planet began to shrink, causing the surface to contract. This compression produced the scarps seen on Mercury's surface and may have prematurely terminated volcanic activity by squeezing shut the cracks and fissures on the surface. Thus Mercury did not experience the extensive volcanic outflows that formed the lunar maria. Despite its larger mass and greater internal temperature, Mercury has probably been geologically inactive for even longer than the Moon.

CHAPTER SUMMARY

Mercury, the closest planet to the Sun, orbits the Sun once every 89 days. Its nearness to the Sun makes it an elusive target for observations, although it is fairly bright to the eye.

Mercury once was believed to rotate synchronously with its revolution around the Sun, but now its rotation rate is known to be exactly two-thirds of the orbital period. This 3:2 spin-orbit resonance in Mercury's motion stems from tidal forces from the Sun acting on the planet. The Mercurian solar day—the time from one noon to the next—is two Mercurian years long. Mercury presents the same face to the Sun not every time around, but every other time.

The best information available about the surface of Mercury came from the flyby of the *Mariner 10* spacecraft in 1974. Mercury is similar in appearance to the Moon, but it lacks extensive lava-flow regions akin to the lunar maria. Scarps, or cliffs, on Mercury probably formed when the crust cooled and shrank long ago.

Mercury's high dayside temperatures and cold nightside temperatures are created in the absence of significant heat conduction or atmospheric blanketing on the planet. Mercury is generally considered to be airless. Any traces of an atmosphere originate in dynamic processes caused by the planet's nearness to the sun. Sunlight strikes Mercury's polar regions at such an oblique angle that temperatures there may be very low, and the planet may have frozen polar caps of water ice.

That Mercury has a magnetic field—even a weak one—is surprising because the planet rotates so slowly. Mercury's magnetic field and large average density imply that the planet is chemically differentiated.

Mercury's evolutionary path was similar to that of the Moon for half a billion years after they both formed. As the planet's large iron core cooled and began to shrink, Mercury's surface contracted and volcanic activity on the planet ended. Mercury's volcanic period probably ended even before that of the Moon.

KEY WORDS

albedo
elongation
hot longitudes
scarp
spin-orbit resonance
transit
weird terrain

REVIEW QUESTIONS

1. Is Mercury a particularly faint planet as seen from Earth? Why is it seldom seen with the eye alone?
2. What happens during a transit of Mercury? How often does a transit of Mercury occur?

3. What early technique was used to lead to the conclusion that Mercury spins on its axis exactly once for each orbit of the planet around the Sun?
4. How fast does Mercury rotate? Explain when and how the planet's true rotation rate was discovered.
5. What does it mean to say that Mercury has a 3:2 spin-orbit resonance? Give a second example of a spin-orbit resonance.
6. Why didn't Mercury settle into a 1:1 spin-orbit resonance with the Sun?
7. To date, what is our best source of information about Mercury's surface?
8. Why is the surface of Mercury often compared to that of the Moon? What are some similarities and some differences between the surfaces of Mercury and the Moon.
9. What are the intercrater plains thought to reveal about Mercury's past?
10. What is a scarp? How are scarps thought to have formed? Why do scientists believe that the scarps formed after most of the meteoritic bombardment ended?
11. Give a brief description of the Caloris basin, and tell how it is different from similar features found on the surface of the Moon.
12. What is weird terrain on Mercury? How is it thought to have formed?
13. In contrast to Earth, Mercury undergoes extremes in temperature. Explain why.
14. What are the hot longitudes on Mercury?
15. In what sense does Mercury have an atmosphere?
16. Why is it so surprising to find a magnetic field associated with Mercury?
17. What do Mercury's magnetic field and its large average density imply about the planet's interior?
18. How is Mercury's evolutionary history like that of the Moon? How is it different?
19. Verify that Mercury's surface gravity is about 0.4 that of Earth's.
20. Assume that Mercury is in its present orbit but that its rotation is in a 5:2 resonance instead of the 3:2 resonance. What would Mercury's rotation rate be?
21. A spacecraft leaves Earth bound for Mercury. Its orbital aphelion just grazes Earth's orbit (at 1 A.U.) and its perihelion just grazes Mercury's orbit (at 0.4 A.U.). Assuming that the two planets are in the phase of their respective orbits that would minimize the length of the journey, how long does the trip to Mercury take? (This orbit is known as the minimum energy orbit for the interplanetary trip.)
22. As seen from Earth, what fraction of the Sun's surface is obscured by Mercury during a transit?
23. What is the angular diameter of the Sun, as seen from Mercury?

DISCUSSION QUESTIONS

1. Explain why Mercury is never seen overhead at midnight in Earth's sky.
2. Mercury used to be called "the Moon of the Sun." Why do you think it had this name? What piece of scientific evidence proved the name inappropriate?
3. Imagine standing on the surface of Mercury, perhaps at the foot of a high scarp, or cliff, on the planet. Describe what you might see.
4. Why might future space travelers want to visit Mercury? What hardships would visitors to the surface of Mercury encounter? Would some landing sites be preferable to others?
5. Mercury is smaller than three moons in the solar system. Why then, is it called a planet instead of a moon? Think about the names and terms for other objects in the solar system. Asteroids orbiting the Sun are even smaller than Mercury. They are sometimes called "minor planets." Do you think this is a good way to describe them? Why or why not?

PROJECTS

1. Try to spot Mercury in morning or evening twilight. (Hint: As seen from the Northern Hemisphere, the best evening apparitions of the planet take place in the spring and the best morning apparitions take place in the fall.)
2. How soon after a conjunction (when Mercury and the Sun are located in the same direction in our sky) can you first see the planet? What conditions might affect your observation? Time of year? City or country? Low clouds or haze?
3. Make a chart or draw a series of pictures depicting Mercury's climb into the twilight sky and subsequent drop back into the Sun's glare with respect to background stars.
4. When you see Mercury shining in the twilight sky, try to

determine whether it has any particular color. Some people call Mercury pinkish. Do you think this is the planet's actual color, or does the pink color appear because we generally see Mercury in a twilight sky?

SUGGESTED READINGS

Chapman, C. R. ''Mercury's Heart of Iron.'' *Astronomy* (November 1988). An article about Mercury's large iron core and possible violent origin.

Cordell, B. M. ''Mercury: The World Closest to the Sun.'' *Mercury* (September/October 1984). A comprehensive look at the innermost planet, by a planetary scientist.

Cowen, R. ''Icy Clues Gleaned from Mercury's Other Half.'' *Science News* (November 9, 1991). The discovery of a polar ice cap for Mercury.

Eberhart, J. ''Mercury's Atmosphere: An Inside Source?'' *Science News* (November 11, 1989). Proposal that Mercury's thin atmosphere might come from gas diffusing up through its crust.

Gingerich, O. ''How Astronomers Finally Captured Mercury.'' *Sky and Telescope* (September 1983). An ''Astronomical Scrapbook'' article about the way that astronomers pinned down Mercury's orbit.

Hartmann, W. K. ''The Significance of the Planet Mercury.'' *Sky and Telescope* (May 1976). How discoveries made by *Mariner 10* about Mercury have increased our knowledge of the planet.

Strom, R. G. *Mercury, the Elusive Planet*. Cambridge: Cambridge University Press, 1987. Short, excellent book on many aspects of the planet Mercury.

Vilas, F., C. R. Chapman, and M. S. Mathews, eds. *Mercury*. Tucson: University of Arizona Press, 1988. Forty-seven contributors give a comprehensive picture of the planet Mercury.

CHAPTER

11

Venus

Earth's Sister Planet

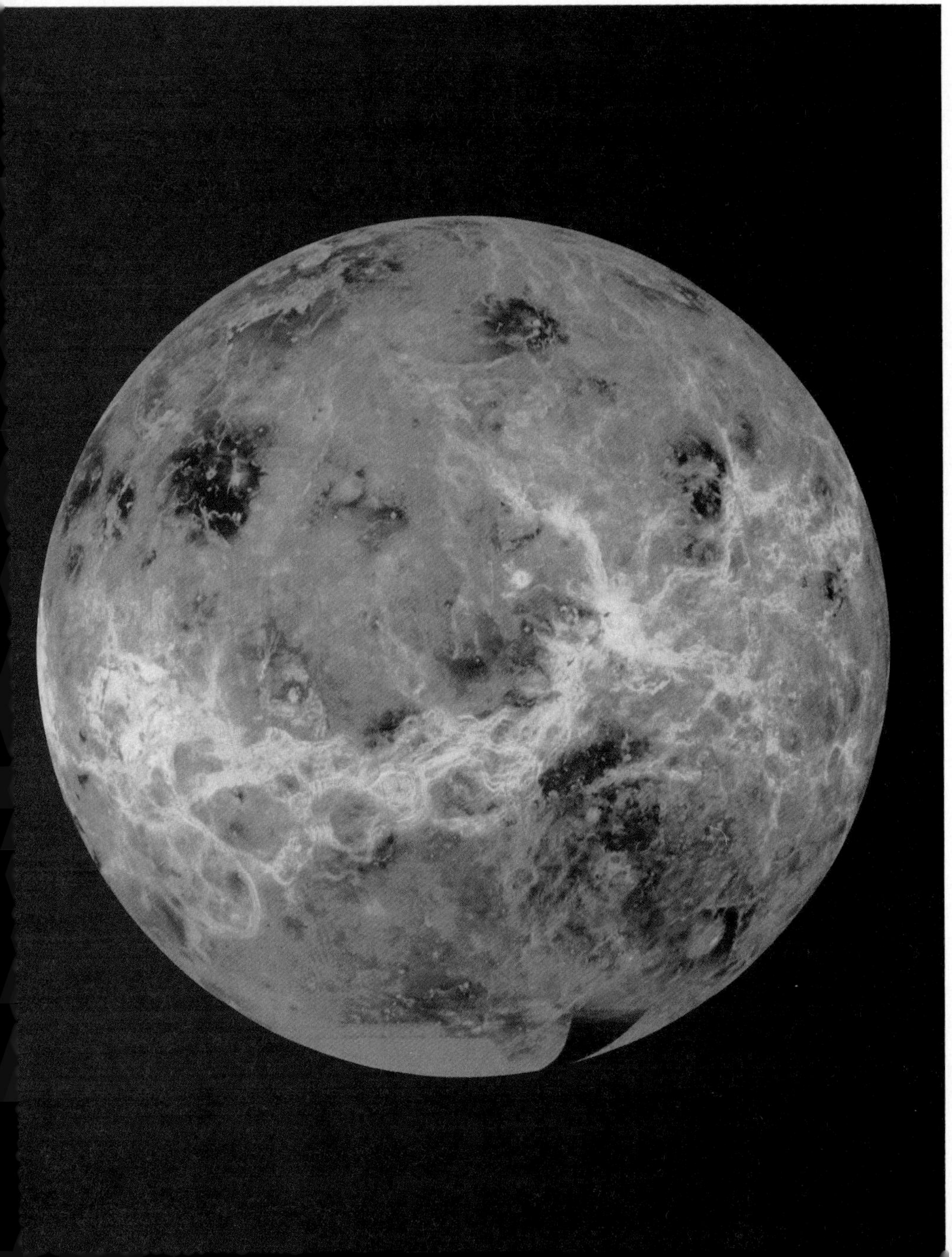

This image of Venus is actually a planetwide mosaic of radio images sent to Earth from the craft Magellan. *The color, which enhances small-scale structure, is based on images recorded by the Soviet* Venera *spacecraft. The roughly triangular black region near the planet's south pole is an area that was not covered on* Magellan's *first mapping pass. The largest "continent" on Venus, Aphrodite Terra, is at the center. (Courtesy NASA)*

R I V U X G

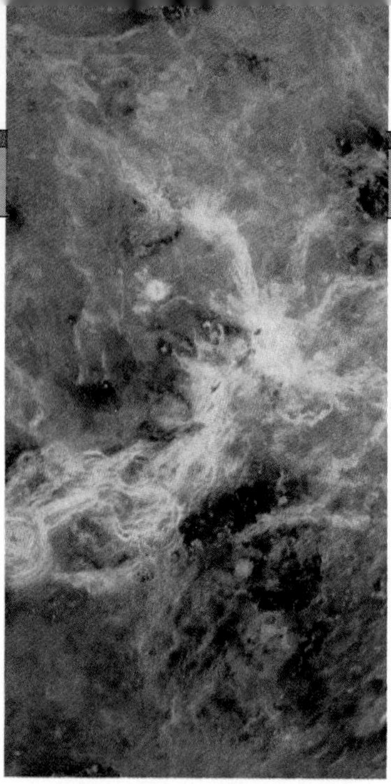

LEARNING GOALS

1 to understand why Venus's orbital location and rotation rate make it hard to observe from Earth

2 to appreciate the importance of unmanned spacecraft in planetary exploration

3 to realize that Venus has an atmosphere, but one quite unlike that of Earth

4 to see why the greenhouse effect ran away on Venus, turning the planet's atmosphere into an inferno

5 to compare the surface and geology of Venus with those on Earth and the Moon

In the previous chapter, we gained insight into the nature and history of Mercury by comparing it with Earth's Moon. Similarly, we can learn about Venus by comparing it with Earth. In its bulk properties at least, Venus is almost a carbon copy of our own world. The two planets are similar in size, density, and chemical composition. They orbit at comparable distances from the Sun. At formation, they must have been almost indistinguishable from one another, yet they are now about as different as two terrestrial planets could be. While Earth is a vibrant world, teeming with life, Venus is an uninhabitable inferno, with a dense, hot atmosphere of carbon dioxide, lacking any trace of oxygen or water.

If Earth and Venus started out alike, then why didn't life emerge and flourish on Venus, too? Somewhere along their respective evolutionary paths, Venus and Earth diverged, and diverged radically. How did this occur? What were the factors leading to Venus's present condition? Why are Venus's surface, atmosphere, and interior so different from Earth's? And is there any chance that Earth could ever evolve into a planet like Venus? In answering these questions, we will discover that a planet's environment, as well as its composition, can play a critical role in determining its future.

GENERAL PROPERTIES OF VENUS

Venus is the second planet from the Sun. Its orbit lies within Earth's, so Venus, like Mercury, is always found fairly close to the Sun in the sky. Because we can see Venus from Earth only just before sunrise or just after sunset, the planet is often called the "morning star" or the "evening star," depending on where it happens to be in its orbit. The early Greek astronomers thought that the morning Venus and the evening Venus were two separate objects (just as they had thought that the morning and evening appearances of Mercury were different planets). By about the sixth century B.C., the truth had become clear, and the planet was named after Aphrodite, the Greek goddess of love. Venus is her Roman name.

Venus's orbital semi-major axis is 0.72 A.U. (108 million km). Its orbital eccentricity is only 0.007, so its path around the Sun is essentially circular. In fact, of all the planets, the orbit of Venus is closest to being a perfect circle—Venus's perihelion and aphelion differ by only 1.5 million km. The planet's sidereal orbital period is 225 days, or about seven and a half Earth months.

Venus is farther from the Sun than is Mercury, so its maximum elongation is correspondingly greater—about 47° (degrees). Given Earth's rotation rate of 15° per hour, this means that Venus is visible above the horizon for at most 3 hours before the Sun rises or after it sets. As with Mercury (see Figure 10.3), there are favorable and unfavorable orientations of Venus's orbit, depending on the observer's latitude and the time of year the observations are made. The most favorable times for viewing Venus from Earth's Northern Hemisphere (assuming that the planet cooperates by being in the right part of its orbit at the time) are during the spring (for morning appearances) and fall (for Venus as the evening star), when the sunward horizon at sunrise or sunset is most nearly perpendicular to the plane of the ecliptic. Because of these constraints, Venus is usually found much less than 47° above the horizon, even at maximum elongation.

Venus is the third brightest object in the entire sky (after the Sun and the Moon), so it very easy to see, even when it is close to the Sun or low in the sky. Although Venus is most definitely *not* a star, it appears more than 10 times brighter than the brightest real star, Sirius. You can see Venus even in the daytime, if you know just where to look. On a moonless night away from city lights, Venus casts a faint shadow. The planet's brightness stems from the fact that, unlike Mercury, Venus is highly reflective, with an albedo of over 0.7. In other words, more than 70 percent of the sunlight reaching Venus is reflected back into space. As we will see, most of the sunlight is reflected from clouds high in the planet's dense atmosphere. Figure 11.1 shows a view of Venus in the western sky just after sunset. The planet appears much brighter than anything else in the photograph, including Sirius and the stars of the constellation Orion.

We might expect Venus to appear brightest when it is "full"—that is, when we can see the entire sunlit side. However, because Venus orbits between Earth and the Sun, Venus is full when it is at its greatest distance from us, 1.7 A.U. away on the other side of the Sun. This configuration—with Venus and the Earth on opposite sides of the Sun, is called **superior conjunction.** (This same term applies also to Mercury, the other planet with an orbit between the Earth and the Sun. A conjunction simply means that two bodies happen to come close together on the sky.) Actually, when Venus is exactly on the other side of the Sun, we can't see it—it is lost in the Sun's glare—but we can see an almost full Venus within a few degrees of superior conjunction.

When Venus is closest to us, at **inferior conjunction,** the planet is at the new phase, lying between Earth and the

R I **V** U X G

Figure 11.1 *Wide-angle view of the western sky just after sunset. The brightest object in the photograph is the planet Venus, which clearly outshines even Sirius, the brightest star in the sky. Also shown here are the stars of the Pleaides star cluster. (S. Westphal)*

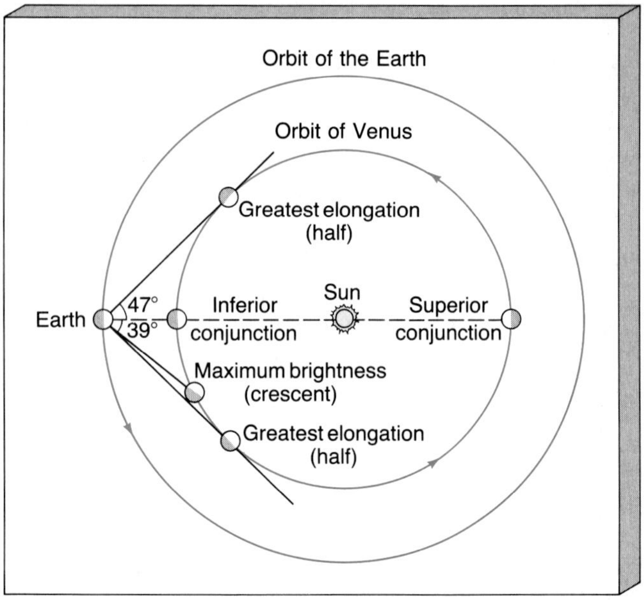

Figure 11.2 *Inferior conjunction occurs when Venus lies between the Sun and Earth. At superior conjunction, Venus is on the opposite side of the Sun from us. Venus's distance increases even as more of its sunlit side becomes visible after inferior conjunction. Venus appears brightest at an elongation of about 39°.*

Sun. At this time we again can't see it, because the sunlit side faces away from us. Only a thin ring of sunlight, caused by refraction in the planet's atmosphere, surrounds the planet. As Venus moves away from inferior conjunction, more and more of it becomes visible. But, of course, as this happens its distance from us also increases. Venus's maximum apparent brightness actually occurs about 36 days before or after inferior conjunction. The elongation of the planet at the time is 39°, and we see it as a rather fat crescent. Figure 11.2 illustrates the relation between Venus's orbit and its brightness.

Like all the planets, Venus orbits the Sun in the same sense as the Earth—counterclockwise, as seen from above the north celestial pole. Its orbital plane is inclined at approximately 3° to the ecliptic. As with Mercury, it is possible for Venus to pass across the face of the Sun, as seen from Earth. However, solar transits of Venus are exceedingly rare events—even rarer than transits of Mercury—primarily because Venus is farther from the Sun and the proper alignment occurs less frequently. The transits occur in pairs, separated by roughly 8 years, and the pairs occur less than once per century. The next pair will occur in 2004 and 2012.

VENUS IN BULK

Radius

We can determine Venus's radius from simple geometry, just as we did for Mercury. At inferior conjunction, when Venus is only 0.28 A.U. from us, its angular diameter is 64″(arc seconds); near superior conjunction it is only 10″ across. From either of these observations, we can determine its radius to be about 6000 km. More accurate measurements from spacecraft lead to a value of 6052 km, or 0.95 Earth radii.

Mass

Like Mercury, Venus has no moon. Before the Space Age, astronomers calculated its mass by indirect means—through studies of its small gravitational effect on the orbits of the other planets, especially Earth. Now that spacecraft have orbited the planet, we know its mass very accurately. It is 4.9×10^{27} g, or 0.82 times the mass of Earth.

Density

From the mass and radius we find that Venus's average density is 5.2 g/cm^3. As far as these bulk properties are concerned, then, Venus appears to be very similar to the Earth. Is Venus's overall composition similar to Earth's? If so, we could reasonably conclude that Venus's internal structure and evolution are basically Earthlike. We will review later in this chapter what little evidence there is on this subject.

VENUS'S ROTATION RATE

✓1 Venus is surrounded by a thick layer of cloud. The same clouds whose reflectivity make Venus so easy to see in the night sky also make it impossible for us to discern any surface features, at least in visible light. Even when viewed through a large optical telescope, the clouds themselves show few features. Until the advent of suitable radar techniques in the 1960s, astronomers did not know the rotation period of Venus. Attempts to determine its rotation by watching motions of the cloud markings as seen from Earth were frustrated by the rapidly changing nature of the clouds themselves. Some astronomers argued for a 25-day period, while others favored a 24-hour cycle. Controversy raged until, to the surprise of all, radar observers announced that the Doppler broadening of their returned echoes implied a sluggish 243-day rotation period. Furthermore, Venus's spin was found to be *retrograde*—that is, opposite to that of the Earth and most other solar system objects, and in the opposite sense to Venus's orbital motion. Figure 11.3 illustrates Venus's retrograde rotation and compares it with the rotation of its neighbors, Mercury and Earth.

If you could stand on the surface of Venus and could see the Sun—which is not possible, by the way—it would rise in the west and then set in the east nearly two Earth months later, rising again in the west two Earth months after that. Because Venus's rotation is so slow, the planet's solar day is quite different from its 243-Earth-day sidereal rota-

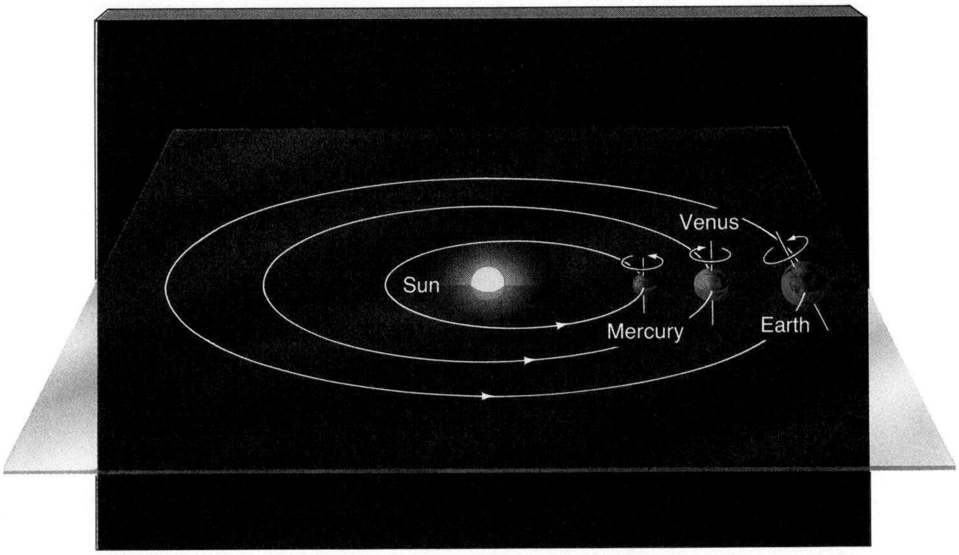

Figure 11.3 *The inner three planets of the solar system—Mercury, Venus, and Earth—display widely differing rotational properties. While all three orbit the Sun in the same sense and in nearly the same plane, Mercury's rotation is slow and prograde (in the same sense as the orbital motion), Earth's is fast and prograde, and Venus's is slow and retrograde. Venus rotates clockwise as seen from above the plane of the ecliptic in the direction of the north celestial pole.*

tion period. In fact, one Venus day is a little more than half a Venus year (225 Earth days). Figure 11.4 depicts the interplay between Venus's orbital and rotational motion.

Venus's slow retrograde rotation presents us with a mystery. *Why* is Venus rotating backwards, and why so slowly? The answer was once thought to involve a resonance between Venus and Earth. However, there are some difficulties with that explanation, as we will see. Let's begin our discussion with a more careful account of synodic periods and solar days.

Recall from Chapter 1 that a *synodic period* is the period as seen from Earth, taking Earth's own motion into account. For example, Venus's synodic period is the time between one inferior (or superior) conjunction and the next. Because Earth moves in the time taken for Venus to orbit the Sun, as illustrated in Figure 11.5, the synodic period is considerably longer than the "true"—that is, the sidereal—period.

The rule for calculating the synodic period is easy: If Earth's sidereal orbital period is E days and Venus's is V days, then the *angular speeds* of the two planets (in degrees per day) are, respectively, $360/E$ (because there are 360° in one complete revolution) and $360/V$. Because Venus moves faster than Earth, it "pulls ahead" of us at a rate of $360/V - 360/E$ degrees per day. The time taken for Venus to outstrip the Earth by exactly one revolution is, by definition, the synodic period, S. It follows that

$$\frac{1}{S} = \frac{1}{V} - \frac{1}{E}.$$

Plugging in the exact numbers, $E = 365.26$ and $V = 224.70$ (the extra decimal digits will turn out to be important in this case), we find that the synodic orbital period of Venus is 583.91 days. By the same computation, from Venus's point of view the Earth's synodic period also is 583.91 days.

Now let's reconsider Venus's rotation period. The sidereal period is -243.08 days (the minus sign conventionally denotes retrograde motion). The same reasoning we used above leads to the following relation between the solar day D (noon to noon), the sidereal rotation period R, and the sidereal year V:

$$\frac{1}{D} = \frac{1}{R} - \frac{1}{V}.$$

The result is a solar day of -116.76 days, almost exactly one-fifth of the synodic period.

The key word in the preceding sentence is *almost*. A resonance, if it existed, would require that the relationship be *exactly* one-fifth. The slight difference—less than 3

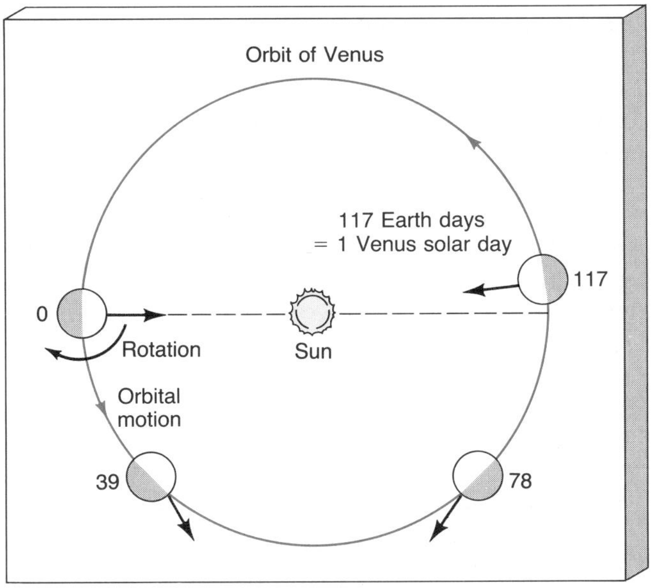

Figure 11.4 *Venus's orbit and retrograde rotation combine to produce a Venus solar day equal to 117 Earth days, or slightly more than half a Venus year. The numbers in the figure mark time in Earth days.*

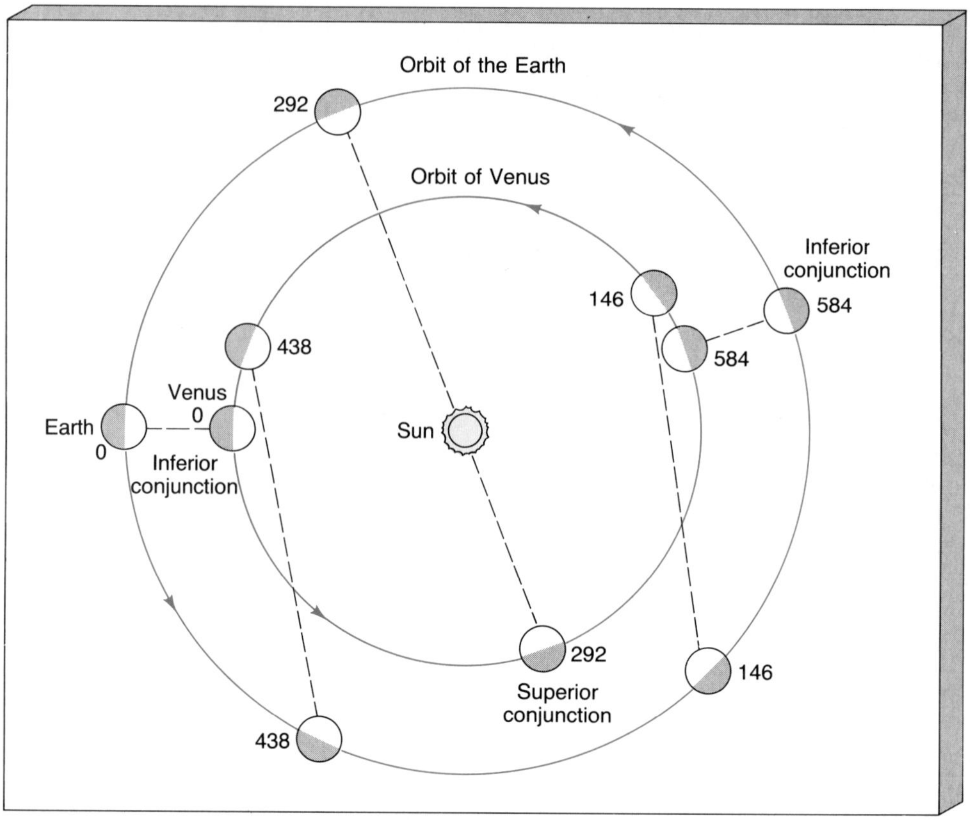

Figure 11.5 *The synodic period of Venus (that is, as seen from the moving Earth) is 584 days. As in Figure 11.4, the numbers in the figure represent time in Earth days.*

hours in 584 days—appears to be real, and, if that is so, no resonance exists. Whether the resonance does or does not exist, the near-coincidence of these periods has the following important observational consequence: In the time between one inferior conjunction and the next, Venus rotates almost exactly five times relative to the Sun. Thus, *Venus always presents nearly the same face to Earth at inferior conjunction*. This means that radar observations of the planet's surface cover one side—the one facing us at closest approach—much more thoroughly than the other side, which we can see only when the planet is close to its maximum distance from Earth.

With or without a resonance, there are still problems in explaining Venus's rotation. If there is no resonance, then we have a complete mystery on our hands. If there is a resonance, and Venus is locked into a 5:1 spin-orbit resonance with Earth, then we lack an explanation of how the resonance arose. Tidal forces, which, as we saw in Chapter 10, explain Mercury's 3:2 resonance, are far too weak between Venus and Earth. Earth's tidal effect on Venus is tiny, and it is much less than the Sun's tidal effect in any case. For now we are compelled to accept this "coincidence" of resonance or near-resonance without any adequate explanation, and the riddle of Venus's slow retrograde rotation remains. Possibly the planet was struck by a large body early on, much like the one that may have hit Earth and formed the Moon, and that impact was sufficient to reduce the planet's spin virtually to zero. For now, we just don't know.

LONG-DISTANCE OBSERVATIONS OF VENUS

General Features

Because Venus of all planets most nearly matches Earth in size, mass, and density, and because its orbit is closest to us, it is often called Earth's sister planet. But Venus has a dense atmosphere that is nearly opaque to visible radiation, making its surface completely invisible from the outside at optical wavelengths. Even cloud patterns are very difficult to detect. Figure 11.6 shows one of the best photographs of Venus taken with a large telescope on Earth. The planet resembles a white-yellow disk and shows rare hints of cloud circulation.

The atmospheric patterns are much more evident when examined with equipment capable of detecting ultraviolet radiation. Some of Venus's atmospheric constituents absorb this high-frequency radiation, thereby increasing the cloud contrast. Figure 11.7 is a photograph—really an ultraviolet image—taken in 1979 by America's *Pioneer Venus* spacecraft at a distance of 200,000 km from the planet. The large, fast-moving cloud patterns resemble Earth's high-altitude jet stream more than the great cyclonic whirls characteristic of Earth's low-altitude clouds. In fact, the upper deck of clouds on Venus rotates around the planet in just 4 days, which is much faster than the underlying surface. The upper-level winds reach speeds of 400 km/hr relative to the planet.

Spectral Analysis of the Cloud Cover

Spectroscopic examination of sunlight reflected from Venus shows the presence of large amounts of carbon dioxide (CO_2), with little evidence for any other atmospheric gases. The first studies to reveal carbon dioxide as a major constituent of Venus's atmosphere were performed in the 1930s. Until the 1950s, astronomers generally believed that observational difficulties alone prevented them from seeing other atmospheric components. For a while, the hope lingered that Venus's clouds were actually predominantly water vapor, like those on Earth, and that below the cloud cover Venus might be a habitable planet similar to our own. Indeed, in the 1930s scientists had measured the temperature of the atmosphere spectroscopically at about 240K, not much different from our own upper atmosphere. Calculations of the surface temperature—taking into account the cloud cover and Venus's proximity to the Sun, and assuming an atmosphere much like our own—suggested that Venus should have a surface temperature only 10 or 20 degrees higher than Earth's.

Radio Measurements below the Clouds

These hopes for an Earthlike Venus were dashed in the 1950s, when radio observations of the planet measured its thermal energy emission. Unlike visible light, radio waves easily penetrate the cloud layer, and they gave the first indication of conditions on or near the surface. Astronomers found that the radiation emitted by the planet has a Planck-curve spectrum characteristic of a temperature near 600K! Almost overnight, the popular conception of Venus changed from that of a lush tropical jungle to an arid, uninhabitable desert.

Radar observations of the surface of Venus are routinely carried out from Earth using the Arecibo radio telescope (see Chapter 6). This instrument can achieve a resolution of a few kilometers, but it can adequately cover only a small fraction (roughly 25 percent) of the planet. This telescope's observation of Venus is limited because of the planet's peculiar near-resonance and because radar reflections from regions near the "edge" of the planet are hard to obtain. However, the results from Arecibo have been combined with information from probes orbiting Venus to build up a detailed picture of the planet's surface. Only very recently, with the arrival of the *Magellan* probe, have more accurate data been obtained.

R I **U** X G

Figure 11.7 *Venus is pictured here by the* Pioneer *spacecraft cameras some 200,000 km away from the planet. This image was made by capturing solar ultraviolet radiation reflected from the planet's clouds, which are probably composed mostly of sulfuric acid droplets, much like the corrosive acid in a car's battery. (In fact, the only kind of precipitation Venus could have is acid rain.) Venus's circulation patterns can be best studied using ultraviolet images. It is possible that these patterns might some day help us better understand Earth's weather. (NASA)*

R I **V** U X G

Figure 11.6 *This photograph, taken from Earth, shows Venus with its creamy yellow mask of clouds. (Lick Observatory)*

SPACECRAFT EXPLORATION

✓2 In all, some 20 spacecraft have visited Venus since the 1970s, far more than have spied on any other planet. It is fair to say that the Soviet space program has taken the lead role in exploring Venus's atmosphere and surface. American spacecraft, however, have performed extensive radar mapping of the planet's surface from orbit. The American *Mariner 2* and *Mariner 5* missions passed within 35,000 km in 1962 and 1967, and the *Mariner 10* craft grazed Venus at a distance of 6000 km in 1974, en route to Mercury. During roughly the same period, the Soviet *Venera* (from the Russian word for Venus) program got under way, and the Soviet *Venera 4* through *Venera 12* probes parachuted into the planet's atmosphere between 1967 and 1978.

The early *Venera* probes were destroyed by enormous atmospheric pressures before reaching the surface. Then in 1970 *Venera 7* became the first spacecraft to soft-land on the planet. During the 23 minutes it survived on the surface, it radioed back information on atmospheric pressure and temperature. The surface atmospheric pressure is about 90 bars—90 times the pressure at sea level on Earth. This is equivalent to an (earthly) underwater depth of about 1 km (3000 feet). Unprotected humans cannot dive much below 100 m. The temperature is a sizzling 750K—about twice as hot as a kitchen oven, and hot enough to melt lead.

Since that time, a number of *Venera* landers have transmitted back to Earth photographs of the surface and have analyzed the atmosphere and the soil. These *Venera* data make up the entirety of our direct knowledge of Venus's surface. Despite their rugged construction (see Figure 11.8), none of the landers survived for more than an hour or so on the surface. They were either cooked by the inferno or crushed by the enormous atmospheric pressure. In 1983, the *Venera 15* and *Venera 16* orbiters sent back detailed radar maps (at about 2-km resolution) of large portions of Venus's northern hemisphere.[1]

On the U.S. side, the *Pioneer Venus* mission in 1978 placed an orbiter at an altitude of some 150 km above Venus's surface and dispatched a "multiprobe," consisting of five separate instrument packages, into the planet's atmosphere. During their hour-long descent to the surface, the multiprobe instruments returned information on the variation of density, temperature, and chemical composition with altitude in the atmosphere. The orbiter's radar produced low-resolution (around 25 km horizontally) images of most of the planet's surface. However, its resolution in the vertical direction was extremely good—around 100 m—so that, on those regions of the planet where its observations overlapped with those from ground-based radars or the *Venera* orbiters, high-quality topographic maps could be constructed.

[1]Normally, a planet's sense of rotation determines which pole is labeled "north" and which "south"—in most cases, the "north" pole is such that the planet rotates from west to east, as on Earth. However, in the case of Venus, it is much more common to regard the pole lying above the ecliptic plane (that is, the same side as Earth's North Pole) as "north," so the planet's rotation is from east to west.

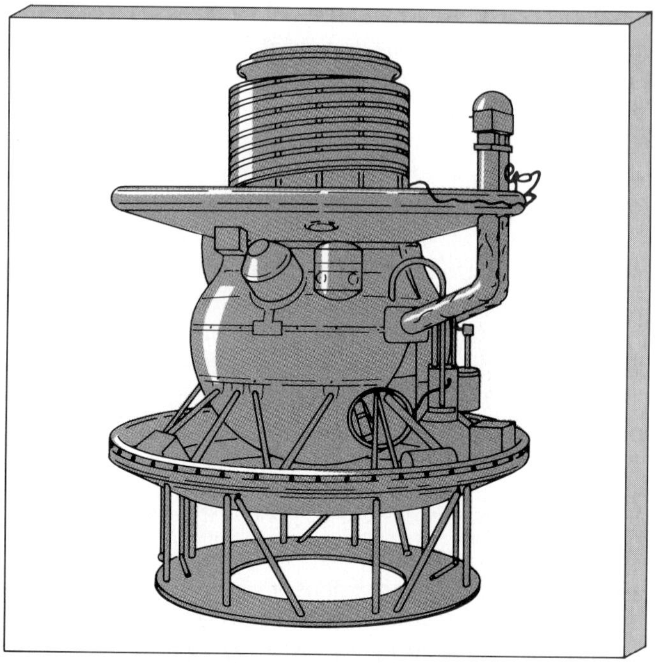

Figure 11.8 *A model of the Soviet* Venera *landers that reached the surface of Venus. The design was essentially similar for all the surface missions. Note the heavily armored construction necessary to withstand the enormous atmospheric pressures at the planet's surface.*

The most recent U.S. mission is the *Magellan* probe (shown in Figure 11.9), which was launched in May 1989 from the space shuttle *Atlantis* and entered orbit around Venus in August 1990. After some initial intermittent problems, the spacecraft began sending back spectacular data in September 1990, and it completed its first 243-day mapping cycle (the time required for Venus to rotate once beneath the probe's orbit) in May 1991, covering 84 percent of the planet's surface with unprecedented clarity. *Magellan*'s spatial resolution is at least 10 times better than the best images previously obtained. It can distinguish objects as small as 120 m across and can measure vertical distances to within a few tens of meters. While planetary scientists are just now analyzing the *Magellan* data, they expect the spacecraft's extraordinary improvement in resolution to answer many key questions concerning the geology and history of Venus. When complete, the probe's coverage will extend to the entire surface of Venus and will render all previous data virtually obsolete. Already, many theories of the processes

Figure 11.9 *The U.S.* Magellan *spacecraft is launched from the space shuttle* Atlantis *in May 1989. (NASA)*

shaping the planet's surface have had to be altered radically or abandoned completely.

THE ATMOSPHERE OF VENUS

Atmospheric Structure

✓3 The data returned from the *Venera* and *Pioneer* spacecraft have allowed us to paint a fairly detailed picture of Venus's atmosphere. Figure 11.10 shows the run of temperature and pressure with height. Compare this figure with Figure 7.6, which gives similar information for Earth. The atmosphere of Venus is much more massive than Earth's, and it extends to a much greater height above the surface. On Earth, 90 percent of the atmosphere lies within about 10 km of sea level. On Venus the corresponding (90 percent) level is found at an altitude of 50 km instead. The surface temperature and pressure of Venus's atmosphere are much greater than Earth's. However, the temperature drops more rapidly with altitude, and the upper atmosphere of Venus is actually colder than our own. The total mass of the atmosphere is about 90 times greater than the Earth's.

The troposphere on Venus extends up to an altitude of nearly 100 km. The reflective clouds that block our view of the surface lie between 50 and 70 km above the surface. The *Pioneer* multiprobe data indicate that the clouds may actually be separated into three distinct layers within that altitude range.

Below the clouds, extending down to an altitude of some 30 km, is a layer of haze. Below 30 km, the air is clear. Above the clouds, a high-speed "jet stream" blows from east to west at about 300–400 km/h, fastest at the equator and slowest at the poles. This high-altitude flow is

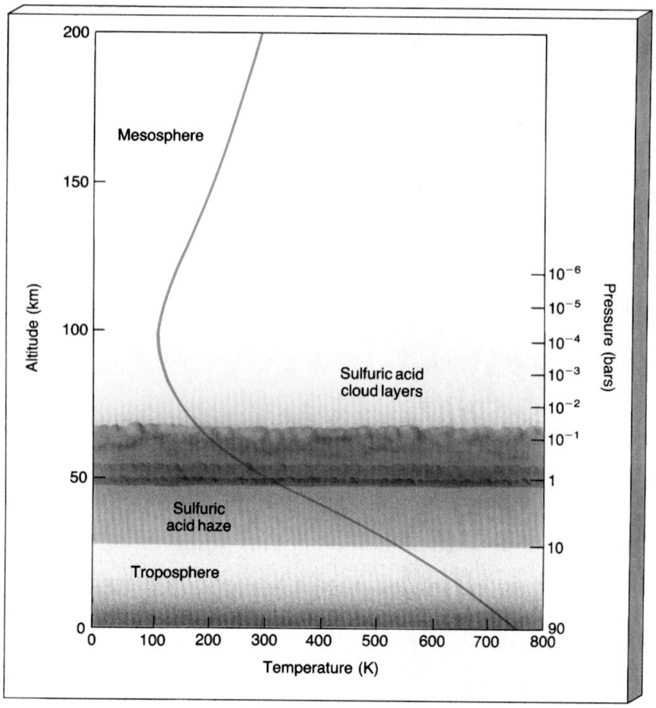

Figure 11.10 *The structure of the atmosphere of Venus, as determined by U.S. and Soviet probes.*

responsible for the rapidly moving cloud patterns seen in ultraviolet light. Figure 11.11 shows a sequence of three ultraviolet images of Venus in which the variations in the cloud patterns can be seen. Note the characteristic V-shaped appearance of the clouds caused by the difference between the equatorial and polar wind speeds. Near the surface, the dense atmosphere moves more sluggishly—indeed, the fluid flow bears more resemblance to Earth's oceans than to its air. Surface wind speeds are typically less than 2 m/s (roughly 4 mph).

Atmospheric Composition

As we have seen, observations from Earth revealed the presence of carbon dioxide in Venus's atmosphere but were inconclusive about other possible constituents. We now know that carbon dioxide is in fact the dominant component of the atmosphere, accounting for 96.5 percent of it by volume. Almost all of the remaining 3.5 percent is nitrogen (N_2). Trace amounts of other gases, such as water vapor, carbon monoxide, sulfur dioxide, and argon, are also present. This composition is clearly radically different from Earth's atmosphere. The absence of oxygen is perhaps not surprising, given the absence of life (recall our discussion of Earth's atmosphere in Chapter 7). However, there is no sign of the water vapor that we might expect to find if a volume of water equivalent to Earth's oceans had evaporated. If Venus started off with Earthlike composition, something has happened to its water—it is now a very dry planet. The lack of water and the dominance of carbon dioxide in the atmosphere of Venus are closely related, as we will see.

For a long time, the chemical makeup of the highly reflective cloud layer surrounding Venus was unknown. At first scientists assumed the clouds were water vapor or ice, as on Earth, but spectroscopic and other studies did not bear out that assumption. The reflectivity of the clouds at different wavelengths didn't match that of water ice. Researchers finally determined, largely by infrared observations carried out in the 1970s, that the clouds (or at least the top layer of clouds) are composed of sulfuric acid (H_2SO_4), created by reactions between water (H_2O) and sulfur dioxide (SO_2). Sulfur dioxide is an excellent absorber of ultraviolet radiation and could be responsible for some of the cloud patterns seen in ultraviolet light. Spacecraft observations have since confirmed the presence of all these compounds in the atmo-

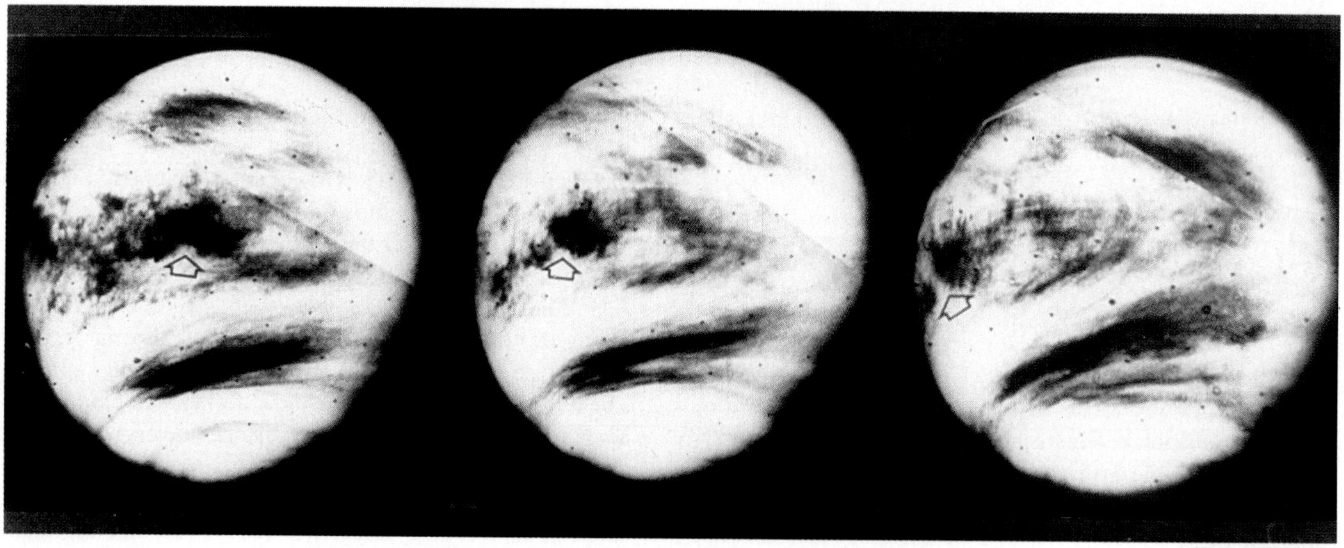

Figure 11.11 *Three ultraviolet views of Venus, taken by the* Pioneer Venus *orbiter, showing the changing cloud patterns in the planet's upper atmosphere. The wind flow is in the direction opposite the* V *in the clouds. Notice the east to west motion of the dark region marked by the arrow. (NASA)*

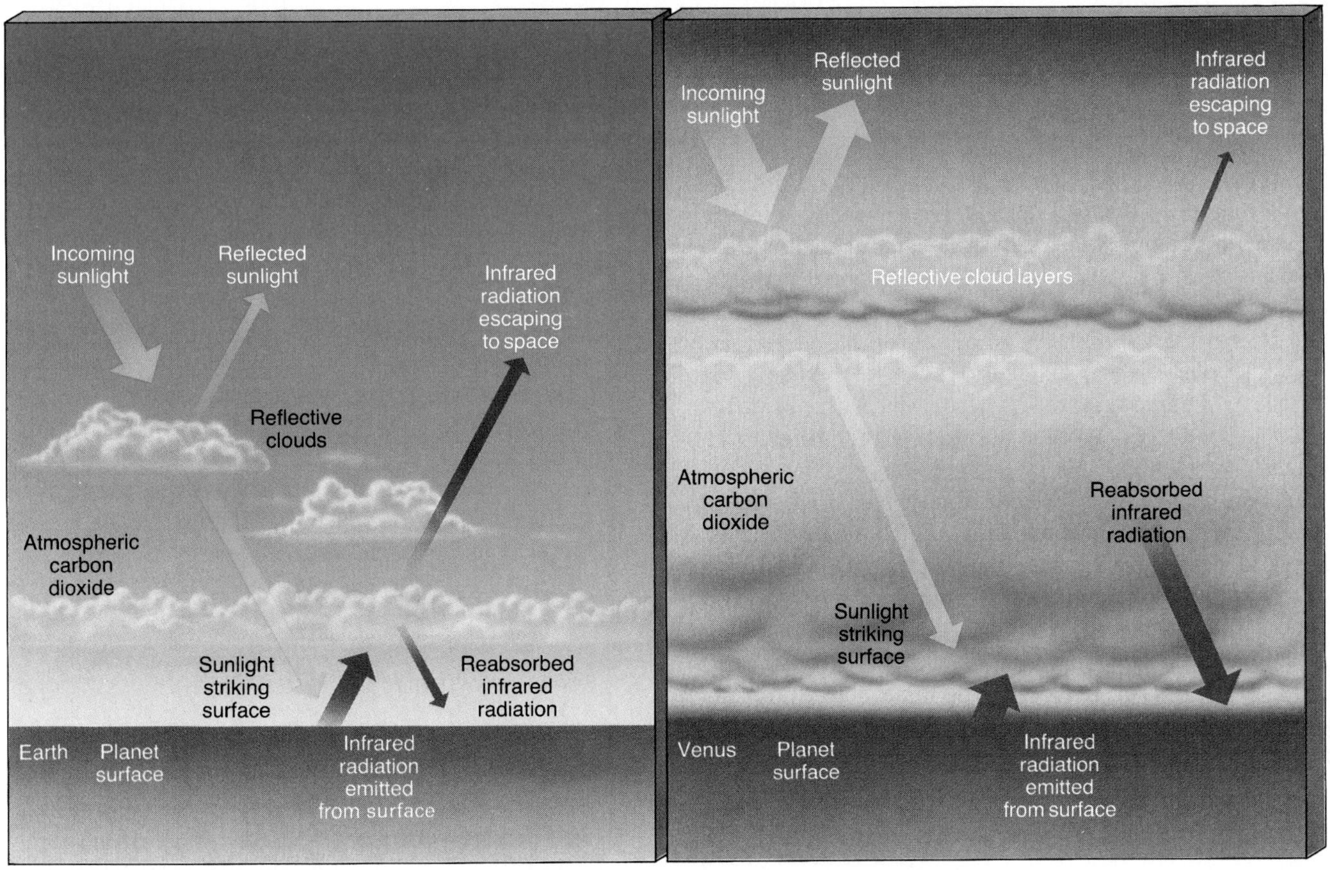

Figure 11.12 *Because Venus's atmosphere is much thicker and denser than Earth's, a much smaller fraction of the infrared radiation leaving the planet's surface actually escapes into space. The result is a much stronger greenhouse effect than on Earth and a correspondingly hotter planet. The outgoing infrared radiation is not absorbed at a single point in the atmosphere. Instead, absorption occurs at all atmospheric levels. The arrows are meant to indicate only that absorption occurs, not that it occurs at one specific level.*

sphere. They also indicate that there may be particles of sulfur suspended in and near the cloud layers, which may account for Venus's characteristic yellowish hue.

The Greenhouse Effect on Venus

Given the distance of Venus from the Sun, the planet was not expected to be such a pressure cooker. Calculations based on Venus's orbit and its surface reflectivity indicated a temperature not much different from Earth's, and early measurements of the cloud temperatures seemed to concur. Certainly, scientists reasoned, Venus could be no hotter than the 500–600K values that characterize the hot side of Mercury, and it should probably be much cooler. This reasoning was obviously seriously in error.

Why is Venus's atmosphere so hot? And if, as we believe, Venus started off like Earth, why is it now so different? The answer to the first question is fairly easy: Given the present composition of its atmosphere, Venus is hot because of the greenhouse effect. Recall from our discussion in Chapter 7 that the "greenhouse gases" in Earth's atmosphere, particularly water vapor and carbon dioxide, serve to trap heat from the Sun. By inhibiting the escape of infrared radiation reradiated from the Earth's surface, these gases serve to increase the planet's equilibrium temperature, in much the same way as an extra blanket keeps you warm on a cold night. Continuing the analogy a little further, the more blankets you place on the bed, the warmer you will become. Similarly, the more greenhouse gases there are in the atmosphere, the hotter the surface will be. (This is precisely the mechanism thought to be responsible for global warming on Earth as atmospheric carbon dioxide levels rise.)

The same effect naturally occurs on Venus, whose dense atmosphere is made up almost entirely of a primary greenhouse gas, carbon dioxide. As illustrated schematically in Figure 11.12, the thick carbon dioxide blanket absorbs 99 percent of all the infrared radiation released from the surface of Venus, and it is the immediate cause of the planet's sweltering 750K surface temperature. Furthermore, the temperature is nearly as high at the poles as at the equator, and there is not much difference between the temperatures on the "night side" (facing directly away from the Sun) and the "day side." The circulation of the atmosphere spreads energy very efficiently around the planet, making it impossible to escape the blazing heat, even at night.

The Runaway Greenhouse Effect

✓4 We have answered the first question about Venus's temperature: Venus is hot because of the abundance of carbon dioxide in its thick atmosphere. Now let us turn to the second question: *Why* is Venus's atmosphere so different from Earth's? Why is there so much carbon dioxide in the atmosphere of Venus, and why is the atmosphere so dense? To address these questions, we must consider the processes that created the atmospheres of the terrestrial planets and then determined their evolution. In fact, we can turn the question around and ask instead, why is there so *little* carbon dioxide in Earth's atmosphere compared with Venus?

We believe that Earth's atmosphere has evolved greatly since it first appeared. Any primary atmosphere escaped soon after the Earth formed and was replaced by a secondary atmosphere outgassed from the interior by volcanic activity 4 billion years ago. Since then, the atmosphere has been reprocessed, in part by living organisms, into its present form. On Venus, the initial stages probably took place in more or less the same way, so that, at some time in the past, Venus may well have had an atmosphere similar to the primitive secondary atmosphere on Earth, containing water, carbon dioxide, sulfur dioxide, and nitrogen-rich compounds. What happened on Venus to cause such a major divergence from subsequent events on our own planet?

On Earth, nitrogen was released into the air by the action of sunlight on the chemical compounds containing it. Meanwhile, the water condensed into oceans, and much of the carbon dioxide and sulfur dioxide eventually became dissolved in them. Most of the remaining carbon dioxide combined with surface rocks. Thus, much of the secondary outgassed atmosphere quickly became part of the surface of the planet. If all the dissolved or chemically combined carbon dioxide were released back into Earth's present-day atmosphere, its new composition would be 98 percent carbon dioxide and 2 percent nitrogen, and it would have a pressure about 70 times its current value. In other words, apart from the presence of oxygen (which appeared on Earth only after the development of life) and water (whose absence on Venus will be explained below), Earth's atmosphere would look a lot like Venus's. The real difference between Earth and Venus is that Venus's greenhouse gases never left the atmosphere.

When Venus's secondary atmosphere appeared, the temperature was higher than on Earth, simply because Venus is closer to the Sun. However, the Sun was probably somewhat dimmer then (see Chapter 22)—perhaps only half its present brightness—so there is some uncertainty as to exactly how much hotter than Earth Venus actually was. If the temperature was already so high that no oceans condensed, the outgassed water vapor and carbon dioxide would have remained in the atmosphere, and the full greenhouse effect would have gone into operation immediately. If oceans did form and most of the greenhouse gases left the atmosphere, the temperature must still have been sufficiently high that a process known as the **runaway greenhouse effect** came into play.

To understand the runaway greenhouse effect, imagine that we took Earth from its present orbit and placed it in Venus's orbit. At its new distance from the Sun, the amount of sunlight hitting the Earth's surface would be almost twice its present level, so the planet would warm up. More water would evaporate from the oceans, leading to an increase in atmospheric water vapor. At the same time more carbon dioxide would escape from the oceans and surface rocks to the atmosphere. This would increase the greenhouse heating, so the planet would warm still further, leading to a further increase in atmospheric greenhouse gasses, and so on. Once started, then, the process would "run away" and lead to the complete evaporation of all the Earth's oceans, in effect restoring all the original greenhouse gases to the atmosphere. Although the details are quite complex, basically the same thing would have happened on Venus long ago.

Venus's atmosphere never lost its greenhouse gases to the surface. The greenhouse effect on Venus was even more extreme in the past, when the atmosphere contained water vapor. By intensifying the blanketing effect of the carbon dioxide, the water vapor helped the surface of Venus reach temperatures perhaps twice as hot as at present. At those high temperatures, the water vapor was able to rise high into the planet's upper atmosphere—so high, in fact, that it was broken up by solar ultraviolet radiation into its components, hydrogen and oxygen. The light hydrogen rapidly escaped, and the reactive oxygen quickly combined with other atmospheric constituents. In this way, essentially *all* of the water in Venus's atmosphere and surface layers was lost forever.

Although it is highly unlikely that global warming will ever send Earth down the path taken by Venus, this episode highlights the relative fragility of the planetary environment. No one knows how close to the Sun Earth could have formed before a runaway greenhouse effect would have occurred. But in comparing our planet with Venus, we have come to understand that there is an orbital limit, presumably between 0.7 and 1.0 A.U., inside of which Earth would have suffered a similar catastrophic runaway. We must consider this "greenhouse limit" when assessing the likelihood that planets harboring life formed elsewhere in the Galaxy.

THE SURFACE OF VENUS

✓5 Although the clouds are extremely thick and the surface totally shrouded, we are by no means ignorant of Venus's topography. As mentioned earlier, radar astronomers have bombarded the planet with radio signals, both from Earth and from the *Pioneer, Venera,* and, most recently,

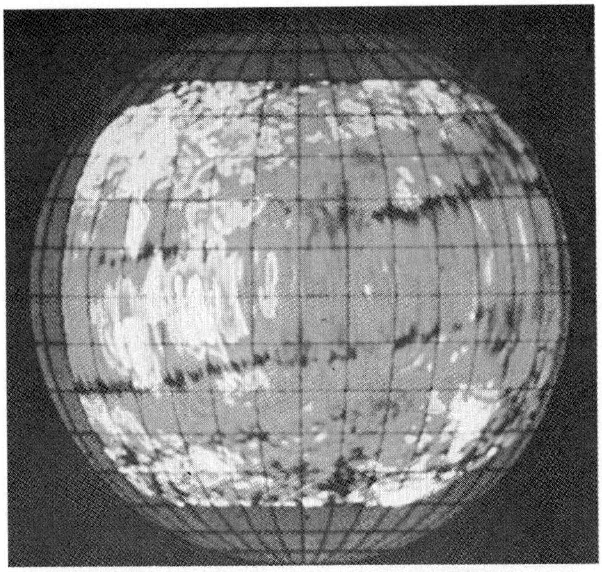

(a) R I V U X G

(b) R I V U X G

(c) R I V U X G

Figure 11.13 *(a) This image of the surface of Venus was made by transmitting radar signals and detecting the returned echos with a large radio telescope on Earth. White represents the highest altitudes, blue the lowest. (b) The vertical resolution of the radar image of Venus was improved dramatically using a transmitter and receiver onboard the* Pioneer *spacecraft, still in orbit about the planet. The two continent-sized landmasses are named Ishtar Terra (upper left) and Aphrodite (lower right). The spatial resolution is about 25 km. (c) At even higher resolution, more surface features are detected. Here the resolution of the Soviet* Venera 15 *orbiter (previously the best available, about 1–2 km) is compared with that of* Magellan *(which can resolve features almost 20 times smaller). The view shows an impact crater named Golubkina, about 34 km in diameter. The lower right side is the* Magellan *image, the upper left came from* Venera. *This was one of the first images radioed back to Earth after* Magellan *became operational in 1990. From the start,* Magellan*'s vastly superior resolution was clearly evident. (NAIC; NASA; Russian Space Agency/NASA)*

Magellan spacecraft. Analysis of the radar echoes yields a map of the planet's surface. As Figure 11.13 illustrates, the early maps of Venus suffered from poor resolution; however, the more recent probes, especially *Magellan,* have provided much sharper views.

Large-Scale Topography

Figure 11.14(a) shows basically the same *Pioneer* data of Venus as Figure 11.13(b), except that it has been flattened out into a more conventional map. The altitude of the surface above the average radius of the planet's surface is indicated by the use of color, with red representing the highest elevations, blue the lowest. For comparison, Figure 11.14(b) shows a map of Earth to the same scale and at the same spatial resolution. Figure 11.14(c) is the same as Figure 11.14(a), except that some of the main features on the planet have been labeled. The planet's surface appears to be mostly smooth, perhaps resembling rolling plains with modest highlands and lowlands. Only two or three continental-sized features adorn the landscape, and these contain mountains comparable in height to those on Earth. The highest peaks rise some 14 km above the level of the deepest surface depressions. For comparison, the highest point on Earth (the summit of Mount Everest), lies about 20 km

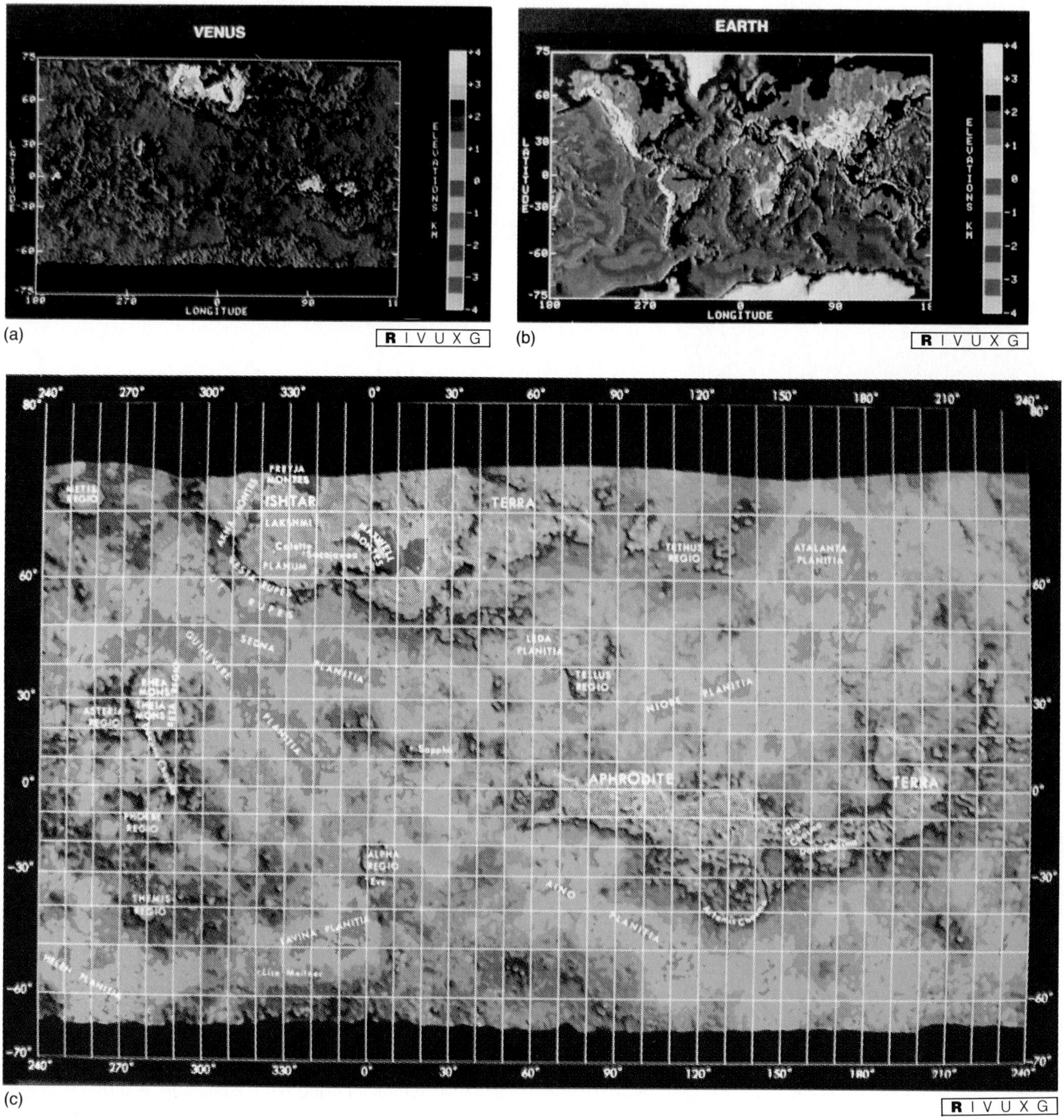

Figure 11.14 *(a) Radar map of the surface of Venus, based on the* Pioneer *Venus data. Color represents elevation according to the scale on the right. (b) A similar map of Earth, at the same spatial resolution. (c) Another version of (a), with the major surface features labeled. (NASA)*

above the deepest section of the ocean floor (Challenger Deep, at the bottom of the Marianas trench on the eastern edge of the Philippines plate).

The *Pioneer* maps allow us to discern the largest-scale features of the surface of Venus. The planet has two elevated continental-sized regions on its surface. In the northern high latitudes, we find an extensive uplifted plateau, called Ishtar Terra (or "Land of Ishtar," after the Babylonian counterpart of the Roman Venus and the Greek Aphrodite). The projection of the map makes Ishtar Terra appear much larger than it really is—it is actually about the same size as Australia. This landmass is dominated by a great plateau known as Lakshmi Planum (see Figure 11.15), which is some 1500 km across at its widest point. This plain is ringed by mountain ranges that include the highest peak on the planet, Maxwell Mons, which reaches an altitude higher than Mount Everest above sea level on Earth. The region also houses a great crater, Cleopatra, about 100 km across. Scientists once thought that Cleopatra was volcanic in origin, but *Magellan* images now indicate that it is actually the result of a meteoritic impact.

Figure 11.15(a) shows a large-scale *Venera* image of Lakshmi Planum, at a resolution of about 2 km. The "wrinkles" are actually chains of mountains, hundreds of kilometers long and tens of kilometers apart. The Maxwell range and the crater Cleopatra appear in the right-hand side of the image. Figure 11.15(b) shows a *Magellan* image of Cleopatra itself. As with all the *Magellan* images, the light areas in Figure 11.15(b) represent regions where the surface is rough and efficiently scatters *Magellan*'s sideways-looking radar beam back to the detector. Smooth areas tend to reflect the beam off into space instead and so appear dark. While close-up views of the crater's structure have led planetary scientists to conclude that it is meteoritic in origin,

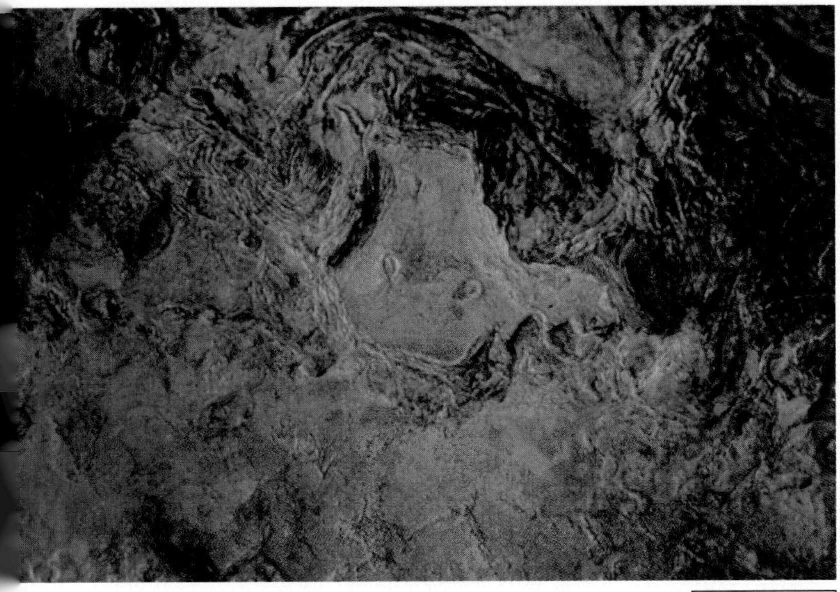

(a) **R** I V U X G

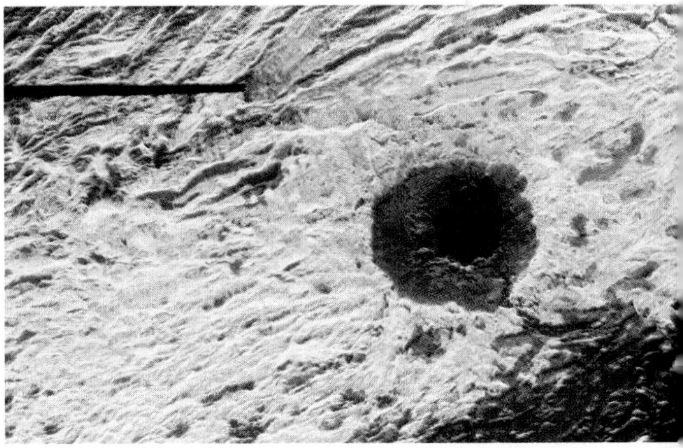

(b) **R** I V U X G

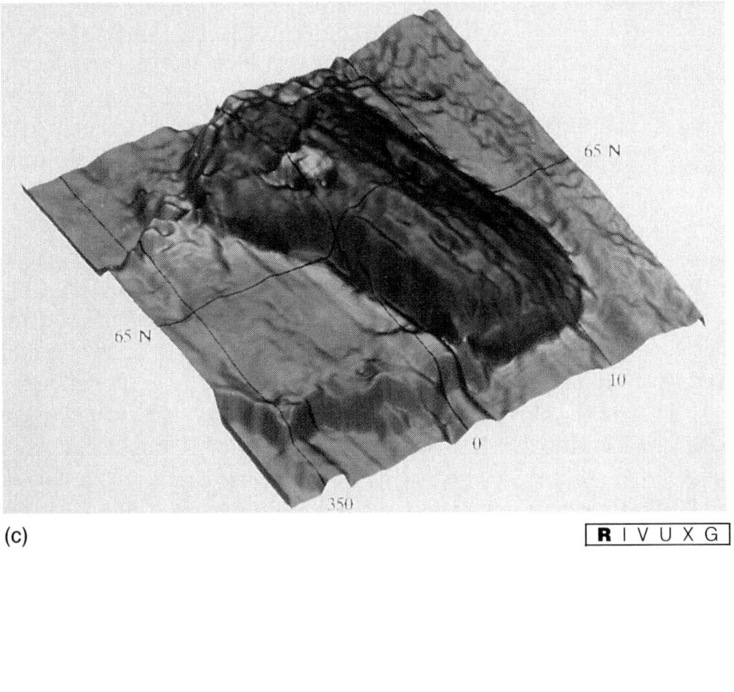

(c) **R** I V U X G

Figure 11.15 *(a) A* Venera *orbiter image of a plateau in Ishtar Terra known as Lakshmi Planum. Maxwell Mons lies on the eastern edge of the plain, near the right-hand edge of the image. The crater Cleopatra is visible at the extreme right, on the eastern slope of the Maxwell mountain range. Note the two other large craters in the center of the plain itself. (b) A* Magellan *image of Cleopatra showing a double-ringed structure that identifies it to geologists as an impact crater. (c) A three-dimensional representation of the Maxwell region, based on* Magellan *data. The height of the mountains is greatly exaggerated, and the colors provide a crude measure of the chemical makeup of the surface. The crater Cleopatra can be seen as the depression on the northeastern slope of the mountain. (Russian Space Agency; NASA)*

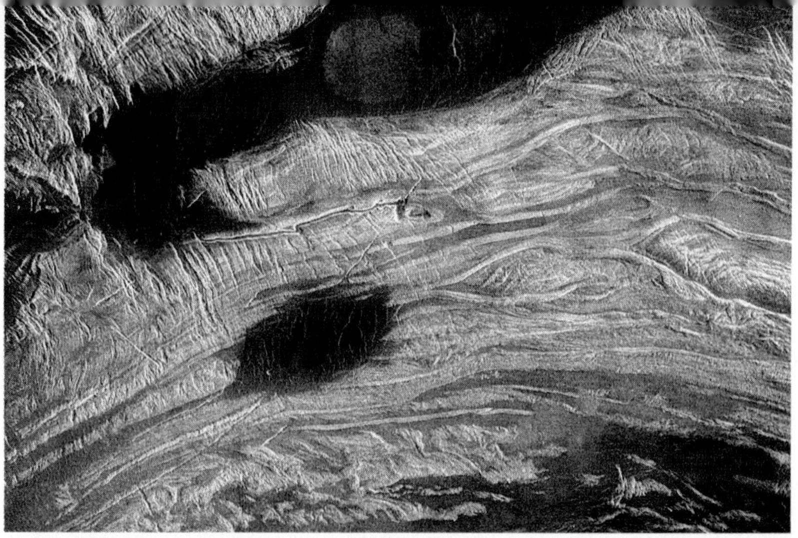

(a)

R I V U X G

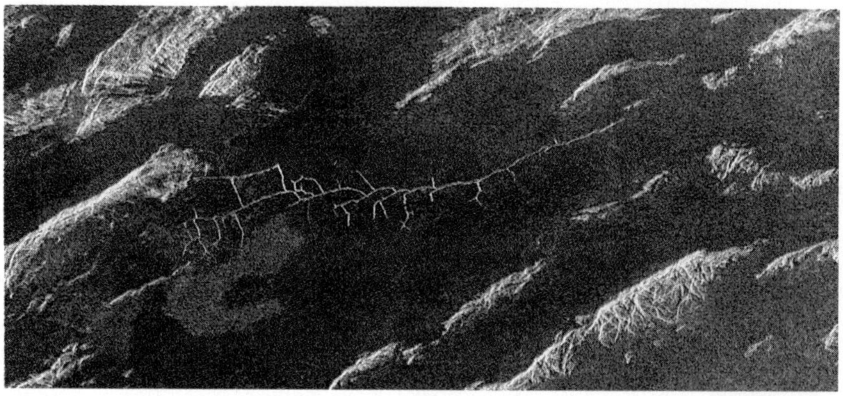

(b)

R I V U X G

Figure 11.16 *(a) A* Magellan *image of Ovda Regio in Aphrodite Terra. The intersecting ridges indicate repeated compression and buckling of the surface. The dark areas represent regions that have been flooded by lava upwelling from cracks like those shown in (b), detected by* Magellan *in another part of Aphrodite Terra. This network of fissures is about 50 km long. (NASA)*

there was some volcanism associated with its formation. Notice the dark lava flow emerging from within the inner ring and cutting across the outer rim at the upper right. Figure 11.15(c) shows a computer-generated three-dimensional view of the Maxwell range. The colors represent a radar property of the surface called its emissivity. Measuring this quantity allows scientists to obtain some information about the chemical makeup of Venus's surface. The purple and blue regions have the lowest emissivity, which may indicate that Maxwell is abnormally rich in iron-bearing minerals relative to the planetary average (represented by orange).

The other continental-sized formation is called Aphrodite Terra. It is located on the planet's equator, south and east of Ishtar Terra, and is comparable in size to Africa. Before *Magellan*'s arrival, some researchers had speculated that Aphrodite Terra might have been the site of something equivalent to seafloor spreading on Earth—a region where two lithospheric plates moved apart and molten rock rose to the surface in the gap between them, forming an extended ridge. With the low-resolution data then available, the issue could not be settled at the time.

The *Magellan* images now seem to rule out any plate tectonic activity on Venus, and the Aphrodite region shows no signs of spreading. The crust appears buckled and fractured, suggesting large compressive forces, and there seem to have been repeated periods when extensive lava flows occurred. Figure 11.16(a) shows a portion of Aphrodite Terra called Ovda Regio. The ridges running in two distinct directions across the image attest to both the magnitude and the variability of the forces compressing and distorting the crust. The dark (smooth) regions are probably solidified lava flows. Some narrow lava channels, akin to rilles on the Moon, also appear. Figure 11.16(b) shows a series of angular cracks in the crust, which are thought to have formed when lava welled up from a deep fissure, flooded the surrounding area, and then retreated back below the planet's surface. As the molten lava withdrew, the thin new crust of solidified material that had formed on top of it was unable to support its own weight, and the surface collapsed, forming the cracks we now see. Even taking into account the temperature and composition differences between Venus's crust and Earth's, this terrain is not at all what we would expect at a spreading site similar to the Mid-Atlantic Ridge.

Two other highland regions, lying close to the equator of Venus and generally south of Ishtar, bear the names Alpha Regio and Beta Regio. These two mountainous areas, which are both somewhat smaller than Ishtar, were identified in the early radar images of Venus made from Earth and designated by the first two letters of the Greek alphabet (alpha and beta; *regio* just means ''region''). It is now conventional to name features on Venus after famous

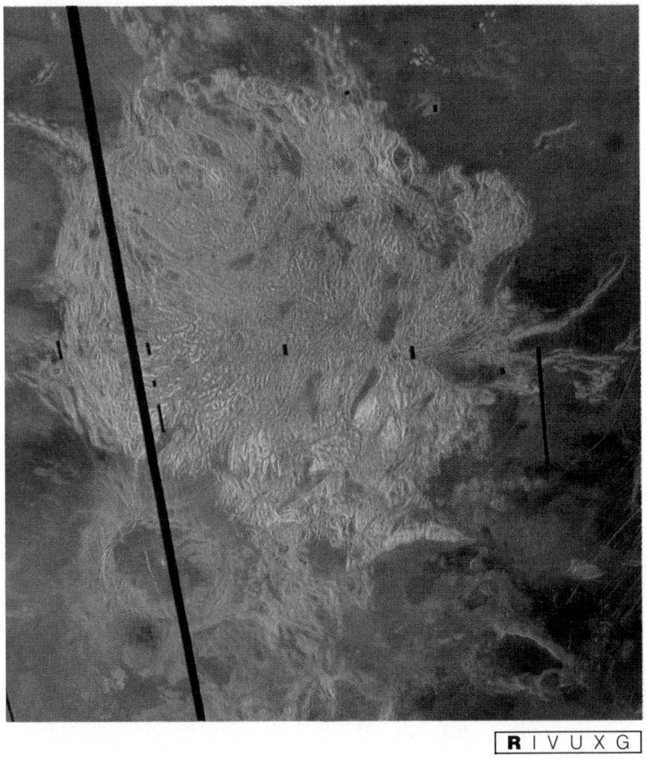

Figure 11.17 Magellan *image of Alpha Regio. The region shown here is about 1300 km across. The image is made up of over 600 separate strips, each about 20 km across, corresponding to consecutive orbits of the spacecraft. The black regions represent periods when data were lost due to technical problems with the spacecraft. The bright dot in the lower left, near the longest black line, is the caldera of the volcano, Eve. (NASA)*

women—Aphrodite, Ishtar, Cleopatra, and so on. However, the early nonfemale names (Maxwell Mons, named after the famous Scottish physicist James Clerk Maxwell, for example) have stuck, and they are unlikely to change.

A large-scale *Magellan* image of Alpha Regio is shown in Figure 11.17. A wide variety of terrains are evident. Alpha Regio itself is the light-colored (that is, rough), somewhat diamond-shaped region in the center of the frame. It consists of mountainous ridges and troughs. Numerous fractures can be seen in the surrounding lava-flooded (dark) lowlands. The large, oval dark region near the southwest edge of Alpha is a volcanic feature known as Eve, which is probably responsible for much of the flooding in the southern portion of the image. Lava flow channels can be seen emerging from its rim. Alpha and Beta Regio are among the many areas of Venus that are now known to have extensive volcanic features. Scientists suspect they are also the sites of currently active volcanos, although no direct evidence has yet been found.

The elevated ''continents'' occupy only 8 percent of Venus's total surface area. (For comparison, continents on Earth make up about 25 percent of the surface.) The remainder of Venus's surface is classified as lowlands (27 percent) or rolling plains (65 percent), although there is probably little geological difference between the two terrains. Although there is no evidence for any large-scale plate tectonic activity on Venus, there is plenty of evidence for volcanism in the recent past. It is likely that the stresses in the crust that led to the large mountain ranges were caused by convective motion within Venus's mantle, the same basic process that drives Earth's plates. Lakshmi Planum, for example, is probably the result of a ''plume'' of upwelling mantle material that raised and buckled the planet's surface.

Volcanism and Cratering

There are numerous craters on Venus's surface. They cannot be seen at the 25-km resolution of the *Pioneer* maps, but more detailed observations by *Venera 15* and *16* and from Arecibo on Earth have revealed their presence. The *Magellan* data allow them to be studied in great detail. Although some craters appear to have arisen from meteoritic impact, most are volcanic in origin.

Figure 11.18 shows an enlargement of part of the southeast portion of Figure 11.17. A series of seven pancake-shaped lava domes, each about 25 km across, can be

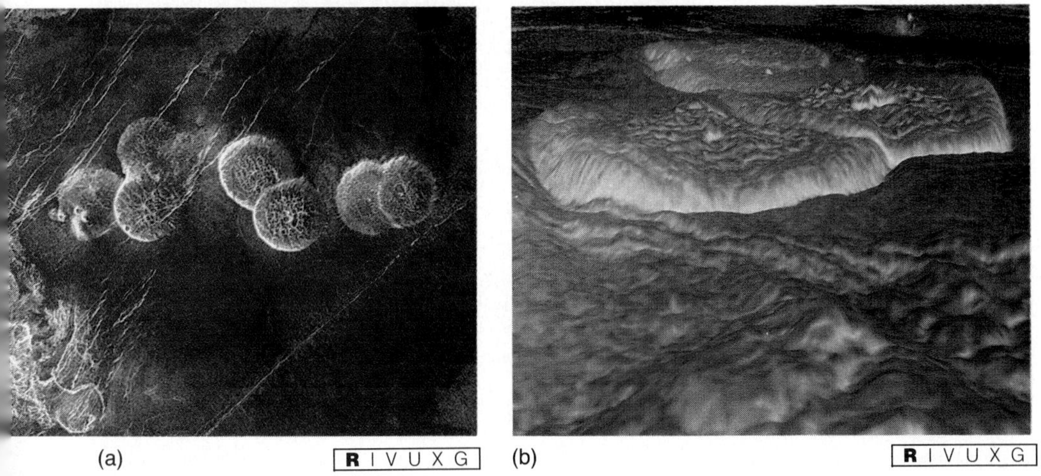

(a) (b)

Figure 11.18 *(a) A series of dome-shaped structures on the southeast edge of Alpha Regio. They are the result of viscous molten rock bulging out of the ground and then retreating, leaving behind a thin solid crust that subsequently cracked and subsided.* Magellan *has found features like this in several locations on Venus. (b) A 3-D representation of four of the domes. The computer view is to the right, near the center of the image in part (a). Color is based on data returned by Soviet landers. (NASA)* seen. They probably formed when lava oozed out of the surface, formed the dome, and then withdrew, leaving the crust to crack and subside. Lava domes such as these are found in several locations on Venus, but the most common volcanos on the planet are of the type known as **shield volcanos.** Two large shield volcanos, called Sif Mons and Gula Mons, are shown (in false color) in Figure 11.19. Shield volcanos on Earth are associated with lava welling up through a "hot spot" in the crust (like the Hawaiian Islands on Earth). They are built up over long periods of time by successive eruptions and lava flows. A characteristic of shield volcanos is the formation of a **caldera,** or crater, at the summit when the underlying lava withdraws and the surface collapses.

The largest volcanic structures on Venus are huge, roughly circular regions known as **coronae.** A large corona, called Heng-O, can be seen in Figure 11.20, another large-scale mosaic of *Magellan* images. The volcanic mountains Sif Mons and Gula Mons can also be seen in the figure, at the top left. Another large volcano, called Sappho, appears at the center right. Coronae are unique to Venus. They appear to have been caused by upwelling mantle material, similar to the uplift that resulted in Lakshmi Planum, but on a somewhat smaller scale. They generally have volcanos both in and around them, and closer inspection of the rims usually shows evidence for extensive lava flows into the plains below.

Not all of the craters found on Venus are volcanic in origin. Some, like the crater Cleopatra on the slopes of Maxwell Mons, were formed by meteoritic impact instead. The largest impact craters in Venus are generally circular, but those less than about 15 km in diameter can be quite asymmetric in appearance. Figure 11.21 shows a *Magellan* image of a relatively small meteoritic impact crater, about 10 km across, in Venus's southern hemisphere. Geologists believe the light-colored region to be the ejecta blanket—material ejected from the crater following the impact. The odd shape may be the result of a large meteoroid that broke up just before impact into pieces that hit the surface near one another. This seems to be a fairly common fate for medium-sized bodies (1 km or so in diameter) plowing through Venus's dense atmosphere. Numerous impact craters, again identifiable by their ejecta blankets, can also be discerned in Figure 11.20.

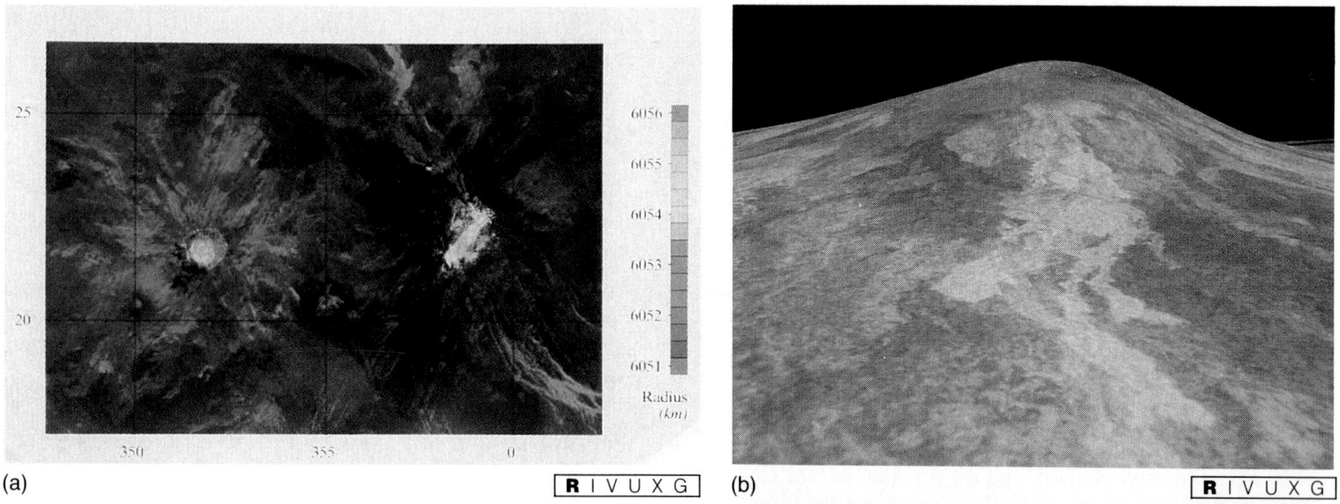

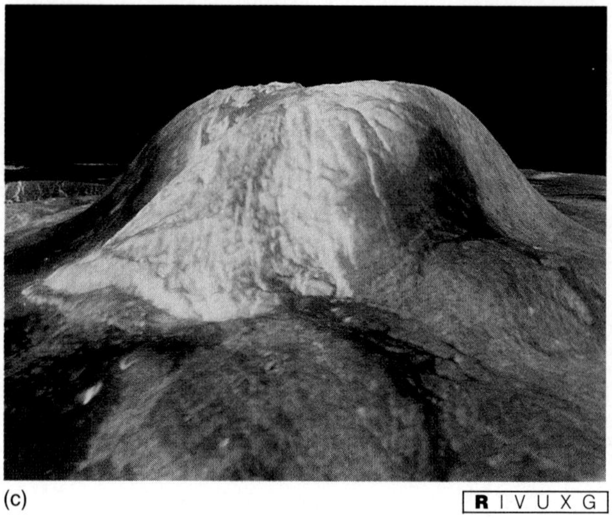

Figure 11.19 *(a) Two larger volcanos, known as Sif Mons (left) and Gula Mons appear in this* Magellan *image of a region east of Beta Regio. Color indicates height, ranging from purple (the level of the surrounding plain) to orange (corresponding to an altitude of about 4 km). The two volcanic calderas at the summits are about 100 km across. (b) A computer-generated view of Sif Mons, as seen from ground level. (c) Gula Mons, as seen from ground level. In (b) and (c), the colors are based on data returned from Soviet landers and the vertical scales have been exaggerated. (NASA)*

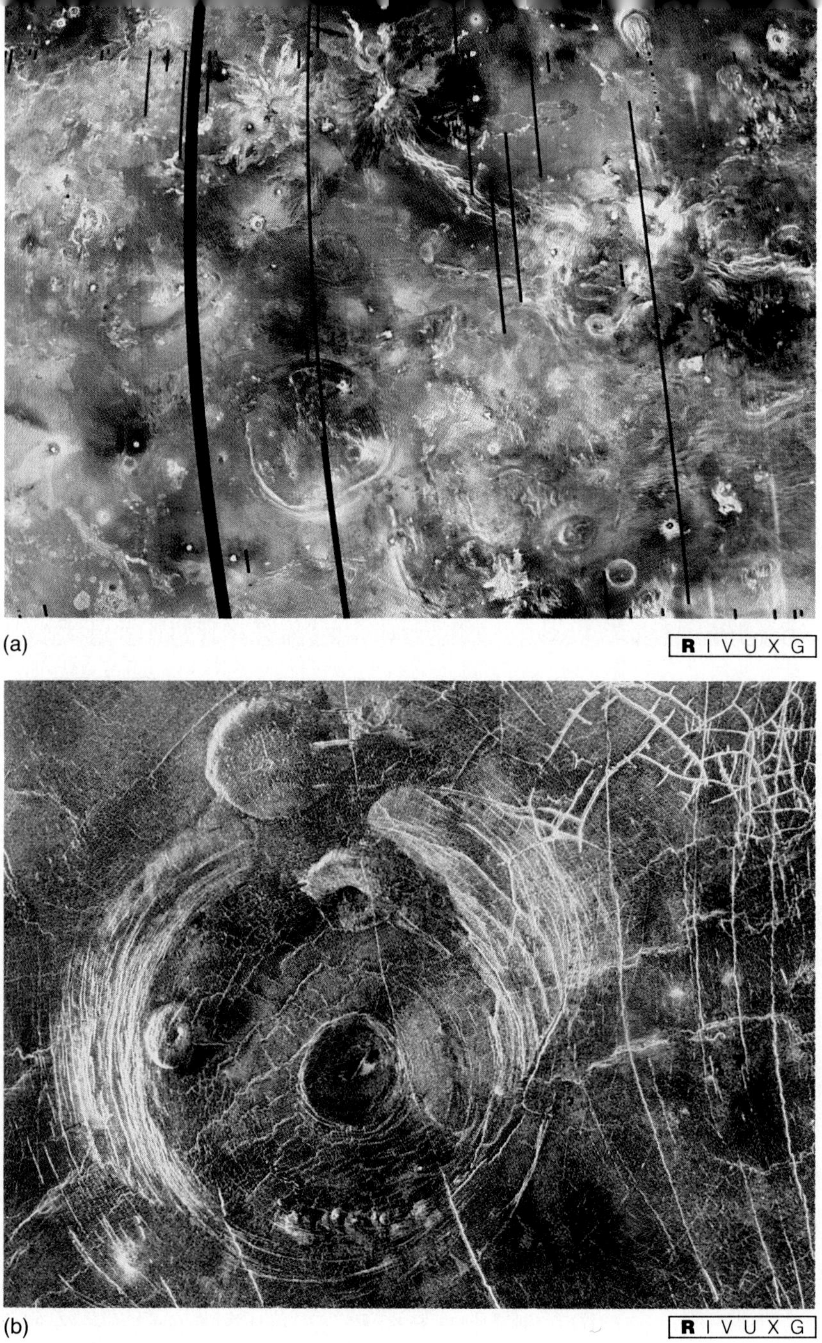

Figure 11.20 (a) *This image shows a region of Venus about 3500 km across—roughly the area of North America. The large circular corona in the lower left is probably the result of mantle material causing the surface to bulge outward. It is larger than the state of California. As usual, the dark regions represent smooth lava plains and the sharp black lines are portions of* Magellan's *orbit where data was lost. Note the light-colored fault line extending between the volcanos Gula Mons and Sappho, the fractures in the crust, and the many large impact craters with their surrounding white (rough) ejecta blankets that stud the region.* (b) *Another corona, called Aine, lies in the plains south of Aphrodite Terra. It is about 300 km across. (NASA)*

Venus's atmosphere is sufficiently thick that small meteorites do not reach the ground, so there are no impact craters smaller than about 3 km across. Atmospheric effects probably also account for the observed deficiency in impact craters less than 25 km in diameter. Overall, the rate of large-diameter craters on Venus's surface seems to be only about 1/10 that in the lunar maria. Applying the same crater-age estimates to Venus as we do to Earth and the Moon suggests that much of the surface of Venus is quite young—around a billion years old. Some planetary scientists have suggested that some areas, such as the region shown in Figure 11.19 are even younger—perhaps as little as 200 or 300 million years. Although erosion by the planet's atmosphere may play some part in obliterating surface features, the main agent is volcanism, which appears to "resurface" the planet every few hundred million years.

Evidence for Ongoing Volcanism

There is now overwhelming evidence for past surface activity on Venus. Has this activity now stopped, or is it still going on? Two pieces of indirect evidence suggest that volcanism continues. First, the level of sulfur dioxide above Venus's clouds shows large and fairly frequent fluctuations. It is quite possible that these variations result from volcanic eruptions on the surface. If so, volcanism may be the primary cause of Venus's thick cloud cover. Second, both the *Pioneer* and the *Venera* orbiters observed bursts of radio energy from the Beta and Aphrodite regions. These bursts are similar to those produced by lightning discharges that often occur in the plumes of erupting volcanos on Earth, again suggesting ongoing activity.

These pieces of evidence may be persuasive, but they

Figure 11.21 *(a) A* Magellan *image of an impact crater in Venus's southern hemisphere. The peculiar kidney shape seems to be the result of a meteorite that fragmented just prior to impact. The dark regions in the crater may be pools of solidified lava. Some more regular looking craters, and their three-dimensional representations, are shown in (b) and (c). (NASA)*

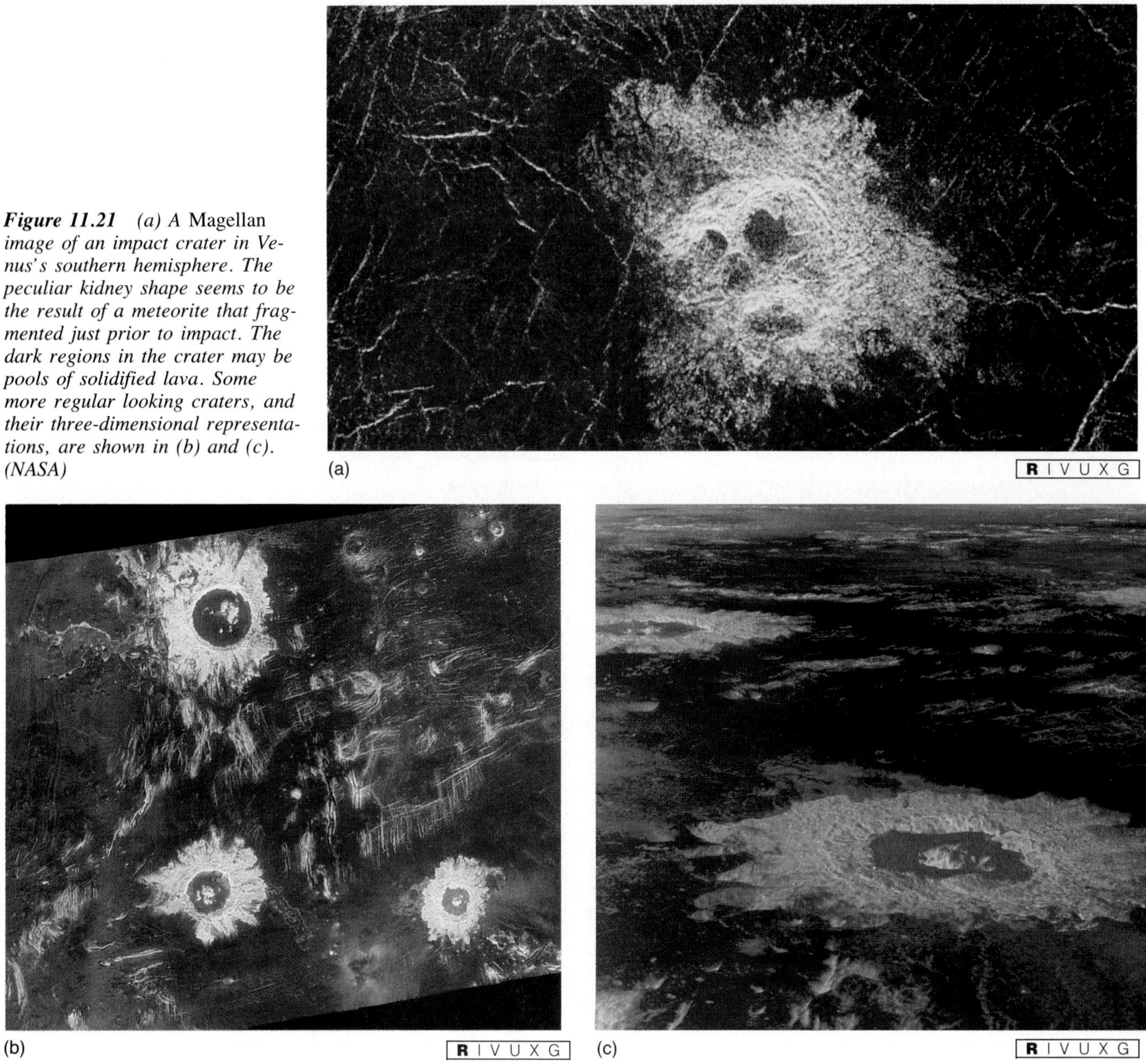

are still only circumstantial. No "smoking gun" (or erupting volcano) has yet been seen, so the case for active volcanism is not yet complete. Perhaps *Magellan* will detect an eruption in progress and settle the issue once and for all.

Data from the Soviet Landers

The 1975 soft-landings of the Soviet *Venera 9* and *Venera 10* spacecraft established directly that Venus's surface is dry and dusty. Figure 11.22 shows two of the first photographs of the surface of Venus radioed back to Earth not long before the spacecrafts' demise. Both crafts lasted about an hour before overheating, their electronic circuitry literally melting in this planetary oven. Typical rocks in the photo measure about 50 cm by 20 cm across—a little like flagstones on Earth. Having sharp edges and a slablike character, they show little evidence of erosion. Apparently they are quite young rocks, again supporting the idea of ongoing surface activity of some kind.

Later *Venera* missions took more detailed photographs, as shown in Figure 11.23. The presence of small rocks and finer material indicates the effects of erosive processes. These later missions also performed simple chemical analyses of the surface of Venus. The samples studied by *Venera 13* and *Venera 14* were predominantly basaltic in nature, again implying a volcanic past. However, not all of the rocks were found to be volcanic in nature. The *Venera 17* and *Venera 18* landers also found surface material resembling terrestrial granite, probably (as on Earth) part of the planet's ancient crust.

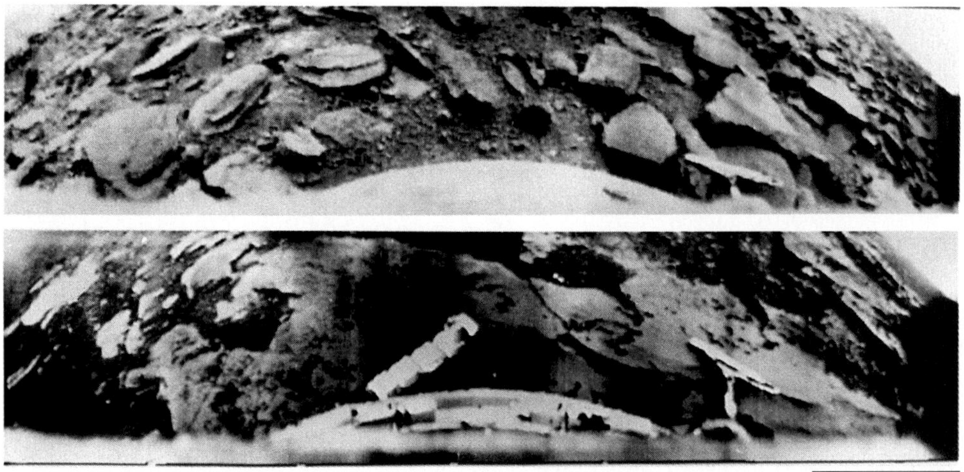

Figure 11.22 *The first direct views of the surface of Venus, radioed back to Earth from the Russian* Venera 9 *and* Venera 10 *spacecraft, which made soft-landings in 1975 in Beta Regio. The amount of sunlight penetrating Venus's cloud cover apparently resembles that on a heavily overcast day on Earth. (Russian Space Agency)*

Figure 11.23 *Another view of Venus, this time from* Venera 14. *Flat rocks like those visible in Figure 11.22 are present, but there are also many smaller rocks and even fine soil on the surface. These craft also landed in Beta Regio, not far from the earlier landing sites. The peculiar filtering effects of whatever light does penetrate the clouds make Venus's air and ground peach-colored. (Russian Space Agency)*

VENUS'S MAGNETIC FIELD AND INTERNAL STRUCTURE

The first probe to reach Venus was *Mariner 2*. In 1962 it flew by the planet, carrying, among other instruments, magnetometers to measure the strength of the planet's venerian magnetic field. None was detected, and subsequent Soviet and U.S. missions, carrying more sensitive detectors, have confirmed this finding. Venus, with an average density similar to Earth's, probably has a similar overall composition and a partially molten iron-rich core. The lack of any detectable magnetic field on Venus, then, is almost surely the result of the planet's extremely slow rotation and consequent lack of dynamo action (see Chapter 7).

Having no magnetosphere, Venus has no protection from the solar wind. Its upper atmosphere is continually bombarded by high-energy particles from the Sun, keeping the topmost layers permanently ionized. However, the great thickness of the atmosphere prevents any of these particles from reaching the surface.

None of the *Venera* landers carried seismic equipment, so no direct measurements of the planet's interior have been made, and theoretical models of the interior have very little hard data to constrain them. However, to many geologists the surface of Venus resembles the young Earth, at an age of perhaps a billion years. At that time, volcanic activity had already begun, but the crust was still relatively thin and the convective processes in the mantle that drive plate tectonic motion were not yet established.

Why has Venus remained in that immature state and not developed plate tectonics like the Earth? That question remains to be answered. Some planetary geologists have speculated that the high surface temperature on Venus has inhibited the planet's evolution by slowing the planet's cooling. Possibly the high surface temperature has made the crust too soft for Earth-style plates to develop. Or perhaps the high temperature and soft crust led to more volcanism, tapping the energy that might otherwise go into convective motion. It may also be that the presence of water plays an important role in lubricating mantle convection and plate motion, so that arid Venus could not evolve along the same path as Earth. Without more detailed data, it is difficult to say. It is hoped that the detailed gravity maps of Venus to be constructed from in-depth studies of *Magellan*'s orbit will shed more light on the planet's internal structure.

As further data are gathered, astronomers will eagerly compare the interior of Venus with that of Earth. The two planets have nearly equal masses and radii but very different environmental conditions. In determining how and why these two near-identical twins diverged in later life, we will surely achieve a much more comprehensive understanding of planetary physics.

CHAPTER SUMMARY

Venus is the second planet from the Sun. Its sidereal orbital period is 225 days. Because its orbit lies inward from that of Earth, Venus is always seen near the Sun—either in the west after sunset or in the east before sunrise. It is one of the brightest objects in Earth's sky—so bright it can be seen even in the daytime.

The extremely thick atmosphere of Venus is nearly opaque to visible radiation, making the planet's surface invisible from the outside. Ninety percent of Venus's atmosphere lies within about 50 km of the surface. On Earth, by comparison, 90 percent of the atmosphere lies within 10 km of the surface. Spectroscopic examination of sunlight reflected from the planet's cloudtops shows the presence of large amounts of carbon dioxide. The temperature of the upper atmosphere is much like that of Earth's upper atmosphere, but the surface temperature is a sizzling 750 K.

Some 20 spacecraft have visited Venus in the last two decades. Soviet spacecraft took the lead role in exploring the planet's atmosphere and surface, while American spacecraft performed extensive radar mapping of the planet's surface from orbit. The most recent U.S. mission is the *Magellan* spacecraft. Its radar images of Venus are at least 10 times better than the best images previously obtained—it can distinguish objects as small as 120 m across.

The evidence for currently active volcanos on Venus includes surface features resembling those produced in earthly volcanism, fluctuating levels of sulfur dioxide in Venus's atmosphere, and bursts of radio energy similar to those produced by lightning discharges that often occur in the plumes of erupting volcanos on Earth. However, no actual eruptions have been seen. Soviet spacecraft that landed on Venus photographed surface rocks with sharp edges and a slablike character. Some rocks on Venus appear predominantly basaltic in nature, implying a volcanic past. Other rocks resemble terrestrial granite and are probably part of the planet's ancient crust.

The lack of any detectable magnetic field on Venus is almost surely the result of the planet's extremely slow rotation and consequent lack of dynamo action. Many geologists think the surface of Venus resembles that of a young Earth. It appears that volcanic activity has begun on Venus but that the crust is still relatively thin. Convective processes in the mantle that drive tectonic motion may not yet be established on Venus.

The total mass of Venus's atmosphere is about 90 times greater than Earth's. The atmosphere of Venus is mostly carbon dioxide, which accounts for 96.5 percent of the atmosphere by volume. Almost all of the remaining 3.5 percent is nitrogen. Trace amounts of other gases, such as water vapor, carbon monoxide, sulphur dioxide, and argon, are also found. The carbon dioxide in the atmosphere of Venus absorbs 99 percent of all the infrared radiation released from the surface of the planet, preventing the radiation from escaping into space. This is the immediate cause of the planet's sweltering surface temperature. Almost all of the water vapor and carbon dioxide present in Earth's early history quickly became part of Earth's oceans or surface rocks. Because Venus orbits closer to the Sun than Earth, and surface temperatures were higher to begin with, its greenhouse gases never left the atmosphere. The result was a runaway greenhouse effect and the extremely hostile conditions found on Venus today.

We know about the surface of Venus thanks to the work of radar astronomers and to U.S. and Soviet spacecraft, especially the *Magellan* probe. The planet's surface is mostly smooth, perhaps resembling rolling plains with modest highlands and lowlands. Two elevated continental-sized regions are called Ishtar Terra and Aphrodite Terra. Some craters on Venus are due to meteoritic impact, but many others are almost certainly volcanic in origin. Most of Venus's volcanos are shield volcanos, associated with lava welling up through "hot spots" in the planet's crust.

KEY WORDS

caldera
coronae
inferior conjunction
runaway greenhouse effect
shield volcano
superior conjunction

REVIEW QUESTIONS

1. Why did early Greek astronomers believe that Venus was two separate objects?
2. Why does Venus appear so bright to the eye? On what two factors does the brightness of Venus depend?
3. Without spacecraft data, how would we know the radius of Venus?
4. Even if Venus were found to be locked in an exact orbital resonance with Earth, what further mystery would remain to be explained?
5. How did radio observations of Venus made in the 1950s change our conception of the planet?
6. What country has taken the lead role in exploring Venus via spacecraft? How long do spacecraft that land on Venus typically survive?
7. What two features of the Venus's atmosphere make Venus extremely hostile to earthly life?
8. How did *Pioneer Venus* increase our knowledge of the planet?
9. Why has *Magellan* revealed so much more than *Pioneer*?
10. Name some ways in which the atmosphere of Venus is different from that of Earth.
11. What do you think a person descending in a spacecraft through the atmosphere of Venus would see? Explain.
12. What are the constituents of Venus's atmosphere? What are clouds in the upper atmosphere made of?
13. What characteristics of the Venus's atmosphere cause Venus to be so hot?
14. Assuming that Venus never had an ocean, explain why there is so much carbon dioxide in the planet's atmosphere.
15. What is the runaway greenhouse effect, and how might it have altered the climate of Venus?
16. How are the ''continents'' of Venus different from earthly continents?
17. Are there craters on Venus? If so, what are they like?
18. What is the evidence for active volcanos on Venus?
19. Given that Venus probably has a partially molten iron-rich core, why doesn't it have a magnetic field?
20. Earth and Venus are nearly alike in size and density. What primary fact caused one planet to evolve as an oasis for life, while the other became a dry and inhospitable inferno?
21. Approximating Venus's atmosphere as a layer of gas 50 km thick, with uniform density 2.1×10^{-2} g/cm^3, calculate its total mass. Compare it with the mass of Earth's atmosphere and with the mass of Venus.
22. According to Stefan's Law (see Chapter 4), how much more radiation—by each square centimeter—is emitted at Venus's surface at 750 K than is emitted at the Earth's surface at 300 K?
23. Given that *HST* orbits 400 km above Earth's surface with a period of 95 minutes, what is the orbit period of the *Magellan* spacecraft if it orbits 500 km above the surface of Venus?

DISCUSSION QUESTIONS

1. Explain why Venus is always found in the same region of the sky as the Sun.
2. If you were standing on Venus, how would Earth look?
3. How might an orbiting spacecraft with sophisticated imaging abilities, such as *Magellan,* be used to discover active volcanos on Venus?
4. If Venus had formed at Earth's distance from the Sun, what do you imagine its climate would be like today? Why do you think so?
5. Might there be life on Venus? Explain your answer.

PROJECTS

1. Is Venus in the morning or evening sky right now? Look for it every few days, over the course of several weeks. Draw a picture of the planet with respect to foreground trees or buildings. If you always observe at the same time every day, you might begin to notice that the planet is getting higher or lower in the sky.
2. Consult an almanac to determine the next time Venus will pass between the Earth and Sun. How many days before and after this event can you glimpse the planet with the eye alone?
3. Consult the almanac again to find out the next time Venus will pass on the far side of the Sun from Earth. How many days before and after *this* event can you glimpse the planet with the eye?
4. When Venus ornaments the predawn sky, try keeping track of the planet with your eye alone until it appears in a blue sky, after sunrise. As always, be careful not to look at the Sun!

SUGGESTED READINGS

Allen, D. A. "Laying Bare Venus' Dark Secrets." *Sky and Telescope* (October 1987). A cloud layer deep within the atmosphere of Venus, revealed by infrared measurements.

Brazilevskiy, A. T. "The Planet Next Door." *Sky and Telescope* (April 1989). Comprehensive article on what was known about Venus shortly before the *Magellan* mission.

Beatty, J. K. "*Magellan* at Venus: First Results." *Sky and Telescope* (December 1990). Early results from a space mission to obtain sophisticated radar images of the surface of Venus.

———. "A New Look at Old Worlds." *Sky and Telescope* (March 1991). What the *Galileo* spacecraft found on its detour through the inner solar system.

Cole, S. "Rediscovering Venus and Jupiter." *Astronomy* (January 1989). On the *Galileo* and *Magellan* missions to Jupiter and Venus.

Eberhart, J. "Venus: The Waters of Yesteryear." *Science News* (December 12, 1981). Evidence for the possibility that Venus once had an ocean.

Eicher, David J. "Ashen Light on Venus." *Astronomy* (August 1988). Modern attempts to see this mysterious glow on the night hemisphere of Venus.

Hunten, D. M., L. Colin, T. M. Donahue, and V. I. Moroz. *Venus*. Tucson: University of Arizona Press, 1983. Part of the Space Science Series. Sixty-nine authors tell what was learned about Venus by spacecraft launched between 1962 and 1978.

Kaula, W. M. "Venus: A Contrast in Evolution to Earth." *Science* (March 9, 1990). Differences in the atmosphere, origin, evolution, geology, and so on of Earth and Venus.

Lemonick, M. "A Restless Venus Unveiled." *Time* (October 8, 1990). Popular article about the *Magellan*'s look at the hidden surface of Venus.

Powell, C. S. "Venus Revealed." *Scientific American* (January 1992). Striking *Magellan* spacecraft radar images of the cloud-covered planet.

Saunders, R. S. "The Surface of Venus." *Scientific American* (December 1990). A portfolio of vivid radar images sent back from Venus by the *Magellan* spacecraft. The accompanying article is written by the mission's project scientist.

CHAPTER

12

Mars

A Near Miss for Life?

This true-color image of Mars is a composite of individual images obtained through red, green, and blue filters using the Planetary Camera aboard the Hubble Space Telescope. *Because about 12 minutes elapsed between exposing the red and blue image, the planet rotated slightly between frames. This is evident as a slightly bluish rim on the eastern (upper right) limb of Mars and a slightly reddish rim on the western limb. The large, dark, ''shark's fin'' feature that dominates this face of Mars is called Syrtis Major Planitia. Arabia Planitia, the bright region to the west of Syrtis, is thought to be bright because of a thin layer of dust deposited on its surface. To the east the bright region is Isidis Planitia, a 1000-kilometer-wide impact basin formed as a result of a large meteoroid collision more than 2 billion years ago. The planet's north polar cap is at upper left, where it was Martian ''winter'' at the time of the observation. The spatial resolution is 0.2 arc second, which yields surface detail as small as 50 kilometers across. (Courtesy NASA)*

R I **V** U X G

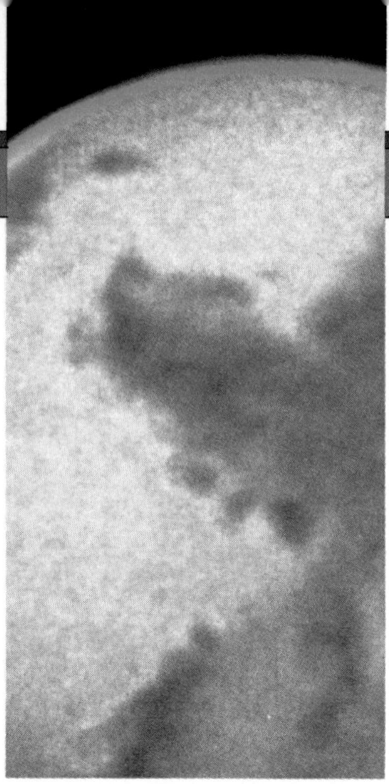

LEARNING GOALS

1 to recognize the existence of seasonal change on another planet

2 to see that the surface of Mars exhibits both cratering like the Moon and evidence of past geology like the Earth

3 to know that Mars has polar caps and once had running water and a thick atmosphere

4 to begin to understand why Mars diverged from Earth early in its evolution

5 to realize that no evidence for life has been found on Mars

As we continue beyond Venus on our outward journey from the Sun, crossing the orbit of the Earth-Moon system, we come to the fourth terrestrial planet—Mars. This red world, named by the ancient Romans for their bloody god of war, is for some people the most intriguing of all celestial objects. Over the years it has inspired much speculation that life—perhaps intelligent, and possibly hostile—may exist there. For much of the first half of the twentieth century, science-fiction writers and the public imagination populated the planet with every imaginable form of life. At the same time, scientists also harbored the belief that life of some sort might indeed be found there.

With the dawn of the Space Age, these notions had to be abandoned. Visits by robot spacecraft have revealed no signs of life of any sort, even at the microbial level, on Mars. Even so, the planet's properties are close enough to Earth's that Mars is still widely regarded as the next most hospitable environment for the appearance of life in the solar system, after Earth itself. At about the same time as Earth's "twin," Venus, was evolving into a searing inferno, the Mars of long ago may have had running water and blue skies. If life ever arose there, however, it must be long extinct. The Mars of today appears to be a dry, dead world.

ORBITAL PROPERTIES OF MARS

Mars is the fourth planet from the Sun and the last of the four terrestrial worlds in the solar system. It lies outside Earth's orbit, as illustrated in Figure 12.1(a), which shows the orbits of both planets drawn to scale. Because of its exterior orbit, Mars ranges in our sky from a position close to the Sun (for example, when Earth and Mars are at the points marked A in the figure) to one far from the Sun (with the planets at points B). Contrast this orbit with the nighttime appearances of Mercury and Venus, whose interior orbits ensure that we never see them far from the Sun. From our Earthly viewpoint, Mars appears to traverse a great circle in the sky, keeping close to the ecliptic and occasionally executing retrograde loops, as we saw in Chapters 2 and 3. Because Mars never comes within Earth's orbit, it can never pass between Earth and the Sun—there can be no Martian transit of the Sun, although a careful Martian astronomer might occasionally glimpse the Earth crossing the solar disk.

Mars revolves around the Sun with an orbital semi-major axis of 1.52 A.U. (228 million km). Its orbital eccentricity is 0.093, much larger than that of most other planets—only the innermost and the outermost planets, Mercury and Pluto, have more elongated orbits. Because of the ellipticity of its orbit, Mars's perihelion distance from the Sun—1.38 A.U. (206 million km)—is substantially smaller than its aphelion distance—1.66 A.U. (249 million km). This difference results in a large variation in the amount of sunlight striking the planet over the course of its year. The intensity of sunlight on the Martian surface is almost 45 percent greater when the planet is at perihelion than when it is at aphelion. As we will see, this has a substantial effect on the Martian climate.

Mars's sidereal orbital period is 687 Earth days. The planet is at its largest and brightest when it is at **opposition**—that is, when the Earth lies between Mars and the Sun (loca-

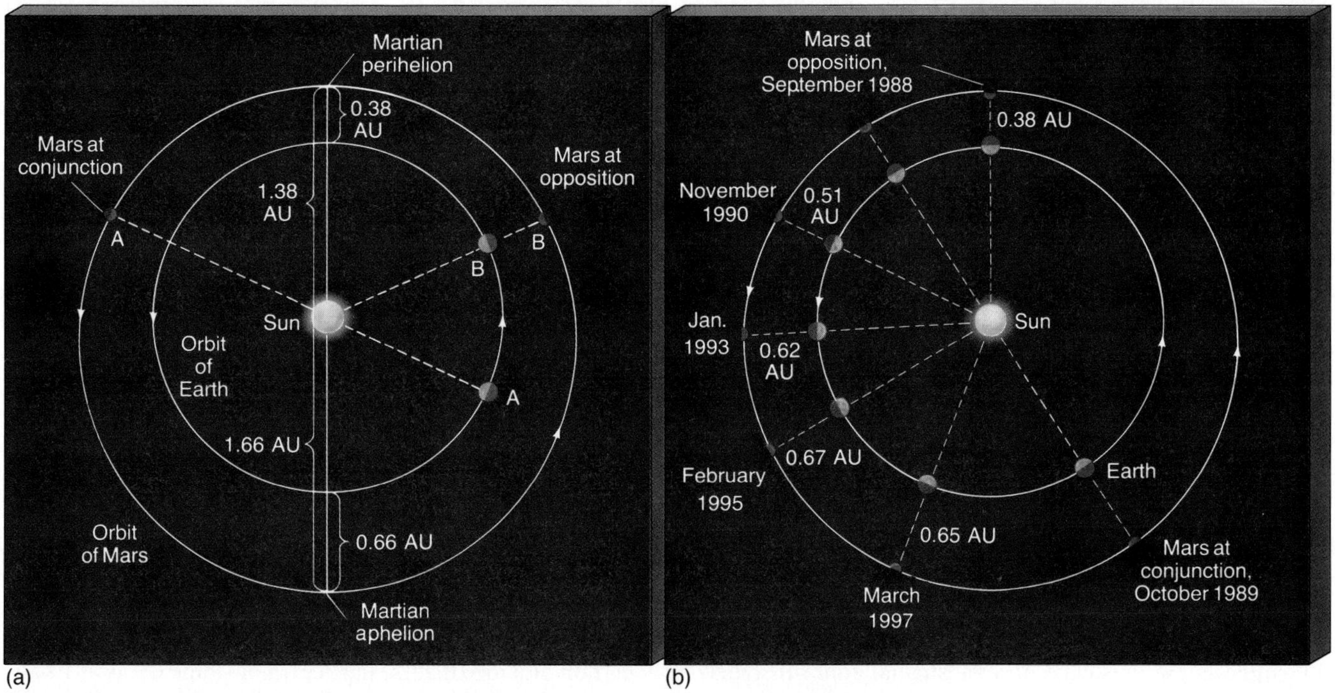

Figure 12.1 *(a) The orbit of Mars compared with that of the Earth. Notice that Mars's orbit is noticeably elliptical, unlike Earth's, whose eccentricity is barely perceptible. When the planets are on opposite sides of the Sun, as at points A, Mars is said to be at conjunction. When Earth and Mars are aligned and on the same side of the Sun, as at points B, Mars is at opposition. (b) Some oppositions of Mars, from the particularly favorable configuration of September 1988 to the unfavorable opposition of March 1997.*

tion B on Figure 12.1). If that configuration happens to occur near a perihelion of Mars, the two planets can come within 0.38 A.U. (56 million km) of each other. Such an opposition is said to be *favorable*. Mars's angular size under those circumstances is about 25″ (arc seconds). Ground-based observations of the planet at those times can distinguish surface features as small as 100 km across—about the same resolution as the unaided human eye can achieve when viewing the Moon. By contrast, when Mars is at **conjunction,** on the opposite side of the Sun from the Earth (configuration A on Figure 12.1), it can be as small as 4″ in diameter. The best observations of Mars, then, are generally made near a favorable opposition.

Using the formula given in Chapter 11, we can easily calculate the time between two successive oppositions (or conjunctions) of Mars. It is 780 days, the synodic period of Mars. The dates and configurations of some oppositions of Mars are illustrated in Figure 12.1(b), which shows the locations of Earth and Mars at five oppositions (at 780-day intervals), starting with the favorable configuration of September 1988 and ending with the unfavorable opposition of March 1997. The locations of the two planets at the October 1989 conjunction, 390 days after the 1988 opposition, are also marked. A favorable opposition is a fairly rare coincidence. It last happened in September 1988 and will next occur in August 2003, although Mars will not be quite so close to perihelion then as it was in 1988.

Although Mars is quite bright and easily seen at opposition, the planet is still considerably fainter than Venus. This faintness results from a combination of three factors. First, Mars is more than twice as far from the Sun as is Venus, so each square centimeter on the Martian surface receives less than one-quarter the amount of light that strikes each square centimeter on Venus. Second, Mars's surface area is only about 30 percent that of Venus, so there are fewer square centimeters to intercept the sunlight. Finally, Mars's albedo is much less than that of Venus—only 0.15, compared with more than 0.7—so each square centimeter that is illuminated by the Sun reflects a far smaller fraction of the light striking it. Still, at its brightest, Mars is brighter than any star.

MARS IN BULK

Radius

As with Mercury and Venus, we can determine Mars's radius from simple geometry. From the data given above for the planet's size and distance, we can calculate a diameter of about 6800 km. More accurate measurements give a result of 6794 km, or 0.53 Earth diameters.

Mass

Unlike Mercury and Venus, Mars has two small moons in orbit around it, which are visible (through telescopes) from Earth. Named Phobos (Fear) and Deimos (Panic) for the horses that drew the Roman war-god's chariot, these moons are little more than large rocks trapped by the planet's gravity. We will return to their individual properties in a moment. The larger of the two, Phobos, orbits at a distance of only 9380 km from the center of the planet once every 459 minutes. Applying the modified version of Kepler's Third Law (which states that the square of a moon's orbital period is proportional to the cube of its orbital semi-major axis divided by the mass of the planet it orbits; see Chapter 3), we find that the mass of Mars is 6.4×10^{26} g, or 0.11 times that of the Earth. Naturally, the orbit of Deimos yields the same result.

Density

From the mass and radius, we find that the average density of Mars is 3.9 g/cm^3, only slightly greater than that of the Moon. If we assume that the Martian surface rocks are similar to those on the other terrestrial planets, this average density suggests the existence of a substantial core of higher-than-average density within the planet. Scientists now believe that this core is composed largely of iron sulfide (a compound about five times denser than surface rock) and has a diameter of about 2500 km.

Rotation

Surface markings easily seen on Mars allow astronomers to track the planet's rotation. Mars rotates once on its axis every 24.6 hours. One Martian day is thus very similar in length to one Earth day. The planet's equator is inclined to the orbit plane at an angle of 25.2° (degrees), again very similar to the Earth's inclination of 23.4°. Thus as Mars orbits the Sun, we find both daily and seasonal cycles, just as on Earth. In the case of Mars, however, the seasons are complicated somewhat by variations in solar heating, due to the planet's eccentric orbit.

EARTH-BASED OBSERVATIONS OF MARS

Surface Features

✓1 At opposition, when Mars is closest to us and most easily observed, it is also full, so the angle of the Sun's rays does not permit us to see any topographical detail, such as craters or mountains. Even through a large telescope Mars appears only as a reddish disk, with some light and dark patches and prominent polar caps. These surface features undergo slow seasonal changes over the course of a Martian year—a consequence of Mars's axial tilt and somewhat eccentric orbit. Figure 12.2 shows some of the best images of Mars ever made from Earth, along with a photograph taken by one of the U.S. *Viking* spacecraft en route to the planet.

Viewed from Earth, the most obvious Martian surface features are its bright polar caps. The caps are mostly frozen carbon dioxide (that is, dry ice), not water ice as at Earth's North and South poles. As shown in Figure 12.3, they grow or diminish according to the seasons, almost disappearing at the time of Martian summer.

The dark surface features on Mars also change from season to season, although their variability probably has little to do with the melting of the polar ice caps. To the more fanciful observers around the start of the twentieth century, these changes suggested the seasonal growth of vegetation on the planet. It was but a small step from seeing polar ice caps and speculating about teeming vegetation to imagining a planet harboring intelligent life, perhaps not unlike us.

But those speculations and imaginings were not to be confirmed. The ice caps contain no water, and the dark markings seen in Figures 12.2 and 12.3, once claimed to be part of a network of "canals" dug by Martians for irrigation purposes (see Interlude 12-1), are actually highly cratered

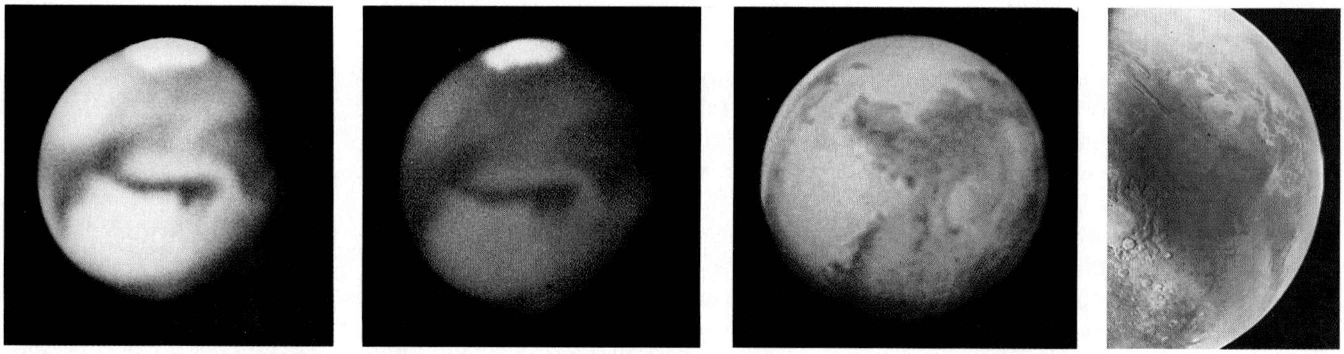

(a) R I **V** U X G (b) R I **V** U X G (c) R I **V** U X G (d) R I **V** U X G

Figure 12.2 *(a) A typical ground-based image of Mars, taken with a professional telescope while the planet was at opposition. One of the planet's polar caps appears at the top. (b) A deep-red (8000 Å) image of Mars, taken in 1991 at Pic du Midi, an exceptionally clear site in the French Alps. (c) A visible-light* Hubble Space Telescope *image of Mars, taken while the planet was near opposition in 1990. (d) A view of Mars taken from a* Viking *spacecraft during its approach in 1976. Some of the planet's surface features are clearly visible at a level of surface detail completely invisible from Earth. (Lick Observatory; Pic du Midi Observatory; NASA; NASA)*

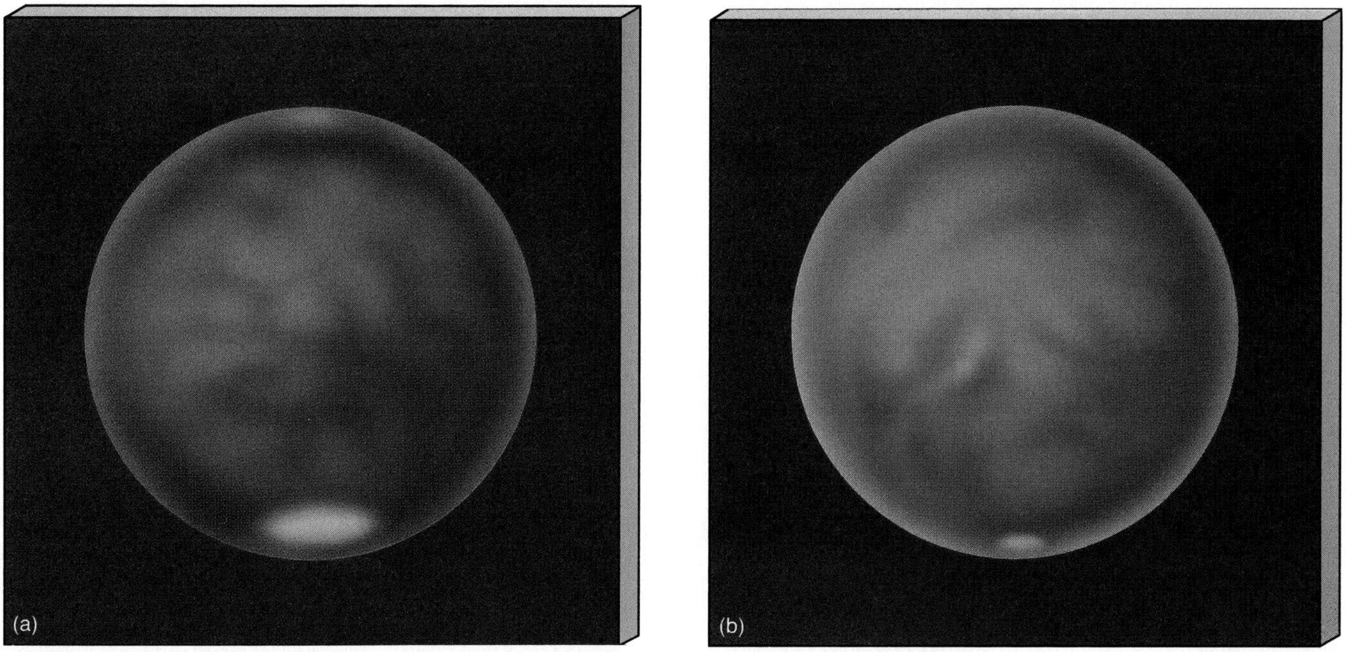

Figure 12.3 *Artist's conception of seasonal variation of the Martian polar caps. The polar caps grow and shrink over the course of a Martian year as both the height of the sun above the horizon and the intensity of sunlight vary. (a) Winter in the southern hemisphere, (b) southern summer.*

and eroded areas around which surface dust occasionally blows. Repeated covering and uncovering of these landmarks gives the impression (from a distance) of surface variability, but it's only the thin dust cover that changes. A powdery material, the surface dust is borne aloft by strong winds that often reach hurricane proportions (that is, hundreds of kilometers per hour). In fact, when the American *Mariner 9* spacecraft went into orbit around Mars in 1971, a planetwide dust storm obscured the entire landscape. Had the craft been on a flyby mission (for a quick look) instead of an orbiting mission (for a longer view), its visit would have been a failure. Fortunately, the storm subsided, enabling the craft to radio home detailed information about the surface.

Atmospheric Composition

Observers in the nineteenth century deduced the existence of a Martian atmosphere, in part from noting the bright clouds that occasionally traversed the face of the planet. In the 1940s, ground-based spectroscopy discovered atmospheric carbon dioxide. Although scientists were unable to detect any other gases, they continued to believe that Mars had a roughly Earthlike atmosphere, composed primarily of nitrogen, and that the observed carbon dioxide was just a small fraction of the total. In the early 1960s, improved observational techniques established that carbon dioxide really is the dominant gas and that the Martian atmosphere is much thinner than had previously been thought. However, it was really only after the planet had been visited by unmanned spacecraft that detailed atmospheric measurements became possible.

SPACECRAFT EXPLORATION OF MARS

Both NASA and the Soviet space agency have Mars exploration programs that extend back to the 1960s. However, the Soviet effort was plagued by a string of technical problems, along with a liberal measure of plain bad luck. As a result, almost all of the detailed planetary data we have on Mars has come from unmanned U.S. probes launched in the 1960s and 1970s.

The first spacecraft to reach the red planet was *Mariner 4*, which flew by Mars in July 1965. This craft was similar in design to *Mariner 2*, which reached Venus in 1962 (see Chapter 11). The images sent back by the craft showed large numbers of impact craters and nothing of the Earthlike terrain some scientists had expected to find. Later flybys in 1969, by *Mariner 6* and *Mariner 7*, confirmed these findings, leading to the conclusion that Mars was a geologically dead planet, with a heavily cratered, old surface. We now know, in fact, that large regions of Mars do not exhibit such a scarred, cratered landscape. It was just a coincidence that this vista was all the first three missions saw. These flybys were also responsible for first determining that the polar caps are composed of frozen carbon dioxide rather than water ice.

Studies of Mars received an enormous boost with the arrival in November 1971 of the *Mariner 9* orbiter. The craft mapped the entire Martian surface at a resolution of about 1 km, and it rapidly became clear that here was a world far more complex than the dead planet imagined only a year or two previously. *Mariner 9*'s maps revealed vast plains, volcanos, drainage channels, and canyons. All of

INTERLUDE 12-1 *Canals on Mars?*

The year 1877 was an important one in the human study of the planet Mars. The red planet came unusually close to Earth, affording astronomers an especially good view. Of particular note was the discovery, by U.S. Naval Observatory astronomer Asaph Hall, of the two moons circling Mars. But most exciting was the report of the Italian astronomer Giovanni Schiaparelli (the man responsible for erroneously reporting the synchronous orbit of Mercury) on his observation of a network of linear markings that he termed *canali,* which in Italian means "grooves" or "channels." The word was translated into English as "canals," suggesting that the grooves had been constructed by intelligent beings. Observations of these features became sensationalized in the world's press (especially in the United States), and some astronomers began drawing elaborate maps of Mars, showing oases and lakes where canals met in desert areas.

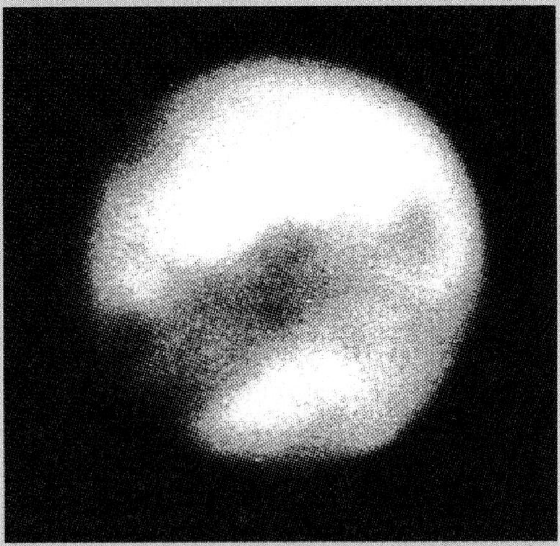

(Naval Observatory)

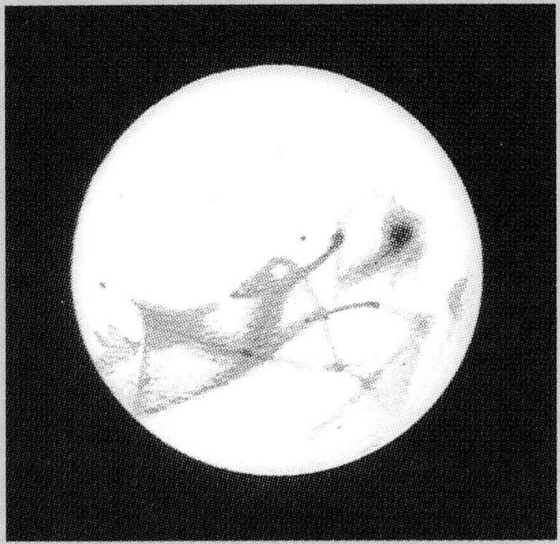

(P. Lowell)

Percival Lowell, a successful Boston businessman (and brother of the poet Amy Lowell and Harvard president Abbott Lawrence Lowell) became fascinated by these reports. He abandoned his business and purchased a clear-sky site at Flagstaff, Arizona, where he built a major observatory. He devoted his life to achieving a better understanding of the Martian "canals." In doing so, he championed the idea that Mars was drying out and that an intelligent society had constructed the canals to transport water from the wet poles to the arid equatorial deserts.

Alas, the Martian valleys and channels photographed by robot spacecraft during the 1970s are far too small to be the "canali" that Schiaparelli, Lowell, and others thought they saw on Mars. The entire episode represents a classic case in the history of science—a case where well-intentioned observers, perhaps obsessed with the notion of life on other worlds, let their personal opinions and prejudices seriously affect their interpretations of reasonable data. The two figures of Mars at the left show how surface features (which were probably genuinely observed by astronomers at the turn of the century) might be imagined to be connected. The figure at top is a photograph of how Mars actually looked in a telescope at the end of the nineteenth century. The sketch at bottom is an interpretation (done at the height of the canal hoopla) of the pictured view. The human eye, under great physiological stress, tends to connect dimly observed yet distinctly separated features. Humans saw patterns and canals where none in fact existed.

The chronicle of the Martian canals illustrates how the scientific method requires scientists to acquire new data to sort out sense from nonsense, fact from fiction. Rather than simply believing the claims about the Martian canals, other scientists demanded further observations to test Lowell's hypothesis. Eventually, improved observations, climaxing in the *Mariner* and *Viking* exploratory missions to the red planet nearly a century after all the fuss began, totally disproved the existence of canals. Although it often takes time, the scientific method does in fact lead to progress toward understanding reality.

---

these features were completely unexpected, given the data provided by the earlier missions. These new findings paved the way for the next step—actual landings on the planet's surface.

The last U.S. spacecraft to visit Mars were the two *Viking* missions, which reached the planet in mid-1976. *Viking 1* and *Viking 2* each consisted of two parts. An orbiter mapped the surface at a resolution of about 100 m (the same as the resolution achieved by *Magellan* on Venus). A lander (shown in Figure 12.4) descended to the surface and performed a wide array of geological and biological experiments. Landing sites had been chosen prior to the mission

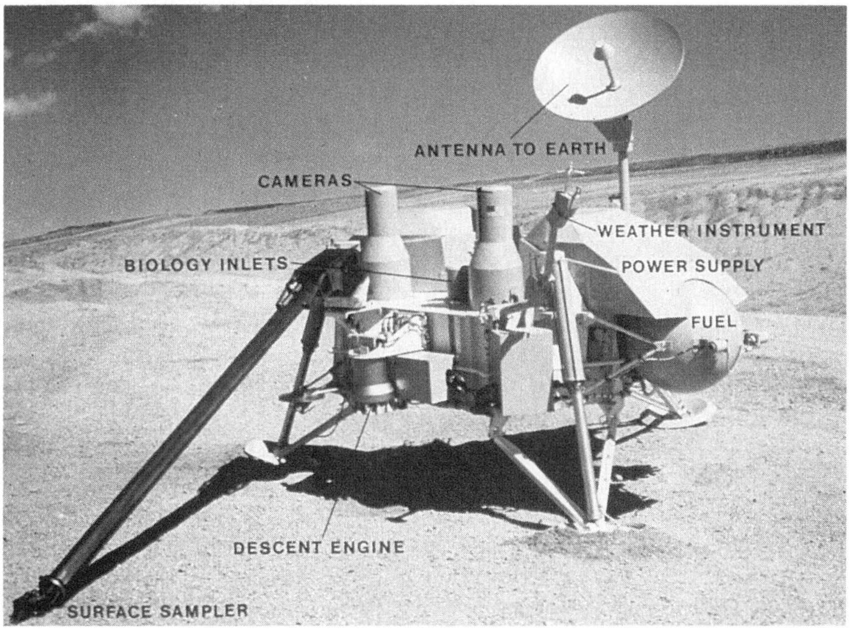

(a)

(b)

Figure 12.4 *(a) A* Viking *lander, here being tested in the Mohave Desert before launch. (b) The descent of the landers to the surface of Mars. Each craft used a heat shield to leave orbit, then parachuted toward the surface, in much the same way as Earth-orbiting spacecraft (before the* Space Shuttle*) return to the ground. The final stage of the* Viking *descent was slowed by retro-rockets, similar to those used by the* Apollo *lunar landers. (NASA)*

on the basis of data returned by *Mariner 9*, and the *Viking* craft went into orbit around the planet ready to release the landers to their preassigned targets. However, the improved resolution of the orbiters quickly revealed that the original sites were nowhere near as smooth and suitable as scientists had thought, and mission controllers spent many days searching through the incoming stream of new data to find alternative sites. *Viking 1* had originally been scheduled to touch down on Mars on July 4, 1976. Instead, it landed on July 20, precisely 7 years after Neil Armstrong first set foot on the Moon. The *Viking 2* lander arrived on the surface in September 1976. Even after the careful search for a new site, the second landing area was found to be strewn with rocks and boulders of all sizes, invisible from space but easily large enough to have destroyed the lander as it descended. *Viking 2* is lucky to have survived its arrival.

The *Viking* orbiters and landers returned a wealth of data on the Martian surface and atmosphere. By any standards, the *Viking* mission was a complete success. Since that time, however, the U.S. Mars program has been on hold. Although most space scientists would agree that more exploration of Mars is needed, they disagree as to the best means of achieving that goal. The Mars *Observer* satellite, expected to be launched sometime in the 1990s, will carry out climatological and geological studies of the planet from orbit, but many scientists believe that another landing—and the return of Martian surface material to Earth—is the next essential step in expanding our knowledge of the planet. Both NASA and the Soviet space agency had ambitious plans for manned missions to Mars, and the idea of a joint venture is raised from time to time. However, the enormous expense of such an undertaking—coupled with the belief of many astronomers that unmanned missions are economically and scientifically preferable to manned missions—make the future of these projects uncertain at best.

THE SURFACE OF MARS

The maps of the surface of Mars returned by the *Mariner* and *Viking* orbiters show a wide range of geological features. Mars has huge volcanos, deep canyons, vast dune fields, and many other geological wonders. The *Viking* data provide no evidence, though, for plate-tectonic activity, as we have on Earth. The orbiters performed large-scale surveys of much of the planet's surface, and the lander data complement these planetwide studies with detailed information on two specific sites.

Large-Scale Topography

✓2 A striking feature of the terrain of Mars is the marked difference between the northern and southern hemispheres. The northern hemisphere is made up largely of rolling volcanic plains, not unlike the lunar maria. These extensive northern lava plains—much larger than those found on Earth or the Moon—were formed by eruptions involving enormous volumes of lava. The plains are strewn with blocks of volcanic rock, as well as with boulders blasted out of impact areas by infalling meteoroids (the Martian atmosphere is too thin to offer much resistance to incoming debris). The southern hemisphere consists of heavily cratered highlands lying several kilometers above the level of the lowland north. Most of the dark regions visible from Earth are mountainous regions in the south. Figure 12.5 contrasts typical terrains in the two hemispheres.

The northern plains are cratered much less than the southern highlands. On the basis of arguments presented in Chapter 8, this smoother surface suggests that the northern surface is younger. Its age is perhaps 3 billion years, compared with 4 billion in the south. In places, the boundary between the southern highlands and the northern plains is quite sharp. The surface level can drop by as much as 4 km in a distance of 100 km or so. Most scientists assume that the southern terrain is the original crust of the planet. How most of the northern hemisphere could have been lowered in elevation and flooded with lava remains a mystery.

The major geological feature on the planet is the Tharsis bulge (shown in Figure 12.6, a large-scale view of the planet). It is a region, roughly the size of North America, on the Martian equator that rises some 10 km higher than the rest of the Martian surface. To the east of Tharsis lies Chryse Planitia (the "Plains of Gold"), and to its west lies a region known as Isidis Planitia (the "Plains of Isis," an Egyptian goddess). These features are wide depressions, hundreds of kilometers across and up to 3 km deep.

Tharsis appears to be even less heavily cratered than the north, making it the youngest region on the planet. It is estimated to be only 2 to 3 billion years old. If we wished to extend the idea of "continents" from Earth and Venus to Mars, we would have to conclude that Tharsis is the only continent on the Martian surface. However, as on Venus, there is no sign of plate tectonics—the continent of Tharsis is not drifting as are its earthly counterparts.

Volcanism

Mars contains the largest known volcanos in the solar system. Four very large volcanos are found on the Tharsis bulge, three of them visible in Figure 12.6. The largest volcano of all is Olympus Mons (shown in Figure 12.7), on the northwestern slope of Tharsis, lying just off the left-hand edge of Figure 12.6. It measures some 700 km in di-

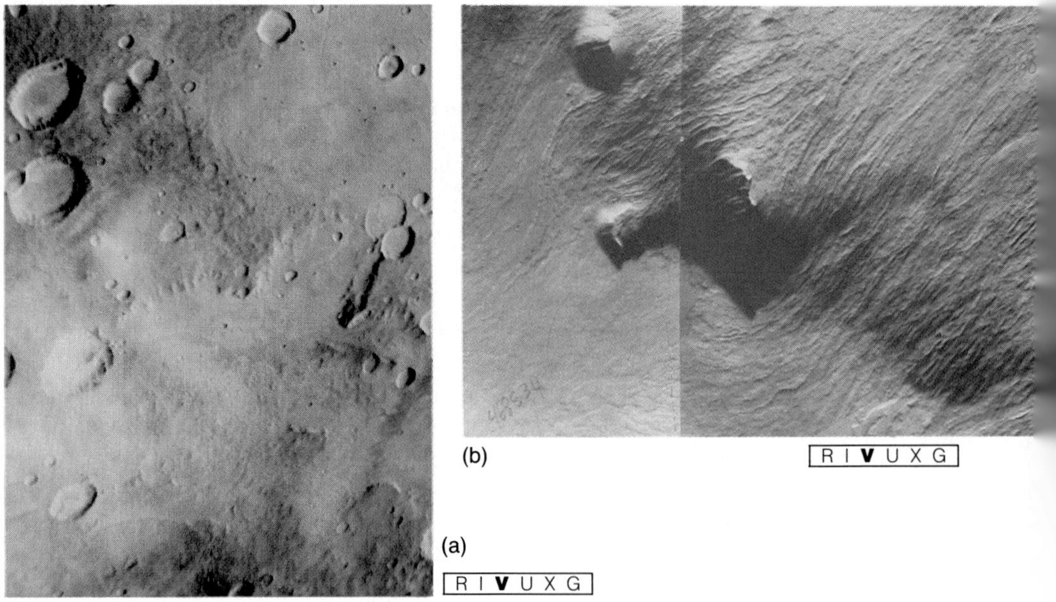

Figure 12.5 *(a) The southern Martian highlands are heavily cratered. (b) The northern hemisphere consists of rolling, volcanic plains. (NASA; NASA)*

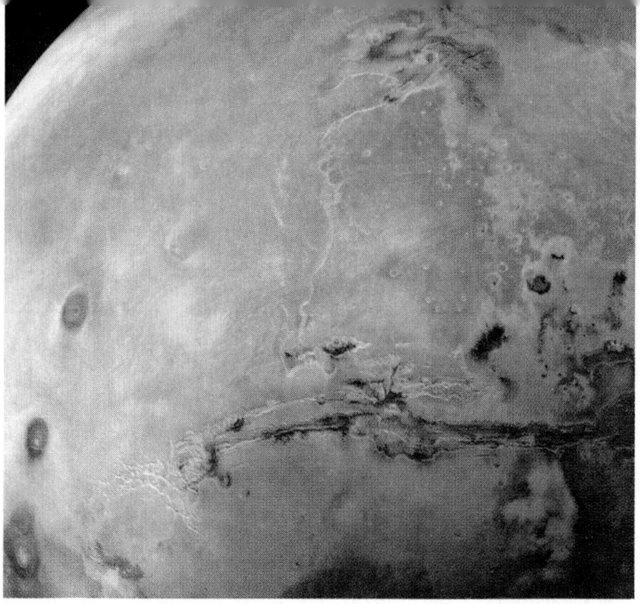

R I V U X G

Figure 12.6 *The Tharsis region of Mars, some 5000 km across, bulges out from the planet's equatorial region, rising to a height of about 10 km. The two large volcanos on the left mark the approximate peak of the bulge. One of the plains flanking the Tharsis bulge, Chryse Planitia, is toward the right. Dominating the center of the field of view is a vast "canyon" known as Valles Marineris. (NASA)*

ameter at its base—only slightly smaller than the state of Texas—and rises to a height of 25 km above the surrounding plains. The caldera, or crater, at its summit, measures 80 km across. The other three volcanos (on the left of Figure 12.6) are a little smaller—a mere 18 km high—and lie near the top of the bulge. Like Maxwell Mons on Venus, these volcanos are not associated with plate motion but instead are shield volcanos (see Chapter 11), sitting atop a hot spot in the underlying Martian mantle. All four volcanos show distinctive lava channels and other flow features very similar to those found on shield volcanos on Earth. The *Viking* images of the Martian surface reveal many hundreds of volcanos. Most of the largest are associated with the Tharsis bulge, but many smaller volcanos are also found in the northern plains.

The great height of Martian volcanos is a direct consequence of the planet's low surface gravity. As a shield volcano forms and lava flows and spreads, its height depends on the new mountain's ability to support its own weight. The lower the gravity, the less the weight and the higher the mountain. It is probably no accident that Maxwell Mons on Venus and the Hawaiian shield volcanos on Earth rise to roughly the same height (about 10 km) above their respective bases—Earth and Venus have similar surface gravity. Mars has surface gravity only 40 percent that of the Earth, and volcanos rise roughly 2.5 times as high.

Are these volcanos still active? Scientists have found no direct evidence for recent or ongoing eruptions. However, if these volcanos have been around since the Tharsis uplift (as the formation of the Tharsis bulge is often known) and have been active as recently as 100 million years ago (an age estimate based on the extent of impact cratering on their slopes), some of them may still be at least intermittently active. Millions of years, though, may pass between eruptions.

Impact Cratering

The *Mariner* series of spacecraft found that the surfaces of Mars and its two moons are pitted with impact craters formed by meteoroids falling in from space. As on our Moon, the smaller craters are often filled with surface matter—mostly dust—confirming that Mars is a dry, desert world. However, Martian craters are filled in considerably faster than their lunar counterparts. On the Moon, ancient craters more than 20 m deep (corresponding to about 100 m in diameter) have been obliterated, primarily by meteoritic erosion, as discussed in Chapter 8. On Mars, there is also a deficit of small craters, but extending to craters 5 km in diameter. The Martian atmosphere is an efficient erosive agent, transporting dust from place to place and erasing surface features much faster than meteoritic impacts alone. As on the Moon, the extent of large impact cratering (that is, craters too big to have been filled in by erosion since they formed) can be used as an age indicator for the Martian surface. Ages derived in this way range from 4 billion years for the southern highlands to a few hundred million years in the youngest volcanic areas.

The detailed appearance of Martian impact craters provides an important piece of information about conditions

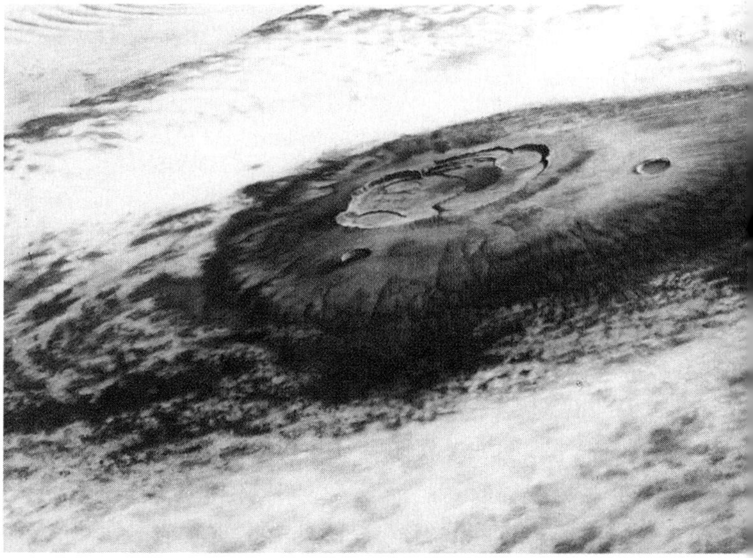

R I V U X G

Figure 12.7 *Olympus Mons, the largest volcano known on Mars or anywhere else in the solar system. Nearly three times taller than Mount Everest on Earth, this Martian mountain measures about 700 km across the base and extends 25 km at the peak. It seems currently inactive and may have been extinct for at least several hundred million years. (By comparison, the largest volcano on Earth, Hawaii's Mauna Loa, measures a mere 120 km across and peaks about 9 km above the Pacific Ocean floor.) (NASA)*

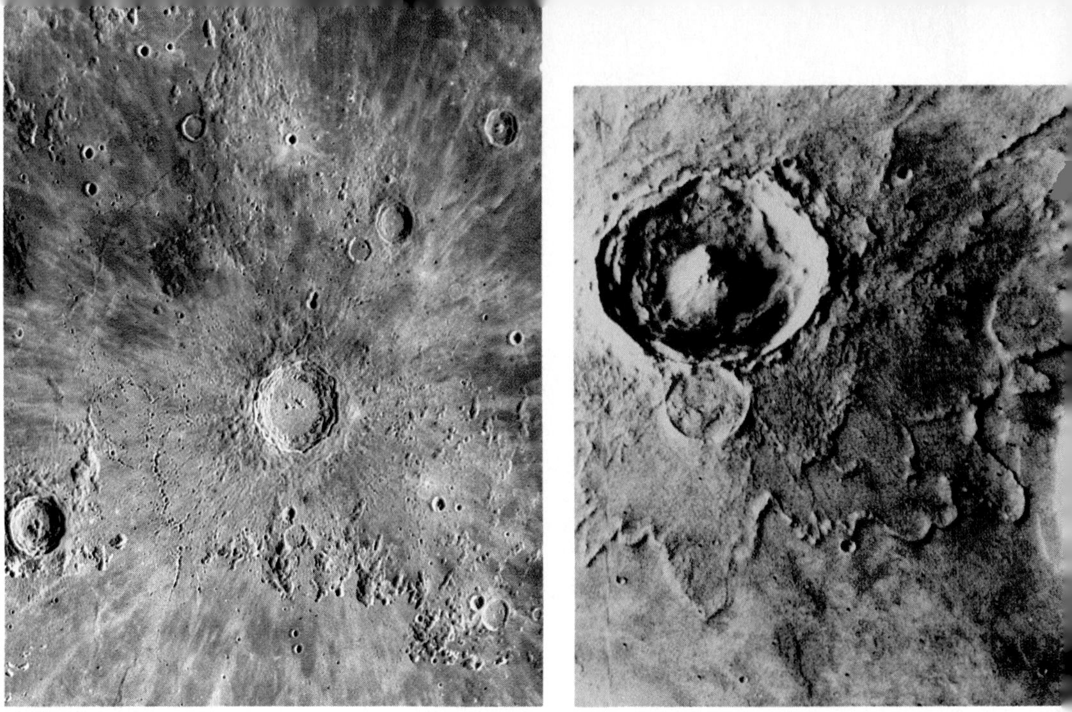

(a) R I **V** U X G (b) R I **V** U X G

Figure 12.8 *(a) The large lunar crater Copernicus is quite typical of those found on Earth's Moon. Its ejecta blanket appears to be composed of dry, powdery material. (b) The ejecta from Mars's crater Yuty (18 km in diameter) evidently was liquid in nature. This type of crater is sometimes called a ''splosh'' crater. (Lick Observatory; NASA)*

just below the planet's surface. The ejecta blankets surrounding many Martian craters look quite different from their lunar counterparts. Figure 12.8 compares the Copernicus crater on the Moon with the (fairly typical) crater Yuty on Mars. The material surrounding the lunar crater is just what one would expect from an explosion ejecting a large volume of dust, soil, and boulders. The ejecta blanket on Mars gives the distinct impression of a liquid that has splashed or flowed out of the crater. Geologists believe that this **fluidized ejecta** crater indicates that a layer of **permafrost,** or water ice, lies just under the surface. The explosive impact heated and liquefied the ice, resulting in the fluid appearance of the ejecta.

The Martian Grand Canyon

Yet another feature associated with the Tharsis bulge is a great ''canyon'' known as Valles Marineris (the Mariner Valley). Shown in its entirety in Figure 12.6 and in more detail in Figure 12.9, it is not really a canyon in the terres-

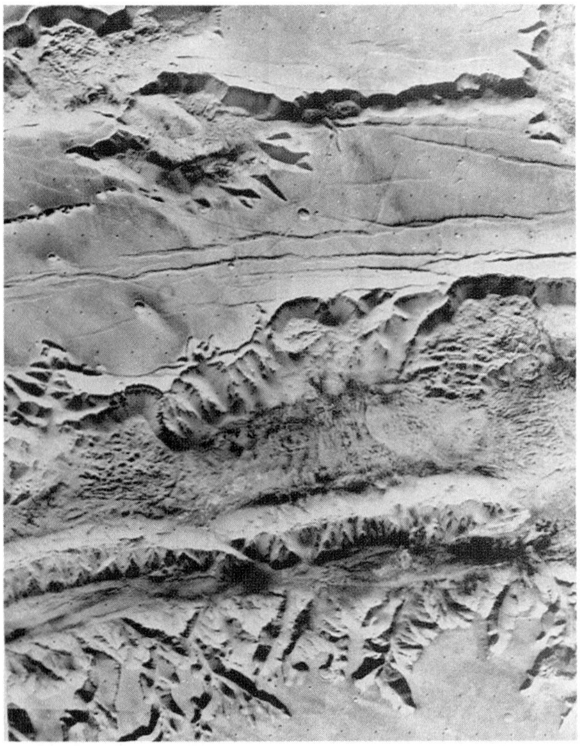

(a) R I **V** U X G

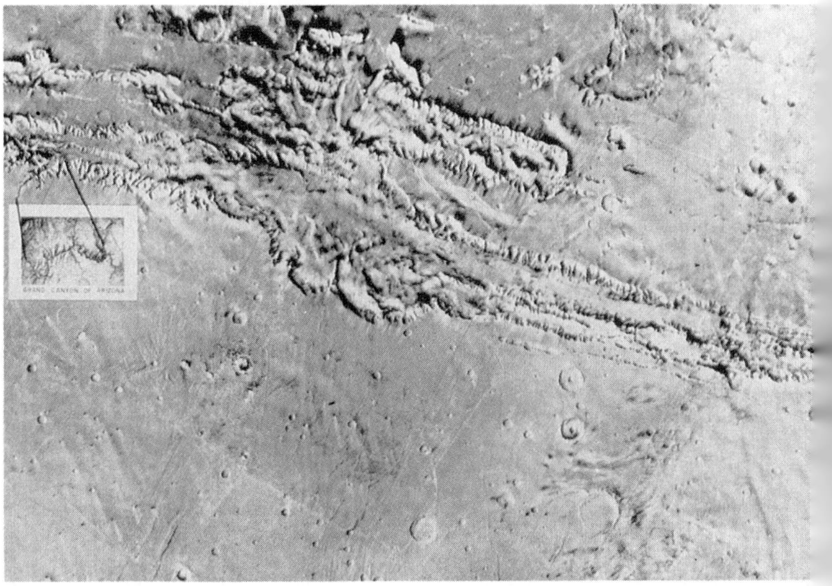

(b) R I **V** U X G

Figure 12.9 *(a) Here we see a closeup of the Mariner Valley, an incredibly large canyon that measures over 4000 km in length, 120 km wide, and 7 km deep. (b) A comparison with Earth's Grand Canyon, which is 20 km wide and 2 km deep.*

trial sense, because running water played no part in its formation. Planetary astronomers believe that it was formed by the same crustal forces that forced the entire Tharsis region to bulge outward, causing the surface to split and crack. These cracks, called **tectonic fractures,** are found all around the Tharsis bulge. The Mariner Valley is the largest of them. Cratering studies suggest that the cracks are at least 2 billion years old. Similar (but smaller) cracks, with similar causes, have been found in the Aphrodite Terra region of Venus, as we saw in Chapter 11.

Valles Marineris runs for almost 4000 km along the Martian equator, about one-fifth of the way around the planet. At its widest, it is some 120 km across, and it is as deep as 7 km in places. Like many Martian surface features, it simply dwarfs the earthly competition. The Grand Canyon in Colorado would easily fit into one of its side "tributary" cracks. It is so large that it can even be seen from Earth—in fact, Valles Marineris was one of the few canals observed by nineteenth-century astronomers (see Interlude 12-1) that actually corresponded to a real feature on the planet's surface (it was known as the Coprates canal). We must reemphasize, however, that this Martian feature was *not* constructed by intelligent beings, nor was it carved by a river, nor is it a result of Martian plate tectonics. For some reason, the crustal forces that formed it never developed into full-fledged plate motion as on Earth.

Evidence for Running Water

✓3 Although the great surface cracks in the Tharsis region are not really canyons and were not formed by running water, photographic evidence reveals that liquid water once existed in great quantity on the surface of Mars. Two types of flow features are seen: the **runoff channels** and the **outflow channels.**

The runoff channels (one of which is shown in Figure 12.10) are found in the southern highlands. They are extensive systems—sometimes hundreds of kilometers in total length—of interconnecting, twisting channels that seem to merge into larger, wider channels. They bear a strong resemblance to river systems on Earth, and it is believed by geologists that that is just what they are—the dried-up beds of long-gone rivers that once carried rainfall on Mars from the mountains down into the valleys. These runoff channels speak of a time 4 billion years ago (the age of the Martian highlands), when the atmosphere was thicker, the surface warmer, and liquid water widespread.

The outflow channels (see Figure 12.11) are probably relics of catastrophic flooding on Mars long ago. They appear only in equatorial regions and generally do not form the extensive interconnected networks that characterize the runoff channels. Instead, they are probably the paths taken by huge volumes of water draining from the southern highlands into the northern plains. The onrushing water arising from these flash floods probably also formed the odd teardrop-shaped "islands" (resembling the miniature versions seen in the wet sand of our beaches at low tide) that have been found on the plains close to the ends of the outflow channels (Figure 12.12). Judging from the width and depth of the channels, the flow rates must have been truly enormous—perhaps as much as a hundred times greater than the 10^5 tons per second carried by the Amazon river, the largest

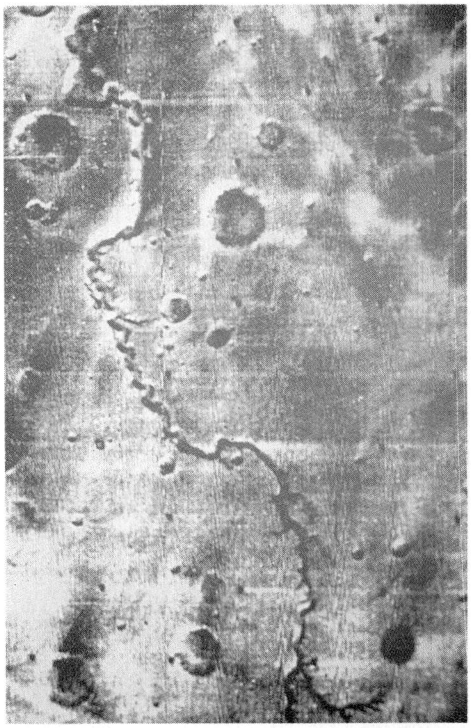

Figure 12.10 *(a) This runoff channel on Mars measures about 400 km long and 5 km wide. Here, we compare it with a photograph of the Red River (b) running from Shreveport, Louisiana, to the Mississippi River. The two differ mainly in that there is currently no liquid water in this, or any other, Martian valley. (NASA; NASA)*

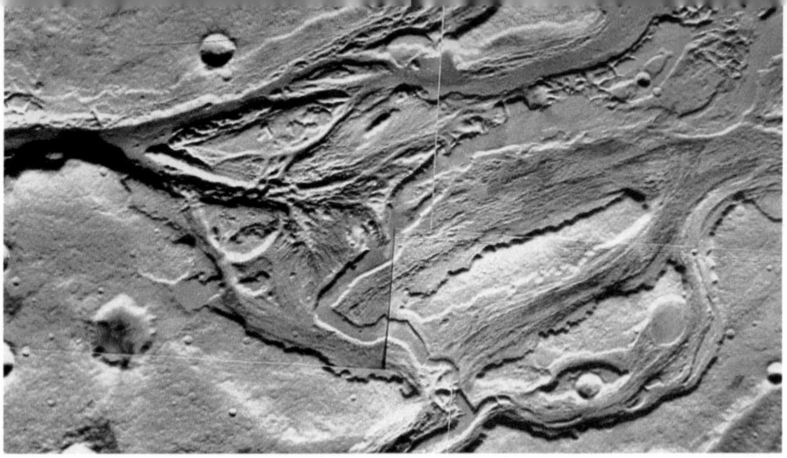

Figure 12.11 *An outflow channel near the Martian equator bears witness to a catastrophic flood that occurred about 3 billion years ago. (NASA)*

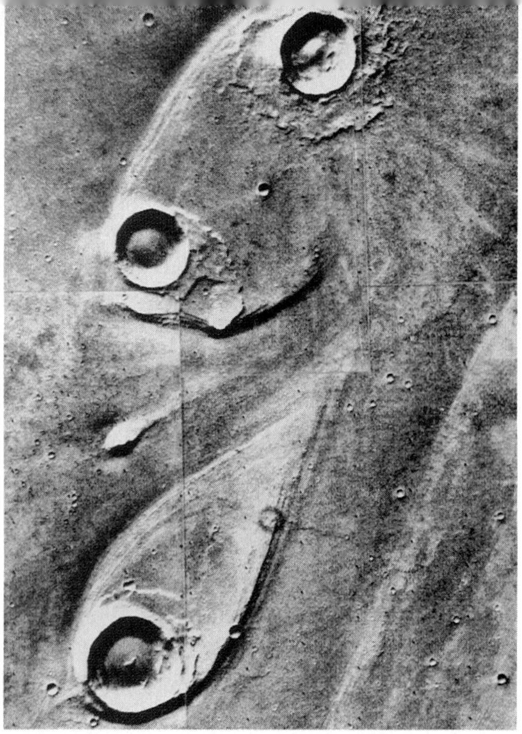

Figure 12.12 *The onrushing water that carved out the outflow channels was responsible for forming these oddly shaped "islands" as the flow encountered obstacles—impact craters—in its path. Each "island" is about 40 km long. (NASA)*

R I V U X G

river system on Earth. Flooding shaped the outflow channels about 3 billion years ago, about the same time as the northern volcanic plains formed.

Planetary astronomers find no evidence for liquid water anywhere on Mars today, and the amount of water vapor in the Martian atmosphere is tiny. Yet the extent of the outflow channels indicates that a huge total volume of water existed on Mars in the past. Where did all that water come from? And where did it all go? The answer may be that virtually all of the water on Mars is now locked in the permafrost layer under the surface, with perhaps a little more contained in the polar caps. Four billion years ago, as climatic conditions changed, the running water that formed the runoff channels began to freeze, forming the permafrost and drying out the river beds. Mars remained frozen for about a billion years, until volcanic (or some other) activity heated large regions of the surface, melting the subsurface ice and causing the flash floods that created the outflow channels. Subsequently, volcanic activity subsided, the water refroze, and Mars once again became a dry world.

Polar Caps

We have already noted that Mars's polar caps are composed predominantly of carbon dioxide frost—dry ice—and show seasonal variations. Each cap in fact consists of two distinct parts—the **seasonal cap,** which grows and shrinks each Martian year, and the **residual cap,** which remains permanently frozen. At maximum size, in southern midwinter, the southern seasonal cap is some 4000 km across, about one-fifth the circumference of the planet. Half a Martian year later, the northern cap is at its largest, reaching a diameter of roughly 3000 km. The seasonal variations in the caps can be clearly discerned in Figure 12.3. Figure 12.13 shows a

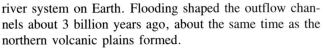

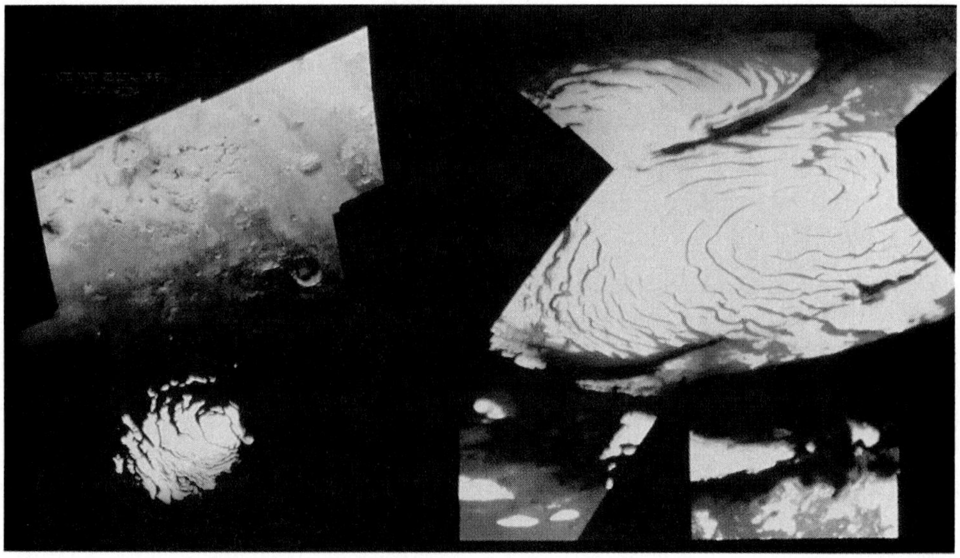

Figure 12.13 *The northern seasonal polar cap of Mars is clearly visible in this mosaic of* Mariner 9 *images. (NASA)*

R I V U X G

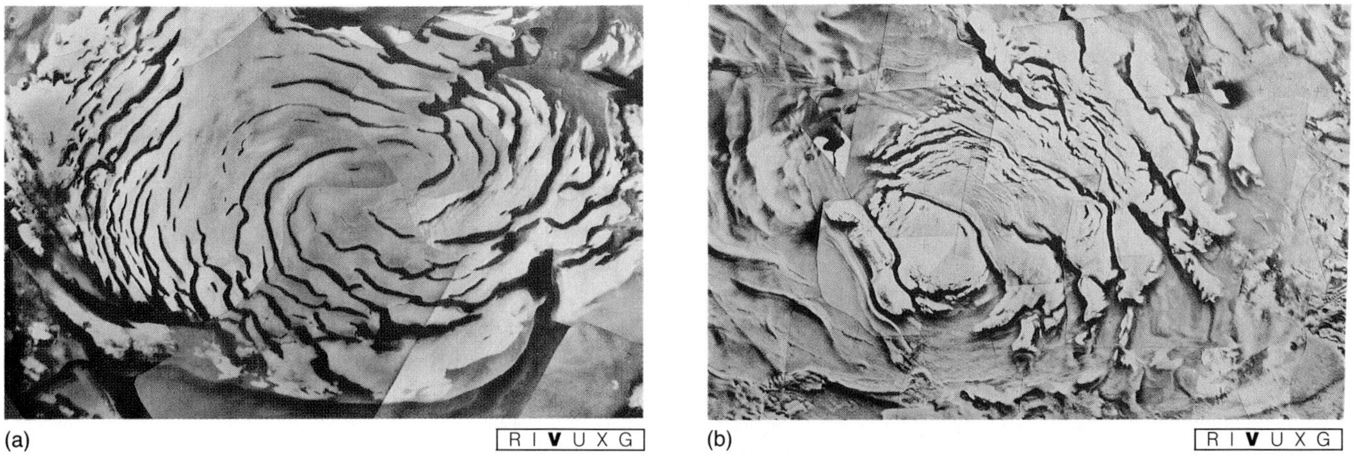

Figure 12.14 *(a) The northern residual polar cap of Mars, seen during northern summer. (b) The southern residual cap, during southern summer, half a Martian year later. Both these images are mosaics made from individual photographs sent back from the* Viking *orbiters. The northern cap is about 1000 km across and composed mostly of water ice. The southern cap is only 350 km across and made up mostly of frozen carbon dioxide. (NASA; NASA)*

Mariner 9 image of the northern cap as it began to recede with the arrival of spring in the northern hemisphere. The two seasonal polar caps do not have the same maximum size because of the eccentricity of Mars's orbit around the Sun. During southern winter, Mars is considerably farther from the Sun than half a year later, in northern winter. Thus, the southern winter season is longer and colder than in the north, and the polar cap grows correspondingly larger.

The seasonal caps are composed entirely of carbon dioxide. Their temperatures are never greater than about 150K (−120°C), the point at which dry ice can form. During Martian summer, when sunlight striking a cap is most intense, carbon dioxide evaporates into the atmosphere, and the cap shrinks. In the winter, atmospheric carbon dioxide refreezes, and the cap reforms. As the caps grow and shrink, they cause substantial variations (up to 30 percent) in the Martian atmospheric pressure. A large fraction of the atmosphere condenses out and evaporates again on a yearly basis. From studies of these atmospheric fluctuations, scientists can estimate the amount of carbon dioxide in the seasonal polar caps. They believe that the maximum thickness of the seasonal caps is about 1 m.

The residual caps, shown in Figure 12.14, are smaller and brighter than the seasonal caps and show an even more marked north-south asymmetry. The southern residual cap is about 350 km across and, like the seasonal caps, is made mostly of carbon dioxide, although it may contain some water ice. Its temperature remains below 150K at all times. The northern residual cap is much larger—about 1000 km across—and warmer, with a temperature that can exceed 200K in northern summertime. Planetary scientists believe that the northern cap is made mostly of water ice, an opinion strengthened by the observed increase in the concentration of water vapor above the north pole in northern summer as some small fraction of its water ice evaporates in the Sun's heat. It is quite possible that the northern residual polar cap is a major storehouse for water on Mars.

We do not know the thickness of the residual caps. As Figure 12.14 shows, however, some frost-free regions seasonally emerge from the ice, suggesting that the caps can be no more than a few kilometers thick.

Why is there such a temperature difference (at least 50K) between the two residual polar caps? The reason seems to be related to the giant dust storms that envelop the planet during southern summer. These storms, which last for a quarter of a Martian year (about six Earth months), tend to blow the dust from the warmer south into the northern hemisphere. As a result, the northern ice cap becomes dusty, reflects less sunlight (ice reflects sunlight much more efficiently than does dust, which absorbs it), and tends to warm up. Over the millennia, this south-north flow has heated the northern polar regions and preferentially concentrated the planet's water ice there. The winds that transport the dust north are probably also responsible for the extensive sand dunes found in the northern polar terrain of Mars.

The View from Viking

Viking 1 landed in Chryse Planitia, the broad depression to the east of Tharsis. The view that greeted its cameras (Figure 12.15) was a windswept, gently rolling, rather desolate plain, littered with rocks of all sizes, not unlike a high desert on Earth. *Viking 2* landed somewhat farther north, in a region of Mars called Utopia. Mission planners chose Utopia in part because they anticipated greater seasonal climatic variations there. The plain on which *Viking 2* landed was flat and featureless. From space, the landing site appeared smooth and dusty. In fact, the surface turned out to be very rocky, even rockier than the Chryse site, and without the dust layer the mission directors had expected. The surface rocks visible in Figure 12.16 are probably part of the ejecta blanket of a nearby impact crater. The views that the two landers recorded may turn out to be quite typical of the low-latitude northern plains.

Figure 12.15 *Panoramic view from the perspective of the* Viking 1 *spacecraft now parked on the surface of Mars. The fine-grained soil and the rock-strewn terrain stretching toward the horizon are reddish. Containing substantial amounts of iron ore, the surface of Mars is literally rusting away. The sky is a pale pink, the result of airborne dust. (NASA)*

Surface Composition

The *Viking* landers performed numerous chemical analyses of the regolith of Mars. One important finding of these studies was the high iron content of the planet's surface. Chemical reactions between the iron-rich surface soil and free oxygen in the atmosphere is responsible for the iron oxide (''rust'') that gives Mars its characteristic color. Although the surface layers are rich in iron relative to the Earth's surface, the overall abundance is similar to the Earth's average iron content. On Earth, much of the iron has differentiated to the center. Chemical differentiation does not appear to have been nearly so complete on Mars.

THE MARTIAN ATMOSPHERE

Composition

As we have already seen, even before the arrival of the *Mariner* and *Viking* spacecraft, astronomers knew from Earth-based spectroscopy that the Martian atmosphere was quite thin and composed primarily of carbon dioxide. In 1964 *Mariner 4* confirmed these results, finding that the atmospheric pressure is only about $\frac{1}{150}$ the pressure of Earth's atmosphere at sea level and that carbon dioxide (CO_2) makes up at least 95 percent of the total atmosphere. With the arrival of *Viking*, more detailed measurements of the Martian atmosphere were made. Its composition is now known to be 95.3 percent carbon dioxide, 2.7 percent nitrogen (N_2), 1.6 percent argon (Ar), 0.13 percent oxygen (O_2), 0.07 percent carbon monoxide (CO), and about 0.03 percent water vapor (H_2O). The level of water vapor is quite variable. While there is some superficial similarity between the atmospheres of Mars and Venus, at least in terms of composition, the two planets obviously have quite different atmospheric histories. Mars's ''air'' is over 10,000 times thinner than that on Venus. Even if Mars's atmosphere were completely compatible with our respiratory system—namely, made of an oxygen-nitrogen mixture—it would still be too thin for earthlings to breathe normally. There is just not enough of it. And, during the seasonal high winds, it is very dusty.

Vertical Structure

As the *Viking* landers descended to the surface, they made measurements of the temperature and pressure at various heights. The results appear in Figure 12.16. The Martian atmosphere contains a troposphere (the lowest-lying atmospheric zone, where convection and ''weather'' occur), which varies both from place to place and from season to season. The variability of the troposphere arises from the

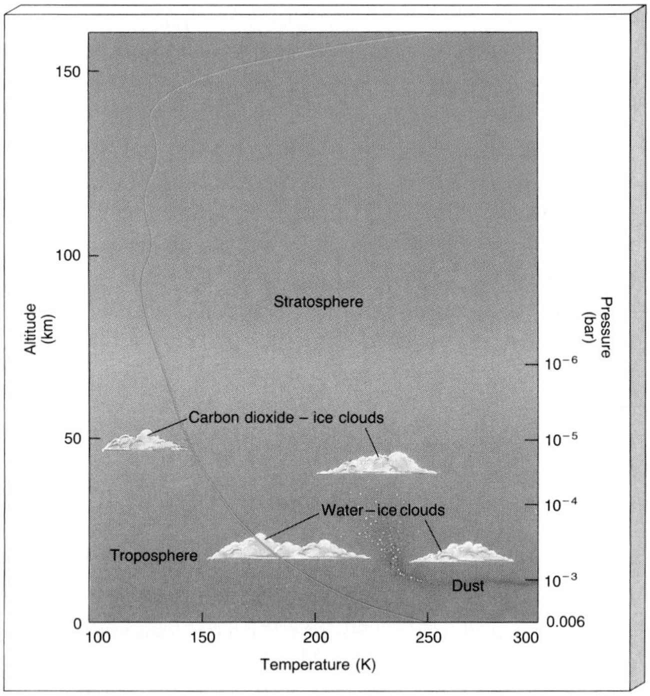

Figure 12.16 *Structure of the Martian atmosphere, as determined by* Viking. *The troposphere rises up to an altitude of about 30 km in the daytime. It occasionally contains clouds of water ice or, more frequently, dust during the planetwide dust storms that occur each year. Above the troposphere lies the stratosphere. Note the absence of a higher-temperature zone in the stratosphere, indicating the absence of an ozone layer. Temperatures in the stratosphere are low enough for carbon dioxide to solidify, giving rise to a high-level layer of carbon dioxide clouds and haze.*

variability of the Martian surface temperature. At noon in the summertime, surface temperatures may reach 300K. Atmospheric convection is strong, and the top of the troposphere can reach as high as 30 km. At night, the atmosphere retains little heat, and the temperature can drop by up to 100K. Convection ceases entirely, and the troposphere vanishes.

Mars does have a stratosphere, but it lacks a stratospheric temperature inversion zone like the one found on Earth (see Figure 7.6). From this fact we can infer that the planet has no significant ozone layer to absorb solar ultraviolet radiation.

Weather

On average, surface temperatures on Mars are about 50K cooler than on Earth. Weather, such as dust storms and most clouds, is confined to the troposphere, as it is on Earth. Only a few thin carbon dioxide clouds are occasionally found in the lower stratosphere. The low early-morning temperatures often produce water-ice "fog" in the Martian canyons, as can be seen in Figure 12.17. For most of the year, there is little day-to-day variation in the weather: The Sun rises, the surface warms up, and light winds blow until sunset, when the temperature drops again. Only in the southern summer does the daily routine change. Strong surface winds (without rain or snow) sweep up the dry dust, carry it aloft, and eventually deposit it elsewhere on the planet. At its greatest fury, a Martian storm floods the atmosphere with dust, making the worst storm we could imagine on Earth's Sahara Desert seem inconsequential by comparison. The dust can remain airborne for months at a time.

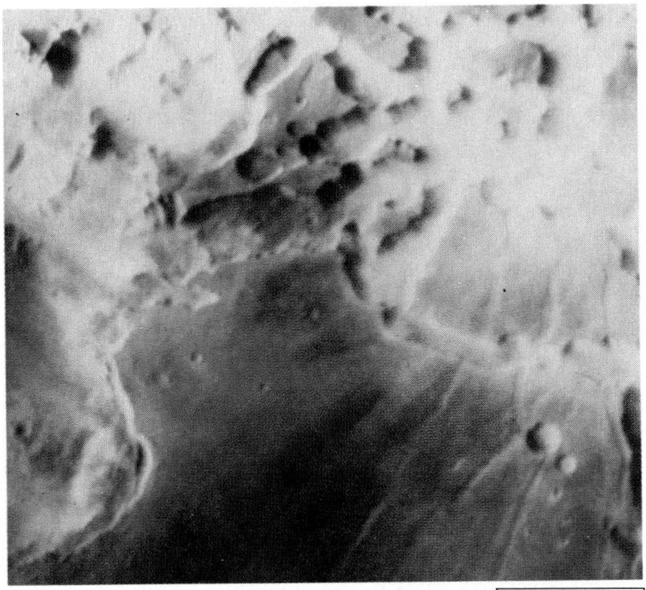

Figure 12.17 *Fog in the Martian canyons, photographed by the* Viking *orbiter. As the Sun's light reaches and heats the canyon floor, it drives water vapor from the surface. When this vapor comes in contact with the colder air above the surface, it condenses again, and a temporary water-ice "fog" results. (NASA)*

In order for the dust to stay aloft for so long, it must be made up of quite fine-grained particles and raised to great heights—probably 20 to 30 km. In addition, the upper-level winds must be very rapid—as high as 150 km/h. Strong local heating of the ground during southern summer likely produces strong updrafts, which inject the dust into the lower stratosphere. With so much dust in the air, some of the incoming sunlight is absorbed, and the whole temperature structure of the troposphere is temporarily modified. The ground is cooled (in a manner similar to the "nuclear winter" scenario on Earth), and tropospheric convection is severely curtailed. We do not fully understand the details, but we do know that the winds continue for the entire summer. Only when the planet moves farther from the Sun does the dust eventually settle and "normal" weather return. The blown dust forms systems of sand dunes similar in appearance to those found on Earth.

Evolution of the Martian Atmosphere

4 As with the other planets we have studied, we can ask *why* the Martian atmosphere is as it is. Presumably, Mars acquired a secondary outgassed atmosphere quite early in its history, just as the other terrestrial worlds did. Around 4 billion years ago, as indicated by the runoff channels in the highlands, Mars may have had a fairly dense atmosphere, complete with blue skies and rain. Despite Mars's distance from the Sun, the greenhouse effect would have kept conditions fairly comfortable, and a surface temperature of over 0°C seems quite possible.

Sometime in the next billion years, most of the Martian atmosphere disappeared. Possibly some of it was lost because of impacts with other large (Moon-sized) bodies in the early solar system. More likely, the Martian atmosphere became unstable, in a kind of reverse runaway greenhouse effect. On Venus, as we have seen, the familiar greenhouse effect ran away to high temperatures and pressures. Planetary scientists theorize that the early Martian atmosphere also ran away, but in the opposite direction. The presence of liquid water would have caused much of the atmospheric carbon dioxide to have dissolved in Martian rivers and lakes (and oceans, if any), ultimately to combine with Martian surface rocks. Recall that, on Earth, most carbon dioxide is presently found in surface rocks. Calculations show that much of the Martian atmospheric carbon dioxide could have been depleted in this way in a relatively short period of time, perhaps as little as a few hundred million years, although some of it may have been replenished by volcanic activity, possibly extending the "comfortable" lifetime of the planet to a half-billion years or so.

As the level of carbon dioxide declined and the greenhouse-heating effect diminished, the planet cooled. The water froze out of the atmosphere, lowering still further the level of atmospheric greenhouse gases, and so accelerating the cooling. Since that time, the overall density of the atmosphere has steadily declined through a steady loss of carbon dioxide, nitrogen, and water vapor. Solar ultraviolet radiation in the upper Martian atmosphere splits the molecules of these gases into their component atoms and provides some with enough energy to escape. The present level of water vapor in the Martian atmosphere is the maximum possible, given the atmosphere's present density and temperature. Estimates of the total amount of water stored as permafrost or in the polar caps are quite uncertain, but it is likely that, if all the water on Mars were to become liquid, it would cover the surface to a depth of several meters.

LIFE ON MARS?

✓5 Even before *Viking* reached Mars, astronomers had abandoned hope of finding intelligent, or even animal, life on the planet. Scientists knew there were no large-scale canal systems, no surface water, almost no oxygen in the atmosphere, and no seasonal vegetation changes. The present lack of liquid water on Mars especially dims the chances for life there now. Might life have existed once? Running water and possibly a dense atmosphere in the past may have created conditions suitable for the emergence of life long ago. In the hope that some form of microbial life might have survived to the present day, the *Viking* landers carried out experiments designed to detect biological activity.

All three *Viking* biological experiments assumed some basic similarity between hypothetical Martian bacteria and those found on Earth. A **gas exchange experiment** offered a nutrient broth to any residents of a sample of Martian soil and looked for gases that would signal metabolic activity. The **labeled release experiment** added compounds containing radioactive carbon to the soil, then waited for results signaling that Martian organisms had either eaten or breathed in this carbon. The **pyrolitic release experiment** added radioactively tagged carbon dioxide to a sample of Martian soil and atmosphere, waited a while, then removed the gas and tested the soil (by heating it) for signs that something had absorbed the tagged gas. In all cases, contamination by terrestrial bacteria was a major concern. Indeed, any release of Earth organisms would have invalidated these and all future such experiments on Martian soil. Both *Viking* landers were carefully sterilized prior to launch, and international agreement presently protects the Martian environment from contamination by future terrestrial probes. (How we will sterilize a manned mission is still a little unclear.)

Initially, all of the experiments appeared to be giving positive signals! However, subsequent careful studies showed that all of the results could be explained by inorganic (that is, nonliving) chemical reactions. Thus, we have no clear evidence for even microbial life on the surface. The *Viking* robots detected peculiar reactions that mimic in some ways the basic chemistry of living organisms, but they did not detect life itself. Despite the lack of clear evidence for life on today's Mars, it is still possible that this most intriguing planet harbors fossilized evidence for a life that flourished during some previous epoch. This possibility rests only on conjecture, however, and remains for future exploration to prove or disprove.

THE MOONS OF MARS

Unlike Earth's moon, Mars's moons are tiny compared with their parent planet and orbit very close to it (in terms of the planet's radius). Discovered by Asaph Hall in 1877, the two Martian moons—Phobos and Deimos—are only a few kilometers across. Their composition is quite dissimilar from that of the planet. Astronomers generally believe that Phobos and Deimos did not form along with Mars but instead are asteroids that were slowed and captured by the outer fringes of the early Martian atmosphere (which, as we have just seen, was probably much denser than the atmosphere today). It is even possible that they are remnants of a single object that broke up during the capture process. They are quite difficult to study from Earth because their proximity to Mars makes it hard to distinguish them from their much brighter parent. The *Mariner* and *Viking* orbiters, however, have studied them in great detail.

Both moons, shown in Figure 12.18, are quite irregularly shaped and heavily cratered. The larger of the two is Phobos (Figure 12.18[a]), which is about 28 km long and 20 km wide and dominated by an enormous 10-km-wide crater named Stickney (after Angelina Stickney, Asaph Hall's wife, who encouraged him to persevere in his observations). The smaller Deimos (Figure 12.18[b]) is only 16 km long by 10 km wide. Its largest crater is 2.3 km in diameter. Both moons have quite dark surfaces, with albedos of 0.06 or less, which contributes to the difficulty in observing them from Earth.

Phobos and Deimos move in circular, equatorial orbits, and they rotate synchronously (that is, they keep the same face permanently turned toward the planet)—all of which are direct consequences of the tidal influence of Mars. Phobos lies only 9378 km (less than three planetary radii) from the center of Mars and, as we saw above, has an orbit period of 7 hours and 59 minutes. This orbit period is much less than a Martian day, so an observer standing on the Martian surface would actually see Phobos move backwards across the Martian sky—that is, in a direction opposite to the apparent daily motion of the Sun. Because the moon moves faster than the observer, it overtakes the planet's rotation, rising in the west and setting in the east, crossing the sky from horizon to horizon in about 5.5 hours. Deimos lies a little farther out, at 23,459 km, or slightly less than seven planetary radii, and orbits in 30 hours and

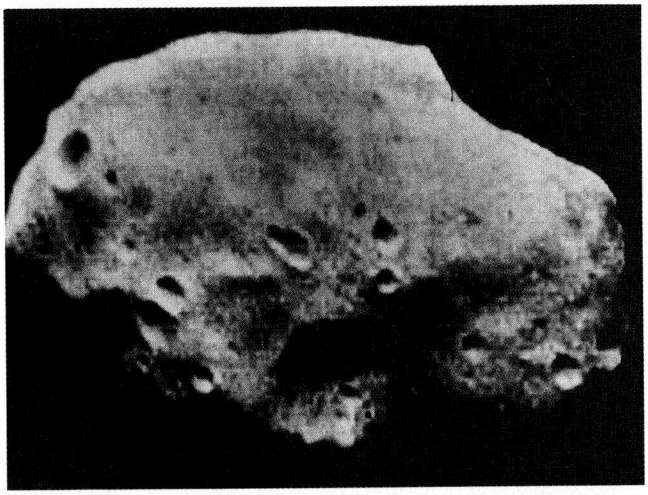

(a) R I **V** U X G

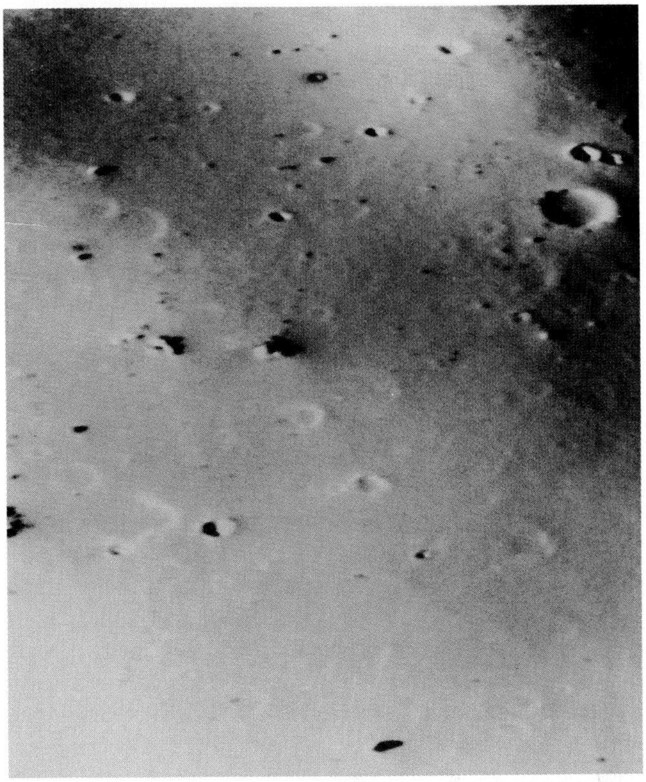

(b) R I **V** U X G

Figure 12.18 *Mars is accompanied in its trek around the Sun by two tiny moons. Deimos (Greek for "panic") and Phobos ("fear") measure about 16 and 28 km long, respectively. Both are shaped like potatoes, with lots of cratering. (a) This is a* Mariner-9 *photograph of the irregularly shaped Phobos, not much larger than Manhattan Island. (b) Like Phobos, the smaller moon Deimos has a composition unlike that of Mars. Both moons are probably captured asteroids. This photograph was taken by the* Viking *orbiter. The field of view is only 2 km across with most of the boulders shown about the size of a house. (NASA; NASA)*

18 minutes. Because it completes its orbit in more than a Martian day, it moves "normally," as seen from the ground (that is, from east to west), taking almost 3 days to traverse the sky. Each moon orbits Mars in the prograde sense—that is, in the same sense (counterclockwise, as seen from above the north celestial pole) as the planet orbits the Sun and rotates on its axis.

On the basis of measurements of their gravitational effect on the *Viking* orbiters, astronomers have estimated the masses of the two moons. Their densities are around 2 g/cm^3, far less than any world we have yet encountered in our outward journey through the solar system. This is an important reason why astronomers do not believe these moons formed along with Mars. If they are indeed captured asteroids, Phobos and Deimos represent material left over from the earliest stages of the solar system. Astronomers study them not to gain insight into Martian evolution, but rather because the moons contain information about the very early solar system, before the major planets had formed.

MARTIAN INTERNAL STRUCTURE

The *Viking* landers carried seismometers to probe the internal structure of the planet, but one failed to work and the other was unable to distinguish seismic activity from the buffeting of the Martian wind. Thus, as yet no seismic studies of Mars's interior have been carried out. On the basis of studies of the stresses that occurred during the Tharsis uplift, astronomers estimate the thickness of the Martian crust to be about 100 km.

Mariner 4 detected no Martian magnetic field. Thus, the strength of Mars's magnetic field is less than $1/1000$ that of the Earth—the level of sensitivity of *Mariner*'s instruments. Because Mars is rotating rapidly, we must conclude that either its core is nonmetallic, or not liquid, or both.

The small size of Mars means that any radioactive (or other internal) heating of its interior would have been less effective at heating and melting the planet than heating on Earth. The heat was able to reach the surface and escape more easily than in a larger planet such as Earth or Venus. The evidence we noted above for ancient surface activity, especially volcanism, suggests that at least parts of the planet's interior must have melted and possibly differentiated at some time in the past. But the lack of current activity, the absence of any significant magnetic field, the relatively low density (4 g/cm^3), and an abnormally high abundance of iron at the surface all suggest that Mars never melted as extensively as did Earth.

The history of Mars appears to be that of a planet where large-scale tectonic activity almost started but was stifled by the planet's rapidly cooling outer layers. The large upwelling of material that formed the Tharsis bulge might have developed on a larger, warmer planet into full-fledged plate tectonic motion, but the Martian mantle became too rigid and the crust too thick for that to occur. Instead, the upwelling continued to fire volcanic activity, almost up to the present day, but geologically much of the planet apparently died 2 billion years ago.

CHAPTER SUMMARY

Mars lies about 1½ times farther from the Sun than does the Earth. Its orbit is more elliptical than Earth's, its distance from the Sun varies more. Mars makes a complete circuit around the sun roughly every 2 Earth years. Because of its axial tilt, Mars has daily and seasonal cycles much like those of Earth. Near opposition, Mars, as seen through a telescope, has a reddish disk with dark and light surface patches. The most obvious Martian surface features are the polar caps of dry ice, which grow and diminish as the seasons change on Mars.

In 1971, *Mariner 9* mapped the entire Martian surface, revealing plains, volcanos, channels, and canyons. *Viking 1* and *Viking 2* landed on Mars in 1976 and returned a wealth of data on the planet's surface and atmosphere.

There is a marked asymmetry between the two Martian hemispheres, with rolling plains in the north and heavily cratered highlands in the south. The lack of craters in the north suggests that this region is younger. The single Martian "continent," called the Tharsis Bulge, is located on the equator. Mars supports the largest known volcanos in the solar system. Their height is a direct consequence of Mars's low surface gravity. No evidence for recent or ongoing eruptions has been found.

Runoff channels and outflow channels indicate that water once existed on the surface of Mars. It is possible that nearly all of the water is now locked in the permafrost layer under the surface, with some contained in the polar caps. The two polar caps on Mars consist of a seasonal cap, which grows and shrinks, and a residual cap, which remains permanently frozen. In contrast to Earth, whose iron has mostly differentiated to the planet's core, Mars has iron-rich surface soil that reacted with free oxygen in the early Martian atmosphere to produce iron oxide (rust). That is why Mars appears reddish.

The atmosphere of Mars consists mostly of carbon dioxide. Its pressure is only 1/150 that of Earth's atmosphere. Mars may have had a dense atmosphere 4 billion years ago, but the atmosphere was lost, partly to space and partly to surface rocks and permafrost. Even today, the thin atmosphere is slowly leaking away. Surface temperatures on Mars average about 50 K cooler than those on Earth. Otherwise, Martian weather is reminiscent of that on Earth, with dust storms, clouds, and fog.

Viking tests for life on Mars initially appeared to give positive results, but all have been explained as inorganic chemical reactions.

Mars's moons, Phobos and Deimos, are probably asteroids captured by Mars early in its history. Their densities are far less than any world in the inner solar system. They may be representative of conditions in the early solar system.

Mars has no magnetic field. Since the planet rotates rapidly, it must have a core that is nonmetallic, or not liquid, or both. The lack of current volcanism, the absence of any significant magnetic field, the relatively low density (4 g/cm^3), and a high abundance of surface iron all suggest that Mars never melted as extensively as Earth.

KEY WORDS

conjunction
fluidized ejecta
gas exchange experiment
labeled release experiment
opposition
outflow channel
permafrost
pyrolitic release experiment
residual cap
runoff channel
seasonal cap
tectonic fracture

REVIEW QUESTIONS

1. Discuss the major characteristics of Mars's orbit. When is the best time to see Mars from Earth?
2. What is the evidence that water once flowed on Mars?
3. Is there water on Mars today?
4. For a century, there was speculation that intelligent life had constructed irrigation canals on Mars. What did the "canals" turn out to be?
5. What evidence from the 1960s showed Mars to be relatively uninteresting geologically? What discoveries changed this perception?
6. Describe the asymmetry between the northern and southern hemispheres of Mars in terms of their surface features.
7. Imagine that you will be visiting the southern hemisphere of Mars during its summer. Describe the atmospheric conditions you might face.
8. How do the Martian polar caps and atmosphere interact?
9. Why is Mars red?
10. Why couldn't you breathe on Mars?

11. Why were Martian volcanos able to grow so large?
12. Did the *Viking* landers discover evidence for life on Mars? Explain.
13. How were the masses of Mars's moons measured, and what did these measurements tell us about their origin?
14. What is the evidence that Mars never melted as extensively as did Earth?
15. Compare and contrast the evolution of the atmospheres of Mars, Venus, and Earth. Include a discussion of the importance of volcanos and water in maintaining a carbon dioxide equilibrium.
16. Verify that the surface gravity on Mars is 40 percent that of Earth.
17. What is the synodic period of the Earth as seen from Mars?
18. What is the maximum elongation of the Earth, as seen from Mars (assuming circular orbits for both planets) (at 1.5 A.U.)?
19. How much less sunlight, on average, does each square centimeter of Martian surface receive than each square centimeter on Earth?
20. What is the angular diameter of the Sun seen from Mars?
21. A certain star, observed from Earth, has a parallax of 0.1″. What would be its parallax as seen from Mars? How might observations from Mars-based telescopes be superior to those made from Earth?

DISCUSSION QUESTIONS

1. How would Earth look from Mars?
2. You are writing a Mars travel guide for future visitors from Earth. Which surface features would you highlight in the guide, and what would you write about them?
3. Do you think that sending humans to Mars in the near future is a reasonable goal? Why or why not?
4. Why do you suppose that people mention the Martian moons as good way stations for future visitors to Mars?

PROJECTS

1. Track the motion of the red planet in front of the stars for several months following its return to the predawn sky. (Consult an almanac to determine where Mars is in the sky this year.) You will see that Mars moves rapidly in front of the stars, crossing many constellation boundaries. The planet is moving eastward in front of the stars, and the stars shift continuously toward the west. Can you explain why Mars's orbit appears as it does?
2. Several months before Earth passes between the Sun and Mars, and Mars appears opposite the Sun in our sky (at opposition), the planet begins retrograde motion. Chart the planet's motion in front of the stars to determine when it stops moving eastward and begins moving toward the west.
3. Notice the increase in Mars's brightness as it approaches opposition. Why is it getting brighter? What other planets appear in the sky now? How do their brightnesses compare with that of Mars?

SUGGESTED READINGS

Beatty, J. K. "Images: A New Perspective on Mars." *Sky and Telescope* (June 1987). Stunning mosaic image, constructed from 102 separate pictures of Mars taken from 20,000 miles up by the *Viking* orbiters.

Chaiken, Andrew. "Four Faces of Mars," *Sky and Telescope* (July 1992). Exciting composite images from the U.S. Geological Survey.

Goldman, S. J. "The Legacy of *Phobos 2*." *Sky and Telescope* (February 1990). Scientific bounty from the Soviet *Phobos* mission to Mars.

Haberle, R. M. "The Climate of Mars." *Scientific American* (May 1986). The past and present climate of the planet next door.

Hartmann, W. K. "What's New on Mars?" *Sky and Telescope* (May 1989). Report from an extremely favorable opposition of Mars, plus new looks at old spacecraft data.

Hoyt, W. G. *Lowell and Mars*. Tucson: University of Arizona Press, 1976. A good look at the process of science from a time earlier in this century and a wonderful tale about Mars.

Kasting, J. F., O. B. Toon, and J. B. Pollack. "How Climate Evolved on the Terrestrial Planets." *Scientific American* (February 1988). Why Mars is too cold, Venus is too hot, and Earth is just right.

Murray, Malin, Greeley, et al. *Earthlike Planets*. San Francisco: W. H. Freeman, 1981. Mercury, Venus, Earth, Mars—how they formed and their characteristics today.

Zakharov, A. V. "Close Encounters with Phobos." *Sky and Telescope* (July 1988). Prelaunch report on the Soviet *Phobos* mission to Mars.

CHAPTER

13

Jupiter
The Giant of the Solar System

This is a true-color image of the giant planet Jupiter, made by accumulating separate red, green, and blue images taken in 1991 with the Hubble Space Telescope. *All features seen are cloud formations in the atmosphere of Jupiter, which contains small crystals of frozen ammonia and traces of colorful chemical compounds of carbon, sulfur, and phosphorus. Most prominent at lower right is Jupiter's famed Great Red Spot, a centuries-old, hurricanelike formation large enough to encompass the whole of planet Earth. Note also the unusual tent-shaped structure on the edge of Jupiter's Equatorial Belt, which is the horizontal dark band just above (north of) the Spot. To the left and below the Spot, there is a so-called white oval, one of several that formed around 1940. This image shows more detail than can be seen with any ground-based telescope, yet it is not as sharp as those seen by the* Voyager *spaceprobes that moved rapidly by the planet in 1979. Detailed comparison with the* Voyager *pictures shows that a totally different cloud structure has formed over the past decade. Accordingly, Jupiter will now be monitored from the* Hubble *spaceborne observatory to allow astronomers to investigate meteorology on an alien world. Studies of such very different atmospheric conditions found on Jupiter might help scientists refine their computer atmospheric models in a way that will help them make more accurate forecasts of weather on Earth. (Courtesy NASA)*

R I **V** U X G

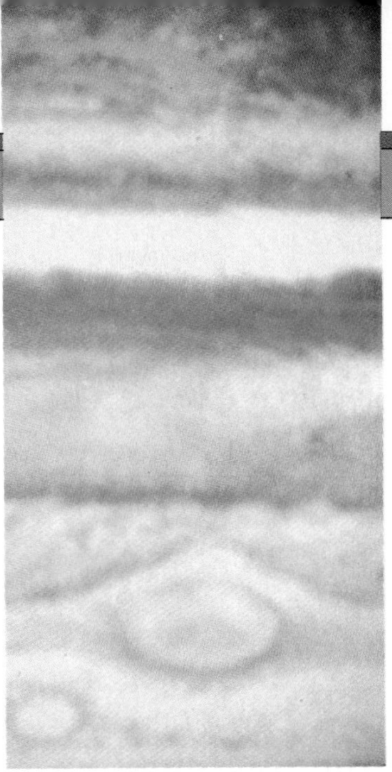

LEARNING GOALS

1. to realize the precision with which the *Voyager* mission skirted the outer planets and changed completely our perception of those worlds
2. to understand some of the processes responsible for the appearance of Jupiter's atmosphere
3. to know why Jupiter radiates more energy than it receives from the Sun
4. to see how Jupiter's internal structure and composition can be inferred from external measurements
5. to appreciate how the Galilean moons form a ''mini solar system'' around Jupiter and exhibit a wide range of properties

Since the 1970s, robotic cameras have taken us into the distant outer realms of our solar system. The U.S. spacecraft that first visited the outer planets revealed in detail worlds that could only have been dreamed of by centuries of Earth-bound astronomers. The spacecraft also found that these distant realms of our planetary system bear little resemblance to our own home or to the other terrestrial planets in our neighborhood. Beyond the orbit of Mars, the solar system is very different from our own backyard.

In sharp contrast to the small, rocky, terrestrial bodies found near the Sun, the outer solar system presents us with a totally unfamiliar environment—huge gas balls, peculiar moons, ringlike structures, and a wide variety of physical and chemical properties, many of which are still only poorly understood. Although the Jovian planets—Jupiter, Saturn, Uranus, and Neptune—differ from each other in many ways, we will find that they have much in common, too. As with the terrestrial planets, we will learn from their differences as well as from their similarities. Our study of these alien places begins with the closest Jovian planet to Earth—Jupiter, the largest planet in the solar system and a model for the other Jovian worlds.

JUPITER'S APPEARANCE FROM EARTH

Named after the most powerful god of the Roman pantheon, Jupiter is by far the largest planet in the solar system. Ancient astronomers could not have known the planet's true size, but their choice of names was very apt. Jupiter is the third-brightest object in the night sky (after the Moon and Venus), making it very easy to locate and study. Like Mars, Jupiter is brightest when it is near opposition. Its angular size is also greatest then—the planet can be up to 50″ (arc seconds) across, and a lot of detail can be discerned through even a small telescope. Figure 13.1 shows two of the best views of Jupiter ever obtained from Earth—one from the ground and one from space. Figure 13.1(a) shows a ground-based photograph of the planet, and Figure 13.1(b) shows a *Hubble Space Telescope (HST)* image taken during the opposition of December 1990. Notice the alternating light and dark bands that cross the planet parallel to its equator and the prominent dark spot at the lower right of Figure 13.1(a).

Jupiter's sidereal orbital period is 11.9 Earth years, so oppositions of Jupiter occur every 399 Earth days (Jupiter's synodic period). The planet's orbit about the Sun has a semi-major axis of 5.20 A.U. (778 million km) and an eccentricity of 0.048, so the closest it can come to Earth is about 3.95 A.U., when opposition happens to occur near perihelion. Again like Mars, Jupiter executes retrograde loops in the sky around the time of opposition, apparently moving backwards relative to the stars as the faster-moving Earth overtakes it (see Chapter 3). Unlike any of the terrestrial planets, Jupiter has many moons, with a wide range of sizes and properties. The four largest are visible from Earth with a small telescope (or even with the naked eye). They are known as the **Galilean moons,** after Galileo Galilei, who first observed them early in the seventeenth century. One of them, Ganymede, can be seen in Figure 13.1(a).

(a) R I V U X G

(b) R I V U X G

Figure 13.1 *(a) Photograph of Jupiter made with a large Earth-based telescope. (b) A* Hubble Space Telescope *image of Jupiter, in true color. Features as small as a few hundred km across are resolved. (Palomar Observatory; NASA)*

JUPITER IN BULK

Mass

The Galilean moons were first observed in the early seventeenth century, so astronomers have been able to study their motion for quite some time. Consequently, Jupiter's mass has long been known. It is 1.9×10^{30} g, or 318 Earth masses. Jupiter has more than twice the mass of all the other planets combined.

Jupiter is such a large planet that many celestial mechanicians—those researchers concerned with the motions of interacting cosmic objects—regard our solar system as containing only two important objects—the Sun and Jupiter. To be sure, in this age of sophisticated and precise spacecraft navigation, the gravitational influence of all the planets must be considered, but in the broadest sense, our solar system is a two-object system with a lot of debris. As massive as Jupiter is, though, it is still some 1000 times less massive than the Sun. This makes studies of Jupiter all the more important, for here we have an object intermediate in size between the Sun and the terrestrial planets.

Radius and Average Density

Knowing Jupiter's distance and angular size, we can easily determine its radius. It is 71,400 km, or 11.2 Earth radii. More dramatically stated, more than 1400 Earths would be needed to equal the volume of Jupiter. From the size and mass, we derive an average density of 1.3 g/cm^3 for the planet. Here (as if we needed it) is yet another indicator that Jupiter is radically different from the terrestrial worlds. It is clear that, whatever Jupiter's composition, it is not made up of the same material as the inner planets (recall from Chapter 7 that the Earth's average density is 5.5 g/cm^3). Studies of the planet's internal structure indicate that Jupiter must be composed primarily of hydrogen and helium. The enormous pressures in the planet's interior greatly compress these light gases, resulting in the high (for hydrogen) average density we observe.

Rotation Rate

As with other planets, we can attempt to determine Jupiter's rotation rate simply by timing a surface feature as it moves around the planet. However, in the case of Jupiter (and, indeed, all the gaseous outer planets), there is a catch. Jupiter has no solid surface. All we see are cloud features in the planet's atmosphere (Figure 13.2). Unlike Earth's atmosphere, different parts of Jupiter's atmosphere, with no solid surface to "tie them down," move independently. Visual observations and Doppler-shifted spectral lines prove that the equatorial zones rotate a little faster (9^h50^m period) than the higher latitudes (9^h56^m period). Thus, Jupiter exhibits **differential rotation**—the rotation rate is not constant from one location to another. Although differential rotation is not possible in solid objects such as the terrestrial planets, it is normal for fluid bodies such as Jupiter.

Observations of Jupiter's magnetosphere provide a more meaningful measurement of the rotation period. The planet's magnetic field is strong and emits radiation at radio wavelengths. Careful studies show a periodicity of 9^h56^m in this radio emission. Scientists assume that this measurement matches the rotation of the planet's interior, where the magnetic field arises. We see that the planet's interior rotates at the same rate as the clouds at its poles. The equatorial zones rotate more rapidly.

A rotation period of 9^h56^m is fast for such a large object. In fact, Jupiter has the fastest rotation rate of *any* solar system object. The strong outward push arising from this spin rate causes a large bulge at the planet's equator. The equatorial radius (71,400 km) exceeds the polar radius (66,800 km) by 6.5 percent. Jupiter is said to have an **oblateness** of 0.065.

The observed oblateness of Jupiter tells us something very important about the planet's deep interior. Careful calculations indicate that Jupiter would be *more* flattened than it actually is if its core were composed of hydrogen and helium alone. To account for the planet's observed oblateness, we must assume that Jupiter has a small, dense, probably rocky core, between 10 and 20 times the mass of the Earth.

The same observations used to determine the rotation period of a planet's surface also let us determine the orientation of its rotation axis. In Jupiter's case, the axis is almost perpendicular to the orbit plane—Jupiter has an axial tilt of only 3°, quite small compared with Earth's tilt of 23½°. With such a small tilt and a nearly circular orbit around the Sun, the planet has a climate that shows very little seasonal change during the course of a year.

Figure 13.2 *We can clearly see Jupiter's belts, bands, and spots in this view taken during* Voyager's *approach in 1979, when the spacecraft was about 50 million km from the planet. (NASA)*

SPACECRAFT EXPLORATION

1 Two pairs of U.S. spacecraft have revolutionized our knowledge of Jupiter and the Jovian planets. The first pair, *Pioneer 10* and *Pioneer 11*, were launched in March 1972 and April 1973, respectively, arriving at Jupiter in December 1973 and December 1974. Among their many important accomplishments, they acted as "scouts" for the upcoming *Voyager* missions. The *Pioneers* showed that spacecraft could travel the long route from Earth to Jupiter without colliding with debris in the solar system. They also discovered—and survived—the perils of Jupiter's extensive radiation belts (somewhat like Earth's Van Allen belts, but on a much larger scale). In addition, *Pioneer 11* used Jupiter's gravity to propel it on the same trajectory to Saturn that the *Voyager* controllers planned for *Voyager 2*'s visit to Saturn's rings.

The *Pioneer* spacecraft also took many photographs and performed measurements not directly related to the upcoming *Voyager* missions and made numerous important scientific discoveries. Their different orbital trajectories also allowed them to observe the polar regions of Jupiter in much greater detail than the *Voyager* missions would achieve.

The two *Voyager* spacecraft (see Figure 13.3) left Earth in 1977, reaching Jupiter in March (*Voyager 1*) and July (*Voyager 2*) of 1979 to study the planet and its major satellites in detail. Each craft carried sophisticated equipment to study the planet's magnetic field and magnetosphere, as well as radio, visible-light, and infrared sensors to analyze its reflected and emitted radiation. Most of the data presented in this and the next two chapters came from *Voyager* sensors.

Both *Voyager 1* and *Voyager 2* used Jupiter's gravity to send them on to Saturn. *Voyager 1* visited Titan, Saturn's largest moon, and so did not come close enough to the planet to receive a gravity-assisted boost to Uranus. However, as we have already seen in Interlude 1-3, *Voyager 2* went on to visit both Uranus and Neptune in a spectacularly successful "Grand Tour" of the outer planets. The data returned by the two craft are still being analyzed today. Like *Pioneer 11*, the two *Voyager* craft are now headed out of the solar system, heading for interstellar space. Figure 13.4 shows the past and present trajectories of the *Voyager* spacecraft.

The most recent mission to Jupiter is the U.S. *Galileo* probe, launched by NASA in 1989. It will arrive at its target

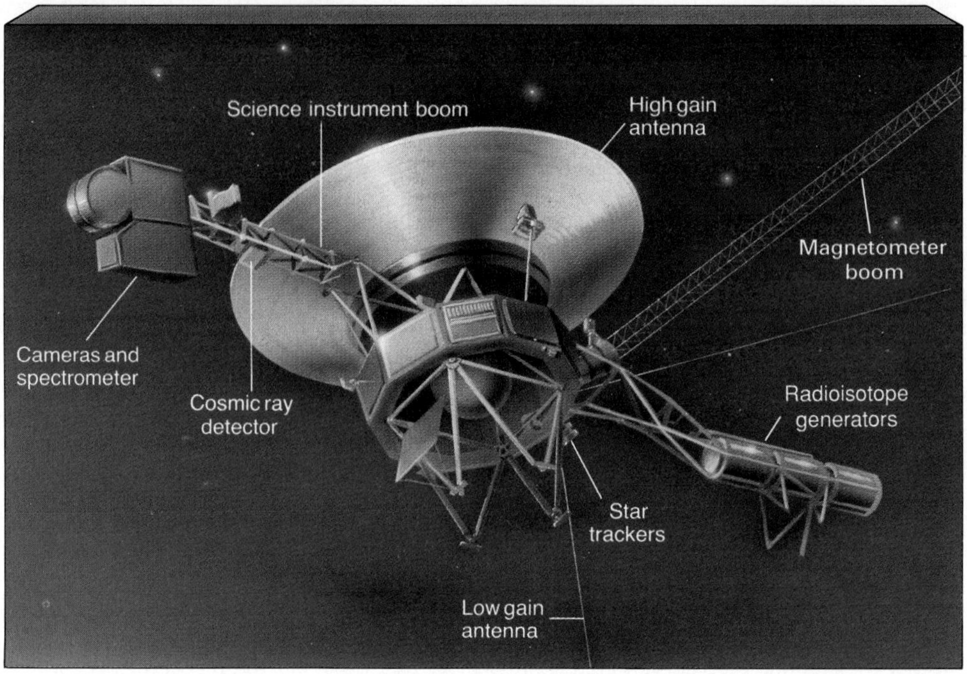

Figure 13.3 The Voyager *spacecraft.* Voyager 1 *and* Voyager 2, *shown here were identical to each other.*

in 1995, after a rather circuitous route (Figure 13.5) involving a gravity assist from Venus and two from Earth itself. The mission consists of an orbiter and an atmospheric probe. The probe will descend into the of Jupiter atmosphere, slowed by a heat shield and a parachute, making measurements and chemical analyses as it goes. The orbiter will execute a complex series of gravity-assisted maneuvers through Jupiter's moon system, returning to some moons already studied by *Voyager* and visiting others for the first time. Scientists around the world eagerly anticipate the results of the mission. Though *Galileo*'s findings may answer many old riddles, they are sure to pose many new ones.

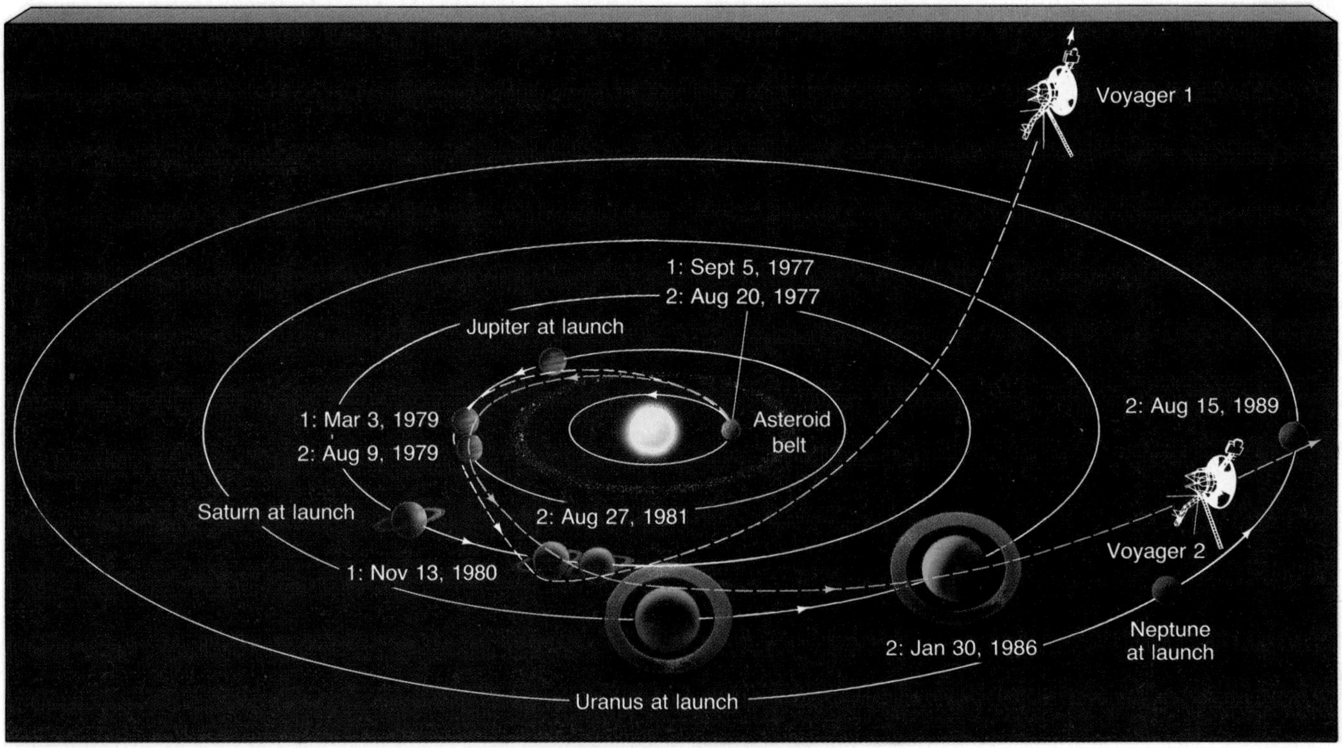

Figure 13.4 The paths followed by the two Voyager *missions to reach the outer planets.* Voyager 1 *is now high above the plane of the ecliptic, having been deflected up and out of the plane of the solar system following its encounter with Saturn.* Voyager 2 *continued on for a "Grand Tour" of the four major outer planets. It is now far beyond the orbit of Neptune.*

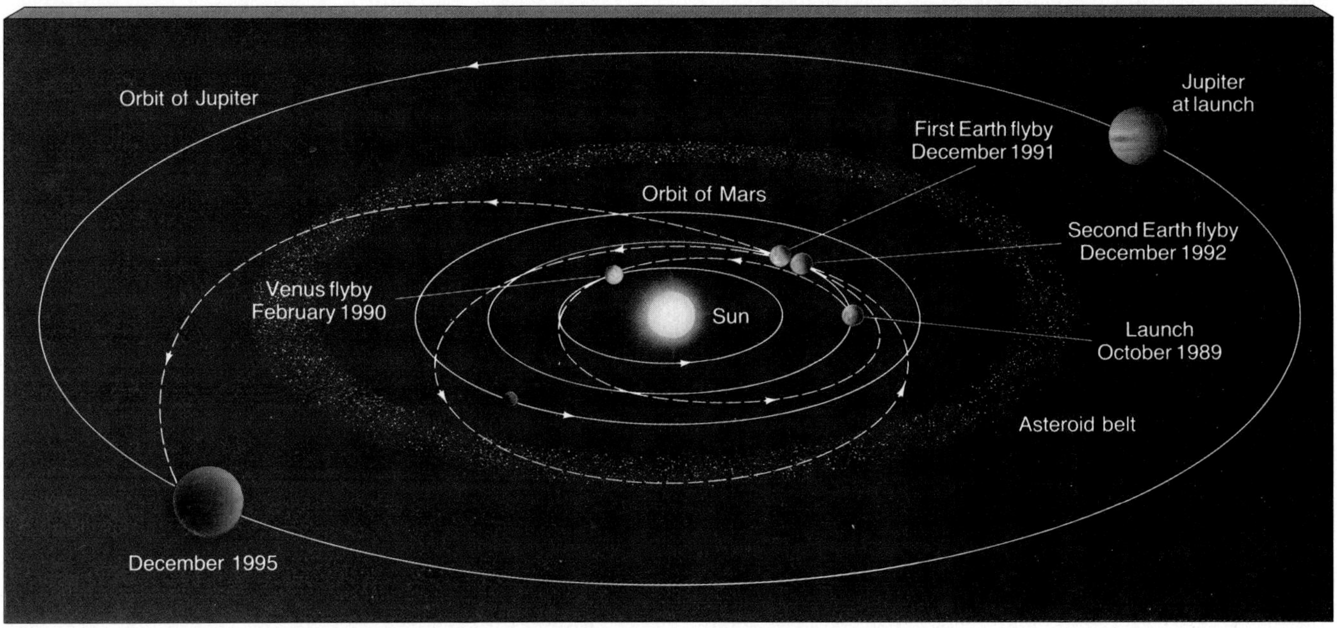

Figure 13.5 Galileo*'s path to Jupiter includes one flyby of Venus, two of Earth, and two trips through the asteroid belt, before arriving at its destination in 1995.*

THE ATMOSPHERE OF JUPITER

✓2 Jupiter is visually dominated by two features, one a series of ever-changing atmospheric bands arranged parallel to the equator and the other an oval atmospheric blob called the **Great Red Spot,** or often just the ''Red Spot.'' The cloud bands, seen clearly in Figure 13.1, display many colors—pale yellows, light blues, deep browns, drab tans, and vivid reds among others. Scientists believe that chemical compounds in Jupiter's atmosphere create these different colors, but the detailed chemistry is still not completely understood. The Red Spot (shown in more detail in Figure 13.6) is one of many features associated with Jupiter's weather. It seems to be an Earth-sized hurricane that has persisted for hundreds of years.

Atmospheric Composition

Spectroscopic studies of sunlight reflected from Jupiter gave scientists their first look at the planet's atmospheric composition. Radio, infrared, and ultraviolet observations later provided more details. The most abundant gas is molecular hydrogen (H_2: 86.1 percent by number), followed by helium (He: 13.8 percent). Together they make up over 99 percent of Jupiter's atmosphere. Scientists generally accept that these two gases also make up the bulk of the planet's interior. This belief is based not on direct evidence of the interior (there is none), but largely on theoretical studies of the internal structure of the planet, such as we have already seen in the discussion of Jupiter's oblateness. Small amounts of atmospheric methane (CH_4), ammonia (NH_3), and water vapor (H_2O) are also found.

Unlike the gravitational pull of the terrestrial planets, the gravity of the larger Jovian planets is strong enough to have retained even hydrogen. Little, if any, of Jupiter's original atmosphere has escaped. Because of the great abundance of hydrogen, all of the common elements other than helium (in particular, carbon, nitrogen, and oxygen) are chemically combined with it.

Atmospheric Structure and Color

None of the atmospheric gases listed above can, by itself, account for Jupiter's observed coloration. For example, frozen ammonia and water vapor would simply produce white

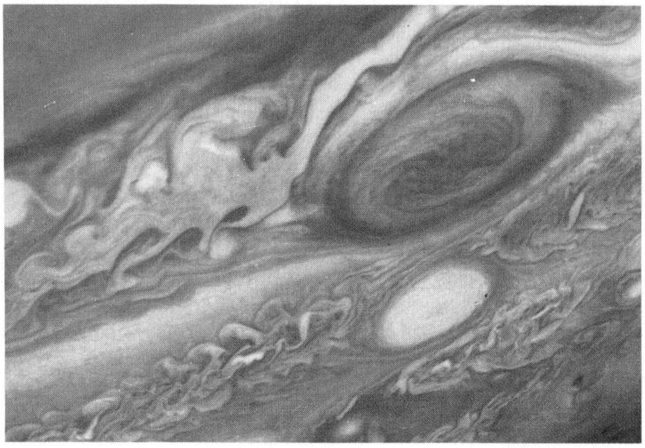

R I **V** U X G

Figure 13.6 Voyager 1 *took this photograph of Jupiter's Red Spot (upper right) from a distance of about 100,000 kilometers. The resolution is about 100 km. Note the complex turbulence patterns to the left of both the Red Spot and the smaller white oval vortex below it. (For scale, planet Earth is about the size of the white oval.) (NASA)* clouds, not the many colors actually seen. We now believe that Jupiter's clouds are arranged in several layers and are the product of complex and continuous chemical processes occurring in the planet's turbulent atmosphere. The various visible clouds lie at different levels—specifically, the white ammonia clouds generally overlie the more brightly colored layers, whose composition we will discuss in a moment. Above the clouds themselves there is a thin, faint layer of haze, created by chemical reactions similar to those that cause smog on Earth. When we observe Jupiter's many colors, we are actually looking down to many different depths in the planet's atmosphere.

Figure 13.7 shows a diagram of Jupiter's atmosphere. The planet lacks a solid surface to use as a reference level for measuring altitude, so by convention scientists take the top of the troposphere to lie at 0 km. As on other planets, the weather on Jupiter is the result of convection in the troposphere, so the colored clouds, which are associated with planetary weather systems, all lie at negative altitudes in the diagram. The haze layer lies at the upper edge of Jupiter's troposphere, at an altitude of zero. The temperature at this level is about 110K. Above the troposphere, as on Earth, the temperature rises as the atmosphere absorbs solar ultraviolet light. Below the haze layer, at a depth of about 40 km (shown as −40 km in Figure 13.7), lie white wispy clouds made up of ammonia ice. At these cloud tops, the temperature is approximately 125–150K. It increases quite rapidly with increasing depth.

A few tens of kilometers below the ammonia clouds, the temperature is a little warmer—over 200K—and the clouds are probably made up mostly of droplets or crystals of ammonium hydrosulfide, produced by reactions between ammonia and hydrogen sulfide in the planet's atmosphere. However, instead of being white (the color of ammonium hydrosulfide on Earth), these clouds are tawny. This is the level at which atmospheric chemistry begins to play its part in determining Jupiter's appearance. Many planetary scientists believe that the element sulfur plays an important role in influencing the cloud colors, particularly the reds, browns, and yellows—all colors associated with sulfur or sulfur compounds. It is also possible that the red and yellow forms of the element phosphorus contribute to the colora-

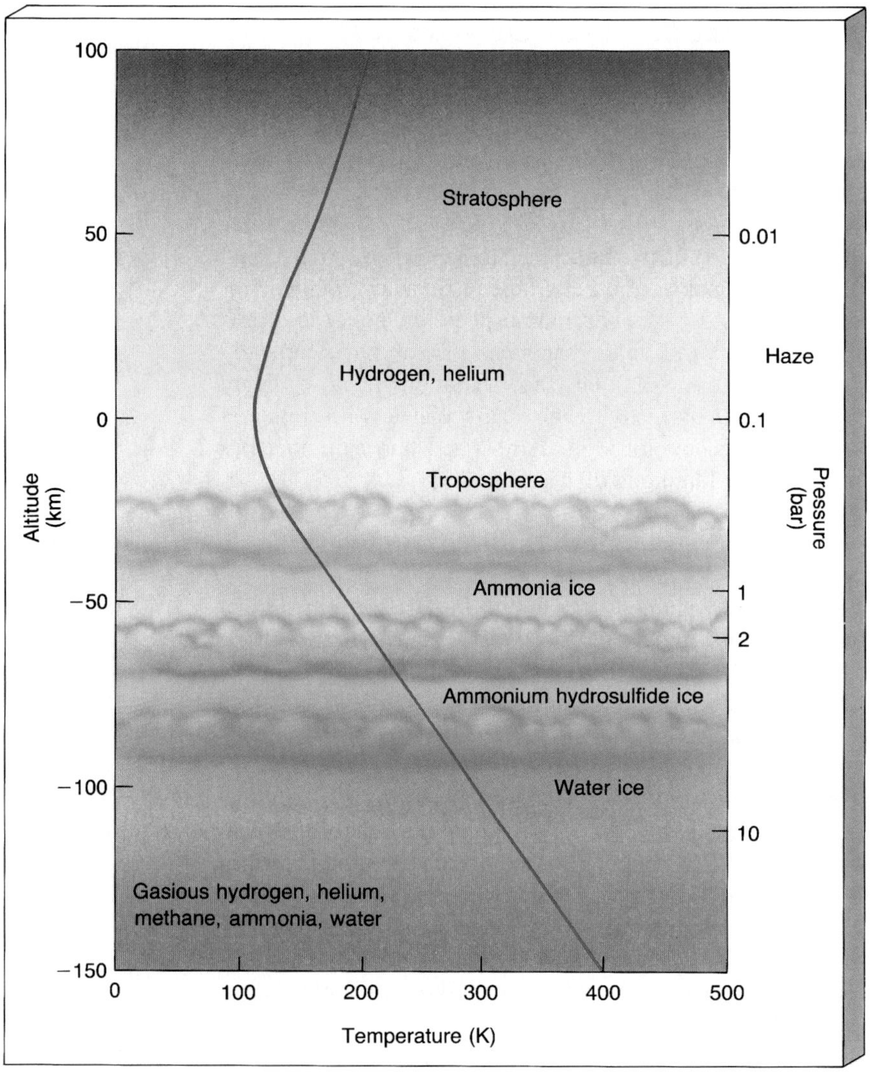

Figure 13.7 The vertical structure of Jupiter's atmosphere. Jupiter's clouds are arranged in three main layers, each with quite different colors and chemistry. The colors we see in photographs of the planet depend on the cloud cover. The white regions are the tops of the upper ammonia clouds. The yellows, reds, and browns are associated with the second cloud layer, which is composed of ammonium hydrosulfide ice. The lowest cloud layer is water ice and bluish in color. However, the overlying layers are sufficiently thick that this level is not seen in visible light.

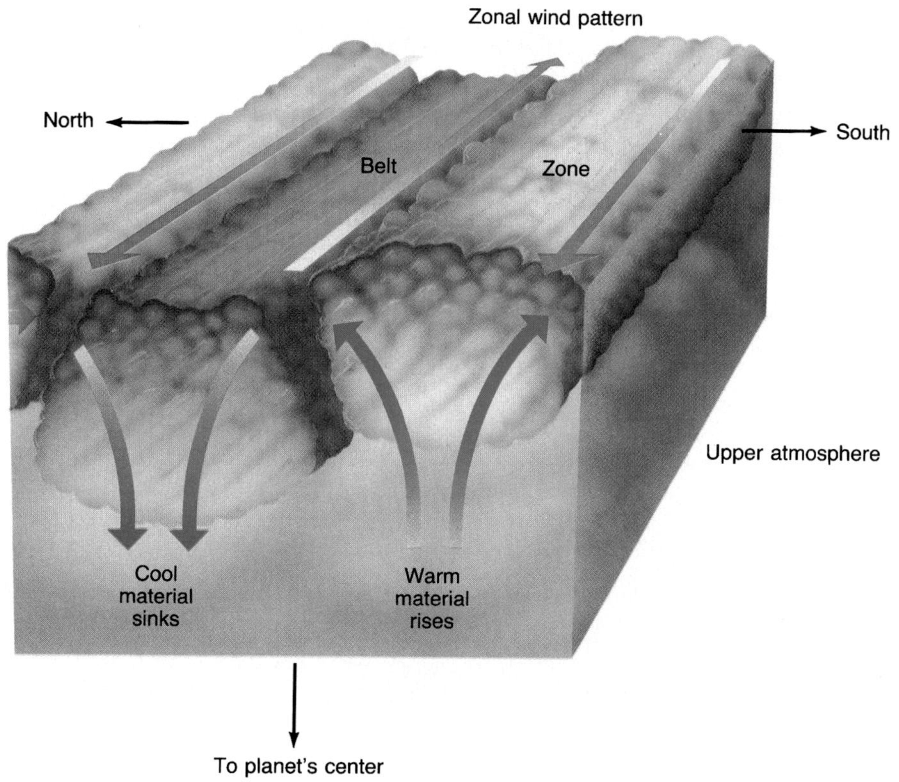

Figure 13.8 *The colored bands in Jupiter's atmosphere are associated with vertical convective motion. Upwelling warm gas results in the lighter-colored zones; the darker bands lie atop lower-pressure regions where cooler gas is sinking back down into the atmosphere. As on Earth, surface winds tend to blow from high- to low-pressure regions. Jupiter's rotation channels these winds into an east-west flow pattern, as indicated.*

tion. At deeper levels in the atmosphere, the ammonium hydrosulfide clouds give way to clouds of water ice or water vapor. This lowest cloud layer, which is not seen in visible-light images of Jupiter, lies some 80 km below the top of the troposphere.

Deciphering the detailed causes of Jupiter's distinctive colors is a difficult task. The cloud chemistry is complex, and it is very sensitive to small changes in atmospheric conditions, such as pressure, temperature, and chemical composition. At the same time, the atmosphere is in incessant, churning motion, causing these conditions to change from place to place and from hour to hour. In addition, the energy that powers the reactions comes in many different forms: the planet's own interior heat, solar ultraviolet radiation, aurorae in the planet's magnetosphere, and lightning discharges within the clouds themselves. All of these factors combine to keep the complete explanation of Jupiter's appearance beyond our present grasp.

An alternative, much more speculative explanation for the clouds' colors has been suggested. Might the colors actually be organic compounds—that is, chemicals produced by living organisms? Jupiter's atmosphere contains all the raw ingredients thought necessary for the appearance of life on Earth, and the frequent lightning discharges on Jupiter could conceivably have powered the same sorts of complex chemical reactions that led to life on our own planet. Furthermore, certain levels in the atmosphere are warm and relatively hospitable, perhaps even providing a suitable environment for living organisms. Fascinating though the possibility of life on Jupiter may be, at present there is simply no hard evidence to support this theory.

Atmospheric Bands

The banded structure of Jupiter's atmosphere is clearly evident in Figure 13.2. Astronomers generally describe its appearance—and, to a lesser extent, the appearance of the other Jovian worlds as well—as a series of bright **zones** and dark **belts** crossing the planet. The zones and belts vary in both latitude and intensity during the year, but the general pattern remains. These variations are not seasonal in nature—Jupiter has no seasons—but instead appear to be the result of dynamic motion in the planet's atmosphere. The light-colored zones lie above upward-moving convective currents in Jupiter's atmosphere. The dark belts are caused by the other part of the convection cycle, representing regions where material is generally sinking downward, as illustrated schematically in Figure 13.8.

Because of the upwelling material below them, the zones are regions of high pressure; the belts, conversely, are low-pressure regions. Thus, the belts and zones are the planet's equivalents of the familiar high- and low-pressure systems that cause our weather on Earth. A major difference is that Jupiter's rapid rotation has caused these systems to wrap all the way around the planet, instead of forming localized circulating storms, as on our own world. Because of the pressure difference, the zones lie slightly higher in the atmosphere than the belts. The temperature difference between the two (recall that the temperature decreases with altitude) is the basic reason for their different colors.

Underlying the bands is an apparently very stable pattern of eastward and westward wind flow, often referred to as Jupiter's **zonal flow.** This zonal flow is evident in Figure

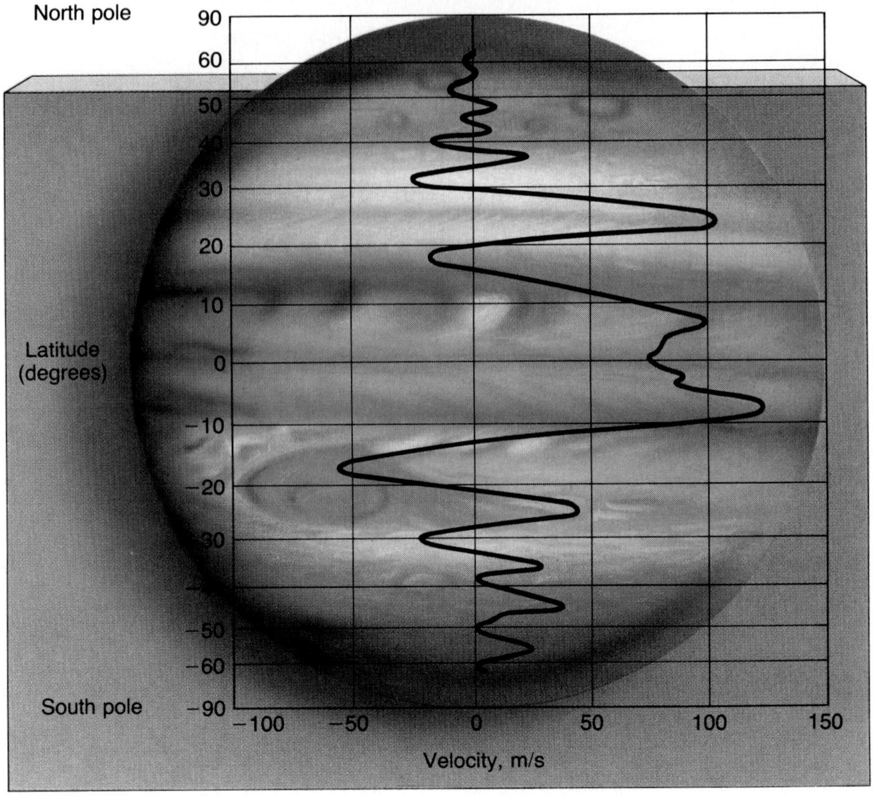

Figure 13.9 *The wind speed in Jupiter's atmosphere, measured relative to the planet's internal rotation rate. The alternations in wind direction are associated with the atmospheric band structure.*

13.9, which shows the wind speed at different planetary latitudes measured relative to the rotation of the planet's interior (determined from studies of Jupiter's magnetic field). As we have already seen, the equatorial regions of the atmosphere rotate faster than the planet, with an average flow speed of some 85 m/s, or about 300 km/h, in the easterly direction. The speed of this equatorial flow is quite similar to that of the jet stream on Earth. At higher latitudes, there are alternating regions of westward and eastward flow, roughly symmetric about the equator, with the flow speed generally diminishing toward the poles.

As Figure 13.9 shows, the belts and zones are closely related to Jupiter's zonal flow pattern. However, closer inspection shows that the simplified picture presented in Figure 13.8, with wind direction alternating between adjacent bands, as Jupiter's rotation deflects surface winds into eastward or westward streams, is really too crude to explain the actual flow. Scientists now believe that the interaction between convective motion in Jupiter's atmosphere and the planet's rapid rotation channels the largest eddies into the observed zonal pattern, but that smaller eddies cause irregularities in the flow. Near the poles, where the zonal flow disappears, the band structure vanishes also.

The Weather on Jupiter

In addition to the zonal flow pattern, Jupiter has many "small-scale" weather patterns. The Great Red Spot, shown in Figure 13.6—a close-up photograph taken as the *Voyager 1* spacecraft glided past in 1979—is a prime example. The Great Red Spot was first reported by the British scientist Robert Hooke in the mid-seventeenth century, so we can be reasonably sure that it has existed in one form or another for more than 300 years, and it might well be much older. *Voyager* observations show the spot to be a region of swirling, circulating winds, rather like a whirlpool or a terrestrial hurricane—a persistent and vast atmospheric storm. The size of the Spot varies, although it averages about twice the diameter of the Earth. Its present dimensions are roughly 25,000 km by 15,000 km. It rotates around Jupiter at a rate similar to the planet's interior, perhaps suggesting that its roots lie far below the atmosphere.

The origin of the Spot's red color is unknown, as is its source of energy, although it is generally supposed that the Spot is somehow sustained by Jupiter's large-scale atmospheric motion. Repeated observations show that the gas flow around the Spot is counterclockwise, with a period of about 6 days. Turbulent eddies form and drift away from its edge. The Spot's center, however, remains quite tranquil in appearance, like the eye of a hurricane on Earth. The zonal motion north of the Spot is westward, while that to the south is eastward (see Figure 13.9), supporting the idea that the Spot is confined and powered by the zonal flow. However, the details of how this occurs are still a matter of conjecture. Computer simulations of the complex fluid dynamics of Jupiter's atmosphere are only now beginning to hint at answers.

Actually, storms, which as a rule are much smaller than the Red Spot, may be quite common on Jupiter. Spacecraft photographs of the dark side of the planet reveal both auroral activity and bright flashes resembling lightning. The *Voyager* mission discovered many smaller light-and-dark-

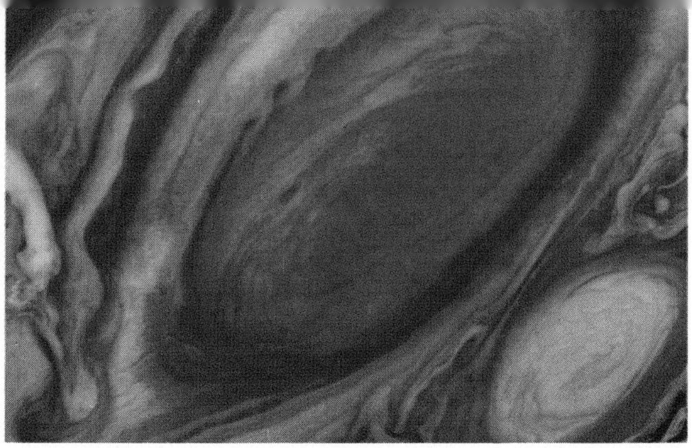

Figure 13.10 *This* Voyager 2 *close-up view of the Red Spot shows clearly the turbulent flow around its edges. The general direction of motion of the gas north of (above) the Spot is westward (to the left), while gas south of the Spot flows east. The Spot itself rotates counterclockwise, suggesting that it is being "rolled" between the two oppositely directed flows. The colors have been exaggerated somewhat to enhance the contrast. (NASA)*

colored spots that are also apparently circulating storm systems. Note the several **white ovals** in Figures 13.6 and 13.10, south of the Red Spot. Like the Red Spot, they rotate counterclockwise. Their high cloud tops give them their color. These particular white ovals are known to be at least 40 years old. Figure 13.11 shows a **brown oval,** a "hole" in the clouds that allows us to look down into the lower atmosphere. For unknown reasons, brown ovals appear

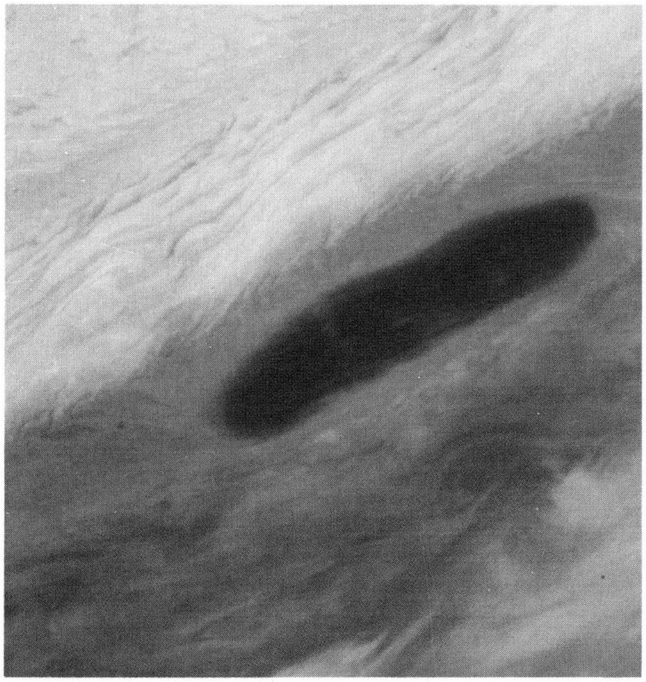

Figure 13.11 *A brown oval in Jupiter's northern hemisphere. Its color comes from the fact that it is actually a break in the upper cloud layer, allowing us to see deeper in. The oval's length is approximately equal to the diameter of the Earth. (NASA)*

only in latitudes around 20°N. Although not as long-lived as the Red Spot, these systems can persist for many years or even decades.

We cannot explain their formation, but we can offer at least a partial explanation for the longevity of these storm systems on Jupiter. On Earth a large storm, such as a hurricane, forms over the ocean and may survive for many days, but it dies quickly once it encounters land. The Earth's continental landmasses disrupt the flow patterns that sustain the storm. Jupiter has no continents, so once a storm is established and has reached a size at which other storm systems cannot destroy it, apparently little affects it. The larger the system, the longer its lifetime.

SURFACE TEMPERATURE AND INTERNAL HEATING

✓3 On the basis of Jupiter's distance from the Sun, astronomers had expected to find that the temperature of the cloud tops was around 105K. At that temperature, they reasoned, Jupiter would radiate back into space exactly the same amount of energy as it received from the Sun. When radio and infrared observations were first made of the planet, however, astronomers found that its Planck spectrum corresponded to a temperature of 125K instead. Subsequent measurements, including those of *Voyager*, as we have just seen, have verified that finding. Although a difference of 20K may seem small, recall from Chapter 5 that the energy emitted by a planet grows as the *fourth* power of the surface temperature (in Jupiter's case, the temperature of the cloud tops). A planet at 125K radiates $(125/105)^4$, or about twice as much energy as a 105K planet. Put another way, Jupiter actually emits about twice as much energy as it receives from the Sun. Thus, unlike any of the terrestrial planets, Jupiter must have its own internal heat source.

What is responsible for Jupiter's extra energy? It is not the decay of radioactive elements within the planet—that must be occurring, but not at nearly the rate necessary to produce the temperature we record. Nor is it the process that generates energy in the Sun, nuclear fusion—the temperature in Jupiter's interior, high as it is, is far too low for that. (See Interlude 13-1.) Instead, astronomers theorize that the source of Jupiter's excess energy is the slow escape of gravitational energy released during the planet's formation. As the planet took shape, some of its energy was converted into heat in the interior. That heat is still slowly leaking out through the planet's heavy atmospheric blanket, resulting in the excess emission we observe. Despite the huge amounts of energy involved—Jupiter's energy emission is about 4×10^{17} watts more than it receives from the Sun—the energy loss is quite slight compared with the planet's total energy. (A typical household light bulb is 75 watts.) A simple calculation indicates that the average temperature of the interior of Jupiter falls by only about a millionth of a kelvin per year.

INTERLUDE 13-1 *Almost a Star?*

Jupiter has a starlike composition—predominantly hydrogen and helium, with a trace of heavier elements. Did Jupiter ever come close to becoming a star itself? Might the solar system have formed as a double-star system? Probably not. Unlike a star, Jupiter is cold. Its central temperature is far too low to ignite the nuclear fires that power our Sun. Jupiter's mass would have to increase eighty-fold before its central temperature rose to the point where nuclear reactions could begin, converting Jupiter into a small, dim star. Even so, it is interesting to note that although Jupiter's present-day energy output is very small (by solar standards, at least), it must have been much greater in the distant past, while the planet was still contracting rapidly toward its present size. For a brief period of time—a few hundred million years—Jupiter may actually have been as bright as a faint star, although its brightness never came within a factor of 100 of the Sun. Seen from Earth at that time, Jupiter would have been about 100 times brighter than the Moon!

What might have happened had our solar system formed as a double-star system? Conceivably, had Jupiter been massive enough, its radiation might have produced severe temperature fluctuations on all the planets, perhaps to the point of making life on Earth impossible. Even if Jupiter's brightness were too low to cause us any problems, its gravitational field (which becomes 1/12 that of the Sun if we increase the mass by a factor of 80) could have made the establishment of stable, roughly circular planetary orbits a fairly improbable event. The size of the ''Jupiter,'' or second-largest body, in a new-born planetary system may be a very important factor in determining the likelihood of the appearance of life.

INTERNAL STRUCTURE

4 Jupiter's clouds, along with their complex chemistry, are probably less than 200 km thick. Below them, the temperature and pressure steadily increase, as the atmosphere becomes the ''interior'' of the planet. Much of our knowledge of Jupiter's interior comes from theoretical modeling. We have no seismographic information on Jupiter, and no prospect of obtaining any, for the very good reason that there is no solid surface to experience a tremor! Instead, we must use all available bulk data on the planet—mass, radius, composition, rotation, temperature, and so on—to construct a model of the interior that agrees with observations. Our statements about Jupiter's interior are, then, really statements about the model that best fits the facts. However, because the interior consists largely of hydrogen and helium—two simple gases whose physics we think we understand well—we can be fairly confident that Jupiter's internal structure is now understood.

Figure 13.12 shows that both the temperature and the density of Jupiter's atmosphere increase with depth below the cloud cover. However, no ''surface'' of any kind exists anywhere inside. Instead, Jupiter's atmosphere just becomes denser and denser, because of the pressure of the overlying layers. At a depth of a few thousand kilometers, the gas makes a gradual transition into the liquid state. By a depth of about 20,000 km, the pressure is about 3 million times greater than atmospheric pressure on Earth. Under those conditions, the hot liquid hydrogen is compressed so much that it undergoes another transition, this time to a ''metallic'' state, with properties in many ways similar to a liquid metal. Of particular importance for Jupiter's magnetic field, this metallic hydrogen is an excellent conductor of electricity.

As we have already mentioned, Jupiter's observed oblateness requires that there be a small, dense core at its center, containing perhaps 15 times the mass of the Earth. The core's exact composition is unknown, but planetary

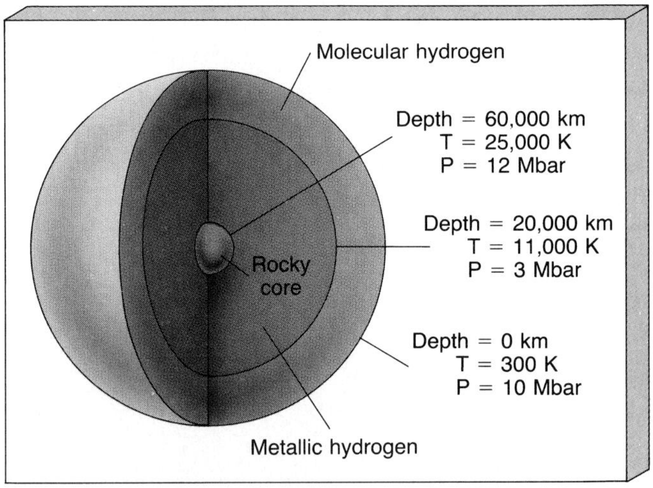

Figure 13.12 *Jupiter's internal structure, as deduced from* Voyager *measurements and theoretical modeling. The outer radius represents the top of the cloud layers, some 70,000 km from the planet's center. The density and temperature increase with depth, and the atmosphere gradually liquefies over the outermost few thousand kilometers. Below a depth of 20,000 km, the hydrogen behaves like a liquid metal. At the center of the planet lies a large rocky core, somewhat terrestrial in composition but much larger than any of the inner planets. Although very uncertain, the temperature and density at the center are probably about 40,000K and 50 Mbar, respectively.*

Figure 13.13 *The* Pioneer 11 *spacecraft took this photograph of convection cells while passing Jupiter in late 1974. The probe was about 1 million km from the cloud tops at the time. (NASA)*

scientists think that it contains heavier materials than the rest of the planet. Present best estimates indicate that it consists of "rocky" materials, similar to those found in the terrestrial worlds. In fact, it now appears that all of the Jovian planets contain large rocky cores and that the formation of such a large "terrestrial" planetary core is a necessary stage in the process of building up a gas giant. Because of the high pressure at the center of Jupiter—approximately 50 million times that on Earth's surface, or 10 times that at its center—the core must be compressed to quite high densities (perhaps twice the core density of the Earth). It is probably not much more than 20,000 km in diameter, and the central temperature may be as high as 40,000K.

This theoretical model does have some direct observational support. In particular, if Jupiter is made up of cold gas lying atop hot gas, we can confidently expect to find evidence for convection, just as in the Earth's atmosphere and interior. Figure 13.13 is a photograph of numerous convection cells near one of Jupiter's poles.

These circular, reddish patches bubble up through the upper atmosphere, essentially enabling the planet to "breathe" or vent its heat. Convection of this sort is probably the source of the day-to-day variations of Jupiter's clouds that we can observe even with a small telescope. Larger-scale convection produces the planet's banded structure and powers the large-scale wind flow.

JUPITER'S MAGNETOSPHERE

For decades, ground-based radio telescopes monitored radiation leaking from Jupiter's magnetosphere, but only when the *Pioneer* and *Voyager* spacecraft reconnoitered the planet in the mid-1970s did astronomers realize the full extent of the planet's magnetic field. Jupiter, as it turns out, is surrounded by a vast sea of energetic charged particles, mostly electrons and protons, somewhat similar to Earth's Van Allen belts but much larger. As in the case of the Earth, the size of the magnetosphere is determined by the interaction between the planet's magnetic field and the solar wind. Outside, the solar-wind particles flow freely away from the Sun, past the planet. Inside, their motions are governed by the planetary magnetic field. On the sunward side, the magnetopause—the boundary of Jupiter's magnetic influence on the solar wind—lies about 3 million km from the planet.

Direct spacecraft measurements show Jupiter's magnetosphere to be almost 30 million km across, roughly a million times more voluminous than Earth's magnetosphere and far larger than the entire Sun. It has a long tail extending away from the Sun at least as far as Saturn's orbit (at a distance of 4 A.U.), as sketched in Figure 13.14. The outer magnetosphere appears to be quite unstable, sometimes deflating in response to "gusts" in the solar wind, then grow-

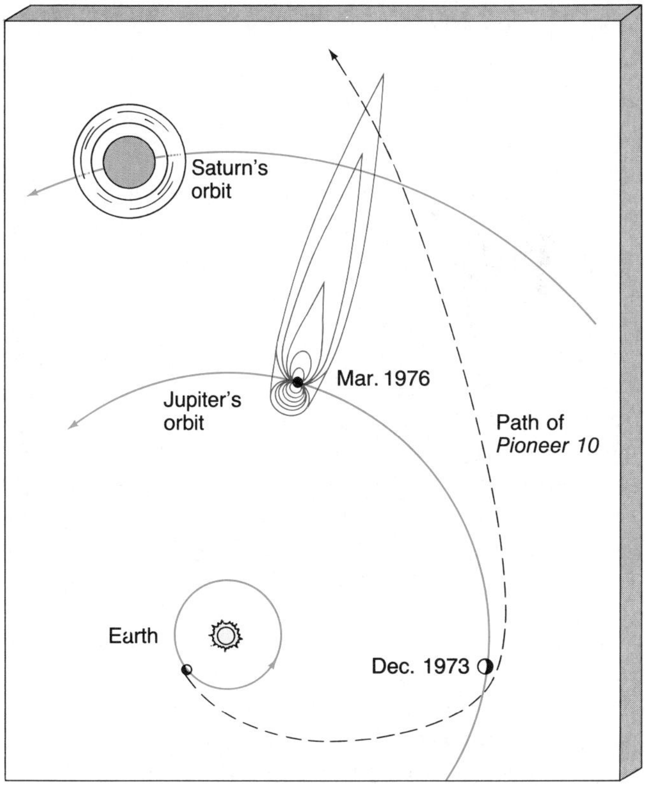

Figure 13.14 *The* Pioneer 10 *spacecraft did not detect any solar particles while moving behind Jupiter. Accordingly, as sketched here, Jupiter's magnetosphere apparently extends beyond the orbit of Saturn.*

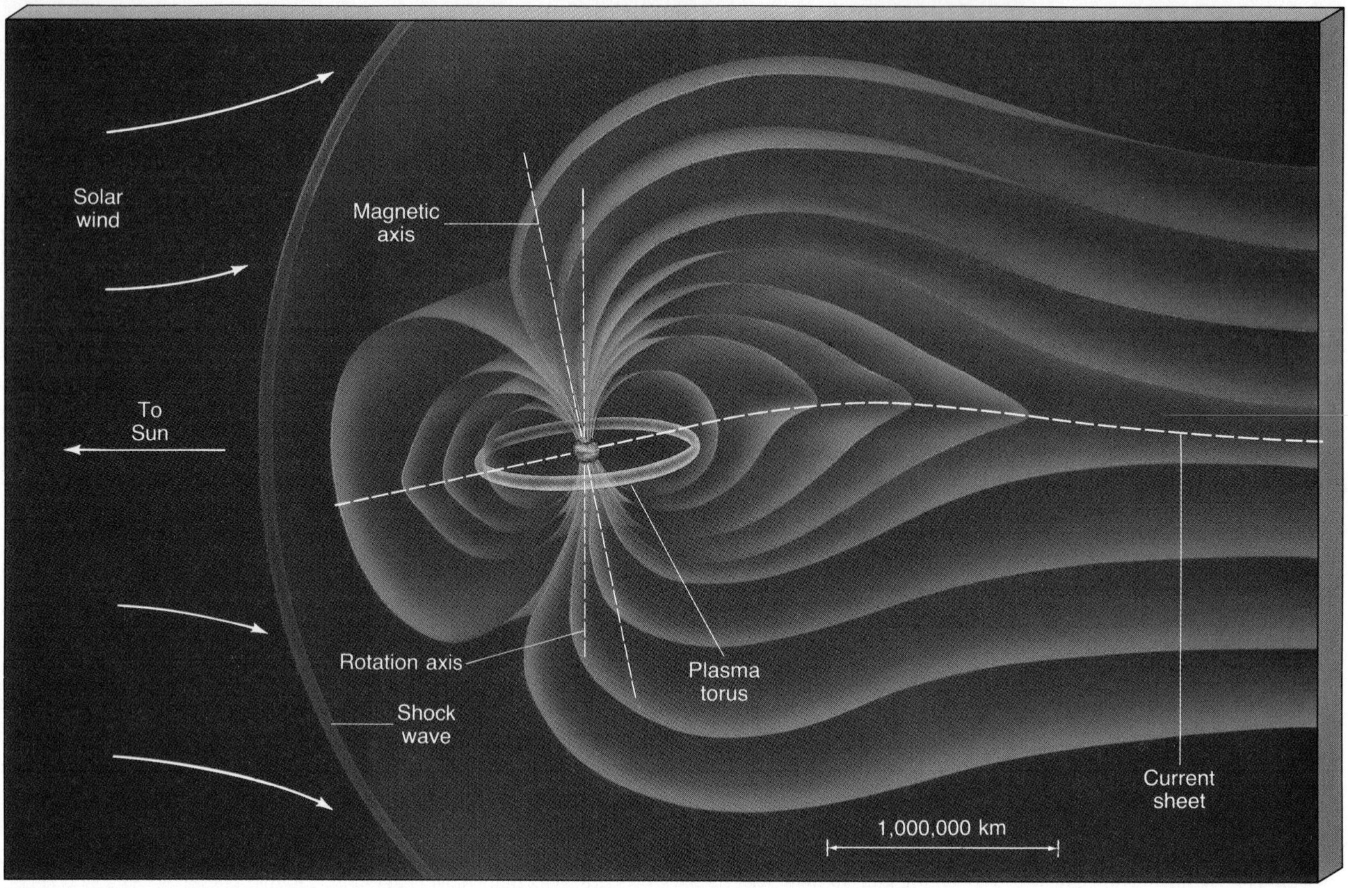

Figure 13.15 *Jupiter's inner magnetosphere is characterized by a flat current sheet, consisting of charged particles squeezed into the magnetic equatorial plane by the planet's rapid rotation.*

ing back. In the inner magnetosphere, Jupiter's rapid rotation has forced most of the charged particles into a flat **current sheet,** lying on the planet's magnetic equator. The inner magnetosphere is sketched in Figure 13.15. Notice that the planet's magnetic axis is not exactly aligned with its rotation axis, but is inclined to it at an angle of approximately 10°.

The radiation associated with Jupiter's magnetosphere is several thousand times more intense than Earth's, creating a tremendous hazard for manned and unmanned space vehicles. Sensitive electronic equipment (not to mention even more sensitive human bodies) would require special protective shielding to operate for long in this hostile environment. The radiation is produced by energetic particles, mostly electrons, that are swept up in Jupiter's powerful magnetic field and accelerated to very high speeds—in fact very close to the speed of light. Under those conditions, the particles radiate with a spectrum quite unlike the Planck spectrum produced by a thermal (''black-body'') emitter, as illustrated in Figure 13.16. For this reason, the radiation is known as **nonthermal radiation.**

Ground-based and space-borne observations of Jupiter's nonthermal radiation imply that the *intrinsic* strength of the planet's magnetic field is nearly 20,000 times greater than Earth's. The existence of such a strong field further supports our theoretical model of the interior; the conducting liquid interior that is thought to make up most of the planet should combine with Jupiter's rapid rotation to produce a large dynamo effect and a strong magnetic field, just as observed.

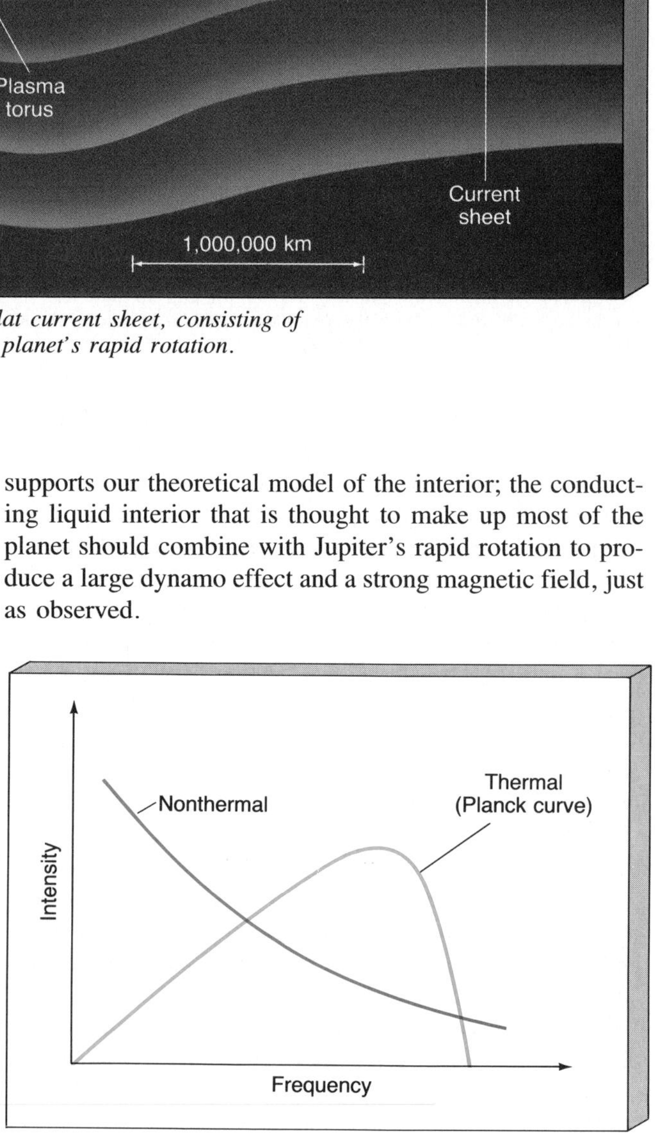

Figure 13.16 *Thermal and nonthermal radiation spectra are usually very easily distinguished. Shown are a thermal Planck spectrum and the nonthermal spectrum of the radiation received from Jupiter's magnetosphere.*

THE MOONS OF JUPITER

✓5 Jupiter has at least 16 moons. In many ways, the entire Jupiter system resembles a miniature solar system. Its four largest moons—the Galilean satellites—are each comparable in size to Earth's Moon. Moving outward from Jupiter, the four are named Io, Europa, Ganymede, and Callisto, after the mythical attendants of the Roman god Jupiter. They move in nearly circular orbits about their parent planet. When the *Voyager 1* spacecraft passed close to the Galilean moons in 1979, it sent some remarkably detailed photographs back to Earth, allowing planetary scientists to discern fine surface detail on each moon and greatly expanding our knowledge of these small, distant worlds. We will consider the Galilean satellites in more detail in a moment.

Within the orbit of Io lie four small satellites, all but one discovered by *Voyager* cameras. The largest of the four, Amalthea, is less than 300 km across and is irregularly shaped. E. E. Barnard discovered it in 1892. It orbits at a distance of 181,000 km from Jupiter's center—only 110,000 km above the cloud tops. Its rotation, like that of most of Jupiter's satellites, is synchronous with its orbit because of Jupiter's strong tidal field. Amalthea rotates once per orbit period, every 11.7 hours.

Beyond the Galilean moons lie eight more small satellites, all discovered in the twentieth century, but before the *Voyager* missions. They fall into two groups of four moons each. The moons in the inner group move in eccentric, inclined orbits, about 11 million km from the planet. The outer four moons lie about 22 million km from Jupiter. Their orbits too are fairly eccentric, but *retrograde*, moving in a sense opposite to all the other moons (and to Jupiter's rotation). It is very likely that each group represents a single body that was captured by Jupiter's strong gravitational field long after the planet and its larger moons originally formed. Both bodys subsequently broke up, either during or after the capture process, resulting in the two families of similar orbits we see today. The masses, and hence the densities, of these small worlds are unknown. However, their appearance and sizes suggest compositions more like asteroids than their larger Galilean companions. Table 13-1 presents the general properties of Jupiter's moons.

The Galilean Moons as a Model of the Inner Solar System

If we think of Jupiter's moon system as a scaled-down solar system, the Galilean moons correspond to the terrestrial planets. Their orbits are direct (that is, in the same sense as Jupiter's rotation), roughly circular, and lie close to Jupiter's equatorial plane. They range in size from slightly smaller than Earth's Moon (Europa) to slightly larger than Mercury (Ganymede). The parallel with the inner solar system continues with the realization that their densities decrease with increasing distance from Jupiter. It is quite likely that the inner two Galilean moons, Io and Europa, have a rocky composition, possibly similar to the crusts of the terrestrial planets. The two outer Galilean moons, Ganymede and Callisto, are clearly deficient in rocky materials. Lighter materials, such as water ice, may account for as much as half their total mass. Figure 13.17 compares the appearances and sizes of the four Galilean satellites. Figure 13.18 shows two of the moons in comparison to their parent planet.

Many astronomers think that the formation of Jupiter and the Galilean satellites mimicked on a small scale the formation of the Sun and the inner planets. For that reason,

TABLE 13-1 ***The Moons of Jupiter***

Name	Distance (km; planet radii)	Orbit Period (days)	Size (longest diameter, km)	Mass (Earth Moon masses)	Density (g/cm^3)	Albedo
Metis	128,000; 1.79	0.29	40			0.05
Adastea	129,000; 1.81	0.30	20			0.05
Amalthea	181,000; 2.54	0.50	200			0.06
Thebe	222,000; 3.11	0.67	90			0.05
Io	422,000; 5.91	1.77	3630	1.22	3.6	0.6
Europa	671,000; 9.40	3.55	3140	0.65	3.0	0.6
Ganymede	1,070,000; 15.0	7.16	5260	2.02	1.9	0.4
Callisto	1,880,000; 26.4	16.7	4800	1.47	1.9	0.2
Leda	11,100,000; 155	239	15			
Himalia	11,500,000; 161	251	180			0.03
Lysithea	11,700,000; 164	259	40			0.05
Elara	11,700,000; 164	260	80			0.03
Ananke	21,200,000; 297	631*	30			
Carme	22,600,000; 317	692*	40			
Pasiphae	23,500,000; 329	735*	70			
Sinope	23,700,000; 332	758*	40			

*Indicates a retrograde orbit

Figure 13.17 *The* Voyager 1 *spacecraft photographed each of the four Galilean moons of Jupiter. Shown here to scale, as they would appear from a distance of about 1 million km, they are, clockwise from upper left, Io, Europa, Callisto, and Ganymede. (NASA)*

R I **V** U X G

studies of the Galilean moon system may provide us with valuable insight into the processes that created our own world. We will return to this parallel in Chapter 17. But let us point out that not all of the properties of the Galilean moons find analogs in the inner solar system. For example, because of Jupiter's tidal effect, all four Galilean satellites are in states of synchronous rotation, so that they all keep one face permanently pointing toward their parent planet. By contrast, of the terrestrial planets, only Mercury is strongly influenced by the Sun's tidal force, and even its orbit is not synchronous. And, of course, the Jupiter system has no analogs of the Jovian planets.

Inspection of Table 13-1 shows a remarkable coincidence in the orbit periods of the three inner Galilean moons: Their periods are almost exactly in the ratio 1 : 2 : 4—a kind of "Bode's Law" for Jupiter. This is most probably the result of a complex, but poorly understood, three-body resonance in the Galilean moon system, something not found among the terrestrial worlds.

R I **V** U X G

Figure 13.18 Voyager 1 *took this photo of Jupiter with ruddy Io on the left and pearl-like Europa toward the right. Note the scale of objects here: Io and Europa are each comparable to the size of our Moon, and the Red Spot (seen here to the left bottom) is roughly twice as big as Earth. (NASA)*

Io

Io, the densest of the Galilean moons, is the most geologically active object in the entire solar system. Its mass and radius are fairly similar to those of Earth's own Moon, but there the resemblance ends. Shown in Figure 13.19, Io's surface is a collage of reds, yellows, and blackish browns—resembling a giant pizza in the minds of some startled *Voyager* scientists. As the spacecraft glided past Io, an outstanding discovery was made: Io has active volcanos! *Voyager 1* photographed eight erupting volcanos, and six were still erupting when *Voyager 2* passed by 4 months later. In Figure 13.20, one volcano is seen ejecting matter to an altitude of over 200 km. The gases are spewed forth at speeds up to 2 km/s, quite unlike the (relatively) sluggish ooze that emanates from Earth's insides.

The orange color immediately surrounding the volcano most likely results from sulfur compounds in the ejected material. In stark contrast to the other Galilean

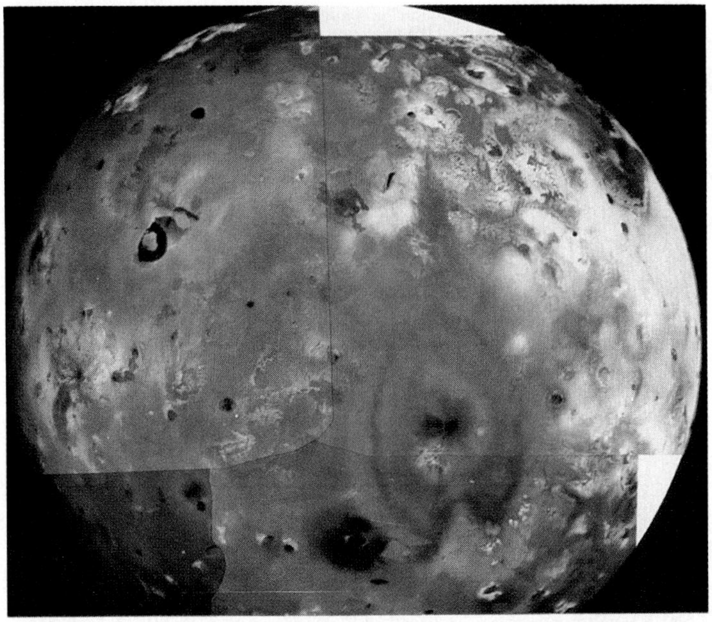

R I V U X G

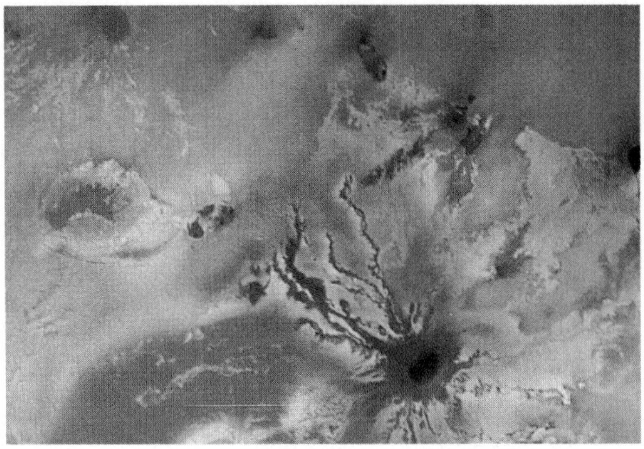

R I V U X G

Figure 13.19 *Jupiter's innermost moon, Io, is quite different in character from the other three Galilean satellites. Its surface is kept smooth and brightly colored by the moon's constant volcanism. The resolution of the photograph on the left is about 7 km. In the more detailed image on the right, features as small as 2 km across can be seen. (NASA; NASA)*

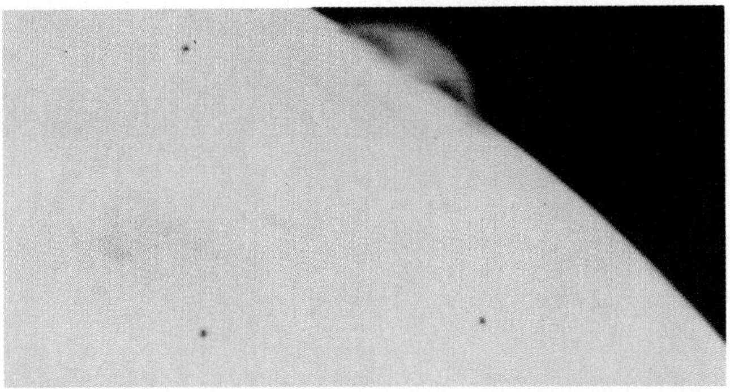

R I V U X G

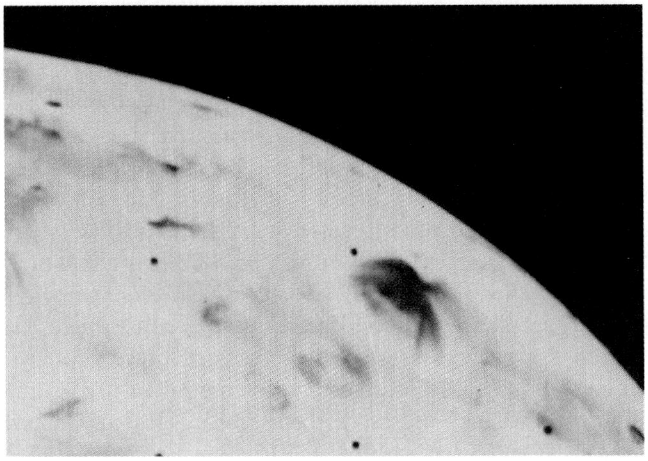

R I V U X G

Figure 13.20 *One of Io's volcanos was caught in the act of erupting while the* Voyager *spacecraft flew past this fascinating moon. Surface features here are resolved to within a few kilometers. At the top, the volcano's umbrella-like profile shows clearly against the darkness of space. The plume measures about 100 km high and 300 km across. At the bottom, several jets of volcanic ejecta (dark against Io's brighter surface) can be discerned as* Voyager *prepares to "overfly" another volcano. (NASA)*

moons, Io's surface is neither cratered nor streaked. (The circular features visible in Figure 13.20 are volcanos.) Its surface is exceptionally smooth, apparently the result of molten matter constantly filling in any "dents and cracks." Accordingly, we can conclude that this remarkable moon has the youngest surface of any known object in the solar system. Of further significance, Io also has a thin, temporary atmosphere made up primarily of sulfur dioxide, presumably the result of gases ejected by volcanic activity.

Io's volcanism has a major effect on Jupiter's magnetosphere. All of the Galilean moons orbit within the magnetosphere and play some part in modifying its properties, but Io's influence is particularly marked. Although many of the charged particles in Jupiter's magnetosphere come from the solar wind, there is strong evidence that Io's volcanism is the primary source of heavy ions in the inner regions. Jupiter's magnetic field continually sweeps past Io, gathering up the particles its volcanos spew into space and accelerating them to high speed. The result is the **Io plasma torus** (shown in Figure 13.21), a doughnut-shaped region of energetic heavy ions that follows Io's orbital track, completely encircling Jupiter. (A plasma is a gas that has been heated to such high temperatures that all of its atoms are ionized.) It is quite easily detectable from Earth, but before *Voyager* its origin was unclear. Spectroscopic analysis shows that sulfur is indeed one of the torus's major constituents, strongly implicating Io's volcanos as its source. As a hazard to spacecraft—manned or unmanned—the plasma torus is formidable. The radiation levels there are lethal.

What causes such astounding volcanic activity on Io? Surely that moon is too small to have geological activity like the Earth. Io should be long-dead, like our own Moon. At one time, some scientists suggested that Jupiter's magnetosphere might be the culprit—perhaps the (then-unknown) processes creating the plasma torus were somehow also stressing the moon. We now know that this is not

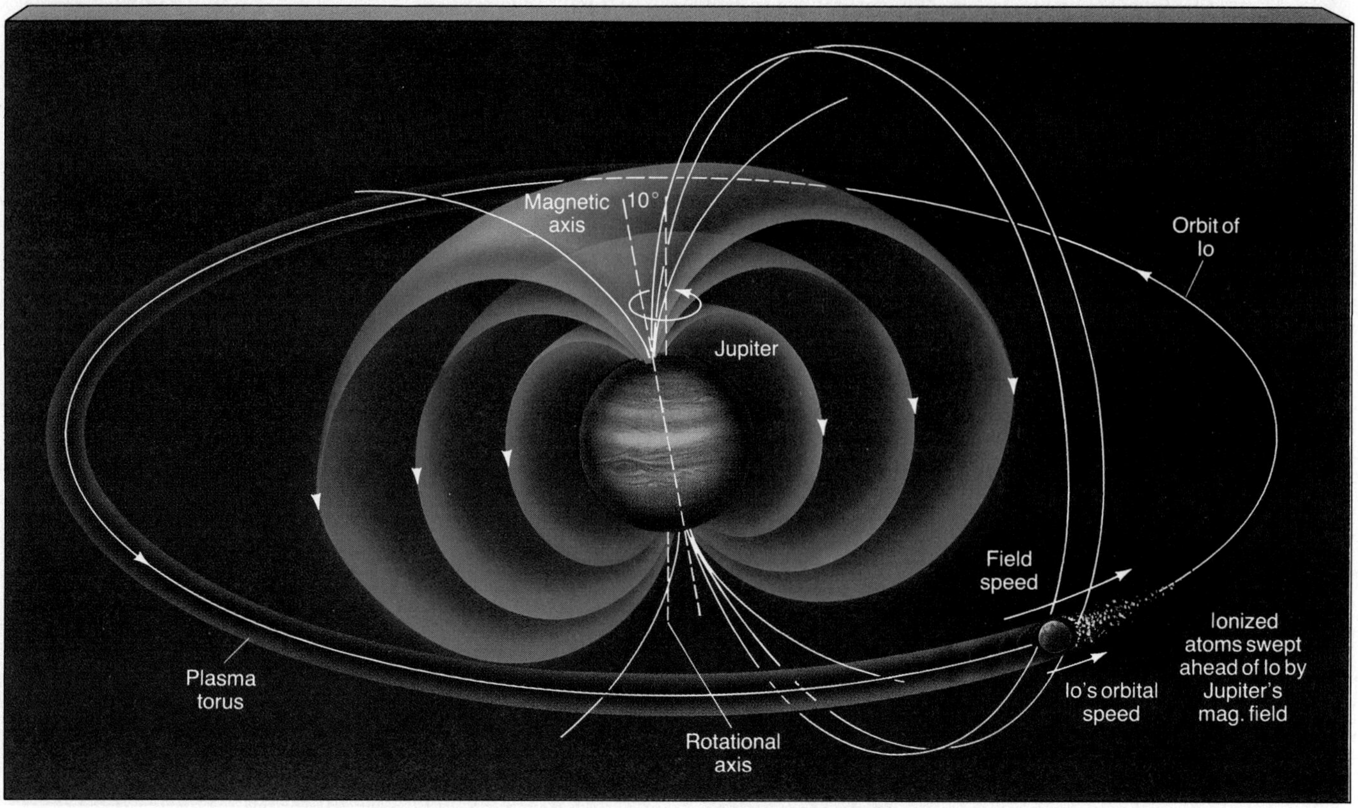

Figure 13.21 *The Io plasma torus is the result of material being ejected from Io's volcanos and swept up by Jupiter's rapidly rotating magnetic field. Because of the 10° inclination between Jupiter's magnetic and rotation axes, the torus does not exactly follow Io's orbit (which is in the equatorial plane). It is instead much fatter than the tubelike region through which the moon itself moves. Spectroscopic analysis indicates that the torus is composed primarily of sodium and sulfur ions.*

the case. The real source of Io's energy is gravity—Jupiter's gravity. Astronomers Stanton Peale, Patrick Cassen, and Ray Reynolds were the first to realize this fact and predicted that Io would have volcanos caused by Jupiter's gravitational field. They published their findings just days before the *Voyager 1* encounter. Rarely does a theoretical prediction receive such immediate and spectacular verification!

Io orbits very close to Jupiter—only 422,000 km, or 5.9 Jupiter radii, from the center of the planet. As a result, Jupiter's huge gravitational field produces strong tidal forces on the moon. If Io were the only satellite in the Jupiter system, it would long ago have come into a state of synchronous rotation with the planet, for the same reasons as discussed for our own Moon in Chapter 8. In that case, Io would move in a perfectly circular orbit, with one face permanently turned toward Jupiter. The tidal bulge would be stationary with respect to the moon, and there would be no internal stresses and hence no volcanism.

However, Io is not alone. As it orbits, it is constantly tugged by the gravity of its nearest large neighbor, Europa. These tugs are small and not enough to cause any great tidal effect in and of themselves, but they are sufficient to make Io's orbit slightly noncircular, preventing the moon from settling into a precisely synchronous state. (The reason for this effect is exactly the same as for the libration of Earth's Moon, discussed in Chapter 8.) Thus, as seen from Jupiter, Io "wobbles" slightly from side to side as it moves. The large tidal bulge, conversely, always points toward Jupiter. These conflicting forces result in enormous tidal stresses that continually flex and squeeze Io's interior. Just as repeated back-and-forth bending of a wire can produce heat through friction, Io is constantly energized by the ever-changing distortion of its interior. This generation of large amounts of heat within Io ultimately causes huge jets of gas and molten rock to squirt out of the surface. It is likely that much of Io's interior is soft or molten, with only a relatively thin solid crust overlying it. In fact, Io's volcanos are probably more like geysers on Earth, but the term volcano has stuck. Researchers estimate that the total amount of heat generated in Io as a result of tidal flexing is about 100 million megawatts. The result is one of the most fascinating objects in our solar system.

Europa

Europa (Figure 13.22) is a very different world from Io. Lying outside Io's orbit, 670,000 km (9.4 Jupiter radii) from Jupiter, it has relatively few craters on its surface, suggesting geologic youth. Recent activity must have erased the scars of ancient meteoric impacts. Europa's surface does display a vast network of lines crisscrossing

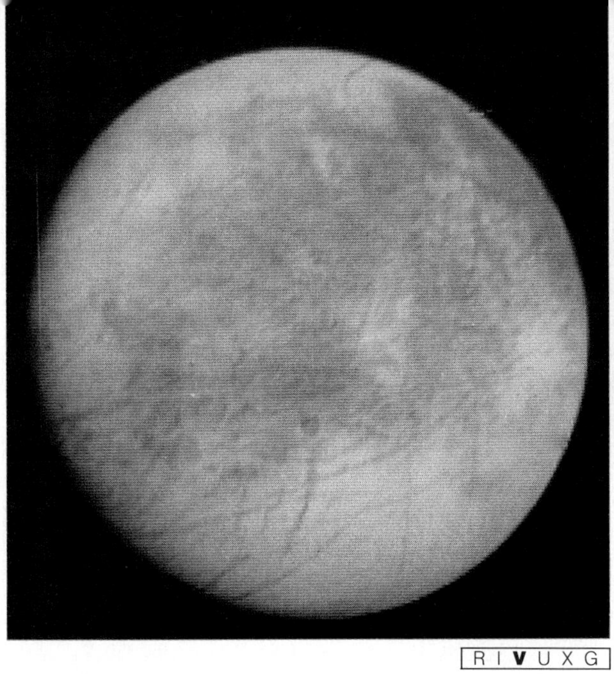

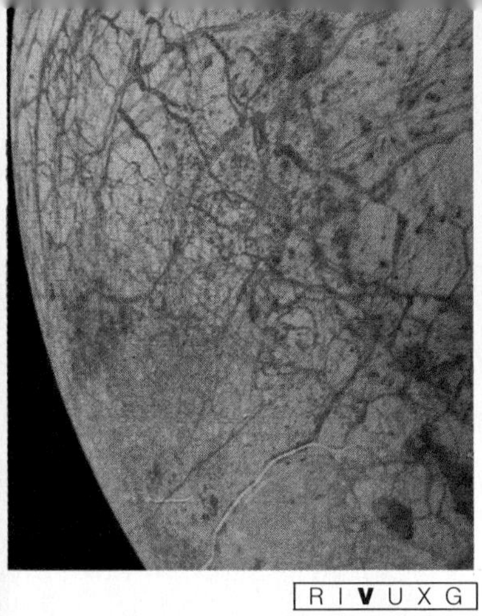

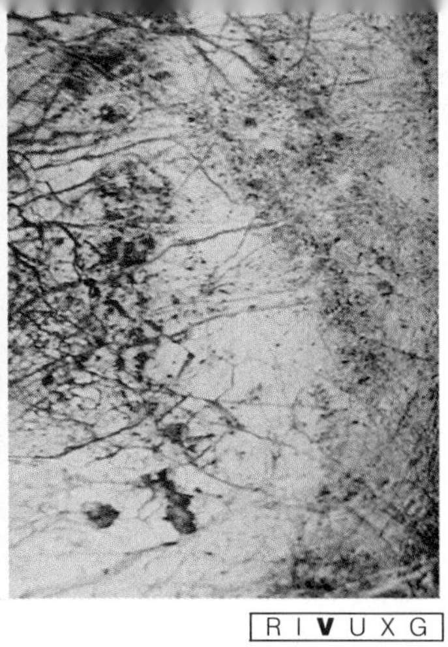

Figure 13.22 *The second Galilean moon is Europa. Its icy surface is only lightly cratered, indicating that some ongoing process must be obliterating impact craters soon after they are formed. The origin of the cracks criss-crossing the surface is uncertain. The resolution of the* Voyager *mosaic on left is about 5 km. The two inserts on the right, from* Voyager 2, *display even finer detail. (NASA)*

bright, clear fields of water ice. Some of these linear "bands," or fractures, appear to extend halfway around the satellite and resemble in some ways the pressure ridges that develop in ice floes on Earth's polar oceans.

Some researchers have theorized that Europa is covered completely by an ocean of liquid water whose top is frozen at the low temperatures that prevail because of this moon's great distance from the Sun. The cracks are attributed to the tidal influence of Jupiter and the gravitational pulls of the other Galilean satellites, although these forces are weaker than those powering Io's volcanic activity. Other planetary scientists suggest that Europa's fractured surface is instead related to some form of tectonic activity, one involving ice rather than rock. If the markings truly are fault lines of ice, then this moon is probably still quite active. If Europa does have a liquid ocean below the ice, it opens up many interesting avenues of speculation into the possible development of life there.

Ganymede and Callisto

The two outermost Galilean moons are Ganymede (at 1.1 million km, or 15 planetary radii, from the center of Jupiter) and Callisto (at 1.9 million km, or 26 Jupiter radii). Their densities are each only about 2 g/cm^3, suggesting that they harbor substantial amounts of ice throughout and are not just covered by thin icy or snowy surfaces. Ganymede, shown in Figure 13.23, is the largest moon in the solar

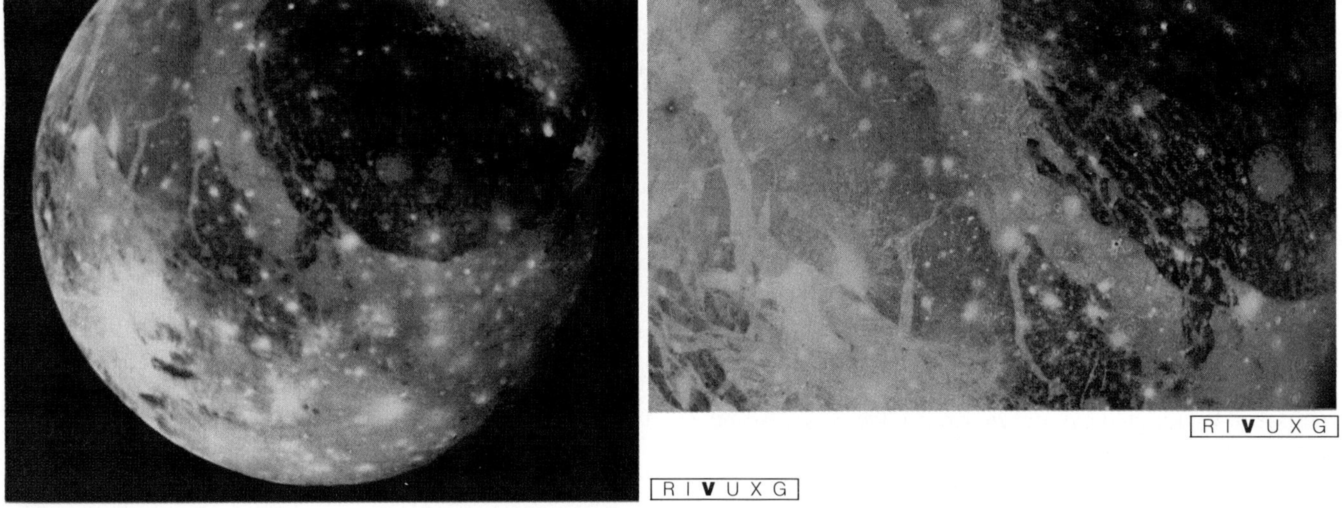

Figure 13.23 *Jupiter's largest moon, Ganymede, is also the largest satellite in the solar system. The dark regions on the surface are the oldest and probably represent the original icy crust of the moon. The largest dark region visible in the* Voyager 2 *image on the left is called Galileo Regio. It spans some 320 km. The lighter, younger regions are the result of flooding and freezing that occurred within a billion years or so of Ganymede's formation. The light-colored spots are recent impact craters. The resolution of the detailed image on the right is about 3 km. (NASA; NASA)*

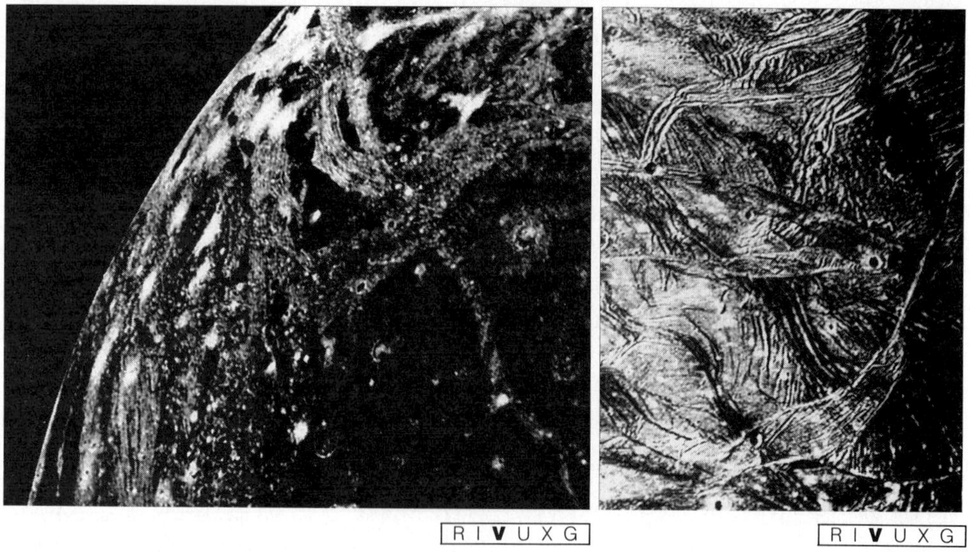

Figure 13.24 *"Grooved terrain" on Ganymede may have been caused by a process similar to plate tectonics on Earth. The resolution of the detailed image on the right is about 3 km. (NASA; NASA)*

system, exceeding not only Earth's Moon but also the planets Mercury and Pluto in size. It has many impact craters on its surface and patterns of dark and light markings that are reminiscent of the highlands and maria on Earth's own Moon. In fact, Ganymede's history has many parallels with that of the Moon (with water ice replacing lunar rock). The large, dark region clearly visible in Figure 13.23 is called Galileo Regio.

As with the inner planets, we can estimate ages on Ganymede by counting craters. We learn that the darker regions, like Galileo Regio, are the oldest parts of Ganymede's surface. These regions are the original icy surface of the moon, just as the ancient highlands on our own Moon are its original crust. The surface darkens with age as micrometeorite dust slowly covers it. The light-colored parts of Ganymede are much less heavily cratered, so they must be younger. They are Ganymede's "maria" and probably formed in a manner similar to the maria on the Moon. Intense meteoritic bombardment caused liquid water—Ganymede's counterpart to our own Moon's molten lava—to upwell from the interior, flooding the impacting regions before solidifying.

Not all of Ganymede's surface features follow the lunar analogy. Ganymede has a system of grooves and ridges (shown in Figure 13.24) that may have resulted from crustal tectonic motion, much as Earth's surface undergoes mountain building and faulting at plate boundaries. Ganymede's large size indicates that its original radioactivity probably helped to heat and differentiate its partly rocky interior, after which the moon cooled and the crust cracked. Ganymede seems to have had some early plate tectonic activity, but the process stopped about 3 billion years ago when the cooling crust became too thick.

Callisto, shown in Figure 13.25, is in many ways similar in appearance to Ganymede, although it has more craters and fewer fault lines. Its most obvious feature is a huge series of concentric ridges surrounding each of two large basins. The larger of the two, on Callisto's Jupiter-facing side, is named Valhalla and measures some 3000 km across. It is clearly visible in Figure 13.25. The ridges resemble the ripples made as a stone hits water, but on Cal-

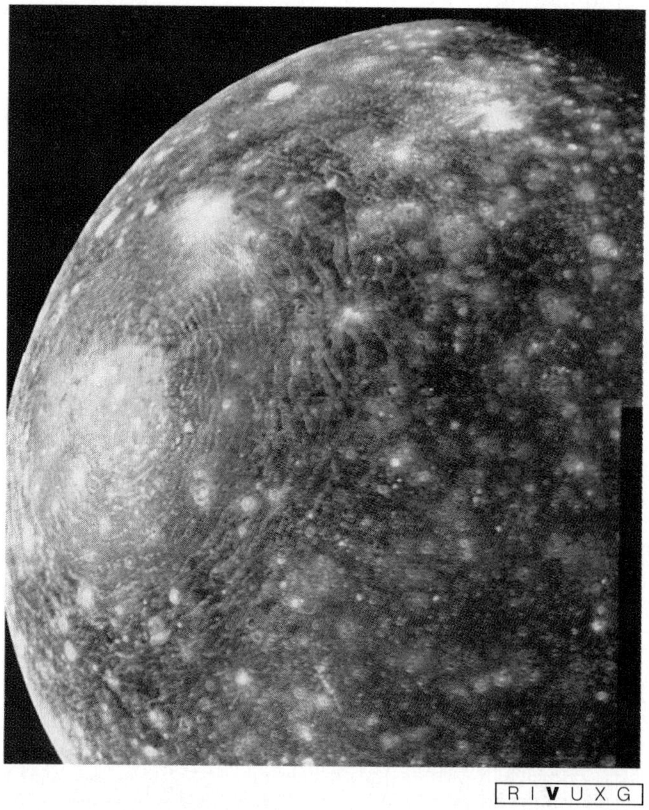

Figure 13.25 *Callisto, the outermost Galilean moon of Jupiter, is similar to Ganymede in composition but is more heavily cratered. The large series of concentric ridges visible on the left of the image is known as Valhalla. Extending nearly 1500 km from the basin center, they formed when "ripples" from a large meteoritic impact froze before they could disperse completely. The resolution here is around 10 km. (NASA)*

listo they probably resulted from a cataclysmic impact with a meteorite. The upthrust ice was partially melted, but it resolidified quickly, before the ripples had a chance to subside. Today, both the ridges and the rest of the crust are frigid ice and show no obvious signs of geological activity (like the grooved terrain on Ganymede). Apparently, Callisto froze before plate tectonic or other activity could start. The density of impact craters on the Valhalla basin indicates that it formed long ago, perhaps 4 billion years in the past.

JUPITER'S RING

Yet another remarkable finding of the 1979 *Voyager* missions was the discovery of a faint ring of matter encircling Jupiter in the plane of the planet's equator (see Figure 13.26). This ring lies roughly 50,000 km above the top cloud layer of the planet, inside the orbit of the innermost moon. A thin sheet of material may extend all the way down to Jupiter's cloud tops, but most of the ring is confined within a region only a few thousand kilometers across. The outer edge of the ring is quite sharply defined. In the direction perpendicular to the equatorial plane, the ring is only a few tens of kilometers thick. The small, dark particles that make up the ring may have been chipped off by meteorite impacts on two small moons—Metis and Adastrea, discovered by *Voyager*—that lie very close to the ring itself. Despite its different appearance and structure, Jupiter's ring can perhaps be best understood by studying the most famous ringed planet—Saturn—and we will postpone further discussion of ring properties until the next chapter.

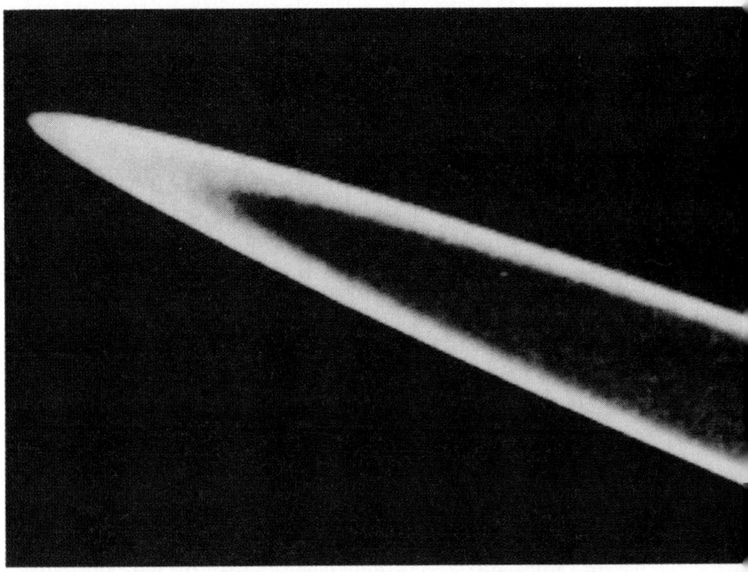

R I **V** U X G

Figure 13.26 *Jupiter's faint ring, as photographed by* Voyager 2. *The ring, made up of dark fragments of rock and dust, possibly chipped off the innermost moons by meteorites, was unknown before the two* Voyager *spacecraft arrived at the planet. It lies in Jupiter's equatorial plane, only 50,000 km above the cloud tops. (NASA)*

CHAPTER SUMMARY

Jupiter is about five times farther from the Sun than Earth is and takes 11.9 years to complete one orbit. To the eye, Jupiter appears as the second-brightest of all planets. Jupiter has more than twice the mass of all the other planets combined, though it is still 1000 times less massive than the Sun. It is composed primarily of hydrogen and helium gas.

Measurements of Jupiter's magnetosphere show that the planet's interior has a rotation period of 9 hours 56 minutes, the fastest rate of any solar system object. Jupiter spins so rapidly that its equator bulges outward. To account for the amount by which Jupiter's poles are flattened, it is necessary to assume that the planet has a dense core.

Pioneers 10 and *11* were the first spacecraft to travel through the asteroid belt and on into the outer solar system. *Voyagers 1* and *2* reached Jupiter in 1979. *Voyager 2* received a gravity-assisted boost to Uranus and later to Neptune. Both *Voyagers* are now heading for interstellar space. NASA's *Galileo* spacecraft is expected to arrive at Jupiter in 1995.

Jupiter's atmosphere is layered with colored clouds thought to be the product of complex chemical processes fueled by the planet's interior heat, solar ultraviolet radiation, auroral phenomena, and lightning. Variable bright zones and dark belts cross Jupiter. The lighter zones are the tops of upwelling, warm currents, and the darker bands are cooler regions where gas is sinking. On Jupiter there are no continents to disrupt weather patterns. The Great Red Spot, a giant hurricane, has existed at least since its discovery three centuries ago. Smaller white ovals and brown ovals can persist for decades.

Jupiter emits about twice as much energy as it receives from the Sun. The most likely explanation is that this energy is the slow escape of heat released during the planet's formation. Jupiter is less massive than a star, and its interior never became hot enough for thermonuclear reactions to begin.

The temperature and density of Jupiter's atmosphere increase with depth below the layers of clouds, and the gas turns into liquid at a depth of a few thousand kilometers. At a depth of about 20,000 km, the liquid hydrogen is compressed into a ''metallic'' state by pressure that is three million times greater than the atmospheric pressure on Earth.

The magnetosphere of Jupiter is about a million times more voluminous than Earth's magnetosphere, and the planet has a long magnetic ''tail'' extending away from the Sun to at least the distance of Saturn's orbit. Energetic particles spiral around magnetic field lines, accelerated by Jupiter's rotating magnetic field, producing radio radiation.

Jupiter and its system of moons resemble a small solar system. Sixteen moons have been discovered. The outermost eight moons resemble asteroids and have retrograde orbits, indicating that they may have been captured by Jupiter's gravity long after the planets and largest moons formed. The four Galilean moons correspond in some ways to the Sun's terrestrial planets and therefore may hold clues to the processes that created our own solar system. As Io orbits Jupiter, it ''wobbles'' because of the gravitational pull of Europa. The ever-changing distortion of its interior energizes this moon, and geyserlike volcanos keep its surface smooth with constant eruptions. Europa's fields of ice are nearly devoid of craters but have extensive fractures attributed either to tidal influence of Jupiter and gravitational effects of other Galilean satellites or to some form of tectonic activity. Ganymede is the largest moon in the solar system. There is evidence of early plate tectonic activity, but Ganymede now is unmoving rock and ice. Callisto apparently froze before tectonic activity could start there. It has an ancient, cratered surface.

The 1979 *Voyager* mission discovered a faint ring of dark particles 50,000 km above Jupiter's cloud layer. The particles could be pieces chipped off of the moons Metis and Adastrea by meteorites.

KEY WORDS

belts
brown oval
current sheet
differential rotation
Galilean moons
Great Red Spot
Io plasma torus
nonthermal radiation
oblateness
white oval
zonal flow
zones

REVIEW QUESTIONS

1. In what sense does our solar system consist of only two important objects?
2. What is differential rotation, and how is it observed on Jupiter?
3. Why doesn't Jupiter have seasons?
4. Describe some of the ways in which the *Voyager* mission changed our perception of Jupiter.
5. What is the Great Red Spot, and what is known about the source of its energy?
6. What is the source of the colors in Jupiter's atmosphere?
7. Briefly describe the weather on Jupiter.
8. Why has Jupiter retained most of its original atmosphere?
9. Explain the theory that accounts for Jupiter's internal heat source.
10. What is Jupiter thought to be like beneath its clouds? Why do we think this?
11. What is responsible for Jupiter's enormous magnetic field?
12. Why might we say that Jupiter was nearly a star?
13. If Jupiter had become a star—even a dim one—what might have been its effect on Earth?
14. In what sense are Jupiter and its moons like a miniature solar system?
15. How do the Galilean moons correspond to the terrestrial planets? How do they differ?
16. What is thought to be the source of Io's volcanic activity?
17. Why is there speculation that the Galilean moon Europa might be an abode for life?
18. Compare the force of gravity at Jupiter's cloudtops with that at Earth's surface. What is the escape velocity from Jupiter? (See Chapter 3.)
19. Calculate the ratio of Jupiter's mass to the total mass of the galilean moons. Compare this with the ratio of Earth's mass to that of the Moon, and the ratio of the Sun's mass to that of the rest of the solar system.
20. Io orbits Jupiter at a distance of six planetary radii in 42 hours. At what distance would a satellite orbit Jupiter in the time taken for Jupiter to rotate exactly once (10 hours, say), so that the satellite would appear ''stationary'' above the planet? (Use Kepler's Third Law; see Chapter 3.)
21. Compare the apparent sizes of the galilean moons, as seen from Jupiter's cloudtops, with the angular diameter of the Sun at Jupiter's distance.

DISCUSSION QUESTIONS

1. What is the possibility of life in the Jovian cloud bands and on Jupiter's moon Europa? On the basis of what you know about both worlds, speculate briefly on the sorts of creatures that might have evolved there.
2. Focus on the idea of seasons on some of the worlds studied so far. Compare and contrast seasonal change on Earth, Mars, and Jupiter.
3. Go to the library and find some recent articles on the *Galileo* mission to Jupiter. Where is the craft now? What is its most recent discovery? Would you expect to ever see a total solar eclipse?

PROJECTS

1. Look in an almanac to find out where Jupiter is now. What constellation is it in?
2. Are there any stars in the night sky that look as bright as Jupiter? What other difference do you notice between Jupiter and the stars?
3. Use binoculars to peer at Jupiter. Be sure to hold them steadily (try propping your arms up on the hood of a car, or sitting down and bracing them against your knees). Can you see any of Jupiter's four largest moons?
4. Through a telescope, you should be able to see the red-and-tan cloud bands of Jupiter, and you can clearly see some moons. If you come back the following evening, the moons' relative positions will have changed. Why? With a telescope, can you see a shadow of a moon on the planet?

SUGGESTED READINGS

Astronomy (May 1979). Special issue on *Voyager* at Jupiter.

Beatty, J. K. ''The Far-Out Worlds of *Voyager 1*: 1.'' *Sky and Telescope* (May 1979). One of the early articles expressing the amazing things found at Jupiter by *Voyager 1*.

Beebe, R. F. ''Queen of the Giant Storms.'' *Sky and Telescope* (October 1990). A detailed look at Jupiter's Great Red Spot.

Bennett, G. L. ''Return to Jupiter.'' *Astronomy* (January 1987). On the *Galileo* mission to Jupiter.

Gehrels, T., ed. *Jupiter*. Tucson: University of Arizona Press, 1976. Scientific aspects of the interior, atmosphere, magnetosphere, and satellites of Jupiter.

Morrison, D. ''The Enigma Called Io.'' *Sky and Telescope* (March 1985). The most dynamic of the four galilean satellites of Jupiter.

———. ''Four New Worlds.'' *Astronomy* (September 1980). The largest moons of Jupiter, the galilean satellites, as seen by the *Voyager* spacecraft.

Soderblom, L. A. ''The Galilean Moons of Jupiter.'' *Scientific American* (January 1980). The Galilean worlds as Earth-like bodies that can be compared with one another to gain insight into how they evolved.

Waldrop, M. M. ''Can Galileo Take the Heat?'' *Science* (September 22, 1989). Can *Galileo* survive its circuitous route to upiter?

CHAPTER

14

Saturn

Spectacular Rings and Mysterious Moons

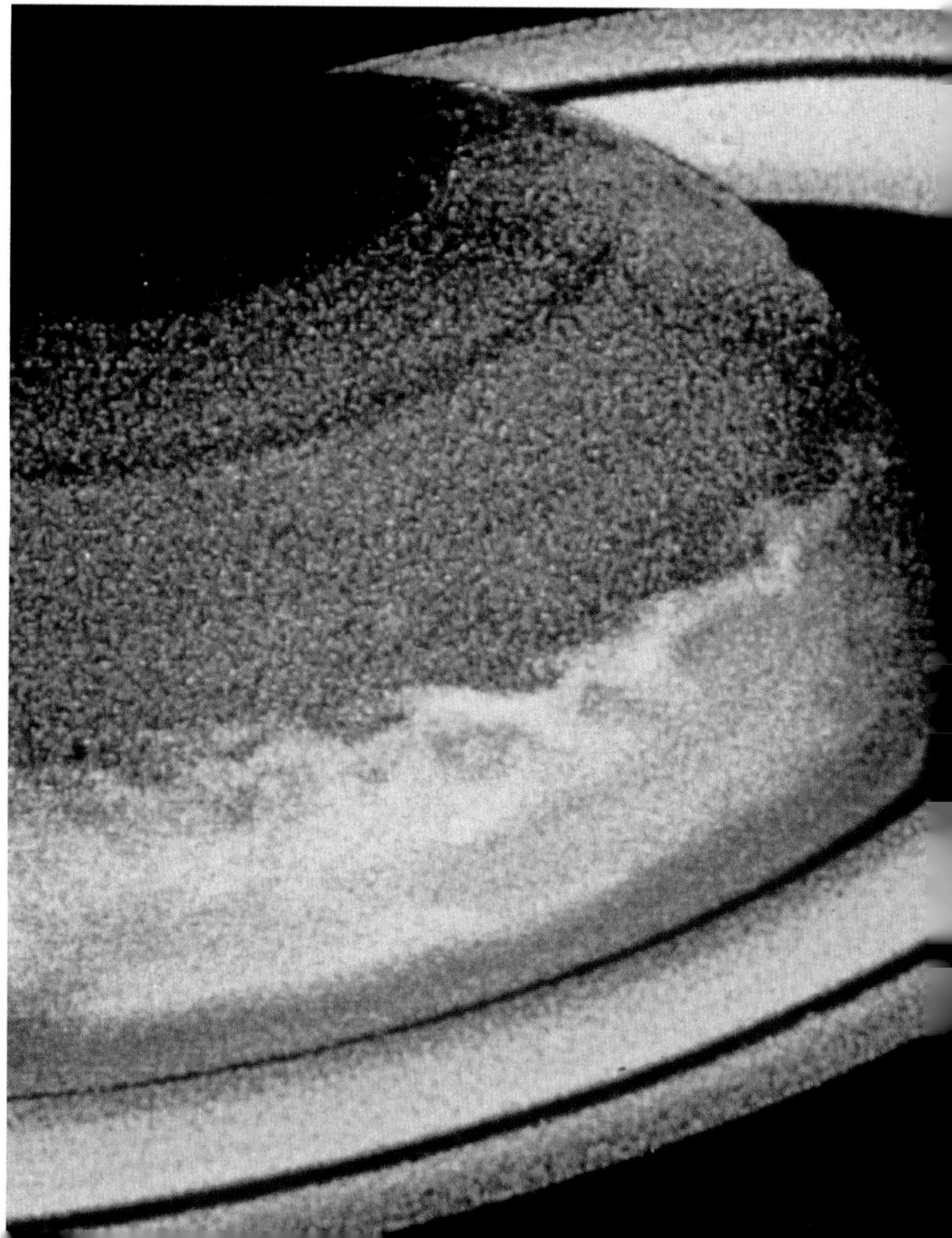

This computer-enhanced image of Saturn, taken by the Hubble Space Telescope, *combines the reflected light from the planet in two ''colors''—blue and infrared—to display novel structure never before seen across the face of the planet. Acquired about a month after a huge storm began brewing on the planet in the fall of 1990, this (false-color) psychedelic cloudscape shows a band of white clouds that the storm has blown across part of the equator. Combining blue and infrared radiation allows the study of the vertical growth of the clouds, which are thought to be mainly ammonia ice crystals. Image resolution is about 700 kilometers. Saturn itself is 120,000 kilometers in diameter. At the time of this observation Saturn was about 1.4 billion kilometers from Earth. (Courtesy NASA)*

R **I** **V** U X G

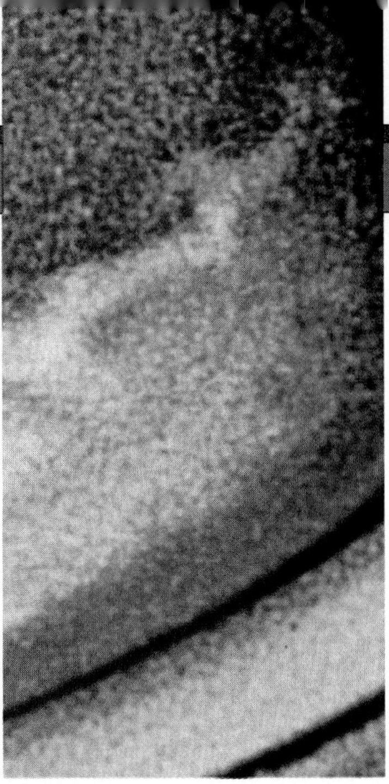

LEARNING GOALS

1. to compare Saturn, the second Jovian world, with Jupiter
2. to understand that Saturn too has an internal heat source, but one that is quite different from Jupiter's
3. to understand the nature of Saturn's rings
4. to realize that Titan, Saturn's largest moon, has an atmosphere with ongoing complex organic chemistry
5. to see that Saturn's smaller moons display a wealth of geological and dynamical detail

To many people, Saturn is the most beautiful and enchanting of all astronomical objects. Saturn's rings are a breathtaking sight when viewed through even a small telescope, and they are probably the planet's best-known feature. Aside from its famous rings, however, Saturn presents us with another good example of a giant gaseous planet in action. Saturn is in many ways similar to its larger neighbor, Jupiter, in terms of composition, size, and structure. Yet when we study the two planets in detail, we find that there are important differences, too. A comparison between Saturn and Jupiter provides us with valuable insight into the structure and evolution of all the Jovian worlds.

Saturn's rings were unknown to ancient astronomers. We now know that all of the Jovian planets have rings of some sort, although there are great differences in ring properties from one world to another. Nevertheless, studies of Saturn's rings have provided scientists with a wealth of information about the way rings work. Saturn's rings, and the planet's extensive collection of moons, exhibit fascinating and almost unbelievable dynamical complexity. The moons themselves are a varied lot and have much to tell us about the planet's formation. The largest of them, Titan, is perhaps the last remaining candidate for study in the search for life elsewhere in the solar system.

ORBITAL PROPERTIES OF SATURN

Saturn was the outermost planet known to ancient astronomers. Named after the father of Jupiter in Greek and Roman mythology, Saturn orbits the Sun at almost twice the distance of Jupiter, with an orbital semi-major axis of 9.54 A.U. (1430 million km). The planet's sidereal orbital period of 29.5 Earth years was the longest natural unit of time known to the ancient world. Saturn's orbital eccentricity is 0.056, so its perihelion is 9.01 A.U. (1350 million km) and its aphelion reaches 10.07 A.U. (1510 million km).

At opposition, when Saturn is at its brightest, it can lie within 8 A.U. of the Earth. However, its great distance from the Sun still makes it considerably fainter than either Jupiter or Mars. In fact, Saturn ranks behind Jupiter, the inner planets, and several of the brightest stars in the sky in terms of apparent brightness. At opposition, Saturn's maximum angular diameter (excluding the rings) is 19.5″ (arc seconds). At conjunction, its minimum diameter is 15″. The time between successive oppositions (or conjunctions), Saturn's synodic period, is 378 days—only slightly more than one Earth year.

SATURN IN BULK

Mass, Radius, and Density

✓1 Less than one-third the mass of Jupiter, Saturn is still an enormous planet, at least by terrestrial standards. As with Jupiter, Saturn's many moons allowed an accurate determination of the planet's mass long before the arrival of the *Pioneer* and *Voyager* missions. Saturn's mass is 5.6×10^{29} g, or 95 times the mass of Earth. From Saturn's distance and angular size, the planet's radius—and hence the average density—quickly follow. Saturn's equatorial radius is 60,000 km, or 9.4 Earth radii. The average density then is 0.7 g/cm^3, which means it is less dense than water (whose density is 1.0 g/cm^3). Here we have a planet that would float in the ocean–if Earth had one big enough! Saturn's low average density indicates that, like Jupiter, it is composed primarily of hydrogen and helium. Saturn's lower mass, however, results in lower interior pressure, so that these gases are less compressed than in Jupiter's case.

Rotation Rate

Saturn, like Jupiter, rotates very rapidly and differentially. The rotation period of the interior (as measured from magnetospheric outbursts, which trace the rotation of the planet's core) and at high planetary latitudes (determined by tracking weather features observed in Saturn's atmosphere) is 10^h40^m. The rotation period at the equator is 10^h14^m, about 26 minutes shorter. Because of Saturn's lower density, this rapid rotation makes Saturn even more oblate (flattened) than Jupiter. The oblateness, as defined in Chapter 13, is 0.11, so that Saturn's polar radius is only 54,000 km, about 6000 km less than the equatorial radius. Careful calculations show that this oblateness is not what would be expected for a planet composed of hydrogen and helium alone. Astronomers believe that Saturn also has a rocky core, perhaps twice the mass of Jupiter's.

Unlike Jupiter, Saturn's rotation axis is significantly tilted with respect to the planet's orbit plane. The axial tilt is 27°, similar to that of both Earth and Mars. We might, then, expect to see substantial seasonal variations on Saturn. However, as we will see, this is not the case—solar heating may not be the dominant part of the planet's surface energy budget.

Rings

Saturn's best known feature is its spectacular *ring system.* Because the rings lie in the equatorial plane, their appearance (as seen from Earth) changes in a seasonal manner, as shown in Figure 14.1. As Saturn orbits the Sun, the angles at which the rings are illuminated, and at which we can view them, vary. When the planet's north or south pole is tipped toward the Sun, in Saturnian summer and winter, the highly reflective rings are at their brightest. In Saturnian spring and fall, the rings are close to being edge-on, both to the Sun and to us, so that they seem to disappear altogether. One important deduction that we can make from this simple observation is that the rings are very thin. In fact, we now know that their thickness is less than a few hundred meters, even though they are over 200,000 km in diameter.

SATURN'S ATMOSPHERE

Saturn is much less colorful than Jupiter. Figures 14.2 and 14.3 show yellowish and tan cloud belts that parallel the equator, but these regions display less atmospheric structure than do the belts on Jupiter. No obvious large ''spots'' or ''ovals'' adorn Saturn's cloud decks. Bands and storms do exist, but the color changes that distinguish them on Jupiter are largely absent on Saturn.

Composition and Coloration

Astronomers first observed methane in the spectrum of sunlight reflected from Saturn in the 1930s, about the same time that it was discovered on Jupiter. However, it was not until the early 1960s, when more sensitive observations became possible, that gaseous ammonia was finally detected. In Saturn's cold upper atmosphere, most ammonia is in the solid or liquid form, and relatively little of it is present as a gas to absorb sunlight and create spectral lines. With continuing improvement in both detector design and observing techniques, astronomers finally made the first accurate determinations of the hydrogen and helium content

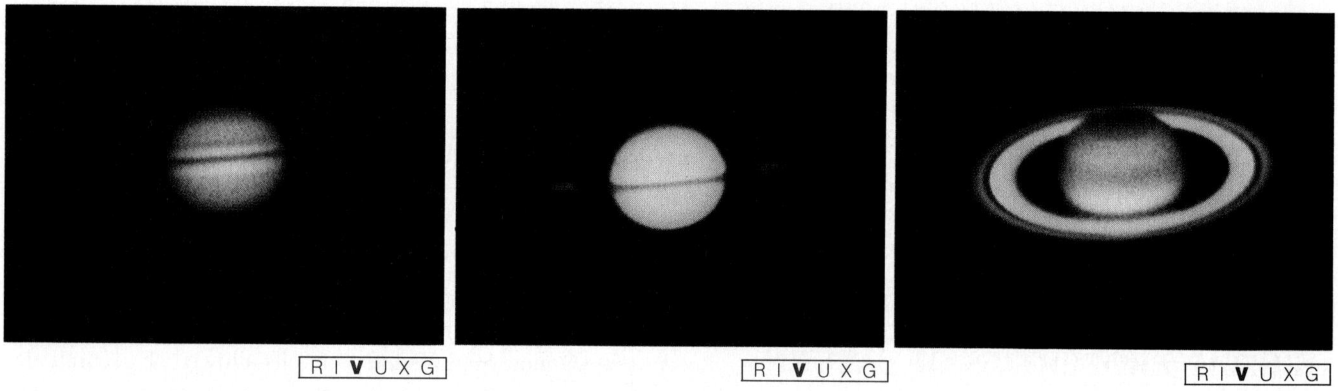

Figure 14.1 *Over a period of several years, Saturn's rings change their appearance to terrestrial observers as the tilted ring plane orbits the Sun. (Lick Observatory)*

R I V U X G

Figure 14.2 Saturn, its rings, and two of its moons (Tethys and Dione) are shown in this photograph taken by the Pioneer 11 *spacecraft as it sped past in 1979 at a distance of 20 million km. Note the absence of the vivid colors that characterize Jupiter's atmospheric cloud layers. (NASA)*

in the late 1960s. The Earth-based measurements were later confirmed with the arrival of the *Pioneer* and *Voyager* spacecraft in the 1970s.

These spectroscopic studies show that the Saturnian atmosphere consists of molecular hydrogen (H_2: 92.4 percent), helium (He: 7.4 percent), methane (CH_4: 0.2 percent), and ammonia (NH_3: 0.02 percent). As on Jupiter, hydrogen and helium dominate. These most abundant elements never escaped from Saturn's atmosphere because of the planet's large mass and low temperature (see Interlude 9-2). However, the fraction of helium on Saturn is far less than is observed on Jupiter (where, as we saw helium accounts for nearly 14 percent of the atmosphere) or in the Sun. It is extremely unlikely that the processes that created the outer planets preferentially stripped Saturn of nearly half its helium or that the missing helium somehow escaped from the planet while the hydrogen remained behind. Instead, astronomers believe that at some time in Saturn's past the heavier helium began to sink toward the center of the planet, reducing its abundance in the outer layers and leaving them relatively hydrogen-rich. We will return to the reasons for this differentiation and its consequences later.

Figure 14.4 illustrates the structure of the Saturnian atmosphere. (Compare this structure with the corresponding diagram for Jupiter, Figure 13.7.) In many respects, Saturn's atmosphere is quite similar to Jupiter's, except that Saturn's temperature is a little lower because of its greater distance from the Sun and because its clouds are somewhat thicker. So Saturn, like Jupiter, lacks a solid surface, we take the tropopause as our reference level and set it to 0 km. The top of the visible clouds lies about 50 km below this level. As on Jupiter, the clouds are arranged in three distinct layers, composed (in order of increasing depth) of ammonia, ammonium hydrosulfide, and water ice. Above the clouds, around the tropopause, lies a layer of haze formed by the action of sunlight on Saturn's upper atmosphere.

The total thickness of the three cloud layers in Saturn's atmosphere is roughly 200 km, compared with a thickness of about 80 km on Jupiter, and each layer is itself somewhat thicker than its counterpart on Jupiter. The reason for this difference is Saturn's weaker gravity. At the haze level, Jupiter's gravitational field is nearly two and a half times stronger than Saturn's, so Jupiter's atmosphere is pulled much more powerfully toward the center of the planet. Thus Jupiter's atmosphere is compressed more than Saturn's, and the clouds are squeezed more closely to-

Figure 14.3 Much more fine structure, especially in the rings, appears in this image taken while the Voyager 2 *spacecraft approached the planet. (Three of Saturn's moons appear at bottom, and a fourth casts a black shadow on the cloud tops.) The banded structure of Saturn's atmosphere is also more evident in this photograph. This is approximately true color—that is, as the human eye sees things. (NASA)*

R I V U X G

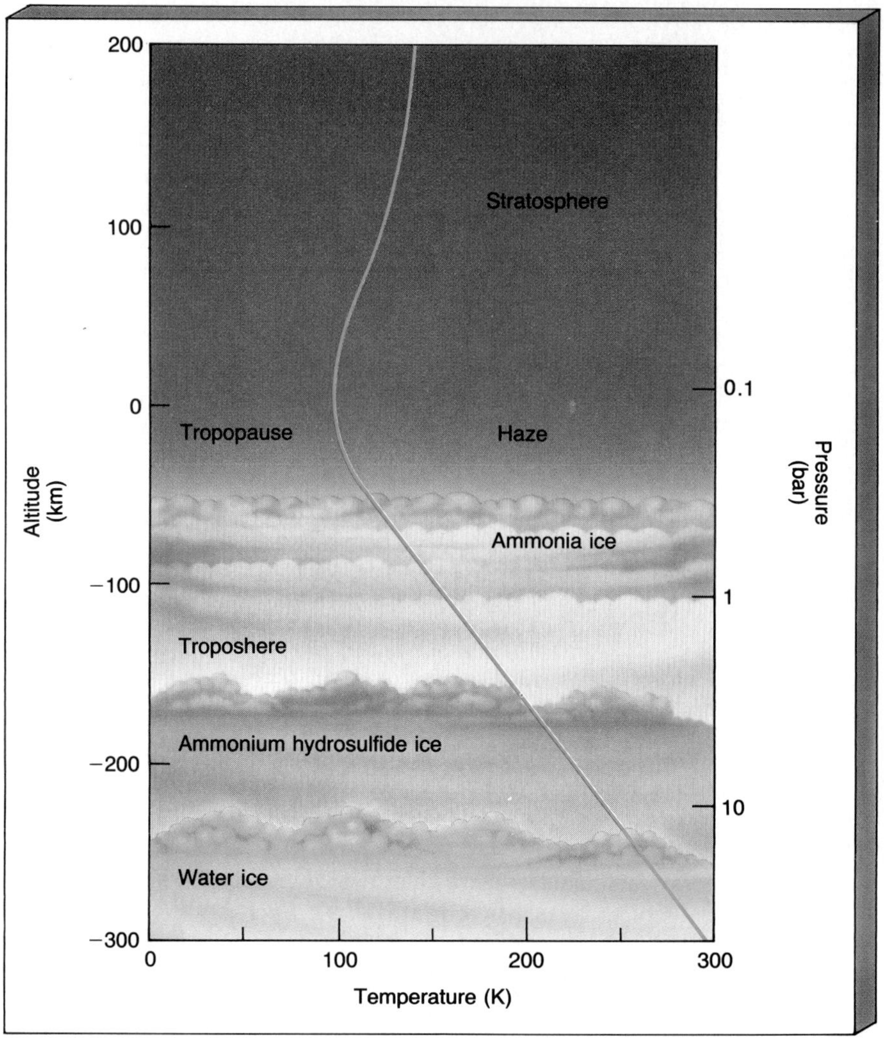

Figure 14.4 The vertical structure of the Saturnian atmosphere. As with Jupiter, there are several cloud layers, but Saturn's weaker gravity results in thicker clouds and a more uniform appearance.

gether. The colors of Saturn's cloud layers, as well as the planet's overall butterscotch hue, are due to the same basic cloud chemistry as on Jupiter. However, because Saturn's clouds are thicker there are few holes and gaps in the top layer, so we rarely glimpse the more colorful levels below. Instead, we see only different levels in the topmost layer, which leads to Saturn's rather uniform appearance.

Weather

Saturn has atmospheric wind patterns that are in many ways reminiscent of those on Jupiter. There is an overall east-west zonal flow, which is apparently quite stable. Computer-enhanced images of the planet that bring out more cloud contrast (see Figure 14.5) clearly show the existence of bands, oval storm systems, and turbulent flow patterns looking very much like those seen on Jupiter. Scientists believe that the cause of Saturn's bands and storms is essentially the same as the cause of Jupiter's weather. Ultimately, the large-scale flows and small-scale storm systems are powered by convective motion in Saturn's interior and the planet's rapid rotation.

The zonal flow on Saturn is considerably faster than on Jupiter and shows fewer east-west alternations, as can be seen from Figure 14.6, which compares the flows on the two planets. The equatorial eastward jet stream, which reaches a speed of about 400 km/h on Jupiter, moves at a brisk 1500 km/h on Saturn, and extends to much higher latitudes. Not until latitudes 40° north and south of the equator are the first westward flows found. Latitude 40°N also marks the strongest bands on Saturn and the most obvious ovals and turbulent eddies. Astronomers still do not fully understand the reasons for the differences between Jupiter's and Saturn's flow patterns.

In September 1990, amateur astronomers detected a large white spot in Saturn's southern hemisphere, just below the equator. In November of that year, when the *Hubble Space Telescope* imaged the phenomenon in more detail, the spot had developed into a band of clouds completely encircling the planet's equator. Some of the *HST* images are shown in Figure 14.7, illustrating the potential of *HST* as an instrument for monitoring atmospheric phenomena on the giant planets. Astronomers believe that the white coloration arose from crystals of ammonia ice. This

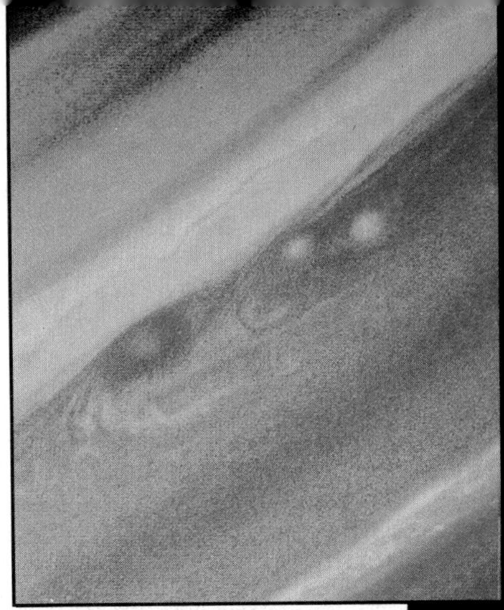

Figure 14.5 *We see more structure in Saturn's cloud cover when computer processing and artificial color is used to enhance the image contrast as in this* Voyager *image. (NASA)*

ice formed when an upwelling plume of warm gas penetrated the cool upper cloud layers. Because the crystals were freshly formed, they had not yet been affected by the chemical reactions that color the planet's other clouds.

Such spots are rare on Saturn. The last one visible from Earth appeared in 1933, but it was much smaller than the 1990 system and much shorter-lived, lasting for only a few weeks. The turbulent flow patterns seen around the 1990 white spot have many similarities to the flow around Jupiter's Great Red Spot. Scientists speculate that these

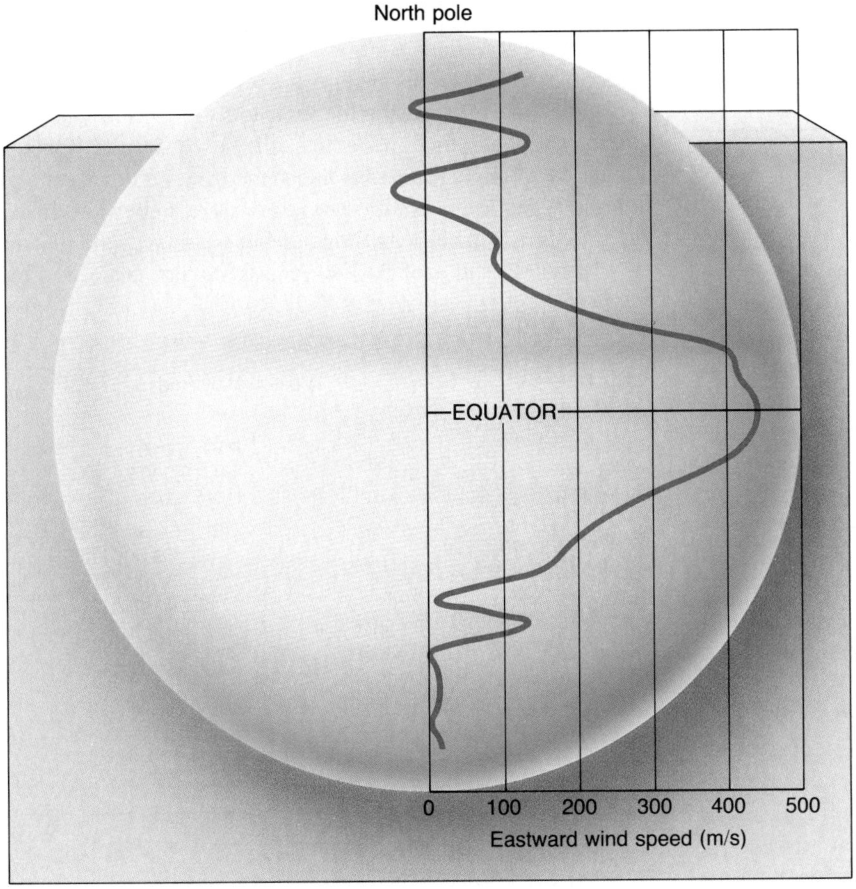

Figure 14.6 *Winds on Saturn reach speeds greater than even the winds on Jupiter. As on Jupiter, the visible bands appear to be associated with variations in wind speed.*

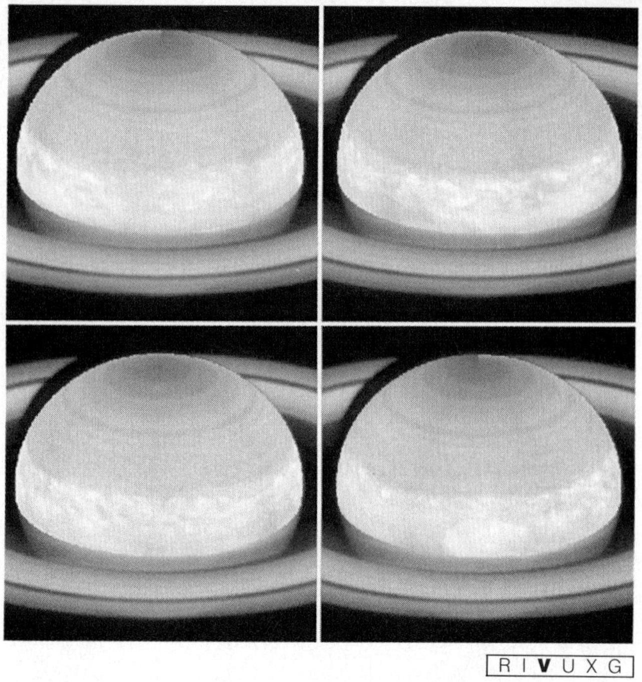

Figure 14.7 *Circulating and evolving cloud systems on Saturn, imaged by the* Hubble Space Telescope *in November 1990. (NASA)*

white spots may represent long-lived Saturnian weather systems and hope that routine observations of such temporary atmospheric phenomena on the outer worlds will enable them to gain greater insight into the dynamics of planetary atmospheres.

SATURN'S INTERIOR AND MAGNETOSPHERE

Internal Heating

✓2 Infrared measurements indicate that Saturn's surface (that is, cloud-top) temperature is 97 K, substantially higher than the temperature at which Saturn would reradiate all the energy it receives from the Sun. In fact, Saturn radiates away almost three times more energy than it absorbs. Thus Saturn, like Jupiter, has an internal energy source. But the explanation behind Jupiter's excess energy—that the planet has a large reservoir of heat left over from its formation—doesn't work for Saturn. Saturn is smaller than Jupiter and so must have cooled more rapidly—rapidly enough that its original supply of energy has long ago been used up. What then is happening inside Saturn to produce this extra heat?

The explanation for this strange state of affairs, first suggested by Ed Salpeter of Cornell and David Stevenson of Caltech, also explains the mystery of Saturn's apparent helium deficit, all in one neat package. At the temperatures and high pressures found in Jupiter's interior, liquid helium *dissolves* in liquid hydrogen. In Saturn, where the internal temperature is lower, the helium doesn't dissolve so easily, and tends to form droplets instead. The phenomenon is familiar to cooks who know that it is generally much easier to dissolve ingredients in hot liquids than in cold ones. Saturn probably started out with a fairly uniform mix of hydrogen and helium, but the helium tended to condense out of the surrounding hydrogen, much as water vapor condenses out of Earth's atmosphere to form a mist. The amount of helium condensation was greatest in the planet's cool outer layers, where the mist turned to rain about 2 billion years ago. A light shower of liquid helium has been falling through Saturn's interior ever since. This **helium precipitation** is responsible for depleting the outer layers of their helium content.

So we account for the unusually low abundance of helium in Saturn's atmosphere—much of it has rained down to lower levels. But what about the excess heating? The answer is simple: As the helium sinks toward the center, the planet's gravitational field compresses it and heats it up. The gravitational energy thus released is the source of Saturn's internal heat. In the distant future, the helium rain will stop and Saturn will cool until its outermost layers radiate only as much energy as they receive from the Sun. When that happens, the temperature at Saturn's cloud tops will be 74 K. As Jupiter cools, it too may someday experience helium precipitation in its interior, causing its surface temperature to rise once again.

Interior Structure

Figure 14.8 (compare Figure 13.12) illustrates Saturn's internal structure. This picture has been pieced together by planetary scientists using the same tools—*Voyager* observations and theoretical modeling—they used to infer Jupiter's inner workings. Saturn has the same basic internal parts as Jupiter, but their relative proportions are somewhat different: Saturn's metallic hydrogen layer is thinner, and its core is larger. Because of its lower mass, Saturn has a less ex-

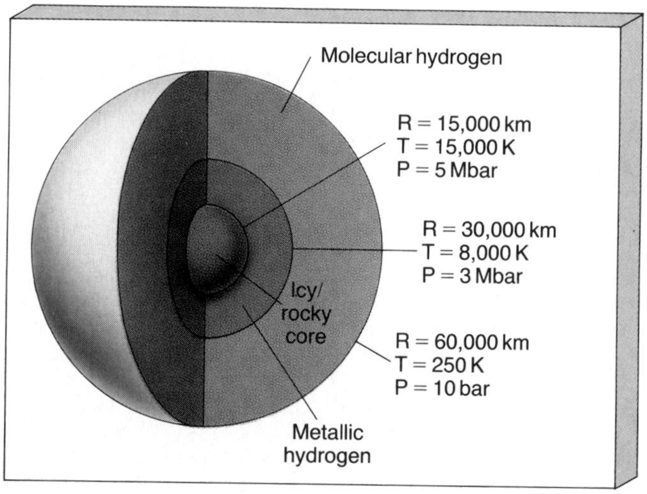

Figure 14.8 *Saturn's internal structure, as deduced from* Voyager *observations and computer modeling.*

treme core temperature, density, and pressure than Jupiter. The central pressure is around a tenth of Jupiter's—not too different from the pressure at the center of the Earth.

Magnetospheric Activity

Saturn's electrically conducting interior and rapid rotation produce a strong magnetic field and an extensive magnetosphere. Probably because of the considerably smaller mass of the Saturnian metallic hydrogen zone, Saturn's basic magnetic field strength is only about 1/20 that of Jupiter, or about 1000 times greater than that of the Earth. The magnetic field at Saturn's cloud tops (roughly 10 Earth radii from the planet's center) is approximately the same as at Earth's surface. *Voyager* measurements indicate that, unlike Jupiter and Earth, whose magnetic axes are slightly tilted, Saturn's magnetic field is not inclined with respect to its rotation axis.

Saturn's magnetosphere extends about 1 million km toward the Sun and is large enough to contain the planet's ring system and the innermost 16 small moons. Its size varies with the strength of the solar wind, which tends to push the sunward side of the magnetosphere closer to the planet. Saturn's largest moon, Titan, orbits about 1.2 million km from the planet, so that it is sometimes found just inside the outer magnetosphere and sometimes just outside, depending on the intensity of the solar wind. Because no major moons lie deep within Saturn's magnetosphere, the details of the magnetospheric structure are different from Jupiter's. For example, there is no equivalent of the Io plasma torus. Saturn emits radio waves but, as luck would have it, they are reflected from the Earth's ionosphere (they lie in the AM band) and were not detected until the *Voyager* craft approached the planet.

Figure 14.9 *(a) One of the best images of Saturn from an Earth-bound telescope. (b) Saturn as seen by the* Hubble Space Telescope *in August 1990. (Palomar Observatory; NASA)*

SATURN'S SPECTACULAR RING SYSTEM

The View from Earth

✓3 The most obvious aspect of Saturn's appearance is, of course, its **planetary ring system.** Astronomers now know that all the Jovian planets have rings, but Saturn's are by far the brightest, the most extensive, and the most beautiful. Galileo saw them first in 1610, but he did not recognize what he saw as a planet with a ring. At the resolution of his small telescope, the rings looked like bumps on the planet, or perhaps a triple planet of some sort. In 1659, the Dutch astronomer Christian Huygens realized what the "bump" was—a thin, flat ring, completely encircling the planet. Figure 14.9(a) shows Saturn as seen from a large Earth-based telescope. A *Hubble Space Telescope* view, from the same distance, but without Earth's atmospheric distortion, is shown in Figure 14.9(b).

In 1675, the French-Italian astronomer Giovanni Domenico Cassini discovered the first ring feature, a dark band about two-thirds of the way out from the inner edge. From Earth, the band looks like a gap in the ring (an observation that is not too far from the truth, although we now know that there is actually some ring material within it). This "gap" is named the **Cassini Division,** in honor of its discoverer. Careful observations from Earth show that the inner "ring" is in reality also composed of two rings. From the outside in, the three rings are known somewhat prosaically as the **A, B,** and **C rings.** The Cassini Division lies between A and B. A smaller gap, known as the **Encke Division,** is found in the outer part of the A ring. Its width is 270 km. These ring features are marked on Figure 14.9(b). No finer ring details are visible from our earthly vantage point. Of the three main rings, the B ring is brightest, followed by the somewhat fainter A ring, and then by the almost translucent C ring. A more complete list of ring properties appears in Table 14-1.

TABLE 14-1 ***The Rings of Saturn***

Ring	Inner Radius (km; planet radii)	Outer Radius (km/planet radii)	Width (km)
D	60,000; 1.00	74,000/1.23	14,000
C	74,000; 1.23	92,000/1.53	18,000
B	92,000; 1.53	117,600/1.96	25,600
A	122,200; 2.04	136,800/2.28	14,600
F	140,500; 2.34	140,600/2.34	100
E	210,000; 3.50	300,000/5.00	90,000

Overall Properties of the Rings

A fairly obvious question—and one that perplexed the best scientists and mathematicians on Earth for centuries—is "What are Saturn's rings?" By the middle of the nineteenth century, various dynamical and thermodynamic arguments had conclusively proved that the rings could not be solid, liquid, or gas! What is left? In 1857, Scottish physicist James Clerk Maxwell, after proving that a solid ring would become unstable and break up, suggested that the rings are composed of a great number of small particles, all independently orbiting Saturn, like so many tiny moons. That inspired speculation was essentially correct. It was verified in 1895, when Lick Observatory astronomer James Keeler measured the Doppler shift of sunlight reflected from the rings and showed that the velocities thus determined were exactly what would be expected from separate particles moving in circular orbits. The orbital velocity decreases with increasing distance from the planet, in accordance with Kepler's Laws and Newton's Law of Gravity. For example, a particle at the inner edge of the B ring circles Saturn in just 8 hours, while at the outer edge of the A ring a particle's orbital period is slightly over 14 hours. Obviously, no solid ring would be possible under these circumstances.

What sort of particles make up the rings? Their high albedo (0.8) suggested to astronomers that they were made of ice. Infrared observations in the 1970s confirmed that water ice is indeed a prime ring constituent. Radar observations and later *Voyager* studies of scattered sunlight showed that the sizes of the particles range from fractions of a millimeter to tens of meters in diameter, with most particles being about the size (and composition) of a large snowball on Earth. We now know that the rings really are truly thin—perhaps only a few tens of meters thick in places. Stars can occasionally be seen through them, like automobile headlights penetrating an open-weave curtain, and, as we have already seen, when viewed edge-on from Earth they disappear altogether.

Why are the rings so thin? The answer is that collisions between ring particles tend to keep them all moving in circular orbits in a single plane. Any particle that tries to stray away from this orderly motion finds itself in an orbit that soon runs into other ring particles. Over long periods of time, the ensuing jostling serves to keep all particles moving in circular, planar orbits. The asymmetric gravitational field of Saturn (the result of its flattened shape) sees to it that the rings lie in the plane of the planet's equator.

The Roche Limit

But why a ring of particles at all? What process produced the rings in the first place? To answer these questions, consider the fate of a small moon orbiting close to a massive planet such as Saturn. The moon is held together by internal forces—its own gravity or, for a small body, its own tensile strength (its resistance to stretching). As we bring our hypothetical moon closer to the planet, the tidal force on it increases. Recall that the effect of such a tidal force is to stretch the moon along the direction to the planet—that is, to create a tidal bulge. Recall also that the tidal force increases rapidly with decreasing distance from the planet. As the moon is brought closer to the planet, it reaches a point where the tidal force tending to stretch it out becomes *greater* than the internal forces holding it together. At that point, the moon is torn apart by the planet's gravity, as shown in Figure 14.10. The pieces of the satellite from that point on pursue their own individual orbits around that planet, eventually spreading all the way around it in the form of a ring.

For any given planet and any given moon, this critical distance, inside of which the moon is destroyed, is known as the **tidal stability limit,** or the **Roche limit,** after the nineteenth-century French mathematician Edouard Roche, who first calculated it. If our hypothetical moon is held together by its own gravity and its average density is the same as that of the parent planet (both reasonably good assumptions for Saturn's larger moons), then the Roche limit is just 2.4 times the radius of the planet. Thus, for Saturn, no moon can survive within a distance of 144,000 km of the planet's center, about 7000 km beyond the outer edge of the A ring. The rings of Saturn occupy the region inside Saturn's Roche limit.

These considerations apply equally well to the other Jovian worlds. Figure 14.11 shows the locations of the ring systems of each of the four Jovian planets relative to the respective planetary Roche limits. Given the approximations in our assumptions, we can conclude that all of the rings are found within the Roche limit of their parent planet. Notice that the calculation of this limit applies only to moons massive enough for their own gravity to be the dominant force binding them together. Sufficiently small moons can survive even within the Roche limit because they are held together mostly by interatomic (electromagnetic) forces, not by gravity.

Two possible origins have been suggested for Saturn's rings. Astronomers estimate that the total mass of ring material is no more than 10^{15} tons—enough to make a satellite about 250 km in diameter. If such a satellite strayed inside Saturn's Roche limit or was destroyed (perhaps by a collision) near that radius, a ring could have resulted. The alternative view is that the rings represent material left over from Saturn's formation stage 4.5 billion years ago. Because of Saturn's tidal field, no moon could form inside the Roche limit, and so the material has remained a ring ever since. Which is correct? One can find respected astronomers on both sides of the issue. However, there does seem to be growing evidence that collisions among ring particles would tend to destroy the rings in much less time than the age of the solar system. If this is so, then either the rings must be replenished, perhaps by fragments of Saturn's moons chipped off by meteorites, or they are the result of a relatively recent, possibly catastrophic event in the planet's

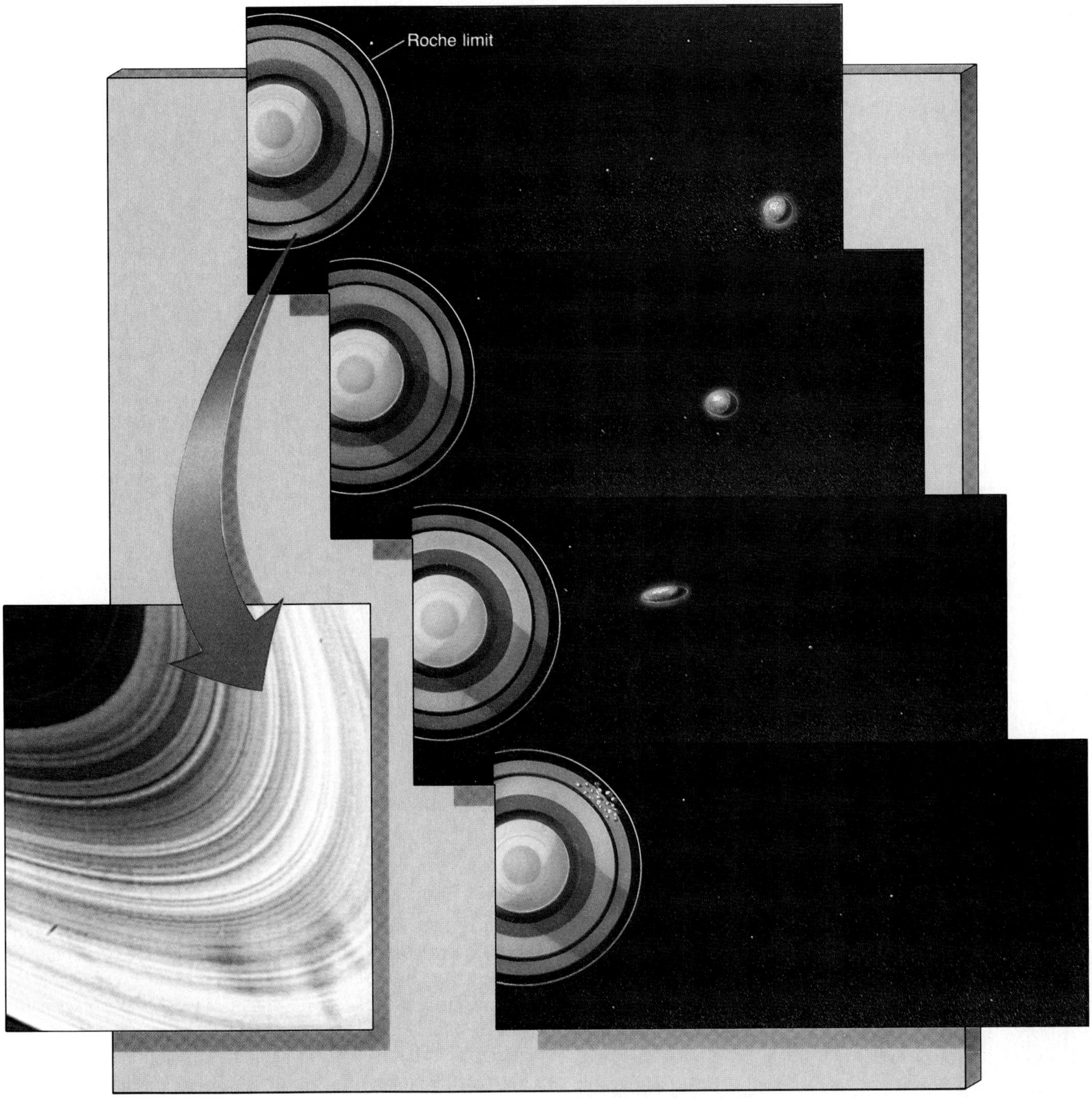

Figure 14.10 *The increasing tidal field of a planet first distorts, then destroys a moon that strays too close. (NASA)*

system. For now, however, the full answer just isn't known.

The View from Voyager

Thus it was, as *Voyager* approached Saturn, that scientists on Earth were fairly confident that they understood the nature (if not the origin) of the rings. However, the rings had many surprises in store. The *Voyager* mission to Saturn changed forever our view of this spectacular region in our cosmic backyard. It revealed the rings to be vastly more complex than astronomers had imagined. Gone is the apparent simplicity of the Saturnian disk observed by generations of Earth-based observers. What a difference a close-up camera can make!

As the *Voyager* probes approached Saturn, it became obvious that the main rings are actually composed of tens of thousands of narrow **ringlets** (shown in Figures 14.12 and

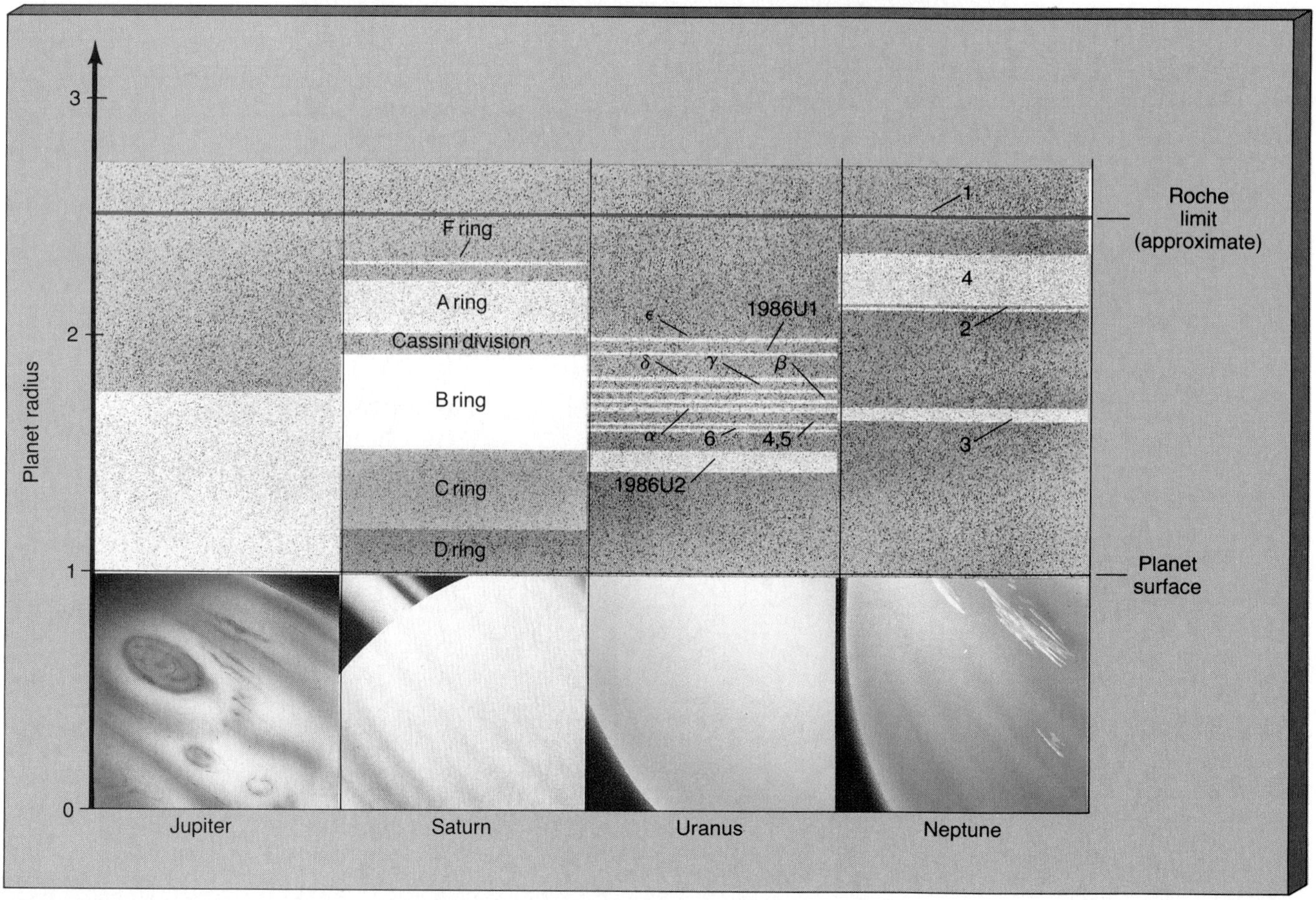

Figure 14.11 The rings of Jupiter, Saturn, Uranus, and Neptune. All distances are expressed in planetary radii. The dashed line represents the Roche limit. In all cases, the rings lie within the Roche limit of the parent planet.

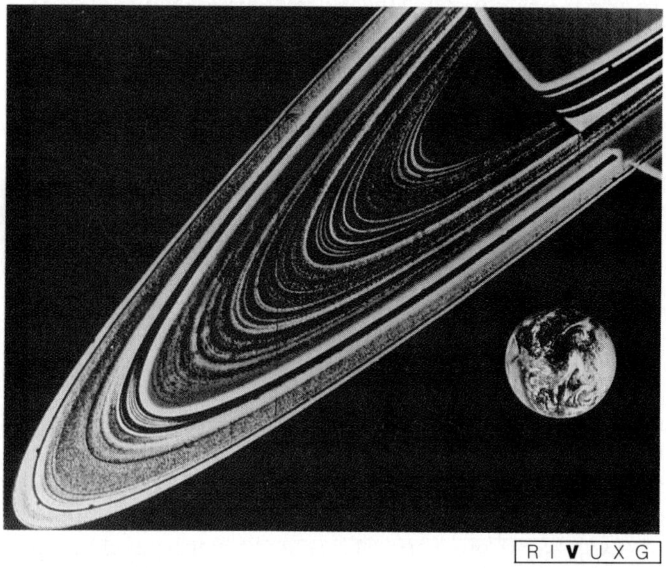

Figure 14.12 The Voyager 2 *cameras took this close-up of the ring structure just before plunging through the tenuous outer rings of Saturn. Earth is superposed, to proper scale, for a size comparison. (NASA)*

14.13). Although *Voyager* cameras did find several new gaps in the rings, the ringlets are generally not separated from each other by empty space. Instead, the rings contain concentric regions of alternating high and low concentrations of ring particles—the ringlets are just the high-density peaks. Most remarkable, this fine structure is not fixed, but varies with both time and position (see Figure 14.14). Although the process is not fully understood, the mutual gravitational attraction of the myriad ring particles (as well as the effects of Saturn's inner moons) enables waves of matter to form and move in the plane of the rings, rather like ripples on the surface of a pond. The wave crests typically wrap around and across the rings, often in the form of spiral patterns resembling grooves in a huge celestial phonograph record. Figure 14.15 illustrates how such **spiral density waves** can lead to the appearance of ringlets.

Although the ringlets are probably the result of spiral waves in the rings, the true gaps are not. The narrower gaps—about 20 of them—are most likely swept clean by the action of small moonlets embedded in them. These moonlets are larger (perhaps 10 or 20 km in diameter) than the largest true ring particles, and they simply ''scoop up''

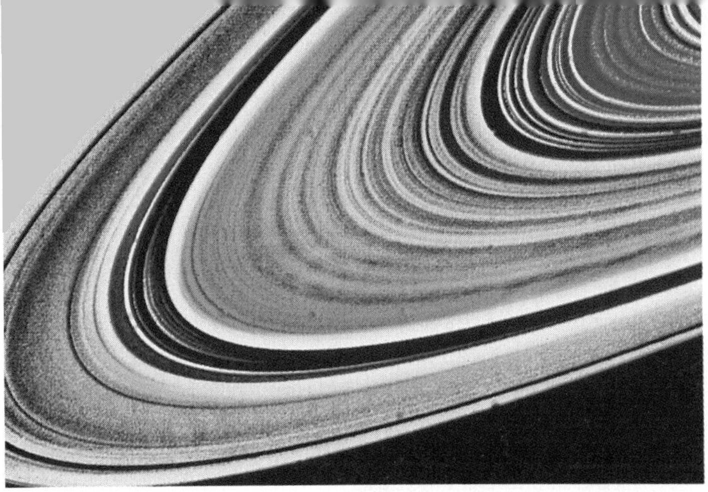

Figure 14.13 Ringlets in the B ring, spread over several thousand kilometers, and resolved here to about 10 km. As Voyager approached Saturn, more and more of these tiny ringlets became noticeable in the main rings. The (false) color variations probably indicate different sizes and compositions of the particles making up the thousands of rings. (NASA)

R I **V** U X G

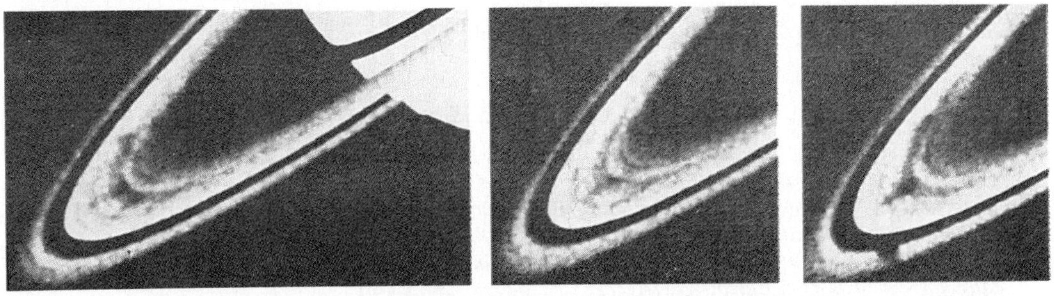

Figure 14.14
Time-dependence of the B ring ringlets. Voyager *took these images at three different times, over a period of several ring orbit times. (NASA)*

R I **V** U X G

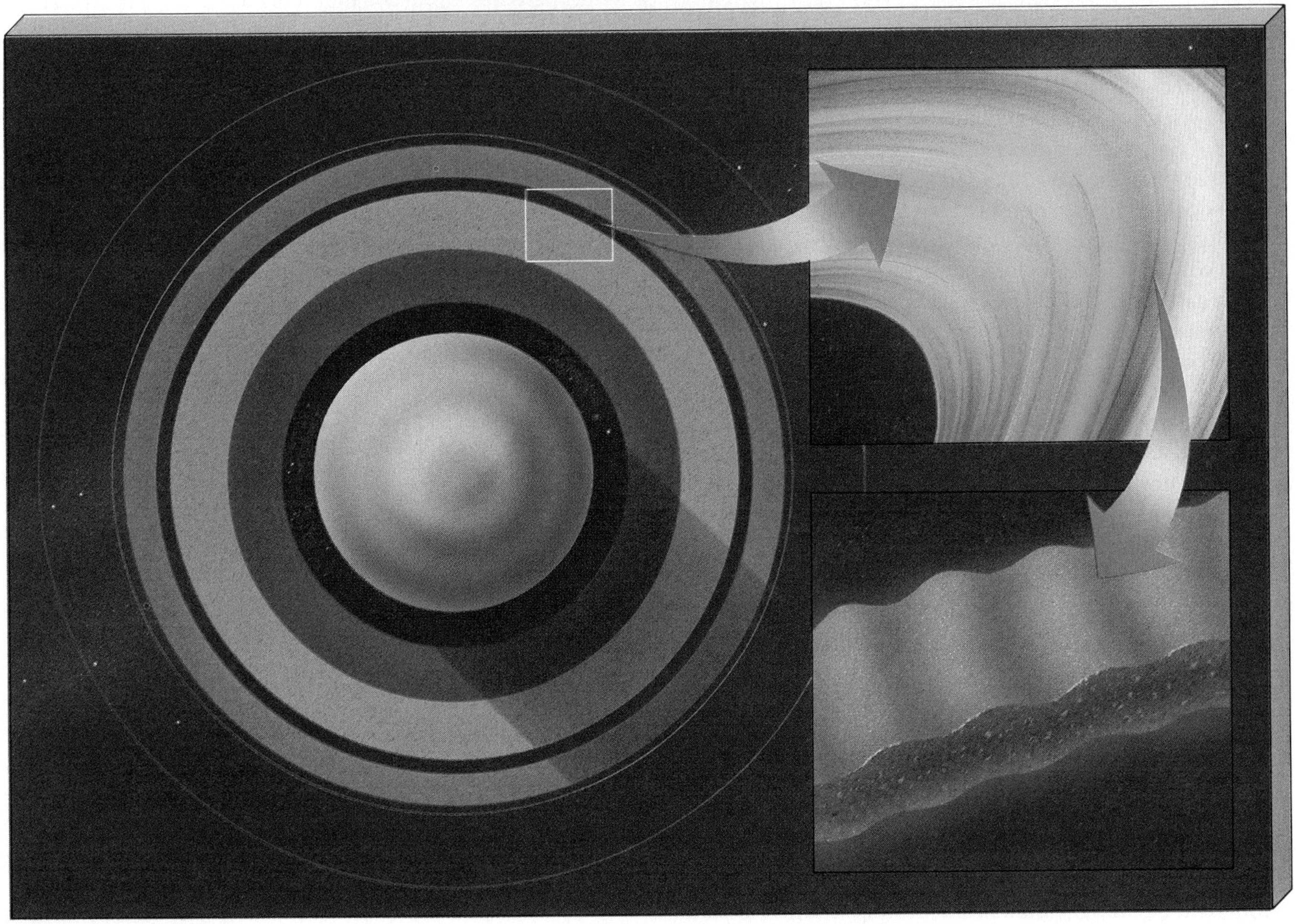

Figure 14.15 Density waves are alternating regions of high and low density in a ring. When the density wave is wrapped into a tight spiral, as is the case in Saturn's rings, the high-density peaks form nearly concentric ringlets.

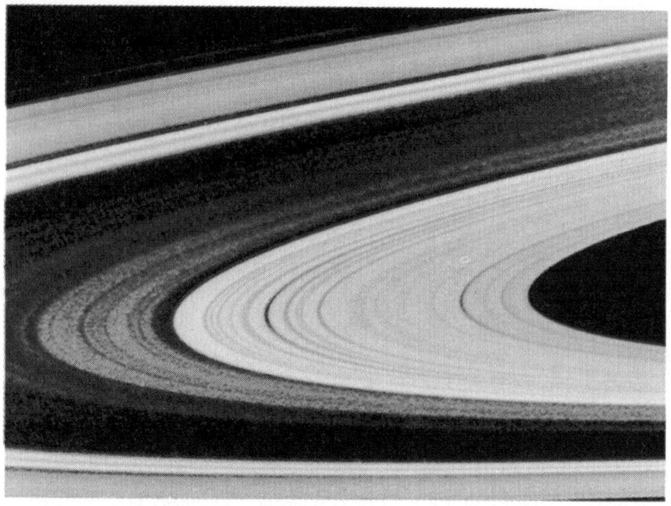

Figure 14.16 *Close inspection by* Voyager *revealed that the Cassini Division (shown here as the darker color) is not completely empty. It contains a series of faint ringlets and gaps, assumed to be caused by unseen embedded satellites. The density of material in the division is very low, accounting for its dark appearance from Earth. (NASA)*

ring material as they go. Despite many careful searches of the *Voyager* images, only one of these moonlets has so far been found—in 1991, after 5 years of exhaustive study, NASA scientists confirmed the discovery of the eighteenth moon of Saturn (tentatively named Pan) in the Encke Division. Astronomers have found indirect evidence for embedded moonlets, in the form of ''wakes'' that they leave behind them in the rings, but no other direct sightings have occurred. Despite their elusiveness, however, moonlets are still regarded as the best explanation for the small gaps.

Voyager images show that the Cassini Division itself is not completely empty of matter. In fact, as shown in Figure 14.16, the division contains a series of faint ringlets and gaps (and, presumably, embedded moonlets too). The overall concentration of ring particles in the division as a whole is, however, much lower than in the A and B rings. Although its small internal gaps probably result from embedded satellites, the division itself does not. It owes its existence to another solar system *resonance*, this time between particles orbiting in the division and Saturn's innermost major moon, Mimas.

A ring particle moving in an orbit within the Cassini Division has an orbital period exactly half that of Mimas. Particles in the division thus complete exactly two orbits around Saturn in the time taken for Mimas to orbit once—a configuration known as a 2:1 resonance. The effect of this resonance is that particles in the division feel a gravitational tug from Mimas at exactly the same location in their orbit every other time around. Successive tugs reinforce one another, and the initially circular trajectories of the ring particles soon get stretched out into ellipses. In their new orbits these particles collide with other particles and eventually find their way into new circular orbits at other radii. The net effect is that the number of ring particles in the Cassini Division is greatly reduced.

Particles in ''nonresonant'' orbits (that is, at radii where the orbital period is not simply related to the period of Mimas) also experience the gravitational pull of Mimas. But the times when the force is greatest are spread uniformly around the orbit, and the tugs cancel out. It's a little like pushing a child on a swing—pushing at the same point in the swing's motion each time produces much better results than do random shoves. Thus, Mimas (or any other moon) has a large effect on the ring at those radii where a resonance exists and little or no effect elsewhere.

In general, the simple resonances—2:1, 3:2, and so on—are the strongest and have the greatest effect on the ring structure. For example, the 3:2 resonance (3 ring orbits in 2 Mimas orbital periods) corresponds to the outer edge of the A ring. Most theories of planetary rings predict that the ring system should spread out with time, basically because of collisions among ring particles. The sharpness of the A ring's outer edge is in large part a consequence of the 3:2 resonance with Mimas. The outer edge of the ring is ''patrolled'' by a small satellite named Atlas, which prevents ring particles from escaping outward and is itself held in place by the resonance with Mimas. Other, weaker resonances, many of them involving moons other than Mimas, also exist. (In fact, a 7:6 resonance—7 ring orbits for every 6 moon orbits—with two tiny moons called Janus and Epimetheus also contributes to the A ring's sharp edge.) However, many of these resonances do not produce gaps but instead are responsible for powering the spiral waves and warping the rings slightly.

Outside the A ring lies the strangest ring of all. The faint, narrow **F ring** (shown in Figure 14.17) was discov-

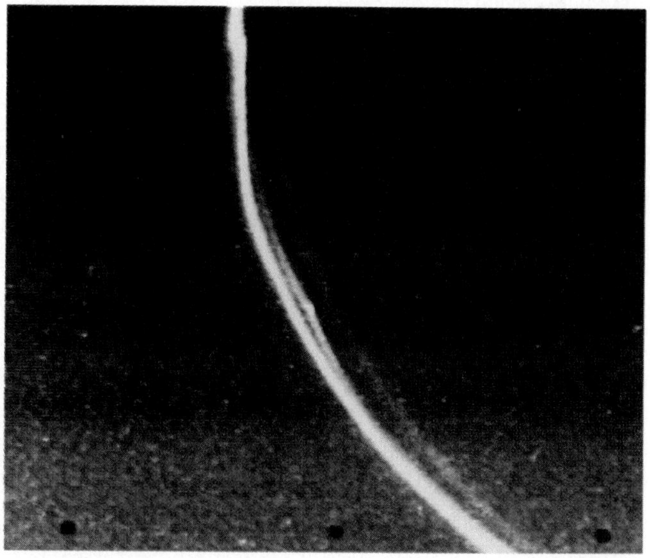

Figure 14.17 *Saturn's narrow F ring appears to contain kinks and braids, making it unlike any of the other Saturnian rings. (NASA)*

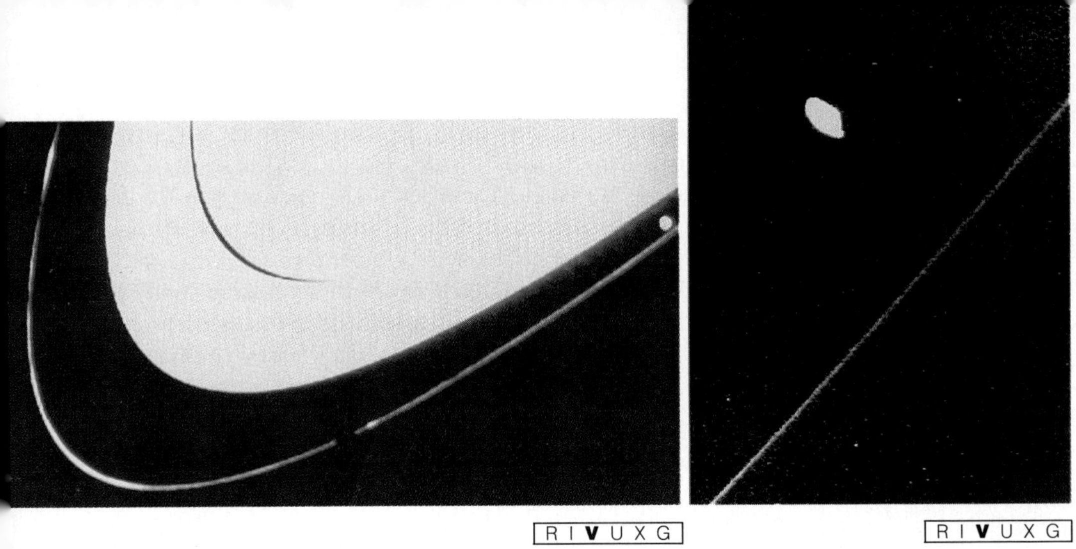

Figure 14.18 *The F ring's thinness, and possibly its other peculiarities too, can be explained by the effects of two shepherd satellites that orbit a few hundred kilometers inside and outside the ring. The photo at right shows one of the shepherding satellites, roughly 100 km in length. (NASA)*

ered by *Pioneer* in 1979, but its full complexity became evident only when *Voyager* took a closer look. Unlike the inner major rings, the F ring is narrow, less than a hundred kilometers wide. It lies just inside Saturn's Roche limit, separated from the A ring by about 3500 km. Its narrowness by itself is unusual, as is its slightly eccentric shape, but its oddest feature is that it looks for all the world as though it is made up of several separate strands braided together! This remarkable discovery sent dynamicists scrambling in search of an explanation. It now seems as though the ring's intricate structure, as well as its thinness, arise from the influence of two small moons, known as **shepherd satellites,** that orbit on either side of it (Figure 14.18).

These two small, dark satellites, each about 50 km in diameter, are called Prometheus and Pandora. They orbit about 1000 km on either side of the F ring, and their gravitational influence on the F ring particles keeps the ring tightly confined in its narrow orbit. As shown in Figure 14.19, any particle straying too far out of the F ring is gently guided back into the fold by one or the other of the moons. (The moon Atlas confines the A ring in a somewhat similar way.) However, the details of how Prometheus and Pandora produce the braids in the F ring and why the two moons are there at all, in such similar orbits, remain unclear. There is some evidence that other eccentric rings found in the gaps in the A, B, and C rings may also result from the effects of shepherding moonlets.

Voyager found a faint series of rings, now known collectively as the **D ring,** inside the inner edge of the C ring, stretching down almost to the Saturnian cloud tops. The D ring contains relatively few particles and is so dark that it is completely invisible from Earth. Another faint ring, also a *Voyager* discovery, lies well outside the main ring structure. Known as the **E ring,** it appears to be associated with volcanism on the moon Enceladus. Finally, the *Voyager* cameras revealed one other completely unexpected feature. A

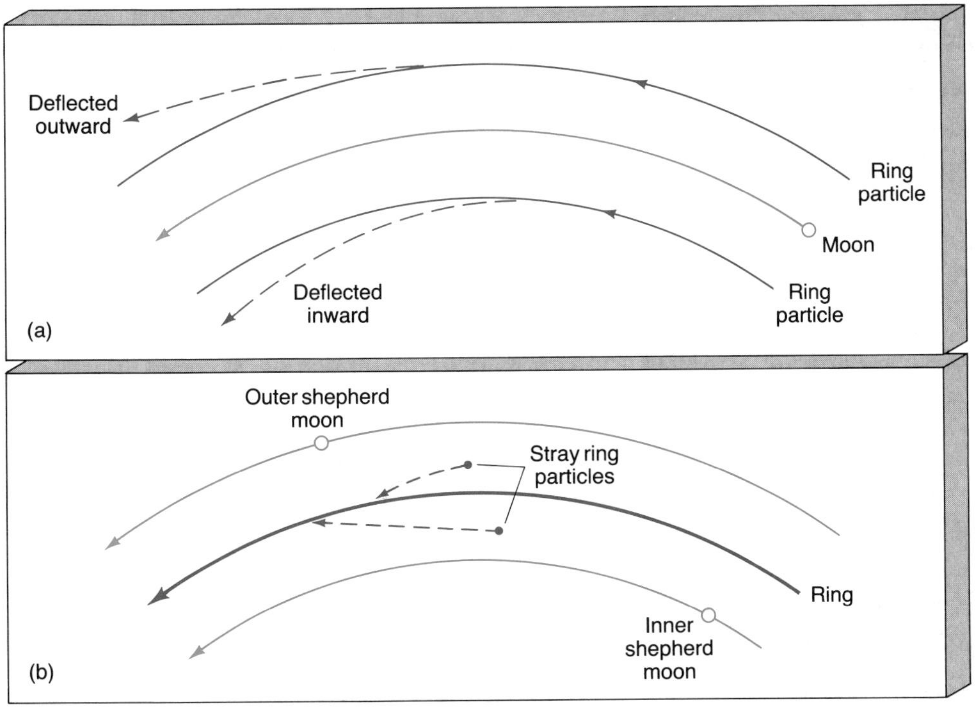

Figure 14.19 *(a) The net effect of the interactions between moon and ring particles is that the moon tends to push those particles* away *from it. (b) The F ring shepherd satellites operate by forcing errant F ring particles back into the main ring. Each moon operates as in part (a) so that the ring is confined between the two moons. As a consequence, the satellites themselves slowly drift away from the ring.*

R I V U X G

Figure 14.20 Saturn's B ring showed a series of dark temporary spokes as Voyager 2 *flew by at a distance of about 4 million km. The spokes were caused by small particles suspended just above the ring plane. (NASA)*

series of dark radial ''spokes'' formed on the B ring, moved around the planet for about one ring orbit period, and then disappeared (Figure 14.20). Careful scrutiny of these peculiar drifters showed that they were composed of very fine (micron-sized) dust hovering a few tens of meters *above* the plane of the rings. Scientists believe that this dust was held in place by electrostatic forces generated in the ring plane, perhaps resulting from particle collisions there. The electrical fields slowly dispersed, and the spokes faded as the ring revolved. We expect that the creation and dissolution of such spokes is a regular occurrence in the Saturn ring system.

All this dynamic activity in Saturn's rings implies to some researchers that the rings are quite young—perhaps no more than 50 million years old, or 100 times younger than the solar system. There is just too much activity for the rings to have remained stable for billions of years. If true, then the ring debris we see today might well be what is left of a small Saturnian moon that was hit by a large comet or perhaps by another moon. Astronomers normally prefer not to invoke catastrophic events to explain observed phenomena, but the more we learn of the universe, the more we realize that catastrophe probably plays an important role.

THE MOONS OF SATURN

General Features

Saturn has the most extensive, and in many ways the most complex, system of natural satellites of all the planets. At least 18 moons, listed in Table 14-2, have been spotted to date. Their reflected light suggests that most are covered with snow and ice. Many of them are probably made almost entirely of water ice. Even so, they are a curious and varied lot.

The satellites fall into three natural groups. First, there are the ''small'' moons—irregularly shaped chunks of ice, all less than 300 km across—which exhibit a bewildering variety of complex and fascinating motion, although they are of little direct importance to the dynamics of the Saturnian system. Second, there are six ''medium-sized'' moons—spherical bodies with diameters ranging from 400 to 1500 km—which offer clues to the past and present state of the environment of Saturn, at the same time presenting

TABLE 14-2 ***The Moons of Saturn***

Name	Distance (km; planet radii)	Orbit Period (days)	Size (longest diameter, km)	Mass (Earth Moon masses)	Density (g/cm^3)	Albedo
Pan	134,000; 2.23	0.58	20	.00000004		
Atlas	138,000; 2.30	0.60	40			0.5
Prometheus	139,000; 2.32	0.61	80			0.5
Pandora	142,000; 2.37	0.63	100			0.5
Janus	151,000; 2.52	0.69	190			0.5
Epimetheus	151,000; 2.52	0.69	120			0.5
Mimas	186,000; 3.10	0.94	394	0.00054	1.2	0.8
Enceladus	238,000; 3.97	1.37	502	0.0011	1.2	1.0
Tethys	295,000; 4.92	1.89	1050	0.010	1.3	0.8
Telesto	295,000; 4.92	1.89	25			0.6
Calypso	295,000; 4.92	1.89	25			0.9
Dione	377,000; 6.28	2.74	1120	0.015	1.4	0.6
Helene	377,000; 6.28	2.74	30			0.6
Rhea	527,000; 8.78	4.52	1530	0.0034	1.3	0.6
Titan	1,220,000; 20.3	16.0	5150	1.83	1.9	0.2
Hyperion	1,480,000; 24.7	21.3	270			0.3
Iapetus	3,560,000; 59.3	79.3	1440	0.026	1.2	0.5
Phoebe	13,000,000; 217	550*	220			0.06

*Indicates a retrograde orbit

many puzzles in their own appearance and history. Finally, there is Saturn's single "large" moon—Titan—which, at 5150 km in diameter, is the second-largest satellite in the solar system (Jupiter's Ganymede is a little bigger) and dominates the Saturnian system. It has an atmosphere denser than Earth's and, some scientists think, surface conditions possibly conducive to life. Notice, by the way, that Jupiter has no "medium" moons, as just defined. The Galilean satellites are large, like Titan, and all of Jupiter's other satellites are small—no more than 200 km in diameter.

Titan

✓4 Perhaps the most intriguing of all Saturn's moons is Titan, discovered by Christian Huygens in 1655. Even through a large Earth-based telescope, Titan is visible only as a barely resolved reddish disk. Long before the *Voyager* missions, astronomers already knew (from spectroscopic observations) that the moon's reddish coloration is caused by something quite special—an atmosphere. So anxious were mission planners to obtain a closer look that they programmed *Voyager 1* to pass very close to Titan, even though that meant the spacecraft could not then use Saturn's gravity to continue on to Uranus and Neptune. (Instead, *Voyager 1* left the Saturnian system on a path taking the craft out of the solar system well above the ecliptic plane.) A *Voyager 1* image of Titan is shown in Figure 14.21.

Scientists believe that Titan's internal composition and structure must be similar to Ganymede and Callisto because these three moons have quite similar masses and radii and hence average densities (Titan's density is 1.9 g/cm^3). Thus, Titan probably contains a rocky core surrounded by a thick mantle of water ice. Despite *Voyager 1*'s close pass, the moon's surface remains a mystery. A thick, uniform haze layer, similar to the photochemical smog found over many cities on Earth, envelops the moon and completely obscured the spacecraft's view.

Voyager 1 was able to provide mission specialists with detailed atmospheric data, however. Titan's atmosphere is thicker and denser even than Earth's, and it is certainly more substantial than that of any other moon. Prior to *Voyager 1*'s arrival in 1980, only methane and a few other simple hydrocarbons had been conclusively detected on Titan, but radio and infrared observations from the spacecraft showed that the atmosphere is actually made up mostly of nitrogen (roughly 90 percent) and possibly argon (at most 10 percent), with only a few percent of methane. In addition, the complex chemistry in Titan's atmosphere maintains steady (but only trace) levels of hydrogen gas (2000 parts per million), ethane (20 ppm), propane (20 ppm), carbon monoxide (100 ppm), and even smaller amounts of other compounds.

In fact, Titan's atmosphere seems to act like a gigantic chemical factory. Powered by the energy of sunlight, it is undergoing a complex series of chemical reactions that ultimately result in the observed smog and trace chemical composition. The upper atmosphere is thick with aerosol haze, and the unseen surface may be covered with organic sediment that has settled down from the clouds. Speculation runs the gamut from oceans of liquid hydrocarbons, especially ethane, to icy valleys laden with petrochemical sludge. Future spacecraft exploration of Titan may present scientists with an opportunity to study the kind of chemistry thought to have occurred billions of years ago on Earth—the prebiotic chemical reactions that eventually led to life on our own planet.

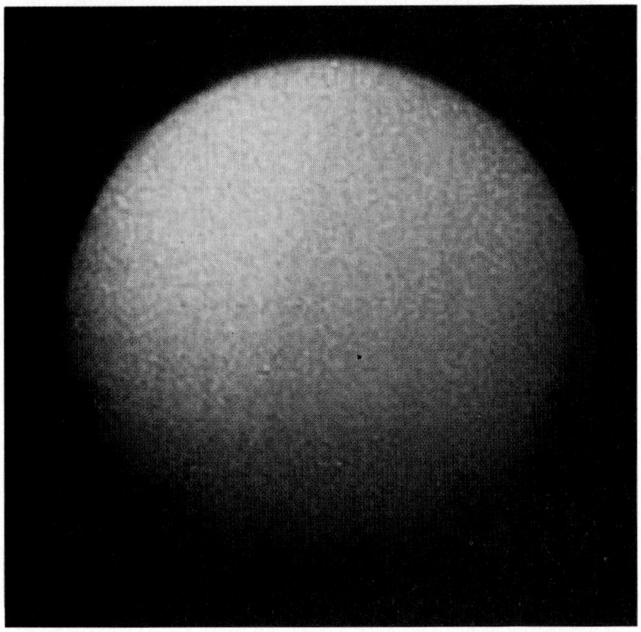

Figure 14.21 *Titan, larger than the planet Mercury and roughly half the size of Earth, was photographed from only 4000 km away as the* Voyager 1 *spacecraft passed by in 1980. All we can see here is Titan's upper cloud deck where the northern hemisphere appears slightly brighter than the southern. (NASA)*

Based largely on *Voyager* measurements, Figure 14.22 shows the structure of Titan's atmosphere. Despite Titan's low mass (a little less than twice that of Earth's Moon) and hence its low surface gravity (one-seventh of Earth's), the atmospheric pressure at ground level is 60 percent *greater* than on Earth. These data imply that Titan's atmosphere contains about 10 times as much gas as Earth's. Also, because of Titan's weaker gravitational pull, the atmosphere extends some 10 times farther into space than does our own. The top of the main haze layer lies some 200 km above the surface, although there are additional layers, seen primarily through their absorption of ultraviolet radiation, at about 300 km and 400 km (Figure 14.23). Below the haze, the atmosphere is reasonably clear, although rather gloomy, because so little sunlight gets through. The surface temperature is a frigid 94K, roughly what we would expect simply on the basis of Titan's distance from the Sun. If there is any greenhouse effect in Titan's atmosphere, it is small. At the temperatures typical of the lower atmosphere, methane and ethane may behave

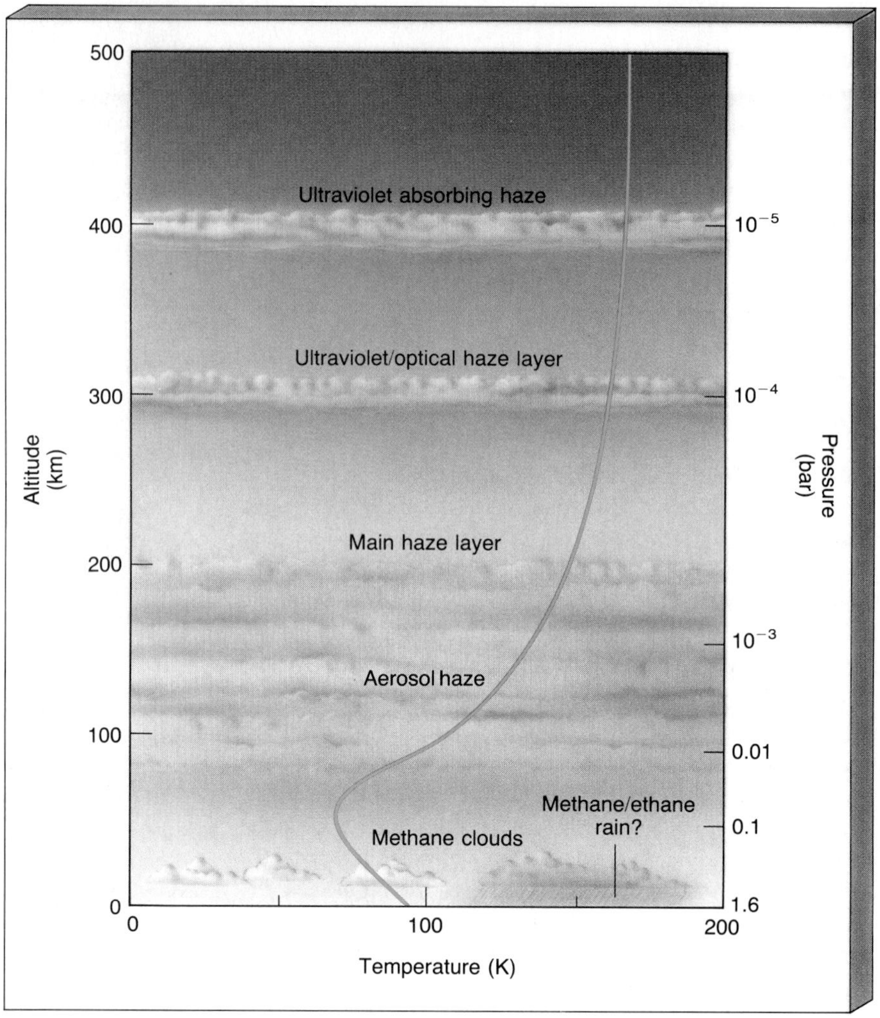

Figure 14.22 *The structure of Titan's atmosphere, as deduced from* Voyager 1 *observations.*

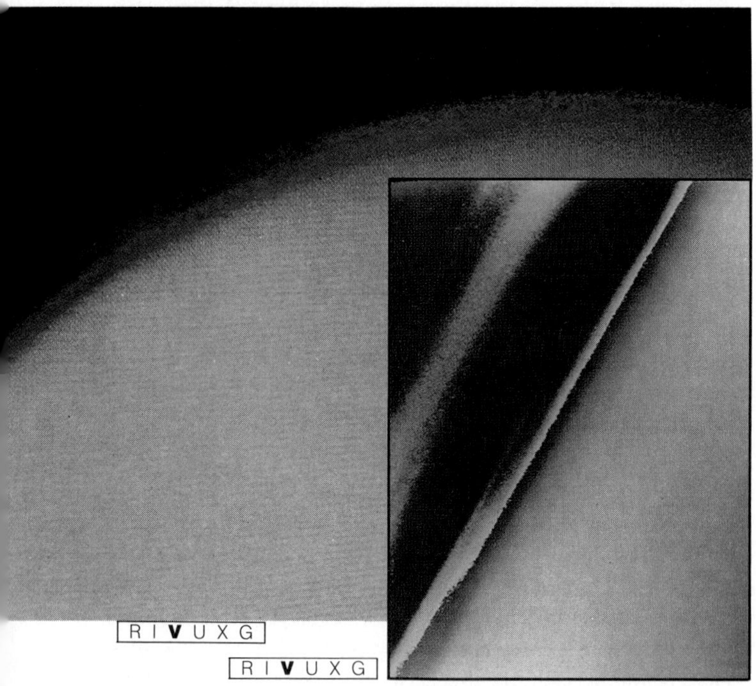

Figure 14.23 *The haze layers (blue) of Titan's atmosphere are visible in this false-color* Voyager 1 *image. (NASA)*

rather like water on Earth, raising the possibility of methane rain, snow, and fog and even ethane oceans! At high altitudes, the temperature rises, the result of photochemical absorption of solar radiation.

Why does Titan have such a thick atmosphere, when similar moons of Jupiter such as Ganymede and Callisto have none? The answer lies largely in Titan's low surface temperature. We have already seen (consult Interlude 9-2) how lower temperatures favor retention of an atmosphere. Just as important for Titan, low temperatures also enhance the ability of water ice to absorb certain light gases, such as methane and ammonia. The moons of Saturn lie farther from the Sun than do their Jupiter counterparts, so they formed at a lower temperature. As a result, the water ice that makes up the bulk of Titan's interior was initially laden with much more methane and ammonia gas than was present in either Ganymede or Callisto. As Titan's internal radioactivity warmed the moon, the ice released the trapped gases, forming a thick methane-ammonia atmosphere. Sunlight split the ammonia into hydrogen, which escaped into space, and nitrogen, which remained in the atmosphere. The methane, which was less easily broken apart, survived intact. Together with argon outgassed from Titan's interior, these gases form the basis of the atmosphere we see today.

Saturn's Medium-Sized Moons

Saturn's complement of mid-sized moons consists (in order of increasing distance from the planet) of Mimas, Enceladus, Tethys, Dione, Rhea, and Iapetus. Cassini found the outermost four in the late seventeenth century. William Herschel discovered Mimas and Enceladus in 1789. The inner five of those moons all move on circular trajectories lying well inside Titan's orbit, and they are tidally locked by Saturn's gravity into synchronous rotation (so that one side always faces the planet). They therefore all have permanently "leading" and "trailing" faces as they move in their orbits, a fact that is important in understanding their often asymmetrical surface markings. The remaining member of the group, Iapetus, orbits far beyond Titan on an inclined, eccentric path.

Unlike the densities of the Galilean satellites of Jupiter, the densities of these six Saturnian moons do not show any correlation with distance from Saturn. Their densities are all about 1.2 g/cm^3. This similarity perhaps indicates that nearness to the central planetary heat source was a less important influence during their formation than it was in the Jupiter system. Scientists believe that the mid-sized moons are composed largely of rock and water ice, as is Titan. Their densities are lower than Titan's primarily because their lower masses produce less compression of their interiors. The largest of the six, Rhea, has a mass only 1/30 that of Earth's Moon.

Rhea, Saturn's second-largest satellite, is shown in Figure 14.24. Its icy surface is very reflective (with an albedo of about 0.6) and heavily cratered. At the low temperatures found on its surface, water ice is very hard and behaves rather like rock on the inner planets. For that reason, Rhea's surface craters look very like craters on the Moon or Mercury. The crater density is similar to that in the lunar highlands, indicating that the surface is old, and there is no evidence of extensive geological activity. Rhea's only real riddle is the presence of so-called **wispy terrain**—prominent light-colored streaks—on its trailing side. The leading face, by contrast, shows no such markings, only craters. Astronomers believe that the "wisps" were caused by some event in the distant past during which water was released from the interior and condensed on the surface. Any similar markings on the leading side have presumably been long since obliterated by cratering, which is expected to be much more frequent on the satellite's forward-facing surface.

Inside Rhea's orbit lie the orbits of Tethys and Dione. These two moons, shown in Figure 14.25, are comparable in size and have masses somewhat less than half the mass of Rhea. Like Rhea, they have reflective surfaces that are heavily cratered, but each shows signs of surface activity, too. Dione's trailing face has prominent bright streaks, which are probably similar to Rhea's wispy terrain. Notice also the "maria" on Dione, where flooding of some sort appears to have obliterated the older craters. The cracks on

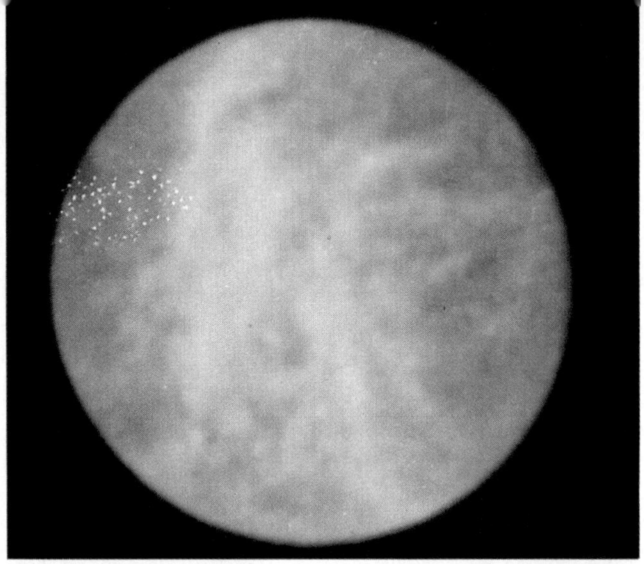

R I V U X G

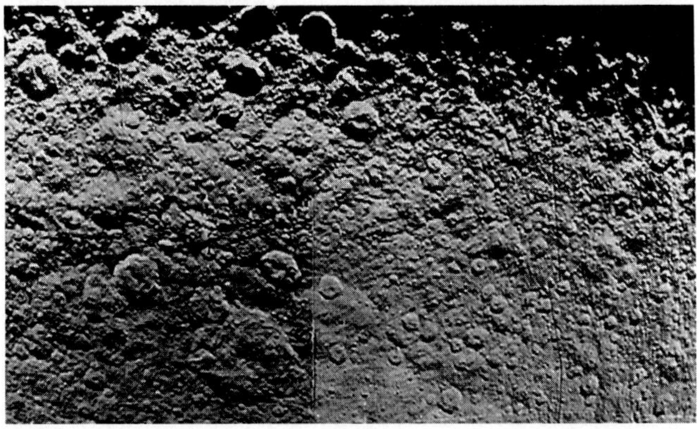

R I V U X G

Figure 14.24 *Saturn's second-largest moon, Rhea, has a heavily cratered, old surface. The light-colored "wisps" on the trailing face of the moon are thought to be water that was released from the moon's interior during some long-ago period of activity and then froze on the surface. The insert shows Rhea's north polar region at a superb resolution of only 1 km. The heavily cratered surface resembled that of the Moon, except that we see here craters in bright ice, rather than dark lunar rock. (NASA)*

Tethys may have been caused by cooling and shrinking of the surface layers or, more probably, by meteoritic bombardment. Tethys contains a huge impact crater, Odysseus, which is two-fifths the diameter of the moon itself. (This makes it the largest crater in the solar system, measured relative to the size of its parent body.) The major surface feature on Tethys, a great canyon known as Ithaca Chasma, is probably associated with the Odysseus impact. It stretches three-quarters of the way around the moon and rivals Mars's Valles Marineris in size.

The innermost, and smallest, medium-sized moon is Mimas, shown in Figure 14.26. Despite its low mass—only 1 percent the mass of Rhea—its closeness to the rings causes resonant interactions with the ring particles, resulting most notably in the Cassini Division, as we have already seen. Mimas is heavily cratered. The moon's chief surface feature, known as Herschel, is an enormous crater on the leading face. Its diameter is almost one-third that of the

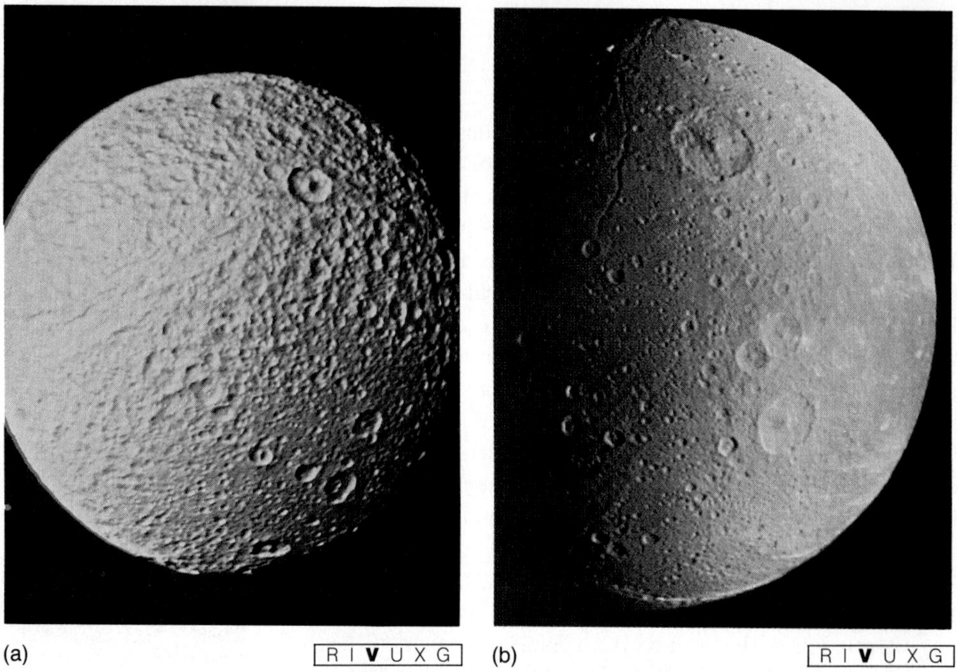

Figure 14.25 Tethys (a) and Dione (b) are similar to each other and are each about 1,000 km across, smaller than Rhea. They show evidence for ancient geological activity of some sort, in addition to extensive meteoritic cratering. Resolution here is just a few kilometers. (NASA)

moon itself. As with Odysseus on Tethys, the impact that formed Herschel must have come very close to destroying Mimas completely. It is quite possible that the debris produced by such impacts is responsible for creating or maintaining the spectacular rings we see.

Enceladus (Figure 14.27) orbits just outside Mimas. Its size, mass, composition, and orbit are so similar to those of Mimas that one might guess that the two moons would be

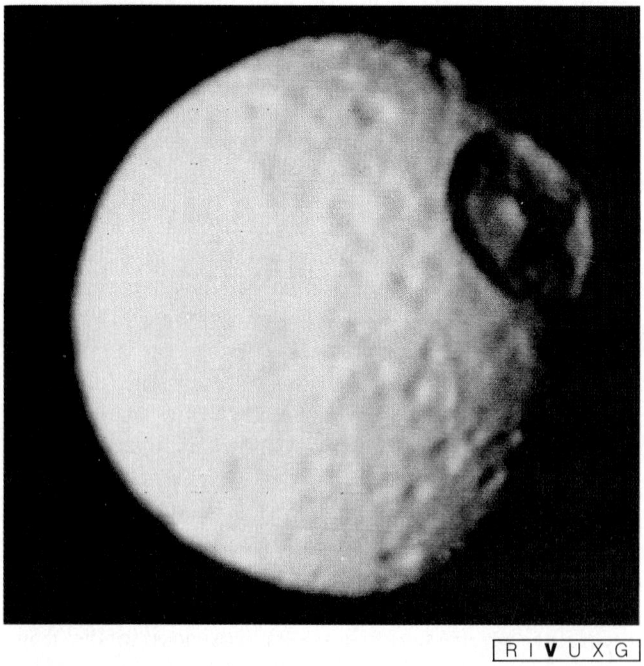

Figure 14.26 *Mimas's main surface feature is the large crater Herschel, plainly visible in this* Voyager *image. The impact that caused Herschel must have come very close to shattering Mimas. The moon is about 400 km in diameter. (NASA)*

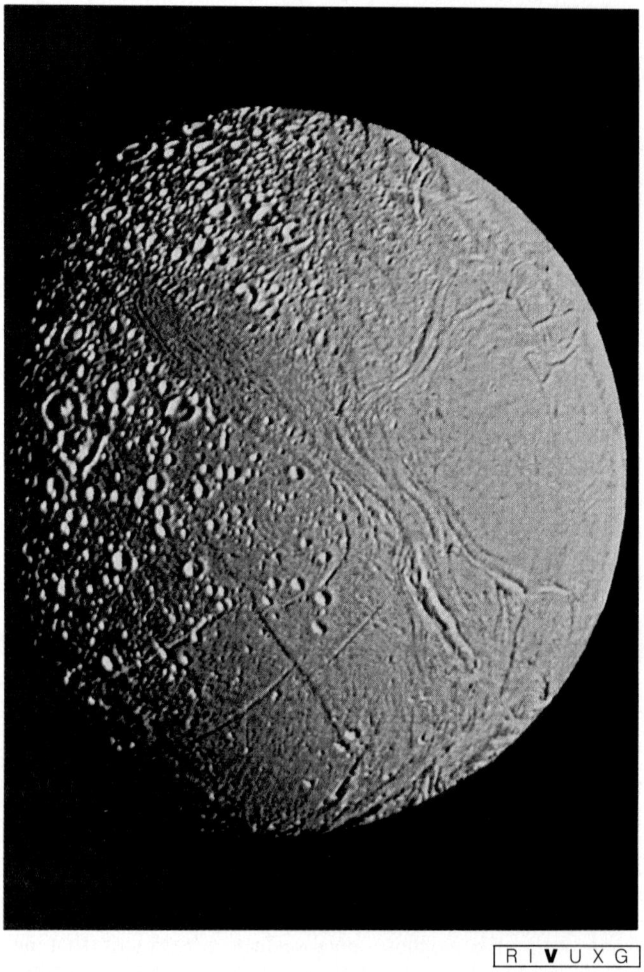

Figure 14.27 *Despite its similarity in size and location to Mimas, Enceladus, with its highly reflective icy surface and apparent water volcanism, looks very different. Apparently, the moon is still active but the cause of the volcanism is unexplained. (NASA)*

very similar to one another in appearance and history. This apparently is not so. Enceladus is so bright and shiny—it reflects virtually 100 percent of the sunlight falling on it—that astronomers believe its surface must be completely coated with fine crystals of pure ice.

These crystals may well be the icy "ash" of water volcanos on Enceladus. Two lines of reasoning lead to this conclusion. First, the nearby thin cloud of small, reflective particles that makes up Saturn's E ring is known to be densest near Enceladus. Are the ring's particles falling onto Enceladus? Or is Enceladus erupting and spewing out particles that become the E ring? The answer is not entirely clear. However, calculations indicate that the E ring is unstable because of the disruptive effects of the solar wind, supporting the view that volcanism on Enceladus continually supplies new particles to maintain the ring. Second, Enceladus bears visible evidence for large-scale volcanic activity. Much of the moon's surface is devoid of impact craters, which seem to have been erased by what look like lava flows, except that the "lava" is water, temporarily liquefied during recent internal upheavals and now frozen again. Although no geysers or volcanos have actually been observed on Enceladus, there seems to be strong circumstantial evidence for volcanism on the satellite.

Why is there so much geological activity on a moon so small? No one knows. Attempts have been made to explain Enceladus's water volcanism in terms of tidal stresses. (Recall the role that Jupiter's tidal stresses play in creating volcanism on Io.) However, Saturn's tidal force on Enceladus is only one-quarter the force exerted by Jupiter on Io, and there are no nearby large satellites to force Enceladus away from a circular trajectory. Mimas is certainly too small to have any major effect, and it is unclear whether other interactions, such as the 2:1 resonance that exists between Enceladus and Dione, can produce a sufficiently eccentric orbit. Thus, the ingredients that power Io's volcanos may not be present on Enceladus. The question of why Mimas, which is even closer to Saturn, doesn't experience similar tidal effects also arises. For now, the mystery of Enceladus's internal activity remains unsolved.

The outermost mid-sized moon is Iapetus. It orbits Saturn on a somewhat eccentric, inclined orbit with a semimajor axis of 3.6 million km. Its mass is about three-quarters that of Rhea. Iapetus's major surface feature is a huge, nearly circular black spot known as Cassini Regio (see Figure 14.28). Iapetus is a two-faced moon. The dark, leading face reflects only about 3 percent of the sunlight reaching it, while the icy trailing side reflects 50 percent. Scientists think similar dark deposits seen elsewhere in the solar system are organic (carbon-containing) in nature; they can be produced by the action of solar radiation on hydrocarbon (for example, methane) ice. Spectral studies of the dark regions seem to indicate that the material originates on Iapetus, so the moon is not simply sweeping up dark material as it orbits. But how the dark markings can adorn only one side of Iapetus is still unknown.

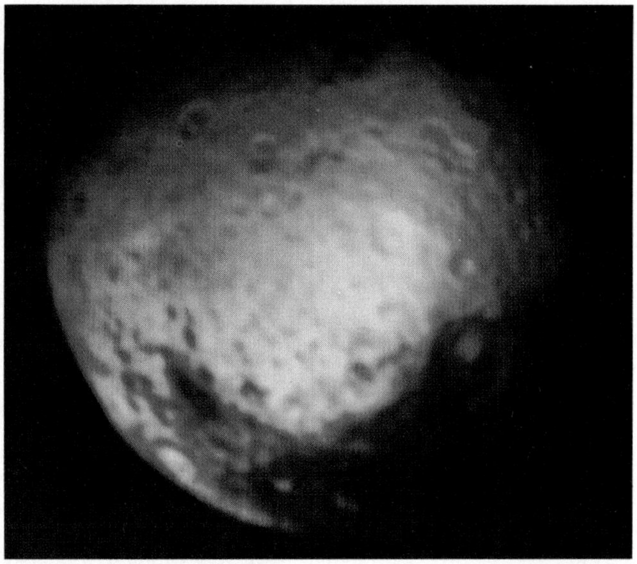

Figure 14.28 *Iapetus is one of the most peculiar worlds known. The contrast is clearly evident here between its light (icy) trailing surface at top and center and the black leading hemisphere at bottom. The large black region at bottom is called Cassini Regio. Its makeup and origin are unknown. (NASA)*

The Small Satellites

✓5 Finally, we come to Saturn's dozen or so small moons. Their masses are mostly unknown, but scientists believe their composition to be similar to the small icy moons of Jupiter. The outermost small moons, Hyperion and Phoebe, were discovered in the nineteenth century, in 1848 and 1898, respectively. The others were first detected in the second half of the twentieth century. Only the moons in or near the rings themselves were actually discovered by *Voyager*. We have already seen how the innermost moons, some of them still unnamed, sweep up ring material to form gaps and how the moon Atlas preserves the sharp outer edge of the A ring. We have also discussed how the shepherd moons Prometheus and Pandora confine the F ring. We now turn to the other members of the small-moon group.

Saturn's outermost moon is Phoebe, lying almost 13 million km from the planet, more than three times farther out than any other moon. Its wide, retrograde orbit (the only retrograde orbit among the Saturnian satellites) suggests to many researchers that it is captured debris and not a true part of Saturn's retinue.

Just 10,000 km beyond the F ring lie the so-called **co-orbital satellites** Janus and Epimetheus. As the name implies, these two satellites "share" an orbit. In fact, ground-based observations once led astronomers to believe they were a single satellite, and their separate motions led to great confusion about orbital properties! The motions of the co-orbital satellites are sketched in Figure 14.29. At almost any instant, both moons are in circular orbits about Saturn, one of them with a slightly smaller orbital radius than the other. In fact, the radii are each about 151,000 km and differ by only 50 km (actually less than the radii of the moons

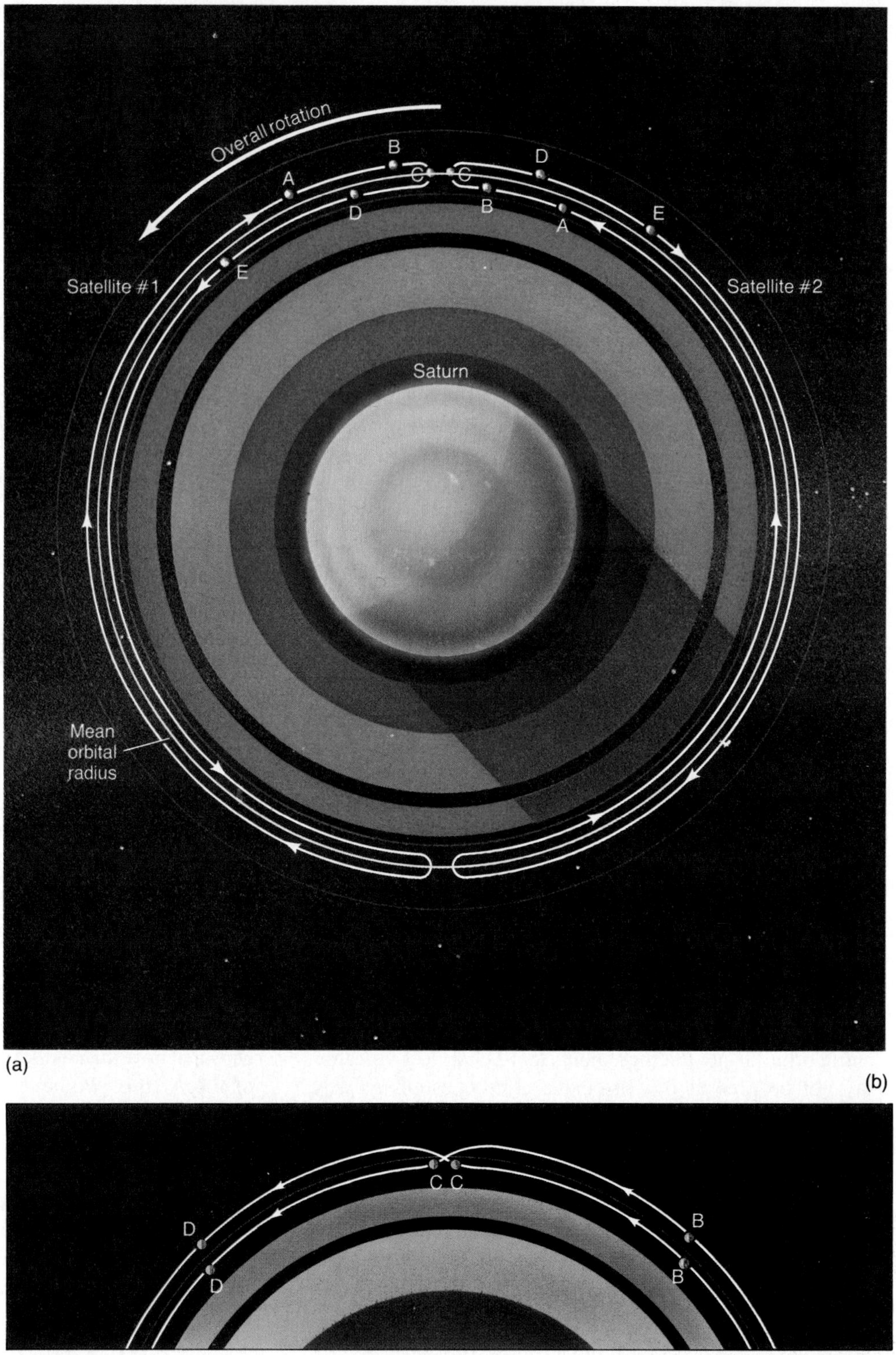

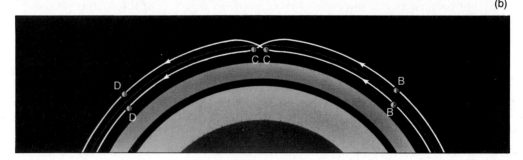

Figure 14.29 *The peculiar motion of Saturn's co-orbital satellites, Janus and Epimetheus. (a) Shown here from the point of view of an observer rotating at the average angular speed of the pair, the two moons play a never-ending game of tag as they move on their horseshoe orbits around the planet. The labeled points represent the locations of the two moons at a few successive times. From A to C, satellite 2 gains on satellite 1. However, before it can overtake it, the two moons swap orbits, and satellite 1 starts to pull ahead of satellite 2 again, through points D and E. The whole process is repeated on the other side of the planet, and so on, ad infinitum. (b) From the point of view of a stationary observer, of course, the moons do not stop and change directions. Instead, they simply swap orbits, as shown.*

themselves). Each satellite obeys Kepler's Laws, so the inner satellite orbits slightly faster than the outer one and slowly catches up to it. The inner and outer orbital periods are 16.664 and 16.672 hours, respectively, so the moons' relative speed is roughly 9 m/s. The inner moon takes about 4 years to "lap" the outer one.

As the inner satellite gains ground on the outer one, a strange thing happens. As depicted in Figure 14.29, when the two get close enough to begin to feel each others' weak gravity, they exchange orbits. The new inner moon (which used to be the outer one) begins to pull away from its companion, and the whole process begins again! If we were to view all this from a frame of reference rotating with the average angular speed of the two moons, they would each move on a horseshoe shaped orbit, forever approaching, swapping orbits, then separating from each other, only to meet again on the other side of the planet. No one knows why the co-orbital satellites are engaged in this curious dance. Possibly they are portions of a single moon that broke up, perhaps after a meteoritic impact, leaving the two pieces in almost the same orbit.

In fact, several of the other small moons also share orbits, this time with larger moons. Two of them, Telesto and Calypso, have orbits that are synchronized with the orbit of Tethys, always remaining fixed relative to the larger moon, lying precisely 60° ahead and 60° behind it as it travels around Saturn (see Figure 14.30). The moon Helene is similarly tied to Dione. These 60° points are known as **Lagrange points,** after the French mathematician Joseph Louis Lagrange, who first studied them. We will see further examples of this special 1:1 orbital resonance in the motion of some asteroids about the Sun, trapped in the Lagrange points of Jupiter's orbit (see Chapter 16).

The strangest motion of all is that of the moon Hyperion, which orbits in the outer part of the Saturnian system, between Titan and Iapetus, at a distance of 1.5 million km from the planet. Unlike most of the other Saturnian moons, its rotation is not synchronous with its orbital motion. Because of the gravitational effect of Titan, Hyperion's orbit is not exactly circular, so synchronous rotation cannot occur. In response to the competing gravitational influences of Titan and Saturn, this irregularly shaped satellite constantly changes both its rotation speed and its rotation axis, in a condition known as **chaotic rotation.** As Hyperion orbits Saturn, it tumbles apparently at random, never stopping and never repeating itself, in a completely unpredictable way. Since the 1970s, the study of chaotic motion on Earth has revealed new classes of unexpected behavior in even very simple systems. Hyperion is one of the few other places in the universe where this behavior has been unambiguously observed.

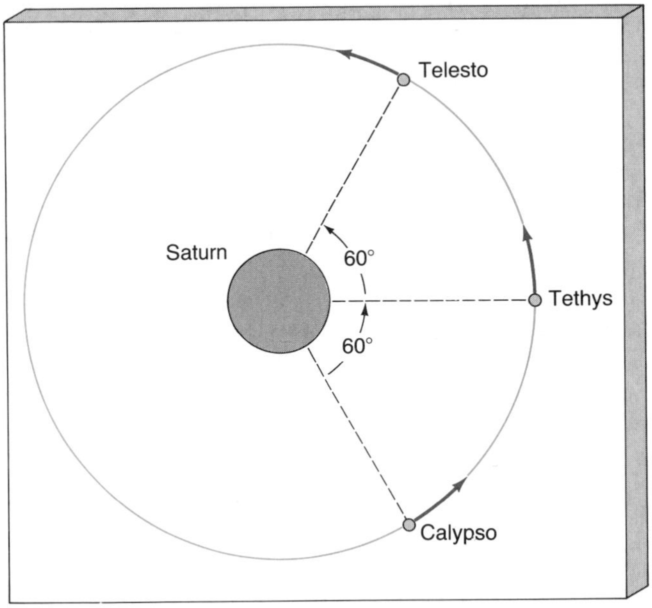

Figure 14.30 *The orbits of the moons Telesto and Calypso are tied to the motion of the moon Tethys. The combined gravitational pulls of Saturn and Tethys keep the small moons exactly 60° ahead and behind the larger moon at all times, so that all three moons share an orbit and never change their relative positions.*

CHAPTER SUMMARY

Saturn is about 9.5 times farther from the Sun than Earth is. Its orbital period is 29.5 Earth years. Like all the outer planets, Saturn is brightest around opposition, but its great distance from the Sun makes it considerably fainter than Jupiter or Mars. Saturn is one-third the mass of Jupiter, about 94 times the mass of Earth. It is composed primarily of hydrogen and helium, though its degree of oblateness suggests a rocky core twice the mass of Jupiter's. Like Jupiter, Saturn rotates differentially. Because its density is lower, its rapid rotation makes it even more oblate than Jupiter. Saturn's axial inclination of 27° seasonally affects our view of the thin, reflective rings.

Saturn has weaker gravity than Jupiter, and therefore it has a more extended atmosphere. The planet's butterscotch hue is probably due to cloud chemistry similar to that on Jupiter. Unlike the atmospheres of the other outer planets, Saturn's atmosphere is deficient in helium, which sank toward the center of the planet at some time during its evolution. Saturn, like Jupiter, has bands, ovals, and turbulent flow patterns powered by convective motion, although zonal flow is faster on Saturn. The strongest Saturnian bands and the most obvious ovals and turbulent eddies are found about 40° from the equator, where westward winds are found.

Some researchers believe that a light shower of helium has been falling through Saturn's interior for 2 billion years, heating up as it loses gravitational energy. The result is a planet with more excess energy than Jupiter, relative to the amount of energy absorbed from the Sun. The interior of Saturn is theoretically the same as that of Jupiter, but it has a thinner layer of metallic hydrogen and a larger core. Its lower mass gives Saturn a less extreme core temperature, density, and pressure than Jupiter. Saturn's conducting interior and rapid rotation produce a strong magnetic field and an extensive magnetosphere. Extending some 1 million km toward the Sun, Saturn's magnetosphere is large enough to contain the planet's ring system and the innermost 16 moons.

Saturn's rings are composed of small icy particles, all orbiting Saturn like tiny moons. These particles range in size from fractions of a millimeter to tens of meters in diameter. The Roche limit, or tidal stability limit, is the critical distance inside of which a moon is destroyed by a planet's tidal forces. Saturn's rings occupy the region inside the planet's Roche limit, 2.4 times the radius of the planet.

The *Voyager* encounter with Saturn (1980–1981) showed unexpected complexity in the planet's rings, caused by the planet's many moons. The A, B, and C rings are actually made up of tens of thousands of narrow ringlets. The Cassini Division, shown to contain faint ringlets and gaps, was formed by the resonance between its particles and the inner moon, Mimas.

Saturn's 18 moons can be grouped into three categories: small, medium, and Titan. Most of the larger moons have nearly circular orbits in the ring plane and are tidally locked by Saturn's gravity into synchronous rotation. Titan is the second-largest moon in the solar system. Its atmosphere contains photochemical smog similar to that found over some Earth cities. Despite Titan's low surface gravity, the atmospheric pressure at ground level is 60 percent greater than on Earth. Titan's atmosphere also extends some 10 times farther into space than our own, a consequence of the moon's weaker gravity. The medium-sized moons show a wide variety of surface terrains. The small moons exhibit many different types of complex motion.

KEY WORDS

A ring
B ring
C ring
Cassini Division
chaotic rotation
co-orbital satellites
D ring
Encke Division
E ring
F ring
helium precipitation
Lagrange point
planetary ring system
ringlet
Roche limit
shepherd satellites
spiral density wave
tidal stability limit
wispy terrain

REVIEW QUESTIONS

1. Why would Saturn float on water?
2. To observers on Earth, Saturn's rings sometimes appear broad and brilliant and sometimes seem to disappear. Why?
3. Why does Saturn have a less-varied appearance than Jupiter?
4. Describe the theoretical model that accounts for Saturn's missing helium and the planet's surprising amount of radiated heat.
5. Give a brief description of Saturn's rings.
6. What is the Roche limit?
7. What are the current views regarding the origin of Saturn's rings?
8. What evidence supports the idea that a relatively recent catastrophic event is responsible for Saturn's rings?
9. What effect does Mimas have on the rings?
10. When *Voyager* passed Saturn in 1980, why didn't it see the surface of Saturn's largest moon, Titan?
11. Why does Titan have a thick atmosphere when other large moons in the solar system don't?
12. What is the evidence for geological activity on Enceladus?
13. What mystery is associated with Iapetus?
14. Describe the behavior of co-orbital satellites.
15. Describe the chaotic rotation of Hyperion.
16. What is the synodic period (that is, the time from opposition to opposition) of Saturn, as seen from Jupiter?
17. Compare the angular sizes of Jupiter (at 9.5 A.U.) and the Sun, as seen from Saturn.
18. How long does it take for Saturn's equatorial flow, moving at 1500 km/h, to encircle the planet?
19. How would Saturn's rings appear from Titan, if the moon's atmosphere were transparent? What would the apparent angular diameter of the rings be as seen from Titan's surface?

DISCUSSION QUESTIONS

1. Compare and contrast Saturn's medium-sized moons with Jupiter's Galilean moons.
2. Compare and contrast the atmospheres of Saturn and Jupiter, describing how the differences affect the appearance of each planet.
3. Do you think it would be worthwhile to send a spacecraft to Titan? Why or why not?
4. Imagine what the sky would look like from Saturn's moon Hyperion. Would the Sun rise and set in the same way it does on Earth? How do you imagine Saturn might look, and on what sort of schedule might it rise and set? Would you ever see stars?

PROJECTS

1. Saturn moves more slowly among the stars than any other visible planet. It crosses a constellation boundary only once every few years. Look in an almanac to see where the planet is now. Find Saturn in the sky. What constellation is it in now? Can you explain why it moves so slowly, in contrast to Mars or Jupiter?
2. How many stars in the night sky are brighter than Saturn? What do you notice about the difference in appearance between Saturn and a star of about equal brightness?
3. If you are observing Saturn in the months around its opposition, can you notice it growing any brighter? Is its color becoming more evident?
4. Binoculars won't reveal the rings of Saturn, but most small telescopes will. Use a telescope to peer at the rings of Saturn. How are they tilted?
5. Can you see a dark line in the rings? It's called the Cassini Division. It once was thought to be a gap in the rings, but the *Voyager* spacecraft discovered that it is filled with tiny ringlets. Can you see the shadow of the rings on Saturn? Can you see Titan or any other of Saturn's moons? Titan would look like a bright star near the planet.

SUGGESTED READINGS

Berry, R. "More Science from Saturn." *Astronomy* (March 1981). *Voyager 1*'s encounter with Saturn.

———. "*Voyager* Science at Saturn." *Astronomy* (February 1981). Long, excellent popular article on *Voyager 1* science at Saturn.

Burnham, R. "Saturn, Lord of the Rings." *Astronomy* (August 1991). Observing the ringed planet.

"Continents on Titan?" *Astronomy* (November 1989). Using radar to penetrate the atmosphere of Saturn's largest moon.

Cowen, R. "Saturn's White Spot: Driven by the Sun?" *Science News* (October 5, 1991). Report on the recent storm on Saturn.

Eberhart, J. "Five-year Hunt Locates Saturn's 18th Moon." *Science News* (August 4, 1990). A tiny moon that apparently creates the Encke Division in Saturn's A ring.

Esposito, L. W. "The Changing Shape of Planetary Rings." *Astronomy* (September 1987). Evidence that planetary rings are young objects that grow and decay.

———. "Ever Decreasing Circles." *Nature* (November 14, 1991). Ideas on the origin of Saturn's rings.

Gehrels, T., and M. S. Matthews. *Saturn*. Tucson: University of Arizona Press, 1984. The planet, its rings, satellites, magnetosphere, and interaction with the interplanetary medium.

Golden, F. "Visit to a Large Planet." *Time* (November 24, 1980). Excellent popular article about *Voyager 1*'s historical encounter with Saturn.

"Iapetus: Saturn's Harlequin Moon." *Astronomy* (November 1989). Theories attempt to explain the dark face of Saturn's moon Iapetus.

Irion, R., A. Hollis, and J. Mitton. "Probing Saturn and Starlight." *Astronomy* (November 1989). Saturn and 28 Sagitarii observations by both amateurs and professionals.

MacRobert, A. M. "Hunting the Moons of Saturn." *Sky and Telescope* (July 1991). How to see and identify Saturn's moons through a small telescope.

Nichols, R. G. "Voyages to Worlds of Ice." *Astronomy* (December 1990). Introduction to CRAF—the Comet Rendezvous/Asteroid Flyby—and the *Cassini* mission to Saturn.

Robertson, D. F. "*Cassini*." *Astronomy* (September 1987). A craft to orbit Saturn and a probe of the atmosphere of its largest moon.

"*Voyager 2*'s Saturn: Still Surprising." *Science News* (August 29, 1991). More on *Voyager 2* at Saturn.

CHAPTER

15

Uranus, Neptune, and Pluto

The Outer Worlds of the Solar System

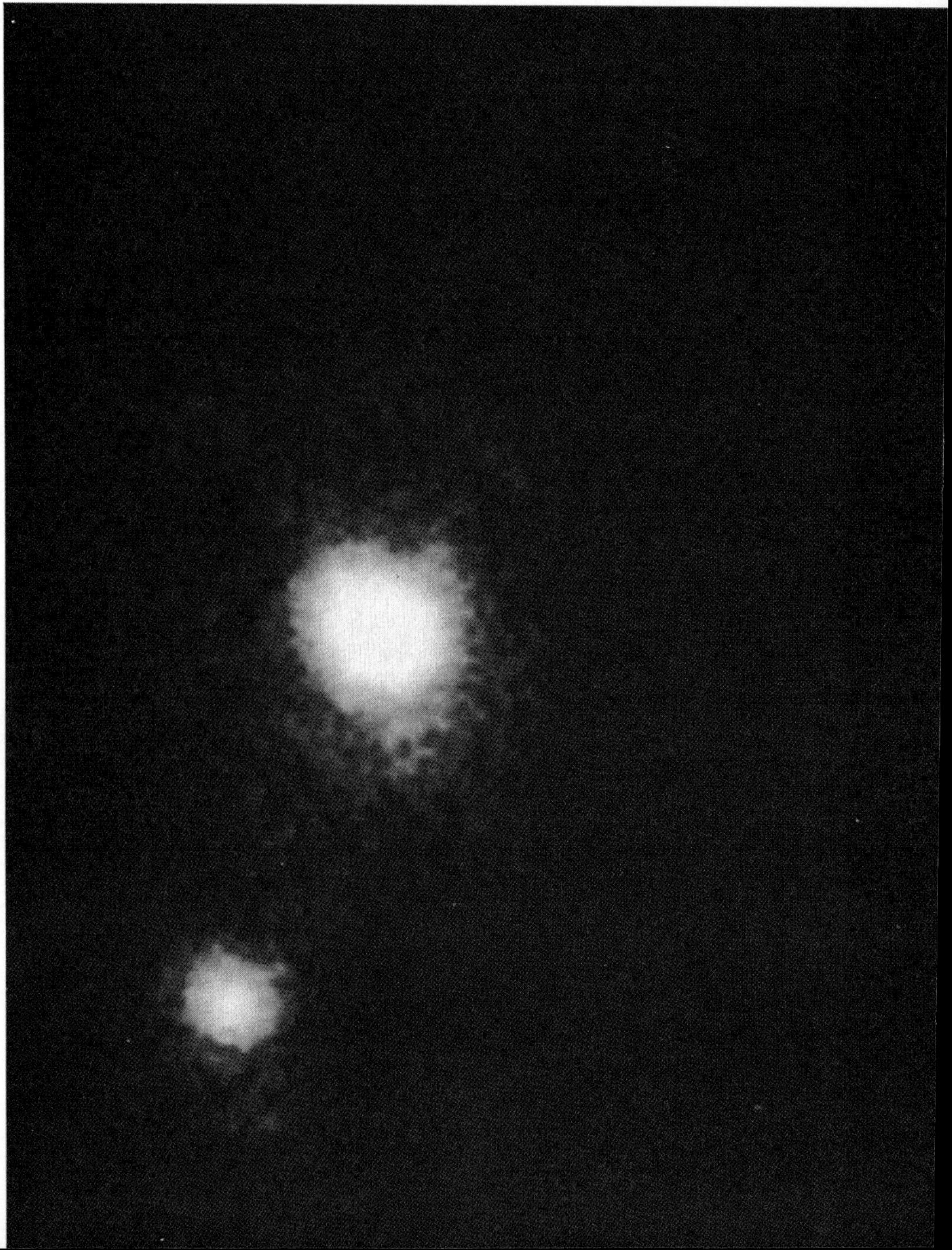

This photograph was taken by the Hubble Space Telescope *and shows Pluto at the center and its fainter moon, Charon, at lower left. For the first time, these two icy worlds are clearly distinguished. At the time of the observation, Charon was near its maximum angular separation of about 1 arc second—equivalent to the angular separation between a pair of car headlights about 500 kilometers away. Charon's orbit has a radius of nearly 20,000 kilometers, a span of about 1.5 times the diameter of Earth. (Courtesy NASA and European Space Agency)*

R I **V** U X G

LEARNING GOALS

1 to see how both calculation and chance played a major role in the discovery of the outer planets

2 to understand the similarities and differences between Uranus and Neptune and the other two Jovian planets

3 to realize that the moons of a planet can tell us much about its past

4 to see how the rings of Uranus and Neptune differ from those of Saturn and Jupiter

5 to understand that the Pluto-Charon system is fundamentally different from all the other planets

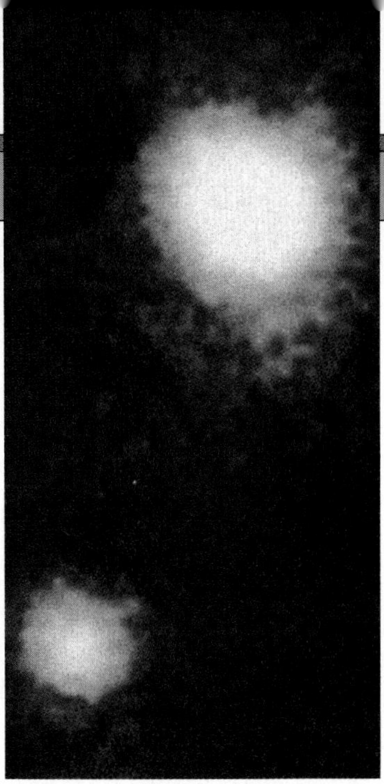

The three outermost planets were unknown to the ancients. All were discovered by telescopic observations—Uranus in 1781, Neptune in 1846, and Pluto in 1930. Uranus and Neptune have very similar masses and radii, so it is natural to consider them together. They are part of the Jovian family of planets (although, as we will see, they are in some ways intermediate between Jupiter and Earth in their properties). Pluto, conversely, is not a Jovian world. It is very much smaller than even the terrestrial planets and generally seems much more moonlike than planetlike in character. Indeed, it may well be a one-time moon that has escaped from one of the outer planets, most likely Neptune. Because of Pluto's similarity to the Jovian moons and its possible (although unproven) connection with Neptune, we study it here along with its larger Jovian neighbors.

Pluto is the only planet in the solar system not visited by unmanned spacecraft, and there is little prospect of such a mission in the foreseeable future. Its great distance from Earth means that little is known about it or its moon Charon, and every new discovery is the result of painstaking observations from Earth or Earth-orbiting instruments. Our knowledge of Uranus and Neptune is far greater, thanks to Voyager 2*, which reached Uranus in 1986 and Neptune in 1989. Virtually all of the detailed information we have on these twin worlds came from that mission. The alignment of the planets that permitted those flybys is a very rare event, occurring only once every 175 years. Every drop of available information must be squeezed from the* Voyager 2 *data, as spacecraft are unlikely to pass that way again for a very long time.*

THE DISCOVERY OF URANUS

1 The British astronomer William Herschel discovered the planet Uranus in 1781. Herschel was engaged in charting the faint stars in the sky when he came across an odd-looking object that he described as "a curious either nebulous star or perhaps a comet." Repeated observations showed that it was neither. The object appeared as a disk in Herschel's 6-inch telescope and moved relative to the stars, but it traveled too slowly to be a comet. Herschel soon realized that he had found the seventh planet in the solar system. Since this was the first new planet discovered in well over 2000 years, the event caused quite a stir at the time. The story goes that Herschel's first instinct was to name the new planet "Georgium Sidus" (Latin for "George's star") after his king, George III of England. The world was saved from a planet named George by the wise advice of another astronomer, Johann Bode. He suggested instead that the tradition of using names from Greco-Roman mythology be continued and that the planet be named after Uranus, the father of Saturn.

Careful observations since its discovery have allowed astronomers to determine the orbital properties of Uranus. Its orbit has a semi-major axis of 19.2 A.U. (2.9 billion km) and an eccentricity of 0.046, so that its perihelion distance from the Sun is 18.3 A.U. (2.7 billion km), and its aphelion is at 20.1 A.U. (3.0 billion km). Its sidereal orbital period is 84.1 years. Since its discovery in 1781, Uranus has completed only two and a half revolutions about the Sun.

Uranus is in fact just barely visible to the naked eye,

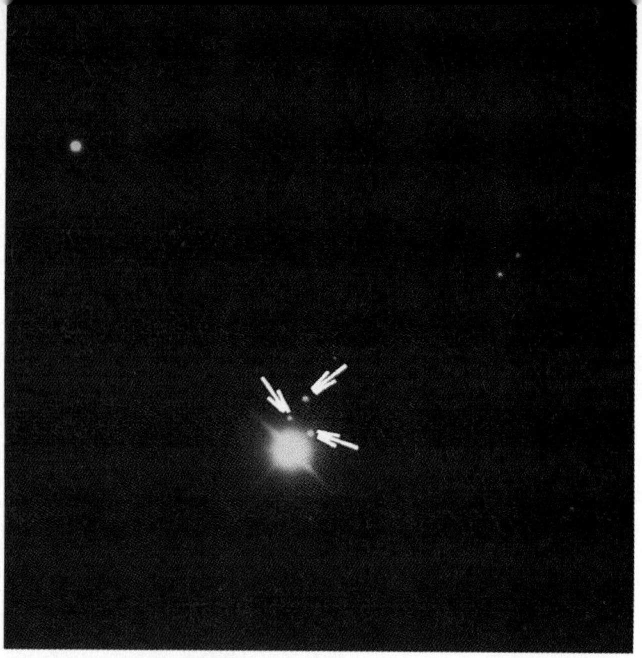

R I **V** U X G

Figure 15.1 Details are barely visible on photographs of Uranus made with large Earth-based telescopes. (Arrows point to three of its moons.) (Lick Observatory)

R I **V** U X G

Figure 15.2 A close-up view of Uranus, sent back to Earth by the Voyager 2 *spacecraft while whizzing past this giant planet at 10 times the speed of a rifle bullet. This montage shows Uranus's blue yet featureless upper atmosphere, taken when* Voyager 2 *was about 100,000 km away. The (real) image of the foreground object is Miranda, one of Uranus's moons. The rings are shown in an artist's concept. (NASA)*

if you know exactly where to look. At opposition, the planet has a maximum angular diameter of 3.7″ and shines just above the unaided eye's threshold of visibility. It looks like a faint, undistinguished star. No wonder it went unnoticed by the ancients. Even today, few astronomers have seen it without a telescope. Through a large Earth-based optical telescope, Uranus appears hardly more than a tiny pale greenish disk, as shown in Figure 15.1. With the flyby of *Voyager 2* in 1986, our detailed knowledge of Uranus increased dramatically. Figure 15.2 is a close-up visible-light image of the planet. Its apparently featureless atmosphere contrasts sharply with the bands and spots visible on all the other Jovian worlds.

THE DISCOVERY OF NEPTUNE

Once Uranus was discovered, astronomers set about charting its orbit. The figures we have just listed do indeed describe Uranus's orbital motion, but eighteenth-century astronomers quickly discovered a small discrepancy between the planet's predicted position and where they actually observed it. Try as they might, astronomers could not find an elliptical orbit that fit the planet's trajectory to within the accuracy of their measurements (a few arc seconds). Half a century after Uranus's discovery, the "discrepancy" had grown to a quarter of an arc minute, far too big to be explained away as observational error.

Uranus seemed to be violating Kepler's laws of orbital motion. Why wasn't the planet's orbit an ellipse? Astronomers turned to Newtonian gravity for the answer. In the mid-nineteenth century, astronomers realized that although the Sun's strong gravitational pull determines Uranus's orbit nearly completely, the deviation meant that an unknown body was exerting a much weaker, but still measurable, gravitational force on Uranus. But what body was this?

Very often in astronomy (and indeed in science in general), the behavior of one object is dominated by the influence of a second, but it is slightly affected, or *perturbed*, by the presence of a third. In large part because of the needs of astronomers, mathematicians in the nineteenth century developed the tools necessary to understand this interaction. These tools are collectively known as **perturbation theory.**

With the aid of this new mathematics, astronomers quickly proved that the deviations of Uranus's orbit from a pure ellipse could not be explained by the gravitational effects of the other known planets. There had to be *another* planet in the solar system perturbing Uranus's motion. In the 1840s, two mathematicians independently solved the difficult problem of determining this new planet's mass and orbit. An Englishman, John Adams, reached the solution in September 1845, after almost 2 years of work, but he was unable to convince any British astronomers to look for the

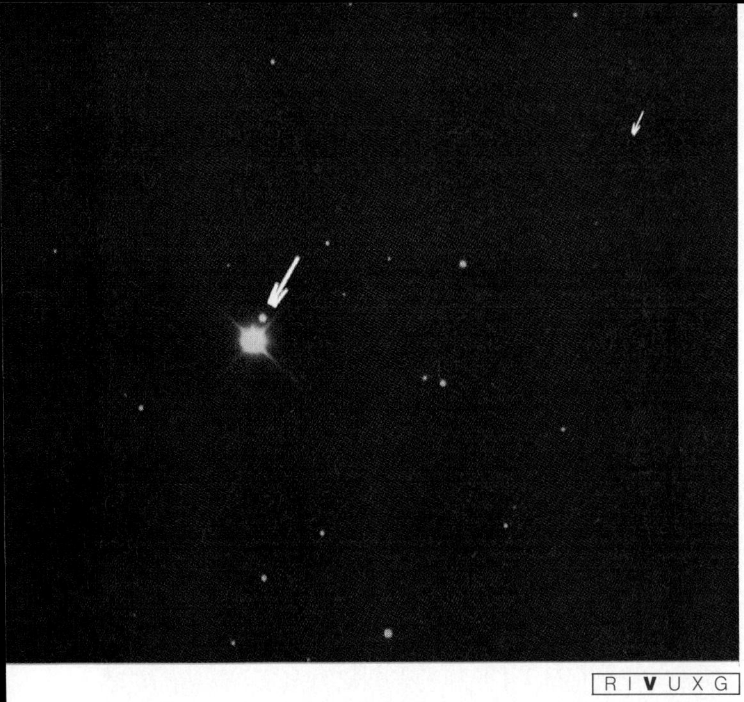

Figure 15.3 *Neptune and one of its moons, Triton (larger arrow), imaged with a large Earth-based telescope. (Lick Observatory)*

new planet. The Astronomer Royal refused even to see him. In June 1846, the French mathematician Urbain Leverrier came up with essentially the same answer, and the search was finally on. British astronomers found nothing during the summer of 1846. In September of that year, a German astronomer named Johann Galle began his own search from the Berlin Observatory, using a newly completed set of more accurate sky charts. He found the new planet within one or two degrees of the predicted position—on his first attempt. After some wrangling over names and credits, the new planet was named Neptune, and Adams and Leverrier (but not Galle!) are now jointly credited with its discovery.

Neptune orbits the Sun with a semi-major axis of 30.1 A.U. (4.5 billion km) and an orbital eccentricity of 0.010. Its minimum and maximum distances from the Sun thus are 29.8 A.U. (4.5 billion km) and 30.4 A.U. (4.5 billion km). Its sidereal orbital period is 164.8 years, so it has not yet completed one revolution since its discovery. Unlike Uranus, Neptune cannot be seen with the naked eye, although it can be seen with a small telescope—Galileo may actually have seen Neptune, although he had no idea what it really was at the time. Through a large telescope, Neptune appears as a bluish disk, with a maximum angular diameter of 2.2″ at opposition.

Figure 15.3 shows a long exposure of Neptune and one of its moons, Triton. Neptune is so distant that surface features are virtually impossible to discern. Even under the best observing conditions, only a few markings can be seen. These are suggestive of multicolored cloud bands—light bluish hues seem to dominate. With *Voyager 2*'s arrival, much more detail emerged, as shown in Figure 15.4. Superficially, at least, Neptune resembles a blue-tinted Jupiter, with atmospheric bands and spots clearly evident.

URANUS AND NEPTUNE IN BULK

Masses and Radii

✓2 Figure 15.5 shows Uranus and Neptune to scale along with the Earth for comparison. The two giant planets are quite similar in their bulk properties, too. Their diameters are: Uranus, 51,100 km (4.0 times that of Earth), and Neptune, 49,500 km (3.9 Earth diameters). Their masses (first determined from terrestrial observations of their larger moons and later refined by *Voyager 2*) are: Uranus, 8.7×10^{28}g (14.6 Earth masses), and Neptune, 1.0×10^{29}g (17.2 Earth masses). Thus, Uranus's average density is 1.2 g/cm^3 and Neptune's is 1.7 g/cm^3. These densities imply that large rocky cores constitute a greater fraction of the planets' masses than do the cores in either Jupiter or Saturn. The

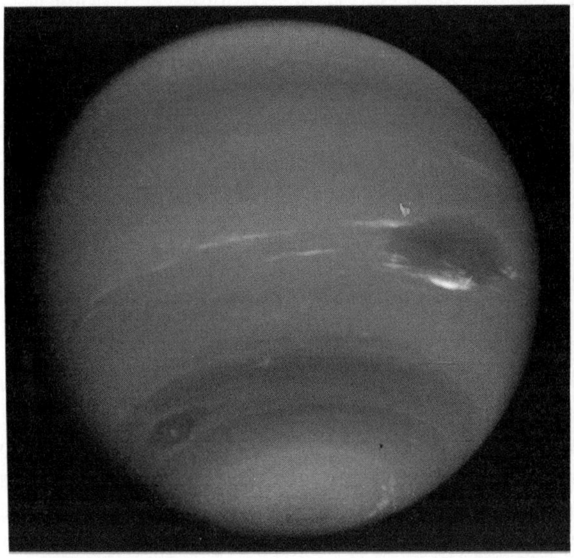

Figure 15.4 *Neptune as seen by* Voyager 2, *from a distance of roughly a million km. The inset, resolved to about 10 km, shows cloud streaks ranging in width from 50 to 200 km. (NASA).*

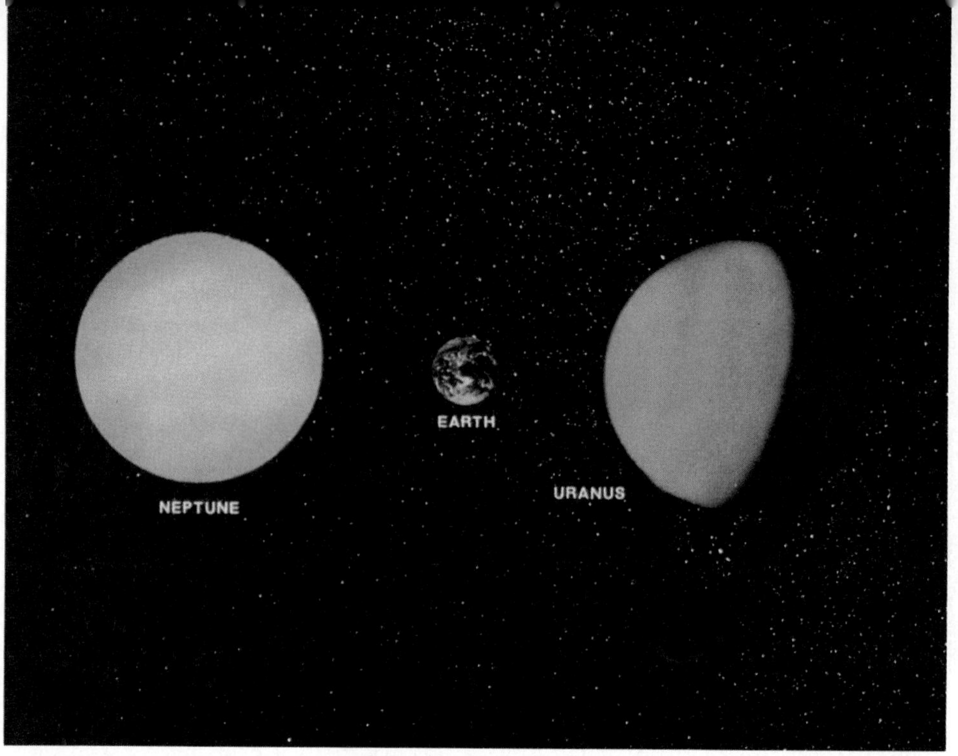

Figure 15.5 *Neptune, Earth, and Uranus, illustrated to scale. Uranus and Neptune are quite similar in their bulk properties. Each probably contains a core about ten times larger than the Earth. (NASA)*

R I **V** U X G

cores themselves are probably comparable in size, mass, and composition to those of the two larger giants.

Rotation Rates

Like the other Jovian planets, Uranus has a short rotation period. Analysis of the Doppler shifts of spectral lines from Earth is difficult, but prior to the *Voyager 2* encounter astronomers knew that Uranus rotates once every 10 to 20 hours. The length of the Uranian ''day'' was determined much more accurately when a radio-astronomy instrument aboard *Voyager 2* timed radio signals associated with the planet's magnetosphere. The Uranian day is 17.2 hours long. The oblateness produced by this rotation is 0.03. Again as with Jupiter and Saturn, the planet's atmosphere rotates differentially, but this time *faster* at the poles (where the period is 14.2 hours) than near the equator (where the period is 16.5 hours).

Each planet in our solar system seems to have some outstanding peculiarity, and Uranus is no exception. Its curious characteristic anomaly concerns its axis of rotation. Unlike all the other planets, which have their spin axes roughly perpendicular to the ecliptic plane, Uranus's rotation axis lies almost within that plane. We might say that Uranus lies on its side—or is tipped over—its axial tilt is 98°. As a result, the ''north'' (spin) pole[1] of Uranus, at some time in its orbit, points almost directly toward the Sun. Half a Uranian ''year'' later, its ''south'' pole faces the Sun, as illustrated in Figure 15.6. When *Voyager 2* encountered the planet in 1986, the north pole happened to be pointing nearly at the Sun, so it was midsummer in the northern hemisphere.

The strange orientation of the Uranian rotation axis produces some extreme seasonal effects (see Figure 15.7). Starting at the height of northern summer, when Uranus's north pole is closest to the Sun, an observer near that pole would see the Sun move in gradually increasing circles in the sky, completing one circuit (counterclockwise) every 17 hours and dipping lower in the sky each day. Eventually the Sun would set, and the nights would grow progressively longer with each passing day. Twenty-one years after the summer solstice, the autumnal equinox occurs, and day and night are each 8.5 hours long. The days would continue to shorten until, one day, the Sun would fail to rise at all. This period of total darkness is equal in length to the earlier period of constant daylight and plunges the northern hemisphere into the depths of winter. Eventually, the Sun rises again, and the planet moves through a vernal equinox, then back into summer in the north. From the point of view of an observer on the equator, summer and winter would be almost equally cold seasons, with the Sun never rising far above the horizon. Spring and fall would be the warmest times of year, as the Sun passes almost overhead each day.

No one knows why Uranus is tilted in this way. Some people have speculated that a catastrophic event, such as a grazing collision between the planet and another planet-sized body, might have altered the planet's spin axis. However there is no direct evidence for such an occurrence, and no theory to tell us how we should seek to confirm it.

Neptune's clouds show more variety and contrast than do those of Uranus, and Earth-based astronomers studying them determined a rotation rate for Neptune even before *Voyager 2's* flyby in 1989. The average rotation period of

[1] For definiteness, we adopt the convention here that a planet's rotation is always counterclockwise as seen from above the north pole (that is, planets always rotate from west to east).

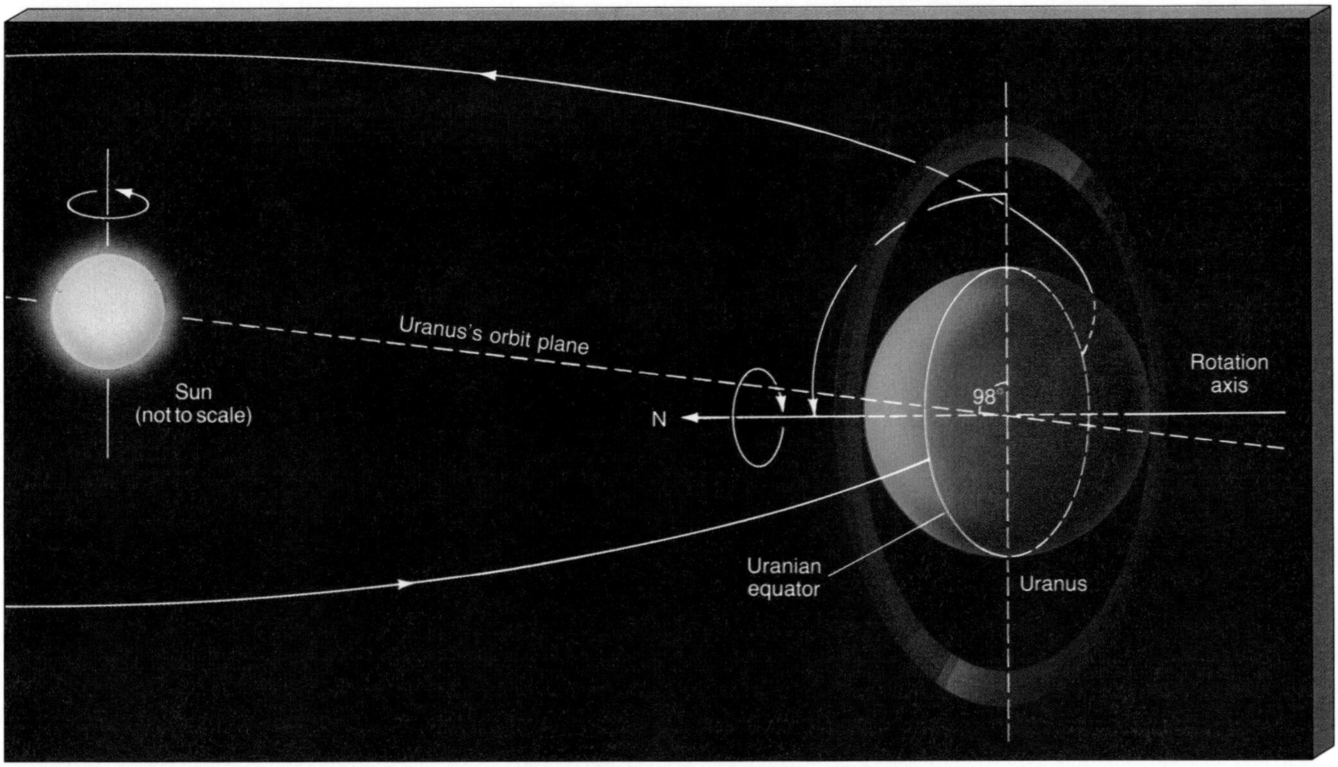

Figure 15.6 Uranus's 98° axial tilt places its equator almost perpendicular to the ecliptic, giving the planet a retrograde rotation.

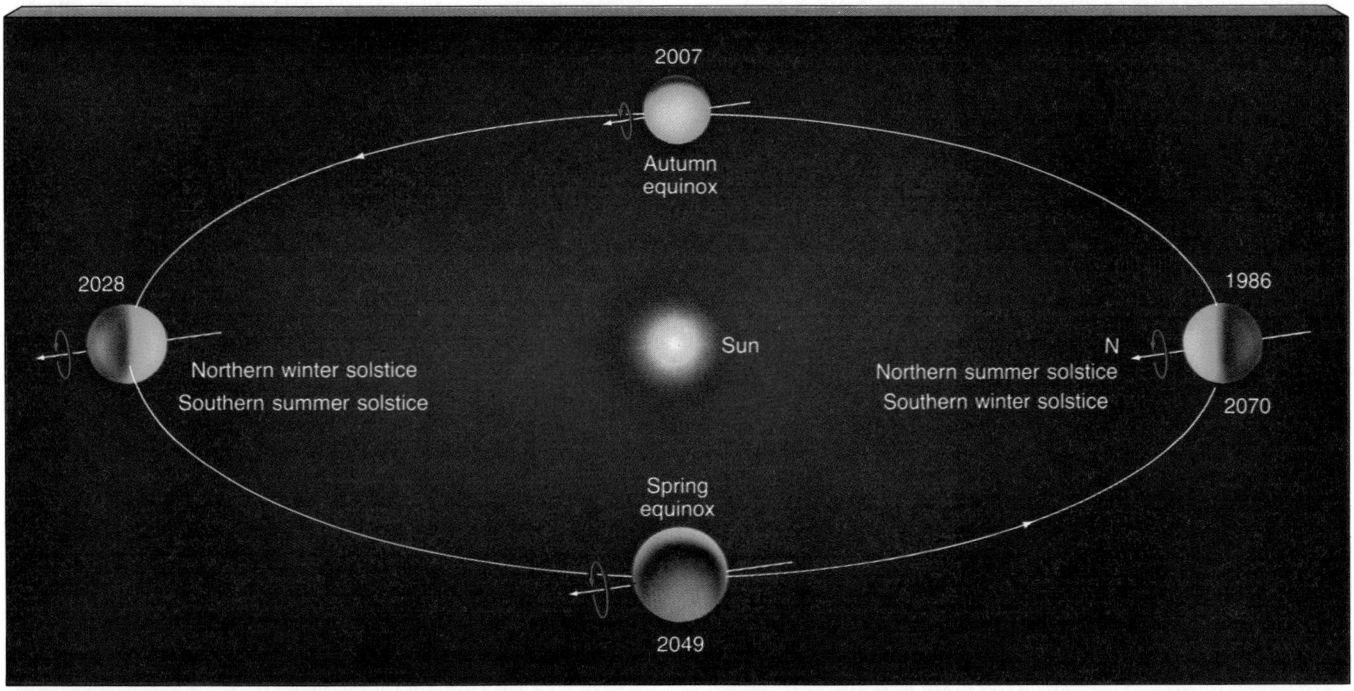

Figure 15.7 Because of Uranus's axial tilt the planet experiences the most extreme seasons known in the solar system. The equatorial regions experience two ''summers'' (around the two equinoxes) and two ''winters'' (at the solstices) each year, and the poles are alternately plunged into darkness for 42 years at a time.

Neptune's atmosphere is 17.3 hours (virtually identical to that of Uranus), and the planet's oblateness is 0.026. Measurements of Neptune's radio emission by *Voyager 2* showed that the magnetic field of the planet, and presumably also its interior, rotates once every 16.1 hours. Thus, Neptune is unique among the Jovian worlds in that its atmosphere rotates *more slowly* than its interior. Neptune's rotation axis is inclined at 29.6° to the perpendicular to the orbit plane, quite similar to the 27° tilt of Saturn.

THE ATMOSPHERES OF URANUS AND NEPTUNE

Composition

Spectroscopic studies of sunlight reflected from Uranus's and Neptune's dense clouds indicate that the two planets' outer atmospheres (the parts we actually measure spectroscopically) are similar to those of Jupiter and Saturn. The most abundant element is molecular hydrogen (84 percent), followed by helium (about 14 percent) and methane, which is more abundant on Neptune (about 3 percent) than on Uranus (2 percent). Ammonia, which plays such an important role in the Jupiter and Saturn systems, is not present in any significant quantity in the outermost Jovian worlds. Helium is not depleted on either planet, so the helium precipitation that produces Saturn's excess internal heat is apparently not occurring on either Uranus or Neptune.

The abundances of gaseous ammonia and methane vary among the Jovian planets. Jupiter has much more gaseous ammonia than methane but, as we proceed outward, we find that the more distant planets have steadily decreasing amounts of ammonia and relatively greater amounts of methane. This variation occurs because of temperature. Ammonia gas freezes into ammonia ice crystals at about 70K. This is cooler than the cloud-top temperatures of Jupiter and Saturn but warmer than those of Uranus (58K) and Neptune (59K). Thus, the outermost Jovian planets have little or no *gaseous* ammonia in their atmospheres, so their spectra (which record atmospheric gases only) show only traces of that gas.

The increasing amounts of methane are largely responsible for the outer Jovian planets' blue coloration. Methane absorbs red long-wavelength visible light quite efficiently, so that sunlight reflected from the planet's atmospheres is deficient in red and yellow photons and appears blue-green or blue. As the concentration of methane increases, the bluer the reflected light should appear. This is just the trend observed: Uranus, with less methane, looks blue-green, while Neptune, with more methane, looks distinctly blue.

Weather

Early observers of Uranus reported dusky atmospheric bands parallel to the equator. Some even claimed to have spotted bright spots reminiscent of the Great Red Spot on Jupiter. These ground-based observations were almost certainly erroneous. *Voyager 2* detected only a few atmospheric features (like the one shown in Figure 15.8), and

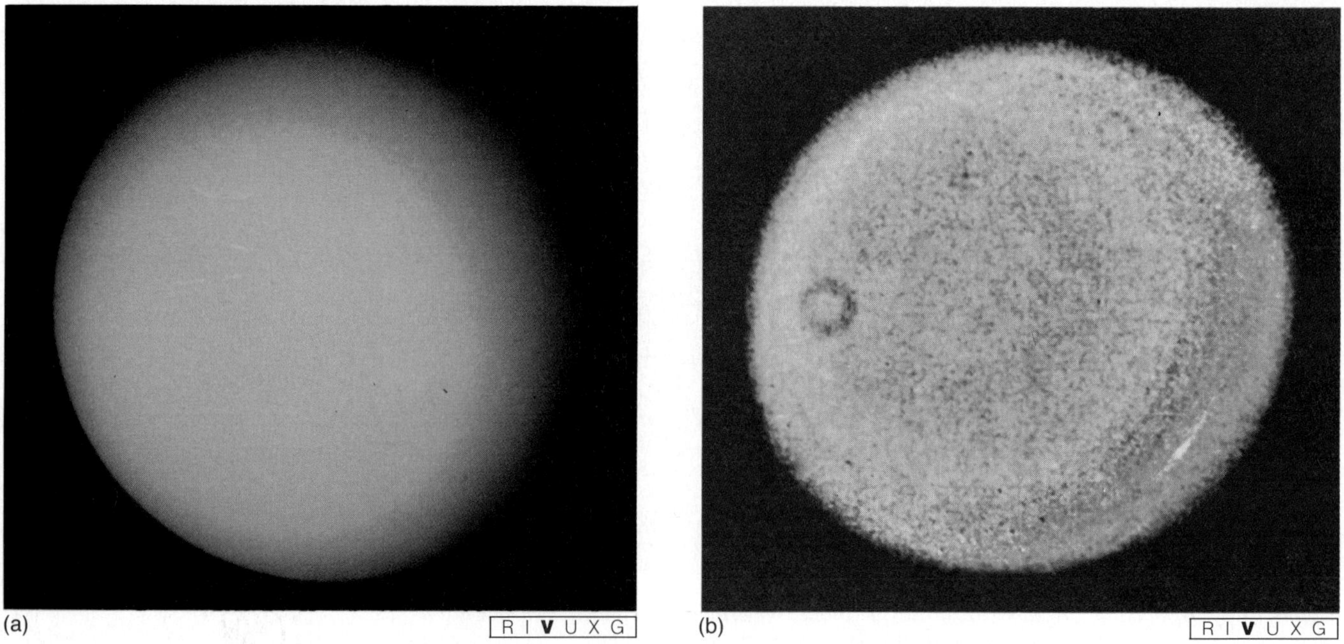

Figure 15.8 *Two* Voyager *views of Uranus, processed by different techniques of computer image enhancement. In the approximately color correct image (a), faint atmospheric bands can just be seen. The false color image (b) brings in subtle details around Uranus's south pole. (NASA)*

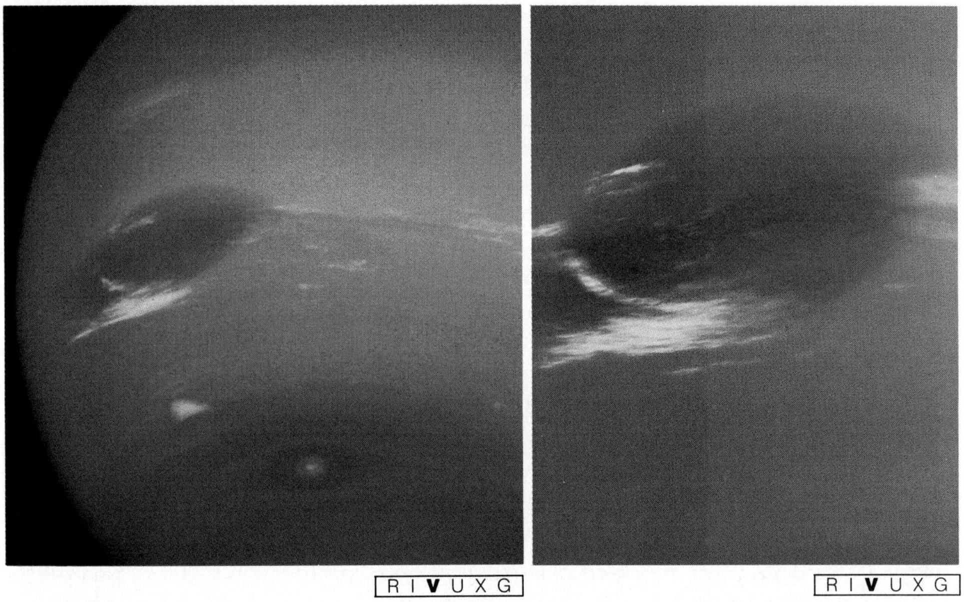

Figure 15.9 Close-up views of the Great Dark Spot of Neptune, which astronomers believe to be a large storm system in the planet's atmosphere, possibly similar in structure to Jupiter's Great Red Spot. Resolution in the photo on the right is about 50 km; the entire dark spot is roughly the size of planet Earth. (NASA)

even they became visible only after extensive computer enhancement. Uranus apparently lacks any significant internal heat source, and because of the planet's low surface temperature its clouds are found only at lower-lying, warmer levels in the Uranian atmosphere. The absence of high-level clouds means that we must look deep into the planet's atmosphere to see any structure, and the bands and spots that characterize flow patterns on the other Jovian worlds are largely "washed out" on Uranus by intervening stratospheric haze. At present, the photochemical haze is particularly thick over Uranus's north pole, which is oriented toward the Sun.

With computer-processed images, astronomers have learned that Uranus has atmospheric clouds and flow patterns that move around the planet in the same sense as the planet's rotation, with wind speeds in the 200-500 km/h range. Tracking these clouds allowed the measurement of the differential rotation mentioned earlier. Despite the odd angle at which sunlight is currently striking the surface (recall that it is just after midsummer in the northern hemisphere), the planet's rapid rotation still channels the wind flow into bands reminiscent of those found on Jupiter and Saturn. The planet's 21-year seasons plunge the planet's poles into darkness for 42 years at a time and lead to wildly uneven solar surface heating. The fact that the north polar region currently receives the greatest amount of solar heating per unit area may explain why wind speeds are greatest there. Even though the predominant wind flow is in the east-west direction, Uranus's atmosphere seems to be quite efficient at transporting energy from the heated north to the unheated southern hemisphere. Although the south is currently in total darkness, the temperature there is only a few kelvin less than in the north.

Neptune's cloud and band structure is much more easily seen. Although it lies at a greater distance from the Sun, Neptune's upper atmosphere is actually slightly warmer than Uranus's. Thus, like Jupiter and Saturn but unlike Uranus, Neptune has an internal energy source—it radiates 2.7 times more heat than it receives from the Sun. The cause of this heating is still uncertain. Recall that Neptune's atmosphere has more methane than does Uranus's. Some scientists have suggested that this methane has helped "insulate" the planet, tending to maintain its initially high internal temperature. If that is so, then Neptune's internal heat has the same basic explanation as Jupiter's—it is energy left over from the planet's formation. The combination of extra heat and less haze may be responsible for the greater visibility of Neptune's atmospheric features, as its cloud layers lie at higher levels in the atmosphere than do Uranus's. *Voyager 2* also detected numerous white methane cirrus clouds (visible in Figure 15.4) lying some 50 km above the main cloud tops.

Neptune's zonal wind flow is unusual in that it is *retrograde* relative to the planet's rotation. As we mentioned, Neptune's interior rotates faster than the visible cloud layers, so the "surface" winds blow from east to west, with speeds of over 2000 km/h—actually close to supersonic! Why such a cold planet should have such rapid winds is still a mystery. Neptune also sports several storm systems similar in appearance to those seen on Jupiter (and assumed to be produced and sustained by the same general processes). The largest such storm is known simply as the **Great Dark Spot,** shown in Figure 15.9. It is about the size of Earth, resides near the planet's equator, and exhibits many of the same general characteristics as the Great Red Spot on Jupiter. The flow around it is counterclockwise, as it is in the Red Spot, and there is probably turbulence where the winds associated with the Great Dark Spot interact with the zonal flow to its north and south. The flow around this and other dark spots may drive updrafts to high altitudes, where methane crystallizes out of the atmosphere to form the high-lying cirrus clouds.

MAGNETOSPHERES

Voyager 2 found that both Uranus and Neptune have fairly strong internal magnetic fields—about 100 times stronger than Earth's field and 1/10 as strong as Saturn's. However, because the radii of Uranus and Neptune are so much larger than the radius of the Earth, the magnetic fields at the cloud tops, spread out over a far larger volume than the field on Earth, are actually comparable in strength to Earth's field. Uranus and Neptune each have a substantial magnetosphere, populated largely by electrons and protons either captured from the solar wind or created from ionized hydrogen gas escaping from the planets themselves.

When *Voyager 2* arrived at Uranus, it discovered that the planet's magnetic field is tilted at some 60° to the axis of rotation. On Earth, such a tilt would put the North (magnetic) Pole somewhere in the Caribbean. Furthermore, the magnetic field lines are *not* centered on the planet. It is as though Uranus's field were due to a bar magnet that is tilted with respect to the planet's rotation axis and displaced from the center by about one-third the radius of the planet. Figure 15.10 compares the magnetic field structures of the four Jovian planets. The locations and orientations of the bar magnets represent the observed planetary fields, and the sizes of the bars indicate magnetic field strength.

Because dynamo theories generally predict that the magnetic axis should be roughly aligned with the rotation axis—as on Earth, Jupiter, Saturn, and the Sun—the misalignment on Uranus suggested to some researchers that perhaps the Uranian field had been caught in the act of reversing. (For more on magnetic field reversals, consult

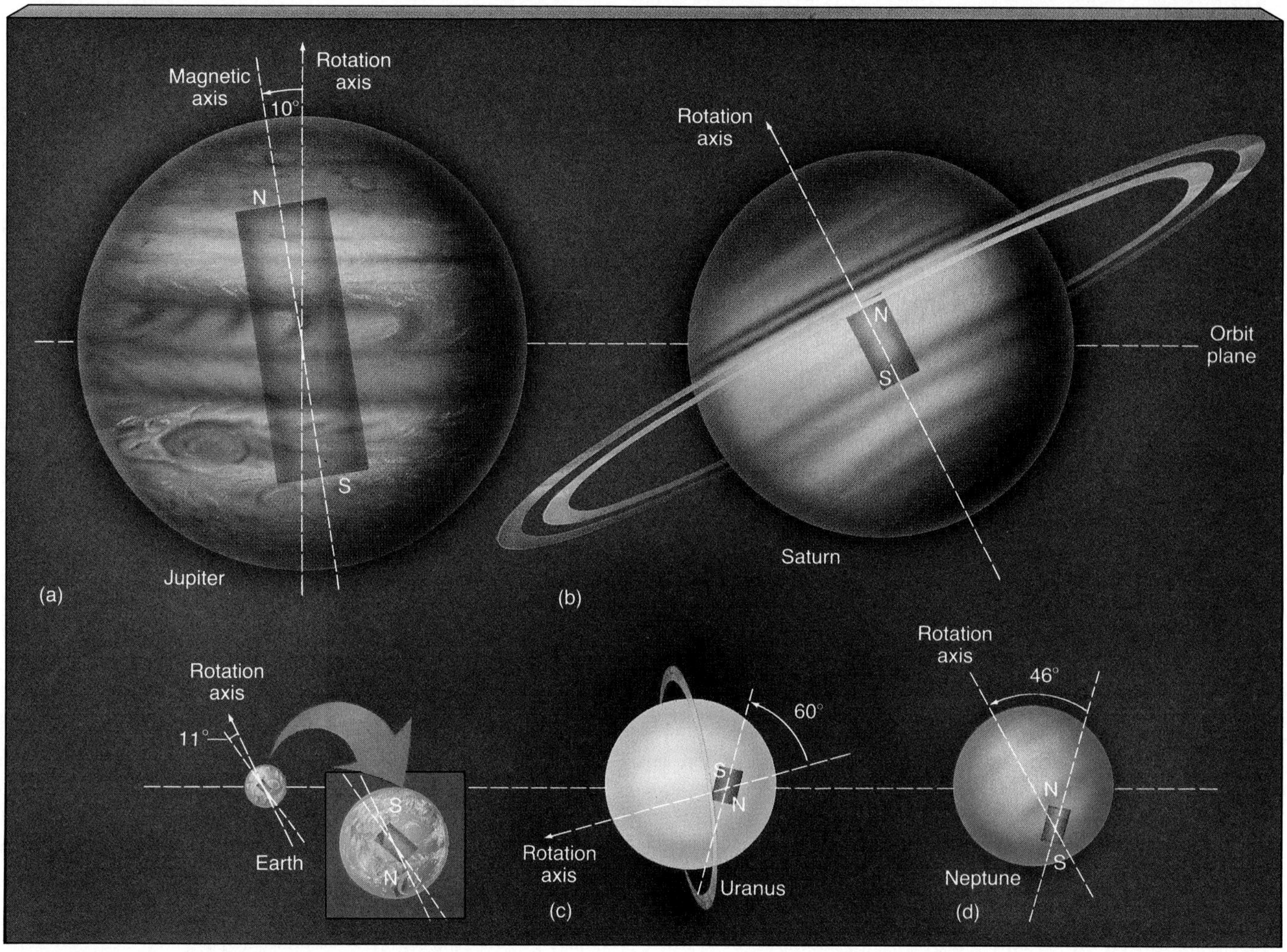

Figure 15.10 A comparison of the magnetic field strengths, orientations, and offsets in the four Jovian planets: (a) Jupiter, (b) Saturn, (c) Uranus, (d) Neptune. The planets are drawn to scale, and in each case the magnetic field is represented as though it came from a simple bar magnet. The size and location of the magnets represent the strength and orientation of the planetary field. Notice that the fields of Uranus and Neptune are significantly offset from the center of the planet and are significantly inclined to the planet's rotation axis. The Earth's magnetic field is shown for comparison.

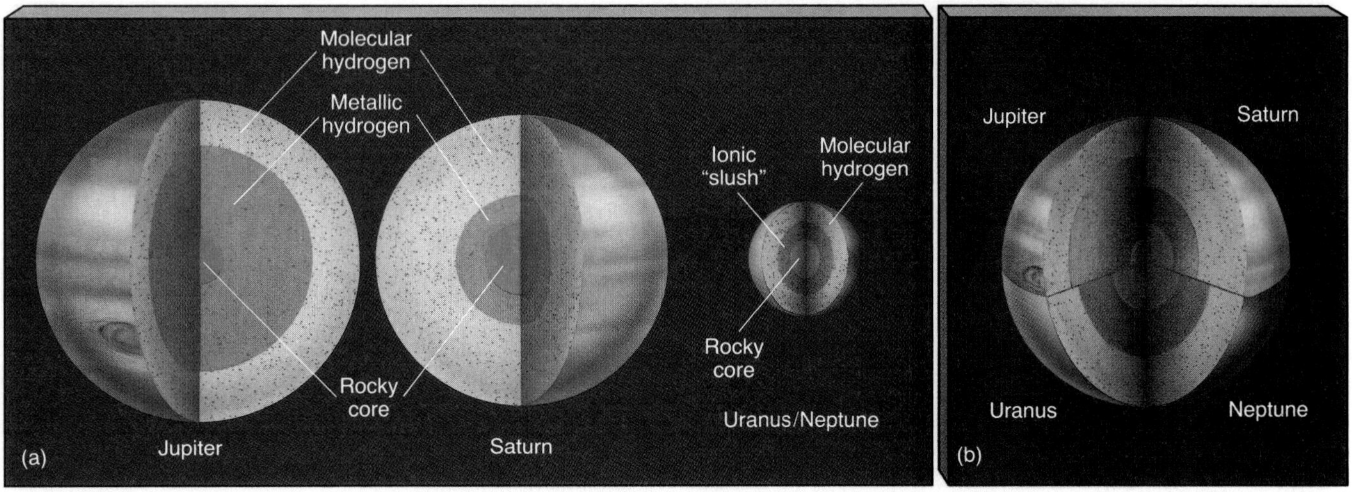

Figure 15.11 *A comparison of the interior structures of the four Jovian planets. (a) The planets drawn to scale. (b) The relative proportions of the various internal zones.*

Interlude 7-3.) Another possibility was that the oddly tilted field was in some way related to the planet's axial tilt—perhaps one catastrophic collision skewed both axes at the same time. Those ideas evaporated in 1989 when *Voyager 2* found that Neptune's field is also inclined to the planet's rotation axis, at an angle of 46° (see Figure 15.10[d]), and also substantially offset from the center. It now appears that the internal structures of Uranus and Neptune are different from those of Jupiter and Saturn, and this difference (or differences) somehow changes the nature of the field-generation process.

INTERIORS

As we have already described in the context of Jupiter and Saturn, studies of a planet's mass, radius, average density, oblateness (and also its detailed gravitational field) allow researchers to construct theoretical models of the interior. Much of what we know about the internal structure of the Jovian planets is based on such studies. These models indicate that Uranus and Neptune have rocky cores similar to those found in Jupiter and Saturn—about the size of Earth, and perhaps 10 times more massive. However, the pressure outside the cores of Uranus and Neptune (unlike the pressure in Jupiter and Saturn) is too low to force hydrogen into the metallic state, so hydrogen stays in its molecular form all the way in to the planets' cores.

Currently, astronomers theorize that deep below the cloud layers Uranus and Neptune may have high-density, "slushy" interiors containing thick layers of water clouds. It is also possible that much of the planets' ammonia could be dissolved in the water, simultaneously accounting for its absence at higher cloud levels and producing a thick, electrically conducting ionic layer that could conceivably explain the planets' magnetic fields. At the present time, however, we don't know enough about the interiors to assess the correctness of this picture. Our current state of knowledge is summarized in Figure 15.11, which compares the internal structures of the four Jovian worlds.

THE MOON SYSTEMS OF URANUS AND NEPTUNE

3 Like Jupiter and Saturn, both Uranus and Neptune have extensive moon systems, each consisting of a few large moons, long known from Earth, and many smaller moonlets, discovered by *Voyager 2*.

Uranus's Moons

The five well-known moons of Uranus (as well as 10 smaller ones discovered by *Voyager 2*) share its skewed rotation plane. In order of increasing distance from the planet, they are Miranda (at 5.1 planetary radii), Ariel (at 7.5), Umbriel (at 10.4), Titania (at 17.1), and Oberon (at 22.8). The smaller moons all lie inside the orbit of Miranda. All the moons revolve in the Uranian equatorial plane, almost perpendicular to the ecliptic, in circular, tidally locked orbits. Because the satellites share Uranus's odd orientation, they experience the same extreme seasons as their parent planet. Many of the small inner moons are intimately related to the Uranian ring system (see Figure 15.12). Uranus's moons are described in more detail in Table 15-1.

Because of Uranus's tilted rotation axis, a minor mystery surrounds the moons. That they share the planet's odd orientation is no surprise—the planet's gravitational forces would see to that. However, the orbital regularity and similarity in composition of the five large moons raise some

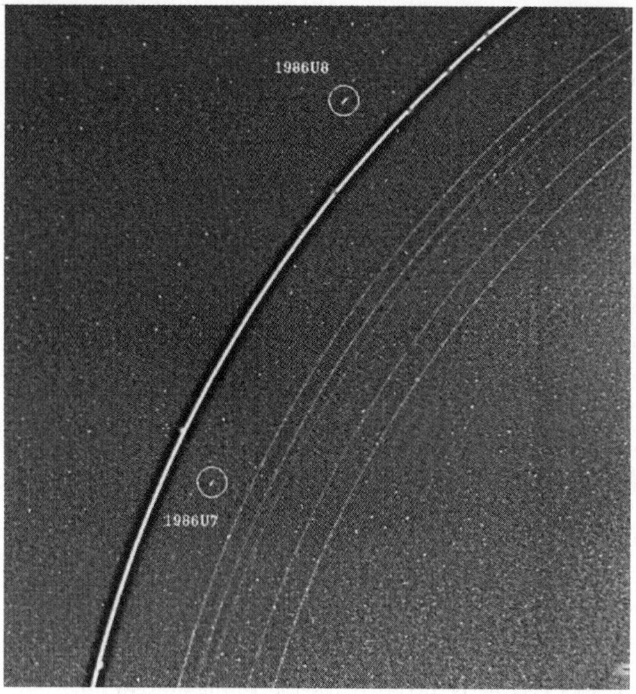

R I V U X G

Figure 15.12 *These two small Uranian moons, discovered by* Voyager 2 *in 1986, tend to "shepherd" one of Uranus's rings, thus keeping it from diffusing away. (NASA)*

questions. If they are the original moons, their presence places quite severe constraints on the event that toppled the planet without ejecting them. Conversely, if they were captured after that event, their regularity is hard to explain.

In 1789 William Herschel discovered and named the two largest moons, Titania and Oberon. British astronomer William Lassell found the next largest, Ariel and Umbriel, in 1851. Gerard Kuiper found the smallest, Miranda, in 1948. These five moons are similar in many respects to the six mid-sized moons of Saturn. Their densities are in the range 1.3–1.6 g/cm^3, suggesting composition of ice and rock, like the Saturnian moons, and their diameters range from 1600 km for Titania and Oberon, to 1200 km for Umbriel and Ariel, to 490 km for Miranda. Uranus has no moons comparable to the Galilean satellites of Jupiter or to Saturn's single large moon, Titan. Figure 15.13 shows the five large Uranian moons to scale. Note the large differences in brightness among them.

The two outermost Uranian moons, Titania and Oberon (shown in Figure 15.14), are heavily cratered and show little indication of any geological activity. Their overall appearance (and quite possibly their history) is comparable to that of Saturn's moon Rhea, except that they lack Rhea's wispy streaks. Also, like all the Uranian moons, they are considerably less reflective than their Saturnian counterparts. In fact, the Uranian moons generally have albedos in the range 0.3–0.4, suggesting that their icy surfaces are quite dirty. Possibly the planetary environment in the outer solar system contains more small "sooty" particles than does the inner solar system (inside the orbit of Saturn). An alternative explanation cites the effects of radiation and high-energy particles striking the surfaces of these moons. These impacts tend to break up the molecules on the moons' surfaces, eventually leading to chemical reactions that slowly build up a layer of dark, organic material. This **radiation darkening** is thought to contribute to the generally darker coloration of many of the moons and rings in the outer solar system. In either case, the longer a moon has been inactive and untouched by meteoritic impact, the darker its surface should be. The moons of the outer Jovian worlds tend to be darker than those of Jupiter and Saturn simply because they contain more carbon-bearing (methane) ice.

The darkest Uranian satellite is Umbriel (Figure 15.15[a]). Its albedo is only 0.2, so its reflectivity is only a little greater than that of Mars. Umbriel displays little evidence for any past surface activity. Its only mark of distinction is a bright spot about 30 km across, of unknown origin,

TABLE 15-1 ***The Moons of Uranus***

Name	Distance (km; planet radii)	Orbit Period (days)	Size (longest diameter, km)	Mass (Earth Moon masses)	Density (g/cm^3)	Albedo
Cordelia	49,700; 1.95	0.34	40			
Ophelia	53,800; 2.11	0.38	50			
Bianca	59,200; 2.32	0.44	50			
Cressida	61,800; 2.42	0.46	60			
Desdemona	62,700; 2.45	0.48	60			
Juliet	64,600; 2.53	0.50	80			
Portia	66,100; 2.59	0.51	80			
Rosalind	69,900; 2.74	0.56	60			
Belinda	75,300; 2.95	0.63	60			
Puck	86,000; 3.37	0.76	170			0.07
Miranda	130,000; 5.09	1.41	485	0.0011	1.3	0.3
Ariel	191,000; 7.48	2.52	1160	0.018	1.6	0.4
Umbriel	266,000; 10.4	4.14	1190	0.018	1.4	0.2
Titania	436,000; 17.1	8.71	1610	0.048	1.6	0.3
Oberon	583,000; 22.8	13.5	1550	0.040	1.5	0.2

R I V U X G

Figure 15.13 The five large moons of Uranus, to scale, in order of increasing distance from the planet. The moons are Miranda, Ariel, Umbriel, Titania, and Oberon. Earth's moon is shown (bottom right) for comparison. (NASA)

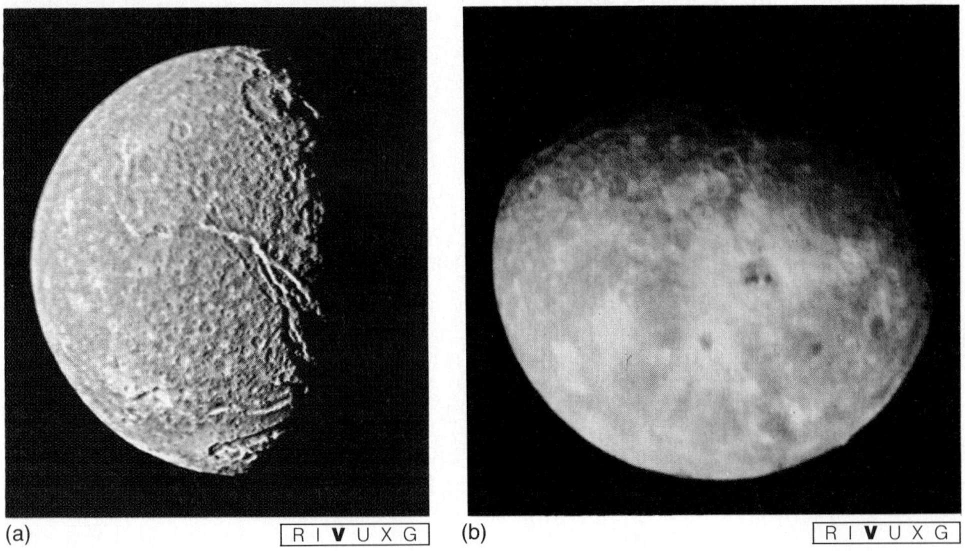

(a) R I V U X G (b) R I V U X G

Figure 15.14 Close-up comparison of the two largest Uranian moons, Titania (a) and Oberon (b). Their appearance, structure, and history may be quite similar to Saturn's moon Rhea. Smallest details visible on both moons are about 15 km across. (NASA)

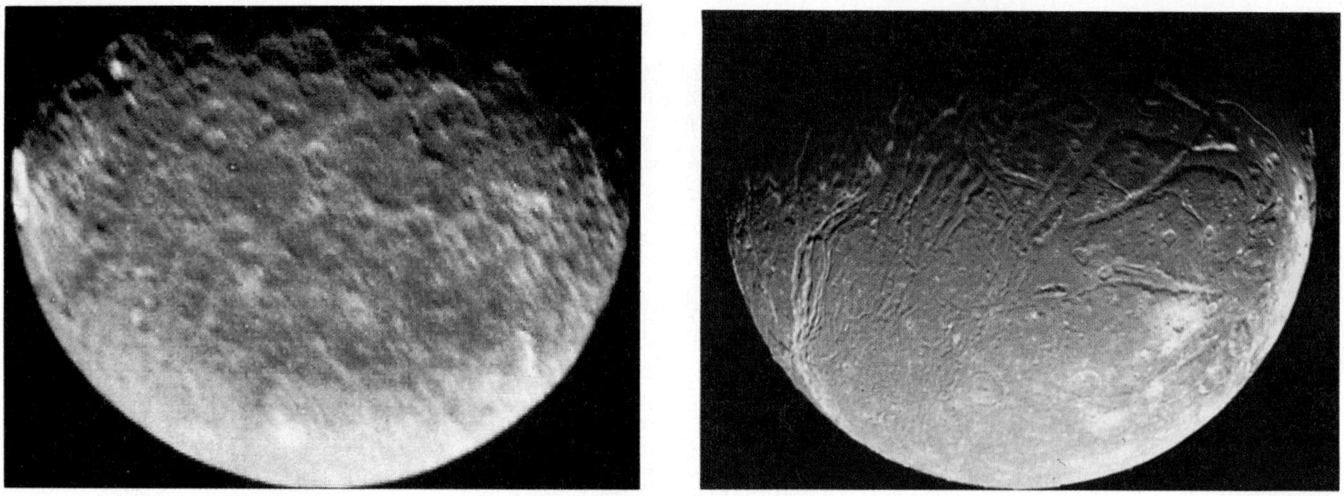

(a) R I **V** U X G (b) R I **V** U X G

Figure 15.15 *(a) The Uranian moon Umbriel is one of the darkest bodies in the solar system. Its most noteworthy feature is a bright white spot on its sunward side. (b) Ariel is similar in size but has a brighter surface. Unlike Umbriel, its surface shows signs of past geological activity. Resolution is approximately 10 km. (NASA)*

on its sunward side. Conversely, Ariel (Figure 15.15[b]), similar in size to Umbriel but closer to Uranus, does appear to have experienced some activity in the past. It shows signs of resurfacing in places and exhibits surface cracks a little like those seen on another Saturnian moon, Tethys. However, unlike Tethys, whose cracks are probably due to meteoritic impact, Ariel's activity probably occurred as internal forces and external tidal stresses (due to Uranus) distorted the moon and cracked its surface.

Strangest of the icy Uranian moons is Miranda, shown in Figure 15.16. Before the *Voyager 2* encounter, astronomers thought that Miranda would most resemble Mimas, the Saturnian moon whose size and location it most closely approximates. However, instead of being a relatively uninteresting cratered, geologically inactive world, Miranda displays a wide range of surface terrains, including ridges, valleys, large oval faults, and many other tortuous geological features. A close-up view of part of this strange landscape is presented in Figure 15.17. In order to explain why Miranda seems to combine so many different types of surface features, some researchers have hypothesized that this baffling object has been catastrophically disrupted several times (from within or without), with the pieces falling back together in a chaotic, jumbled way. Certainly, the frequency of large craters on the outer moons suggests that destructive impacts may once have been quite common in the Uranian system. It will be a long time, though, before we can obtain more detailed information to test this theory.

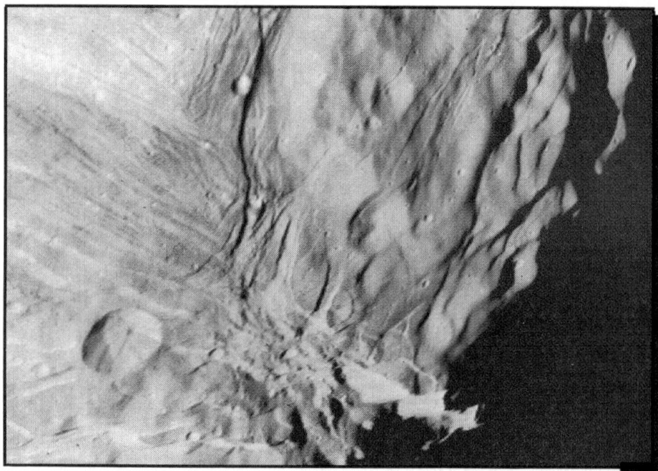

R I **V** U X G

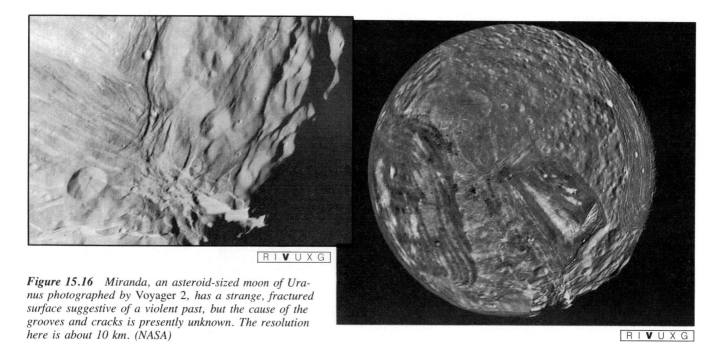

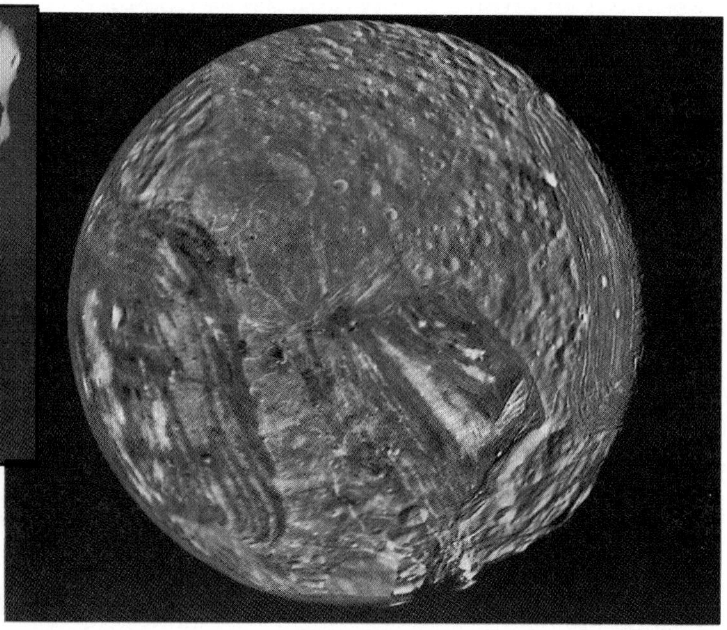

Figure 15.16 *Miranda, an asteroid-sized moon of Uranus photographed by* Voyager 2*, has a strange, fractured surface suggestive of a violent past, but the cause of the grooves and cracks is presently unknown. The resolution here is about 10 km. (NASA)*

R I **V** U X G

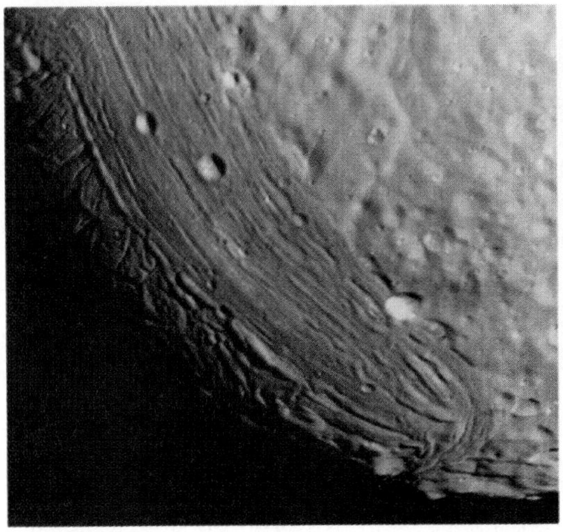

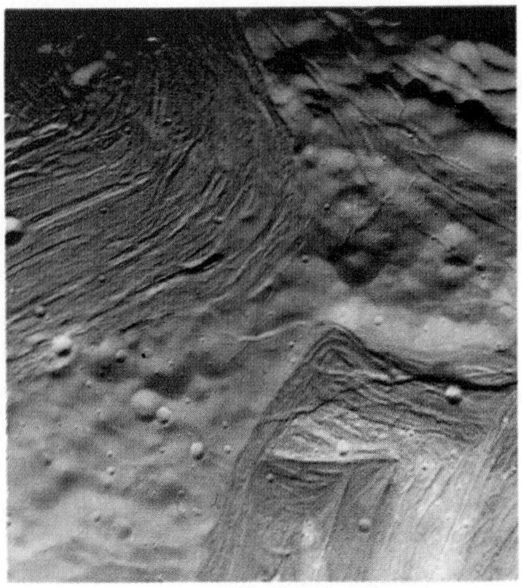

Figure 15.17 *Closer views of Miranda's surface, showing an elevated feature, of unknown origin, that resembles a giant racetrack. The insert shows details as small as 1 km across. (NASA)*

Neptune's Moons

From Earth, we can see only two moons orbiting Neptune. William Lassell discovered the inner moon, Triton, in 1846. The outer moon, Nereid, was located by Gerard Kuiper in 1949. *Voyager 2* discovered an additional six moons, all less than a few hundred kilometers across, and all lying within Nereid's orbit. Neptune's known moons are listed in Table 15-2.

In its moons we find Neptune's contribution to our list of solar system peculiarities. Unlike the other Jovian worlds, Neptune has no regular moon system. The larger moon, Triton, is 2800 km in diameter and occupies a circular retrograde orbit 354,000 km (14.2 planetary radii) from the planet, inclined at about 20° to Neptune's equatorial plane. It is the only large moon in our solar system to have a retrograde orbit. The other moon visible from Earth, Nereid, is only 200 km across. It orbits Neptune in the prograde sense, but on an elongated trajectory (with an orbital eccentricity of 0.75) with a semi-major axis of 5.5 million km. It comes as close as 1.4 million km to the planet and ventures as far away as 9.7 million km. Nereid is probably similar in both size and composition to Neptune's small inner moons.

Voyager 2 approached to within 24,000 km of Triton's surface, providing us with essentially all that we now know about that distant, icy world. Astronomers redetermined the moon's radius (which was corrected downward by about 20 percent) and measured its mass for the first time. Along with Saturn's Titan and the four Galilean moons of Jupiter, Triton is one of the six large moons in the outer solar system. Triton is the smallest of them, with about half the mass of the next smallest, Jupiter's Europa. Lying 4.5 billion km from the Sun, and with a fairly reflective surface, Triton has a surface temperature of just 37 K.

On the basis of spectroscopic observations from Earth, astronomers once thought that Triton's surface might be covered with a shallow ocean of liquid nitrogen. The *Voyager 2* images show this is not the case. Although Triton has a tenuous nitrogen atmosphere, perhaps a hundred thousand times thinner than Earth's, the surface itself is solid and frozen and probably consists primarily of water ice.

A *Voyager 2* mosaic of Triton's south polar region is shown in Figure 15.18. The moon's low temperatures produce a layer of nitrogen frost that forms and evaporates over the polar caps, a little like the carbon dioxide frost responsible for the seasonal caps on Mars. The frost is visible as the pinkish region on the left of Figure 15.18. Overall, there is a marked lack of cratering on Triton, presumably indicating that surface activity has obliterated the evidence of most

TABLE 15-2 ***The Moons of Neptune***

Name	Distance (km; planet radii)	Orbit Period (days)	Size (longest diameter, km)	Mass (Earth Moon masses)	Density (g/cm^3)	Albedo
Naiad	48,200; 1.94	0.30	60			0.06
Thalassa	50,100; 2.02	0.31	80			0.06
Despina	52,500; 2.12	0.33	180			0.06
Galatea	62,000; 2.50	0.43	150			0.05
Larissa	73,600; 2.97	0.55	190			0.06
Proteus	118,000; 4.76	1.12	415			0.06
Triton	354,000; 14.3	5.88*	2760	0.017	2.1	0.8
Nereid	5,520,000; 223	360.2	200	0.0000034	2.0	0.14

*Indicates a retrograde orbit

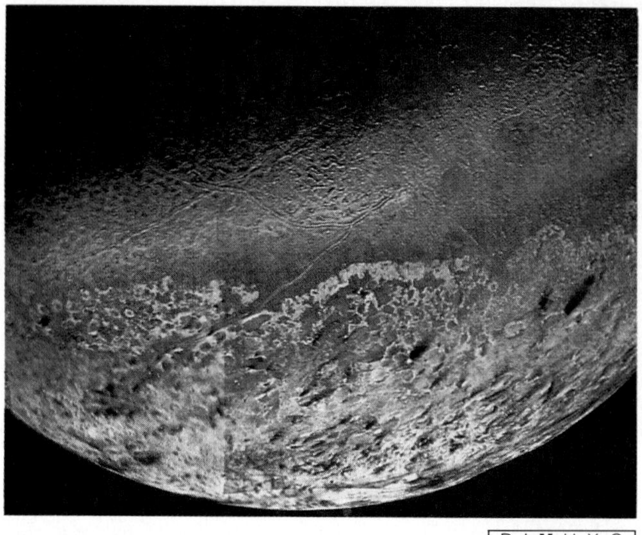

Figure 15.18 *The south polar region of Triton, showing a variety of terrains, ranging from deep ridges and gashes to what appear to be frozen water lakes, all indicative of past surface activity. The pinkish region on the left is nitrogen frost, forming the moon's polar cap. Resolution is about 4 km. (NASA)*

impacts. There are many other signs of an active past. Triton's face is scarred by large fissures similar to those seen on Ganymede, and the moon's odd cantaloupe-like terrain may indicate repeated faulting and deformation over the moon's lifetime. In addition, Triton has numerous frozen "lakes" of water ice (Figure 15.19), which are believed to be volcanic in origin.

Triton's surface activity is not just a thing of the past. As *Voyager 2* passed the moon, its cameras detected two great jets of nitrogen gas erupting from below the surface, rising several kilometers into the sky. It is thought that these "geysers" result when liquid nitrogen below Triton's surface is heated and vaporized by some internal energy source, or perhaps even by the Sun's feeble light. Vaporization produces high pressures, which force the gas through cracks and fissures in the crust, creating the displays *Voyager 2* saw. Scientists conjecture that nitrogen geysers may be very common on Triton and are perhaps responsible for much of the moon's thin atmosphere.

The event or events that placed Triton on a retrograde orbit and Nereid on such an eccentric path are unknown, but they are the subject of considerable speculation. Triton's peculiar orbit and surface features suggest to some astronomers that the moon did not form as part of the Neptunian system but instead was captured by Neptune, perhaps not too long ago. Other astronomers, basing their views on Triton's chemical composition, maintain that it formed as a "normal" moon but was later kicked into its abnormal orbit by some catastrophic event, such as an interaction with another similar-sized body. It has even been suggested that the planet Pluto may have played a role in this process, although no really convincing demonstration of such an encounter has ever been presented. The surface deformations on Triton certainly suggest fairly violent and relatively recent events in the moon's past. However, they were most likely caused by the tidal stresses produced in Triton as Neptune's gravity circularized its orbit and synchronized its spin, and they give little indication of the processes leading to the orbit in the first place. We will return to the possibility of a Pluto-Triton connection in a moment.

Whatever its past, Triton's future is fairly clear. Because of its retrograde orbit, the tidal bulge Triton raises on Neptune tends to make the moon spiral *toward* the planet rather than away from it (as our Moon moves away from Earth). Thus, Triton is doomed to be torn apart by Neptune's tidal gravitational field, probably in no more 100 million years or so, the time required for the moon's inward spiral to bring it inside Neptune's Roche limit (see Chapter 14). By that time, it is conceivable that Saturn's ring system may have disappeared, so that Neptune will then be the planet in the solar system with spectacular rings!

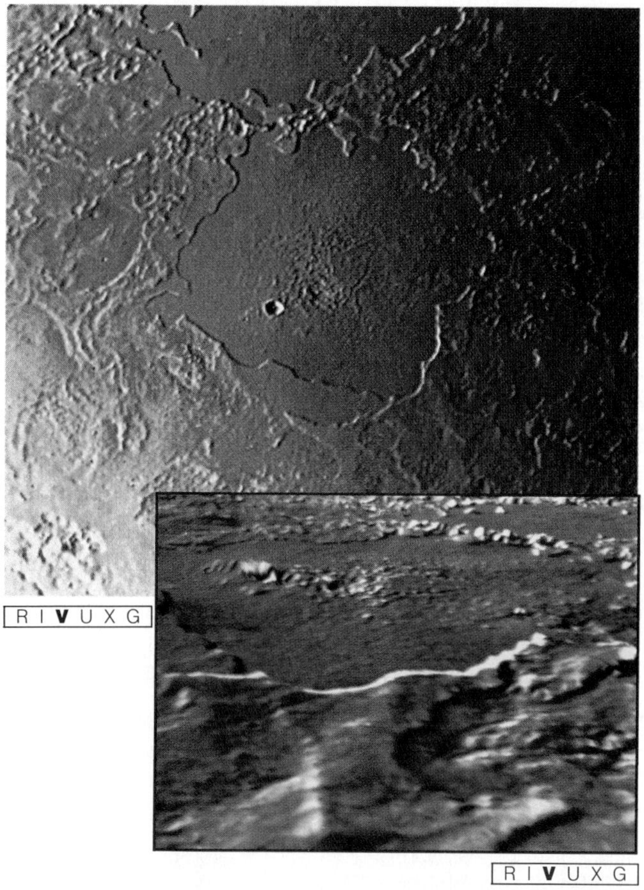

Figure 15.19 *Scientists believe that this lakelike feature on Triton may have been caused by the eruption of an ice volcano. The water "lava" has since solidified, leaving a smooth surface. The absence of craters indicates that this eruption was a relatively recent event in Triton's past. The nearly circular feature at the center of this image spans some 200 km in diameter; its details are resolved to a remarkable 1 km. The insert is a computer-generated view illustrating the topographic relief of the same area. (NASA)*

THE RINGS OF THE OUTERMOST JOVIAN PLANETS

✓4 All of the Jovian planets have rings. The ring system surrounding Uranus was discovered in 1977, when astronomers observed it passing in front of a bright star, momentarily dimming its light. Such a **stellar occultation** (see Figure 15.20) happens a few times per decade and allows astronomers to measure planetary structures that are too small and faint to be detected directly. The 1977 observation was actually aimed at studying the Uranian atmosphere by watching how it absorbed starlight. However, 40 minutes before and after the planet itself occulted the star, the flickering starlight revealed the presence of a set of rings. The discovery was particularly exciting because, at the time, only Saturn was known to have rings. Jupiter's rings went unseen until *Voyager 1* arrived there in 1979, and those of Neptune were unambiguously detected only in 1989, by *Voyager 2*.

The ground-based observations revealed the presence of a total of nine thin rings. The main rings, in order of increasing radius, are named Alpha, Beta, Gamma, Delta, and Epsilon, and they range from 44,000 to 51,000 km from the planet's center. For reference, the Roche limit of Uranus is about 62,000 km. A fainter ring, known as the Eta ring, lies between the Beta and Gamma rings, and three other faint rings, known as 4, 5, and 6, lie between the Alpha ring and the planet itself. In 1986, *Voyager 2* discovered two more even fainter rings, one between Delta and Epsilon, and one between ring 6 and Uranus. The main Uranian rings are shown in Figure 15.21. More details on the rings are provided in Table 15-3.

The Uranian rings are quite different from those of Saturn. While Saturn's rings are bright and wide with relatively narrow dark gaps between them, the rings of Uranus are dark, narrow, and widely spaced. With the exception of the Epsilon ring, which is about 100 km wide, the Uranian rings are all less than 10 km wide, and the spacing between them ranges from a few hundred to about a thousand kilometers. However, like Saturn's rings, all of the Uranian rings are less than a few tens of meters thick (that is, measured in the direction perpendicular to the ring plane).

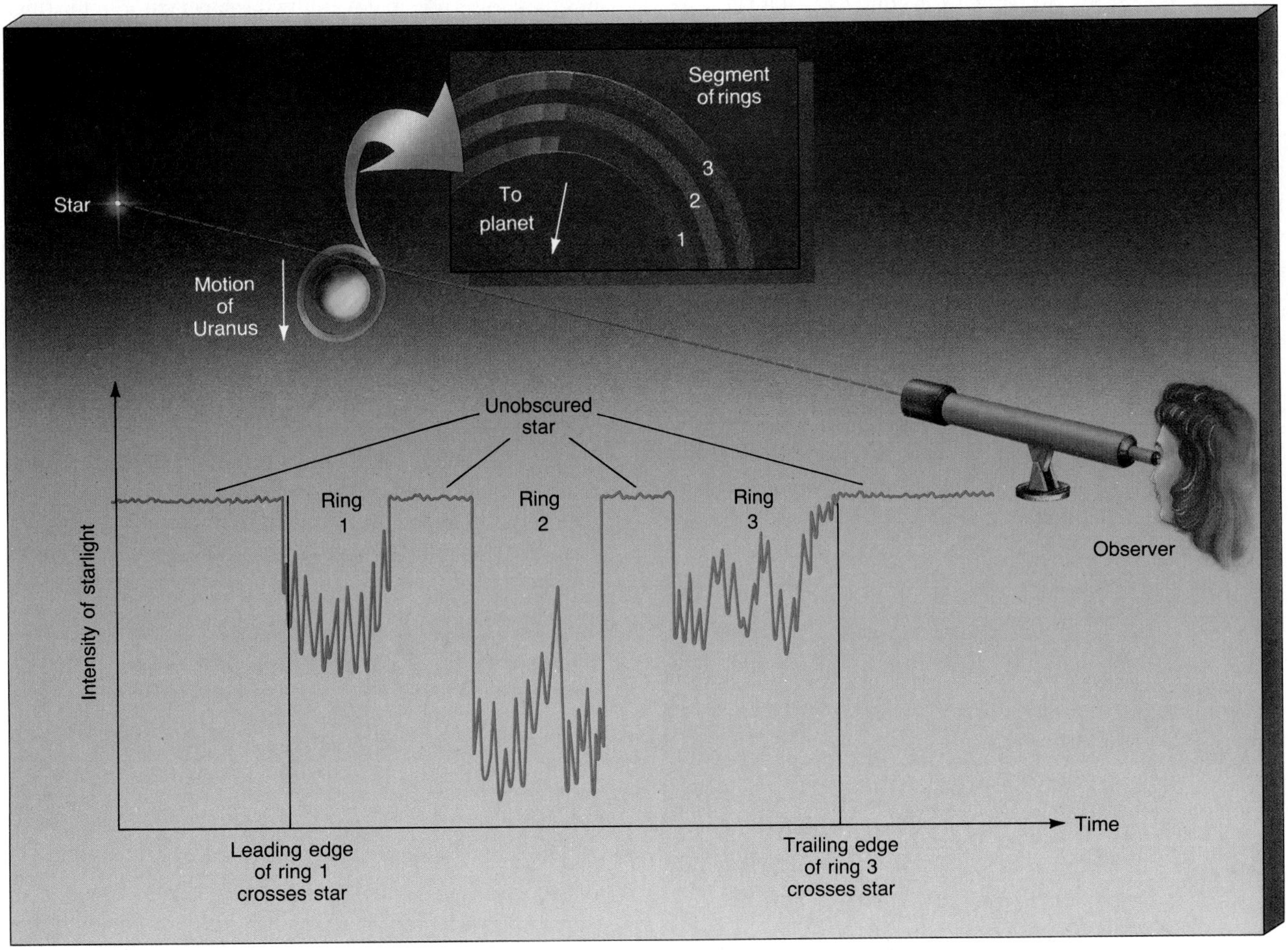

Figure 15.20 *How occultation of starlight allows astronomers to detect fine detail on a distant planet. The rings of Uranus were discovered using this technique.*

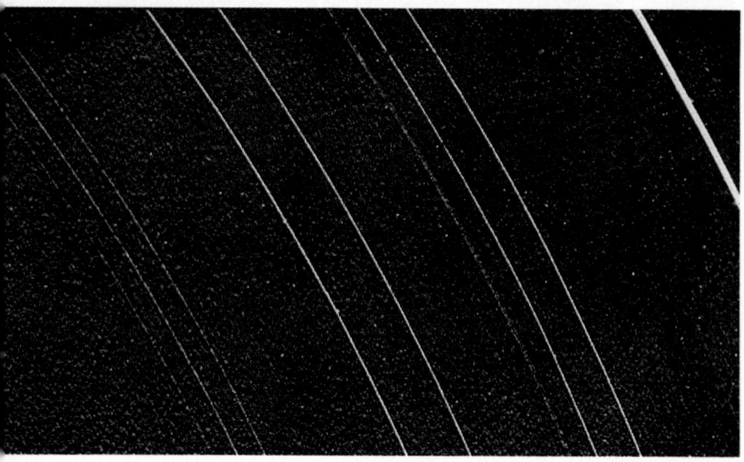

R I V U X G

Figure 15.21 The main rings of Uranus, as imaged by Voyager 2. *All of the rings known before* Voyager 2's *arrival can be seen in this photo. From the inside out, they are 6, 5, 4, Alpha, Beta, Eta, Gamma, Delta, and Epsilon. Resolution is about 10 km, which is just about the width of most of these rings. The two rings discovered by* Voyager 2 *are too faint to be seen here. (NASA)*

TABLE 15-3 ***The Rings of Uranus***

Ring	Inner Radius (km; planet radii)	Outer Radius* (km; planet radii)	Width (km)
1986U2R	37,000; 1.45	39,500; 1.55	2,500
6	41,900; 1.64		2
5	42,200; 1.65		2
4	42,600; 1.67		2
alpha	44,700; 1.75		10
beta	45,700; 1.79		10
eta	47,200; 1.85		1
gamma	47,600; 1.86		3
delta	48,300; 1.89		6
1986U1R	50,000; 1.96		2
epsilon	51,200; 2.00		20–90

*Most of Uranus's rings are so thin that there is little difference between their inner radius and outer radius.

The density of particles within the Uranian rings themselves is comparable to that found in Saturn's A and B rings. The particles that make up Saturn's rings range in size from dust grains to boulders, but the Uranian particles show a much smaller spread—few if any are smaller than a centimeter or so in diameter. The Uranian ring particles are also considerably less reflective than Saturn's ring particles, possibly because they are covered with the same dark material as the Uranian moons. The Epsilon ring (shown in detail in Figure 15.22), the widest of the 11, exhibits properties a little like Saturn's F ring. It is slightly eccentric (its eccentricity is 0.008), and it is of variable width, although no braids are found. It also appears to be composed of ringlets.

Like the F ring of Saturn, the narrow Uranian rings require shepherding satellites to keep them from diffusing away. In fact, the theory of shepherd satellites was first worked out to explain the rings of Uranus, which had been detected by stellar occultation even before *Voyager 2's* Saturn encounter. Thus, the existence of the F ring did not come as quite such a surprise as it might otherwise have done! Presumably many of the small inner satellites of Uranus play some role in governing the appearance of the rings. *Voyager 2* detected the shepherds of the Epsilon ring, Cordelia and Ophelia. Many other, undetected shepherd satellites must also exist.

Before 1989, astronomers knew that three Jovian planets had rings, and it was widely assumed that Neptune had rings too. There were some indications, again based on occultation measurements, that some strange ''partial rings'' existed, but it was the *Voyager 2* flyby that finally confirmed the rings' existence. As shown in Figure 15.23 and presented in more detail in Table 15-4, Neptune is surrounded by four dark rings. Three are quite narrow, like the Uranian rings, and one is quite broad and diffuse, more like Jupiter's ring. The dark coloration probably results from radiation darkening, as discussed above in the context of the moons of Uranus. All the rings lie within Neptune's Roche limit. The outermost ring (now known as the Adams ring) is noticeably clumped in places. From Earth we see not a complete ring, but only partial arcs—the unseen parts of the ring are simply too thin (unclumped) to be detected. The

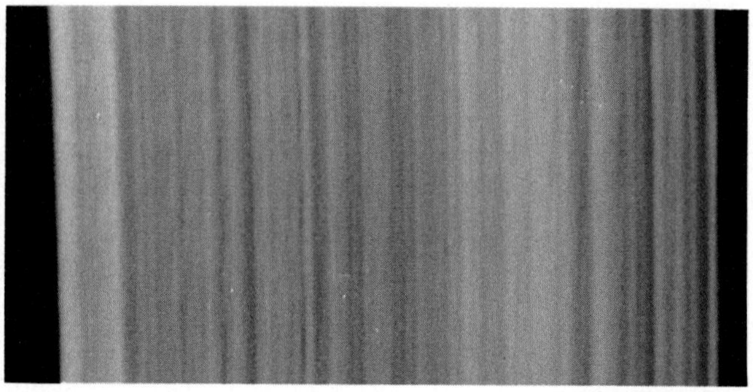

R I V U X G

Figure 15.22 A close-up of Uranus's Epsilon ring, showing some of its internal structure. The width of the rings averages 30 km; special image processing has magnified the resolution to about 100 meters. (NASA)

TABLE 15-4 ***The Rings of Neptune***

Ring	Inner Radius (km; planet radii)	Outer Radius (km; planet radii)	Width (km)
1989N3R (Galle)*	41,900; 1.69	41,900; 1.69	15
1989N2R (Leverrier)*	53,200; 2.15	53,200; 2.15	30
1989N4R	53,200; 2.15	59,000; 2.38	5,800
1989N1R (Adams)*	62,900; 2.54	62,900; 2.54	50

*In July 1991, the International Astronomical Union renamed Neptune's three brightest rings after the three nineteenth-century astronomers who played major roles in the planet's discovery.

Figure 15.23 Neptune's faint rings. In this long-exposure image, the planet (center) is heavily overexposed and has been artificially blotted out to make the rings easier to see. One of the two fainter rings lies between the inner bright ring and the planet. The other lies between the two bright rings. (NASA)

R I **V** U X G

connection between the rings and the planet's small inner satellites has not yet been firmly established, although many astronomers now believe that the clumping is caused by shepherd satellites.

All the Jovian worlds have ring systems, but the rings themselves differ widely from planet to planet. Is there some ''standard'' way in which rings form around a planet? And is there a standard manner in which ring systems evolve? Or do the processes of ring formation and evolution depend entirely on the particular planet in question? These questions are central to any understanding of planetary rings. If, as now appears to be the case, ring systems are relatively short-lived, their formation must be a fairly common event. Otherwise, we would not expect to find rings around all four Jovian planets at once. There are also many indications that the individual planetary environment plays an important role in determining a ring system's appearance and longevity. Although many aspects of ring formation and evolution are now understood, it must be admitted that no comprehensive theory yet exists.

THE DISCOVERY OF PLUTO

By the end of the nineteenth century, observations of the orbits of Uranus and Neptune suggested that Neptune's influence was not sufficient to account for all of the irregularities in Uranus's motion. Further, it seemed that Neptune itself might be affected by some other unknown body. Following the success of perturbation theory in guiding observers to the discovery of Neptune, astronomers hoped to pinpoint the location of this new perturber using similar techniques. One of the most ardent searchers was Percival Lowell, a capable and persistent observer and one of the best-known astronomers of his day. Recall that he was the leading proponent of the theory that the ''canals'' on Mars were constructed by an intelligent race of Martians (see Interlude 12-1).

Basing his investigation primarily on the motion of Uranus (Neptune's orbit was still relatively poorly determined at the time), Lowell set about calculating where the supposed ninth planet should be. He searched for it, without success, during the decade preceding his death in 1916. The region of the sky in which Lowell expected to find the new planet lay in the Milky Way, a region of high star density, making it very difficult to detect the slow night-to-night motion of any one small speck of light. Not until 14 years after Lowell's death did the American astronomer Clyde Tombaugh, working with improved equipment and photographic techniques at the Lowell Observatory, finally succeed in finding Lowell's ninth planet, only 6° away from Lowell's predicted position. The new planet was named Pluto for the Roman god of the dead who presided over eternal darkness (and also because its first two letters and its astrological symbol ♇ are Lowell's initials). Its discovery was announced on March 13, 1930, Percival Lowell's birthday.

On the face of it, the discovery of Pluto looks like another spectacular success for perturbation theory and Newtonian gravity. Unfortunately, it now appears that the supposed irregularities in the motions of Uranus and Neptune did not exist and that the mass of Pluto, not measured accurately until the 1980s, is far too small to have caused them anyway. The discovery of Pluto owed much more to dumb luck than to elegant mathematics!

Pluto's orbital semi-major axis is 39.5 A.U. (5.9 billion km). But unlike the paths of the other outer planets, Pluto's orbit is quite elongated, with an eccentricity of 0.25. It is inclined at 17.2° to the plane of the ecliptic. Already we have some indications that Pluto is unlike its Jovian neighbors. Because of its substantial orbital eccentricity, Pluto's distance from the Sun varies considerably. At perihelion,

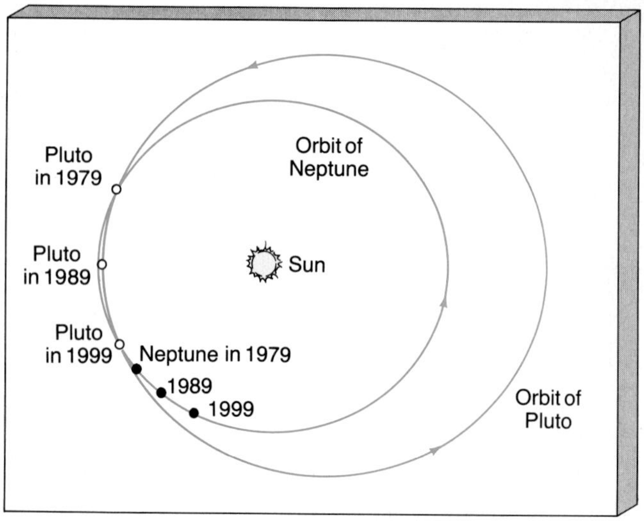

Figure 15.24 *The orbits of Neptune and Pluto cross, although Pluto's orbital inclination and a 3:2 resonance prevents the planets from actually coming close to one another. Between 1979 and 1999, Pluto is inside Neptune's orbit, making Neptune the most distant planet from the Sun.*

Pluto lies 29.7 A.U. (4.4 billion km) from the Sun, inside the orbit of Neptune. At aphelion, the distance is 49.3 A.U. (7.4 billion km), outside Neptune's orbit. Pluto last passed perihelion in 1989, and it will remain inside Neptune's orbit until 1999. Its sidereal period is 248.6 years, so the next perihelion passage will not occur until the middle of the twenty-third century. Pluto's orbital period is apparently exactly 1.5 times that of Neptune—the two planets are locked into a 3:2 resonance (2 orbits of Pluto for every 3 of Neptune) as they orbit the Sun. The orbits of these two outer planets are sketched in Figure 15.24.

At nearly 40 A.U. from the Sun, Pluto is often hard to distinguish from the background stars. Shown in the two photographs of Figure 15.25, the planet is actually considerably fainter than many stars in the sky. Like Neptune, it is never visible to the naked eye. Unlike the other outer planets, there is no present or proposed space mission that will suddenly and radically improve our knowledge of this distant world.

PLUTO IN BULK

√5 Pluto is so far away that little is known of its physical nature. Until the late 1970s, studies of its reflected light variations suggested a rotation period of nearly a week, but measurements of its mass and diameter were very uncertain. Astronomers thought that Pluto's mass might be as much as a few tenths the mass of the Earth—less than Lowell's original estimate of six Earth masses, but still fairly substantial.

All this changed in July 1978, when James Christy and Robert Harrington, of the U.S. Naval Observatory, discovered that Pluto has a satellite. It is now named Charon, after the mythical boatman who ferried the dead across the river Styx into Hades, Pluto's domain. The discovery photograph of Charon is shown in Figure 15.26. Charon is the small bump near the top of the image. Using the moon's orbital period, astronomers now know the mass of Pluto to much greater accuracy. It is 0.0025 Earth masses (1.5×10^{25} g), far smaller than any earlier estimate, and more like

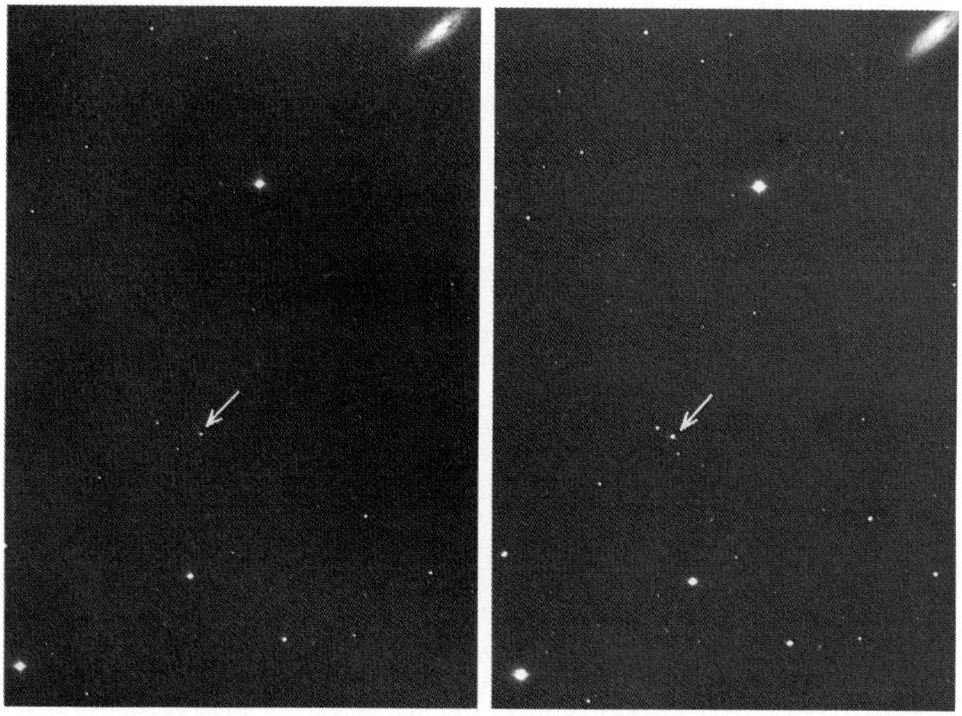

Figure 15.25 *These two photographs, taken one night apart, show motion of the planet Pluto (arrow) projected against a field of much more distant stars. Most of Pluto's apparent motion in these two frames is actually due to the orbital motion of the Earth. (Lick Observatory)*

R I **V** U X G

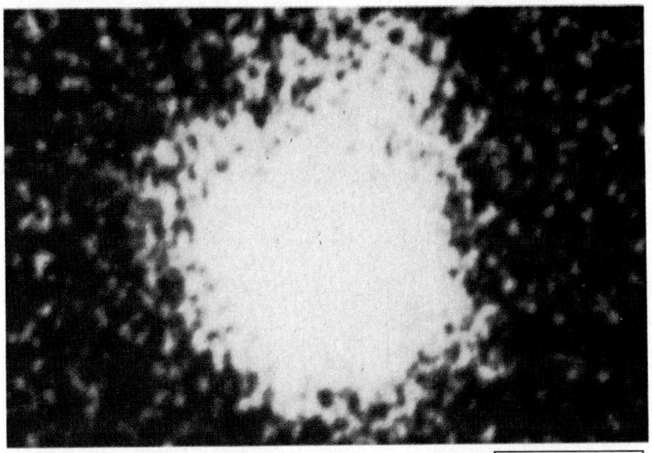

Figure 15.26 The discovery photograph of Pluto's moon, Charon. The moon is the small bump on the top right portion of the image. (U.S. Naval Observatory)

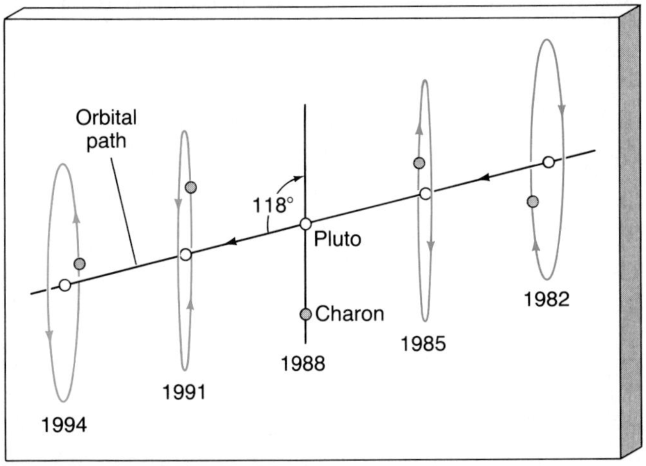

Figure 15.28 The orbital orientation of Charon produced a series of eclipses between 1985 and 1991. Observations of eclipses of Charon by Pluto and of Pluto by Charon have provided detailed information about both bodies' sizes and orbits.

the mass of a moon than of a planet. In 1990, the *Hubble Space Telescope* imaged the Pluto-Charon system (Figure 15.27). The improved resolution of that instrument clearly separates the two bodies and should allow even more accurate measurements of planetary properties.

Before Charon was discovered, Pluto's radius was also poorly known. Pluto's angular size is much less than 1″, so its true diameter is blurred by the effects of Earth's turbulent atmosphere. But Charon's orbital orientation has given astronomers new insight into the system. By pure chance, Charon's orbit over the 6-year period from 1985 to 1991 (less than 10 years after the moon was discovered) has

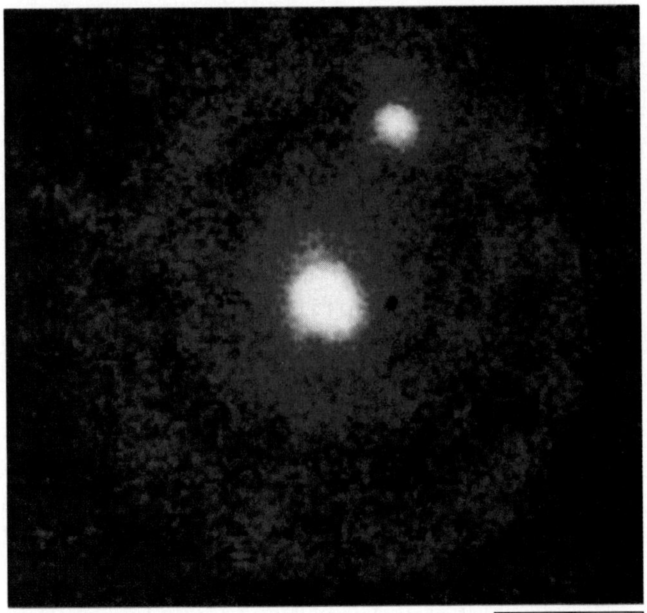

Figure 15.27 The Pluto-Charon system, as seen by the Hubble Space Telescope. *The angular separation of the planet and its moon is about 0.9 arc seconds. (NASA)*

produced for Earth viewers a series of eclipses. Pluto and Charon repeatedly passed in front of one other, as seen from our vantage point. Figure 15.28 sketches this orbital configuration. With more good fortune, these eclipses took place while Pluto was closest to the Sun, making for the best possible Earth-based observations.

Basing their calculations on the variations in light as Pluto and Charon periodically hid each other, astronomers have computed the masses and radii of each of them and have determined their orbit plane. Additional studies of sunlight reflected from Pluto's surface indicate that the two objects are tidally locked as they orbit each other. Pluto's diameter is 2300 km, about one-fifth the size of the Earth. Charon is about 1300 km across and orbits at a distance of 19,700 km from Pluto, with an orbital period of 6.4 days, the same as the rotation period of each body. If planet and moon have the same composition (probably a reasonable assumption), Charon's mass must be about one-sixth that of Pluto, giving the Pluto-Charon system by far the largest satellite-to-planet mass ratio in the solar system. As shown in Figure 15.29, Charon's orbit and the spin axes of both planet and moon are inclined at an angle of 118° to the plane of the ecliptic. Thus, Pluto is the third planet in the solar system found to have retrograde rotation (Venus and Uranus are the other two).

The known mass and radius of Pluto allow us to determine its average density, which is 2 g/cm^3. This density is too low for a terrestrial planet and far too high for a mixture of hydrogen and helium of that mass. Instead, the mass, radius, and density of Pluto are just what we would expect for one of the icy moons of a Jovian planet. In fact, Pluto is quite similar in mass and radius to Neptune's large moon, Triton. As we will see in a moment, this similarity has fueled much speculation, both about Pluto's origin and the strange state of the Neptunian moon system.

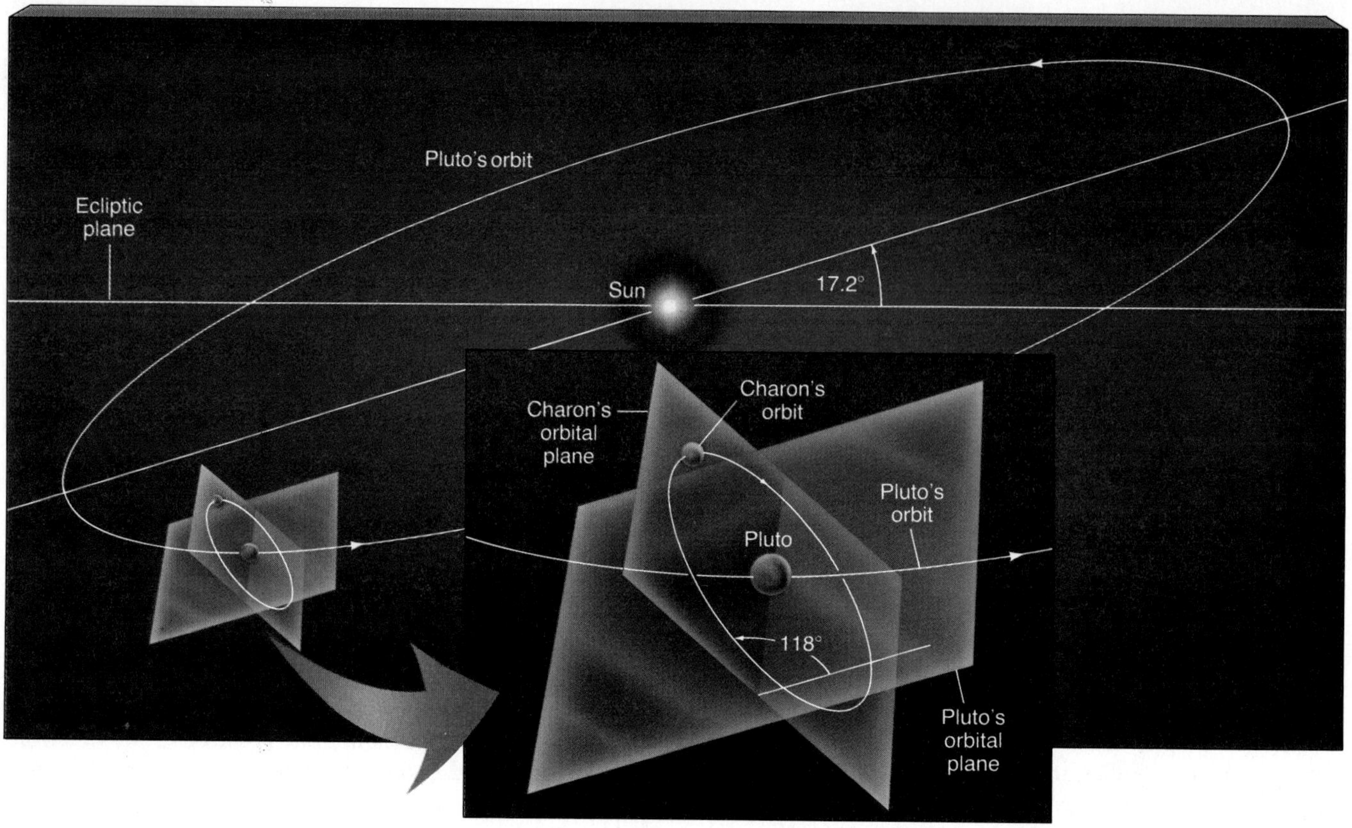

Figure 15.29 *Charon's path around Pluto is circular, synchronous, and inclined at 118° to the orbit plane of the Pluto-Charon system about the Sun. The Pluto-Charon orbit plane is itself inclined at 17° to the plane of the ecliptic.*

THE COMPOSITION OF PLUTO

Pluto's relatively high albedo of 0.6 and its similarities to the icy Jovian moons suggest that the planet is almost certainly made up mostly of water ice. In addition, spectroscopy reveals the presence of *frozen* methane as a major surface constituent. Pluto is the only planet in the solar system on which methane exists in the solid state, implying that the surface temperature on Pluto is no more than 50K. Based on Pluto's distance from the Sun, its surface temperature should vary between 50K at aphelion and 60K at perihelion. Pluto may also have a thin methane atmosphere, associated with the methane ice on its surface. Recent computer studies indicate that Charon may have bright polar caps, but their composition and nature are as yet unknown.

THE ORIGIN OF PLUTO

Because Pluto is neither terrestrial nor Jovian in its makeup and because of its similarity to the ice moons of the outer planets, some researchers suspect that Pluto is not a "true" planet at all. Pluto may be an escaped planetary moon or a large icy chunk of debris left over from the formation of the solar system. This idea is bolstered by Pluto's eccentric, inclined orbit, which is quite unlike the orbits of the other known planets. Since 1978, the explanation of Pluto's origin has been greatly complicated by the presence of Charon. It was much easier to suppose that Pluto was an escaped moon before we learned that it had a moon of its own. There is still no clear or easy answer to the puzzle of Pluto's origin.

Pluto may be just what it seems—a planet that formed in its current orbit, possibly even with its own moon right from the outset. Because we know so little about the environment in the outer solar system, we cannot rule out the possibility that planets beyond Neptune should simply look like Pluto. There is evidence for large chunks of ice circulating in interplanetary space beyond the orbits of Jupiter or Saturn (see Chapter 15), and some researchers have even suggested that there might have been thousands of Pluto-sized objects initially present in the outer solar system. The capture of a few of these objects by the giant planets would explain the strange moons of the outer worlds, especially Triton. And if there were enough moon-sized chunks originally orbiting beyond Neptune, it is quite plausible that Pluto could have captured Charon following a collision (or near-miss) between the two. At present, our scant knowledge of the compositions of the two bodies does not allow us to confirm or disprove either the coformation or the capture theory of the Pluto-Charon system.

CHAPTER SUMMARY

The outer planets Uranus, Neptune, and Pluto were unknown to ancient stargazers. Today we know the giant planets Uranus and Neptune mainly through data taken by *Voyager 2*. Small, remote Pluto has not been visited by a spacecraft, and our knowledge of it stems from painstaking observations.

Uranus has circled the Sun only two and a half times since its discovery in 1781. At opposition this planet is barely visible to the unaided eye. It appears as a pale green disk through a telescope. Deviations in the orbit of Uranus led astronomers to believe that another planet lay even farther from the Sun. In 1846, Neptune was found within one or two degrees of the predicted position. Neptune cannot be seen with the naked eye, but a telescope shows it as a tiny bluish disk.

Uranus and Neptune have similar bulk properties. Their densities imply large, rocky cores making up a greater fraction of the planets' masses than in either Jupiter or Saturn. Models indicate that Uranus and Neptune have rocky cores like those theorized for Jupiter and Saturn—with a similar size and perhaps 10 times the mass of the Earth. With computer-enhanced images, *Voyager 2* revealed atmospheric clouds and flow patterns moving beneath Uranus's haze. Neptune, though farther away, has atmospheric features that are clearer because of warmer temperatures and less haze. Both Uranus and Neptune have a substantial magnetosphere. *Voyager 2* discovered that the magnetic field of Uranus is tilted some 60° from the planet's axis of rotation. Neptune's field is inclined to that planet's rotation axis by 47°.

All of the uranian moons revolve in the equatorial plane of Uranus, almost perpendicular to the ecliptic, in circular, synchronous orbits. The strangest uranian moon is Miranda, with geological features that suggest a number of internal or external catastrophic events. Neptune's largest moon, Triton, has a solid, frozen surface with features that differ from those found on Neptune's other moons. Triton orbits in a retrograde fashion and is spiraling inward toward Neptune. Eventually, it will be torn apart by Neptune's gravity. The rings of Uranus are dark, narrow, and widely spaced. Neptune now is known to have four dark rings. Two are narrow, and two are broad and diffuse.

Pluto was discovered in 1930, after a determined search for a ninth planet. Pluto's orbit is highly elongated and inclined at 17° to the plane of the ecliptic. Pluto is now closer to the Sun than Neptune and will remain so until 1999. The planet is so tiny and remote that it is hard to distinguish from background stars. Pluto is too small for a terrestrial planet. It is more like a Jovian ice moon, and it may be an escaped moon, or an icy asteroid.

After the discovery of Pluto, searches continued for a planet that would explain discrepancies in the orbits of Uranus and Neptune. A survey of the infrared sky by the *IRAS* satellite in 1983 revealed no evidence for a tenth planet.

KEY WORDS

Great Dark Spot
perturbation theory
radiation darkening
stellar occultation

REVIEW QUESTIONS

1. How was Uranus discovered?
2. Why did astronomers suspect an eighth planet beyond Uranus?
3. What is the most unusual feature about the rotation of Uranus?
4. Why do Uranus and Neptune have a bluish color? Why is Uranus more blue-green than Neptune?
5. How are the interiors of Uranus and Neptune thought to differ from those of Jupiter and Saturn?
6. Name an unusual feature of the winds on Neptune.
7. What is so unusual about the magnetic fields of Uranus and Neptune?
8. What observation weakens the idea that the sideways rotation of Uranus was caused by a catastrophic event?
9. What is unique about Miranda? What explains this uniqueness?
10. How does Neptune's system of moons differ from those of other Jovian worlds? What do these differences suggest about their origins?
11. What is the predicted fate of Triton?
12. How do the rings of Uranus differ from the rings of Saturn?
13. How do the rings of Neptune differ from those of Uranus or Saturn?
14. How was Pluto discovered?
15. How were the masses and radii of Pluto and Charon determined?
16. In what respect is Pluto more like a moon than like a Jovian or terrestrial planet?
17. Compare the amount of sunlight reaching Uranus (at 19.2 A.U.) and Neptune (at 30.1 A.U.) with the amount of sunlight reaching Earth. What is the amount

of sunlight per square centimeter reaching Uranus and reaching Neptune as a percentage of the amount of sunlight reaching Earth?

18. What is the angular diameter of the Sun, as seen from Neptune? Compare it with the angular diameter of Triton. Would you expect solar eclipses to be common on Neptune? Why or why not?

19. How close is Charon to Pluto's Roche limit?

20. What is the round-trip travel time of light from Earth to Pluto? (Assume a distance of 40 A.U.) How far would a spacecraft orbiting Pluto at a velocity of 0.5 km/s travel in that time?

DISCUSSION QUESTIONS

1. Why do you suppose the discovery of Uranus in 1781 was so surprising? Do you think that similar surprises are in store for today's astronomers?
2. Select a moon each from Uranus, Neptune, and Pluto. Explain what each moon reveals about its planet's past.
3. Give a brief description of the outer solar system, as we would know it *without* the *Voyager* mission.
4. Do you think Pluto should be called a planet? Why or why not?
5. Which planet or moon in the solar system would you most like to have a future spacecraft visit? Why?

PROJECTS

1. The major astronomy magazines *Sky and Telescope* and *Astronomy* generally print charts showing the whereabouts of the planets in their January issues. Consult the charts in one of these magazines, and locate Neptune and Uranus in the sky. Uranus may be visible to the naked eye, but binoculars make the search much easier. Hint: Uranus shines more steadily than the background stars. With the eye alone, can you detect a color to Uranus? Does any color become more intense through binoculars?
2. The search for Neptune requires a much more determined effort! A telescope is best, but high-powered binoculars mounted on a steady support also reveal the planet. If you have an opportunity to see both planets through a telescope—and they will remain close together on the sky's dome for the rest of this century—contrast their colors. Which planet appears more blue? Through a telescope, can you see that Uranus shows a disk? Can you see that Neptune shows a disk, or does it look more like a pinpoint of light?
3. Pluto does not show a disk, even through powerful telescopes. It always looks like a star. To see Pluto at all, you must have at least an 8-inch telescope. Then you must locate the field of stars in which it currently resides (check the January issue of *Astronomy* or *Sky and Telescope* for this year's chart). Draw a picture of all the stars you see in that field of view. Come back a few nights later, and draw a picture of the field again. The "star" that has moved is Pluto!

SUGGESTED READINGS

Beatty, J. K. "A Place Called Uranus." *Sky and Telescope* (April 1986). Report on the Voyager 2 spacecraft encounter with Uranus.

———. "Getting to Know Neptune." *Sky and Telescope* (February 1990). Additional insight from the Voyager 2 Neptune encounter.

———. "Pluto and Charon: The Dance Goes On." *Sky and Telescope* (September 1987). New discoveries made possible by the decade-long series of mutual eclipses by Pluto and its moon.

Binzel, R. "Pluto." *Scientific American* (June 1990). What was learned about Pluto and its moon Charon during the 1980s series of mutual eclipses.

Chiles, J. R. "For Voyager 2: From There to Eternity." *Smithsonian* (September 1988). Popular article on the journey of Voyager 2 through the outer solar system, written just before the Neptune encounter.

Cuzzi, J. N., and L. W. Esposito. "The Rings of Uranus." *Scientific American* (July 1987). Uranian rings.

Ingersoll, A. P. "Uranus." *Scientific American* (January 1987). Report from the Voyager 2 encounter with Uranus.

Johnson, T. V., R. H. Brown, and L. A. Soderblom. "The Moons of Uranus." *Scientific American* (April 1987). Uranian moons.

Kinoshita, J. "Neptune." *Scientific American* (November 1989). The Voyager 2 encounter with Neptune.

Littmann, M. *Planets Beyond: Discovering the Outer Solar System.* New York: Wiley, 1988. The stories of Uranus, Neptune, and Pluto, the only major planets to have been discovered in recorded history.

CHAPTER 16

Solar System Debris

Keys to Our Origin

This is an artist's conception of a Comet Rendezvous and Asteroid Flyby *mission that the U.S. space program is considering sending to visit a comet and to gather information about its chemical makeup by directly sampling its gases. In this version, which is actually a single frame grabbed from a video animation, we see that the comet is in the midst of outgassing, or sublimating. Accordingly, huge plumes of steam are rising toward the Sun out of fissures in the comet's crust. (Courtesy D. Berry)*

LEARNING GOALS

1 to understand the origin and nature of asteroids
2 to understand the origin and nature of comets
3 to understand the origin and nature of meteoroids

According to classical definitions, there are only nine planets in the solar system. But several thousand other celestial bodies are also known to revolve around the Sun in well-determined orbits. These minor objects—called asteroids and comets—are small and of negligible mass compared with the planets and their major moons. Yet each is a separate world, with its own story to tell about the early solar system. On the basis of statistical deductions, astronomers estimate that there are more than a billion such objects still to be discovered. These objects may seem to be only rocky and icy "debris," but, more so than the planets themselves, they hold a record of the formative stages of our planetary system. Many are nearly "pure," unevolved bodies that may one day teach us much about our local origins.

ASTEROIDS

1 **Asteroids** are relatively small, rocky objects that revolve around the Sun. Their name literally means "starlike bodies," but asteroids are definitely not stars. They are too small even to be classified as planets. Astronomers often refer to them as "minor planets" or sometimes "planetoids." The largest known asteroid, Ceres, is just 1/10,000 the mass of Earth and measures only 940 km across. Together, the 4000 or so known asteroids amount to less than 1/10 the mass of the Moon, so they do not contribute significantly to the total mass of the solar system.

Asteroids differ from planets first by their size and second by their orbits. Few are larger than 300 km in diameter, and most are far smaller—as small as a tenth of a kilometer across. They generally move on quite eccentric trajectories between Mars and Jupiter, quite unlike the almost circular paths of the major planets. To the extent that astronomers can determine their compositions, asteroids have been found to differ not only from the nine known planets and their many moons, but also among themselves.

Orbital Properties

European astronomers discovered the first asteroids early in the nineteenth century as they searched for an additional planet in the region between Mars and Jupiter, where the Titius-Bode "Law" (see Interlude 9-1) suggested one might be found. Italian astronomer Giuseppe Piazzi was the first to discover an asteroid. He detected Ceres in 1801 and measured its orbital semi-major axis to be 2.8 A.U., exactly as predicted by the "Law." Within a few years, three more asteroids—Pallas (at 2.8 A.U.), Juno (at 2.7 A.U.), and Vesta (at 3.4 A.U.)—were discovered. But astronomers soon realized that these bodies are not "true" planets. Together, they do not even come close to the mass of the Moon, let alone a planet. Moreover, the eccentricities of the asteroids' orbits are generally larger than those of the planets (although most are still less than 0.25 or so). Figure 16.1 shows an Earth-based photograph of the asteroid Icarus. This body is actually somewhat atypical, as its orbit brings it close to the Sun and never far beyond the orbit of Mars. However, because of its occasional close encounters with Earth it is relatively easy to study.

By the start of the twentieth century, astronomers had cataloged several hundred asteroids with well-determined orbits. In this last decade in the twentieth century, the list

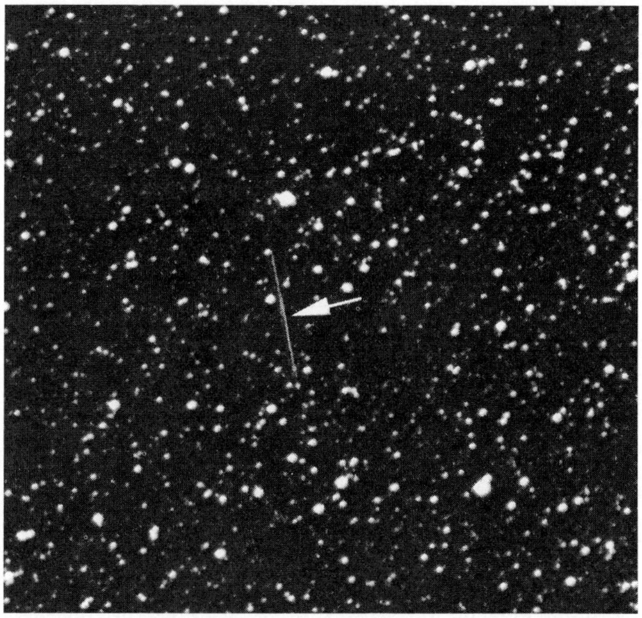

Figure 16.1 *The asteroid Icarus has an orbit that passes within 0.2 A.U. of the Sun, well within the orbit of Earth. Icarus occasionally comes close to our own planet, making it one of the best-studied asteroids in the solar system. (Palomar)*

numbers close to 4000. The great majority are found in a region of the solar system known as the **asteroid belt,** located between 2.1 and 3.3 A.U. from the Sun—roughly halfway between the orbits of Mars (at 1.5 A.U.) and Jupiter (at 5.2 A.U.).

Such a compact concentration of asteroids in a well-defined belt suggests that they are either the fragments of a planet broken up long ago or primal rocks that never managed to accumulate into a genuine planet. On the basis of the best evidence currently available, researchers favor the latter view. There is far too little mass in the belt to constitute a planet, and the marked chemical differences between individual asteroids strongly suggest that the asteroids could not all have originated in a single planet. Instead, astronomers believe that the strong gravitational field of Jupiter continuously disturbs the motions of these chunks of primitive matter, nudging and pulling at them, thereby prohibiting them from aggregating into a planet. The existence and composition of the asteroid belt joins the general properties of the planets and their moons on our list of features that any theory of solar system formation must explain.

The orbits of most asteroids have eccentricities lying in the range 0.05–0.3, ensuring that they always remain between the orbits of Mars and Jupiter. Very few asteroids have eccentricities greater than 0.4. Those that do, of course, are of great interest to us, as their orbits may intersect that of the Earth, leading to the possibility of a collision with our planet. Parts of the asteroid belt are also highly inclined—by as much as 30°—to the plane of the solar system. (Recall that none of the large planets deviates from this plane by more than 3°. Only Mercury and Pluto, the innermost and outermost worlds—and also the smallest—fall outside this range.) Despite these differences, all but one of the known asteroids revolve about the Sun in prograde orbits, in the same sense as the planets.

The few stray asteroids with very elliptical orbits have probably been influenced by the gravitational fields of nearby Mars and especially Jupiter. These planets can disturb the normal asteroid orbits, deflecting them into the inner solar system. Asteroids whose paths cross the orbit of Earth are termed **Apollo asteroids** (after the first-known Earth-crossing asteroid, Apollo). Those crossing only the orbit of Mars are known as **Amor asteroids** (again, named after the first of that type to be discovered). Although we are currently aware of only a few dozen Apollo asteroids, they are among the most famous, because of their occasional close encounters with Earth. For example, the Apollo asteroid Icarus (see Figure 16.1) periodically comes to within 0.2 A.U. of the Sun. On its way past Earth in 1968, it missed our planet by ''only'' 6 million km—a close call by cosmic standards. More recently, in 1989 an unnamed asteroid (designated 1989FC) came even closer, passing only 800,000 km from Earth, only twice the distance to the Moon. In 1991, asteroid 1991BA missed us by a mere 170,000 km.

The potential for collision with Earth is real. Some astronomers argue that, during any given million-year period, Earth is struck by about three Apollo asteroids. Because our planet is largely covered with water, on average two of the impacts should occur in the ocean, the one other on land. Several dozen large land basins and eroded craters on our planet are suspected to be sites of ancient asteroid collisions, the Barringer Crater, shown in Figure 8.10, being the premier example. The many large impact craters on Venus (see Chapter 11) are direct evidence of similar events on that world. Calculations imply that most Apollo asteroids will eventually collide with Earth.

Most known Apollo asteroids are relatively small—about 1 km in diameter (although one of 10 km in diameter has been identified). However, a visit of even a kilometer-sized asteroid to Earth could be catastrophic by human standards. Such an asteroid packs enough kinetic energy to devastate an area some 100 km in diameter. The explosive power would be equivalent to about a million 1-megaton nuclear bombs, a hundred times more than all the nuclear weapons currently in existence on Earth. A fatal blast wave would doubtless affect a much larger area still. Should asteroids hit our planet hard enough, they might even cause the extinction of entire species (see Interlude 30-2, which describes a theory that explains the extinction of dinosaurs through an asteroid impact). That the Amor and Apollo asteroids eventually hit Mars or Earth also accounts for their rarity. Once an asteroid finds itself on a planet-crossing orbit, its days are numbered. Some scientists take the prospect of an asteroid impact sufficiently seriously that they advocate an ''asteroid watch''—cataloging and monitoring all Earth-crossing asteroids in order to maximize our warning time of any impending collision. This program, called Spaceguard, would cost $15 million a year—a small fraction of the cost of even a small impact—but funding is currently uncertain.

Asteroid Families

The overall layout of the asteroid belt is sketched in Figure 16.2. Many Apollo and other asteroids move in very similar orbits. These "families" of asteroids are known as **Hirayama families,** after the Japanese astronomer Kiyotsugu Hirayama, who first discovered them in 1918. Astronomers believe that each family, containing perhaps 5 to 10 members, represents a single asteroid that broke up into smaller pieces after colliding (at a typical impact velocity of a few kilometers per second) with another asteroid. The individual fragments are now spread out all along the orbit of the original body.

Given the relative congestion of the asteroid belt, collisions may be quite common, providing one source of interplanetary dust and smaller asteroids. In addition, one or both of the bodies involved may be deflected onto eccentric, Earth-crossing orbits. Less violent collisions may be responsible for the curious **binary asteroids,** which consist of a large asteroid orbited by a smaller satellite body. At least two such systems are known—in fact, the second-largest (and second-discovered) asteroid, Pallas, is a binary. The strangely shaped asteroid Hektor, which looks like two small asteroids glued together, may also be the result of a low-velocity collision.

Although most asteroids orbit in the main belt, between 2 and 3 A.U. from the Sun, an additional class of asteroids, known as the **Trojan asteroids,** orbit at the distance of Jupiter. These asteroids are locked into a 1:1 orbital resonance with Jupiter by that planet's strong gravity, just as some of the small moons of Saturn share orbits with the medium-sized moons Tethys and Dione (see Chapter 14). Calculations first performed by the French mathematician Lagrange show that there are exactly five places in the solar system where a small body can orbit the Sun in synchronism with Jupiter, subject to the combined gravitational influence of both large bodies. These places are known as

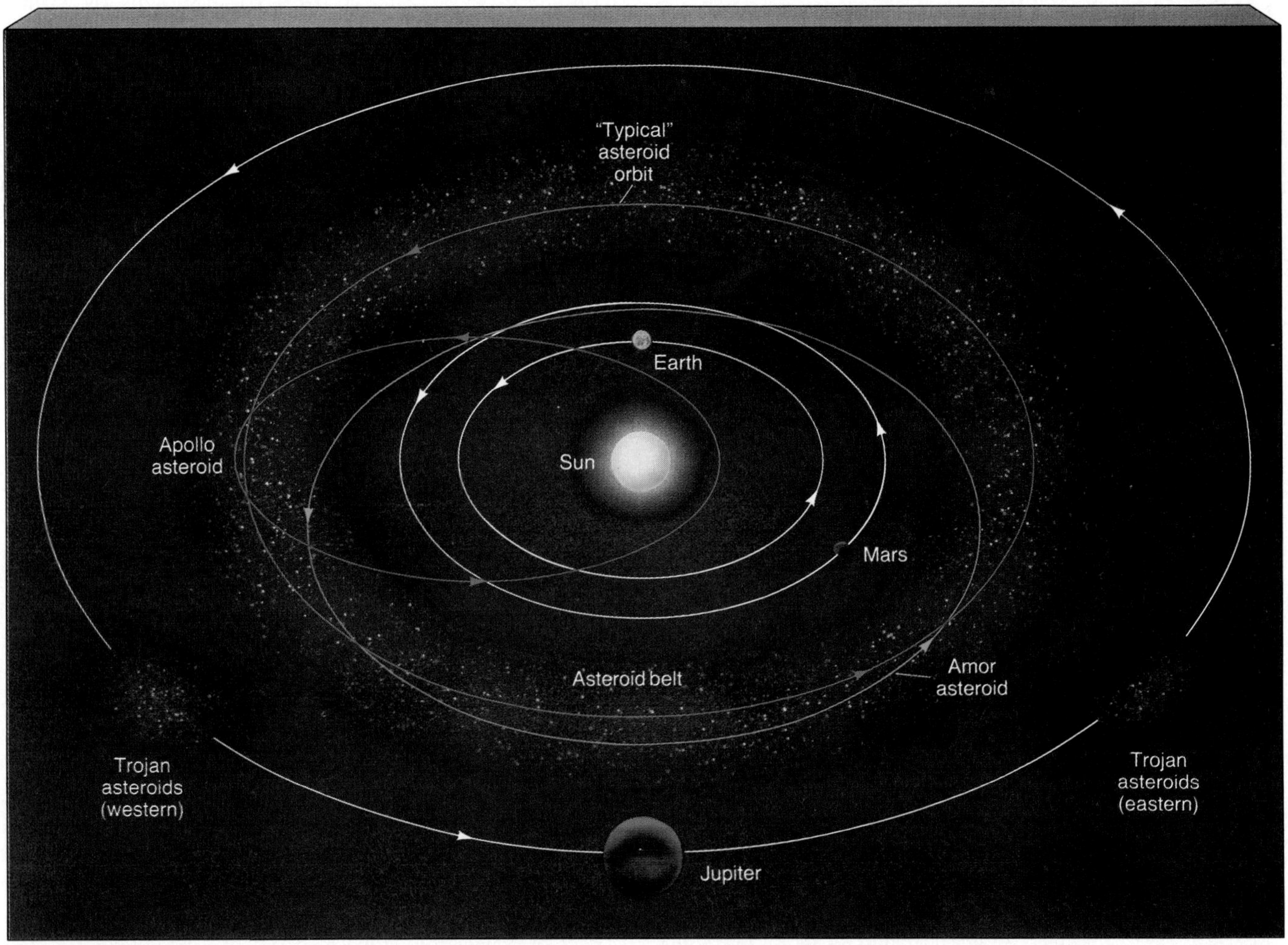

Figure 16.2 The asteroid belt, along with the orbits of Earth, Mars, and Jupiter. The main belt, the Trojan asteroids, and some Apollo and Amor orbits are shown.

the *Lagrange points*. Three of them, all lying on the line joining Jupiter to the Sun (or along its extension) are *unstable*—a body placed near one of these points will tend to drift slowly away from (not toward) it. Thus, matter does not accumulate in those regions. However, the other two Lagrange points, designated $\mathbf{L_4}$ and $\mathbf{L_5}$, are both *stable*—asteroids placed near them tend to remain nearby. These $\mathbf{L_4}$ and $\mathbf{L_5}$ points are located on Jupiter's orbit exactly 60° ahead of and behind the planet, and they revolve around the Sun at the same rate as Jupiter (see Figure 16.3). The 50 or so Trojan asteroids are roughly equally divided between the leading ($\mathbf{L_4}$) and trailing ($\mathbf{L_5}$) Lagrange points. Astronomers know of no asteroids orbiting in the Lagrange points of any other planet.

The main asteroid belt contains substructure, not as obvious as in the Saturnian ring system, with its prominent gaps, but nonetheless of great dynamical significance. A graph of the number of asteroids having various orbital semi-major axes (Figure 16.4) shows that there are several prominent underrepresented regions in the distribution. These "holes" are known as the **Kirkwood gaps,** after their discoverer, the nineteenth-century American astronomer Daniel Kirkwood.

We have just seen how the Trojan asteroids share an orbit with Jupiter—they orbit in 1:1 resonance with that planet. The Kirkwood gaps result from other, more complex orbital resonances with Jupiter. For example, an asteroid with a semi-major axis of 3.3 A.U. would (by Kepler's Third Law) orbit the Sun in exactly half the time taken by Jupiter. The gap at 3.3 A.U., then, corresponds to the 2:1 resonance. An asteroid at a resonance feels a regular, periodic "kick" from Jupiter at the same point in every orbit. The cumulative effect of those kicks is to deflect the asteroid into an elongated orbit—one that crosses the orbit of Mars or Earth. Eventually, the asteroid collides with one of

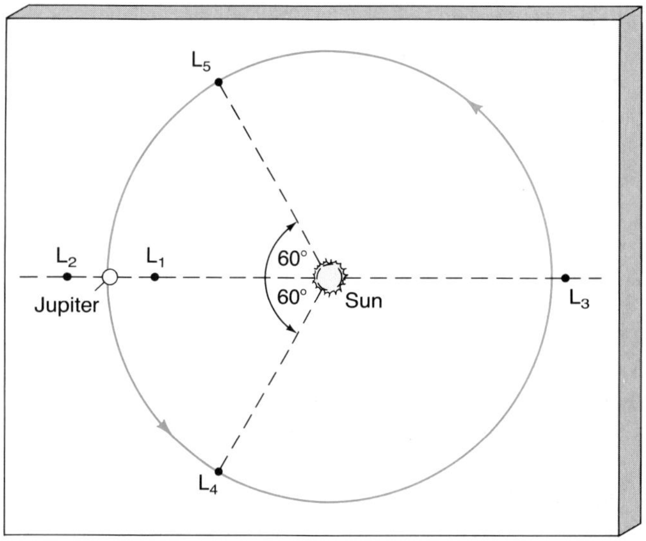

Figure 16.3 *The Lagrange points of the Jupiter-Sun system, where a third body could orbit in synchronism with Jupiter on a circular trajectory. Only the L_4 and L_5 points are stable. They are the locations of the Trojan asteroids.*

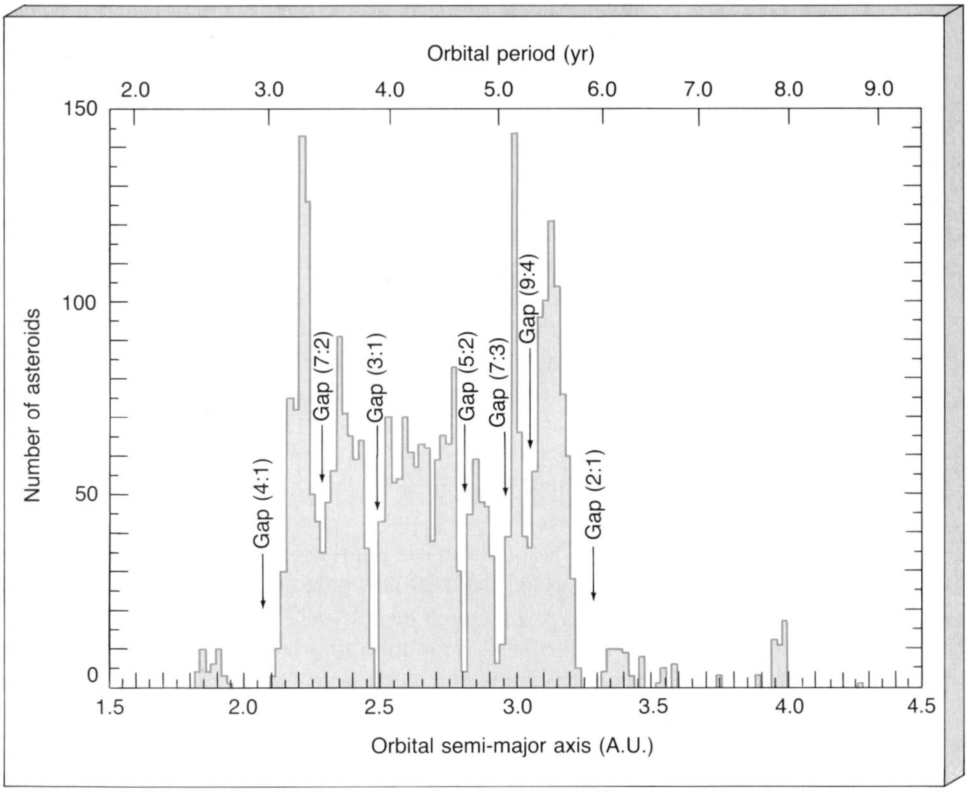

Figure 16.4 *The distribution of asteroid semi-major axes shows some prominent gaps caused by resonances with Jupiter's orbital motion. Note, for example, the prominent gap at 3.3 A.U., which corresponds to the 2:1 resonance—the orbital period is 5.9 years, exactly half that of Jupiter.*

those two planets or comes close enough that it is pushed onto an entirely different trajectory. In this way, Jupiter's gravity creates the Kirkwood gaps, while some of the cleared-out asteroids become Apollo or Amor asteroids.

Notice that there are many similarities between this mechanism and the resonances that produce the gaps in Saturn's rings (see Chapter 14). However, unlike Saturn's rings, where eccentric orbits are rapidly circularized by collisions among ring particles, there are no *physical* gaps in the asteroid belt. The in-and-out motion of the belt asteroids as they travel in their eccentric orbits around the Sun means that no orbit is actually empty. Only when we look at semi-major axes (or, equivalently, at orbital *energies*) do the gaps become apparent.

Physical Properties

With few exceptions, asteroids are too small to be resolved by Earth-based telescopes. We must rely on indirect methods to find their sizes, shapes, and compositions. Consequently, only a few of their physical and chemical properties are accurately known.

One established fact is that all asteroids reflect a varying amount of sunlight. This observation is attributed to rotation or tumbling of a nonspherical objects or one with a nonuniform surface. This explanation has been verified directly in the case of the asteroid Eros, which made a relatively close approach (about 1.5 A.U.) to Earth in 1975. Infrared observations of Eros implied a highly porous, rocky surface, covered with "dust" of some sort. Radar observations showed its surface to be much rougher than the surface of any planet or moon. If Eros is a typical asteroid, then they must all be pocked with surface holes and peppered with embedded chunks—not a terribly surprising description, given that asteroids normally reside in what is surely the most congested part of our solar system.

Asteroids are generally classified in terms of their spectroscopic properties. For example, astronomers found spectroscopic evidence for iron and silicate material on Eros. The darkest, or least reflective, asteroids, with typical albedos around 0.05, contain a large fraction of carbon in their makeup. They are known as **C-type** (or **carbonaceous**) asteroids. The more reflective asteroids, with albedos of about 0.15 to 0.25, contain silicate, or rocky, material. They are **S-type** asteroids. Generally, S-type asteroids predominate in the inner portions of the asteroid belt, and the fraction of C-type bodies steadily increases as we move out. Overall, 15 percent of all asteroids are S-type, 75 percent are C-type, and 10 percent are other types (such as the M-type asteroids, which contain large fractions of iron). The C-type asteroids may consist of primitive material representative of the early solar system that has not suffered significant heating or chemical evolution since its formation.

Astronomers generally estimate the sizes of asteroids from the amount of sunlight they reflect and the amount of heat they radiate. Making these observations is difficult, but we now have measurements of albedos and sizes for more than 1000 asteroids. In rare cases, astronomers witness an asteroid occulting a star, which allows them to determine the asteroid's size and shape with great accuracy. The largest asteroids are roughly spherical, but the smaller bodies can be highly irregular. Astronomers have measured masses only for the largest asteroids. The computed densities are consistent with the rocky or carbonaceous compositions just described. As mentioned in Chapter 12, many astronomers think that the Martian moons Phobos and Deimos are actually captured asteroids. If so, the images obtained by the *Viking* missions (see Figure 12.22) may be the best asteroid photographs to date.

The largest three asteroids, Ceres, Pallas, and Vesta, have diameters of 940, 580, and 540 km, respectively. Only two dozen or so asteroids are more than 200 km across, and most are much smaller. Almost assuredly, there exist hundreds of thousands more asteroids awaiting discovery. However, observers estimate that they are mostly small. Probably 99 percent of all asteroids larger than 100 km are known and cataloged, and at least 50 percent of those asteroids larger than 10 km are accounted for. In general, the number of asteroids increases with decreasing size. There are roughly 30 times as many 10-km asteroids as there are 100-km bodies, roughly 30 times as many 1-km bodies as there are 10-km, and so on, but the scaling is only approximate. The total number of potentially visible asteroids (that is, visible through telescopes if we knew just where and when to look) may exceed 100,000. Most asteroids are probably less than a few kilometers across. However, most of the mass in the asteroid belt resides in objects greater than a few tens of kilometers in diameter.

The Jupiter probe *Galileo* has recently provided scientists with the first close-up view of an asteroid. As mentioned in Chapter 13, *Galileo* is taking a rather roundabout path to the giant planet, receiving one gravity assist from Venus, followed by two from Earth (see Figure 13.5). By the time it reaches Jupiter in 1995, *Galileo* will have passed twice through the asteroid belt, having made close encounters with asteroid Gaspra in October 1991 and asteroid Ida in August 1993. Technical problems limited the amount of data that could be sent back from the spacecraft during the Gaspra flyby, so scientists have to wait until *Galileo*'s next close approach to Earth (in December 1992) to receive high-resolution images. As illustrated in Figure 16.5, the data show far more detail than has ever before been seen on an asteroid.

Gaspra is an irregularly shaped body with a maximum diameter of about 20 km, pitted with craters ranging in size from a few hundred meters to 2 km across and covered with a layer of dust (regolith) of variable thickness. It appears to be a fragment of a much larger object, suggesting to scientists that it is the product of a series of violent collisions in which several large bodies have been broken down into many smaller pieces. Figure 16.5 shows a false-color image

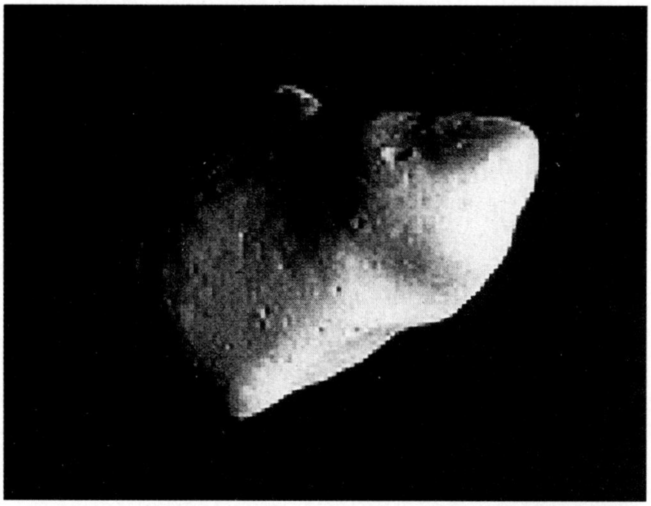

Figure 16.5 The asteroid Gaspra, as seen at a distance of 1600 km by the satellite Galileo *en route to Jupiter. This true-color image shows the asteroid's surface to be a fairly uniform shade of gray. However, sensors on board the spacecraft indicated that the amount of infrared radiation absorbed by the surface varies from place to place. The differences in absorption, astronomers believe, arise from variations in the thickness of the regolith on the asteroid's surface. The smallest features visible in this image are about 160 m across. The asteroid itself is some 20 km long. (NASA)*

designed to accentuate variations in Gaspra's surface properties. The asteroid's true color is a rather uniform gray, hinting that it may be chemically quite unevolved and relatively unchanged since the formation of the solar system. *Galileo*'s asteroid encounters will likely yield important new information on these primitive constituents of our solar system.

COMETS

✓2 Another resident of the solar system is the **comet.** Known for their long wispy tails, comets derive their name from the Greek *kome*, meaning ''hair.'' They are usually discovered as faint, fuzzy patches of light on the sky while still several astronomical units away from the Sun, in whose direction they are racing. Far from the Sun, comets are visible only by reflected sunlight, but as they near the Sun, they can emit radiation of their own, or at least reprocess the Sun's light into other forms. Traveling in a highly elliptical orbit with the Sun at one focus, a comet brightens and develops a tail as its icy matter becomes heated and sublimes[1] away, as depicted in Figure 16.6.

[1] *Sublimation* is the process by which a solid changes directly into a gas without passing through the liquid phase. Frozen carbon dioxide—dry ice—on Earth is an example of a solid that undergoes sublimation rather than melting and subsequently evaporating. In space, sublimation is the rule, rather than the exception, for the behavior of ice when exposed to heat.

Comet Orbits

As a comet departs from the Sun's vicinity, its brightness and its tail diminish until it becomes once again a faint point of light receding into the distance. Comets that survive a close encounter with the Sun—some break up entirely—continue their outward journey to the edge of the solar system. Their highly elliptical orbits take many comets far beyond Pluto, perhaps even as far as 50,000 A.U., where, in accord with Kepler's Second Law, they spend most of their time. Most take hundreds of thousands, some even millions, of years for a round trip, although a few ''short-period'' comets return for another encounter within a human life span. (In this context, the dividing line between ''short'' and ''long'' periods is conventionally taken to be 200 years.) The short-period comets do not venture far beyond the orbit of Pluto at aphelion.

Unlike the orbits of the solar system objects we have studied so far, the orbits of comets are *not* confined to within a few degrees of the ecliptic plane. They come in all

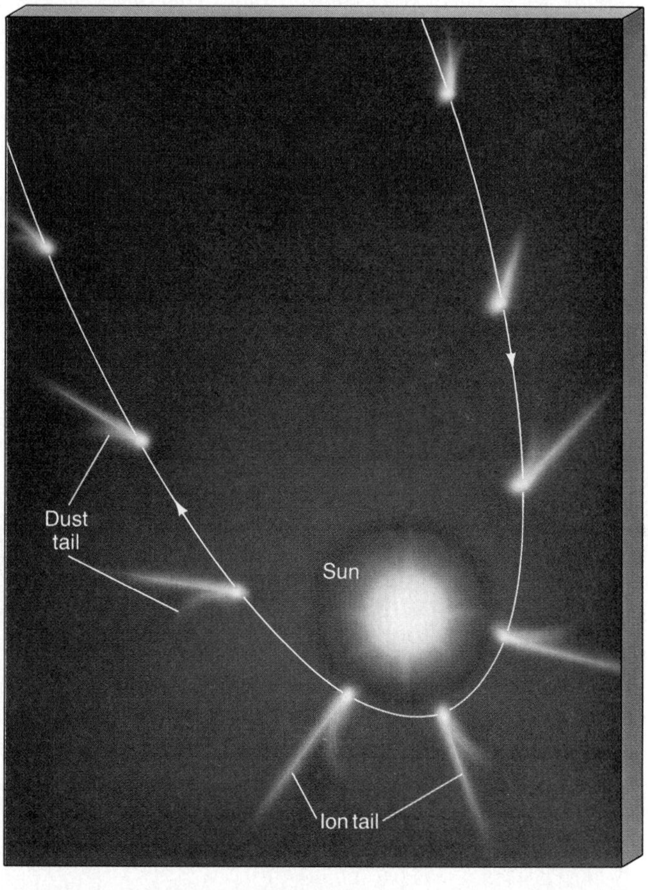

Figure 16.6 Schematic diagram of part of the orbit of a typical comet. As the comet approaches the Sun, it develops an ion tail, which is always directed away from the Sun. Closer in, a curved dust tail, also directed generally away from the Sun, may also appear. Notice that although the ion tail always points directly away from the Sun on both the inbound and the outgoing portions of the orbit, the dust tail has a marked asymmetry, always tending to ''lag behind'' the ion tail.

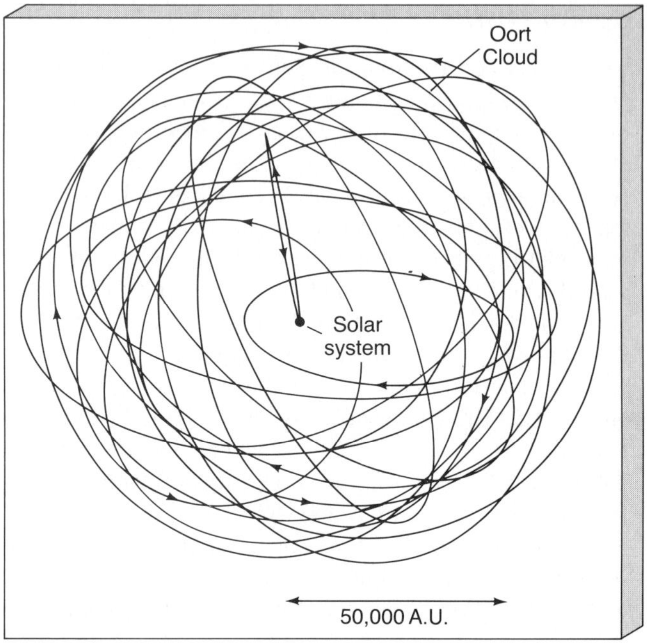

Figure 16.7 *Diagram of the Oort Cloud, showing some cometary orbits. Most Oort Cloud comets never come close to the Sun. Of all the orbits shown, only the most elongated ellipse represents a comet that will actually enter the solar system (which is smaller than the dot at the center of the figure on this scale) and possibly become visible from Earth.*

inclinations and all orientations, both prograde and retrograde, so that comets are uniformly distributed in all directions from the Sun. Only a tiny portion of a typical cometary orbit lies within the inner solar system, so it follows that for every comet we see, there must be many more at great distances from the Sun. Many astronomers reason that there is a huge ''cloud'' of comets far beyond the orbit of Pluto, completely surrounding the Sun. It is named the **Oort Cloud** after the Dutch astronomer Jan Oort, who first wrote (in the 1950s) of the possibility of such a vast and distant reservoir of inactive, frozen comets. Despite their great distances and long periods, the Oort Cloud comets are still gravitationally bound to the Sun. Their orbits are governed by the same laws of motion that control the planets.

Astronomers now believe that the comets whose orbits bring them near the Sun are themselves only a small fraction of the total. Most comets in the Oort Cloud remain at very large distances from the Sun throughout their entire orbits, never approaching even the orbit of Pluto, let alone that of the Earth. During all the time they spend far from the Sun's warmth, the Oort Cloud comets are nothing more than small icy chunks. Occasionally, however, the gravitational field of a passing star kicks a comet into an orbit that happens to take it through the inner solar system. Only when it enters the vicinity of the Sun does it acquire its spectacular tail. The orbits of some comets in the Oort Cloud are sketched in Figure 16.7.

Perhaps the most famous comet is Halley's Comet. In 1705, the British astronomer Edmund Halley realized that the 1682 appearance of this comet was not a one-time event. Basing his work on previous sightings of the comet, Halley calculated its path and found that the comet orbited the Sun with a period of 76 years. He predicted its reappearance in 1758. His successful determination of the comet's trajectory and his prediction of its return was an early triumph of Newton's laws of motion and gravity. Although Halley did not live to see his calculations proved correct, the comet was named in his honor.

Once astronomers knew the comet's period, they traced its appearances back in time. Historical records show that Halley's Comet has been observed at every passage since 240 B.C. A spectacular show, the tail of Halley's Comet can reach almost a full astronomical unit in length, stretching many tens of degrees across the sky. Figure 16.8(a) shows Halley's Comet as seen from Earth in 1910. Its most recent visit, in 1986 (see Figure 16.8[b]), was not ideal for terrestrial viewing, but the comet was closely scru-

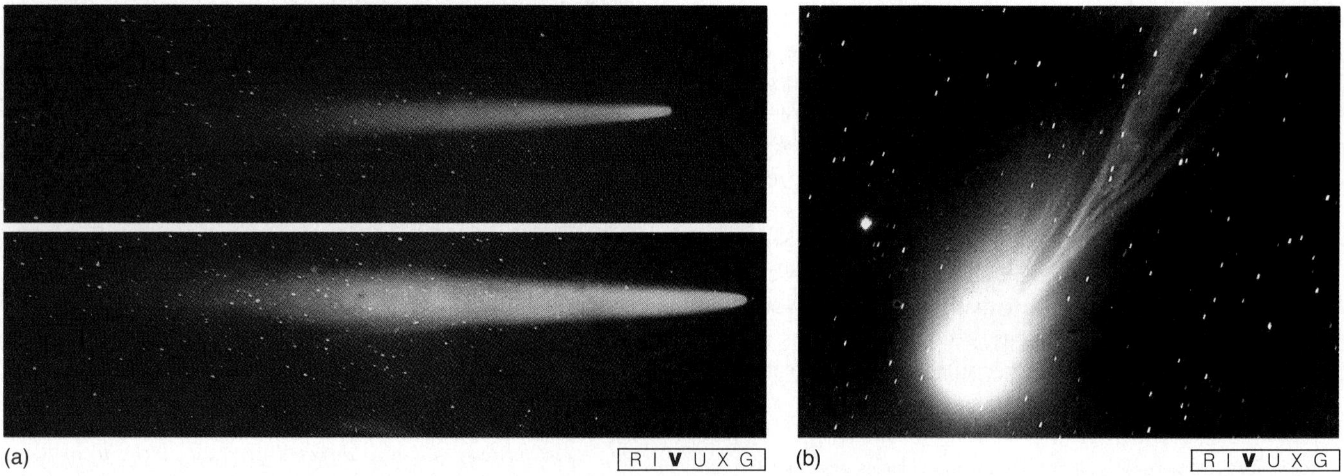

Figure 16.8 *(a) A series of photographs of Halley's Comet as it appeared in 1910; top, on May 10, with a 30° tail, bottom, on May 12, with a 40° tail. (b) Halley's Comet in 1986, about one month before perihelion. (Palomar Observatory; Lick Observatory).*

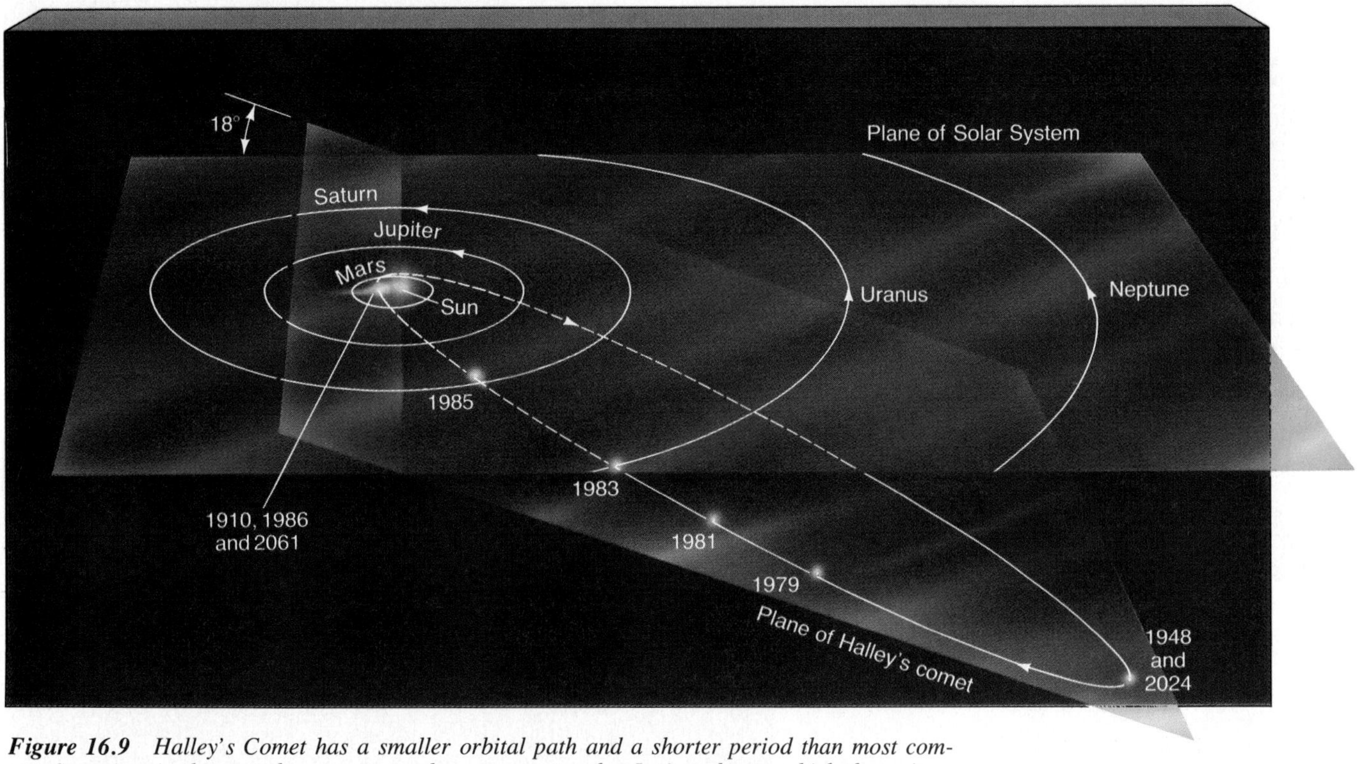

Figure 16.9 *Halley's Comet has a smaller orbital path and a shorter period than most comets. Sometime in the past the comet must have encountered a Jovian planet, which threw it into a tighter orbit that extends not to the Oort Cloud but merely a little beyond Neptune. Halley applied Newton's Law of Gravity to predict this comet's return.*

tinized by spacecraft. As indicated in the sketch of its orbit shown in Figure 16.9, the comet's next scheduled visit is in 2061.

Comet Structure

When a comet comes within a few astronomical units of the Sun, its icy surface becomes too warm to remain stable. Part of it becomes gaseous and expands into space, ultimately forming the exciting display we see from Earth. Even through a large telescope, the **nucleus,** or main solid body, of a comet is no more than a minute point of light. A typical cometary nucleus is extremely small—only a few kilometers in diameter. During most of the comet's orbit, far from the Sun, only this frozen nucleus exists. However, as the comet responds to the Sun's heat, a diffuse **coma** (or ''head,'' or ''halo'') of dust and evaporated gas begins to form, surrounding the nucleus and growing in size as the comet nears the Sun. At maximum size, the coma can measure as much as 100,000 km in diameter—almost as large as Saturn or Jupiter. Engulfing the coma, an invisible **hydrogen envelope,** usually distorted by the solar wind, stretches across millions of kilometers of space. The comet's **tail,** most pronounced when the comet is closest to the Sun and the rate of evaporation from the nucleus is greatest, is much larger still, sometimes spanning as much as an astronomical unit. From Earth, only the coma and tail of a comet are visible to the naked eye. Despite the size of the tail, most of a comet's light comes from the coma.

Comets generally come in two types, distinguished mostly by the appearance and material composition of their tails. **Type I** tails are approximately straight, often made of glowing, linear streamers like those seen in Figure 16.10(a). Their spectra show emission lines of numerous *ionized* molecules, including carbon monoxide, nitrogen, and water among many others. The type I tail is often called the **ion,** or **plasma, tail. Type II** tails are usually broad, diffuse, and gently curved (Figure 16.10[b]). They are rich in microscopic dust particles that reflect sunlight, making the tail visible even though it emits no light of its own. Type II tails are also known as **dust tails.**

Some comets' tails are mixtures of types I and II. Comet Kohoutek (Figure 16.10[c]), which appeared in 1975, is a typical example. This comet caused great excitement, as astronomers predicted it would be very bright in the twilight sky. To everyone's dismay, however, Kohoutek was barely visible to the naked eye despite its abundance of ionized matter and emitted radiation (its type I tail). Its disappointing appearance was due to an unusually large amount of dust (its type II tail) that scattered the light from the tail of the comet and thus dimmed it.

The tails are in all cases directed *away* from the Sun by the solar wind (the invisible stream of matter and radiation escaping the Sun). Consequently, as depicted in Figure 16.6, the tail always lies outside the comet's orbit and actually *leads* the comet during the portion of the orbit that is outbound from the Sun. Every tiny particle in space in our solar system—including those in comet tails—follows an orbit determined by gravity and the solar wind. If gravity alone were acting, the particle would follow the same

(a) R I V U X G

(b) R I V U X G

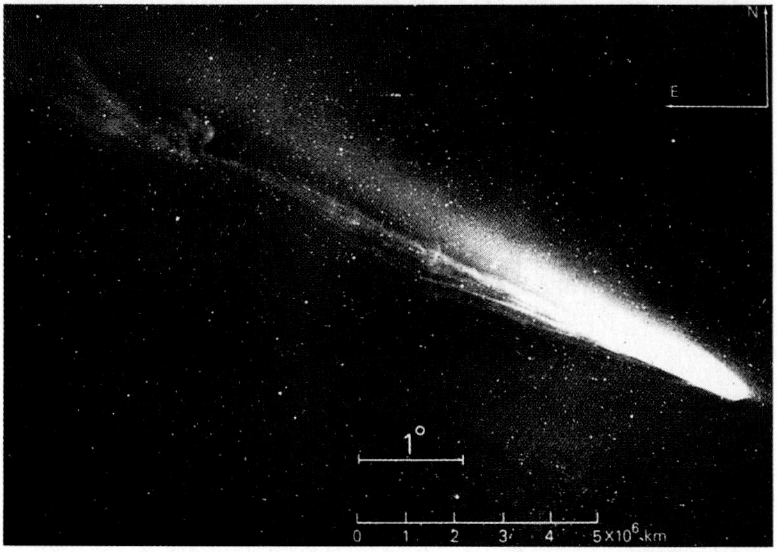

(c) R I V U X G

Figure 16.10 *(a) Comet with a primarily type I tail. Called Comet Giacobini-Zimmer and seen here in 1959, its coma measures 70,000 km across and its tail well over 500,000 km long. (b) Photograph of a comet having (mostly) a type II tail, showing both its gentle curvature and inherent fuzziness. This is Comet West, in 1976, whose tail stretches 13° across the sky. (c) Photograph of Comet Kohoutek, whose tail in 1975 was a mixture of type-I and type-II categories. The "tail" is not a sudden streak in time across the sky, as in the case of meteors or fireworks. Instead, it travels sedately along with the head of the comet. (U.S. Naval Observatory; NOAO)*

curved path as its parent comet, in accord with Newton's laws of motion. If the solar wind were the only influence, the tail would be swept up by it and would trail radially outward from the Sun. The ions that make up the type I tails are much more strongly influenced by the solar wind than by the Sun's gravity, so those tails always point directly away from the Sun. The inertia of the heavier dust particles gives them more of a tendency to follow the comet's orbit, giving rise to the slightly curved tail of the type II variety.

A Visit to Halley's Comet

In 1986, when Halley's Comet last rounded the Sun, a small armada of spacecraft launched by the USSR, Japan, and a consortium of western European countries examined the comet up close. One of the Soviet craft, *Vega 2*, traveled through the comet's coma, coming to within some 8000 km of the nucleus. Using positional knowledge of the comet gained from the Soviet craft encounters, the European *Giotto* spacecraft (named after the Italian artist who painted Halley Comet's image not long after its appearance in the year A.D. 1301) was navigated to within 600 km of the nucleus—a daring trajectory that in fact damaged *Giotto*'s camera, but not before it sent home a wealth of data. At 70 km/s, the speed of the craft relative to the comet, a colliding dust particle becomes a devastating bullet. Figure 16.11 shows two images of the comet's nucleus radioed back to Earth by these spacecraft.

The results of the Halley encounters were somewhat surprising. Halley's nucleus is an irregular, potato-shaped object and larger than astronomers had estimated. Spacecraft measurements showed it to be 15 km long by as much as 10 km wide. Also, the nucleus appeared almost jet black—as dark as finely ground charcoal or carbon black. This solid nucleus was enveloped by a cloud of dust, which

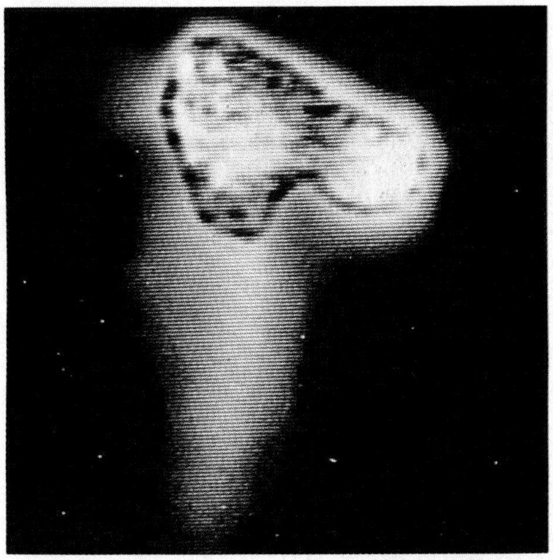

Figure 16.11 *(a)* Vega 2's *camera imaged Halley's nucleus, coming within a mere 8000 km of the spinning bundle of dirty ice. (b) The* Giotto *spacecraft better resolved the comet, showing its nucleus to be very dark, although heavy dust in the area obscured any surface features. Resolution here is about 50 m—half the size of a football field. At the time of both pictures, March of 1986, and within days of perihelion, the sun was toward the bottom. Hence, the brightest parts of the images are jets of evaporated gas and dust spewing from the comet's nucleus. (Russian Space Agency; Eurospace Center)*

scattered light throughout the coma of the comet. Partly because of this scattering and partly because of direct dimming by the dust, no spacecraft was able to discern much surface detail on the nucleus.

Figure 16.12 is a composite of several findings concerning Halley's nucleus. The visiting spacecraft found direct evidence for several jets of matter streaming from the nucleus. Instead of evaporating uniformly from the whole surface to form the comet's coma and tail, gas and dust apparently vent from small areas on the sunlit side of Halley's nucleus. The force of these jets may be largely responsible for the comet's 53-hour rotation period. Like maneuvering rockets on a spacecraft, such jets can cause a comet to change its rotation rate and even to veer away from a perfectly elliptical orbit. Astronomers had supposed the existence of these nongravitational forces for comets in general on the basis of the slight deviations from Kepler's laws observed in some cometary trajectories. However, only on the Halley encounter did astronomers actually see these jets at work.

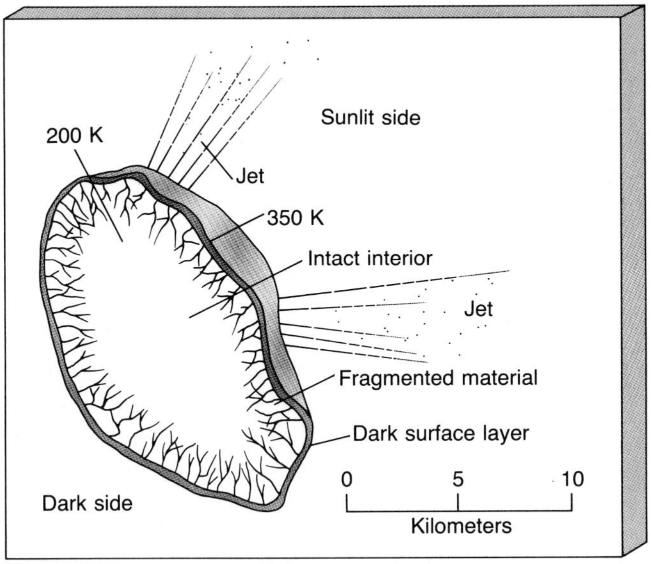

Figure 16.12 *A drawing of Halley's nucleus, depicting its size, shape, jets, and other physical and chemical properties.*

Physical Properties of Comets

The mass of a comet, most of which resides in its nucleus, can sometimes be estimated. Researchers can watch how the comet interacts with other solar system objects or use more indirect means, such as determining the size of the nucleus and assuming a density characteristic of icy composition. These methods yield typical cometary masses ranging from 10^{15} to 10^{19} g, comparable to the masses of small asteroids.

Actually, a comet's mass decreases with time because some of it is lost each time it rounds the Sun. For some comets that travel within an astronomical unit of the Sun, this evaporation rate can reach as high as 10^{30} molecules per second or, equivalently, 10^7 g/s. This is a loss of 10 tons of cometary material for *every second* the comet spends near the Sun. However, this rapid evaporation does not last long because comets, like all other bodies orbiting the Sun, obey Kepler's Second Law and speed up in their orbits when near the Sun. But comets do not live forever. Astronomers have estimated that this loss of material will destroy Halley's Comet in about 5000 orbits, or 40,000 years.

A few comets break up while near the Sun. In later years, they can be seen to return as pairs or swarms of smaller comets. Some comets have even disappeared entirely after an especially close approach to the Sun, and a few are known to have fallen into the Sun. In the act of breaking up, these **Sun-grazing comets** often emit prominent spectral lines that give important clues about comet composition. Figure 16.13 shows Comet West, which later broke apart as it rounded the Sun.

In seeking the physical makeup of a cometary body itself, astronomers are guided by the observation that comets have dust that reflects light, as well as gas that emits spectral lines of the abundant elements hydrogen, nitrogen,

Figure 16.13 *Comet West is seen here approaching the Sun in 1976. Shortly thereafter, it shattered into several fragments while passing close to the Sun. (NASA)*

R I **V** U X G

carbon, and oxygen. Even as the atoms, molecules, and dust particles boil off, creating the coma and tail, the nucleus itself remains a cold mixture of gas and dust, hardly more than a ball of loosely packed ice having a density of about 0.1 g/cm^3 and a temperature of only a few tens of kelvins. Experts now consider cometary nuclei to be largely made of dust particles trapped within a mixture of methane, ammonia, and ordinary water ice. (By now, these constituents should be fairly familiar to you as the main components of most of the small moons in the outer solar system.) Comets are often called ''dirty snowballs,'' a term first coined by Fred Whipple, of Harvard University, whose name has become virtually synonymous with cometary studies.

Puzzlingly, though, infrared heat sensors aboard the *Vega* craft indicated that the nucleus of Halley's Comet was as hot as 400K in places. This is too hot for ice, although an icy core could be surrounded by a mantle of warmer material. Possibly the instruments measured the heat of ejected dust just outside the nucleus and not the nucleus itself. Whatever the source of the heat, instruments confirmed that Halley is mostly ice by determining that roughly 80 percent of the molecules in the tail are water. Halley's classification as a moderately dusty comet was also confirmed, because about a tenth of the mass of matter in its tail was found to be dust.

METEOROIDS

✓3 On a clear night, it is possible to see a few **meteors,** or ''shooting stars,'' every hour. As a small piece of interplanetary debris enters Earth's atmosphere, it heats and excites air molecules through friction. These molecules emit light as they return to their ground states, and a sudden bright streak results. The object responsible for the display is known by several names, depending on circumstances. The bright streak is a meteor. Before encountering the atmosphere, the chunk of debris is known as a **meteoroid,** in analogy with the word *asteroid*. If any part of a meteoroid survives its fiery passage through our atmosphere, the piece that finds its way to the ground is known as a **meteorite.** Notice that the sudden flash of light produced by a meteor is in no way similar to the broad, steady swath of light associated with a comet's tail. A meteor is a fleeting event in our own atmosphere, while a comet tail is truly enormous, possibly an astronomical unit or more long, and visible in the sky for weeks or even months.

Meteoroids and asteroids are both interplanetary fragments. What separates them is size. The dividing line between them is a little fuzzy, but meteors are usually taken to be less than 100 m in diameter. The distinction applies only in space—if an asteroid strikes the Earth's atmosphere, we call it a meteor, and if it strikes the Earth, its remnant is called a meteorite. Some meteoroids are the rocky remains of broken-up comets. Others seem to be stray asteroids or pieces left over from asteroid collisions. Like comets and asteroids, many meteoroids are composed of ancient material and provide us with evidence of conditions in the early solar system.

Cometary Fragments

On each pass of a comet near the Sun, some cometary fragments dislodge from the main body. The fragments initially travel in a tight-knit group, called a **meteoroid swarm,** of dust or pebble-sized objects moving in nearly the same orbit as the parent comet. Over the course of time, the swarm gradually disperses along the orbit, so that eventually the **micrometeoroids,** as these small meteoroids are known, become more or less smoothly spread all the way around the parent comet's orbit. Whenever Earth intersects the orbit of a young, relatively undispersed cluster of meteoroids, a spectacular **meteor shower** can result. Earth's motion takes it across a given comet's orbit only once or twice a year (depending on the precise orbit of each body). Intersection occurs at the same time each year (see Figure 16.14), so the appearance of certain meteor showers is a regular and (fairly) predictable event.

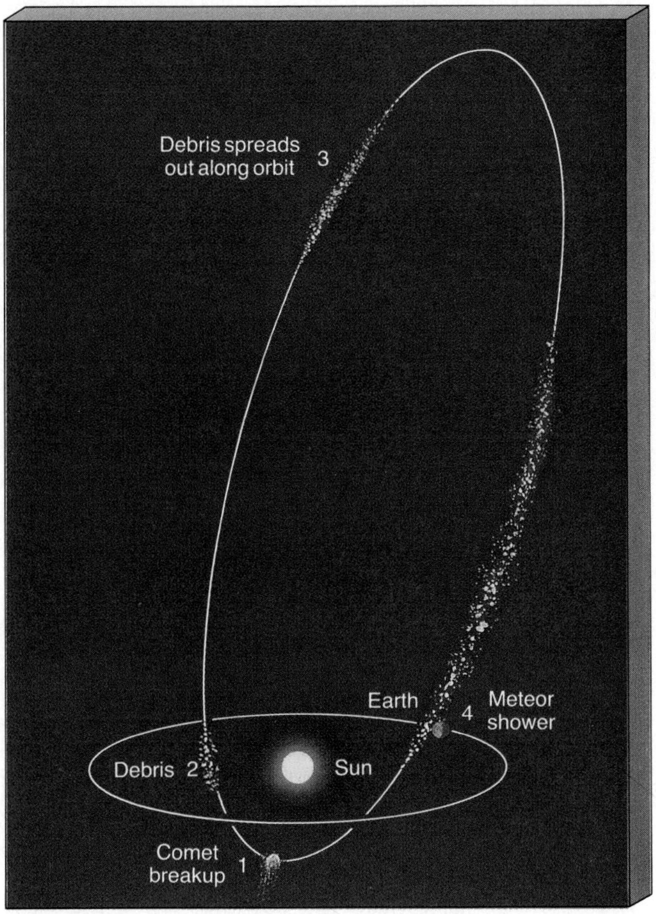

Figure 16.14 *A meteor swarm associated with a given comet intersects the Earth's orbit at specific locations, giving rise to meteor showers at specific times of the year. We imagine that a portion of the comet breaks up near aphelion, at the point marked [1]. The fragments continue along the original comet orbit, gradually spreading out as they go (points [2] and [3]). The rate at which the debris disperses is actually much slower than depicted here—it takes many orbits for the material to spread out as shown. Eventually, the fragments will extend all around the orbit, more or less uniformly. If the orbit happens to intersect Earth's orbit, a meteor shower is seen each time Earth passes through the intersection (point [4]).*

Meteor showers, like the one shown in Figure 16.15, are composed of large numbers of meteors burning their way through the atmosphere. The showers are usually named for their **radiant,** the constellation from whose direction they appear to come. For example, the Perseid shower is seen to emanate from the constellation Perseus. It can last for several days, but reaches maximum every year on the morning of August 12, when upward of 50 meteors per hour can be observed.

Astronomers can use the velocity and direction of a meteor's flight to compute its interplanetary trajectory. This is how certain meteoroid swarms have come to be identified with well-known comet orbits. For example, the Perseid shower shares the same orbit as Comet 1862III, the third comet discovered in the year 1862 (also known as Comet Swift-Tuttle). Meteoroids travel around the Sun in elliptical orbits just as comets, asteroids, and planets do. All these objects obey Kepler's laws, as they all move under the gravitational guidance of the Sun.

Table 16-1 lists some prominent meteor showers, the dates they are visible from Earth, and the comet from which

R I **V** U X G

Figure 16.15 *This photograph shows several streaks in Earth's atmosphere caused by micrometeoroids of the Leonid meteor shower (which appears to come from the constellation Leo). (Harvard-Smithsonian Center for Astrophysics)*

TABLE 16-1 ***Some Prominent Meteor Showers***

Morning of Maximum Activity	Shower Name	Rough Hourly Count	Parent Comet
Jan. 3	Quadrantid	40	—
Apr. 21	Lyrid	10	1861I (Thatcher)
May 4	Eta Aquarid	20	Halley
June 30	Beta Taurid	25	Encke
July 30	Delta Aquarid	20	—
Aug. 11	Perseid	50	1862III (Swift-Tuttle)
Oct. 9	Draconid	up to 500	Giacobini-Zinner
Oct. 20	Orionid	30	Halley
Nov. 7	Taurid	10	Encke
Nov. 16	Leonid	up to 100,000	1866I (Tuttle)
Dec. 13	Geminid	50	3200 Phaeton?

they are thought to originate. Notice in the table that the last "parent comet," Phaeton, is actually an asteroid. This object shows no sign of cometary activity, but its orbit matches the meteoroid paths very well.

Most micrometeoroids are rather tenuous, fragile objects, varying in overall density from 0.01 to about 1 g/cm^3. With an average velocity of about 25 km/s, a typical small (10^{-4} cm) micrometeoroid has little chance of reaching Earth's surface before burning up. In fact, as far as we know, no particle in *any* meteor shower has ever been large enough or dense enough to survive the atmospheric encounter and reach the ground. However, some have been collected during high-altitude U-2 aircraft flights. Figure 16.16 shows a typical low-density, irregularly shaped micrometeoroid collected at an altitude of about 20 km. With a frothy, nonmetallic structure, micrometeoroids often resemble bits of burned newspaper or charred toast.

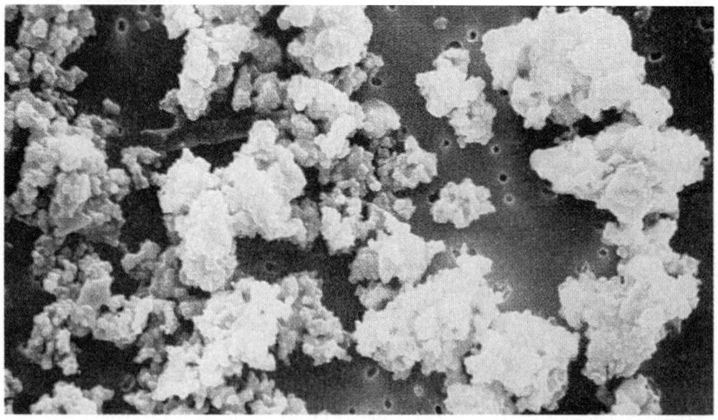

Figure 16.16 *This cluster of micrometeoroids was collected by a reconnaissance (U-2) aircraft at an altitude of about 20 km. The field of view shown in this photograph is 10^{-2} cm across. If reconstituted into a ball, these fluffy structures would make an object about 10^{-3} cm in diameter. (University of Washington)*

Stray Asteroids

Larger meteoroids—more than a few centimeters in diameter—are generally *not* associated with swarms of cometary debris. Regarded as stray asteroids or as the result of asteroid collisions, these objects have produced most of the cratering on the surfaces of the Moon, Mercury, Venus, Mars, and some of the moons of the jovian planets. When these large meteoroids enter Earth's atmosphere with a typical velocity of nearly 20 km/s, they produce energetic shock waves, or "sonic booms," as well as a bright sky streak and a dusty trail of discarded debris. Such large meteors are sometimes known as *fireballs*. The greater the velocity of the incoming object, the hotter its surface becomes and the faster it burns up. A few large meteors enter the atmosphere with such great velocity (about 75 km/s) that they either fragment or disperse entirely at high altitudes.

More massive meteors (each at least a ton in mass and a meter across) do make it to the surface, converting their kinetic energy (motion) into mechanical (damage), thermal (heat), and acoustical (sound) energy. This combination is inevitably explosive and produces a crater such as the kilometer-wide Barringer Crater of Figure 8.10. From the size of this crater, we can estimate that the meteoroid responsible must have had a mass of about 50,000 tons. Only 25 tons of iron meteorite fragments have been found at the crash site. The remaining mass must have been scattered by the explosion at impact, broken down by subsequent erosion, or buried in the landscape.

Currently, Earth is scarred with nearly 100 craters larger than 0.1 km in diameter. Most of these are so heavily eroded by weather and distorted by crustal activity that they can be identified only in satellite photography, as shown in Figure 16.17. Fortunately, such major collisions between Earth and large meteoroids are thought to be rare events

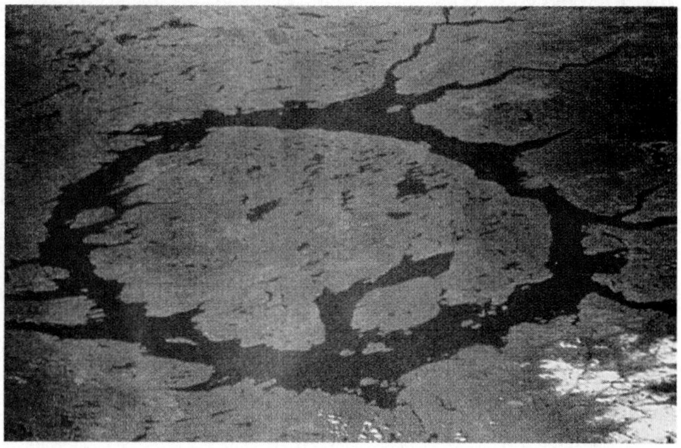

R I V U X G

Figure 16.17 *This photograph, taken from orbit by the U.S.* Skylab *space station, clearly shows the ancient impact basin that forms Quebec's Manicouagan Reservoir. A large meteorite must have landed there long ago. The most ancient terrestrial meteoroid crater is nearly 2 billion years old. (Canadian Institute for Theoretical Astrophysics)* now. Researchers believe that, on average, they occur only once every few hundred thousand years.

Major impacts with the Earth are not confined to meteoroids. One of the most recent occurred in central Siberia on June 30, 1908. The presence of only a shallow depression as well as a complete lack of fragments implies that this Siberian event was probably a collision with an icy comet. The comet apparently exploded several kilometers above the ground, leaving a blasted depression at ground level but no well-formed crater. The explosion, estimated to have been equal in energy to a 1-megaton nuclear detonation, was heard hundreds of kilometers away and produced measurable increases in atmospheric dust levels all across the Northern Hemisphere.

Orbits of large meteorites that survive their plunge through Earth's atmosphere can be reconstructed in a manner similar to that used to determine the orbits of meteor showers. There is one problem, however: Accurate observations of large meteors falling through Earth's atmosphere are quite rare. In fact, astronomers have been able to determine the orbits for only a few recovered meteorites. Figure 16.18 diagrams the reconstructed orbits of three of them: the Pribram fall (Czechoslovakia, April 7, 1959), the Lost City fall (Oklahoma, January 3, 1970), and the Innisfree fall (Alberta, February 5, 1977). As shown, the most distant part of each of their orbits intersects the asteroid belt. This reconstructed orbital information for these three fallen meteorites is the strongest evidence that large meteorites were probably once part of the asteroid belt before being redirected, probably by a collision with another asteroid, into an Earth-crossing orbit that led to the impact with our planet.

Finds and Falls

In order to study a meteorite in detail, we must retrieve it from its landing site and take it to a laboratory. Several thousand meteorites have been recovered over the years. Their acquisition falls into two categories. Some meteorites are accidental **finds,** and others are discovered because of an observed **fall.** A find is largely a matter of chance, but when a fall is observed, a deliberate search is made at the point of impact. Many meteorites look similar to Earth rocks—for that reason, finds tend to be "unusual" meteorites that stand out in some way from their surroundings (for example, the metallic meteorites discussed below). Antarctica is an especially good place to seek relatively clean me-

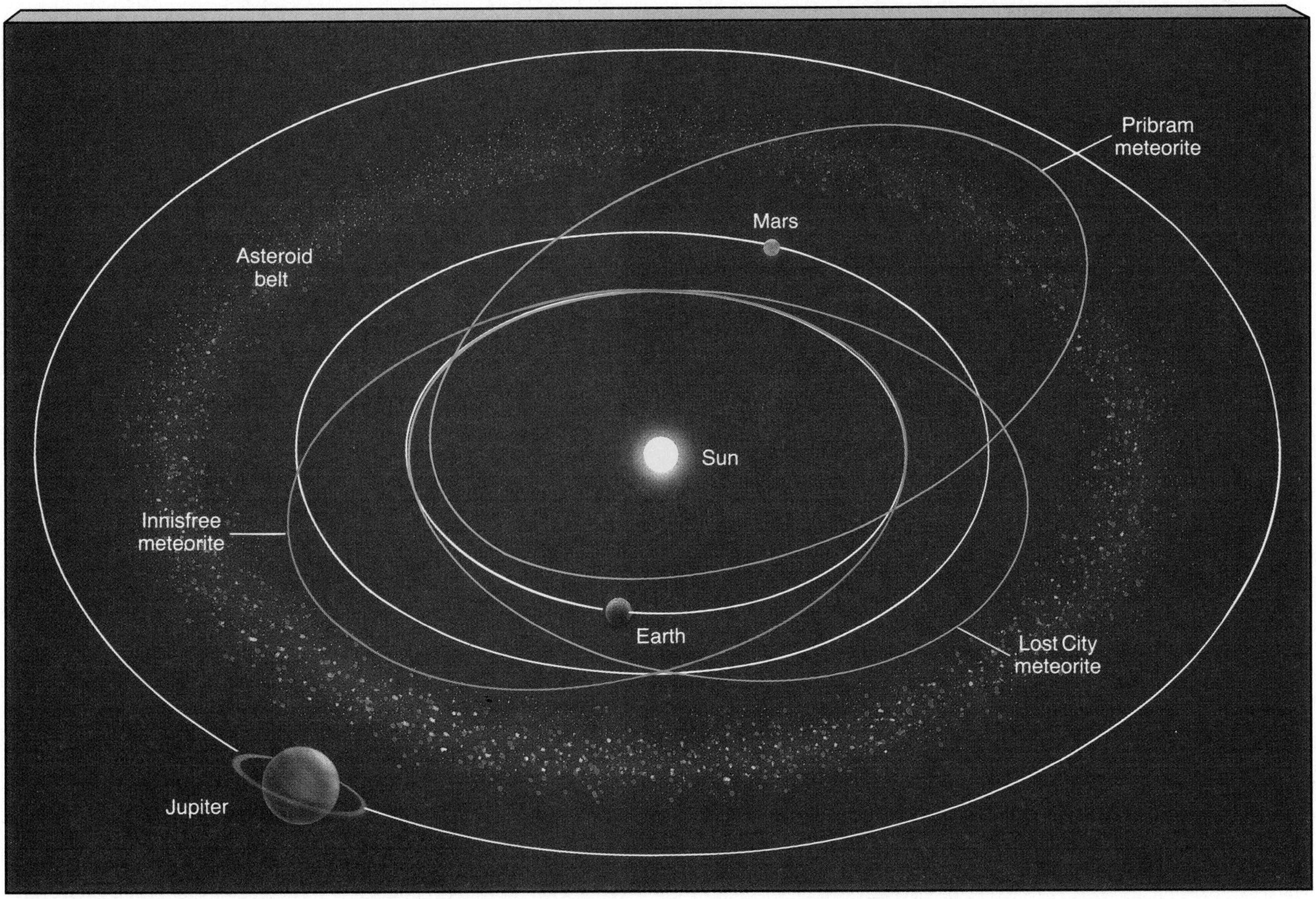

Figure 16.18 *Reconstructed orbits of three large meteorites that collided with Earth during the past few decades. All three orbits pass through the asteroid belt, consistent with the view that large meteorites originate there.*

teorites because the subfreezing temperatures and snow cover protect them from erosion and contamination that would otherwise alter their properties. Every so often, strong winds remove the snow cover, enabling researchers to locate and collect the meteorite fragments. The snow also makes the fragments especially easy to see.

Meteoritic Composition

One feature that distinguishes the small micrometeoroids that burn up in Earth's atmosphere from the large meteoroids that manage to reach the ground is their composition. The average density of meteoritic fireballs too small to reach the ground is about 0.5 g/cm^3. Such a low density is typical of comets, which are made of loosely packed ice and dust. By contrast, the meteorites that reach Earth's surface are much denser—around 5 g/cm^3—suggesting a composition more like that of the asteroids.

Meteorites like the one shown in Figure 16.19 have received close scrutiny in terrestrial laboratories. Prior to the Space Age, meteorites were the only type of extraterrestrial matter to which we had access. When meteorites are heated and broken down into their component elements, we learn that their basic composition is much like the stony inner planets and the Moon. They do exhibit, however, an underabundance of some of the lighter elements—such as hydrogen and oxygen—that may have boiled away into space when the meteoroids were molten long ago.

All meteorites, then, are made up of fairly dense material. They are classified into three major types. The **irons** are composed mostly of nickel-iron and are of correspondingly high density (over 7 g/cm^3). Pure iron is not found on Earth, making it clear that these meteorites are extraterrestrial in origin. The **stones** are meteorites that more closely resemble terrestrial rock, with densities around 3 g/cm^3. The relatively rare **stony-irons** contain a mixture of rocks and metal, as the name suggests, and have intermediate density. Most meteorites (over 95 percent) are stones. However, almost half of all meteorite finds are irons or stony-irons, simply because these types look unlike rocks native to the Earth and so are more readily found. We can determine the chemical and geological history of meteorites from detailed scrutiny of their structure. Some **differentiated meteorites** show clear evidence of strong heating at some time in their past, indicating that they originated on a larger body that either experienced some geological activity or was partially melted during the collision that liberated the meteorite. The **primitive meteorites,** which show no such evidence, date back to the formation of the solar system. The primitive meteorites are all stones.

Figure 16.19 This blackened rock is the "Lost City meteorite," which was detected in 1970 by a network of cameras as it plunged through Earth's atmosphere. It was recovered in Oklahoma 6 days later. (The tape measures inches.) (NASA)

The most primitive meteorites of all are the **carbonaceous meteorites,** so called because of their relatively high carbon content. They are black or dark gray, and they may well be related to the C-type asteroids that populate the outer asteroid belt. (Similarly, silicate-rich stones are associated with the inner, S-type asteroids.) Many, but not all, carbonaceous meteorites contain **chondrules,** which are rocky spheres embedded in their interiors. They are known as **carbonaceous chondrites** and are believed to be the most primitive of all known meteorites. Carbonaceous chondrites often contain significant amounts of ice and other volatile substances, and they are usually rich in organic molecules. Figure 16.20 compares a carbonaceous chondrite and a typical iron meteorite.

Meteoritic Ages

Finally, almost all meteorites are *old.* Direct radioactive dating shows most of them to be between 4.4 and 4.6 billion years old—roughly the age of the oldest lunar rocks. The primitive meteorites are generally older than the differentiated meteorites, by about 100 million years. Consequently, meteorites, along with some lunar rocks, comets, and perhaps the planet Pluto, provide important clues about the original state of matter in the solar neighborhood.

Actually, some meteorites found in Antarctica may well have come from the Moon and even Mars. At least one meteorite is identical to rocks returned by the *Apollo* astronauts from the lunar highlands. It was probably chipped off the face of the Moon by a giant meteor that struck with such force that surface material was hurled out into space. This material fell toward Earth and finally landed as a meteorite in the Antarctic ice. Another class of meteorites, very few in number, are distinctly younger than all the rest. Only about 1.3 billion years old (compared with the usual 4.6 billion years), their composition appears to be similar to the crust of Mars. Accordingly, some researchers speculate that a small fraction of all meteorites might well have come from Mars, providing us with a direct means of examining part of that planet's surface.

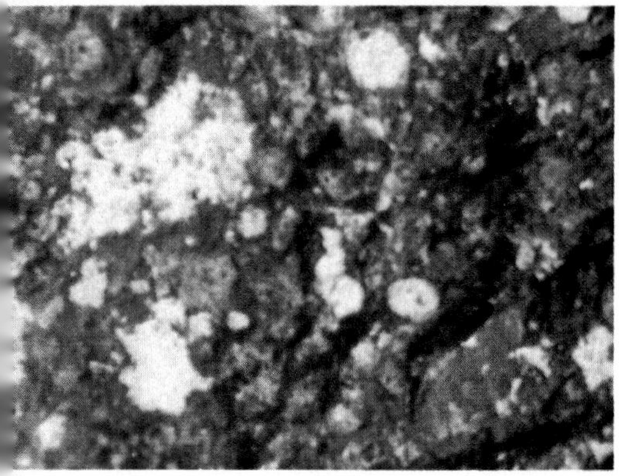

(a)

(b)

Figure 16.20 Exposed slabs of a carbonaceous meteorite (a) and an iron meteorite (b). Note the dark, charcoal-like color and texture of the carbonaceous meteorite and the interlocking crisscross pattern in the iron, caused by intersecting regions of different metals within the meteorite. The pattern is made visible by polishing the exposed surface, then etching it with dilute nitric acid. Meteorites do not generally look so clean and shiny! (Harvard-Smithsonian Center for Astrophysics)

CHAPTER SUMMARY

In addition to the nine known planets, the solar system contains perhaps billions of smaller bodies. Many asteroids and comets are pristine, containing clues to the origin of the solar system.

The word asteroid means "starlike," but asteroids are really small chunks of rock orbiting the Sun. More than 4000 asteroids have been cataloged. Their combined mass is less than 1⁄10 the mass of the Moon. Most asteroids orbit in the asteroid belt between Mars and Jupiter. They are probably primal rocks that never clumped together to form a planet. Some asteroids cross the orbit of Earth. Many asteroids are categorized as belonging to different "families," with each family moving in very similar orbits. Collisions may have created some binary asteroids. Trojan asteroids orbit at the distance of Jupiter, exactly 60° ahead of or behind the planet. The Kirkwood gaps in the main asteroid belt have been cleared by Jupiter's gravity. Asteroids are generally classified according to the properties of their reflected light. Brighter S-type (silicate) asteroids dominate the inner asteroid belt, while darker C-type (carbonaceous) asteroids are more plentiful in the outer region.

Comets are chunks of ice that grow long wispy tails when they approach the Sun. They can swing close by the Sun before heading again toward the far outer reaches of the solar system. Their orbits come in all inclinations and orientations. It is generally believed that a vast reservoir of comets, called the Oort Cloud, completely surrounds our solar system. Comets can be classified by their tails. Type I tails are approximately straight and glow in the light of ionized molecules. Type II tails are rich in microscopic dust particles that reflect sunlight. They are usually broad, diffuse, and gently curved, and they can be millions of miles long. Many comets have both kinds of tails. The core, or nucleus, of a comet may be only a few kilometers in diameter. This solid nucleus may be surrounded by a tenuous coma, or halo, of dust and evaporated gas. Comets are icy, dusty bodies, sometimes called "dirty snowballs."

Meteors, or "shooting stars," are bright streaks of light seen in Earth's upper atmosphere. Meteoroids are the same objects before they encounter the atmosphere. Meteorites are the same objects when they strike the ground. The major difference between meteoroids and asteroids is their size. The dividing line between them is sometimes taken to be around 100 m. Comet fragments form a meteoroid swarm, which orbits along a path similar to that of the parent comet. When Earth crosses the orbit of such a swarm, we see a meteor shower. There are a dozen or so major meteor showers every year. No meteor associated with an annual shower has ever been known to reach the ground.

Comets and stray asteroids are responsible for most of the cratering on the various worlds in the solar system. Earth is still subject to these sorts of collisions. The most recent one occurred in 1908, when a comet apparently exploded several miles above Siberia. Several thousand meteorites have been found, either accidentally or after an observed fall. Antarctica is a good place to find meteorites because subfreezing temperatures and snow protect them from erosion and contamination. Meteorites are classified into three major types: irons, stones, and stony-irons. Differentiated meteorites show evidence of heating, but primitive meteorites do not. Most meteorites are between 4.4 and 4.6 billion years old. They can therefore provide clues about matter in the parent cloud of gas from which our solar system formed. Some meteorites found in Antarctica may have been chipped off the Moon and Mars.

KEY WORDS

Amor asteroid
Apollo asteroid
asteroid
asteroid belt
binary asteroid
carbonaceous asteroid
carbonaceous chondrite
carbonaceous meteorite
chondrule
coma
comet
C-type asteroid
differentiated meteorite
dust tail
fall
find
Hirayama family
hydrogen envelope
ion tail
iron meteorite
Kirkwood gaps
meteor
meteor shower
meteorite
meteoroid
meteoroid swarm
micrometeoroid
nucleus
Oort cloud
plasma tail
primitive meteorite
radiant
stone meteorite
stony-iron meteorite
S-type asteroid
Sun-grazing comet
Trojan asteroid
type I tail
type II tail

REVIEW QUESTIONS

1. What clues about the origin of asteroids may be taken from their masses, compositions, and orbits?

2. What is the evidence of a high potential for collisions of asteroids with Earth?

3. What are Trojan asteroids?

4. Since asteroids are generally too small to be resolved by Earth-based telescopes, how do we know anything about their sizes, shapes, and compositions?

5. What are comets like when they're far from the Sun? What happens when they come into the inner solar system?

6. What are the two types of comet tails?

7. What are some possible fates of comets?

8. Explain the difference between a meteor, a meteoroid, and a meteorite.

9. What causes a meteor shower?

10. What are the various types of meteorites?

11. What are the most primitive meteorites?

12. What have meteorites revealed about our solar system?

13. The largest asteroid, Ceres, has a radius 0.073 times the radius of the Earth and a mass of 0.0002 Earth masses. How much would a 100 kg astronaut weigh on Ceres?

14. (a) Using Kepler's Laws of Planetary Motion (see Chapter 3), calculate the orbital period of a comet with a perihelion distance of 0.5 A.U. and aphelion in the Oort cloud, at a distance of 50,000 A.U. from the Sun. (b) A short-period comet has a perihelion of 1 A.U. and an orbit period of 125 yr. What is its maximum distance from the Sun?

15. A particular comet has a total mass of 10^{16} g, 95 percent of which is ice and dust. The remaining 5 percent is in the form of rocky fragments of average mass 1 kg. How many meteoroids would you expect to find in the swarm formed by the breakup of this comet?

DISCUSSION QUESTIONS

1. Asteroids were called ''the vermin of the skies'' by astronomers a century ago. Yet in the late 1970s, they were widely discussed as having a practical use. Look in the library for articles about the possible uses of asteroids, and write a report on the subject. What are your thoughts on the usefulness of asteroids as resources?

2. Explain why comets can approach the Sun from any direction, while asteroids orbit generally in the plane of the solar system. Since asteroids and comets are both part of the Sun's family, what might this difference tell us about the way in which the solar system formed?

3. Why do meteorites contain information about the early solar system, yet Earth does not?

4. The United States did not send its own spacecraft to Halley's Comet. Do you think we should have? Why or why not?

5. Write a report about the Tunguska event. What is the evidence that it was caused by the collision of a small comet with Earth?

PROJECTS

1. Asteroids came to be called by a name that means ''starlike'' because, through earthly telescopes or binoculars, they look exactly like stars. The only way to tell an asteroid from a star is to watch it over several nights.

Then you can easily detect its movement in front of the star background. The astronomy magazines *Sky and Telescope* and *Astronomy* often publish charts for especially prominent asteroids. Use the chart to locate the appropriate star field. Aim a telescope or binoculars at that location in the sky, and make a rough drawing of the entire field. Come back a night or two later, and look again. The "star" that has moved is the asteroid. (Hint: Asteroids move more quickly in front of the stars than the planet Pluto, which was described in the projects for Chapter 15. They are often much brighter than Pluto! This makes them easier to see than Pluto, if you watch for several nights.)

Why do asteroids move apart from the stars?

Why do asteroids exhibit a faster motion in front of the stars than Pluto?

2. Although a spectacular naked-eye comet comes along only about once a decade, fainter comets can be seen with binoculars and telescopes in the course of every year. *Sky and Telescope* often runs a "Comet Digest" column announcing the whereabouts of comets. A comprehensive list of periodic comets expected to return in a given year can be found in Guy Ottewell's *Astronomical Calendar*. This calendar contains a wealth of other sky information as well, including monthly star charts. At this writing, it costs $15 a year and can be purchased from:

Astronomical Workshop
Furman University
Greenville, South Carolina 29613
(803) 294-2208

3. There are a number of major meteor showers every year, but, if you plan to watch one, be sure to notice the phase of the Moon. Bright moonlight or city lights can obliterate a meteor shower. A common misconception about meteor watching is that most meteors are seen in the direction of the shower's radiant point. It's true that if you trace the paths of the meteors backward in the sky, they all can be seen to come from the radiant. But most meteors don't become visible until they are 20 or 30 degrees from the radiant. Meteors can appear in all parts of the sky! Just relax and let your eyes rove among the stars. You will generally see many more meteors in the hours before dawn than in the hours after sunset.

Why do you suppose meteors have different brightnesses?

Can you detect their variety of colors?

Watch for meteors that appear to "explode" as they fall.

Watch for ghostly vapor trails that linger after the meteor itself has disappeared.

SUGGESTED READINGS

Alvarez, W., and F. Asaro. "What Caused the Mass Extinction? An Extraterrestrial Impact." *Scientific American* (October 1990). Debate between those who say it was caused by an extraterrestrial impact and those who claim the cause was a period of extreme Earth vulcanism.

Balsiger, H., H. Fechtig, and J. Geiss. "A Close Look at Halley's Comet." *Scientific American* (September 1988). Data sent by European, Japanese, and Soviet probes to Halley's Comet.

———. "Killer Crater in the Yucatan?" *Sky and Telescope* (July 1991). On the possibility that a crater in the Yucatan is the long-sought site of the impact that triggered the demise of the dinosaurs.

Brandt, J. C., and M. B. Niedner, Jr. "The Structure of Comet Tails." *Scientific American* (January 1986). Interrelations of the plasma tail of Halley's Comet with the solar wind.

Chapman, R., and J. Brandt. *The Comet Book*. Boston: Jones and Bartlett, 1984. Popular review of comet history and science from two leaders in the field.

Delsemme, A. H. "Whence Come Comets?" *Sky and Telescope* (March 1989). The mysterious origin of comets.

Dodd, R. T. *Thunderstones and Shooting Stars*. Cambridge, MA: Harvard University Press, 1986. All about meteorites: types, origins, and the evidence they bring of conditions beyond Earth.

Grieve, R. A. F. "Impact Cratering on the Earth." *Scientific American* (April 1990). What happens when a large meteoroid strikes Earth.

Levy, D. H. "How to Discover a Comet." *Astronomy* (December 1987). Advice from a successful comet hunter.

Morrison, D., and C. R. Chapman. "Target Earth: It Will Happen." *Sky and Telescope* (March 1990). On the realization that comets and small asteroids still occasionally collide with the Earth.

Spratt, C. E. "On the Trail of a Meteorite." *Astronomy* (August 1989). Excellent article on hunting for meteorites.

Weissman, P. "Are Periodic Bombardments Real?" *Sky and Telescope* (March 1990). Article discussing the idea that Earth is regularly bombarded by comets, which cause periodic mass extinctions.

Whipple, F. L. "The Black Heart of Comet Halley." *Sky and Telescope* (March 1987). The originator of the "dirty snowball" theory discusses Halley's nucleus.

———. *The Mystery of Comets*. Washington, D.C.: Smithsonian Institution Press, 1985. A comet specialist since the 1930s tells an insider's tale of a half-century of cometary science.

CHAPTER

17

The Formation of the Solar System

The Birth of Our World

This is an artist's conception drawn to illustrate the peculiar star Beta Pictoris and its surrounding environment. The star was long believed to have a disk around it, in which new planets might be forming. But new spectroscopic observations with the Hubble *telescope suggest that this illustration is more the case. Here, we see a stable disk with huge, comet-like objects falling in toward the central star. (Courtesy D. Berry)*

LEARNING GOALS

1 to appreciate the architecture of our solar system

2 to realize the role played by dust in the currently accepted model of solar system formation

3 to understand how planets form as natural by-products of star formation

4 to understand the reasons for the differences between the terrestrial and the Jovian planets

Completing these chapters on the planets, you might be struck by the vast range of physical and chemical properties found in the solar system. The planets present a long list of interesting features and bizarre peculiarities, and the list grows even longer when we consider the characteristics of their moons. Every object has its idiosyncrasies, some of them due to particular circumstances, many others the result of planetary evolution. Each time a new discovery is made, we learn a little more about the properties and history of our planetary system. Our astronomical neighborhood may seem more like a great junkyard than a smoothly running planetary system. Can we really make any sense of the entire collection of solar system matter? Is there some underlying principle that unifies the knowledge we have gained? The answer, as we will see, is yes.

It is part of the job of the planetary scientist to distinguish between those properties of the solar system that are inherent—that is, that were imposed at formation—and those that must have evolved since the solar system formed. In this chapter, we will draw together all the planetary data we have amassed and show how the regularities—and the irregularities—of the solar system can be explained by a single comprehensive theory.

MODELING THE ORIGIN OF THE SOLAR SYSTEM

✓1 The origin of the planets and their moons is a complex and as yet incompletely solved puzzle, although the basic outline of the process is by now fairly well understood. Most of our knowledge of the solar system's formative stages has emerged from studies of interstellar gas clouds, fallen meteorites, and Earth's Moon, as well as from the various planets observed with ground-based telescopes and planetary space probes. Ironically, studies of Earth itself do not help much because information about our planet's early stages eroded away long ago. Meteorites provide perhaps the most useful information, for nearly all have preserved within them traces of the solid and gaseous matter from the earliest solar system.

Astronomers still have no firm evidence of the existence of planets anywhere beyond our solar system. For that reason, we must concentrate in this chapter on the origin of the planetary system in which we live. Bear in mind, however, that no part of the scenario we will describe below is in any way unique to our own system. The same basic processes could have occurred and (many astronomers believe) probably did occur during the formative stages of most of the stars in our Galaxy.

Model Requirements

Any theory purporting to explain the origin and architecture of our planetary system must adhere to the known facts. We know of nine outstanding properties of our solar system as a whole. They can be summarized as follows.

1. *Each planet is relatively isolated in space.* The planets exist as independent entities at progressively larger distances from the central Sun; they are not bunched together. In very rough terms, each planet tends to be twice as far from the Sun as its next inward neighbor.

Although the Titius-Bode rule does not have the force of law, it certainly describes a general planetary configuration that must have been established by the forces that shaped the early solar system.

2. *The orbits of the planets are nearly circular.* In fact, with the exception of Mercury and Pluto, each planetary orbit closely describes a perfect circle. The slight orbital eccentricity of Mercury, the innermost planet, is perhaps related to the influence of the Sun's intense gravity. As for the outermost Pluto, some researchers regard this planet as a cometlike body or an escaped moon of Neptune, thereby accounting for its large orbital eccentricity (see Figure 15.24). But this is only conjecture. In any case, there is sufficient cause to believe that the innermost and outermost planets have experienced "special circumstances" causing them to depart from accurately circular motion. Thus, we will explain why planetary orbits are nearly circular, then argue that Mercury and Pluto are special cases.
3. *The orbits of the planets all lie in nearly the same plane.* The planes swept out by the planets' orbits are accurately aligned to within a few degrees. Put another way, the solar system has the shape of a very thin disk. Again, Mercury and Pluto are slight exceptions, probably for the same reasons as noted above.
4. *The direction of the planets' revolution in their orbits about the Sun (counterclockwise as viewed from Earth's north) is the same as the Sun's rotation on its axis.* Virtually all the large-scale motions in the solar system—such as the planets' orbits and the Sun's spin—are in the same plane and in the same sense. The plane is that of the Sun's equator, and the sense is that of the Sun's rotation.
5. *The direction in which most planets rotate on their axes also mimics that of the Sun's spin.* This property is a little less general, as three planets (Venus, Uranus, and Pluto) do not share it. Venus's slow spin is opposite (retrograde) to that of the Sun and the other planets. Uranus apparently lies on its side, with its poles almost in the plane of its orbit. The Pluto-Charon system also rotates almost on its side. Pluto, as we have seen, is an exception to almost all our rules. Theorists usually cite catastrophic events occurring early in the solar system's history as the reasons for the anomalous spins of Venus and Uranus.
6. *Most of the known moons revolve about their parent planets in the same direction that the planets rotate on their axes.* Some moons, like those of Jupiter, resemble miniature solar systems, revolving about their parent planet in roughly the same plane as the planet's equator. This again suggests uniformity in our planetary system. In the Jovian moon systems, revolution in the equatorial plane is probably forced by the planets' oblateness. However, the prograde motion is not forced but must have been determined by the environment in which the planets and moons formed. Again, the few retrograde moons in the solar system must be explained as special cases.
7. *Our planetary system is highly differentiated.* The inner terrestrial planets are characterized by high densities, moderate atmospheres, slow rotation rates, and few or no moons. By contrast, the outer Jovian planets (as usual, except Pluto) have low densities, thick atmospheres, rapid rotation rates, and many moons.
8. *The asteroids are very old and exhibit a range of properties not characteristic of either the inner or the outer planets or their moons.* As we have seen, the asteroid belt appears to be made of primitive, unevolved material, and the meteorites that strike the Earth are the oldest rocks known.
9. *The comets are primitive, icy fragments that do not orbit in the ecliptic plane and probably reside primarily at large distances from the Sun.* The probable existence of the Oort Cloud, surrounding the Sun at tens of thousands of astronomical units, is thus directly related to the processes that formed the solar system long ago.

All these observed facts, when taken together, strongly suggest a high degree of order within our solar system. The whole system is not a random assortment of objects spinning or orbiting this way or that. Consequently, it hardly seems possible that our solar system could have formed by the slow accumulation of already-made interstellar "planets" casually captured by our Sun over the course of billions of years. The overall architecture of our solar system is too neat, and the ages of its members too uniform, to be the result of random chaotic events. The overall organization points toward a single formation, an ancient but one-time event, 4.6 billion years ago. A convincing theory that explains all of the nine features listed above has been a goal of astronomers for many centuries.

Additional Considerations

It is important to recognize what our theory of the solar system does *not* have to explain. There is plenty of scope for planets to evolve after their formation, so circumstances that have developed since the initial state of the solar system was established need not be included in our list. Examples are Mercury's 3:2 spin-orbit coupling, Venus's runaway greenhouse effect, the Moon's synchronous rotation, the emergence of life on Earth and its absence on Mars, the Kirkwood gaps in the asteroid belt, and the rings and atmospheric appearance of the outer planets. There are many more. Indeed, all of the properties of the planets for which we have already provided an *evolutionary* explanation need not be included as items that our theory must account for at the outset.

In addition to its many regularities, our solar system also has many notable *irregularities,* some of which we

INTERLUDE 17-1 *The Concept of Angular Momentum*

It seems that most celestial objects rotate. Planets, moons, stars, galaxies—virtually all have some *angular momentum,* which we can define as the tendency of a body to keep spinning or moving in a circle. Angular momentum is as basic to any object as its mass or its energy.

Consider first a simpler motion—*linear momentum,* which is the tendency of an object to keep moving in a straight line in the absence of external forces. Consider a truck and a bicycle rolling equally fast down a street. Each has some linear momentum, but you would obviously find it easier to halt the less massive bicycle. Although the two vehicles have the same speed, the truck has more momentum. We see that the linear momentum of an object depends on the mass of that object. It also depends on the velocity. If two bicycles were rolling down the street at different speeds, the slower one could be stopped more easily. Linear momentum is defined as the product of mass and velocity:

$$\textit{linear momentum} = \textit{mass} \times \textit{velocity}$$

Angular momentum relates in an analogous fashion to objects having some rotation or revolution. However, in addition to mass and velocity, the angular momentum also depends on the size of the object:

$$\textit{angular momentum} = \textit{mass} \times \textit{velocity} \times \textit{size}$$

Notice that although the ''velocity'' for linear momentum is velocity in a straight line, for angular momentum ''velocity'' refers to the circular velocity of spinning or orbital motion.

Both of these types of momentum—linear and angular—must be conserved at all times. In other words, both linear and angular momentum must remain constant before, during, and after a physical change in any object (again, as long as no external forces act). For example, if a spherical object having some spin begins to contract, the relationship above demands that it spin faster. After all, the object's mass does not change during the contraction, yet the size of the object clearly decreases. The circular velocity of the spinning object must therefore increase in order to keep the total angular momentum unchanged.

Figure skaters use the principle of angular-momentum conservation. They spin faster by drawing in their arms and slow down by extending them. Here, the mass of the human body remains the same, but its lateral size changes, causing the body's circular velocity to change in order to conserve angular momentum.

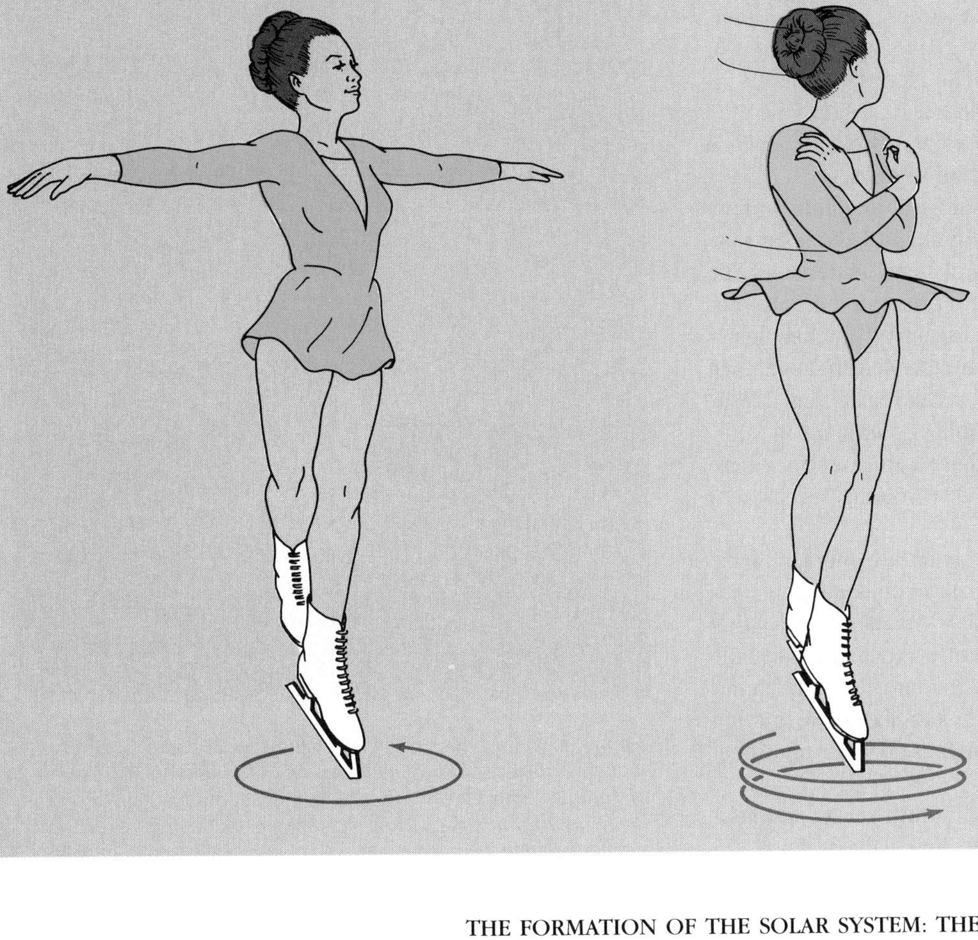

have already mentioned. Far from threatening our theory, however, these irregularities are important facts for us to consider in shaping our explanations. For example, it is necessary that the explanation for the solar system not insist that *all* planets rotate in the same sense or have *only* prograde moons, because that is not what we observe. Instead, the theory of the solar system should provide strong reasons for the observed planetary characteristics, yet be flexible enough to allow for and explain the deviations, too. And, of course, the existence of the asteroids and comets that tell us so much about our past must be an integral part of the picture. That's quite a tall order, yet many researchers now believe that we are close to that level of understanding.

NEBULAR CONTRACTION

Laplace's Nebular Theory

One of the earliest heliocentric models of solar system formation is termed the **nebular theory.** It is often attributed to the eighteenth-century German philosopher Immanuel Kant, but he merely elaborated upon a proposal made a century earlier by the French philosopher René Descartes, in 1644. In this model, a large cloud of interstellar gas began to collapse under its own gravity. As it contracted, it became denser and hotter, eventually forming a star—the Sun—at its center. While all this was going on, the outer, cooler, parts of the cloud formed a giant swirling region of matter, creating the planets and their moons essentially as by-products of the star-formation process. We have called this swirling mass "the primitive solar system," but it is more usually referred to as the **solar nebula.**

The nebular theory is an example of an **evolutionary theory,** which describes the development of the solar system as a series of gradual and natural steps, understandable in terms of well-established physical principles. Evolutionary theories may be contrasted with **catastrophic theories**—theories that invoke accidental or unlikely celestial events in order to interpret observations. Generally speaking, scientists do not like to invoke catastrophes to explain the universe. However, as we will see, there are instances where pure chance has played a role in determining the present state of the solar system.

In 1796 the French mathematician-astronomer Pierre Simon de Laplace tried to develop the nebular model in a quantitative way. He was able to show mathematically that the conservation of angular momentum (consult Interlude 17-1) demands that the contracting matter of an interstellar fragment spin faster. A decrease in the size of a rotating mass must be balanced by an increase in its rotational speed.

The increase in rotation speed in turn caused the nebula's *shape* to change as it collapsed. In Chapters 7 and 13, we saw how a spinning body tends to develop a bulge around its middle. The rapidly spinning nebula behaved in exactly this way. As shown in Figure 17.1, the fragment eventually flattened into a pancake-shaped primitive solar system. If we now suppose that planets formed out of this spinning material, we can already begin to understand the origin of some of the architecture observed in our planetary system today, such as the circularity of the planets' orbits and the fact that they move in nearly the same plane.

Laplace imagined that as the spinning nebula contracted, it left behind a series of concentric rings, each of which would eventually become a planet orbiting a central **protosun.** Each ring then clumped into a **protoplanet**—a forerunner of a genuine planet. Figure 17.2 is an artist's illustration of this scenario. In this diagram, several outer planets have already begun to develop, while the interior of the primitive solar system is still contracting to shape the inner planets and the central Sun.

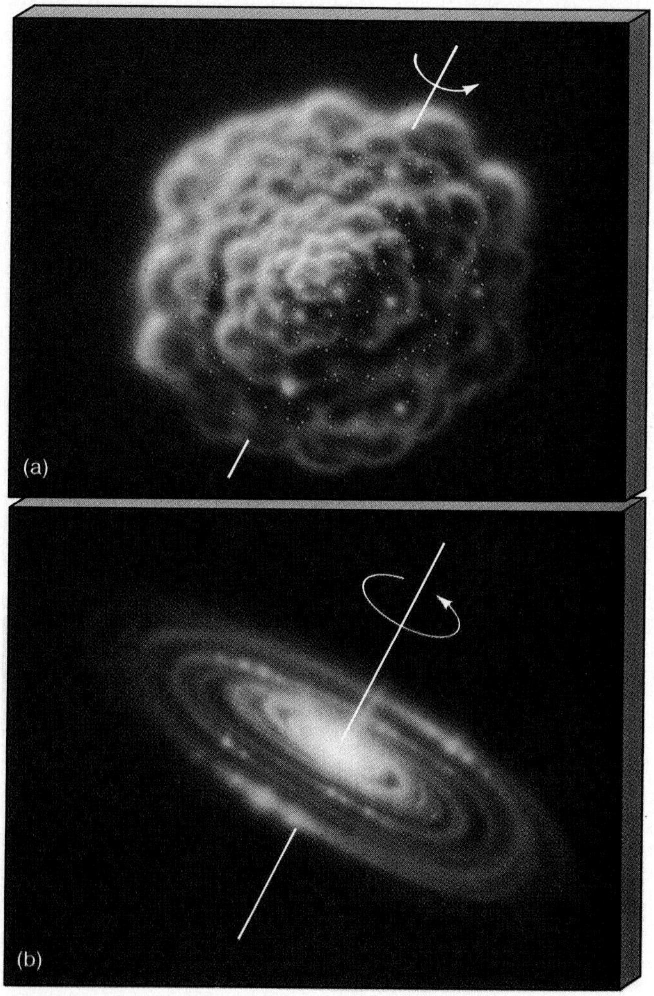

Figure 17.1 *Conservation of angular momentum demands that a contracting, rotating cloud (a) gradually speed its rate of rotation. Eventually (b), the primitive solar system resembled a giant pancake. The large blob at the center would ultimately become the Sun.*

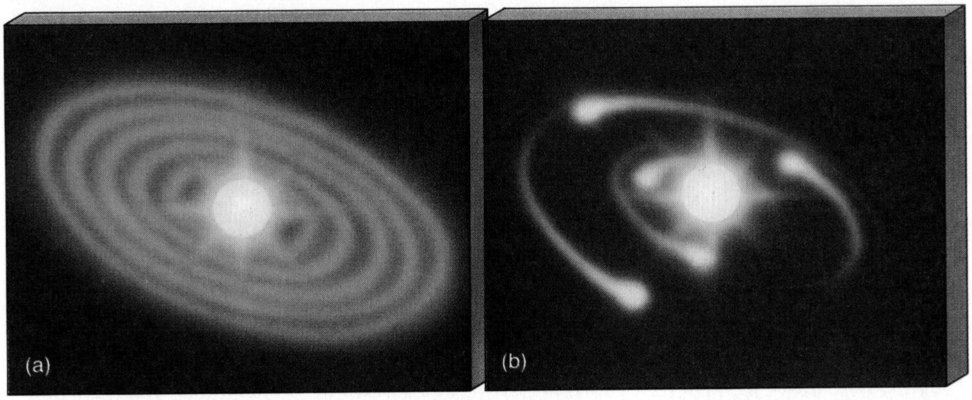

Figure 17.2 *Laplace's nebular theory envisioned the formation of (a) rings of gaseous matter at various distances from the central protosun. Eventually, (b) the rings clump into planets.*

Fatal Problems in the Nebular Theory

Our description of the changes experienced by a shrinking interstellar gas fragment is rather straightforward. It follows directly from the obedience of a parcel of gas to the known laws of physics. Two centuries ago, Laplace got the description of the collapse and flattening of the nebula essentially correct. Yet in modern times, as we use computers to study more subtle aspects of the problem, some fatal flaws have been found in the simple nebular theory.

The first difficulty stems from gas dynamics. Detailed computations show that a ring of the sort assumed in the theory would probably not form and would not necessarily accumulate to form a planet in any case. In fact, computer calculations predict just the opposite: The rings would tend to disperse. The protoplanetary matter is too warm, and, moreover, no one ring would have enough mass to bind its own matter into a ball.

A second difficulty with the nebular theory is known as the **angular momentum problem.** Although our Sun contains about 1000 times more mass than all the planets combined, it possesses a mere 0.3 percent of the total angular momentum of the solar system. Jupiter, for example, has a lot more angular momentum than does our Sun. Because of its large mass and great distance from the Sun, Jupiter possesses about 60 percent of the solar system's angular momentum. All told, the four Jovian planets account for well over 99 percent of the total angular momentum of the solar system. By comparison, the lighter (and closer) terrestrial planets have negligible angular momentum.

The problem here is that the nebular theory predicts that the Sun should command most of the solar system's angular momentum, precisely because it contains most of the mass. However, as we have just seen, the reverse is true. In fact, if all the planets, with their large amounts of orbital angular momentum, were placed inside the Sun, it would spin on its axis about 100 times as fast as it does at present. Instead of rotating roughly once a month, the Sun would spin once every few hours (which would be fast enough to deform the whole Sun into a thin disk). Thus, within the nebular theory, the slow spin rate of our Sun is a mystery. As we will see, this problem plagues not only the nebular theory but also other models proposed for the origin of our solar system.

A final difficulty that caused the nebular theory to lose support was the growing realization during the nineteenth century that there are exceptions to the ordered architecture demanded in the nebular theory's planetary system. The observation that did greatest damage to the theory at the time was the discovery of Triton's retrograde motion around Neptune. There was no known way for the nebular theory to explain such irregularity in the planetary system. As a result, other theories of the solar system began to gain ground.

CLOSE ENCOUNTERS

The Collision Hypothesis

The difficulties with the nebular theory caused researchers to consider alternatives. One competing model is the **collision theory** or, more popularly, the **close-encounter theory.** In this picture, a near collision (or close encounter) between our Sun and another passing star tore streamers of hot matter from the solar surface. Some of these streamers, the theory holds, remained gravitationally bound to our Sun and went into orbit around it (Figure 17.3). The matter cooled and eventually condensed into planets. If the encounter also produced the Sun's spin, this scheme neatly accounts for the alignment of the planets' orbital planes with the Sun's equator and the fact that all planets revolve in the same sense as the Sun rotates.

The French mathematician Georges Louis de Buffon proposed the close-encounter model in the eighteenth century. Enthusiasm for it grew about 100 years ago as the shortcomings in the nebular theory became evident. For a time, catastrophic models like the collision theory were preferred over the more ordered evolutionary nebular model. However, this ascendency was short-lived.

Fatal Problems in the Collision Theory

Although the collision theory has some qualitative points in its favor, it has serious shortcomings as well. Foremost among them is that a near collision between two stars is

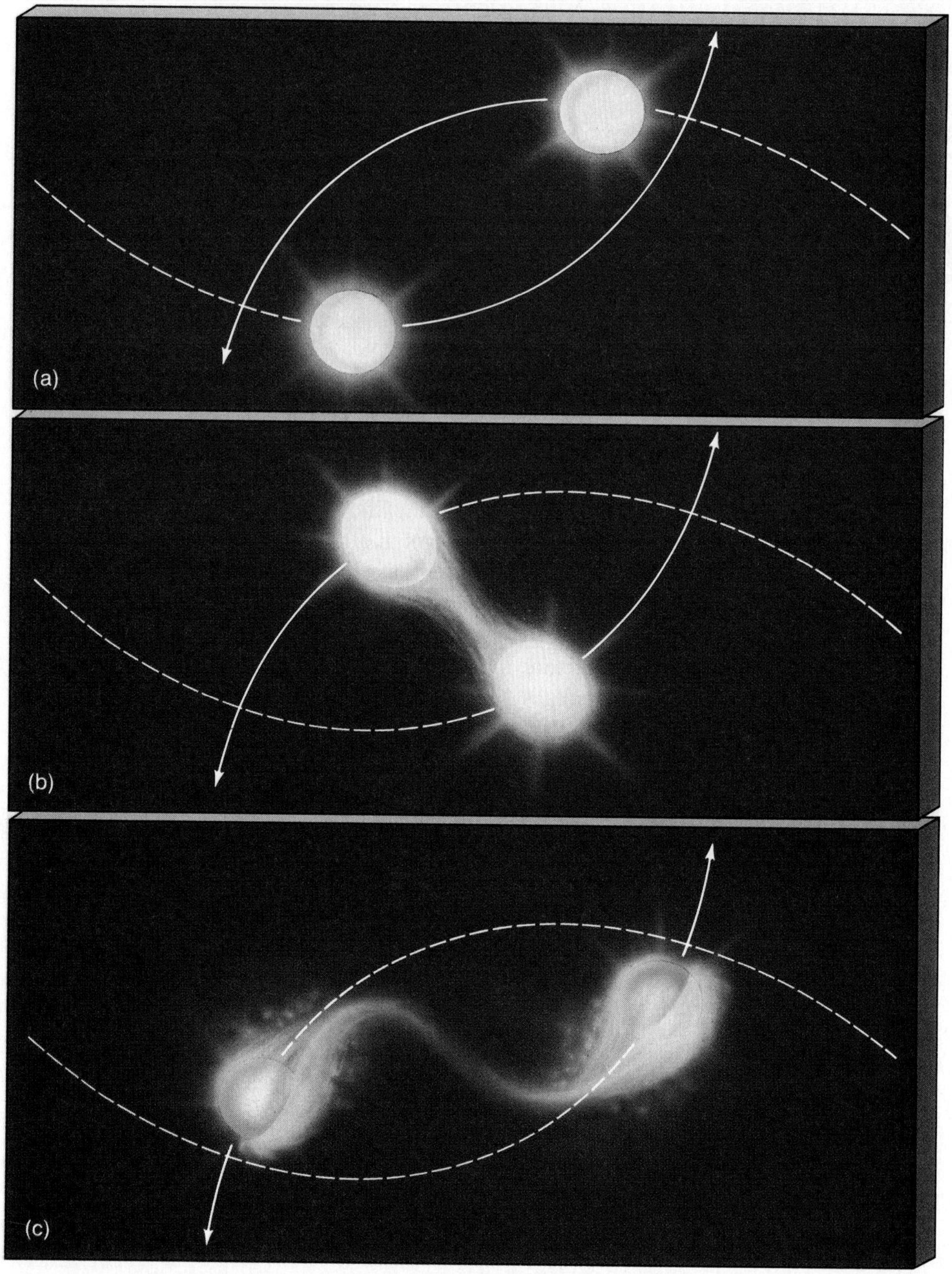

Figure 17.3 *A sufficiently close encounter between two stars would cause flaming matter to be torn from each object. According to the collision theory, some of these streamers eventually formed planets around each of the stars involved. The three frames show various stages in such a hypothetical near miss. (a) The two stars approach each other on an unbound (hyperbolic) orbit. (b) Tidal forces pull large streamers of gas from the surface of each star. Finally, (c) the stars continue on their separate paths through the Galaxy, and the streamers condense into planets.*

highly improbable. Stars are large by terrestrial standards, but still minute compared with the typical distances that separate them. For example, the Sun is about 1.4 million km in diameter, but the distance to the nearest star (Alpha Centauri) is about 4×10^{13} km, more than 25 million times greater. Probability theory suggests that, given the number of stars and their sizes, velocities, and typical separations, not more than a handful of such close encounters are likely throughout the entire expanse and history of the Milky Way Galaxy. Galaxy collisions are frequent (see Interlude 26-2 for more information), but stellar collisions are extremely rare.

The improbability of such collisions does not, of course, prove that the collision theory is wrong. After all, our solar system could be the foremost—even the only—example of this extraordinarily uncommon phenomenon. Should this theory be correct, we can justifiably conclude that our planetary system is a rare type of astronomical region. Very few stars would be expected to have planets, and therefore the chances for extraterrestrial life diminish accordingly.

However, besides the small chance of the collision itself, other difficulties confront the collision theory. First, it is hard to understand how hot solar gas torn from the Sun could condense into planets—hot gases usually disperse before they cool. Consequently, although such a near collision between two stars might occasionally happen, the hot fragments are very unlikely to form planets. Some of the hot streamers would surely fall back into the Sun. Others, because of their high temperatures, would disperse even more quickly than the cooler gases in the rings of the nebular theory. A second quandary concerns the nearly circular orbits of the planets. There is simply no way to explain how clumps of matter ripped from the Sun to form the planets should end up orbiting the central Sun in near-perfect circles. The collision theory cannot explain this observed fact even qualitatively.

Figure 17.4 *Interstellar gas and dark dust lanes mark this region of star formation. The dark cloud known as Barnard 86 (left) flanks a cluster of young blue stars called NGC 6520 (right). The former might be part of an interstellar cloud that gave rise to the latter. (Anglo-Australian Observatory)*

THE CONDENSATION THEORY

The model currently favored by most astronomers is really just a more sophisticated version of the nebular theory. Known as the **condensation theory,** it combines the good features of the old nebular theory with new information about interstellar chemistry to avoid most of the old theory's problems.

The Role of Dust

✓2 The key new ingredient in the modern picture is the presence of **interstellar dust** in the solar nebula. Astronomers now recognize that the space between the stars is strewn with microscopic dust grains, an accumulation of the ejected matter of many long-dead stars (see Chapter 22). These dust particles probably formed in the cool atmospheres of old stars, then grew by accumulating more atoms and molecules from the interstellar gas within the Milky Way Galaxy. The end result is that our entire Galaxy is littered with miniature chunks of icy and rocky matter having typical sizes of about 10^{-5} cm. Figure 17.4 shows one of many such dusty regions found in the vicinity of the Sun.

Dust grains play an important role in the evolution of any gas. Dust helps to cool warm matter by efficiently radiating its heat away, allowing it to collapse more easily. Furthermore, the dust grains greatly speed up the process of collecting enough atoms to form a planet. They act as **condensation nuclei**—microscopic platforms to which other atoms can attach, forming larger and larger balls of matter. (This is similar to the way that raindrops form in Earth's atmosphere; dust and soot in the air act as condensation nuclei around which water molecules cluster.) Put another way, the presence of dust virtually guarantees that gaseous matter will clump, both because the dust "seeds" it with sites for the condensation process to start and because it cools the matter below the point at which outward-pushing pressure (which is just proportional to the gas temperature) can effectively compete with inward-pulling gravity.

Modern models trace the formative stages of our solar system along the following broad lines. Imagine a dusty interstellar cloud fragment measuring about a light-year across. Intermingled with the preponderance of hydrogen and helium atoms, the cloud harbors some heavy-element gas and dust. Some external influence, such as the passage of another interstellar cloud or perhaps the concussion from a nearby exploding star (a *supernova*—see Chapter 23), start the fragment contracting, down to a size of about 100 A.U. At this time turbulent gas eddies form at various locations throughout the primitive solar system. As the cloud collapses, it rotates faster and begins to flatten (just as described in the old nebular theory). By the time it has contracted to 100 A.U., the solar nebula has already formed an extended, rotating disk.

Astronomers are fairly confident that our own solar nebula formed such a disk because similar disks have been observed (or inferred) around other stars. Figure 17.5 shows a visible-light image of the region around a star known as Beta Pictoris, lying about 15 pc from the Sun. When the light from the star itself is suppressed and the resulting image enhanced by a computer, a faint disk of warm matter (viewed almost edge-on here) can be seen. This particular disk is roughly 1000 A.U. across—about 10 times the diameter of Pluto's orbit. Astronomers believe that Beta Pictoris is a very young star, perhaps only 100 million years old, and that we are witnessing it pass through an evolutionary stage similar to that our own Sun experienced some 4.6 billion years ago.

Unlike the nebular theory sketched above, the condensation theory predicts no rings. As depicted in Figure 17.6, turbulent gas eddies would have naturally appeared,

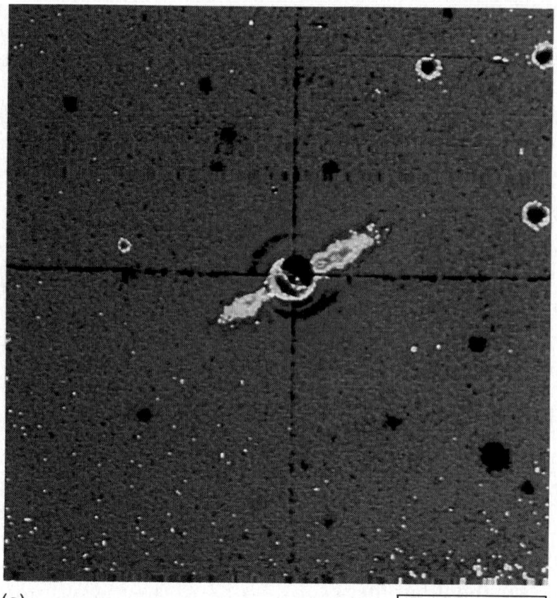

Figure 17.5 (a) A computer-enhanced photograph (taken from Las Campanas Observatory in Chile) of a disk of warm matter surrounding the star Beta Pictoris. Most of the light from the star itself is blocked by an instrument called a coronagraph, designed to detect faint halos around bright objects. The full extent of the disk, seen almost edge-on here, is about 1000 A.U. (b) An artist's conception of the disk of clumped matter. (NASA; D. Berry)

disappeared, and often reappeared at various places throughout the primitive, rotating solar system.

Recall that a major problem with the nebular theory was the inability of the loose, ringed matter to cluster into a tight-knit protoplanet. The gas in each of the rings possessed too little mass and too much heat to initiate gravitational contraction. Without the dust, a similar problem would occur in the condensation model. Although eddies would continually form and swirl, the chance of one of them sweeping up enough mass to overcome the gas pressure and begin to contract into a planet is exceedingly small—so small, in fact, that planets would not have formed in the time available. However, by introducing dust into our interstellar cloud, we can be sure that dust-grain cooling and condensation occurred before the matter had a chance to disperse. This way the solar nebula could contract into planets, instead of dispersing into interstellar space.

Accretion and Fragmentation

√3 The process of planet formation took place in three stages. Early on in the solar nebula, dust grains formed condensation nuclei around which small clumps of matter began to grow. This vital step greatly hastened the critical process of forming the first small clumps of matter. Once these clumps formed, they grew quite rapidly by sticking to other clumps or dust grains they encountered. (Imagine a snowball thrown through a fierce snowstorm, growing bigger as it encounters more snowflakes.) As the clumps grew larger, their surface areas increased, and the rate at which they accumulated new material accelerated. They gradually grew into objects of pebble size, baseball size, basketball size, and larger.

Eventually, this process of **accretion**—this gradual growth by collision and sticking—created objects a few hundred kilometers across. By that time, their gravity was strong enough to sweep up material that would otherwise not have collided with them, and their rate of growth became faster still. At the end of this first stage, the solar system was made up of hydrogen and helium gas and millions of **planetesimals**—objects the size of small moons, whose gravitational fields were just strong enough to affect their neighbors.

In the second phase of the accretion process, gravitational forces between the planetesimals caused them to collide and merge, again forming larger and larger objects. Because larger objects have stronger gravity, the rich became richer in the early solar system, and eventually almost all the planestesimal material was swept up into a few large protoplanets—the accumulations of matter that would eventually evolve into the planets we know today. Figure 17.7 shows a computer simulation of accretion in the inner solar system. Notice how, as the number of bodies decreases, the orbits of the remainder become more widely spaced and more nearly circular.

The four largest protoplanets became large enough that they were able to enter a third phase of planetary development, sweeping up large amounts of gas from the solar nebula to form what would ultimately become the Jovian planets. The smaller, inner protoplanets never reached that point, so their masses remained relatively modest.

There is an alternative explanation for the significant difference in size between the Jovian and the terrestrial planets. The eddies that arose in the cool outer solar nebula may have been large enough for protoplanets to form directly, skipping the initial accretion stage. These first protoplanets had gravitational fields strong enough to scoop up more of the remaining gas and dust in the solar nebula, allowing them to grow into the gas giants we see today. Their large size reflects the "head start" they obtained in

Figure 17.6 Planetesimals and planets forming in a collapsing, dusty cloud. Dust grains act as condensation nuclei, forming clumps of matter that collide, stick, and grow. Large eddies also come and go as the gas in the disk swirls around. According to the condensation theory, the larger eddies and clumps of matter will eventually become planets. The large blob in the center will become the Sun.

the accretion process. The smaller terrestrial planets did not form directly in this way, however, but had to go through the stages described above.

As the protoplanets grew, another process became important. The strong gravitational fields produced many high-speed collisions between planetesimals and protoplanets. These collisions led to **fragmentation,** as small objects broke into still smaller chunks, which were then swept up by the protoplanets. Not only did the rich get richer, but the poor were mostly driven to destruction!

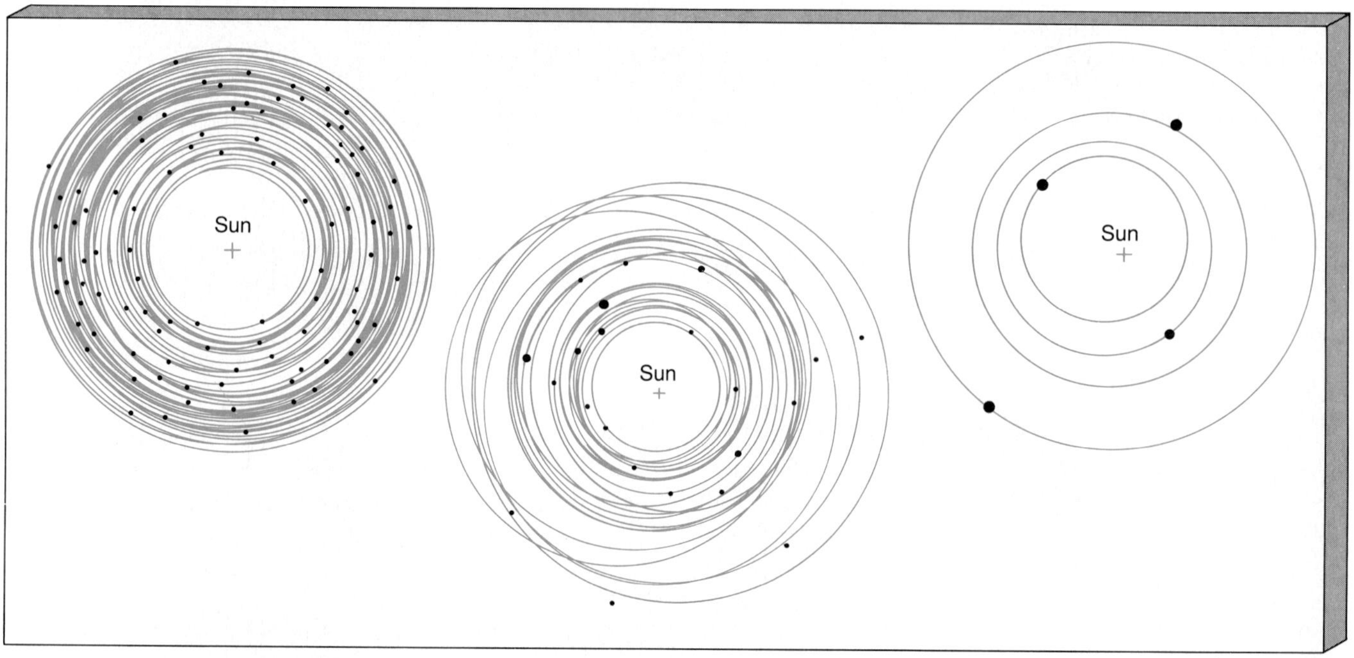

Figure 17.7 *Accretion in the inner solar system: Initially, many moon-sized planetesimals orbited the Sun. Over the course of a hundred million years or so, they gradually collided and coalesced, forming a few large planets in roughly circular orbits.*

Some of these fragments produced the intense meteoritic bombardment we know occurred during the early evolution of the planets and moons, as we have seen in the last few chapters. Only a relatively small number of 10–100 km fragments escaped capture by a planet or a moon and became the asteroids and comets. Assuming that the accretion process was reasonably efficient throughout the disk, we can thus understand how our present solar system has come to exist as a collection of rather small planets orbiting throughout an otherwise empty region of space.

Many of the moons of the planets (but not our own—see Chapter 8) presumably also formed through accretion but on a smaller scale, in the gravitational field of their parent planets. Once the nebular gas began to accrete onto the large Jovian protoplanets, conditions probably resembled a miniature solar nebula, with condensation and accretion continuing to occur. The large moons of the outer planets almost certainly formed in this way. Some of the smaller moons may have been "chipped off" their parent planets during collisions with asteroids; others may be captured asteroids themselves.

Mathematical modeling, like the computer simulation shown in Figure 17.7, indicates that, after about 100 million years, the primitive solar system had evolved into nine protoplanets, dozens of protomoons, and the big protosolar mass at the center. Computer simulations generally reproduce the increasing spacing between the planets ("Bode's Law"), although the reasons for the regularity seen in the actual planetary spacing remain unclear. Roughly a billion years more were required to sweep the system reasonably clear of interplanetary trash. This was the billion-year period that saw the heaviest meteoritic bombardment, tapering off as the number of planetesimals decreased.

THE DIFFERENTIATION OF THE SOLAR SYSTEM

✓4 We can understand the basic differences in content and structure between the terrestrial and the Jovian planets using the condensation theory of the solar system's origin. Indeed, it is in this context that the adjective *condensation* derives its true meaning. To see why the planets' composition depends on location in the solar system, it is necessary to consider the temperature structure of the solar nebula.

The Role of Heat

As the primitive solar system contracted under the influence of gravity, it heated up as it flattened into a disk. The density and temperature were greatest near the central protosun and much lower in the outlying regions. Detailed calculations indicate that the gas temperature near the core of the contracting system was several thousand kelvins. At a distance of 10 A.U., out where Saturn now resides, the temperature was only about 100 K.

In the warmer regions of the cloud, dust grains broke apart into molecules, and they in turn split into excited atoms. Because the extent to which the dust was destroyed depended on the temperature, it also depended on location in the solar nebula. Most of the original dust in the inner solar system disappeared at this stage, but the grains in the outermost parts probably remained largely intact.

The destruction of the dust in the hot inner portion of the solar nebula introduced an important new ingredient into the theoretical mix, one that we omitted from our earlier account of the accretion process. With the passage of time, the temperature decreased at all locations, except in the very core, where the Sun was forming. Everywhere beyond the protosun, new dust grains began to condense (or crystallize) from their hotter gas phase to their cooler solid phase, much as raindrops, snowflakes, and hailstones condense from moist, cooling air here on Earth. It may seem strange that although there was plenty of interstellar dust early on, it was mostly destroyed, only to form again later. However, a critical change had occurred. Initially, the nebular gas was uniformly peppered with dust grains. When the dust reformed later, the distribution of grains was very different.

Figure 17.8 plots the temperature gradient across the primitive solar system just prior to the onset of the accretion stage. At any given location, the only materials to condense out were those able to survive the temperature there. As marked on the figure, in the innermost regions, around Mercury's present orbit, only metallic grains could form. It was simply too hot for anything else to exist. A little farther out, at about 1 A.U., it was possible for rocky, silicate grains to form, too. Beyond about 3 or 4 A.U., water ice could exist, and so on, with the condensation of more and more material possible at greater and greater distances from the Sun. The composition of the material that could condense out at any given radius would ultimately determine the types of planets that formed there. In a sense, the contractive heating "sterilized" much of the solar nebula, setting the stage for a highly diversified planetary system.

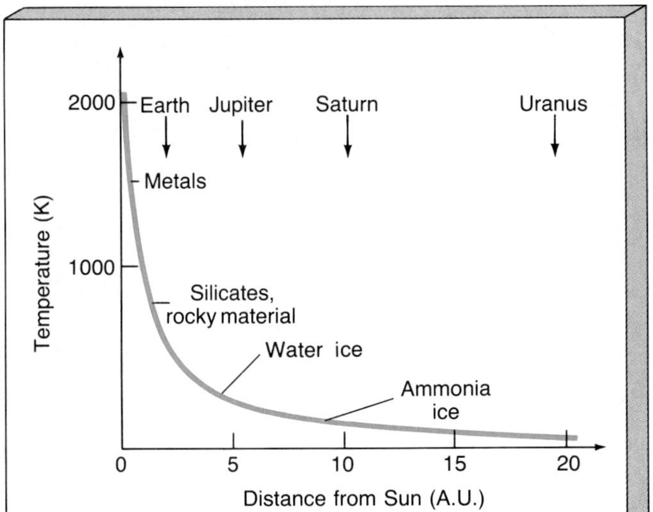

Figure 17.8 *Theoretically computed variation of temperature across the primitive solar nebula. In the hot central regions, only metals could condense out of the gaseous state to form grains. At greater distances from the central protosun, the temperature was lower, so rocky and icy grains could also form. The labels indicate the minimum radii at which grains of various types could condense out of the nebula.*

The Jovian Planets

In the middle and outer regions of the primitive planetary system, beyond about 5 A.U. from the center, the temperature was low enough for the condensation of several abundant gases into solid form. After hydrogen and helium, the most common materials in the solar nebula (as they are today in the universe as a whole) were the elements carbon, nitrogen, and oxygen. The most common chemical compounds were those containing those elements—specifically, water vapor, ammonia, and methane. As we have seen, these compounds are still the primary constituents of Jovian atmospheres. At temperatures of a few hundred kelvins or less, these gases condensed out of the nebula. Consequently, the ancestral fragments destined to become the cores of the Jovian planets were formed under cold conditions out of low-density, icy material. The planetesimals that formed at these distances were predominantly composed of ice. Because more material could condense out of the solar nebula at these radii than in the inner regions near the protosun, accretion began sooner, with more resources to draw on. The outer planets grew rapidly to the point where they could accrete nebular gas, not just grains, and eventually formed the hydrogen-rich Jovian worlds we see today.

With the formation of the four giant Jovian planets, the remaining planetesimals were subject to those planets' strong gravitational fields. Over a period of hundreds of millions of years and after repeated "gravity assists" from the giant planets, many of the interplanetary fragments in the outer solar system were flung into orbits taking them far from the Sun. Astronomers believe that those fragments now make up the Oort Cloud, whose members occasionally visit the inner solar system as comets. During this period, many icy planetesimals were also deflected into the inner solar system, where they played an important role in the evolution of the inner planets.

The Terrestrial Planets

In the inner regions of the primitive solar system, condensation from gas to solid began when the average temperature was about 1000 K. The environment there was too hot for ices to survive. Many of the abundant heavier elements, such as silicon, iron, magnesium, and aluminum, combined with oxygen to produce a variety of rocky materials. Planetesimals in the inner solar system were therefore rocky in nature, as were the protoplanets and planets they ultimately formed.

These heavier materials condensed into grains in the outer solar system, too, of course. However, they would have been vastly outnumbered by the far more abundant light elements there. The outer solar system is not deficient in heavy elements. The inner solar system is underrepresented in light material. Here we have another reason why the Jovian planets grew so much bigger than the terrestrial worlds. The inner regions of the nebula had to wait for the temperature to drop so that a few rocky grains could appear and begin the accretion process, but the outer regions may not have had to wait at all. The accretion process in the outer solar system began almost with the formation of the disk itself.

Very abundant light elements such as hydrogen and helium, as well as any other gases that failed to condense into solids, would have escaped from the terrestrial protoplanets, or, more likely, they were simply never accreted from the solar nebula. The inner planets' surface temperature was too high, and their gravity too low, to capture and retain those gases. Where then did Earth's volatile gases, particularly water, come from? The answer seems to be that icy fragments—comets—from the outer solar system, deflected into eccentric orbits by the Jovian planets' gravity, participated in the meteoritic bombardment of the newly born inner planets, supplying them with water *after* their formation.

Why the myriad rocks of the asteroid belt between Mars and Jupiter failed to accumulate into a planet remains a small mystery. Perhaps nearby Jupiter's huge gravitational field caused the early rocks to collide too destructively to coalesce. Strong Jupiter tides on the planetesimals in the belt would have also hindered the development of a protoplanet. The result is a band of planetesimals, still colliding and occasionally fragmenting, but never coalescing into a larger body—surviving witnesses to the birth of the planets.

Cleaning Up the Debris

Most of the planetesimals left over after the major planets formed eventually collided with a planet or were ejected into the Oort Cloud. Little solid material remained. But what of the gas that made up most of the original cloud? Why don't we see it today throughout the planetary system? In the outer solar system, some (but not all) of that gas was swept up into planets. But that did not occur in the inner regions, where the terrestrial protoplanets never became massive enough to accrete such light material. Instead, the newly formed Sun took a hand. All young stars apparently experience a highly active evolutionary stage known as the *T-Tauri* phase (Figure 17.9; see also Chapter 21, during which their radiation and stellar winds are very intense. Any gas remaining between the planets was blown away into interstellar space by the solar wind and the Sun's radiation pressure when the Sun entered this phase, just before nuclear burning started at its center. Afterwards, all that remained were protoplanets and planetesimal fragments, ready to continue their long evolution into the solar system we know today.

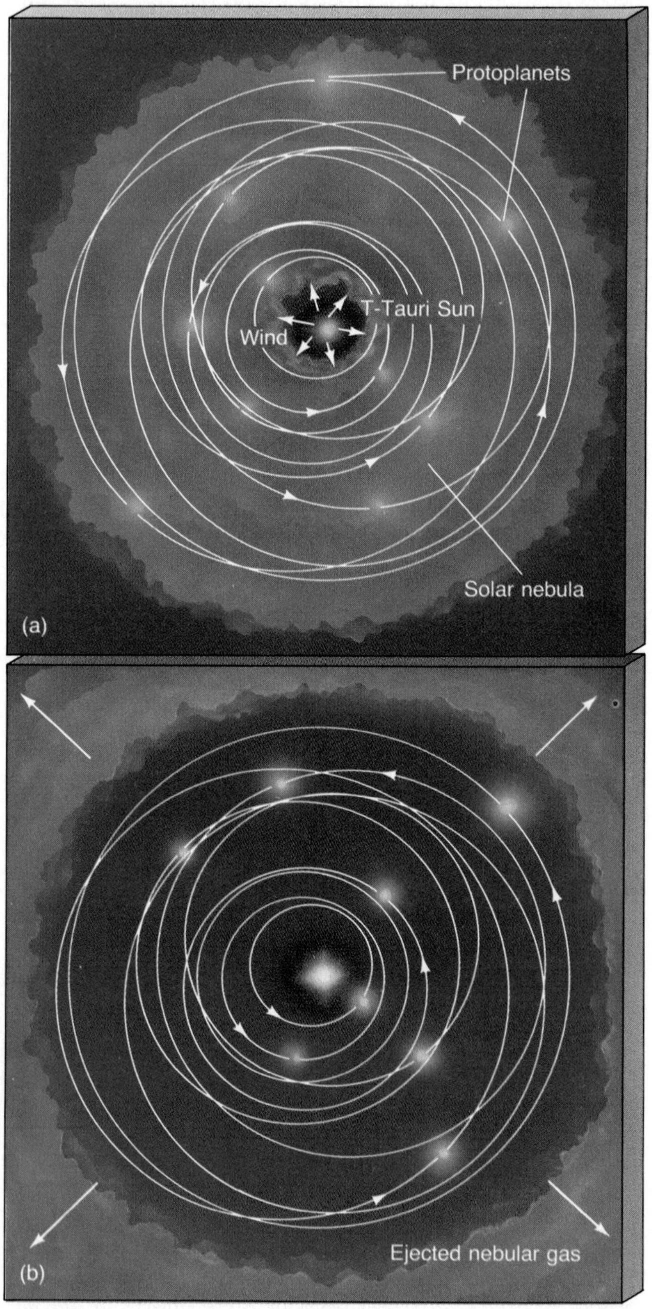

Figure 17.9 *(a) Strong stellar winds from newly born stars are responsible for sweeping away any dust and gas left over from the star formation process, (b) leaving only planets and planetesimals behind.*

THE ROLE OF CATASTROPHES

The condensation theory accounts for the nine "characteristic" points listed at the start of this chapter. Specifically, the circular (2), coplanar (3), prograde (4) orbits are a direct consequence of the nebula's shape and rotation. The rotation of the planets (5) and the orbits of the moon systems (6) are due to the tendency of the smaller-scale eddies to inherit the nebula's overall sense of rotation. The growth of planetesimals throughout the nebula, with each protoplanet ultimately sweeping up the material near it, accounts for point (1), which is the fact that the planets are widely spaced (even if the theory does not explain the regularity of the spacing). The heating of the nebula and the Sun's ignition resulted in the observed differentiation (7), while the debris from the accretion-fragmentation stage naturally accounts for the asteroids (8) and comets (9).

We stressed above that an important aspect of any solar system theory is its ability to allow for the possibility of imperfections—deviations from the otherwise well-ordered scheme of things. In the condensation theory, that capacity is provided by the randomness inherent in the encounters that ultimately combined the planetesimals into protoplanets. As the numbers of large bodies decreased and their masses increased, individual collisions acquired greater and greater importance. The effects of these collisions can still be seen today in many parts of the solar system—the large craters on many of the moons we have studied thus far are a case in point.

Having started with nine regular points to explain, we end with seven irregular solar system features that still fall within the theory's scope. It is impossible to test any of these assertions directly, but it is reasonable to suppose that some (or even all) of the following "odd" aspects of the solar system can be explained in terms of collisions late in the formative stages of the protoplanetary system. Not all astronomers believe all of these explanations; however, most would accept at least some.

1. Mercury's exceptionally large nickel-iron core may be the result of a collision between two partially differentiated protoplanets. The cores may have merged, and much of the mantle material may have been lost.
2. Two large bodies could have merged to form Venus, giving it its abnormally low rotation rate.
3. The Earth-Moon system may have formed from a collision between the proto-Earth and a Mars-sized object.
4. A late collision with a large planetesimal may have caused Mars's curious north-south asymmetry and ejected much of the Mars atmosphere.
5. The tilted rotation axis of Uranus might have been caused by a grazing collision with a sufficiently large planetesimal or by a merger of two smaller planets.
6. The Uranian moon Miranda may have been almost destroyed by a planetesimal collision, accounting for its bizarre surface terrain.
7. Interactions between the proto-Neptune and one or more planetesimals might account for Triton's retrograde motion, Nereid's eccentric orbit, and perhaps even the Pluto-Charon system.

THE ANGULAR MOMENTUM PROBLEM

A weak link in the condensation theory is, as we have seen in other theories, the very small angular momentum of the Sun. All mathematical modeling requires the Sun to have been spinning very fast in the earliest epochs of the solar system. Somehow, the Sun must have lost most of its angular momentum. Although the precise way it did so is unknown, we can surmise that the Sun probably transferred much of its spin angular momentum to the orbital angular momentum of the planets.

Many researchers speculate that the solar wind, moving away from the Sun into interplanetary space, carried away much of the Sun's initial angular momentum. The early Sun probably produced more of a dense solar gale than the relatively gentle "breezes" now measured by our spacecraft. High-velocity particles leaving the Sun followed the solar magnetic field lines. As the rotating magnetic field of the Sun tried to drag those particles around with it, they acted as a brake on the Sun's spin. Although each particle boiled off the Sun carries only a minute amount of the Sun's angular momentum with it, over the course of nearly 5 billion years the vast numbers of escaping particles have robbed the Sun of most of its initial spin momentum. Even today, our Sun's spin continues to slow.

Other researchers prefer to solve the Sun's momentum problem by assuming that the primitive solar system was much more massive than the present-day system. They argue that the accretion process was not entirely successful during the system's formative stages. Matter not captured by the Sun or the planets may well have transported much angular momentum while escaping back into interstellar space. This proposal is difficult to test, because the escaped matter would be well beyond the range of our current robot space probes. Perhaps the remote Oort Cloud of innumerable comets is the "escaped" matter.

Despite some minor controversy as to how this angular momentum quandary can best be resolved, nearly all astronomers agree that some version of the condensation theory is correct. The details have yet to be fully worked out, but the broad outlines of the processes involved are quite firmly established. Our planet is a by-product of the formation of the Sun. We might reasonably now ask what preceded the Sun, and what circumstances led to the collapse of the solar nebula in the first place. To answer these questions, we must widen the scope of our studies. We will find it necessary to understand the workings not only of the stars and the gas between them, but also of the Galaxy in which they reside. In the next chapter, we will begin this expanded inquiry with a closer look at our own parent star—the Sun.

CHAPTER SUMMARY

Most of our knowledge about the solar system's formative stages comes from studies of interstellar gas clouds, fallen meteorites, Earth's moon, and other planets. Meteorites are useful because they hold traces of the solid and gaseous matter from the early solar system.

Our solar system is an orderly place, making it unlikely that the planets were captured by the Sun. The overall organization points toward formation as the product of an ancient, one-time event, 4.6 billion years ago. An ideal theory of the solar system should provide strong reasons for the observed characteristics of the planets, yet be flexible enough to allow for deviations. The nebular theory correctly maintained that planets and their moons were by-products of the star formation process. However, refined calculations show that the rings of matter envisaged in the nebular theory would tend to disperse instead of accumulate into a planet.

According to the collision theory, the Sun had a close encounter with another star. Hot debris torn from the Sun would have provided the raw material for the planets. The main problems with this theory are that collisions between stars are extremely rare and that the observed characteristics of the planets are not easily explained.

The modern condensation theory combines the good features of the old nebular theory with new information about interstellar chemistry. Interstellar dust grains are thought to act as condensation nuclei, to which other atoms can attach. Once clumps form, they grow rapidly by sticking to other clumps. Finally, gravity takes over to complete the process of planet formation. The condensation theory can explain the basic differences between the jovian and terrestrial planets because the temperature of the solar nebula would be expected to decrease with increasing distance from the Sun. At any given location, the temperature would determine which materials could condense out of the nebula and so control the composition of any planets forming there.

The ancestral fragments destined to become the cores of the jovian planets were formed under cold conditions out of low-density, icy material. After the four giant planets formed, leftover planetesimals were subject to their strong gravitational fields. Many were flung into what became the Oort Cloud and now occasionally revisit our part of the solar system as comets. The terrestrial planets formed in the inner, hotter regions out of rocky or metallic materials. Light elements such as hydrogen and helium would have escaped into space. It is possible that Earth's water was carried to our world by comets deflected from the outer solar system. The myriad rocks of the asteroid belt failed to accumulate into a planet, possibly because the strong gravity of nearby Jupiter prevented them from coalescing.

Soon after the Sun ''turned on'' as a star, a brief period of intense radiation and stellar winds would have cleared the solar system of any remaining gas, setting the stage for the evolution of the planets as we know them today. Many ''odd'' aspects of the solar system may be explained in terms of collisions late in the formation stages of the protoplanetary system.

KEY WORDS

accretion
angular momentum problem
catastrophic theories
close-encounter theory
collision theory
condensation nuclei
condensation theory
evolutionary theory
fragmentation
interstellar dust
nebular theory
planestesimal
protoplanet
protosun
solar nebula

REVIEW QUESTIONS

1. Why does it seem unlikely that the planets of our solar system were captured by the Sun?
2. Explain the difference between evolutionary theories and catastrophic theories of the solar system's origin.
3. Describe the basic features of the nebular theory of the solar system's origin.
4. Give an example of how the nebular theory explains some features of the present-day solar system.

5. Name two basic flaws in the nebular theory.
6. Explain the difference between angular momentum and linear momentum.
7. Describe the basic features of the collision, or close-encounter, theory of the solar system's origin.
8. Give three reasons why it seems unlikely that the planets formed following a close encounter between the Sun and another star.
9. What is the key ingredient in the modern condensation theory of the solar system's origin?
10. Explain the role of accretion in the formation of the planets, in light of condensation theory.
11. Give two possible scenarios explaining why the Jovian planets are so much larger than the terrestrial planets.
12. What was the role of fragmentation in the process by which the planets formed?
13. What influence did Earth's location in the solar system have on its composition?
14. What was the role of heat in the formation of the Jovian planets?
15. What is the modern explanation for the Oort Cloud?
16. Why would Earth have been unable to retain volatile gases, particularly water, in the process of its formation? How might Earth have gotten its water?
17. What happened in the early solar system when the Sun became a T Tauri star?
18. How do modern astronomers attempt to explain the angular momentum problem in light of modern theories of solar system formation?
19. The orbital angular momentum of a planet in a (roughly) circular orbit is the product of its mass, its orbital velocity, and its distance from the Sun. Compare the orbital angular momenta of Jupiter, Saturn, and the Earth.
20. A typical comet contains some 10^{16} g of water ice. How many comets would have to strike the Earth in order to account for the roughly 2×10^{24} g of water presently found on our planet? If this amount of water accumulated over a period of 0.5 billion years, how many comets per year, on average, would have to have hit our planet during that time?
21. Consider a planet growing by accretion of material from the solar nebula. As it grows, its *density* remains roughly constant. Do you expect the force of gravity at its surface to increase, decrease, or stay the same? Specifically, what would happen to the surface gravity as the radius of the planet doubled? Give reasons for your answer.

DISCUSSION QUESTIONS

1. Give at least two reasons why we can't simply look to a distant solar system that is in the process of formation to understand how our solar system came to be.
2. Discuss the possibility that other stars have planets in light of two different theories of solar system formation: collision theory and condensation theory.
3. From what you now know about the origin of our solar system, how likely is it that other stars have planets much like Earth?
4. Describe the possible history of a single comet, starting with its birth somewhere near the planet Jupiter.

SUGGESTED READINGS

Barnes-Svarney, P. "The Chronology of Planetary Bombardments." *Astronomy* (July 1988). Collisions in the early solar system.

Black, D. C., and E. H. Levy. "A Profusion of Planets." *The Sciences* (May/June 1989). Excellent article outlining the various theories of solar system formation.

Boss, A. "The Origin of the Moon." *Science* (January 24, 1986). On the theory that the Moon formed following the impact of a Mars-sized object on the protoearth.

Cowen, R. "New Evidence of Budding Solar Systems." *Science News* (March 10, 1990). Good basic article on the ongoing search for newly forming solar systems.

"From Ring Particles to Whole Planets." *Science News* (January 19, 1991). How small particles may coagulate and large particles may grow in size.

Kerridge, J. F., and M. S. Matthews, eds. *Meteorites and the Early Solar System*. Tucson: University of Arizona Press, 1988. The broad sweep of meteorite studies. Includes summaries of current astrophysical models of prenebula and solar nebula evolution as well as of models of nebula accretion processes.

Pollard, W. G. "The Prevalence of Earthlike Planets." *American Scientist* (November–December 1979). How many Earth-like planets might actually exist in the universe?

"Protoplanetary Disks Are Common." *Science News* (July 11, 1987). Report on searches for newly forming solar systems.

CHAPTER

18

The Sun

Our Parent Star

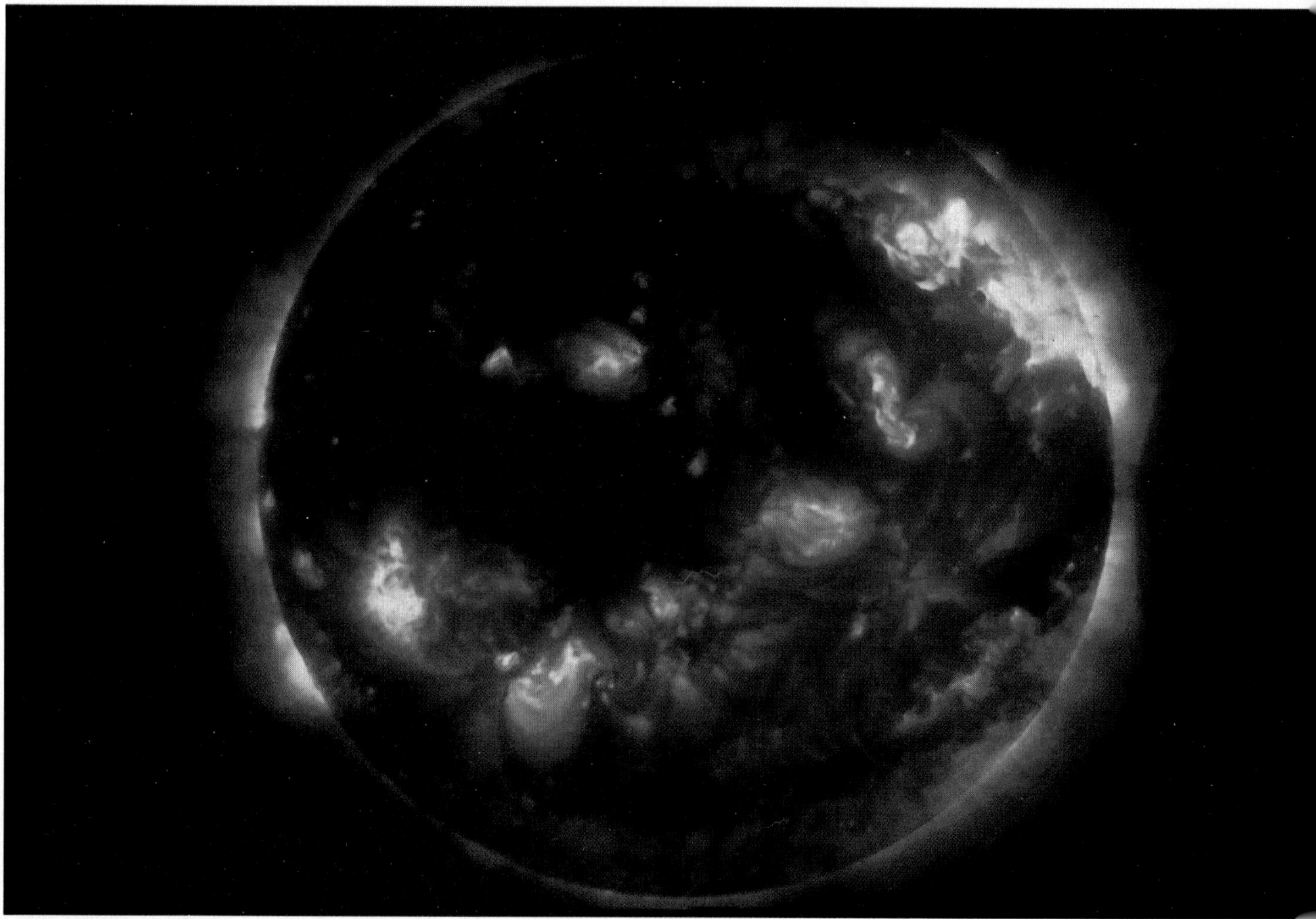

R I V U **X** G

This spectacular image of the Sun was taken from a rocket flight that lasted only a few minutes on July 11, 1991, just prior to a total solar eclipse. The payload of the rocket contained a special, high-resolution (~0.7 arc second) telescope that captured unique details of the bright arches connecting solar flares and other regions of intense activity on the Sun. This false-colored image is not an optical photograph; rather, it shows x-ray emission, mostly from the solar corona. The brightest regions of x-rays (intense white areas) denote the hottest gases in the Sun's atmosphere, some a fiery 3 million degrees Kelvin. Note the intricate complexity of the emitting gases in and around locations of solar activity, implying an extremely tangled magnetic structure that controls the gases at those locations. Note also Earth's Moon approaching from the west, about to eclipse the Sun and thereby bathe parts of the Earth in its long and eerie shadow. (Courtesy IBM and Smithsonian Astrophysical Observatory)

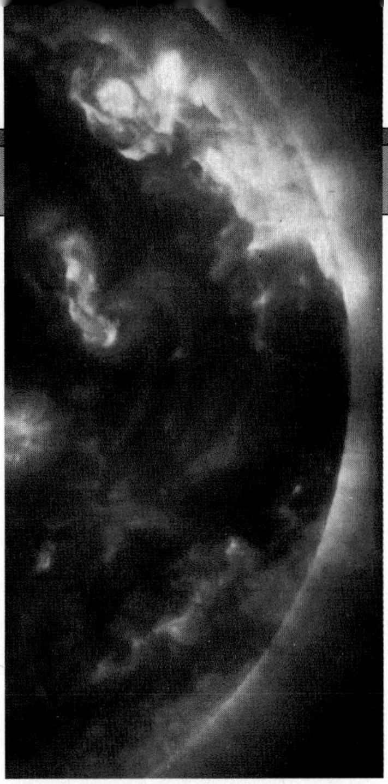

LEARNING GOALS

1 to appreciate the properties of the nearest star, our Sun
2 to understand how energy travels from the solar core, through its interior, and out into space
3 to realize what the structure of the Sun's outer layers is and what those layers tell us about the Sun's surface
4 to understand the various types of solar activity
5 to know how stars shine
6 to realize that observations of the Sun's core challenge our present understanding of the Sun

*Living in the solar system, we have the chance to study, at close range, perhaps the most common type of cosmic object—a star. Our Sun is a star, and a fairly average star at that, but with one unique feature: It is very close to us—some 300,000 times closer than our next nearest neighbor, Alpha Centauri. While Alpha Centauri is 4.3 light-years distant, the Sun is only 8 light-*minutes *away from us. Consequently, astronomers know far more about the properties of the Sun than about any of the other distant points of light in the universe. A good fraction of all our astronomical knowledge is based on modern studies of the Sun. Just as we studied our parent planet, the Earth, to set the stage for our exploration of the solar system, we now study our parent star, the Sun, as the next step in our exploration of the universe.*

THE SUN IN BULK

✓1 The Sun is the sole source of light and heat for the maintenance of life on Earth. It is a **star,** a glowing ball of gas held together by its own gravity and powered by nuclear fusion at its center. In its physical and chemical properties, the Sun is very similar to most other stars, regardless of when and where they formed. Our Sun appears to be a rather "typical" star, lying right in the middle of the observed ranges of stellar mass, radius, brightness, and composition. Far from detracting from the interest in the Sun, this very mediocrity is one of the main reasons that astronomers study the Sun—they can apply knowledge of solar phenomena to so many other stars in our Galaxy.

Makeup

The Sun has a surface of sorts—not a solid surface (the Sun contains no solid material), but the part of the sunny gas ball we see with our eyes or through a heavily filtered telescope—the so-called **solar disk.** This "surface" is known as the **photosphere.** Just above the photosphere is the Sun's lower atmosphere, called the **chromosphere.** Stretching far beyond that is a tenuous (thin) outer atmosphere, the solar **corona.** At still greater distances, the corona turns into the solar wind, which flows away from the Sun and permeates the entire solar system. Figure 18.1 shows the approximate dimensions of each of these regions.

Figure 18.1 also shows three more regions, all lying within the photosphere. Just below the surface, extending down to a depth of some 200,000 km, lies the **convection zone,** a region where the material of the Sun is in constant convective motion. The region at the heart of the Sun, about 400,000 km in diameter, is known as the **solar core** and is the site of powerful nuclear reactions that generate the Sun's enormous energy output. In between lies a region of gas usually simply referred to as the **solar interior.** Often the term *interior* is used to mean everything—including the convection zone—between the core and the photosphere.

Size

We have already seen that the Earth-Sun distance—the astronomical unit—is about 1.5×10^8 km. Knowing that distance, we can find the physical size of the solar photosphere. The solar disk is about 0.5° across, so we can compute the diameter of the Sun to be about 1.4×10^6 km,

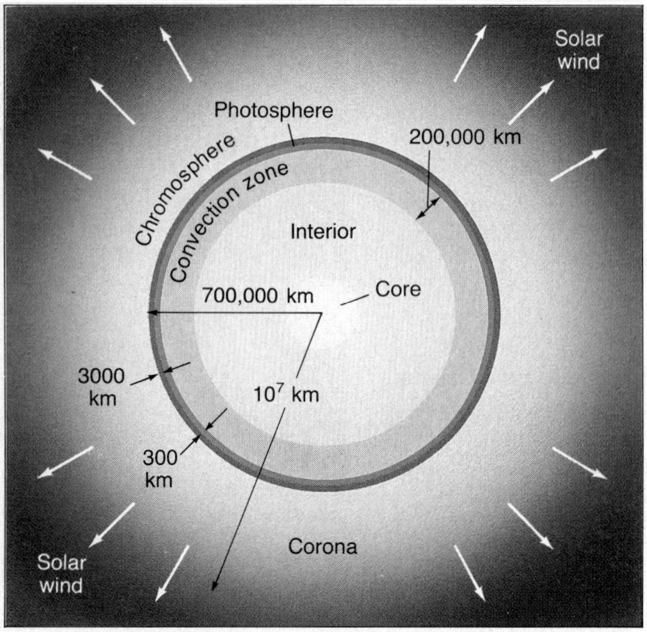

Figure 18.1 *Six main regions of the Sun, not drawn to scale, with physical dimensions labeled.*

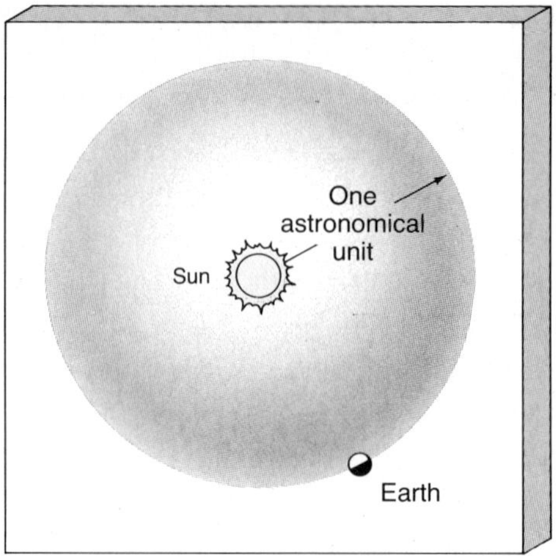

Figure 18.2 *We can draw an imaginary sphere around the Sun so that the sphere's edge passes through Earth's center. The radius of this imaginary sphere equals one astronomical unit. By multiplying the sphere's surface area by the solar constant, we can measure the Sun's luminosity, the amount of energy it emits each second.*

a little over 100 times the diameter of the Earth. The Sun's volume is therefore more than 100^3, or 1 million times that of the Earth. The Sun is truly vast by terrestrial standards.

Mass and Density

By studying the motion of objects influenced by the Sun's gravity—such as planets and spacecraft (see Chapter 3)—we can determine the Sun's mass. It is about 2.0×10^{33} g, about 300,000 times greater than the mass of the Earth. However, the Sun is not especially massive considering its huge volume. In fact, the average solar density—about 1.4 g/cm^3—is quite similar to that of the Jovian planets and only about one-quarter the average density of the Earth.

Temperature

In order to measure the Sun's temperature, we can apply the ideas discussed in Chapter 4. The distribution of solar radiation with respect to wavelength has the shape of a Planck curve for an object at about 6000K. This is the average temperature—sometimes called the **effective temperature**—of the Sun's photosphere. (A more careful calculation yields a temperature of 5800K, but we will use the rounded-off value here.) Notice, by the way, that this measurement tells us nothing whatever about the temperature in the solar interior. Later in this chapter we will consider conditions deeper within the Sun.

Luminosity

The properties of size, mass, density, and temperature are familiar from our study of the planets. But the Sun has an additional property, perhaps the most important of all from the point of view of life on Earth—it *radiates* a great deal of energy into space, uniformly (we assume) in all directions.

What is the total energy output of the Sun? If we were to hold a light-sensitive device—perhaps a solar cell of known dimensions—perpendicular to the Sun's rays, we could answer this question by measuring the amount of solar energy it received per square centimeter per second. The amount of solar energy reaching the Earth per unit area per unit time is known as the **solar constant.** Its value is approximately 1.4×10^6 $erg/cm^2/s$. Although an erg is a very small amount of energy—less than 1/10 the energy expended by a jumping flea—1.4 million ergs received by every square centimeter on Earth every second quickly adds up. A sunbather's body, with a total surface area of about 0.5 m^2 (5,000 cm^2), receives solar energy at a rate of about 7.0×10^9 erg/s, or 700 watts. This amount is roughly equivalent to the output of a typical electric room heater or about 10 household light bulbs.[1]

Let us now ask about the *total* amount of energy radiated in all directions from the Sun, not just the small fraction intercepted by the Earth. Figure 18.2 shows how this can be measured. Imagine a three-dimensional sphere centered on the Sun. The surface of this sphere intersects the Earth, so the sphere's radius is 1 A.U. and its surface area is approximately 2.8×10^{27} cm^2. Multiplying the amount of solar energy falling on each square centimeter (that is, the solar constant) by the total surface area of our imaginary sphere, we can determine the total *rate* at which energy leaves the Sun's surface. It turns out to be about 4×10^{33} erg/s. (A more precise measurement gives $3.9 \times$

[1] Although we use CGS units in this book, the *watt* (W) is probably a more familiar unit of power to most readers. Power is the rate of energy expenditure. One watt of power is the same thing as 10^7 erg/s.

10^{33} erg/s, but we will use the rounded-off figure here.) This quantity is known as the **luminosity** of the Sun. The solar luminosity is the total energy radiated by the Sun each second. As much as mass and radius, luminosity is one of the basic properties we use to characterize stars.

The Sun is a very powerful source of energy. An energy of 4×10^{33} erg is equivalent to the detonation of about 100 billion 1-megaton nuclear bombs, and the Sun produces this every *second*. Expressed in units of power, the solar luminosity is 4×10^{26} W, which is equivalent to 4 trillion trillion 100-watt light bulbs shining simultaneously. It also equals about 10^{19} U.S. dollars' worth of energy radiated per second (at current rates).

THE SOLAR INTERIOR

Modeling the Structure of the Sun

✓2 Lacking any direct measurements of the solar interior, astronomers must use more indirect means to probe the inner workings of our parent star. To accomplish this, they construct mathematical models of the Sun. The physical processes that are believed to be important in determining the Sun's internal structure are incorporated into a computer program, and guesses are made about any unknown quantities. The program then calculates the internal properties of this model star and predicts how it would look from the outside. These predictions are compared with observations, the guesses are modified, and the cycle is repeated until the solar model agrees with reality. By requiring our theoretical model to agree with observations when comparisons can be made, and trusting our knowledge of the laws of physics when comparisons are impossible, we arrive at a self-consistent picture of the Sun. The model that has gained widespread acceptance is often referred to as the **Standard Solar Model.**

In the 1960s, astronomers discovered that the surface of the Sun vibrates like a complex set of bells. These vibrations, illustrated in Figure 18.3, are the result of internal pressure (''sound'') waves that reflect off the photosphere and repeatedly cross the solar interior. Because these waves can penetrate deep inside the Sun, analysis of their surface patterns allows scientists to study conditions far below the Sun's surface. This process is similar to the way in which seismologists study the interior of the Earth by observing the P- and S-waves produced by earthquakes (see Chapter 7). For this reason, study of solar surface patterns is usually called **helioseismology,** even though solar pressure waves have nothing whatever to do with solar seismic activity—there is no such thing.

The observed surface vibrations can be extremely complex. Many separate patterns are superimposed on one another, so that deciphering the observations is a difficult task. The most recent and most extensive study of solar oscillations is the ongoing **GONG** (short for **G**lobal **O**scillations **N**etwork **G**roup) project. By making continuous observations of the Sun from many clear sites around the Earth, solar astronomers can obtain uninterrupted high-quality solar data spanning many days and even weeks—almost as though the Earth were not rotating and the Sun never set. Analysis of these data provides important additional information about the Sun, refining the Standard Solar Model. The temperature, rotation, and convective state of the solar interior have all been studied in this way, allowing direct comparisons between theory and reality to be made. As the GONG network expands during the 1990s, we can expect more and more detailed information about the heart of the Sun to become available.

Figure 18.3 *The Sun has been found to vibrate in a very complex way. By observing the motion of the solar surface, scientists can determine the wavelength and the frequencies of the individual waves and deduce information about the solar interior not obtainable by other means. The alternating red and blue patches represent gas moving down (red) and up (blue). (National Solar Observatory)*

Figure 18.4 shows the solar density and temperature plotted as functions of distance from the core of the Sun. Notice how the density drops rather precipitously at first and then decreases more slowly near the solar photosphere, some 700,000 km from the center. This variation of density is large, ranging from a core value of about 150 g/cm^3, 20 times the density of iron, to an intermediate value (at 350,000 km) of about 1 g/cm^3, the density of water, to an extremely small photospheric value of $10^{-7} g/cm^3$, about 10,000 times less dense than air at the surface of the Earth. The *average* density, as we have seen, of the entire Sun is 1.5 g/cm^3. Because the density is so high in the core, roughly 90 percent of the Sun's mass is contained within the inner half of its radius.

The solar density continues to decrease out beyond the photosphere, reaching values as low as 10^{-26} g/cm^3 in the far corona. Gas of such low density is about as thin

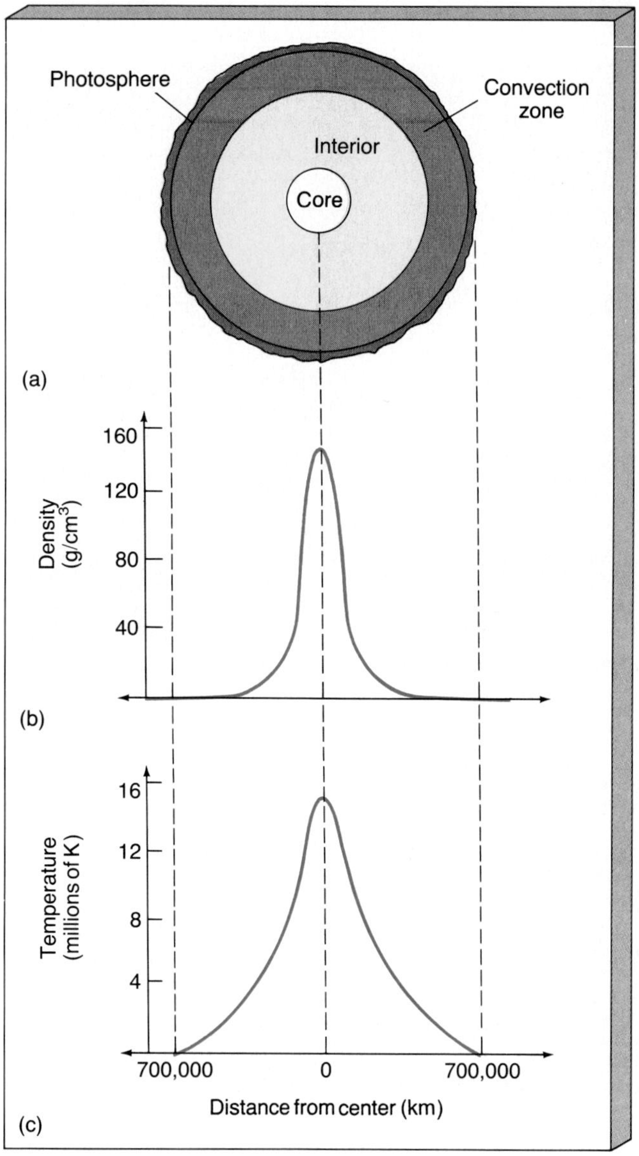

Figure 18.4 *Theoretically modeled profiles of density (b) and temperature (c) for the interior of the Sun, presented for perspective in (a). All three parts describe a cross-sectional cut through the center of the Sun.*

as the best vacuum that physicists can create in laboratories on Earth.

As shown in Figure 18.4(c), the solar temperature also decreases with increasing radius, but not as rapidly as the density. Computer models indicate a temperature of about 15 million K at the core, consistent with the minimum 10 million needed to initiate the nuclear reactions known to power all stars, decreasing to the observed value of about 6000K at the photosphere.

Convection

Computer models of the solar interior predict the existence of an extensive convection zone lying just below the photosphere. As in the Earth's interior and atmosphere (as discussed in Chapter 7), and elsewhere in the solar system, convection can occur whenever cooler material overlies warmer material. The result is a rather well-defined circulation pattern that strives to even out the temperature. Because the process of solar convection is critically important for transporting energy to the Sun's surface, let's study it in a little more detail.

The very hot solar interior ensures violent and frequent collisions among gas particles. Particles move in all directions with high velocities, bumping into one another unceasingly. In and near the core, the extremely high temperatures guarantee that the gas is completely ionized. Recall that as atoms absorb photons, their electrons are boosted to more excited states. With no electrons left on atoms to capture the photons, the deep solar interior is quite transparent to radiation. Only occasionally does a photon encounter and scatter off a free electron or proton. As a result, the energy produced by nuclear reactions in the core travels outward toward the surface in the form of radiation with relative ease. This part of the interior is known as the **radiation zone.**

As we move outward from the core, the temperature falls, atoms collide less frequently and less violently, and more and more electrons manage to remain bound to their parent nuclei. With more and more ions to absorb the outgoing radiation, the gas in the interior changes from being relatively transparent to being almost totally opaque. By the outer edge of the radiation zone, 200,000 km below the photosphere, *all* of the photons produced in the Sun's core have been absorbed. Not one of them reaches the surface. But what happens to the energy they carry?

The photons' energy must travel beyond the Sun's interior. If it did not, the Sun would have exploded long ago. That we see sunlight—visible energy—proves that some energy escapes. That energy reaches the surface by convection. Hot solar gas, beginning at the top of the radiation zone, *physically* moves upward, while cooler gas above it sinks, creating a characteristic pattern of convection cells. This action is the solar analog of the warm, rising bubbles in a saucepan of boiling soup. All through the upper interior, energy is transported to the surface by physical motion of the solar gas. Figure 18.5(a) is a diagram of solar convection cells, where columns of hot gas rise, cool, and descend. Convection and radiation are fundamentally different mechanisms of energy transport.

In reality, the zone of convection is much more complex than we have just described. There is a hierarchy of convection cells, organized in tiers at different depths, as illustrated in Figure 18.5(b). The deepest tier, about 200,000 km below the surface, is thought to contain large cells some tens of thousands of kilometers in diameter. Heat is then successively carried upward through a series of progressively smaller-sized cells, stacked one upon another until, at a depth of about 1000 km, the individual cells are about 1000 km across. The top of this uppermost tier of convection is the visible surface, where astronomers can directly observe the cell sizes. Information about convec-

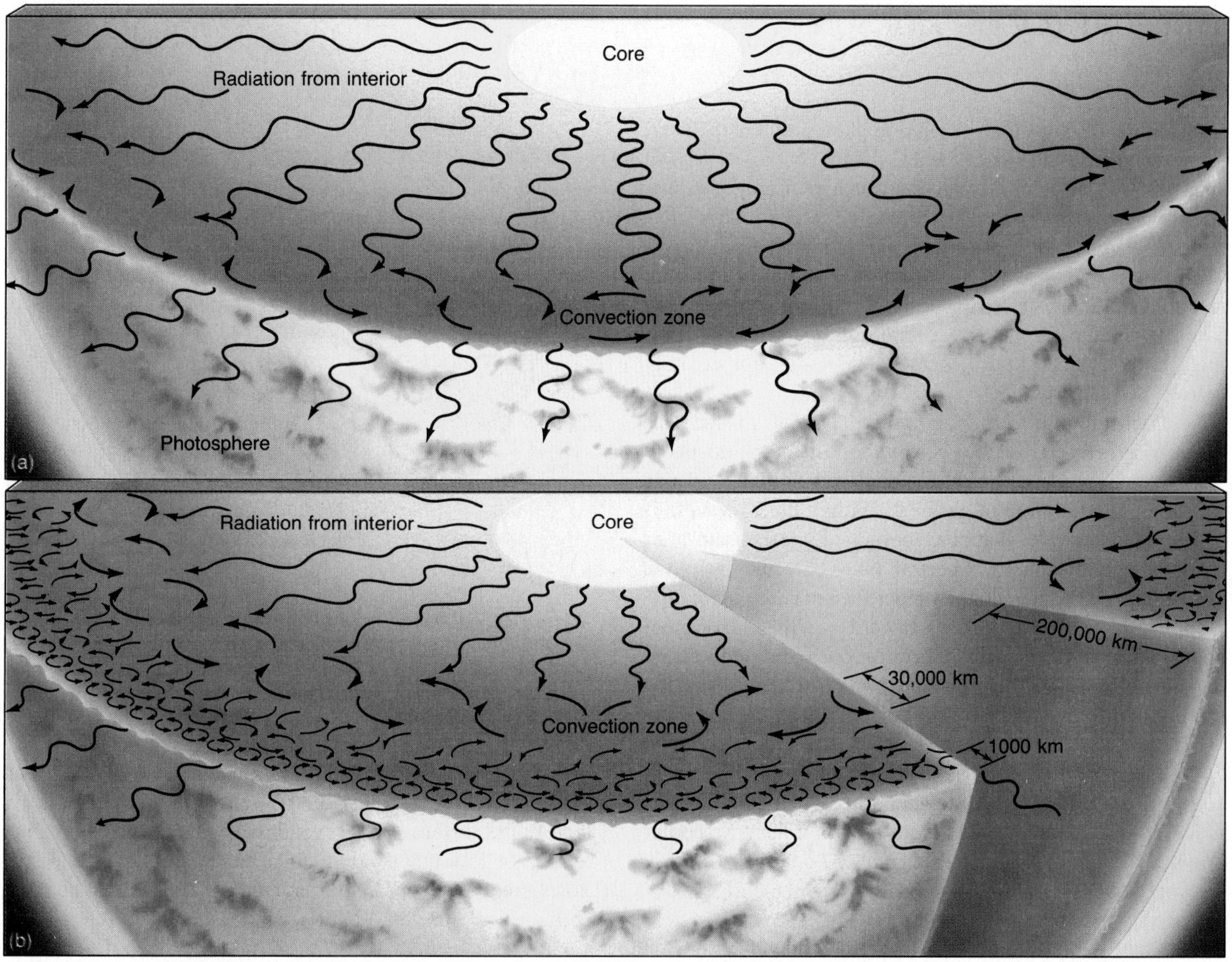

Figure 18.5 (a) Simplified diagram of the physical transport of energy in the Sun's convection zone. We can visualize the upper interior as a boiling, seething sea of gas. Each convective loop is about 1000 km across. (b) In reality, the convective cells are arranged in tiers containing cells of progressively smaller size as the surface is neared. This is still a highly simplified diagram, however. There are many different cell sizes, and they are not so neatly arranged.

tion below that level is inferred mostly from computer models of the solar interior.

At some distance from the core, the solar gas becomes too thin to sustain further upwelling by convection. Theory suggests that this distance roughly coincides with the photospheric surface we see. Convection does *not* proceed into the solar atmosphere. There is simply not enough gas there. The density is so low that the gas becomes transparent once again, and we return to radiation as the mechanism of energy transport. Photons reaching the photosphere escape more or less freely into space, and the photosphere emits thermal radiation, like any other hot object.

Energy Transport in the Sun

Here then is the history of a typical ''parcel'' of energy created in the Sun's core. Produced in the form of gamma-ray photons (as we will see later in this chapter), the energy repeatedly interacts with the free electrons that populate the inner radiation zone. As photons are continually scattered in regions of lower and lower temperature, the energy is steadily ''downgraded'' into lower- and lower-energy photons as it progresses outward. Eventually, after a few tens of thousands of years, it reaches the base of the convection zone in the form of x-rays. At that point, the photons are absorbed, and the energy is carried upward by convection, arriving at the photosphere a few hundred thousand years later. Once above the convection zone, the energy travels outward in the form of visible radiation without further interruption, reaching Earth in about 8 minutes. Thus, the sunlight we see starts its journey to Earth *from the photosphere*. The solar interior effectively transforms copious amounts of lethal high-energy radiation into the lower-energy light we actually see.

THE SURFACE OF THE SUN

Granulation

☑3 Figure 18.6 is a high-resolution photograph of the solar surface taken with instruments aboard NASA's *Skylab* space station as it orbited above much of Earth's atmosphere in the mid-1970s. The visible surface is highly mottled, or *granulated,* with bright and dark gas. Each bright **granule** measures about 1000 km across—comparable in size to a continent on Earth—and has a lifetime of between 5 and 10 minutes. Together, several million granules constitute the top layer of the convection zone, immediately below the photosphere.

Each granule forms the topmost part of a solar convection cell. Spectroscopic observation of the photosphere within and around the bright regions shows direct evidence for the upward motion of gas as it "boils" up from within. This evidence proves that convection really does occur at or below the photosphere. Spectral lines detected from the bright granules appear slightly bluer than normal, indicating Doppler-shifted matter coming toward us with a velocity of about 1 km/s. Conversely, spectroscopes focused on the darker portions of the granulated photosphere show the same spectral lines to be redshifted, indicating matter moving away.

The brightness variations of the granules result strictly from differences in temperature. The upwelling gas is hotter and therefore emits more radiation than the cooler downwelling gas. The adjacent bright and dark gases appear to contrast considerably, but in reality their temperature difference is less than about 500K.

Careful measurements also reveal a much larger-scale flow on the solar surface. **Supergranulation** is a flow pattern quite similar to granulation except that supergranulation cells measure some 30,000 km across. As with granulation, material upwells at the center of the cells, flows across the surface, then sinks down again at the edges. Scientists believe that supergranules are the imprint on the photosphere of a deeper tier of large convective cells, like those depicted in Figure 18.5(b).

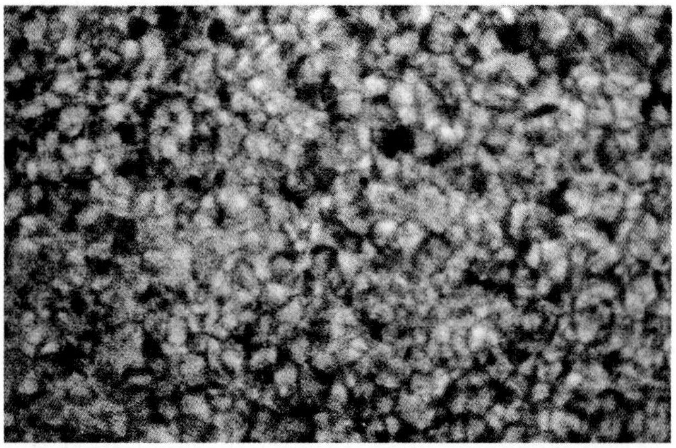

Figure 18.6 Skylab *photograph of the granulated solar photosphere. Typical solar granules are comparable in size to Earth's continents. The bright portions of the image are regions where hot material is upwelling from below. The dark regions correspond to cooler gas that sinking back down into the interior. (NASA)*

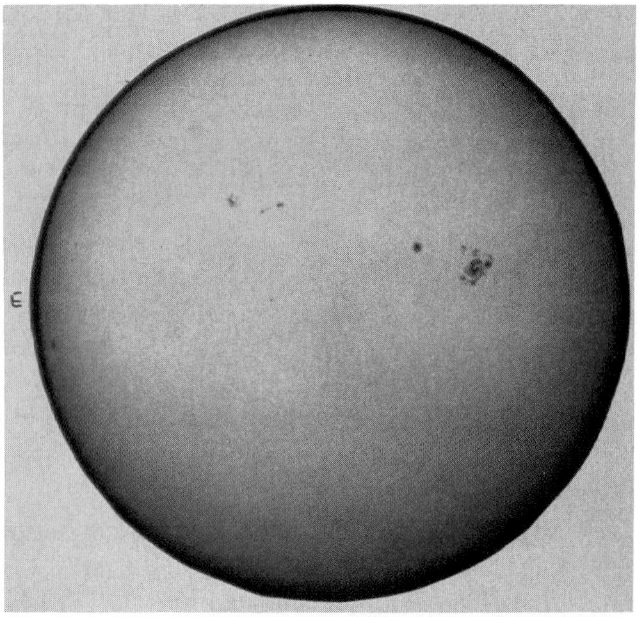

Figure 18.7 *This photograph of the Sun taken through a hydrogen-alpha filter shows a sharp solar limb, although our star, like all stars, is made of a gradually thinning gas. Notice also the phenomenon of limb darkening—the edge of the solar disk is noticeably darker than the center. (NOAO)*

The "Edge" of the Sun

Figure 18.7 shows the entire Sun's disk photographed through a heavily filtered telescope. Despite the steady decrease in density and temperature from the interior to the atmosphere, the Sun displays a reasonably sharp edge, or **limb.** The limb is sharp because the overwhelming majority of visible photons arise in the extremely shallow photosphere. Slightly below the photosphere, the gas is still convective, and the radiation does not reach us directly. Slightly above, the gas is too thin to emit appreciable amounts of radiation. Recent estimates suggest that the depth of the photosphere is no more than 500 km, very small compared with the size of the Sun. The photosphere's thickness is less than 1/10 of 1 percent of the solar radius, which is why we perceive the Sun as having a well-defined edge.

There is one further observation that allows us to check a prediction of the Standard Solar Model—namely, that the temperature increases with depth. In Figure 18.7, the disk of the Sun is noticeably darker near the edge than it is at the center. This phenomenon is known as **limb darkening.** It is a direct consequence of the rise in temperature

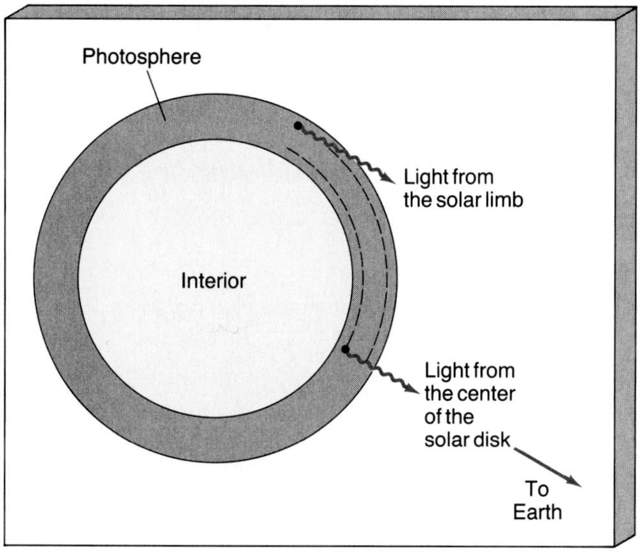

Figure 18.8 *Limb darkening occurs because the radiation we receive from the center of the solar disk comes from deeper, and hence hotter, regions of the photosphere (whose thickness is exaggerated here) than the radiation coming from the limb. Limb darkening is direct evidence that solar temperature increases with depth, at least in the photosphere.*

with increasing depth below the solar surface. Let us imagine that an ''average'' photon of sunlight traverses a fixed distance in the photosphere before leaving the Sun on its way to Earth. As illustrated in Figure 18.8, photons coming from the center of the solar disk (as seen from Earth) are emitted from a deeper level than photons coming from the limb. Because temperature increases with depth, the photons coming from the center of the disk come from a hotter (deeper) part of the photosphere. The hotter the gas, the more radiation it emits. Therefore, the center of the Sun appears brighter than the limb.

THE SOLAR ATMOSPHERE

Since the 1960s, there has been a surge of interest in the regions beyond the Sun's photosphere. This interest has been stimulated largely by the Space Age discovery that Earth is engulfed by the solar wind, whose influence on the magnetosphere and atmosphere of our planet was described in Chapter 7. Scientists have discovered that the solar atmosphere is a violent, volatile place. We are only now beginning to understand the complexities of its behavior.

The Chromosphere

The photosphere of the Sun is the region that actually emits the light we see from Earth. Above the photosphere lies the cooler chromosphere, the inner part of the solar atmosphere. This region emits very little light of its own and cannot be observed visually under normal conditions. The photosphere is just too bright, dominating the chromosphere's radiation. The relative dimness of the chromosphere results from its low density. Large numbers of photons simply cannot be emitted (or absorbed) by a tenuous gas containing very few atoms per unit volume. Although the chromosphere is not normally seen, astronomers have long been aware of its existence.

If we could shut off the photosphere's intense light, we could clearly observe the radiation emitted by the Sun's atmosphere. This is precisely what happens during a total solar eclipse (see Chapter 1), when the Moon blocks the Sun's disk from view. Figure 18.9 shows the Sun during an eclipse in which the photosphere, but not the chromosphere, is obscured by the Moon. The chromosphere's characteristic reddish hue is plainly visible. This coloration is due to the Hα (read ''hydrogen alpha'') emission line of hydrogen, which dominates the chromospheric spectrum. (Recall from Chapter 5 that the wavelength of this line is 6563 Å, right in the middle of the red portion of the spectrum.)

The chromosphere is far from tranquil. Every few minutes, small solar storms erupt, expelling jets of hot matter known as spicules into the Sun's upper atmosphere (see Figure 18.10). These long, thin spikes of flaming matter leave the Sun's surface at typical velocities of about 100 km/s, reaching several thousand kilometers above the photosphere. Spicules are not spread evenly across the solar surface. Instead, they cover only about 1 percent of the total area, tending to accumulate around the edges of supergranules. The Sun's magnetic field is also known to be

R I **V** U X G

Figure 18.9 *This photograph of a total solar eclipse shows the solar chromosphere, a few thousand kilometers above the Sun's surface. Note the true-color prominences on the east and west limbs. (G. Schneider)*

Figure 18.10 *Solar spicules, short-lived narrow jets of gas that typically last mere minutes, can be seen sprouting up from the solar chromosphere in this Hα image of the Sun. (NOAO)*

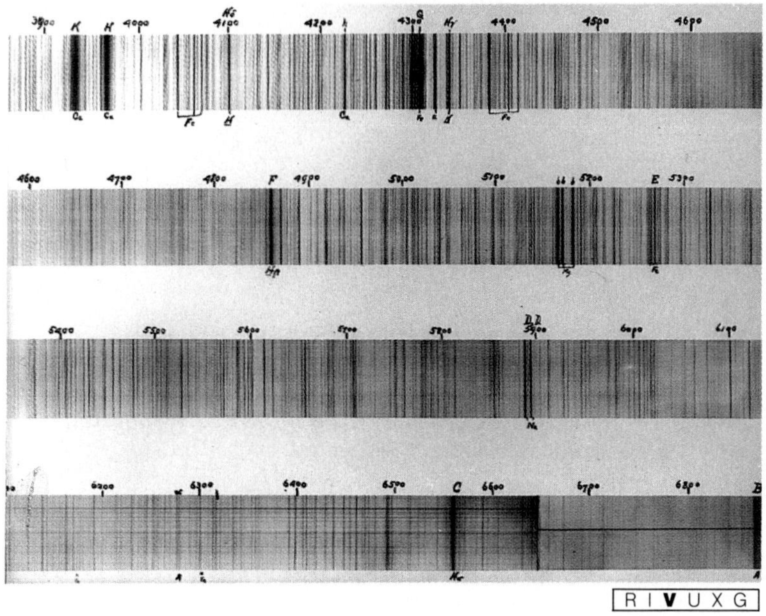

Figure 18.11 *A detailed spectrum of our Sun in the visible domain shows thousands of Fraunhofer spectral lines, which indicate some 67 different elements in various stages of excitation and ionization in the lower solar atmosphere. This spectrum has been cut into four strips to fit on the page, but it is really a single spectrum stretching from long-wavelength red at bottom right to blue at upper left. (Palomar Observatory)*

somewhat stronger than average in those regions. Scientists speculate that the downwelling material there tends to strengthen the solar magnetic field and that spicules are the result of magnetic disturbances in the Sun's churning outer layers.

The Spectrum of the Sun

Astronomers can also glean an enormous amount of information about the solar atmosphere from an analysis of the *absorption* lines that arise there. Figure 18.11 (see also Figure 5.4) is a detailed spectrum of the Sun, obtained for the full range of visual wavelengths, from 4000 to 7000 Å. Recall from Chapter 5 that photons radiated from the surface are absorbed by *cooler* overlying gas. The result is an absorption spectrum characteristic of the density and temperature of the region containing the absorbing atoms and ions. Most of the absorption lines seen in the solar spectrum are produced in the upper regions of the photosphere, or in the chromosphere. Because these lines are seen in absorption rather than in emission, we know that the temperature beyond the photosphere continues to decrease with increasing distance from the center of the Sun.

Thousands of spectral lines have been observed and cataloged in the solar spectrum, although there are not nearly this many elements in the Sun. Some of the heavier, multi-electron elements can absorb a variety of photons having many different energies. For example, hundreds of lines are attributed to just the element iron in various stages of excitation and ionization, as we discussed in Chapter 5. In all, some 67 elements have been identified in the Sun. More elements probably exist there, but they are present in such small quantities that our instruments are simply not sensitive enough to detect them. Table 18-1 lists the 10 most common elements in the Sun. Notice that hydrogen is by far the most abundant element, followed by helium. This distribution is just what we saw on the Jovian planets, and it is what we will find for the universe as a whole.

The Corona

During the brief moments of an eclipse, if the Moon's angular size is large enough that both the photosphere and the chromosphere are blocked, the ghostly solar corona can be seen, as in Figure 18.12. With the photospheric light removed, the pattern of spectral lines changes dramatically. The intensities of the usual lines alter, suggesting changes

TABLE 18-1 ***The Composition of the Sun***

Element	Abundance (percent of total number of atoms)	Abundance (percent of total mass)
Hydrogen	91.2	71.0
Helium	8.7	27.1
Oxygen	0.078	0.97
Carbon	0.043	0.40
Nitrogen	.0088	0.096
Silicon	.0045	0.099
Magnesium	.0038	0.076
Neon	.0035	0.058
Iron	.0030	0.14
Sulfur	.0015	0.040

in elemental abundances or gas temperature or both. Most lines shift from absorption to emission because the usual hot background source is now blocked, and entirely new spectral lines not seen in the photospheric absorption spectrum suddenly appear. These new coronal (and in some cases chromospheric) lines were first observed during eclipses in the 1920s, and for many years afterward researchers attributed them to a nonterrestrial element called "coronium."

We now recognize that these new spectral lines do not indicate any new kind of atom. Coronium does not really exist. Rather, the new lines arise because atoms in the corona have lost several more electrons than atoms in the photosphere—that is, the coronal atoms are much more highly ionized. For example, astronomers have identified coronal lines corresponding to iron ions with as many as 13 of their normal 26 electrons missing. In the photosphere, most iron atoms have lost only 1 or 2 of their electrons.

The cause of this extensive electron stripping is the high coronal temperature. The degree of ionization inferred from spectra observed during solar eclipses tell us that the gas temperature of the upper chromosphere exceeds that of the photosphere. Furthermore, the temperature of the solar corona, where even more ionization is seen, is still higher.

Based on many observations of conditions at different distances from the limb of the Sun, from the photosphere outward into the corona, Figure 18.13 plots the variation of gas temperature with altitude. The temperature decreases to a minimum of about 4500K some 500 km above the photosphere, after which it rises steadily. About 1500 km above the photosphere, the gas temperature begins to rise rapidly, reaching more than 1 million K at an altitude of 10,000 km. Thereafter, it remains roughly constant. Using this temperature profile, we can draw a clear distinction between the chromosphere and the corona: The chromosphere extends from the top of the photosphere for approximately 1500 km. The region in which the temperature rises rapidly—from about 1500 km to 10,000 km—is called the **transition zone.** At 10,000 km, the corona begins.

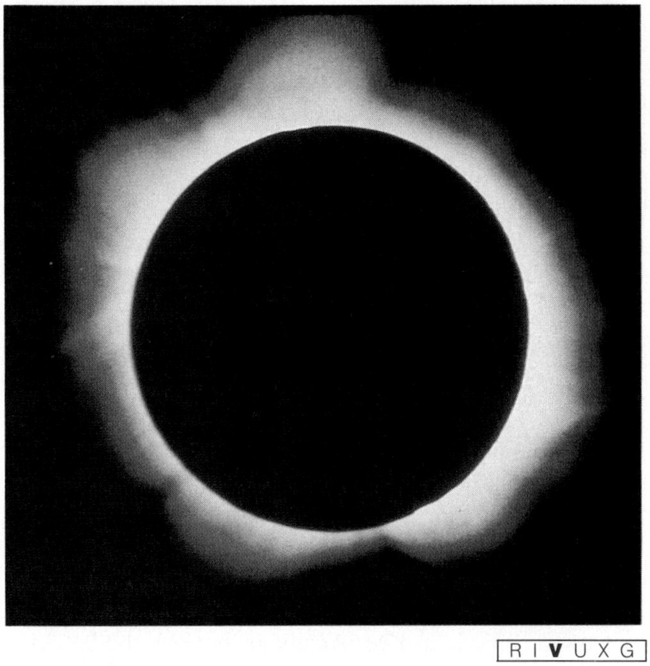

Figure 18.12 *When both the photosphere and the chromosphere are obscured by the Moon during a solar eclipse, the faint corona becomes visible. This photograph shows clearly the emission of radiation from the solar corona. (NOAO)*

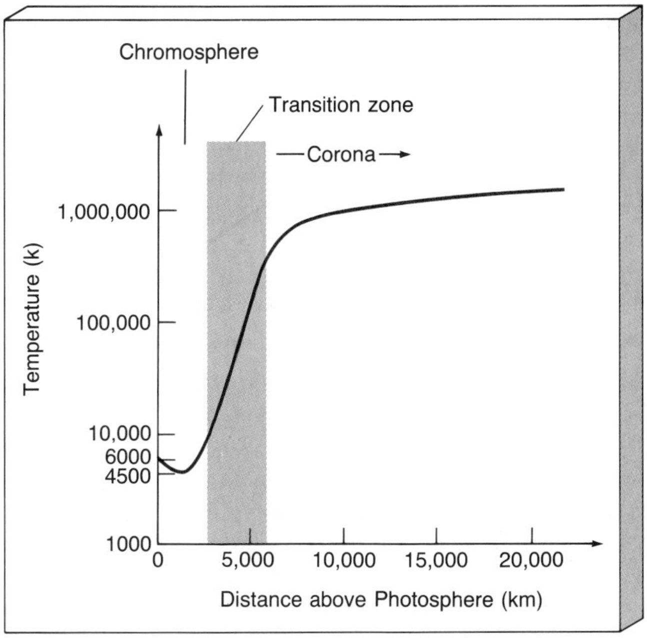

Figure 18.13 *The change of gas temperature in the lower solar atmosphere is dramatic. The minimum temperature marks the outer edge of the chromosphere. Beyond that, the temperature rises sharply in the transition zone, finally leveling off at about 1 million K in the corona.*

The cause of this rapid temperature rise is not fully understood. The temperature profile runs contrary to intuition—moving away from a heat source, we would normally expect the heat to diminish, but this is not the case for the Sun. The corona must have another energy source. Astronomers now believe that magnetic disturbances in the solar photosphere—a little like spicules, but on a much larger scale—are ultimately responsible for heating the corona. We will return to these disturbances in more detail in the next section.

The Solar Wind

Electromagnetic radiation and fast-moving particles—mostly protons and electrons—escape from the Sun all the time. The radiation moves away from the photosphere at the speed of light, taking 8 minutes to reach Earth. The particles travel more slowly, although at the still considerable speed of about 500 km/s, reaching Earth in a few days. This constant stream of escaping solar particles is the solar wind.

The solar wind results from the high temperature of the solar corona. About 10 million km above the photosphere, the coronal gas is hot enough to escape the Sun's gravity, and it begins to flow outward into space. At the same time, the solar atmosphere is continually replenished from below. If that were not the case, the corona would disappear in about a day. The Sun is, in effect, ''evaporating''—constantly shedding mass through the solar wind. However, the wind is an extremely thin medium. Even though it blows all the time, carrying away about a million tons of solar matter each second, less than 0.1 percent of the Sun has been lost in this way since its formation billions of years ago. Our star is evaporating, but it is losing only a negligible fraction of its huge bulk.

The Sun in X-Rays

What sort of radiation is emitted by a gas of 1 million K? Unlike the 6000K photosphere, which emits most strongly in the visible part of the electromagnetic spectrum, the hotter coronal gas radiates at much higher frequencies—primarily in x-rays. For this reason, x-ray telescopes have become important tools in the study of the solar corona. Figure 18.14 shows an x-ray image of the Sun. The full corona extends well beyond the regions shown, but the density of coronal particles emitting the radiation diminishes rapidly with distance from the Sun. The intensity of x-ray radiation farther out is too dim to be seen here.

In the mid-1970s, instruments aboard NASA's *Skylab* space station revealed that the solar wind escapes mostly through solar ''windows'' called **coronal holes.** The dark area to the right side of Figure 18.14 shows a coronal hole. Not really holes, such structures are simply deficient in matter—vast regions of the Sun's atmosphere where the density is about 10 times lower than the already tenuous, normal corona. Coronal holes are underabundant in matter because the gas there is able to stream freely into space at particularly high speeds, driven by disturbances in the Sun's atmosphere and magnetic field. In other regions of the corona, the solar magnetic field tends to inhibit the outward flow of the solar wind, just as the Earth's magnetic field tends to prevent the incoming solar wind from striking the Earth, and the density remains (relatively) high. The largest coronal holes can be hundreds of thousands of kilometers across. Structures of this size are seen only a few times each decade. Smaller holes—perhaps only a few tens of thousand kilometers in size—are much more common and seem to appear every few hours.

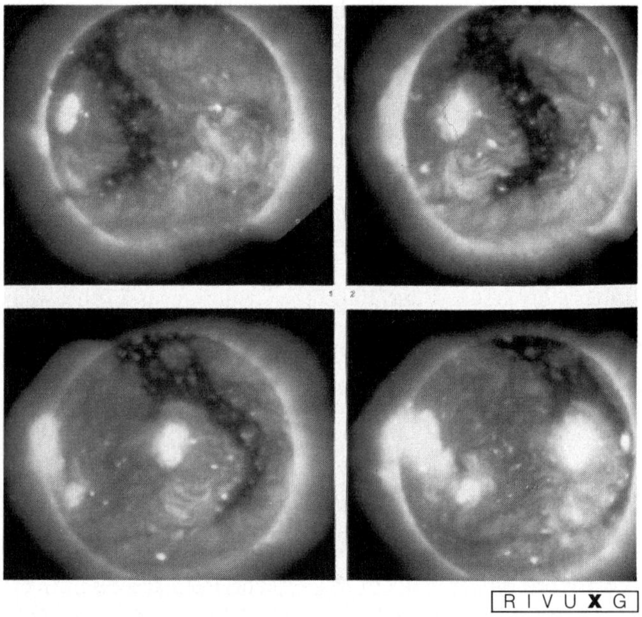

Figure 18.14 *An image of the x-ray emission from the Sun observed by the* Skylab *space station. Note the dark, boot shaped coronal hole traveling to the right, where the x-ray observations outline in dramatic detail the abnormally thin regions through which the high-speed solar wind streams forth. These frames, taken from left to right and top to bottom are spaced at one-day intervals. (NASA)*

THE ACTIVE SUN

✓4 Most of the Sun's luminosity results from continuous emission from the photosphere. This radiation arises from what we call the **quiet Sun,** which is the underlying predictable condition of the Sun that blazes forth on a daily basis. This steady behavior contrasts with the sporadic, unpredictable radiation of the **active Sun,** a much more irregular condition in which the star is prone to sudden explosive outbursts, some of them intense enough to affect us directly here on Earth. The size and duration of coronal holes are strongly influenced by the level of solar activity, as is the strength of the solar wind.

Despite its violence, the active component of solar radiation contributes little to the Sun's total luminosity, and it has little effect on the evolution of the Sun as a star. Indeed, these transitory disturbances would be very difficult to detect at all if the Sun were at the distance of even the next nearest star. A little like thunderstorms and tornadoes on Earth, these solar ''storms'' are usually localized in their effect and almost impossible to predict with any certainty.

Sunspots

Figure 18.15 is an optical photograph of the entire Sun. Notice the numerous dark blemishes on its surface. First studied in detail by Galileo, these ''spots'' provided one of the first clues that the Sun was not a perfect unvarying creation, but a place of constant change. These dark areas are called **sunspots.** They typically measure about 10,000 km across, about the size of the Earth. As shown in the figure, they often occur in groups. At any given time, the Sun may have hundreds of sunspots, or it may have none at all.

Studies of sunspots show an **umbra,** or dark center, surrounded by a grayish **penumbra.** The close-up view of a pair of sunspots in Figure 18.16 shows each of these dark

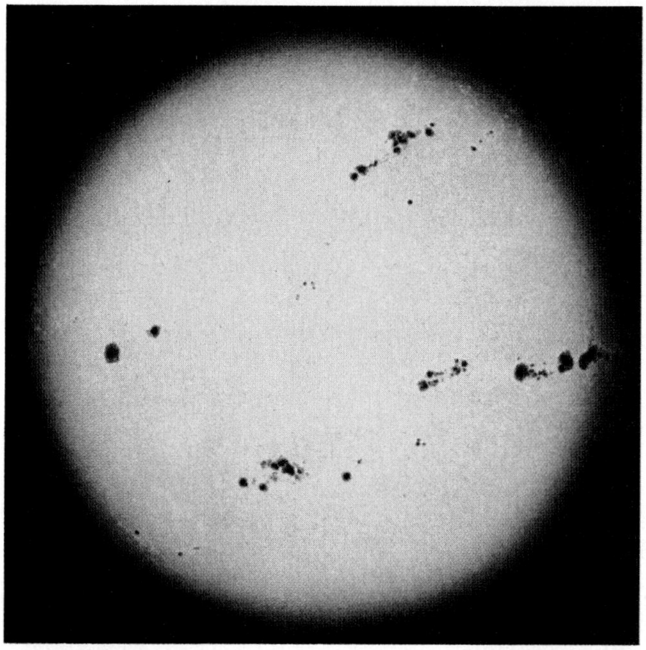

R I V U X G

Figure 18.15 *This photograph of the entire Sun shows several groups of sunspots, taken during the 1957 period of solar maximum activity. The largest spots in this image are about 20,000 km across—twice the diameter of the Earth. Typical sunspots are only about half this size. (Palomar Observatory)*

areas and the brighter undisturbed photosphere nearby. This gradation in darkness is really a gradual change in photospheric temperature—sunspots are simply *cooler* regions of the photospheric gas. The temperature of the umbra is about 4500K, and the penumbra measures about 5500K. The spots, then, are certainly composed of hot gases. They seem dark only because they appear against an even brighter background (the hotter, 6000K photosphere). If we could magically remove a sunspot from the Sun (or just block out the rest of the Sun's emission), the spot would glow brightly, just like any 4000–6000K emitter.

Sunspots are not steady. Most change their size and shape, and all come and go. Figure 18.17 shows a time sequence in which several spots vary—sometimes growing, sometimes dissipating—over a period of several days. Individual spots may last anywhere from 1 to 100 days. A large group of spots typically lasts 50 days.

All sunspots are observed to move across the face of the Sun. This motion is part of the general rotation of the Sun on its axis. Observations also indicate that the Sun does not rotate as a solid body. Instead, it spins *differentially*—faster at the equator and slower at the poles, like Jupiter and Saturn. Spot movements imply that the photosphere rotates once every 27 days at the equator, but only once in 31 days at the poles.

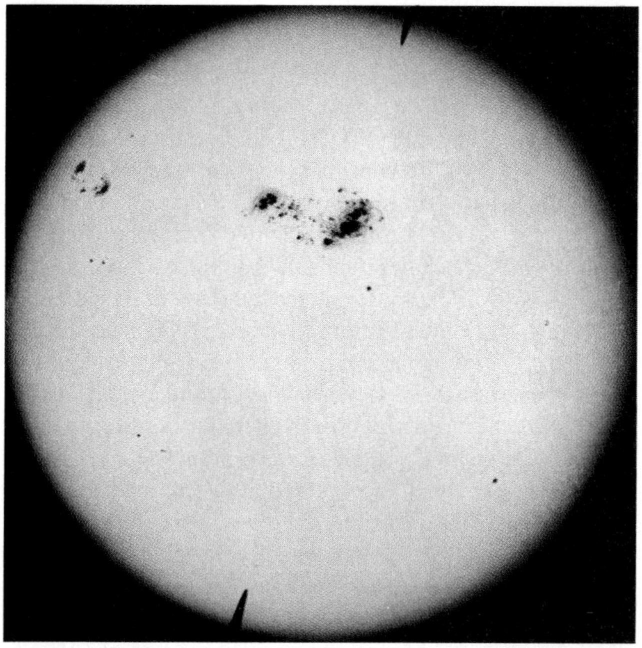

R I V U X G

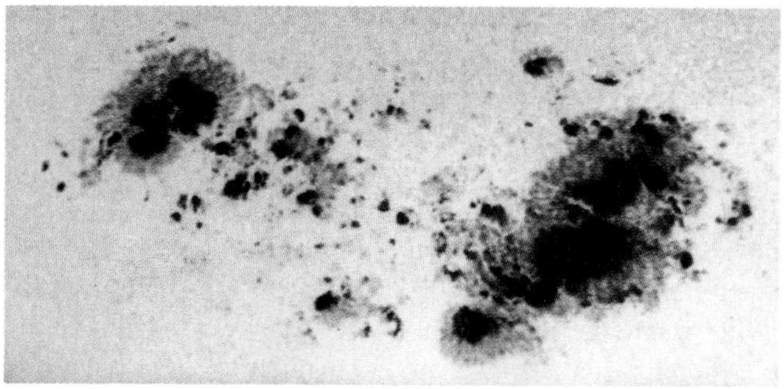

R I V U X G

Figure 18.16 *Enlarged photograph of a pair of sunspots (bottom) that can be seen in their broader context (top). Each spot consists of a cool, dark inner region called the umbra surrounded by a warmer, brighter region called the penumbra. The spots appear dark because they are slightly cooler than the surrounding photosphere. (Palomar Observatory)*

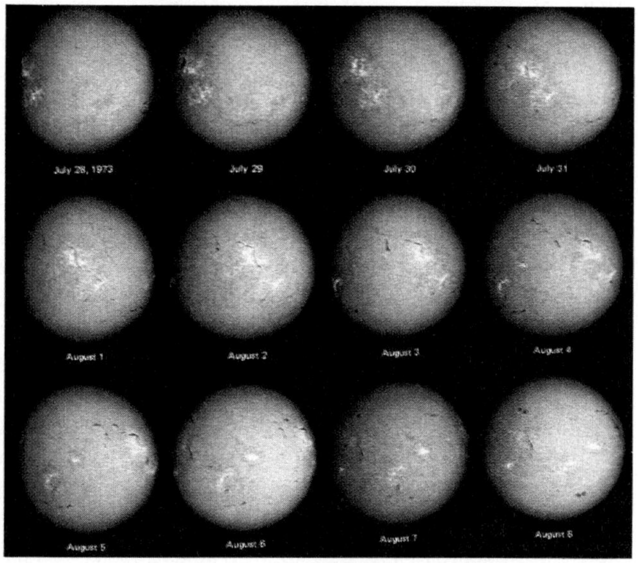

Figure 18.17 *The evolution of some sunspots and lower chromospheric activity over a period of nearly two weeks. A hydrogen-alpha fitter was used to make the observations from the Skylab space station. (NASA)*

R I V U X G

Solar Magnetism

What causes a sunspot? Why is it cooler than the surrounding photosphere? The answers to these questions involve the Sun's magnetism. Absorption lines in a sunspot are significantly wider than those observed anywhere else on the Sun. Careful analysis shows that this additional line width is caused by a strong magnetic field (see Chapter 5). The magnetic field in a typical sunspot is about 1000 times greater than the field in neighboring, undisturbed photospheric regions (which is itself several times stronger than Earth's field). No other naturally occurring magnets of comparable power are known anywhere in the solar system.

Sunspots, then, are sites of concentrated magnetic fields. Scientists believe that they are cooler than their surroundings because these abnormally strong fields tend to block (or redirect) the normal convective flow of hot gas toward the surface of the Sun.

Another indicator of the magnetic nature of sunspots is their grouping. They almost always come in pairs, and the magnetic fields observed in the two members of any pair are always opposite to one another—that is, the members of the pair have opposite magnetic *polarity*. As illustrated in Figure 18.18(a), magnetic field lines emerge from the interior through one member of a sunspot pair, loop through the solar atmosphere, then reenter the solar surface through the other spot. What's more, *all* the sunspot pairs in the same solar hemisphere (north or south) at any instant have the same magnetic configuration—if the magnetic field lines are directed into the Sun in one leading spot (measured in the direction of the Sun's rotation), they are inwardly directed in all leading spots in that hemisphere, as shown in Figure 18.18(b). What's more, in the other hemisphere at the same time, all sunspot pairs have the *opposite* polarity. Despite the irregularity of the sunspots themselves, these correlations suggest a high degree of order in the solar magnetic field.

The Sun's magnetic field is much stronger than Earth's. In addition, the Sun is gaseous and rotates differentially, and these facts radically affect the character of solar magnetism. The differential rotation distorts the solar magnetic field, wrapping it around the solar equator. Figure 18.19 shows an idealized sketch of this wrapping. The Sun's nonuniform rotation at the poles and equator eventually causes the original north-south magnetic field to reorient itself in an east-west direction. Convection then causes the magnetized gas to upwell toward the surface, twisting and tangling the magnetic field pattern. In some places, the field becomes kinked like a knot in a garden hose, causing it to increase in strength. Occasionally, the field strength becomes so great that it overwhelms the Sun's gravitational field and a "tube" of field lines bursts out of the surface and loops through the lower atmosphere, forming a sunspot pair. The general east-west organization of the underlying solar field accounts for the observed polarities of the pairs in each hemisphere.

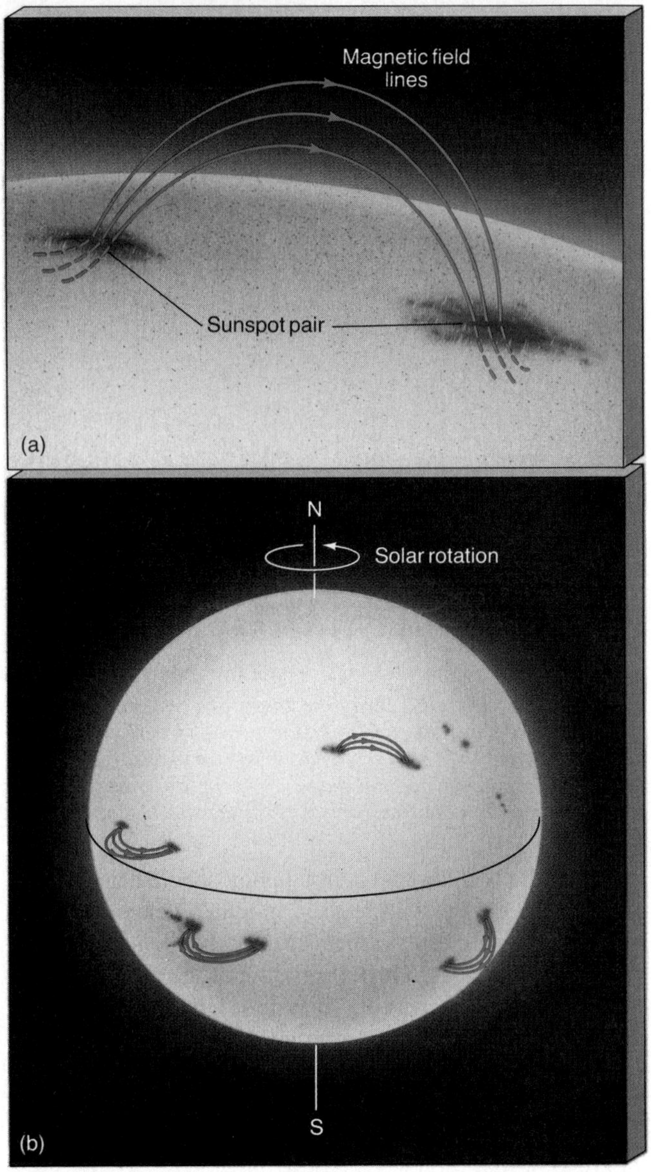

Figure 18.18 *(a) Sunspot pairs are linked by magnetic field lines. The Sun's magnetic field emerges from the surface through one member of the pair and reenters the Sun through the other. (b) The leading members of all sunspot pairs in the solar northern hemisphere have the same polarity—if the magnetic field lines are directed into the Sun in one leading spot, they are inwardly directed in all leading spots in that hemisphere. The same is true in the southern hemisphere, except that the polarities are always opposite to those in the north.*

The Solar Cycle

Not only do sunspots come and go with time, but their numbers and distribution across the face of the Sun also change in a fairly regular fashion. Centuries of observations have established a clear **sunspot cycle.** Figure 18.20 shows the number of sunspots observed each year during the twentieth century. The smooth line shows the 5-year average of the data, to make long-term trends more evident. The average number of spots reaches a maximum every 11 or so years, then falls off almost to zero before the cycle begins afresh. The repetition is not exact, however. While the *av-*

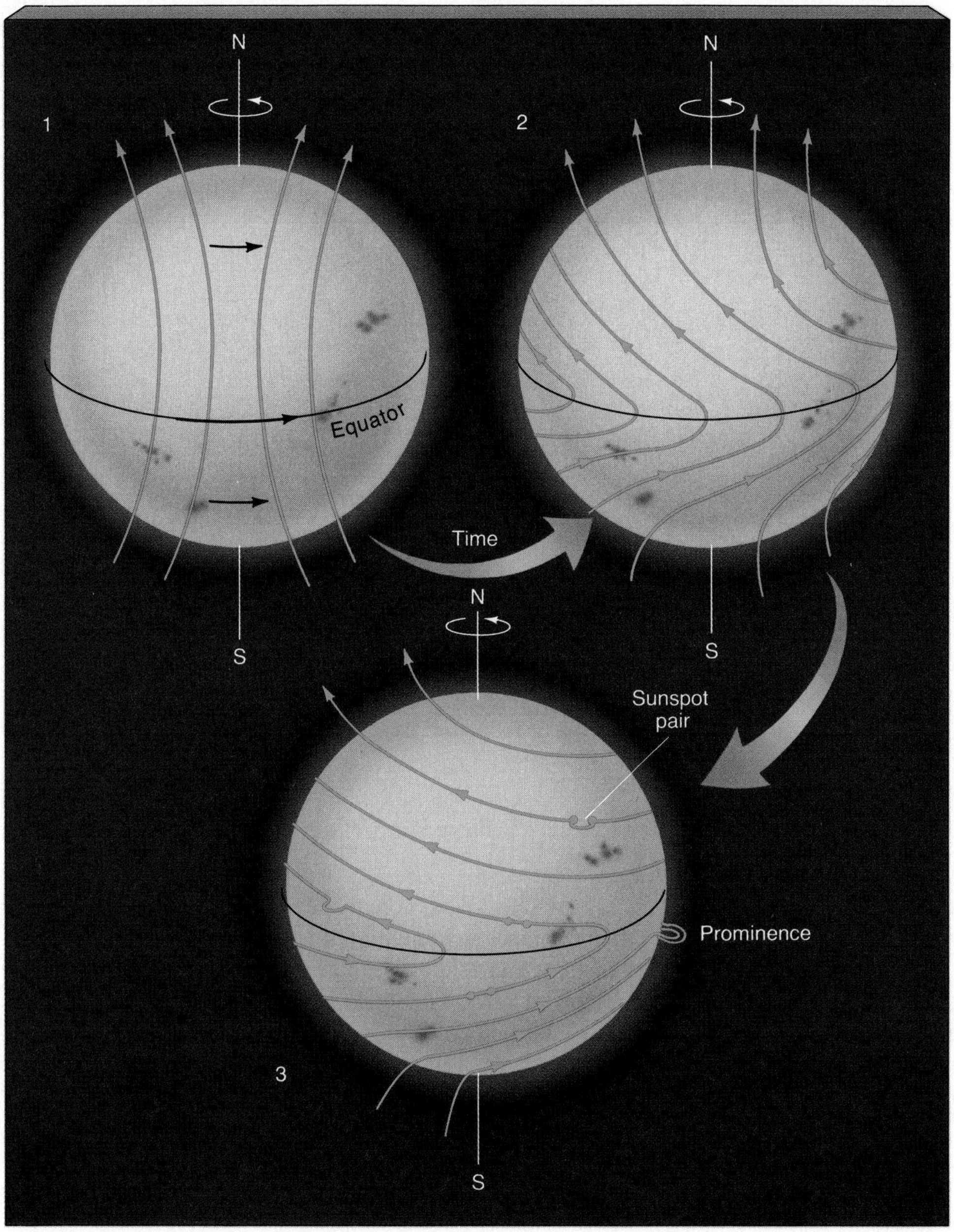

Figure 18.19 This diagram illustrates how the Sun's differential rotation wraps and distorts the solar magnetic field. Occasionally, the field lines burst out of the surface and loop through the lower atmosphere, thereby creating a sunspot pair. The underlying pattern of the solar field lines explains the observed pattern of sunspot polarities.

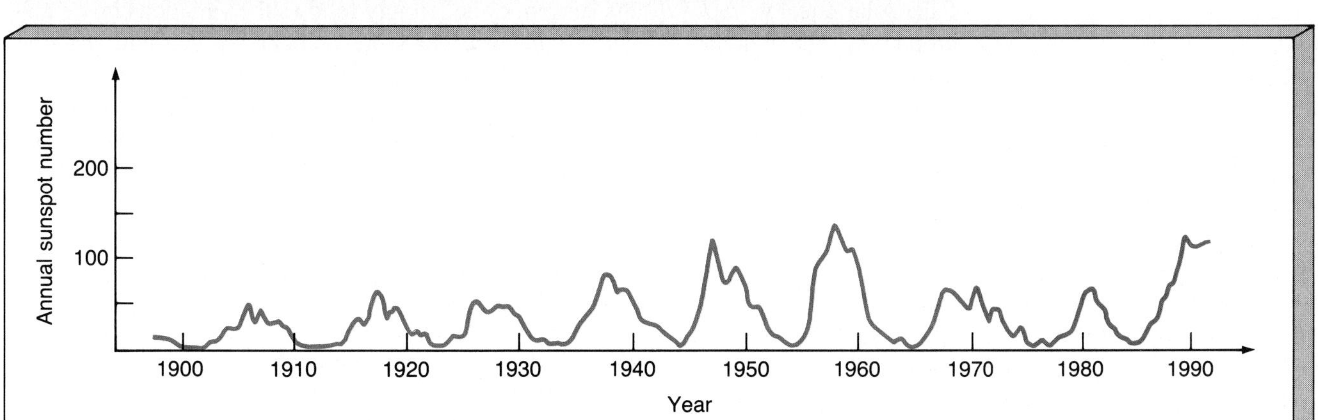

Figure 18.20 This graph presents the annual number of sunspots throughout this century. The red line shows the 5-year average of the annual data. The (roughly) 11-year solar cycle is clearly visible. At the time of solar minimum, virtually no sunspots are seen. About 4 years later, at solar maximum, as many as 100–200 spots are observed per year.

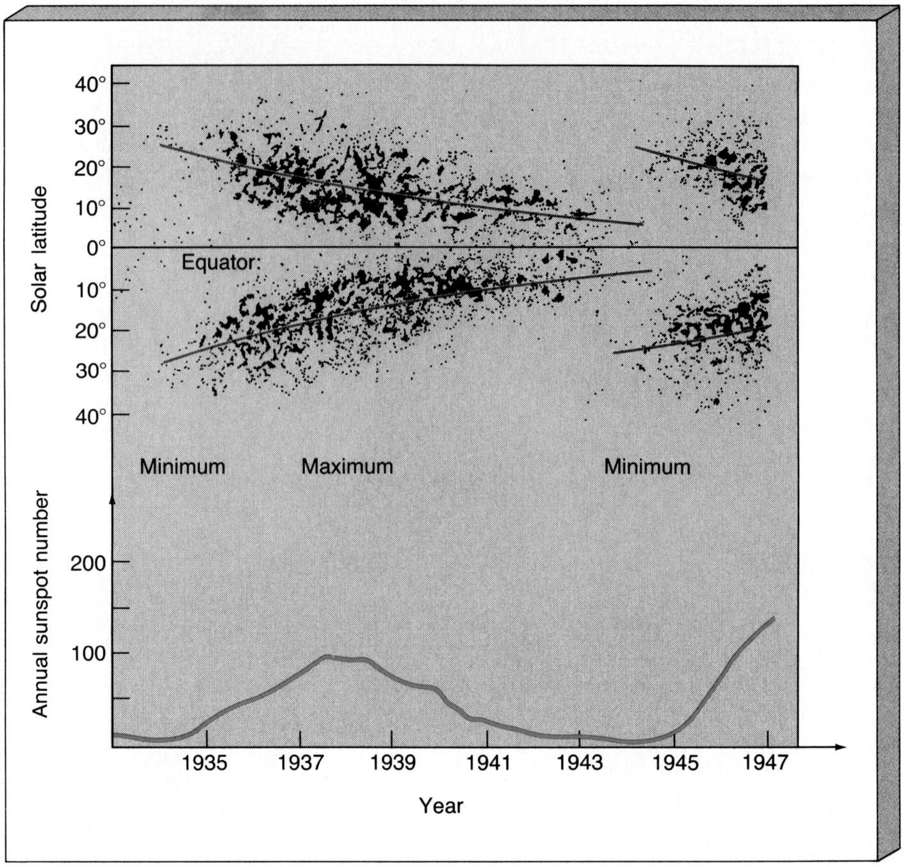

Figure 18.21 Sunspots cluster at high latitudes when solar activity is at a minimum. They appear at lower latitudes as the number of sunspots peaks. Finally, they are prominent near the Sun's equator as solar minimum is again approached. The most recent solar maximum occurred in 1990.

erage time from one maximum to the next is 11 years, there is quite a spread in individual cycles, which can vary in length from 7 to 15 years.

The latitudes at which sunspots appear vary as the sunspot cycle progresses. Individual sunspots do not move up or down in latitude, but new spots appear closer to the equator as older ones at higher latitudes fade away over the course of the 11-year cycle. Figure 18.21 is a plot of observed sunspot latitude as a function of time. At the start of each cycle, at the time of **solar minimum,** only a few spots are seen. They are generally confined to two narrow zones about 25° to 30° north and south of the solar equator. Approximately 4 years into the cycle, around the time of **solar maximum,** the number of spots has increased markedly. They are found within about 15° to 20° of the equator. Finally, by the end of the cycle, at solar minimum again, the total number has fallen again, and most sunspots lie within about 10° of the solar equator. The beginning of each new cycle appears to overlap the end of the last.

Complicating this picture further, the 11-year sunspot cycle is actually only half of a longer 22-year **solar cycle.** During any given sunspot cycle, the leading spots of all the pairs in the northern hemisphere have the same polarity, while spots in the southern hemisphere have the opposite polarity (see Figure 18.18[b]). However, these polarities reverse their signs on successive 11-year cycles. The full cycle takes 22 years to repeat, when we take the Sun's magnetism into account. Between one 11-year segment and the next, the *entire* solar magnetic field reverses itself.

Astronomers now believe that the Sun's magnetic field is generated and amplified by the constant stretching, twisting, and folding of the field lines that results from the combined effects of differential rotation and convection. This theory is essentially the same dynamo theory that accounts for the repeated reversals of the Earth's magnetic field over the past few tens of millions of years (see Chapter 7), except that the solar dynamo operates much faster. Dynamo theory predicts that the Sun's magnetic field should rise to a maximum, then fall to zero and reverse itself in a more-or-less periodic way. Solar surface activity, such as the sunspot cycle, simply follows the variations in the magnetic field. The changing numbers of sunspots and their migration to lower latitudes are both consequences of the strengthening and eventual decay of the field lines as they become more and more tightly wrapped around the solar equator.

Figure 18.22 below plots the full extent of all the sunspot data recorded since the invention of the telescope. It is a simple extension of the data given in Figures 18.20 and 18.21, and it shows the average number of sunspots observed at any one time. As can be seen, the 11-year "periodicity" of the solar sunspot cycle is far from regular. Not only does the period vary from 7 to 15 years, but the sunspot cycle has disappeared entirely in the relatively recent past. In honor of the British astronomer who drew attention to these historical records, the lengthy period of solar inactivity that extended from 1645 to 1715 is called the **Maunder minimum.**

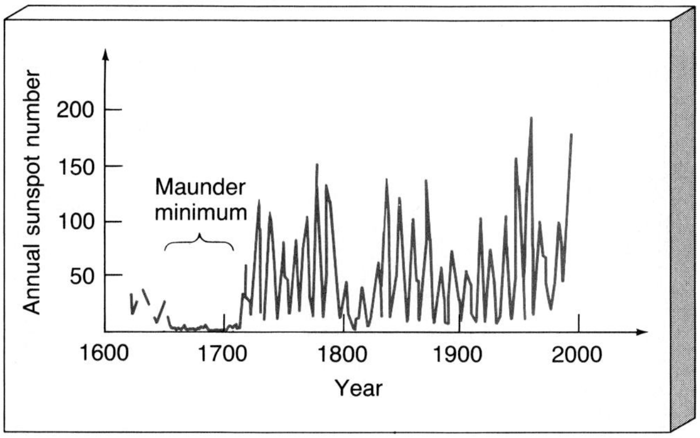

Figure 18.22 *This graph plots the number of sunspots occurring each year. Note the approximate 11-year "periodicity," and the absence of spots during the late seventeenth century.*

Active Regions

Sunspots are relatively quiescent aspects of solar activity. However, the photosphere surrounding them occasionally erupts violently, spewing forth large quantities of energetic particles into the surrounding corona. The sites of these explosive events are known simply as **active regions.** Most pairs or groups of sunspots have active regions associated with them. Like all aspects of solar activity, these phenomena tend to follow the solar cycle and are most frequent and violent around the time of solar maximum.

Active regions are generally somewhat hotter than their surroundings. As a result, even though sunspots are cooler and darker than the rest of the photosphere, the additional luminosity produced by the active regions around them makes the Sun slightly *brighter* than average at times of sunspot maximum and slightly *fainter* than average at solar minimum. Observations made by the *Solar Maximum Mission* (an Earth-orbiting satellite launched in 1980 to study solar activity, and repaired in 1984 by a NASA shuttle astronaut) showed variations in the Sun's luminosity of a few tenths of 1 percent over the course of one sunspot cycle.

Figure 18.23 shows two solar **prominences.** Prominences are loops or sheets of glowing gas ejected from an active region on the solar surface, moving through the inner parts of the corona under the influence of the Sun's magnetic field. Magnetic instabilities in the strong fields found in and near sunspot groups may cause the prominences, although the details are still not completely understood. Many observations, as in Figure 18.23(a), clearly show streams of hot ionized gas soaring high into the solar atmosphere, following the arching magnetic field lines between members of a sunspot pair.

Quiescent prominences persist for days or even weeks, hovering high above the photosphere, suspended by the Sun's magnetic field. **Active prominences** come and go much more erratically, changing their appearance in a matter of hours or surging up from the solar photosphere, then immediately falling back on themselves. A typical solar prominence measures some 100,000 km in extent, nearly 10 times the diameter of planet Earth. Prominences as large as that shown in Figure 18.23(b) (which traversed almost half a million kilometers of the solar surface) are less common and usually appear only at times of greatest solar activity. The largest prominences can release up to 10^{32} ergs of energy, counting both particles and radiation—not much compared with the total solar luminosity of 4×10^{33} erg/s, but still enormous by terrestrial standards.

Flares are another type of solar activity observed low in the Sun's atmosphere near active regions. Also the result of magnetic instabilities, flares, like that shown in Figure 18.24, are even more violent (and even less well understood) than prominences. They often flash across a region of the Sun in minutes, releasing enormous amounts of energy as they go. Observations made by the *Solar Maximum* satellite demonstrated that x-ray and ultraviolet emissions are especially intense in the extremely compact hearts of flares, where temperatures can reach 10^8 K. So energetic are these cataclysmic explosions that some researchers have likened flares to bombs exploding in the lower regions of the Sun's atmosphere. A major flare can release as much energy as the largest prominences, but in a matter of minutes or hours rather than days or weeks. Unlike the gas that

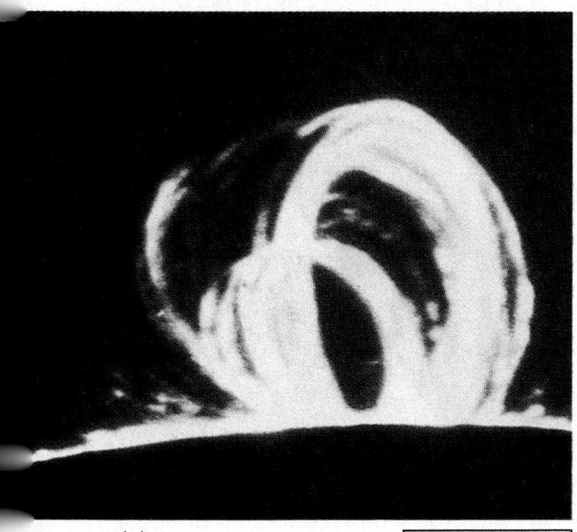

(a) R I **V** U X G

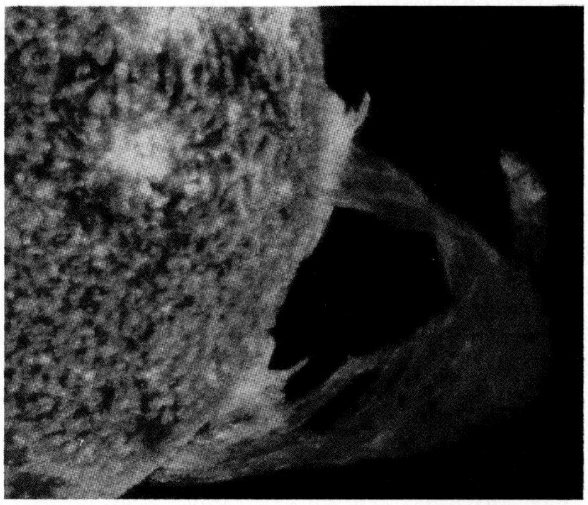

(b) R I V **U** X G

Figure 18.23 *(a) The looplike structure of this prominence clearly reveals the magnetic field lines connecting the two members of a sunspot pair. (b) This image of a particularly large solar prominence was observed by ultraviolet detectors aboard the* Skylab *space station in 1973. (NOAO; NASA)*

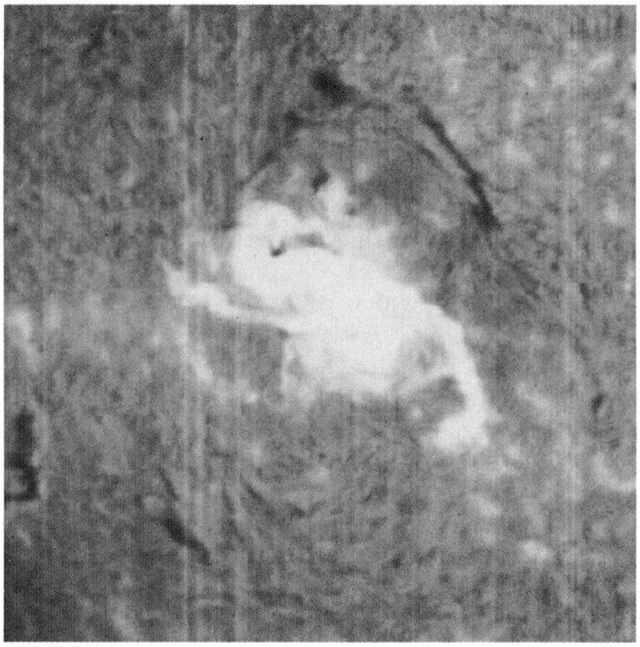

R I **V** U X G

Figure 18.24 *Much more violent than a prominence, a solar flare is an explosion on the Sun's surface that sweeps across an active region in a matter of minutes, accelerating solar material to high speeds and blasting it into space. (Palomar)*

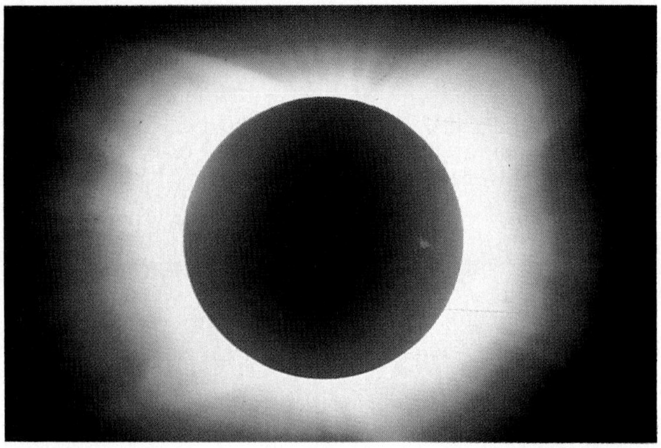

R I **V** U X G

Figure 18.25 *The solar corona at the peak of the sunspot cycle. At these times, the corona is much less regular and much more extended than at sunspot minimum (compare Figure 18.12). Astronomers believe that coronal heating is caused by surface activity on the Sun. The changing shape and size of the corona are the direct result of variations in prominence and flare activity over the course of the solar cycle. (G. Schneider)*

makes up the characteristic loop of a prominence, the particles produced by a flare are so energetic that the Sun's magnetic field is unable to hold them and shepherd them back to the surface. Instead, the particles are simply blasted into space by the violence of the explosion.

The Changing Solar Corona

The solar corona also varies in step with the sunspot cycle. The photograph of the corona in Figure 18.12 shows the quiet Sun, at sunspot minimum. The corona is fairly regular in appearance and appears to surround the Sun more or less uniformly. Compare this image with Figure 18.25, which was taken in 1991, close to the most recent peak of the sunspot cycle. The active corona is much more irregular in appearance and appears to extend farther from the solar surface. The "streamers" of coronal material pointing away from the Sun are characteristic of this phase of the solar cycle.

Astronomers now believe that the corona is heated primarily by solar surface activity, particularly prominences and flares, which can inject large amounts of energy into the upper solar atmosphere, greatly distorting its shape. Extensive disturbances often move through the corona above an active site in the photosphere, distributing the energy throughout the coronal gas. Given this connection, it is hardly surprising that both the appearance of the corona and the strength of the solar wind are closely correlated with the solar cycle.

THE HEART OF THE SUN

√5 Compared with the daily activity on Earth—storms, floods, geysers, even violent volcanos and earthquakes—the spots and flares of the active Sun are enormously energetic. The much greater steady emission from the inactive quiet Sun simply staggers the imagination. The Sun is somehow able to produce huge amounts of energy, and, according to Earth's fossil record, it has been doing so for the last several billion years. What powers the Sun? What forces are at work in the Sun's core to produce such energy? By what process does the Sun shine, day after day, year after year, eon after eon? Answers to these questions are central to all of astronomy. Without them, we can understand neither the physical existence of stars and galaxies in the universe nor the biological existence of life on Earth.

Solar Energy Production

As we saw earlier, the Sun's luminosity is 4×10^{33} erg/s, and its mass is 2×10^{33} g. We can quantify how efficiently the Sun generates energy by dividing the solar luminosity by the solar mass:

$$\frac{\textit{solar luminosity}}{\textit{solar mass}} = 2 \text{ erg/g/s}$$

This simply means that every gram of solar material yields about 2 ergs of energy every second. A couple of ergs is not much energy. A piece of burning wood generates about a million times more energy per unit mass per unit time than does our Sun. But there is an important difference: The wood will not burn for billions of years.

In order to appreciate the magnitude of the energy generated by our Sun, we must consider not the ratio of the solar luminosity to the solar mass, but instead the *total* amount of energy generated by each gram of solar matter *over the entire lifetime of the Sun as a star*. This is easy to do. We simply multiply the rate at which the Sun generates energy by the age of the Sun, about 5 billion years. We obtain a value of 3×10^{17} erg/g. This is the average amount of energy radiated by every gram of solar material since the Sun formed. It represents a *minimum* value for the total energy radiated by the Sun, for more energy will be needed for every additional day the Sun shines. Should the Sun endure for another 5 billion years (as is predicted by theory), we would have to double this value.

This energy-to-mass ratio value is very large. Nearly a billion billion ergs of energy must arise from *every* gram of solar matter to power the Sun throughout its lifetime. But the generation of energy is not explosive, releasing large amounts of energy in a short period of time. Instead, it is slow and steady, providing a *uniform* and long-lived rate of energy production. Only one known energy-generation mechanism can conceivably power the Sun in this way. That process is nuclear **fusion**—the combining of light nuclei into heavier ones.

Nuclear Fusion

We can represent a typical fusion reaction symbolically as

$$\text{nucleus 1} + \text{nucleus 2} \longrightarrow \text{nucleus 3} + \text{energy}$$

For powering the Sun, the most important piece of this equation is the energy produced. Let's see what gives rise to this energy.

The key point is that during a fusion reaction, the total mass *decreases*: The mass of nucleus 3 is *less* than the combined masses of nuclei 1 and 2. To understand the consequences of this, we can use a very important law of modern physics—the law of **conservation of mass and energy.** Albert Einstein showed at the beginning of the twentieth century that matter and energy are interchangeable. One can be converted into the other using Einstein's famous equation, $E = mc^2$. To determine the amount of energy corresponding to a given mass, simply multiply it by the square of the speed of light (c in the equation). For example, the energy equivalent of 1 g of matter is $1 \times (3 \times 10^{10})^2$, or 9×10^{20} ergs. The speed of light is so large that small amounts of mass translate into enormous amounts of energy.

The law of conservation of mass and energy states that the *sum* of mass and energy must always remain constant in any physical process. There are no known exceptions. According to this law, an object can literally disappear, provided that some energy appears in its place. If magicians really made rabbits disappear, the result would be a flash of energy equaling the product of the rabbit's mass and the square of the speed of light—enough to destroy the magician, everyone in the audience, and most of the surrounding city! In the case of a fusion reaction, the lost mass is converted into energy, primarily in the form of electromagnetic radiation. The light energy we see coming from the Sun means that the Sun's mass must be slowly decreasing

The Proton-Proton Chain

All atomic nuclei are positively charged, so they repel one another. Furthermore, the closer two nuclei come to one another, the greater is the repulsive force between them. How then do nuclei—two protons, say—ever manage to fuse into anything heavier? If they collide at high enough speeds, one proton can momentarily plow deep into the other, eventually coming within the exceedingly short range of the strong nuclear force (see Interlude 5-1). At distances less than about 10^{-13} cm, the attraction of the nuclear force overwhelms the electromagnetic repulsion, and fusion occurs. Speeds in excess of a few hundred kilometers per second, corresponding to a gas temperature of 10^7 K or more, are needed to slam protons together fast enough to initiate fusion. Such conditions are found in the core of the Sun and at the center of all stars.

At these temperatures, two protons can interact to produce another proton, a neutron, and two additional elementary particles. Figure 18.26 is a diagram of this event. We can also represent it by the following equation:

$$\text{proton 1} + \text{proton 2} \longrightarrow \text{proton 3} + \text{neutron} + \text{positron} + \text{neutrino}.$$

The **positron** particle in this reaction is a positively charged electron. Its properties are identical to those of a normal negatively charged electron, except for its positive charge. Scientists call the electron and the positron a "matter antimatter pair"—the positron is said to be the *antiparticle* of the electron. These newly created positrons find themselves in the midst of a sea of electrons, with which they interact immediately and violently. The particles and antiparticles annihilate one another, producing pure energy in the form of gamma-rays.

The final product of the reaction is a particle known as a **neutrino,** a word derived from the Italian for "little neutral one." Neutrinos are virtually massless and chargeless. They move at (or nearly at) the speed of light and interact with hardly anything. They can penetrate, without stopping, several light-years of lead. Despite their elusiveness, neutrinos can be detected with carefully constructed instruments. At the end of this chapter we will see an example of a rudimentary neutrino "telescope" and the important contribution it has made to solar astronomy.

The neutron and proton produced in the collision merge to form a **deuteron,** the nucleus of a special form of hydrogen—deuterium, also known as "heavy hydrogen."

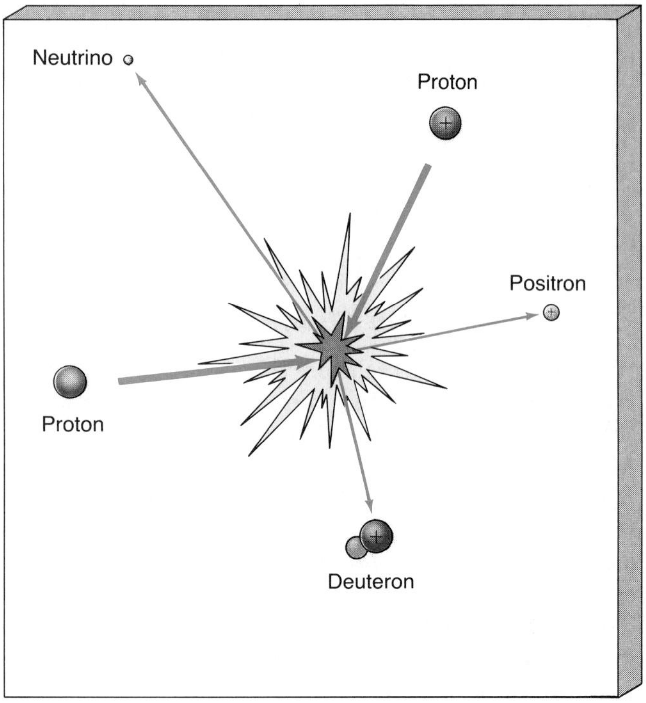

Figure 18.26 Diagram of two protons colliding violently, initiating the chain of nuclear fusion that powers the Sun.

Deuterium (D) differs from ordinary hydrogen by virtue of an extra neutron. Nuclei containing the same number of protons but different numbers of neutrons represent different forms of the same element—they are known as **isotopes** of that element. Usually, there are about as many neutrons in a nucleus as protons, but the exact number of neutrons can vary, and most elements can exist in a number of forms.

In order to avoid confusion when talking about isotopes of the same element, nuclear physicists attach a number to the symbol representing the element. This number indicates the total number of particles (protons plus neutrons) in its nucleus. Ordinary hydrogen is denoted by ^{1}H, deuterium as ^{2}H (or sometimes ^{2}D), tritium is ^{3}H, normal helium (2 protons plus 2 neutrons) is ^{4}He (also referred to as helium-4), and so on. We will adopt this convention for the remainder of this book. We can now write the net effect of the proton-proton reaction, including the formation of the deuteron, as

$$^1\text{H} + {}^1\text{H} \longrightarrow {}^2\text{D} + \text{positron} + \text{energy} + \text{neutrino}. \quad (1)$$

This equation is labeled (I) because the production of a deuteron by the fusion of two protons is the first step in the fusion process powering most stars. This type of reaction is called the **proton-proton chain.** Gargantuan quantities of protons are fused within the core of the Sun each and every second.

The next step in solar fusion is the formation of an isotope of helium. It is produced in another nuclear process, symbolized by the equation

$$^2\text{D} + {}^1\text{H} \longrightarrow {}^3\text{He} + \text{energy}. \quad \text{(II)}$$

A proton, of which there are many in the solar core, interacts with the deuteron particle produced in step (I). Step (II) begins as soon as deuterons appear. The main product is an isotope of helium—helium-3—lacking one of the neutrons contained in the normal (two proton and two neutron) helium-4 nucleus. Energy is also emitted, again in the form of gamma-ray photons.

The third and final step in the proton-proton chain, also verified by direct laboratory experiments, involves the production of nuclei of helium-4. Helium-4 (^{4}He) comes about most often through the fusion of two of the helium-3 nuclei created in step (II):

$$^3\text{He} + {}^3\text{He} \longrightarrow {}^4\text{He} + {}^1\text{H} + {}^1\text{H} + \text{energy}. \quad \text{(III)}$$

A helium-4 nucleus plus two more protons arise. Alternative reactions leading to the same final result exist, but they are relatively rare in stars like the Sun.

The net effect of steps (I) through (III) is this: Four hydrogen nuclei (protons) combine to create one helium-4 nucleus, plus some gamma-ray radiation, and two neutrinos. The whole process is illustrated in Figure 18.27. Symbolically, we have

$$4 \times {}^1\text{H} \longrightarrow {}^4\text{He} + \text{energy} + 2 \text{ neutrinos}.$$

As we have seen, the gamma-ray photons are slowly degraded in energy as they pass through the solar interior, and the energy eventually escapes in the form of visible light. The neutrinos escape unhindered into space at the speed of light. The helium stays put in the core.

Energy Generated by the Proton-Proton Chain

Let us now calculate the energy produced in the fusion process and compare it with the 3×10^{17} erg/g needed to account for the Sun's luminosity. Careful laboratory experiments have determined the masses of all of the particles involved in the conversion of 4 protons into a helium-4 nucleus: The total mass of the protons is 6.6943×10^{-24} g, the mass of the helium-4 nucleus is 6.6466×10^{-24} g, and the neutrinos are virtually massless. The difference between the total mass of the protons and the helium nucleus is lost in the reaction. This is not much mass, but it is measurable. This difference, 0.048×10^{-24} g, is transformed into energy. Multiplying the vanished mass by the square of the speed of light yields 4.3×10^{-5} erg. This is the energy produced by the fusion of 6.7×10^{-24} g (the rounded-off mass of the 4 protons) of hydrogen into helium. It follows that fusion of 1 g of hydrogen would generate 6.4×10^{18} erg—more than enough to power the Sun. In order to fuel the Sun's present energy output, hydrogen must be fused into helium in the core at a rate of 600 million

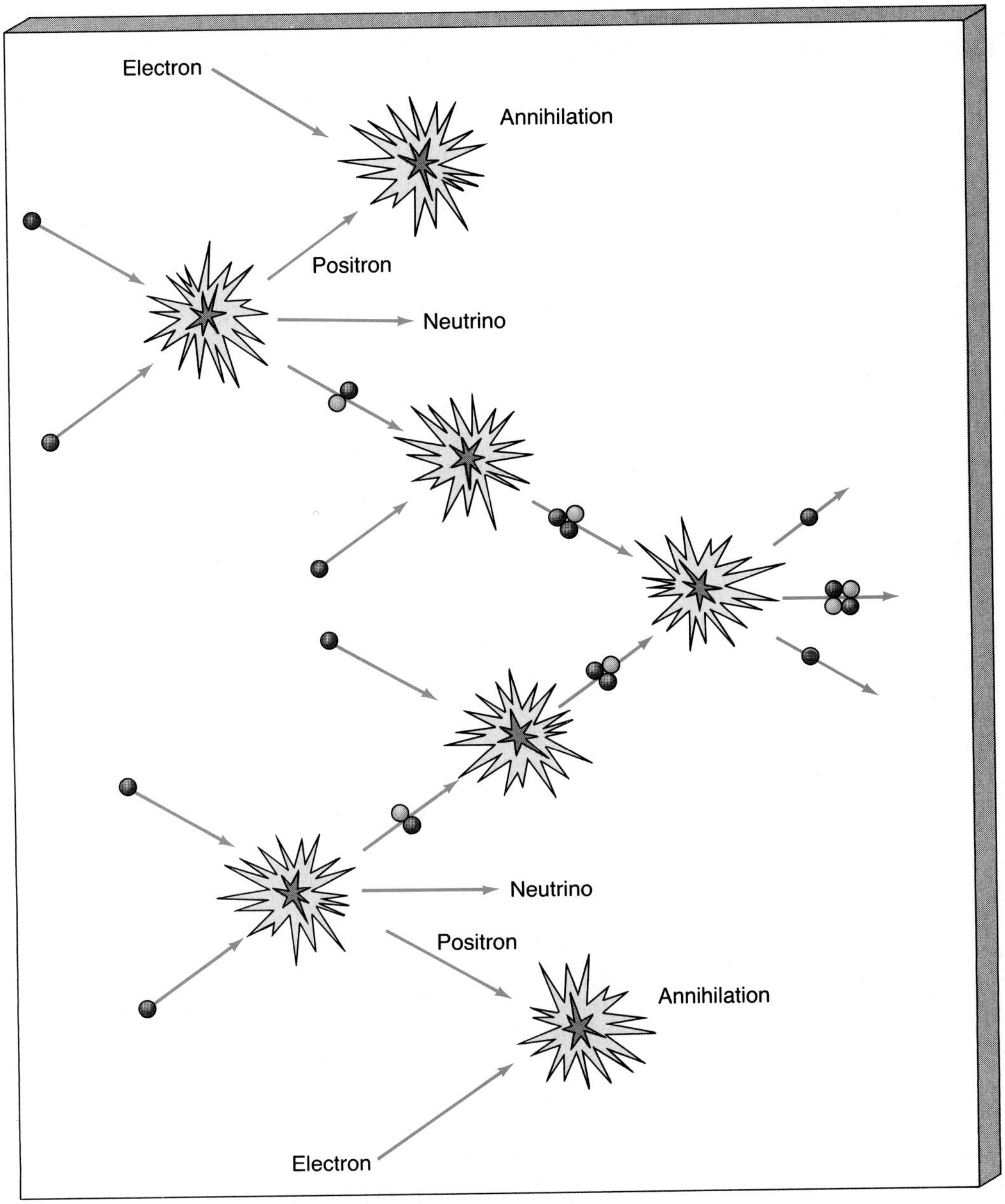

Figure 18.27 Diagram of the entire proton-proton chain. A total of 6 protons (and 2 electrons) are converted into 2 protons, 1 helium-4 nucleus, and 2 neutrinos. The 2 leftover protons are available as fuel for new proton-proton reactions, so the net effect is that 4 protons are fused to form 1 helium-4 nucleus. Energy, in the form of gamma-rays, is produced in each reaction. (Red particles are protons and green particles are neutrons)

tons per second. This same basic process (see Interlude 18-1 for a variation on the theme) is responsible for the light emitted by almost all the stars we see.

OBSERVATIONS OF SOLAR NEUTRINOS

Neutrino Astronomy

√6 Electromagnetic radiation cannot escape directly from the Sun's interior. The powerful gamma-ray photons created in the solar core are absorbed and reemitted many times, bouncing their way through the interior for a few hundred thousand years before they finally leave the solar surface as visible or infrared photons. As they bounce around, the photons heat the gas of the solar interior.

By contrast, the neutrinos that arise as by-products of the proton-proton cycle *do* travel cleanly out of the Sun, interacting with virtually nothing. They leave at or near the speed of light, escaping into space a few seconds after being created at the core. The neutrinos do not heat the solar gas at all. On the contrary, they serve to cool it by swiftly carrying energy away.

Theorists are quite sure that the proton-proton chain operates in the Sun. Based on detailed "numerical experiments"—that is, simulations—on large computers, their models of the Sun's interior temperature, density, composition, and nuclear burning rates predict bulk properties that agree well with observations. The observations, however, are almost exclusively confined to the solar exterior—the photosphere, chromosphere, and corona. Astronomers have little *direct* evidence of the nuclear reactions occurring at the center of the Sun.

One experiment does test the inner workings of the

INTERLUDE 18-1 *The CNO Cycle*

The proton-proton chain is not the only nuclear process operating in the Sun and other late-generation stars. Another fusion mechanism capable of converting hydrogen into helium, starting from carbon-12 (^{12}C), proceeds according to the following six steps. Nitrogen (N) and oxygen (O) nuclei are created as intermediate products:

1. $^{12}C + {}^{1}H \longrightarrow {}^{13}N$ + energy,
2. $^{13}N \longrightarrow {}^{13}C$ + positron + neutrino,
3. $^{13}C + {}^{1}H \longrightarrow {}^{14}N$ + energy,
4. $^{14}N + {}^{1}H \longrightarrow {}^{15}O$ + energy,
5. $^{15}O \longrightarrow {}^{15}N$ + positron + neutrino,
6. $^{15}N + {}^{1}H \longrightarrow {}^{12}C + {}^{4}He$.

These six steps are termed the *CNO cycle*. Aside from the radiation and neutrinos, notice that the sum total of these six reactions is

$$^{12}C + 4(^{1}H) \longrightarrow {}^{12}C + {}^{4}He.$$

Thus, the *net* result is the fusion of four protons into a single helium-4 nucleus, just as in the proton-proton chain. The carbon-12 acts merely as a *catalyst*, a stimulant of change that is not itself consumed in the reaction.

The repulsive electromagnetic forces operating in the CNO cycle are greater than in the proton-proton chain because the charges of the heavy-element nuclei are larger. Accordingly, higher temperatures are required to coerce the heavy nuclei into the realm of the strong nuclear force to ignite fusion. The following figure presents a numerical estimate of the energy released in the Sun by the proton-proton chain and the CNO cycle, each as a function of gas temperature. The proton-proton chain dominates at lower temperatures, up to about 16 million K. Above this temperature, the CNO cycle is the more important fusion process.

According to our theoretical models of the Sun, the temperature of the solar core is 15 million K, so these curves indicate that the proton-proton cycle is the dominant source of solar energy (notice that each step on the vertical scale corresponds to a *factor* of 100 in energy generation). The CNO cycle contributes no more than 10 percent of the observed solar radiation. However, stars more massive than our Sun often have core temperatures much higher than 20 million K, making the CNO cycle the dominant energy emission mechanism.

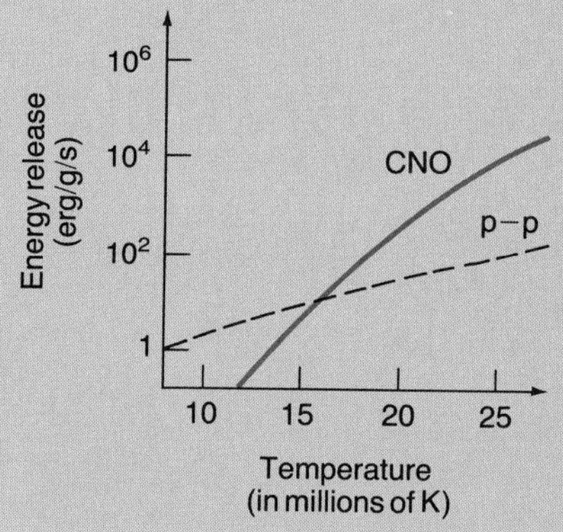

solar core. Unfortunately, its results do not agree with theoretical predictions. Let's examine it in a little more detail.

Neutrinos offer, at least in principle, the possibility of probing directly the conditions at the heart of the Sun. Of course, the fact that they can pass through the entire Sun without interacting also makes neutrinos fairly difficult to detect on Earth. Nevertheless, they do interact a little more strongly with some elements than with others, and this knowledge can be used in the construction of Earth-based neutrino-detection devices. Some of the neutrinos produced by the Sun happen to interact more strongly with chlorine than with most materials. Occasionally, a neutrino will encounter a chlorine-37 nucleus and convert it into a nucleus of argon-37.

In the late 1960s, a team of researchers from Brookhaven National Laboratory, led by physicist Ray Davis, built a large tank near the bottom of the Homestake gold mine in South Dakota and filled it with 400,000 liters (100,000 gallons) of a chlorine-containing chemical—the common cleaning fluid used by dry cleaners. Figure 18.28 shows a photograph of the apparatus. At 1.5 km below

Figure 18.28 *This swimming-pool-sized detector is a "neutrino telescope" of sorts, buried underground in a South Dakota gold mine. (R. Davis)*

ground level, the experimenters could be reasonably sure of avoiding interference from other sources, as most subatomic particles are unable to penetrate the Earth to such a depth. They left their tank in the mine for months at a time, periodically checking to see if any of the chlorine had been converted into argon, which would signal the absorption of a neutrino. Given the size of the detector and the expected physical conditions at the Sun's core, theory predicts that about one solar neutrino of the roughly 10^{16} that stream through the tank each day should be detected.

The Solar Neutrino Problem

For the first few years the experiment ran (in the late 1960s), no neutrinos were found. More sensitive apparatus installed in the 1970s succeeded in detecting *some* neutrinos, but not as many as predicted. Neutrinos are currently detected about two or three times per week, not once per day. Apparently, the only way we have of peering directly into the core of the Sun presents us with a problem. It is generally known as the **solar neutrino problem.**

Where are the missing neutrinos? Is the proton-proton chain operating as we think? Do we *really* know what processes are at work deep in the hearts of stars? Astronomers are now wrestling with these questions as they attempt to understand stellar fusion. What was once regarded as a fully understood phenomenon—the proton-proton production of solar energy—is once again under intensive study by theoretical and experimental astronomers alike.

Although the detection of solar neutrinos is an incredibly exacting task, it is unlikely that the experimental apparatus is at fault. The neutrino deficit has persisted over two decades of almost continuous monitoring and, though scores of technicians have examined every facet of the Homestake instrument for instrumental flaws, they have found none. The experimental gear is well designed and well built. The conclusion that the Homestake detector works properly is bolstered by a more recent experiment (with a quite different detector design) conducted at Kamioka, Japan, which reports a similar neutrino deficit.

A drawback of the Homestake and Kamioka experiments is that they are sensitive to only a tiny fraction of the neutrinos actually produced by fusion in the Sun's core. The particular reaction creating the neutrinos they can detect is not reaction (I) above, but a much less probable sequence of events—a chain of reactions involving unstable nuclei of the light elements beryllium and boron—which occurs only about 0.25 percent of the time. Uncertainties in the exact rate at which these reactions occur once led astronomers to suspect that the theoretical calculation of "Homestake-detectable" neutrinos might simply be wrong. Because the beryllium-boron reaction chain accounts for only a tiny fraction of the Sun's total energy output, it has a negligible effect on the solar luminosity and surface temperature. As a result, modifications to its reaction rate might have made the Homestake and Kamioka results consistent with the Standard Model without significantly affecting any other aspect of the Sun's appearance.

That possibility has been all but eliminated by the two most recent neutrino detectors—the *Soviet-American Gallium Experiment* (or SAGE, for short) and the U.S.-European GALLEX collaboration—each of which uses the element gallium to capture solar neutrinos. Unlike Homestake and Kamioka, both SAGE and GALLEX can detect the neutrinos produced by reaction (I)—the initial step in the proton-proton chain—so they provide a much more direct probe of energy generation in the solar core. In 1991 and 1992, the first results were reported from these experiments, conducted in underground laboratories in Russia and in Italy. Like Homestake and Kamioka, each of them indicates a substantial shortfall of neutrinos below the predicted number.

The four neutrino-detection experiments we have just described disagree somewhat in their measurements of the precise extent of the deficit, but *each* of them sees considerably less than the expected number of solar neutrinos. The unavoidable conclusion is that there is a real discrepancy between the Sun's theoretical neutrino output and the neutrinos we actually observe on Earth. How can we explain this contradiction? If, as we think, the detectors are working correctly, there are really only two possibilities. Either neutrinos are not produced as frequently as we think, or not all of them make it to Earth. Let us now consider these alternatives in turn.

If the temperature in the solar core were lower, the number of neutrinos predicted by theory would be lower. If the center of the Sun were about 10 percent cooler than in the Standard Solar Model—about 13.5 million K—helium-4 would still be produced, but it would be accompanied by fewer neutrinos detectable by the Homestake experiment. However, lowering the temperature would also lower the Sun's luminosity, and most theorists agree that the numerical models could not be in error by as much as 1.5 million K while remaining consistent with all other solar observations. In addition, observations by the GONG group (discussed earlier) seem to rule out a central temperature below 15 million K. Most astronomers regard it as very unlikely that the resolution of the solar neutrino problem will be found in the nuclear physics of the Sun's interior.

Having discarded the most obvious explanation, scientists have sought other solutions to the solar neutrino problem. Some have suggested that perhaps the Sun is pulsating slightly, changing its size and luminosity in a regular way. If so, the solar core would experience alternating periods of higher and lower temperatures. If the Sun is currently experiencing a low core temperature and is shrinking, part of its present luminosity could be generated by the release of gravitational energy due to shrinkage, not by nuclear fusion, and the neutrino observations could be reconciled with the Standard Solar Model.

Another possibility is that the Sun's internal structure might be influenced by material not included in the Standard Solar Model. A current favorite among theorists is a class of hypothetical objects known as **W**eakly **I**nteracting **M**assive **P**articles—or **WIMPs** for short. Cosmologists "invented" WIMPs during their studies of questions concerning the very large-scale structure of the universe. WIMPs have mass but interact hardly at all with matter or radiation, making them extremely difficult to detect with conventional telescopes. As we will see in Chapter 26, these properties make them an ideal way to "hide" large amounts of mass in the universe, as many cosmologists contend is necessary to bring theory into agreement with observations. In the context of the Sun, it turns out that if enough WIMPs accumulated in the solar core, they might modify the density and temperature structure of the interior to the point where the observed neutrino emission could be consistent with theory.

The properties of the neutrinos themselves might also provide the answer. Although neutrinos are usually assumed to be massless, like photons, theorists have proposed that they might perhaps have a minute amount of mass—about 10,000 times less mass than electrons, which are themselves some 2000 times less massive than protons. This tiny mass could allow neutrinos to change their properties, even to transform into other particles, during their 8-minute flight from the solar core to Earth, through a process generally known as **neutrino oscillations.** In this picture, neutrinos are produced in the Sun at the rate required by the Standard Solar Model, but some of them turn into something else (oscillate) on their way to Earth and so go undetected. Proposed experiments near neutrino-producing nuclear reactors on Earth may be able to test this idea within the next few years.

At present, there is no conclusive observational evidence in favor of any of these theories. No solar pulsations have been observed that could account for the required energy generation. WIMPs remain an unproven theoretical possibility, and no firm experimental evidence yet exists that the neutrino mass is anything other than zero. For now, the mystery of the solar neutrinos remains unsolved; although most physicists favor the neutrino-oscillation explanation. With continued observations of the neutrinos from the main reaction in the proton-proton chain, astronomers are hopeful that the solution will be in hand by the end of the century. Virtually all researchers concur—or at least hope—that the correct interpretation of the solar neutrino problem will not tear apart the theoretical fabric of the proton-proton chain. Most believe that the description of solar fusion we have presented is this chapter is basically right; our understanding of neutrino physics just needs to be fine-tuned. But should drastic measures be needed to solve the solar puzzle, we may yet have to return to the drawing board to answer one of the most fundamental scientific questions of all: How does a star shine?

CHAPTER SUMMARY

Astronomers use surface observations and a theoretical understanding of the Sun's interior to formulate models of the interior. Information about the solar interior can also be obtained from observation of solar oscillations. The Standard Solar Model indicates that 90 percent of the Sun's mass is contained within the inner half of its radius, and the core temperature is about 15 million K. Energy created in the Sun's core travels toward the surface, first in the form of radiation, then by convection in the outer layers.

The photosphere, or visible surface of the Sun, has several million bright and dark gas granules that form the upper part of convection cells. A typical granule is about the size of an Earth continent. The photosphere's thickness is less than 1⁄10 of 1 percent of the solar radius, so we perceive the Sun as having a well-defined edge. Beyond the photosphere and the chromosphere—the Sun's lower atmosphere—the temperature rises sharply, reaching about 1 million K in the corona. The corona eventually merges into the solar wind. X-ray images reveal coronal holes, which are vast regions of lower-than-average density in the Sun's atmosphere. The solar wind from coronal holes streams forth at particularly high speeds.

Hydrogen and helium are the most abundant elements in the Sun, although many other elements have been detected there. Spectral lines observed from the Sun's chromosphere give information about temperature, density, and elemental abundances. Theory suggests that the same proportions of atoms are present everywhere in the Sun, with the exception of the core. These atoms may be in many different stages of excitation and ionization.

Small dark areas called sunspots are disturbed regions in an otherwise well-behaved photosphere. They appear dark because they are cooler than their surroundings. The number of spots on the Sun varies in a fairly regular 11-year cycle. Explosive events called prominences often display well-defined looplike structures, nearly 10 times the size of Earth. Solar flares are even more violent events occurring in the Sun's lower atmosphere. The Sun's differential rotation is thought to distort the solar magnetic field, creating sunspots, flares, and prominences. Activity on the Sun is thought to play a role in Earth's climate and magnetic activity.

Earth's fossil record shows that the Sun has produced enormous amounts of energy for several billion years. Of all possible energy sources, only thermonuclear fusion is capable of powering the Sun.

In the proton-proton chain, by which 4 protons are fused into a helium nucleus, some mass is lost and transformed into energy. In the Sun, the proton-proton chain is the dominant energy source, though the CNO cycle contributes up to 10 percent. In more massive stars with higher core temperatures, the CNO cycle dominates.

Seconds after their creation during the proton-proton chain, neutrinos escape into space at the speed of light. Neutrinos offer a glimpse of conditions at the solar core. However, experiments on Earth report neutrino deficits that are not easily explained by any current theories of solar energy production.

KEY WORDS

active prominences
active region
active Sun
chromosphere
conservation of mass and energy
convection zone
corona
coronal hole
deuteron
effective temperature
flare
fusion
GONG
granule
helioseismology
isotope
limb
limb darkening
luminosity
Maunder minimum
neutrino
neutrino oscillation
penumbra
photosphere
positron
prominence
proton-proton chain
quiescent prominences
quiet Sun
radiation zone
solar constant
solar core
solar cycle
solar disk
solar interior
solar maximum
solar minimum
Standard Solar Model
star
sunspot
sunspot cycle
supergranulation
transition zone
umbra
WIMP

REVIEW QUESTIONS

1. How nearby is the Sun compared to the next-nearest star?
2. Name and briefly describe the main regions of the Sun.
3. How massive is the Sun, compared with the Earth?
4. How hot is the solar surface? The solar core?
5. Describe how scientists construct models of the Sun.
6. Describe what is thought to happen to a parcel of solar energy, from the time it is created inside the Sun until it arrives at Earth.
7. Why does the Sun appear to have a sharp edge?
8. Give the history of "coronium," and tell how it increased our understanding of the Sun.
9. What is the solar wind?
10. What is a sunspot?
11. What was the Maunder minimum?
12. What is thought to be the cause of sunspots, flares, and prominences?
13. What fuels the Sun's enormous energy output?
14. Briefly describe the law of conservation of mass and energy.
15. As mass decreases in the proton-proton chain, where does it go?
16. Why are modern scientists trying so hard to detect solar neutrinos?
17. What are some possible explanations for the scarcity of detected neutrinos?
18. Use Wien's Law (peak wavelength = 0.29 cm/T, in Kelvins; see Chapter 4) to determine the wavelength corresponding to the peak of the Planck curve (a) in the core of the Sun, where the temperature is 10^7 K, (b) in the solar convection zone, where the temperature is 10^5 K, and (c) just below the solar photosphere, where the temperature is 10^4 K. What form (visible, infrared, x-ray, and so on) does the radiation take in each case?
19. Granules represent the top of convective cells in the Sun. If the convected material moves at 1 km/s, how long does it take to flow across the 1000-km expanse of a typical granule? Compare this with the roughly 10-minute lifetimes observed for most solar granules.
20. Use Stefan's Law (Flux = σT^4, in Kelvins; see Chapter 4) to calculate how much less energy (as a fraction) is emitted per unit area of a 4500-K sunspot than from the surrounding 6000-K photosphere.
21. How long does it take for the Sun to convert 1 Earth mass of hydrogen into helium?

DISCUSSION QUESTIONS

1. What would happen to us on Earth if the Sun's internal energy source suddenly shut off?
2. What is the status of the attempt to create controlled fusion on Earth?
3. What results have come from the *Ulysses* mission to explore the region of space above the Sun's poles?
4. What ancient cultures have worshipped the Sun? Is there a modern-day equivalent of Sun worship or—with the world's ozone layer problems—has the Sun's light come to be more feared than revered?

PROJECTS

1. An appropriately filtered telescope will show you all manner of spots on the Sun, if you are looking near the peak of the 11-year sunspot cycle. **NEVER LOOK DIRECTLY AT THE SUN WITHOUT A SPECIAL FILTER!** Count the number of dark spots you see on the Sun's surface. Notice that sunspots often come in pairs. Come back and look again a few days later, and you'll see that the Sun's rotation has caused spots to move.
2. Activity on the Sun can seriously disrupt shortwave broadcasting and other radio communications. Shortwave listeners know to tune to station WWV—which also provides accurate minute-by-minute time signals—for up-to-date information about solar activity. The "solar activity reports" come at 18 minutes after every hour on WWV. They are full of jargon and may sound incomprehensible at first, but if you keep listening day after day, you'll begin to notice trends. Listen particularly for the A Index, a daily report of geomagnetic activity, which is tied to activity on the Sun.
3. If you want to see a star in the night sky that's similar to our Sun, look for Capella in the constellation Auriga. This bright star is visible in the evening in the fall, winter, and spring. Like the Sun, it is a G star.

SUGGESTED READINGS

Bahcall, J. N. "The Solar-Neutrino Problem." *Scientific American* (May 1990). Solar neutrinos and how they complicate our ideas about the Sun.

Bennett, G. L. "Rendezvous with a Star." *Sky and Telescope* (November 1990). Article about *Ulysses,* the mission to study the Sun's polar regions.

Foukal, P. V. "The Variable Sun." *Scientific American* (February 1990). How the Sun does not shine as steadily as was once believed.

Giampapa, M. S. "The Solar-Stellar Connection." *Sky and Telescope* (August 1987). What studies of other stars may reveal about the Sun.

Giovanelli, R. *Secrets of the Sun*. Cambridge: Cambridge University Press, 1984. Excellent, very clear book explaining many aspects of solar science.

"Great Balls of Fire." *Sky and Telescope* (May 1989). A spectacular sunspot, flares, and auroras seen in March 1989.

Harvey, J. W., J. R. Kennedy, and J. W. Leibacher. "GONG: To See Inside Our Sun." *Sky and Telescope* (November 1987). A telescope network for solar studies.

Kaler, J. B. "Cousins of Our Sun: The G Stars." *Sky and Telescope* (November 1986). Comprehensive article about stars similar to the Sun.

Maran, S. P. "Do Solar Fireworks Bring Stormy Weather?" *Smithsonian* (March 1990). Written near the peak of a solar cycle, this article focuses on how the Sun's activity may affect the Earth.

McIntosh, P. S. "Did Sunspot Maximum Occur in 1989?" *Sky and Telescope* (January 1991). About the most recent solar maximum.

McIntosh, P. S., and H. Leinbach. "Watching the Premier Star." *Sky and Telescope* (November 1988). How observations by amateurs can help scientists understand how the Sun works.

Menzel, D. *Our Sun*. Cambridge, MA: Harvard University Press, 1959. Classic book about the Sun.

Robinson, L. J. "The Sunspot Cycle: Tip of the Iceberg." *Sky and Telescope* (June 1987). New discoveries suggesting that the 11-year sunspot cycle is only part of the story of the Sun's cycle.

Smith, D. H. "The Solar-Neutrino Mystery Deepens." *Sky and Telescope* (October 1990). The solar neutrino saga.

Williams, G. E. "Solar Cycle in Precambrian Time." *Scientific American* (August 1986). Thin layers in an Australian rock formation reveal the Sun's activity 680 million years ago.

Zirin, H. "Heading toward Solar Maximum." *Sky and Telescope* (October 1988). Solar astronomers view the Sun's recent activity outburst.

CHAPTER 19

Measuring the Stars

Giants, Dwarfs, and the Main Sequence

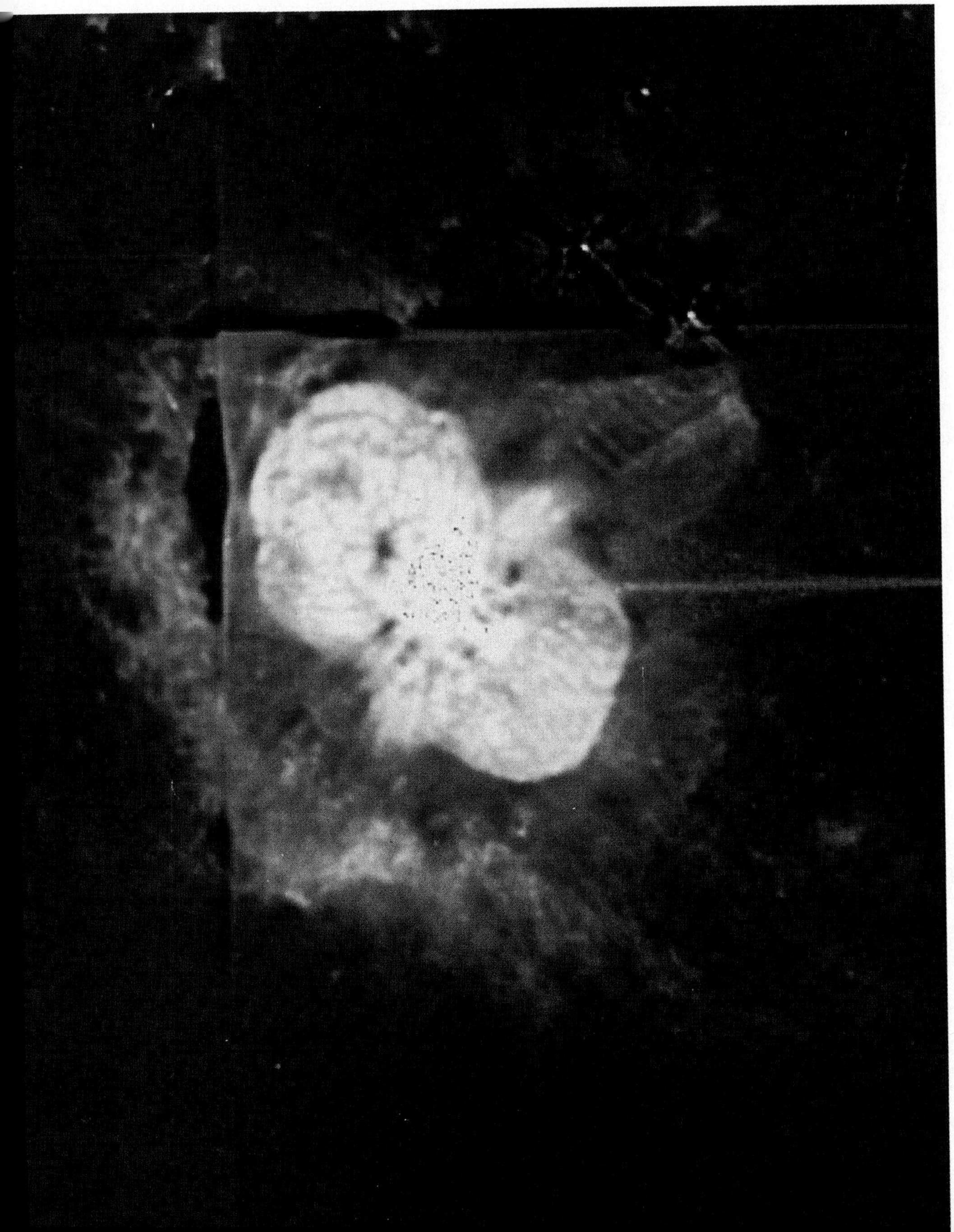

This Hubble Space Telescope *image has captured one of the most bizarre objects in the sky. Called Eta Carinae, this southern-hemisphere nebula is the result of a star having expelled much of itself during an outburst in 1843. Until a few years ago, when the central star was spotted by infrared means, astronomers had thought that the entire star had been blown to smithereens. Having a well-defined edge stretching over a couple of light-years, the surrounding peanut-shaped nebula is probably a thin shell of matter, rather than a filled volume. Knots and filaments trace the locations of shock fronts within the nebula; individual clumps as small as ten times the size of our solar system can be seen in this high-resolution image. Jets and ladderlike ''rungs'' also appear. No one has ever seen structures quite like this in space before. The sharp edges at top and left, as well as the spike at right, are artifacts of the CCD chip. (NASA)*

R I **V** U X G

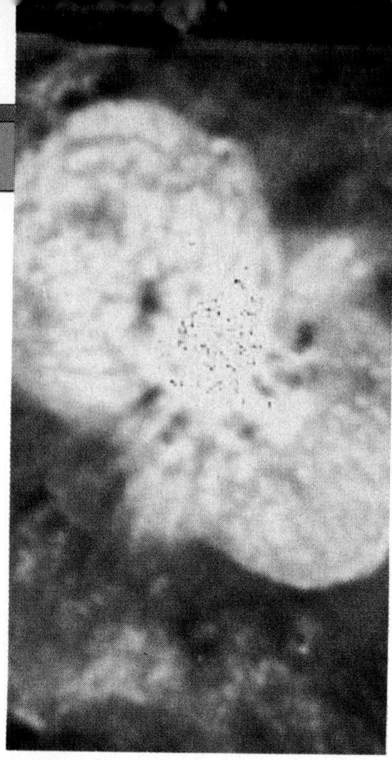

LEARNING GOALS

1. to understand how stellar distances are determined
2. to recognize that stars move through space
3. to understand why the sizes of stars, with few exceptions, must be estimated indirectly from the laws of physics
4. to appreciate the difference between true and apparent brightness
5. to know that stars are often categorized according to their surface temperatures
6. to understand how the masses of stars can be measured and how mass is related to other stellar properties

Up to this point, we have studied the Earth, the Moon, the solar system, and the Sun itself. To continue our inventory of the contents of the universe, we must move away from our local environment, into the depths of space. In this chapter we take a great leap in distance and consider stars in general. Our primary goal is to comprehend the nature *of the stars that make up the constellations as well as the myriad more distant stars we cannot perceive with our unaided eyes. Rather than studying their individual peculiarities, we will concentrate on determining the physical and chemical properties they share. There is order to be found in the legions of stars scattered across our skies. Like comparative planetology in the solar system, comparing and cataloging the stars plays a vital role in furthering our understanding of the galaxy and the universe we inhabit.*

THE DISTANCES TO THE STARS

Stellar Parallax

✓1 In Chapter 2, we studied how we can use *parallax* to measure distances to terrestrial and solar system objects. Recall that parallax is an object's apparent shift relative to some more distant background as the observer's point of view changes. In order to measure parallax, we must observe the object from either end of some baseline and measure the angle through which the line of sight to the object shifts. In astronomical contexts, the parallax is determined by comparing photographs made from the two ends of the baseline. As the distance to the object increases or the baseline shrinks, the parallax becomes smaller and harder to measure. Accordingly, a large baseline is essential for measuring the distance to a very remote object.

The stars are so far away that even Earth's diameter is too short to use as a baseline in determining their distance. Their apparent shift, as seen from different points on Earth, is just too small to measure. However, by comparing observations made of a star at different times of year, as shown in Figure 19.1, the baseline is effectively extended to the diameter of the Earth's orbit around the Sun, 2 A.U. Only with this enormously longer baseline do some stellar parallaxes become measurable. Other more distant stars do not reveal any apparent shifts, even with this 2-A.U. baseline.

The parallactic angle is always very small. Even for the closest stars, it is less than 1 arc second, so the imaginary triangle formed by the Earth, the Sun, and the star is actually much longer and narrower than suggested by Figure 19.1. Astronomers generally find it convenient to measure parallax in arc seconds, rather than in degrees. If we ask at what distance a star must lie in order for its parallax to measure exactly 1″ (arc second), we get an answer of 206,265 A.U., or 3.1×10^{18} cm. Astronomers call this distance 1 **parsec** (1 pc), the name being a contraction of "*par*allax in arc *sec*onds." Because parallax decreases as distance increases, we can relate the parallactic angle to a star's distance by the following simple formula:

$$\textit{distance} \text{ (in parsecs)} = \frac{1}{\textit{parallax} \text{ (in arc seconds)}}.$$

Thus, a star with a measured parallax of 1″ (arc second) lies at a distance of 1 pc from the Sun. The parsec is

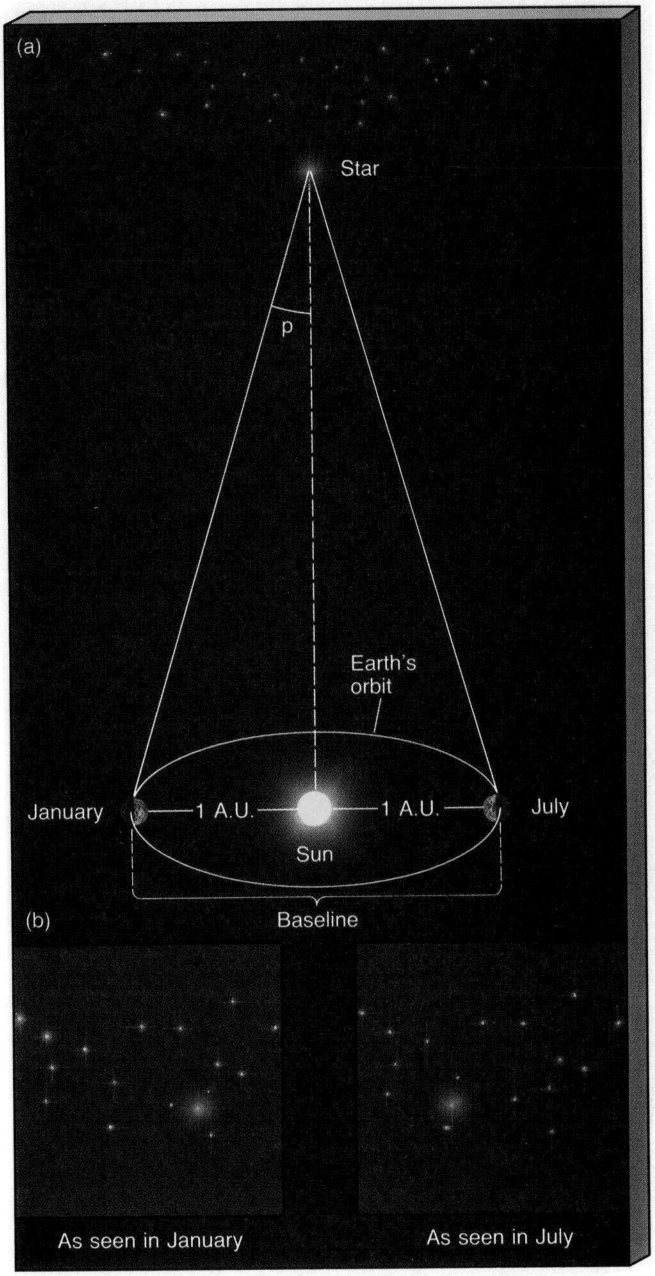

Figure 19.1 *(a) The geometry of stellar parallax. For observations made 6 months apart, the baseline is twice the Earth-Sun distance. (b) The parallactic angle* p *is usually measured photographically.*

constructed to make the conversion between distance and parallactic angle easy—an object with a parallax of 0.1″ lies at a distance of 10 pc, one with a parallax of 10″ lies at 0.1 pc, and so on. One parsec is approximately equal to 3.3 light-years.

We saw in Chapter 2 how Aristotle and other geometers of ancient Greece used the apparent absence of stellar parallax to argue that the Earth was stationary. They were eventually proved wrong, but only in the nineteenth century, when a German astronomer named Friedrich Bessel (1784–1846) finally succeeded in measuring the parallax of a nearby star. The observation of stellar parallax proves that Earth moves about the Sun. However, the lack of easily observed stellar parallax proves that even "nearby" stars lie at great distances from us.

Our Nearest Neighbors

The closest star to Earth (excluding the Sun) is called Proxima Centauri, a member of a triple-star system (three separate stars orbiting one another, bound together by gravity) known as the Alpha Centauri complex. Proxima Centauri displays the largest known stellar parallax, 0.76″, which means that it is about 1.3 pc away—about 270,000 A.U., or 4.3 light-years. That's the *nearest* star to Earth—at almost three hundred thousand times the distance from Earth to the Sun! This is a fairly typical interstellar distance in the Milky Way Galaxy.

Vast distances can sometimes be grasped through analogies. Here is one that might be helpful: Imagine Earth as a grain of sand orbiting a golfball-sized Sun at a distance of about 1 m. The nearest star, also a golfball-sized object, is then more than 100 *kilo*meters distant. Except for the other planets in our solar system, themselves ranging in size from grains of sand to small marbles and all lying within 50 m of the "Sun," nothing else of consequence exists in the 100 km separating the two stars. Such is the void of interstellar space.

The next nearest neighbor to the Sun beyond the Alpha Centauri system is called Barnard's Star. Its parallax is 0.55″, so it lies at a distance of 1.8 pc, or 6.0 light-years. Increasing numbers of stars are found at greater distances, as the volume of space surveyed rises, but only about 30 stars are known to lie within 4 pc of Earth. Figure 19.2 is a map of our nearest galactic neighbors.

Although ground-based images of stars are generally smeared out into a disk of radius 1″ or so by turbulence in Earth's atmosphere, astronomers have special equipment that can routinely measure stellar parallaxes as small as 0.03″ and often even smaller. Thus, stars within about 30 pc (100 light-years) of Earth have distances we can measure by this technique. Several thousand stars lie within this range. The majority of them are dimmer than the Sun and invisible to the naked eye. Most of the bright stars in the night sky are really much brighter than our Sun, so they are visible from Earth despite their great distance. The vast majority of stars in our Galaxy are far more distant than 30 pc.

In Chapter 3 we introduced the first "rung" on a ladder of distance-measurement techniques that will ultimately carry us to the edge of the observable universe. That rung is radar-ranging of the inner planets. It establishes the scale of the solar system to great accuracy and, in doing so, defines the astronomical unit. The second rung on our ladder is stellar parallax. We will discuss one more rung in this chapter and several more in later chapters as we continue to expand our cosmic field of view (see Figure 19.13).

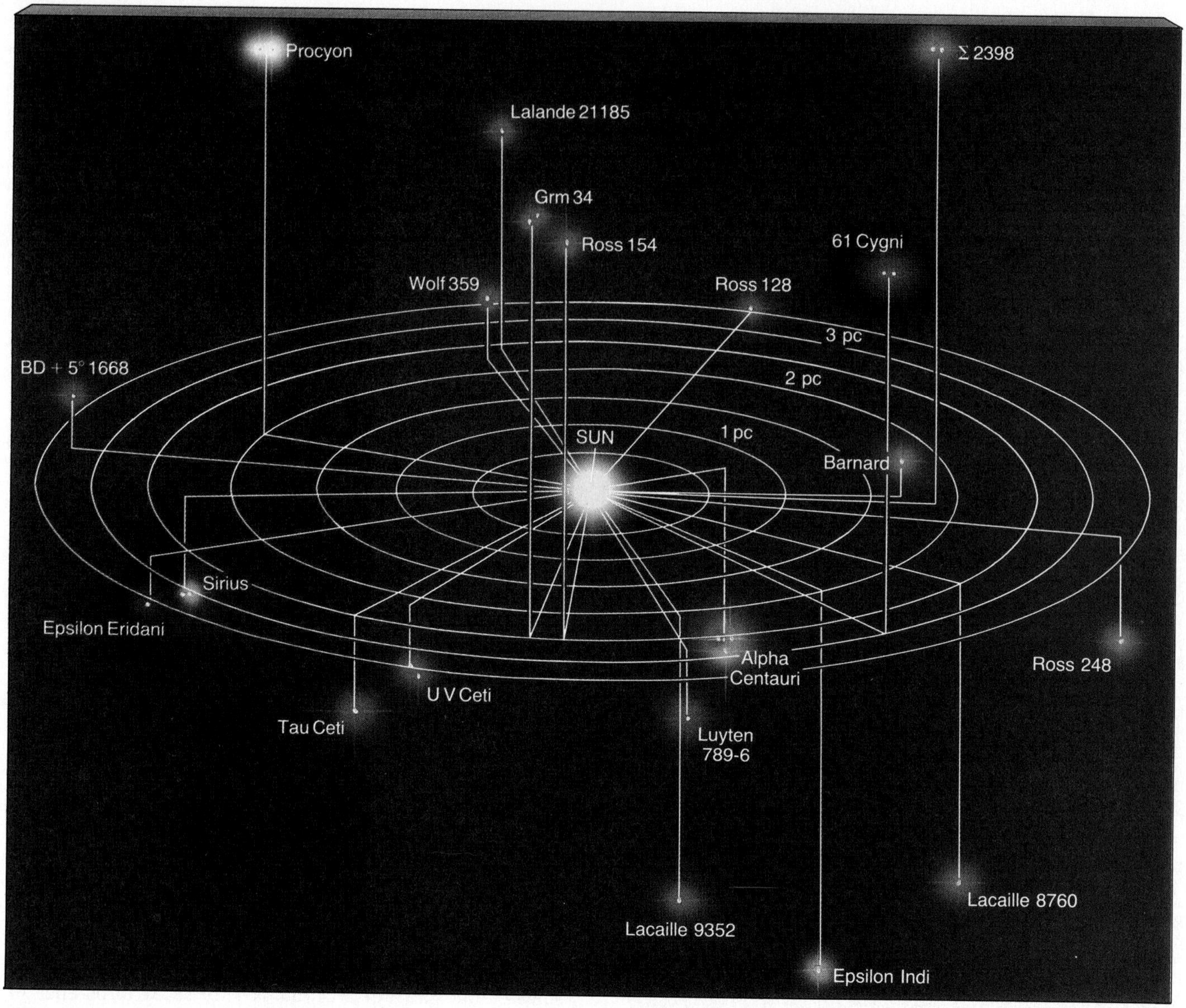

Figure 19.2 *A plot of the 30 closest stars to the Sun, projected in such a way as to reveal their three-dimensional relationships. Note that many are members of multiple-star systems, and most are cooler than the Sun. All lie within 4 pc (about 13 light-years) of Earth.*

Notice that the distances measured by stellar parallax depend on accurate knowledge of the Earth-Sun distance. An error in the measurement of the astronomical unit would immediately translate into errors in all distances obtained by parallax. (Observe, however, that the astronomical unit is actually known to an accuracy of about 1 part in 10^8—which is millions of times more accurately than parallactic angles can be measured—but the principle applies to all distance measurements.) We will find that every new rung on our ladder depends on the accuracy of all lower rungs, and any errors accumulate. Therefore, it is important that we fully understand each new measurement technique, for uncertainties at each level have ramifications for all studies of the universe on larger scales.

STELLAR MOTION

✓2 In addition to the apparent motion caused by parallax, stars have real motion, too. In other words, stars really do travel through space. The annual movement of a star across the sky, as seen from Earth (and corrected for parallax), is called **proper motion.** Like parallax, it is measured in terms of angular displacement, and the angles involved are typically very small. Proper motion is most commonly expressed in arc seconds per year. Stars' velocities can be quite large, but because of their great distances many years are needed before we can discern their movement in most cases.

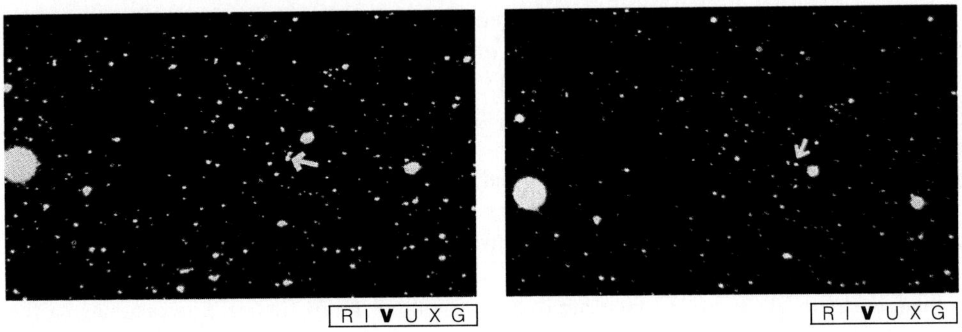

Figure 19.3 *Comparison of two photographic plates taken 22 years apart shows evidence of real space motion for Barnard's star (denoted by an arrow). (Harvard-Smithsonian Center for Astrophysics)*

Figure 19.3 compares two photographs of the sky around Barnard's Star. They were made on the same day of the year, but 22 years apart. As can be seen, Barnard's Star moved during this interval. If the two photographs were superposed, the two images of the star would not coincide. Because Earth was at the same point in its orbit when these photographs were taken, the displacement cannot be due to parallax caused by Earth's motion around the Sun. We conclude that the observed displacement indicates real space motion of Barnard's Star relative to the Sun.

Careful measurements show that Barnard's Star moved 227″ in the 22-year interval. The proper motion—*the annual angular displacement*—of Barnard's Star is 227″/22 years, or 10.3″/yr. This is the largest known proper motion of any star. Only a few hundred stars have proper motions greater than 1″/yr.

Proper motion does not account for the total space motion of a star. It is only one part of the real motion—namely, the component called the **transverse velocity,** which is projected onto the plane of the sky perpendicular to the line of sight. The other component of motion, along the line of sight, is called the **radial velocity.**

A star's transverse velocity is easily calculated if its proper motion and its distance are known. Figure 19.4 is a sketch of the Alpha Centauri star system in relation to the solar system. Its proper motion has been measured, relative to more distant background stars, at about 3.5″/yr. At Alpha Centauri's distance of 1.3 pc, an angle of 3.5″ corresponds to a physical displacement of 0.00002 pc, about 700 million km. Alpha Centauri takes a year to travel this distance, so its transverse velocity is (700 million km)/(3.2×10^7 seconds/year), or 22 km/s.

More generally, if a star's distance is d pc and its proper motion is μ″/yr (μ is the Greek letter mu, the symbol conventionally used by astronomers to denote proper motion), its transverse velocity V_t (in kilometers per second) can be shown to be

$$V_t = 4.7\mu d$$

For example, the transverse velocity of Barnard's Star, whose distance is 1.8 pc and whose proper motion is 10.3″/yr, can be calculated with the help of this equation to be 87 km/s.

We can determine the other component of a star's motion, the radial velocity, using the Doppler effect, as discussed in Chapter 4. Spectral lines from Alpha Centauri are slightly blueshifted, allowing astronomers to measure the star system's radial velocity as 20 km/s in the direction of Earth.

What is the true space motion of Alpha Centauri? Will this alien system collide with our own at some time in the future? The answer is no—Alpha Centauri's transverse velocity will steer it well clear of the Sun. We can combine the transverse and radial velocities according to the theorem of Pythagoras. The total velocity is $\sqrt{22^2 + 20^2}$, or about 30 km/s, in the direction shown by the color horizontal arrow in Figure 19.4. As indicated in the diagram, Alpha Centauri will get no closer to us than about 1 pc, and that won't happen for another 280 centuries.

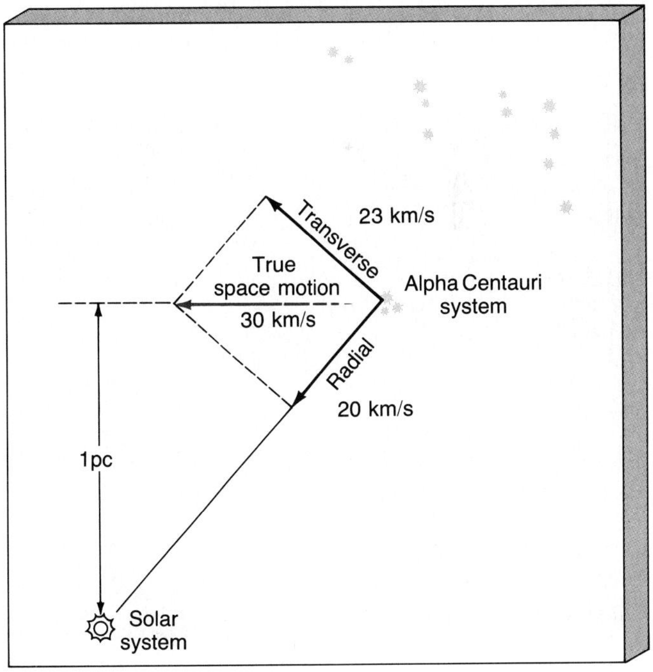

Figure 19.4 *The motion of the Alpha Centauri star system drawn relative to our solar system. The transverse component of the velocity has been determined by observing the system's proper motion. The radial component is measured using the Doppler shift of lines in Alpha Centauri's spectrum. The true space velocity, indicated by the red arrow, results from the combination of the two.*

STELLAR SIZES

Direct Measurement

✓3 Most stars are unresolvable points of light in the sky, even when viewed through the largest telescopes. Still, a few are big enough and close enough to allow *direct* measurement of their sizes. The best-known example is shown in Figure 19.5. The star Betelgeuse, in the constellation Orion, is hundreds of times larger than our Sun. It is a bright, reddish, swollen star, clearly visible in the winter sky. Using a special optical technique known as **speckle interferometry,** astronomers piece together many short-exposure images of the star. Each exposure is too brief for Earth's turbulent atmosphere to smear it out into a seeing disk. By combining many such faint images, astronomers can construct a very high-resolution map of the star's surface. The result is a composite image of Betelgeuse detailed enough to show individual surface features.

Using this technique, optical astronomers have directly measured the size of a few dozen stars. By measuring a star's angular size and knowing its distance from Earth, astronomers can determine its physical radius by simple geometry. These exceptions aside, however, the sizes of most stars must be inferred by more indirect means.

Indirect Estimates

Recall from Chapter 4 that the radiation emitted by a hot body is governed by *Stefan's Law*, which states that the energy emitted per unit area per unit time—a quantity known as the **energy flux**—by a hot body increases proportionally to the fourth power of the temperature. This law applies to a hot piece of metal, a glowing light bulb, or a star.

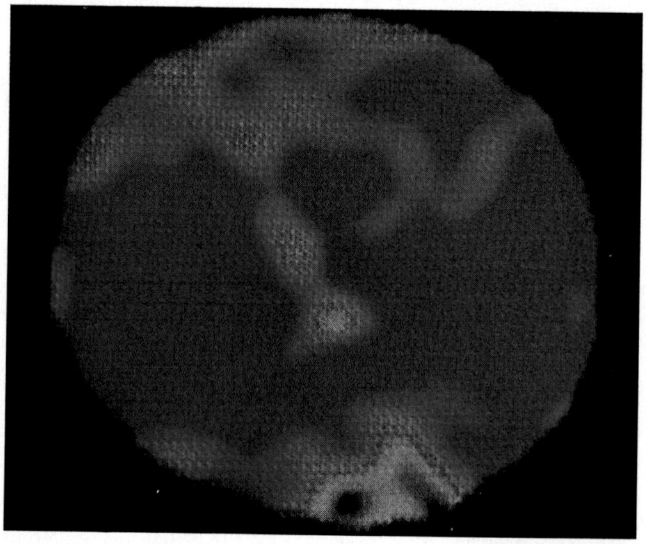

Figure 19.5 *The swollen star Betelgeuse (shown here in false color) is close enough that we can directly resolve its size, as well as some surface features thought to be storms similar to those occurring on the Sun. Betelgeuse is such a giant star (some 300 times the size of our Sun) that its photosphere roughly spans the size of Mars's orbit. Most of the surface features visible here are much larger than the entire Sun. (NOAO)*

Note that the energy flux is measured per square centimeter. To extend this measure of energy emission to incorporate the entire surface—in other words, to determine the *luminosity* of the star—we must multiply by the star's surface area. Because the energy flux is proportional to the fourth power of the stellar surface temperature and the area is proportional (the symbol for proportionality is $\propto$) to the square of the stellar radius, it follows that

$$luminosity \propto radius^2 \times temperature^4.$$

The proportionality may also be rewritten as

$$radius \propto \frac{\sqrt{luminosity}}{temperature^2}.$$

This **radius-luminosity-temperature relationship** is important because it demonstrates that knowledge of a star's luminosity and temperature can yield an estimate of its radius—an *indirect* determination of stellar size.

Giants and Dwarfs

Let's consider some examples to clarify the above ideas. The star Omicron Ceti (also called Mira) has a surface temperature of about 3000K and a luminosity of 1.6×10^{36} erg/s. Thus, the surface temperature is half, and the luminosity about 400 times, the corresponding quantities for our Sun. The above proportionality then implies that the star's radius is $\sqrt{400/0.5^2} = 80$ times that of our Sun. If our Sun were this large, its photosphere would extend as far as the orbit of Mercury. Omicron Ceti is a large object—in fact, such a star, whose radius is much greater than that of the Sun, is known as a **giant.** Since the color of any 3000K object is red, Omicron Ceti is a **red giant.** More precisely, giants are stars with radii between 10 and 100 times that of the Sun. Even larger stars, ranging up to 1000 solar radii in size, are known as **supergiants.**

Now consider Sirius B, a faint companion to Sirius A, the brightest star in the night sky. Sirius B's surface temperature is nearly 12,000 K, twice that of the Sun. Its luminosity is 10^{31} erg/s, about 0.002 times the solar value. Substituting these quantities into our equation, we find a radius of $\sqrt{0.002/2^2} = 0.01$ times the solar radius. Sirius B is much hotter but smaller and dimmer than our Sun. In fact, it is roughly the size of the Earth. Such a star is known as a **dwarf.** Because any 12,000K object glows white, Sirius B is an example of a **white dwarf** star. In astronomical terminology, any star of radius comparable to or smaller than the Sun (including the Sun itself), is called a dwarf.

Stellar sizes determined in this way range from $0.01\ R_\odot$ to $100\ R_\odot$ (recall that $\odot$ is the symbol for Sun, so $R_\odot$ is the radius of the Sun, 7.0×10^{10} cm). We will encounter some exceptions to this statement as we proceed through this text; however, it is valid for the vast majority of stars.

We see that to determine a star's size, we need know its luminosity and temperature. How do we determine these quantities? Let us now consider in turn each of these basic stellar properties. By studying them, we will learn a lot more about the general properties of stars.

LUMINOSITY AND BRIGHTNESS

✓4 Luminosity is the total amount of energy radiated into space each second from a star's surface. It is the *rate* of energy emission—not just the energy of visible light, but also the energy of any type of electromagnetic radiation, from radio waves to gamma-rays. For the majority of stars, the Planck spectrum peaks in or near the visible part of the spectrum, so most of the energy is in fact emitted in the form of visible light. Luminosity is an intrinsic property of a star. It does not depend in any way on the location or motion of the observer.

What makes one star appear brighter and another star dimmer? That is, what explains the **apparent brightness** of a star, which is how bright a star appears to an observer on Earth? Two factors determine a star's apparent brightness—its luminosity and its distance. A bright star is a powerful emitter of radiation, is near Earth, or both. A dim star is a weak emitter, is far from Earth, or both.

When we see streetlights at night, we can tell which are most distant because we can assume that they all have the same design and hence the same intrinsic brightness, or luminosity. When we look at the scene in the daytime, we have plenty of other points of reference to guide our eyes and provide visual cues. In astronomy, however, there usually are no other points of reference, and stellar luminosities can vary widely. As a result, the connection between apparent brightness and distance is not so easy to make. Two identical stars will have the same apparent brightness in the sky if (and only if) they lie at the same distance from Earth, but two different stars can also appear equally bright if the more luminous one happens to lie farther away.

Another Inverse-Square Law

How does apparent brightness diminish with distance? Figure 19.6 shows light leaving a star. Moving outward, the radiation passes through imaginary spheres of ever-increasing radius surrounding the emitting source. The radiation leaving the star is constant, and so the farther the light from the source, the less energy it carries per unit area. Think of the energy as being "diluted" as it expands into space. Because the area of a sphere grows as the square of the radius, the energy per unit area—the star's apparent brightness—is *inversely* proportional to the square of the distance from the star. Doubling the distance from a star makes it appear 2^2, or 4, times dimmer. Tripling the distance makes the apparent brightness 3^2, or 9, times less, and so on.

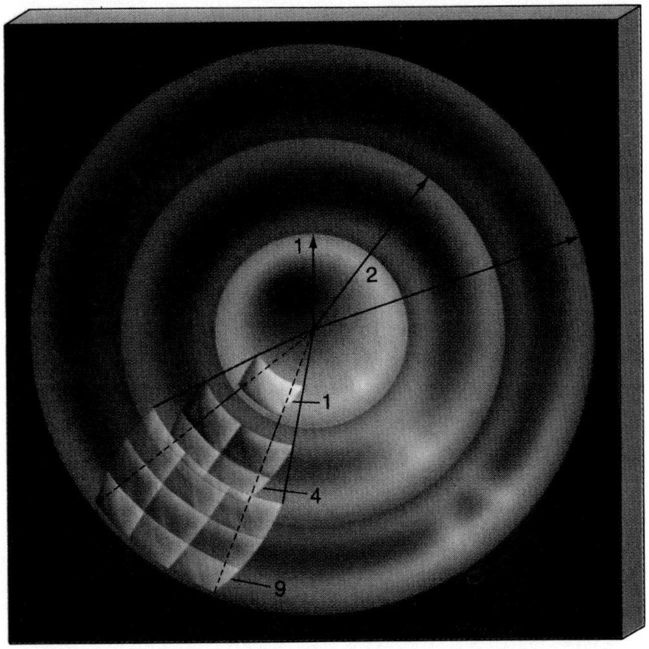

Figure 19.6 *The amount of detectable radiation (apparent brightness) varies inversely as the square of the distance from an emitting object. As radiation moves away from a source, it is steadily diluted as it spreads over progressively larger surface areas (depicted here as sections of spherical shells).*

If we were to change the luminosity of the source in Figure 19.6, we would also change its apparent brightness. Doubling its energy output would double the energy crossing each spherical shell, so the apparent brightness (the energy per unit area per unit time) would double. We can therefore make a fuller statement about apparent brightness: The apparent brightness of a star is *directly* proportional to the star's luminosity and *inversely* proportional to the square of its distance.

Absolute and Apparent Brightness

To compare sources of light—be they streetlights or starlight—we need some common ground. In order to understand the differences between stars, we must consider their *intrinsic*, not their *apparent*, properties.

If we were to look at all streetlights from the same distance, then differences in apparent brightness would reflect differences in the power of the bulbs used, not in our location relative to the bulbs. When comparing the luminosities of stars, astronomers do exactly the same thing. They imagine looking at all stars from a standard distance of 10 pc. There is no particular reason to use 10 pc; it is simply convenient. A star's **absolute brightness** is the apparent brightness the star would have if it were placed at a distance of 10 pc from Earth. Because the distance is fixed in this definition, the absolute brightness of a star depends *only* on its luminosity.

Finding a star's absolute brightness is a twofold task.

First, the astronomer determines the star's apparent brightness by measuring the flux of energy detected through a telescope. Second, the distance must be measured—by parallax for nearby stars, by other means (to be discussed later) for more distant stars. Using the inverse-square law, the astronomer can then mathematically place the star at a distance of 10 pc and so calculate its absolute brightness.

Notice that a star whose true distance is greater than 10 pc will have an absolute brightness greater than its apparent brightness as the astronomer "pulls" it closer, to 10 pc. Conversely, a star within 10 pc will have an absolute brightness less than its apparent brightness as the astronomer "pushes" it out to 10 pc. Consider for example the Sun. Because of its closeness to Earth, it appears extremely bright. However, if it were moved out to a distance of 10 pc, it would be barely visible to the naked eye of an Earth-based observer. See Interlude 19-1.

TEMPERATURE AND COLOR

Astronomers can obtain the surface temperature of a star from measurements of its brightness (radiation intensity) at different frequencies. Figure 19.7 shows three Planck curves describing the emission of radiation from three different stars. As we saw in Chapter 18, the surface ("effective") temperature of the Sun is found by measuring its radiation at many different frequencies, then matching the observations to the appropriate Planck curve. The theoretical Planck curve best fitting the Sun's emission describes a 5800K emitter.

We can use the same technique for any star, regardless of its distance. Actually, we need not measure every wavelength. Because the basic *shape* of the Planck curve is so well understood, we need only a few data points to determine the correct curve and thus determine the surface tem-

INTERLUDE 19-1 *The Magnitude Scale*

Not all stars in the night sky have the same apparent brightness—some are bright, others faint. This fact is clear even to the casual observer, without any specialized astronomical measuring equipment. Astronomers working at any wavelength speak of luminosity, flux, and brightness, but optical astronomers also describe brightness in terms of an ancient notion called the *magnitude* of an object.

In the second century B.C., the Greek astronomer Hipparchus ranked the naked-eye stars into six groups. The brightest stars were placed in the first group: He categorized them as *first magnitude*. The next brightest stars were labeled *second magnitude*, and so on, down to the faintest stars visible to the naked eye, which were classified as *sixth magnitude*. The range 1 (brightest) through 6 (faintest) spanned all the stars known to the ancients. Notice that a *large* magnitude means a *faint* star!

When modern astronomers began using telescopes with sophisticated detectors to measure the light received from the stars, they quickly discovered two important facts about the magnitude scale. First, the 5-magnitude range (from 1 to 6) defined by Hipparchus spans about a factor of 100 in energy received per unit area per unit time—energy flux—from stars: a first-magnitude star is approximately 100 times brighter than a star of the sixth magnitude. Second, the physiological characteristics of the human eye are such that each 1-magnitude change corresponds to a decrease in flux by a factor of about 2.5. A second-magnitude star is roughly 2.5 times fainter than a first-magnitude star, a third-magnitude star is roughly 2.5 times fainter than a second-magnitude star, and so on. By combining factors of 2.5, we confirm that a sixth-magnitude star is indeed $(2.5)^5 \approx 100$ times fainter than a first-magnitude star.

Astronomers have found it convenient to retain the ancient magnitude scale. They have formalized it by *defining* 5 magnitudes to be equivalent to a factor of exactly 100 in apparent brightness. One magnitude, then, is equivalent to a factor of the fifth root of 100, or approximately 2.512. A star of magnitude 2 is 2.512 times fainter than a star of magnitude 1, while a star of magnitude 3 is 6.31 ($= 2.512^2$) times fainter than a star of magnitude 1. More generally, the ratio of the apparent brightnesses of any two objects can be calculated simply by raising 2.512 to the power of the magnitude difference between them. For example, the planet Jupiter is about 10 magnitudes fainter than the Moon, as seen from Earth—in other words, we receive 10,000 (2.512^{10}) times less light from Jupiter than from the Moon.

Astronomers have broadened the magnitude scale in two ways. First, it no longer includes only whole numbers. A star intermediate in brightness between magnitude 4 and magnitude 5 has magnitude 4.5. The rule for comparing brightnesses still holds. A star of magnitude 4.5 is about 1.6 ($2.512^{0.5}$) times fainter than a star measuring magnitude 4.0 in brightness.

The second extension of Hipparchus's system (already hinted at by our example involving Jupiter and the Moon) is that magnitudes outside the range 1–6 are allowed. Very bright objects have magnitudes less than 1.0, and very faint objects can have magnitudes far greater than 6.0. Consider the star Sirius, the brightest star in our sky. Its energy flux, as measured by modern telescopes, actually places it brighter than magnitude 1. Because increases in flux are associated with decreases in magnitude, Sirius must have a magnitude less than 1. In fact, its magnitude is -1.5, 2.512 magnitudes brighter (10 times more flux) than the magnitude assigned to it by Hipparchus. Other very bright objects also have negative magnitudes—the Sun at perature. In practice, an astronomer observes a star's intensity at only a few selected wavelengths. This is done with telescope filters that block out all radiation except that within specific wavelength ranges. For example, a B (for blue) filter rejects all radiation except for a certain range of blue light. Defined by international agreement to extend from 3800 to 4800 Å, this range corresponds to wavelengths to which photographic film happens to be most sensitive. Similarly, a V (for visual) filter passes only radiation within the 4900 to 5900 Å range, corresponding to that part of the spectrum to which human eyes are particularly sensitive. Many other filters are also in routine use—a U (for ultraviolet) filter covers the near ultraviolet, and infrared filters span longer-wavelength parts of the spectrum.

Temperature can be determined with as few as two filters. Figure 19.7 shows how the B and V filters admit different amounts of light for objects of different temperatures. These measurements, or those made in any two suitably separated wavelength ranges, are enough to specify the Planck curve and thus yield the surface temperature. This is a completely general technique. It works for all objects that emit radiation through heat, be they glowing metals, flashlight bulbs, burning fires, or distant stars.

Let's consider a few examples. Suppose that we observe a particular star and find that the light passing through the B filter is almost twice as intense as that passing through the V filter, as shown in curve (a) of Figure 19.17. We can construct the entire Planck curve based only on those two measurements—no other Planck curve could be drawn through *both* measured points. By determining the temperature associated with that particular curve, we can measure the temperature of the star to be 30,000K. Such a hot star would have a distinct bluish tint when observed without any restricting filters.

−26.8, the full Moon at −12.5, Venus at −4.4 (at its brightest), and Jupiter at −2.7.

Modern telescopes can collect and focus light from stars much dimmer than were visible to the ancient Greeks. For example, the large Hale telescope on Mount Palomar can detect objects as faint as magnitude 25—equivalent to seeing a candle at a distance of about 15,000 km—from which we receive only about 1 photon per square centimeter per *hour*. The orbiting *Hubble Space Telescope* can study even fainter objects—as dim as magnitude 30. The figure below illustrates the magnitudes of some astronomical objects.

So far, the magnitude we have been discussing is equivalent to radiation flux, or apparent brightness. It is more usually known as *apparent magnitude*. Apparent magnitude denotes a star's brightness when viewed at its actual distance from Earth. *Absolute magnitude,* in analogy to absolute brightness, is the apparent magnitude of a star placed at the "standard" distance of 10 pc. It is a measure of the intrinsic brightness, or luminosity, of the star itself.

We can now recast our earlier discussion of absolute and apparent brightness in terms of magnitudes. A star with a distance of less than 10 pc would be diminished in apparent brightness and so *increased* in apparent magnitude, if moved to 10 pc from us. Stars less than 10 pc from Earth therefore have apparent magnitudes that are less than their absolute magnitudes. For stars more distant than 10 pc, the reverse is true. An extreme example is our own Sun. Because of its proximity to Earth, it appears very bright and thus has a large negative apparent magnitude. However, the Sun's absolute magnitude is 4.8—if the Sun were moved to a distance of 10 pc, it would be only slightly brighter than the faintest stars visible in the night sky.

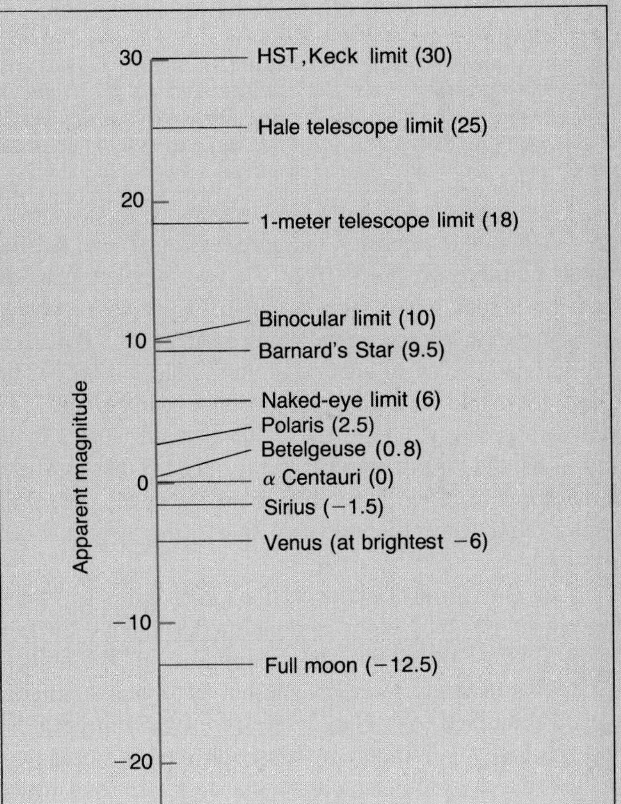

The magnitude system is a quaint and, to many students (scientists and nonscientist alike), highly confusing heirloom of traditional astronomy. Nevertheless, in its modern form it is deeply embedded in optical astronomy and is unlikely to be replaced in the foreseeable future.

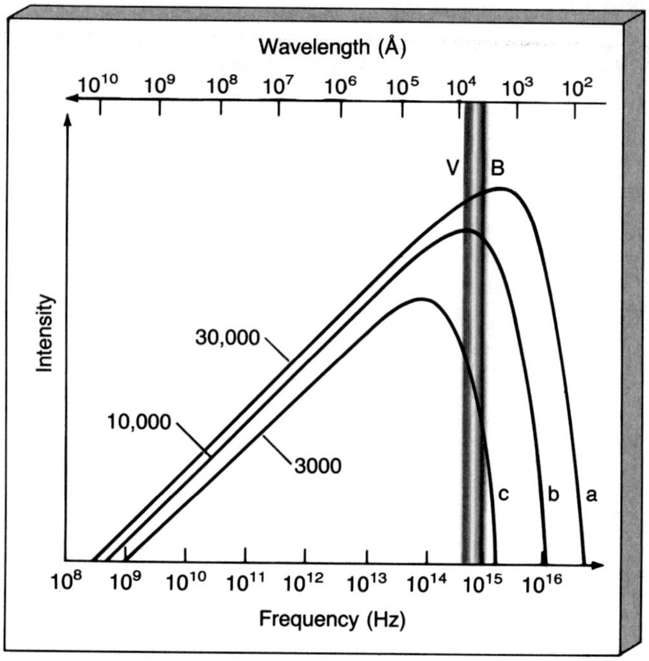

Figure 19.7 *A set of Planck curves for different temperatures, along with the locations of the B and V filters. Star (a) is very hot—30,000K—so its blue (B) intensity is considerably greater than its visual (V) intensity. Star (b) has B and V readings about the same and appears white. Its temperature is about 10,000K. Star (c) is red. Its V intensity greatly exceeds the B value. Its temperature is 3000K.*

When the intensity measured through the B filter equals that through the V filter, the two measurements define the Planck curve labeled (b) in Figure 19.17, corresponding to a surface temperature of about 10,000K. The characteristic color of such a glowing object is white, because all visible wavelengths contribute nearly equally. Finally, curve (c) of Figure 19.17 shows a star whose intensity measured through the B filter is only 1/5 that measured through the V filter. These two intensity values describe a Planck curve for a relatively cool, and distinctly red, 3000K emitter.

In astronomical parlance, the **color index** (or sometimes even just the "color") of a glowing object is the ratio of its B to V intensities. It is equivalent to the object's surface temperature to the extent that the object's radiation is well described by a Planck spectrum (which it often is), and it is easily measurable by telescopic means. Looking up at the night sky, you can tell at a glance which stars are hot and which are cool. In Figure 19.8, which shows the constellation Orion as it appears through a small telescope, the colors of the cool red star Betelgeuse and the hot blue star Rigel are plainly evident. However, to obtain the actual temperatures (3000K for Betelgeuse and 20,000K for Rigel), at least two intensity measurements at different wavelengths are required. (Notice, by the way, that these colors are intrinsic properties of the stars and have *nothing* whatever to do with Doppler red or blue shifts.)

Table 19-1 lists a more detailed description of the color index, the surface temperature derived from it, the dominant color perceived in the absence of any filters, and some stellar examples. This type of non-spectral-line analysis, where a star's intensity is measured through each of a set of standard filters, is known in the trade as **photometry** (literally meaning "light measurement"). In a particularly common photometric system—the **UBV system**—astronomers use U, B, and V filters to characterize stellar properties.

Figure 19.8 *The constellation Orion, as it would appear through a small telescope or binoculars. The different colors of the member stars are easily distinguished. The bright red star at the upper left is Betelgeuse; the bright blue-white star at the lower right is Rigel. (S. Westphal)*

TABLE 19-1 ***Stellar Colors and Temperatures***

B intensity / V intensity	Surface Temperature (K)	Color	Familiar Examples
1.7	30,000	electric blue	
1.3	20,000	blue	Rigel
1.0	10,000	white	Vega, Sirius
0.8	8,000	yellow-white	Canopus
0.6	6,000	yellow	Sun, Alpha Centauri
0.4	4,000	orange	Arcturus, Aldebaran
0.2	3,000	red	Betelgeuse, Barnard's Star

THE CLASSIFICATION OF STARS

✓5 Astronomers can use color and temperature to classify stars reasonably well, but they often use a more detailed classification scheme. This scheme incorporates additional knowledge of stellar physics obtained through spectroscopy, the study of spectral-line radiation.

Detailed Spectra

Figure 19.9 compares the spectra of several different stars, arranged in order of decreasing surface temperature (as determined from measurements of their colors). The spectra all extend from 4000 to 7000 Å, and each shows a series of dark absorption lines superposed on a background of continuous color, like those of the Sun shown earlier in Chapters 5 and 18. However, the precise patterns of lines show many differences. Some stars display strong lines in the long-wavelength part of the spectrum. Other stars have their strongest lines at short wavelengths. Still others show strong absorption lines spread across the whole visible spectrum. What do these differences tell us?

Although spectral lines of many elements are present with widely varying strengths, the differences among the spectra in Figure 19.9 are not due to differences in composition. Detailed spectral analysis indicates that the seven stars shown all have very similar elemental abundances—all are more or less solar in makeup. Instead, the differences are due almost entirely to the stars' *temperatures*. The spectrum at the top of Figure 19.9 is exactly what we would expect from a star with solar composition and a surface temperature of about 30,000K, the second from a 20,000K star, and so on, down to the 3000K star at the bottom of the figure. Let's now discuss these spectra in a little more detail.

The spectra of stars with surface temperatures exceeding 25,000K usually show intense absorption lines of singly-ionized helium and multiply-ionized heavier elements, such as oxygen, nitrogen, and silicon. These lines are not seen in the spectra of cooler stars. Only very hot (blue) stars have surface temperatures high enough to excite and ionize these tightly bound atoms. By contrast, hydrogen produces only weak absorption lines in these stars. The reason is not a lack of hydrogen. It is still by far the most abundant element, but at these high temperatures it is mostly ionized, so there are few intact hydrogen atoms to produce a characteristic spectrum. What's more, the few hydrogen atoms that have managed to retain their single electron are mostly in such highly excited states that their spectral lines are invisible (see Interlude 5-3).

In cooler stars, hydrogen lines are more intense. In (whitish) stars, with surface temperatures of about 10,000K, hydrogen is responsible for the strongest absorption feature. This temperature is just right for electrons to move frequently between hydrogen's second and third orbitals, producing the characteristic H-α (hydrogen alpha) line at 6563 Å. Tightly bound atoms—elements such as helium, oxygen, and nitrogen, which need lots of energy for excitation or ionization—are hardly observed at all, while ions of more loosely bound atoms—such as calcium and titanium—are fairly common. The spectrum of a yellow star like the Sun, with a surface temperature of about 6000K, shows few strong lines from ionized elements—the

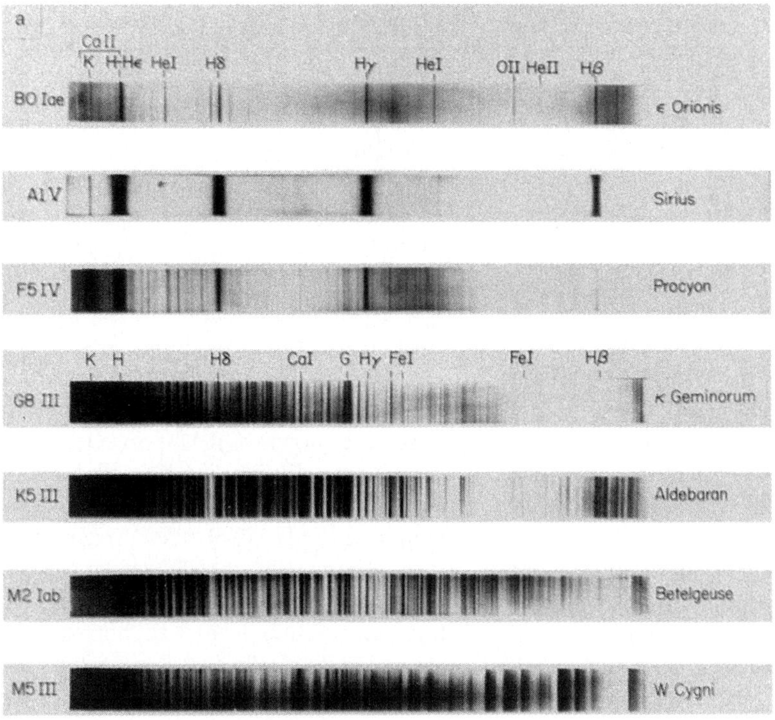

Figure 19.9 *Comparison of spectra observed for seven different stars having a range of surface temperatures. The hottest stars, at the top, show lines of helium and multiply-ionized heavy elements. In the coolest stars, helium lines are not seen, but lines of neutral atoms and molecules are plentiful. At intermediate temperatures, hydrogen lines are strongest. The compositions of all seven stars are about the same. In this figure, blue is at left, red at right.* (*J. Kaler*)

Sun is too cool for that. The H-α is no longer the most intense feature because, as with the hottest stars, the gas in the Sun does not have much electron traffic between the second and the third atomic orbitals. But unlike the hottest stars, where most hydrogen atoms are highly excited or ionized, the Sun's hydrogen lines are weak because most of the electrons reside in the ground orbital.

Cool red stars, with surface temperatures of only a few thousand kelvins, show extremely weak hydrogen lines. The most intense features in their spectra are due to neutral heavy atoms—and weakly excited ones at that. Astronomers observe no lines from ionized elements. In fact, the average energy of the photons leaving the surface of the coolest stars is less even than that needed to destroy some molecules, and many absorption lines observed in red stars are molecular in origin.

The differences among stellar spectra, then, stem from differences in temperature, not in composition. Most stars are made from the same elements, in roughly the same proportion. Observing the spectrum of a star provides us with another means of measuring its surface temperature. Spectroscopy generally requires more telescope time than photometry, because the available photons must be spread out over the entire visible spectrum instead of just being divided into a few wide bands, as discussed earlier. Nevertheless, so much more information is available from a complete spectrum than from a simple color index that astronomers prefer to work with spectra whenever possible. We now discuss the method actually used by astronomers to classify the stellar spectra they observe.

Spectral Classification

Stellar spectra like those shown in Figure 19.9 had been obtained for numerous stars even before the start of the twentieth century. Observatories around the world amassed spectra for several hundred thousand stars, in both hemispheres of the sky. Early researchers (1880 to 1920) correctly identified some of the observed spectral lines on the basis of comparisons between stellar lines and those obtained in the laboratory. Those workers, though, had no firm understanding of how the lines were produced. Modern atomic theory had not yet been developed, so the correct interpretation of the line strengths, as described in this chapter and in Chapter 5, was impossible at the time. Lacking full understanding of the atom, early workers classified stars primarily according to their hydrogen-line intensities. They adopted an A, B, C, D, . . . scheme, thinking that A stars, with the strongest hydrogen lines, had more hydrogen than did B stars, and so on. The classification extended as far as the letter P.

In the 1920s, scientists discovered the intricacies of atomic structure and the causes of spectral lines, and astronomers quickly realized that stars could be more meaningfully classified according to their surface temperature. Instead of adopting an entirely new scheme, however, they chose to shuffle the existing alphabetical categories—those based on the strength of hydrogen lines—into a new sequence based on temperature. In the modern scheme, the hottest stars have the letter designation O, because they have very weak absorption lines of hydrogen and they were classified toward the end of the original scheme. When listed in order of decreasing temperature, the original letters now run O, B, A, F, G, K, M. (The other letter classes have long since been dropped.) These stellar designations are called **spectral classes.** They can be remembered, in the correct order, by the mnemonic, "**O**h, **Be** **A** **F**ine **G**uy, **K**iss **M**e."

Nowadays astronomers subdivide each lettered spectral classification into 10 subdivisions, denoted by the numbers 0–9. By convention, the lower the number, the hotter the star. Thus, for example, our Sun is classified as a G2 star (a little cooler than G1 and a little hotter than G3), Vega is a type A0, Barnard's Star is M5, Betelgeuse is M2, and so on. Table 19-2 lists the main properties of each stellar spectral type, extending the data presented in Table 19-1.

We should not underestimate the importance of the early work in classifying stellar spectra. Even though the original classification was based on erroneous assumptions, the painstaking accumulation of large quantities of accurate data paved the way for rapid improvements in understanding once a theory came along to explain the observations.

THE HERTZSPRUNG-RUSSELL DIAGRAM

We have now studied the two most important, basic properties of any star: its luminosity (or absolute brightness, or absolute magnitude) and its surface temperature (or color, or spectral type). Astronomers use these two quantities to classify stars in much the same way that height and weight serve to classify the bulk properties of human beings. We know that people's height and weight are well-correlated—tall persons are usually heavier, short persons lighter. We might naturally wonder if the two basic stellar properties are also related in some way. In the second decade of the twentieth century, Danish astronomer Ejnar Hertzsprung and the U.S. astronomer Henry Norris Russell independently discovered just such a relationship.

Figure 19.10 shows the way that Hertzsprung and Russell originally plotted the temperatures and luminosities of stars. In honor of these two scientists, such a plot is now known as a Hertzsprung-Russell diagram, or **HR diagram** for short. The vertical scale, expressed in units of the solar luminosity ($L_\odot = 3.9 \times 10^{33}$ erg/s), extends over a large range, from 10^{-4} to 10^4 (so the Sun appears right in the middle of the luminosity range, at a luminosity of 1). Surface temperature is plotted horizontally, although in the unconventional sense of temperature increasing to the *left* (so the spectral sequence O, B, A, . . . reads left to right).

TABLE 19-2 ***Stellar Spectral Types***

Spectral Class	Surface Temperature (K)	Prominent Absorption Lines	Familiar Examples
O	30,000	Ionized helium strong; multiply-ionized heavies; hydrogen faint	
B	20,000	Neutral helium moderate; singly-ionized heavies; hydrogen moderate	Rigel (B8)
A	10,000	Neutral helium very faint; singly-ionized heavies; hydrogen strong	Vega (A0), Sirius (A1)
F	8,000	Singly-ionized heavies; neutral metals; hydrogen moderate	Canopus (F0)
G	6,000	Singly-ionized heavies; neutral metals; hydrogen faint	Sun (G2), Alpha Centauri (G2)
K	4,000	Singly-ionized heavies; neutral metals strong; hydrogen faint	Arcturus (K2), Aldebaran (K5)
M	3,000	Neutral atoms strong; molecules moderate; hydrogen very faint	Betelgeuse (M2), Barnard's Star (M5)

To change the horizontal scale so that temperature increases conventionally to the right would play havoc with historical precedent.

As we have just seen, astronomers often use a star's *color* to measure its temperature. Indeed, the spectral classes plotted along the horizontal axis of spectral class in Figure 19.10 are equivalent to the B/V color index. Also, since astronomers commonly express a star's luminosity as an absolute magnitude (see Interlude 19-1), stellar *magnitude* could be plotted vertically instead of stellar luminosity. For these reasons, many astronomers refer to diagrams such as Figure 19.10 as **color-magnitude diagrams.** In this book, however, we will use temperature and luminosity measurements.

Constructing an HR Diagram

The first step in making an HR diagram for any given collection of stars is to determine the surface temperature (or spectral class) of each. We can do this in one of two ways. Either we can measure the star's intensity in the B and V bands and fit a Planck curve to those measurements, or we can observe the star's spectrum and so determine its spectral type. As we have seen, both of these methods provide a means of determining stellar temperature. Neither requires any knowledge of the star's distance.

The second step is to determine each star's luminosity. This is either an easy or an impossible task! It is easy if the star's distance is known, because we can then easily convert the measured apparent brightness (flux) into absolute brightness (luminosity) using the inverse-square law. If the distance is unknown, however, the star's luminosity cannot be measured, so it cannot be used in the construction of the diagram. Thus it is important to start by choosing stars with known distances.

If the only distance-measurement technique we had at our disposal were stellar parallax, we would be able to construct HR diagrams only for stars lying within a few tens of parsecs of the Sun. In a moment, we will see how the HR diagram *itself* can be used to estimate the distances to stars that are much farther away. But before we can do that, we must study the properties of the HR diagram in detail.

The Main Sequence

As Hertzsprung and Russell plotted more and more stellar temperatures and luminosities, they found that stars are *not* uniformly scattered across the HR diagram. Instead, most are confined to a fairly well-defined band stretching diagonally across Figure 19.10 from the top left (high temperature, high luminosity stars) to the bottom right (low temperature, low luminosity stars). In other words, cool stars tend to be faint, and hot stars tend to be bright. This band of stars spanning the HR diagram is known as the **main sequence.**

The HR diagram in Figure 19.10 shows just a few stars, chosen for the sake of illustration from among the

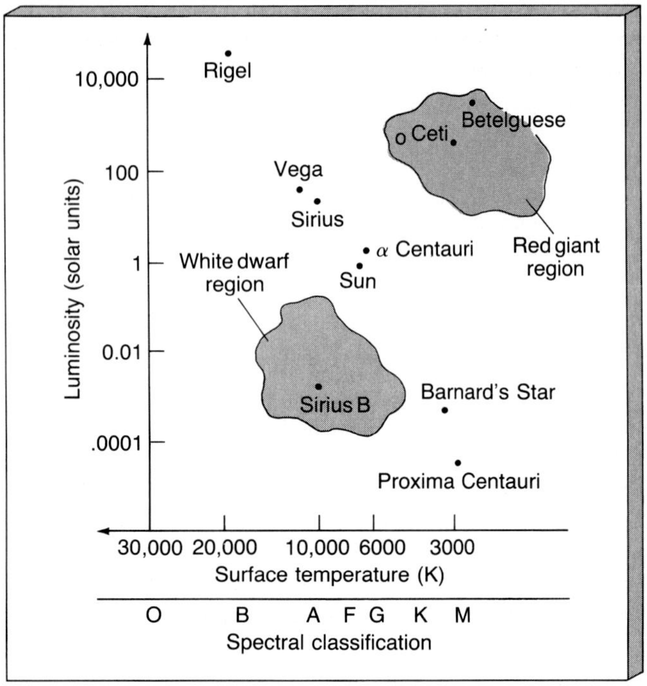

Figure 19.10 *A plot of luminosity and surface temperature (or spectral classification), known as an HR diagram, is a useful way to compare stars. Plotted here are the locations of some stars mentioned earlier in the text.*

examples used earlier in this chapter. Figure 19.11 shows a more systematic study of stellar properties, covering the 40 or so stars that lie within 5 pc of the Sun. As more points are included in the diagram, the main sequence "fills up," and the pattern becomes more evident. Most stars lie within the shaded band. The vast majority of stars in the immediate vicinity of the Sun lie on the main sequence.

Main sequence stars have surface temperatures ranging from about 3000K (spectral type M) to 30,000K (spectral type O). This relatively small temperature range—only a factor of 10—is determined mainly by the rates at which nuclear reactions occur in stellar cores. By contrast, the observed range in luminosities is very large, covering some eight orders of magnitude (that is, a factor of 100 million), ranging from 10^{-4} to 10^4 times the luminosity of the Sun.

Astronomers can use the radius-luminosity-temperature relation ($L \propto R^2 T^4$) studied earlier in this chapter to estimate the radii (R) of main-sequence stars from their temperatures (T) and luminosities (L). They find that in order to account for the observed range in luminosities, stellar radii must also vary along the main sequence. The faint, red M-type stars in the bottom right of the HR diagram are only about 1/10 the size of the Sun, while the bright, blue O-type stars in the upper left are about 10 times larger. The oblique dashed lines in Figure 19.11 represent stars with the same radii (so $L \propto T^4$ along these lines). By including such lines on this and subsequent HR diagrams, we can indicate stellar temperatures, luminosities, and radii on a single plot.

We see a very clear trend as we traverse the main sequence from top to bottom. At one end, the stars are large, hot, and bright. Because of their size and color, they are referred to as **blue giants.** The very largest are called **blue supergiants.** At the other end, stars are small, cool, and faint. They are known as **red dwarfs.** Our Sun lies right in the middle.

Figure 19.12 shows an HR diagram for a different group of stars—the 100 stars of known distance having the greatest apparent brightness, as seen from Earth. Compare this with Figure 19.11, which shows only the closest stars, lying within 5 pc of Earth. Notice the much larger number of very luminous stars toward the upper end of the main sequence in Figure 19.12. The reason for this excess of blue giants is quite simple—we can see very luminous stars a long way off. The stars shown in Figure 19.12 are scattered through a much greater volume of space than those in Fig-

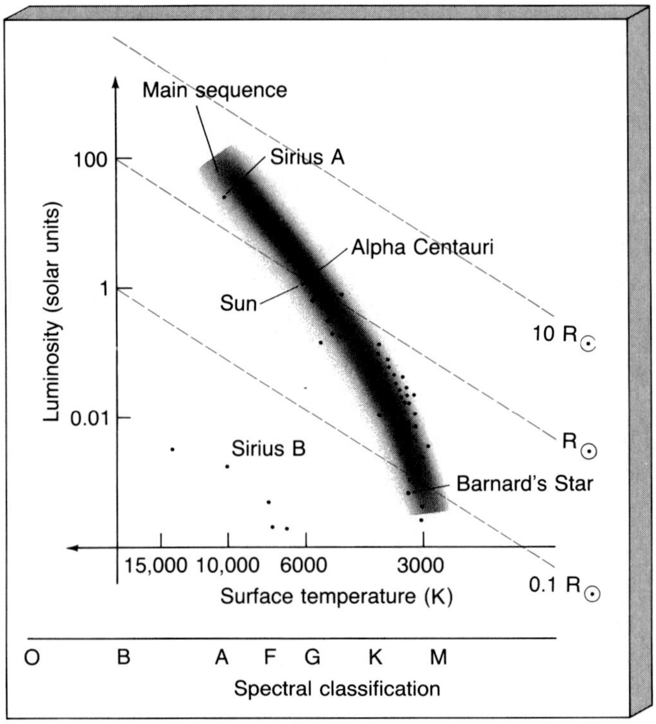

Figure 19.11 *Most stars have properties within the shaded region known as the main sequence. The points plotted here are for stars lying within about 5 pc of the Sun. The diagonal lines correspond to constant stellar radius, as discussed in the text, so that stellar size can be represented on the same diagram as luminosity and temperature. (Recall that ☉ stands for the Sun.)*

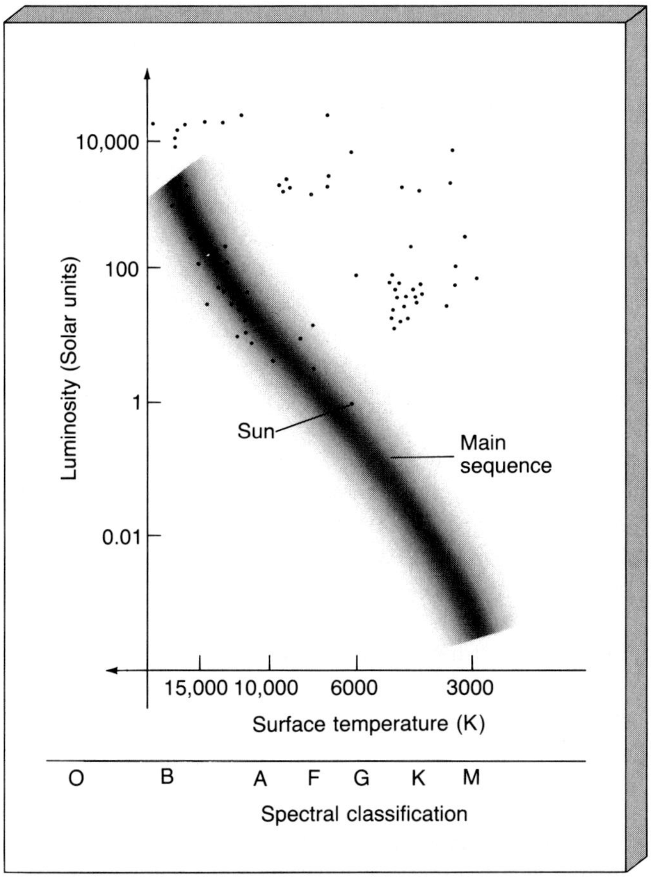

Figure 19.12 *An HR diagram for the 100 brightest stars in the sky. Such a plot is biased in favor of more luminous stars—which appear toward the upper left—because we can see them more easily than we can faint stars.*

ure 19.11, but the sample is heavily biased toward the brightest objects. In fact, of the 20 brightest stars in the sky, only 6 lie within 10 pc of us; the rest are visible, despite their large distances, because of their large luminosities.

If very luminous blue giants are overrepresented in Figure 19.12, low-luminosity red dwarfs are surely underrepresented. In fact, no dwarfs are present on that diagram. This absence is not surprising, because such low-luminosity stars are very difficult to observe from Earth. They are just too faint, and they radiate much of their energy in the invisible, infrared part of the electromagnetic spectrum. However, in the 1970s astronomers began to realize that they had greatly underestimated the number of red dwarfs in the Galaxy. As hinted at by the HR diagram in Figure 19.11, which shows an unbiased sample of stars in the solar neighborhood, red dwarfs are actually the most common type of star in the sky, probably accounting for upward of 80 percent of all stars in the universe.

Red Giants and White Dwarfs

Most stars lie on the main sequence. However, some of the points plotted in Figures 19.10–19.12 clearly do not. One such point in Figure 19.10 represents Mira (Omicron Ceti), whose surface temperature (3000K), as we saw earlier, is about half that of the Sun and whose luminosity is some 400 times greater than the Sun's. Another point represents Betelgeuse (Alpha Orionis), the ninth brightest star in the sky, a little cooler than Mira but more than 30 times brighter. The upper right-hand corner of the HR diagram, where these stars reside, is called the **red giant region.** Notice that no red giants are found within 5 pc of the Sun (Figure 19.11), but many of the brightest stars seen in the sky are in fact red giants (Figure 19.12). Red giants are relatively rare, but they are so bright they are visible to very great distances. The red giant region in the HR diagram represents a class of stars quite distinct from the stars of the main sequence.

Also shown in Figure 19.10 is Sirius B, a white dwarf whose surface temperature (12,000K) is about double that of the Sun and whose luminosity is about 0.002 L. A few more such faint A-type stars can be seen in Figure 19.11 in the bottom left-hand corner of the HR diagram. This region is known as the **white dwarf region.** White dwarfs form a third distinct class of stars in the HR diagram, very different in their properties from both main-sequence stars and red giants.

Dwarfs and giants give some feeling for the extreme properties of stars, but, as we have seen, most stars have properties much more like our Sun and lie on the main sequence in the HR diagram. About 90 percent of all stars in our solar neighborhood, and probably a similar percentage elsewhere in the universe, are main-sequence stars. About 9 percent of stars are white dwarfs, and 1 percent are red giants.

Spectroscopic Parallax

We have discussed the connections between absolute brightness (luminosity), apparent brightness (energy flux), and distance. If we know a star's apparent brightness and distance, we can determine its absolute brightness using the inverse-square law. Suppose we know a star's absolute and apparent brightnesses. Can we use this information to determine the star's distance? Can we add a new rung to the cosmic distance ladder begun in Chapters 2 and 3 with radar ranging and parallax? The answer is yes. If both the absolute and apparent brightnesses of a star are known, the distance can be determined. This new rung, indicated in Figure 19.13, is known as **spectroscopic parallax.**

How is this accomplished operationally? The first step is to measure the apparent brightness of a star as it appears on a photograph or through a telescope. The second step is to obtain the star's luminosity or absolute brightness by some means. The final step is to calculate the distance using the inverse-square law.

Consider for a moment another analogy. Most of us have a rough idea of the approximate brightness and size of a red traffic signal. Suppose we are driving down an unfa-

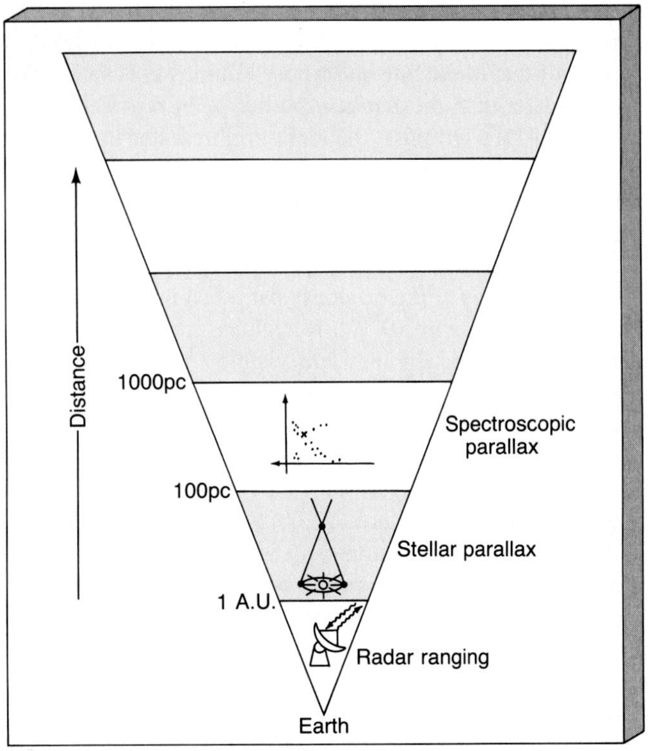

Figure 19.13 *Knowledge of a star's absolute and apparent brightnesses can yield an estimate of its distance. Astronomers can use this third "rung" in our distance ladder, called spectroscopic parallax, to measure distances as far out as individual stars can be clearly discerned—a few thousand parsecs.*

miliar street and see a red traffic light in the distance. Our knowledge of the intrinsic luminosity of the light often enables us to immediately make a mental estimate of its distance. A normal traffic light that is relatively dim must be quite distant (assuming it's not just dirty); a bright one must be relatively nearby.

The important point here is that a measurement of the apparent brightness of a light source, combined with some knowledge of its intrinsic properties, can yield an estimate of the source's distance. For a star, the trick is to find an independent measure of the luminosity without knowing the distance. The HR diagram can provide just that.

The main sequence is a close correlation between temperature and luminosity for most stars, with the exception of a few giants and dwarfs. As such, the main sequence can be used to specify the *average* properties of most stars. Let's imagine for a moment that the main sequence really is a *line* in the HR diagram, rather than a somewhat fuzzy band, and that *all* stars lie on the main sequence. From a star's spectrum, we can determine its surface temperature or spectral type. If the star lies on the main sequence, there is then only one possible absolute brightness corresponding to that temperature. We can read the star's luminosity directly off a graph like Figure 19.12, then determine its distance by measuring the energy flux at Earth and using the inverse-square law. The existence of the main sequence allows us to make a connection between an easily measured quantity (temperature) and the star's luminosity, which would otherwise be unknown.

Spectroscopic parallax can be used out to distances of several thousand parsecs. Beyond that, spectra and colors of individual stars are difficult to obtain. The "standard" main sequence is obtained from HR diagrams of stars whose distances can be measured by (geometric) parallax, so that the method of spectroscopic parallax is calibrated using nearby stars. In using this method, we are applying the "principle of mediocrity" that we discussed in Chapter 1. Specifically, we are assuming (without proof) that distant stars are basically similar to nearby stars—in particular, that they *fall on the same main sequence as nearby stars*. Only by making this assumption can we expand the boundaries of our distance-measurement techniques. We will see many more applications of the principle of mediocrity in subsequent chapters, as we continually extend our "local" knowledge to larger and larger spatial domains.

Of course, the main sequence is not really a line in the HR diagram. It has some thickness to it. For example, the luminosity of a main-sequence G2-type star (like the Sun) can actually range from about 0.5 to 1.5 $L_\odot$. The main reason for this range is the variation in stellar composition and age from place to place in the Galaxy. The effect is that there is an uncertainty in the luminosity obtained using this method and so some uncertainty in the distance. Distances obtained by spectroscopic parallax are probably accurate to no better than 25 percent. While this may not seem very accurate—a cross-country traveler in the United States would hardly be impressed to be told that the best estimate of the distance between Los Angeles and New York is somewhere between 3000 and 5000 km—it illustrates that in astronomy even something as simple as the distance to another star can be very difficult to measure. Still, an estimate with an uncertainty of ± 25 percent is far better than no estimate at all.

Luminosity Class

If, by chance, the star in question happens to be a giant or a dwarf, the determination of distance by spectroscopic parallax will be incorrect. But roughly 90 percent of all stars are on the main sequence, so we could argue that the assumption that a star is a main-sequence star will be valid 9 out of 10 times. In fact, astronomers can do much better than that. Recall from Chapter 5 that the width of a spectral line can provide information on the *density* of the gas where the line formed. The atmosphere of a red giant is much less dense than that of a main-sequence star, and this in turn is much less dense than the atmosphere of a white dwarf. By studying the width of a star's spectral lines, astronomers can usually tell with a high degree of confidence whether or not a star is on the main sequence.

Over the years, astronomers have developed a system for classifying stars according to the width of their spectral

INTERLUDE 19-2 *Stacks and Stacks of Photographs*

Several of the world's observatories have stacks not only of books, filed by author and title, but also of celestial photographs, filed by sky location and date. The largest collection of stellar photographs is housed at the Harvard College Observatory, where there are now nearly a million glass plates cataloged and stored in a building suspended on springs to guard against earthquakes.

Prior to the invention of photography in the mid-nineteenth century, astronomical observations were made by visual impressions and hand sketches. Viewed by most astronomers at its inception as an idle diversion from the real study of the skies, photography soon became a major tool for recording and quantifying observations. In effect, photography transformed observational astronomy from an art into a science. Foresight on the part of E. C. Pickering, director of the Harvard Observatory in the 1880s, who assigned funds and staff to this new pursuit, resulted in the large Harvard collection, which includes photographs regularly taken with several telescopes in both hemispheres for the past century. Pickering is shown in the accompanying photograph making his way to the summit of El Misti in Arequipa, Peru, the site of one of the most advanced observatories south of the equator around the turn of the century.

A small fraction of all stars in the Galaxy are known to vary in brightness over a period of days or weeks. Early photographic surveys concentrated mostly on the luminosity fluctuations of these *variable stars*. The cataloged plates are the best way to monitor long-term changes in stellar brightness. Toward the end of the nineteenth century, researchers realized that the spectroscopy of stars contains even more information than luminosity alone. Detailed spectral observations were made for tens of thousands of stars by 1900 and for millions of stars since then.

The Harvard Photographic Collection is a rich source of astronomical lore, only part of which has yet been tapped. Astronomers from around the world apply to analyze the plates, much as they routinely request observing time on telescopes or orbiting satellites. The cataloged plates allow astronomers to study the brightness of almost any visible cosmic object and the spectra of the brighter ones. Even in these days of CCDs (charge-coupled devices) and orbiting observatories, the long time span of the Harvard archival data makes the plate collection invaluable.

(Harvard-Smithsonian Center for Astrophysics)

lines. Because the line width is particularly sensitive to density in the stellar photosphere, and the atmospheric density in turn is well correlated with luminosity, the category in which a star falls has come to be known as its **luminosity class.** The point of the classification is this: It provides a means for astronomers to distinguish supergiants from giants, giants from main-sequence stars, and main-sequence stars from subdwarfs by measuring a single spectral property—the line broadening—of the radiation received.

The standard stellar luminosity classes are as follows:

Class	Description
Ia	Bright supergiants
Ib	Supergiants
II	Bright giants
III	Giants
IV	Subgiants
V	Main-sequence stars/dwarfs

The locations of these classes on the HR diagram are shown in Figure 19.14. Now we have a way of specifying a star's location in the diagram in terms of properties that are measurable by purely spectroscopic means—spectral type and luminosity class locate a star just as surely as do temperature and luminosity. The full specification of a star's spectral properties includes its luminosity class. For example, the Sun, on the main sequence, is of class G2V, Vega is A0V, the red dwarf Barnard's Star is M5V, the red supergiant Betelgeuse is M2Ia, and so on.

Consider, for example, a K0-type star (see Table 19-3). If the star lies on the main sequence (that is, it is a K0V star), its luminosity is about 0.1 $L_\odot$. If its spectral lines are observed to be narrower than normally found in main-sequence stars, the star may be recognized as a giant, with a luminosity of 10 $L_\odot$. If the lines are very narrow, the star might instead be classified as a supergiant, brighter by a further factor of 100, at 1000 $L_\odot$. In this way, the observed width of the star's spectral lines translates directly into a measure of the physical state of the star.

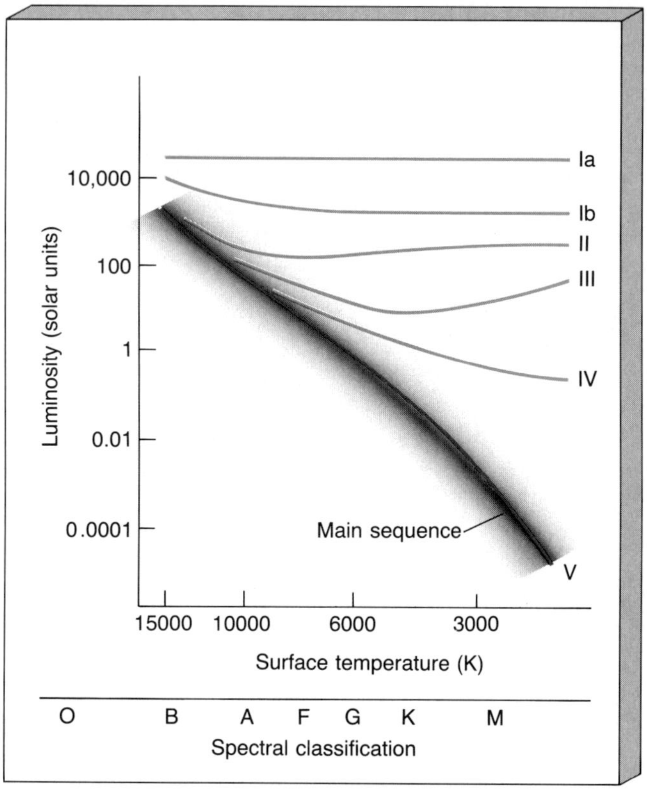

Figure 19.14 Stellar luminosity classes in the HR diagram. Note that a star's location could be specified by giving its spectral type and luminosity class instead of by its temperature and luminosity.

TABLE 19-3 ***Variation in Stellar Properties within a Spectral Class***

Surface Temperature (K)	Luminosity ($L_\odot$)	Radius ($R_\odot$)	Object
4000	0.1	0.7	K main-sequence star
4000	10	7	K giant star
4000	1000	70	K supergiant star

STELLAR MASS

✓6 What ultimately determines a star's position on the main sequence? The answer is its *mass*. Mass and composition are fundamental properties of any star. They are set once and for all at the time of a star's birth. Together, they uniquely determine both the star's internal structure and its external appearance and even (as we will see in Chapter 22) its future evolution. The ability to measure these two key stellar properties is of the utmost importance if we are to understand how stars work. We have already seen how spectroscopy is used to determine a star's composition. Now we turn to the problem of finding a star's mass.

The mass of an isolated star is practically impossible to determine. As with all other objects, we must measure a star's mass by observing its gravitational influence on some other nearby body—another star, perhaps, or a planet. If the distance between the two bodies is known, Newton's laws may be used to calculate the masses involved. In this way, we can derive the mass of the Earth by watching the Moon or artificial satellites orbit it and the mass of the Sun by studying the orbital motions of the planets. Unfortunately, astronomers have not yet been able to detect planets around any other star, much less measure their orbits. Even so, there is a way to estimate the masses of many stars.

Binary Stars

Most stars are members of multiple-star systems—groups of two or more stars in orbit around one another. The majority of stars are found in **binary-star systems,** which consist of two stars in orbit about their common center of mass, held together by their mutual gravitational attraction. Other stars are members of triple, quadruple, and even more complex systems. Most complex are the rich star clusters discussed below. The Sun is not part of a multiple-star system, putting it in the minority among stars. If our Sun has anything at all uncommon about it, it may be its lack of stellar companions.

Astronomers classify binaries according to their appearance from Earth and the ease with which they can be observed. **Visual binaries** have widely separated members that are bright enough to be observed and monitored separately. Other binaries, known as **spectroscopic binaries,** are too distant to be resolved into separate stars, but they can be indirectly perceived by monitoring the back-and-forth Doppler shifts of their spectral lines as the stars orbit one another. Recall that motion toward an observer blueshifts the lines, and motion away from the observer redshifts them. In a *double-line* spectroscopic binary, two distinct sets of spectral lines—one for each component star—shift back and forth as the stars move. Because we see particular lines alternately approaching and receding, we know that the objects emitting the lines are in orbit. In the more common *single-line* systems, one star is too faint for its spectrum to be distinguished, so that only one set of lines is observed to shift back and forth. Still, we realize that this shifting means that the detected star must be in orbit around another star, even though the companion cannot be directly observed.

In the much less common **eclipsing binaries,** the orbital plane of the pair of stars is almost edge-on to our line of sight, and we observe a periodic decrease of starlight as one component passes in front of the other. By studying the variation of the light from the binary system—the binary's **light curve**—astronomers can derive detailed information not only about the stars' orbits but also about their radii.

Figure 19.15 illustrates how we can discover and study each of the above types of binary systems. Notice, by the way, that these categories are not mutually exclusive.

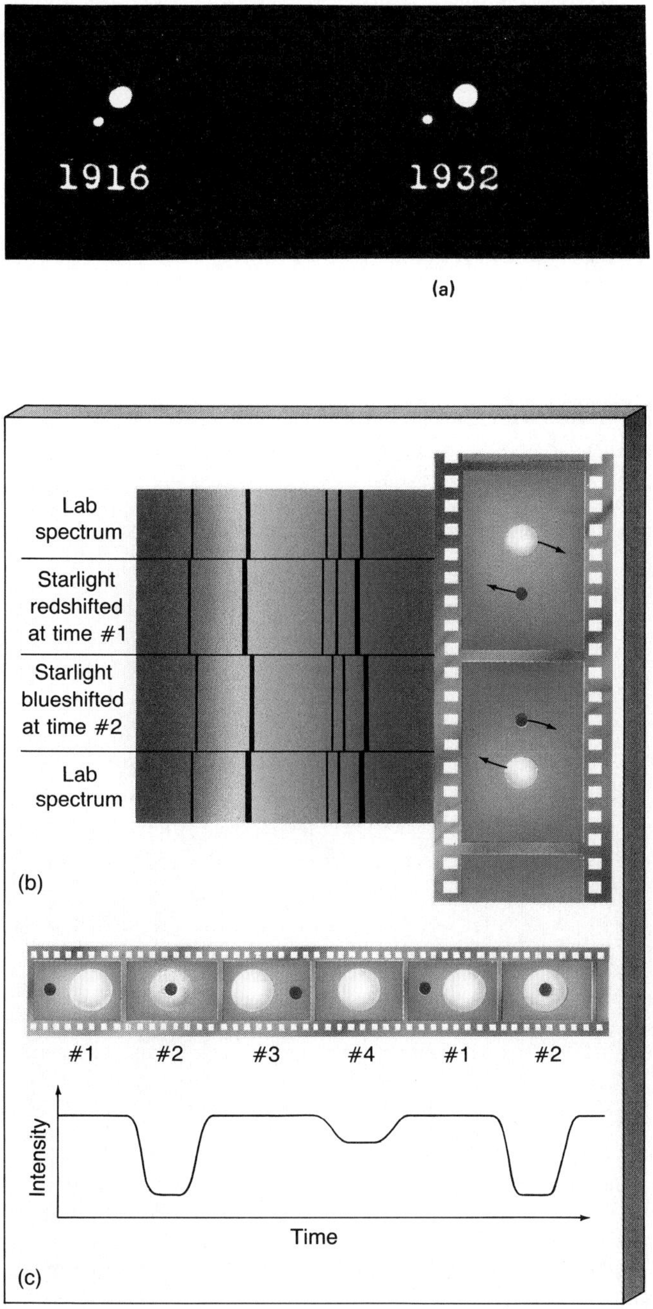

Figure 19.15 (a) The periods and separations of binary stars can be observed directly if each star is clearly seen. This binary star is known as 70 Ophiuchus. (b) Binary properties can also be found indirectly by measuring the periodic Doppler shift of one star relative to the other as they move in their orbit. This shows a single-line system, where only one spectrum (from the brighter component) is visible. The observer is situated left of the diagram. (c) If the two stars happen to eclipse one another, their radii and masses can be determined by observing the periodic decrease in starlight as one passes in front of the other. (Harvard College Observatory)

For example, a single-line spectroscopic binary may also happen to be an eclipsing system. In that case astronomers can use the eclipses to gain extra information about the fainter member of the pair.

Occasionally, two unrelated stars just happen to lie close together in the sky, even though they are actually widely separated. These **optical doubles** are just chance superpositions and carry no useful information about stellar properties.

A great deal of data can be obtained from repeated observations of a binary system. By observing the actual orbit of the stars, or the back-and-forth motion of the spectral lines, or the dips in the light curve—whatever information happens to be available—we can measure the binary's orbital period. Observed periods span a broad range—from hours to centuries. Doppler-shift measurements give us information on the orbital velocities of the member stars. In addition, if the distance to a visual binary is known, the size (semi-major axis) of its orbit can be determined directly, by simple geometry.

Knowledge of the binary period and orbit size is all we need to determine the combined mass of the component stars, using the modified form of Kepler's Third Law (as discussed in Chapter 3). Additional observations are needed to determine the individual masses of the components. For example, in any system of orbiting objects, each orbits the common center of mass. Measuring the distance from each star to the center of mass of a visual binary yields the ratio of the stellar masses. Knowing both the sum of the masses and the ratio of the masses, we can then find the mass of each star. The individual component masses of a single-line spectroscopic binary system generally cannot be determined (unless the binary happens also to be an eclipsing system, in which case the extra information allows the mass ratio to be inferred). If a binary is too distant or if only one component is visible, it is not possible to determine the component masses, only their sum. Nevertheless, for many nearby systems, individual masses can be obtained. Virtually all we know about the masses of stars is based on those observations.

Consider, for example, the nearby double-star system made up of the bright star Sirius A and its faint companion Sirius B. Observations of their orbit show that the sum of their masses is three times the mass of the Sun—3 $M_{\odot}$. Further observations show Sirius A to have roughly twice the mass of its companion. It follows that the mass of Sirius A is 2 $M_{\odot}$ and that of Sirius B is 1 $M_{\odot}$.

Dependence of Stellar Properties on Mass

Now that we know how to determine stellar masses, at least for stars found in binary systems, we can ask how these masses are correlated with the other properties of stars—temperature, luminosity, and radius—discussed earlier. Figure 19.16 is an HR diagram showing how stellar mass varies along the main sequence. Note the clear progression from low-mass red dwarfs to high-mass blue giants. With few exceptions, main-sequence stars range in mass from about 0.1 to 20 $M_{\odot}$. The hot O- and B-type stars are generally about 10 to 20 times more massive than our Sun. The

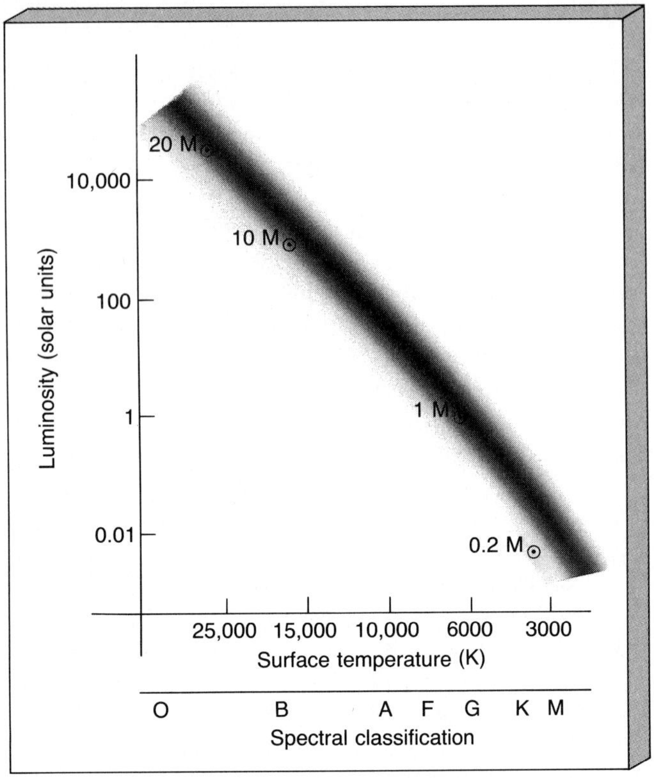

Figure 19.16 *Mass, more than any other stellar property, determines a star's position on the main sequence. If a star forms with a low mass, it will be cool and faint, lying at the bottom of the main sequence. Very massive stars, conversely, are hot and bright, placing them at the top of the main sequence.*

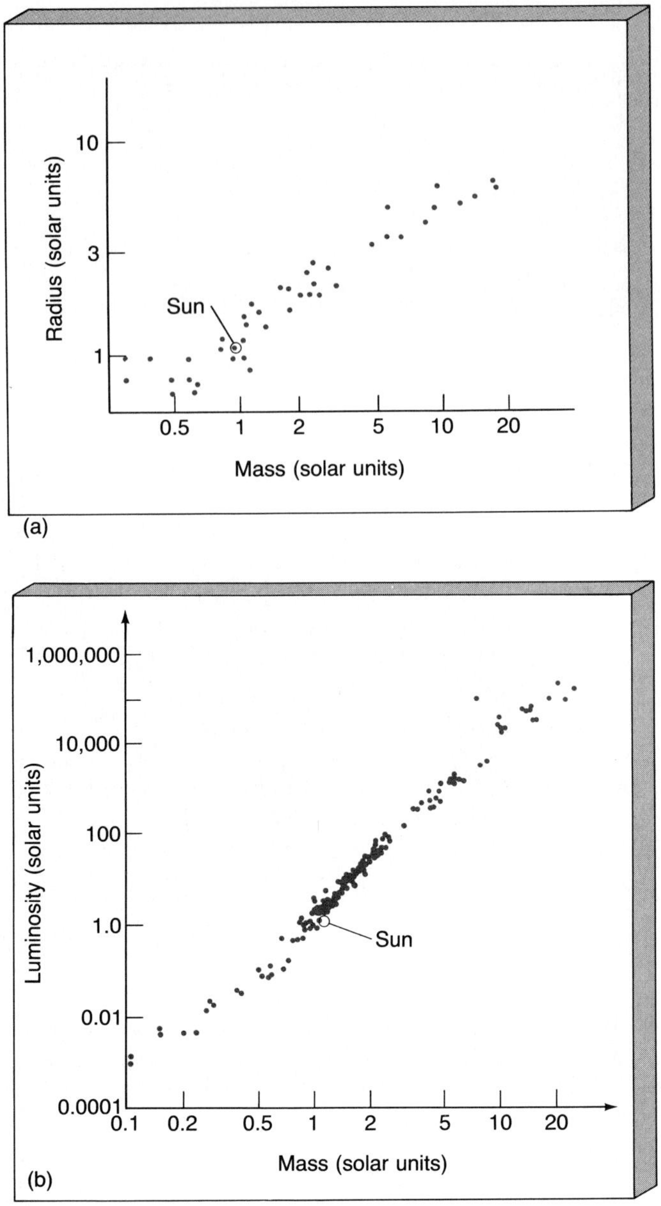

Figure 19.17 *(a) Dependence of stellar radius on mass, for main-sequence stars. The radius increases roughly in proportion to the mass over much of the range. (b) Dependence of luminosity on mass. Notice that the luminosity increases much faster than the mass.*

coolest K- and M-type stars contain only a few tenths of a solar mass. Because all other stellar properties are set once a star's mass is known, we can say that *the mass of a star at the time of formation determines its location on the main-sequence.*

Figure 19.17 illustrates in a little more detail how a main-sequence star's radius and luminosity depend on its mass. The two curves are based on observations of binary-star systems. The curves show the **mass-radius** and **mass-luminosity** relations for main-sequence stars. Along the main sequence, both radius and luminosity increase with mass. As a rough rule of thumb, radius rises in proportion to mass, while luminosity increases much faster—more like the *cube* (a number raised to the third power) of the mass. For example, a 2 $M_\odot$ main-sequence star has a radius twice that of the Sun and a luminosity of 8 $L_\odot$ (= 2^3); a 0.2 $M_\odot$ main-sequence star has a radius of about 0.2 $R_\odot$ and a luminosity of 0.008 (= 0.2^3) $L_\odot$, and so on.

Stellar Lifetimes

Mass determines a star's position on the main sequence because mass, more than any other stellar property, governs a star's ability to radiate. Very massive stars have so much matter pushing on their cores that the gas there becomes more compressed than the gas in the core of our Sun. As a result, the core nuclei collide more vigorously, raising the gas temperature and greatly increasing the rate of nuclear fusion. The opposite is true of the least massive stars. Their core particles collide less frequently than the core particles in our Sun, and their central temperatures and nuclear fusion rates are correspondingly smaller.

The rapid rate of nuclear burning deep inside a star releases vast amounts of energy per unit time and creates high surface temperatures and luminosities. How long can the fire continue to burn? To estimate a star's lifetime, we simply divide the amount of fuel available—the mass of the star—by the rate at which the fuel is being consumed—the star's luminosity:

TABLE 19-4 ***Key Properties of Some Well-Known Stars***

Star	Spectral Type	Mass ($M_\odot$)	Central Temperature ($\times 10^6$ K)	Luminosity ($L_\odot$)	Estimated Lifetime ($\times 10^6$ years)
Rigel	B	10	30	44,000	20
Sirius	A	2.3	20	23	1,000
Alpha Centauri	G	1.1	17	1.4	7,000
Sun	G	1.0	15	1.0	10,000
Proxima Centauri	M	0.1	6	0.00006	>1,000,000

$$\textit{stellar lifetime} \propto \frac{\textit{stellar mass}}{\textit{stellar luminosity}}$$

For example, O and B stars, as we have seen, have masses 10 to 20 times that of the Sun and luminosities thousands of times higher than the solar luminosity. Accordingly, these massive stars can endure only for short times. Their nuclear reactions proceed so rapidly that their fuel is quickly depleted, despite their large mass. They might be likened to large, inefficient gas-guzzling automobiles. We can be sure that all the O and B stars now observable in the sky are relatively young objects. Most of them are less than 20 million years old. Massive stars older than that have already exhausted their fuel and no longer emit large amounts of energy. They have, in effect, died.

At the opposite end of the main sequence, the cooler K- and M-type stars have less mass than our Sun. With their low core densities and temperatures, their proton-proton reactions churn away rather sluggishly, much more slowly than in the Sun's core. The small energy release per unit time leads to low luminosities and surface temperatures for these stars. They endure for long times; their low fuel consumption likens them to the more efficient compact cars. Many of the K- and M-type stars now seen in the sky are expected to shine for at least another trillion years!

Table 19-4 compares some key properties of several well-known main-sequence stars, arranged in order of decreasing mass. Notice how little the central temperature differs from one star to another and how large is the spread in stellar luminosities and lifetimes.

INTERLUDE 19-3 *Stellar Activity*

Our Sun's activity is not unique. Astronomers have recently found similar activity in all types of stars. Advances in ultraviolet and x-ray space-borne astronomy and in radio and infrared ground-based astronomy have revealed that many stars have surface activity and strong stellar winds. For example, in the late 1970s and early 1980s, the *Einstein* x-ray observatory (*HEAO-B*; see Chapter 6) quite unexpectedly detected x-rays from nearly all types of stars, thereby revealing that they are surrounded by coronas having temperatures of 1 million K or more.

Apparently all stars have active regions, including spots, flares, and prominences much like those on our own Sun. Some stars exhibit "starspots"—what we would call sunspots on our own Sun—so large that an entire face of the star is darkened. Others display flare activity thousands of times more intense than that on the Sun. As in the case of our Sun, however, virtually all such activity is invisible, detectable only at wavelengths to which the human eye is insensitive.

In retrospect, we should hardly be surprised at discovering stellar activity. By now the doctrine that our Sun is a typical star is firmly entrenched in astronomical thought. Still, it is exciting to see this philosophically reasonable viewpoint backed up by direct observation.

What *is* surprising is the strength of the stellar winds from the most massive stars. Ultraviolet observations made by both the early ultraviolet satellite *Copernicus* and the satellite *International Ultraviolet Explorer* have shown key stellar absorption features to be shifted in wavelength by the Doppler effect—the result of the rapid motion of the outflowing gas. For some O stars, wind speeds often reach 3000 km/s. Such stars lose mass almost a billion times faster than the Sun. They shed an entire solar mass in the relatively short span of 100,000 years. The winds of the most luminous stars must have profound effects not only on the stars' immediate environment, including any planets, but also on the evolution of the stars themselves.

All things considered, one of the most fascinating features of stars is their remarkable repertoire of activity. Magnetic fields, photospheric violence, flarelike effects, coronal holes, and stellar winds are part of the normal daily routine of essentially every type of star. Clearly, the atmospheres of most stars are a far cry from the steady visible images shining forth in the evening sky.

STAR CLUSTERS

While an HR diagram can be drawn for any group of stars, astronomers are usually interested in comparing the properties of stars that have something in common. Very often, this common feature is that the stars all lie in the same region of space, as shown in Figure 19.11. The HR diagram then becomes a means of describing the average properties of stars in a given part of the Galaxy. The diagram might then be compared with similar plots made for stars found elsewhere in the Galaxy, and astronomers could use the results of these comparisons to test their theories about the way stars (and the Galaxy itself) formed and evolved.

When trying to obtain an HR diagram for a distant region of the Galaxy, however, astronomers face a problem. In order to plot the diagram we must know luminosities, and to know luminosities we must know distances. It would seem impossible to construct HR diagrams for stars more distant than 100 pc or so, the maximum distance measurable by stellar parallax. Notice that we can't use the method of spectroscopic parallax because that already *assumes* that the stars lie on the main sequence.

In some circumstances, however, it is possible to plot an HR diagram for very distant stars even though their distances may not be known. If we could find and recognize a group of stars that we knew were all at the same distance from us, then comparing *apparent* brightnesses would be equivalent to comparing *absolute* brightnesses. Why? Because as the radiation traveled toward Earth, the brightness of every star in the group would be diminished by the same amount by the inverse-square law. By measuring and plotting apparent brightnesses, we could create a ''relative'' HR diagram for the group that would look (apart from the numbers on the vertical axis) exactly the same as a real HR diagram based on absolute brightnesses. Such an easily recognizable group of distant stars is called a **star cluster.**

Star clusters can house anywhere from a few dozen to a million stars in a region a few parsecs across. Astronomers believe that all the stars in a given cluster formed at the same time, out of the same cloud of interstellar gas, under the same conditions. Thus when we look at a star cluster, we are looking at a group of stars that all have the same age, similar composition, and lie in the same region of space, at essentially the same distance from Earth. Unlike the stars plotted in Figure 19.14, which differ in mass, age, and (to a lesser extent) in chemical composition, the only factor distinguishing one cluster star from another is its mass.

Clusters are therefore almost ideal ''laboratories'' for stellar studies—not in the sense that astronomers can perform experiments on the stars in them, but because the properties of the stars are very tightly constrained, so that theoretical models of star formation and evolution can be compared with reality without the major complications introduced by broad spreads in age, composition, and place of origin. We will see in Chapter 22 that star clusters are of central importance to astronomers who wish to understand how stars evolve in time.

Open Clusters

Figure 19.18(a) shows a rather loose cluster—the Pleiades, or Seven Sisters—a well-known naked-eye object. Individual stellar colors provide an estimate of the surface temperature of each star. The luminosities follow directly from measurement of the apparent brightness and the cluster's distance (which in this case is known to be about 120 pc). Figure 19.18(b) shows the cluster HR diagram obtained from these data. This type of cluster, found mainly in the

(a) R I V U X G

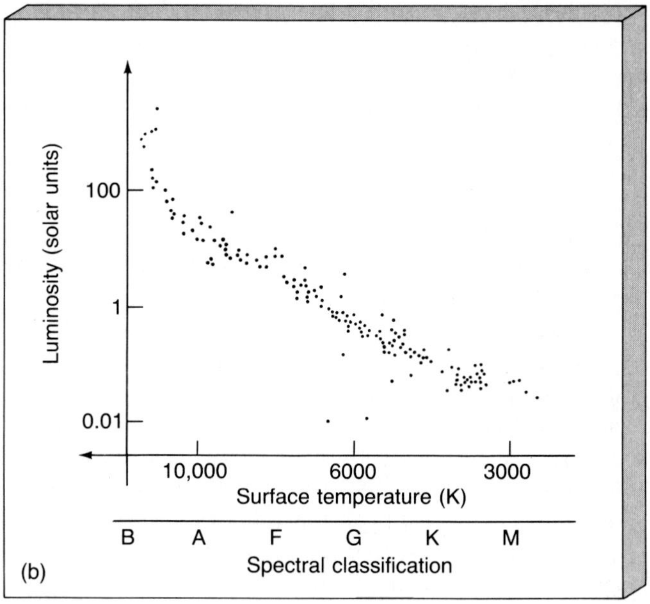

Figure 19.18 (a) This is the Pleiades cluster (also known as M45), about 120 pc from the Sun. The naked eye can see only its brightest stars. (b) The stars of this well-known open (or galactic) cluster yield an HR diagram. (NOAO)

strip across the sky known as the Milky Way, is known as an **open cluster** (or, sometimes, a *galactic cluster*). Open clusters typically contain from a few tens to a few hundred stars and are a few parsecs across.

The HR diagram in Figure 19.18(b) shows stars throughout the main sequence—stars of all colors are represented. The blue stars must be relatively young, for, as we have seen, they burn their fuel rapidly. If all the stars in the cluster formed at the same time, then the red stars must be young too. Thus, even though we have no direct evidence of the cluster's birth, we can estimate its age as less than 20 million years, the lifetime of an O star. Other factors also hint at the cluster's youth. It contains a large amount of interstellar gas and dust not yet processed into stars or lost from the cluster, and it is abundant in heavy elements that (as we will see in Chapter 22) can have been cooked only within the cores of many generations of ancient stars long since perished.

Globular Clusters

A second type of stellar swarm, of which a representative is shown in Figure 19.19(a), is called a **globular cluster.** Globular clusters are much more tightly knit than the loose groups of stars that make up open clusters. All globular clusters are roughly spherical (which accounts for their name) and contain tens of thousands, and sometimes millions, of stars spread out over about 50 pc. As with open clusters, the entire assemblage is held together by gravity.

Figure 19.19(b) shows an HR diagram for this cluster. Notice its many differences from Figure 19.18(b)—globular clusters are a very different stellar environment from open clusters like the Pleiades. The distance to this cluster has been determined by a variation on the method of spectroscopic parallax, but applied to the entire main sequence rather than to individual stars. By calculating the distance at which the apparent brightnesses of the cluster's main-sequence stars taken as a whole best matches a "standard" main sequence, the cluster is found to lie about 10,000 pc from Earth.

The most outstanding feature of all globular clusters is their lack of O- and B-type stars. This deficit is clear from Figure 19.19(b), which is an HR diagram for the accompanying cluster. Low-mass red stars and intermediate-mass yellow stars abound, but high-mass white or blue stars are nearly absent—in fact, there are no main-sequence stars with masses greater than about 0.8 $M_\odot$. (The blue supergiants in this plot are actually stars at a much later stage in their evolution that just happen to be "passing through" the location of the upper main sequence. Their internal structure is quite different from main-sequence giants.) Apparently, globular clusters formed long ago—the more massive O- through G-type stars have long since exhausted their nuclear fuel and disappeared from the main sequence. Other factors confirm that globular clusters are old. Their spectra show few if any heavy elements, implying that these stars formed much earlier in the universe, when heavy elements were much less abundant than they are today.

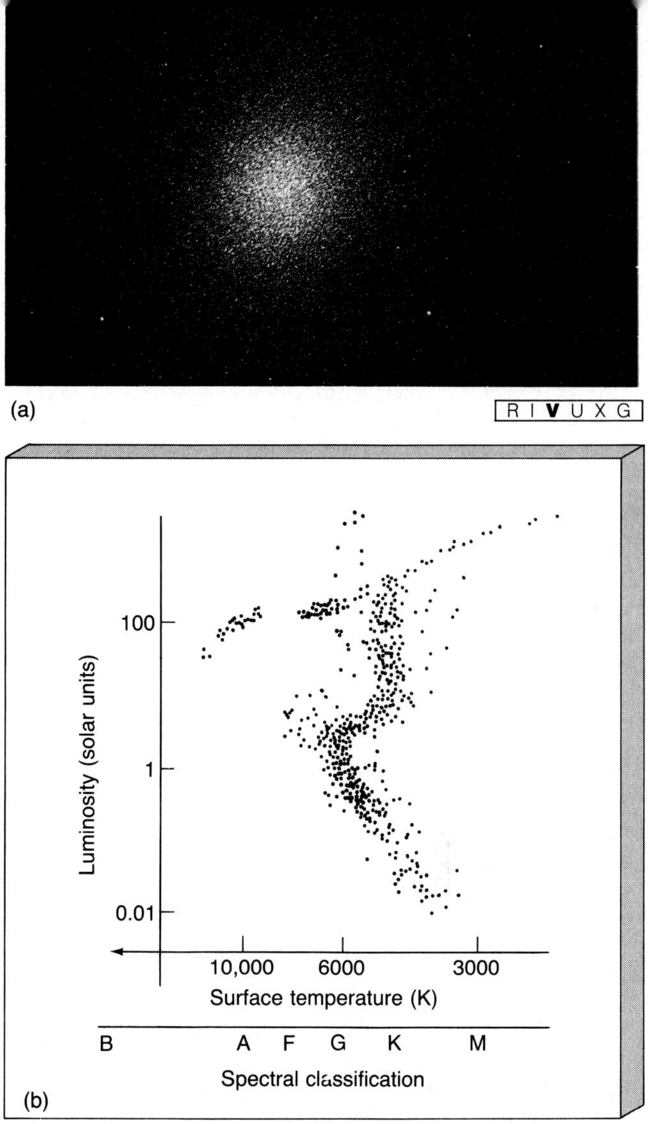

Figure 19.19 *(a) Photograph of a globular cluster and (b) an HR diagram for many (but not all) of its stars. This cluster is called Omega Centauri. It lies approximately 5000 pc from Earth and spans some 40 pc in diameter. (NOAO)*

On the basis of these and other observations, astronomers estimate that globular clusters are all at least 10 billion years old. They contain the oldest known stars in our Milky Way Galaxy. As such, globular clusters are considered to be remnants of the earliest stages of our Galaxy's existence. We can therefore conclude that the age of our Milky Way Galaxy is at least 10 billion years. We will return to this question in Chapter 22, when we discuss the evolution of stars in much greater detail.

We will never be able to watch a single star move through all its evolutionary phases. The lifetimes of humans—probably even of human civilizations—are far too short compared with the lifetimes of even the shortest-lived O and B stars. Instead, we must observe stars as they presently exist—through snapshots taken at specific moments in their life cycles. The HR diagram is just such a snapshot.

By studying stars of different ages or, even better, by studying stars in clusters, where the ages are known to be the same, we can patch together an understanding of a star's "life story" without having to follow a few individuals from birth to death. Such evolutionary studies will be the subject of the next few chapters.

CHAPTER SUMMARY

Astronomers can determine the distance to a star by measuring its parallax, observing the star from either end of a baseline defined by the diameter of Earth's orbit. The annual shift in position of a star relative to other stars, after any effects of parallax have been removed, is called its proper motion. Knowledge of a star's proper motion and its distance reveals its true motion through space.

Most stars are so far away that even the largest telescopes cannot resolve their sizes. Speckle interferometry produces images of stars that help astronomers measure stellar diameters, but the sizes of most stars must be measured indirectly, by using the laws of physics to interpret the radiation they emit. A star's radius can be estimated once its luminosity and temperature are known. Stars with radii between 10 and 100 times that of the Sun are called giants. Stars with larger radii, as big as 1000 solar radii, are called supergiants. A star the size of our Sun or smaller is known as a dwarf.

Luminosity, or absolute brightness, is the total amount of energy that a star's surface radiates into space each second. Apparent brightness, determined by luminosity and distance, is how bright a star appears to us on Earth. As light leaves a star, it spreads out and becomes diluted. The amount of light we see varies inversely as the square of the distance from the star. A star's absolute brightness is the apparent brightness it would have if it resided at a distance of 10 pc from Earth.

The colors of stars reveal which are hot and which are cool. Astronomers can determine a star's surface temperature precisely by observing its brightness at various frequencies and fitting the measurements to a Planck curve. A more detailed classification of stars employs a knowledge of stellar physics and spectroscopy. The spectrum of a cool star shows lines of neutral atoms and molecules. Warmer stars show hydrogen lines. The hottest stars show lines of helium and multiply ionized heavy elements. Spectral classes are designated by the letters O, B, A, F, G, K, M, with O specifying the hottest stars.

The Hertzsprung-Russell diagram plots the spectral type (or temperature) and luminosity of a star. Most stars, like our Sun, lie on the "main sequence," a well-defined diagonal band running across the HR diagram from hot bright giants at the upper left to cool faint dwarfs at the lower right. Low-luminosity stars are difficult to see, radiating much of their energy in the invisible part of the spectrum. For this reason, red dwarfs are often underrepresented on HR diagrams.

A star can be located on the HR diagram by spectral type and luminosity class as well as by temperature and luminosity. Astronomers determine luminosity class by studying the width of stellar spectral lines. Knowledge of a star's absolute and apparent brightnesses yields an estimate of its distance. This "spectroscopic parallax" can be used to measure distances as far out as individual stars can be clearly discerned—a few thousand parsecs.

Most stars come in binary or multiple systems. These systems can provide astronomers with information about the masses of the component stars. It is the mass of a star that determines its position on the main sequence. More than any other stellar property, mass governs a star's ability to radiate. Massive stars burn brightly and quickly. Smaller stars are less luminous and burn for a longer time. A star's radius rises roughly proportionally to its mass, while its luminosity increases much faster. The increase in luminosity with mass helps determine how long main-sequence stars can sustain their energy output. Every type of star has activity such as magnetic fields, photospheric violence, flarelike effects, coronal holes, and stellar winds.

Star clusters are good "laboratories" for stellar studies because their members have the same age and similar composition and lie at essentially the same distance from Earth. Open clusters, like the Pleiades, are loosely grouped and usually contain from a few tens to a few hundred stars. They are generally a few parsecs across, and most are less than a few billion years old. Globular clusters are tightly knit spheres of tens of thousands to millions of stars. They may span up to 50 pc across. Globular clusters are thought to be at least 10 billion years old. They are not forming stars today, so their most massive stars are absent, having burned themselves out long ago.

KEY WORDS

absolute brightness	binary-star system	blue supergiant	color-magnitude diagram
apparent brightness	blue giant	color index	dwarf

eclipsing binary
energy flux
giant
globular cluster
HR diagram
light curve
luminosity class
main sequence
mass-luminosity relation
mass-radius relation
open cluster
optical double
parsec
photometry
proper motion
radial velocity
radius-luminosity-temperature relation
red dwarf
red giant
red giant region
speckle interferometry
spectral class
spectroscopic binary
spectroscopic parallax
star cluster
supergiant
transverse velocity
UBV system
visual binary
white dwarf
white dwarf region

REVIEW QUESTIONS

1. What is parallax, and how is it used to measure the distances to the stars?
2. What is a parsec?
3. Explain two ways in which a star's real space motion translates into motion observable from Earth.
4. How is the space motion of a star determined?
5. How big are giant stars? How small are dwarf stars? Name some other characteristics of these two sorts of stars.
6. What do astronomers mean when they speak of a star's luminosity?
7. What is the difference between the absolute and apparent brightnesses of stars?
8. What is the inverse-square law, and how does it affect our ability to see distant stars?
9. How do astronomers measure the temperatures of stars?
10. Briefly describe the system of the classification of stars according to their spectral characteristics.
11. Explain how stars are plotted on the Hertzsprung-Russell diagram.
12. What is the main sequence?
13. Why does an HR diagram tend to be biased?
14. What is the most important factor influencing a star's ability to radiate?
15. What is a method that astronomers use to create an unbiased HR diagram?
16. What is the difference between an open star cluster and a globular star cluster?
17. What is the distance to the star Spica, whose observed parallax is 0.013″?
18. A certain star has a temperature twice that of the Sun and a luminosity 64 times greater than the solar value. What is its radius, in units of the solar radius?
19. Two stars—A and B, of luminosities 0.5 and 4.5 times the luminosity of the Sun—are observed to have the *same* apparent brightness. Which one is more distant, and how much farther away is it than the other?
20. Given that the Sun's lifetime is about 10 billion years, estimate the life expectancy of (a) a 0.2-solar mass, 0.01-solar luminosity red dwarf, (b) a 3-solar mass, 30-solar luminosity star, (c) a 10-solar mass, 1000-solar luminosity blue giant.

DISCUSSION QUESTIONS

1. Contrast what you now know about the real motions of stars with what earlier cultures believed about the "fixed" stars. Draw your contrast in light of early and modern philosophical notions about the nature of space, time, and the universe.
2. Go to the library and find some references to "blue stragglers." What are they? Has the mystery of blue stragglers been solved?
3. Discuss the HR diagram as an astronomical tool. Why is it so useful?

PROJECTS

1. Every winter, you can find an astronomy lesson in the evening sky. The Winter Circle is an asterism—or pattern of stars—made up six bright stars in five different constellations. The stars are Sirius, Rigel, Betelgeuse, Aldebaran, Capella, and Procyon. These stars span nearly the entire range of colors (and therefore temperatures) possible for normal stars. Rigel is a B star. Sirius is an A. Procyon is an F star. Capella is a G star. Aldebaran a K star. Betelgeuse is an M star. It's easy to see the color difference of these stars in the Winter Circle.

   What do the colors of these stars tell us about their temperatures?

   Why do you suppose there is no O star in the Winter Circle?

2. Summer is a good time to search with binoculars for open star clusters. Open clusters are generally found in the plane of the Galaxy. If you can see the hazy band of the Milky Way arcing across your night sky—in other words, if you are far from city lights and looking at an appropriate time of night and year—you can simply sweep with your binoculars along the Milky Way. Numerous "clumps" of stars will pop into view. Many will turn out to be open star clusters.
3. Globular star clusters are harder to find. They are intrinsically larger, but they are also typically much farther away and therefore appear smaller in our sky. The most famous globular cluster visible from the Northern Hemisphere is M13 in the constellation Hercules, visible on spring and summer evenings. This cluster contains half a million or so of the Galaxy's most ancient stars. It may be glimpsed in binoculars as a little ball of light, located about one-third of the way from the star Eta to the star Zeta in the Keystone asterism of the constellation Hercules. Telescopes reveal this cluster as a magnificent, symmetrical grouping of stars.

   Another fairly easy-to-find globular cluster is up in the evening during the summer months. M4 is located 1.3° directly west of the bright star Antares in the constellation Scorpius. It can be seen with the eye under ideal conditions. It's easy to pick out with binoculars! From a dark location, just aim toward Antares, and M4 will probably be in the same binocular field.

SUGGESTED READINGS

Boyko, A. "Inside a Globular Cluster." *Sky and Telescope* (November 1964). Calculations show how the night sky might appear from inside a globular star cluster.

Fortier, E. "Touring the Stellar Cycle." *Astronomy* (March 1987). Using a small telescope to observe different objects in various stages of stellar evolution.

Griffin, R. "Radial-Velocity Revolution." *Sky and Telescope* (September 1989). A new era in the study of stars' motions toward and away from Earth.

Kaler, J. B. "The B Stars: Beacons of the Skies." *Sky and Telescope* (August 1987). Hot, young objects that shine brightly and mark regions of recent star formation.

———. "Cousins of Our Sun: The G Stars." *Sky and Telescope* (November 1986). Stars of the same spectral class as our Sun.

———. "Extraordinary Spectral Types." *Sky and Telescope* (February 1988). Stars that can't be pigeonholed into any of the usual spectral classes.

———. "Journeys on the H-R Diagram." *Sky and Telescope* (May 1988). How stars change as they age.

———. "The K Stars: Orange Giants and Dwarfs." *Sky and Telescope* (August 1986). Some are among the most familiar sights in the heavens, while others are the least luminous stars visible with the unaided eye.

———. "M Stars: Supergiants to Dwarfs." *Sky and Telescope* (May 1986). The biggest and the coolest of stars, some famous variables, and feeble red dwarfs.

———. "Origins of the Spectral Sequence." *Sky and Telescope* (February 1986). How astronomers classify spectra and interpret them to unlock the secrets of all kinds of stars.

———. "The Spectacular O Stars." *Sky and Telescope* (November 1987). Hottest, bluest, brightest, rarest, and most massive, the O stars form the blazing tip of the main sequence.

———. "The Temperate F Stars." *Sky and Telescope* (February 1987). F stars mark the transition from hot to cool stars and include the pulsating Cepheid variables.

———. "White Sirian Stars: Class A." *Sky and Telescope* (May 1987). The spectral class of two of the most famous stars in the sky: brilliant Sirius and its faint white-dwarf companion.

Sneden, C. "Reading the Colors of the Stars." *Astronomy* (April 1989). Using stellar spectra to understand stars.

Trimble, V. "White Dwarfs: The Once and Future Suns." *Sky and Telescope* (October 1986). The whys and hows of white dwarf stars.

White, R. E. "Globular Clusters: Fads and Fallacies." *Sky and Telescope* (January 1991). How much of the conventional wisdom concerning these clusters may not be so certain after all.

CHAPTER

20

The Interstellar Medium

Gas and Dust between the Stars

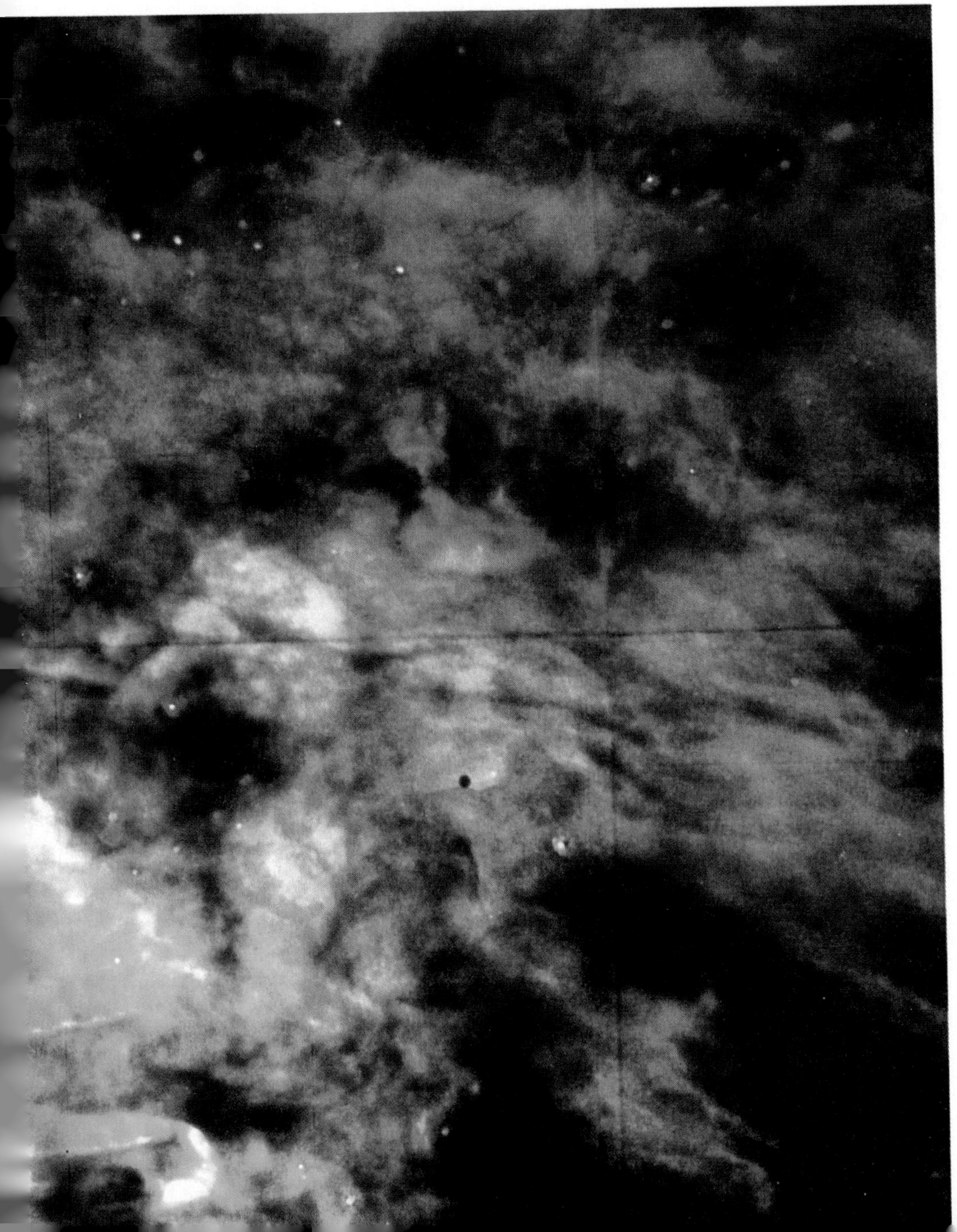

This computer-enhanced image taken with the Hubble Space Telescope *shows a highly magnified view of a very small part of the Orion Nebula, a bright region of star formation some 1500 light-years distant. Red light is sulfur emission, blue is oxygen, and green is hydrogen. Note the jetlike structure evident at the 9 o'clock position and the bright ''elephant trunk'' to the lower left. The scale of this image is about 0.1 arc second, which at Orion's distance translates into a dimension of about 6 light-hours, or roughly the radius of our solar system. (Courtesy NASA)*

R I **V** U X G

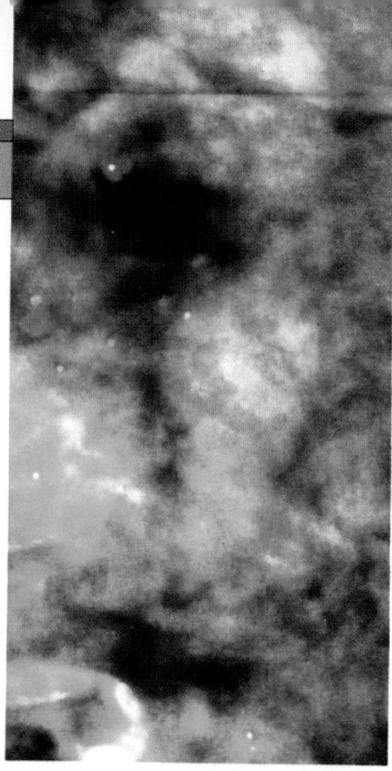

LEARNING GOALS

1 to appreciate that the vast regions among the stars harbor gargantuan quantities of interstellar gas and dust

2 to recognize emission nebulae as the signposts of stellar birth

3 to appreciate the role of interstellar dust in dark clouds

4 to understand the radio techniques used to probe the nature of interstellar matter

5 to understand the nature and significance of interstellar molecules

We have now studied the basic properties of stars and planets, well-defined dense collections of matter that are visible either because they emit light of their own or because they reflect light from their surfaces. Space, however, is populated not only with stars and planets. It also harbors matter throughout the invisible regions of interstellar space, in the dark voids between the stars in the evening sky.

The density of interstellar matter is extremely low—approximately a trillion trillion times less dense than matter in either stars or planets, far more tenuous than the best vacuum attainable on Earth. Only because the volume of interstellar space is so vast does its mass amount to anything at all. So why bother to study this near-perfect vacuum? We do so for three important reasons. First, there is as much mass in the ''voids'' between the stars as there is in the stars themselves. Second, interstellar space is the region out of which new stars are born. Third, it is the region into which some old stars explode at death. It is one of the most significant crossroads through which matter passes in the history of our universe.

INTERSTELLAR MATTER

✓1 Figure 20.1 shows a large region of space, a much greater expanse of universal real estate than anything we have studied thus far. The bright regions are congregations of innumerable stars, some of whose properties we have just studied in Chapter 19. However, the dark areas are not simply ''holes'' in the stellar distribution. They are regions of space that obscure or extinguish light from the stars beyond. These regions house **interstellar matter,** consisting of great clouds of gas and dust. Their very darkness means that they cannot be easily studied by the optical methods used for stellar matter. There is, quite simply, very little to see!

Figure 20.1 shows that the dark interstellar matter is rather patchy. It is spread very irregularly through space. In some directions, the obscuring matter is largely absent, and astronomers can study objects literally billions of parsecs from Earth. In other directions, the obscuration is moderate, prohibiting visual observation of objects beyond more than a few thousand parsecs, but still allowing us to study nearby

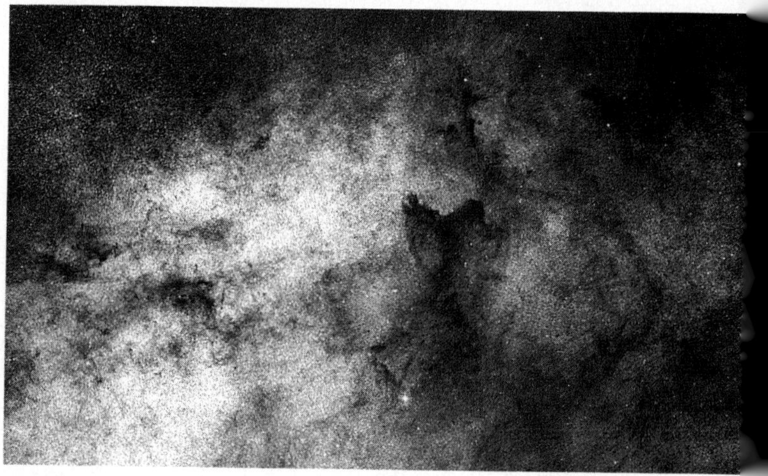

R I **V** U X G

Figure 20.1 *A wide-angle photograph of a great swath of space, showing regions of brightness (vast fields of stars) as well as regions of darkness (obscuring interstellar matter). (Palomar Observatory)*

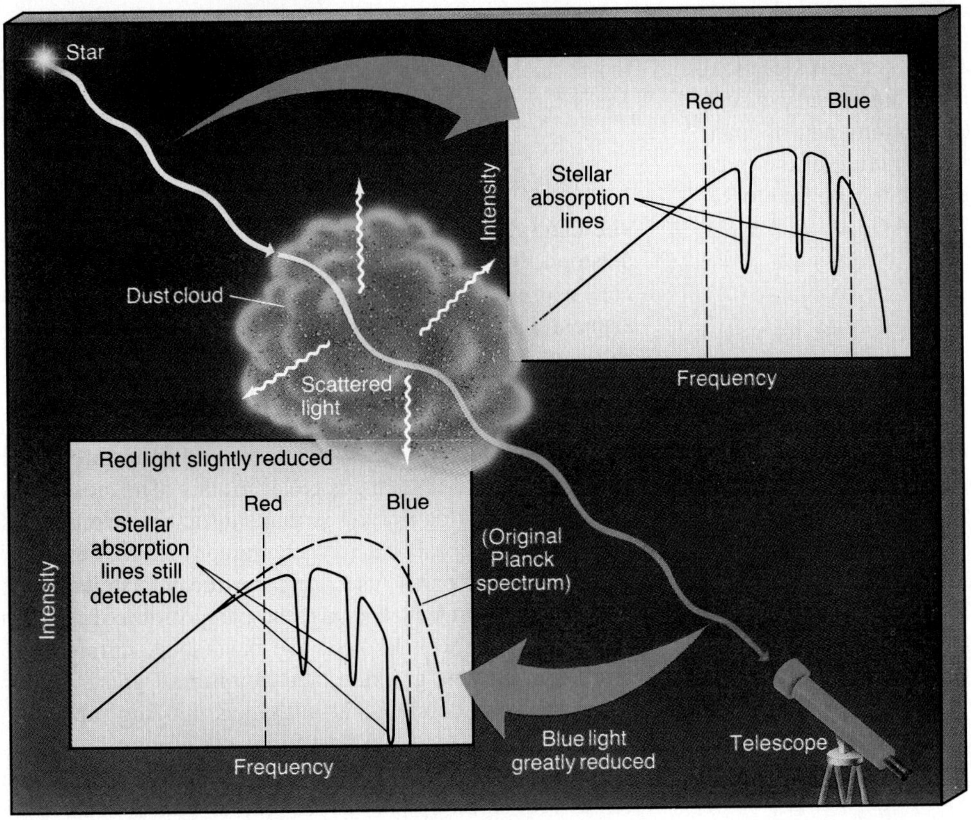

Figure 20.2 *Starlight passing through a dusty region of space is both dimmed and reddened, but spectral lines are still recognizable in the light that reaches Earth.*

stars. Still other regions are so heavily obscured that starlight from even relatively nearby stars is completely absorbed before reaching Earth.

Gas and Dust

The matter between the stars—the **interstellar medium**—is made up of two components, gas and dust, intermixed throughout all of space.

Interstellar gas is made up mainly of individual atoms, of average size 10^{-8} cm (1 Å) or so. The gas also contains some small molecules, no larger than about 10^{-7} cm across. Regions with such small particles are transparent to nearly all types of radiation, including ultraviolet, visible, infrared, and radio waves. Apart from narrow atomic and molecular absorption lines, gas alone does not block radiation to any great extent.

Interstellar dust is more complex. It consists of larger clusters of atoms—really mixtures of millions of molecules—not unlike chalk dust and the microscopic particles that make up smoke, soot, or fog. Light from distant stars cannot penetrate the densest accumulations of interstellar dust any more than a car's headlights can illuminate roadside objects in a thick Earth-bound fog. Comparisons of how starlight is diminished in interstellar space with the scattering of humanmade light in terrestrial fog indicate that the typical size of an interstellar dust particle—or **dust grain**—is about 10^{-5} cm—comparable in size to the wavelength of visible light. Typical interstellar dust particles are thus about 1000 times larger in diameter than interstellar gas particles.

The ability of a particle to scatter light depends strongly on both the size of the particle and the wavelength of the radiation involved. As a rule of thumb, we can say that radiation of a given wavelength—λ, say—is significantly affected only by particles comparable to or larger than λ in size. Because the wavelength of radio waves greatly exceeds the size of the dust grains, dusty interstellar regions are completely transparent to long-wavelength radio radiation. These regions are also partially transparent to infrared radiation. Conversely, interstellar dust is a very effective blocker of high-frequency optical, ultraviolet, and x-ray radiation.

In astronomical parlance, this general dimming of starlight by interstellar matter is called **extinction.** In addition, because the interstellar medium is more opaque to radiation of shorter wavelengths, light from distant stars is preferentially robbed of its higher-frequency (''blue'') components, so that stars also tend to appear redder than they really are. This effect is known as **reddening.** The process is basically similar to the one that makes sunsets on Earth appear red (see Interlude 7-1).

As illustrated in Figure 20.2, extinction and reddening change a star's apparent brightness and color, but they have no effect on its spectral type. Absorption lines in the star's spectrum are largely unaffected by interstellar dust. Astronomers can use this fact to study the interstellar medium. By determining a main-sequence star's spectral type, astronomers first learn its true luminosity and color (see Chapter 19). They can then measure the degree to which the starlight has been affected by extinction and reddening en route to Earth, and this in turn allows them to estimate both the numbers and the sizes of interstellar dust particles along the line of sight to the star. By repeating these measurements for stars in many different directions and at many different distances from Earth, a rough picture of the interstellar medium in the surrounding solar neighborhood can be built up.

Temperature and Density

The temperature of the interstellar gas and dust ranges from a few kelvins to a few hundred kelvins, depending on its proximity to a star or some other source of radiation. Generally, we can take 100K as an average temperature of a typical region of interstellar space. Compare this with 273K, at which water freezes, and 0K, at which atomic and molecular motions virtually cease. Interstellar space is very cold.

The density of interstellar matter is extremely low. It averages roughly 1 *atom* per cubic centimeter, but densities as great as 1000 atoms/cm^3 and as small as 0.01 atom/cm^3 have been found. Matter of this low density is far more tenuous (thin) than the best vacuum—about 10^4 molecules/cm^3—that we can make in laboratories here on Earth. Interstellar matter is about a trillion trillion times less dense than water.

Interstellar space is populated with gas so thin that harvesting all the matter in an interstellar region the size of Earth would yield barely enough matter to make a pair of dice. How can matter so thinly distributed play a significant role compared with stars and planets that are so much denser? The key is size—interstellar space is vast. The typical distance between stars—1 pc or so in the vicinity of the Sun—is far, far greater than the typical size—around 10^{-7} pc—of the stars themselves. Stellar and planetary sizes pale in comparison to the vastness of interstellar space. Thus matter can accumulate, regardless of how thinly spread.

Interstellar dust is even rarer than interstellar gas. On average there are only about 10^{-12} dust particles per cubic centimeter—that is, 1000 per cubic *kilo*meter. How can such fantastically thin matter diminish light radiation so effectively? The reason, again, is that interstellar dust is spread throughout enormous volumes of space—and that means great distances between Earth and astronomical sources of light. For example, an imaginary cylinder 1 cm^2 in cross-section and extending from Earth to Alpha Centauri would contain more than a million dust particles. Over huge distances, dust particles accumulate slowly but surely, to the point where they can effectively block visible light and other short-wavelength radiation.

Despite their rarity, dust particles make interstellar space a *relatively* dirty place. Earth's atmosphere, by comparison, is about a million times cleaner. Our air is tainted by only one dust particle for about every billion billion (10^{18}) atoms of atmospheric gas. If we could compress a typical parcel of interstellar space to equal the density of air on Earth, this parcel would contain enough dust to make a fog so thick that we would be unable to see our hand held at arm's length in front of us.

Composition

The composition of interstellar gas is reasonably well understood. Spectroscopic studies of the light it absorbs provide astronomers with comprehensive information on its elemental abundances. Most of it—about 90 percent of all particles—is atomic and molecular hydrogen, some 9 percent is helium, and the remaining 1 percent is heavier elements. The abundances of several of the heavy elements (at least in gaseous form) are much lower in the interstellar medium than in our solar system or in stars. The most likely explanation for this finding is that substantial quantities of elements such as carbon, oxygen, silicon, magnesium, and iron have been partly used to form the interstellar dust.

By contrast with the gas, the composition of the dust is currently not well known. We have some infrared evidence for silicates, graphite, and iron in dust grains. These materials are underabundant in the gas, lending support to the theory that interstellar dust forms out of interstellar gas. The dust probably also contains some "dirty ice," a frozen mixture of ordinary water-ice contaminated with trace amounts of ammonia, methane, and other chemical compounds—composition quite reminiscent of cometary nuclei in our own solar system (see Chapter 16).

Dust Shape

Curiously, astronomers know the *shape* of interstellar dust particles better than their composition. Although the minute atoms in the interstellar gas are basically spherical, the dust particles are not. Individual dust grains are apparently elongated or rodlike, as shown in Figure 20.3. We can infer this because the light emitted by stars is dimmed and partially **polarized** by the dust.

Recall from Chapter 4 that light consists of electromagnetic waves composed of vibrating electric and magnetic fields. Normally, these waves are randomly oriented, and we say the radiation is unpolarized. Stars emit *unpolarized* radiation from their photospheres. Under some circumstances, the electric fields can become aligned—all vibrating in the same plane as the radiation moves through space. We then say the radiation is *polarized*. Polarization of starlight does not occur by chance. If the light detected by our telescope is polarized, it is because some interstellar matter lies between the emitting object and our equipment. The polarization of starlight, then, provides a means of studying the interstellar medium.

On Earth we can produce polarized light by passing unpolarized light through a Polaroid filter, which has specially aligned elongated molecules that pass only those waves with their electric fields oriented in some specific direction (see Figure 20.4[a]). Other waves are absorbed and do not pass through the filter. The alignment of the

CHAPTER SUMMARY

The dark areas between the stars contain clouds of gas and dust that obscure starlight from beyond. Interstellar gas does not block radiation to any great extent, but interstellar dust is an effective blocker of optical, ultraviolet, and x-ray radiation. The matter density between the stars is extremely low, but outer space is so vast that interstellar space contains about as much total mass as the stars themselves.

Interstellar space is bitterly cold and of extremely low density. Even the densest interstellar regions are more tenuous than the best vacuum we can make in earthly laboratories. Interstellar gas is made up of atomic and molecular hydrogen, helium, and about 1 percent heavier elements. Interstellar dust is less well understood, but it is thought to be composed of silicates, graphite, iron, and "dirty ice." Interstellar dust particles are apparently elongated or rodlike. The polarization of starlight provides a means of studying them.

Emission nebulae are regions of glowing, ionized gas. At or near the center of each is at least one hot O- or B-type star, whose radiation causes the surrounding gas to glow. Accordingly, the photons detected at Earth provide direct information about the nebular gas, as well as indirect information about the embedded hot stars.

Some excited atomic states take so long to emit a photon that their lines are never seen in terrestrial laboratories, where the gas is very dense by interstellar standards and collisions always knock the atom into another excited state before it can emit any radiation. As a result, the spectral lines associated with these transitions are extremely rare, or "forbidden," on Earth.

Cold, dark interstellar clouds along the line of sight from Earth to a star can be studied by measuring their absorption of optical light. These clouds cannot be seen directly at optical wavelengths because they are too cold to emit any visible light. The radio signature of cold, atomic hydrogen is a wavelength of 21 centimeters. Radio telescopes can observe any interstellar region having enough gas to produce a detectable signal.

The *International Ultraviolet Explorer* found evidence for speeding clouds and "sheets" of interstellar matter. The *IUE* also showed that some regions of interstellar space are much thinner and hotter than expected. Some of the space between clouds and nebulae may contain superheated interstellar "bubbles."

Dwarfing the emission nebulae are the molecular clouds—dense regions containing gas, mostly hydrogen, in molecular form. Their emitted radiation is usually in the radio range. The dust in interstellar clouds probably both protects molecules and acts as a catalyst to help them form. Radio maps of interstellar gas and infrared maps of interstellar dust reveal that molecular clouds are part of huge molecular cloud complexes. Embedded in these are cloud cores where stars and star clusters are forming. Molecules in space are likely formed from smaller atoms and molecules already present. Some approach the complexity of the precursors of life on Earth.

KEY WORDS

dark dust cloud	extinction	interstellar matter	nebula
dust grain	forbidden line	interstellar medium	polarization
dust lane	HI region	molecular cloud	reddening
emission nebula	HII region	molecular cloud complex	21-centimeter line

REVIEW QUESTIONS

1. Give a brief description of the interstellar medium.
2. What is the composition of interstellar gas? What about interstellar dust?
3. How much matter resides between stars? How dense is this material?
4. If space is a near-perfect vacuum, then how can there be enough dust in it to block light?
5. Is interstellar matter spread uniformly through space?
6. What are some ways that astronomers use to study interstellar dust?
7. What is an emission nebula?
8. What determines the size of an emission nebula?
9. Eventually an emission nebula stops growing in size. What are the factors that may keep it from expanding indefinitely?
10. How do emission nebulae enable astronomers to study stars?
11. Give a brief description of a dark dust cloud.
12. What is 21-cm radiation, and why is it so profoundly important to astronomers?
13. How does a molecular cloud differ from other interstellar matter?

14. What is observed to be happening in the cores of molecular clouds?

15. Of what significance are complex molecules in interstellar clouds?

16. Calculate the total mass of interstellar matter (of density 10 hydrogen atoms, each of mass 1.7×10^{-24} grams, per cubic centimeter, say) contained in a volume equal to the volume of the Earth.

17. In order to carry enough energy to ionize a hydrogen atom, a photon must have a wavelength less than 912 Å or 9.12×10^{-6} cm. Using Wien's Law (Chapter 4), calculate the temperature a star must have for the peak wavelength of its Planck curve to equal this value.

18. Calculate the frequency of 21-cm radiation.

19. A beam of light shining through a dense molecular cloud is diminished in intntensity by a factor of 2.5 for every 3 parsecs it travels. By what total factor is it reduced if the total thickness of the cloud is 60 pc?

20. Calculate the radius of a spherical molecular cloud whose total mass equals the mass of the Sun. Assume a cloud density of 10^6 hydrogen atoms per cubic centimeter.

DISCUSSION QUESTIONS

1. If our Sun were surrounded by a cloud of gas, would this cloud be an emission nebula? Why or why not?
2. Compare the reason for the reddening of stars by interstellar dust with that for the reddening of the setting Sun.
3. Explain what it means for a star's light to be polarized. How does the polarization of starlight provide a means of studying the interstellar medium?
4. Suppose that life *can* exist in interstellar clouds. Speculate on what sort of life it might be. What link might there be between life in the clouds and the origin of life on Earth?

PROJECTS

1. The constellation Orion the Hunter is prominent in the evening in winter. Its most noticeable feature is a short, straight row of three medium-bright stars: the famous Belt of Orion. A curved line of stars extends from the Belt. This line represents Orion's Sword. The second from the bottom in this line of stars is the sky's most famous emission nebula, M42, the Orion Nebula.

   The Orion Nebula appears to the eye as a misty star. Binoculars show more haze, and a telescope reveals not only a complex region of haze, but also a little grouping of at least four young hot stars. These stars are known as the Trapezium. They are O, B, and A stars. Their energy causes the Orion Nebula to glow. Why does the Orion Nebula appear greenish?

   The Orion Nebula is like the visible tip of a vast celestial iceberg. It marks the location in our sky of a huge dark cloud in which many new stars are forming. These stars cannot be seen in visible light, but they have been studied at other wavelengths. No one knows how many new stars may eventually emerge from this region of the sky, but they may number in the thousands. If you could return to Earth in 50 million years and gaze toward the Orion Nebula, what might you see?
2. Can you find in the sky any of the other emission nebulae mentioned in the text?

SUGGESTED READINGS

Gillett, F. C., I. Gatley, and D. Hollenbach. "Infrared Astronomy Takes Center Stage." *Sky and Telescope* (August 1991). The 1990s should see major new infrared telescopes for use on the ground, in the air, and in space.

Kanipe, J. "Inside Orion's Stellar Nursery." *Astronomy* (August 1989). How an infrared image revealed hundreds of stars forming inside the Orion Nebula.

Malin, D. "In the Shadow of the Horsehead." *Sky and Telescope* (September 1987). Photographic feature about a famous dark nebula.

Malin, D., and D. Allen. "Echoes of the Supernova." *Sky and Telescope* (January 1990). The detailed, three-dimensional structure of interstellar dust sheets in the Large Magellanic Cloud, as revealed by studies of Supernova 1987A.

Paresce, F. and S. Bowyer. "The Sun and the Interstellar Medium." *Scientific American* (September 1986). The Sun in the larger context of the Galaxy.

Schorn, R. "The Strange Case of NGC 6164-6165." *Sky and Telescope* (September 1986). A bizarre S-shaped nebula, possibly the work of an evolved O star.

CHAPTER 21

Star Formation
A Traumatic Birth

This is an artist's conception of a gas cloud that has contracted and begun to spin up. As matter rushes in, the density and temperature in the center mount, creating high pressure that ejects excess material along its north and south poles. (Courtesy D. Berry)

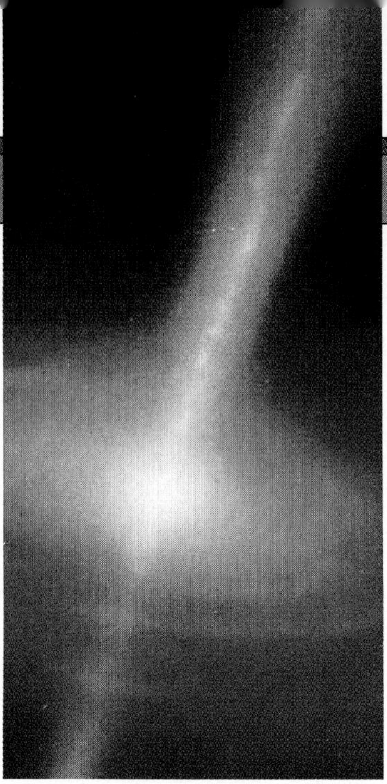

LEARNING GOALS

1 to recognize the many factors, such as heat, rotation, and magnetism, that compete against gravity in the process of star formation

2 to know the sequence of events that led to the formation of our Sun

3 to understand that stars of different masses form in similar ways but at very different rates

4 to see some of the observational evidence supporting the modern theory of star formation

5 to realize that interstellar space is so complex that the details of prestellar evolution are very hard to unravel

We now move from the interstellar medium—the gas and dust between the stars—to the stars themselves. The next four chapters discuss the formation and evolution of stars. We have already seen in Chapter 19 that stars must necessarily evolve as they consume their fuel supply, and we have extensive observational evidence of stars at many different evolutionary stages. With the help of these observations, scientists have developed a good understanding of stellar evolution—the complex changes experienced by stars as they form, mature, grow old, and die.

Stars take so long to evolve that we cannot track one from birth to death. Nevertheless, by observing many different stars, we can build up a picture of their evolution. Astronomical studies of the changes stars experience are a little like biological studies of the life cycles of trees. Trees live longer than humans, but we can understand their evolution by taking a census of the forest and applying our knowledge of the trees' environment to infer how different species grow and die. Likewise, we take a census of the stars—the HR diagram—and apply our knowledge of physics and the interstellar medium to understand how they evolve.

We begin by studying the process of star formation, through which interstellar clouds of gas and dust are transformed into the myriad stars we see in the night sky.

GRAVITATIONAL COMPETITION

✓1 How do stars form? What factors determine the masses, luminosities, and distribution of stars in our Galaxy? In short, what basic processes are responsible for the appearance of our nighttime sky? Simply stated, star formation begins when part of the interstellar medium—one of the cold dark clouds discussed in Chapter 20—starts to collapse under its own weight. The cloud fragment heats up as it shrinks, and eventually its center becomes hot enough for nuclear burning to begin. At that point, the contraction stops, and a star is born. But what determines which interstellar clouds collapse? For that matter, because all clouds exert a gravitational pull, why didn't they all collapse long ago? To begin to answer these questions, let us consider a small portion of a large cloud of interstellar gas. Concentrate first on just a few atoms, as shown in Figure 21.1.

Each atom has some random motion because of the

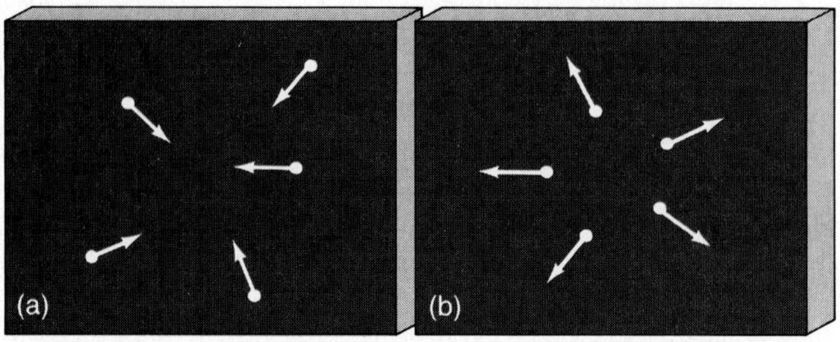

Figure 21.1 *Motions of a few atoms within an interstellar cloud are influenced by gravity so slightly that their paths are hardly changed before (a) and after (b) an accidental, random encounter.*

cloud's heat, even if the cloud's temperature is very low. Each atom is also influenced by the gravitational attraction of all its neighbors. The gravitational force is not large, however, because the mass of each atom is so small. Even when a few atoms accidentally cluster for an instant, as shown in Figure 21.1(a), their combined gravity is insufficient to bind them into a lasting, distinct clump of matter. This accidental cluster would disperse as quickly as it formed. The effect of heat—the random motion of the atoms—is much stronger than the effect of gravity.

Now let's concentrate on a larger group of atoms. Imagine, for example, 50, 100, 1000, even a million atoms, each gravitationally pulling on all the others. The force of gravity is now stronger than before. Would this many atoms exert a combined gravitational attraction strong enough to prevent the clump from dispersing again? The answer—at least under the conditions found in interstellar space—is still no. Gravity is still far too weak to overcome the effect of heat.

How many atoms must be brought together in order for gravity to hold them together? We cannot answer this all-important question by considering just gravity alone. Given enough atoms, gravity *will* eventually be able to bind our interstellar fragment together and make it contract, ultimately to form one or more stars, but other physical agents, such as heat, rotation, and magnetism, all strongly influence the fragment's evolution. Although these factors should not be regarded in any sense as "antigravity," they do compete with gravity in determining the fate of our clump of matter. Let's consider each in turn.

The Effect of Heat

We have already seen numerous instances of the competition between heat and gravity (see, for example, Interlude 9-1). Recall that the temperature of a gas is simply a measure of the average speed of the atoms or molecules in it. The higher the temperature, the greater the average speed, and the greater this speed, the higher the pressure within the gas. A hot gas exerts pressure that competes with the effects of gravity—heat is the main reason that the Sun and other stars don't collapse. The outward pressure of heated gases exactly balances gravity's inward pull.

Analysis of the radiation emitted by interstellar clouds indicates that they contain some heat, but not much. Their temperatures are low by solar, and even terrestrial, standards. Observations of dark interstellar regions show that most clouds have temperatures below about 100K. Some have temperatures as low as 10K; a few are known to be even colder. Thermal effects are barely able to prevent these clouds from collapsing under their own gravity. Once a cloud crosses the threshold to where its gravity overcomes thermal pressure and begins to contract as enough atoms clump together, heat becomes a relatively unimportant factor as the collapse proceeds. Only when the cloud has shrunk to a small fraction of its original size and its temperature has risen to several thousand kelvins do thermal effects again begin to play an important role.

The Effect of Rotation

Rotation—that is, spin—can also compete with the inward pull of gravity. As we saw in Chapter 17, a contracting cloud having even a small spin tends to develop a bulge around its midsection. As the cloud contracts, it must spin faster (to conserve its angular momentum), and the bulge grows—material on the edge tends to fly off into space. Figure 21.2 illustrates this important feature of rotation. (Consider as an analogy mud flung from a rapidly rotating bicycle wheel.) Eventually, the cloud forms a flattened, rotating disk.

For material to remain part of the cloud and not be spun off into space, a force must be applied—in this case, the force of gravity. The more rapid the rotation, the greater is the tendency for the gas to escape, and the greater is the gravitational force needed to retain it. It is in this sense that we can regard rotation as opposing the inward pull of gravity. Should the rotation of a contracting gas cloud overpower gravity, the cloud would simply disperse. The upshot: Rapidly rotating interstellar clouds need more mass for contraction into stars than clouds having no rotation at all.

The Effect of Magnetism

Magnetism can also hinder a cloud's contraction. Just as Earth, most of the other planets, and the Sun all have some magnetism, magnetic fields permeate most interstellar

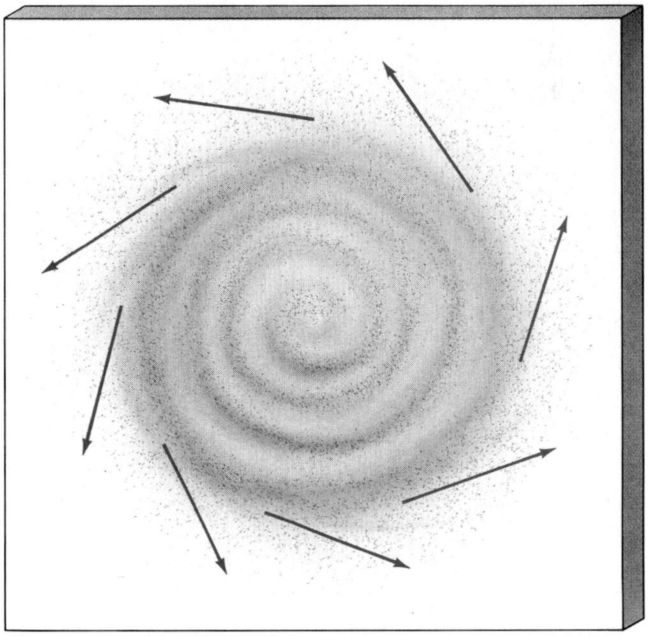

Figure 21.2 *A rapidly rotating gas cloud tends to resist contraction. The spin tends to fling matter from the cloud, like mud spinning off the rim of a bicycle wheel. Spin, then, competes with the inward pull of gravity.*

clouds. As a cloud contracts, it heats up, and atomic encounters become violent enough to ionize the gas. As we noted in Chapter 7 when discussing Earth's Van Allen belts, magnetic fields can exert electromagnetic control over charged particles. In effect, the particles tend to become "tied" to the magnetic field—free to move *along* the field lines but inhibited from moving *perpendicular* to them. If the magnetism is strong enough, the ions in the gas will be influenced more by the magnetic field than by gravity. It is in this sense that magnetism can resist gravity.

This tug-of-war between gravity and magnetism often causes interstellar clouds to contract in distorted ways. Because the charged particles and the magnetic field are tied together, the field itself follows the contraction of a cloud, as indicated in Figure 21.3. The charged particles literally pull the magnetic field toward the cloud's center in the direction perpendicular to the field lines. As the field lines are compressed, the magnetic field strength increases. In this way, the strength of magnetism in a cloud can become much larger than that normally permeating general interstellar space. The primitive solar nebula may have contained a strong magnetic field created in just this way.

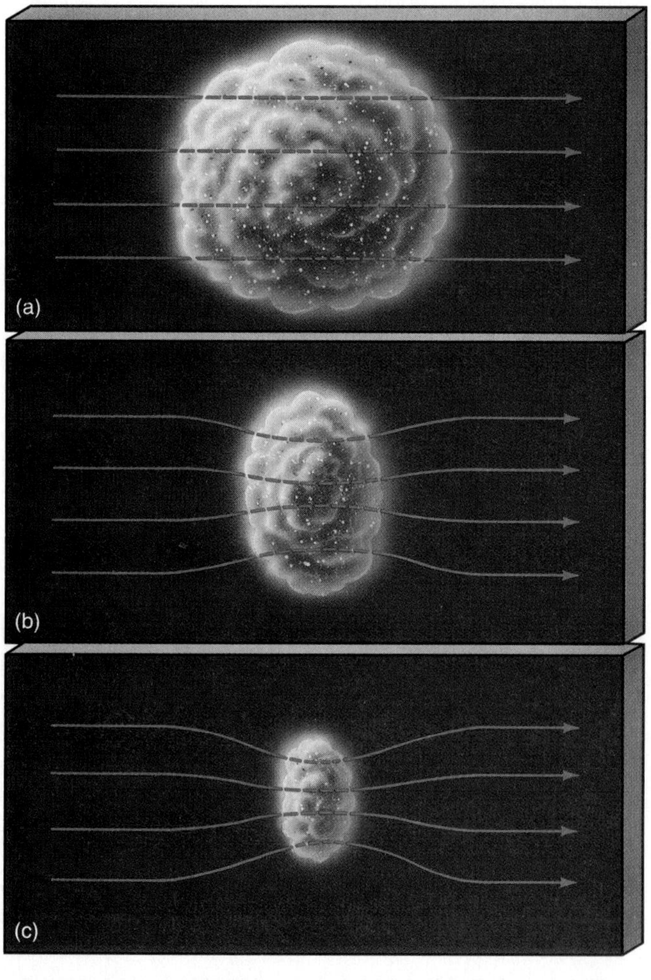

Figure 21.3 *Magnetism can hinder the contraction of a gas cloud, especially in directions perpendicular to the magnetic field (solid lines). Frames (a), (b), and (c) trace the evolution of a slowly contracting interstellar cloud having some magnetism.*

The Simplest Case

Observations made during the past decade show that real interstellar clouds are not very hot, spin only slowly, and are only slightly magnetized. Nevertheless, theory suggests that even small quantities of any of these agents can compete quite effectively with gravity and can greatly alter the evolution of a typical gas cloud. The existence of the solar system illustrates the importance of rotation, for example—without rotation, no planetary system would exist at all. Unfortunately, the interplay of these factors is not well understood. Both rotation and magnetism can lead to very complex behavior as a cloud contracts, and the combination of the two is extremely difficult to study theoretically. We can gain an appreciation for the broad outlines of the star-formation process by neglecting these two complicating factors, but bear in mind that both are probably important in determining the details.

We now return to our original question: How many atoms need to be accumulated for their collective pull of gravity to prevent them from dispersing back into interstellar space? The answer, even for a typical cool (100K) cloud having no rotation or magnetism, is a truly huge number. Nearly 10^{57} atoms are required—much larger than the 10^{25} grains of sand on all the beaches of the world, even larger than the 10^{51} elementary particles that constitute all the atomic nuclei in the entire Earth. There is simply nothing on Earth comparable to a star.

The total number of atoms that make up the mass of the Sun is about 10^{57}. It is no coincidence that that is also the number of atoms required to build up the gravitational field necessary to make a star. Our Sun is a very typical star. As we have seen, most stars in our Galaxy have masses between 0.1 and 10 times that of our Sun or, equivalently, contain between 10^{56} and 10^{58} atoms. The more massive stars probably formed in interstellar regions where heat, rotation, or magnetism competed strongly with gravity, so that more mass was needed to overcome them. Stars less massive than our Sun presumably formed in regions having relatively little heat, rotation, or magnetism.

THE FORMATION OF STARS LIKE THE SUN

✓2 We can best study the specific steps of star formation by considering the Hertzsprung-Russell (HR) diagram studied earlier in Chapter 19. Recall that an HR diagram is a plot of two key stellar properties—surface temperature increasing to the left and luminosity increasing upward. The luminosity scale in Figure 21.4 is expressed in terms of the solar luminosity ($L_\odot \approx 4 \times 10^{33}$ erg/s). Our G2-type Sun

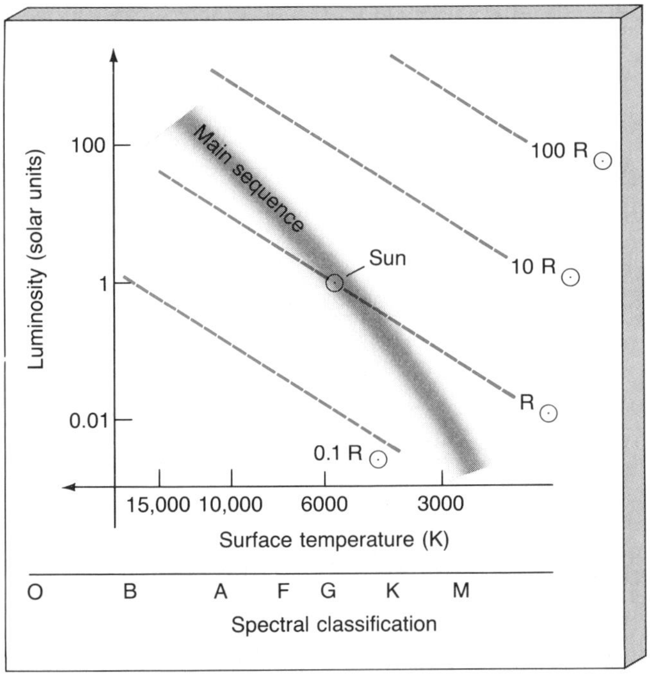

Figure 21.4 *The HR diagram is a useful way to summarize the observed properties of both stars and prestellar objects. The diagonal lines correspond to stars having the same radius (so $L \propto T^4$).*

is plotted at a temperature of 6000K and a luminosity of 1 unit. As we saw in Chapter 19, we can also indicate the *size* of a star represented by any point on the diagram because of the radiation law connecting radius, temperature, and luminosity. In all the HR diagrams used in this and the following chapter, the dashed diagonal lines mark stellar radius, allowing us to follow the changes in a star's size as it evolves.

As we saw in Chapter 19, most stars plotted on the HR diagram fall along the main sequence. For roughly 90 percent of their lifetimes, stars burn rather quietly and their physical properties do not change much. Data points representing such stable, full-fledged stars remain almost stationary on the HR diagram.

Near the beginning and end of its existence, however, a star's properties do change drastically, and the HR diagram is a useful aid in describing these phases of its life. At each phase of a star's evolution, its surface temperature and luminosity can be represented by a point on the HR diagram. The motion of that point about the diagram as the star evolves is known as the star's **evolutionary track.** It is a graphical representation of a star's life.

Table 21-1 lists seven evolutionary stages that an interstellar cloud goes through prior to becoming a main-sequence star like our Sun. These stages are characterized by varying central temperatures, surface temperatures, central densities, and radii of the prestellar object. They trace its progress from a quiet interstellar cloud to a genuine star. The numbers given in Table 21-1 and the following discussion are valid *only* for stars having approximately the same mass as the Sun. In the next section we will relax this restriction and consider the formation of other stars.

Stage 1

The first stage in the star-formation process is just an ordinary dense interstellar cloud, like those studied in Chapter 20. Many of these clouds are truly vast, spanning tens of parsecs (10^{14}–10^{15} km) across. Typical temperatures are about 10 K throughout, and densities are usually not much more than 10^3 particles/cm^3. Stage-1 clouds contain thousands of times the mass of the Sun in the form of cold atomic and molecular gas.

If such a cloud is to be the birthplace of stars, it must become unstable and eventually break up into smaller pieces. The initial collapse occurs when a pocket of gas becomes gravitationally unstable. Perhaps it is squeezed by some external event, such as the pressure wave produced when a nearby O- or B-type star forms and ionizes its surroundings. Or possibly its supporting magnetic field leaks away as charged particles slowly drift across the confining field lines. Whatever the specific cause, theory suggests that once the collapse begins, fragmentation into smaller and smaller clumps of matter naturally follows, as gravitational instabilities continue to operate in the gas. As illustrated in Figure 21.5, a typical cloud can break up into tens, hundreds, even thousands, of fragments, each imitating the shrinking behavior of the parent cloud and contracting ever faster. The whole process, from a single quiescent cloud to many collapsing fragments, takes a few million years.

TABLE 21-1 ***Prestellar Evolution of a Solar-type Star***

Stage	Approximate Time to Next Stage (yr)	Central Temperature (K)	Surface Temperature (K)	Central Density (particles/cm^3)	Diameter (km)	Object
1	2×10^6	10	10	10^3	10^{14}	Interstellar cloud
2	3×10^4	100	10	10^6	10^{12}	Cloud fragment
3	10^5	10,000	100	10^{12}	10^{10}	Cloud fragment/protostar
4	10^6	1,000,000	3,000	10^{18}	10^8	Protostar
5	10^7	5,000,000	4,000	10^{22}	10^7	Protostar
6	3×10^7	10,000,000	4,500	10^{25}	2×10^6	Star
7	10^{10}	15,000,000	6,000	10^{26}	1.5×10^6	Main-sequence star

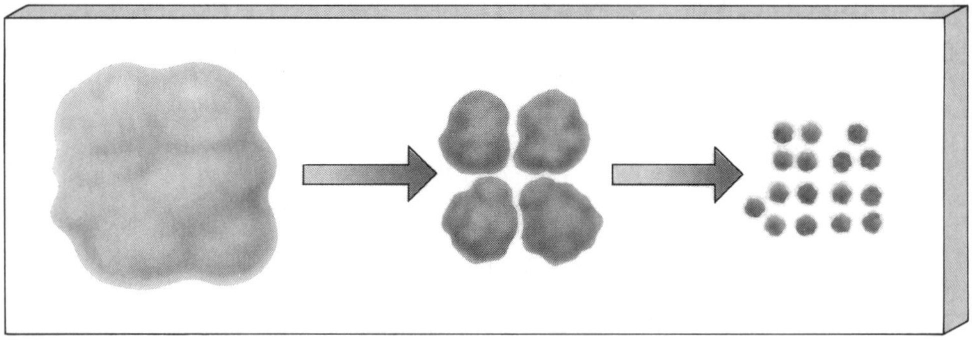

Figure 21.5 As an interstellar cloud collapses, gravitational instabilities cause it to fragment into smaller pieces. The pieces themselves continue to collapse and fragment, eventually to form many tens or hundreds of separate stars.

In this way, depending on the precise conditions under which fragmentation takes place, an interstellar cloud can produce either a few dozen stars, each much larger than our Sun, or a whole cluster of hundreds of stars, each comparable to or smaller than our Sun. There is *little* evidence for stars born in isolation, one star from one cloud. Most stars—perhaps even all stars—appear to originate as members of multiple systems or star clusters. The Sun, which is now found alone and isolated in space, probably escaped from the multiple-star system where it formed after an encounter with another star or some much larger galactic object (such as a molecular cloud).

Stage 2

The second stage in our evolutionary scenario represents the physical conditions in just one of the many fragments that develop in a typical interstellar cloud. A fragment destined to form a star like the Sun contains between 1 and 2 solar masses of material at this stage. Estimated to span a few hundredths of a parsec across, this fuzzy, gaseous blob is still about 100 times the size of our solar system. Its central density is now some 10^6 particles/cm^3.

Even though it has shrunk substantially in size, the fragment's average temperature is not much different from that of the original cloud. The reason for this is that the gas constantly radiates large amounts of energy into space. The material of the fragment is so tenuous that photons produced within it easily escape, without being reabsorbed by the cloud, so that virtually all of the energy released in the collapse is radiated away and does not cause any significant increase in temperature. Only at the center, where the radiation must traverse the greatest amount of material in order to escape, is there any appreciable temperature increase—the gas there might be as warm as 100K by this stage. For the most part, however, the fragment stays cold as it shrinks.

The process of continued fragmentation is eventually stopped by the increasing density within the shrinking cloud. As stage-2 fragments continue to contract, they eventually become so dense that radiation cannot get out easily. The trapped radiation causes the temperature to rise, the pressure to increase, and the fragmentation to cease.

Stage 3

Several tens of thousands of years after it first began contracting, a typical stage-2 fragment has shrunk by the start of stage 3 to a gaseous sphere with a diameter roughly the size of our solar system (still 10,000 times the size of our Sun). As noted in Table 21-1, the inner regions of the fragment have just become opaque to their own radiation and so have started to heat up. The central temperature has reached about 10,000K—hotter than the hottest steel furnace on Earth. However, the temperature at the fragment's periphery has not increased much. It is still able to radiate its energy into space and so remains cool. The density increases much faster in the core of the fragment than at its periphery, so the outer portions of the cloud are both cooler and thinner than the interior. The central density by this time is approximately 10^{12} particles/cm^3 (still only 10^{-12} g/cm^3 or so).

For the first time, our fragment is beginning to resemble a star. The dense, opaque region at the center is called a **protostar**—an embryonic object perched at the dawn of star birth. Its mass grows as more and more material rains down on it from outside, although its radius continues to shrink because its pressure is still unable to overcome the relentless pull of gravity. After stage 3, we can distinguish a "surface" on the protostar—its *photosphere*. Inside the photosphere, the protostellar material is opaque to the radiation it emits. (Notice this is exactly the same definition of "surface" as we used for the Sun in Chapter 18.) From now on, the surface temperatures listed in Table 21-1 refer to the photosphere and not to the "periphery" of the collapsing fragment, whose temperature remains low.

Stage 4

As the protostar continues to evolve, it shrinks, its density grows, and its temperature rises, both in the core and at the photosphere. Some 100,000 years after the fragment began to form, it reaches stage 4, where its center seethes at about 1,000,000K. The electrons and protons ripped from atoms whiz around at hundreds of kilometers per second, yet the temperature is still short of the 10^7 K needed to ignite the proton-proton nuclear reactions that fuse hydrogen into helium. Still much larger than the Sun, our gaseous heap at this stage is about the size of Mercury's orbit. Heated by the material falling on it from above, its surface temperature has risen to a few thousand kelvins.

By the time stage 4 is reached, the surface temperature has become high enough for our protostar's physical properties to be plotted on the HR diagram, as shown in Figure 21.6. Knowing the protostar's radius and surface temperature, we can calculate its luminosity. Surprisingly, it turns out to be several thousand times the luminosity of the Sun. Even though the protostar has a surface temperature only about half that of the Sun, it is hundreds of times larger, making its total luminosity very large indeed—in fact, much greater than the luminosity of most main-sequence stars. Because nuclear reactions have not yet begun, this luminosity is due entirely to the release of gravitational energy as the protostar continues to shrink in size and nebular material continues to rain down on its surface.

Figure 21.6 depicts the approximate path followed by our interstellar cloud fragment since it became a protostar at stage 3 (which itself lies off the right-hand side of the figure). This early evolutionary track is known as the **Kelvin-Helmholtz contraction phase,** after two European physicists (Lord Kelvin and Hermann von Helmholtz) who first studied the theory of contracting clouds in the late nineteenth century. Figure 21.7 is a series of artist's sketches of an interstellar gas cloud proceeding along the evolutionary path outlined so far.

Our protostar is still not in equilibrium. Even though

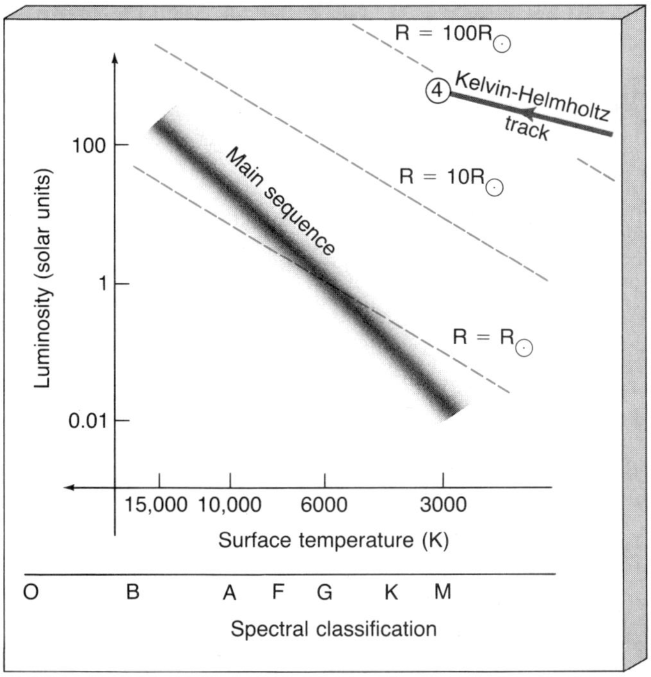

Figure 21.6 *Diagram of the approximate evolutionary track followed by an interstellar cloud fragment prior to arriving, as a stage-4 protostar, at the end of the Kelvin-Helmholtz contraction phase.*

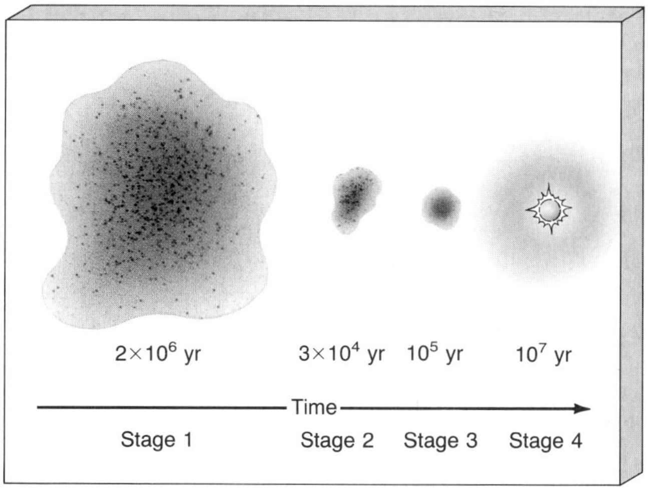

Figure 21.7 *Artist's conception of the changes in an interstellar cloud during the early evolutionary stages outlined in Table 21-1. Shown are a stage-1 interstellar cloud, a stage-2 fragment, a smaller, hotter stage-3 fragment, and a stage-4 protostar. (Not drawn to scale.) The duration of each stage, in years, is also indicated.*

its temperature is now so high that outward-directed pressure has become a powerful countervailing influence against gravity's continued inward pull, the balance is not yet perfect. The protostar's internal heat gradually diffuses out from the hot center to the cooler surface, where it is radiated away into space. As a result, the overall contraction slows, but it does not stop completely. From our earthly perspective, this is quite fortunate: If the heated gas were somehow able to counteract gravity completely before the star reached the temperature and density needed to start nuclear burning in its core, the star would simply radiate away its heat and never become a true star. The night sky would be abundant in faint protostars, but completely lacking in the genuine article. Of course, there would be no Sun either, so it is unlikely that we, or any other intelligent life form, would exist to appreciate these astronomical subtleties.

After stage 4, the protostar's internal structure changes. At a surface temperature of 3000K or so, the outer layers become so opaque to the radiation from within that the entire protostar becomes convective. Radiation has such difficulty penetrating the protostar's outer layers that energy is much more easily carried through the protostellar interior by bulk motion of the gas. One result of this is that the protostar's surface temperature remains almost constant during the next phase of its evolution. With little or no increase in temperature, the protostar becomes less luminous as it shrinks. The protostar on the HR diagram during this phase moves down (toward lower luminosity) and slightly to the left (toward higher temperature), as shown in Figure 21.8.

This portion of our protostar's evolutionary path is called the **Hayashi track,** after C. Hayashi, a twentieth-century Japanese researcher who has made major contributions to the theory of protostars. The Hayashi track runs

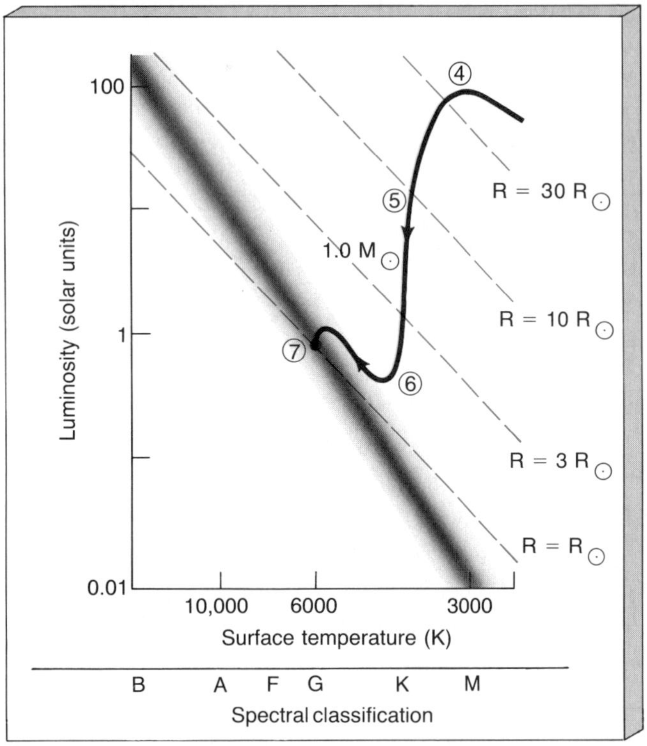

Figure 21.8 *The changes in a protostar's observed properties are shown by the path of decreasing luminosity, from stage 4 to stage 6, often called the Hayashi track.*

from point 4 to point 6 in Figure 21.8. Protostars on the Hayashi track often exhibit violent surface activity. As a consequence, they can have extremely strong protostellar winds, much denser than that of our own Sun. The **T Tauri stars** discussed in Interlude 21-1 may well be direct observational evidence of this phase of stellar evolution.

Stage 5

By stage 5 on the Hayashi track, the protostar has shrunk to about 10 times the size of the Sun, its surface temperature is about 4000K, and its luminosity has fallen to about 10 $L_\odot$. At this point, the central temperature has reached about 5,000,000K. Brutal collisions in the core have completely ionized the gas particles by now. However, the protons still do not have enough thermal energy to overwhelm their mutual electromagnetic repulsion and enter the realm of the nuclear binding force. The core is still too cool for nuclear burning to begin.

Events in a protostar's development happen more slowly as it approaches the main sequence. The initial contraction and fragmentation of the interstellar cloud occurred quite rapidly, but as the protostar nears the status of a full-fledged star, its evolution slows. The cause of this slowdown is heat—even gravity must struggle to compress a hot object. The contraction is governed largely by the rate at which the protostar's internal energy can be radiated away into space. The greater this radiation of internal energy—that is, the more energy that moves through the star to escape from its surface—the faster the contraction occurs. As the luminosity decreases, so too does the contraction rate.

Stage 6

Some 10 million years after its first appearance, the protostar finally becomes a genuine star. By the bottom of the Hayashi track, at stage 6, when our roughly 1-solar-mass object has shrunk to a radius of about 1,000,000 km, the contraction has raised the central temperature to 10,000,000 K, enough to ignite nuclear burning. Protons begin fusing into helium nuclei in the core, and a star is born. As shown in Figure 21.8, the star's surface temperature at this point is about 4500K, still a little cooler than the Sun. Even though the newly formed star is slightly larger in radius than our Sun, its lower temperature means that its luminosity is somewhat less than (actually, about two-thirds) the solar value.

Stage 7

Over the next 30 million years or so, the stage-6 star contracts a little more. In making this slight adjustment, the central density rises to about 10^{26} particles/cm^3 (more conveniently expressed as 100 g/cm^3), the central temperature increases to 15,000,000K, and the surface temperature reaches 6000K. During this time, the star once again becomes radiative, rather than convective, throughout most of its interior, and both its temperature and luminosity increase. At stage 7, it finally reaches the main sequence just about where our Sun now resides. Pressure and gravity are finally balanced, and the rate at which nuclear energy is generated in the core exactly matches the rate at which energy is radiated from the surface. The star's location on the HR diagram will remain virtually unchanged for the next 10 billion years. Our Sun is presently about halfway through this next stage of its evolution.

All the evolutionary events just described occur over the course of some 40–50 million years. While this is a long time by human standards, it is still less than 1 percent of the Sun's lifetime on the main sequence. Once an object begins fusing hydrogen and establishes a "gravity-in/pressure-out" equilibrium, it burns steadily for a very long time.

STARS OF DIFFERENT MASSES

✓3 *The Zero-Age Main Sequence*

The numerical values and the evolutionary track described above are valid only for the case of a 1-solar-mass star. The temperatures, densities, and radii of prestellar objects of other masses exhibit similar trends, but the numbers differ, in some cases quite considerably. Not surprisingly, the most massive fragments within interstellar clouds tend to produce

INTERLUDE 21-1 *Evolution Observed*

Astronomical objects generally evolve over enormously long intervals of time, making it almost impossible to study their changes during a human lifetime. Even the relatively short span of 30 million years needed to form a star like our Sun is roughly a million human generations—far more than have yet occurred.

Some stages of a star's evolution are nonetheless expected to occur extraordinarily rapidly by cosmic standards. One such stage is the sudden explosion of a massive star near death, to be studied in Chapter 23. Another occurs as a newly formed protostar approaches the main sequence. Given the huge number of stars in the sky, we can detect a few objects in this evolutionary stage, despite the relatively short time they spend there.

T Tauri stars are a class of young protostars on the verge of reaching the main sequence. Their peculiar name derives from the star labeled T in the constellation Taurus, whose unpredictable variations in brightness have long marked it as an unusual object. (As is common in astronomy, the first known of a particular object gives its name to the entire class.) In fact, the term *star* is somewhat misleading—the average T Tauri star lies on the Hayashi track (somewhere around stage 5 or 6 in Figure 21.8), so T Tauri stars are really protostars—but the name has stuck. During the past half-century, astronomers have watched some T Tauri stars brighten greatly over the course of a few years, then remain at that increased level of brightness. We do not know how long this phase lasts, however, because many of these young stars have been discovered only since the 1970s, and most are still bright. The accompanying photographs record the change of one such T Tauri star.

The image at the top shows a region of interstellar space containing a fan-shaped gaseous nebula and a faint star located at the nebula's tip. The nebula, labeled NGC2261, is filled with dust and gas and reflects the light of the star, which is called R Monocerotis. The same field of view, photographed 3 years apart, is shown at the bottom. The star has brightened considerably, as has the nebula (reflecting the star's increased brightness). The star, a T Tauri variable, has retained this brightness ever since.

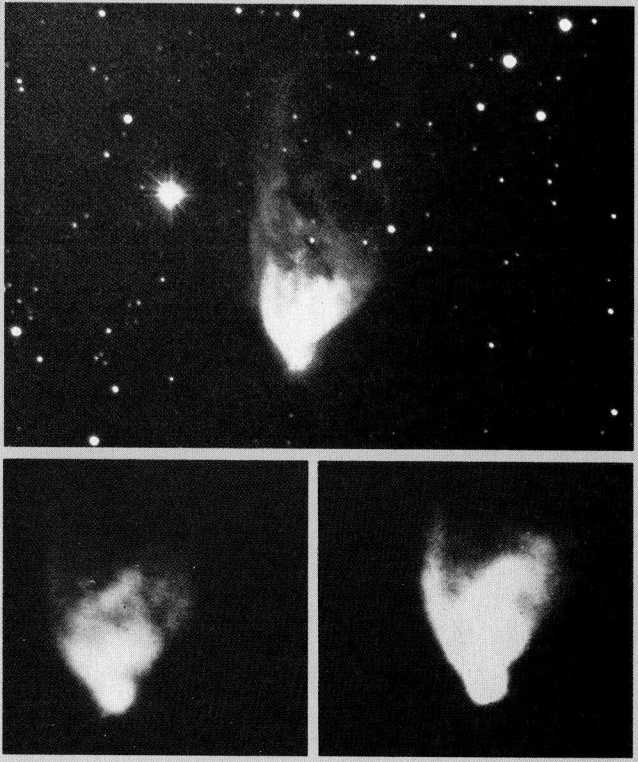

(NOAO)

R I **V** U X G

Currently, we have no complete explanation for the sudden brightening of young stars like the one in this photograph. Is it caused by interstellar matter falling onto the newly formed star? Or by flares on the star's surface driven by the protostar's extensive convection zone and strong magnetic field? Or possibly by surrounding gas and dust being blown away from the new star, making its true brightness more clearly visible to us? Or are there internal changes in the star's nuclear burning rate? Whatever the cause, several T Tauri stars have undergone very definite changes in appearance on time scales shorter than a human lifetime. Astronomers around the world are monitoring them closely, hoping that further observations will reveal the full reason for their extraordinarily rapid evolution.

the most massive protostars and eventually the most massive stars; similarly, low-mass fragments give rise to low-mass stars.

Figure 21.9 compares the paths taken by two prestellar objects—one of 0.3 and one of 3 solar masses—with the pre-main-sequence track followed by the Sun. Notice how all objects traverse the HR diagram in the same general manner, but their luminosities and temperatures differ. Cloud fragments that eventually form more massive stars approach the main sequence along a higher track on the diagram; those destined to form less massive stars take a lower track. More massive objects generally have larger luminosities and surface temperatures at any given evolutionary stage—Kelvin-Helmholtz contraction, the Hayashi track, or the main sequence—than their less massive counterparts.

The time required for an interstellar cloud to become a main-sequence star depends strongly on its mass. Large cloud fragments have more gas particles and more frequent particle collisions than smaller fragments. They heat up to

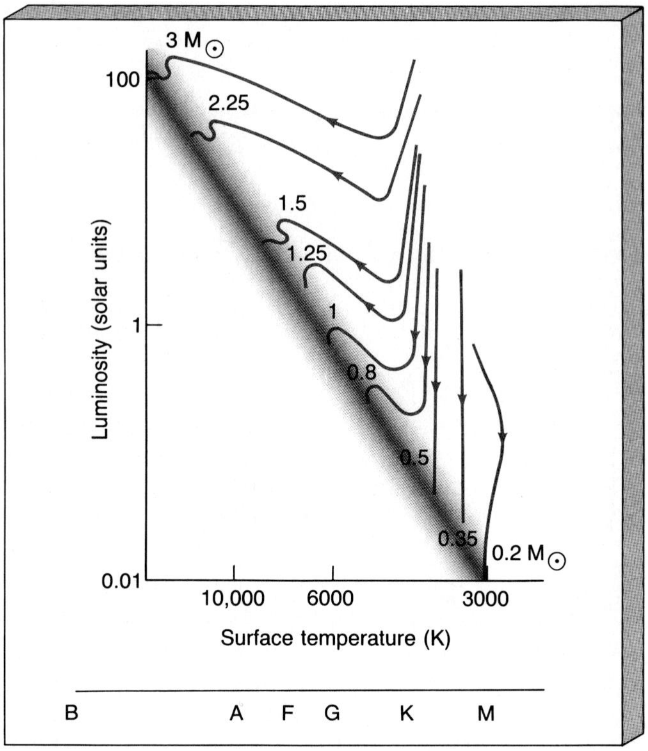

Figure 21.9 *Prestellar evolutionary paths for stars more massive and less massive than our Sun differ markedly from those sketched in Figures 21.6 and 21.8.*

the required 10,000,000K more rapidly than do less massive objects. The most massive fragments contract into stars in a mere million years, roughly 1/50 the time taken by the Sun. This rapid evolution parallels the way in which the most massive stars also race through their main-sequence lifetimes in the shortest times, as described in Chapter 19. As we will see in the next three chapters, these stars speed through all parts of their life cycle, eventually to die in a catastrophic explosion.

The opposite is the case for stars and prestellar objects having masses less than our Sun. Cloud fragments that evolve into low-mass stars are smaller and cooler, and particle encounters occur less frequently. Not only do these fragments take a long time to become protostars, but the protostars also take their time changing into full-fledged stars. A typical M-type star, for example, requires nearly a billion years just to form, some 20 times longer than the time taken by the Sun.

Whatever the mass, the end-point of the prestellar evolutionary track is the main sequence. The main sequence predicted by theoretical models, where stellar properties finally settle down to stable values and an extended period of steady burning ensues, is usually called the **zero-age main sequence** (or ZAMS for short). It agrees quite well with main sequences actually observed in the vicinity of the Sun and in more distant star clusters. Let us stress that the main sequence itself is *not* an evolutionary track—stars do not evolve along it. Rather, it is just a ''way station'' on the HR diagram where stars stop and spend most of their lives, low-mass stars at the bottom, high-mass stars at the top.

Together, mass and composition uniquely determine a newly formed star's appearance. Once the star's total mass and chemical makeup are specified, all other stellar properties follow. If all gas clouds contained precisely the same elements in exactly the same proportions, mass would be the sole determinant of a newborn star's location on the HR diagram, and the zero-age main sequence really would be a well-defined line. However, the composition of a star affects its internal structure (mainly by changing the opacity of its outer layers), and this in turn affects both its temperature and its luminosity on the main sequence. Stars with more heavy elements tend to be cooler and slightly less luminous than stars of the same mass containing fewer heavies. As a result, differences in composition between stars ''blur'' the zero-age main sequence into a broad band instead of a narrow line. When composition differences are taken into account, theoretical models provide an excellent explanation of observed main sequences.

Dark Clinkers

Some cloud fragments are too small ever to become stars at all. The giant planet Jupiter is a good example. Jupiter contracted under the influence of gravity, and the resultant heat is still detectable, as we saw in Chapter 13. But the planet did not have enough mass for gravity to crush its matter to the point of nuclear ignition. It became stabilized by heat, rotation, and possibly also magnetism, all of which opposed the pull of gravity, before the central temperature became hot enough to fuse hydrogen. Jupiter never evolved beyond the protostar stage.

If Jupiter, or the other Jovian planets, had continued to accumulate gas from the solar nebula, they would eventually have become stars. But virtually all the matter present during the formative stages of our solar system is gone now, swept away by the solar wind. Like the Jovian worlds, low-mass interstellar gas fragments simply lack the mass needed to initiate nuclear burning. Rather than turning into stars, they will continue to cool, eventually becoming compact, dark ''clinkers''—cold fragments of unburnt matter—in interstellar space. On the basis of theoretical modeling, astronomers believe that the minimum mass of gas needed to generate core temperatures high enough to begin nuclear fusion is about 0.08 $M_\odot$ (solar masses).

Vast numbers of Jupiter-like objects may well be scattered throughout the universe—fragments frozen in time somewhere along the Kelvin-Helmholtz contraction phase. Small, faint, and cool (and growing ever colder), they are known collectively as **brown dwarfs.** Our technology has great difficulty currently of detecting them, be they planets associated with stars or interstellar cloud fragments far away from any star. (We can telescopically detect stars and spectroscopically infer atoms and molecules, but astronomical objects of intermediate size outside our solar system are hard to see.) Interstellar space *could* contain many cold, dark Jupiter-sized objects without our knowing it. They might even account for more mass than we observe in the form of stars and interstellar gas combined.

OBSERVATIONS OF CLOUD FRAGMENTS AND PROTOSTARS

✓4 The evolutionary stages we have just described are derived from numerical experiments performed on high-speed computers. Table 21-1 and the evolutionary paths described in Figures 21.6, 21.8, and 21.9 are mathematical predictions of a multifaceted problem incorporating gravity, heat, rotation, magnetism, nuclear reaction rates, elemental abundances, and a few other physical conditions specifying the state of contracting interstellar clouds. Computer technology has enabled theorists to construct these models, but their accuracy is only partly known, because it is difficult to test them observationally.

How then can we verify the theoretical predictions just outlined? Even the total lifetime of our entire civilization is much, much shorter than the time needed for a cloud to contract and form a star. We can never observe individual objects proceed through the full panorama of star birth. We can, however, observe many different objects—interstellar clouds, protostars, young stars approaching the main sequence—as they appear today at different stages of their evolutionary cycles. Each observation is like part of a jigsaw puzzle. When properly oriented relative to all the others, the pieces can be used to build up a picture of the full life cycle of a star. By observing pre-main-sequence objects at many sites in our Galaxy, astronomers have directly verified most of the prestellar stages just described.

Let us now consider in more detail some of the observational pieces that make up the modern picture of prestellar evolution.

Evidence of Cloud Contraction

Prestellar objects at stages 1 and 2 are not yet hot enough to emit much infrared radiation, and certainly no optical radiation arises from their dark, cool interiors. The best way to study the early stages of cloud contraction and fragmentation is to use radio telescopes to detect the radiation emitted or absorbed by one or more interstellar molecules. Only long-wavelength radiation can escape from these clouds to our telescopes on or near Earth.

Figure 21.10 shows M20, the splendid emission nebula studied in the previous chapter, along with some of its surroundings. The brilliant region of glowing, ionized gas is not our main interest here, however. Instead, the youthful O- and B-type stars that energize the nebula serve to alert us to the general environment where stars are forming. Emission nebulae are signposts of star birth.

The region surrounding M20 contains galactic matter that seems to be contracting. The presence of (optically) invisible gas there was illustrated in Figure 20.21, which showed a contour map of the abundance of the formaldehyde (H_2CO) molecule. These and many other kinds of molecules are widespread in the vicinity of M20, especially throughout the dusty regions below and to the right of the nebula. Their radio emission shows that they are especially

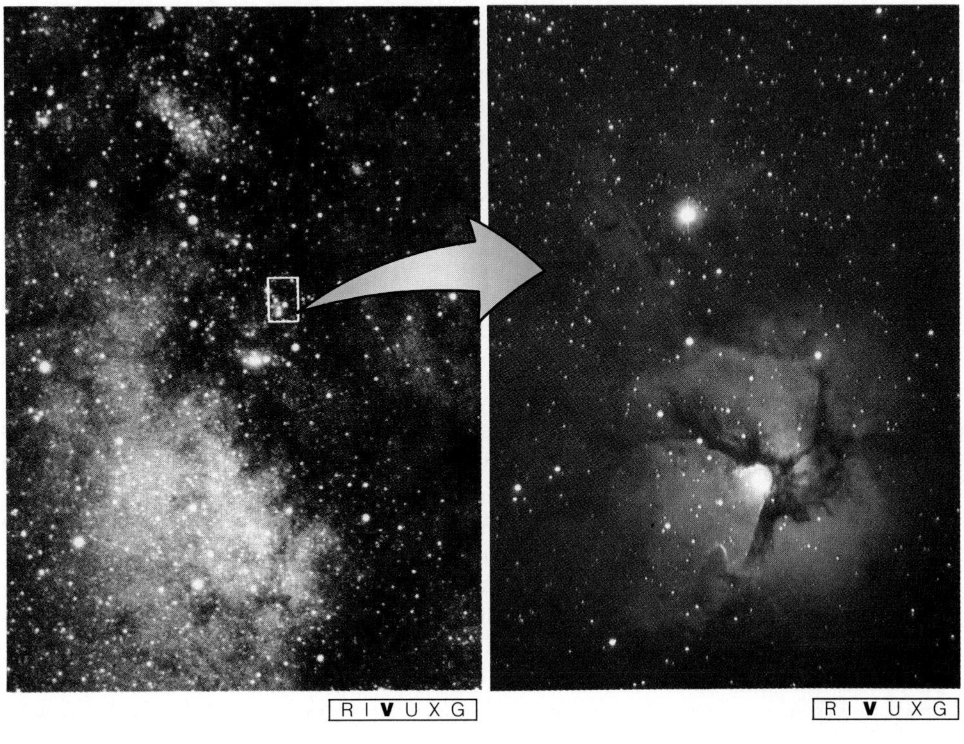

Figure 21.10 *The beautiful emission nebula M20 (right) and its dark surroundings (left) provide examples of many phases of star formation. The nebula itself is an HII region, glowing because of the energy of the hot young stars embedded in it. The surrounding dark regions show evidence of cloud collapse and fragmentation. (Anglo-Australian Telescope Board; Palomar Observatory)*

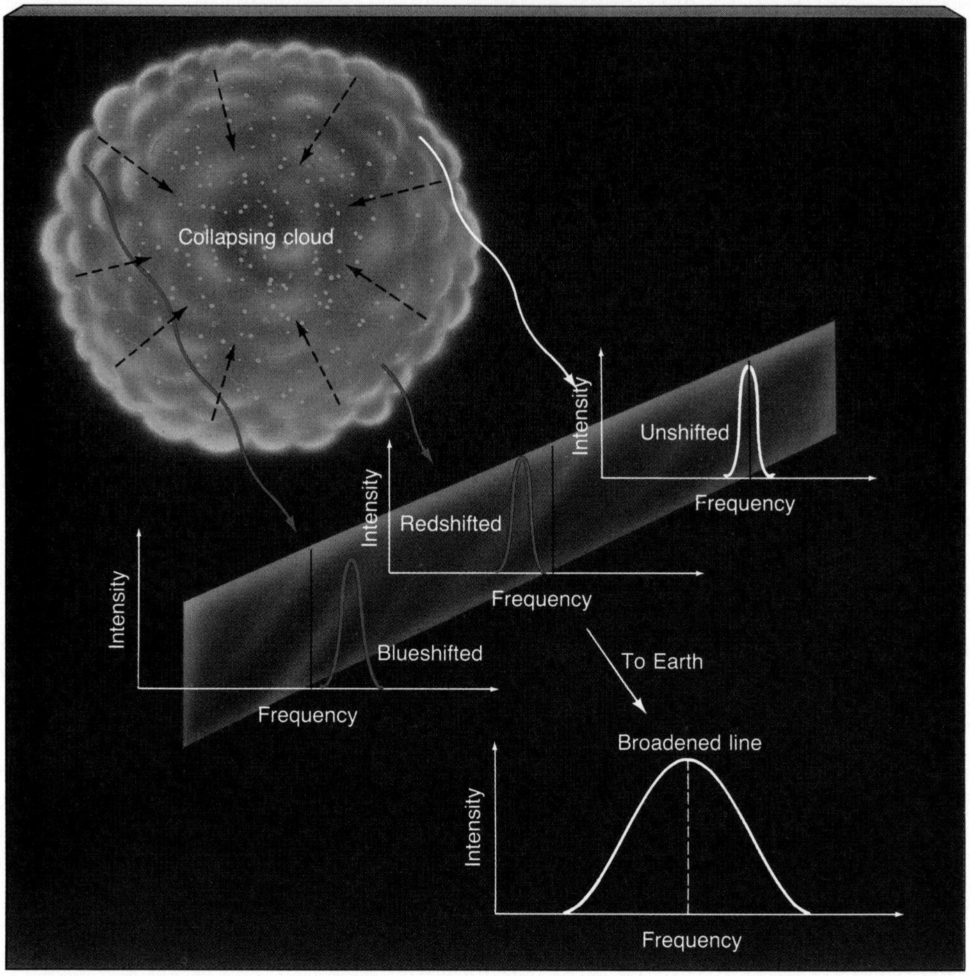

Figure 21.11 *Spectral lines observed from a collapsing gas cloud are broadened by the motion of the infalling gas. Lines from the front of the cloud are redshifted as the gas falls away from the observer, while lines from the back are blueshifted. When these Doppler shifts are superposed at Earth, the net effect is a broadening of the original line. Notice that the radio radiation from the back of the cloud passes unhindered through the cloud itself, allowing us to observe the entire collapsing region at once.*

abundant near the completely opaque dark region below the nebula. Further analysis of the observations suggests that this region of greatest molecular abundance is also contracting and fragmenting, well on its way toward forming a star or, more likely, a star cluster.

We can tell that this portion of the dark cloud outside M20 is contracting because of the widths of the radio spectral lines observed there. As the cloud shrinks, radio radiation from the side nearest us is redshifted, as that part of the cloud falls away from us (see Figure 21.11). At the same time, radiation from the far side (which passes right through the cloud without hindrance) is blueshifted. The net effect is that a radio line is broadened by an amount that depends on the rate at which the cloud is contracting. Figure 21.12 shows a map of the width of a formaldehyde line observed in the vicinity of M20. This map was made by measuring the line width at various locations across the cloud. Contours were then drawn connecting places having spectral lines of equal width.

At the periphery of the region, we observe narrow spectral lines because the motion of the infalling matter is mainly perpendicular to our line of sight, causing little Doppler broadening of the spectral-line profiles. Toward the center of the dark region, however, infalling gas at the front and back of the cloud is moving mostly parallel to our line of sight, producing Doppler-broadened line profiles of much greater width. Significantly, this line-width map peaks in roughly the same place as the molecular abundance map—in the totally opaque region below M20 itself. The similarity between the maps of molecular motion and molecular abundance strongly suggests that this portion of the interstellar cloud surrounding M20 is currently collapsing.

The interstellar clouds in and around M20 thus provide tentative evidence for three distinct phases of star formation, as shown in Figure 21.13. The huge dark molecular cloud surrounding the visible nebula—the "stage-1" cloud, in the language of Table 21-1—is sometimes called the *quiescent* phase of the evolution. It is characterized by low densities and temperatures, in the range of about 100 particles/cm^3 and 20K.

Greater densities and temperatures typify smaller regions within this huge cloud. The totally obscured regions where the molecular abundances peak represent such denser, warmer fragments. Here, the total gas density is observed to be at least 1000 particles/cm^3, and the temperature is about 100K. Less than a light-year across, the region in the *collapsing* phase shown in Figure 21.13 has a total mass of over 1000 $M_\odot$, considerably more than the mass of M20 itself. This second broad phase of the star-formation sequence represents a cloud somewhere between stages 1 and 2 of Table 21-1.

The third phase shown in Figure 21.13 is the *nebular*

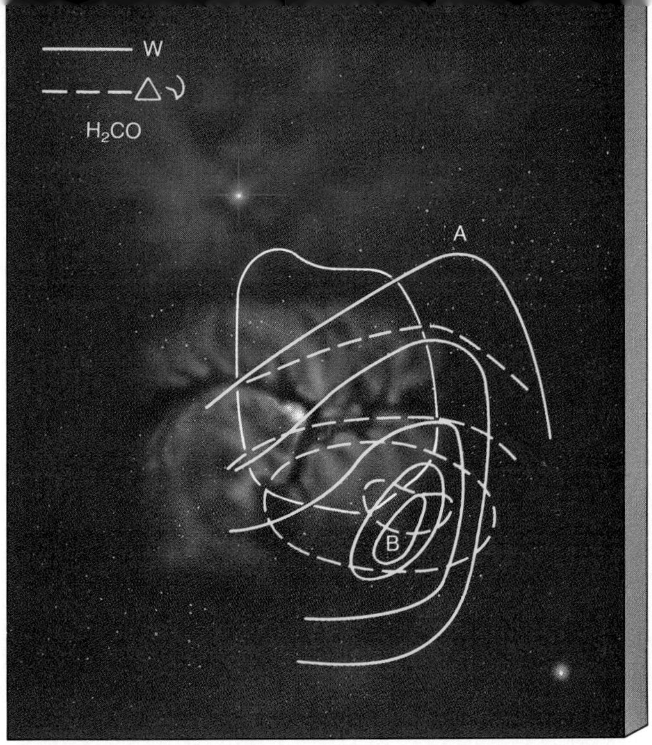

Figure 21.12 A map of the spectral-line width of the formaldehyde molecule's radiation toward the south of M20 reveals evidence for a contracting interstellar cloud. The line-width contours increase in uniform steps from the outside to the inside of the cloud, and are indicated by dashed lines here. They approximately coincide with the map of formaldehyde abundance (displayed here as solid contours).

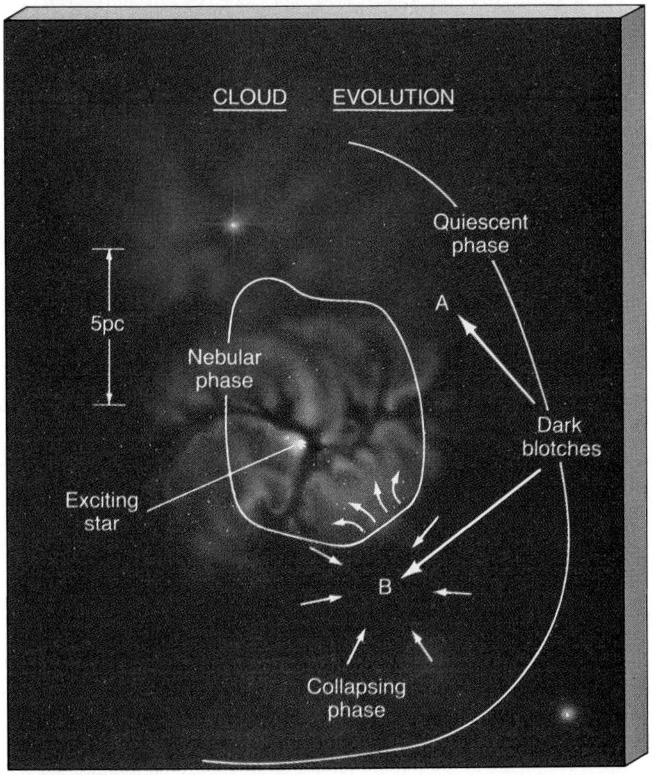

Figure 21.13 The M20 region shows observational evidence of three broad phases in the birth of a star: a quiescent phase (the parent cloud), followed by a collapsing phase (a contracting fragment), and last, a nebular phase (M20 itself).

phase. It is M20 itself. The glowing region of ionized gas results directly from a massive O-type star that formed there within the past million years or so. Because the central star is already fully formed, this final phase corresponds to stage 6 or 7 of our earlier evolutionary scenario.

Evidence of Cloud Fragments

Other parts of our Milky Way Galaxy provide sketchy evidence for prestellar objects in stages 3 through 5. The Orion complex, shown in Figure 21.14, is one such region. Lit from within by several O-type stars, the bright Orion Nebula is partly surrounded by a vast molecular cloud that extends well beyond the roughly 5 × 10 pc region bounded by the photograph in Figure 21.14(b).

Astronomers often study the Orion complex and other molecular clouds by means of radio radiation emitted by so-called tracer molecules in the gas. Although these tracers represent only a tiny fraction of the cloud's mass, they are always found in regions where other, more abundant, molecules exist. When we detect tracers, then, we have indirectly detected these regions of more abundant molecules as well. What makes tracers useful is that they happen to radiate at frequencies that are particularly convenient for radio astronomers to observe. As a result, tracers are widely used to map out regions where molecules are present. Carbon monoxide (CO) and formaldehyde (H_2CO) are especially useful probes of interstellar gas of moderate density—around 1000 particles/cm^3.

Other molecules, such as hydrogen cyanide (HCN) and carbon monosulfide (CS), are tracers of even denser regions. As shown in Figure 21.14, these molecules extend over only a small part of the molecular cloud, just behind the bright nebula itself. They delineate a small fragment of the larger molecular cloud where the average density is approximately 100,000 particles/cm^3. The measured extent of the fragment is somewhat less than 1 pc, and its temperature is about 200K. We can thus identify it as an object around stage 2 of Table 21-1.

The Orion molecular cloud also harbors several smaller sites of intense radiation emitted by molecules under very special conditions (see Interlude 21-2). Molecules such as hydroxyl (OH) and water vapor (H_20) have been found by radio techniques to be buried within the core of the cloud fragment. Their extent, also shown in Figure 21.14, measures about 10^{10} km, or 1/1000 of a light-year, about the diameter of our solar system. The gas density of these smaller regions is about 10^9 particles/cm^3, much denser than the surrounding cloud. Although the temperature cannot be estimated reliably, many researchers regard these regions as objects well on their way toward stage 3. We cannot determine if these regions will eventually form stars like the Sun, but it does seem certain that these intensely emitting regions are on the threshold of becoming protostars.

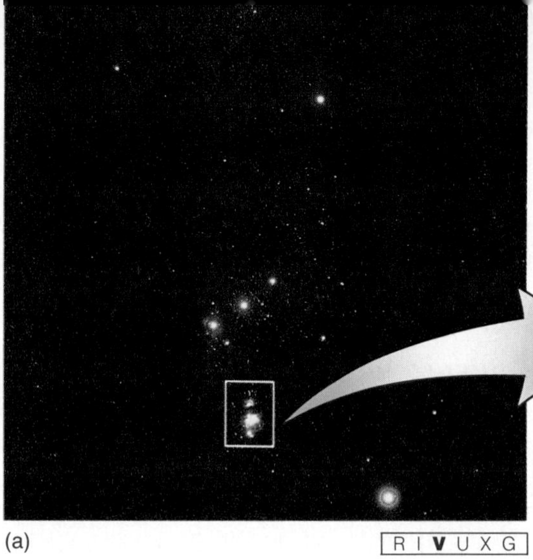

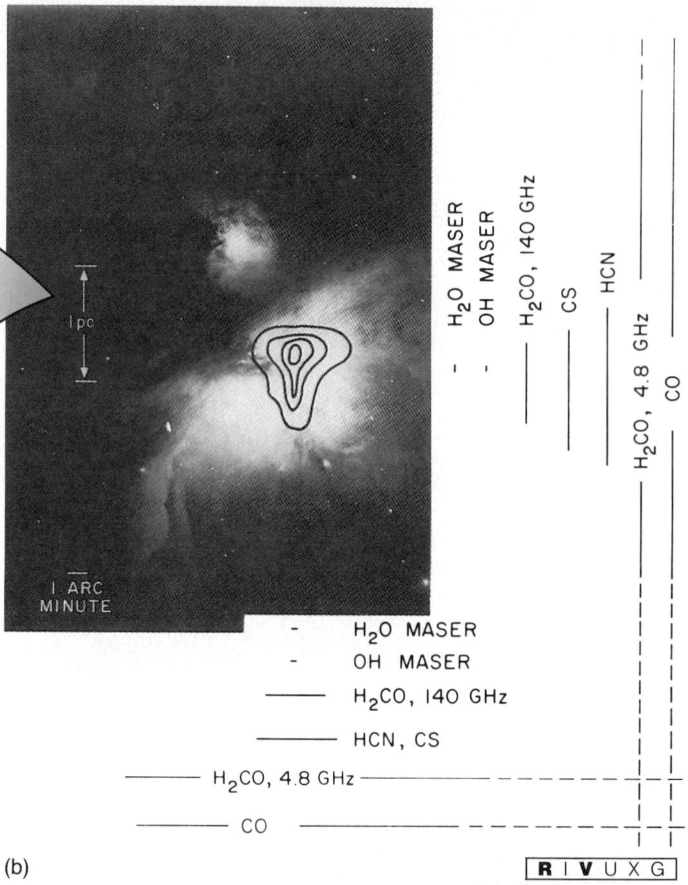

Figure 21.14 *The Orion Nebula is a stunning gaseous nebula within the small rectangle in (a). Visible in this large-scale image are the stars of Orion's belt (top). This framed region is enlarged in (b), showing how Orion is partly surrounded by a vast molecular cloud. Various parts of this cloud are probably fragmenting and contracting, with even smaller sites forming protostars. The lines to the right and bottom of (b) depict the extent of molecular emissions from a dark cloud* behind *the bright nebula. (Harvard-Smithsonian Center for Astrophysics; NOAO)*

Evidence of Protostars

In the hunt for and study of objects at more advanced stages of star formation, radio techniques become less useful because stages 4, 5, and 6 are expected to display increasingly high temperatures. As the Planck curve of thermal emission from warm protostars and young stars shifts toward shorter wavelengths, these objects should be observable largely in the infrared.

A most interesting object within the core of the Orion molecular cloud was detected by infrared astronomers in the 1970s. Known as the Kleinmann-Low Nebula after its discoverers, it is a strong infrared emitter with a luminosity of some 10^3 $L_\odot$ and lies behind the visible nebula. This compact source is outlined by contours in Figure 21.14. Most astronomers agree that this warm, dense blob is a genuine protostar, poised on the verge of stardom near the end of the Kelvin-Helmholtz contraction phase, probably around stage 4.

Until the *Infrared Astronomy Satellite* was launched in the early 1980s, astronomers were aware only of giant stars forming in clouds far away. But *IRAS* showed that stars are forming much closer to home, and some of these protostars have masses comparable to our Sun's mass. Figure 21.15 shows a premier example of a solar-mass protostar—Barnard 5. Its infrared heat signature is that expected of an object on the Hayashi track, around stage 5.

The energy sources for some infrared objects seem to be luminous hot stars that are hidden from optical view by

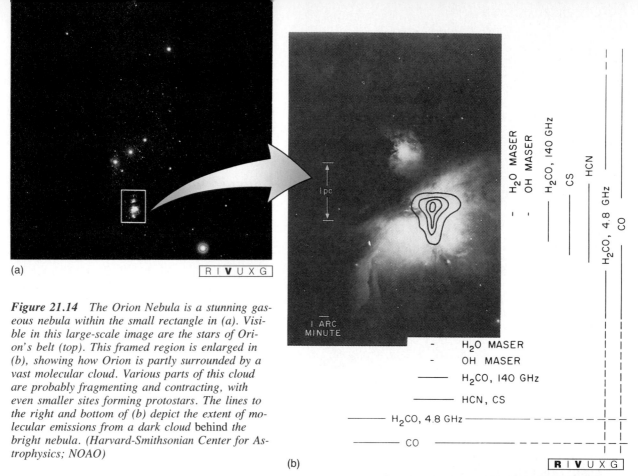

Figure 21.15 *An infrared image of the nearby region containing the source Barnard 5 (indicated by the arrow). Based on its temperature and luminosity, Barnard 5 appears to be a protostar on the Hayashi track in the HR diagram. (NASA)*

INTERLUDE 21-2 *Interstellar Masers*

The word *laser* has become a common everyday term. It is an acronym for ''light amplification by stimulated emission of radiation.'' Lasers are devices that emit a concentrated stream of optical radiation in a very narrow beam. They operate by using radiation to excite atoms and molecules in a gas or solid, after which the same radiation stimulates the gas back into a lower-energy state. Provided that all the atoms or molecules return quickly and simultaneously to the lower state, a powerful packet of light is emitted from the gas. Although laser emission is much more intense than the emission from, say, a light bulb, the wavelength or frequency of the emitted radiation is still uniquely characteristic of the particular atom or molecule excited. The original beam of radiation is amplified enormously by the operation of the laser.

Masers are similar to lasers, except that they produce *m*icrowave (radio) radiation rather than optical radiation. Some parts of interstellar space are naturally suited to the production of amplified microwave radiation. We know this because some interstellar regions emit extremely intense radiation—much more intense than is possible from a normal collection of molecules energized only by random collisions with one another. In fact, some of the early observations of radio radiation from the hydroxyl (OH) molecule in the 1960s were so mysterious that a few puzzled researchers even began calling the emitter ''mysterium'' (recall ''coronium'' and ''nebulium'' from earlier generations of confused astronomers). Later identified as OH, the radiation emitted by these molecules is amplified by the same maser process as described above.

Apparently, some sites within interstellar clouds have the special conditions required first to excite some molecules and then to stimulate them into intense emission. In addition to OH, water vapor (H_2O) and silicon monoxide (SiO) also emit microwave radiation in a maserlike fashion.

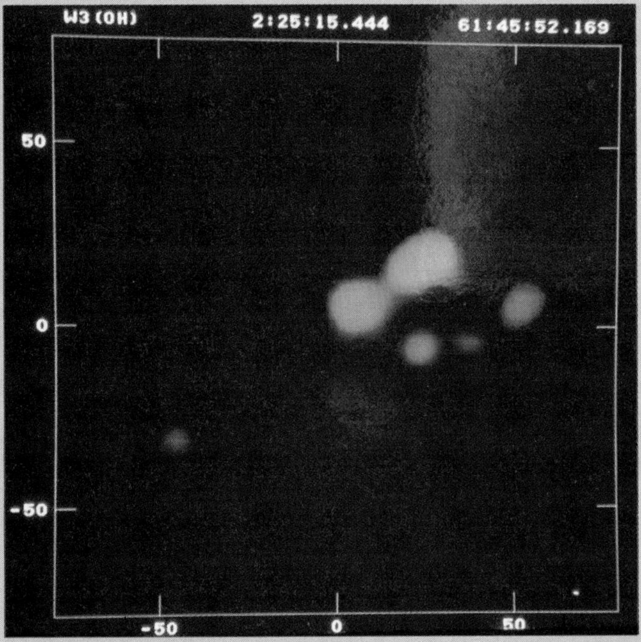

(Harvard-Smithsonian Center for Astrophysics)

R I V U X G

The special physical conditions needed for maser radiation —warm temperatures of about 1000K and densities of about 10^{12} particles/cm^3—exist naturally near protostars.

The above figure is a very-high-resolution map of several radio-emitting maser sources in a star-forming region known as W3. Each of these regions has roughly solar system dimensions and has been resolved here by the technique of very-long-baseline-interferometry discussed in Chapter 6. The left and bottom scales refer to ± 50 milli-arc seconds, or 0.05″ of angular measure. In all probability, these molecular masers are protostars or, at the least, about to become protostars.

surrounding dark clouds. Apparently, some of the stars are already so hot that they emit large amounts of ultraviolet radiation, which is mostly absorbed by a ''cocoon'' of dust surrounding the central star. The absorbed energy is then reemitted by the dust as infrared radiation. These bright infrared sources are known as **cocoon nebulae.**

Intense OH and H_2O line radiation (which escapes the region) enables the clouds forming the cocoon to cool despite their continual heating by the stars within. Some of the clouds are so massive that they may themselves be on the verge of gravitational collapse. The cooling provided by the escape of the line radiation may be an important factor in controlling the overall cloud contraction process. Two considerations support the idea that the hot stars responsible for the clouds' heating have only recently ignited. (1) These dust cocoons are predicted to disperse quite rapidly once their central stars form, and (2) they are invariably found in the dense cores of molecular clouds. The central stars probably lie near stage 6 in Table 21-1.

Protostellar Winds

Protostars often exhibit strong winds. Radio and infrared observations of hydrogen and carbon monoxide molecules, again in the Orion cloud, have revealed gas expanding outward at velocities approaching 100 km/s. High-resolution interferometric observations have disclosed expanding knots of water emission within the same star-forming region and have linked the strong winds to the protostars themselves. As mentioned earlier, these winds may be related to the violent surface activity associated with many protostars (see also Interlude 21-1).

Early in a protostar's life, it may still be embedded in an extensive disk of nebular material in which planets are forming, as discussed in Chapter 17. When the protostellar wind begins to blow, it encounters less resistance in the directions perpendicular to the disk than in the disk plane. The result is known as a **bipolar flow**—two "jets" of matter are expelled in the directions of the poles of the protostar, as illustrated in Figure 21.16. As the protostellar wind gradually destroys the disk, blowing it away into space, the jets widen until, with the disk gone, the wind flows away from the star equally in all directions. Figure 21.17 shows an example of a real bipolar flow.

Figure 21.16 *When a protostellar wind encounters the disk of nebular gas surrounding the protostar, it tends to form a bipolar jet, preferentially leaving the system along the line of least resistance, which is perpendicular to the disk, as shown in (a). (b) As the disk is blown away by the wind, the jets fan out, eventually (c) merging into a spherical wind.*

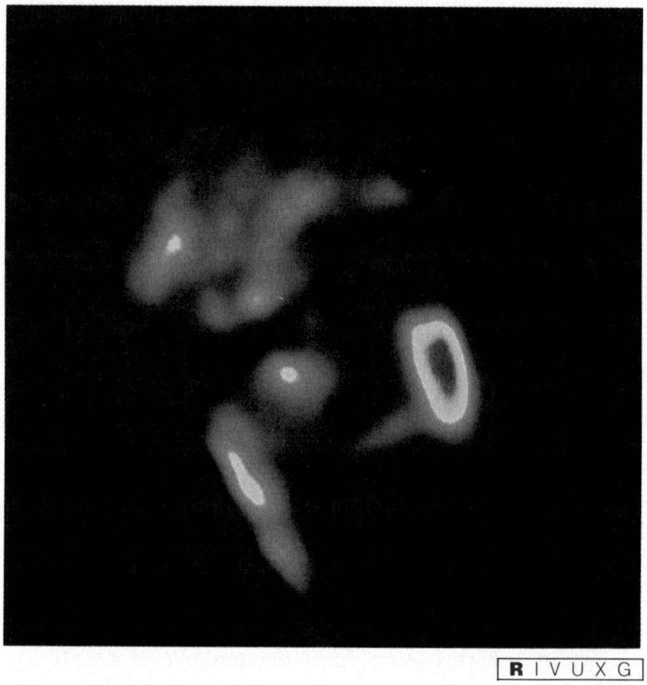

Figure 21.17 *This false-colored radiograph has captured a bipolar outflow from the emission nebula S106. (The red blob is a hot spot in the nebula and is unrelated to the outflow.) The outflows are seen as irregular (blue) patches of gas to the north and south of a weak source (not visible here) embedded in the dark diagonal lane bisecting the lobes. (NRAO)*

THE COMPLEXITIES OF REALITY

✓5 The subject of star formation is actually much more complicated than the previous discussion suggests. Interstellar space is populated with many kinds of clouds, fragments, protostars, stars, and nebulae. They all interact in a complex fashion, and each type of object undoubtedly affects the behavior of the others.

For example, the presence of an emission nebula in or near a molecular cloud probably influences the evolution of the entire region. For any of the nebulae studied so far, we can easily visualize expanding waves of matter driven by the pressure of stellar ultraviolet radiation in the nebula. As the waves push outward into the surrounding molecular cloud, interstellar gas tends to pile up and become compressed. Such a shell of gas, rushing rapidly through space, is known as a **shock wave.** It can push ordinarily thin matter into dense sheets, just as a plow pushes snow.

Many astronomers regard the passage of a shock wave through interstellar matter as the triggering mechanism needed to initiate star formation in a galaxy. Calculations show that when a shock wave encounters an interstellar cloud, it races around the thinner exterior of the cloud more rapidly than it can penetrate its thicker interior. Thus, shock waves do not blast a cloud from only one direction. They effectively squeeze it from many directions, as illustrated in Figure 21.18. Atomic bomb tests have experimen-

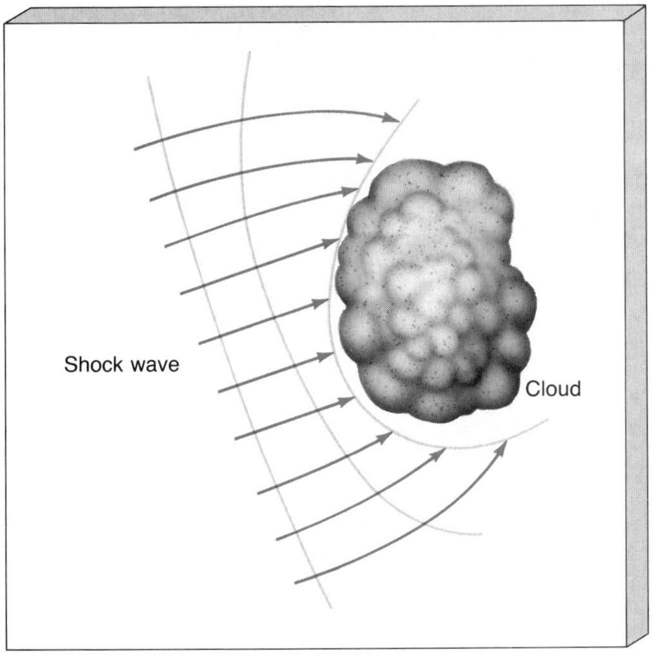

Figure 21.18 Shock waves tend to wrap around interstellar clouds, compressing them to greater densities, and thus possibly triggering star formation.

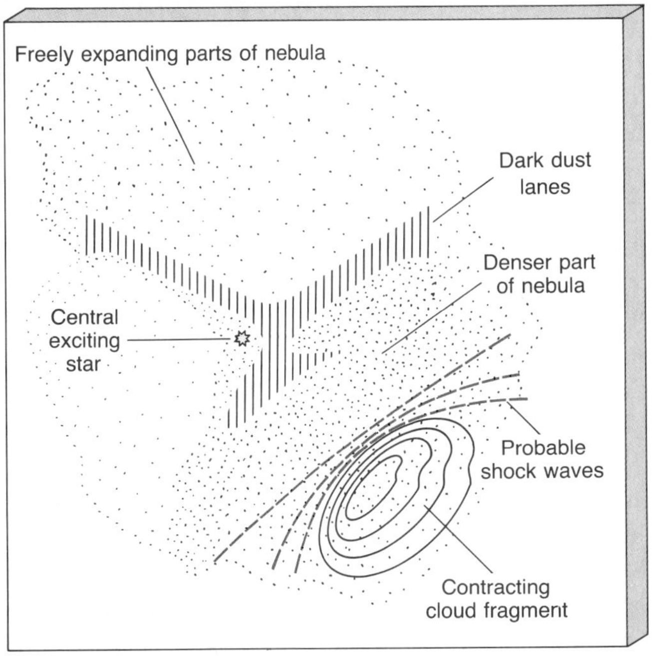

Figure 21.19 A cloud fragment is shown undergoing compression on the southerly edge of M20 as shock waves from the nebula penetrate the surrounding interstellar cloud.

tally demonstrated this squeezing—shock waves created in the blast surround buildings, causing them to be blown together (imploded) rather than apart (exploded). After shock waves cause the initial compression of an interstellar cloud, natural gravitational instabilities may divide it into the fragments that eventually form stars. Figure 21.19 suggests how this mechanism might be at work near M20.

HII regions are not the only generators of shock waves. At least two other sources are available—the spiral-arm waves that plow through the Milky Way (to be discussed in Chapter 25), and the remnants of old stars (planetary nebulae and supernovae, to be discussed in Chapters 22 and 23). Supernovae are probably the most efficient way to pile up matter into dense clumps, but they are relatively few and far between, so the other mechanisms may be more important in triggering star formation.

The photograph in Figure 21.20 shows a semicircular band of glowing gas in the region of the sky known as Canis Major. The bright interstellar matter along the arc is almost certainly part of a three-dimensional (spherical) shell, although we cannot be absolutely certain from just this optical view. Radio observations of the region reveal that other parts of the expanding shell are made mostly of invisible neutral hydrogen gas. The arc in Figure 21.20 is visible because of the star-formation activity it has already triggered. The gas making up the shell is thought to be the remnant of a stellar explosion that occurred long ago—the size and expansion velocity of the shell's gas indicate that the star blew up about half a million years ago. Ever since, a shock wave has been moving away from this point, piling up matter and forming new stars as it goes. The arrows in the figure indicate the locations of some young stars that probably formed in this way.

Although the evidence is somewhat circumstantial, the presence of young (and thus quick-forming) O- and B-type stars in the vicinity of this remnant does suggest that the birth of stars is often initiated by the violent, explosive deaths of others. In some cases at least, the demise of old stars is the trigger needed to conceive new ones.

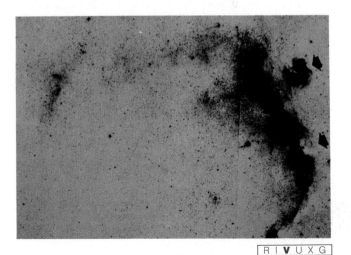

Figure 21.20 This arc of glowing gas is part of a nearly complete shell of interstellar matter that was probably ejected by a massive star that exploded nearly 600,000 years ago. Young stars are found on the inside edge of the shell, while additional stars are probably forming on the outer edge (marked by arrows) as the shell's shock wave piles up matter before it. (Harvard College Observatory)

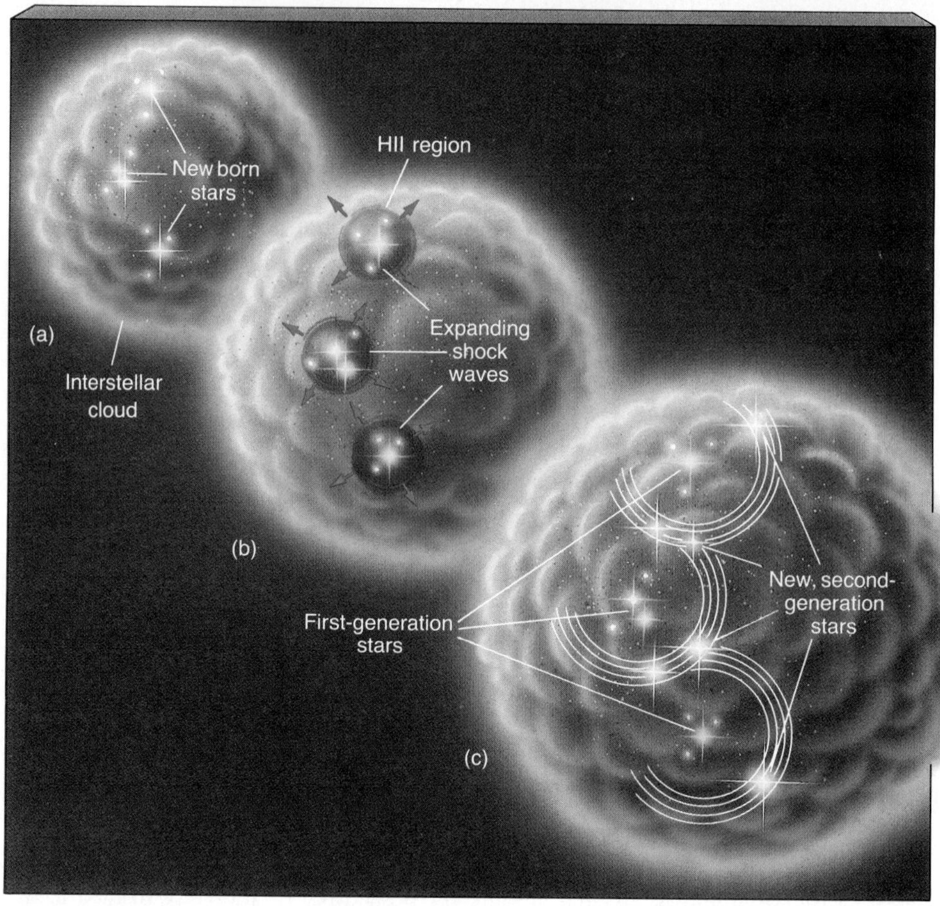

Figure 21.21 *Star birth (a) and shock waves (b) lead to more star births and more shock waves (c) in a continuous cycle of star formation in many areas of our Galaxy. Like a chain reaction, old stars trigger the formation of new stars ever deeper into an interstellar cloud.*

Wherever O- and B-type stars have recently formed, we can assume that less massive stars are still in the process of formation. It takes longer for the less massive stars to form, and thus we should not expect to find many A-, F-, G-, K-, or M-type stars in the Canis Major shell, provided that the star formation mechanism really was triggered less than 1 million years ago. The whole Canis Major region, and many others like it, is probably a vast stellar nursery—the site of many invisible interstellar cloud fragments and protostars, as well as the young, massive stars we see.

This scenario of shock-induced star formation is even more complicated because O- and B-type stars form quickly, live briefly, and die explosively (as we will see in later chapters). These massive stars, themselves born of a passing shock wave, may in turn create new shock waves, either through their expanding nebular gas or by their explosive deaths. These new shock waves may produce "second-generation" stars, which in turn will explode and give rise to still more shock waves, and so on, as depicted in Figure 21.21. Star formation thus resembles a chain reaction. Other, lighter, stars are also formed in the process, of course, but they are largely "along for the ride." It is the O- and B-type stars that drive the star-formation wave through the cloud. Observational evidence lends some support to this chain-reaction model. Groups of stars nearest molecular clouds appear to be the youngest, while those farther away appear to be older, just as we would expect on the basis of this picture.

HII REGIONS AND STAR CLUSTERS

We have seen how a portion of an interstellar cloud can become unstable, collapse, and fragment into stars. Let us take a moment to ask what happens next—not to the newborn stars themselves (that will be the subject of the next three chapters), but to the galactic environment in which they have just formed.

The end result of the collapse process is a group of stars, all formed from the same region of the parent cloud, along with a certain amount of unused gas and dust. How many stars form, and of what type? How much gas is left over? What does the collapsed cloud look like once the star formation process has run its course? At present, while the main stages in the formation of an individual star (stages 3–7) are becoming clearer, the answers to these more general questions (involving stages 1 and 2) are still rather sketchy and await a more thorough understanding of the cloud-collapse process.

In general, the more massive the collapsing region, the more stars are likely to form there. On the basis of observed HR diagrams, we also know that low-mass stars are much more common than high-mass ones. However, the precise number of stars of any given mass or spectral type depends in a complex (and poorly understood) way on conditions within the parent cloud. The same is true of the *efficiency* of star formation—that is, the fraction of the total

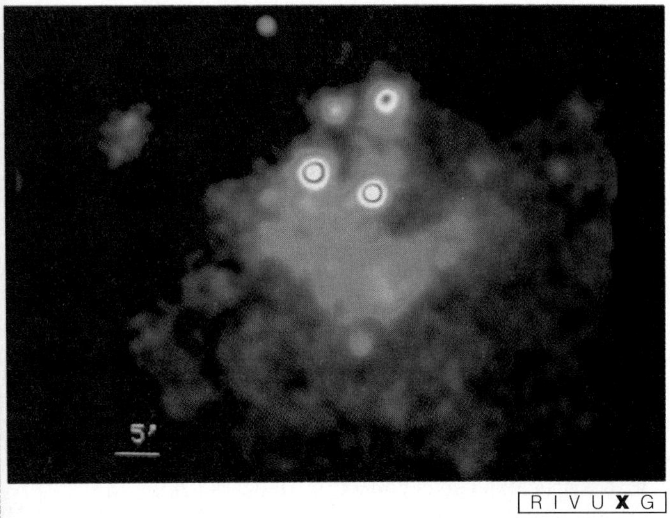

Figure 21.22 The somewhat open cluster and HII region Eta Carinae is seen here in true-color visual light. The region is about 2,700 parsecs away and extends across some 30 parsecs. The insert is an x-ray image of the hottest (O-type) stars clustered near the core of the nebula. (NOAO; NASA)

mass that actually finds its way into stars—which determines the amount of leftover gas. Composition, rotation, and magnetism in the cloud all seem to play important roles in determining the eventual outcome of the collapse.

If, as is usually the case, one or more O- or B-type stars form, an emission nebula, or HII region, results. Ultraviolet radiation from the hot stars heats and ionizes the surrounding gas, causing it to glow, as described in Chapter 20. If enough hot stars are present, the HII region may extend well beyond the region containing the newborn stars, into the surrounding dark cloud. The heat of the newborn stars causes the HII region to expand into its surroundings, possibly triggering new star formation, as we saw in the previous section. Eventually, the intense radiation and winds from the stars will cause the surrounding gas to disperse, leaving behind a clump of young stars—a star cluster. Figure 21.22 shows an open cluster in which the gas dispersal process is almost complete.

Until recently, the existence of star clusters within HII regions was largely conjecture. The stars themselves could not be seen optically because of obscuration by dust. Infrared observations have now clearly demonstrated that stars really are found within star-forming regions! Figure 21.23 compares optical and infrared views of the central

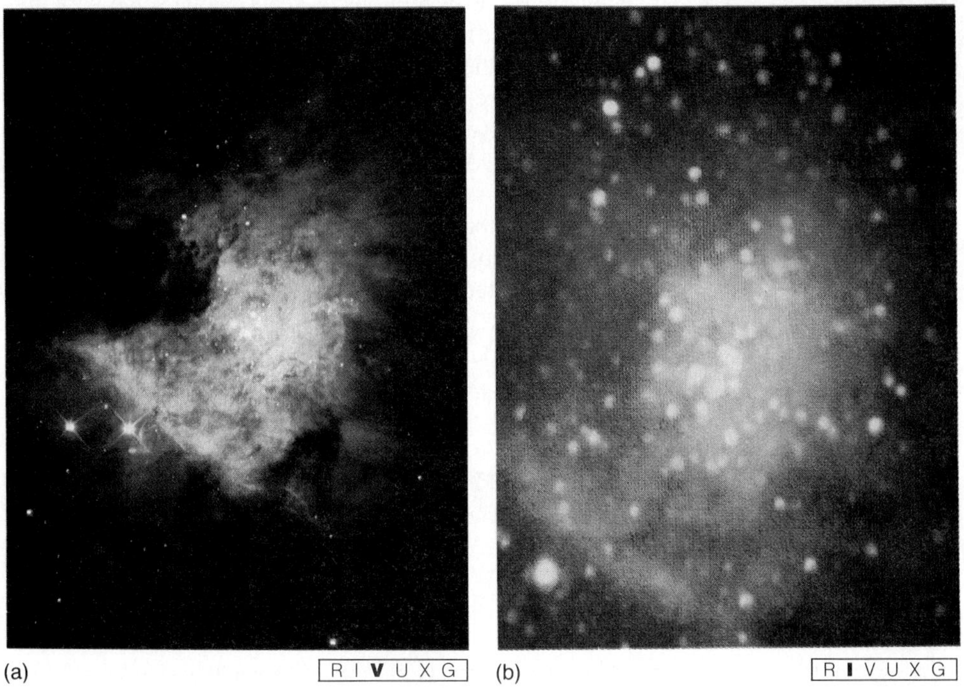

(a) (b)

Figure 21.23 The central regions of the Orion nebula seen (a) in a short exposure visible light image (using an oxygen-line filter) and (b) in the infrared to roughly the same scale. The visible image shows the nebula itself and four bright O stars, known as the Trapezium. The infrared view, seen here in false color, where red is coolest and white is warmest, shows many cool stars within the nebula that are undetectable in visible light. The reddish region near the center of the image is the Kleinmann-Low Nebula, discussed earlier in the text. (Lick Observatory; NASA) regions of the Orion Nebula. The optical image in Figure 21.23(a) shows only the Trapezium, the group of four bright stars responsible for ionizing the nebula. However, the false-color infrared image in Figure 21.23(b) also reveals an extensive cluster of stars (shown here in blue) within and behind the visible nebula. The Kleinmann-Low Nebula, discussed earlier, can be seen here as the orange- and red-colored region close to the center of the infrared image. The bright spot within it (known as the Becklin-Neugebauer object) is thought to be a dust-shrouded B star just beginning to form its own HII region around it. This one image shows many separate stages of star formation.

For every O or B giant, tens or even hundreds of G, K, and M dwarfs may form, so even a modest HII region can give rise to a fairly extensive collection of stars. A typical open star cluster in the plane of our Galaxy, like that shown in Figure 21.24, may measure 10 pc across and contain 1000 or more stars. Less massive, but more extended, clusters are usually known as **associations.** These typically contain no more than 100 stars but may span many tens of parsecs. Associations tend to be rich in very young stars. Those containing many pre-main-sequence T Tauri stars are known as *T associations,* while those with prominent O and B stars, like the Trapezium in Orion, are called *OB associations*. It is quite likely that the main difference between associations and open clusters is simply the efficiency with which stars formed from the parent cloud.

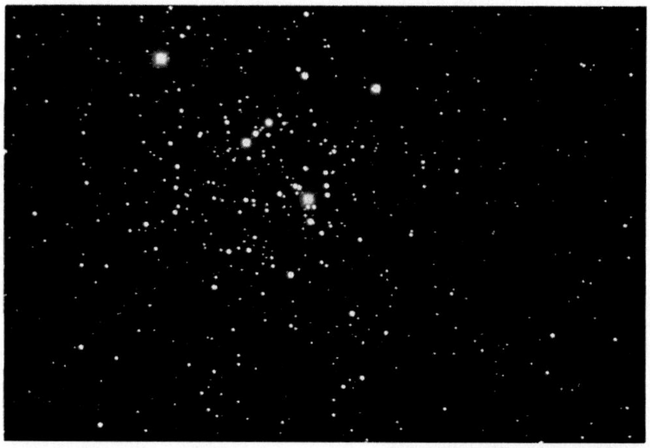

R I **V** U X G

Figure 21.24 *The Jewel Box cluster is a relatively young open cluster in the southern sky. While many bright stars appear in this image, the cluster contains many more low-mass, less-luminous stars. Because some red giants appear among the cluster's blue main-sequence stars, we can estimate the age of the cluster to be about 10 million years. (NOAO)*

R I **V** U X G

Figure 21.25 *M67, one of the oldest known open clusters, has survived for almost 5 billion years, an unusually long time for a star system in the plane of the Milky Way Galaxy. (NOAO)*

Eventually, star clusters dissolve into individual stars. One reason is that stellar encounters tend to eject the lightest stars from the cluster, just as the gravitational slingshot effect (see Interlude 3-2) can propel spacecraft around the solar system. At the same time, the tidal gravitational field of the Milky Way Galaxy slowly strips outlying stars from the cluster. Occasional distant encounters with giant molecular clouds also tend to remove cluster stars; a near miss may even disrupt the cluster entirely. As a result of all these influences, most open clusters break up in a few hundred million years, although the actual lifetime depends on the cluster's mass. Loosely bound associations may survive for only a few tens of millions of years, while some very massive open clusters like M67, shown in Figure 21.25, are known from their HR diagrams to be almost 5 billion years old. In a sense, only when a star's parent cluster has completely dissolved is the star-formation process really complete. The road from a gas cloud to a single, isolated star like the Sun is long and tortuous indeed!

Take another look at the nighttime sky. Ponder all that cosmic activity while gazing upward some clear, dark evening. After studying this chapter, you may find your view of the night sky greatly changed. Even the seemingly quiet nighttime darkness is dominated by continual change.

CHAPTER SUMMARY

Heat, rotation, and magnetism all compete with gravity to influence the evolution of a star-forming cloud. Interstellar clouds contain very little heat, so once gravity has overcome thermal pressure and a cloud has begun to contract, thermal effects become relatively unimportant. Only later, when the cloud has shrunk to about the size of the solar system and its temperature has risen to several thousand kelvins, does heat again start to play an important role.

A cold interstellar cloud containing a few thousand solar masses of gas can fragment into tens or hundreds of smaller clumps of matter, from which stars eventually form. As each cloud fragment shrinks, its density increases. The result is an increase in temperature and pressure, which eventually causes the fragmentation to stop. The opaque center of the fragment forms a protostar, whose mass grows as more material rains down on it from outside, while its radius continues to shrink. A protostar's luminosity is much greater than that of most main-sequence stars, due to the release of gravitational energy.

The protostar's luminosity decreases as it shrinks. After a few million years, the protostar becomes a full-fledged star, with a core temperature high enough to fuse hydrogen into helium. Nuclear energy generation in the core matches the rate at which energy is radiated from the surface. At every phase of the star's evolution, its surface temperature and luminosity can be represented by a point on the Hertzsprung-Russell diagram.

Mass is the key property for determining a star's characteristics and life span. The most massive stars have the shortest formation times and main-sequence lifetimes. At the other extreme, some low-mass objects never even reach the point of nuclear ignition. The universe may be populated with a vast number of brown dwarfs—objects that are not massive enough to fuse hydrogen to helium in their interiors.

Emission nebulae are signposts of star birth. Dark interstellar regions near them often provide evidence for cloud fragmentation and protostars. Radio telescopes are used for studying the early phases of cloud contraction and fragmentation; infrared observations allow us to see later stages of the process. Many well-known emission nebulae, lit by several O-type stars, are partly engulfed by a molecular cloud, parts of which are probably fragmenting and contracting, with smaller sites forming protostars.

Protostellar winds encounter less resistance in the directions perpendicular to a protostar's disk. Thus they create two jets of matter, expelled in the directions of the protostar's poles. As the protostellar wind gradually destroys the disk, the jets widen until, with the disk gone, the wind flows away from the star equally in all directions. Shock waves generated by emission nebulae, by galactic spiral-arm waves, and by late stellar evolutionary stages compress other interstellar clouds and trigger further star formation. Star birth and the production of shock waves are thought to produce a chain reaction of star formation in molecular cloud complexes.

KEY WORDS

association
bipolar flow
brown dwarf
cocoon nebula
evolutionary track
Hayashi track
protostar
Kelvin-Helmholtz contraction phase
shock wave
T Tauri star
zero-age main sequence

REVIEW QUESTIONS

1. What is the role of heat in the process of stellar birth?

2. What is the role of rotation in the process of stellar birth?

3. What is the role of magnetism in the process of stellar birth?

4. Roughly how many atoms are needed to make a star? How much mass is this?

5. What is an evolutionary track?

6. Why do stars tend to form in groups?

7. At what point does a star-forming cloud become a protostar?

8. At what point does a protostar become a full-fledged star?

9. What are brown dwarfs?

10. What are T Tauri stars?

11. Because stars live much longer than we do, how do astronomers test the accuracy of theories of star formation?

12. Why do astronomers use radio and infrared radiation to study prestellar objects?

13. Describe the typical evolution of protostellar winds.

14. What is a maser?

15. What is a shock wave? What phenomena are thought to produce shock waves in the instellar medium?

16. Of what significance are shock waves in star formation?

17. A certain interstellar cloud contains 10^{60} atoms. Hydrogen (mass per atom $= 2 \times 10^{-24}$ g) accounts for 90 percent of the atoms, and the remainder are helium (each helium atom has four times more mass than an atom of hydrogen). What is the cloud's mass? Express your answer in solar masses ($M_\odot = 2 \times 10^{33}$ g).

18. Use the radius-luminosity-temperature relation ($L = R^2T^4$; see Chapter 19) to explain how a protostar's luminosity changes as it moves from stage 4 ($T =$ 3000K, $R = 2 \times 10^8$ km) to stage 6 ($T =$ 4500K, $R = 4 \times 10^6$ km).

19. What is the luminosity, in solar units, of a brown dwarf star whose radius is 0.1 solar radii and whose surface temperature is 600K (0.1 times that of the Sun)?

20. The average distance between stars in a newborn open cluster is about 0.1 pc (3×10^{17} cm). Compare this with (a) the diameter of the Sun (1.4 million km), (b) the diameter of an O-type main-sequence star (20 solar diameters), and (c) the diameter of a stage-4 protostar (10^8 km). Show your comparisons as ratios.

DISCUSSION QUESTIONS

1. Explain the usefulness of the Hertzsprung-Russell diagram to astronomy.

2. Write a report on brown dwarfs. Have any objects been discovered yet that most astronomers agree are brown dwarfs?

3. What is the likelihood that the most massive stars have life-bearing planets? Explain your answer.

4. Astronomers who study star formation are limited by the fact that stars live much longer than we do. What other limitations exist in this field of study? How have astronomers managed to overcome these limitations in developing a theory of star formation?

PROJECT

1. The Trifid Nebula, otherwise known as M20, is a place where new stars are forming. It has been called a "dark night revelation, even in modest apertures." An 8- to 10-inch telescope is needed to see the triple-lobed structure of the nebula. Ordinary binoculars reveal the Trifid as a hazy patch located in the constellation Sagittarius. This nebula is set against the richest part of the Milky Way, the edgewise projection of our own Galaxy around the sky. It is one of many wonders in this region of the heavens. What are the dark lanes in M20? Why are other parts of the nebula bright? There have been reports of large-scale changes occurring in this nebula in the last century and a half. The reports are based on old drawings, which show M20 looking slightly different from how it appears today. Is it possible that a cloud in space might undergo a change in appearance on a time scale of years, decades, or centuries?

SUGGESTED READINGS

Byrd, D. "Do Brown Dwarfs Really Exist?" *Astronomy* (April 1989). The search for brown dwarfs, and why some astronomers think they will not be found.

———. "Bad News for Brown Dwarfs." *Sky and Telescope* (October 1990). More on the struggle to discover brown dwarfs.

Fienberg, R. T. "Brown Dwarfs Coming and Going." *Sky and Telescope* (November 1989). More on why brown dwarfs should exist, despite the fact that no one has yet discovered one.

Fortier, E. "Touring the Stellar Cycle." *Astronomy* (March 1987). Stellar evolution explained.

Hartley, K. "How a Star Is Born." *Astronomy* (December 1989). The birth of stars.

Lada, C. J. "Star in the Making." *Sky and Telescope* (October 1986). How astronomers observed an interstellar cloud of gas and dust collapsing to form a protostar.

Norris, R. "Cosmic Masers." *Sky and Telescope* (March 1986). Cosmic radio amplifiers and what they tell us about comets, dying stars, and active galactic nuclei.

Sagdeev, R. Z., and C. F. Kennel. "Collisionless Shock Waves." *Scientific American* (April 1991). Shock waves in space, transmitted through the tenuous plasma by electric and magnetic fields, may help explain some of the most violent phenomena in the universe.

Schild, R. E. "A Star Is Born." *Sky and Telescope* (December 1990). Why the birth of a star is one of the least understood events in astronomy.

Stahler, S. W. "The Early Life of Stars." *Scientific American* (July 1991). The complex life cycle of stars, beginning when clouds of interstellar gas coalesce into protostars detectable only in the infrared and becoming optically visible stars.

Stahler, S., and N. Comins. "The Difficult Birth of Sunlike Stars." *Astronomy* (September 1988). How stars like the Sun come to exist.

CHAPTER

22

Stellar Evolution

From Middle-Age to Death

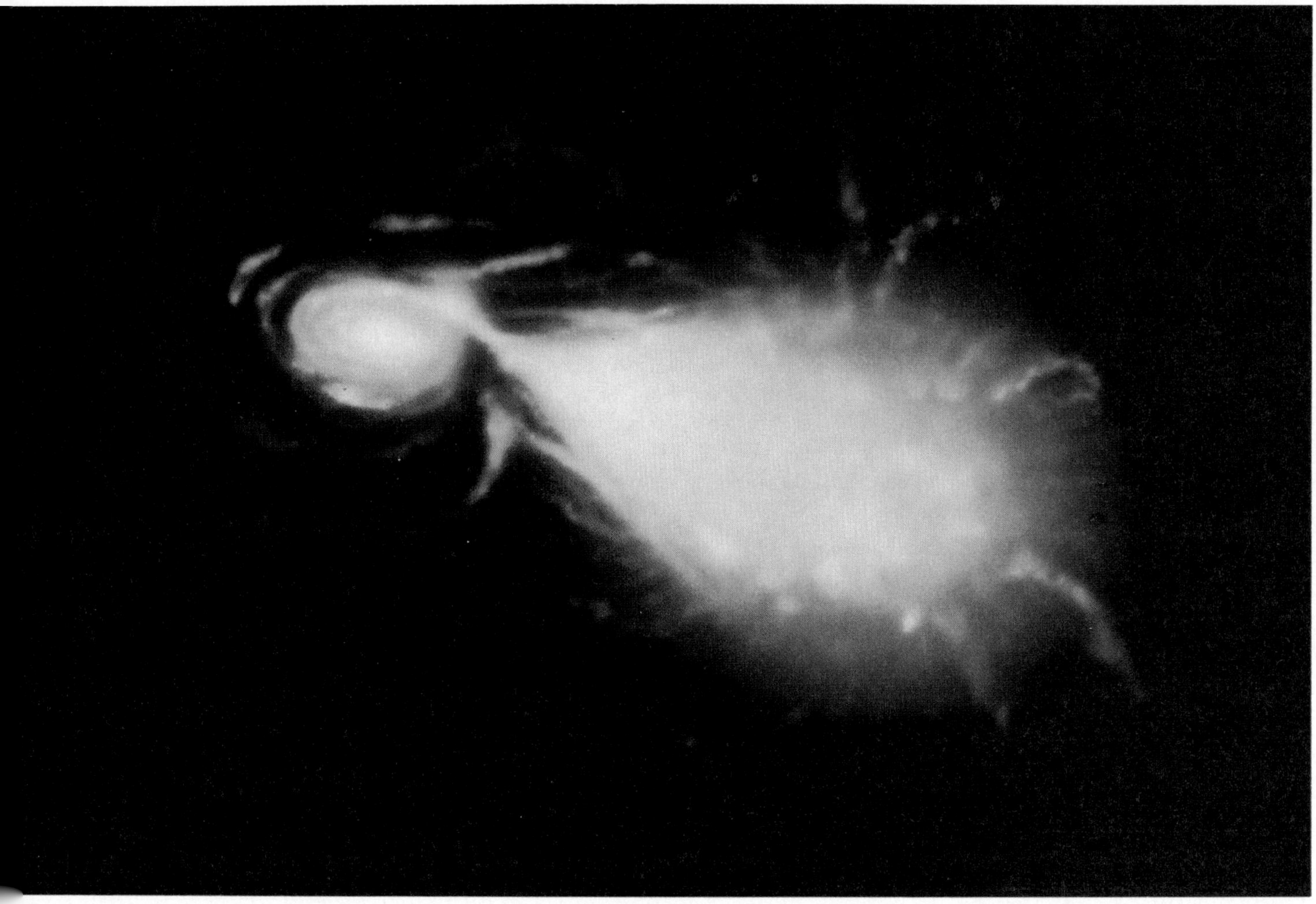

This artist's conception depicts a 4-solar-mass star that has swollen and is now spilling over onto its dwarf companion star (at top left). As the star swells, its shape is defined by the Roche lobe, structured by the gravitational field surrounding the two stars. As matter streams across to the companion, an accretion disk begins to form. This is clearly an unstable, even violent stage in the evolution of a binary-star system. (Courtesy D. Berry)

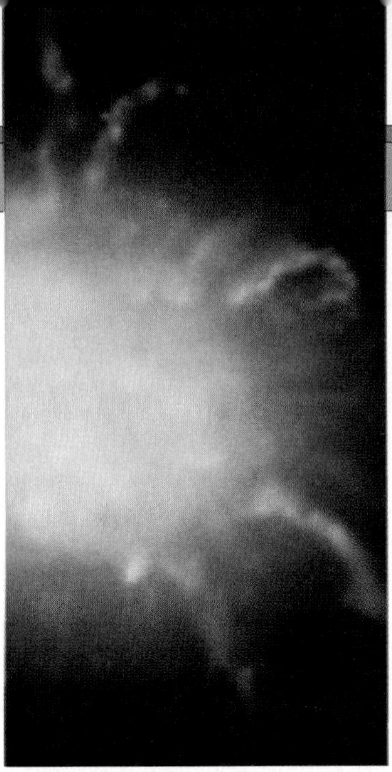

LEARNING GOALS

1 to understand why stars evolve

2 to understand what happens when a Sun-like star exhausts the supply of hydrogen at its center

3 to understand the reasons for the gentle death of low-mass stars

4 to grasp the essential evolutionary differences between high-mass and low-mass stars

5 to examine some of the many observations that help verify the theory of stellar evolution

6 to realize that stars in binary systems may evolve quite differently from stars in isolation

The process of star formation produces great changes in the physical properties of a prestellar object. After traveling the Hertzsprung-Russell diagram as a protostar, a newborn star eventually reaches the main sequence, where it is destined to remain, almost unchanging in outward appearance, for more than 90 percent of its entire stellar lifetime. At the end of this period, as the star begins to run out of fuel and die, its properties once again change greatly. Aging stars travel along evolutionary tracks that take them far from the main sequence as they head toward death.

In this and the next two chapters, we will study the evolution of stars during and after their main-sequence burning stages. We will find that the ultimate fate of a star depends primarily on its mass, although interactions with other stars can also play a vital role, and that the final states of stars can be strange indeed. By continually comparing theoretical calculations with detailed observations of stars of all types, astronomers have refined the theory of stellar evolution into a precise and powerful tool for understanding the universe.

EVOLUTION OFF THE MAIN SEQUENCE

✓1 The main sequence of the HR diagram is the evolutionary stage where most stars spend most of their lives. For example, a star like the Sun, after spending a few tens of millions of years in formation, is expected to reside on or near the main sequence for 10 *billion* years before evolving into something else. That "something else" is the main topic of this chapter. As we will see, once a star leaves the main sequence, its days are numbered. To all intents and purposes, evolving off the main sequence represents the beginning of the end for any star.

Virtually all of the low-mass stars that ever formed still exist as stars. M-type dwarfs burn their fuel so slowly that not one of them has yet left the main sequence. Some of them will burn steadily for a trillion years or more. Conversely, the most massive O- and B-type stars evolve away from the main sequence after only a few tens of millions of years of burning. Most of the massive stars that have ever existed perished long ago. Between these two extremes, many stars are observed in advanced stages of stellar evolution, their properties quite different from when they first arrived on the main sequence.

All theoretical models suggest that the final stages of stellar evolution depend critically on the mass of the star. As a rule of thumb, low-mass stars die gently, and high-mass stars die catastrophically. Depending on which astronomer you ask, the dividing line between "low-mass" and "high-mass" is anywhere from 3 to 8 times the mass of the Sun. We will take 4 solar masses as a compromise value—that is, we will consider stars of 4 solar masses or more to be high-mass stars.

We will begin by considering the evolution of a fairly low-mass star like the Sun. The stages described in the next few sections will pertain to the Sun at some future time, as it nears the end of its fusion cycle 5 billion years from now. In fact, most of the qualitative features of the discussion apply to any low-mass star, although the exact numbers vary considerably. Later, we will broaden our discussion to include all stars, large and small.

Main-Sequence Equilibrium

Gravity is always present wherever matter exists. Only some counteracting phenomenon prevents an astronomical object from completely collapsing under its own weight. In the case of stars, that competing phenomenon is gas pressure, caused by the heat of the raging inferno at the stellar core. Figure 22.1 diagrams the balance between gravity's inward pull and pressure's outward push. You should keep this simple picture in mind while studying the various stages of stellar evolution described in this chapter. Also bear in mind a simple maxim that summarizes the eventual outcome of the struggle between gravity and heat in almost all phases of a star's life: Sooner or later, gravity wins.

Provided that a star remains in this equilibrium state, nothing spectacular happens to it. While the star is a resident of the main sequence, its hydrogen fuel slowly burns into helium, its surface occasionally erupts in flares and spots, and its atmosphere ejects copious amounts of particles and photons. But by and large, main-sequence stars do not experience sudden, large-scale changes in their properties. Their average temperatures and luminosities remain fairly constant (in fact, the luminosity increases very slowly—the Sun is now some 30 percent brighter than it was 5 billion years ago). We might expect them to release energy indefinitely unless something drastic were to occur. Eventually something drastic does occur.

Depletion of Hydrogen in the Core

After approximately 10 billion years of steady **core hydrogen burning,** a Sun-like star begins to run out of fuel. Hydrogen becomes depleted, at least in a small central region about 1/100 of the star's full size. The depletion of hydrogen is slow and steady, but the consequences are severe. It is a little like an automobile cruising effortlessly along a highway at a constant speed of 55 mph for many hours, only to have the engine cough and sputter as the gas gauge reaches empty. Unlike automobiles, though, stars are not so easy to refuel.

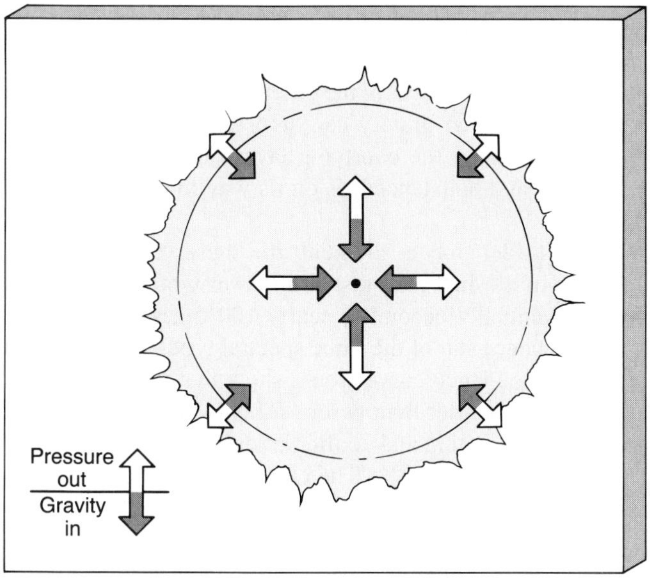

Figure 22.1 *In a steadily burning star on the main sequence, the outward pressure of hot gas counterbalances the inward pull of gravity. This is true at every point within the star, guaranteeing its stability.*

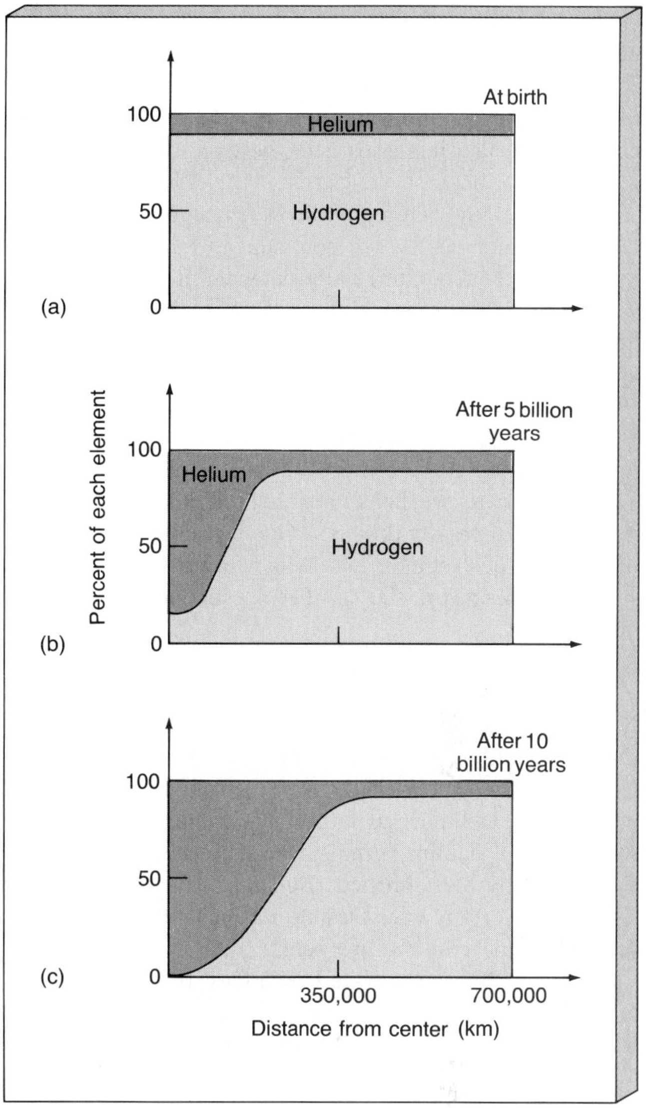

Figure 22.2 *Theoretical estimates of the changes in a solar-type star's composition. Hydrogen (yellow) and helium (orange) abundances are shown (a) at birth, on the zero-age main sequence, (b) after 5 billion years, and (c) after 10 billion years. At stage (b) only about 5 percent of the star's total mass has been converted from hydrogen into helium. This change speeds up as the nuclear burning rate increases with time.*

As the nuclear burning proceeds, the composition of the star's interior changes. Figure 22.2 illustrates the increase in helium abundance and the corresponding decrease in hydrogen in the stellar core as the star ages. Three cases are drawn—(a) as the chemical composition of the original core, (b) the composition after 5 billion years, and (c) the composition after 10 billion years. Case (b) represents approximately the present state of our Sun.

The star's helium content increases fastest at the center, where temperatures are highest and the burning is fastest. Helium also increases near the edge of the core, but more slowly, because the burning rate is less rapid there. The inner helium-rich region becomes larger and more hydrogen-deficient as the star continues to shine. Eventually, hydrogen becomes completely depleted at the center, the nuclear fires there cease, and the location of principal burning moves to higher layers in the core. An inner core of nonburning pure helium starts to grow.

After about 10 billion years, a serious problem arises. While hydrogen burning continues in the outer core, the lack of burning at the center leads to an unstable situation. The gas pressure weakens in the helium inner core, but the inward pull of gravity does not. Gravity never lets up. Once the outward push against gravity is relaxed—even a little—structural changes in the star become inevitable.

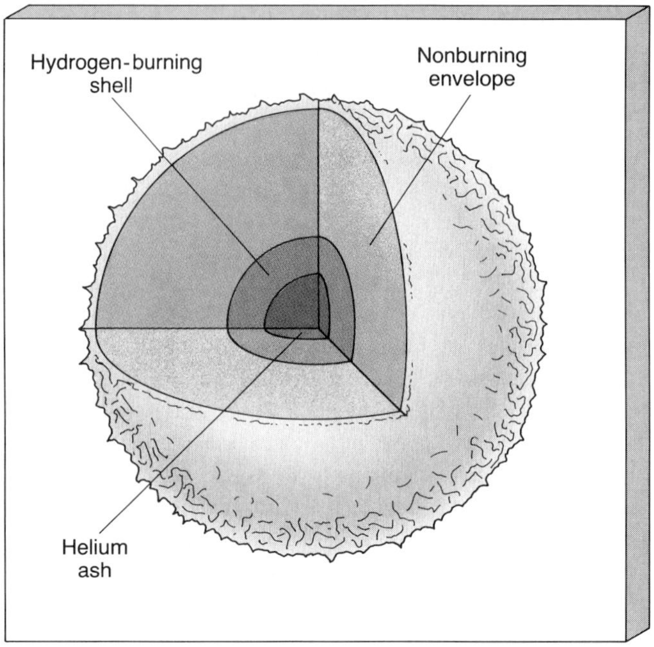

Figure 22.3 *As a star's core becomes progressively depleted of hydrogen, hydrogen continues to burn in the layers surrounding the nonburning helium ash.*

Contraction of the Helium Core

If more heat could be generated, the star might possibly return to equilibrium. For example, were helium in the core to begin fusing into some heavier element such as carbon, all would be well once again. Energy would be created as a by-product of helium burning, and the necessary gas pressure could be reestablished. But the helium there cannot burn—not yet anyway. Despite its high temperature, the core is far too cold to fuse helium into anything heavier.

Recall that a temperature of at least 10^7 K is needed to burn hydrogen into helium. Above that temperature, colliding hydrogen nuclei—that is, protons—have enough speed to overwhelm the repulsive electromagnetic force between them. With helium, however, even 10^7 K is insufficient for fusion. Each helium nucleus, composed of two protons and two neutrons, has a net positive charge twice that of the hydrogen nucleus. As a result, the repulsive electromagnetic force between two helium nuclei is also larger, and more violent collisions are needed to fuse helium. Tremendously high temperatures are required—about 10^8 K.

A core composed of helium at 10^7 K thus cannot generate energy through fusion. As soon as the hydrogen fuel becomes substantially depleted, the helium core begins to contract because the pressure there—without nuclear burning—is too low to counteract gravity. Just as in earlier phases of stellar evolution, this shrinkage releases gravitational energy, driving up the central temperature.

The increasingly hot core heats the overlying layers of the star's core. The higher temperatures—now well over 10^7 K—cause hydrogen nuclei to fuse even more rapidly than before. Figure 22.3 depicts this situation, where hydrogen is burning at a fantastic rate in a shell surrounding the nonburning helium "ash" in the center. This phase is usually known as the **hydrogen shell-burning** stage. The hydrogen shell generates energy at a rate greater than the original main-sequence star's hydrogen-burning core, and its energy production continues to grow. Strange as it may seem, the star's response to the disappearance of the fire at its center is to get brighter!

Conditions in the aging star have clearly changed from the equilibrium that once characterized it as a main-sequence object. The helium core is unbalanced and shrinking, on its way to becoming hot enough for helium fusion. The rest of the core is also unbalanced, fusing hydrogen into helium at a growing rate. The gas pressure exerted by this enhanced hydrogen burning increases, forcing the intermediate layers and especially the outermost layers of the star to expand. Not even gravity can stop them. Even while the core is shrinking, the overlying layers are expanding! The star, aged and unbalanced, is on its way to becoming a red giant.

Consider for a moment the observational consequences of all this. An outside observer would see the star swell, eventually becoming nearly 100 times larger than a main-sequence star of the same spectral type, while analysis of the star's Planck curve would show that the surface was about 2000K cooler than before. This is not to say that the period of ballooning and cooling of an aged star could actually be observed directly. The change from a normal main-sequence star to an elderly red giant takes about 100 million years to complete.

We can trace these large-scale changes in an aged 1-solar-mass star on the HR diagram. Figure 22.4 shows the path away from the main sequence. The luminosity of the giant at the point marked 8 on the figure (recall from Chapter 21 that stage 7 was the star's arrival on the main se-

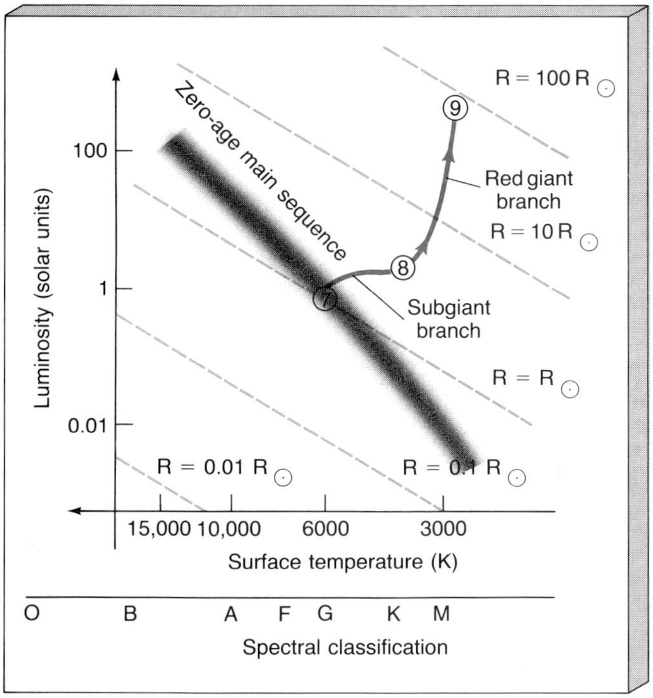

Figure 22.4 *As the core of helium ash shrinks and the intermediate stellar layers expand, the star leaves the main sequence. At stage 8, the star is on its way to becoming a red giant star. The star continues to brighten and grow as it ascends the red-giant branch to stage 9, the top of the red-giant branch. As in Chapter 21, the diagonal lines correspond to stars of constant radius, allowing us to gauge the changes in the size of our star.*

quence) is about 10☉. (Remember that 1 L☉ is the Sun's luminosity.) It exceeds 100 L☉ at point 9. The surface temperature at stage 8 has fallen to the point where much of the interior becomes opaque to the radiation from within. Beyond this point, convection carries the core's enormous energy output to the surface. One consequence of this is that the star's surface temperature remains nearly constant between stages 8 and 9.

Given the rising luminosity and the falling surface temperature, it follows that the star's radius must increase, to about 20 R☉ (1 R☉ is the Sun's radius) at stage 8 and eventually to about 70 R☉ at stage 9. The roughly constant-luminosity path from stage 7 (the main sequence) to stage 8 is often called the **subgiant branch.** The nearly vertical path followed by the star between stages 8 and 9 is known as the **red-giant branch** of the HR diagram.

Red Giants

Figure 22.5 compares the relative sizes of our Sun and a stage-9 red giant star. The red giant is huge, having swollen to about 70 times its main-sequence size, about the size of Mercury's orbit. By contrast, its helium core is surprisingly small, probably about 1000 times smaller than the entire star, making it only a few times larger than the Earth.

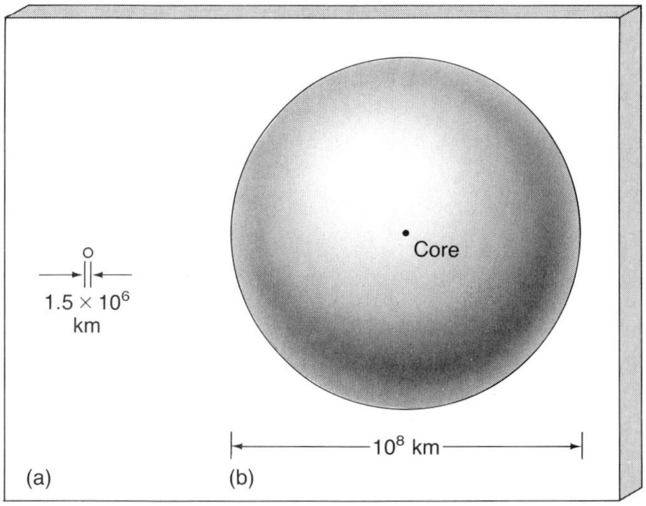

Figure 22.5 *Diagram of the relative sizes of (a) a normal G-type star (this is our Sun) on the main sequence and (b) a normal G-type star after it has ballooned to become a red giant. The difference in size is approximately a factor of 70. The core of the red giant at this stage is about 15 times smaller than the main-sequence star (a) and is barely discernable on the scale of this figure.*

The density in the core is now enormous. Continued shrinkage of the red giant's core has compacted its helium gas to approximately 10^5 g/cm^3. Contrast this with the 10^{-6} g/cm^3 in the outermost layers of the red giant, with the 5 g/cm^{3}average density of the Earth, and with the 150 g/cm^3 in the present core of the Sun. About 25 percent of the mass of the entire star is packed into its small core.

Perhaps the most famous red giant is the naked-eye star Betelgeuse in the constellation Orion (shown in Figures 19.7 and 19.11). Despite its great distance—about 150 pc—its enormous luminosity of 10^4 L☉ makes it one of the brightest stars in the night sky.

HELIUM FUSION IN LOW-MASS STARS

✓2 Should the unbalanced state of a red giant star continue, the core would eventually collapse, and the rest of the star would slowly drift into space. The forces and pressures at work inside a red giant would literally pull it apart. However, this simultaneous shrinking and expanding does not continue indefinitely. A few hundred million years after a solar-mass star leaves the main sequence, something else happens—helium begins to burn in the core.

By the time the central density has risen to about 10^5 g/cm^3, the temperature in the core has reached the 10^8 K needed for helium fusion. Helium nuclei then collide with one another, fusing into carbon nuclei and igniting the central fires once again. The reaction that transforms helium into carbon occurs in two steps. First, two helium nuclei come together to form a nucleus of beryllium-8 (^{8}Be). This is a very unstable isotope that would normally break up into two helium nuclei in about 10^{-12} s. However, at the densities in the core of a red giant, it is very likely that the beryllium-8 nucleus will encounter another helium nucleus before this occurs, fusing with it to form carbon-12 (^{12}C). This is the second step of the helium-burning reaction. In part, it is because of the electrostatic repulsion between beryllium-8 (containing 4 protons) and helium-4 (containing 2) that the temperature has to rise to 10^8 K before this reaction can take place.

Symbolically, we can represent this next stage of stellar fusion as follows:

$$^4He + {}^4He \longrightarrow {}^8Be + \text{energy}$$

$$^8Be + {}^4He \longrightarrow {}^{12}C + \text{energy}$$

Helium-4 nuclei are traditionally known as **alpha particles.** The term dates from the early days of nuclear physics, when the true nature of these particles was unknown. Because three alpha particles are required to get from helium-4 to carbon-12, the above reaction is usually called the **triple-alpha process.**

The Helium Flash

For low-mass stars, there is a major complication in the helium burning process. At the high densities found in the core, the gas has entered a new state of matter whose properties are governed by the laws of quantum mechanics rather than by those of classical physics. Up to now, we have been primarily concerned with the nuclei—protons, alpha particles, and so on—that make up virtually all of the star's mass and participate in the reactions that generate its energy. However, the star contains another important constituent—a vast sea of electrons stripped from their parent nuclei by the ferocious heat in the stellar interior. At this stage in our story, these electrons play a critical role in determining the star's evolution.

Given the conditions in the stage-9 red-giant core, a rule of quantum mechanics known as the **Pauli exclusion principle** (after Wolfgang Pauli, one of the founding fathers of quantum physics) prohibits the electrons in the core from being squeezed too close together. In effect, the exclusion principle tells us we can think of the electrons as tiny rigid spheres, which can be squeezed relatively easily up to the point of contact but become virtually incompressible thereafter. This condition is known (for historical reasons) as **electron degeneracy,** and the pressure associated with the contact of the tiny electron spheres is called **electron degeneracy pressure.** It has nothing whatever to do with the thermal pressure (due to the star's heat) we have been studying up to now. In a red-giant core, the pressure resisting the force of gravity is supplied almost entirely by degenerate electrons. Hardly any of the core's support results from "normal" thermal pressure.

The importance of electron degeneracy to the onset of helium burning in the core of a red giant is this: Under normal ("nondegenerate") circumstances, the core can react to and accommodate the onset of helium burning, but in its degenerate state the burning becomes unstable, with explosive consequences. In a normal star, the increase in temperature produced by the onset of helium fusion would lead to an increase in pressure. The gas would then expand and cool, reducing the burning rate and reestablishing equilibrium. In the degenerate case, however, the pressure is largely *independent* of the temperature. When burning starts and the temperature increases, there is no corresponding rise in pressure, no expansion of the gas, no drop in the temperature, and no stabilization of the core. Instead, the pressure remains more or less unchanged while the nuclear reaction rates increase and the temperature continues to rise. The temperature increases rapidly in a runaway explosion called the **helium flash.**

For a period of a few hours, the helium burns ferociously, like an uncontrolled bomb. Despite its brevity, this period of uncontrolled fusion releases a flood of new energy, enough to expand the core, lowering its density and ultimately returning it to a stable, nondegenerate state. This expansive adjustment of the core halts its gravitational collapse, returning it to equilibrium—an equilibrium reached once again between the inward pull of gravity and the outward push of gas pressure. The core, now stable, begins to burn helium into carbon at temperatures well above 10^8 K.

The helium flash terminates the giant star's ascent on the red-giant branch of the HR diagram. Despite the explosive detonation of helium in the core, the flash does *not* increase the star's luminosity. On the contrary, the helium flash produces a rearrangement of the core that ultimately results in a *reduction* in the energy output. On the HR diagram, the star jumps from stage 9 to stage 10, a stable state with steady helium burning in the core. As indicated in Figure 22.6(a), at this stage the surface temperature is higher than it was on the red-giant branch, while the luminosity is considerably less than at the helium flash. This adjustment in the star's properties occurs quite quickly—in about 100,000 years.

The Horizontal Branch

At stage 10 our star is now stably burning helium in its core and fusing hydrogen in a shell surrounding it. It resides in a well-defined region of the HR diagram known as the **horizontal branch**—a "helium main sequence" of sorts, where core-helium-burning stars remain for a time before resuming their journey around the HR diagram. The star's specific position within this region is determined mostly by its mass—not its original mass, but whatever mass remains after its ascent of the giant branch. The two masses differ because during the red-giant stage, strong stellar winds eject large amounts of matter from a star's surface (see Interlude 22-1). As much as 20–30 percent of the original stellar mass may escape during this period. It so happens that more massive stars have lower surface temperatures at this stage,

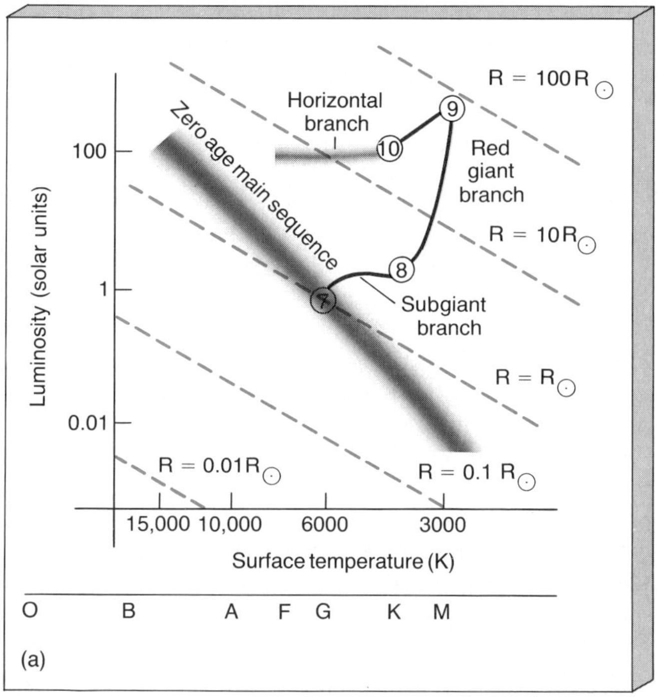

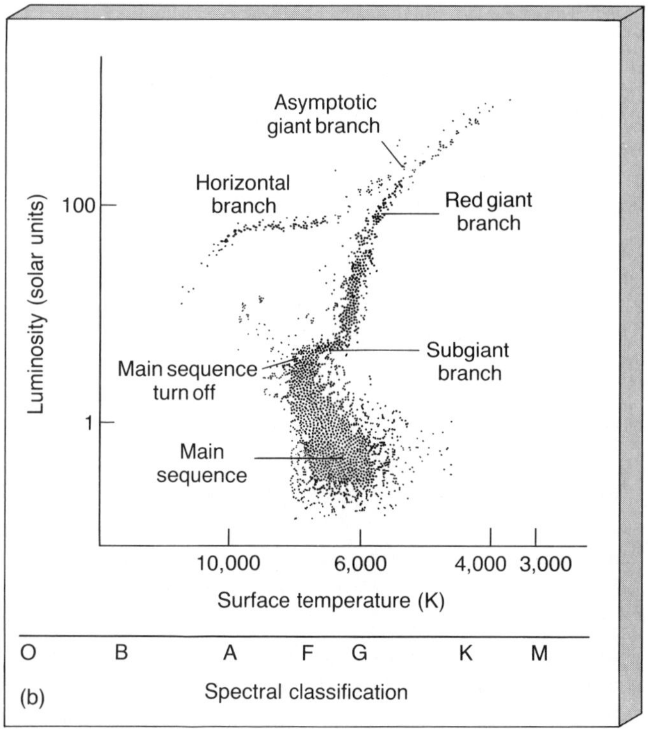

Figure 22.6 *(a) After its large increase in luminosity ascending the red-giant branch, our star settles down into another equilibrium state at stage 10, on the horizontal branch. (b) The horizontal branch is clearly visible in this HR diagram of an old star cluster (the globular cluster M 5). (Courtesy F. Renzini and A. Fusi Pecci)*

but all stars have roughly the same luminosity after the helium flash. As a result, stage-10 stars tend to lie along a horizontal line on the HR diagram, with more massive stars to the right, less massive ones to the left. Figure 22.6(b) shows how this next stage of stellar evolution appears in a real HR diagram (in this case, for the old globular cluster M 5).

Table 22-1 summarizes the evolution of a 1-solar-mass star. It is a continuation of the compilation listed in Table 21-1 except that the density units have been changed from particles per cubic centimeter to the more convenient grams per cubic centimeter. The previous table ended with stage 7, a main-sequence object fusing hydrogen into helium over the course of some 10 billion years. The table here begins with stage 7, then moves on to stage 8 on the subgiant branch, as the star evolves away from the main sequence. Stage 9 is the helium flash, at the tip of the red-giant branch. Stage 10 describes an established horizontal-branch star stably fusing helium into carbon at its core.

The Carbon Core

The nuclear reactions in a star's helium core burn on, but not for long. Whatever helium exists in the core is rapidly

TABLE 22-1 ***Evolution of a Sun-like Star***

Stage	Approximate Time to next Stage (yr)	Central Temperature (10^6 K)	Surface Temperature (K)	Central Density (g/cm^3)	Diameter (km)	Object
7	10^{10}	15	6,000	10^2	1.5×10^6	Main-sequence star
8	10^8	50	4,000	10^4	2×10^7	Subgiant branch
9	10^5	100	4,000	10^5	10^8	Helium flash
10	5×10^7	200	5,000	10^4	2×10^7	Horizontal branch
11	10^4	250	4,000	10^5	10^9	Asymptotic giant branch
12	10^5	300	100,000	10^6	10^5	Carbon core
		—	3,000*	10^{-20}	10^8	Planetary nebula
13	—	100	50,000	10^7	10^4	White dwarf star
14	—	Close to 0	Close to 0	10^7	10^4	Black dwarf star

*envelope

INTERLUDE 22-1 *Mass Loss from Giant Stars*

As noted in Interlude 19-3, astronomers now know that stars of all spectral types are active and have stellar winds. Consider for a moment the highly luminous, hot, blue O- and B-type stars that have by far the strongest winds. Satellite and rocket observations of their ultraviolet spectra have shown that their wind speeds often reach gale force—as much as 3000 km/s. The corresponding mass-loss rates approach and sometimes exceed 10^{-6} $M_\odot$/yr. (Recall that 1 $M_\odot$ is the Sun's mass.) This is about a tenth of the total mass of the O- or B-type star—an entire solar mass or more—carried off into space in the relatively short span of 1 million years. These powerful stellar winds hollow out vast cavities in the interstellar gas, pushing outward expanding shells of galactic matter.

Observations made by the *International Ultraviolet Explorer* satellite have shown that the pressure of hot gases in the corona is not sufficient to produce such strong winds. Instead, the winds of these luminous hot stars must be driven directly by the pressure of the ultraviolet radiation emitted by these stars. The details of the process are not well understood. Whatever is going on, it is quite complex. The ultraviolet spectra of the stars are observed to vary with time, implying that the wind is not steady. Apparently, instabilities of one kind or another are at the heart of the issue.

Observations made with radio, infrared, and optical telescopes have shown that luminous cool stars (for example, K- and M-type red giants) also lose mass at rates comparable to the luminous hot stars. Red-giant wind velocities, however, are much lower, averaging "merely" 30 km/s. They carry roughly as much mass into space as O-star winds because their densities are generally much greater. Because luminous red stars are inherently cool objects (with surface temperatures of only about 3000K), they emit virtually no ultraviolet radiation, so the mechanism driving the winds must differ from that in luminous hot stars. We can only surmise that gas turbulence or magnetic fields or both in the atmospheres of these stars are somehow responsible. The surface conditions in red giants are in some ways similar to those in Hayashi-track protostars, which are also known to exhibit strong winds. Possibly the same basic mechanism—violent surface activity—is responsible for both.

Unlike winds from hot stars, winds from these cool stars are rich in dust particles and molecules. Because nearly all stars eventually evolve into red giants, these winds provide a major source of new gas and dust to interstellar space. These stellar winds provide a vital link in the cycle of star formation and the evolution of the interstellar medium.

The accompanying figure shows imaging and spectroscopic data acquired by the *Hubble Space Telescope* toward the massive star called Melnick 42. At left the star is shown in false-color to the north of the 30 Doradus nebula in the Large Magellanic Cloud. At right is a three-dimensional graph of the star. At center is an ultraviolet spectrum showing a wide carbon-line in absorption, followed by an emission peak—a telltale sign of significant mass loss from this star.

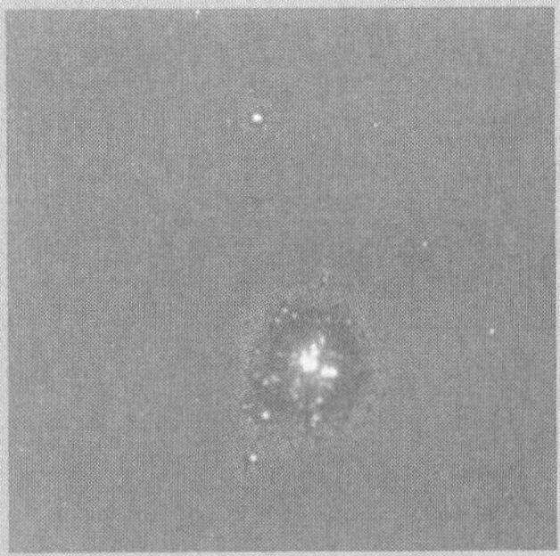

(NASA) R I V **U** X G

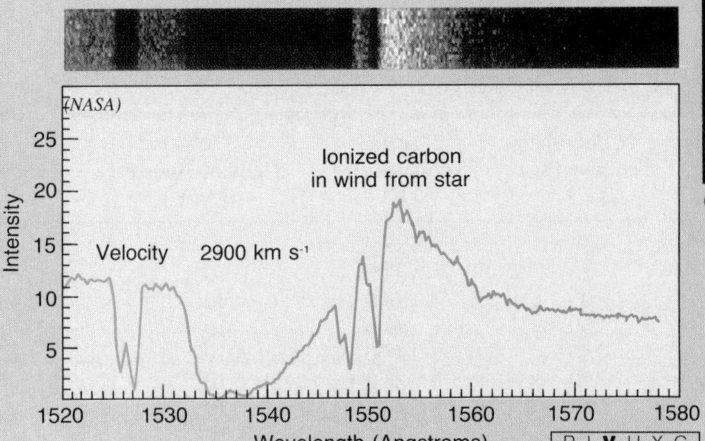

R I **V** U X G

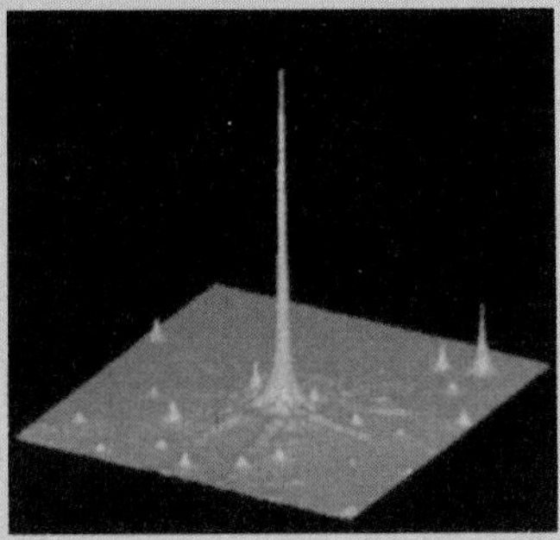

(NASA) R I V **U** X G

consumed. The triple-alpha helium-to-carbon fusion reaction—like the proton-proton and CNO-cycle hydrogen-to-helium reactions before it—proceeds at a rate that increases very rapidly with temperature. At the extremely high temperatures found in the horizontal-branch core, the helium fuel doesn't last long—no more than a few tens of million years after the initial flash.

As the helium burns, a new inner core of carbon ash forms, and phenomena familiar to us from the earlier buildup of helium ash begin to occur. Now helium becomes depleted at the very center, and eventually fusion ceases there. In response, the carbon core shrinks and heats up a little as gravity pulls it inward, causing the hydrogen and helium burning rates in the overlying layers of the core to increase. The star now contains a shrinking carbon core surrounded by a helium-burning shell, which is in turn surrounded by a hydrogen-burning shell. The outer envelope of the star expands, much as it did earlier in the first red-giant stage, and at stage 11 of Table 22-1 the star becomes a swollen red giant for a second time. Figure 22.7 depicts the star's interior structure during this time.

The star's second ascent of the giant branch is shown in Figure 22.8. To distinguish this second track from the first red-giant stage, this phase is sometimes known as the **asymptotic giant branch.** The burning rates at the center are much fiercer this time around, and the star's radius and luminosity increase to values even greater than those reached at the helium flash on the first ascent. Our star is now a **red supergiant.** The carbon core continues to shrink, which drives the hydrogen-burning and helium-burning shells to higher and higher temperatures and luminosities.

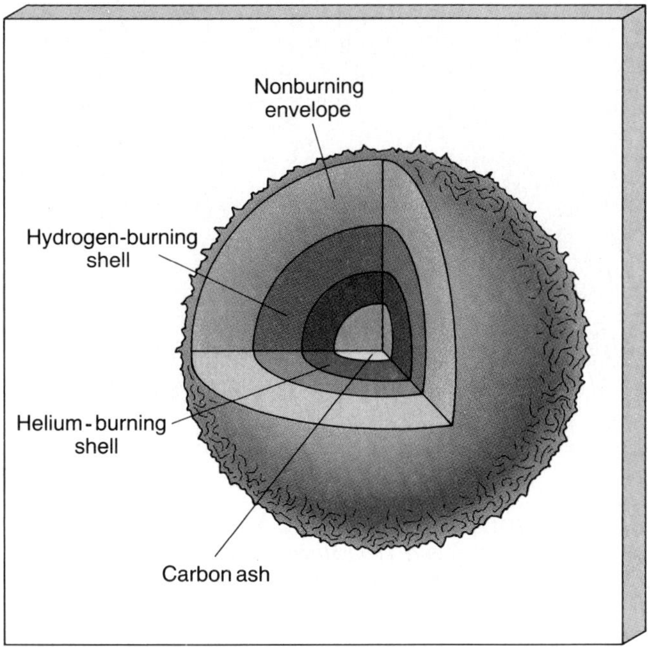

Figure 22.7 *Within a few million years after the onset of helium burning, carbon ash accumulates in the inner core of a star, above which hydrogen and helium are still burning in concentric shells.*

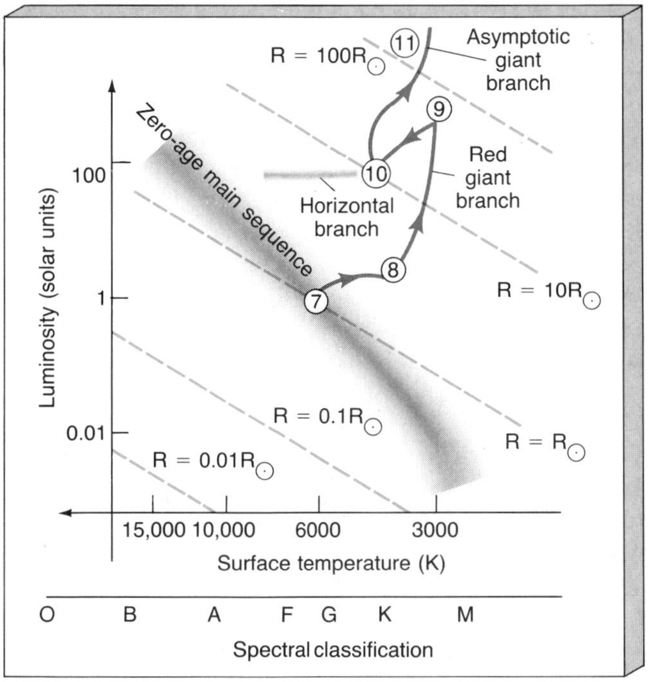

Figure 22.8 *A carbon-core star reascends the giant branch of the HR diagram—this time on a track called the asymptotic giant branch—for the same reason it evolved there the first time around: Lack of nuclear burning at the core causes contraction of the core and expansion of the overlying layers.*

If the core temperature could become high enough for the fusion of carbon nuclei, or even a mixture of carbon and helium nuclei, still heavier products could be synthesized, and the newly generated energy would again support the star, restoring for a while the previous equilibrium between gravity and heat. For the case of our solar-type star, however, this does not occur. As we will see in a moment, the temperature never reaches the point at which new nuclear reactions can occur. The red supergiant is very close to the end of its nuclear-burning lifetime.

THE DEATH OF A LOW-MASS STAR

✓3 How do stars die? To answer this question, we must rely partly on mathematical modeling and partly on what is observable in the sky. The problem is that in the four centuries since the invention of the telescope, no one has actually witnessed the death of a star in our Galaxy. Guided by theoretical predictions of how stars behave near death, astronomers search the universe for evidence of stars at or near the end of their lives.

We said at the end of the previous section that carbon nuclei could begin to burn in the core *if* the temperature were to reach a high enough value. However, the temperature necessary for carbon burning is so high—600 million K—that low-mass stars can never attain it. Here's why.

As the carbon core shrinks in response to the pull of gravity, its temperature and density increase. However, before it can attain the incredibly high carbon-ignition temperature, its density reaches a point beyond which it cannot be compressed further. At about 10^7 g/cm^3, the electrons in the core once again become degenerate. This point represents the maximum compression that the star can achieve, even in its core. There is simply not enough matter in the overlying layers to bear down any harder. At this point, the contraction of the core ceases, and its temperature stops rising.

The core density at this stage is extraordinarily high. A single cubic centimeter of core matter would weigh 1000 kg on Earth—a ton of matter compressed into a volume about the size of a grape. Yet even at these high densities, collisions among nuclei are neither frequent nor violent enough to fuse carbon into any of the heavier elements. Consequently, oxygen, nitrogen, iron, gold, uranium, and the other heavy elements are not synthesized in low-mass stars. The central fires go out after the formation of carbon.

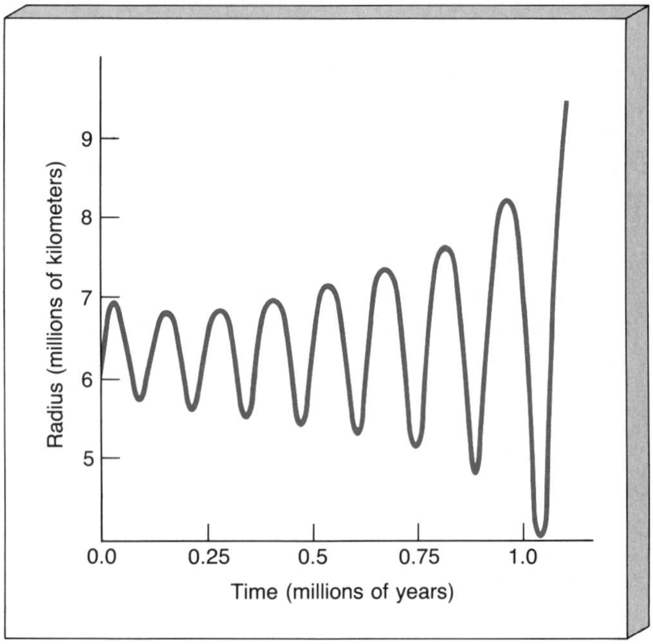

Figure 22.9 *Buffeted by helium shell flashes from within and subject to the destabilizing influence of recombination, the outer layers of a red giant become unstable and enter into a series of growing pulsations. Eventually the envelope is ejected and forms a planetary nebula.*

Planetary Nebulae

Our aged star is now in quite a predicament. Its inner carbon core is, for all intents and purposes, dead. The outer-core shells continue to burn hydrogen and helium and, as more and more of the inner core reaches its final, high-density state, the zone of nuclear burning increases in intensity. Meanwhile, the outermost layers of the star continue to expand. Eventually, the surface is so far from the central nuclear burning that it cools to the point where electrons can recombine with nuclei, forming neutral atoms once again.

Around this time, the star becomes very unstable. The helium-burning shell is subject to a series of explosive **helium shell flashes** caused by the enormous pressure there and the extreme sensitivity of the triple-alpha burning rate to small changes in temperature. These flashes produce large fluctuations in the intensity of the radiation reaching the star's outermost layers, causing them to pulsate. At the same time, these outer layers are themselves unstable. As the temperature drops to the point where electrons can recombine with nuclei to form atoms, each recombination produces additional photons, which tend to push the outer envelope to greater and greater distances from the core. As shown in Figure 22.9, the radius of the star oscillates more and more violently. To cut a long story short, the net effect of all these processes is to eject most of the envelope of the star at speeds of a few tens of kilometers per second.

In time, a rather unusual-looking object results. We say unusual because the "star" now has two distinct parts, both of which constitute stage 12 of Table 22-1. At the center, there is a small well-defined core of mostly carbon ash. Hot and dense, only the outermost part of this core still burns helium into carbon. Well beyond the core, there is a spherical shell of cooler and thinner matter spread over a volume roughly the size of our solar system—the ejected envelope of the giant. Such an object is called a **planetary nebula.** A typical example is shown in Figure 22.10. In all, some 1000 planetary nebulae are known in our Galaxy.

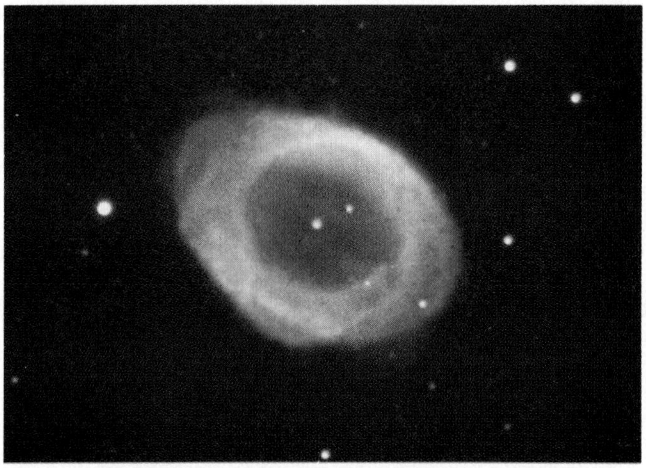

R I **V** U X G

Figure 22.10 *A planetary nebula is an object with a small dense core (central blue star) surrounded by an extended shell (or shells) of glowing matter. This one, the Ring Nebula,resides about 1500 pc away in the constellation Lyra. Its apparent size is about $\frac{1}{1000}$ that of the Moon, although it is actually much larger than our solar system. (In fact, it is about 0.2 pc in diameter.) Even so, the nebula is too dim to see well with the naked eye. All of the other stars shown are either foreground or background objects unrelated to the planetary nebula. (NOAO)*

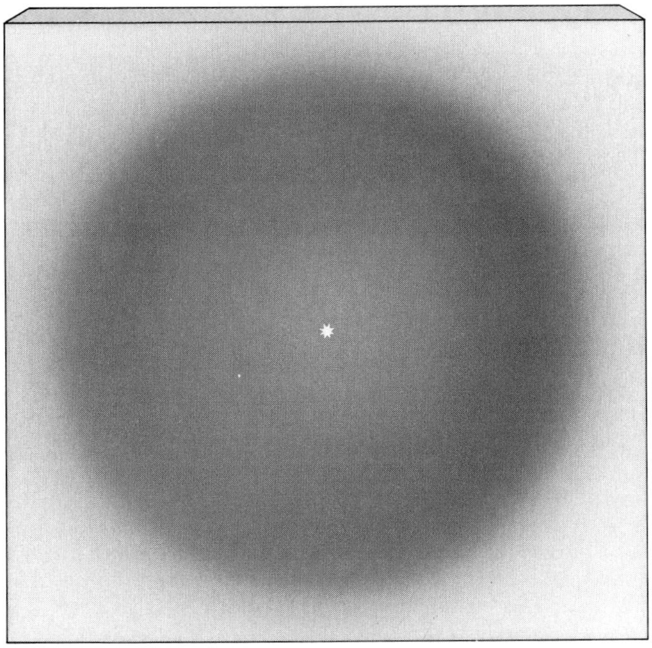
(a)

(b)

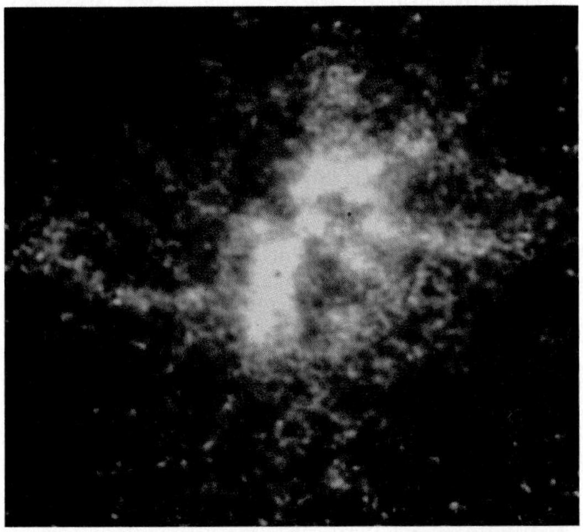
(c)

Figure 22.11 *A planetary nebula (a), with a spherical outer shell (a three-dimensional envelope), appears to the eye as a small star with a halo around it (b). The object in (b) is called the Helix Nebula and is only about 140 pc away; its apparent size in the sky is roughly half that of the full Moon. (c) Not all planetaries are symmetrical. This one (called N 66) was found in the Large Magellanic Cloud by the* Hubble Space Telescope. *(Anglo-Australian Telescope Board; NASA)*

The term *planetary nebula* is very misleading. These objects have no association with planets in any way. The name originated in the eighteenth century, when optical astronomers could barely distinguish between the myriad faint, fuzzy patches of light in the nighttime sky. With poor resolution, some of these patches did not appear as points, like stars, but instead looked more like disks—in other words, like planets. However, later observations have clearly demonstrated that the planetary nebula's fuzzy circular shape results from a shell of warm, glowing gas.

The term *nebula* is itself a little confusing because it suggests kinship with the various gaseous nebulae studied in Chapter 20. Although in some ways planetary nebulae resemble some emission nebulae, and both even undergo similar ionization-recombination processes, these two types of objects are very different. Not only are planetary nebulae much smaller than emission nebulae, they are also much older. Emission nebulae are the signposts of recent stellar birth. Planetary nebulae indicate impending stellar death.

The "ring" of the nebula is really a three-dimensional shell completely surrounding the core. Its halo-shaped appearance is only an illusion. The shell is a complete envelope that has been expelled from atop the core. It is so thin, however, that we can see it only at the edges, where the emitting matter has accumulated along our line of sight. As shown in Figures 22.11 and 22.12, the shell is virtually invisible in the direction of the core.

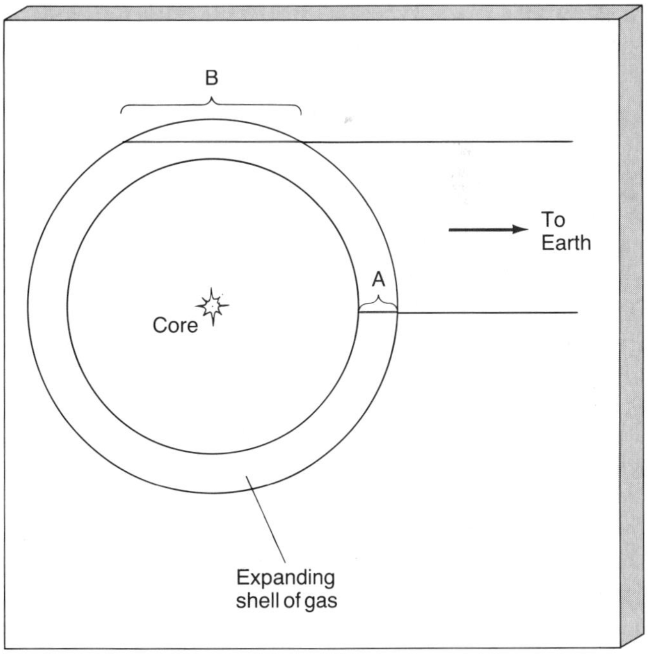

Figure 22.12 *The appearance of the planetary nebula can be explained once we realize that the shell of glowing gas around the central core is actually very thin. There is very little gas along the line of sight between the observer and the central star (path A), so that that part of the shell is invisible. Near the edge of the shell, however, there is more gas along the line of sight (path B), so the observer sees a glowing ring.*

White Dwarfs

The expanding envelope of a planetary nebula continues to spread out with time, becoming more diffuse and cooler, gradually merging with interstellar space. This is one way in which interstellar space becomes enriched with additional helium atoms and possibly some carbon atoms as well, dredged up from the depths of the core into the envelope by convection during the star's final years.

The carbon core, the stellar remnant at the center of the planetary nebula, continues to evolve. Formerly concealed by the atmosphere of the red giant star, the core appears as the envelope recedes. The core is very small. By the time the envelope is ejected as a planetary nebula, it has shrunk to about the size of the Earth. In some cases, it may be even smaller than our planet. Shining only by stored heat, not by nuclear reactions, this small star has a white-hot surface when it first becomes visible, although it appears rather dim because of its small size. The core's heat and size give rise to its new label—*white dwarf*. This is stage 13 of Table 22-1.

Part of the dashed line in Figure 22.13 depicts the star's evolutionary change from horizontal branch to red giant to white dwarf. The trek across the HR diagram between stages 11 and 13 describes the steady transformation of a large, cool red giant into a small, hot white dwarf.

Not all white dwarf stars are found as the cores of planetary nebulae (although most at one time did have a planetary nebula around them). Several hundred have been discovered "naked" in our Galaxy, their envelopes expelled to invisibility long ago. Figure 22.14 shows an example of a typical white dwarf, Sirius B. This star is a faint binary companion of the much brighter star Sirius A (see Chapter 19). Detailed observations show it to have the properties listed in Table 22-2.

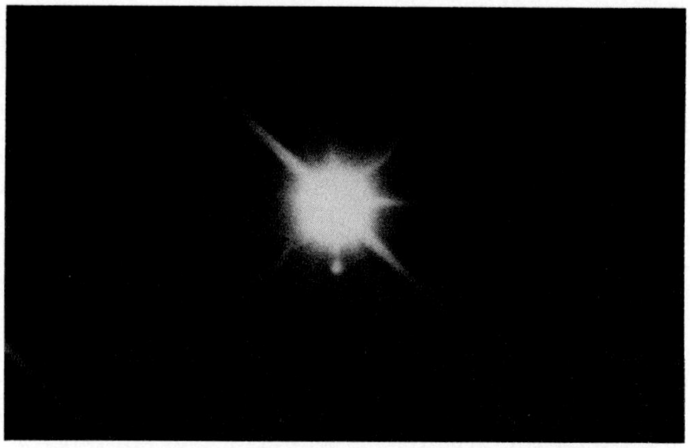

Figure 22.14 *Sirius B (the speck of light at bottom) is a white dwarf star, a companion to the much larger and brighter star Sirius A. (The hexagonal shape of the image of Sirius A is not real. The "spikes" are artifacts caused by the support struts of the telescope.) (NASA)*

Our planet, with a radius of 0.009 $R_\odot$, is actually larger than the star Sirius B. This white dwarf has more than the mass of the Sun packed into an volume smaller than the Earth. Its density is about a million times larger than anything familiar to us in the solar system.

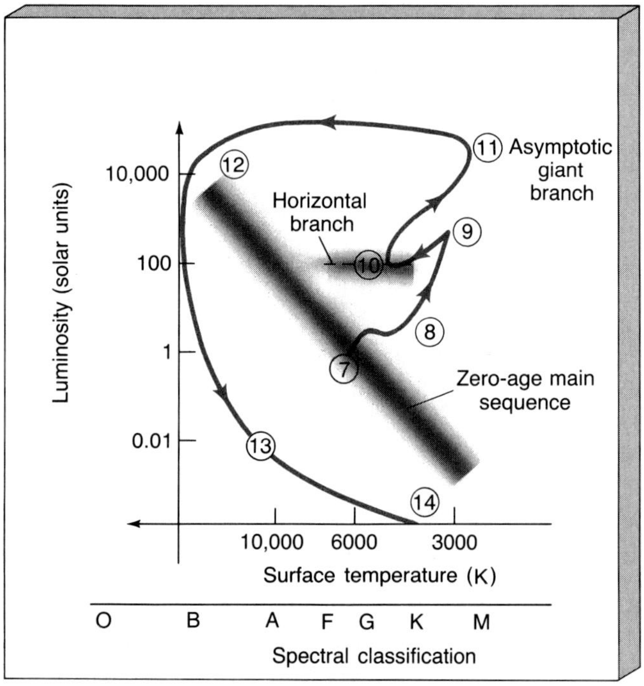

Figure 22.13 *A star's passage from the horizontal branch (stage 10) to the white dwarf stage (stage 13) by way of the asymptotic giant branch creates an evolutionary path that cuts across the entire HR diagram.*

Dark Clinkers Again

Astronomers can identify red giants, planetary nebulae, and white dwarfs in the nearby cosmos. These objects match fairly well the expectations of the theoretical calculations for aged low-mass stars. Of course, we should not expect to witness the actual expulsion of the envelope during the course of a human lifetime. Several tens of thousands of years are needed for the white dwarf to appear from behind the veil of expanding gas.

Once a star becomes a white dwarf, it is, for all practical purposes, dead. It continues to cool and dim with time,

TABLE 22-2 ***Sirius B—A Nearby White Dwarf Star****

Mass	= 1.1 $M_\odot$
Radius	= 0.008 $R_\odot$
Luminosity	= 0.002 $L_\odot$
Surface temperature	= 10,000 K
Average density	= 3×10^6 g/cm^3

*Recall that the symbol $\odot$ refers to our Sun. For example, 1 $M_\odot$ means 1 solar mass.

INTERLUDE 22-2 *Learning Astronomy from History*

Sirius A, the brighter of the two objects shown in Figure 22.14, appears twice as luminous as any other visible star, excluding the Sun. Its absolute brightness is not very great, but because its distance from us is small (less than 3 pc), its apparent brightness is very large. Sirius has been prominent in the nighttime sky since the beginning of recorded history. Ancient cuneiform texts of the Babylonians refer to the star as far back as 1000 B.C., and historians know that the star strongly influenced the agriculture and religion of the Egyptians of 3000 B.C.

Because recorded observations of this star go back several thousand years, we might have a chance to observe a slight evolutionary change in it, despite the long time thought necessary for evolution. The chances for success are improved in this case because Sirius is so bright that even the naked-eye observations of the ancients should be reasonably accurate. Interestingly, recorded history does suggest that Sirius A has changed its appearance, but the observations are confusing. Every piece of information about Sirius recorded between the years 100 B.C. and A.D. 200 claims that this star was *red.* (No earlier records of its color are known.) By contrast, modern observations now show it to be white, or bluish white, but definitely *not* red.

Sirius has apparently changed from red to blue-white in the intervening years. The problem is this: According to the theory of stellar evolution, no star should be able to change its color in this way in such a short time. Any change of this sort should take at least several tens of thousands of years, and perhaps a lot longer. It should also leave some evidence of its occurrence.

Astronomers have offered several explanations for the rather sudden change in Sirius A. These include the suggestions that (1) some ancient observers were wrong and other scribes copied them; (2) a galactic dust cloud passed between Sirius and Earth some 2000 years ago, reddening the star much as Earth's dusty atmosphere often reddens our Sun at dusk; (3) the companion to Sirius A, namely Sirius B, was a red giant and the dominant star of this double-star system 2000 years ago but has since expelled its planetary nebular shell to reveal the white dwarf star that we now observe.

Each of these explanations presents problems. How could the color of the sky's brightest star be incorrectly recorded for hundreds of years? Where is the intervening galactic cloud now? Where is the shell of the former red giant? We are left with the uneasy feeling that the sky's brightest star doesn't seem to fit particularly well into the currently accepted scenario of stellar evolution.

---

following the dashed line near the bottom of the HR diagram of Figure 22.13. Such an elderly "star" slowly transforms from a white dwarf to a yellow dwarf and then to a faint red dwarf, finally becoming a black dwarf—a cold, dense, burned-out ember in space. This is stage 14 of Table 22-1, the graveyard of stars.

The cooling dwarf does not shrink much as it fades away, however. Even though its heat is leaking away into space, gravity does not compress it further. Why not? Because at the enormously high densities in the star (from the white dwarf stage on), the resistance of electrons to being squeezed together—the same electron degeneracy that prevailed in the red-giant core around the time of the helium flash—holds the star up, even as its temperature drops almost to absolute zero. As the dwarf cools, it remains about the size of the Earth.

HIGH-MASS STARS

✓4 As we saw in Chapter 19, high-mass stars evolve much faster than their low-mass counterparts. The more massive a star, the more ravenous is its fuel consumption and the shorter is its main-sequence lifetime. For example, the Sun will spend a total of some 10 billion years on the main sequence, but a 5-solar-mass B-type star will remain there for only a few hundred million years. A 10-solar-mass O-type star will depart in only 20 million years or so. This trend toward much faster evolution for more massive stars continues even after the main sequence. All evolutionary changes happen much more rapidly for high-mass stars because their larger mass generates more heat, which speeds up all phases of stellar evolution.

Stars leave the main sequence for one basic reason: They run out of hydrogen in their cores. As a result, the early stages of stellar evolution are qualitatively the same in all cases: Main-sequence core hydrogen burning eventually gives way to the formation of a nonburning, collapsing helium core surrounded by a hydrogen-burning shell. A high-mass star leaves the main sequence on its journey toward the red-giant region with an internal structure quite similar to that of its low-mass cousin. However, its subsequent evolution soon begins to deviate from the low-mass scenario just described.

After the main sequence, two major divergences between low-mass and high-mass stars occur. First, when a high-mass star reaches the point at which helium begins to burn, its core is less dense than the core in a low-mass star. In fact, the density is so low that the core is *non*degenerate when helium fusion starts. As a result, the burning begins smoothly and stably, not explosively. There is no helium flash. The red giant remains a red giant as helium fuses into carbon. Second, the core *is* subsequently able to attain the 600 million K needed to burn carbon, so the evolution does not end with a carbon white dwarf. Instead, evolution continues smoothly as the star creates heavier and heavier elements, burning ever faster as it goes.

Large mass is the key here. Massive stars have stronger gravitational fields than solar-type stars. The added gravity ensures higher temperatures in the core, which causes more frequent and violent collisions among the particles there. Consequently, carbon nuclei can fuse to form heavier, more complex nuclei within the cores of these stars. (The heavier the nucleus, the greater its charge and the higher the temperature needed for it to fuse into creation.) Massive stars are able to produce the temperatures necessary for the fusion process to continue.

Figure 22.15 compares evolutionary tracks for several stars of different masses, from the point where they leave the main sequence to their arrival in the red-giant region. Notice how the higher-mass stars move nearly horizontally across the HR diagram from the upper main sequence to the giant branch. Their luminosities stay roughly constant as their radii increase and their surface temperatures drop. Low-mass stars ascend the giant branch almost vertically, as described earlier. Like many of the figures in this chapter, these tracks are based on computer calculations performed by Icko Iben, Jr., and coworkers at the University of Illinois.

The sudden "loops," or changes in direction of a star's motion, in Figure 22.15 are associated with the onset of new burning stages in the core. With no helium flash, there is no sudden jump to the horizontal branch and no subsequent reascent of the asymptotic giant branch. Instead, the star simply loops back and forth at the top of the HR diagram at roughly constant luminosity, creating heavier and heavier nuclear ash in its core. Some of the burning stages are marked on the figure, along with the times taken to get there from the zero-age main sequence. Notice that the most massive stars (those more massive than about 15 $M_\odot$, in fact) don't even reach the red-giant region before they start to fuse helium in their cores. They achieve a central temperature of 10^8 K while still quite close to the main sequence, and their evolutionary track continues smoothly across the HR diagram, apparently unaffected by each successive phase of burning.

With heavier and heavier elements forming at an ever-increasing rate, the high-mass stars shown in Figure 22.15 are very close to the ends of their lives. We will return to the details of their fate in Chapter 23, but suffice it to say here that they are destined to die in a violent explosion soon after carbon and oxygen begin to fuse in their cores. These stars evolve so rapidly that we are unlikely to "catch one in the act" of leaving the main sequence and traversing the HR diagram. For all practical observational purposes, high-mass stars explode and die as soon as they leave the main sequence.

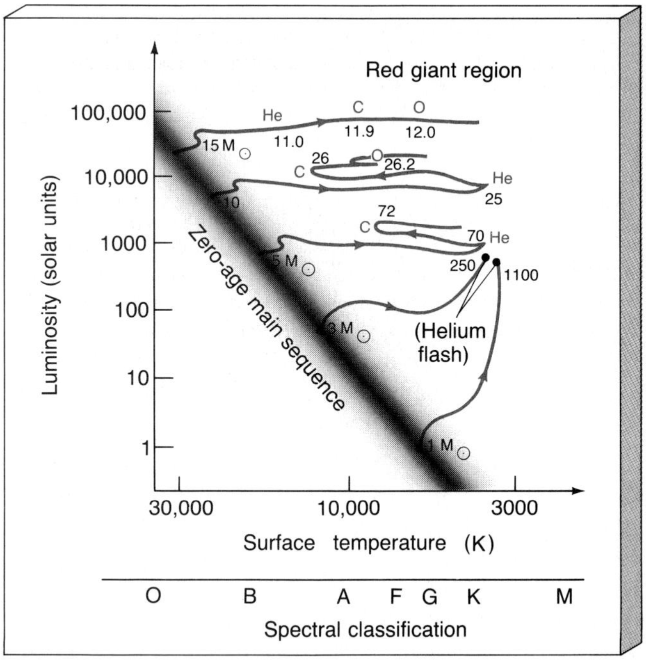

Figure 22.15 *Evolutionary tracks for stars of 1, 3, 5, 10, and 15 solar masses (shown only up to the point of the helium flash in the low-mass cases). Notice how low-mass stars ascend the giant branch almost vertically, while the higher-mass stars move roughly horizontally across the HR diagram from the main sequence into the red giant region. The most massive stars experience smooth transitions into each new burning stage. No helium flash occurs for stars more massive than about 4 solar masses. The loops in the tracks generally indicate the point at which a new burning stage begins. Some points are marked with the element that has just started to fuse in the inner core and with the time (in millions of years) at which this occurs (He = helium, C = carbon, O = oxygen).*

OBSERVING STELLAR EVOLUTION IN STAR CLUSTERS

✓5 Star clusters provide excellent test sites for the theory of stellar evolution. Every star in a given cluster formed at the same time, from the same interstellar cloud, with virtually the same composition. Only the mass varies from one star to another. This allows us to check the accuracy of our theoretical models in a fairly straightforward way. Having studied in some detail the evolutionary tracks of individual stars, we now consider how their collective appearance changes in time.

In Chapter 19, we saw how astronomers estimate the ages of star clusters by determining which of their stars have already left the main sequence. In fact, the main-sequence lifetimes that go into those age measurements represent only a tiny fraction of the data obtained from theoretical models of stellar evolution. Using the information presented in the preceding sections, it is possible, starting from the zero-age

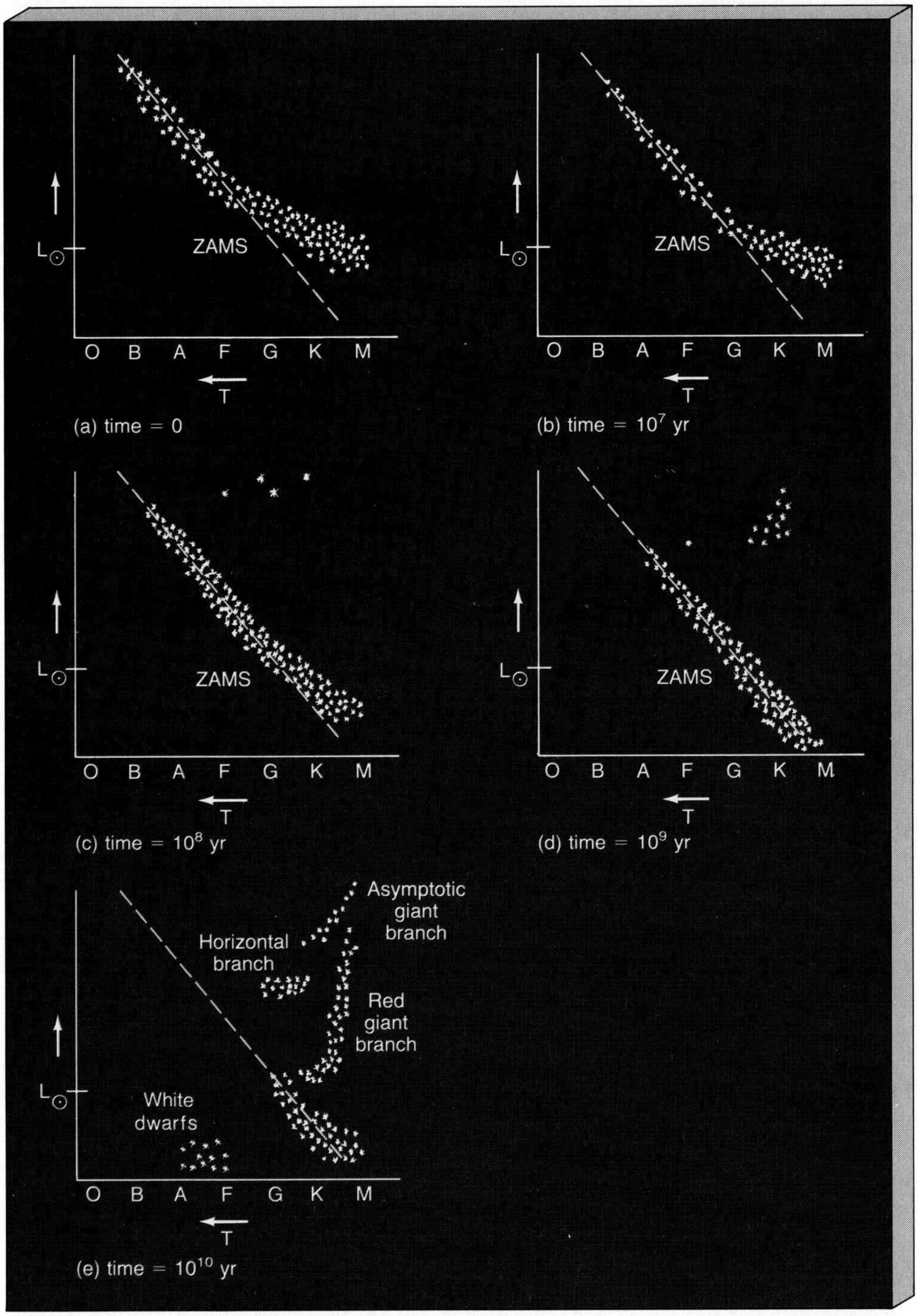

Figure 22.16 *The changing HR diagram of a hypothetical star cluster. (a) Initially, stars on the upper main sequence are already burning steadily, while the lower main sequence is still forming. (b) At 10^7 years, O-type stars have already left the main sequence, and a few red giants are visible. (c) At 10^8 years, the main sequence turnoff has moved into spectral type B. More red giants are visible, and the lower main sequence is almost fully formed. (d) At 10^9 years, the main sequence is cut off at about spectral type A. The subgiant and red-giant branches are just becoming evident, and the formation of the lower main sequence is complete. A few white dwarfs may be present. (e) At 10^{10} years, only stars less massive than the Sun still remain on the main sequence. The cluster's subgiant, red-giant, horizontal, and asymptotic giant branches are all discernible. Many white dwarfs have now formed.*

main sequence, to predict exactly how a newborn cluster should look at any later time. Ever since high-speed computers began to become available in the 1960s, this is precisely what astronomers have done. While they cannot see into the interiors of stars to test their models, they can compare stars' outward appearances with theoretical predictions. The agreement—in detail—between theory and observation is remarkably good.

To illustrate this point, let us consider the evolution of a hypothetical star cluster somewhere in the Galaxy. We begin our study shortly after the cluster's formation, with the upper main sequence already fully formed and burning steadily, and lower-mass stars just beginning to arrive on the main sequence, as shown in Figure 22.16(a). The appearance of the cluster at this early stage is dominated by its most massive stars—the bright blue supergiants. Using the evolutionary tracks described above, let us follow the cluster forward in time and ask how its HR diagram evolves.

Figure 22.16(b) shows the appearance of our cluster's HR diagram after 10 million years. The most massive O-type stars have evolved off the main sequence. Most have already exploded and vanished, as just discussed, but one or two may still be visible as red giants. The remaining cluster stars are largely unchanged in appearance—their evolution is slow enough that little happens to them in 10^7 years. The cluster's HR diagram shows the main sequence slightly cut

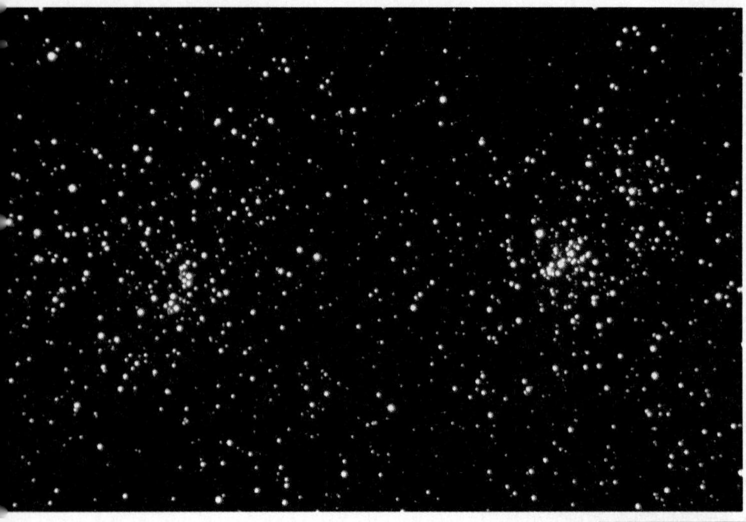

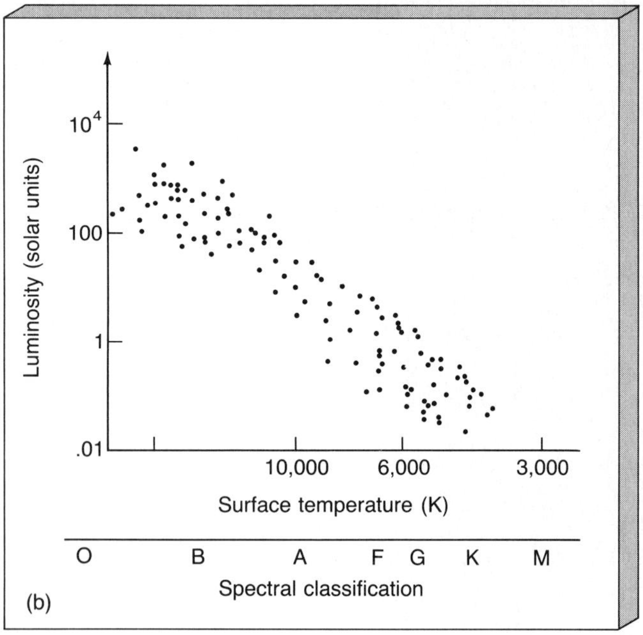

Figure 22.17 *(a) The "double cluster" h and χ Persei. (b) The HR diagram of the pair indicates that they are very young—probably only about 10 million years old. (NOAO)*

off, along with a rather poorly defined red-giant region. Figure 22.17 shows the twin open clusters h and χ (the Greek letter chi) Persei, along with their combined HR diagram. Comparing this figure with diagrams like Figure 22.16(a), astronomers estimate the age of this pair of clusters to be about 10 million years.

At any time during the evolution, the cluster's original main sequence is intact up to some well-defined stellar mass, corresponding to the stars that are just leaving the main sequence at that instant. We can imagine the main sequence being "peeled away" from the top down, with fainter and fainter stars turning off and heading for the giant branch as time goes on. Astronomers refer to the high-luminosity end of the observed main sequence as the **main-sequence turnoff.** The mass of the star that is just evolving off the main sequence at any moment is known as the **turnoff mass.**

After 100 million years (Figure 22.16[c]), stars brighter than type B5 or so (about 4-5 $M_{\odot}$) have left the main sequence, and a few more red supergiants are visible. By this time, most of the cluster's low-mass stars have finally arrived on the main sequence, although the dimmest M stars may still be in their contraction phase. The appearance of the cluster is now dominated by bright B stars and brighter red giants.

At 1 billion years, the main sequence turnoff mass is around 2 $M_{\odot}$, corresponding roughly to spectral type A2. The subgiant and giant branches associated with the evolution of low-mass stars are just becoming visible, as indicated in Figure 22.16(d). The formation of the lower main sequence is now complete. In addition, the first white dwarfs have just appeared, although they are generally too faint to be observed at the distances of most clusters. Figure 22.18 shows the Hyades open cluster, again with its HR diagram alongside. Its similarity to Figure 22.16(d) suggests its age is about 5×10^8 years.

At 10 billion years, the turnoff point has reached solar-mass stars, of spectral type G2. The subgiant and giant branches are now clearly discernible (see Figure 22.16[e]), and the horizontal and asymptotic giant branches appear as distinct regions in the HR diagram. Many white dwarfs are also present in the cluster. Although stars in all these evolutionary stages are also present in the 1-billion-year-old cluster shown in Figure 22.16(d), they are few in number—typically only a few percent of the total number of stars in the cluster. Also, because they evolve so rapidly, they spend very little time in these regions. Lower-mass stars are much more numerous and evolve more slowly, so their evolutionary tracks are more easily detected.

Figure 22.19 shows the globular cluster 47 Tucanae. By carefully adjusting their theoretical models until the cluster's main sequence, subgiant, red-giant, and horizontal branches are all well matched, astronomers have determined its age to be roughly 14 billion years, a little older than our hypothetical cluster in Figure 22.16(e). In fact, as mentioned in Chapter 19, globular cluster ages determined this way show a remarkably small spread. All the globular clusters in our Galaxy appear to have formed between about 12 and 15 billion years ago.

Cluster age measurements allow us to chart not only the development of stars, but also the evolution of the interstellar medium. Stellar evolution theory indicates that a star's outer layers remain largely unchanged in composition from its birth almost up to the moment the star dies. By analyzing the spectra of stars of different (but known) ages, then, we are in effect probing what the composition of the interstellar medium was in the distant past (at the time of the stars' births). In this way, the steady buildup of heavy elements in the Galaxy can be studied. Just how this buildup occurs will be the subject of the next chapter.

R I **V** U X G

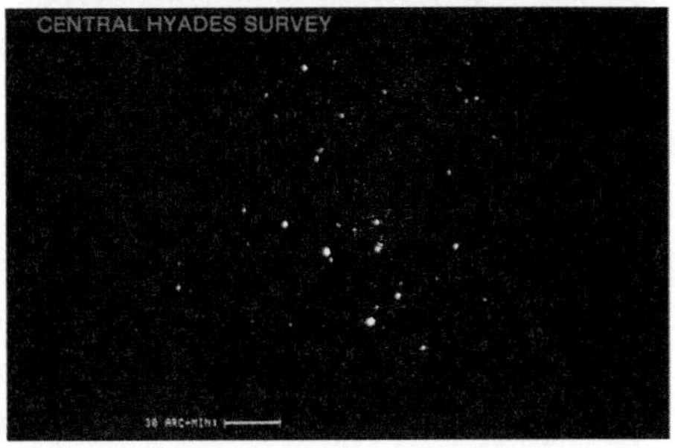

(a) R I V U **X** G

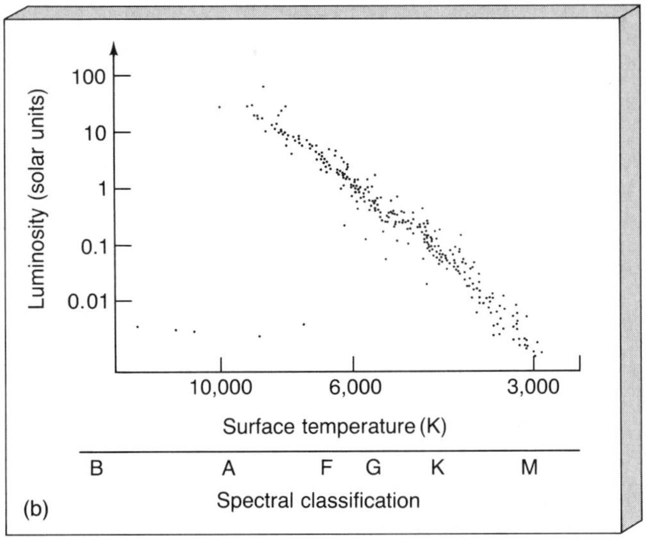

Figure 22.18 *(a) The Hyades cluster, a relatively young group of stars visible to the naked eye. The insert below is the same region, imaged in x-ray. (b) Its HR diagram is cut off at about spectral type A0, implying an age of about 500 million years. (Harvard-Smithsonian Center for Astrophysics)*

R I **V** U X G

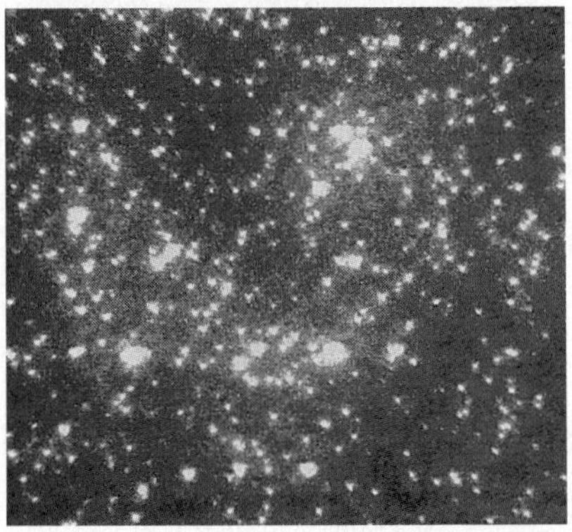

(a) R I V **U** X G

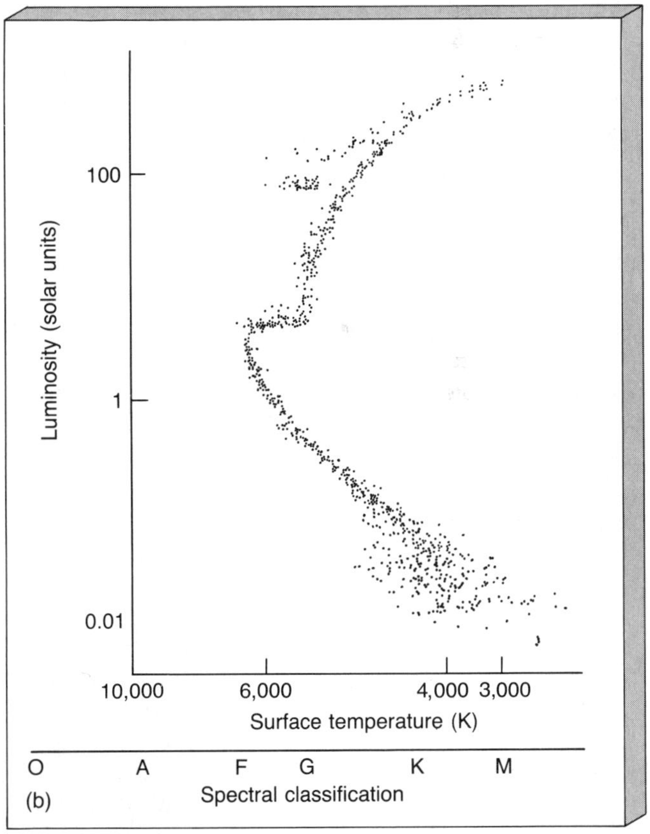

Figure 22.19 *(a) The southern globular cluster 47 Tucanae. (b) Fitting its main-sequence turnoff and its giant and horizontal branches to theoretical models gives 47 Tucanae an age of about 14 billion years, making this one of the oldest known objects in the Milky Way Galaxy. The insert is a high-resolution ultraviolet image of 47 Tucanae's core region, taken with the* Hubble Space Telescope *and showing unexpected evidence for "blue stragglers"—massive stars resulting perhaps from a merging of binary star systems—that lie on the main sequence above the cutoff point. (European Southern Observatory; NASA; DAO)*

Stellar evolution is one of the great success stories of astrophysics. Like all good scientific theories, it makes definite testable predictions about the universe, while at the same time remaining flexible enough to incorporate new discoveries as they occur. Theory and observation have advanced hand in hand. At the start of the twentieth century, many scientists despaired of ever knowing even the compositions of the stars, let alone why they shine and how they change. Today, the theory of stellar evolution is a cornerstone of modern astronomy.

THE EVOLUTION OF BINARY-STAR SYSTEMS

✓6 We have noted that most stars in our Galaxy are not isolated objects, but are actually members of binary-star systems. However, our discussion of stellar evolution has so far focused exclusively on isolated stars. This prompts us to ask the following obvious question. How does membership in a binary-star system change the evolutionary tracks we have just described? Since nuclear burning occurs deep in the core, does the presence of a stellar companion have any significant effect at all? Not surprisingly, the answer depends on the distance between the two stars in question.

For a binary system whose component stars are very widely separated—by which we mean that the distance between the stars is greater than perhaps a thousand stellar radii—the answer is that the two stars evolve more or less independently of one another, each following the track for an isolated star of its particular mass. However, if the two stars are closer, the gravitational pull of one may strongly influence the envelope of the other, and the physical properties of both may deviate greatly from those calculated for isolated single stars.

As an example, consider the star Algol (also known as Beta Persei, the second brightest star in the constellation Perseus). By studying its spectrum and the variation in its light intensity, astronomers have determined that Algol is actually a binary (in fact, an eclipsing double-lined spectroscopic binary—see Chapter 19), and they have measured its properties very accurately. Algol consists of a 3.7 $M_\odot$ main-sequence star of spectral type B8 with a 0.8 $M_\odot$ subgiant companion moving in a circular orbit around it. The stars are 4 million km apart, and the subgiant companion has an orbital period of about 3 days.

A moment's thought reveals that there is something odd about these findings. On the basis of our earlier discussion, the more massive main-sequence star should have evolved *faster* than the less massive component. If the two stars formed at the same time (as is assumed to be the case), there should be no way that the 0.8 $M_\odot$ star could have reached the giant stage first. Either our theory of stellar evolution is seriously in error or something has modified the evolution of the Algol system. Fortunately for theorists, the latter is the case.

To understand Algol, we must consider binary systems in a little more detail. As sketched in Figure 22.20, each star is surrounded by its own tear-shaped "zone of influence," inside of which its gravitational pull dominates the effects of both the other star and the overall rotation of the binary. Any matter within that region "belongs" to the star. It cannot easily flow onto the other component or out of the system. Outside the two regions, it is possible for gas to flow toward either star relatively easily. The two tear-shaped regions are usually called **Roche lobes,** after Edouard Roche, the French mathematician who first studied the problem in the nineteenth century and whose work we have already encountered in the context of planetary rings (see Chapter 14). The Roche lobes of the two stars meet at a point on the line joining them—the inner Lagrange point (L_1), which we discussed in Chapter 16 when discussing

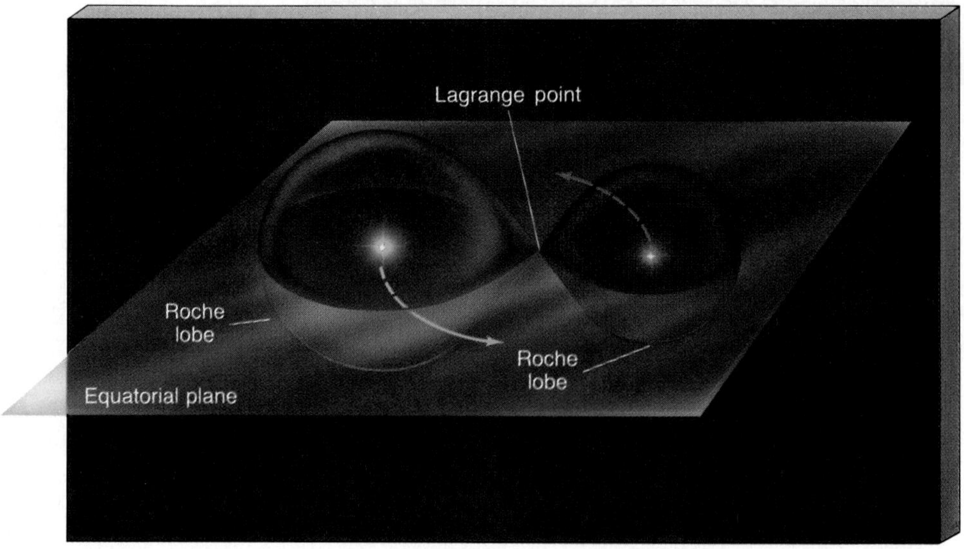

Figure 22.20 Each star in a binary system can be pictured as being surrounded by a "zone of influence," or Roche lobe, inside of which matter may be thought of as being "part" of that star. The two teardrop-shaped Roche lobes meet at the Lagrange point between the two stars. Outside the Roche lobes, matter may flow onto either star with relative ease.

asteroid motions in the solar system. This Lagrange point is a place where the gravitational pulls of the two stars exactly balance the rotation of the binary system. The greater the mass of one component, the larger is its Roche lobe and the closer is the Lagrange point to the other star.

Normally, the two stars each lie well within their respective Roche lobes, and such a binary system is said to be **detached,** as in Figure 22.21(a). However, as a star evolves off the main sequence and moves toward the giant branch, it is possible for its radius to become so large that it overflows its Roche lobe. Its gas begins to flow onto the companion through the Lagrange point. The binary in this case is said to be **semi-detached** (Figure 22.21[b]). Because matter is flowing from one star onto the other, semi-detached binaries are also known as **mass-transfer binaries.** If, for some reason, the other star also overflows its Roche lobe (either because of stellar evolution or because so much extra material is dumped onto it), the surfaces of the two stars merge, and a new configuration results. The binary system then consists of two nuclear-burning stellar cores surrounded by a single continuous envelope—a **common-envelope binary,** or **contact binary,** shown in Figure 22.21(c).

In a very close binary system, neither star has to evolve far off the main sequence before it overflows its Roche lobe and mass transfer begins. In a wide binary, both stars may evolve all the way up the giant branch without either surface ever reaching the Lagrange point, and they evolve just as though they were isolated. Depending on the stars involved and their orbital separations, there are many different possibilities for the eventual outcome of the evolution. Let's make these ideas more definite by returning to the question of how the binary star Algol reached its present state.

Astronomers believe that Algol started off as a detached binary. For reference, let us label the component that is now the 0.8 $M_\odot$ subgiant as star 1 and the star that is now on the main sequence, with mass 3.5 $M_\odot$, as star 2. Initially, star 1 was the more massive of the two—perhaps 3 $M_\odot$ or so—so that it evolved off the main sequence first. As it ascended the giant branch, it overflowed its Roche lobe, and gas began to flow onto star 2. This had the effect of reducing the mass of 1 and increasing that of 2, which in turn caused the Roche lobe of star 1 to shrink as its gravity decreased. As a result, the rate at which 1 overflowed its Roche lobe increased, and a period of *rapid mass transfer* ensued, transplanting most of star 1's envelope onto star 2. Eventually, the mass of star 1 became less than that of star 2. Detailed calculations show that the rate of mass transfer dropped sharply at that point, and the stars entered the relatively stable state we see today. These changes in Algol's components are illustrated in Figure 22.22.

We see that being part of a binary system has radically altered the evolution of both stars in the Algol system. The original high-mass star 1 is now a low-mass red giant, while the roughly solar-type star 2 is now a massive main-sequence B star. The removal of mass from the envelope of star 1 may prevent it from ever reaching the helium flash. Instead, its naked core may eventually be left behind as a *helium white dwarf.* In a few tens of millions of years, star 2 will itself begin to ascend the giant branch and fill its own Roche lobe. If star 1 is still a giant at that time, a contact binary system will result. If star 1 is not a giant but a white dwarf instead, a new mass-transferring period—with matter streaming from star 2 back onto star 1—will begin. In that case, as we will see, Algol may have a very active and violent future in store.

Just as molecules exhibit few of the physical or chem-

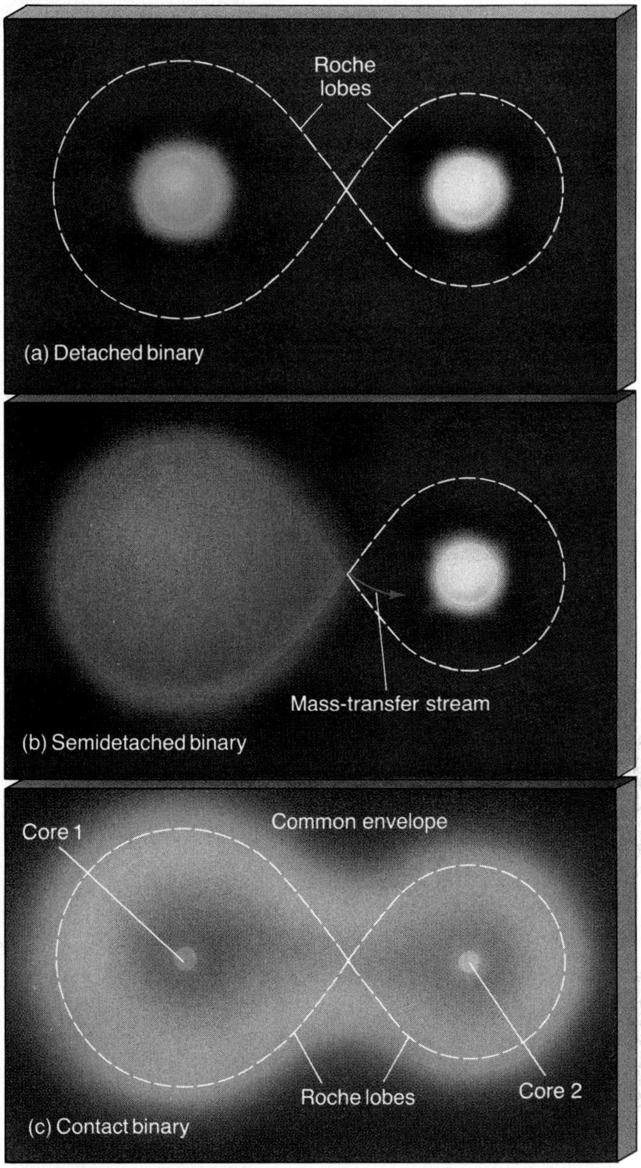

Figure 22.21 *(a) In a detached binary, each star lies within its respective Roche lobe. (b) In a semi-detached binary, one of the stars fills its Roche lobe and transfers matter onto the other, which still lies within its own Roche lobe. (c) In a contact, or common-envelope, binary, both stars have overflowed their Roche lobes, and a single star with two distinct nuclear-burning cores results.*

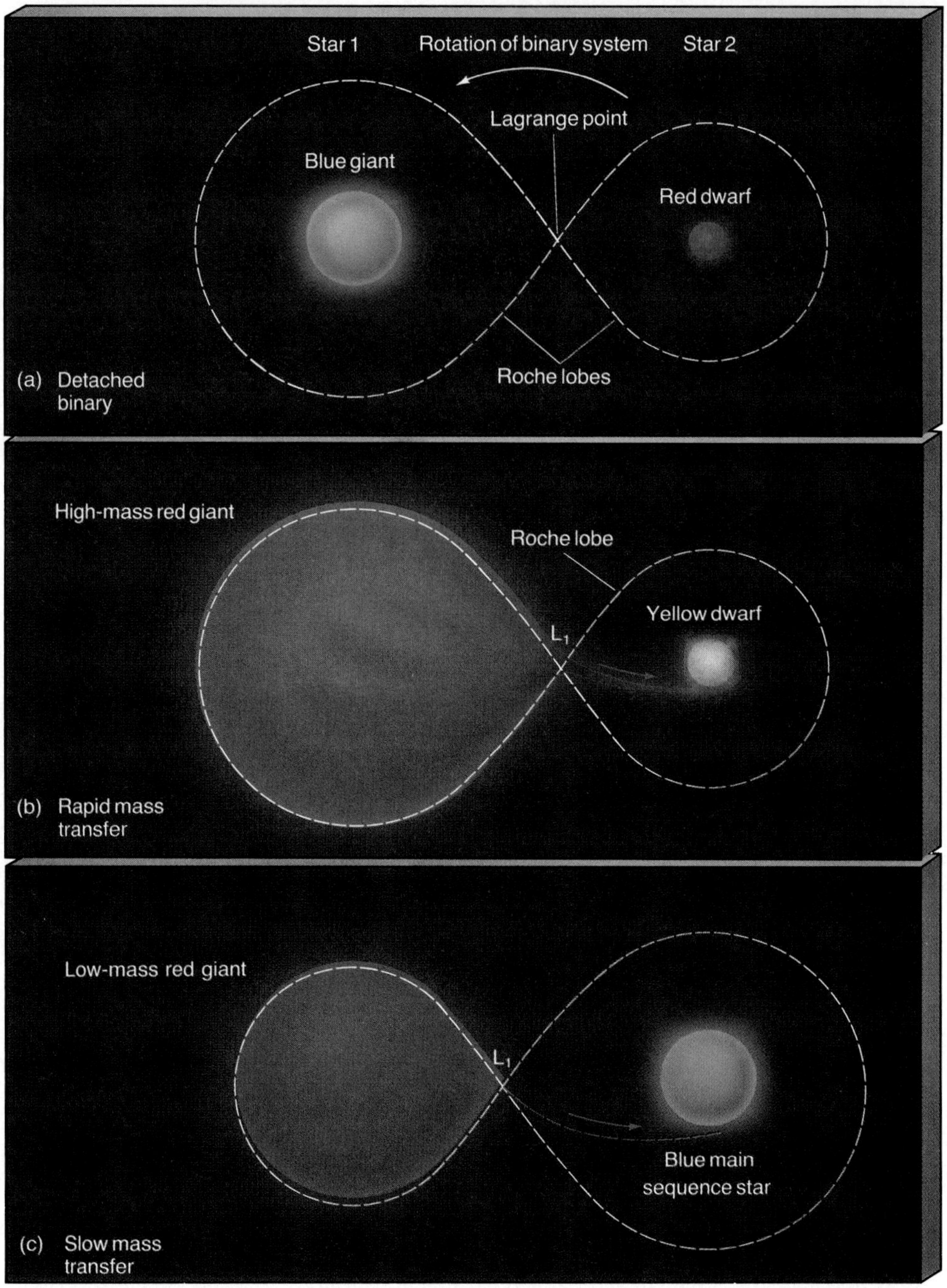

Figure 22.22 *The evolution of the binary star Algol. (a) Initially, Algol was a detached binary, made up of two main-sequence stars. (b) As the more massive component (star 1) evolved off the main sequence, it expanded to fill and eventually overflow its Roche lobe, transferring large amounts of matter onto its smaller companion (star 2). (c) Today, star 2 is the more massive of the two, but it is still on the main sequence. Star 1 is still in the giant phase and fills its Roche lobe, causing a steady stream of matter to pour onto its companion.*

ical properties of their constituent atoms, binaries can display types of behavior that are quite different from either of their component stars. The Algol system is a fairly simple example of binary evolution, yet it gives us an idea of the sorts of complications that can arise when two stars evolve in an interdependent way. A substantial fraction of all the binary stars in the Galaxy will pass through some sort of mass-transfer or common-envelope phase. In this chapter, we have seen one possible result of mass transfer involving main-sequence stars. We will return to this subject in the next two chapters, when we continue our discussion of stellar evolution and the strange states of matter that may result.

CHAPTER SUMMARY

Stars spend most of their lives on the main sequence. As a star runs out of fuel and begins to die, its properties change greatly. Of all the low-mass stars that ever formed, nearly all still exist. Most massive stars that ever existed perished long ago. Some intermediate stars that have evolved away from the main sequence can be observed in advanced stages of stellar evolution.

In stars, the force of gravity is balanced by gas pressure created in the stellar core. While this equilibrium is maintained, hydrogen fuel slowly changes into helium, and the star burns steadily. When the star runs out of hydrogen in its core, the lack of nuclear burning at the core results in weaker gas pressure against the force of gravity. The core of nonburning helium ash contracts, releasing gravitational energy. This energy in turn heats overlying layers, where hydrogen burning increases, forcing the outer layers of the star to expand. For a star like the Sun, this change from a main-sequence star to a red giant takes about 100 million years.

In low-mass stars, by the time the core temperature has risen high enough for helium fusion to occur, the density is so great that electron degeneracy pressure dominates. Under these conditions, the onset of helium burning causes a runaway explosion—a helium flash—that expands the core, bringing about stable helium burning. The star's position on the horizontal branch marks the point where it resides once helium burning becomes stabilized. Its location there is determined largely by its mass. After remaining on the horizontal branch for a time, the star reascends the giant branch, now with a shrinking carbon core surrounded by a helium-burning shell and a hydrogen-burning shell. The star is now a red supergiant.

The giant's bloated outer layers eventually become unstable and drift off into space. The glowing remains of the giant envelope are known as a planetary nebula. The star's small, hot core is left behind as a white dwarf. As the white dwarf cools, it evolves into a yellow dwarf, then a red dwarf, and finally a black dwarf, a cold object held up purely by the pressure of degenerate electrons. Our Galaxy has not existed long enough for many low-mass stars to have died and cooled to this final state.

Evolutionary changes happen more rapidly for high-mass stars because their large mass enables them to generate more heat. The core of a high-mass red giant can reach temperatures high enough to fuse heavier elements. Each new burning stage moves toward the production of an iron core. Nuclear fusion cannot continue in a core of iron, and the stellar core collapses in seconds at this point, reaching nuclear densities and rebounding violently. A shock wave blasts through the star and into space in one of the most energetic events known in the universe—a supernova.

In a close binary system, a star evolving off the main sequence can overflow its Roche lobe and begin mass transfer, greatly influencing the evolution of both components. In a wide binary, both stars can evolve as though they were isolated.

KEY WORDS

alpha particle
asymptotic giant branch
common-envelope binary
contact binary
core hydrogen burning
detached binary
electron degeneracy
electron degeneracy pressure
helium flash
helium shell flash
horizontal branch
hydrogen shell burning
main-sequence turnoff
mass-transfer binary
Pauli exclusion principle
planetary nebula
red-giant branch
red supergiant
Roche lobe
semi-detached binary
subgiant branch
triple-alpha process
turnoff mass

REVIEW QUESTIONS

1. What types of stars live the longest, and why?
2. What is main-sequence equilibrium?
3. Why don't stars live forever?
4. How long can a star like our Sun keep burning hydrogen in its core?
5. Why is the depletion of hydrogen at the core of a star such an important event?
6. What makes an ordinary star become a red giant?
7. How big is a red giant?
8. How long does it take for a star like the Sun to evolve from the main sequence to the top of the red-giant branch?
9. What is the helium flash?
10. What is a red supergiant?
11. Describe an important way in which winds from red giant stars are linked to the interstellar medium.
12. How do stars of low mass die? How do stars of high mass die?
13. What are yellow dwarfs, red dwarfs, and black dwarfs?
14. What is the ultimate fate of a black dwarf?
15. Do many black dwarfs exist in the Galaxy?
16. What happens when the inner core of a massive star begins to change into iron?
17. What determines whether the stars in a binary system will influence one another's evolution?
18. The main-sequence lifetime of the Sun is about 10^{10} years. What would be the lifetimes of (a) a 10-solar mass, 1000-solar luminosity blue giant, and (b) an 0.1-

solar mass, 0.001-solar luminosity red dwarf? Recall that the lifetime of a star is proportional to its mass divided by its luminosity—that is, M/L.

19. (a) Use the radius-luminosity-temperature relation ($L \propto R^2T^4$; see Chapter 19) to calculate the radius of a red supergiant with temperature 3000K (one-half the solar value) and luminosity 10,000 solar luminosities. How many planets of our solar system would this star engulf? (b) Repeat your calculation for a 12,000K (twice the temperature of the Sun), 0.0004 solar luminosity white dwarf. To which terrestrial planet is this star closest in size?

20. The Sun will reside on the main sequence for 10^{10} years. If the luminosity of a main-sequence star is roughly proportional to the cube of the star's mass, what mass star is just now leaving the main sequence in a cluster that formed 400 million years ago?

DISCUSSION QUESTIONS

1. Suppose you were not limited by the speed of light in your ability to travel in space. What sorts of stars and other objects might you see as you journeyed through our Milky Way Galaxy? How many, very roughly, of each type would you see?

2. Suppose a new technology were developed that enabled astronomers to discover huge numbers of black dwarfs in our Galaxy. Which astronomical theories would be affected by this discovery? Say how they would be affected.

PROJECT

1. You can tour the Galaxy without ever leaving Earth, just by looking up. In the winter sky, you'll find the red supergiant Betelgeuse in the constellation Orion. It's easy to see because it's one of the brightest stars visible in our night sky. Betelgeuse is a variable star, with a period of about 6.5 years. Its brightness changes as the star expands and contracts. At its maximum size, Betelgeuse fills a volume of space that—if centered on our Sun—would extend beyond the orbit of Jupiter. Betelgeuse is thought to be about 10 to 15 times more massive than our Sun. It is probably between 4 and 10 million years old—and in the final stages of its evolution. A similar star can be found shining prominently in midsummer. This is the red supergiant Antares in the constellation Scorpius. Depending on the time of year, can you find one of these stars? Why are these stars red? What will happen to them next?

SUGGESTED READINGS

Balick, B. ''The Shaping of Planetary Nebulae.'' *Sky and Telescope* (February 1987). New images help explain the varied and intricate forms of planetary nebulae.

Kaler, J. B. ''Journeys on the H-R Diagram.'' *Sky and Telescope* (May 1988). How stars may change dramatically as they age, taking on an amazing variety of characters.

Kawaler, S. D., and D. E. Winget. ''White Dwarfs: Fossil Stars.'' *Sky and Telescope* (August 1987). Why many stars will end their lives as crystals.

MacRobert, A. ''Epsilon Aurigae.'' *Sky and Telescope* (January 1988). New light on a long-mysterious star.

Schorn, R. A. ''Goodbye to Supermassive Stars.'' *Sky and Telescope* (January 1986). A star thousands of times heavier than the Sun was suspected to lie in the heart of the Tarantula Nebula. But it's not there.

Tomkin, J., and D. L. Lambert. ''The Strange Case of Beta Lyrae.'' *Sky and Telescope* (October 1987). This familiar eclipsing binary star offers a rare glimpse of an exposed stellar core.

Trimble, V. ''White Dwarfs: The Once and Future Suns.'' *Sky and Telescope* (October 1986). White dwarf stars.

CHAPTER

23

Stellar Explosions

Novae, Supernovae, and the Formation of the Heavy Elements

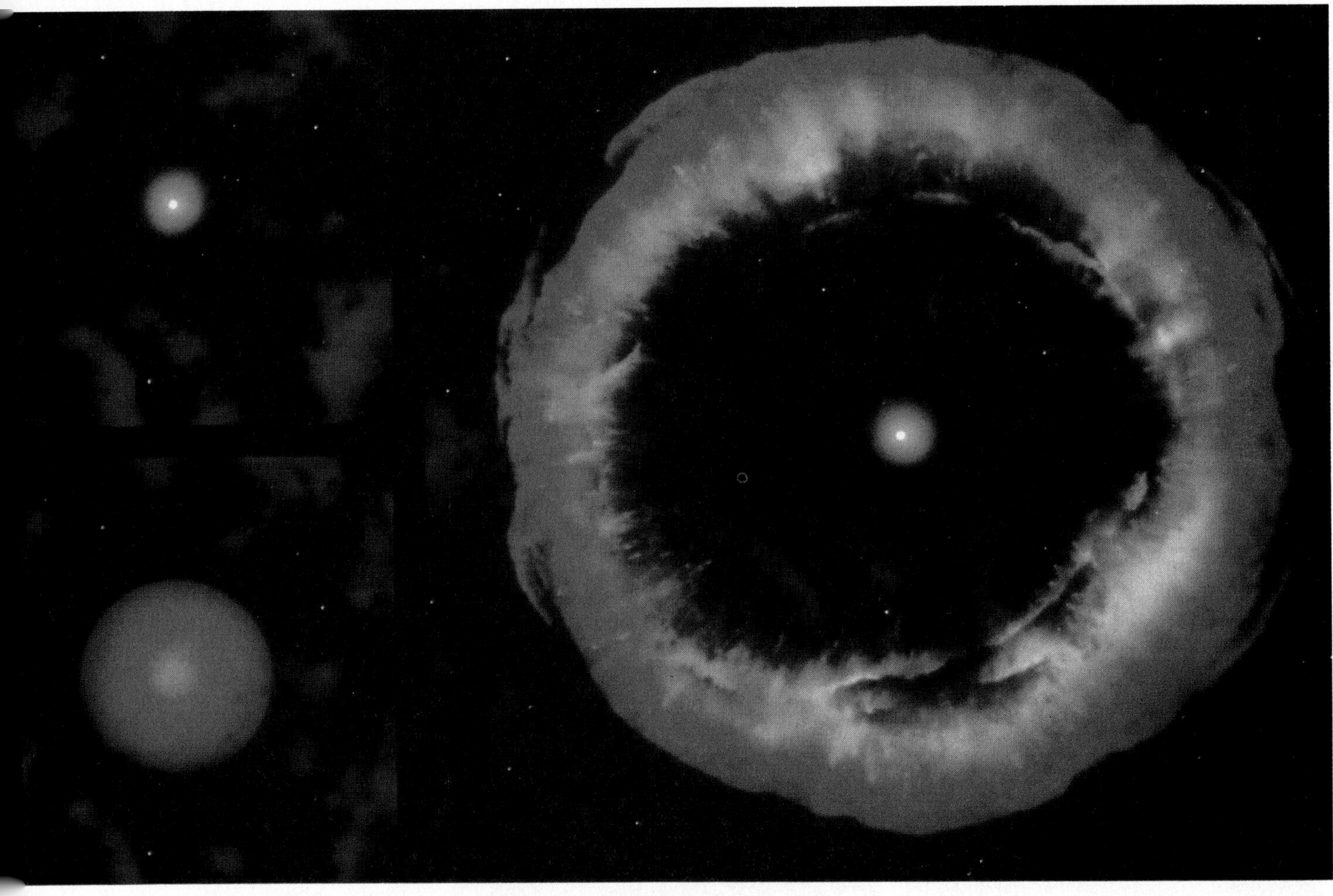

This is an artist's conception of the evolution of a nova. The model in the sky on which this drawing is based is known technically as V605 Aquiae and colloquially as the "Lazarus Star." Among novae, this one is rather strange, for it might well be self-induced by a single white-dwarf star. In this three-frame illustration, the dwarf itself is initially seen in the top left. At the bottom left, the dwarf has swelled back into the red-giant stage. At right, after the star gravitationally collapsed to a dwarf—for the second time—the still-expanding outer atmosphere remains as a planetary nebula. (Courtesy D. Berry)

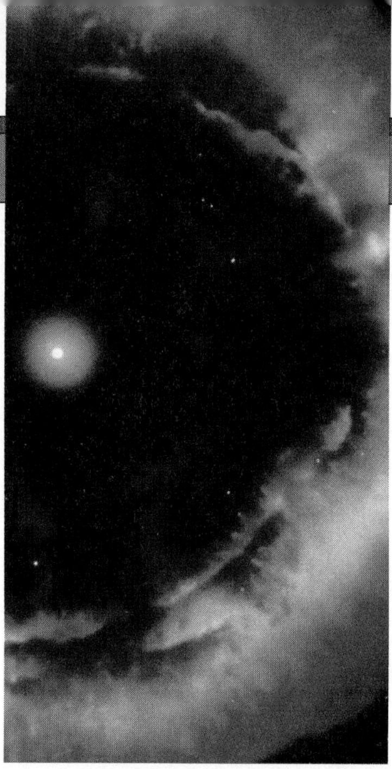

LEARNING GOALS

1 to understand how white dwarfs in binary systems can become explosively active as novae or supernovae

2 to see the sequence of events leading to the explosive death of a massive star

3 to recognize that there are two distinct types of supernova, triggered in very different ways

4 to realize that of all the common elements, only hydrogen and helium existed in the early universe—the rest formed in stars

5 to understand how the heavy elements formed during supernova explosions

In the previous chapter, we saw how low-mass stars evolve off the main sequence, pass through the red-giant stage, and eventually die. The white-dwarf stage is not necessarily the end of the story for all such stars, however. The potential exists for further violent activity if a neighboring star can provide additional fuel.

We have also seen how massive stars race through all of their evolutionary stages, creating heavier and heavier elements in their cores at an ever-increasing rate. What fate awaits them when they run out of fuel? In contrast to the gentle death of their low-mass companions, these stars are destined to die explosively, releasing vast amounts of energy and scattering their matter throughout interstellar space. The explosion leads to the creation of many heavy elements and their expulsion into the interstellar medium. It may also trigger the formation of new stars, continuing the cycle of stellar birth and death.

In this chapter we will study in more detail the processes leading to these stellar explosions and the mechanisms that create heavy elements during all phases of stellar evolution.

NOVAE: LIFE AFTER DEATH FOR WHITE DWARFS

✓1 A **nova** is a star that dramatically increases in brightness, often by as much as a factor of 10,000, in a very short period of time. The word *nova* means "new" in Latin, and to ancient observers these stars did indeed seem new, as they suddenly appeared in the night sky. Astronomers now recognize that a nova is not really a new star at all. It is instead a white dwarf—a faint star—undergoing an explosion on its surface that results in a rapid, temporary increase in luminosity. Figure 23.1 illustrates the brightening of a

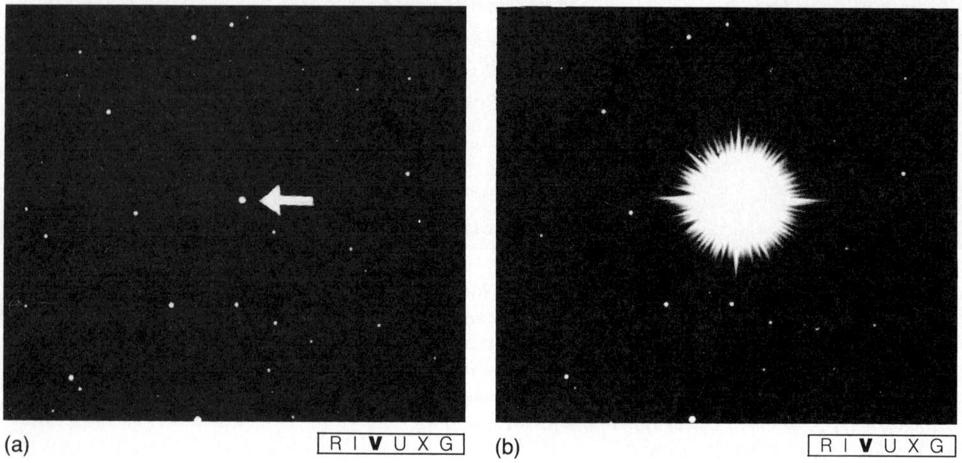

Figure 23.1 *A nova is a star that suddenly increases enormously in brightness, then slowly fades back to its original luminosity. Novae are the result of explosions on the surfaces of faint white dwarf stars, caused by matter falling onto their surfaces from the atmosphere of a larger binary companion. Shown is Nova Herculis 1934, in (a) March, 1935, and (b) in May, 1935, after brightening by a factor of 60,000. (Lick Observatory)*

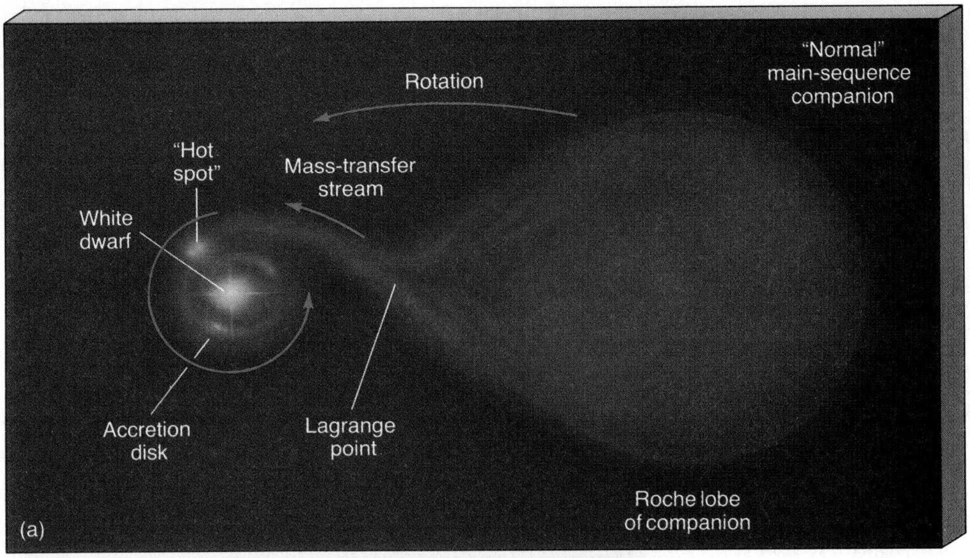

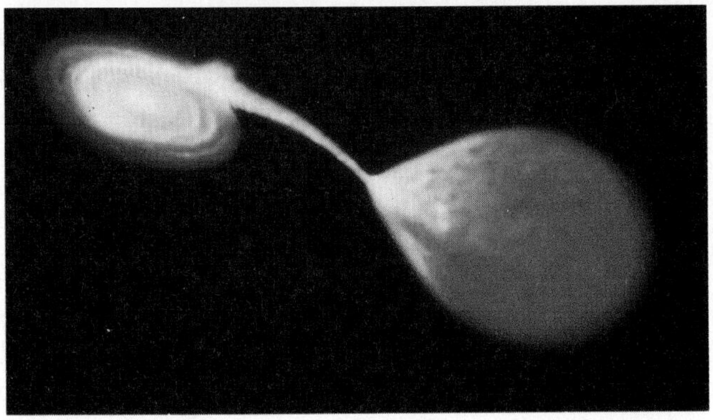
(b)

Figure 23.2 *A white dwarf in a binary system may be close enough to its companion that its gravitational field can tear material from the companion's surface. (a) A diagram representing the various processes occurring. (b) An artist's conception of such interacting binaries. (D. Berry)*

typical nova over a period of 3 days. Novae eventually fade back to normal, usually after a few weeks or months. On average, two or three novae are observed each year. Astronomers also know of many **recurrent novae**—stars that have been observed to "go nova" several times over the course of a few decades.

What could cause such an explosion on a faint, dead star? The energy involved is far too great to be explained by flares or other surface activity, and, as we have seen, there is no nuclear activity in the dwarf's interior. To understand what happens, we must reconsider the fate of a low-mass star after it enters the white-dwarf phase.

We remarked in the previous chapter that the white-dwarf stage represents the end-point of a star's evolution. Subsequently, the star simply cools, eventually becoming a black clinker in interstellar space. This picture is quite correct for an *isolated* star, like our Sun. However, should the star be part of a *binary* system, an important new possibility exists. If the distance between the two stars is small enough, the dwarf's tidal gravitational field can pull matter—primarily hydrogen and helium—away from the surface of its main-sequence or giant companion, as illustrated in Figure 23.2. A stream of gas leaves the companion through the Lagrange point (see Chapter 22) and flows onto the dwarf.

As material builds up on the white dwarf's surface, the newly accreted gas becomes hotter and denser. Eventually its temperature exceeds 10^7 K, and the hydrogen ignites, fusing into helium at a furious rate. This surface burning stage is as brief as it is violent. The star suddenly flares up in luminosity, then fades away, as some of the fuel is exhausted and the remainder is blown off into space. If the event happens to be visible from Earth, we see a nova. Figure 23.3 is a photograph of a nova apparently caught in the act of expelling mass from its surface.

The initial flareup of luminosity from a nova steadily declines, and eventually the star returns to its normal, pre-explosion appearance. The luminosity and temperature of a nova are not usually plotted on an HR diagram, however. Instead, the change in luminosity is plotted in the form of a *light curve,* like that shown in Figure 23.4. Such curves show the dramatic rise in luminosity over a few days, followed by the much slower decay over the course of several months. The decline in brightness results from the expansion and cooling of the dwarf's surface layers as they are blown into space. Studies of the details of these curves provide astronomers with a wealth of information, both on the dwarf and about its binary companion.

The manner in which the matter reaches the dwarf's

Figure 23.3 Material accumulates on the surface after being accreted from a companion, then ignites in hydrogen fusion as a nova outburst, ejecting part of the surface gas into space, as shown here. This nova is called Nova Persei (1901). (Palomar Observatory)

surface provides an important piece of observational evidence in support of this general picture. Because of the binary's rotation, the stream does not fall directly onto the dwarf. Instead, it misses the compact star and goes into orbit around it, forming a disk of matter known as an **accretion disk** (shown in Figure 23.2). Friction causes the gas to spiral slowly down onto the dwarf's surface, heating up as it goes. The hot disk is observable as a source of x-rays. The point where the infalling stream of matter strikes the accretion disk may form a turbulent "hot spot," which can cause observable fluctuations in the light emitted by the binary system between nova outbursts. Both the x-ray emission and the optical light variations are routinely observed in galactic novae.

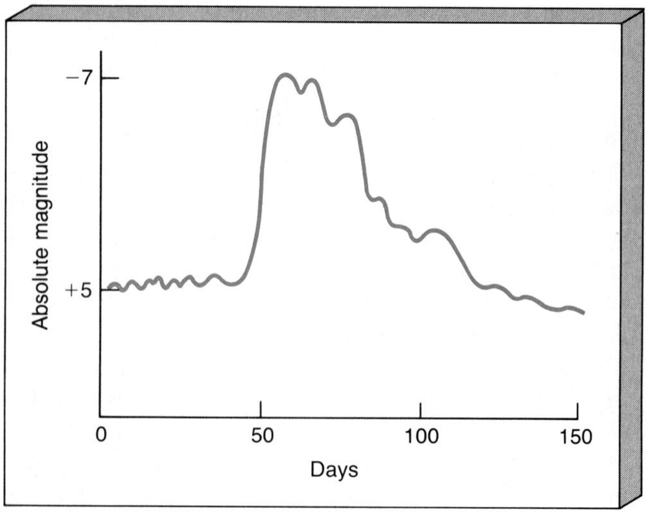

Figure 23.4 The light curve of a typical nova. The rapid rise and slow decline in the light received from the star, as well as the maximum brightness attained, are all in good agreement with the explanation of the nova as a nuclear flash on a white dwarf's surface.

THE DEATH OF A HIGH-MASS STAR

✓2 A low-mass star—a star with mass less than about 4 $M_\odot$ (4 solar masses)—never becomes hot enough to burn carbon in its core. It ends its life as a carbon white dwarf. A high-mass star, however, can fuse not just hydrogen and helium, but also carbon, oxygen, and other even heavier elements as its inner core continues to contract and its central temperature continues to rise. The burning rate accelerates as the core evolves. In short, the star is in a highly unstable state and headed for disaster. Can anything stop this runaway process? Is there a stable "white-dwarf-like" state at the end of the evolution of a high-mass star? What is its ultimate fate? To answer these questions, we must look more carefully at fusion in massive stars.

Heavy Element Fusion

Figure 23.5 is a cutaway diagram of the interior of a highly evolved star of large mass. Note the numerous layers where various nuclei burn. As the temperature increases with depth, the ash of each burning stage becomes the fuel for the next stage. At the relatively cool periphery of the core, hydrogen fuses into helium. In the intermediate layers, shells of helium, carbon, and oxygen burn to form heavier nuclei. Deeper down reside neon, magnesium, silicon, and other heavy nuclei, all produced by nuclear fusion in the layers overlying the core. The core itself is composed of iron nuclei, complex pieces of matter each containing 26

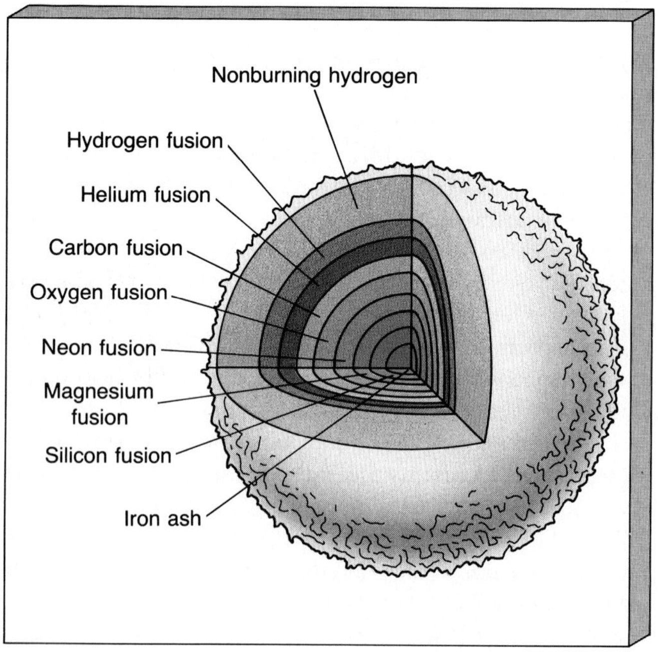

Figure 23.5 Cutaway diagram of the interior of a highly evolved star of mass greater than 5 $M_\odot$. The interior resembles the layers of an onion, with shells of progressively heavier elements burning at smaller and smaller radii and at higher and higher temperatures.

protons and 30 neutrons. We will study the key reactions in this burning chain in more detail later in this chapter.

As each element is burned to depletion at the center, the core contracts, heats up, and starts to fuse the ash of the previous burning stage. A new inner core forms, contracts again, heats again, and so on. Through each period of stability and instability, the star's central temperature increases, the nuclear reactions speed up, and the newly released energy supports the star for ever-shorter periods of time. For example, in approximate numbers, a star of 20 $M_\odot$ burns hydrogen for 10 million years, helium for 1 million years, carbon for 1000 years, oxygen for 1 year, and silicon for a week. Its iron core grows for less than a day.

Collapse of the Iron Core

Once the inner core begins to change into iron, our massive star is in trouble. Nuclear fusion involving iron does not produce energy. Iron nuclei are so compact that energy cannot be extracted by combining them into heavier elements. In effect, iron plays the role of a fire extinguisher, damping the stellar inferno in the core. As substantial quantities of iron appear, the central fires cease for the last time.

This situation is most unstable. A highly evolved star of large mass can no longer be supported by nuclear burning in its core. The star's foundation has been effectively destroyed, and its equilibrium is gone forever. Even though the temperature in the iron core has reached several billion kelvins by this stage, the enormous inward gravitational pull of matter ensures catastrophe in the very near future.

Once gravity overwhelms the pressure of the hot gas, the star implodes, falling in on itself. Two factors act to speed this contraction. First, the core temperature rises to nearly 10 billion K. At such temperatures, individual photons, according to Wien's Law, have tremendously high energies—high enough to split iron into lighter nuclei and to break those lighter nuclei apart in turn until only protons and neutrons remain. This process is known as **photodisintegration** of the heavy elements in the core. In less than a second, the collapsing core undoes all of the effects of nuclear fusion that occurred during the previous 10 million years! But to split iron and lighter nuclei into smaller pieces requires a lot of energy. After all, this is just the opposite of the fusion reactions that generated the star's energy during earlier times. The process of photodisintegration *absorbs* some of the core's thermal energy—in other words, it tends to cool the core and so reduces the pressure. Thus, as nuclei are destroyed, the core of the star becomes even less able to support itself against its own gravity. As a result, the collapse accelerates.

Now the core consists entirely of simple elementary particles—electrons, protons, neutrons, and photons—at enormously high densities, and it is still shrinking. Soon the second destabilizing factor comes into play. As the core density continues to rise, the protons and electrons are crushed together, forming neutrons and neutrinos:

$$p + e \longrightarrow n + \text{neutrino}.$$

This process is sometimes called the **neutronization** of the core. Recall from our discussion in Chapter 18 that the neutrino is an extremely elusive particle and interacts hardly at all with matter. Even though the central density by this time may have reached 10^9 g/cm^3 or more, most of the neutrinos produced by neutronization pass through the core as if it weren't there and escape into space, carrying away energy as they go.

The absence of the electrons and the escape of the neutrinos make matters even worse for the core's stability. There is now nothing to prevent it from collapsing all the way to the point where the neutrons themselves come into contact, at the incredible density of about 10^{12} g/cm^3. At that stage, **neutron degeneracy pressure,** similar to the electron degeneracy pressure that operates in red giants and white dwarfs (see Chapter 22), finally begins to slow the collapse. By the time the collapse is actually stopped, however, the core has overshot its point of equilibrium, and it may reach densities as high as 10^{14} or 10^{15} g/cm^3 before it turns around and begins to reexpand. Like a fast-moving ball hitting a brick wall, the core becomes compressed, stops, then bounces back.

The events just described do not take long. Only about a second elapses from the start of the collapse to the "bounce" at nuclear densities. At that point, the core rebounds—with a vengeance! An enormously energetic shock wave sweeps through the star at high speed, blasting all of the overlying layers—including the heavy elements outside the iron inner core—into space. Although the details of how the shock reaches the surface and destroys the star are still uncertain, the end result is not. The star literally explodes, in a manner vastly more violent than the expulsion of matter in planetary nebulae that marks the end of a low mass star. The explosion is one of the most energetic events known in the entire universe. For a period of a few days, as shown in Figure 23.6, the exploding star may well

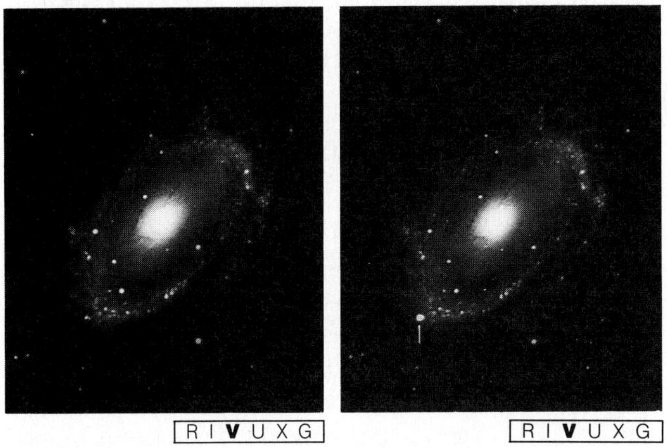

Figure 23.6 A supernova was exploding in this far away galaxy (NGC 4725) at the moment the photograph on the right was taken. The photograph on the left is the normal appearance of the galaxy. (Palomar Observatory)

outshine the rest of the trillion-star galaxy in which it resides. This spectacular death rattle of a high-mass star is known as a **supernova.**

SUPERNOVA EXPLOSIONS

Novae and Supernovae

✓3 Let's consider for a moment the observational appearance of a supernova explosion and compare it with a nova, the phenomenon from which it derives its name. A supernova, like a nova, is a star that suddenly increases dramatically in brightness, then slowly dims again, eventually fading from view. The exploding star is commonly called the supernova's **progenitor.** Supernova light curves can appear very similar to those of novae—so much so, in fact, that the difference between the two was not fully appreciated until the 1920s—but novae and supernovae are known to be quite distinct phenomena. Supernovae are vastly more violent events, driven by quite different underlying physical processes.

Before they understood the causes of either novae or supernovae, astronomers knew of observational differences between them, the most important being that supernovae are about a million times brighter than novae. A supernova produces a flash equaling nearly a billion solar luminosities, reaching that brightness within just a few hours of the start of the outburst. The total amount of energy radiated by a supernova during the few months it takes to brighten and fade away is roughly 10^{51} erg—as much energy as the Sun will radiate during its *entire* 10^{10}-year lifetime!

A second important difference is that the same star may become a nova many times, but a star can become a supernova only once. This is easily understood once we know why novae and supernovae occur. The nova accretion-explosion cycle described earlier can occur over and over again, but a supernova destroys the star involved, with no possibility of a repeat performance.

Observational Classification of Supernovae

Observationally, there are major differences among supernovae, too. Some contain very little hydrogen, according to their spectra, while others contain a lot. In addition, the light curves of the hydrogen-poor supernovae are qualitatively different from those of the hydrogen-rich ones, as illustrated in Figure 23.7. On the basis of these observations, astronomers divide supernovae into two classes: type-I and type-II. **Type-I supernovae,** the hydrogen-poor kind, have a light curve similar in shape to that of a typical nova. **Type-II supernovae,** whose spectra show lots of hydrogen, have a characteristic "bump" in the light curve a few months after the maximum.

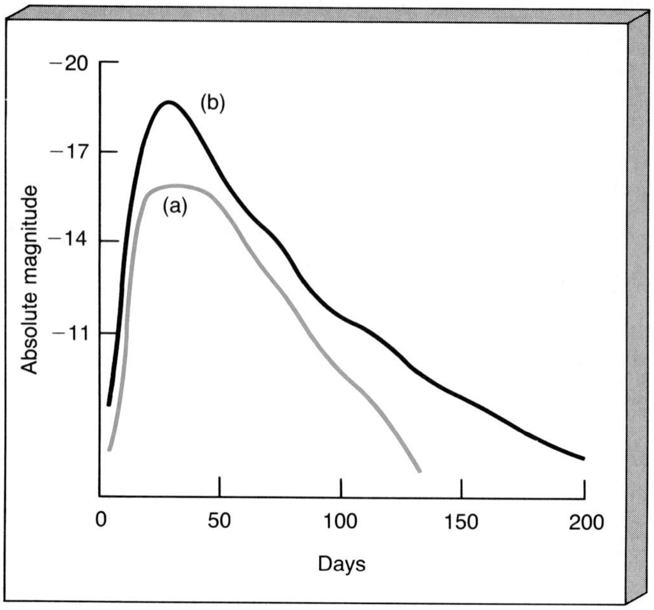

Figure 23.7 *The light curves of typical (a) type-I and (b) type-II supernovae. In both cases, the maximum brightness or intensity can sometimes reach that of a billion suns, but there are characteristic differences in the fall-off of the luminosity after the initial peak. Type-I light curves resemble those of novae (Figure 23.4). Type-II curves have a characteristic bump in the declining phase.*

Carbon Detonation Supernovae

What is responsible for these differences between supernovae? Is there more than one way in which a supernova explosion can occur? The answer is yes. To understand the alternate supernova mechanism, we must return to the processes that cause novae and consider the long-term consequences of their accretion-explosion cycle.

Nova explosions eject matter from a white dwarf's surface, but they do not necessarily expel or burn *all* of the material that has accumulated since the last outburst. In other words, there is a tendency for the dwarf to increase slowly in mass with each new nova cycle. As the star's mass grows and the internal pressure required to support its weight rises, it can enter into a new period of instability—with disastrous consequences.

Recall that a white dwarf is held up not by thermal pressure (heat) but by the pressure of electrons that have been squeezed so close together that they have effectively come into contact with one another. Is there a limit to the pressure that even these electrons can exert? Is there, therefore, a limit to the mass of a white dwarf beyond which electrons cannot provide the pressure needed to support the star? The answer to both questions is yes. Detailed calculations show the maximum mass of a white dwarf is about 1.4 $M_\odot$, a mass often called the **Chandrasekhar mass,** after the Indian astronomer Subramanyan Chandrasekhar, whose work in theoretical astrophysics earned him a Nobel Prize in 1983.

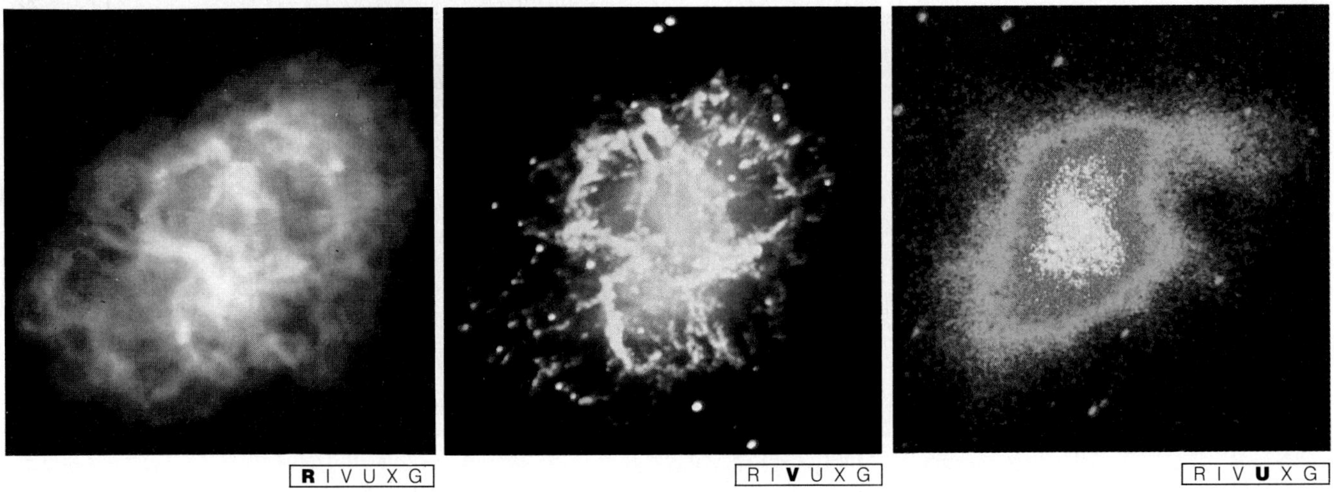

Figure 23.8 *This remnant of an ancient supernova is called the Crab Nebula (or M1 in the Messier catalog). It resides about 1800 pc from Earth and has an angular diameter about one-fifth that of the full Moon. Because its debris is scattered over "only" a 2-pc extent, the Crab is considered to be a young supernova remnant. In A.D. 1054 Chinese astronomers observed the supernova explosion itself. The left and right inserts, to the same scale, show the Crab Nebula in the radio and ultraviolet domains, respectively. (NRAO, NOAO, NASA)*

If an accreting white dwarf exceeds the Chandrasekhar mass, the pressure of degenerate electrons in its interior becomes unable to withstand the pull of gravity, and the star immediately starts to collapse. Its internal temperature rapidly rises to the point where carbon can fuse into heavier elements. Carbon fusion begins everywhere throughout the white dwarf almost simultaneously, and the entire star explodes in another type of supernova—a so-called **carbon detonation supernova**—comparable in violence, but very different in cause, to the "implosion" supernova associated with the death of a high-mass star. In an alternative scenario, two white dwarfs in a binary system may collide to form a massive, unstable star, again with explosive results.

Type-I and Type-II Supernovae Explained

The explosion resulting from the detonation of a carbon white dwarf, the descendant of a low-mass star, is a type-I supernova. Because this conflagration stems from a system containing virtually no hydrogen, we can readily understand why the spectrum of a type-I supernova shows little evidence for that element. The fall-off in light results from the expanding and cooling gases produced in the explosion, just as in a nova. The implosion-explosion of the core of a massive star, described earlier, produces a type-II supernova. Its strong hydrogen lines come from the expanding unburned stellar envelope being blown off into space. The bump in the light curve, as we will see below, results from the radioactive decay of unstable heavy elements—notably ^{56}Ni (nickel-56)—produced in the explosion itself.

Supernova Remnants

We have plenty of evidence that supernovae have occurred in our Galaxy. Occasionally, the explosions themselves are visible from Earth (see Interlude 23-1). In many other cases we can detect their glowing remains, or **supernova remnants.**

One of the best-studied supernova remnants is known as the Crab Nebula, shown in Figure 23.8. Its brightness has greatly dimmed now, but the original explosion was so brilliant that ancient manuscripts of Chinese and Middle Eastern astronomers claim that its brightness greatly exceeded that of Venus and, according to some (possibly exaggerated) accounts, even rivaled that of the Moon in the year A.D. 1054. For nearly a month, this exploded star reportedly could be seen in broad daylight. Native Americans also left engravings of the event in the rocks of what is now the southwestern United States.

Certainly, the Crab Nebula has the appearance of exploded debris. In fact, astronomers have proved that this matter was ejected from some central explosion. Figure 23.9 is a superposition of a positive image of the Crab Nebula taken in 1960 and a negative image taken in 1974. If the gas were not in motion, the positive and negative images would overlap perfectly, but they do not. The gas moved outward in the intervening 14 years. Knowing the total distance traveled by the gas in this time, astronomers have computed a velocity of several thousand kilometers per second for the expelled debris. Running the motion backward in time, astronomers have found that the explosion must have occurred about nine centuries ago. This date is consistent with the Chinese observations of the Crab.

The nighttime sky harbors many relics of stars that blew up long ago. Figure 23.10 is another example. It shows a Milky Way supernova remnant, known as the Gum Nebula, whose expansion velocities imply that its central star exploded around 9000 B.C. It lies only 500 pc away from Earth. Given its close proximity, we can only speculate what impact such a bright supernova might have had on

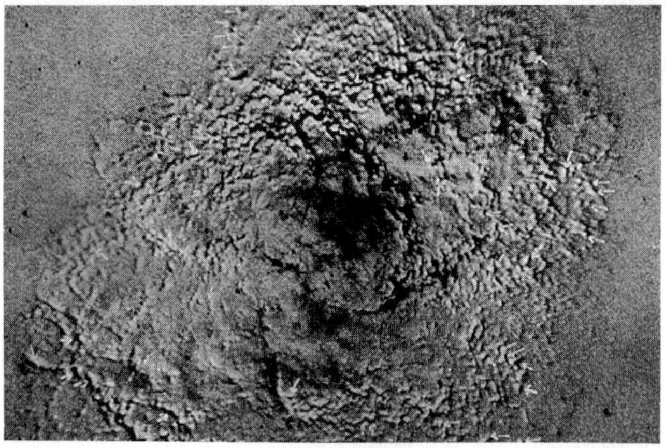

Figure 23.9 *Positive and negative photographs of the Crab Nebula taken 14 years apart do not superpose exactly, indicating that the gaseous filaments are still moving away from the site of the explosion. (Harvard-Smithsonian Center for Astrophysics)*

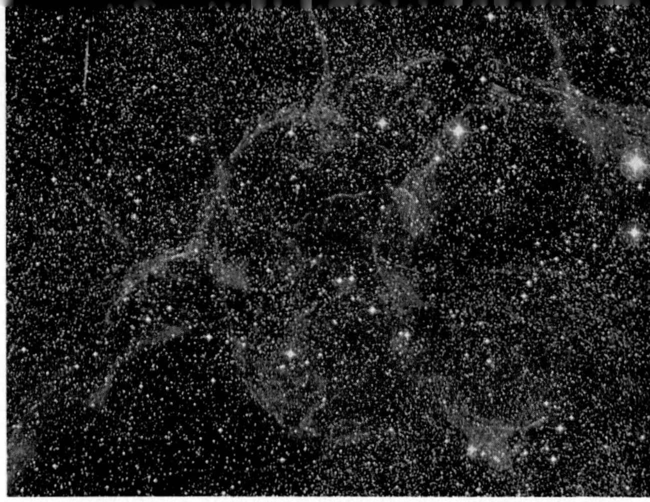

Figure 23.10 *The glowing gases of the Gum Nebula supernova remnant are spread across an amazingly large 60° of the sky. The closest point of the expanding shell is only 100 pc away from Earth. (European Southern Observatory)*

the myths, religions, and cultures of Stone Age humans when it first appeared in the sky.

Astronomers patrolling the skies with telescopes occasionally notice a sudden brightening of a small part of some distant galaxy (see, for example, Figure 23.6). Drawing a light curve, like those shown in Figure 23.7, of this sharp rise in luminosity and studying the supernova's development, astronomers refine their theoretical models. Hundreds of supernovae have been observed in other galaxies during the twentieth century. However, no one has ever observed with modern equipment a supernova in our own Galaxy (see also Interlude 23-1). A viewable Milky Way star has not exploded since Galileo first turned his telescope to the heavens almost four centuries ago. The last supernova observed in our Galaxy (now known as Tycho's supernova) caused a worldwide sensation in Renaissance times. The sudden appearance and subsequent fading of a very bright object in the year 1604 helped shatter the Aristotelian idea of an unchanging universe.

Why doesn't our Galaxy have more supernovae? Knowing the rates at which stellar evolutionary stages occur and estimating the number of high-mass stars in our Galaxy, we would expect a galactic supernova to occur in an observable location every 100 years or so. The brilliance of a nearby supernova might rival the brightness of a full Moon. It is unlikely that astronomers could have missed any since

INTERLUDE 23-1 *Nearby Supernovae*

Only six galactic supernovae have been recorded in the past 1000 years. This figure shows their positions within our Milky Way. They are labeled by the year in which they first appeared. The supernova Cassiopeia A apparently went unnoticed optically, although modern radio studies suggest that the first light from the explosion should have reached Earth midway through the seventeenth century. The following combined radio/optical/x-ray image is dramatic proof that this supernova remnant, although invisible optically, makes quite an impact at other wavelengths.

Most astronomers assume that many more stars than these six have blown up in our Galaxy. Why haven't we seen them? Possibly because they were too distant to be detected by the naked eye, or perhaps dark clouds in the galactic plane kept them from our view. All the Milky Way supernovae mapped above are confined to our quadrant of the Galaxy, and each of them is at least 100 pc above or below the galactic plane. Studies of the rate at which supernovae occur suggest that we can expect one within 100 pc of our Sun only every 500,000 years. Thus, a truly "nearby" supernova would be a rare event indeed. Humanity may be destined to see all supernovae from a distance.

(Harvard-Smithsonian)

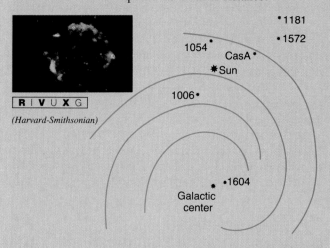

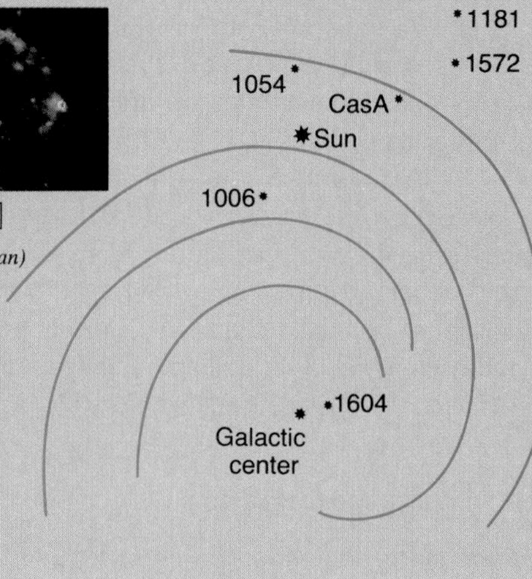

the last one nearly four centuries ago. The Milky Way seems long overdue for a supernova. Unless massive stars explode much less frequently than predicted by the theory of stellar evolution, we should be treated to a nearby version of nature's most spectacular cosmic event any day now.

Supernovae as Distance Indicators

Astronomers are often concerned with the measurement of the distances to astronomical objects. Knowing the size of the universe is essential for understanding how it works. Astronomers are especially interested in **standard candles**—cosmic objects of known absolute brightness. Once a standard candle is recognized, its absolute brightness is immediately known, and so a measurement of its apparent brightness yields its distance. Supernovae are good examples of standard candles. Astronomers have found that nearly all supernova of a given type (I or II) have the same peak luminosity (but see Interlude 23-2 for a notable exception).

The reason for this similarity in peak luminosities is that the explosion occurs only when a steadily growing stellar core reaches a well-defined, specific critical mass. The conditions immediately preceding the explosions are much less varied than the objects in which they occur. For example, regardless of how quickly or slowly a white dwarf is pushed over the Chandrasekhar mass, a type-I supernova results from the explosion of a 1.4 $M_{\odot}$ carbon star. These luminosities and light curves are almost always the same. Similarly, the iron core of a massive star implodes and produces a type-II supernova only when its mass reaches a critical value (in fact also the Chandrasekhar mass, because the iron core is supported by electron degeneracy pressure as it grows). So, once again, the conditions producing the supernova are very similar, even in the cores of progenitor stars of quite different masses.

When astronomers recognize a supernova, they plot its light curve. By comparing the peak apparent brightness against the theoretical peak luminosity, they compute the supernova's distance. The great advantage of these objects is that they are very bright, so they can be seen at great distances. Observations of supernovae are invaluable in determining cosmic distances well beyond the Milky Way Galaxy.

THE FORMATION OF THE ELEMENTS

[✓4] The determination of the origin of the elements is an important problem in modern astronomy. Up to now, we have studied nuclear reactions for their role in stellar energy generation. Now let's consider them again, but this time as the processes responsible for creating much of the world we live in. The evolution of the elements, combining nuclear physics with astronomy, is a complex subject. Let us begin by taking inventory of the composition of the universe.

Types of Matter

We currently know of 104 different elements, ranging from the simplest—hydrogen, containing 1 proton—to the most complex—kirchotovium, with 104 protons in its nucleus. Elements having 105, 106, 107, and 108 protons may also have been discovered in experiments, but they have not yet been confirmed by other researchers. All elements exist in several different isotopic forms, each isotope having the same number of protons but a different number of neutrons. We often think of the most common or stable isotope as being the "normal" form of an element. These isotopes are listed in Appendix Table 2. Some elements, and many isotopes, are radioactively unstable, which means that they eventually decay into more stable nuclei.

The 81 stable elements found on Earth make up the overwhelming bulk of matter in the universe. In addition, 10 radioactive elements also occur naturally on our planet. Well-known examples are radon and uranium. Like all radioactive elements, they have definite half-lives (recall that the half-life is the time required for half the nuclei of an element to decay into something else). Even though their half-lives are very long (millions or billions of years, typically), their steady decay means that they are scarce on Earth, in meteorites, and in lunar samples. They are not observed in stars—there is just too little of them to produce detectable spectral lines.

Besides these 10 naturally occurring radioactive elements, 11 more radioactive elements have been artificially produced under special conditions in nuclear laboratories on Earth. The debris collected after nuclear weapons tests also contains traces of these elements. Unlike the naturally occurring radioactive elements, these artificial ones decay into other elements quite quickly (in much less than a million years). Consequently, they too are extremely rare in nature. Two other elements round out our list. Promethium is a stable element that is found on our planet only as a by-product of nuclear laboratory experiments. Technetium is an unstable element that is found in stars but does not occur naturally on Earth.

Abundance of Matter

How and where did all these elements form? Were they always present in the universe, or were they created after the universe formed? Since the 1950s, astronomers have come to realize that the hydrogen and most of the helium in the universe are *primordial*—that is, these elements date back to the very earliest times. The other elements in our universe result from **stellar nucleosynthesis**—that is, they were formed by nuclear fusion in the hearts of stars. For the remainder of this chapter, we will delve more deeply into the details of this important subject.

A key point in understanding the creation of heavy elements is that large nuclei can be built from smaller ones by nuclear fusion. We might naturally theorize that all the heavy elements have been created in this way. In this picture, the ultimate source of the heavy elements is the lightest and simplest of all—hydrogen. To test this idea, we must consider not just the list of different *kinds* of elements and isotopes but also their observed *abundances,* shown in Figure 23.11. This curve is derived largely from spectroscopic studies of stars, including the Sun. The essence of the figure is summarized in Table 23-2. (All isotopes of all elements are included in both Figure 23.11 and Table 23-2.) Any theory proposed for the creation of the elements must reproduce these observed abundances. The most obvious feature is that the heavy elements are much less abundant than most light elements.

Let us now study the specific reactions leading to heavy-element production at different stages of stellar evolution. Some of the following material simply repeats the stages of burning already studied in Chapters 18 and 22. The rest expands upon reactions mentioned previously but not discussed in detail. Why repeat all this? Because we now want to discuss these reactions not from the point of view of the star and its energy needs, but instead from the perspective of building the heavy elements that make up the universe we inhabit.

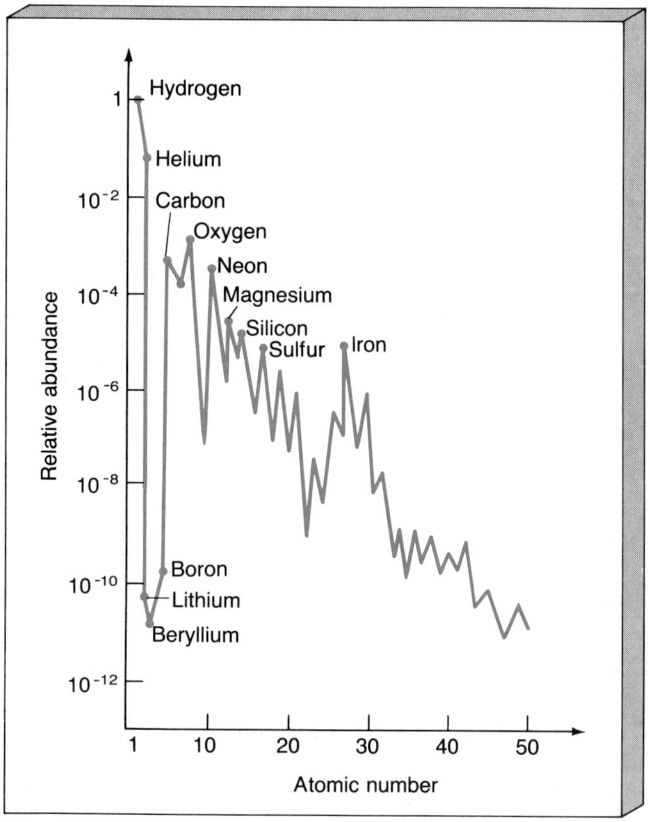

Figure 23.11 *A summary of the cosmic abundances of the elements and their isotopes. The horizontal axis shows atomic number—the number of protons in the nucleus. Notice how many common terrestrial elements are found on "peaks" of the distribution, surrounded by elements that are tens or hundreds of times less abundant. Notice especially the large peak around the element iron.*

TABLE 23-2 ***Cosmic Abundances of the Elements***

Elemental Group	Percent Abundance by Number of Particles*
Hydrogen (1 nuclear particle	90
Helium (4 nuclear particles)	9
Lithium group (7–11 nuclear particles)	.000001
Carbon group (12–20 nuclear particles)	.2
Silicon group (23–48 nuclear particles)	.01
Iron group (50–62 nuclear particles)	.01
Middle-weight group (63–100 nuclear particles)	.00000001
Heaviest-weight group (over 100 nuclear particles)	.000000001

*The total does not equal 100% because of uncertainties in the helium abundance. All isotopes of all elements are included.

Hydrogen Burning

Stellar nucleosynthesis begins with the proton-proton chain studied in Chapter 18. Provided that the temperature is high enough—at least 10^7 K—a series of nuclear reactions occurs, ultimately forming a nucleus of ordinary helium (^{4}He) from four protons (^{1}H):

$$4\ (^1\text{H}) \longrightarrow {}^4\text{He} + 2 \text{ positrons} + 2 \text{ neutrinos} + \text{energy}.$$

Recall that the positrons immediately interact with nearby free electrons, producing high-energy gamma-rays through matter-antimatter annihilation. The neutrinos rapidly escape, carrying energy with them but playing no direct role in nucleosynthesis. The validity of these reactions has been directly confirmed in nuclear experiments conducted in laboratories around the world during the past decades. In massive stars, the CNO cycle (see Interlude 18-3) may greatly accelerate the hydrogen-burning process, but the basic 4-proton-to-1-helium-nucleus reaction, shown in Figure 23.12, is unchanged.

Helium Burning

As helium builds up in the core of a star, the burning ceases, and the core contracts and heats up. When the temperature exceeds about 10^8 K, helium nuclei can overcome their mutual electrical repulsion, leading to the *triple-alpha reaction*, as we saw in Chapter 22:

$$3\ (^4\text{He}) \longrightarrow {}^{12}\text{C} + \text{energy}.$$

The net result of this reaction is that three helium-4 nuclei are combined into one carbon-12 nucleus, as shown in Figure 23.13, releasing energy in the process.

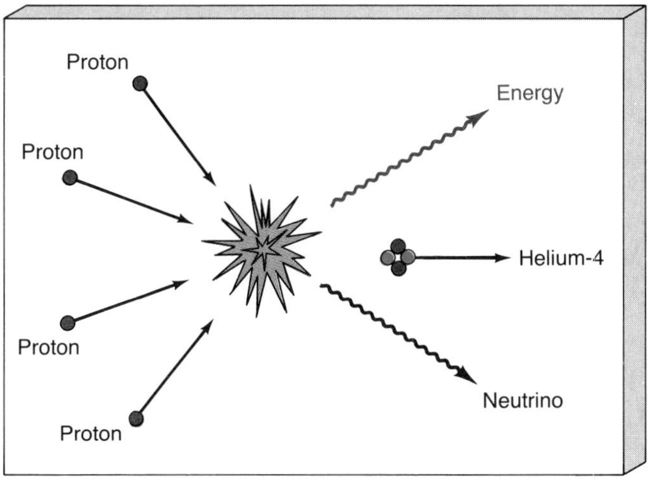

Figure 23.12 Diagram of the basic proton-proton hydrogen-burning reaction. Four protons combine to form a nucleus of helium-4, releasing energy in the process.

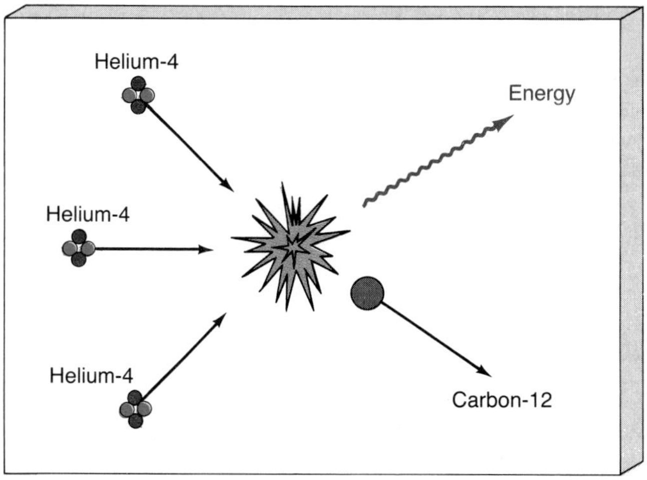

Figure 23.13 Diagram of the triple-alpha helium-burning reaction, when three helium-4 nuclei combine to form carbon-12.

Carbon Burning and Helium Capture

At higher and higher temperatures, heavier and heavier nuclei can gain enough energy to overcome the electrical repulsion between them. At about 10^9 K, carbon nuclei can fuse into magnesium, as depicted in Figure 23.14(a):

$$^{12}C + ^{12}C \longrightarrow ^{24}Mg + \text{energy}.$$

Because of the rapidly mounting nuclear charges—that is, the increasing number of protons in the nuclei—fusion reactions between any nuclei larger than carbon require such high temperatures that they are quite uncommon in stars. Instead, an easier path leading to the formation of heavier elements is found. For example, the repulsive force between two carbon nuclei is three times greater than the force between a nucleus of carbon and one of helium. Thus carbon-helium fusion occurs at a lower temperature than carbon-carbon fusion. At temperatures above 6×10^8 K, a carbon-12 nucleus colliding with a helium-4 nucleus can produce oxygen-16:

$$^{12}C + ^{4}He \longrightarrow ^{16}O + \text{energy}.$$

If any helium-4 is present, this reaction, shown in Figure 23.14(b), is much more likely to occur than the carbon-carbon reaction.

Similarly, the oxygen-16 thus produced may fuse with other oxygen-16 nuclei at a temperature of about 1.2 billion K to form sulfur-32:

$$^{16}O + ^{16}O \longrightarrow ^{32}S + \text{energy}.$$

However, it is much more likely that an oxygen-16 nucleus will capture a helium-4 nucleus (if one is available) to form neon-20:

$$^{16}O + ^{4}He \longrightarrow ^{20}Ne + \text{energy}.$$

This reaction is more likely because it requires lower temperatures than the oxygen-oxygen reaction.

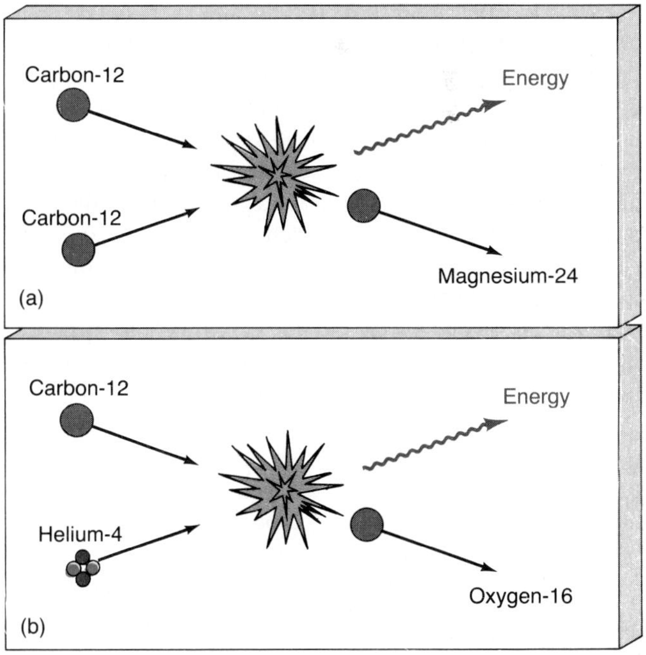

Figure 23.14 Carbon can form heavier elements (a) by fusion with other carbon nuclei or, more commonly, (b) by fusion with a helium nucleus.

As the star evolves, heavier elements tend to form through **helium capture** rather than by fusion of like nuclei. Because these helium-capture reactions are so much more common, elements with nuclear masses of 4 units (that is, helium itself), 12 units (carbon), 16 units (oxygen), 20 units (neon), 24 units (magnesium), and 28 units (silicon) stand out as prominent peaks in our chart of cosmic abundances, Figure 23.11. Each element is built by combining the preceding element and a helium-4 nucleus as the star evolves.

Proton and Neutron Capture

Helium capture is not the only type of nuclear reaction occurring in evolved stars. As many nuclei of different kinds accumulate, a great variety of reactions becomes possible. In some reactions protons and neutrons are freed from their parent atom. In turn they are absorbed by other nuclei, so that proton-capture and neutron-capture reactions begin to play a role in the construction of new elements. The result is a family of many nuclei having masses intermediate between the nuclei formed by helium capture. Laboratory studies confirm that common nuclei such as fluorine-19, sodium-23, phosphorus-31, and many other nuclei are created in this way. Their abundances, however, are not as great as those produced directly by helium capture, simply because the helium-capture reactions are much more common in stars. For this reason, many of these elements—and note that they are not divisible by 4, the mass of a helium nucleus—reside in the troughs of Figure 23.11.

Some Complications

With the appearance of silicon-28 in the core of a star, another process begins to complicate the helium-capture scheme. A competitive struggle begins between the continued capture of helium to produce even heavier nuclei and the tendency of the heavier nuclei to break down into simpler ones. The cause of this breakdown is heat. By the point at which silicon-28 has been produced, the star's core temperature has reached the unimaginably large value of 3 billion K, and the gamma-rays associated with that temperature have enough energy to break a nucleus apart, as illustrated in Figure 23.15. This is the same process of photodisintegration that will ultimately accelerate the star's iron core in its final collapse toward a type-II supernova.

Under the intense heat, some silicon-28 nuclei break apart into 7 helium-4 nuclei. Other nearby nuclei that have not yet photodisintegrated may capture some or all of these helium-4 nuclei, leading to the formation of still heavier elements. The process of photodisintegration provides raw material that allows the helium-capture process to proceed to greater masses. The process continues, with some heavy nuclei being destroyed and others increasing in mass. In succession the star forms sulfur-32, argon-36, calcium-40, titanium-44, chromium-48, iron-52, and nickel-56. The chain of reactions building from silicon-28 up to nickel-56 is:

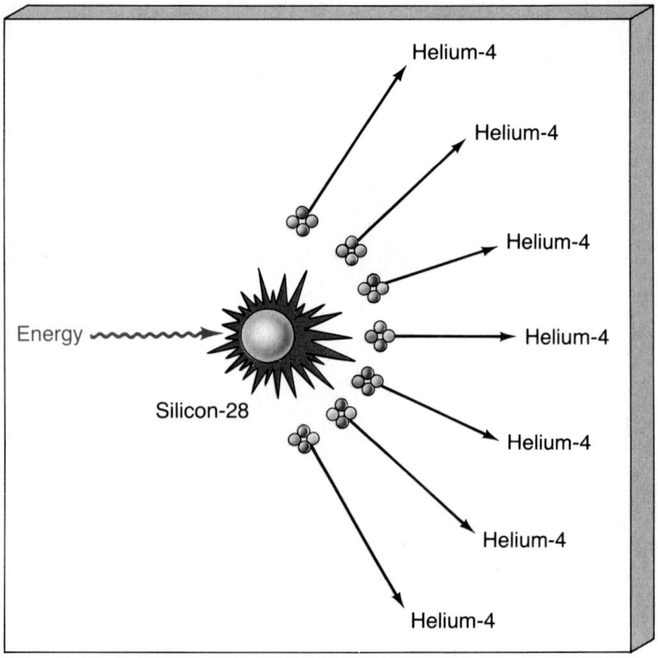

Figure 23.15 *At high temperatures, heavy nuclei (such as silicon, shown here) can be broken apart into helium nuclei by high-energy photons. Other nuclei can capture the helium nuclei—or alpha particles—thus produced, forming heavier elements by the so-called alpha process. This process continues all the way to the formation of iron.*

$$^{28}\text{Si} + 7\ (^{4}\text{He}) \longrightarrow\ ^{56}\text{Ni} + \text{energy}.$$

This two-step process—photodisintegration followed by the direct capture of some or all the resulting helium-4 nuclei (or alpha particles)—is called the **alpha process.**

A further complication enters the picture here: Nickel-56 is unstable. It rapidly decays, first into cobalt-56, then into a normal iron-56 nucleus. Any unstable nucleus will continue to decay until stability is achieved—we know this from experiments in nuclear physics laboratories—and iron-56 is the stablest of all nuclei. Thus, the alpha process leads inevitably to the buildup of iron in the stellar core.

Ironclad Stability

Iron's 26 protons and 30 neutrons are bound together more strongly than the particles in any other nucleus. Iron is said to have the greatest *nuclear binding energy* of any element. Any nucleus with more or fewer protons or neutrons has less nuclear binding energy and is not quite as stable as the iron-56 nucleus. This enhanced stability of iron explains why some of the heavier nuclei in the iron group are more abundant than many lighter nuclei (see Table 23-2 and Figure 23.13).

Because of the unusual stability of the iron-56 nucleus, it cannot fuse with other nuclei to produce more energy. In fact, any fusion process involving iron *consumes* energy. As we have seen, nuclear fusion in a highly evolved massive star inevitably produces an iron core that not only fails to produce energy but also begins to rob the star of some of its much needed pressure support. Without this support, the massive star begins its collapse to a supernova.

Making Elements beyond Iron

The mechanism of helium capture does not create nuclei heavier than iron. But if heavier nuclei are to be made, then some other nuclear process must operate. This other process involves the capture of neutrons. Deep in the interiors of highly evolved stars, conditions are ripe for neutron capture. Neutrons are produced as "by-products" of many nuclear reactions, and there are many neutrons present to interact with other nuclei. Neutrons have no charge, so they have no repulsive barrier to overcome in combining with positively charged nuclei. As more and more nuclei join an iron nucleus, its mass continues to grow beyond that of "normal" iron.

Adding neutrons to a nucleus—iron, for example—does not change the element. Rather, a more massive isotope of iron is produced. Eventually, however, so many neutrons are added to the nucleus that an unstable iron isotope is formed, which then decays radioactively to form a stable nucleus of some other element, and the process continues. For example, an iron-56 nucleus can capture a single neutron (n) to form a relatively stable isotope, iron-57:

$$^{56}\text{Fe} + \text{n} \longrightarrow {}^{57}\text{Fe}.$$

This may then be followed by another neutron capture,

$$^{57}\text{Fe} + \text{n} \longrightarrow {}^{58}\text{Fe},$$

producing another relatively stable isotope, iron-58, which can capture yet another neutron to produce an even heavier isotope of iron:

$$^{58}\text{Fe} + \text{n} \longrightarrow {}^{59}\text{Fe}.$$

Iron-59 is known from laboratory experiments to be radioactively unstable. It decays in about a month into cobalt-59, which is stable. The neutron-capture process then resumes: Cobalt-59 captures a neutron to form the unstable cobalt-60, which in turn decays to nickel-60, and so on. In this way, heavier and heavier nuclei form. This process is the origin of the copper and silver in the coins in our pockets, the lead in our car batteries, and the gold (or the zirconium) in the rings on our fingers. These reactions continue all the way up to bismuth-209, the heaviest known nonradioactive nucleus. Any attempt to form elements heavier than bismuth-209 by neutron capture in stars fails because the new nuclei decay back to bismuth as fast as they form.

Each successive capture of a neutron by a nucleus typically takes about a year, so that most unstable nuclei have plenty of time to decay before the next neutron comes along. Researchers refer to this "slow" neutron-capture mechanism as the **s-process** (the *s* stands for "slow").

Making the Heaviest Elements

✓5 The s-process explains the synthesis of stable nuclei up to and including bismuth-209, but it cannot account for the heaviest nuclei, such as thorium-232, uranium-238, or plutonium-242. There must be yet another nuclear mechanism that produces the very heaviest nuclei. This process is called the **r-process,** where *r* stands for "rapid," in contrast to the "slow" s-process we just described. The r-process operates very quickly, occurring literally during the supernova explosion that signals the death of a massive star.

For about the first 15 minutes of the supernova blast, the number of free neutrons increases dramatically as heavy nuclei are broken apart by the violence of the explosion. Unlike the s-process, which stops when it runs out of stable nuclei, the neutron-capture rate during the supernova is so great that even unstable nuclei can capture many neutrons before they have time to decay. Jamming neutrons into light- and middle-weight nuclei, the r-process is responsible for the creation of the heaviest known elements. The heaviest of the heavy elements, then, are actually born *after* their parent stars have died. Because the time available for synthesizing these heaviest nuclei is so brief, they never become very abundant. Elements heavier than iron (see Table 23-2) are a billion times less abundant than hydrogen and helium.

Observational Evidence for Stellar Nucleosynthesis

As we have seen, the modern picture of element formation involves many different types of nuclear reactions occurring at many different stages of stellar evolution, from main-sequence stars all the way to supernovae. Light elements—through iron—are built first by fusion, then by alpha capture, with proton and neutron capture filling in the gaps. Elements beyond iron form by neutron capture and radioactive decay. How do we know that stars really produce heavy elements in this way? Can we be sure that this scenario is correct? We are assured of the soundness of our theories by three convincing pieces of evidence.

First, the rate at which various nuclei are captured and the rate at which they decay are known from laboratory experiments. When these rates are incorporated into detailed computer models of the nuclear processes occurring in stars and supernovae, the resulting elemental abundances agree extremely well, point by point, with the observational data presented in Figure 23.11 and Table 23-2. The match is remarkably good for elements up through iron and is still fairly close for heavier nuclei. Thus, although no one has ever directly observed the formation of heavy nuclei in stars, we can be reasonably confident that the theory of stellar nucleosynthesis makes good sense in the context of nuclear physics and stellar evolution. While the reasoning is indirect, the agreement between theory and observation is so striking that most astronomers regard it as strong evidence in support of the entire theory of stellar evolution.

Second, the presence of one nucleus—technetium-99—provides direct evidence that heavy-element formation really does occur in the cores of stars. Laboratory measurements show that the technetium nucleus has a radioactive half-life of about 200,000 years. This is a very short time astronomically speaking. No one has ever found even traces of naturally occurring technetium on Earth because it all decayed long ago. The observed presence of technetium in the spectra of many red giant stars implies that it must have been synthesized through neutron capture—the only known way that technetium can form—within the past few hundred thousand years. Otherwise, we wouldn't observe it. Many

INTERLUDE 23-2 *Supernova 1987A*

In 1987, astronomers were treated to a spectacular supernova in the Large Magellanic Cloud (usually abbreviated to LMC), the Milky Way's nearest companion galaxy, some 50 kpc away from Earth. Observers in Chile first saw the explosion on February 24, and within a few hours practically all Southern Hemisphere telescopes and every available orbiting spacecraft were focused on the object. It was officially named SN1987A (the SN stands for "supernova," 1987 gives the year, and A identifies it as the first supernova seen that year). This was one of the most dramatic changes observed in the universe in nearly 400 years. A 15 $M_\odot$ B-type supergiant star with the catalog name of SK-69°202 detonated, outshining for a few weeks all the other stars in the LMC combined, as shown in the images below.

Because the LMC is relatively close to Earth and because the explosion was detected so soon after it occurred, SN1987A provided scientists with an enormous volume of detailed information on supernovae, allowing astronomers to make key comparisons between theoretical models and observational reality. By and large, the theory of stellar evolution described in the text has held up very well. Still, SN1987A did have some surprises.

According to its hydrogen-rich spectrum, the supernova was of type II—the iron core implosion-explosion type—as expected for a high-mass parent star like SK-69°202. But a glance at Figure 22.15 (which was computed for stars in our own Galaxy) shows that, according to theory, the parent star should have been a *red* supergiant at the time of the explosion, not blue, as was actually observed. This unexpected finding caused theorists to scramble in search of a reason, and many possibilities were considered before an explanation fitting the facts emerged. It seems that the progenitor's envelope was deficient in heavy elements compared with young stars in the Milky Way. This deficiency had little effect on the evolution of the core and on the supernova explosion itself, but it did change the star's evolutionary track on the HR diagram. Unlike a Milky Way star of the same mass, once helium ignited in the core of SK-69°202, the star shrank in size and looped back toward the main sequence. The star had just begun to return to the right on the HR diagram following the ignition of carbon, with a surface temperature of around 20,000K, when the rapid chain of events leading to the supernova occurred.

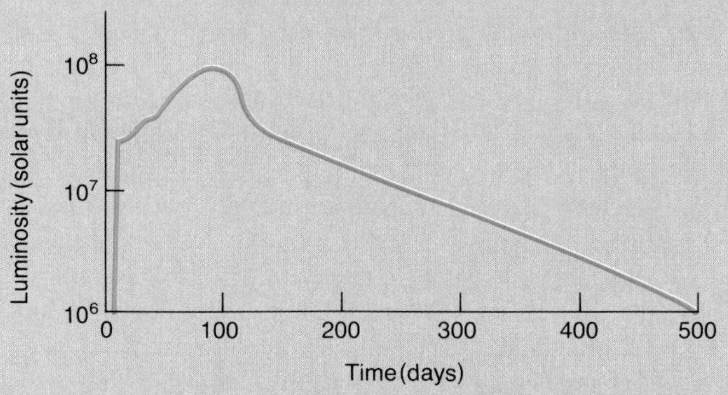

The light curve of SN1987A, shown above, also differed somewhat from the "standard" type-II shape (see Figure 23.7[b]), and the peak brightness was only about 1/10 the expected value. For a few days after its initial detection, the supernova faded as it expanded and cooled rapidly. After about a week the surface temperature had dropped to about 5000K, at which point electrons and protons near the expanding surface recombined into atomic hydrogen. This recombination made the surface layers less opaque and allowed more radiation from the interior to leak out. The supernova brightened rapidly as it grew. The temperature of the expanding layers reached a peak in late May, by which point the radius of the expanding photosphere was about 2×10^{10} km—a little larger than our solar system. Subsequently, the photosphere cooled as it expanded, and

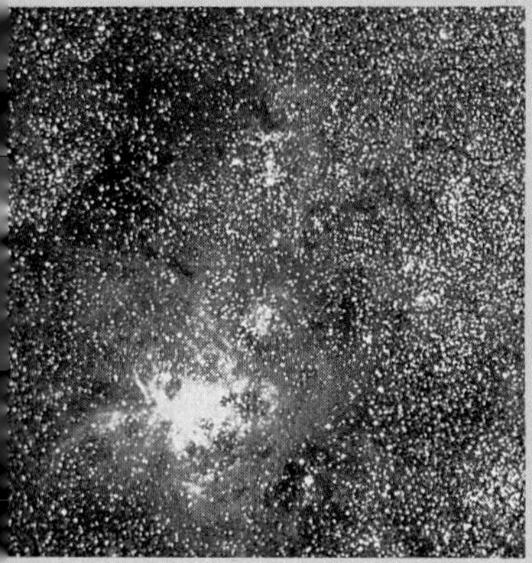

R I V U X G

R I V U X G

(both photos European Southern Observatory)

astronomers consider the spectroscopic evidence for technetium as proof that the s-process really does operate in evolved stars.

Third, the study of typical light curves from type-I supernovae indicates that radioactive nuclei form as a result of the explosion. Figure 23.16 (see also Figure 23.7[a]) displays the dramatic rise in luminosity at the moment of explosion and the characteristic slower decrease in brightness. Depending on the initial mass of the exploded star, the luminosity takes from several months to many years to decrease to its original value, but the *shape* of the decay curve is nearly the same for all exploded stars. These curves have

the luminosity dropped as the internal supply of heat from the explosion leaked away into space.

Much of the preceding description would apply equally well to a type-II supernova in our own Galaxy. The differences between the SN1987A light curve shown above and the light curve in Figure 23.7(b) are the result of the (relatively) small size of the progenitor star. Incidentally, astronomers observing extragalactic supernovae, on which Figure 23.7 is based, do not normally catch the initial bright flash, which occurs very shortly after the explosion and dims rapidly. Instead, the supernova is usually detected as its luminosity grows again, so the left-most point of the figure in the text probably corresponds to SN1987A at some time in early March.

The peak luminosity of SN1987A was less than that of a "normal" type-II supernova because the progenitor SK-69°202 was small and quite tightly bound by gravity. A lot of the energy emitted in the form of visible radiation (Figures 23.7[b] and 23.16) was expended in expanding SN1987A's stellar envelope, so that far less was left over to be radiated into space. Thus, SN1987A's luminosity during the first few months was lower, and the early peak evident in Figure 23.7(b) did not occur. The peak in the SN1987A light curve at about 80 days actually corresponds to the "knee" in Figure 23.7(b). As discussed in the text, this peak was produced by the radioactive decay of unstable heavy elements formed in the seconds following the explosion (especially cobalt-56, whose 77-day half-life closely matches the fall-off in the light curve after the peak). The energy contained in the initial shock wave itself had been dissipated perhaps a month earlier. Because of the lower early luminosity of SN1987A, this extra source of energy—the radioactive decay—actually dominated its light output, appearing as a peak in the luminosity rather than as a secondary bump on a declining curve.

That stellar evolution theory could extend to fit the facts of SN1987A is of course very reassuring. However, the unexpected color and size of the supernova's progenitor star underscore the importance of observational tests and checks in hammering out different theoretical models. It is just as important to know *which* model—in this case, which parent star—to use in the calculations as it is to perform the calculation correctly!

About 20 hours before the supernova was detected optically, a brief (about 13-second) burst of neutrinos was simultaneously recorded by underground detectors in Japan and the United States. As discussed in the text, the neutrinos are predicted to arise as electrons and protons in the star's collapsing core merge to form neutrons. The neutrinos preceded the light because they escaped during the collapse, while the first light of the explosion was emitted only after the supernova shock had plowed through the body of the star to the surface. In fact, theoretical models, consistent with these observations, suggest that vastly more energy was emitted in the form of neutrinos than in any other form—the supernova's neutrino luminosity was many tens of thousands of times greater than its optical energy output. Despite some unresolved details in SN1987A's behavior, detection of this neutrino pulse is considered to be a brilliant confirmation of theory. This singular event—the detection of neutrinos—may well herald a new age of astronomy. For the first time, astronomers have received information from beyond the solar system by radiation outside the electromagnetic spectrum.

Theory predicts that the expanding remnant of SN1987A will be large enough to be resolvable by optical telescopes in a few years. The photograph below was taken by the *Hubble Space Telescope* in late 1990. It shows the remnant itself (in red) at the center, surrounded by a much larger ring of glowing gas (in yellow). Scientists believe that the progenitor star expelled this ring during its red-giant phase, some 40,000 years before the explosion. The image we see results from the initial flash of ultraviolet light from the supernova hitting the ring and causing it to glow brightly. In about 10 years, the fastest moving debris from the remnant will strike this ring, making it a temporary but intense source of x-rays.

Buoyed by the success of stellar evolution theory and armed with firm theoretical predictions of what should happen next, astronomers eagerly await future developments in the story of this remarkable object.

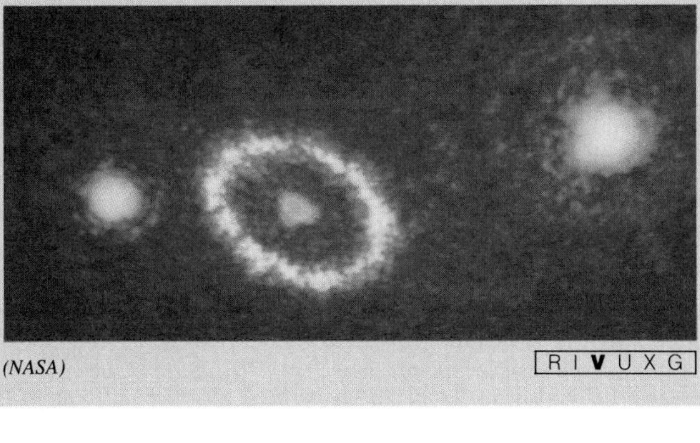

(NASA) R I **V** U X G

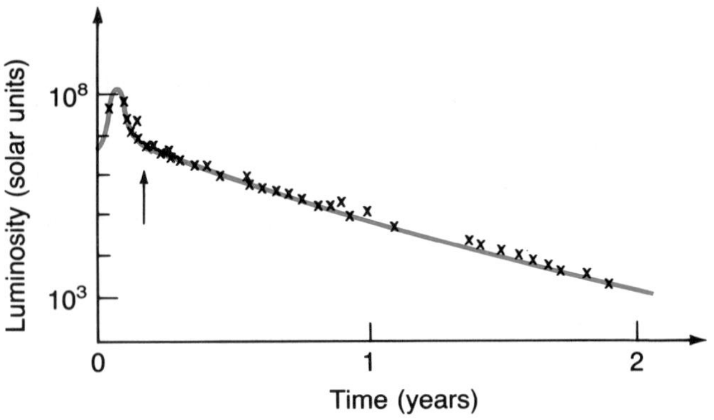

Figure 23.16 The light curve of a type-I supernova, showing not only the dramatic increase and slow decrease in luminosity, but also the characteristic change in the rate of decay about 2 months after the explosion (after the time indicated by the arrow). This particular supernova occurred in the faraway galaxy IC4182 in 1938. The crosses are the actual observations of the supernova's light.

two distinct features. After the initial peak, the luminosity first declines rapidly, then begins to decrease at a slower rate. This change in the luminosity decay invariably occurs about 2 months after the explosion, regardless of the intensity of the outburst.

We can explain the two-stage decline of the luminosity curve in Figure 23.16 in terms of the radioactive decay of unstable nuclei, notably nickel-56 and cobalt-56, produced in abundance during the early moments of the supernova explosion. From theoretical models of the explosion, we can calculate the amounts of these elements expected to form, and we know their half-lives from laboratory experiments. Because each radioactive decay produces a known amount of visible light, we can then determine how the light emitted by these unstable elements should vary in time. The result is in very good agreement with the observed light curve in Figure 23.17. More direct evidence for the presence of these unstable nuclei was obtained in the 1970s, when a gamma-ray spectral feature of decaying cobalt-56 was first identified in a supernova observed in a distant galaxy.

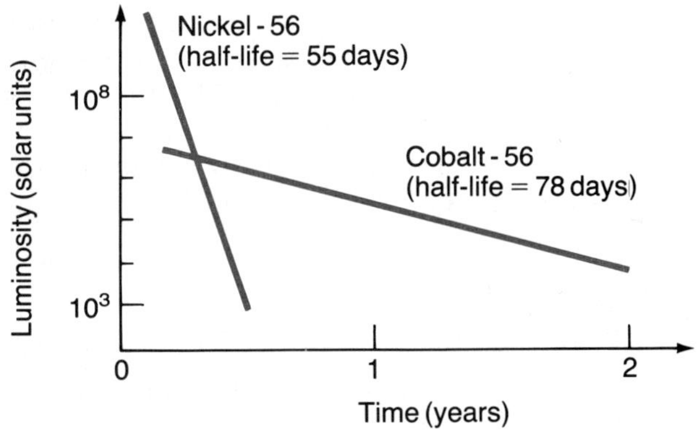

Figure 23.17 Theoretical calculations of the light emitted by the radioactive decay of nickel-56 and cobalt-56 produce a light curve very similar to those actually observed in real supernova explosions (compare Figure 23.16), adding strong support to the theory of stellar nucleosynthesis.

THE CYCLE OF STELLAR EVOLUTION

The evidence in favor of the theory of stellar nucleosynthesis is overwhelming. Theoretical calculations of stellar evolutionary paths predict that heavy elements are created deep inside stars, and spectroscopic studies of giants and stellar remnants confirm this idea. Theory likewise predicts the observed distinct differences in heavy-element abundance between the old globular cluster stars and the younger galactic cluster stars. The youngest stars contain the most heavy elements, which suggests that these elements are slowly produced over time and that each new generation of stars increases the concentration of heavy elements in the interstellar clouds from which the next generation forms. In the past three chapters, we have seen all of the ingredients that make up the complete cycle of star formation and evolution in our Galaxy. Let us now briefly summarize that process, which is illustrated in Figure 23.18.

Stars form when part of an interstellar cloud is compressed beyond the point where it can support itself against its own gravity. The cloud collapses and fragments, forming a cluster of stars. The hottest stars heat and ionize the surrounding gas, sending shock waves through the surrounding cloud, possibly triggering new rounds of star formation.

Within the cluster, stars evolve. The most massive stars evolve fastest, creating heavy elements in their cores and spewing them forth into the interstellar medium in supernova explosions. The lighter stars take longer to evolve,

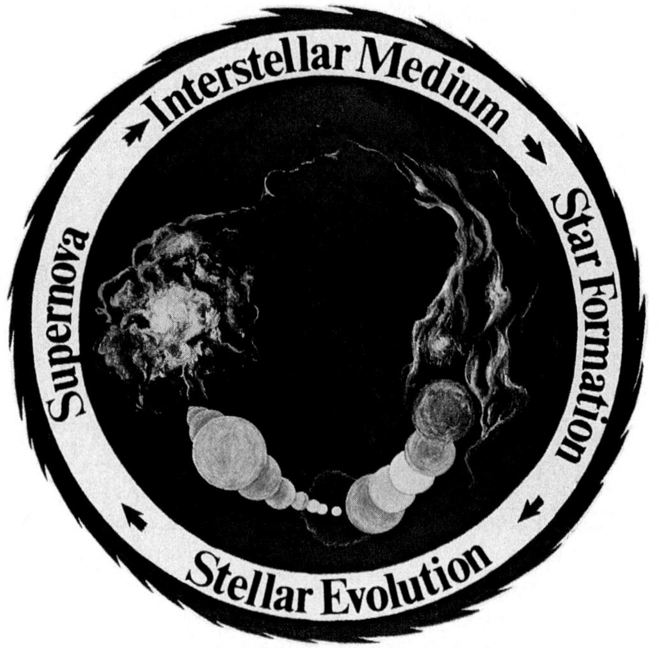

Figure 23.18 The cycle of star formation and evolution continuously replenishes the Galaxy with new heavy elements and provides the driving force for the creation of new generations of stars. (L. J. Chaisson)

but they too can create heavy elements and may contribute to the ''seeding'' of interstellar space when they shed their envelopes as planetary nebulae. The creation and dispersal of new heavy elements are accompanied by further shock waves. Their passage simultaneously enriches the interstellar medium and compresses it into further star formation.

In this way, although some material is used up in each cycle—turned into energy or locked up in low-mass stars—the Galaxy continuously recycles its matter. Each new round of formation creates stars with more heavy elements than the preceding generation had. From the old, metal-poor globular clusters to the young, metal-rich open clusters, we observe this enrichment process in action. Our Sun is the product of many such cycles. We ourselves are another. Without the heavy elements synthesized in the hearts of stars, life on Earth would not exist.

CHAPTER SUMMARY

Novae are stars that increase markedly, but temporarily, in brightness. They are explosions of the surface of a white dwarf that is accreting matter from a companion star.

Type-I supernovae are hydrogen-poor and have a light curve similar in shape to that of a nova. These explosions result from the detonation of a carbon white dwarf. Type-II supernovae are hydrogen-rich and have a characteristic bump in the light curve a few months after maximum, caused by the decay of radioactive elements created during the early moments of the supernova. They result from the implosion-explosion of the iron core of a massive star. Only six galactic supernovae have been recorded in the past 1000 years, but studies suggest that a supernova can be expected within 10 pc of our Sun roughly every 500 million years. Supernova remnants provide ample evidence that stars have exploded in our Galaxy. A star can become a nova many times, but a supernova only once.

Because the conditions producing supernovae are very similar even in the cores of stars of different masses, supernovae of a given type tend to release very similar amounts of energy. As a result, astronomers can use them as distance indicators.

The nuclear reactions within stars produce energy and synthesize heavy elements. Stars ultimately are responsible for much of the matter that we see around us, though hydrogen and most helium date back to the beginning of the universe. As a star evolves, elements form first by fusion of like nuclei, then by helium capture at later stages. In evolved stars, proton capture and neutron capture also play a role in the formation of some nuclei. These nuclei are not as plentiful as those produced by helium capture because helium-capture reactions are more common.

At a core temperature of 3 billion K, photodisintegration breaks apart some heavy nuclei, providing many helium-4 nuclei for synthesis of even more massive elements, ultimately leading to a buildup of iron-56 in the stellar core. Iron is the most stable of all elements. It cannot fuse with other nuclei to produce energy, so an iron core robs a star of its pressure support. It is then that the core begins to collapse.

Neutrons are produced as by-products of many nuclear reactions. With no repulsive electromagnetic barrier to overcome—neutrons carry no charge—neutrons can easily combine with nuclei, pushing their masses far beyond that of iron. If the resultant nucleus is unstable, radioactive decay follows, and no heavier nuclei can be formed. During a supernova explosion rapid neutron capture occurs, producing the heaviest nuclei of all. The neutron capture rate is so great at supernovae that even unstable nuclei can capture many neutrons before they have time to decay. Laboratory experiments support the theory of stellar nucleosynthesis.

Supernova 1987A was detected in the Large Magellanic Cloud during February of 1987. Although the color and size of the supernova's progenitor star were surprising, stellar evolutionary theory has been able to account for the supernova's appearance.

KEY WORDS

accretion disk
alpha process
carbon detonation supernova
Chandrasekhar mass
helium capture
neutron degeneracy pressure
neutronization
nova
photodisintegration
progenitor
recurrent nova
r-process
s-process
standard candle
stellar nucleosynthesis
supernova
supernova remnant
type-I supernova
type-II supernova

REVIEW QUESTIONS

1. What makes a nova?

2. What is a light curve?

3. What is a supernova?

4. How often can we expect to see a supernova?

5. What evidence is there that many supernovae have occurred?

6. According to historical accounts, how did the explosion creating the Crab Nebula appear to observers on Earth?

7. How do supernovae work as "standard candles"?

8. Which elements existed in the early universe?

9. How were all of the other elements in the universe formed?

10. What proof do astronomers have that heavy elements are formed in stars?

11. As a star evolves, why do heavier elements tend to form by helium capture rather than by fusion of like nuclei?

12. Why do stars' cores evolve into iron?

13. How are nuclei heavier than iron formed?

14. What is the r-process?

15. What makes a massive star collapse?

16. The Crab Nebula is now about 1 pc in radius. It was observed to explode in A.D. 1054, so roughly how fast is it expanding? (Assume constant expansion velocity. Is this a reasonable assumption?)

17. A certain telescope could just detect the Sun at a distance of 10,000 pc. What is the maximum distance at which it could detect a nova with a peak luminosity of 10^5 solar luminosities? Repeat the calculation for a supernova with a peak luminosity 10^{10} times that of the Sun.

18. At what distance would the supernova in question 17 look as bright as the Sun? Would you expect a supernova to occur that close to us?

DISCUSSION QUESTIONS

1. Why was the appearance of a naked-eye nova or supernova so amazing to people a few centuries ago? What differences might there be between current views of novae and supernovae and those of the ancients?

2. Write a report on Supernova 1987A. What are some of the major discoveries associated with it? Has the neutron star created by the supernova been discovered yet?

PROJECT

1. In 1758, the comet hunter Charles Messier discovered the sky's most legendary supernova remnant, now called M1, or the Crab Nebula. The Crab can be detected with a 3- or 4-inch telescope. It appears as a whitish patch, located northwest of Zeta Tauri, the star that marks the southern tip of the horns of Taurus the Bull. A 6-inch telescope reveals the Crab's oval shape. A 10-inch or larger telescope reveals some of its famous filamentary structure. What elements were created by the star that exploded to form the Crab? What elements may have been created in the explosion? When you look at the Crab Nebula, are you gazing at a collection of elements that may one day be incorporated into planets or even living beings?

SUGGESTED READINGS

Lattimer, J. M., and A. S. Burrows. "Neutrinos from Supernova 1987A." *Sky and Telescope* (October 1988). Scientists' first ringside seat at the formation of a neutron star.

Malin, D., and D. Allen. "Echoes of the Supernova." *Sky and Telescope* (January 1990). How light from Supernova 1987A revealed the detailed, three-dimensional structure of interstellar dust sheets in the Large Magellanic Cloud.

Maran, S. P. "In Our Backyard, a Star Explodes." *Smithsonian* (April 1988). Popular articles about the first supernova visible to the naked eye since 1604.

———. "A Supernova in Our Backyard." *Sky and Telescope* (April 1987). An early report on the discovery of a naked-eye supernova.

———. "Supernova Shines On." *Sky and Telescope* (May 1987). More on SN 1987A.

———. "Supernova 1987A's Changing Face." *Sky and Telescope* (July 1988). The supernova, a year and a half later.

Seward, F. "Neutron Stars in Supernova Remnants." *Sky and Telescope* (January 1986). The discovery of many more pulsars and neutron stars in supernova remnants.

Woosley, S., and T. Weaver. "The Great Supernova of 1987." *Scientific American* (August 1989). Comprehensive article on Supernova 1987A.

CHAPTER

24

Neutron Stars and Black Holes

Strange States of Matter

This artist's illustration depicts one possible "central engine" at the core of an active galaxy—double jets of matter protruding from a flattened accretion disk surrounding a supermassive black hole. Although the central hole is too small to be seen in this view, the broad pancake-shaped accretion disk is clearly visible. Superheated matter escapes away along the rotation axis of the black hole in the form of jets that often extend for thousands of light-years beyond the galaxy. (Courtesy D. Berry)

LEARNING GOALS

1. to understand how neutron stars and black holes fit into the overall theory of stellar evolution
2. to realize that neutron stars can be observed and studied through their regular pulsed energy emission
3. to appreciate how black holes can trap matter and radiation
4. to recognize that the bizarre phenomena near black holes result from the severe warping of the space around them
5. to understand various implications of space travel near black holes
6. to know some of the regions in the Galaxy where researchers suspect black holes are hiding

Stars spend most of their lives quietly, steadily burning hydrogen on the main sequence. Subsequently, a star changes very rapidly, passing through a series of unstable phases as it evolves toward its death. Just how a star ends its life depends strongly on its mass. Our study of stellar evolution has led us to some very unusual and unexpected objects. Red giants, white dwarfs, and supernova explosions surely represent extreme states of matter completely unfamiliar to us here on Earth. Yet stellar evolution can have even more extreme consequences. The strangest states of all result from the catastrophic implosion-explosion of stars much more massive than our Sun.

Even the almost unimaginable violence of a supernova explosion may not represent the end of the road for a massive star. A small, incredibly dense remnant can remain behind—a neutron star, so tightly squeezed by its own gravity that its component particles are literally squashed against one another. Many neutron stars have been observed in our Galaxy, and observational methods exist for probing their interior structure, allowing us to refine our theories of how they work. But even the remarkable properties of those exotic, ultracompressed objects pale in comparison with the fate that may await the most massive stars. Black holes are objects so extreme in their behavior that they require us to reconsider some of our most hallowed laws of physics. They open up a science-fiction writer's dream of bizarre phenomena, and may one day force scientists to construct a whole new theory of the universe.

NEUTRON STARS

✓1 What remains after a supernova explosion? Is the entire progenitor star just blown to bits and ejected into interstellar space, or does some portion of it survive? For a type-I (carbon detonation) supernova, most astronomers regard it as quite unlikely that any central remnant is left after the explosion. The entire star is shattered by the blast. However, for a type-II supernova, involving the implosion and subsequent rebound of a massive star's iron core, theoretical calculations indicate that part of the star may survive. Just as the expulsion of a red giant's envelope into space as a planetary nebula leaves the star's compact core behind as a white dwarf, the much more violent type-II supernova explosion destroys the parent star but leaves a tiny compressed remnant at its center. Even by the high-density standards of a white dwarf, however, the matter within this severely compressed core is in a very strange state, unlike anything we are ever likely to find (or create) on Earth.

Core Collapse

Recall from Chapter 23 that during the moment of implosion of a massive star—just prior to the supernova explosion itself—the electrons in the core violently smash into the protons there, forming neutrons and neutrinos. The free electrons were there all along; the protons were freed when the heavy nuclei disintegrated under the bombardment of high-energy photons as the core collapsed and its temperature rose. The neutrinos leave the scene at (or nearly at) the speed of light, accelerating the collapse of the neutron core, which continues to contract until its particles come into contact. At that point, neutron degeneracy pressure causes the central portion of the core to rebound, creating a powerful shock wave that races outward through the star, violently expelling matter into space. At these very high densities, even the neutrinos become trapped, and they too help to drive the explosion.

The key point here is that the shock wave does not start at the very center of the collapsing core. The innermost part of the core—the region that bounces—remains intact as the shock wave it causes destroys the rest of the star. After the violence of the supernova has subsided, this ball of neutrons is all that is left. Researchers colloquially call this core remnant a **neutron star,** although it is not a star in any true sense of the word—all of its nuclear reactions have ceased forever.

Neutron stars are extremely small and very massive. Composed purely of neutrons packed together in a tight ball about 20 km across, a typical neutron star is not much bigger than a small asteroid or a terrestrial city (see Figure 24.1), yet its mass is greater than that of the Sun. With so much mass squeezed into such a small volume, neutron stars are incredibly dense. Their average density can reach 10^{14} or even 10^{15} g/cm^3, nearly a billion times denser than a white dwarf. A single thimbleful of neutron-star material would weigh 100 million tons—about as much as a good-sized terrestrial mountain. Even the density of a normal atomic nucleus is "only" 10^{14} g/cm^3. In a sense, we can think of a neutron star as a single enormous nucleus, with an "atomic mass" of around 10^{57}.

Neutron stars are solid objects. Provided that a sufficiently cool one could be found, you might even imagine standing on it. However, this would not be easy, as a neutron star's gravity is extremely powerful. A 150-pound (70-kg) human being would weigh the Earth-equivalent of about 1 million tons (1 billion kg). The severe pull of a neutron star's gravity would squeeze you much thinner than this piece of paper!

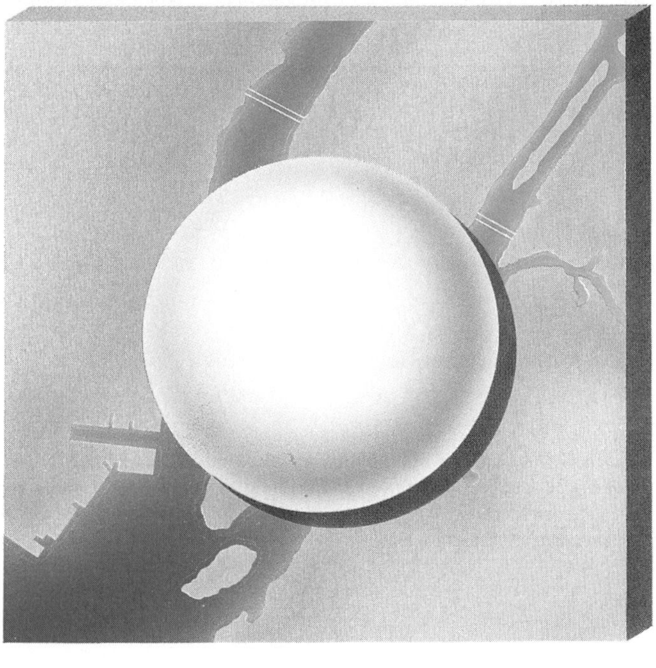

Figure 24.1 *Neutron stars are not much larger than many of Earth's major cities. In this fanciful comparison, a typical neutron star sits atop Manhattan Island.*

Rotation and Magnetism

Apart from large mass and small size, neutron stars have two other very important properties. First, newly formed neutron stars rotate extremely rapidly, with periods measured in fractions of a second. This is a direct result of the law of conservation of angular momentum (see Chapter 17), which tells us that any rotating body must spin faster as it shrinks. The collapsing iron (and later neutron) core of an evolved star is no exception to this law. Even if the core of the progenitor star were initially rotating quite slowly (once a month, say), it would be spinning about once a second by the time it had reached a radius of 10 km.

The second important property of a young neutron star is its strong magnetic field. The original field of the progenitor (parent) star is amplified by the collapse of the core because the contracting material squeezes the field lines closer together, increasing the field strength to a value on the order of a *trillion* times that of Earth's field (millions of times stronger than the fields found even in the hearts of the most violent solar flares).

These processes of spin-up—a speeding up of the rotation—and magnetic field strengthening are similar to those discussed at the start of Chapter 21 in the context of interstellar cloud collapse prior to star formation. The main difference here, in the context of neutron stars, is that these processes operate on a much more energetic scale. In time, theory indicates, our neutron star will spin more slowly as it radiates its energy into space, and its magnetic field will diminish. However, for a few million years after its birth, these two properties combine to provide the primary means by which this strange object can be detected and studied, as we now discuss.

Pulsars

✓2 Can we be sure that objects as strange as neutron stars really exist? The answer is a confident yes. The first observation of a neutron star occurred in 1967, when Jocelyn Bell, a graduate student at Cambridge University, made a remarkable discovery. She observed an astronomical object emitting radio radiation in the form of rapid pulses lasting about 0.01 s (second) apiece. Each pulse contained a burst of radiation, after which there was nothing. Then, after 1.34 s, another pulse arrived. The time interval between pulses was astonishingly uniform—so accurate, in fact, that the repeated emissions could be used as a very precise clock. Figure 24.2 is a recording of the periodic radio radiation from the pulsating object Bell discovered.

Many hundreds of these pulsating objects are now known in our Milky Way Galaxy. They are called **pulsars.** Each has its own characteristic pulse period and duration that repeat indefinitely. The pulse periods of some pulsars are so stable that they are by far the most accurate natural clocks known in the universe—more accurate even than the best atomic clocks on Earth. In some cases, the period is

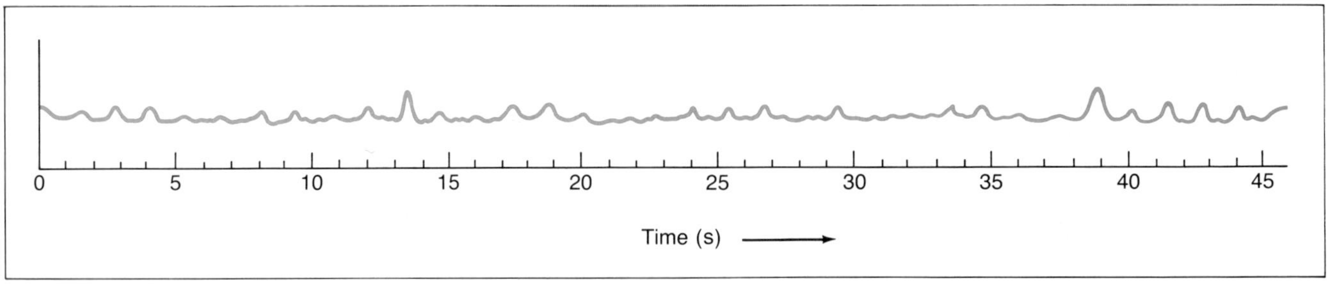

Figure 24.2 *Pulsars emit periodic bursts of radiation. This recording shows the regular change in the intensity of the radio radiation emitted by the first such object known. It was discovered in 1967. The pulse period is 1.34 seconds.*

predicted to change by only a few seconds in a million years.

A few pulsars are directly associated with supernova remnants, although not all such remnants have a detectable pulsar within them. The pulsar whose radio "light curve" is shown in Figure 24.3 resides close to the center of the Crab Nebula remnant (discussed in Chapter 23). By observing the velocity and direction of travel of the Crab's ejected matter, we can work backward to pinpoint the location in space at which the explosion must have occurred and where the supernova core remnant should be located. That is precisely the region of the Crab Nebula from which the pulsating signals arise. The Crab pulsar is evidently all that remains of the once-massive star whose supernova was observed in 1054.

Most pulsars emit their pulses in the form of radio radiation. Some have been observed to pulse in the visible, x-ray, and gamma-ray parts of the spectrum as well. Whatever types of radiation are produced, all of these electromagnetic flashes are synchronized—that is, occurring at regular, repeated time intervals—as we would expect if they arose from the same astronomical object. The period of most pulsars is usually short—ranging from about 0.03 to 0.3 s, corresponding to a flashing rate of between 3 and 30 times per second. The human eye is insensitive to such rapid flashes, making it impossible to observe the flickering of a pulsar with the naked eye or even using a large telescope. Fortunately, instruments can record pulsations of light that the human eye cannot. Figure 24.4 shows a series of optical photographs of the Crab pulsar. In some frames, the pulsar is on; in others, it is off.

When Jocelyn Bell made her discovery in 1967, she did not know what she was looking at. Indeed, no one at the time knew what a pulsar really was. The explanation of pulsars as spinning neutron stars won Bell's thesis advisor, Anthony Hewish, the 1974 Nobel Prize in physics. Hewish reasoned that the only physical mechanism consistent with such precisely timed pulsations is a small, rotating source of radiation. Only rotation can cause the high degree of regularity of the observed pulses, and only a small object can account for the sharpness of each pulse. Radiation emitted from different regions of an object larger than a few tens of kilometers across would arrive at Earth at slightly different times, blurring the pulse profile. The best current model

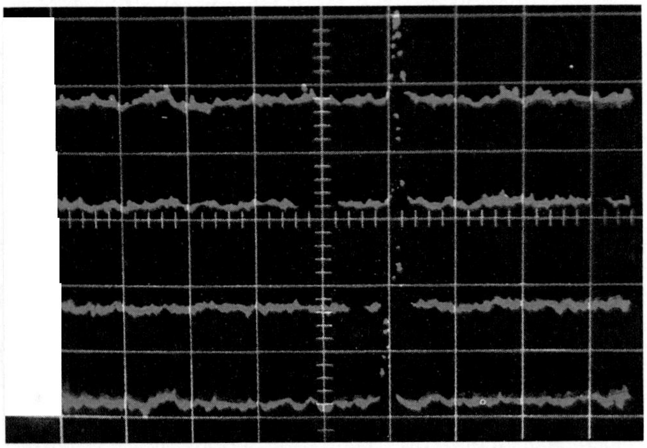

Figure 24.3 *This graph, taken from a video screen, shows the variation in the intensity of the radio radiation emitted by the pulsar at the center of the Crab Nebula.*

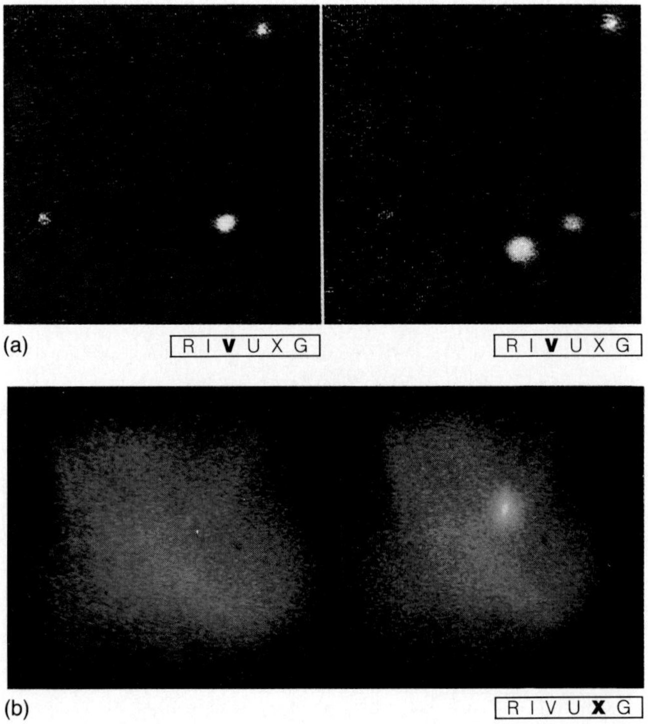

Figure 24.4 *The pulsar in the core of the Crab Nebula blinks on and off about 30 times each second. (a) In this pair of closely spaced optical images, the pulsing can be seen clearly. (b) The same phenomenon is also detected in x-rays. (Lick Observatory; NASA)*

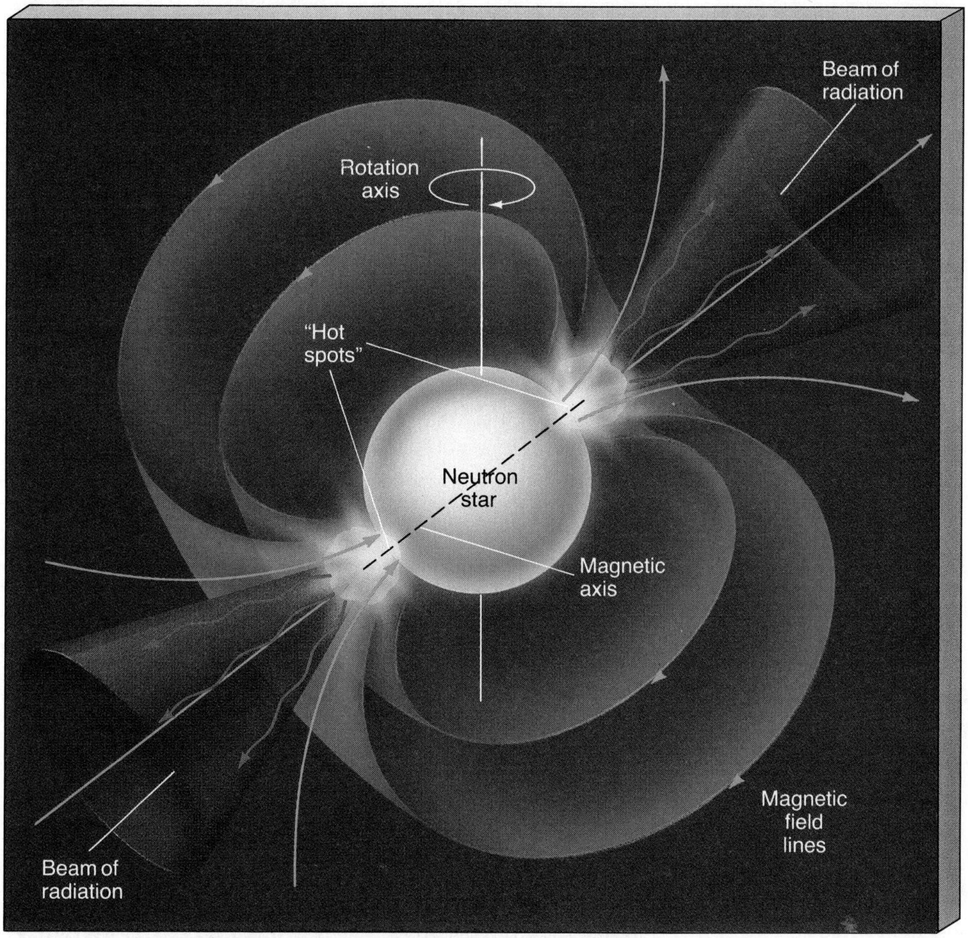

Figure 24.5 This diagram of the "lighthouse model" of neutron-star emission accounts for many of the observed properties of pulsars. As depicted, charged particles, accelerated by the magnetism of the neutron star, flow along the magnetic field lines producing radio radiation that beams outward.

describes a pulsar as a compact, spinning neutron star that periodically flashes radiation toward Earth.

Figure 24.5 outlines the important features of this pulsar model. A "hot spot" on the surface of a neutron star, or in the magnetosphere above it, continuously emits radiation in a narrow "searchlight" pattern. This spot is most likely a localized region near one of the neutron star's magnetic poles, where charged particles, accelerated to extremely high energies by the star's rotating magnetic field, emit radiation along the star's magnetic axis. The hot spot radiates more or less steadily, and the resulting beam sweeps through space, like a revolving lighthouse beacon, as the neutron star rotates. Indeed, this pulsar model is often known as the **lighthouse model.** The beam is observed as a series of rapid pulses—each time the beam sweeps past Earth, a pulse is seen. The period of the pulses is the star's rotation period.

All pulsars are neutron stars, but not all neutron stars are pulsars—at least, as viewed from Earth. The reason for this is simple. The pulsar beam depicted in Figure 24.5 is relatively narrow—perhaps only a few degrees across. Unless the neutron star happens to be oriented in just the right way, the beam never sweeps across Earth, and we never see a pulsar. However, given our current knowledge of star formation, stellar evolution, and neutron stars, pulsar observations are quite consistent with the idea that *every* high-mass star dies in a supernova, leaving a neutron star behind, and that *all* neutron stars emit beams of radiation, just like the pulsars we actually see.

Pulsar "Glitches" and Neutron Starquakes

We mentioned above that pulsars are very accurate clocks. In fact, their spin rate decreases very slowly as they radiate their energy into space and slow down. However, occasionally some pulsars are observed to suffer a sudden small drop, or **"glitch,"** in the pulse period. The pulsar rotates at a (nearly) constant rate for a long time, then abruptly changes its rotation speed and continues to spin at that new rate. Far from being an annoying imperfection in an otherwise perfectly predictable system, however, these pulsar glitches allow astronomers to probe the interior structure of neutron stars.

Based on theoretical calculations and observations of pulsar periods and glitches, a model of the interior structure of a neutron star has been developed. Researchers believe that a neutron star has a fluid interior, as illustrated in Figure 24.6. Despite the enormously high density, the neutrons are free to move around inside the star. Based on our present understanding of the properties of degenerate matter, the interior is believed to be a **superfluid**—that is, able to flow without friction—and possibly also a **superconductor**—able to conduct electricity without any resistance. At the surface lies a 1-km thick solid "crust" made up of

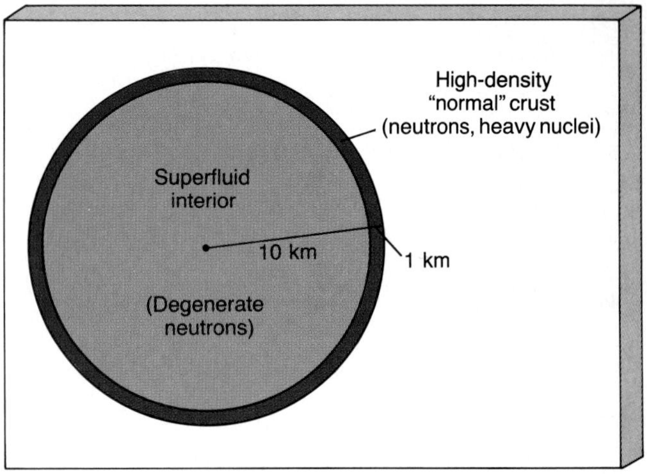

Figure 24.6 *The interior structure of a neutron star, according to theoretical models and pulsar observations. A thin, dense, solid crust about 1 km thick sits atop an even denser superfluid interior.*

nondegenerate neutrons and high-density "normal" material, such as iron nuclei and electrons.

Occasionally, under the influence of the neutron star's immense gravity, the crust cracks and settles slightly. This rearrangement of the crust is often called a **starquake.** It is far more violent (by a factor of at least a trillion) than any terrestrial earthquake. A starquake slightly reduces the radius of the neutron star. As a result, the star's spin increases slightly, to conserve angular momentum, and a small drop in the pulse period—a pulsar glitch—is the result. Just as seismology on Earth and helioseismology on the Sun allow scientists to understand more fully the interiors of planets and stars, observations of glitches in pulsars allow us to study the interiors of neutron stars.

Neutron-Star Binaries

In Chapter 19, we saw that most stars are not single but instead are members of binary systems. While many pulsars are known to be isolated (that is, not part of any binary), there is strong evidence that at least some neutron stars do have binary companions.

The late 1970s saw several important discoveries relating to neutron stars in binary-star systems. First, intense bursts of x-rays were discovered from sources near the central regions of our Galaxy and also near the centers of rich star clusters. These x-ray emitting systems, known as **x-ray bursters,** radiate thousands of times more energy than our Sun, in rapid bursts that last only a few seconds. A typical burst is shown in Figure 24.7.

These bursts are thought to arise on neutron stars that are members of binary systems. Matter torn from the surface of the (main-sequence) companion by the neutron star's strong gravitational pull accumulates on the neutron star's surface. As the gas builds up, its temperature rises, due to the pressure of overlying material. Eventually, it becomes hot enough to fuse hydrogen. The result is a sudden period of rapid nuclear burning that releases a huge amount of energy in a brief but intense flash of x-rays—an x-ray burst. After several hours of renewed accumulation, a fresh layer of matter produces the next burst. Thus, an x-ray burst is just like a nova explosion on a white dwarf (discussed in Chapter 23), but on a far more violent scale because of the neutron star's much stronger gravity.

At around the same time as the first x-ray bursts were seen, military satellites detected **gamma-ray bursters**—bright, irregular flashes of gamma-rays lasting only a few seconds. Most astronomers believe that gamma-ray bursters are "scaled up" versions of x-ray bursters, in which matter from the binary companion experiences particularly violent nuclear burning, accompanied by the release of gamma-rays.

SS433

Another important discovery of the 1970's is one of the most remarkable objects in the entire Milky Way Galaxy. It is known as SS433, a name that simply identifies it as the 433rd entry in a particular catalog of stars with strong optical emission lines. SS433 emits both radio waves and

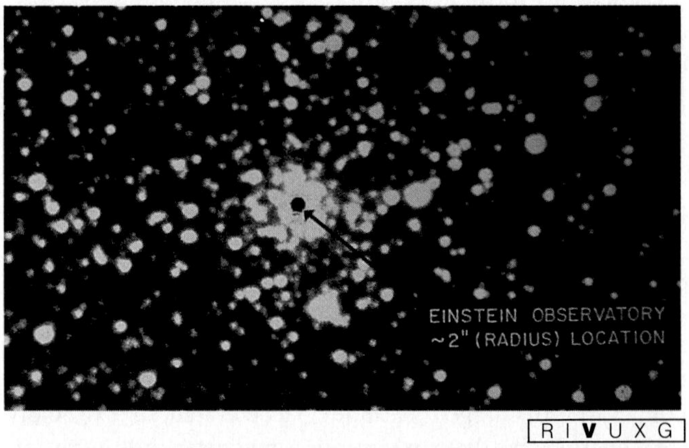

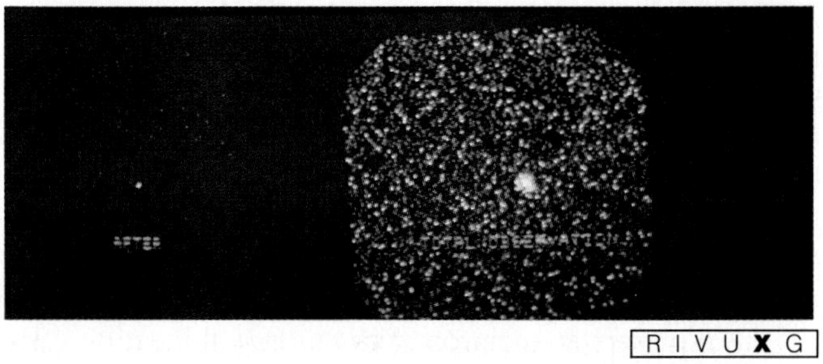

Figure 24.7 *An x-ray burster produces a sudden, intense flash of x-rays, followed by a period of relative inactivity lasting as long as several hours. Then another burst occurs. The bursts are thought to be caused by explosive nuclear burning on the surface of an accreting neutron star, similar to the explosions on a white dwarf that give rise to novae. The upper image is an optical photograph of the star cluster Terzan 2, showing a 2″ dot at the center where the x-ray bursts originate. The lower photo is an X-ray image taken during the outburst. (Harvard-Smithsonian Center for Astrophysics; NASA)*

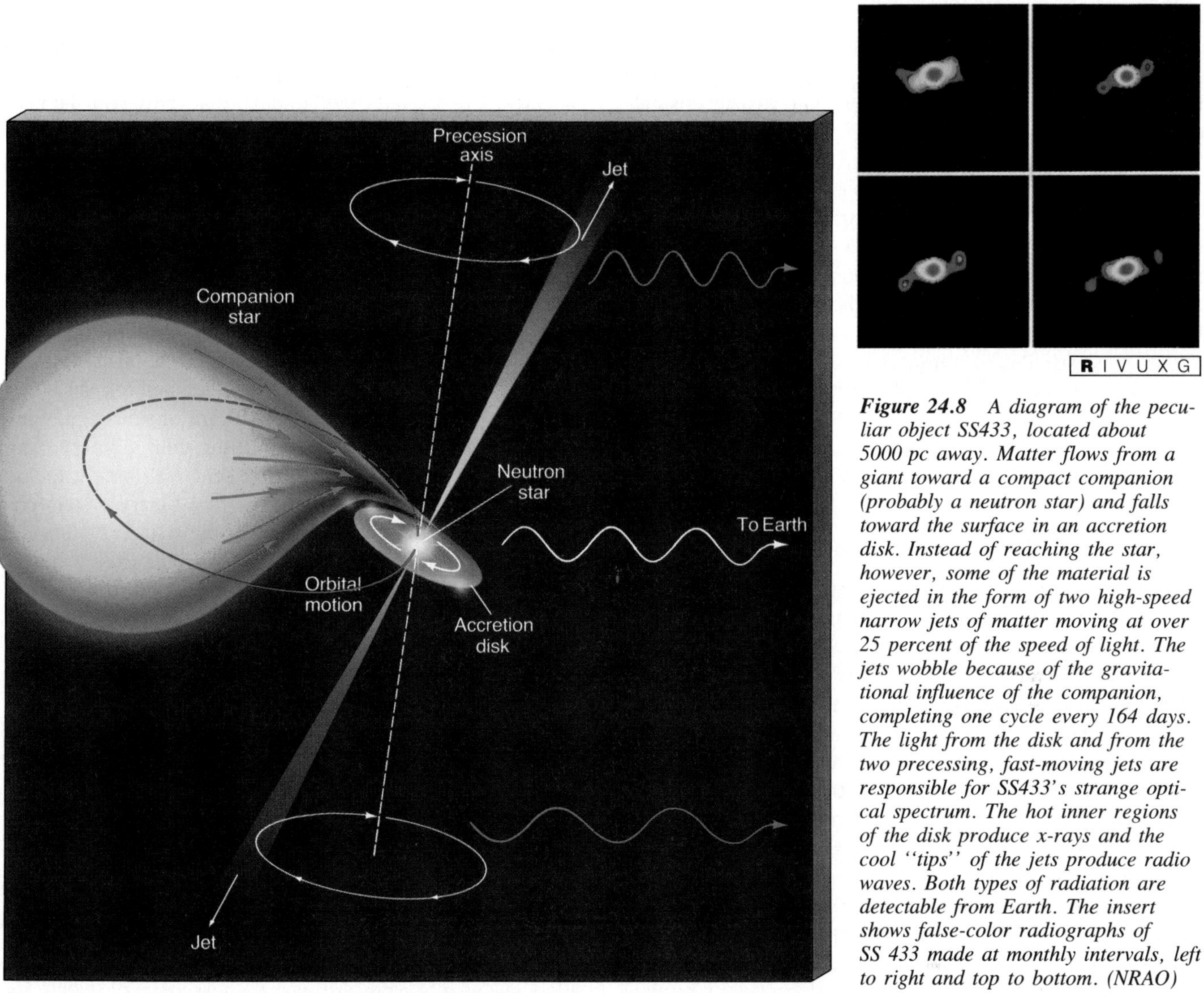

Figure 24.8 A diagram of the peculiar object SS433, located about 5000 pc away. Matter flows from a giant toward a compact companion (probably a neutron star) and falls toward the surface in an accretion disk. Instead of reaching the star, however, some of the material is ejected in the form of two high-speed narrow jets of matter moving at over 25 percent of the speed of light. The jets wobble because of the gravitational influence of the companion, completing one cycle every 164 days. The light from the disk and from the two precessing, fast-moving jets are responsible for SS433's strange optical spectrum. The hot inner regions of the disk produce x-rays and the cool "tips" of the jets produce radio waves. Both types of radiation are detectable from Earth. The insert shows false-color radiographs of SS 433 made at monthly intervals, left to right and top to bottom. (NRAO)

x-rays, but its main claim to fame is its very unusual optical spectrum. It shows *three* distinct sets of optical emission lines. One set has a fairly small Doppler shift, corresponding to a velocity of a few tens of kilometers per second, that alternates back and forth (that is, from approach to recession and back) every 13 days. The other two sets of spectral lines each have very large Doppler shifts, indicating velocities on the order of 10 to 20 percent of the speed of light—tens of *thousands* of kilometers per second. One set is redshifted, the other is blueshifted, and the two sets shift back and forth every 164 days. According to its spectrum, SS433 is simultaneously racing away from us, racing toward us, and virtually standing still! For a few months in 1978, astronomers scrambled to find an explanation for this bizarre behavior.

After an extensive series of optical observations, Bruce Margon (University of Washington) and his collaborators came up with the following model, now generally accepted, that explains SS433's peculiar spectrum. SS433 is a binary system, with a neutron star and an ordinary companion orbiting each other once every 13 days. Matter from the companion falls onto the neutron star, spiraling down onto the surface in an accretion disk (just as discussed in Chapter 23, in the case of white-dwarf novae), heating up as it goes. The low-velocity spectral lines (the ones with the 13-day period) come from the disk. Their Doppler shifts reflect the orbital motion of the binary.

Not all of the infalling material is accreted onto the neutron star, however. SS433 also ejects more than an Earth mass of material every year in the form of two oppositely directed narrow jets of gas moving roughly perpendicular to the disk at just over 25 percent of the speed of light, as shown in Figure 24.8. (The velocities quoted above are the components of this motion along our line of sight.) Jets of this sort are apparently quite common in astronomical systems where an accretion disk surrounds a compact object (such as a neutron star or a black hole). They are believed to be produced by the intense radiation and magnetic fields near the inner edge of the disk, although the details of their formation are still uncertain. While SS433 is the only stellar object known to produce jets, we will see examples of similar phenomena on much larger scales in later chapters. One of the most important aspects of SS433 is that we can actually study both the disk and the jets, instead of simply having to assume their existence, as in more distant cosmic objects.

The gravitational pull of the companion on the neutron star and the accretion disk causes the two high-speed jets to wobble like a spinning top, tracing out a complete cone every 164 days. This motion is similar to the precession of the Earth's rotation axis caused by the combined gravitational pulls of the Moon and the Sun, as discussed in Chapter 1. The precession of the hot gas in the jets explains the variability of SS433's strongly Doppler-shifted emission lines. As illustrated in Figure 24.8, when the jets happen to be almost perpendicular to our line of sight, only a small shift is seen; when they lie most nearly along the line of sight, the shift (red or blue) is much larger.

Millisecond Pulsars

In the mid-1980s an important new category of pulsars was found—a class of very rapidly rotating objects called **millisecond pulsars.** Several dozen millisecond pulsars are currently known in the Milky Way Galaxy. These objects spin hundreds of times per second (that is, their pulse period is a few 1 milliseconds, 0.001 s). This speed is about as fast as a typical neutron star can spin without flying apart. This fact suggests a phenomenon bordering on the incredible—a cosmic object of kilometer dimensions, more massive than our Sun, spinning almost at breakup speed, making nearly 1000 complete revolutions *every second*—yet the observations and their interpretation leave little room for doubt.

The story of these remarkable objects is further complicated because many of them (over 40 at last count) are found in globular clusters. This is odd because globular clusters are known to be very old—10 billion years, at least. Yet type-II supernovae (the kind that create neutron stars) are associated with massive stars that explode within a few tens of *millions* of years after their formation, and no stars have formed in any globular cluster since the cluster itself came into being. Thus, no new neutron star has been produced in a globular cluster in a very long time. But, as we have mentioned, the pulsar produced in a supernova explosion is expected to slow down in only a few million years. After 10 billion years, its rotation should have all but ceased. The rapid rotation of the pulsars found in globular clusters cannot be a relic of their birth. These objects must have been spun up—that is, had their rotation rates increased—by some other, much more recent, mechanism.

The most likely explanation for the high rotation rate of these objects is that the neutron star has been spun up by drawing in matter from a companion star. As we have seen, the infalling stream of matter does not fall directly onto the neutron star, but instead goes into orbit around it, forming an accretion disk (see Figure 24.8). As matter from the disk spirals down onto the star's surface, it provides the "push" needed to make the neutron star spin faster. Theoretical calculations indicate that this process can spin the star up to breakup speed in about a hundred million years. Subsequently, an encounter with another star may eject the neutron star from the binary, or the pulsar may destroy its companion, so that an isolated millisecond pulsar results. This general picture is supported by the finding that, of the 40 or so millisecond pulsars seen in globular clusters, 10 are known currently to be members of binary systems. These numbers are quite consistent with the rate at which binaries can be broken up by encounters with other cluster members.

Thus, although a pulsar like the Crab is the direct result of a supernova explosion, millisecond pulsars are the product of a two-stage process. The neutron star was formed in an ancient supernova, billions of years ago. Only relatively recently, through interaction with a binary companion, has it achieved the rapid spin we observe today. Once again, we see how members of a binary system can evolve in ways quite different from single stars. Notice that the picture of accretion onto a neutron star from a binary companion is the same picture that we just used to explain the existence of x-ray bursters. In fact, the two phenomena are very closely linked. Many x-ray bursters may be on their way to one day becoming millisecond pulsars.

The way in which a neutron star can come to be a member of a binary system is the subject of active research, because the violence of a supernova explosion would be expected to blow the binary apart in many cases. Only if the supernova progenitor lost a lot of mass before the explosion would the binary be likely to survive. Alternatively, by interacting with an existing binary and displacing one of its components, a neutron star may become part of a binary system *after* it is formed, as depicted in Figure 24.9. Only further observations will tell whether this scenario is possible, and astronomers are eagerly searching the skies for more millisecond pulsars to test their ideas.

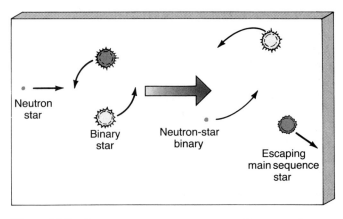

Figure 24.9 *A neutron star can encounter a binary made up of two low-mass stars, ejecting one of them and taking its place. This mechanism provides a means of forming a binary system with a neutron-star component (which may later evolve into a millisecond pulsar) without having to explain how the binary survived the supernova explosion that formed the neutron star.*

Pulsar Planets?

In January 1992, radio astronomers at the Arecibo Observatory found that the pulse period of a recently discovered millisecond pulsar lying some 500 pc from Earth varied in an unexpected but quite regular way. Careful analysis of the data has revealed that the period fluctuates on two distinct time scales, one of 67 days, the other of 98 days. The changes in the pulse period are small—less than one part in 10^7—but repeated observations seem to confirm their reality.

The leading explanation for these fluctuations holds that they are caused by the Doppler effect as the pulsar wobbles back and forth in space. But what causes the wobble? The Arecibo group believes it is the result of the combined gravitational pulls of not one, but *two* planets, each about three times the mass of the Earth! One orbits the pulsar at a distance of 0.4 and the other at a distance of 0.5 A.U. The 67-day and 98-day fluctuations are the planets' orbital periods.

If this explanation holds up, it will be the first definite evidence for planets outside the solar system. However, even if the results are confirmed, it is unlikely that these planets formed in the same way as our own. Any planetary system that orbited the pulsar's progenitor star was almost certainly destroyed in the supernova explosion that created the pulsar. As a result, scientists are still uncertain about how these planets came into being. One possibility involves the binary companion, which provided the matter necessary to spin the pulsar up to millisecond speeds. The pulsar's intense radiation and strong gravity may have destroyed the companion, spreading its matter out into a disk (rather like the solar nebula), in whose cool outer regions the planets condensed. These measurements are difficult and their interpretation still controversial. The first highly publicized claim (by a British group) of the discovery of a planet orbiting a pulsar was publicly withdrawn at the same meeting where the Arecibo researchers announced these results.

Astronomers have been searching for decades for planets orbiting main-sequence stars like our Sun, on the assumption that planets are a natural by-product of star formation (see Chapter 17). It would be ironic indeed if the first planets to be found outside the solar system orbited a dead star and had little or nothing in common with our own world.

Ultracompression

Neutron stars are very peculiar objects. Nevertheless, theory predicts that they are in equilibrium, just like most other stars. For neutron stars, however, equilibrium does not mean a balance between the inward pull of gravity and the outward pressure of hot gas. Instead, as we have seen, the outward force is provided by the pressure of tightly packed neutrons. Squeezed together, the neutrons form a hard ball of matter that not even gravity can compress further. Or do they? Is it possible that, given enough matter packed into a small enough volume, the collective pull of gravity can eventually crush *any* opposing pressure? Can gravity continue to compress a massive star into an object the size of a planet, a city, a pinhead—even smaller? The answer, apparently, is *yes*.

DISAPPEARING MATTER

The Final Stage of Stellar Evolution

√3 We have seen that the eventual evolution of a star depends critically on its mass. Low-mass stars leave behind a compact remnant known as a white dwarf. High-mass stars can also produce a compact remnant, in the form of a neutron star. The laws of physics make specific predictions about the masses of these core remnants. A white dwarf must be less than about 1.4 $M_\odot$ (solar masses)—the so-called Chandrasekhar mass, beyond which the electrons cannot support the core against its own gravity. Similarly, a neutron star resulting from a supernova must have a mass between about 1.4 and 3 solar masses.[1] The lower limit of 1.4 $M_\odot$ stems from the theory of stellar evolution: The iron core of an evolved star must exceed the Chandrasekhar mass for core collapse to begin and a supernova to occur. The upper limit of 3 $M_\odot$ is the neutron-star equivalent of the Chandrasekhar mass—beyond 3 $M_\odot$, not even tightly packed neutrons can withstand the star's gravitational pull.

In fact, we know of *no* force that can counteract gravity beyond the point where neutron degeneracy pressure is overwhelmed. If enough material is left behind after a supernova, as may happen in the case of an extremely massive progenitor star, gravity finally wins once and for all, and the central core collapses forever. As the core shrinks, the gravitational pull in its vicinity eventually becomes so great that even light itself is unable to escape. The resultant object therefore emits no light, no radiation, no information whatsoever. Astronomers call this bizarre end-point of stellar evolution, where a massive core remnant collapses in on itself and vanishes forever, a **black hole.**

Can an entire star simply shrink to a point and vanish? Doesn't this violate some law of physics? Does it really make sense to talk about black holes? These questions bring us to some very fundamental issues that are presently at the forefront of modern physics. Without some agent to com-

[1]The dividing lines at 1.4 and 3 $M_\odot$ are somewhat uncertain because they ignore the effects of magnetism and rotation, both of which are surely present in the cores of evolved stars. Because these effects can compete with gravity (as discussed in Chapter 21), they influence the evolution of stars. In addition, we do not know for certain how the basic laws of physics might change in regions of very dense matter that is both rapidly spinning and strongly magnetized. However, we expect that these dividing lines will shift generally upward when magnetism and rotation are included, because even larger amounts of mass will then be needed for gravity to compress stellar cores into neutron stars or black holes.

pete against gravity, the present laws of gravitational physics predict that a massive core remnant *will* collapse all the way to a point, where both its density and its gravitational field become infinite—a so-called **singularity.** However, we should not take this prediction of infinite density too literally. Singularities always signal the breakdown of the theory producing them. In other words, the present laws of physics are simply inadequate to describe the final moments of the star's collapse.

As it stands today, the theory of gravity is incomplete because it does not incorporate a proper (that is, a quantum-mechanical) description of matter on very small scales. As our collapsing stellar core shrinks to smaller and smaller radii, we eventually lose our ability to predict, or even describe, its behavior. However, we can at least estimate how small the core can get *before* quantum effects must become important. It turns out that by the time this stage is reached the core is already much smaller than any elementary particle. Although the correct description of the end-point of the collapse may well require a major overhaul of the laws of physics, for all practical purposes the prediction of collapse to a point is valid.

A complete analysis of Einstein's **theories of relativity** (see Interludes 24-1 and 24-2), which form the complex mathematical framework needed to understand the true nature of black holes, is far beyond the scope of this book. However, we can still usefully discuss many qualitative aspects of these strange regions of space. We can understand the essence of black holes by using the following two key facts from relativity: Nothing can travel faster than the

INTERLUDE 24-1 *Einstein's Theories of Relativity*

Albert Einstein won a Nobel Prize in 1921 for his explanation of the photoelectric effect, as described in Chapter 4. However, he is probably best known for his two theories of relativity, the successors to Newtonian mechanics (see Chapter 3) that form the foundation of twentieth-century physics.

The *special theory of relativity* (or just *special relativity*), proposed by Einstein in 1905, deals with the preferred status of the velocity of light. We have noted at numerous points in this text that the speed of light, c, is the maximum speed attainable in the universe. But there is more to it than that. In 1887, a fundamental experiment carried out by two American physicists, A. A. Michelson and E. W. Morley, demonstrated a further important and unique aspect of light—the measured speed of a beam of light is *independent* of the motion of the observer. No matter what our velocity may be, we always measure precisely the same value for c—299,792.458 km/s.

A moment's thought leads us to the conclusion that this is a decidedly nonintuitive statement. For example, if we were traveling in a car moving at 100 km/h and we fired a bullet forward with a velocity of 1000 km/h relative to the car, an observer standing at the side of the road would see the bullet pass by at 100 + 1000 = 1100 km/h, as illustrated below. However, if we were traveling in a rocket ship at 1/10 the speed of light, 0.1 c, and we shone a searchlight beam ahead of us, the Michelson-Morley experiment tells us that an outside observer would measure the speed of the

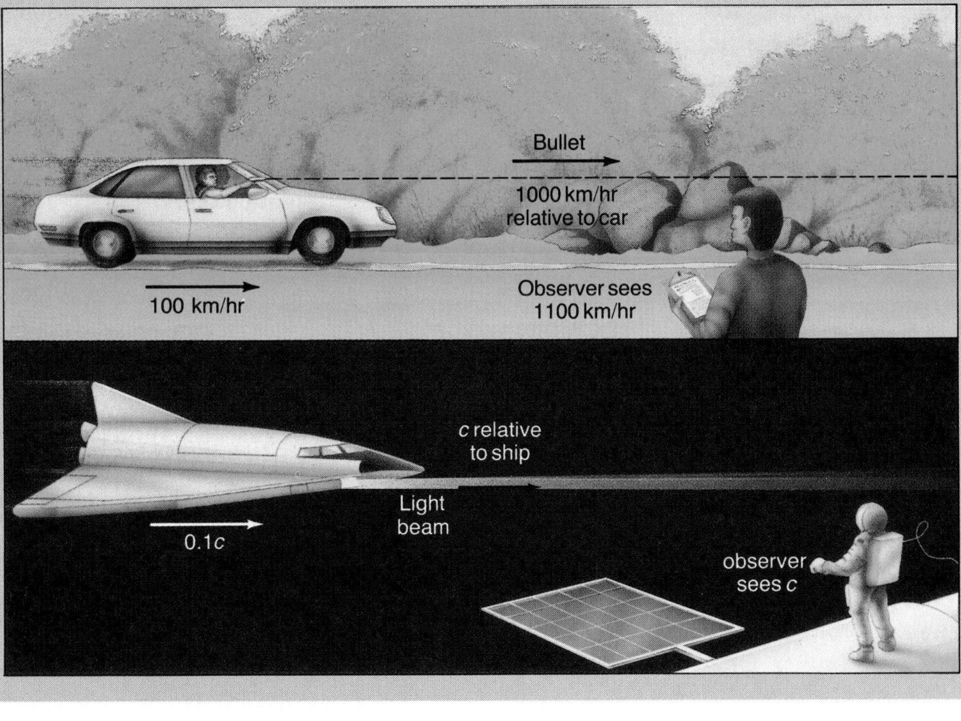

speed of light, and all things, *including light,* are attracted by gravity.

Escape Velocity

To explore how gravity attracts even light, let's consider the concept of escape velocity—the velocity needed for one object to escape from the gravitational pull of another. In Chapter 9, we noted that escape velocity is proportional to the square root of a body's mass divided by the square root of its radius. On Earth, with a radius of about 6500 km, the escape velocity is nearly 11 km/s. In order for any object—molecule, baseball, rocket, whatever—to be launched away from Earth, it must move faster than 11 km/s.

Consider now a hypothetical experiment in which Earth is squeezed on all sides by a gigantic vise. As our planet shrinks under the pressure, its mass remains the same. The escape velocity increases because the radius is decreasing. Suppose that Earth were compressed to one-fourth its present size. The proportionality just mentioned for escape velocity then predicts that the escape velocity would double (since $1/\sqrt{1/4} = 2$). Any object escaping from this hypothetically compressed Earth would need a velocity of about 22 km/s.

Imagine compressing the Earth still more. Squeeze it, for example, by an additional factor of 1000, making its radius hardly more than a kilometer. The escape velocity increases dramatically. In fact, a velocity of about 630 km/s would be needed to escape from an object having a radius of 1 km and a mass equal to that of the Earth.

beam not as 1.1 *c*, as the above example would suggest, but *c*. The rules that apply to particles moving at or near the speed of light are different from those we are used to in everyday life.

Special relativity is the mathematical framework that allows us to extend the familiar laws of physics from low speeds (that is, speeds much less than *c*, which are often referred to as *nonrelativistic*) to very high (or *relativistic*) speeds, comparable to *c*. Relativity is equivalent to Newtonian mechanics when objects move much more slowly than light, but it differs greatly in its predictions at relativistic velocities. For example, special relativity predicts that a rapidly moving spacecraft will appear to contract in the direction of its motion, its clocks will appear to run slow, and its mass will appear to increase. All of the theory's predictions have been repeatedly verified to very high accuracy. Today special relativity is at the heart of all physical science. No scientist seriously doubts its validity.

General relativity is what results when gravity is included in the framework of special relativity. In 1915, Einstein made the connection between special relativity and gravity with the following famous "thought experiment." Imagine that you are enclosed in an elevator with no windows, so that you cannot directly observe the outside world, and that the elevator is floating in space. You are weightless. Now suppose that you begin to feel the floor press up against your feet. Weight has apparently returned. There are two possible explanations for this, shown in the diagram at the right. A large mass could have come nearby, and you are feeling its downward gravitational attraction, *or* the elevator has begun to accelerate upward and the force you feel is accelerating you at the same rate. The crux of Einstein's argument is this: There is *no* experiment that you can perform within the elevator, without looking outside, that will let you distinguish between these two possibilities.

Thus, Einstein reasoned, there is no way to tell the difference between a gravitational field and an accelerated frame of reference (which would be the rising elevator in the thought experiment). Gravity can therefore be incorporated into special relativity as a general acceleration of all particles. However, another major modification to the theory of special relativity must be made. Central to relativity is the notion that space and time are not separate quantities, but instead must be treated as a single entity—*spacetime*. In order to incorporate the effects of gravity, the mathematics forces us to the conclusion that spacetime has to be *curved*.

In general relativity, then, gravity is a manifestation of curved spacetime. There is no such thing as a "gravitational field," in the Newtonian sense. Instead, objects move as they do because they follow the curvature of spacetime, and this curvature of spacetime is determined by the amount of matter present. We will explore some of the consequences of this view of gravity in more detail in the text.

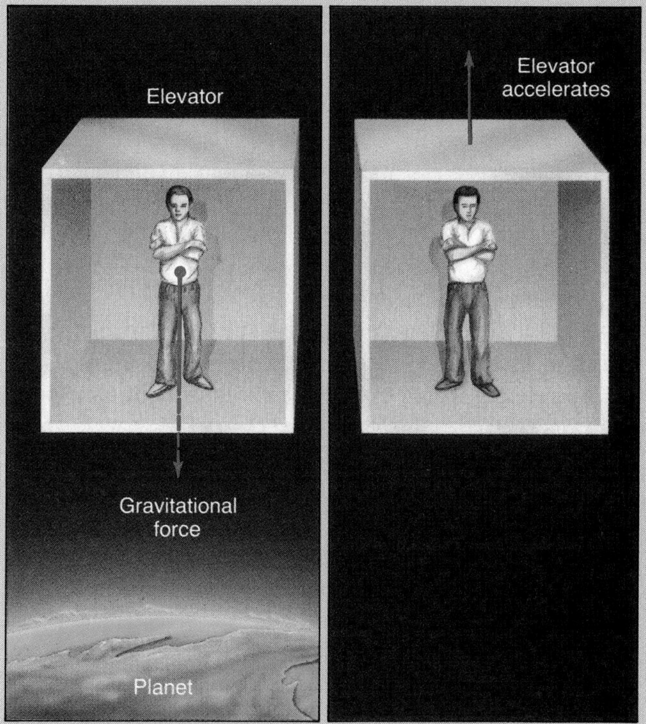

INTERLUDE 24-2 *Tests of General Relativity*

Special relativity is the most thoroughly tested and most accurately verified theory in the history of science. General relativity, however, is on less firm experimental ground than is special relativity.

The problem with verifying general relativity is that its effects on Earth and in the solar system—the places where we can most easily perform tests—are very small. Just as special relativity produces major departures from Newtonian mechanics only when velocities approach the speed of light, general relativity predicts large departures from Newtonian gravity only when extremely strong gravitational fields are involved—in effect, when orbit speeds and escape velocities become relativistic.

We will encounter other experimental and observational tests of general relativity elsewhere in this chapter. In this interlude, we will consider just two ''classical'' tests of the theory. These tests are solar system experiments that helped ensure acceptance of Einstein's theory. Bear in mind, however, that there are no known tests of general relativity in the ''strong-field'' regime—that part of the theory that predicts black holes, for example—so the full theory has never been experimentally tested.

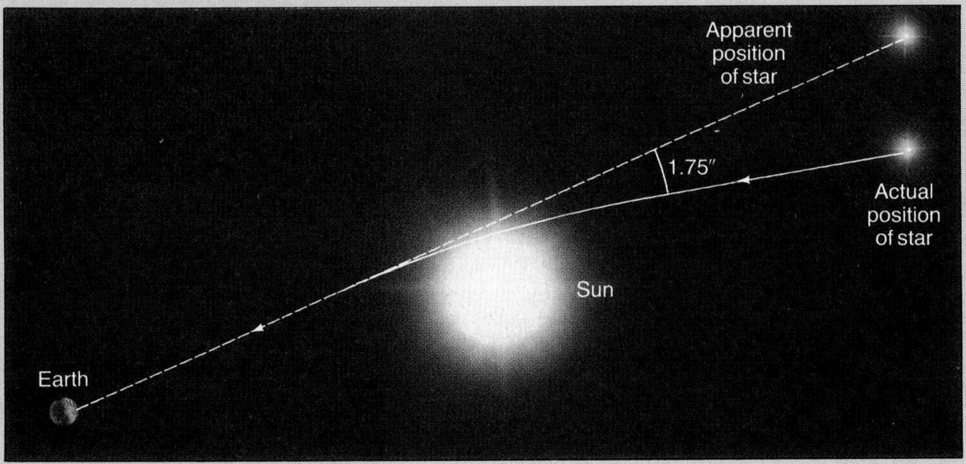

This then is the trend: As an object of any given mass contracts, the force of gravity grows stronger at its surface, and the escape velocity grows. If we could imagine Earth being further compacted, the escape velocity would continue to rise. If our hypothetical vise were to squeeze Earth hard enough to crush its radius to about a centimeter, the velocity needed to escape its surface would reach 300,000 km/s. But this is no ordinary velocity. It is the speed of light, the fastest velocity allowed by the laws of physics as we now know them.

If by some fantastic means the entire planet Earth could be compressed to less than the size of a grape, the escape velocity would exceed the speed of light. And since nothing can exceed that speed, the compelling conclusion is that nothing—absolutely nothing—could escape from the surface of such a compressed body. Even radiation—radio waves, visible light, x-rays, photons of all wavelengths—would be unable to escape the intense gravity of our reshaped Earth. With no photons leaving, our planet would be invisible and uncommunicative, for no signal of any sort could be sent to the universe beyond. The origin of the term *black hole* becomes clear. For all practical purposes, such a supercompact Earth could be said to have disappeared from the universe! Only its gravitational field would remain behind, betraying the presence of its mass, now shrunk to a point.[2]

The Event Horizon

Astronomers have a special name for the critical radius at which the escape velocity from an object would equal the speed of light and within which the object could no longer be seen. It is the **Schwarzschild radius,** after Karl Schwarzschild, the German scientist who first studied its properties. The Schwarzschild radius of any object is simply proportional to its mass. For Earth, it is 1 cm; for Jupiter, at about 300 Earth masses, it is about 3 m; for the Sun, at 300,000 Earth masses, it is 3 km. For a 3-$M_\odot$ stellar core

[2] In fact, we now know that, regardless of the composition or condition of the object that formed the hole, only three physical properties can be measured from the outside—the hole's mass, charge, and angular momentum. All other information is lost once the infalling matter crosses the event horizon. Thus, only three numbers are required to completely describe a black hole's outward appearance. In this chapter, we will consider only holes that formed from nonrotating, electrically neutral matter. Such objects are completely specified once their masses are known.

At the heart of general relativity is the premise that everything, including light, is affected by gravity because of the curvature of spacetime. Shortly after he published his theory in 1915, Einstein noted that light from a star should be deflected by a measurable amount as it passes the Sun. The closer to the Sun the light comes, the more it is deflected. Thus, the maximum deflection should occur for a ray that just grazes the solar surface. Einstein calculated that the deflection angle should be 1.75″ (arc seconds)—a small, but detectable, amount. Of course, it is normally impossible to see stars close to the Sun. During a solar eclipse, however, when the Moon blocks the Sun's light, the observation becomes possible, as illustrated at left.

In 1919, a team of observers, led by the British astronomer Sir Arthur Eddington, succeeded in measuring the deflection of starlight during an eclipse. The results were in excellent agreement with the prediction of general relativity. Virtually overnight Einstein became world famous. His previous major accomplishments notwithstanding, this single prediction assured him a permanent position as the best-known scientist on Earth!

A second prediction of general relativity is that planetary orbits should deviate slightly from the perfect ellipses of Kepler's laws. Again, the effect is greatest where gravity is strongest—that is, closest to the Sun. Thus, the largest relativistic effects are found in the orbit of Mercury. Relativity predicts that Mercury's orbit is not a closed ellipse. Instead, its orbit should rotate slowly, as shown in the (highly exaggerated) diagram at right. The amount of rotation is very small—only 43″ per century—but Mercury's orbit is so well charted that even this tiny effect is measurable.

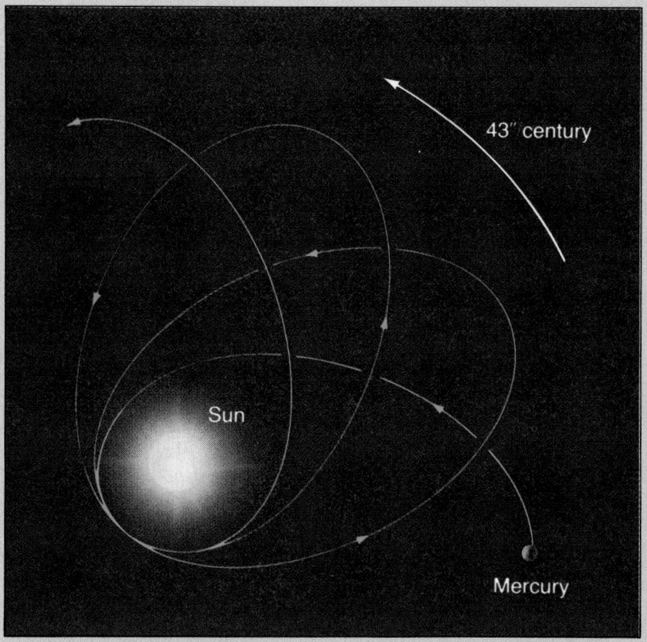

In fact, the observed rotation rate is 574″ per century, much greater than that predicted by relativity. However, when other (nonrelativistic) gravitational influences, primarily the perturbations due to the other planets, are taken into account, the rotation is in complete agreement with the above prediction.

remnant, the Schwarzschild radius is about 9 km. As a convenient rule of thumb, the Schwarzschild radius of an object is simply 3 km multiplied by the object's mass, measured in solar masses. Notice that *every* object has a Schwarzschild radius. It is simply the radius to which the object would have to be compressed for it to become a black hole. Put another way, a black hole is an object that happens to lie within its own Schwarzschild radius.

The surface of an imaginary sphere with radius equal to the Schwarzschild radius and centered on a collapsing star is called the **event horizon.** It defines the region within which no event can ever be seen, heard, or known by anyone outside. Even though there is no matter of any sort associated with it, we can think of the event horizon as the "surface" of a black hole.

A 1.4-$M_\odot$ neutron star has a radius of about 10 km and a Schwarzschild radius of 4.2 km. If we were to keep increasing the star's mass, the star's Schwarzschild radius would grow, although its actual physical radius would not. In fact the radius of a neutron star *decreases* slightly with increasing mass. By the time our neutron star's mass exceeded about 3 $M_\odot$, it would lie just within its own event horizon, and it would collapse of its own accord. It would not stop shrinking at the Schwarzschild radius—the event horizon is not a physical boundary of any kind, just a communications barrier. The remnant would shrink right past it to ever-diminishing size on its way toward becoming a point singularity.

Thus, provided that at least 3 $M_\odot$ remain behind after a supernova explosion, the remnant core will collapse catastrophically, the whole core diving below the event horizon in less than a second. The core simply "winks out," disappearing and becoming a small dark region from which nothing can escape—a literal black hole in space.

Photon Orbits

An alternative way of seeing the significance of the Schwarzschild radius is to consider what happens to rays of light emitted at different distances from the event horizon of a black hole. Imagine moving a light source closer and closer to the hole, as shown in Figure 24.10. Let's suppose that the source emits radiation uniformly in all directions. At large distances from the black hole, essentially all of the radiation eventually escapes into space. Only that portion of the beam that is aimed directly at the hole is captured. As

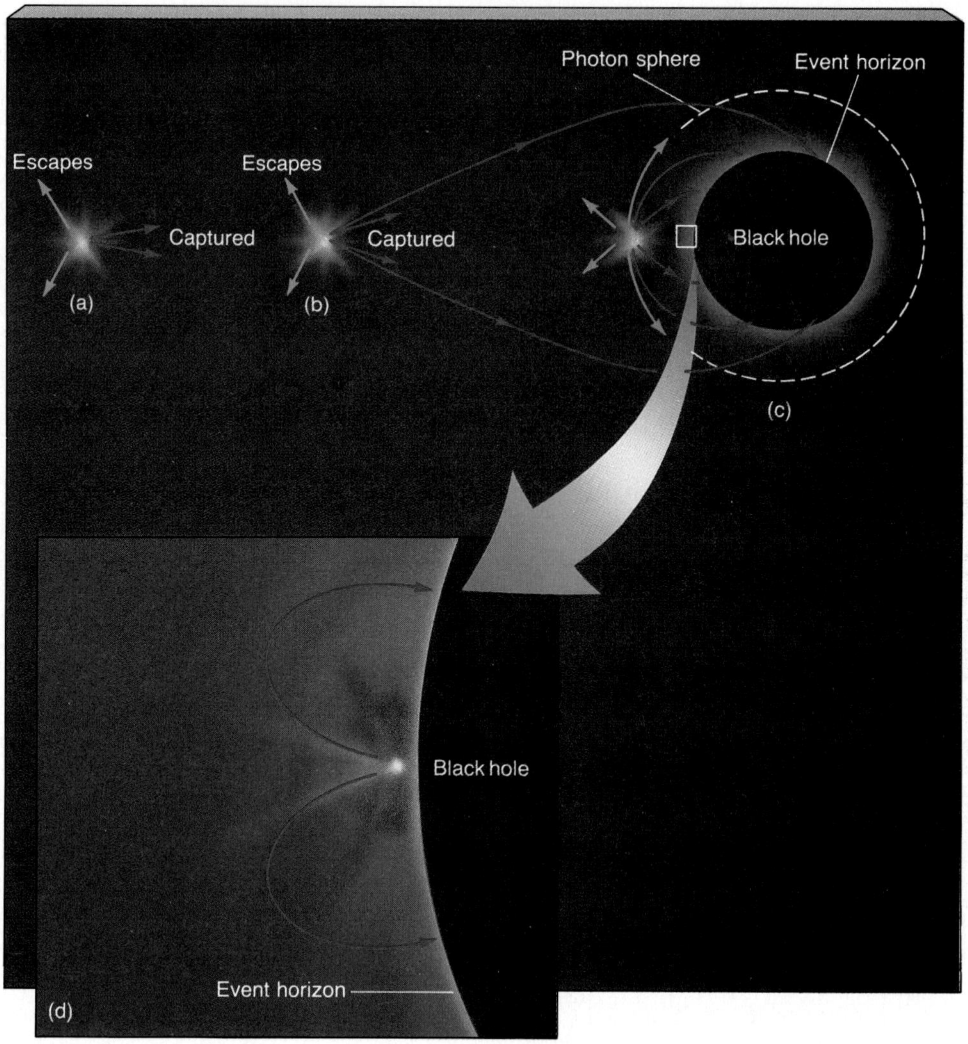

Figure 24.10 *This light source, moving closer and closer to a black hole, emits radiation in all directions. (a) At large distances from the hole, most of the light (marked in blue) escapes into space. (b) As the distance decreases, more and more of the radiation is deflected by the black hole's gravity onto paths that intersect the event horizon, and this radiation is trapped (shown in red). (c) At a distance of 1.5 Schwarzschild radii, photons can orbit the hole on circular trajectories, which mark the photon sphere. (d) Eventually, when the source reaches the horizon, all of the light it emits is destined to enter the hole.*

the light source moves closer, however, the effect of the hole's strong gravity becomes evident, and some photons that would have missed the hole if light traveled in straight lines are instead deflected onto paths that cross the event horizon, and these photons are trapped by the black hole.

At 1.5 Schwarzschild radii, exactly half of the radiation emitted by our light source escapes into space. Photons emitted perpendicular to the line joining the source to the center of the hole move in circular orbits at this radius, never escaping the hole's gravity but never crossing the event horizon. The surface on which these photons move as they travel on their circular paths is called the **photon sphere.** Closer still, the amount of deflection continues to increase, and the fraction of the radiation that escapes into space steadily decreases until, close to the event horizon, only a thin sliver of our beam is able to escape. At the event horizon itself, the sliver vanishes completely. All light rays from that surface, whatever their initial direction, are destined to enter the hole. This example clearly shows how the black hole's immense gravitational field dominates the trajectories of all particles, even photons, in its vicinity.

PROPERTIES OF BLACK HOLES

Warped Space

✓4 Modern notions about black holes rest squarely on the theory of relativity. Though white dwarfs and neutron stars can be adequately described by the classical Newtonian theory of gravity, only the modern Einsteinian theory of relativity can properly describe the bizarre physical properties of black holes.

A central concept of general relativity (see Interlude 24-1) is this: Matter—all matter—tends to "warp" or curve space in its vicinity. Objects such as planets and stars react to this warping by changing their paths. In the Newtonian view of gravity, particles move on curved trajectories because they feel a gravitational force. In Einsteinian relativity, those same particles move on curved trajectories because they are following the curvature of space produced by some nearby massive object. The more mass, the greater the warping. Close to a black hole, the gravitational field becomes overwhelming and the curvature of space extreme.

At the event horizon itself, the curvature is so great that space "folds over" on itself, causing objects within to become trapped and disappear.

Rubber Sheets and Curved Space

Some props may help us visualize the curvature of space near a black hole. Bear in mind, however, that these props are not "real," but only tools to help us grasp some exceedingly strange concepts.

First, imagine a pool table with the tabletop made of a thin rubber sheet rather than the usual hard felt. As suggested by Figure 24.11, such a rubber sheet becomes distorted when a heavy weight, such as a rock, is placed on it. The otherwise flat rubber sheet sags or warps (or curves), especially near the rock. The heavier the rock, the larger the curvature. Trying to play billiards, you would quickly find that balls passing near the rock are deflected by the curvature of the tabletop.

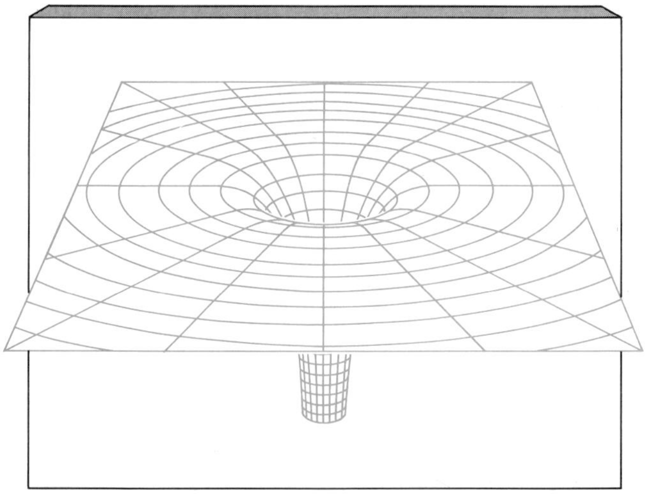

Figure 24.11 *A pool table made of a thin rubber sheet will sag with a weight on it. Likewise, space is bent, or warped, in the vicinity of an astronomical object.*

In much the same way, both matter *and radiation* are deflected by the curvature of space near a star. For example, Earth's orbital path is governed by the relatively gentle curvature of space created by our Sun. In the case of a very massive star, space is more severely curved. In the extreme case, a black hole curves space more than any other object.

Let's consider another analogy. Imagine a large family of people living on a huge rubber sheet—a sort of gigantic trampoline. Deciding to hold a reunion, they converge on a given place at a given time. As shown in Figure 24.12, one person remains behind, not wishing to attend. He keeps in touch with his relatives by means of "message balls" rolled out to him (and back from him) along the surface of the sheet. These message balls are the analog of radiation carrying information through space.

As the people converge, the rubber sheet sags more and more. Their accumulating mass creates an increasing amount of space curvature. The message balls can still reach the lone person far away in essentially flat space, but they arrive less frequently as the sheet becomes more and more warped and stretched—as shown in Figure 24.12(b) and (c)—and the balls have to climb out of a deeper and deeper well. Finally, when enough people have arrived at the appointed spot, the mass becomes too great for the rubber to support. As illustrated in Figure 24.12(d), the sheet pinches off into a "bubble," compressing the people into oblivion and severing their communications with the lone survivor outside. This final stage represents the formation of an event horizon around the party.

Right up to the end—the pinching off of the "bubble"—two-way communication is possible. Message balls can reach the outside from within (but at a slower and slower rate as the rubber stretches), and messages from outside can get in without difficulty. Once the event horizon (the bubble) forms, balls from the outside can still fall in,

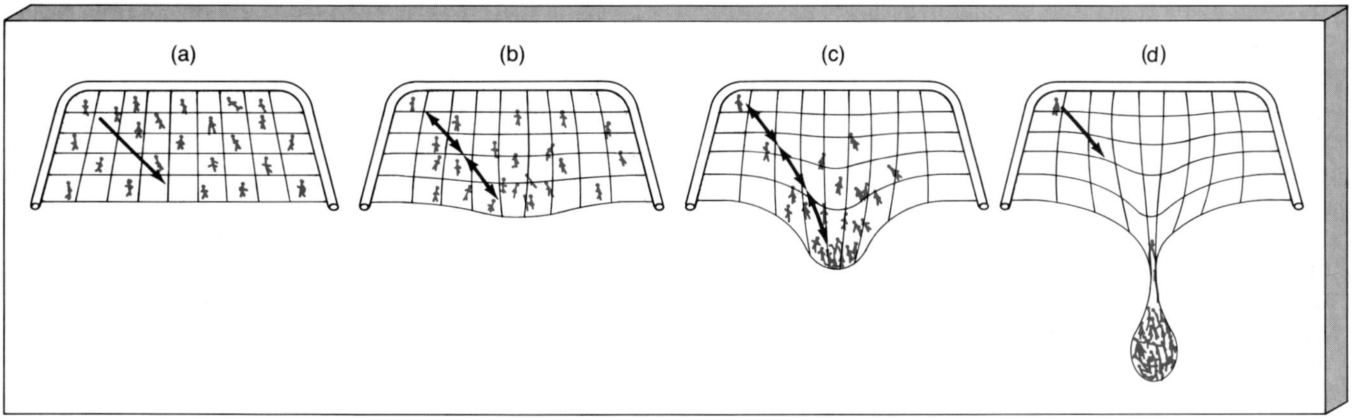

Figure 24.12 *Any mass causes the rubber sheet (space) to be curved. As people assemble at the appointed spot on the sheet, the curvature grows progressively larger, as shown in frames (a), (b), and (c). The people are finally sealed inside the bubble (d), forever trapped and cut off from the outside world.*

but they can no longer be sent back out to the person left behind, no matter how fast they are rolled. They cannot make it past the "lip" of the bubble in Figure 24.12(d). This analogy (very) roughly depicts how a black hole warps space completely around on itself, isolating its interior from the rest of the universe. The essential ideas—the slowing down and eventual cessation of outward-going signals and the one-way nature of the event horizon once it forms—all have clear parallels in the case of stellar black holes.

Cosmic Cleaners, No

Black holes are *not* cosmic vacuum cleaners. They do not cruise around interstellar space, sucking up everything in sight. The orbit of an object near a black hole is essentially the same as its orbit near a star of the same mass. Only if the object happens to pass within a few Schwarzchild radii (perhaps 50 or 100 km for a typical black hole formed in a supernova explosion) of the event horizon is there any significant difference between its actual orbit and the one predicted by Newtonian gravity and described by Kepler's laws. From a distance, the main observational difference is that an object orbiting a black hole would appear to orbit a dark, empty region of space. Neither emitted nor reflected radiation would emerge from the black hole itself.

Black holes, then, do not go out of their way to drag in matter. However, if some matter does happen to fall into one—if its orbit happens to take it too close to the event horizon—it will be unable to get out. Black holes are like turnstiles, permitting matter to flow in only one direction—inward. A black hole's mass never decreases (but see Interlude 24-3 for a modification to this statement). Because a black hole will accrete at least a little material from its surroundings, its mass tends to increase over time. The black hole's size is proportional to its mass, so the radius of the event horizon grows with time.

Cosmic Heaters, Yes

Matter flowing into a black hole is subject to great tidal stress. An unfortunate person falling feet first into a solar-mass black hole would find herself stretched enormously in height and squeezed unmercifully laterally. She would be torn apart even before she reached the event horizon, for the pull of gravity would be much stronger at her feet (which are closer to the hole) than at her head. The tidal forces at work in and near a black hole are the same phenomenon responsible for ocean tides on Earth and the spectacular volcanos on Io. The only difference is that the tidal forces near a black hole are far stronger than any force we know in the solar system.

As shown in Figure 24.13, a similar fate awaits any kind of matter falling into a black hole. Whatever falls in—gas, people, space probes—is vertically stretched and horizontally squeezed, in the process being accelerated to high speeds. The net result of all this stretching and squeezing is numerous and violent collisions among the torn-up debris, causing a great deal of frictional heating of the infalling matter. Material is simultaneously torn apart and heated to high temperatures as it plunges into the hole.

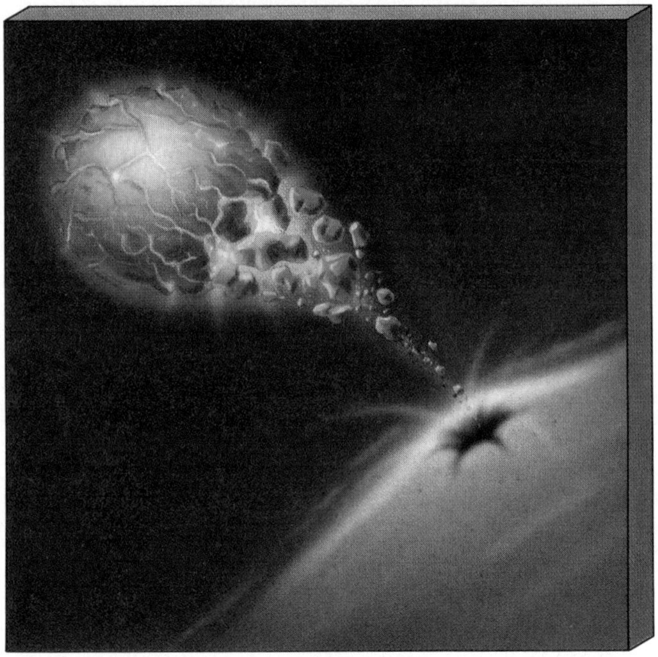

Figure 24.13 *Any matter falling into the clutches of a black hole will become severely distorted and heated. In this sketch, an imaginary planet is shown being pulled apart by the gravitational tides of the black hole.*

The rapid heating of matter by tides and collisions is so efficient that, prior to submersion below the hole's event horizon, the matter emits radiation on its own accord. For a black hole of the mass of the Sun, the energy is expected to be emitted in the form of x-rays. In effect, the gravitational energy of matter outside the black hole is converted into heat while that matter falls toward the hole. Once the hot matter falls below the event horizon, its radiation is no longer detectable, as it never leaves the hole. Contrary to what we might expect from an object whose defining property is that nothing can escape from it, the region surrounding a black hole is expected to be a *source* of energy. Interlude 24-3 presents another, quite different, way in which a black hole can produce energy.

SPACE TRAVEL NEAR BLACK HOLES

✓5 How close to a black hole can we travel? Could we actually observe the event horizon at close range? One reasonably safe way to study a black hole would be to go into orbit around it, safely beyond the disruptive influence of the hole's strong tidal forces. After all, Earth and the other planets of our solar system all orbit the Sun without falling into it and without being torn apart. The gravity field around a black hole is basically no different.

However, even from a stable circular orbit, a close investigation of the hole would be unsafe for humans. Human Endurance tests conducted on astronauts of the United States and the former Soviet Union suggest that the human body cannot withstand stress greater than about 10 times the pull of gravity on the Earth's surface. This break-

INTERLUDE 24-3 *Black Hole Evaporation*

Some attempts to understand gravity on a microscopic scale suggest that black holes may not be entirely black after all. Applying what they know of subatomic physics, scientists now believe it possible that some matter and radiation can escape from a black hole. Here's how. The laws of quantum physics allow a process known as *pair creation* to occur anywhere in space: A particle and its antiparticle—an electron and a positron, say—can come into being spontaneously, literally formed out of nothing. This, of course, violates one of the most cherished laws of physics, the law of conservation of mass and energy (recall that mass and energy are equivalent, related to one another by Einstein's famous equation $E = mc^2$), but this violation is permitted so long as the "books are balanced" by the disappearance of the particles (by mutual annihilation) within a short enough period of time. In effect, the rules can be broken, so long as they are repaired before anyone notices.

Most of the time, pairs of particles appear and disappear so rapidly that energy is conserved on all macroscopic scales. However, should pair creation happen near a black hole, as illustrated in the diagram at right, it is possible for one of the two particles to cross the event horizon *before* it meets and annihilates its partner. The other particle would then be free to leave the scene, making the black hole appear to the outside world as a source of matter or radiation. The energy required to create the new particle ultimately comes from the black hole. Because energy and mass are equivalent, this means that the hole must decrease in mass as it radiates. Thus, black holes do not last forever—they slowly "evaporate." This possibility was first realized by Cambridge University mathematician Stephen Hawking. The radiation resulting is known as *Hawking radiation.*

A remarkable result, also discovered by Hawking's group, is that the spectrum of Hawking radiation is described by a Planck curve—exactly the same curve that characterizes emission from any hot body. Black holes emit black-body radiation! The temperature of the radiation turns out to be inversely related to the mass of the hole. Big black holes are very cold, while small black holes are hot. A hole the mass of the Sun would emit radiation at a temperature of 10^{-6} K, while one the mass of a mountain—about 10^{15} g—would have a temperature of some 10^{12} K. Knowing a black hole's temperature T and surface area A, we can calculate its luminosity L in exactly the same way as for stars ($L \propto AT^4$—see Chapter 19).

A black hole radiates energy (and hence mass) into space. As the hole radiates, its mass drops and its temperature increases. That is, the black hole's temperature is inversely proportional to its mass and its area decreases as the square of the mass. The black hole, then, increases its luminosity as it evaporates. The increased luminosity in turn leads to a faster loss of mass. This runaway situation eventually ends violently, and the black hole explodes in a burst of gamma-rays.

The lifetime of a hole depends on its mass. For a 1-$M_\odot$ black hole, the explosion is predicted to occur after about 10^{70} years! Astronomers today hardly expect to observe such an event. Thus, the issue of evaporation is moot for the black holes described in this chapter; astronomers do not expect ever to observe either their slow decay (which begins the moment they form) or their eventual explosion.

By contrast, very small black holes, with masses of about 10^{15} g, should have lifetimes roughly equal to the current age of the universe. While we know of no way in which such objects could be created in the universe today, it is conceivable that conditions in the very earliest epochs of the universe might have been just right to compress pockets of matter into miniature black holes. Such black holes would have Schwarzchild radii of about 10^{-13} cm, comparable to the size of a subatomic particle. If they exist, very small black holes should be exploding right now. Attempts have been made to observe the resultant gamma-rays, so far without success.

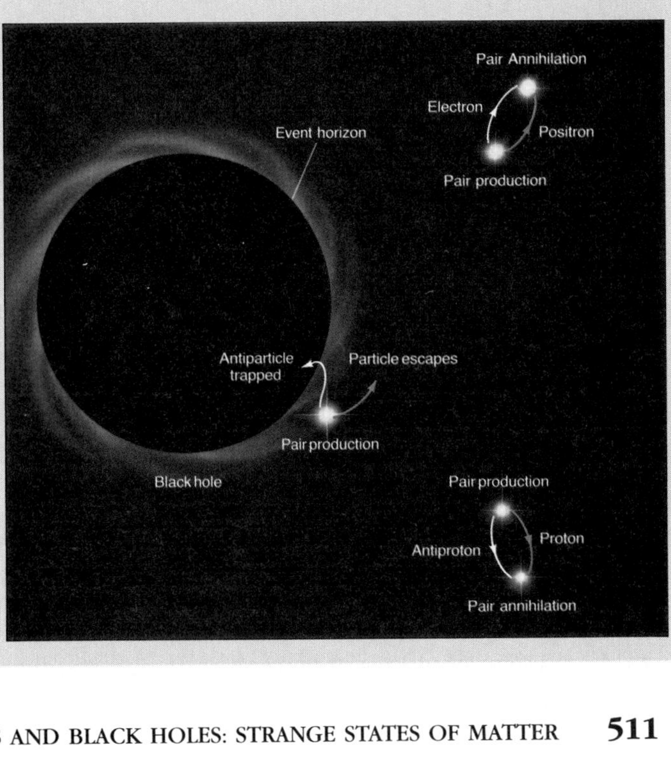

ing point would occur about 3000 km from a 10-$M_\odot$ black hole (which, recall, would have a 30 km event horizon). Closer than that, the tidal effect of the hole would tear a human body apart.

An Indestructible Robot

Let's imagine using only an indestructible astronaut—a mechanical robot. We will send our robot toward the center of the hole, as illustrated in Figure 24.14. Watching from a safe distance in our orbiting spacecraft, we can then examine the nature of space and time near the hole. Our robot will be a useful probe of theoretical ideas, at least down to the event horizon. After that, there is no way for it to return any information about its findings.

Suppose, for example, our robot has an accurate clock and a light source of known frequency mounted on it. From our safe vantage point far outside the event horizon, we could use telescopes to read the clock and measure the frequency of the light we receive. What might we discover?

We would find that the light we receive from the robot would become more and more redshifted as it neared the event horizon. Even if the robot used rocket engines to remain motionless, the red shift would still be detected. The red shift is *not* caused by motion. It is not the result of the Doppler effect arising as the robot falls into the hole. Rather, it is a red shift induced by the black hole's gravitational field and is a clear prediction of Einstein's general theory of relativity. It is known as **gravitational red shift.**

We can explain the gravitational red shift as follows. According to general relativity, photons are attracted by gravity. As a result, in order to escape from a source of gravity, photons must expend some energy: They have to work to get out of the gravitational field. They don't slow down at all—photons always move at the speed of light—they just lose energy. Because a photon's energy is proportional to the frequency of its radiation, light that loses energy must have its frequency reduced (or, conversely, its wavelength lengthened). Radiation coming from the vicinity of a gravitating object will be redshifted by an amount depending on the strength of the gravitational field.

Most astronomical objects' gravitational fields are far too weak to shift radiation much toward the red, although the effect can still be measured. Exceedingly delicate laboratory experiments on Earth have succeeded in detecting the tiny gravitational red shift produced by even our own planet's weak gravity. Sunlight is redshifted by only about a hundredth of an angstrom. A few white dwarf stars do show some significant reddening of their emitted light, however. Their smaller radii means that their surface gravity is much stronger than the Sun's. Neutron stars should show an appreciable shift in their radiation, but it is currently impossible to disentangle the effects of gravity and magnetism on the observed signals. Only near black holes is the gravitational pull so great that the red shift should be large and easily measurable, at least in principle.

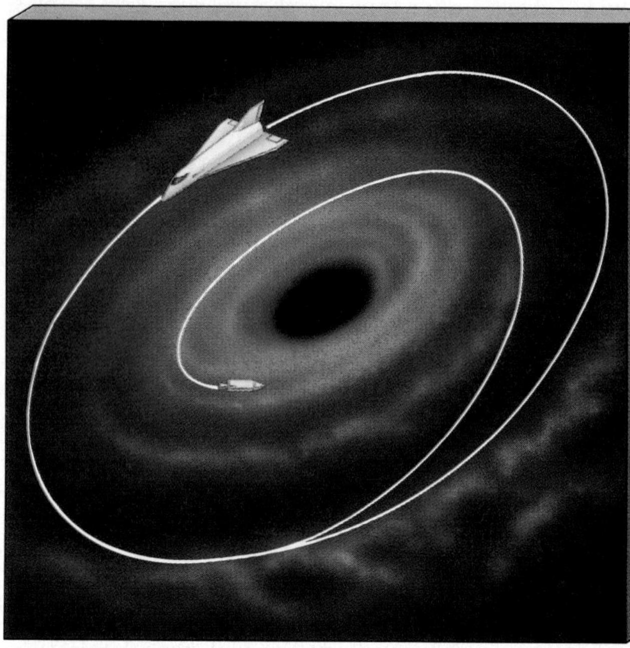

Figure 24.14 *Robot astronauts can travel toward a black hole while performing experiments that humans, farther away, can monitor in order to learn something about the nature of space near a black hole.*

As photons traveled from the robot's light source to the orbiting spacecraft, they would become gravitationally redshifted. From the standpoint of the orbiting humans, a green light, say, would become yellow and then red as the robot astronaut neared the black hole. From the robot's perspective, the light would remain green. As the robot got closer to the event horizon, the radiation from its light source would become undetectable with optical telescopes. The radiation reaching the humans in the orbiting spacecraft would by then be lengthened so much that infrared and then radio telescopes would be needed to detect it. Closer still to the event horizon, the radiation emitted as visible light from the robot probe would be shifted to wavelengths even longer than conventional radio waves by the time it reached the human observers.

Light emitted *from the event horizon itself* would be gravitationally redshifted to infinitely long wavelengths. In other words, each photon would use all of its energy trying to escape from the edge of the hole. What was once light (on the robot) has no energy left upon arrival at the safely orbiting spacecraft. Theoretically, this radiation makes it to us—still moving at the velocity of light—but with zero energy. The light radiation originally emitted has become redshifted beyond our perception.

What about the robot's clock? Assuming that the distant observers in the safely orbiting spacecraft can read it, what time does it tell? Is there any observable change in the rate at which the clock ticks while moving deeper into the hole's gravitational field? We would find that, from the safely orbiting spacecraft, any clock close to the hole would appear to tick more *slowly* than an equivalent clock on board the spacecraft. The closer the clock came to the hole, the slower it would appear run. The clock closest to the hole would operate slowest of all. Upon reaching the event horizon, the clock would seem to stop altogether. It would be as if the robot astronaut had found immortality! All action would become virtually frozen in time. Consequently, an external observer will never actually witness an infalling astronaut sink below the event horizon. Such a process would appear to take forever.

This apparent slowing down of the robot's clock is known as **time dilation.** It is another clear prediction of general relativity, and in fact it is closely related to the gravitational red shift. To see this connection, imagine that we use our light source as a clock, with the passage of a wave crest (say) constituting a "tick." The clock thus ticks at the frequency of the radiation. As the wave is redshifted, the frequency drops, and fewer wave crests pass the distant observer each second—the clock appears to slow down. This thought experiment demonstrates that the red shift of the radiation and the slowing of the clock are essentially one and the same thing.

From the point of view of the indestructible robot, however, relativity theory predicts no strange effects at all. To the infalling robot, the light source hasn't reddened and the clock keeps perfect time. In the robot's frame of reference, everything is normal. Nothing prohibits it from approaching within the Schwarzschild radius of the hole. No law of physics constrains an object from passing through an event horizon. There is nothing like a brick wall at the event horizon, and no sudden lurch as it is crossed; it is only an imaginary boundary in space. Travelers passing through the event horizon of a sufficiently massive hole (such as might lurk in the heart of our own Galaxy, as we will see) might not even know it—at least until they tried to get out!

Deep Down Inside

No doubt, you are wondering what lies within the event horizon of a black hole. The answer is simple: No one really knows.

Some researchers maintain that the inner workings of black holes are irrelevant. Experiments could conceivably be done by robots sent "down under" to study conditions within the event horizon, but that information could never reach the rest of us outside the black hole. Theories of the insides of black holes can never be put to the experimental test. From a purely observational perspective at least, anyone's theory is as valid as anyone else's.

Other researchers point out, as we noted earlier, that relativity must be incorrect, or at least incomplete, when applied to the centers of black holes. The current laws of physics are inadequate in the vicinity of a singularity, because they lose their predictive power there. Perhaps matter trapped in black holes never actually reaches the singularity. Perhaps it just approaches this bizarre state, in a manner that we will someday understand as the subject of *quantum gravity*—the merger of general relativity with quantum mechanics—develops. However, even if this new theory succeeds in doing away with the singularity, it is unlikely that the external appearance of the hole, or the existence of its event horizon, will change. Any modifications to general relativity are expected to occur only on submicroscopic scales, not on the macroscopic (kilometer-sized) scale of the Schwarzschild radius.

Singularities are places where the rules break down, and some very strange things may occur near them. Many possibilities have been envisaged—gateways into other universes, time travel, the creation of new states of matter—but none of them has been proved, and certainly none of them has been observed. Because these regions are places where science fails, their presence causes serious problems for many of our cherished laws of physics, from causality (the idea that cause should precede effect, which runs into immediate problems if time travel is possible), to energy conservation (which is violated if material can hop from one universe to another through a black hole). Disturbed by the possibility of such chaos in science, some relativists, notably Roger Penrose of Oxford University, have proposed the **Principle of Cosmic Censorship:** Nature always hides *any* singularity, such as that found at the center of a black hole, inside an event horizon. In that case, even though physics fails, that breakdown cannot affect us outside, so we are safely insulated from any harmful (or embarrassing) effects the singularity might have. What would happen if we one day found a so-called *naked singularity* somewhere, a singularity uncloaked by an event horizon? Would relativity theory still hold there? For now, we just don't know.

What sense are we to make of black holes? Do black holes, and all the strange phenomena that go on in and around them, really exist? The basis for understanding these weird phenomena is the relativistic concept that mass warps space—which has already been found to be a surprisingly good representation of reality, at least for the weak gravitational fields produced by stars and planets (see Interlude 24-2). The larger the mass concentration, the greater the warping, and thus, apparently, the stranger the observational consequences. Although general relativity is not proven, there is presently no reason to disbelieve it, and black holes are one of its most striking predictions. As long as general relativity stands as the correct theory of gravity in the universe, black holes are real.

OBSERVATIONAL EVIDENCE FOR BLACK HOLES

✓6 Theoretical ideas aside, is there any observational evidence for black holes? More to the point, can we prove that these invisible objects really do exist?

Stellar Occultation?

One way in which we might detect a black hole, at least in principle, would be to observe it passing in front of something else. Figure 24.15 is a sketch of a black hole passing in front of a much larger visible companion star. We would expect to see a minute black dot glide across the otherwise bright star. Unfortunately, this would be extremely hard to see; the 12,000-km planet Venus is barely noticeable when transiting the Sun, so a kilometer-sized object moving across the image of a faraway star would be completely invisible with current equipment or any equipment available in the foreseeable future.

Actually, this observation is not even as clearcut as just suggested. Even if we were close enough to the star that we could resolve the 10- (or so) km disk of the black hole, the observable effect would not really be a black dot superposed on a bright background. The background starlight would be deflected as it passed the black hole on its way to Earth, as indicated in Figure 24.16. The effect is the same as the bending of distant starlight around the edge of the Sun, a phenomenon that has been repeatedly measured during solar eclipses throughout the last several decades (see Interlude 24-2). With a black hole, much larger deflections would occur. As a result, our perception of a black hole in front of a bright companion star would not show a neat, well-defined black dot. Instead, the bending of light around it would make its image fuzzy. Recent studies have shown that such a blurred image would be virtually impossible to observe, even from nearby.

Black Holes in Binary Systems

A better way to find black holes is to look for their effects on other objects. Although black holes are invisible, they are so massive that astronomers can test for a hole's existence by studying its associated gravitational field. For example, the motion of a spacecraft, a planet, or a star could conceivably be used to probe the vicinity of a suspected hole. The hole would have the same effect on a nearby body as would a visible object of the same mass.

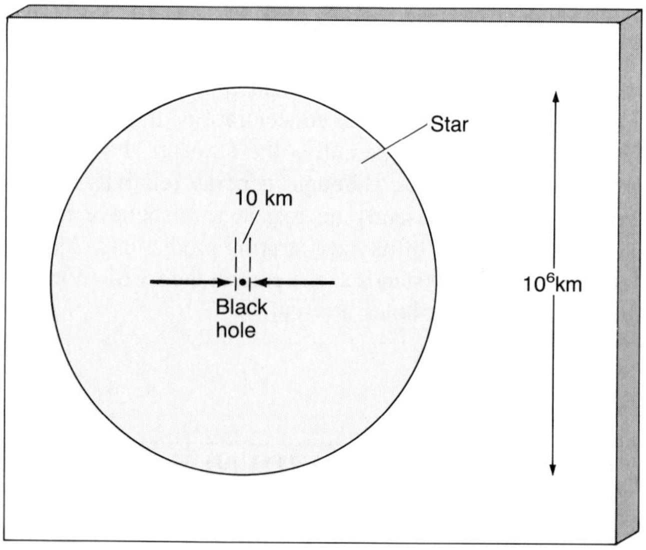

Figure 24.15 *We might try to observe a black hole pass in front of a stellar companion, but the small size of the hole makes this effort impractical.*

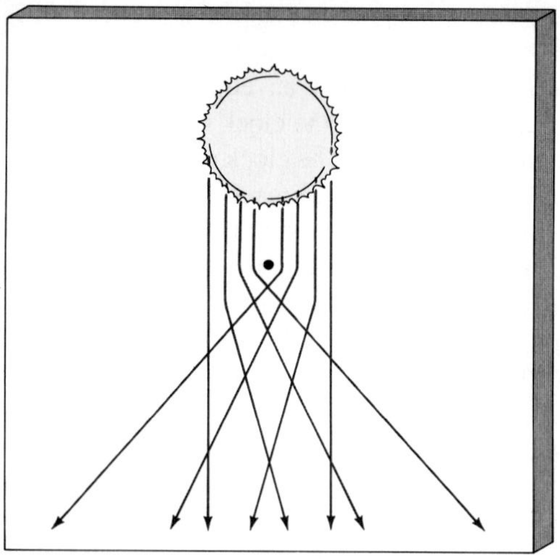

Figure 24.16 *The gravitational bending of light around the edges of a small, massive black hole makes it impossible to observe the hole as a black dot superposed against the bright background of its stellar companion.*

Our Milky Way Galaxy harbors many binary-star systems where only one object can be seen. Recall from our study of double stars in Chapter 19 that we need observe the motion of only one star to infer the existence of an unseen companion and measure some of its properties. In the majority of cases, the invisible companion is probably simply small and dim, nothing more than an M-type star hidden in the glare of an O- or B-type partner. In other cases, dust or other circumstellar debris probably shroud one object, making it invisible to even the best available equipment. In either case, the invisible object is not a black hole.

A few binary systems, however, have peculiarities that suggest that one of their members may indeed be a black hole. Some of the most interesting observations, made during the 1970s and 1980s by Earth-orbiting satellites, reveal binary systems in which the invisible member emits large amounts of x-rays. The mass of the emitting object is measured as several solar masses, so we know it is not simply a small, dim star, and radiation pressure from the binary members makes circumstellar debris an unlikely explanation for its invisibility. One particular binary system drawing much attention lies in the constellation Cygnus.

A Black Hole Candidate

Figure 24.17 shows the area of the sky in Cygnus, where astronomers have reasonably good evidence for a black hole. The rectangle outlines the celestial system of interest, some 2000 pc from Earth. The black hole candidate is an x-ray source called Cygnus X-1, discovered by the *Uhuru* satellite in the early 1970s. Its visible companion was iden-

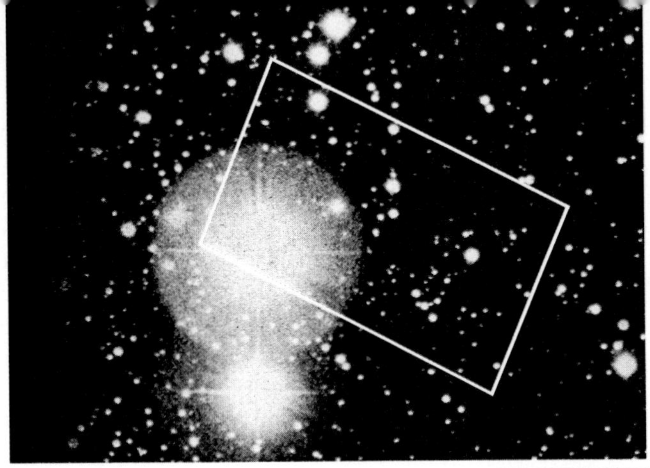

Figure 24.17 *The brightest star in this photograph is a member of a binary system whose unseen companion, called Cygnus X-1, is thought to be a good candidate for a black hole. (The rectangle outlines the field of view illustrated in the next figure.) (Harvard-Smithsonian Center for Astrophysics)*

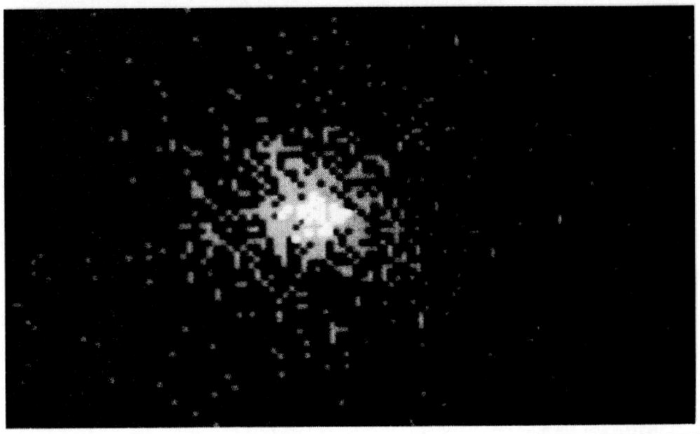

Figure 24.18 *X-rays emitted by the Cygnus X-1 source can be analyzed by changing them into electronic signals, which can then be viewed on a video screen, from which this picture is taken. (The field of view here is outlined by the rectangle in the previous figure.) (NASA)*

tified a few years later. The main observational features of this binary system are:

1. Spectroscopic observations of visible radiation show that the bright object—a blue B-type supergiant with the catalog name of HDE226868—is only one member of a binary-star system whose orbital period (5.6 days) and size (20 million km in diameter) are both well determined. Assuming that the visible component lies on the main sequence, we know its mass must be around 30 $M_{\odot}$.
2. Knowledge of the binary's size and period allows a limit to be placed on the combined mass of the two components. Knowing the mass of the visible component (see 1 above) and subtracting it from the total, we can estimate the mass of Cygnus X-1 to be between 10 and 20 $M_{\odot}$.
3. Other spectroscopic studies suggest that hot gas is flowing from the bright star toward an unseen companion.
4. X-ray radiation emitted from the immediate neighborhood of Cygnus X-1 suggests the presence of high-temperature gas, perhaps as hot as several million K (see Figure 24.18).
5. Rapid time variations of this x-ray radiation imply that the size of the x-ray-emitting region of Cygnus X-1 itself must be less than a few hundred kilometers across. The reasoning goes as follows. If the emitting region were, say, 300,000 km—1 light-second—across, even an instantaneous change in intensity at the source would be smeared out over a time interval of 1 s as seen from Earth, because light from the far side of the object would take 1 s longer to reach us than light from the near side. X-rays from Cygnus X-1 have been observed to vary in intensity on time scales as short as a millisecond. For this variation not to be blurred by the travel time of light across the source, Cygnus X-1 cannot be more than 1 light-millisecond, or 300 km, in diameter.

These general properties suggest that the invisible x-ray-emitting companion could be a black hole. The x-ray-emitting region is likely an accretion disk formed as matter drawn from the visible star spirals down onto the unseen component. The rapid variability of the x-ray emission indicates that the unseen component must be compact—a neutron star or a black hole. The mass limit on the dark component argues for the latter, as astronomers believe that neutron stars' masses cannot exceed about 3 $M_{\odot}$. Figure 24.19 is an artist's conception of the intriguing Cygnus X-1. As shown, most of the gas drawn from the visible star ends up in a Life-Saver-shaped accretion disk of matter. As the

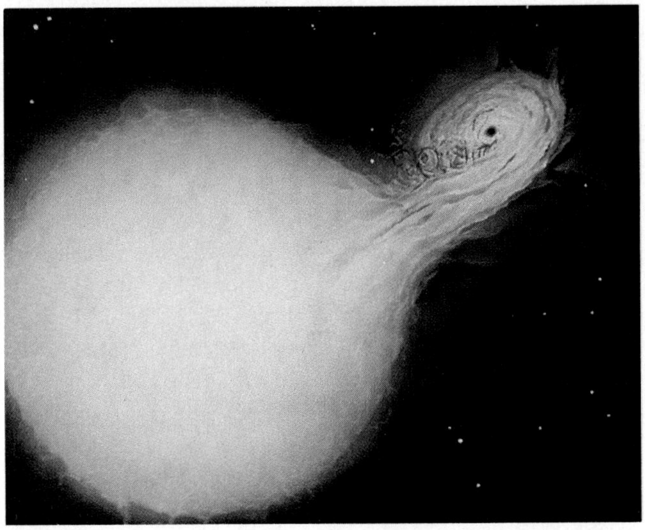

Figure 24.19 *Artist's conception of Cygnus X-1 and its environment. This binary system contains a large, bright visible star (left) and an invisible black hole (upper right) surrounded by a thin disk of accreted, x-ray-emitting matter torn from the giant star. The colors of the various gases are derived from thermodynamic considerations. Note the disposition of the expected turbulence caused by matter in the accretion disk colliding with newly stolen gas to the left of the suspected hole. (L. J. Chaisson)*

gas flows toward the black hole, it becomes superheated and emits x-rays, just before being trapped forever below the event horizon.

Have Black Holes Been Detected?

A few other black hole candidates are known. For example, the third x-ray source ever discovered in the Large Magellanic Cloud—called LMC X-3—is an invisible object that, like Cygnus X-1, orbits a bright companion star. (The Large Magellanic Cloud is a small companion galaxy to our own; we will discuss it in Chapter 26.) LMC X-3's visible companion seems to be distorted into the shape of an egg by the unseen object's intense gravitational pull. Reasoning similar to that applied to Cygnus X-1 leads to the conclusion that the compact object LMC X-3 has a mass of nearly 10 $M_\odot$, making it too massive to be anything but a black hole. The X-ray binary system A0620-00 has been found to contain a compact object of mass 3.8 $M_\odot$.

In total, there are perhaps half a dozen known objects

INTERLUDE 24-4 *Gravity Waves*

Electromagnetic waves are common everyday phenomena. As we saw in Chapter 4, these waves move through space and transport energy. Whether they are radio, infrared, visible, ultraviolet, x-ray, or gamma-ray radiation, all electromagnetic waves involve periodic changes in the strengths of electric and magnetic fields. Any accelerating charged particle, such as an electron within a broadcasting antenna or on the surface of a star, generates electromagnetic waves. Measurements of these waves agree perfectly with the theory of electromagnetism that predicts them.

The modern theory of gravity—Einstein's theory of relativity—also predicts waves that move through space. A *gravity wave* is the gravitational analog of an electromagnetic wave. A gravity wave, or *gravitational radiation*, results from a change in the strength of gravitational field. In principle, any time an object of any mass accelerates, a gravity wave should be emitted at the speed of light. The passage of a gravity wave should produce small distortions in the space through which it passes. Gravity is an exceedingly weak force compared with electromagnetism, so these distortions are expected to be very small—in fact, much smaller than the diameter of an atomic nucleus for the waves that might be produced by galactic sources. Yet many researchers believe that they should be measurable. So far, however, no one has yet succeeded in detecting gravitational radiation. However, since detection of gravity waves, like an unambiguous discovery of a black hole, would provide very strong support for the theory of relativity, scientists are eager to search for them.

Theorists are still arguing about what kinds of astronomical objects should produce gravity waves detectable on Earth. Leading candidates include (1) the merger of a binary-star system, (2) the collapse of a star into a black hole, and (3) the collision of two black holes. Each of these possibilities involves the acceleration of huge masses, so the strength of the gravitational fields should change drastically and rapidly in each case. Other astronomical objects are also expected to emit gravity waves, but only changes involving large masses will produce waves intense enough to be observed.

Of the three candidates, the first one probably presents the best chance to detect gravity waves, at least for the present. Binary-star systems should emit gravitational radiation as the component stars orbit one another. As energy escapes in the form of gravity waves, the two stars slowly spiral toward one another, orbiting more rapidly and emitting even more gravitational radiation. This runaway situation can lead to the decay and eventual merger of close binary systems in a relatively short period of time (which, in this case, means tens or hundreds of millions of years).

Such a slow but steady decay in the orbit of a binary system has in fact been detected. In 1974, Joseph Taylor and his colleagues at the University of Massachusetts discovered a very unusual galactic binary system. Both components are neutron stars, and one is observable from Earth as a pulsar. This system has become known as the *binary pulsar*. Measurements of the periodic Doppler shift of the pulsar's radiation prove that its orbit is slowly shrinking in size. Furthermore, the rate at which the orbit is shrinking is exactly what would be predicted by relativity theory if the energy were being carried off by gravity waves. Even though the gravity waves themselves have not yet been detected, the binary pulsar is regarded by most astronomers as a very strong piece of evidence in favor of general relativity.

We need to develop experimental tools to detect gravity waves from astronomical objects. Radiation is energy, and energy is information. Gravity waves should contain a great deal of information about the physical events in some of the most exotic regions of space. In 1992, funding was approved for an ambitious gravity-wave observatory called **LIGO**—short for **L**aser **I**nterferometric **G**ravity-wave **O**bservatory. This detector, which will use laser beams to measure the extremely small distortions of space produced by gravitational radiation, should be capable of detecting gravity waves from many galactic and extragalactic sources. If successful, the discovery of gravity waves could herald a new age in astronomy, in much the same way that invisible electromagnetic waves, unknown a century ago, revolutionized classical astronomy and led to modern astrophysics.

that may turn out to be black holes, although Cygnus X-1, A0620-00, and LMC X-3 have the strongest cases. But can we be sure that these objects are black holes? Cygnus X-1 and the other suspected black holes in binary systems all have masses relatively close to the neutron star-black hole dividing line. When the effects of rotation and magnetism are someday fully included in the theory, and the uncertainties in the binary mass estimates are taken into account, there may still be an outside chance that the dark objects in question could turn out to be quite "ordinary." In short, they might conceivably be merely dim and dense neutron stars and not black holes at all (although most astronomers do not regard this as a likely possibility).

The real problem here is that there is presently no observational test that can unambiguously distinguish a 10-$M_\odot$ black hole from, say, a 10-$M_\odot$ neutron star. Both objects would affect a companion star's orbit in the same way; both would tear mass from its surface, and both would form an accretion disk around themselves that would emit intense x-rays. The radiation from the surface of the neutron star itself, which would distinguish it from a black hole, might in some cases be so weak that it would be impossible to detect against the emission from the disk. Thus, we have to rule out the neutron star on purely theoretical grounds, on the basis of our understanding of neutron star masses. The argument really proceeds by elimination. Loosely stated, it goes: "Object X is compact and very massive. We don't know of anything else that can be that small and that massive. Therefore object X is a black hole." We will see several other instances later in this book of the same reasoning being used to infer the existence of very massive black holes in the hearts of galaxies. However, some astronomers are troubled that the black hole category has in some ways become a "catch-all" for things that have no other reasonable explanation in terms of present-day physics.

So have stellar black holes really been discovered? The answer is probably yes. Skepticism is healthy in science, but only the most stubborn astronomer would take serious issue with the above lines of reasoning supporting the case for black holes. Can we guarantee that future modifications to the theory of compact objects will not invalidate our arguments? No, but similar statements could be made about many areas of astronomy. We conclude that, strange as they are, black holes have been detected in our Galaxy. Perhaps some day, future generations of space travelers will visit Cygnus X-1 or LMC X-3 and (carefully!) verify these conclusions first hand. Until then, we will have to continue to rely on improving theoretical models and observational techniques to guide our discussions of these most mysterious objects.

CHAPTER SUMMARY

The most massive stars end their lives as neutron stars or black holes. During a type-II supernova explosion, the innermost part of the core remains intact as the shock wave it causes destroys the rest of the star. This core remnant is called a neutron star. Newly formed neutron stars rotate extremely rapidly, with periods measured in fractions of a second. A young neutron star's magnetic field is increased by the collapse of the core to a strength trillions of times greater than Earth's field.

The best current model describes a pulsar as a compact, spinning neutron start that periodically flashes radiation toward Earth. Occasionally, the crust of a neutron star may crack and settle slightly in a "starquake." The star's spin then increases slightly to conserve angular momentum, and a small drop in the pulse period—called a glitch—may result. These pulsar glitches allow astronomers to probe the interior structure of neutron stars. The fastest pulsars are the millisecond pulsars, which have kilometer dimensions, more mass than our Sun, and spin almost at breakup speed. Many are found in globular clusters. They are thought to have been "spun up" by drawing in huge amounts of matter from a companion star.

Observed fluctuations in one pulsar's pulse period may be caused by the Doppler effect as the neutron star wobbles back and forth in space. The wobble may be caused by the combined gravitational pulls of two or more planets, each with about three times the mass of Earth.

An object having more than three times the Sun's mass will collapse beyond the neutron star stage to form a black hole. As an object of any given mass contracts, the force of gravity grows stronger at its surface, and the escape velocity increases. When the object's escape velocity exceeds the speed of light, absolutely nothing can escape its surface. Such an object emits no light, no radiation, no information whatsoever (except via the quantum-mechanical process known as Hawking radiation). The Schwarzchild radius is the name for the critical radius at which the escape velocity from an object would equal the speed of light and within which the object could no longer be seen. The surface of an imaginary sphere with radius equal to the Schwarzchild radius and centered on a collapsing star is called the event horizon. Photons passing too close to a black hole are deflected onto paths that cross the event horizon and become trapped.

Einstein's special theory of relativity deals with the preferred status of the velocity of light. General relativity is what results when gravity is included in the framework of special relativity. In general relativity, gravity is a result of curved spacetime. Matter tends to "warp," or curve, space in its vicinity. The more mass, the greater the warping. All particles—including photons—respond to that warping by moving along curved paths.

Whatever falls into a black hole is vertically stretched and horizontally squeezed as it accelerates to high speeds, causing a great deal of frictional heating. In this way, infalling matter can emit large amounts of radiation before crossing the event horizon.

A few binary systems have peculiarities that suggest that one of their members may be a black hole. An x-ray source in the constellation Cygnus is a long-standing black hole candidate. This object, called Cygnus X-1, appears to be very massive, yet small in volume. There is presently no observational test that can distinguish a 10-$M_\odot$ black hole from, say, a 10-$M_\odot$ neutron star. Stellar black holes probably have been discovered, but it is still not possible to say for sure.

KEY WORDS

black hole	millisecond pulsar	pulsar	superconductor
event horizon	neutron star	pulsar glitch	superfluid
gamma-ray burster	photon sphere	Schwarzchild radius	theory of relativity
gravitational red shift	Principle of Cosmic Censorship	singularity	time dilation
lighthouse model		starquake	x-ray burster

REVIEW QUESTIONS

1. How does the way in which a neutron star forms determine some of its most basic properties?
2. Use your knowledge about how gravity works to explain what would happen to a person standing on a neutron star. Tell why it happens.
3. How can we observe objects as small as neutron stars?
4. Why aren't all neutron stars seen as pulsars?
5. What are x-ray bursters?
6. What is the favored explanation for the fast spin rates of millisecond pulsars?
7. How might some neutron stars have come to be members of binary systems?
8. How might a pulsar acquire planets?
9. What does it mean to say that the measured speed of a light beam is independent of the motion of the observer?
10. Use your knowledge of escape velocity to explain why black holes are said to be "black."
11. Why is it so difficult to test the predictions of general relativity?
12. Describe two tests of general relativity.
13. What would happen to someone falling into a black hole? Why?
14. Explain how a black hole can evaporate.
15. What makes Cygnus X-1 a good black hole candidate?
16. The angular momentum of a body (see Chapter 17) is proportional to its angular velocity times the *square* of its radius. Using the law of conservation of angular momentum, estimate how fast a collapsed stellar core would spin if its initial spin rate was 1 revolution per day and its radius decreased from 100,000 km to 10 km.
17. Given that the Schwarzchild radius for the Sun (mass = 2×10^{33} g) is 3 km and the Schwarzchild radius is proportional to mass, calculate the Schwarzchild radius for (a) a proton, with mass 1.7×10^{-24} g, (b) a 100-kg human, and (c) a 10^{11} solar mass galaxy.
18. A 10-km radius neutron star is spinning 1000 times per second. Calculate the speed of a point on its equator, and compare it with the speed of light. (Consider the equator as the circumference of a circle, and recall that circumference = $2\pi r$.)
19. Using the Newtonian formula given in Chapter 3, calculate the acceleration due to gravity at the event horizon of (a) a 1-$M_\odot$ black hole and (b) a trillion solar-mass black hole. Compare this with the acceleration due to gravity at the surface of the Earth (G = 981 cm/s^2).

DISCUSSION QUESTIONS

1. Imagine that you had the ability to travel at will through the Galaxy. Explain why you would discover many more neutron stars than those known to observers on Earth. Where would you be most likely to find these objects?

2. Go to the library and look for recent articles about gamma-ray bursters. Are they any less mysterious now than when they were first discovered?
3. Do you believe that planet-sized objects discovered in orbit around a pulsar should be called planets? Why or why not?
4. What is the most bizarre thing you have ever heard concerning black holes? Ask your instructor, or go to the library, to find out if what you heard has any relationship to known facts.

PROJECT

Many amateur astronomers enjoy turning their telescopes on the ninth-magnitude companion to Cygnus X-1, the sky's most famous black hole candidate. Because none of us can see in x-rays, no sign of anything unusual can be seen. Still, it is fun to gaze toward this region of the heavens and contemplate Cyg X-1's powerful energy emission and profound strangeness.

Even without a telescope, it is easy to locate the region of the heavens where Cyg X-1 resides. The constellation Cygnus contains a recognizable star pattern, or asterism, in the shape of a large cross. This asterism is called the Northern Cross. The star in the center of the crossbar is called Sadr. The star at the bottom of the cross is called Albireo. Approximately midway along an imaginary line between Sadr and Albireo lies the star Eta Cygni. Cygnus X-1 is located slightly less than 0.5 degrees from this star. With or without a telescope, sketch what you see.

SUGGESTED READINGS

Backer, D., and S. Kulkarni. "A New Class of Pulsars." *Physics Today* (March 1990). How binary pulsars, pulsars with millisecond periods, and pulsars in globular clusters are providing tools for fundamental tests of physics.

Chaisson, E. *Relatively Speaking: Relativity, Black Holes, and the Fate of the Universe*. New York: W. W. Norton, 1988.

Dolan, J. F. "Placing Faith in the Masses?" *Nature* (February 13, 1992). Detecting black holes in space.

Flam, F. "How to Find a Black Hole." *Science* (February 14, 1992). Uses a new black hole candidate, V404 Cygni, as an example while discussing the black hole search strategy of looking for an x-ray signature, then trying to measure masses.

Graham-Smith, F. "Pulsars Today." *Sky and Telescope* (September 1990). Current research on pulsars.

Greenstein, G. *Frozen Star*. New York: Freundlich Books, 1983. A historical perspective on neutron stars and black holes, along with many fine explanations.

Hawley, K, "What Are Gamma Ray Bursters?" *Sky and Telescope* (August 1990). Discusses these high-energy phenomena.

Shapiro, S. L. "Black Holes, Naked Singularities and Cosmic Censorship." *American Scientist* (July–August 1991). Comprehensive article concerning some of the amazing concepts surrounding the subject of black holes.

Shipman, H. L. *Black Holes, Quasars and the Universe*. Boston: Houghton Mifflin, 1976. Excellent basic explanations of black holes and other energetic phenomena.

CHAPTER

25 The Milky Way Galaxy

A Grand Design

This image displays an extraordinary view of the core of the globular cluster, 47 Tucanae. It was captured with an ultraviolet-sensing camera aboard the Hubble Space Telescope. *The scale of the image is 20 arc seconds across, which at a distance of 5000 pc, equals about 0.6 pc. Resolution is a superb 0.08 arc second. Surprisingly, the image reveals a very high concentration of a unique class of stars called "blue stragglers." Such star systems might actually be evolving "in reverse," from old-age back to a hotter and brighter youth, their new lease on life made possible as stars in the cluster's core experience close encounters, possibly capturing one another, forming tight binary systems, or even colliding or merging into one single star. Their "blueness" results from reinvigorated burning, and the term "straggler" suggests they have been left behind by all the other evolving stars in the cluster. (Courtesy NASA and European Space Agency)*

R I V **U** X G

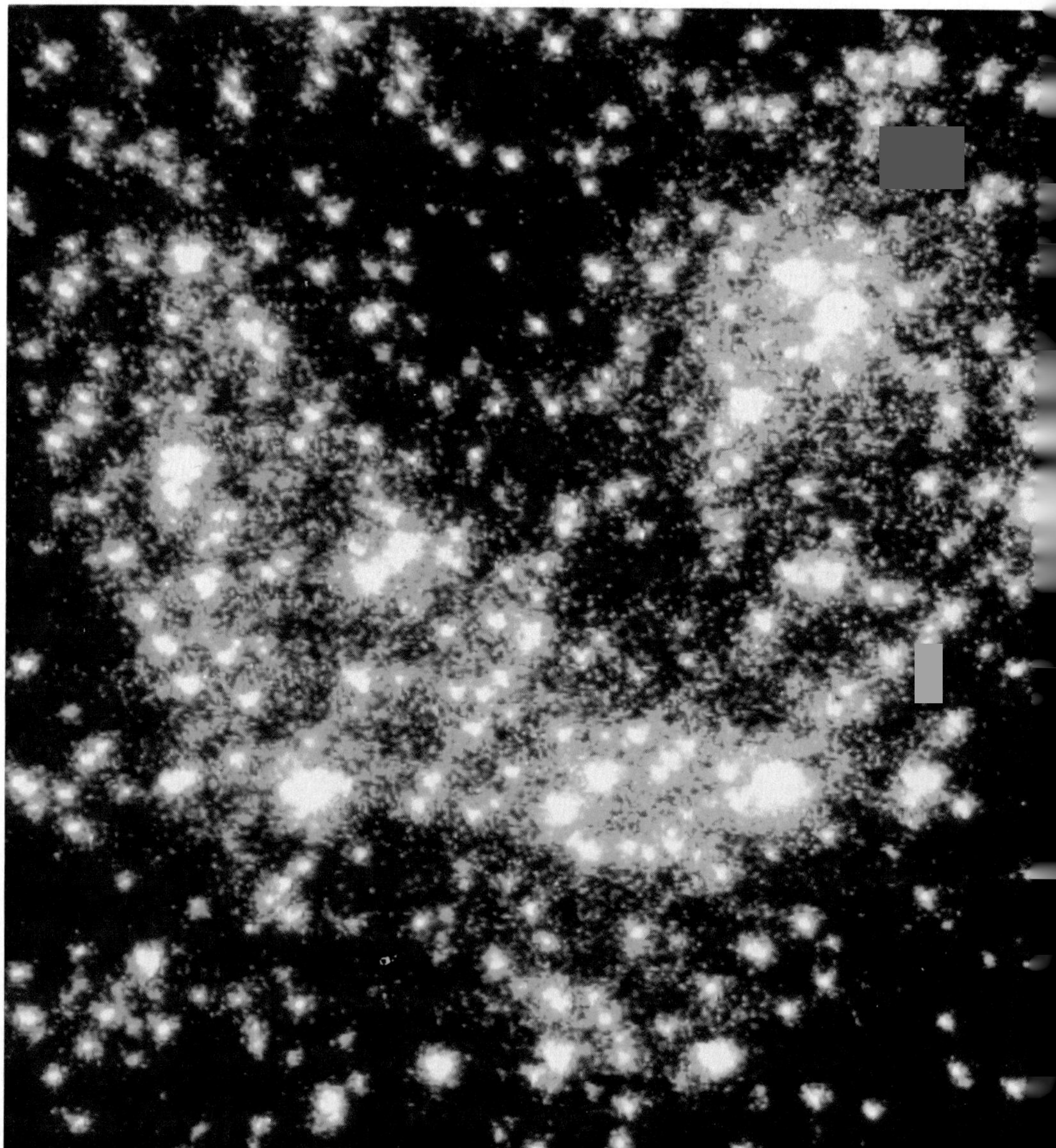

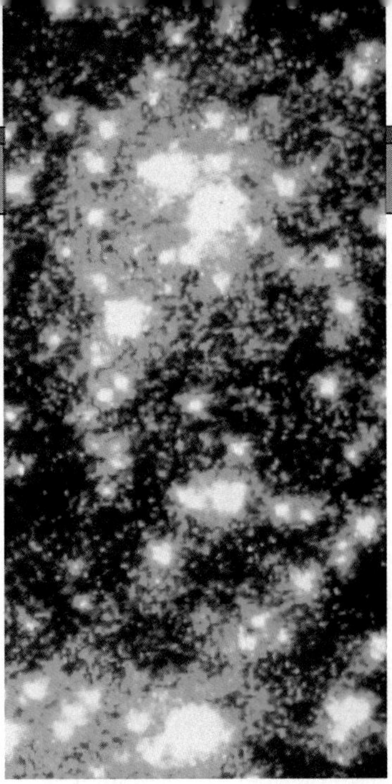

LEARNING GOALS

1 to appreciate how our knowledge of the Milky Way has evolved

2 to understand the use of variable stars in determining distances within the Milky Way Galaxy

3 to understand how astronomers use radio astronomy to map the Galaxy

4 to know the overall structure of the Milky Way Galaxy

5 to understand some of the strange phenomena observed at the center of our Galaxy

Looking out from Earth, beyond the Moon and the planets of our solar system, we are struck by two aspects of the night sky. The first is a fuzzy band of light—the Milky Way—that stretches across the heavens. From the Northern Hemisphere, it is most easily visible in the summertime, when it arcs high above the horizon. Although the Milky Way is brightest in the general direction of the constellation Sagittarius, its full extent forms a great circle that encompasses the entire celestial sphere.

Our second impression is that away from this glowing band the nighttime sky seems more or less the same in all directions. Bunches of stars cluster here and there, but overall our immediate locale seems fairly homogeneous. Regardless of where they happen to lie, equal volumes of space in the neighborhood of the Sun seem to contain pretty much the same number of stars and the same amount of interstellar gas and dust. In short, apart from the Milky Way itself, the rest of the evening sky appears fairly uniform.

But this is only a local impression. Ours is a rather provincial view. When we consider much larger volumes of space, with dimensions much, much larger than the distances between neighboring stars, the spread of stellar and interstellar matter changes. It becomes patchy and irregular. Eventually, it thins to essentially nothing. There is a definite distance beyond which the spread of matter is so thin that stars, gas, and dust are virtually nonexistent. But within this boundary lies a huge collection of stellar and interstellar matter—the Milky Way Galaxy.

OUR PARENT GALAXY

✓1 A **galaxy** is a gargantuan collection of stellar and interstellar matter—planets, stars, gas, dust, brown dwarfs, black holes—isolated in space and held together by its own gravity. A fairly typical galaxy—not the one we inhabit, but another, similar system—is shown in Figure 25.1. This is the Andromeda Galaxy, which we have encountered several times already in this text. It lies about 700 kpc (700 kiloparsecs, or 700,000 pc, about 2 million light-years) from Earth. Despite its enormous distance, Andromeda is the nearest major galaxy to our own. We show it here in place of our own because our domain—the Milky Way Galaxy, or just ''the Galaxy,'' with a capital *G*—is too large for us to see in its entirety or to photograph. We live inside the Milky Way Galaxy, and we cannot move away from it to a distance suitable for a snapshot.

Andromeda's apparent elongated shape is just a consequence of the angle at which we happen to view it. In fact, this galaxy, like our own, consists of a flattened, circular disk of matter that fattens to a bulge at the center and is embedded in an extended, roughly spherical ball of faint old stars. (Andromeda's disk and bulge are clearly visible in Figure 25.1.) This basic knowledge allows us to understand and interpret the appearance of our Galaxy from Earth, as illustrated in Figure 25.2. From our perspective within the disk, the Galaxy is seen as a band of light stretching across the sky—the Milky Way. In this chapter, we will study in some detail the various parts that make up our parent Galaxy.

The above description represents the modern view of

Figure 25.1 *The Andromeda Galaxy probably resembles fairly closely the overall layout of our own Milky Way Galaxy. The disk and bulge are clearly visible in this image. The inserts at top left and bottom right show more detailed views of the galaxy's spiral arms and central bulge. (NOAO; Palomar Observatory)*

our Galaxy. Before the early part of the twentieth century, however, astronomers had a markedly different view of the stellar system we inhabit. The fact that we live in just one of many enormous ''islands'' of matter separated by even larger tracts of apparently empty space was unknown, and the clear distinction between ''our Galaxy'' and ''the universe'' did not exist. Like the Copernican Revolution before them, the twin ideas—that (1) the Sun is not at the center of the Galaxy, and (2) the Galaxy is not the center of the universe—required both time and hard observational evidence before they gained widespread acceptance.

SPIRAL NEBULAE AND ISLAND UNIVERSES

Nineteenth-century astronomers were hampered in their efforts to probe and understand the Galaxy by their inability to determine reliable distances to astronomical objects. Prior to the discovery of the main sequence in 1911 and the development of the distance-measurement technique now known as spectroscopic parallax (see Chapter 19), the locations of objects lying more than a few hundred parsecs from Earth were all but unknown.

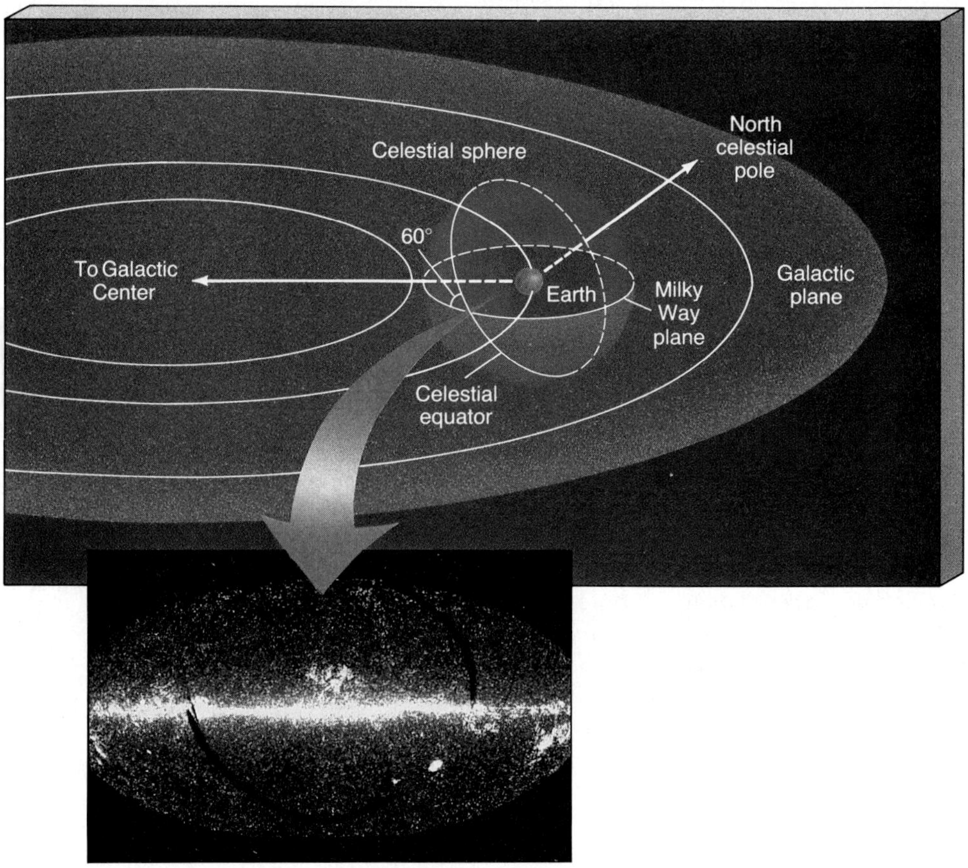

Figure 25.2 *Seen from within, the flattened disk of our Galaxy appears as a band of light across the heavens, known as the Milky Way. When we gaze at the Milky Way, we are looking in the plane of our Galaxy's disk; in other directions, our line of sight is out of the plane. This is (mostly) the Milky Way's plane as detected by* IRAS, *the* Infrared Astronomy Satellite. *(NASA)*

R I V U X G

As an example of the difficulties that resulted from the lack of a reliable distance scale, let's briefly reconsider the Andromeda Galaxy of Figure 25.1, or the Andromeda *spiral nebula,* as it was known in the mid-nineteenth century. With the observational techniques available at the time, it appeared only as a rather fuzzy, indistinct patch of light in the sky, with hints of a swirling, spiral structure in it. Astronomers had no means of determining its distance, and they simply assumed that Andromeda and other spiral nebulae were located in our Galaxy, perhaps somehow similar to the emission nebulae (which we discussed in Chapter 20).

By the end of the nineteenth century, improved telescopes and photographic techniques had allowed astronomers to obtain better images, showing detail comparable to that visible in Figure 25.1. However, the "nebula's" distance was still unknown. In 1888, when these images were first presented publicly, they caused great excitement among astronomers, who thought they were seeing the formation of a star from a swirling gaseous disk! Comparing Figure 25.1 with the figures in Chapter 17 (see especially Figure 17.6), we can perhaps understand how such a mistake could be made—*if* we believed we were looking at a relatively close, star-sized object. Far from demonstrating that Andromeda was distant and large, the observations seemed to confirm that it was just a small part of our own Galaxy.

Further observations soon made it clear that the interpretation of Andromeda as a star-forming region was incorrect. For example, Andromeda's parallax is too small to measure, indicating that it must be at least a few hundred parsecs from Earth. Even at 100 pc (which we now know, of course, is vastly less than the true distance), an object the size of the solar nebula would be impossible to resolve and simply would not look like Figure 25.1.

By the early 1900s, the questions of the size of our Galaxy and the distance to Andromeda and the other spiral nebulae were being hotly debated in astronomical circles. One school of thought maintained that they were galactic nebulae much smaller than, and contained within, our own Galaxy. Other astronomers held that they were "island universes" outside the Milky Way Galaxy and comparable to it in size. However, with no firm distance information on which to base them, both sides' arguments were quite inconclusive. It was not until the late 1920s, with the discovery of the next rung in our cosmic distance "ladder," that the issue was finally settled in favor of the island universe theory. The "spiral nebulae" are now known as *spiral galaxies*.

This story of Andromeda and the spiral nebulae illustrates how important it is to know the distance to an astronomical object before we can really decipher its nature. The growth in our understanding the architecture of our Galaxy, and the realization that there are many other distant galaxies comparable in size to our own, have gone hand in hand with the development of the cosmic distance scale.

VARIABLE STARS AS DISTANCE INDICATORS

✓2 Studies of nearby stars, gas, and dust, as well as comparisons with other galaxies such as Andromeda, can give us a general idea of the overall distribution of matter in our Galaxy. But the resulting picture is crude, for most stars are obscured from our view and interpretations of the gas and dust are tricky, as we saw in Chapter 20. Better observational tools are needed if we are to fully appreciate the size and shape of our Galaxy. We now consider one of the most important of these tools.

Pulsating Stars

One by-product of the laborious effort to catalog stars around the turn of the twentieth century was the systematic study of **variable stars.** These are stars whose luminosity changes with time, some quite erratically, others more regularly. Only a small fraction of all stars fall into this category, but those that do are of great astronomical significance.

We have encountered numerous examples of variable stars in earlier chapters. Often, the variability is the result of membership in a binary system. Eclipsing binaries, novae, and type-I supernovae are cases in point. Novae and supernovae are often collectively called **cataclysmic variables** because of their sudden, large changes in brightness. In other cases, the variability is a basic trait of the star—such as in the T-Tauri phase of star formation, or in so-called **flare stars,** which are low-mass main-sequence stars exhibiting violent surface activity.

As far as the study of galactic structure is concerned, the most important intrinsic variables are the **pulsating variable** stars, whose luminosity varies in a smooth and predictable way. Figure 25.3 shows the light curve of one such star, whose brightness rises and falls by about a factor of 2 every 3 days. This type of pulsating star is known as a **Cepheid variable** (or simply a "Cepheid"), after the first star of the type to be discovered—Delta Cephei, the fourth brightest star in the constellation Cepheus. Cepheid variables are recognizable by the characteristic shape of their light curves (the rapid rise followed by the slower decline in Figure 25.3). Different Cepheids have different pulsation periods, ranging from about 1 to 100 days, but the period of any given Cepheid is essentially constant from one cycle to the next.

Closer study reveals that, as the luminosity of a Cepheid varies, both its radius *and* its surface temperature change too. The temperature is greatest at times of maximum luminosity and least when the luminosity is at a minimum. The radius also changes, but by a relatively small amount—only about 10 or 20 percent—so the Cepheid's luminosity variations are primarily due to changes in its surface temperature (recall from Chapter 19 the radius-luminosity-temperature relation $L \propto R^2T^4$).

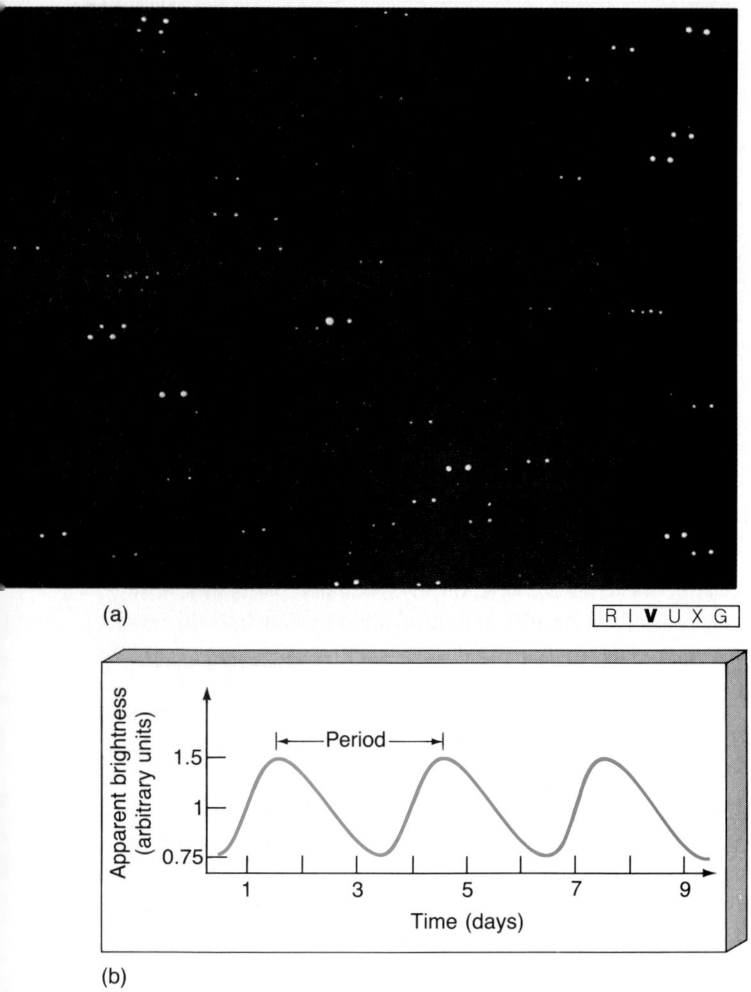

Figure 25.3 A Cepheid variable star (center) appears on these two nearly superposed photographs (a) taken on successive nights. The star is called WW Cygni, and is shown at its maximum and minimum brightness. Monitoring the star over the course of a week or so (b) yields a record of the star's changing brightness. (Harvard Observatory)

A related class of pulsating variables are the **RR Lyrae** stars (again named after the first known of their kind, in this case the variable star labeled RR in the constellation Lyra). Like the Cepheids, they have a characteristic, easily recognizable light curve (shown in Figure 25.4), which accurately repeats itself and which is the result of variations in the temperature and radius of the star. Unlike the Cepheids, however, there is little difference in the period between one RR Lyrae variable and another.

Why do Cepheids and RR Lyrae variables pulsate? What is responsible for the continuous heating and cooling of their outer layers? The basic mechanism was first suggested by the British astrophysicist Sir Arthur Eddington in 1941 and has subsequently been borne out by detailed calculations. The structure of a star is determined in part by the ability of radiation to travel from the core to the photosphere—that is, by the *opacity* of the interior (which, recall from Chapter 4, is the degree to which the gas hinders the

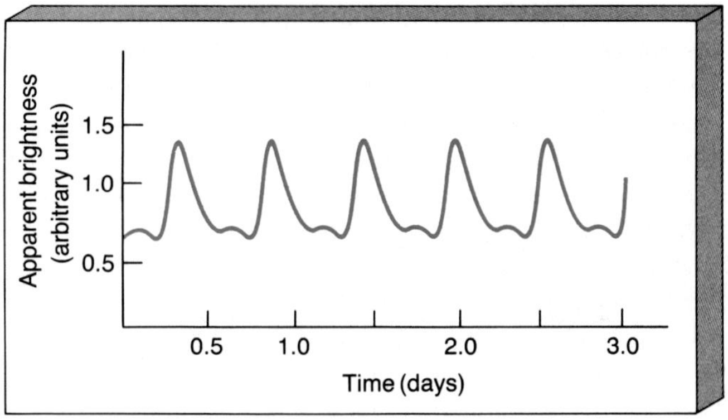

Figure 25.4 Light curve of the pulsating variable star RR Lyrae. All RR Lyrae-type variables have essentially similar light curves, with a period of about half a day.

passage of light through it). If the opacity rises, the radiation becomes trapped, the internal pressure increases, and the star "puffs up." If the opacity falls, radiation can escape more easily, and the star shrinks. Under certain circumstances, it is possible for the star to become slightly unbalanced and enter a condition where the flow of radiation causes the opacity to alternately rise and fall, leading to pulsations.

The conditions necessary to cause pulsations are not found in main-sequence stars. However, they *do* occur in evolved stars as they pass through a region of the Hertzsprung-Russell diagram known as the **instability strip,** shown in Figure 25.5. Cepheids are found in the higher-

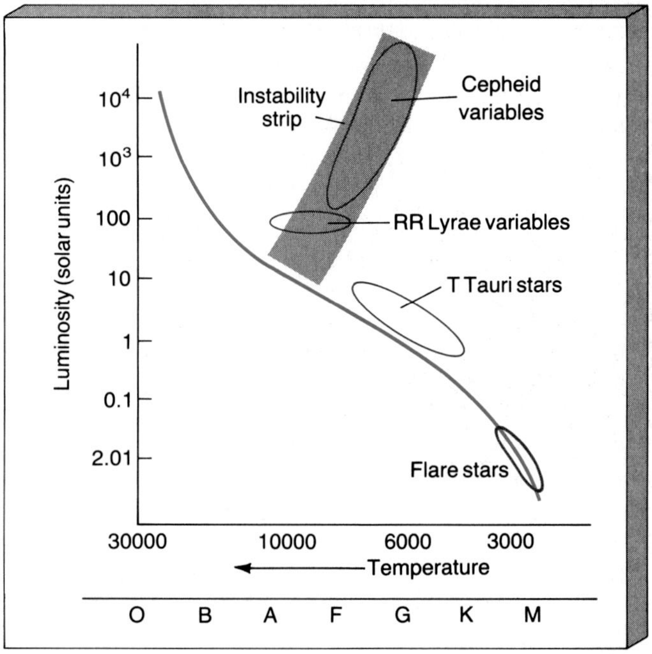

Figure 25.5 Many pulsating variable stars are found in the so-called instability strip of the HR diagram. As a high-mass star evolves through the strip, it becomes a Cepheid variable. Low-mass horizontal branch stars in the instability strip are RR Lyrae variables.

luminosity regions of the instability strip and make up an evolutionary stage of high-mass stars as they evolve across the HR diagram. RR Lyrae variables, in contrast, are low-mass horizontal branch stars that happen to lie in the lower portion of the strip. Similar unstable conditions can occur within stars in other regions of the HR diagram, particularly on the red-giant branch, and other classes of variable stars are indeed associated with giants. However, we will confine ourselves here to the Cepheids and RR Lyrae variables discussed above.

A New Yardstick

Let's not get lost in the detailed nature of Cepheids and other variable stars. While they are certainly interesting objects in their own right, our primary goal here is to use them to obtain information about the large-scale distribution of stars in our Galaxy.

The key point about Cepheid variables is this: Their *absolute* brightnesses and their pulsation periods are quite tightly correlated. Cepheids that vary slowly—that is, that have long periods—have large absolute brightnesses (averaged over an entire pulsation cycle). The converse is also true: Short-period Cepheids have small absolute brightnesses. Figure 25.6 illustrates this relationship for Cepheids found close to Earth. Astronomers can plot this diagram for nearby Cepheids because they know the distances to these stars and hence their absolute brightnesses (or luminosities). Measuring the pulsation period is straightforward. This link between period and brightness is known as the **period-luminosity relation,** as shown in Figure 25.6. It was discovered in 1908 by Henrietta Leavitt, at Harvard University (see Interlude 25-1). We know of no exceptions to this relation, which is also consistent with theoretical calculations of pulsations in evolved stars. Consequently, we assume that the relation holds for all Cepheids, near and far. The roughly constant period of the RR Lyrae variables is also marked in Figure 25.6. For our purposes, it is convenient to think of them simply as short-period Cepheids.

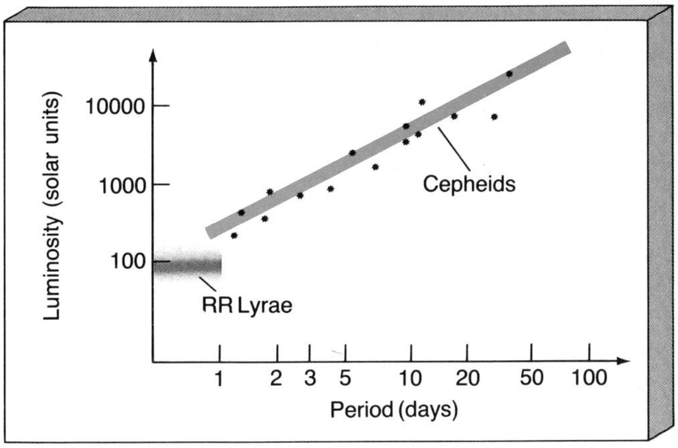

Figure 25.6 *A plot of pulsation period versus average absolute brightness (that is, luminosity) for a group of Cepheid variable stars. The two properties are quite tightly linked. The pulsation periods of some RR Lyrae variables are also marked on this diagram.*

The beauty of the period-luminosity relation is that a simple measurement of a Cepheid variable's pulsation period is enough to determine its distance. From observations of the star's light curve, we can measure its period. We then read its luminosity off the plot in Figure 25.6 (accurate to within the scatter in the graph). Because the apparent brightness is usually easy to measure, we can find the distance to the star by comparing its apparent and absolute brightnesses, as discussed in Chapter 19. The absolute brightness of an RR Lyrae star is even easier to determine. All such stars have basically the same luminosity—about 100 solar luminosities. Once the light curve of a variable star is recognized as belonging to a star of the RR Lyrae type, the star's approximate luminosity is immediately known, and the distance calculation based on apparent and absolute brightnesses can be performed.

Variable stars, then, make up an important new rung in our cosmic distance ladder. The technique works well provided the star can be clearly identified and its period of variability measured. With Cepheids, this method allows astronomers to estimate distances out to several million parsecs, well beyond the range of either stellar parallax (around 100 pc) or spectroscopic parallax (about 1000 pc)—in fact, far enough to include the nearest galaxies. Being less luminous, RR Lyrae stars are not as easily seen as Cepheids, so their useful "range" is not as great. However, they are much more common, so within their limited range they are actually more useful than Cepheids. Figure 25.7 extends our inverted pyramid to include Cepheid variables as a fourth method of determining distance. The period-luminosity relation is calibrated using nearby stars, so this latest rung inherits any and all uncertainties and errors present in the lower levels, and it also has the uncertainties caused by the scatter shown in Figure 25.6.

Harlow Shapley and the Size of Our Galaxy

Many variable stars—specifically, of the RR Lyrae type—are found in globular clusters, those tightly bound swarms of old, reddish stars that we first met in Chapter 19. Early in the twentieth century, the American astronomer Harlow Shapley used observations of RR Lyrae stars to make two very important discoveries. First, he proved that most globular clusters reside at great distances—many thousands of parsecs—from the Sun. Second, he showed that the globular clusters are distributed nearly spherically in space. However, the center of that sphere lies nowhere near our Sun—in fact, it is located nearly 8 kpc away from us in the direction of the constellation Sagittarius. (Shapley originally estimated that the Sun lies 15 kpc from the center; more recent measurements give us the 8 kpc value.) In a brilliant intellectual leap, Shapley realized that the globular clusters

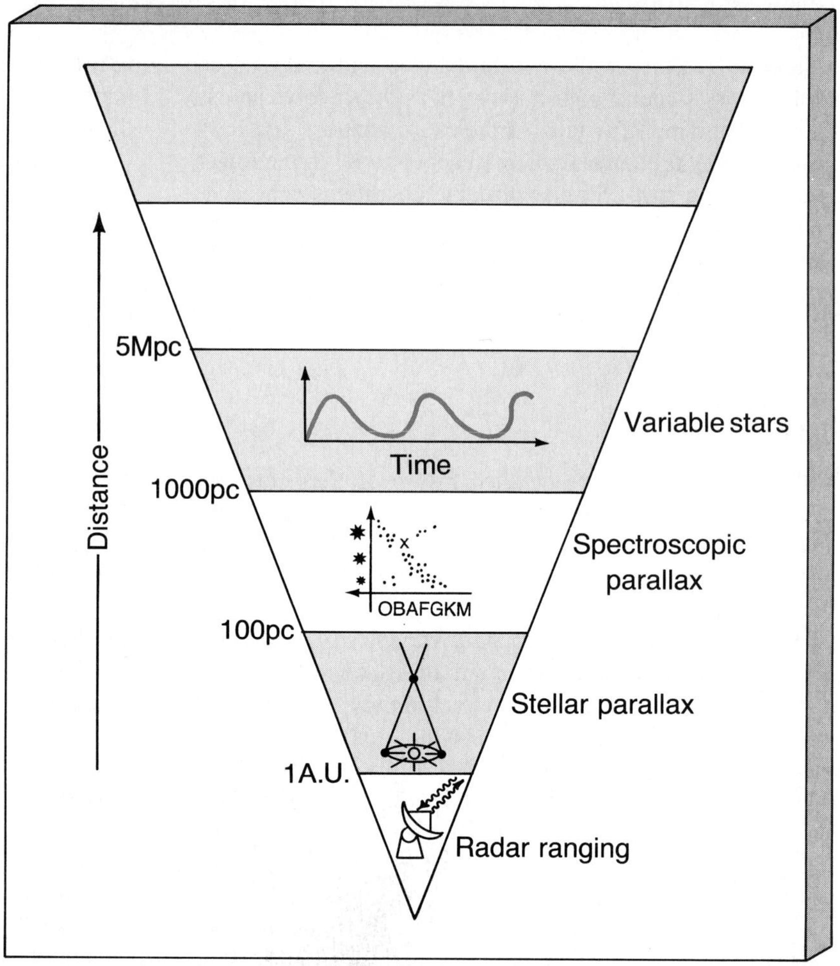

Figure 25.7 Application of the period-luminosity law for (Cepheid) variable stars allows us to determine distances out to about 5 Mpc with reasonable accuracy.

outline the true extent of stars in the Milky Way Galaxy, and he concluded that our Sun does *not* reside in the middle of this vast collection of stars. It is not even close to the center.

Figure 25.8 illustrates how the globular clusters map out what is now known as the **galactic halo** of the Milky Way Galaxy. The whole galactic system of stars spans approximately 30 kpc. Its hub—the **galactic center**—lies about 8 kpc from our Sun. We live in the suburbs of this truly huge ensemble of matter, in the thin sheet of young stars, gas, and dust, usually known as the **galactic disk,** or **galactic plane** which cuts through the center of the halo.

The halo and the disk are quite distinct components of the Milky Way Galaxy. The ancient stars and globular clusters that make up the halo extend far above and below the disk. They are not confined to the galactic plane, and their properties differ in many ways from those of younger disk stars like the Sun. The precise shape of the halo is not well known. There is evidence that it is somewhat flattened in the direction perpendicular to the disk, but the degree of flattening is still uncertain. Notice, by the way, the jump in scale of our units. When talking about stars and "nearby" nebulae, we generally measured distances in parsecs. Now, on a galactic scale, the kiloparsec (kpc) is more appropriate. Soon, as we leave our Galaxy behind, megaparsecs (Mpc) will become the norm.

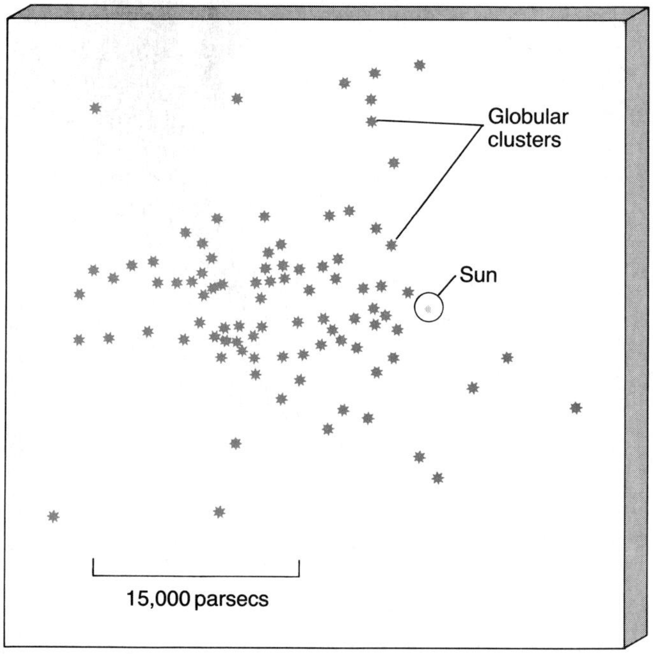

Figure 25.8 Our Sun (denoted by the symbol ⊙) does not coincide with the center of the very large collection of globular clusters (denoted by asterisks). Instead, more globular clusters are found in one direction than in any other. The Sun resides closer to the edge of the collection, which measures some 30 kpc across. We now know that the globular clusters outline the true distribution of stars in the galactic halo.

INTERLUDE 25-1 *Early Computers*

A large portion of the early research in observational astronomy focused on monitoring stellar luminosities and analyzing stellar spectra. Much of this pioneering work was done using photographic methods. What is not so well known is that most of the labor was accomplished by women. Around the turn of the century, a few dozen dedicated women—assistants at the Harvard College Observatory—created an enormous data base by observing, sorting, measuring, and cataloging photographic information that helped form the foundation of modern astronomy. Some of them went far beyond their duties in the lab to make several of the basic astronomical discoveries often taken for granted today.

This 1910 photograph shows several of those women carefully examining star images and measuring variations in luminosity or wavelengths of spectral lines. Their data yielded information on hundreds of thousands of stars through direct eye inspection in the cramped quarters of the Harvard Observatory. Note the plot of stellar luminosity changes pasted on the wall at the left. The cyclical pattern is so regular that it likely belongs to a Cepheid variable. Known as "computers" (for there were no electronic devices then), these women were paid 25 cents an hour.

These workers, beginning in 1880, started a survey of the skies that would be carried on for a century. Their first major accomplishment was a catalog of the brightnesses and spectra of tens of thousands of stars, published in 1890 under the direction of Williamina Fleming. On the basis of this compilation, several of these women made fundamental contributions to astronomy. In 1897, Antonia Maury undertook the most detailed study of stellar spectra to that time, enabling Hertzsprung and Russell independently to develop what is now called the HR diagram. In 1898, Annie Cannon proposed the Harvard Spectral Classification System described in Chapter 19 and now adopted as the international standard for categorizing stars. In 1908, Henrietta Leavitt discovered the period-luminosity relation for Cepheid variable stars, enabling Harlow Shapley (her boss) later to recognize our Sun's position near the edge of our Galaxy, as described in the text.

(Harvard College Observatory)

Shapley's bold interpretation of the globular clusters as defining the overall dimensions of our Galaxy was an enormous step forward in human understanding of our place in the universe. Five hundred years ago, Earth was considered the center of all things. Copernicus argued otherwise, banishing our planet to an undistinguished place removed from the central Sun. Yet even in the early twentieth century, the prevailing view was that our Sun was the center of not only the Galaxy, but also of the universe. Shapley showed otherwise. With his observations of globular clusters, he simultaneously increased the size of our Galaxy by almost a factor of 10 over earlier estimates and banished our parent Sun to its periphery, virtually overnight!

Curiously, Shapley's dramatic revision of the size of the Milky Way Galaxy and our place in it served only to strengthen his own erroneous opinion that the spiral nebulae discussed above were part of our Galaxy and that our Galaxy was essentially the entire universe. He regarded the idea that there could be other structures as large as our Galaxy as simply beyond belief. Only in the late 1920s was the Copernican principle extended to the Galaxy itself, when American astronomer Edwin Hubble observed Cepheids in the Andromeda Galaxy and finally succeeded in measuring its distance.

Thus, astronomers of the first three decades of the twentieth century learned that our Sun is part of the giant system of stars, dust, and gas that makes up our Milky Way Galaxy and recognized that the Earth lies far from the galactic center. Yet not until the 1950s did scientists develop the tool needed to explore the full grandeur of our Galaxy. That new tool was spectroscopic radio astronomy.

RADIO STUDIES OF THE MILKY WAY

Optical Limitations

✓3 In the late eighteenth century, long before the distances to any stars were known, the English astronomer William Herschel tried to estimate the size and shape of our Galaxy simply by counting how many stars he could see in different directions in the sky. Assuming that all stars were of roughly equal brightness, he concluded that the Galaxy was a roughly lozenge-shaped collection of stars lying in the plane of the Milky Way, with the Sun at the center. These observations and arguments were improved and extended over the years and reached their peak of refinement in a model of the Galaxy proposed around 1920 by the Dutch astronomer J. C. Kapteyn. Sketched in Figure 25.9, this so-called **Kapteyn universe** was about 10 kpc in diameter and 2 kpc thick. Like Herschel's Galaxy, its midplane was the plane of the Milky Way, and the Sun was at the center.

Models of the Galaxy based solely on observations made in the visible part of the electromagnetic spectrum suffer from a fatal flaw: They ignore the absorption of visible light by interstellar gas and dust. As we saw in Chapter 20, objects more than a few kiloparsecs away in the plane of the Milky Way are hidden from our view by the effects of interstellar absorption. The central regions of our Galaxy cannot be studied by optical techniques. By contrast, radiation coming to us from above or below the plane of the Galaxy, where there is less obscuring matter, arrives on Earth relatively unscathed. The apparent fall-off in stars with distance in the galactic plane observed by Herschel and Kapteyn is not a real thinning of their numbers in space but simply a consequence of the relatively murky environment in the galactic disk. Because the obscuration occurs in all directions in the disk, the fall-off is roughly similar no matter which way we look, so the Sun appears to be at the center. Perpendicular to the disk, the fall-off in density is real (although there is still some patchy obscuration), and optical measurements reflect more accurately the true disk thickness. The Sun happens to reside at a location where our line of sight perpendicular to the disk is relatively unimpeded by interstellar dust and gas.

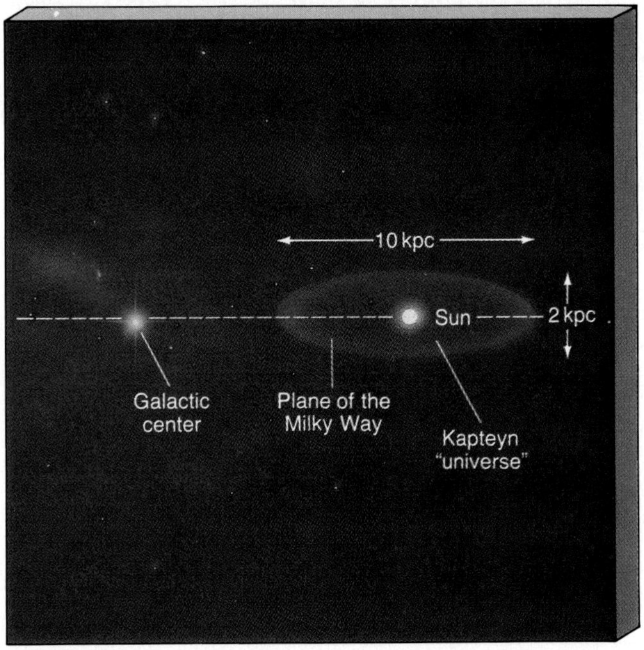

Figure 25.9 *The "Kapteyn universe" was a model of the Milky Way Galaxy based on optical observations of stellar brightnesses and motions. Because it failed to take into account the absorption of light by interstellar matter, the model seriously underestimated the extent of the Galaxy. The Kapteyn model is here compared with the much larger Galaxy revealed by Shapley's observations of globular clusters (Figure 25.8).*

Exploring Galactic Clouds

The key to peering into the center of our Galaxy and observing the full extent of the galactic disk is the 21-cm radio emission line from neutral hydrogen, which falls in the radio range. This long-wavelength radiation is unaffected by the intervening dust. Also, hydrogen is by far the most abundant element, so the 21-cm signals are usually strong enough to measure and let us "see" into the heart of the Galaxy. By studying the 21-cm radio lines emitted by interstellar clouds that lie in many different directions, we can map out the large-scale distribution and motion of the Galaxy's hydrogen gas.

The 21-cm technique works well only for regions abundant in *atomic* gas. However, as noted in Chapter 20, *molecular* clouds are also widespread throughout the Galaxy. Because these clouds contain little or no atomic hydrogen, they are not so easily detectable at this wavelength. In fact, as we noted in Chapter 20, the hydrogen is quite difficult to detect at all because it is mainly bound up in the molecular form H_2, which has no characteristic rotational, vibrational, or electronic transitions in the accessible part of the radio spectrum.

Other molecules, fortunately, do emit radio radiation that we can use to study the spread of molecular gas throughout the Galaxy. The carbon monoxide (CO) molecule is most appropriate for this purpose, mainly because it is abundant in all molecular clouds (although still a million times less abundant than H_2) and also because one of its characteristic spectral lines occurs in a part of the radio domain (at about 0.3-cm wavelength) that is easily observed. Carbon monoxide observations complement neutral hydrogen studies. By measuring these two gases, we can probe most of the disk of the Milky Way.

Mapping the Milky Way

Years of radio observations and detailed analyses have given researchers an appreciation for the true spread of atomic and molecular gas in our Galaxy. The work is still under way, but the big picture has already emerged. The upshot is that the distribution of interstellar clouds in the galactic disk displays an organized pattern on a grand scale.

The neutral hydrogen and carbon monoxide studies show that the gas in our Galaxy is centered far from our Sun. In fact, the center of the gas distribution coincides roughly with the center of our Galaxy, as determined from observations of globular clusters. Unlike the globular clus-

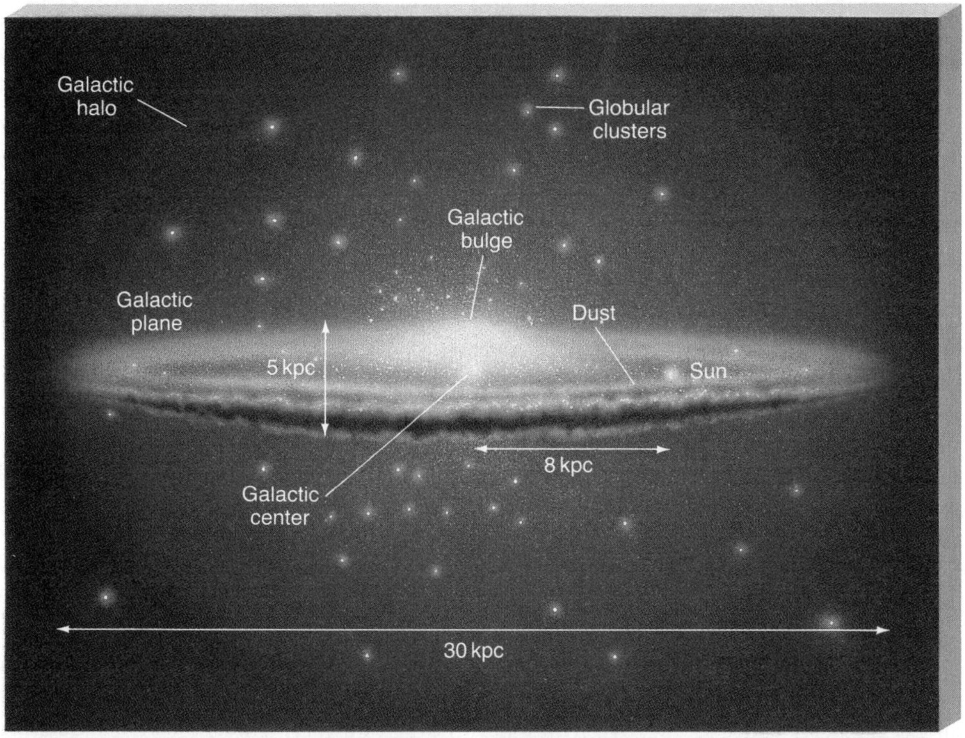

Figure 25.10 *An edge-on diagram of the interstellar gas and dust (shaded area), shown in relation to the globular and galactic star clusters. The numerical values give the primary dimensions of our Milky Way Galaxy in units of kiloparsecs. The bottom left is an edge-on view of a similar, but far away galaxy, NGC 4565. The bottom right frame is a wide-angle radio image of the plane and bulge of our Milky Way, as observed by the cosmic background explorer (COBE) satellite. (NASA; NOAO)*

ters and the halo stars, however, the radio-emitting gas in our Galaxy is not spread roughly spherically through space. Instead, the interstellar clouds are rather flatly distributed, looking a bit like a double Frisbee clamped together at the edges. As shown in Figure 25.10, the gas is confined to the galactic plane. It is this gas, along with the associated dust, young stars, and star clusters, that forms the band of the Milky Way that we see from Earth, shown in Figure 25.2.

The thickness of the galactic disk depends on what types of objects you measure. Young stars and gas are more tightly confined to the plane than stars like the Sun, and solar-type stars in turn are more narrowly defined than older K and M-type dwarfs. The main reason for this appears to be that stars have a tendency to drift up out of the disk over time, due to their interactions with other stars and interstellar gas clouds. Thus, as stars age, their distribution slowly becomes fatter.

At the location of the Sun, some 8 kpc from our Galaxy's center, the plane is quite thin, at least by galactic standards—perhaps 300 pc thick. As we saw earlier, the full diameter of the visible part of our Galaxy is about 30 kpc, so the average diameter of our Galaxy's disk is some 100 times larger than its thickness. Don't be fooled, though. Even if you could travel at the speed of light, it would take you 1000 years to traverse the thickness of the galactic plane. The disk may be thin relative to the diameter of the Galaxy, but it is huge by human standards.

In addition, there is good observational evidence for an intermediate category of galactic stars, midway between the old halo stars and the younger disk stars in age and spatial distribution. Consisting of stars with estimated ages in the range of 7–10 billion years, this so-called **thick disk** component of the Milky Way Galaxy measures some 2–3 kpc from top to bottom. Its thickness is too great to be explained by the slow drift just described. Instead, the presence of the thick disk may have much to tell us about the way the Galaxy formed, as we will see later. To try to avoid confusion, we will adopt the convention that the term "disk" refers to the thin, star-forming plane of the Milky Way. When we mean specifically the thick disk, we will refer to it explicitly.

The gas in the disk fattens markedly toward the center, as shown in Figure 25.10, reaching a maximum thickness of about 2 kpc above and below the plane. The distribution of disk stars (including the thick disk) also fattens there, to about the same thickness. This "puffed up" region of stars and warm gas around the galactic center is known as the **galactic bulge.** The high gas density in the inner part of the bulge makes it the site of vigorous ongoing star formation. To some extent, the stars in the bulge can be thought of as the dense central portion of the halo—both very old and very young stars intermingle there. The bulge is flattened in shape, perhaps even more so than the halo—it measures some 6 kpc across in the plane of the disk, but only about 4 kpc from top to bottom. At larger distances from the galactic center, the bulge merges more or less smoothly with the halo. Recent detailed studies of the motion of gas and stars in and near the bulge indicate that it may really be football-shaped, with the long axis of the football lying in the galactic plane.

Spiral Structure

Figure 25.11 shows a sample of some of the known interstellar clouds in our Galaxy. Together, the clouds show clear evidence for pinwheel-type structures called **spiral arms.** Each arm seems to originate close to the galactic bulge. One of them, as best we can tell, encompasses our Sun. We apparently reside in a vast spiral arm that begins near the galactic center and wraps around a large part of the entire disk. Studies of spiral arms in our Galaxy and especially in other galaxies indicate that spiral arms are made up of much more than just interstellar clouds. Young objects—such as O- and B-type stars, open clusters, and emission nebulae—all reside in the arms, too. Because these young objects are generally not found outside the spiral arms and because the arms are the sites of the greatest concentration of interstellar gas and dust, we conclude that the spiral arms are the scene of recent and ongoing star formation.

As discussed in more detail in Interlude 25-2, a leading explanation for the existence of spiral arms holds that they are **spiral density waves**—coiled waves of gas compression that move through the galactic disk. The squeezing of interstellar gas clouds by the passage of these density waves may well trigger cloud collapse and begin the process of star formation, as described in Chapter 21. The spiral arms we see are outlined by dense clouds of gas and new stars, formed in response to the spiral wave's presence.

An alternative possibility is that the formation of stars drives the waves. Imagine a row of newly formed massive stars somewhere in the disk. When these stars form, the HII regions (the emission nebulae discussed in Chapters 20 and 21) that appear around them send shock waves through the surrounding gas, possibly triggering new star formation. Similarly, when the stars explode in supernovae, more shocks are formed. The formation of one group of stars provides the mechanism for the creation of more stars. Detailed simulations suggest that it is possible for the "wave" of star formation thus created to take on a spiral form—the spiral shape is favored in disk galaxies—and for the spiral pattern to persist for some time. This process, sometimes called **self-propagating star formation,** can produce pieces of spirals, as are in fact seen in some galaxies. However, it apparently cannot produce the galaxy-wide spiral arcs seen in others. It may well be that there is actually more than one process at work in creating the spectacular spirals we see around us.

As best we can tell, our Galaxy looks very much like the Andromeda Galaxy. It has an overall shape and size like Andromeda's and apparently contains similar proportions of stars, gas, and dust. Just how similar are the two galaxies? We'll probably never know for sure. We live embedded within the Milky Way's galactic plane, and we can map the Galaxy from only one vantage point. Imagine trying to unravel the layout of paths, bushes, and trees on Boston Common from the vantage of just one park bench. Of course, we could more easily examine the park by walking around it or by looking at it from some distant vantage point—a skyscraper or a helicopter, say. But we will probably never be

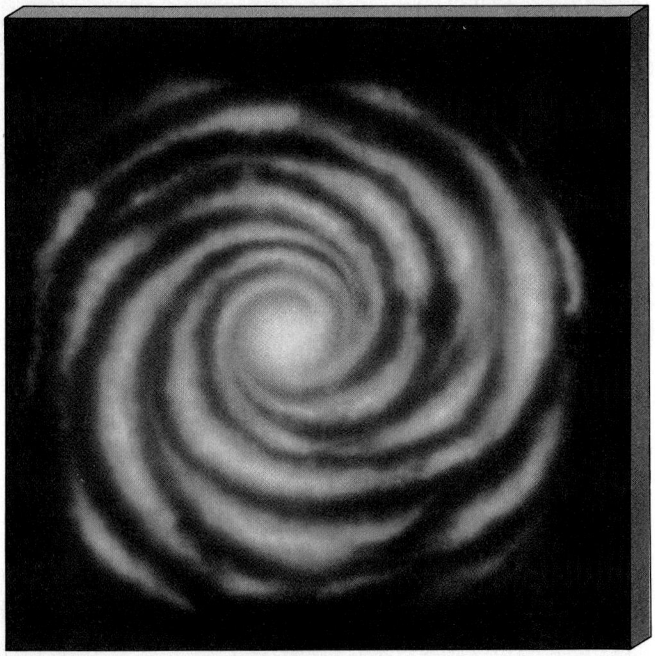

Figure 25.11 *A map of many of the Milky Way's interstellar clouds suggests huge "spiral arms" growing out from near the galactic center.*

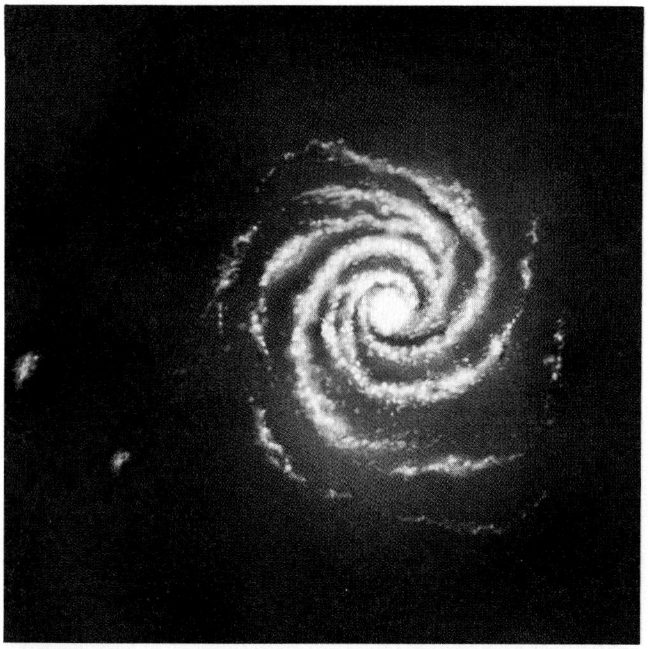
(a)

(b)

R I **V** U X G

Figure 25.12 (a) An artist's conception of our Milky Way Galaxy seen face-on. This illustration is based on data accumulated by legions of astronomers during the past few decades. Painted from the perspective of an observer 100 kpc above the galactic plane, the spiral arms are at their best-determined positions, and all the features are drawn to scale (except for the oversized dot near the top, which represents our Sun). The two small blotches to the left are dwarf galaxies, called the Magellanic Clouds. We will study them in the next chapter. (b) An image of the galaxy NGC891, whose properties are believed to be quite similar to those of the Milky Way. (L. J. Chaisson; Palomar)

able to achieve such a direct, clear view of the Galaxy. The notion of traveling far enough away to look back at our Galaxy from some distant vantage point is destined to remain science fiction for a good long time, perhaps forever. Instead, we must be content with artists' conceptions like Figure 25.12(a), based on the view from within, and images of other galaxies, such as Figure 25-12(b).

RED AND BLUE STARS

Now let's return to the optical appearance of galaxies like the Milky Way. Spiral galaxies imaged in visible light, like those shown in Figures 25.1 and 25.12(b) often exhibit the same basic features we have described for our own Galaxy: clearly defined disks with roughly spherical bulges at their centers and spiral arms that grow out from the bulge region. In addition, faint halos of stars and globular clusters surround the bulge. The thick disks of spiral galaxies beyond our own are generally too faint to be seen (assuming they exist). We will return in Chapter 26 to a more detailed study of the similarities and differences among galaxies, spiral and otherwise.

Within a spiral galaxy, a significant difference exists between stars in the galactic disk and those in the (outer) bulge and halo: Bulge and halo stars appear distinctly *redder* than stars found in the disk. This difference—the bluish tint of the disk and the yellowish or reddish coloration of the halo—is clearly evident in Figures 25.1 and 25.11. In distant galaxies, only the brightest stars can be distinguished as individuals—the remainder merge into a continuous blur of light—so we can say only that, *on average,* their disk stars tend to be bluer in appearance than stars in their bulges and halos. In our own Galaxy, we cannot see the big picture (because of interstellar absorption), but careful studies of individual stars near the Sun indicate a similar distribution. Stars in spiral galaxies are not a homogeneous mixture. Different kinds of stars are found in different regions of our Galaxy.

Figure 25.13 sketches this twofold distribution for our own Galaxy. The local blue stars, observable to distances of a few thousand parsecs, are generally confined to the galactic plane, as are the young open star clusters. Because these O- and B-type objects burn their fuel very rapidly, it follows that the galactic plane must be the site of current (or very recent) star formation. By contrast, the redder stars—including those found in the old globular clusters—inhabit a more nearly spherical region extending well above and below the galactic plane. They seem to be more uniformly distributed throughout the disk, bulge, and halo. (The spiral arms appear blue simply because main sequence O- and

Figure 25.13 *Oblique view of a galaxy, like our own Milky Way or Andromeda, showing the distributions of young blue stars, open clusters, old red stars, and globular clusters.*

B-type blue supergiants are very much brighter than the G, K, and M dwarfs, even though the dwarfs dominate the total mass.)

The generally accepted explanation for this striking difference in stellar population between the halo and the disk takes us all the way back to the birth of our Galaxy, some 15 billion years ago. It is thought that when the first stars and globular clusters formed, the gas in our Galaxy had not yet accumulated into a thin disk. Instead, it was spread out over a rather irregular, and quite extended, region of space, spanning many tens of kiloparsecs in all directions. When the first stars formed, they were distributed throughout this volume. Their distribution today (in the halo) reflects that fact—it is an imprint of their birth. It is also possible that the earliest stars formed in smaller systems that later *merged* to create our Galaxy—the present-day halo would look the same in either case. We will return to the subject of galaxy formation in Chapter 26.

During the past 10–15 billion years, rotation has flattened the gas in the Galaxy into a relatively thin disk, which contains virtually all of the gas, dust, and young stars in the Milky Way. Star formation in the halo ceased billions of years ago when the raw materials fell to the plane. Ongoing star formation in the disk gives the plane its bluish tint, but the halo's short-lived blue stars have long since burned out, leaving only the long-lived red stars that give it its characteristic pinkish glow. The galactic halo is ancient; the disk is full of youthful activity. The thick disk, with its intermediate-age stars may represent an intermediate stage of star formation that occurred while the gas was still flattening into the plane. (However, the process is poorly understood, and this is not the only possible explanation.)

THE MASS OF THE GALAXY

The Galactic Traffic Pattern

✓4 What about the dynamics of the matter in our Galaxy? Can we learn anything by studying the motions of the clouds and stars? Are the internal motions of our Galaxy's members chaotic and random, or are they part of a gigantic "traffic pattern"? The answer depends on our perspective. The motion of stars and clouds we see on small scales (a few tens of parsecs, say) seems random, but the motion we perceive at larger scales (hundreds or thousands of parsecs) appears much more orderly.

To appreciate this orderly movement in our Galaxy's

system, let us consider the galactic disk a little further. We see the gas and stars in the spiral arms, of course, only from our vantage point on Earth as we orbit the Sun. Yet as we look around the galactic disk in different directions, a clear pattern of motion emerges, as summarized in Figure 25.14. All the spectral lines emitted from the interstellar gas in the upper-right quadrant and the lower-left quadrant of Figure 25.14 are blueshifted. All the interstellar regions sampled in the upper-left quadrant and the lower-right quadrant are redshifted. In short, some regions in the Galaxy are approaching the Sun, and others are receding. The important point is that they are moving in a systematic fashion.

Careful study of this fourfold pattern of Doppler-shifted interstellar gas leads to the following important conclusion: The entire galactic disk is rotating about the galactic center. Furthermore, the disk does not rotate at a uniform rate. It spins *differentially,* by which we mean that stars and gas at different distances from the center take different lengths of time to complete one orbit. (Compare this with *solid-body rotation,* where every piece of the disk would move with the same angular speed and so would have the same orbit period, like a record spinning on a turntable.) Specifically, radio observations demonstrate that the inner regions of the galactic disk take less time to orbit the galactic center than do the outer parts. Similar differential rotation is observed in Andromeda (and, in fact, in all other spiral galaxies). The stars in the disk of any galaxy (our own

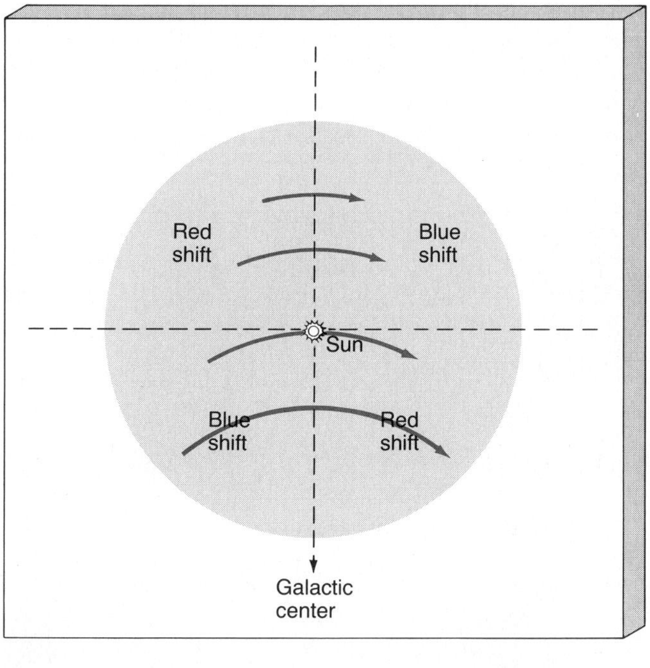

Figure 25.14 *Diagram of the four galactic quadrants in which interstellar clouds show systematic Doppler motions. This information tells us that the disk of the Galaxy is spinning in a well-ordered way. Note that these quadrants are drawn (as dashed lines) to intersect at the Sun, not at the galactic center, for it is from the viewpoint of our own planetary system that the observations are made. The length of the arrows represent the angular velocity of the disk material.*

included) do not move smoothly together, but ceaselessly change their positions relative to one another as they orbit the galactic center.

We can apply many of the same principles we used to look at planetary motions to our observations of galactic dynamics. Recall from Chapter 3 that Kepler's Third Law (as modified by Newton) relates the orbital period, orbit size, and masses of any two objects in orbit around one another. We expressed this law as follows:

$$\textit{total mass (in } M_\odot) = \frac{[\textit{orbit size (in A.U.)}]^3}{[\textit{orbital period (in years)}]^2}.$$

In the solar system, the total mass includes both the mass of the Sun and the mass of an orbiting planet. Because the mass of any planet is small compared with that of the Sun, we can safely ignore the planet. The mass we compute is the mass of the Sun, to good accuracy. Similarly, in working Kepler's Law to weigh the Galaxy, we can neglect the Sun's mass. Once we know the orbit of the Sun around the galactic center, we have all the quantities we need to apply the equation. We can consider the result to be a measure of the mass of the Galaxy.

There is one important clarification to make at this point. In the case of a planet orbiting the Sun, there is no ambiguity about *what* mass is being measured—it is the Sun's. However, the case of the Sun orbiting the center of the Galaxy is more complicated. The Galaxy's matter is distributed over a large volume of space and is not concentrated at the galactic center (as the Sun's mass is concentrated at the center of the solar system). Some of the Galaxy's mass lies inside the Sun's orbit (that is, within 8 kpc of the galactic center), and some lies outside, at large distances from both the Sun and the galactic center. The question naturally arises: "Which portion of the Galaxy's mass controls the Sun's orbit?" Isaac Newton answered this three centuries ago: The Sun's orbital period is determined by the portion of the Galaxy that lies *within the orbit of the Sun.* This is the mass resulting from the above equation.

Neutral hydrogen and carbon monoxide observations provide both of the quantities needed to calculate the mass of the Galaxy in this way. Radio observations of the distribution of gas in the galactic disk have been important in refining our measurement of the distance from the Sun to the galactic center. (In fact, the distance of 8 kpc stated earlier is based largely on radio studies). Measurements of gas velocities in the solar neighborhood show that the Sun, and everything in its vicinity, orbits the galactic center at a speed of about 220 km/s, taking some 225 million years—an interval of time sometimes known as 1 *galactic year*—to complete the trip. Substituting these quantities into our new version of Kepler's Law, we find that the mass of the Milky Way Galaxy within the Sun's orbit is about 10^{11} solar masses—100 *billion* times the mass of our Sun.

Most stars in our Galaxy are obscured from our vantage point on Earth. But astronomers estimate, on the basis of studies of stellar and interstellar matter near the Sun, that

INTERLUDE 25-2 *Spiral Density Waves*

The plane of our Galaxy is shaped like a thin disk with a bulge at the center. Within the disk lie spiral arms, where gas, dust, and young stellar objects tend to reside. This gargantuan collection of billions of stars and vast quantities of gas and dust is not static. All of this stellar and interstellar matter is rotating. At our distance of 8 kpc from the galactic center, the Sun takes about 220 million years to complete one orbit.

This rate of movement implies that our 4.5 billion-year-old Sun has cycled around the galactic center some 20 times. But the Galaxy existed long before our Sun's birth. The presence of the globular clusters implies an age for our Galaxy of some 15 billion years. Over this period of time, the galactic matter at a distance of 8 kpc has orbited the galactic center at least 60 times. Because of differential rotation, the inner parts of our Galaxy have rotated many more times than that, while the outer parts have completed far fewer circuits.

These observations suggest a problem: If the Galaxy is a continuously rotating collection of stars and gas, how do the spiral arms retain their structure? In other words, how do the arms survive differential rotation? We might expect the pinwheel-like arms either to wrap up or to stretch out into thinner sheets of stars, gas, and dust after a few revolutions (see the diagram below). Yet the arms *do* exist, and the prevalence of spiral arms in other disk galaxies suggests that they have existed for quite some time.

To begin to address this problem, let's examine the types of objects found in the spiral arms. The most obvious objects are the young blue stars. These O- and B-type stars, with typical lifetimes of 10 million years or less, do not live long enough to complete one full circuit around the Galaxy. We might then expect the arms to disappear in a single stellar generation. Because the grand spiral pattern seemingly persists, we must conclude that the arms must be continuously replenished with new stars and matter.

In the late 1960s, the American astrophysicists C. C. Lin and Frank Shu proposed a way in which this replenishment could occur. They argued that the arms themselves contain no "permanent" matter. A spiral arm should not be viewed as an assemblage of stars, gas, and dust moving intact through the disk—that would quickly be destroyed by differential rotation. Instead, the arm should be envisaged as a spiral *density wave*—a wave of compression and expansion that rotates through the Galaxy. In this picture, the wave moves *relative* to the stars and gas that make up the galactic disk, just as a sound wave moves through air or an ocean wave passes through water. The problem of differential rotation is avoided because the wave pattern is not tied to the stars and gas of the disk. The disk material rotates at differing rates at different radii, but Lin and Shu showed that the wave itself remains intact. The wave defines the Galaxy's spiral arms.

A wave in water builds up the material temporarily in some places (crests) and lets it down in others (troughs). Similarly, as the spiral density wave encounters galactic matter, the gas is compressed into a crest of slightly higher-than-normal density. Over much of the inner part of the Milky Way Galaxy (within about 15 kpc of the center), the wave pattern actually rotates *slower* than the stars and gas,

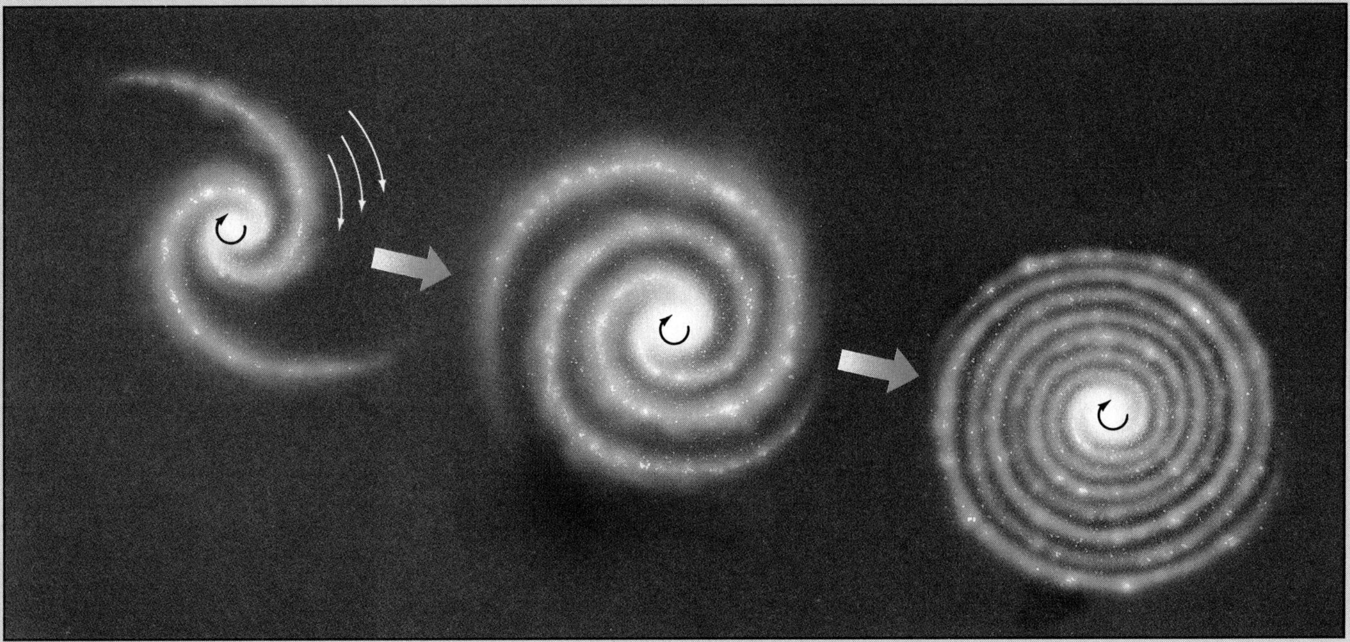

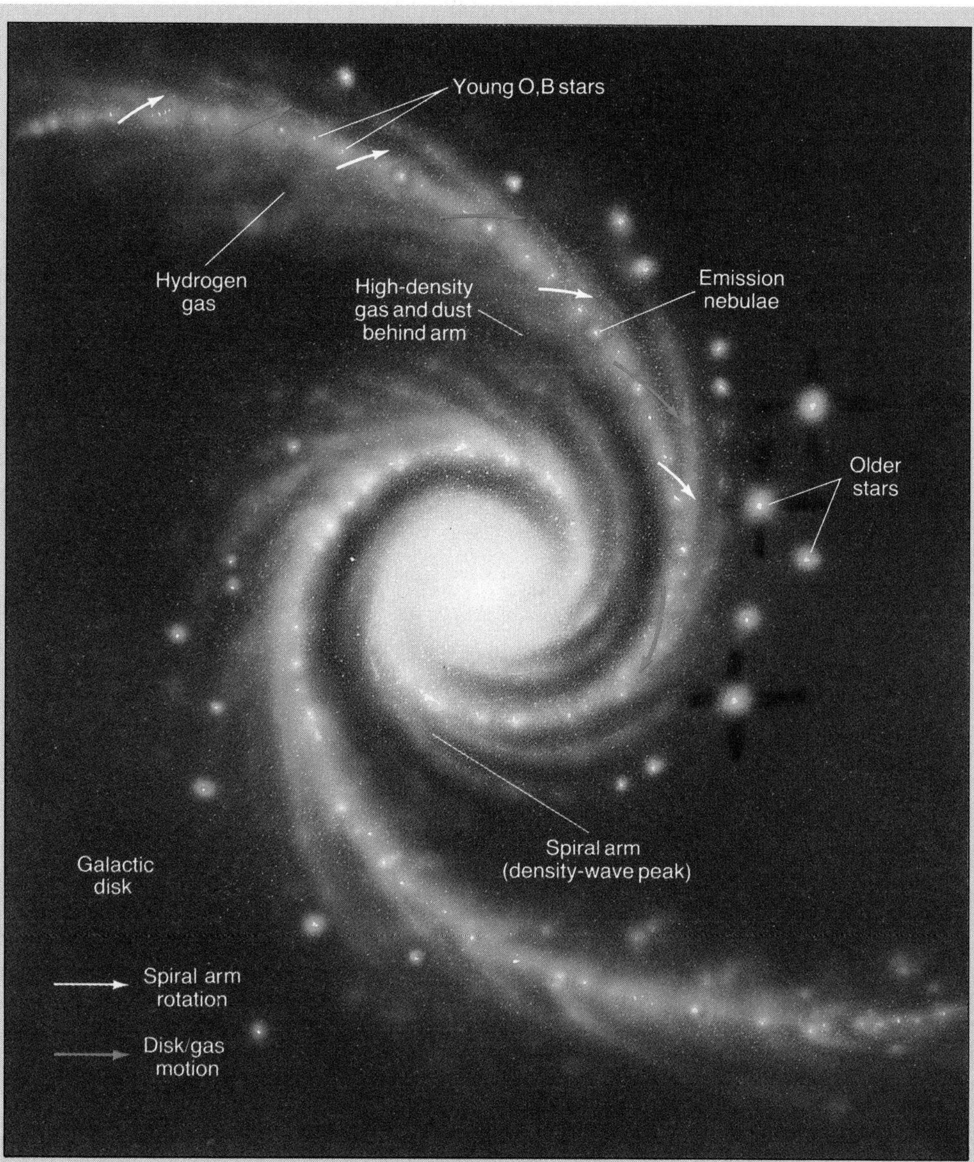

so galactic material catches up with the wave, is temporarily slowed down and compressed as it passes through, and then continues on its way. This compression is thought to trigger the formation of new stars and nebulae. In this way, the spiral arms are formed and reformed repeatedly, without wrapping up. As shown in the above illustration, the (slowly) moving spiral density wave (short dashed arrows) is outrun by the (faster) rotation of the matter (longer arrows).

As gas enters a spiral arm from behind, it is compressed and forms stars. The most prominent stars—the bright O and B giants—live for only a short time, so OB associations, young star clusters, and emission nebulae are found only near the site of their birth, emphasizing the spiral structure. Further downstream, ahead of the newborn stars, older stars and star clusters are seen. These have had time to move ahead of the wave since they formed. As older objects move downstream, their original spiral pattern becomes smeared out and is eventually lost—they become part of the general disk population. Dust lanes, which mark the regions of high-density gas, blue stars, and HII regions—appear in the spiral galaxy shown on the right in the above figure.

Where did these spiral waves come from? The answer to that question is unfortunately not provided by the theory. Scientists speculate that (1) instabilities in the gas near the galactic bulge, (2) the gravitational effects of our satellite galaxies (the Magellanic Clouds), or (3) the asymmetry in the bulge itself might have had a big enough influence on the disk to get the process going. The truth of the matter is, we don't know for sure.

about half of the Galaxy's mass exists in the form of stars. If every star has an average mass of 1 $M_\odot$ (a few big stars and many little stars combine to give us this average), our Galaxy must house at least 100 billion stars. The Milky Way Galaxy is truly vast, both in size and in mass.

Halo Stars

The picture of orderly circular orbital motion about the galactic center applies only to the galactic disk. Stars in the galactic halo (as well as in the thick disk and galactic bulge) are not so well-behaved. The old globular clusters and the faint, reddish individual stars that make up the halo do *not* share in the overall rotation of the disk. Instead, their orbits are largely random. Although they do in fact orbit the galactic center, halo objects move in all directions, their paths filling the entire halo. Halo stars in the vicinity of the Sun move at speeds comparable to the disk's 220 km/s rotation rate, but in every direction, not just one—their orbits carry them repeatedly through the disk plane and out the other side. Figure 25.15 illustrates this motion and contrasts it with the more regular motion in the disk. Halo stars do have a net rotation around the galactic center—about 100 km/s at the location of the Sun—but it is overwhelmed by the larger random component of their motion.

The chaotic orbits of the halo stars are explained by the same theory that accounts for their age and color. When the halo developed, the Galaxy was quite irregular in shape and was rotating only very slowly, so there was no strongly preferred direction in which matter tended to move. As a result, halo stars were free to travel along nearly any path once they formed (or when their parent systems merged). As the disk formed, however, conservation of angular momentum caused the disk to spin. Stars forming from the gas and dust of the galactic disk inherit the disk's rotational motion and so move on well-defined, circular orbits.

Stars in the bulge and the thick disk are intermediate in their orbital properties, having both relatively large random motions and a fairly strong overall rotation. The bulge rotates about 80 km/s more slowly than the disk (at the bulge's outer edge), while the thick disk rotates at about 180 km/s in the vicinity of the Sun. Again, the thick disk's properties suggest (but do not prove) that it formed while gas was still sinking to the Galaxy's mid-plane and had not yet reached its final (present-day) rotation rate.

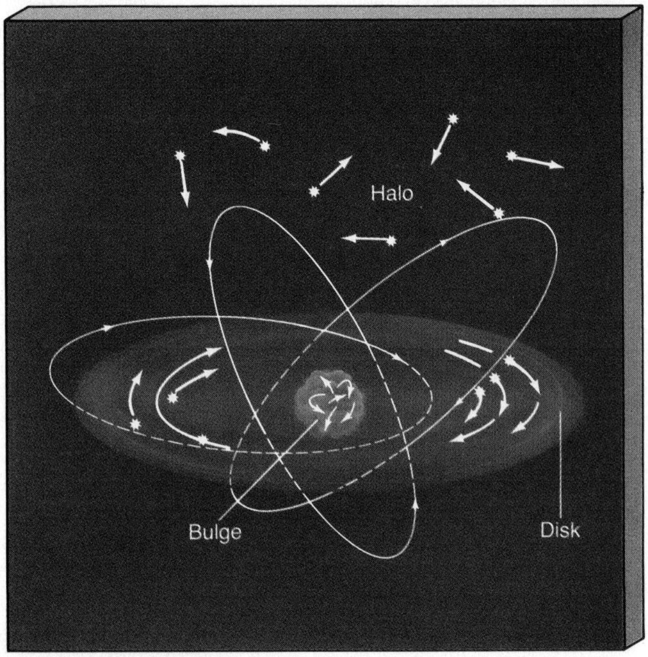

Figure 25.15 *Stars in the Galactic disk move on orderly, circular orbits about the Galactic center. By contrast, halo stars have orbits with largely random orientations and eccentricities. The orbit of a typical halo star takes it high above the Galactic plane, through the disk, then far below the plane on the other side of the Galaxy.*

The Galaxy's Corona

The mass we computed for the Galaxy is the mass of stars, gas, dust, and everything else residing *inside* the Sun's orbit around the galactic center—virtually everything within the spiral structure sketched in Figure 25.12(a). But the Galaxy outlined by the globular clusters and by the spiral arms is merely the "tip of the galactic iceberg." There is strong evidence that the Galaxy is actually much larger.

The same basic techniques that allow astronomers to measure the rotation speed of the galactic disk in the vicinity of the Sun have also been used to measure the Galaxy's rotation at other distances from the galactic center. The resulting plot of our Galaxy's rotation rate at various distances from the center, known as the galactic **rotation curve,** is shown in Figure 25.16. Conventionally, rotation curves show how the orbital speed (in kilometers per second) of disk material varies with distance from the galactic center. On these graphs, solid-body rotation would then have speed proportional to distance (since gas at larger radii would have to move farther, and hence faster, to complete one orbit in a given time). Any rotation curve other than a straight line through the origin therefore represents differential rotation, as discussed earlier. Knowing the rotation speed and the distance, we can determine the period and use our earlier formula to compute the mass inside any given radius. As we extend our measure to include distances farther from the Galaxy's center, we find that our estimate of the Galaxy's mass is too small by *at least* a factor of 2 or 3, and possibly by much more.

If most of the matter in the Galaxy ended at the edge of the globular cluster distribution or the visible spiral structure illustrated in Figure 25.12(a)—at a distance of about 15 kpc from the galactic center—Newton's laws of motion would predict that the orbital speed of stars and gas beyond that radius should decrease outward, just as the orbital speeds of the planets diminish with increasing distance from the Sun. The dashed line in Figure 25.16 indicates how the rotation curve should look in that case. However, the true

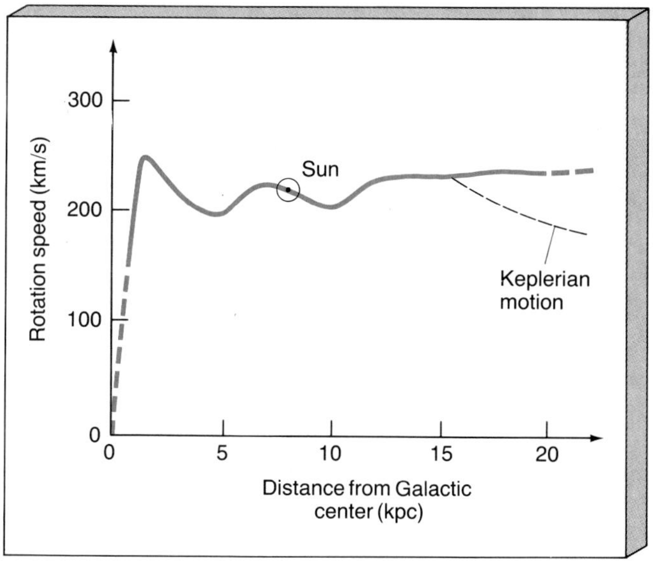

Figure 25.16 *The rotation curve for the Milky Way Galaxy plots rotation speed against distance from the Galactic center. We can use this curve to compute the mass of the Galaxy lying within any given radius. The dashed curve is the rotation curve that would be expected if the Galaxy "stopped" abruptly at a radius of 15 kpc, the limit of most of the known spiral structure and the globular cluster distribution. The fact that the curve does not fall off, but in fact seems to be rising beyond that point indicates that there must be additional matter outside of that radius.*

rotation curve is quite different. Far from falling off at larger distances, it appears to rise slightly out to the limits of our measurement capabilities. This implies that the amount of mass contained within spheres of successively larger radii is continuing to grow beyond the orbit of the Sun, at least out to a distance of 40 or 50 kpc. The amount of mass within 50 kpc is at least 3×10^{11} solar masses and could easily be much greater.

These observations indicate that our Galaxy is surrounded by an extensive, invisible **outer halo,** or **galactic corona,** that seems to dwarf the inner halo of stars and globular clusters. The corona reaches well beyond the 30-kpc diameter once thought to be the limit of our Galaxy. As we will discuss in Chapter 26, astronomers have found evidence for similar coronae in many other galaxies, leading them to suspect that most galaxies are actually much larger than is suggested by their optical images. But where is this mass? We do not detect enough stars or interstellar matter to account for the mass that our computations tell us must be there. We are inescapably drawn to the conclusion that most of the mass in our own and other galaxies exists in the form of invisible **dark matter** that we presently do not understand. Notice that the term *dark* here does not just refer to matter not detectable in visible light. The material has (so far) escaped detection at *all* wavelengths, from radio to gamma-rays. Only by its gravitational pull do we know of its existence. Dark matter is not hydrogen gas (atomic or molecular), nor is it made up of stars. Given the amount of matter that must be present, we would have been able to detect it by now with present-day equipment if it were in either of those forms. Its nature and its consequences for the evolution of galaxies and the universe are among the most important questions in astronomy today.

Many candidates have been suggested for this dark matter, although none is proven. Among the strongest contenders are the *brown dwarfs* discussed in Chapter 21, failed low-mass stars that never reached the point of nuclear burning in their cores. As we remarked earlier, these objects could exist in great numbers throughout the Galaxy yet be exceedingly hard to see. The fact that we see so many low-mass stars strongly suggests that the even lower-mass brown dwarfs may be a very common by-product of star formation. *Black dwarfs,* the end result of low-mass stellar evolution (see Chapter 22) are another possibility. Again, they would be very hard to detect, although it seems unlikely that many stars have actually had time to reach this advanced evolutionary stage. *Black holes* also might supply the unseen mass, although their very existence is still debated, and very few candidates exist. Given that they are the evolutionary product of (relatively rare) massive stars, it is hard to believe that there could be enough of them to hide large amounts of galactic matter.

All of the above candidates are *stellar* in nature—associated with star formation or evolution. A radically different alternative is that the dark matter may be made up of (as yet undetected) exotic subatomic particles that may pervade the entire universe. The **W**eakly **I**nteracting **M**assive **P**articles (WIMPs), introduced in Chapter 18 in the context of the solar neutrino problem, are a prime example.* Whether or not they have any bearing on solar neutrinos (and it appears increasingly likely that they do not), many astrophysicists believe that they could well be present in sufficient numbers to account for all the dark matter. Their very elusiveness makes this idea hard to test, however.

Bear in mind that the identity of the dark matter is not necessarily an "all or nothing" determination. It is perfectly conceivable that more than one type of dark matter exists. For example, it is quite possible that most of the dark matter in the inner (visible) parts of galaxies could be in the form of brown dwarfs, while the dark matter on larger scales might be primarily in the form of massive particles. We will return to this perplexing problem in later chapters.

THE CENTER OF OUR GALAXY

✓5 Theory predicts that the galactic bulge, and especially the region close to the galactic center, should be densely populated with billions of stars tightly concentrated in and

*The term "WIMPs" was coined in the early 1980s, at least in part as a joke. Toward the end of that decade, the generic term "**MA**ssive **C**ompact **H**alo **O**bject," or **MACHO,** was invented to describe the stellar dark-matter candidates. While the preferred general term is "dark matter," the more light-hearted expressions WIMPs and MACHOs are very widely used.

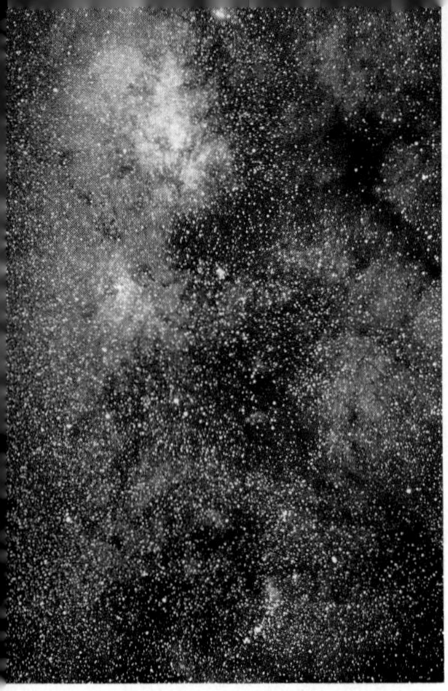

Figure 25.17 A photograph of stellar and interstellar matter in the direction of the galactic center. Because of heavy obscuration, even the largest optical telescopes can see no farther than 1/10 the distance to the center. This photograph is a southerly continuation of Figure 20.6. The M8 nebula can be seen at very top center. (European Southern Observatory)

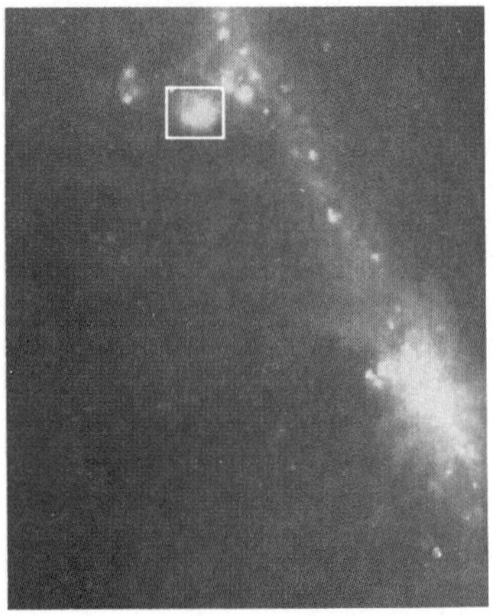

Figure 25.18 An infrared image of the region around the center of our Galaxy shows many bright stars packed into a relatively small volume. The average density of matter in this region is estimated to be about a million times that in the solar neighborhood. The size of the boxed region near top is a few hundred parsecs across. (NASA)

around our Galaxy's midsection. However, we are unable to see this region of our Galaxy—the dark dust clouds within a few thousand parsecs of the Sun effectively shroud what otherwise would be a stunning view. Figure 25.17 shows the (optical) view we do have of the region of the Milky Way toward the galactic center, in the general direction of the constellation Sagittarius.

With the help of infrared and radio techniques, we can peer more deeply into the central regions of our Galaxy. Infrared observations (see Figure 25.18) indicate that the Galaxy's core harbors roughly 50,000 stars per cubic parsec. That's a stellar density about a million times greater than in our solar neighborhood. Had any planets formed along with these galactic-center stars, they would probably have been ripped from their orbits and obliterated, as the stars must experience frequent close encounters and even collisions. Infrared radiation has also been detected from what appear to be huge clouds rich in dust.

High-resolution radio observations show more structure on small scales. Figure 25.19 shows the bright radio source Sagittarius A, which lies inside the box in Figure 25.17—and, we think, at the center of our Galaxy. On a scale of about 100 pc, extended filaments can be seen, suggesting to many astronomers that strong magnetic fields

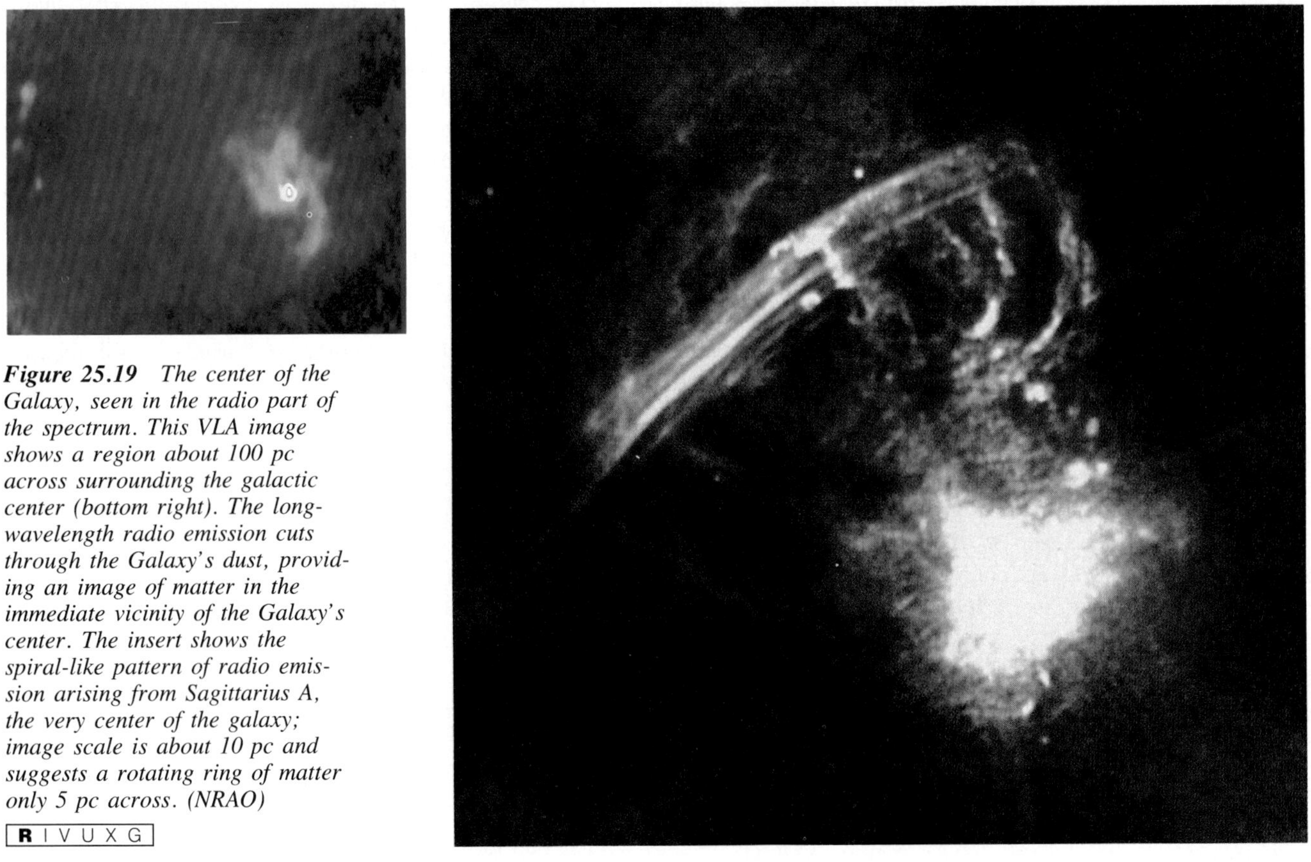

Figure 25.19 The center of the Galaxy, seen in the radio part of the spectrum. This VLA image shows a region about 100 pc across surrounding the galactic center (bottom right). The long-wavelength radio emission cuts through the Galaxy's dust, providing an image of matter in the immediate vicinity of the Galaxy's center. The insert shows the spiral-like pattern of radio emission arising from Sagittarius A, the very center of the galaxy; image scale is about 10 pc and suggests a rotating ring of matter only 5 pc across. (NRAO)

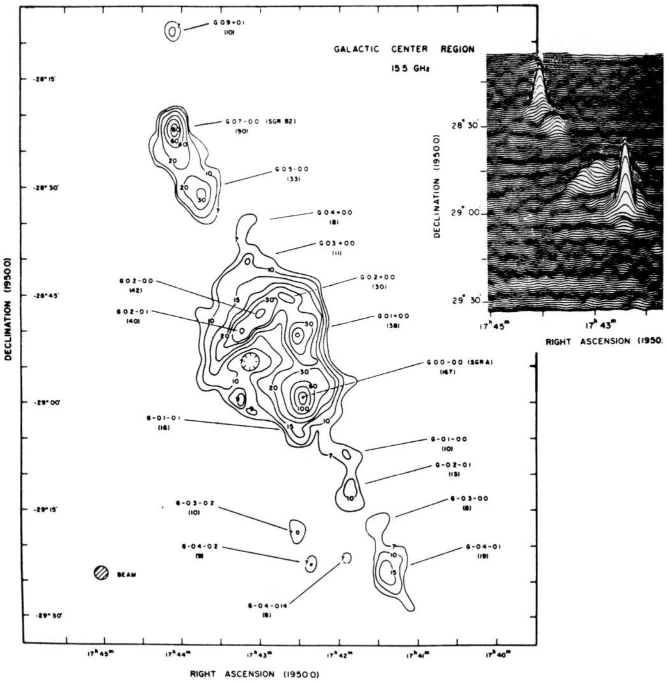

R I V U X G

Figure 25.20 A map of radio signals emitted by matter within the superposed box surrounding the Galaxy's center in Figure 25.18. This image shows a region a few hundred parsecs across. This is the innermost 0.5 percent of our galaxy. (MIT)

operate in the vicinity of the center, creating structures similar in appearance (but much larger in size) to those observed on the active Sun. Figure 25.20 depicts the radio emission near Sagittarius A, viewed on a scale of 1000 pc or so. These observations indicate that a ring of molecular gas 500 pc across surrounds the central source. The ring contains about 30,000 solar masses of material. The nature of Sagittarius A itself is still uncertain.

Figure 25.21 places these findings into a simplified perspective. In a series of six images, an artist has captured the important results of long-wavelength radio and infrared studies of the Milky Way's heart. Each painting is centered on the Galaxy's core, and each increases in resolution by a factor of 10.

Frame (a) renders the Galaxy's overall shape, as painted in Figure 25.12(a). The scale of this frame measures about 100 kpc from top to bottom. Frame (b) spans a distance of 10 kpc from top to bottom and is nearly filled by the great circular sweep of the innermost spiral arm. Moving another 10 times closer, frame (c) depicts a ring of matter made mostly of giant molecular clouds and gaseous nebulae. This entire flattened, circular feature, about 1 kpc in diameter, is rotating at about 100 km/s. The origin of this ring of gas is still unclear. In frame (d), at 100 pc, a pinkish region of ionized gas surrounds the reddish heart of the

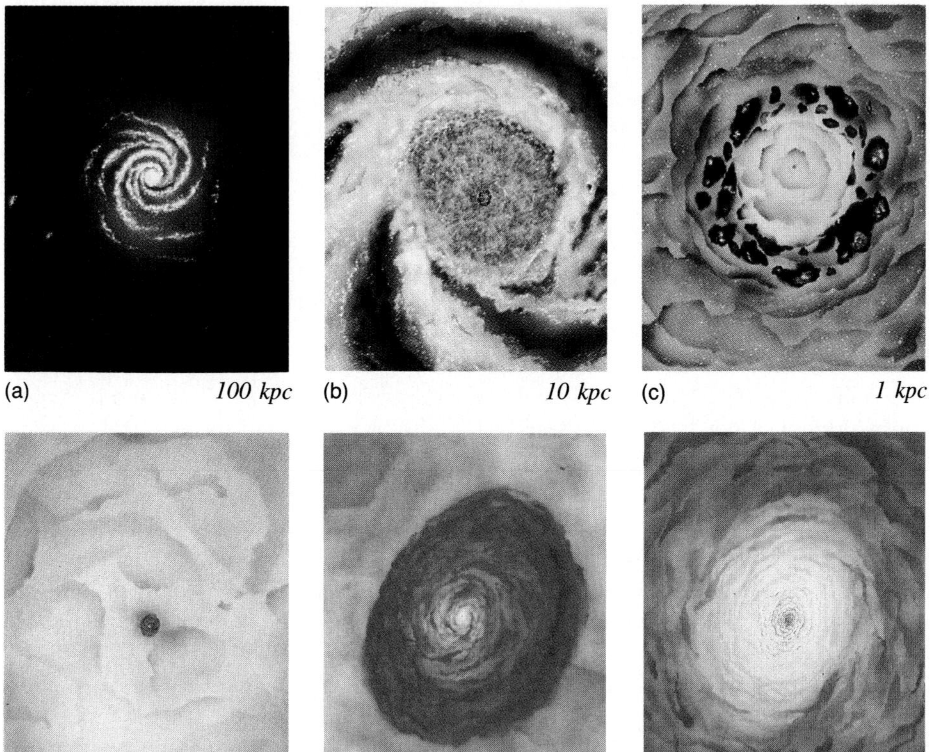

Figure 25.21 Six artist's conceptions, each centered on the galactic core and each increasing in resolution by a factor of ten. Frame (a) shows the same scene as Figure 25.12(a). Frame (f) is an artist's rendition of a vast whirlpool within the innermost parsec of our Galaxy. (L. J. Chaisson)

INTERLUDE 25-3 *Cosmic Rays*

In addition to stars, gas, dust, and possibly dark matter, one other type of matter populates our Galaxy. These are the *cosmic-ray particles,* or cosmic rays for short. Cosmic rays continuously collide with Earth, and they make up our only sample of matter from outside the solar system. In recent decades, high-altitude balloons, rockets, and satellites have enabled astronomers to discover the chemical composition and the energies of cosmic-ray particles. The figure at right shows an actual track of some cosmic rays photographed while colliding with a balloon-borne detector. When a galactic cosmic-ray particle hits our atmosphere, it creates a series of cascades or "showers" of many lower-energy particles. These lower-energy "secondary" particles in turn collide with objects, including humans, on the Earth's surface virtually all the time. Our bodies are being peppered with them right now.

Cosmic rays are actually not "rays" at all. They are subatomic particles. Nearly 90 percent are protons, the nuclei of hydrogen atoms. The nuclei of heavier atoms—helium and a long list of others—make up 9 percent. Electrons amount to 1 percent. The abundances of elements that make up the cosmic rays are close to the abundances that we find throughout the Galaxy.

Cosmic-ray particles are very energetic. By some unknown mechanism (or mechanisms), cosmic rays are accelerated to extremely high velocities, much higher than any velocities we can achieve in terrestrial laboratories. Virtually all cosmic rays—even the heaviest—travel very close to the speed of light. On the basis of the numbers of cosmic rays hitting the Earth, astronomers estimate the number density of cosmic rays in interstellar space to be on the order of 10^{-9} particle/cm^3. That makes them rare, although still more numerous than interstellar dust particles.

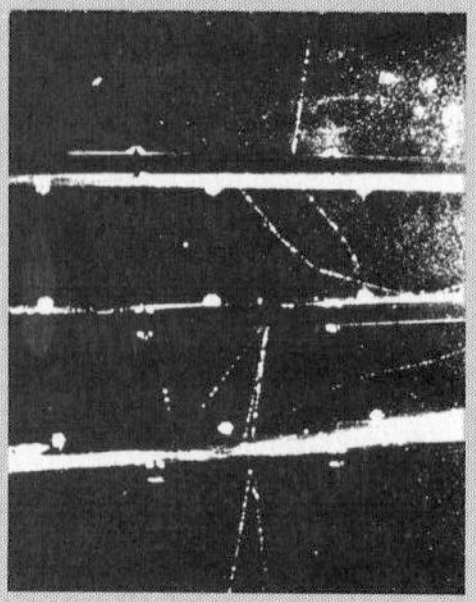

What is the source of our Galaxy's cosmic rays? The answer is currently uncertain. The circuitous and complex paths taken by the cosmic-ray particles while traversing the Galaxy, with its tangled magnetic field, prevent us from pinpointing their source by observing their direction of arrival at Earth. Indeed, careful searches have failed to show any preferential direction for cosmic rays. They seem to arrive from all directions with equal likelihood. Candidates for the origin of cosmic rays include (1) violent events at the galactic center, where strong electromagnetic forces accelerate these particles to high energies, (2) supernova explosions, described in Chapter 23, and (3) the vicinity of neutron stars and black holes, as discussed in Chapter 24.

Most researchers currently favor the second of these explanations. In reality, each of these candidate events probably contributes some particles to the observed hodgepodge of cosmic rays in our Galaxy. Some of the highest-energy cosmic rays may even originate far beyond our own Galaxy, as we will discuss in the next two chapters.

Galaxy. (Of course, we cannot see these regions. Astronomers infer these colors on the basis of the gases' temperatures and densities.) The source of energy producing this vast ionized cloud is unknown, although it is presumed to be related to activity in the galactic center itself. Frame (e), spanning 10 pc, depicts a tilted, spinning whirlpool of hot (10^4 K) gas that marks the core of our Galaxy. The innermost sanctum of this gigantic whirlpool is painted in frame (f), where a swiftly spinning, white-hot disk of superheated (million-kelvin) gas nearly engulfs an enormously massive object too small in size to be pictured (even as a minute dot) on this scale.

What is the cause of all this activity? An important clue comes from the Doppler broadening of infrared spectral lines from the central swirling whirlpool (frame [e] of Figure 25.21). This broadening indicates that in order to keep the gas in orbit, whatever is at the center must be very massive—a million solar masses or more. Given the twin requirements of large mass and small size, a leading contender for the cause of the violence is a supermassive (million-solar-mass) black hole. The hole itself is not the source of the energy, of course. Instead, the vast accretion disk of matter being drawn toward the hole by its enormous gravity emits the energy as it falls in. We will see later that astronomers have reason to suspect that similar events are occurring at the centers of many other galaxies.

If our knowledge of the Galaxy's center seems sketchy, that's because it is sketchy. Astronomers are still learning to decipher the clues hidden within its invisible radiation. We are only beginning to appreciate the full magnitude of this entirely novel realm deep in the heart of the Milky Way. In some respects, our research should not yet even be called mature science. Rather, it's exploration—but absolutely fascinating exploration, enabling us to return from our telescopes with tales of new wonders at the core of our Galaxy's system.

CHAPTER SUMMARY

The Milky Way Galaxy is a huge collection of stellar and interstellar matter. From our perspective within the galactic disk, the Milky Way is seen as a band of light stretching across the sky. Improved observational tools have enabled astronomers to better appreciate the size and shape of the Milky Way. The realization that there are many other distant galaxies has gone hand in hand with the development of the cosmic distance scale.

Variable stars change in luminosity over time. The most important variables to the study of galactic structure are those that pulsate regularly, such as Cepheids and RR Lyrae stars. Cepheid variables are critical to establishing the cosmic distance scale because their absolute brightnesses and their pulsation periods are quite tightly linked. A simple measurement of a Cepheid's pulsation period allows astronomers to estimate its distance. By observing RR Lyrae stars in globular clusters, Harlow Shapley obtained the first observational evidence that our Sun does not reside at the center of the Milky Way Galaxy.

Models of the Galaxy based solely on observations made in visible light are flawed because they ignore the absorption of light by interstellar gas and dust. In the 1950s, spectroscopic radio astronomy was used to begin exploring the full grandeur of our Galaxy. Studies of the 21-cm radio waves emitted by interstellar clouds enable astronomers to map the large-scale distribution and motion of galactic hydrogen gas. The center of the gas distribution coincides roughly with the galactic center, as determined from observations of globular clusters. The gas is confined mainly to the galactic plane. There is clear evidence for pinwheel like structures called spiral arms. A spiral arm should be envisaged not as an assemblage of stars, gas, and dust moving intact through the Galactic disk, but as a spiral wave of star formation rotating through the Galaxy.

Some parts of the galactic disk are approaching the Sun, and others are receding. The entire disk is rotating differentially. Studies of galactic rotation reveal that the mass of the Galaxy within the Sun's orbit is about 100 billion times the mass of our Sun. The old globular clusters and faint, reddish individual stars that make up the halo do not share in the overall rotation of the galactic disk. Instead, their orbits are quite random. Observations indicate that our Galaxy is surrounded by an extensive, invisible outer halo that seems to dwarf that inner halo of stars and globular clusters. The nature of the dark matter that makes up the outer halo is unknown. It might take the form of brown dwarfs, black holes, or much more exotic subatomic particles.

Violent events at the center of the Galaxy may indicate the presence of a million solar-mass black hole there. The energy may come from a vast accretion disk of matter being drawn toward the hole by its enormous gravity.

Cosmic rays continually collide with Earth, and they make up our only sample of matter from outside the solar system. They are subatomic particles, generally the nuclei of hydrogen atoms or other atoms. The abundances of elements in the cosmic rays are close to the abundances that we find throughout the Galaxy. Cosmic rays may originate in violent events at the galactic center, in supernova explosions, or in the vicinity of neutron stars and black holes.

KEY WORDS

cataclysmic variable
Cepheid variable
dark matter
flare star
galactic bulge
galactic center
galactic corona
galactic disk
galactic halo
galactic plane
galaxy
instability strip
Kapteyn universe
MACHO
outer halo
period-luminosity relation
pulsating variable
rotation curve
RR Lyrae star
self-propagating star formation
spiral arm
spiral density wave
thick disk
variable star
WIMP

REVIEW QUESTIONS

1. What are spiral nebulae? When and how did they get that name?
2. What is the basic mechanism that causes pulsating variable stars to rise and fall in luminosity?
3. How are Cepheid variables used to measure distances?
4. How far out into space can we use Cepheids to measure distances?
5. What important discoveries were made possible earlier in this century by the RR Lyrae variables?
6. Why can't we study the central regions of the Galaxy with optical telescopes?
7. What is the thickness of the plane of the Galaxy? Contrast this to its diameter.
8. Explain why the spiral arms are believed to be regions of recent and ongoing star formation.
9. What is self-propagating star formation?
10. What do the red stars in the galactic halo tell us about the history of the Milky Way?

11. What is differential rotation? What does the rotation curve of our Galaxy tell us about its total mass?

12. How many times has the Sun orbited the center of the Galaxy? What line of reasoning reveals this?

13. Contrast the motion of stars in the disk with that of stars in the halo.

14. Why do astronomers believe that a supermassive black hole inhabits the center of the Milky Way?

15. What are cosmic rays?

16. A Cepheid variable is 100 times brighter than a typical RR Lyrae star. How much farther away could the Cepheid be used as a distance-measuring tool?

17. Estimate the volume of the Galactic disk (taking it to be a cylinder of radius 15 kpc and thickness 300 pc), and compare it with the volume of the halo (approximating it as a sphere of radius 15 kpc). (Recall that the volume of a sphere = $\frac{4}{3}\pi r^3$ and that the volume of a cylinder = $\pi r^2 h$, where h is height.)

18. Calculate the total mass of the Galaxy within 20 kpc of the Galactic center if the rotation speed at that radius is 240 km/s.

19. A density wave made up of two spiral arms opposite each other is moving through the galactic disk. At the orbit of the Sun, the wave's speed is 120 km/s. Assuming that the Sun's speed is 220 km/s, calculate how many times the Sun has passed through a spiral arm since the solar system formed 4.5 billion years ago.

DISCUSSION QUESTION

1. In 1907, G. E. Mitton wrote, "Such wonderful discoveries have already been made that it is not too much to say that perhaps some day we may be able to establish some sort of communication with Mars, and if it be inhabited by any intelligent beings, we may be able to signal to them; but it is almost impossible that any contrivance could bridge the gulf of airless space that separates us." In this chapter, we assert that we will probably never be able to gaze at our own Milky Way Galaxy from outside its boundaries. Are we just as likely to be proven wrong as Mitton? Why or why not?

PROJECTS

1. If you're far from city lights on a late summer evening, look for a hazy band of stars arching overhead. This is the edgewise view into our own Milky Way Galaxy. The center of the Galaxy is located in the direction of the constellation Sagittarius. Can you find it? Can you see why Sagittarius is said to have a teapot-shaped arrangement of stars? What lies along your line of sight toward the Galaxy's heart?

2. In the spring, look for Arcturus in the constellation Bootes. It now appears as one of the brightest stars in the sky, but it's only a temporary visitor to our region of space. Arcturus became visible to the eye only half a million years ago. It's one of the stars belonging to the halo of the Milky Way. It orbits the center of the Galaxy, but its orbit carries it up and down through the Galactic disk. Arcturus has a large velocity relative to our Sun, and its proper motion is very large. It moves an apparent distance of about 1⁄10 of a full Moon about every 70 years. It's fun to contemplate Arcturus—a great yellow-orange globe of a star—rushing past us in the night. Only the vastness of space and time, prevent us from observing its motion. How different would the night sky be to an observer on Arcturus today from what it was half a million years ago?

SUGGESTED READINGS

Bok, B. J. *The Milky Way*. 5th ed. Cambridge, MA: Harvard University Press, 1981. Timeless exposition of the Milky Way, as it was known to one of the first experts on the subject.

Ferris, T. *Coming of Age in the Milky Way*. New York: Doubleday, 1988. Insightful and lyrical account of the history of our understanding of our home galaxy.

Harrington, P. "Journey to the Center of the Galaxy." *Astronomy* (July 1991). What it might be like to go there.

Kaufmann, W. J. "Our Galaxy, Part I." *Mercury* (May–June 1989). First part of a popular article on the Milky Way.

———. "Our Galaxy, Part 2." *Mercury* (July–August 1989). Second part of a popular article on the Milky Way.

Rees, M. J. "Black Holes in Galactic Centers." *Scientific American* (November 1990). On the possibility of black holes at the hearts of many galaxies.

Tucker, W. "A Massive Halo, Part 1," *Mercury* (January–February 1989). First part of a comprehensive article on dark matter in our Galaxy.

———. "A Massive Halo, Part 2." *Mercury* (March–April 1989). Second part of a comprehensive article on dark matter in our Galaxy.

Waldrop, M. M. "Heart of Darkness." *Science* (February 19, 1988). Evidence that the Andromeda Galaxy harbors an ultramassive black hole at its center, with implications for our Galaxy and others.

CHAPTER

26

Normal Galaxies

The Local Group and Beyond

This is an artist's conception of interacting galaxies. Such galaxy collisions and mergers might play a significant role in galaxy formation and evolution. Bypassing galaxies pull great tails of loose matter and even whole stars because of gravitational tidal effects. This painting is based on two specific galaxies well studied in the sky, NGC4038 and NGC4039. (Courtesy D. Berry)

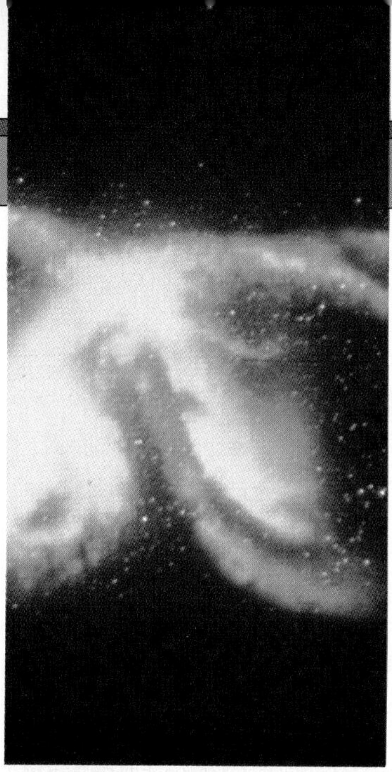

LEARNING GOALS

- 1 to understand the basic properties of normal galaxies
- 2 to learn about some of the distance-measurement techniques that enable astronomers to probe realms of the universe far beyond our Milky Way
- 3 to understand that there is order in the distribution of galaxies across the universe
- 4 to appreciate the techniques used to determine the masses of faraway galaxies
- 5 to realize that most of the matter in the universe is invisible
- 6 to know Hubble's law and how it is used to derive distances to the most remote objects in the observable universe
- 7 to study some theories of how galaxies form and evolve

Perhaps one day our descendants will become sufficiently advanced to journey beyond the Milky Way Galaxy and see its full grandeur from afar. For now, however—and for the foreseeable future—the big picture of our patch of starlight floating silent in the near-void of intergalactic space eludes us. Much of our knowledge of the workings of our own Galaxy is based on observations of the myriad other galaxies in the universe.

We know of literally millions of galaxies beyond the Milky Way. Large telescopes reveal these blurry beacons as collections of matter comparable to our own galactic home. Some galaxies are smaller than our own. Others are much larger. All are vast star systems separated from us by almost incomprehensibly large distances, each a gravitationally bound assemblage of stars, gas, dust, dark matter, and radiation. Even a modest-sized galaxy harbors more stars than people who have ever lived on Earth. The light we receive tonight from the most distant galaxies was emitted long before the Earth even existed.

By observing the skies perpendicular to the plane of the Milky Way (to avoid our own Galaxy's obscuring light and dust), we can study the full extent of these other galaxies. By comparing and classifying their properties, astronomers have begun to understand their complex dynamics. By mapping out their distribution in space, they have traced out the immense realms of the universe. The galaxies remind us that our position in the universe is no more special than that of a boat adrift at sea.

GALAXY CLASSIFICATION

✓1 Galaxies look quite different from stars. Many have a fuzzy, lens-shaped appearance, often resembling a disk rather than the clear, bright, spherical image usually associated with a star (see Figure 26.1). The eighteenth-century German philosopher Immanuel Kant was the first to conceive of these fuzzy objects as individual "island universes" lying far beyond the confines of our Milky Way Galaxy. While the label "universe" was inaccurate, Kant was absolutely correct in arguing that these nonstellar patches of light reside outside our Galaxy. However, as we

Figure 26.1 *A cluster of many galaxies, each galaxy housing hundreds of billions of stars. This one, called the Coma Cluster, lies over 100 million parsecs away. (Palomar Observatory)*

(a) R I V U X G

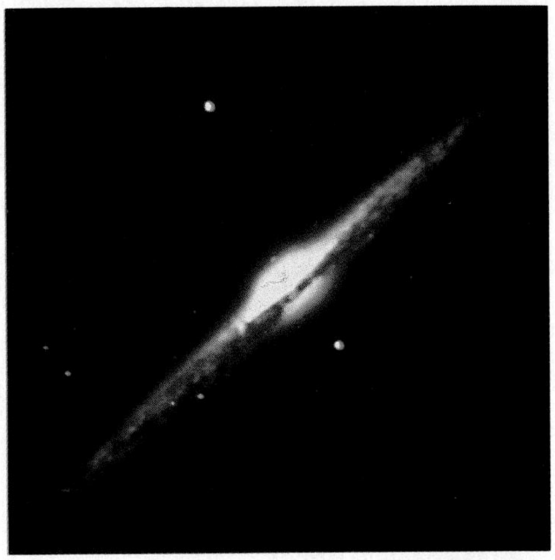

(b) R I V U X G

Figure 26.2 These two galaxies are good examples of the architecture of our own Milky Way Galaxy, of the Andromeda Galaxy, and of many of the other objects shown in Figure 26.1. (a) The NGC 4565 galaxy is a nearly face-on spiral system. (b) NGC 3184 is an edge-on spiral system. (AURA)

saw in the last chapter, only in the third decade of the twentieth century that this fact was conclusively proven.

With one exception (the object with the four "spikes" near top center), *every* patch or point of light in Figure 26.1 is a separate galaxy. A total of several hundred galaxies reside in the distant region of space shown here. Although it is rather difficult to tell from the photograph, many of the fuzzy "blobs" of light are in fact spiral galaxies, like the Milky Way. Two somewhat closer (and more easily discerned) spirals are shown in Figure 26.2.

Figure 26.1 is only one of many photographs showing legion upon legion of remote galaxies. Naturally, we wonder what kinds of objects they are. Do they all have roughly the same size and shape as our Milky Way? Or do galaxies display an array of shapes and sizes? Can we divide galaxies into well-defined categories, much as we classify stars? The American astronomer Edwin Hubble was the first to categorize galaxies in a comprehensive way. Working with the then recently completed 2.5-m optical telescope on Mount Wilson in California in 1924, he classified the galaxies he saw into four basic types—*spirals, barred spirals, ellipticals,* and *irregulars*—based purely on their appearance. He arranged them into the "tuning fork" diagram shown in Figure 26.3. Many modifications to the scheme have been proposed over the years, but the basic **Hubble classification scheme** described here is still widely used today.

Spirals

We have already seen numerous examples of **spiral galaxies.** All contain a flattened galactic disk, where spiral arms are found, and a central galactic bulge. Our own Milky Way Galaxy and our neighbor Andromeda are good examples of this type of system. Within this general description, however, spiral galaxies can exhibit a variety of shapes. Some spiral galaxies have a large central bulge and fairly tightly wrapped spiral arms. Others have a more open pattern spiral and an intermediate-sized central region. Still others have a rather small bulge and long, stringy arms, sometimes making it difficult even to recognize these galaxies as spirals at all. Figure 26.4 illustrates this variation.

In fact, the size of the bulge and the tightness of the spiral pattern are quite well correlated. Tightly wrapped spirals tend to have large central bulges, and spirals with large central bulges often have tightly wrapped arms. More open spirals exhibit smaller bulges, and spirals with smaller bulges generally have less tightly-wound arms. The arms also tend to become more "knotty" in appearance as the spiral pattern becomes more open. In Hubble's scheme, spiral galaxies are denoted by the letter *S* and subdivided according to the form of the spiral arms and the size of the central bulge. *Type Sa* spiral galaxies have large bulges and tight, almost circular, spiral arms. *Type Sb* galaxies have smaller bulges and more open spiral arms. *Type Sc* spirals have small bulges and a loose, often poorly defined spiral pattern. The correlation between bulge size and spiral appearance is far from perfect, but it still provides the basis for a useful classification of galaxy properties. In the 1950s, one additional class, Sd, with even more poorly defined arms, was added to Hubble's scheme.

Detailed photographs of typical spiral galaxies clearly show obscuring dust, and the 21-cm radio radiation emitted by spirals is reasonably intense, signaling the presence of large amounts of interstellar gas. Stars are still forming within the spiral arms, where the interstellar medium is densest. Type Sc galaxies contain the most interstellar gas, Sa galaxies the least. The Sc galaxy M51 shown in Figure 26.5 clearly shows the preponderance of interstellar dust and young blue stars tracing the spiral arms. Spirals are not necessarily young galaxies, however. Rather, like our own Galaxy, they are simply rich enough in interstellar gas to provide for continued stellar birth.

Much of our description of the large-scale structure of the Milky Way Galaxy in Chapter 25 applies to spiral galaxies in general: Our Galaxy is probably a fairly typical type Sb (or Sbc—midway between Sb and Sc) spiral. The four basic regions of our Galaxy—the *disk,* the *bulge,* the *halo,*

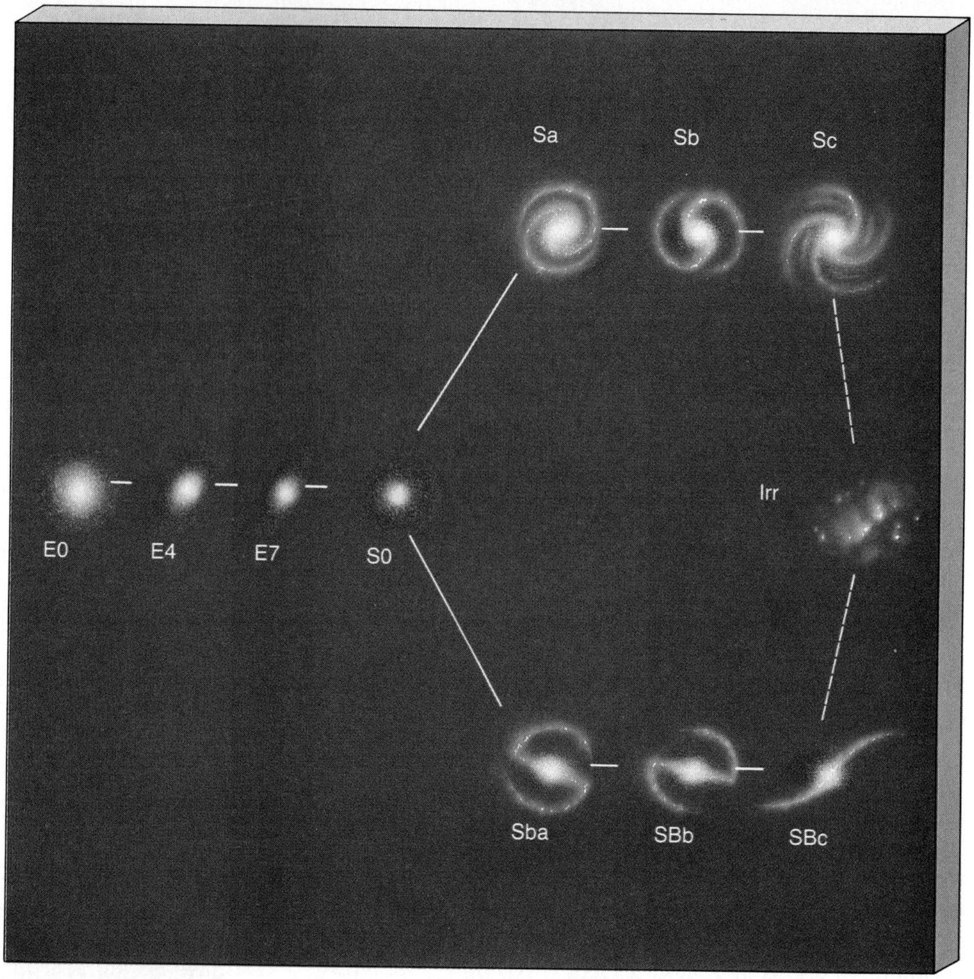

Figure 26.3 Hubble's "tuning fork" diagram, showing the basic galaxy classification scheme he developed. The placement of the four basic galaxy types—ellipticals, spirals, barred spirals, and irregulars—in the diagram is suggestive, but no "evolutionary track" along the sequence (in either direction) is proven.

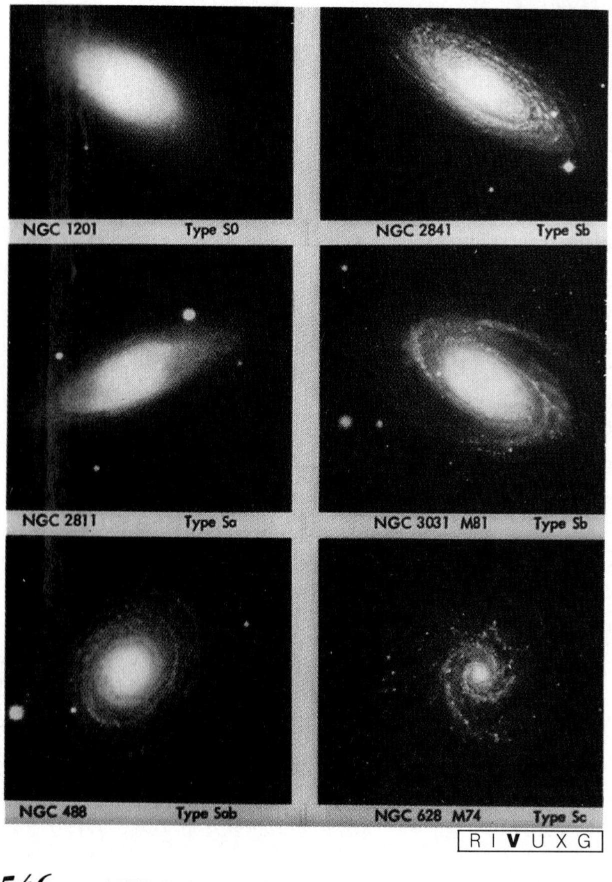

Figure 26.4 Variation in shape among different spiral galaxies. As we progress from type Sa to Sb to Sc, the bulges tend to get smaller, while the spiral arms become less tightly-wound. (Palomar Observatory)

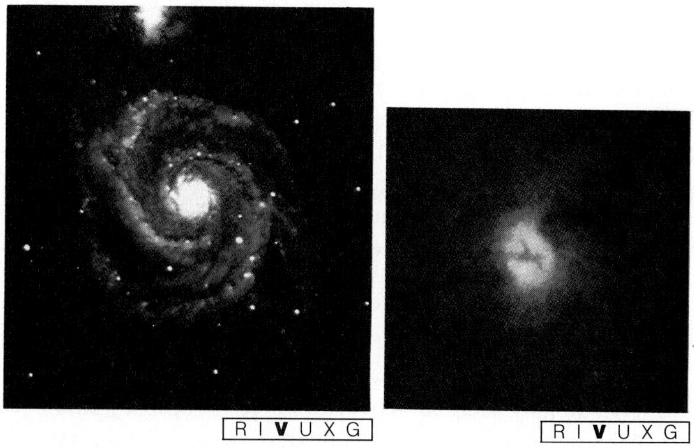

Figure 26.5 The type Sc galaxy M51 clearly shows young (blue) stars spread along its spiral arms. This galaxy, also known as the "Whirlpool Galaxy," is actually a pair of interacting galaxies approximately 10 Mpc away from Earth. (One megaparsec [Mpc] equals 10^6 parsecs.) The insert is a recent Hubble Space Telescope view of the innermost core of M 51, showing an "X" shaped dark feature which is probably an accretion disk of matter girdling a black hole. (AURA; NASA)

Figure 26.6 *The Sombrero Galaxy, a spiral system seen edge-on. Officially cataloged as M104, this galaxy has a dark band composed of interstellar gas and dust. The large size of this galaxy's central bulge marks it as type Sa, even though its spiral arms cannot be seen. (AURA)*

and the dark *corona*—are found in all spirals. (We assume that thick disks exist, too, but their faintness makes this hard to confirm—the thick disk in the Milky Way contributes only a percent or so of our Galaxy's total light.) The flat galactic disks of spiral galaxies are rich in gas and dust, and the spiral arms contain numerous emission nebulae and newly formed blue O- and B-type stars. The bulges and halos contain large numbers of reddish K- and M-type stars, and the halos are often found to contain many old globular clusters, similar to those observed in our own Galaxy and in Andromeda. Most of the light from spiral galaxies, however, comes from fairly average A-, F-, G-, and K-type stars in the disk, giving these galaxies an overall whitish-yellowish glow.

Many spirals are observable only at sharp angles, not face-on as in Figure 26.5. However, we do not need actually to see spiral structure to classify a galaxy as a "spiral." The presence of the disk, with its gas, dust, and newborn stars, is sufficient. For example, the angles at which we see Andromeda (see Figure 25.1) and the edge-on NGC 4565 (Figure 26.2[b]), prohibit a clear view of their arms, but both are nonetheless believed to be spiral galaxies. (NGC is the abbreviation for *New General Catalogue*. NGC 4565 is the 4565th item listed in that collection of astronomical objects.) Similarly, the galaxy shown in Figure 26.6—the so-called Sombrero Galaxy—is thought to be of the spiral variety because of the clear line of obscuring dust that can be seen along its midplane.

Barred Spirals

A variation of the spiral category in Hubble's classification scheme is the **barred-spiral galaxy.** The barred spirals differ from ordinary spirals mainly by the presence of an elongated "bar" of stellar and interstellar matter passing through the center and extending beyond the bulge into the disk. The spiral arms project from near the ends of the bar rather than from the bulge as in normal spirals. Barred spirals are designated by the letters *SB* and are subdivided, like the ordinary spirals, into SBa, SBb, and SBc, depending on the size of the bulge and the tightness of the spiral pattern. Figure 26.7 shows the variation of these types of galaxies, from the tight-knit SBa to the looser SBc. In the case of the SBc category, it is often hard to tell where the bar ends and the spiral arms begin.

Astronomers often cannot distinguish between spirals and barred spirals, especially when a galaxy happens to be oriented with its galactic plane nearly edge-on toward Earth (as in Figures 26.2[b] and 26.6). Because of the physical and chemical similarities of spiral and barred-spiral galaxies, some researchers do not even bother to distinguish between them. Others, however, regard the differences in their structures as very important, arguing that the differences suggest basic dissimilarities in the conditions that led to the formation of the galaxies eons ago. The recent findings that the bulge of our own Galaxy is elongated (see Chapter 25) suggests (to some astronomers, at least) that the Milky Way itself may be a barred spiral, of type SBc.

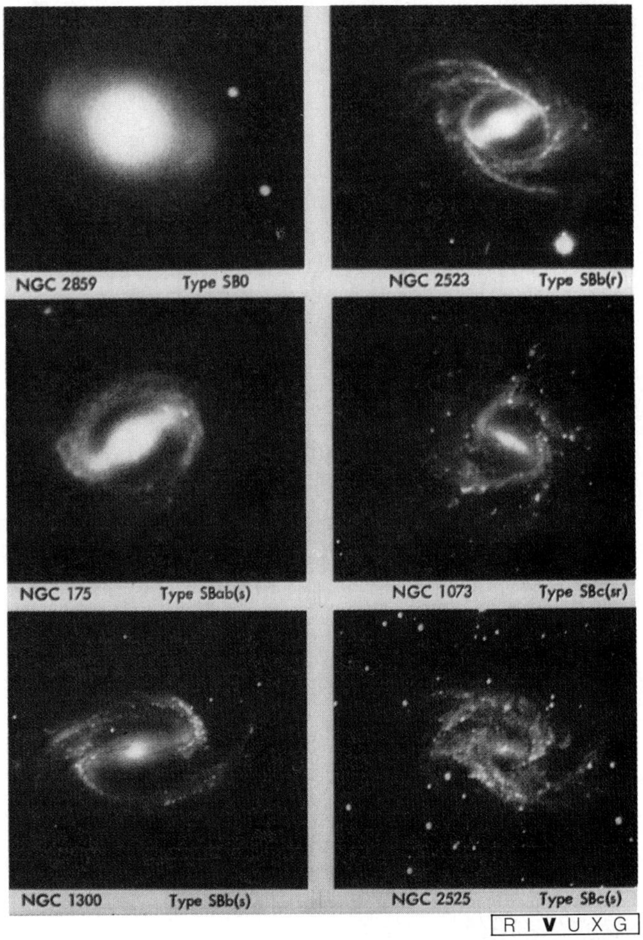

Figure 26.7 *Variation in shape among different barred-spiral galaxies. The variation from SBa to SBc is similar to that for the spirals in Figure 26.4, except that now the spiral arms begin at either end of a bar through the galactic center. (Palomar Observatory)*

Ellipticals

The next major category in the Hubble scheme contains the **elliptical galaxies,** which range in shape from highly elongated to nearly circular in appearance. Denoted by the letter E, these systems are further classified by how elliptical they are. The most circular are designated E0, the flattest E7; intermediate elongations are classified El through E6.[1] Figure 26.8 displays this variation in shape. Unlike the spirals, ellipticals have no spiral arms and, in most cases, no flattened galactic disk—in fact, they often exhibit little internal structure of any kind. It is presently uncertain whether these galaxies are actually *oblate* (lozenge-shaped) or *prolate* (cigar-shaped). Both shapes could appear elliptical, depending on their orientation, but it is difficult to distinguish between the two on the basis of observations alone.

There is a large range in both the size and the number of stars contained in elliptical galaxies. The largest ellipticals are much larger than our own Milky Way Galaxy. The so-called **giant ellipticals** can range up to a few megaparsecs (1 Mpc = 10^6 pc) across and contain trillions (10^{12}) of stars. At the other extreme, **dwarf ellipticals** may be as little as a kiloparsec in size and contain fewer than a million (10^6) stars. The substantial observational differences between giant and dwarf ellipticals have led many astronomers to conclude that they are really separate classes of galaxies, with quite different formation histories and stellar content. The dwarfs are by far the most common type of ellipticals, outnumbering their brighter cousins by about 10 to 1. However, most of the *mass* that exists in the form of galaxies is contained in the larger systems.

Lack of spiral arms is not the only difference between spirals and ellipticals. Most ellipticals also contain little or no gas and dust. No obscuring dust lanes are seen, and 21-cm radio emission from neutral hydrogen gas is, with few exceptions, completely absent. In most cases, there is no evidence for any young stars, or for any ongoing star formation, and there is no sign of star formation in the recent past. Like the halo of our own Galaxy, ellipticals are made up mostly of old, reddish, low-mass stars. Indeed, with no disk, dust, or gas, elliptical galaxies are, in a sense, "all halo." Some E3 or E4 ellipticals may even resemble quite closely how our Galaxy would look if the disk were magically removed. Again like the halo of our Galaxy, the orbits of stars in ellipticals are disordered, with objects moving in all directions, not in regular, circular paths as in our Galaxy's disk. Apparently all, or nearly all, of the interstellar gas within elliptical galaxies was swept up into stars (or out of the galaxy) long ago, before a disk had a chance to form, leaving no loose gas and dust for the continued formation of future generations of stars.

[1]More precisely, if the long axis of the ellipse has length a and the short axis has length b, the n in the designation En is defined to be 10 (1 − b/a). Thus, a spherical galaxy, with $a = b$, is of type E0, a somewhat elongated galaxy with $b = 0.7a$ is classified E3, and so on. No galaxies flatter than E7 are known.

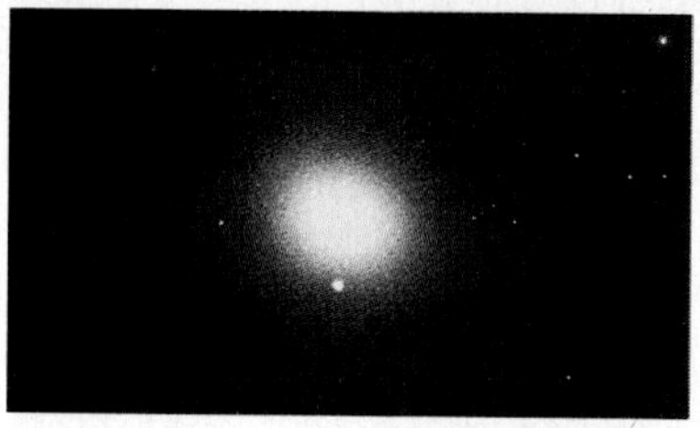

(a) R I V U X G

(b) R I V U X G

Figure 26.8 *Variation in shape among different elliptical galaxies. (a) The El galaxy M49 is nearly circular. (b) M84 is a slightly more elongated elliptical galaxy. It is classified as E3. Both these galaxies lack spiral structure, and neither shows evidence for interstellar matter. (AURA)*

Some giant ellipticals are exceptions to many of the above statements, as they have been found to contain disks of gas and dust in which stars are forming. Astronomers speculate that these galaxies may really be otherwise "normal" ellipticals that have collided and merged with a companion spiral system (although this explanation is not universally accepted by workers in the field).

Intermediate between the E7 ellipticals and the Sa spirals in the Hubble classification is a class of galaxies that show evidence for a thin disk and a flattened bulge but that contain no gas and no spiral arms. Two such objects are shown in Figures 26.4 and 26.7. These galaxies are known as **S0 galaxies** if no bar is evident and **SB0 galaxies** otherwise. They look a little like spirals whose dust and gas have been stripped away, leaving behind a stellar disk. Here too there is a potentially important modification to Hubble's original scheme. Observations in recent years have shown that many "ellipticals" actually have faint disks within them, like the S0 galaxies. Again, the origin of these disks is uncertain, but some researchers suspect that the *SOs* and ellipticals actually form a continuous sequence, along which the bulge-to-disk ratio varies smoothly.

Irregulars

The final galaxy class identified by Hubble is a catchall category—**irregular galaxies.** They are named irregular galaxies largely because their visual appearance does not allow us to place them into any of the other categories just discussed. They are divided into two subclasses. The Irr I galaxies generally look like rather misshapen spirals. The much rarer Irr II galaxies, in addition to having a rather irregular shape, are also downright odd-looking. Figure 26.9 shows some examples of these strangely shaped galaxies, many of which are known to be rich in interstellar matter. Their appearance once led astronomers to suspect that violent, explosive events had occurred within them. However, it now seems possible that in some (but *not* all) cases we are actually seeing the result of a close encounter, or even a collision, between two previously "normal" galaxies.

Irregular galaxies tend to be smaller than spirals but somewhat larger than dwarf ellipticals. They typically contain between 10^8 and 10^{10} stars. The smallest are called

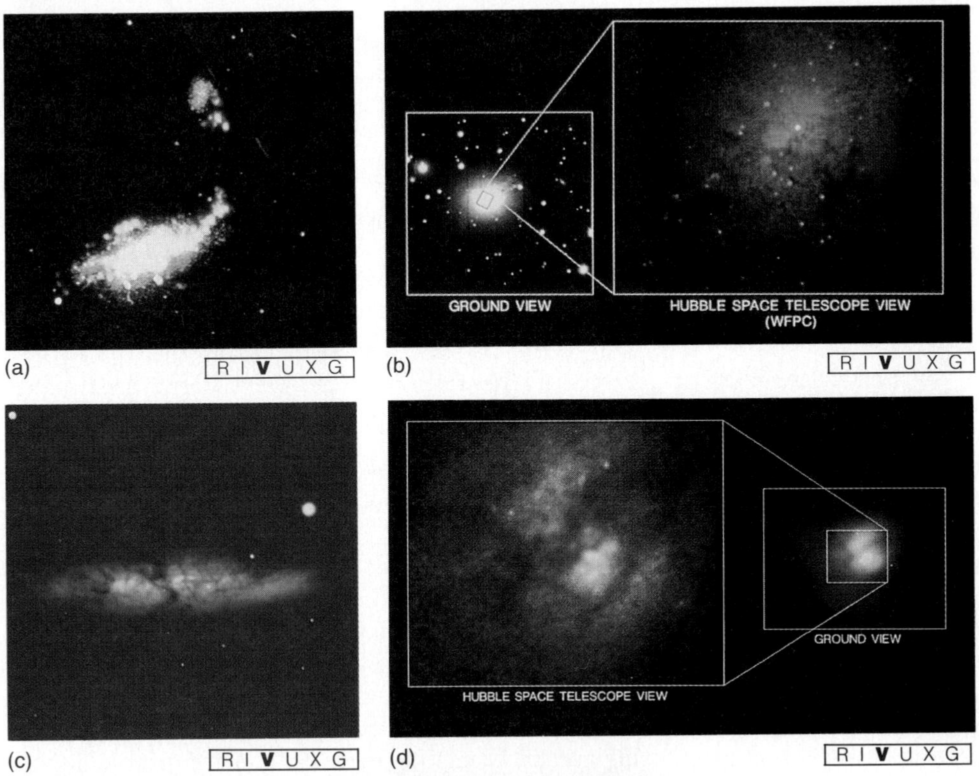

Figure 26.9 *Photographs of some irregular and peculiar galaxies. (a) The oddly shaped galaxies NGC 4485 and NGC 4490 may actually be physically close and be interacting with one another gravitationally. (b) The peculiar galaxy NGC 1275 depicts what seems to be a system of long filaments exploding outward into space. Its blue blobs, as revealed by the* Hubble Space Telescope, *are probably young globular clusters formed by the collision of two galaxies. (c) The galaxy M82 likewise seems to show an explosive appearance, although interpretations remain uncertain. (d) The peculiar galaxy, Arp 220, as seen from the ground and via the* Hubble *telescope, again perhaps two galaxies in collision. (AURA; NASA; Palomar Observatory)*

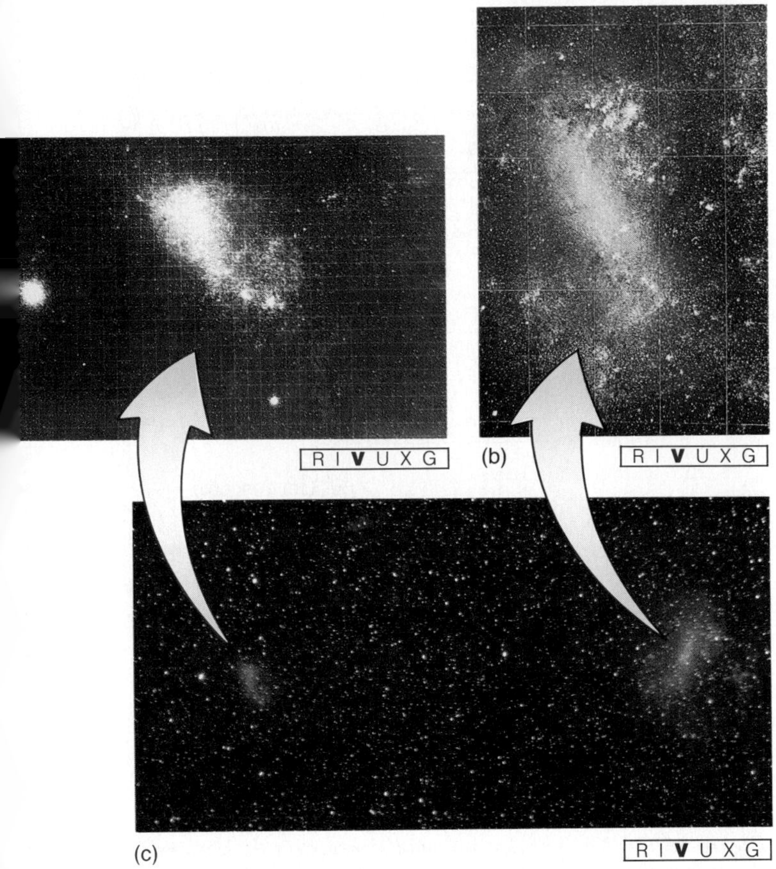

Figure 26.10 *These irregular galaxies, called the Large (a) and Small (b) Magellanic Clouds, are dwarf "satellite" galaxies gravitationally bound to our own Milky Way Galaxy. (c) Wider view of the two Magellanic Clouds, showing their relation to one another in the southern sky. (Harvard College Observatory; AURA)*

dwarf irregulars. As with elliptical galaxies, the dwarf type is most common. Dwarf ellipticals and dwarf irregulars occur in approximately equal numbers and together make up the vast majority of galaxies in the universe. They are often allied with a larger "parent" galaxy from one of the other three categories. Figure 26.10 shows the **Magellanic Clouds,** a famous pair of irregular galaxies that orbit the Milky Way. (They are shown to proper size and scale on the left of Figure 25.12 and discussed further in Interlude 26-1.)

THE DISTRIBUTION OF GALAXIES

✓2 Having studied the different types of galaxies beyond our Milky Way, we naturally wonder about their distribution in space. How are the galaxies spread through the expanse of the universe beyond our Milky Way? Do they reside everywhere, scattered throughout the virtual void of **intergalactic space** all the way out to the very limits of the observable universe? Or is there, as with the Earth, the solar system, and our Galaxy, some terminal point, in this case a boundary beyond which galaxies are no longer observed? To answer these questions, we must first know the distances to the galaxies.

Reviewing the Distance Scale

Let's review the techniques we have applied so far in determining distances, as summarized in the bottom four levels

INTERLUDE 26-1 *The Clouds of Magellan*

Far to the south, out of viewing range of most of the Northern Hemisphere, reside two nearby galaxies called the *Magellanic Clouds*. These "Clouds of Magellan" are dwarf irregular galaxies, gravitationally bound to our own Milky Way. They not only orbit our Galaxy, but also accompany it on its trek through the cosmos.

The Large Magellanic Cloud and its companion, the Small Magellanic Cloud, are visible to the naked eye from any location south of Earth's equator. Looking much like dimly luminous atmospheric clouds, they are named for the sixteenth-century Portuguese explorer Magellan, whose round-the-world expedition first brought word of these giant fuzzy patches of light to the European civilizations.

Studies of Cepheid variables within the Magellanic Clouds show them to be approximately 50,000 pc from the center of our Galaxy. The larger cloud contains about 6 billion solar masses of material and is a few kiloparsecs across. Figure 25.12 shows the position, size, and scale of the two Clouds relative to the Milky Way. Close-ups of the Large Magellanic Cloud in Figure 26.10 illustrate its distorted, irregular shape, although some observers claim they can discern a single spiral arm. Whatever its structure, direct observations show that this irregular galaxy contains lots of gas, dust, and blue stars (not to mention a recent well-known supernova, discussed in Interlude 23-2), indicating youthful activity and presumably current star formation. However, both Clouds also contain many old stars (and several old globular clusters), so we know that star formation has been going on there for a very long time.

Over the years, radio studies have hinted at a possible bridge of hydrogen gas connecting our Milky Way to the Magellanic Clouds, but more observational research is needed to establish this link beyond doubt. It is possible that the tidal force of the Milky Way tore this stream of gas from the Clouds the last time their orbits brought them close to our Galaxy. Of course, gravity works both ways, and the forces exerted by the Clouds also distort the Galaxy, warping and thickening the outer parts of the galactic disk.

of Figure 26.11. We used radar first, then extended our measurements confidently out to about 100 pc using stellar parallax (as discussed in Chapters 2 and 19). However, 100 pc is but a small step compared with the 30 kpc size of the Milky Way Galaxy. To study even our own Galaxy, let alone others, more powerful methods are needed.

Our third distance-measurement technique, comparing the absolute (intrinsic) and apparent (observed) brightnesses of main-sequence stars, moves us beyond the immediate vicinity of the Sun. As described in Chapter 19, the operational procedure for measuring the absolute brightness of a star is twofold. First we determine its spectral class (or surface temperature), and second, assuming it to be a main-sequence star, we read the absolute brightness from the HR diagram. Knowing that the intensity of radiation diminishes as the square of the distance the radiation has traveled, we can calculate the star's distance. This method enables us to estimate distances to stars as far as several thousand parsecs from Earth—about the greatest distance at which we can identify and study individual stars in our Galaxy's plane, but still well within our local niche in the Milky Way.

A fourth technique for distance determination uses variable stars—RR Lyrae variables and Cepheids—as discussed in Chapter 25. This method also requires a comparison of absolute and apparent brightnesses. Variable stars provide a welcome advantage, however, for they enable us to determine a star's absolute brightness with reasonable accuracy without having either to know its spectral classification or to assume that it is located on the main sequence.

We can extend this distance yardstick to include far-

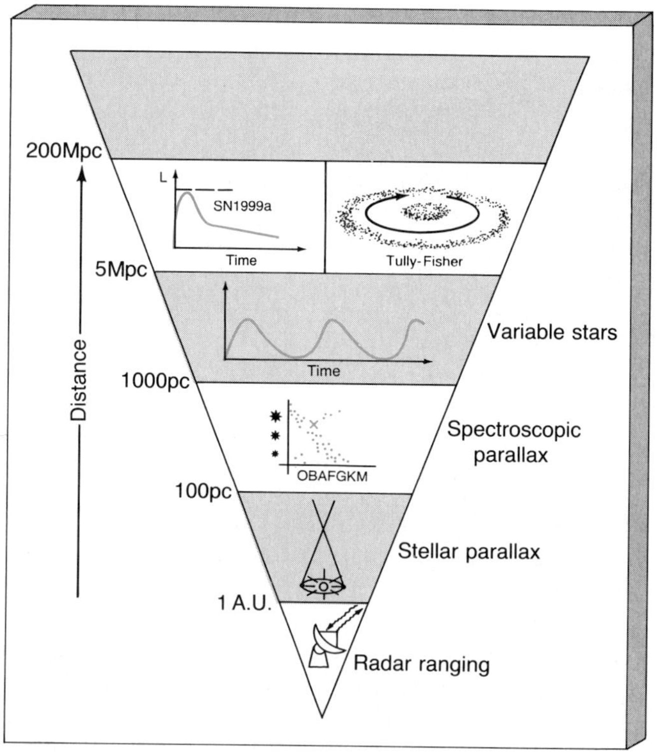

Figure 26.11 *The inverted pyramid summarizes the various distance techniques used to study different realms of the universe. Radar ranging, stellar parallax, spectroscopic parallax, and variable stars take us as far as the nearest galaxies in our study of the universe. To go farther, new techniques must be employed, each based on distances known by techniques at lower levels. The top level shown here will be described later in this chapter.*

away galaxies provided that we can identify an individual Cepheid variable star and measure its period. Note that, in doing this, we assume that the period-luminosity law derived for the Cepheids in our Galaxy holds true for Cepheids in other galaxies. We have no reason to doubt this assumption. The Cepheid technique is a powerful tool, enabling us to find distances as far as the best telescopes can detect individual Cepheids and measure their apparent brightness and pulsation period. At present, we can see the brightest Cepheids out to a distance of approximately 5 Mpc.

The Local Group

Now we can reframe the questions posed earlier. On studying the Cepheid variable stars within other galaxies, what does the spread of galaxies look like? How is matter distributed within a few million parsecs of our Milky Way?

Figure 26.12 sketches all the known major astronomical objects within about 1 Mpc of the Milky Way. Our Galaxy appears with its two satellite galaxies, the Magellanic Clouds. The Andromeda Galaxy, lying some 700 kpc from the Milky Way, is also shown. Two of the larger satellite galaxies of Andromeda—one spiral, one elliptical—can easily be seen in Figure 25.1.

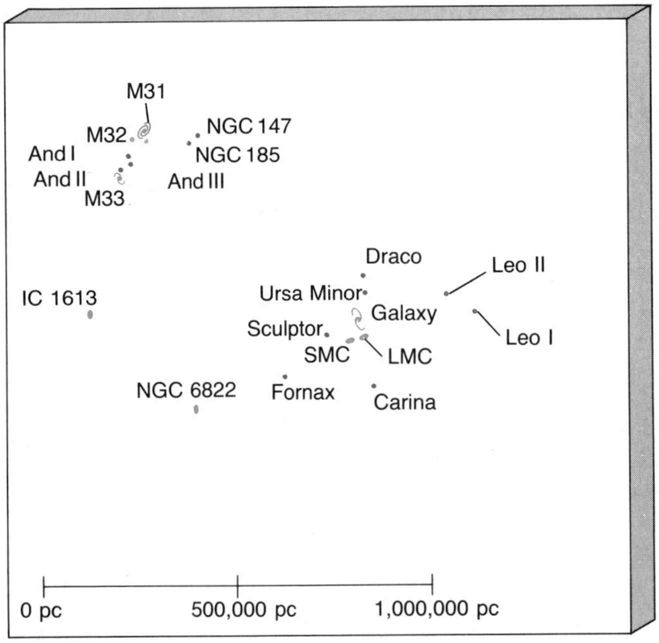

Figure 26.12 *Diagram of the Local Group of some twenty galaxies within approximately 1 Mpc of our Milky Way Galaxy. Only a few are spirals; most of the rest are dwarf elliptical and irregular galaxies.*

Observations have pinpointed several more galaxies within a few million parsecs of us. All told, some 20 galaxies populate our Galaxy's neighborhood, including three spiral galaxies (ourselves, Andromeda, and M33, visible at the lower right of Figure 25.1) and with many dwarf irregular and elliptical systems (such as M32, also visible in Figure 25.1).

A group of galaxies held together by their mutual gravitational attraction is called a **galaxy cluster** (or **cluster of galaxies**). The galaxy cluster that includes our Milky Way is known as the **Local Group.** The diameter of this well-defined galaxy cluster is roughly 1 Mpc. The words *vast, huge,* and *gargantuan*—even the slang adjective *astronomical*—begin to take on their real meanings when describing galaxy clusters.

Let's slow down for a moment to comprehend the content of this section. We have just made a great jump in spatial dimensions, from the 30-kpc size of our own Galaxy to the 1-Mpc size of the Local Group—a factor of 30 in scale. Notice also that the Copernican Principle (or the principle of mediocrity) continues to apply. Not only is Earth not the center of our solar system and the Sun not the center of our Galaxy, but our Galaxy is also not the center of the much larger Local Group.

Beyond the Local Group

Our Local Group contains 20 or so galaxies, but we know that many more galaxies exist in the universe. The single photograph in Figure 26.1 reveals several hundred galaxies, and thousands of similar photos have been taken for different regions of the sky. In all, astronomers estimate that some *100 billion* other galaxies roam the observable universe.

The next large concentration of galaxies lies well beyond our local galaxy cluster. Although some galaxies do reside between 1 and 5 Mpc away, most known galaxies lie well beyond 5 Mpc—too far for the Cepheid technique to work. Cepheid stars in very distant galaxies simply cannot be observed well enough, even through the world's largest telescopes, to permit measurement of their luminosity and period. When the Hubble telescope is repaired, it may be able to see Cepheids as far away as 20–30 Mpc, but that is still in the future.

To extend our distance-measurement ladder, we must find some new object to study. We cannot use ordinary stars or Cepheid variables beyond 5 Mpc because we cannot distinguish their spectral types or pulsation periods. What individual objects are bright enough for us to observe and study at great distances?

Researchers have tackled this problem by using a variety of **standard candles**—easily recognizable astronomical objects whose intrinsic brightnesses are believed to be well known. The basic idea is very simple. Once an object is identified as a standard candle—simply by its appearance or by the shape of its light-curve, say—an estimate of its luminosity can be made. Measurement of its *apparent* brightness and application of the inverse-square law then allow its distance to be determined. To be most useful, a standard candle must (1) have a narrowly defined luminosity, so the uncertainty in estimating its brightness is small, and (2) be bright enough to be seen at large distances.

Astronomers have used many different standard candles over the years. Novae, HII regions, star clusters (especially globular clusters), supernovae, even entire galaxies have been employed, with varying degrees of success. In fact, virtually every "bright" object discussed in this book has been considered at one time or another, so eager have astronomers been to expand their cosmic horizon. Just as we did with the lower rungs on our distance ladder, we calibrate the properties of each new standard-candle hung using distances measured by more "local" techniques. In this manner, the distance-measurement process "bootstraps" its way to greater and greater distances—many hundreds of megaparsecs, by the time entire galaxies are used as standard candles. At the same time, the errors and uncertainties in each step continue to accumulate, so the distances to the farthest objects are the least well known.

An important alternative technique was discovered in the 1970s, when astronomers R. Brent Tully and J. Richard Fisher discovered a clear correlation between the rotational velocity (as measured by the broadening of spectral lines) and the total luminosity of spiral galaxies within a few tens of megaparsecs of the Milky Way. Because (as we saw in Chapter 25) the rotation speed is a measure of a galaxy's total mass, we should perhaps not be surprised that it is related to total luminosity. What *is* surprising is how tight the correlation is. The spectral line normally used in these studies is the 21-cm line of cold, neutral hydrogen in a galactic disk. By measuring its width, we can very reliably determine a spiral galaxy's absolute brightness and then compare it with the galaxy's apparent brightness to yield its distance.

The **Tully-Fisher relation,** as it is now known, can be used to measure distances to spiral galaxies out to about 200 Mpc (beyond which the line broadening becomes increasingly difficult to measure accurately). A somewhat similar connection exists for elliptical galaxies, linking the broadening of a galaxy's spectral lines (a measure of the average random velocity of the stars in the galaxy—see Chapter 4) and the galaxy's *size*. By measuring the broadening, astronomers determine the true size of the galaxy, which is then compared with the apparent size to give the distance. The important point here is that both of these methods bypass many of the standard candles just described and so provide an independent means of determining distances to faraway objects.

To emphasize the point that they are conceptually different approaches, both Tully-Fisher and standard candles share the fifth ring of our distance ladder in Figure 26.11. In a moment, we will see what happens when these two parallel techniques are compared.

Other Galaxy Clusters

✓3 The nearest large cluster of galaxies to the Local Group is known as the Virgo Cluster. It resides some 20 Mpc from the Milky Way. Like the Local Group, the Virgo Cluster is held together by the mutual gravitational attraction of its member systems. But the Virgo Cluster does not contain a mere 20 individual galaxies; it houses approximately 2500 galaxies, bound together in a tight-knit group about 3 Mpc across, each galaxy containing 100 billion or so individual stars. Figure 26.13 shows a view of the central portion of the Virgo Cluster.

Figure 26.13 *Part of the Virgo Cluster of galaxies, about 16 Mpc away. Several large spiral and elliptical galaxies can be seen. The galaxy near the center is a bright elliptical known as M86. (Royal Observatory)*

The Virgo Cluster is not the only "nearby" group of galaxies. Figure 26.14 illustrates several well-defined clusters sprinkled across our cosmic neighborhood, within about 50 Mpc of the Milky Way. Such a "map" clearly demonstrates that galaxies are not evenly spread throughout space—at least on scales of 50 Mpc or so. Galaxy clusters come in many shapes and sizes. The largest, "rich" clusters contain many thousands of individual galaxies distributed fairly smoothly in space. The smallest clusters, like the Local Group, contain only a few galaxies and are quite irregular in shape.

Not all galaxies are members of clusters. A few are apparently isolated systems, moving alone through intercluster space. Apart from these galaxies, is there any gaseous matter of any kind outside clusters? The answer seems to be no. Astronomers have never found even a hint of an intergalactic medium beyond any of the well-defined galaxy clusters. Evidently, when the clusters formed eons ago, they did so very efficiently, sweeping up all the matter within any given region of the universe. Matter definitely exists between the individual galaxies *within* many clusters,

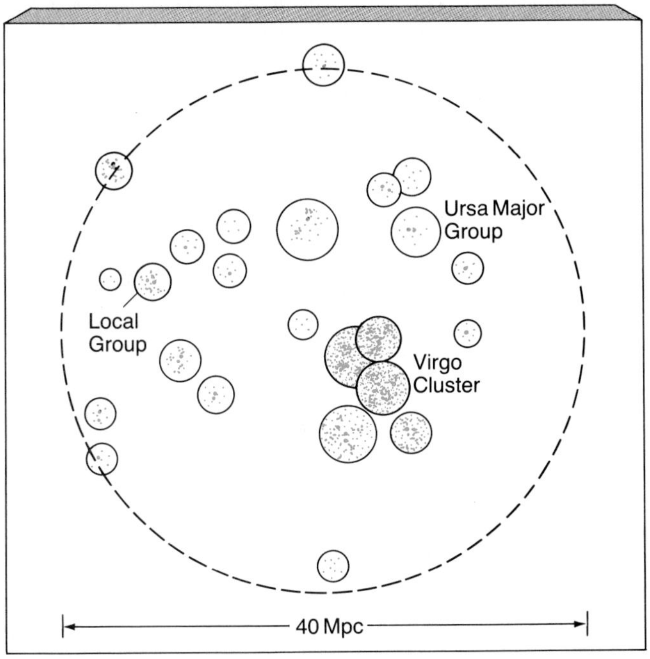

Figure 26.14 *The locations of several galaxy clusters in our part of the universe. Our Milky Way is only one of these dots and our Local Group only one of the clusters of dots.*

but the space between the galaxy clusters is apparently empty of luminous material.

Clusters of Clusters

Does the universe have even greater groupings of matter, or do galaxy clusters top the cosmic hierarchy? Most astronomers now believe that the galaxy clusters themselves are clustered, thereby shaping titanic agglomerations of matter known as **superclusters.** Figure 26.15 shows the so-called *Local Supercluster,* containing the Local Group, the Virgo Cluster, and most of the other clusters shown in Figure 26.14. (The direction of view is different from that shown in the earlier figure, so some of the clusters noted there are difficult to discern here.) Each point represents a separate galaxy, and the diagram is centered on the Milky Way. The perspective is such that the disk of our Galaxy is seen edge-on and is oriented in an up-down direction. The two empty V-shaped regions at the top and bottom are not really devoid of galaxies—they are simply obscured from view by the dust in our own Galaxy's plane. The total mass of the entire supercluster—which is centered in the Virgo Cluster (the large "blob" of points on the right)—is about 10^{15} $M_\odot$ (solar masses). Perhaps by now it should come as no sur-

INTERLUDE 26-2 *Colliding Galaxies*

Contemplating the congested confines of a rich galaxy cluster like Virgo, with its thousands of members, we might expect that collisions among galaxies would be common. Gas particles collide in our atmosphere and hockey players collide in the rink—do galaxies in clusters collide? The answer is yes.

Direct observational evidence shows that galaxies in clusters apparently collide quite often. In the smaller groups, the galaxies' velocities are low enough that they tend to "stick together," and *mergers* are a common outcome of the encounter. The left-hand image below is a computer-enhancement of the pair of galaxies NGC 4676 A and B (also known as "The Mice"), which seem to show streams of connecting gas apparently generated by the encounter between the two galaxies. Whether these galaxies are genuinely colliding or only experiencing a close encounter cannot easily be determined. No human being could witness an entire collision, for it would last many millions of years. However, computer simulations of these systems (right-hand frame) show formations remarkably similar to the real thing. This particular simulation began with two spiral galaxies, but the details of the original structure have been largely obliterated by the collision. Ultimately, the calculations show the two galaxies merge into one.

Since the early 1980s, it has become increasingly clear, on the basis of both observations and numerical simulations (such as that shown below), that collisions can have very large effects on the galaxies involved. The stellar and interstellar contents of each galaxy are rearranged, and the merged interstellar matter very likely experiences shock waves that trigger bursts of star formation (see Figure 26.29). Some researchers would go so far as to suggest that *most* galaxies have been strongly influenced by collisions, in many cases in the relatively recent past. We discuss some of the possible effects of galaxy collisions in more detail later in the text.

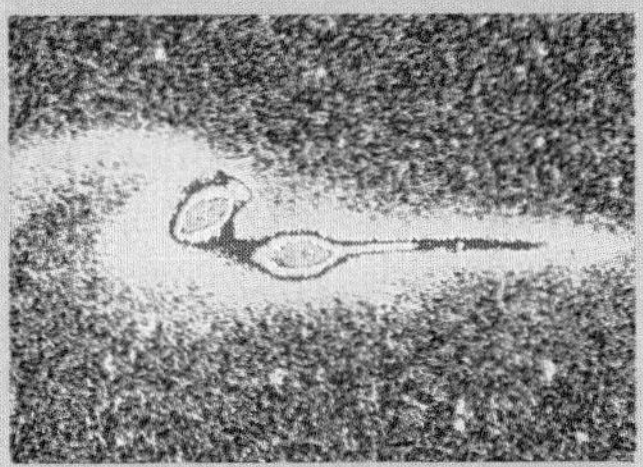

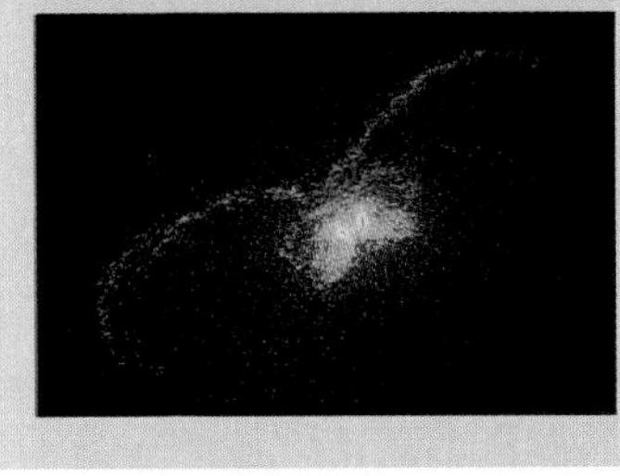

(D. H. Hubel/J. Robbins; J. Barnes/L. Hernquist)

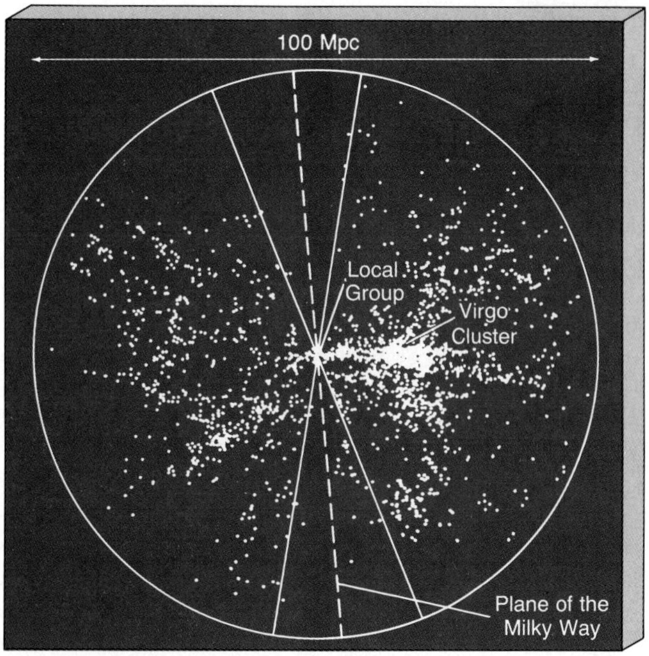

Figure 26.15 The Local Supercluster. Each of the 2200 points shown represents a galaxy, and the Sun is at the center of the diagram. The Virgo Cluster and the plane of our own Galaxy are marked. (Our Galaxy is seen edge-on. Its dust obscures our views to the top and the bottom, and two empty V-shaped regions on the map result.) The radius of the circle shown here is about 50 Mpc.

prise that the Local Group is not found at the heart of the Local Supercluster. Instead, it lies far off in the periphery, some 20 Mpc from the center.

Deep Space

The Local Supercluster contains a huge number of individual galaxies—perhaps several tens of thousands. Yet the great majority of known galaxies exist far beyond its edge. Figure 26.16 is a long-exposure photograph of one such remote cluster. Called the Corona Borealis Cluster, this gravitationally bound aggregate is much farther away than the 50 Mpc shown in Figure 26.14. This rich cluster is only one of many other large and distant groups of galaxies scattered throughout the observable universe. The Coma Cluster shown in Figure 26.1 is another. On and on, the picture is much the same. The farther we peer into deep space, the more galaxies, clusters of galaxies, and superclusters we see. Is there structure on even larger scales? The answer is still yes!

But we are moving too fast in our intellectual rush toward the limits of the universe. We are missing some interesting and important information about the nature of our extended cosmic neighborhood—within a billion or so parsecs, if we stretch our distance-measurement techniques to their limits. Let's pause for a moment to reconsider the properties of galaxies and galaxy clusters before making our final leap in distance.

Figure 26.16 The Corona Borealis Cluster contains huge numbers of galaxies and resides roughly half a billion parsecs away from Earth. (Palomar Observatory)

GALAXY MASSES

✓4 We have quoted several galaxy masses above without giving any indication of how they were determined. Because mass is such an important physical property of any object, let us correct that oversight now and describe the methods used to weigh galaxies and galaxy clusters. How can we find the masses of such huge systems? Surely, we can neither count all their stars nor estimate their interstellar content very well. Galaxies are just too complex to take direct inventory of their material makeup. Instead, we must rely on indirect techniques. As usual, they involve the Law of Gravity.

Individual Galaxies

Two techniques for measuring the mass of an individual galaxy employ Kepler's Third Law, in much the same way as our estimation of the mass of our own Galaxy (in Chapter 25) depends on that law. By observing different locations in a galaxy with a telescope and spectrometer, we can measure the Doppler shift of spectral lines and determine the galaxy's *rotation curve,* the plot of rotational velocity versus distance from the center of the galaxy. The mass within any given radius then follows directly, just as the mass for our own Milky Way was computed. To use this technique, a galaxy must appear large enough on the sky to permit us to observe each end separately, as shown in Figure 26.17(a). The rotation curves for some nearby spiral galaxies are shown in Figure 26.18. Notice, incidentally, how *none* of them shows much decline in rotational velocity at large distances, even though the entire visible disk of the galaxy is included in the plots. We will return to this very important point in a moment. For now, however, we note only that these curves imply masses ranging from about 10^{11} to $5 \times 10^{11}\ M_{\odot}$ within about 25 kpc of the center.

In many cases, galaxies are too distant and therefore

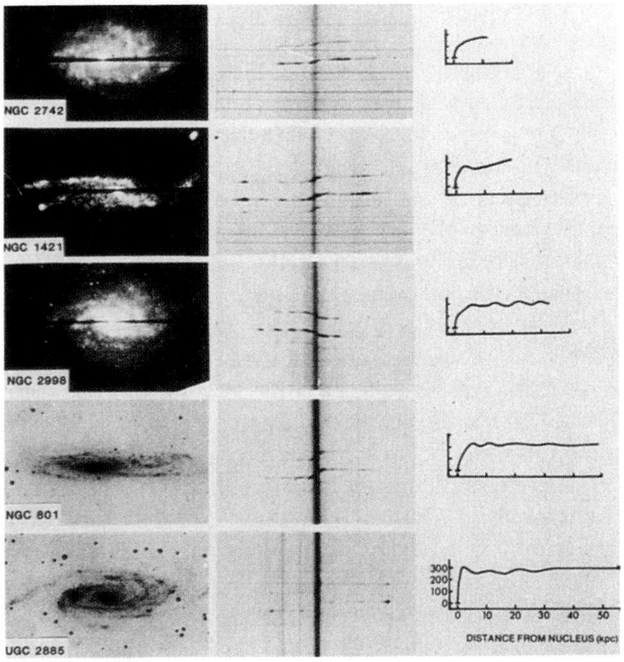

Figure 26.18 *Rotation curves for some nearby spiral galaxies indicate masses of a few hundred billion times the mass of the Sun. Compare this figure with Figure 25.16, which plots a similar curve for our own Galaxy. (Carnegie Institution)*

appear too small for such detailed measurements to be made. Nevertheless, if the entire galaxy is observed at once, the galaxy's spectral lines are *broadened* by the overall rotation in much the same way as lines from a rotating planet or star. This is precisely the broadening that forms the basis of the Tully-Fisher distance-measurement technique described earlier. Thus it is possible to measure the overall rotation (although not the rotation curve), again allowing us to determine the galactic mass.

We can also use (the modified version of) Kepler's Third Law to estimate the masses of binary galaxies—two galaxies orbiting each other—in a manner quite similar to the way that binary-star masses were determined in Chapter 19. The orbital size and the rotational period of one galaxy around the other are needed to infer the mass, as sketched in Figure 26.17(b). Of course, we cannot watch the galaxies travel even a small fraction of their entire orbit. Usually, the orbital period is simply estimated on the basis of the measured orbital speed and the present distance between the two galaxies. Masses obtained in this way, then, are usually fairly uncertain. However, we can combine many such measurements to obtain quite reliable *statistical* information about galaxy masses.

The results of investigations applying these methods indicate that individual galaxies have masses much like our Milky Way. Most spirals and large ellipticals contain between 10^{11} and 10^{12} $M_\odot$, although the irregulars often contain less, about 10^8 to 10^{10} $M_\odot$. Dwarf ellipticals (and irregulars) can contain as little as 10^6 to 10^7 $M_\odot$ of material.

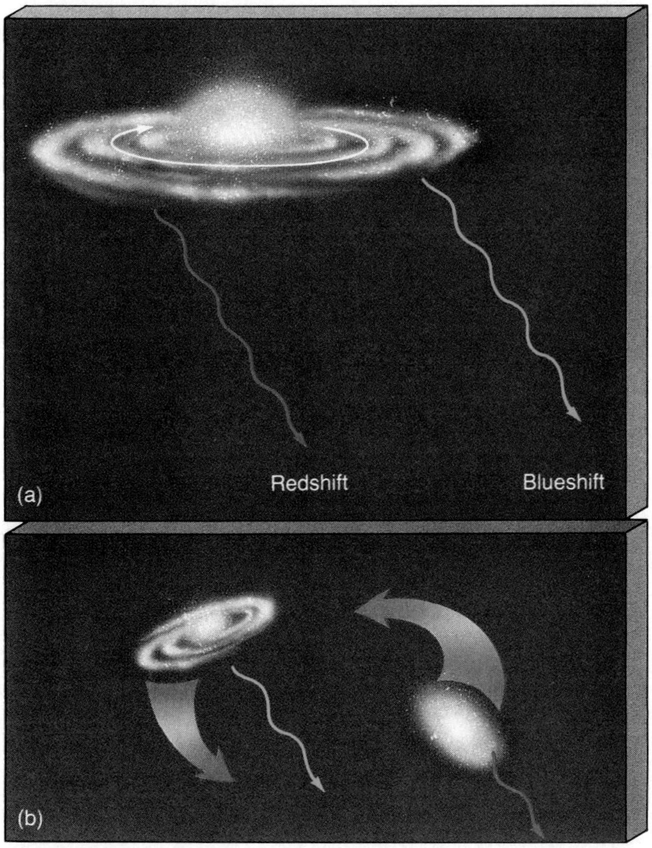

Figure 26.17 *Galaxy masses are derived by observing either (a) two ends of a galaxy of known size or (b) the orbital period of one galaxy about another.*

Galaxy Clusters

We can use another technique to derive the combined mass of all the galaxies within a galaxy cluster. This technique, known as the **Virial method,** is a statistical procedure useful when the previous methods cannot be applied (either because a galaxy is too distant for its line broadening to be measured, or because it is not part of a binary system). The Virial method mathematically ties the speed of the galaxies in a cluster to the total mass of the cluster required to keep the galaxies in orbit.

Consider the gas particles that make up Earth's atmosphere. Because the air is heated (usually about 300 K at sea level), its atoms and molecules have some motion. On warm days, these particles move more rapidly than on cool days. In either case, the atmosphere remains intact—it doesn't disperse or drift away. What holds it to the Earth? Gravity, of course. As we saw in Interlude 9-2, gravity keeps our atmosphere from escaping by constantly pulling back on each of the gas particles. The particles in our atmosphere do not move fast enough to escape the pull of Earth's gravity. Gravity depends on mass, and the Virial method, as applied to our planet, tells us how massive the Earth must be to retain its atmospheric gas particles.

Each galaxy in a cluster, like each gas particle in our atmosphere, moves. Most galaxy clusters are well-defined, tight-knit groups, suggesting that they are indeed gravitationally bound together. If the clusters were not bound, their member galaxies would long ago have dispersed into intergalactic space. We can ask how massive a galaxy cluster must be in order to retain its individual particles—to bind its galaxies gravitationally. In this way, we arrive at an estimate of the cluster's mass. Notice that the Virial method gives us no information about the masses of individual galaxies. It informs us only about the *total* mass of the entire cluster. Typical cluster masses obtained by this method are in the range of 10^{13}–10^{14} $M_\odot$.

Let us stress that the Virial method of mass determination is only a *statistical* technique. It yields a trustworthy answer only when there are enough galaxies (say, a dozen or more) to provide a sufficiently large statistical sample. Furthermore, it assumes that the galaxy cluster is *bound*. If, for example, the cluster were actually breaking up and flying apart, the method would greatly overestimate the total mass.

Dark Matter in the Universe

✓5 These methods of mass determination sound very reasonable—until we apply them and compare the results with what we see. Radio observations indicate that spiral galaxy rotation curves remain flat (that is, do not decline) far beyond the visible image of the galaxy. We conclude that these spiral galaxies—perhaps all spiral galaxies—must have invisible dark halos similar to that surrounding the Milky Way (see Chapter 25). Depending on how far these coronae extend (and observations appear to indicate that they reach at least as far as 50 kpc from the center of a Milky Way-sized galaxy), spiral galaxies may contain anywhere from 3 to 10 times more mass than can be accounted for in the form of visible matter. Studies of elliptical galaxies suggest similarly large dark coronae surrounding them too, but the observations are still far from conclusive.

When the Virial method is applied to clusters, the mass needed to bind the galaxy clusters is once again found to be much larger than the mass estimated on the basis of the total light emitted by the cluster members. The calculated mass is some 10 to 100 times larger than that found by adding together the masses of all the stars visible in the galaxies. Stated another way, a good deal more mass is needed to bind galaxy clusters than we can see. The problem of dark matter exists, then, not just in our own Milky Way Galaxy, but also in other galaxies and, to an even greater degree, in galaxy clusters as well. It most likely applies to the entire universe. In that case, we must come to terms with the fact that *upward of 90 percent of the universe is composed of inherently dark material*. As we noted in Chapter 25, this matter is not just dark in the visible portion of the spectrum—it is invisible at *all* electromagnetic wavelengths.

What could the dark matter within the galaxy clusters be? Where could it hide? At present, we don't know. As discussed in the previous chapter, many solutions have been suggested, from MACHOs (stellar remnants of various sorts) to WIMPs (exotic subatomic particles). Whatever the answer, the dark matter in clusters apparently cannot be simply the accumulation of smaller amounts of dark matter within individual galaxies—even including the galaxies' dark coronae, we cannot account for all of the dark matter in galaxy clusters. As we look on larger and larger scales, we find that a larger and larger fraction of the matter in the universe is dark.

Intracluster Gas

Could the dark matter be diffuse intergalactic matter existing among the galaxies within the clusters—that is, intracluster gas? Until the late 1970s, astronomers had no observational evidence for intergalactic matter, either inside or outside of galaxy clusters. Then satellites orbiting above Earth's atmosphere detected substantial amounts of x-ray radiation in the direction of some galaxy clusters. Figure 26.19 shows a false-color x-ray image of a large galaxy cluster. The x-ray-emitting region is centered on, and com-

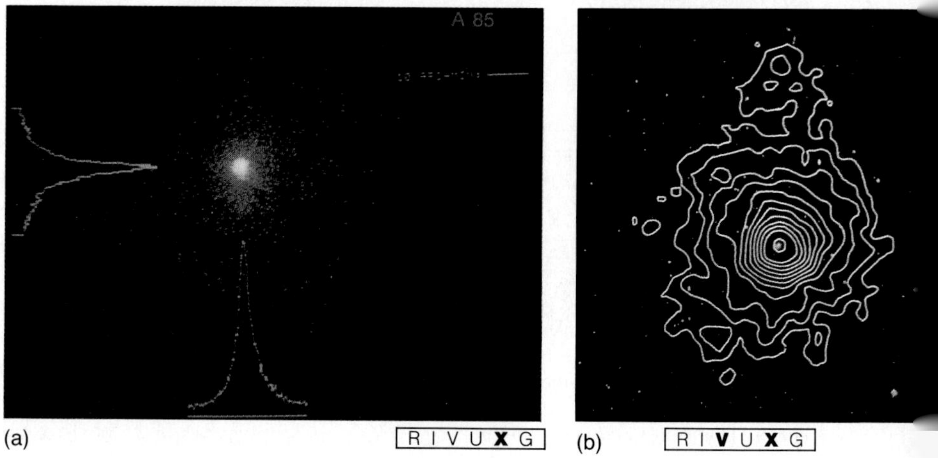

(a) R I V U X G (b) R I V U X G

Figure 26.19 *X-ray image of an old distant cluster of galaxies having few spirals, taken with the* Einstein *x-ray satellite observatory. Called A85, its x-ray emission is shown here in (a) as an orange map and green graphs displaying a smooth, peaked intensity profile centered on the cluster but not associated with individual galaxies. The contour map of x-rays (b) is superposed on an optical photo showing its x-rays peaked on Abell 85's central super-giant galaxy. Images like these demonstrated for the first time that the space between the galaxies within galaxy clusters is filled with superheated gas. (NASA; AURA)* parable in size to, the visible image of the cluster. These x-ray observations demonstrated for the first time the existence of large amounts of invisible hot gas within clusters. The radiation is thermal in nature, with the Planck curve indicating temperatures of about 1 million K. However, *no* gas has been observed outside the clusters—no "extracluster" matter has ever been found.

How much matter have the x-ray satellites found? The observations suggest that roughly as much matter exists outside the galaxies in the form of gas as is visible within them in the form of stars. This is a substantial amount of material, but it still only doubles the total amount of mass accounted for in the cluster. In order to solve the dark-matter problem, we would have to find from 10 to 100 times more gas. We will return to the question of the nature of dark matter in Chapter 28, when we consider the dynamics of the universe on the largest scales.

HUBBLE'S LAW

✓6 We chose our earlier analogy of galaxies in a cluster to molecules in an atmosphere to stress the fact that the motion of individual galaxies within clusters is essentially random. You might expect that, on the largest possible scales, the clusters themselves would also be found to have random, disordered motion—some clusters moving this way, some that. This is not the case at all, however. On the largest scales, unclustered galaxies and galaxy clusters alike display very ordered motion.

Universal Recession

In 1912, the American astronomer Vesto M. Slipher, working under the direction of Percival Lowell, discovered that virtually every spiral nebula (that is, spiral galaxy) he observed had a redshifted spectrum—it was apparently *receding* from our Galaxy. In fact, with the exception of only a few nearby systems, *every* known galaxy is part of a general motion away from us in all directions. Individual galaxies not part of galaxy clusters are steadily receding. Galaxy clusters too have an overall recessional motion, although their individual member galaxies move randomly with respect to one another. (As an analogy, consider a jar full of fireflies that has been thrown into the air. The fireflies within the jar, like the galaxies, have random motions due to their individual whims, but the jar as a whole, like the galaxy cluster, has some directed motion as well.)

Figure 26.20 shows the optical spectra of several galaxies. Interpreted as a Doppler effect, their red shifts indicate that the galaxies are steadily receding. Furthermore, the extent of the red shift increases progressively from top to bottom in the figure. Because the distance also increases from top to bottom, we conclude that there is a connection between Doppler shift and distance. This trend of greater red shifts for objects farther away holds for nearly all galaxies in the universe. (Two galaxies within our Local Group, including Andromeda, and a few galaxies in the Virgo Cluster display blue shifts and so have some motion toward us, but this results from their random motions within their parent clusters.)

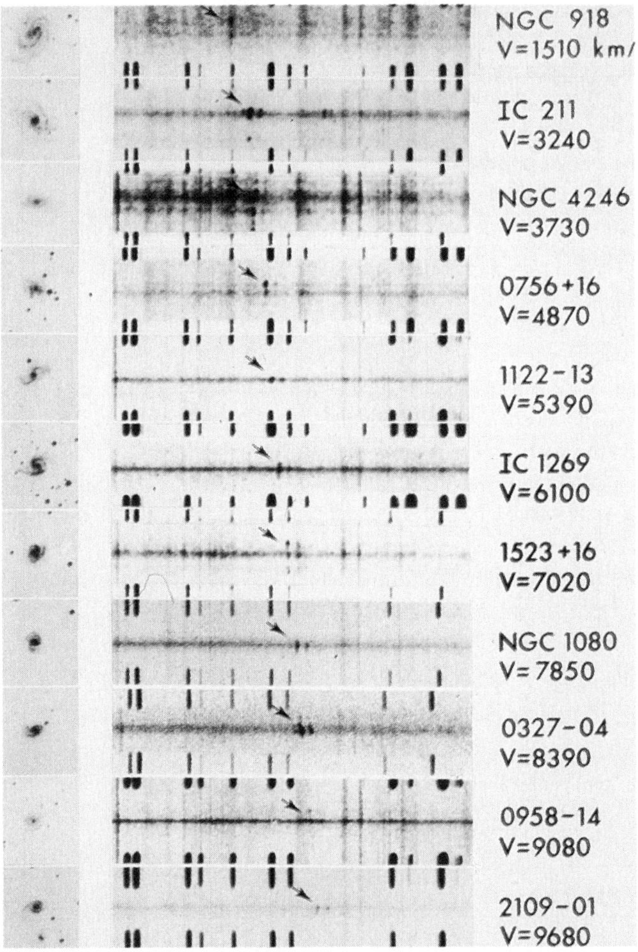

Figure 26.20 *Optical spectra (at center) of several different galaxies (shown on the left). The extent of the red shift (denoted by the arrows at center) and the distance to each galaxy increase from top to bottom. Here, wavelength increases to the right. (Carnegie Institute)*

Not only are the galaxies receding, but they recede with velocities that are proportional to their distances. A simple linear relationship connects velocity and distance: The farther an object is from us, the faster it recedes. Figure 26.21(a) shows a diagram of recessional velocity plotted against distance for the galaxies of Figure 26.20. Figure 26.21(b) is a similar plot for numerous galaxies within 1 billion parsecs of Earth. Diagrams like these were first made by Edwin Hubble in the 1920s, and now bear his name—*Hubble diagrams*. The resultant straight-line fit to the data is called **Hubble's law.** We could construct such a diagram for any group of galaxies, provided that we could determine their distances using one of the previously discussed techniques and that we could measure their radial velocities using spectroscopy. The universal recession described by the Hubble diagram is often referred to as the **Hubble flow.**

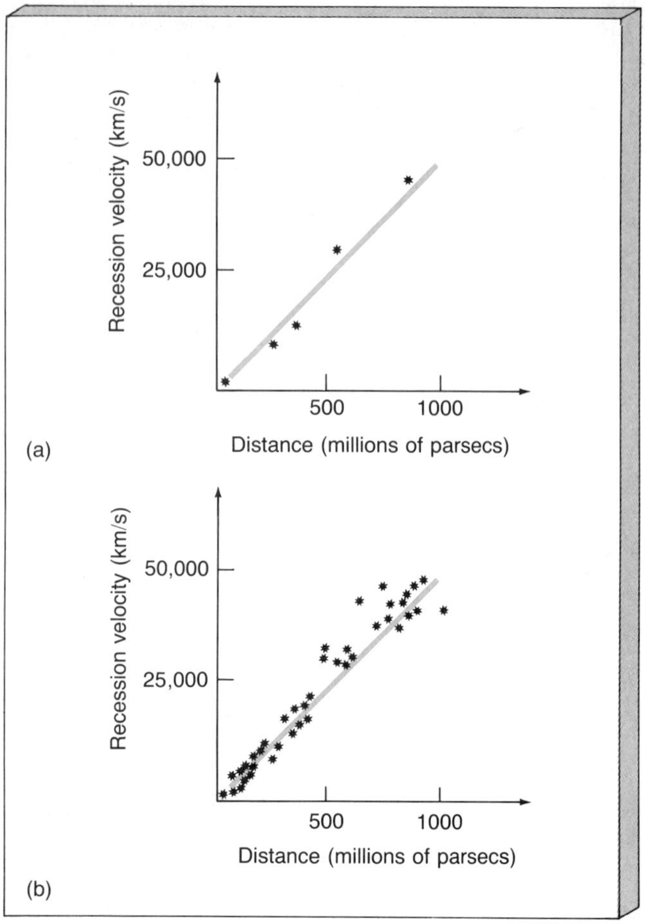

Figure 26.21 *Plots of recessional velocity against distance (a) for some of the galaxies shown in Figure 26.20 and (b) for numerous other galaxies within about 1 billion parsecs.*

An Observational Finding

Hubble's law is an *empirical* discovery—that is, a discovery based strictly on observational results. Its central relationship—a statistical correlation between recessional velocity and distance—is well documented as far as galaxy distances can be reliably determined, but no law of nature demands that all galaxies recede, and no law of physics requires that a link exist between velocity and distance. In that sense, Hubble's "Law" is not really a law at all. It is strictly a convenient way of noting the *observational* fact that any galaxy's recessional velocity is directly proportional to its distance from us.

The recessional motions of the galaxies prove that the cosmos is not steady and unchanging on the largest scales. Its contents are in constant relative motion. Nor do they move entirely randomly. The universe is expanding, and expanding in a directed fashion. In short, it is evolving. But before going any further, let's be clear on *what* is expanding and what is not. The Hubble law does *not* mean that humans, the Earth, the solar system, or even the galaxies are physically increasing in size. These groups of atoms, rocks, planets, and stars are held together by their own internal forces and are not themselves getting bigger. Only the largest framework of the universe—the ever-increasing distances separating the galaxies and the galaxy clusters—is expanding.

Although we will use it for the remainder of this chapter merely as a convenient distance-measuring tool, Hubble's law has some fairly obvious and dramatic implications for the universe as a whole. Specifically, if nearly all galaxies show recessional velocity according to Hubble's law, then doesn't that mean that they all started their journey from a single point? If we could run time backwards, wouldn't all the galaxies fly back to this one point, perhaps the site of some explosion in the remote past? In Chapter 28 we will explore some of the implications of the Hubble flow for the past and future evolution of our universe.

Hubble's Constant

We can quantify Hubble's law to make it more useful. The constant of proportionality between recessional velocity and distance is known as **Hubble's constant** and is usually denoted by the symbol H_0. The data of Figure 26.22 obey the following equation:

$$\textit{recessional velocity} = \textit{Hubble's constant} \times \textit{distance}.$$

or

$$V = H_0 \times D$$

The value of Hubble's constant is the slope of the dashed line in Figure 26.22(b). Reading the numbers off the graph, this comes to roughly 50,000 km/s divided by 1000 Mpc, or 50 km/s/Mpc (kilometers per second per megaparsec, the most commonly used unit for H_0).

Astronomers continually strive to refine the accuracy of the Hubble diagram and the resulting estimate of H_0 because Hubble's constant is one of the most fundamental quantities of nature—it specifies the rate of movement of the entire cosmos.

Observational Uncertainties

The value for the Hubble constant just quoted is essentially that obtained in the 1970s by astronomers Allan Sandage and Gustav Tammann, whose distance measurements relied on a chain of standard candles (as described above) extending out to large distances. In the early 1980s, when the Tully-Fisher technique had become fairly well established, Marc Aaronson, John Huchra, and Jeremy Mould used it to obtain a measurement of the Hubble constant that was largely *independent* of the methods used by Sandage and Tammann. (Aaronson, Huchra, and Mould used infrared luminosities to avoid absorption problems caused by our Galaxy's dust.) From observations of galaxies within about 150 Mpc, they deduced a value of $H_0 = 90$ km/s/Mpc, a result definitely inconsistent with that of Sandage and Tammann, given the estimated uncertainties in both sets of measurements. For some reason, distances obtained using the infrared Tully-Fisher method are only about half those determined by using standard candles.

It is presently unknown which of the two values (if either) is correct—no one has yet succeeded in finding a fatal flaw in either of these two approaches. Measurements of the Hubble constant made by other groups of researchers, using different galaxies and alternate distance-measurement techniques, generally also give results within the range of 50–100 km/s/Mpc. Thus, for now, in the absence of any good explanation for the discrepancy, astronomers must simply live with this factor of 2 uncertainty in the expansion rate of the universe. Some like to "split the difference" and just take H_0 = 75 km/s/Mpc. We will adopt this compromise value as the best current value for Hubble's constant in the remainder of the text.

Distance Once Again

One of the most useful aspects of Hubble's law is its potential for determining distances of very distant astronomical objects. Using Hubble's law, we can derive distances to the most remote objects simply by measuring their recessional velocities. Even within the distances accessible by other methods, Hubble's law is often the most convenient means of distance measurement.

Operationally, the method works like this. An astronomer measures the red shift of an object's spectral lines. The extent of the shift is then converted to velocity using the Doppler relationship of Chapter 4. Knowing the object's velocity, the astronomer then finds its distance by using the plot of Figure 26.21(b). Notice, however, that the factor of 2 uncertainty in the Hubble constant translates directly into a factor of 2 uncertainty in the distance determined by this method.

Using Hubble's law in this way tops our inverted pyramid of distance-measurement techniques. Sketched (for the last time) in Figure 26.22, this sixth method simply assumes that Hubble's law holds. If this assumption is correct, Hubble's law enables us to measure great distances in the universe—so long as we can obtain an object's spectrum, we can determine how far away it is.

Many redshifted objects have recessional motions that are a substantial fraction of the speed of light. Some move faster than 50 percent of the velocity of light. The most distant object thus far observed in the universe has the peculiar catalog name of QO051-279. Its extremely high red shift implies a recessional velocity 93 percent that of light. At that speed, ultraviolet spectral lines are shifted all the way into the far infrared! Hubble's law predicts that QO051-279, solely on the basis of its observed red shift, lies more than 4000 Mpc away. This object resides as close to the edge of the observable universe as astronomers have yet been able to probe.

The velocity of light is finite. It takes time for light or any kind of radiation to travel from one point in space to another. The radiation that we now see from distant objects originated long ago. Incredibly, the radiation that astronomers now detect from QO051-279 left that object some 14 billion years ago, well before our planet, our Sun, or even our Galaxy came into being.

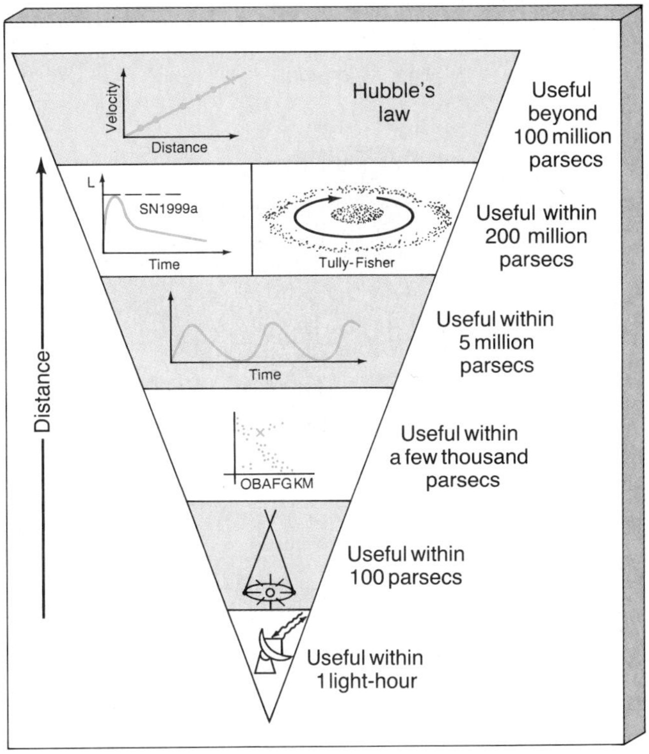

Figure 26.22 *Hubble's law tops the inverted pyramid of distance techniques. This last method is used to find the distances of astronomical objects all the way out to the limits of the observable universe.*

The Great Wall and Gaping Voids

Finally, we can complete our census of the large-scale distribution of galaxies. One of the most extensive surveys of the universe is being carried out by astronomers at the Center for Astrophysics (CfA) at Harvard University. Using Hubble's law as their distance indicator, these researchers are compiling a catalog of the positions and red shifts of all galaxies within about 300 Mpc of the Milky Way. This is an extremely painstaking task—even with a large telescope, detailed spectra of distant galaxies are time-consuming to obtain. Rather than trying to cover the entire sky at once, the team instead has elected to map the universe in a series of wedge-shaped "slices," each 6° (degrees) wide and spanning the northern sky. The first slice, covering a region of the sky containing the Coma Cluster (Figure 26.1), which happens to lie in a direction almost perpendicular to our Galaxy's plane, is shown in Figure 26.23.

The most striking feature of such maps is that the distribution of galaxies on large scales is decidedly nonrandom. The galaxies appear to be arranged in a network of strings, or filaments, surrounding large, relatively empty regions of space known as **voids.** The biggest voids measure some 100 Mpc across. For a time, they were the largest objects in the universe that astronomers knew about. The most likely ex-

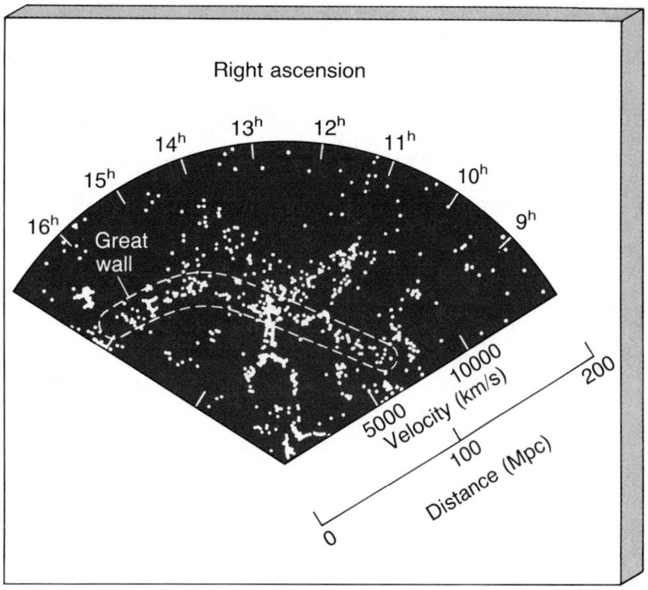

Figure 26.23 *The first slice of a survey of the universe covering 1057 galaxies out to an approximate distance of 300 Mpc. The distances shown assume H_0 = 75 km/s/Mpc. Each dot represents a galaxy. (CFA)*

planation for the filamentary appearance is that the galaxies and galaxy clusters are spread across the surfaces of vast bubbles in space. The voids are the interiors of these gigantic bubbles. The galaxies only seem to be distributed like beads on strings because of the way we view them on the rims of the bubbles. Like suds on soapy water, these bubbles fill the entire universe. The densest clusters and superclusters lie in regions where several bubbles meet.

The notion that the filaments are just the intersection of the survey slice with much larger sheets of galaxies (the bubble surfaces) was confirmed when the second and third slices of the survey, lying above and below the first, were completed. The region of Figure 26.23 indicated by the dashed line was found to continue through both the other slices, so we know that it covers *at least* 36° on the sky perpendicular to the outline in the figure. This extended sheet of galaxies, which has come to be known as the **Great Wall,** measures at least 70 Mpc (out of the page) by 200 Mpc (across the page). For now at least, it is one of the largest known structures in the universe.

Figure 26.24 combines all of the available CfA data with results of other surveys of the southern sky. (The large empty regions are obscured by our Galaxy or are simply not yet mapped.) The Great Wall can be seen arcing around the left side of the figure. It is difficult to avoid the impression that it will join up with the large ''blob'' of galaxies on the right (known as the Perseus-Pisces chain) when all the data are in. If that impression proves to be correct, the true dimensions of the Great Wall are even greater than those quoted above.

Are there still-larger structures in the universe? Only with more extensive surveys will we know for sure. Geller and Huchra are quick to point out that all large-scale surveys performed to date have seen structures comparable in size to the region surveyed—in other words, they have found objects as large as the largest object they could have hoped to find. Observers and theorists alike are eager to find out how much farther this trend continues. (In fact, as we will see in Chapter 28, there is some indication that the CfA survey may actually be very close to the largest scale on which structure exists.)

What might be the origin of this ''sudsy'' or ''frothy'' distribution of galaxies? For a time, theorists tried to explain the voids as gigantic cavities hollowed out by the explosions of extremely massive stars early in the history of the universe. Unfortunately, this explanation does not work. The energy requirements are just too great, and the explosions would have had other observable consequences (mostly related to the radiation they would have released) that simply are not seen. At present, there is no clearcut explanation for the walls and voids seen in this and other surveys. Most theorists believe that all large-scale structure in the universe (on scales larger than a few megaparsecs) is the result of small ''ripples'' (density fluctuations) in the early universe that became unstable and grew with time, eventually forming the large inhomogeneities we see today. In the next section, we will see how some of the smaller fluctuations formed the galaxies. We will return to the subject of structure on larger scales in Chapters 28 and 29, when we study the evolution of the universe as a whole.

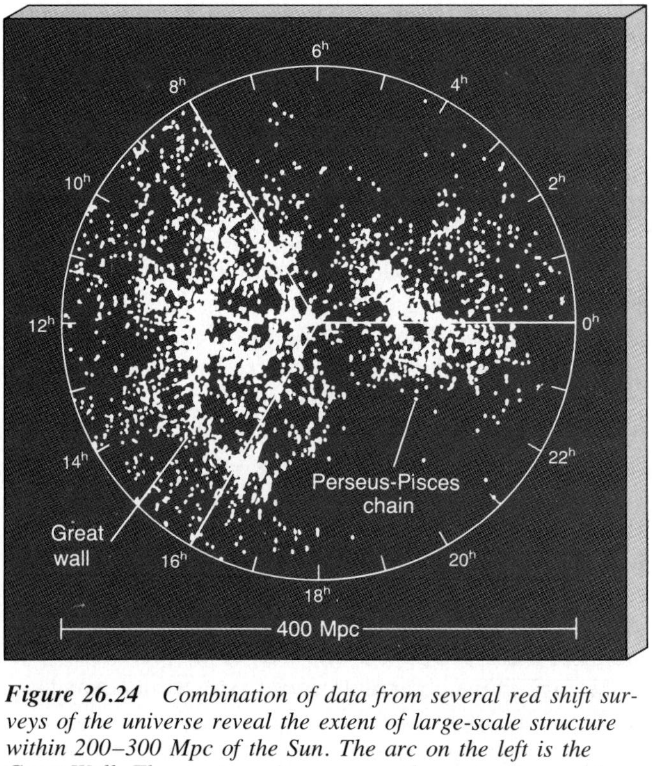

Figure 26.24 *Combination of data from several red shift surveys of the universe reveal the extent of large-scale structure within 200–300 Mpc of the Sun. The arc on the left is the Great Wall. The empty regions are mostly areas obscured by our Galaxy. Positions for more than 4500 galaxies are plotted here. If we assume a Hubble constant of 75 km/s/Mpc, the diameter of the region of space covered by this figure is 400 Mpc.*

GALAXY FORMATION AND EVOLUTION

An HR Diagram for Galaxies?

√7 Having earlier found a convenient classification scheme for stars—namely, the Hertzsprung-Russell diagram discussed in Chapter 19—we naturally wonder if there might be a similar overall pattern, or evolutionary scheme, relating the various types of galaxies. In other words, is there anything that links together the different galaxies in the same manner as the HR diagram relates red stars to blue stars and dwarfs to giants? The answer, as best we can tell, is no. Astronomers have so far failed to discover any underlying physical process that would enable us to relate one type of galaxy to any other.

Some decades ago, not long after the Hubble classification was developed, some astronomers did propose an evolutionary progression of galaxies. (Incidentally, Hubble himself never subscribed to this interpretation of his classification scheme.) The idea was that all galaxies originate with a more-or-less spherical E0 shape. As they grow older, they tend to flatten as a result of their rotation, first becoming more elliptical (E1–E7), then developing spiral arms (Sa), which slowly open (Sb, Sc), prior to their eventually breaking up as aged irregular galaxies.

This scenario has many serious shortcomings. We will mention only two here. First, elliptical galaxies are *not* necessarily elliptical because of rotation. In fact, some (especially the larger ones) have only a very slight overall rotation—their elliptical shape is actually a consequence of the flattened orbits of their stars. Second, this evolutionary hypothesis requires that all ellipticals be young and all irregulars old. But this is not the case at all. Observationally, elliptical galaxies are *old,* nearly depleted of interstellar gas and dust and displaying little or no evidence of current star formation. There is no way they could predate the gas-rich, star-forming irregulars.

We might instead argue that because ellipticals are clearly old galaxies, the evolutionary scheme progresses in the opposite direction. Maybe irregulars are young and, once formed, gradually evolve into ellipticals. Perhaps it is easier to imagine loose spiral galaxies wrapping up into tighter spirals and eventually into elliptical galaxies. But there are problems here, too.

First, we are hard pressed to understand how the beautiful spiral galaxies might have grown from the distorted irregulars. Second, we cannot reconcile this idea with the abundance of old (halo and globular cluster) stars in the irregular and loose spiral galaxies. Many of those stars are known to lack heavy elements, proving their old age. Simply put: If irregular galaxies and loose spirals are the starting point in any scheme of galaxy evolution, all of them should be young. But they are not. Virtually all irregulars and spirals house a mixture of old and new stars.

The bottom line is that normal galaxies probably do not evolve from one type to another. Spirals do not seem to be ellipticals with arms, nor do ellipticals appear to be spirals without star-forming disks. In short, astronomers know of no parent-child relationship among normal galaxies. The various types of galaxies are more like cousins who trace their birth to the same ancestor—the galaxy formation process.

Galaxy Formation

Many millions of galaxies can be found in the night sky, and they have been the object of careful study by astronomers for well over half a century. Having seen in this chapter a little of the galaxies' overall properties and distribution, we might reasonably wonder where they come from. Do we understand galaxy formation as well as, say, star formation? The answer is a resounding *no!* The theory of galaxy formation is still very much in its infancy. It is presently unable to answer completely even a basic question like why spirals and ellipticals exist at all, let alone offer an explanation of the Hubble sequence or predict the different evolutionary tracks that galaxies might take.

There are several good reasons for this lack of understanding of galaxy formation: Galaxies are much more complex than stars, they are harder to observe, and the observations are harder to interpret. We have only a partial understanding, and no observations, of the conditions in the universe immediately preceding galaxy formation, quite unlike the corresponding situation for stars (see Chapters 20 and 21). Finally, although stars evolve in isolation, galaxies may suffer many collisions and mergers during their lives, making it much harder to decipher their pasts. Despite all these difficulties, some general ideas have begun to gain widespread acceptance, and we can offer a few insights into the processes whereby the galaxies we see came into being.

The seeds of galaxy formation were sown in the very early universe, when small density fluctuations in the primordial matter began to grow. We will study this process in more detail in Chapter 29. For now, let us simply begin our discussion with these "pregalactic" blobs of gas already formed. The masses of these fragments were quite small—perhaps only a few million solar masses—comparable, in fact, to the masses of the smallest present-day dwarf galaxies, which may in fact be remnants of this early time. The key point in our current understanding is the realization that although stars form when a large cloud *fragments* into many smaller pieces, large galaxies grow by repeated *merging* of smaller objects, as illustrated in Figure 26.25. Strong support for this picture comes from recent observations by Lennox Cowie of the University of Hawaii and others, which indicate that, at large distances (well over 1000 Mpc away, or 4 billion years ago), galaxies appear distinctly smaller and more irregular than those found nearby. Furthermore, these galaxies show definite signs of recent and ongoing merging. Figure 26.26 shows one of Cowie's images. The rather vague patches are separate small galaxies, each containing only a few percent of the mass of the Milky Way.

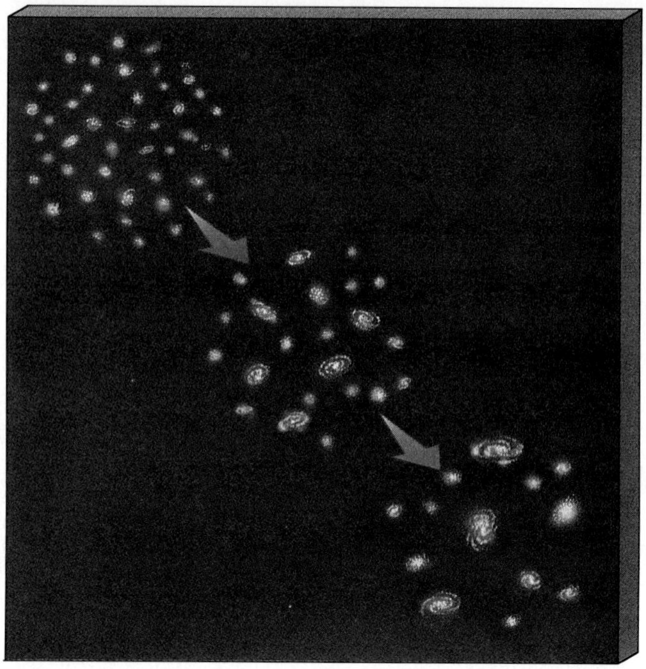

Figure 26.25 *The present view of the formation of galaxies holds that large systems were built up from smaller ones through collisions and merges.*

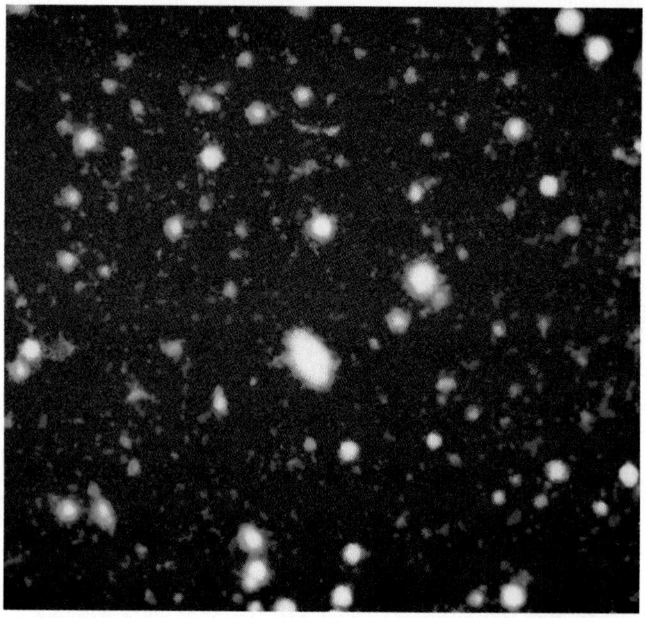

R I V U X G

Figure 26.26 *Numerous small, irregularly-shaped young galaxies can be seen in this very deep (that is, long-exposure) optical image. Redshift measurements indicate that the galaxies lie well over 1000 Mpc from Earth. Their size and appearance support the theory that galaxies grew by merger and were smaller and less regular in the past. (L. Cowie)*

Additional theoretical evidence is provided by numerical simualtions of the early universe, which clearly show this agglomeration process in action.

Given that galaxies form by repeated mergers, how might we account for the differences between spirals and ellipticals? The answer is unclear. One apparently very important factor is just when and where stars first appeared—in the original blobs, during the merger process, or later—and how much gas was used up, or ejected from the young galaxy, in the process. If many stars formed early on and little gas was left over, an elliptical galaxy would be a likely outcome, with many old stars on random orbits and no gas to form a central disk. Alternatively, if a lot of gas remained, it would tend to sink to a central plane and form a rotating disk—in other words, a spiral galaxy would result. However, it is not known what determines the time, the place, or the rate of star formation, so whether spirals and ellipticals can form in basically the same environment, or if they tend to form in different places, is still an open question. For example, it is known that spirals are relatively rare in regions of high galaxy density. Is this because they tended not to form there, or is it because their disks are so fragile that they are easily destroyed by collisions and mergers, which are more common in dense regions? For now, we just don't know.

Mergers and Acquisitions

Another important question is, When did the formation process *stop?* Astronomers are divided on this issue. Some maintain that there was a fairly well-defined time in the past—given by the age of the globular clusters, for example—by which most formation was over. Others point out that many galaxies show evidence for repeated mergers and the accumulation of smaller satellite galaxies over an extended period of time—even up to the present. These astronomers suggest that the many galaxy interactions we observe today (see for example Interlude 26-2) are really just part of the same process begun when the first fragments merged. In the latter view, galaxy formation is still occurring today. In either case, astronomers have ample evidence that galaxies evolve in response to external environmental factors, even long after they first formed.

We now know that spiral galaxies have huge, invisible coronae surrounding them, and we strongly suspect that *all* galaxies have similar dark halos. Consider two galaxies orbiting one another—a binary galaxy. As they orbit, the galaxies interact with each other's halos, one galaxy stripping halo material from the other by tidal forces. The freed matter is either redistributed within a common envelope or is entirely lost from the binary system. This interaction between the halos changes the orbits of the galaxies themselves. They tend to spiral toward one another, eventually coalescing in a **galaxy merger.** If one galaxy of the pair happens to have a lower mass than the other, the process is colloquially termed **galactic cannibalism.** Such cannibalism might explain why supermassive galaxies are often found at the cores of rich galaxy clusters. Having dined on their companions, they now lie at the center of the cluster, waiting for more food to arrive. Figure 26.27 is an astonishing combination of images that has apparently captured this process at work.

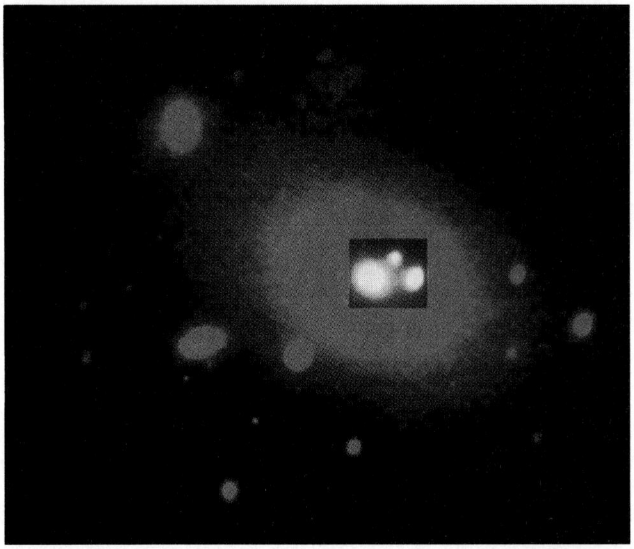

R I V U X G

Figure 26.27 This composite (computer-enhanced and false colored) optical photograph of the galaxy cluster known as A2199 is thought to show an example of galactic cannibalism. The central galaxy of the cluster is displayed with a superposed "window." (This results from a shorter time exposure, which shows only the brightest objects that fall within the frame.) Within the core of the cluster are several smaller galaxies (the three bright yellow images at center) apparently already "eaten" and now being "digested" (that is, being torn apart and becoming part of the larger galaxy). Other small galaxies swarm on the outskirts of the swelling galaxy, almost certainly to be eaten too. (Harvard-Smithsonian Center for Astrophysics)

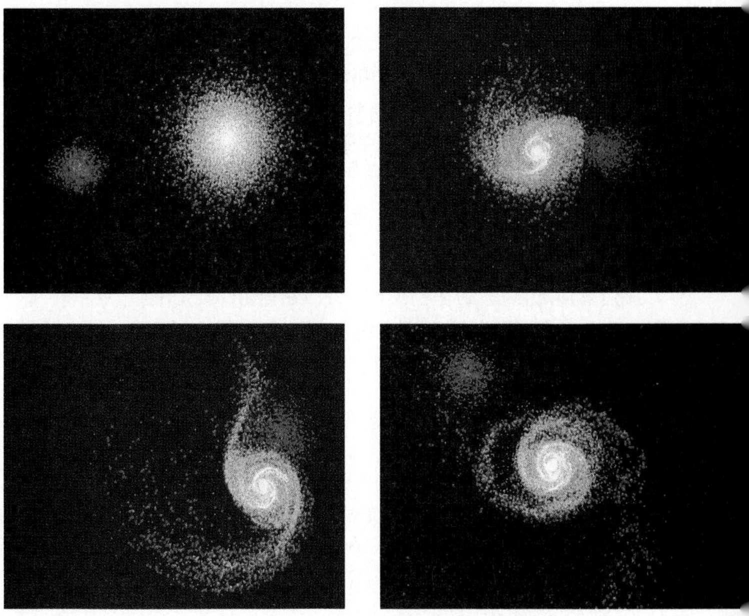

Figure 26.28 Galaxies might change their shapes long after their formation. In this computer-generated reenactment, two galaxies closely interact over several hundred million years left to right and top to bottom. The smaller galaxy in red has gravitationally disrupted the larger galaxy in blue, changing it into a spiral galaxy. Compare the result of this supercomputer simulation with a photograph (Figure 26.5) of M51 and its small companion. (J. Barnes/L. Hernquist)

Now consider two interacting disk galaxies, one a little smaller than the other but each having a mass comparable to our Milky Way. As shown in the computer-generated frames of Figure 26.28, the smaller galaxy can substantially distort the larger one, causing spiral arms to appear where none existed before. The entire event requires several hundred million years—a span of evolution that supercomputers can model in minutes.

The final frame of Figure 26.28 looks remarkably similar to the double galaxy shown in Figure 26.5(a). Shown there are two galaxies with sizes, shapes, and velocities corresponding very closely to those in the computer simulation. The magnificent spiral galaxy is M51, popularly known as the Whirlpool Galaxy. Its smaller companion is an irregular galaxy that may have drifted past M51 millions of years ago. Did this smaller galaxy cause the spiral structure we see in M51? Does the model mirror reality? Perhaps. We need more evidence from other galaxies to confirm the accuracy of these and similar simulations. Still, the computer rendition does demonstrate a plausible way that two galaxies might have interacted millions of years ago and how spiral arms might be created or enhanced as a result.

The M81-M82 system is another example of a galactic interaction that probably rearranged much matter in at least one of the two galaxies. Shown in Figure 26.29, the two galaxies seem separated by several hundred thousand

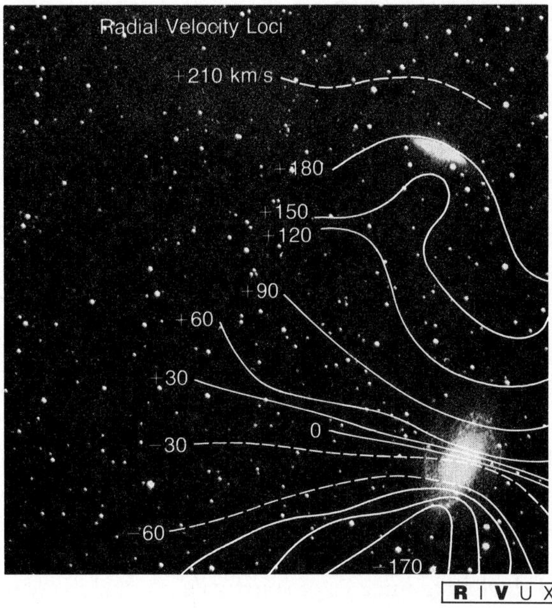

R I V U X G

Figure 26.29 An optical photograph of the M81-M82 system, approximately 3 Mpc away, overlaid by a radio map of the system showing that M81 (bottom) and M82 (top) are wrapped within a common envelope of invisible hydrogen gas. (The numbers on the contours are the relative velocities of the connecting gas in kilometers per second.) (NOAO; NRAO) parsecs of space. But note the overlaid radio contours of neutral hydrogen. The gas seems to link the two galaxies. Their relative motions, along with the motion of the intervening gas, suggest that M81 swept past the smaller M82 about 200 million years ago and severely affected it. Strong gravitational tides doubtless rearranged M82's internal structure, triggering bursts of new stars and generally causing violent activity in M82's central regions.

These systems support the notion that some of the most spectacular changes that occur in galaxies result from interactions with other galaxies. Astronomers know of numerous **starburst galaxies,** where violent events, possibly a near-collision with a neighbor, appear to have caused a sudden, intense burst of star formation in the recent past. Such close encounters are random events and do not seem to represent any genuine evolutionary sequence linking all spirals to all ellipticals and irregulars. However, it *is* clear that many galaxies and galaxy clusters have evolved greatly since they first formed long ago.

CHAPTER SUMMARY

Galaxies can be classified into four basic types—spirals, barred spirals, ellipticals, and irregulars. Spiral galaxies can exhibit a variety of shapes, but the four basic regions of our Galaxy—the disk, the bulge, the halo, and the corona—are found in all spirals. The barred spirals differ from ordinary spirals mainly by the presence of an elongated "bar" of stellar and interstellar matter passing through the center and extending beyond the bulge into the disk. Elliptical galaxies range in shape from highly elongated to nearly circular in appearance. There is no evidence for any young stars or any recent star formation in most ellipticals. Irregular galaxies are so named largely because their visual appearance does not allow them to be placed into any of the other three categories. Most galaxies in the universe fall into the categories of dwarf elliptical or dwarf irregular.

Four techniques—radar, stellar parallax, comparing the absolute and apparent brightnesses of main-sequence stars, and the use of variable stars—enable astronomers to measure distances out to about 5 Mpc. The fifth rung in our cosmic distance ladder is shared by a variety of standard candles and an independent distance-measurement technique known as the Tully-Fisher method. Standard candles are objects that are easily recognized and have a fairly well-known luminosity, which allows their distance to be estimated. The Tully-Fisher method is based on a correlation between the rotational velocity and the total luminosity of spiral galaxies within a few tens of megaparsecs of the Milky Way. Measuring the broadening of galactic spectral lines yields a galaxy's rotational velocity, and this in turn gives its luminosity.

Some 20 galaxies populate our Galaxy's neighborhood, including three large spiral galaxies and many dwarf irregular and dwarf elliptical systems. The galaxy cluster that includes our Milky Way is known as the Local Group. The nearest large galaxy cluster, the Virgo Cluster, houses approximately 2500 galaxies in a tight-knit group about 3 Mpc across. Galaxy clusters come in many shapes and sizes. The largest, "rich" clusters contain many thousands of individual galaxies distributed fairly smoothly in space. The smallest clusters, like the Local Group, contain only a few galaxies and are quite irregular in shape. Galaxy clusters themselves are clustered, forming titanic agglomerations of matter known as superclusters. On even larger scales, galaxies seem to lie in great sheets or filaments, leaving vast regions devoid of luminous matter. An extended sheet of galaxies is known as the Great Wall.

Two techniques for measuring the masses of individual galaxies employ Kepler's Third Law. Many galaxies have been found to have masses comparable to that of our own Milky Way. We can also ask how massive a galaxy cluster must be in order to bind its galaxies gravitationally, thereby obtaining an estimate of the cluster's mass. This method, known as the Virial method, gives no information about the masses of individual galaxies. Rotation curves of spiral galaxies and the movements of galaxies within clusters suggest that upward of 90 percent of the universe is composed of dark matter. X-ray observations have demonstrated the existence of large amounts of hot gas within clusters of galaxies, but not nearly enough to account for the unseen material.

On the largest scales, unclustered galaxies and galaxy clusters alike display very ordered motion. With the exception of only a few nearby systems, every known galaxy is part of a general expansion away from us in all directions. The farther an object is from us, the faster it recedes. This correlation between velocity and distance is called Hubble's law. The constant of proportionality between recessional velocity and distance is known as Hubble's constant. Hubble's constant is one of the most fundamental quantities of nature because it specifies the rate expansion of the entire cosmos. Hubble's law allows us to derive distances to the remotest objects simply by measuring their recessional velocities. However, the Tully-Fisher relation yields distances to galaxies that disagree somewhat with distances obtained using standard candles, so there is some uncertainty in the value of the Hubble constant.

KEY WORDS

barred-spiral galaxy
dwarf elliptical
elliptical galaxy
galactic cannibalism
galaxy cluster
galaxy merger
giant elliptical
Great Wall
Hubble classification scheme
Hubble flow
Hubble's constant
Hubble's law
intergalactic space
irregular galaxy
Local Group
Magellanic Clouds
S0 galaxy
SB0 galaxy
spiral galaxy
standard candle
starburst galaxy
supercluster
Tully-Fisher relation
Virial method
void

REVIEW QUESTIONS

1. When and how did galaxies come to be classified into four basic types?

2. In what sense are elliptical galaxies ''all halo''?

3. Briefly review the four techniques used to measure distances out to about 5 Mpc from our Sun.

4. Describe the contents of the Local Group. How much space does it occupy in contrast to the Milky Way?

5. What is a standard candle, and why is it useful? Give some examples of standard candles.

6. What is the Tully-Fisher relation, and how is it used?

7. What lies in the space between galaxy clusters?

8. What is the Virgo Cluster?

9. Describe two techniques for measuring the mass of a galaxy that involve Kepler's Third Law.

10. Why do astronomers believe that galaxies in clusters are associated with more mass than we can see?

11. What is Hubble's law?

12. Explain how Hubble's law is used to measure distances to galaxies.

13. Why is the value of the Hubble constant uncertain?

14. In what way does the distribution of galaxies resemble soap bubbles?

15. Describe the role of collisions in the formation and evolution of galaxies.

16. A certain telescope could just see the Sun at a distance of 10 kpc. What is the maximum distance at which it could detect a supernova with a luminosity 1 billion times that of the Sun?

17. According to Hubble's law, with $H_0 = 75$ km/s/Mpc, what is the recessional velocity of a galaxy at a distance of 200 Mpc? How far away is a galaxy whose recessional velocity is 4000 km/s? How do these answers change if $H_0 = 50$ km/s/Mpc instead?

18. Two galaxies are orbiting one another at a separation of 500 kpc, and the orbit period is (estimated to be) 30 billion years. Use Kepler's Law (as stated in Chapter 25) to find the total mass of the pair.

DISCUSSION QUESTIONS

1. One of the first steps toward understanding the galaxies was to classify them according to appearance. Why was this step so important?

2. Briefly describe what your ''personal cosmology''—your own conception of the universe—was before taking this class. How has that picture changed, in light of what you now know?

SUGGESTED READINGS

Burbidge, G. ''The Cult of the Missing Mass.'' *Sky and Telescope* (June 1990). An outspoken extragalactic astronomer presents an alternative view.

Freeman, M. ''Galaxies.'' *Smithsonian* (January 1989). Popular article about the variety of galaxies.

Geller, M. J., and J. P. Huchra. ''Mapping the Universe.'' *Sky and Telescope* (August 1991). Researchers explain how they probe the three-dimensional structure of the universe.

Hodge, P. W. *Galaxies*. Cambridge, MA: Harvard University Press, 1986. Good basic information about most of the topics discussed in this chapter.

Shapley, H. *Galaxies*. Cambridge, MA: Harvard University Press, 1943, 1961, 1972. Engaging, readable, and classic book by one of the world's foremost galaxy pioneers.

CHAPTER

27

Active Galaxies and Quasars

Limits of the Observable Universe

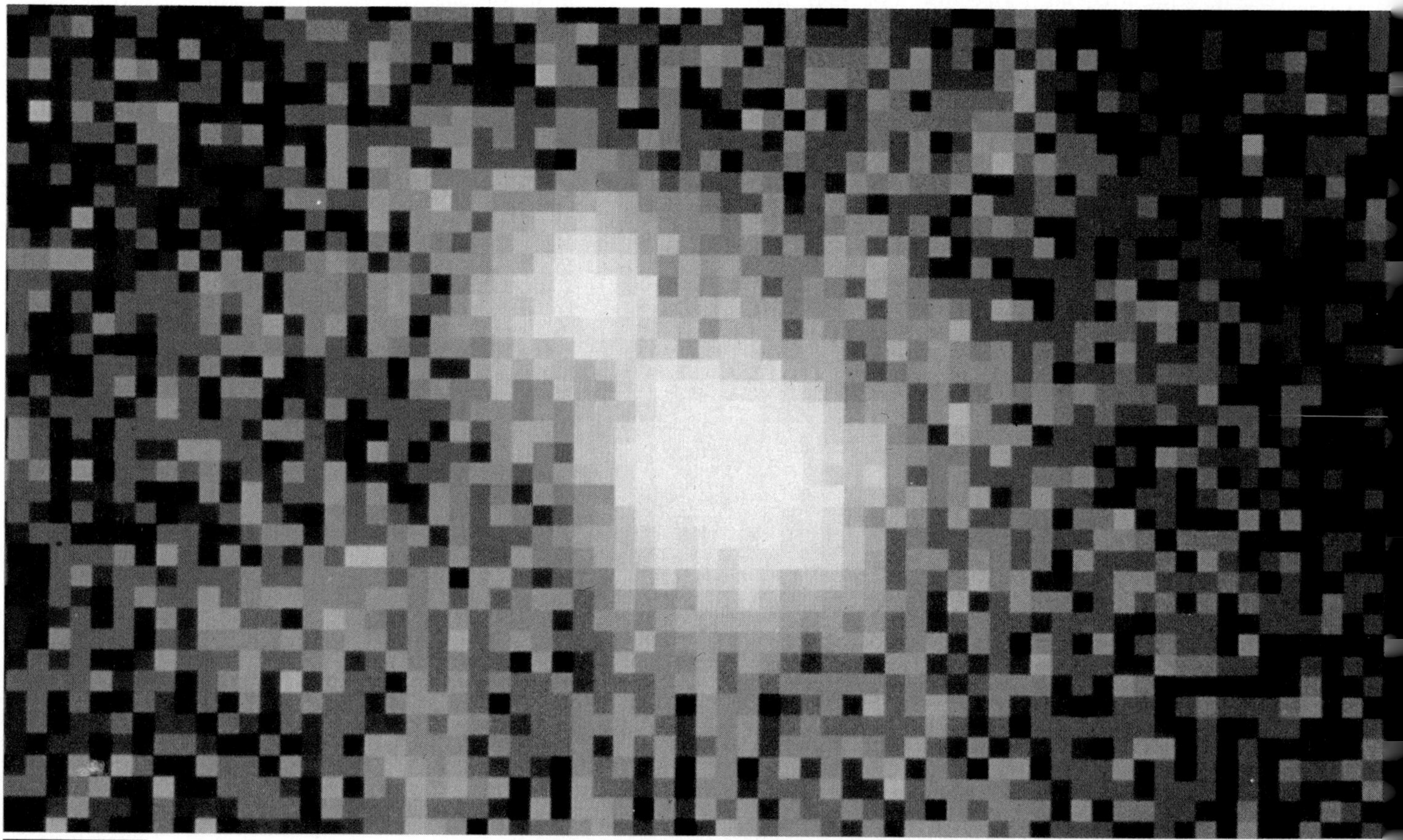

R I **V** U X G

Quasars remain among the most mysterious objects in the known universe. Through even the best ground-based telescopes, these powerful celestial denizens show hardly more than a point of light smeared over an interval of about 1 arc second. Even to orbiting spacecraft above our atmosphere, they seem point-like. In this image taken by the spaceborne Hubble *telescope, we see a magnified view of a typical quasar with the name 1208+101. This image fascinatingly suggests that the quasar actually consists of two images—two images of the same object, much like a mirage. Astronomers believe, but have not yet proved, that a gravitational lens is producing the multiple images of the one object. Only further observations will confirm if this is really a picture of a single quasar imaged twice or a chance superposition of a star in our own galaxy and a distant, unrelated quasar. (Courtesy NASA)*

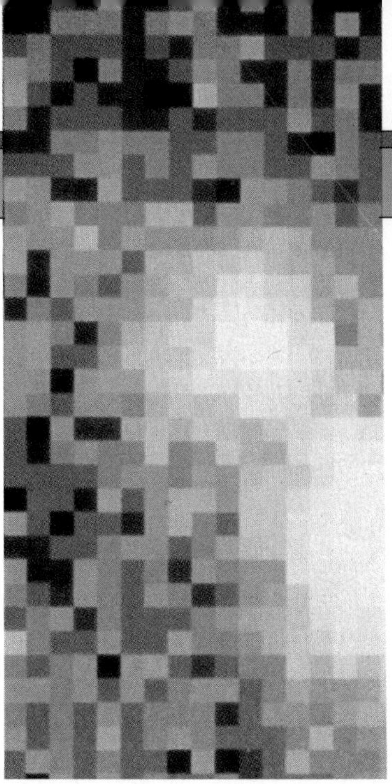

LEARNING GOALS

1 to appreciate the basic differences between active and normal galaxies
2 to realize that active galaxies emit tremendous amount of energy
3 to recognize the important features of Seyfert and radio galaxies
4 to see what drives the central engine that powers all active objects
5 to recognize that the powerful, distant quasars may simply be an early stage of galaxy evolution

In Chapter 25 we gained an appreciation of our location within our own Milky Way Galaxy. In Chapter 26 we expanded our view, to reveal our place within the larger universe. We found that our Galaxy is just one normal galaxy among many others that together form the small cluster known as the Local Group. Beyond, other galaxy clusters abound, each a gravitationally bound unit, separated from neighboring clusters by vast tracts of intergalactic space. Galaxy clusters themselves are grouped into superclusters. On still larger scales, we find great sheets of galaxies, separated by enormous empty voids. Our journey from the Milky Way to the Great Wall in the past two chapters has widened our cosmic field of view by a factor of 10,000, yet the galaxies that make up the structures we see show remarkable consistency in their properties.

The overwhelming majority of galaxies fit neatly into the Hubble Classification Scheme, showing few, if any, ''unusual'' characteristics. However, sprinkled through the mix of normal galaxies, even fairly close to the Milky Way, are galaxies that are decidedly abnormal in their properties. Although their optical appearances are often quite ordinary, these abnormal galaxies emit huge amounts of energy—far more than a normal galaxy—mostly in the invisible part of the electromagnetic spectrum. Observing such objects at great distances, we may be seeing some of the formative stages of our own galactic home.

BEYOND THE LOCAL REALM

✓1 Astronomers have recognized and cataloged spiral and elliptical galaxies as far away as several hundred megaparsecs from Earth. Beyond this distance, normal galaxies appear so faint that it is difficult to discern their characteristic shapes, so their types are largely unknown. Nevertheless, according to their observed red shifts (and Hubble's law), we know that many galaxies lie well beyond this distance, in the outer reaches of the observable universe. But what kinds of objects are they? Are they spiral, elliptical, and irregular galaxies—close relatives of the material conglomerations that populate the local universe—or are they somehow different? The answer is that they often seem to be *different* from the galaxies found in our cosmic backyard.

By and large, very distant objects are more active—more violent—than those found closer to home. Their absolute brightnesses (that is, their luminosities) can be much greater than those of normal spiral and elliptical galaxies in our vicinity. The most energetic objects, which can emit hundreds or thousands of times more energy per second than the entire Milky Way, are known collectively as **active galaxies.** Although they are certainly more common at greater distances, some active galaxies are also found locally, scattered among the normal galaxies that make up most of our cosmic neighborhood. Not all active galaxies are distant, nor are all distant galaxies active—many faraway ''normal-luminosity'' galaxies are known. As a general rule, however, the most active objects are found at the largest red shifts and so lie the farthest from Earth.

We might wonder if this predominance of energetic objects at large red shifts is just an observational effect, resulting from our inability to detect relatively faint normal galaxies at great distances. After all, the apparent brightness of any astronomical object decreases as the square of its distance from us, so even with the very best telescopes we would expect to preferentially observe the more energetic and powerful galaxies in remote regions of space. Although this observational bias does play a role, it turns out that the weakness of signals from faraway normal galaxies can only partly explain the apparent predominance of energetic objects at great distances. We are led to conclude that bright active galaxies really are more common at large redshifts—that is, long ago.

In addition to their great brightness, there is also something basically different about active galaxies. It seems that their *radiative character* differs fundamentally from that of normal galaxies. A normal galaxy's radiation extends from the radio to the gamma-ray part of the electromagnetic spectrum, in much the same way as radiation from ordinary stars does. In fact, to a large extent the light we see from a normal galaxy *is* just the accumulated light of its many component stars. Most of a normal galaxy's radiated energy is visible because the Planck curves for most stars peak in the visible part of the spectrum. For example, our entire Milky Way has a luminosity of about 10^{44} erg/s at optical frequencies—20 billion Suns' worth of radiation—but only 10^{38} erg/s at radio frequencies, which is a million times less. By contrast, the radiation observed from active galaxies does not peak at optical frequencies—far more energy is emitted at longer wavelengths than in the visible range. The emission of radiation from active galaxies is *inconsistent* with what the combined emission of myriad stars would be: The radiation is *nonstellar*. Figure 27.1 shows schematically the differences in the energy emission of active and normal galaxies.

Physical conditions were undoubtedly different at earlier times than they are now. Perhaps, then, we should not be surprised that remote astronomical objects, which emitted long ago the radiation we observe today, differ from nearby objects, which emitted their radiation much more recently. What *is* surprising—in fact astounding—is the *amount* of energy radiated from some of the most luminous objects. Their total energy production stretches scientific theory to its limits, as we will see. The abnormal power and nonstellar character of the light from active galaxies suggest to astronomers that the universe was once a much more violent place than it is today.

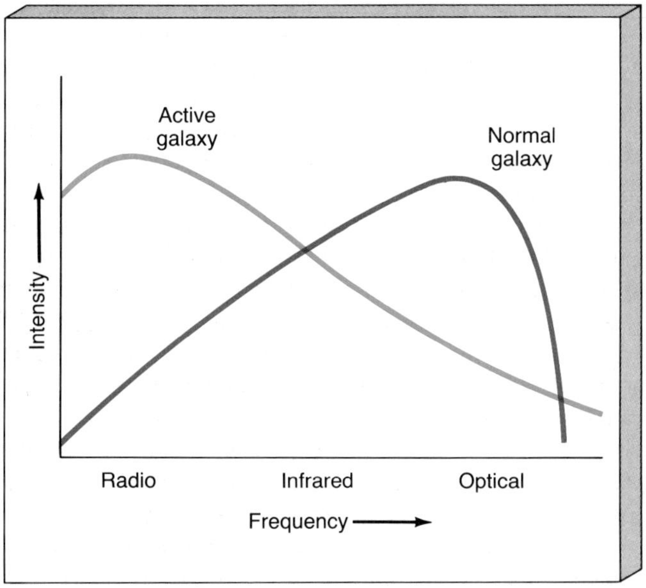

Figure 27.1 *The nature of the energy emitted from a normal galaxy is different from that of an active galaxy. (This plot illustrates the general run of intensity for all galaxies of a particular type and does not represent any one individual galaxy.)*

COSMIC VIOLENCE

✓2 We consider in this section the energies of several well-known sources of radiation. The most powerful of these will serve as benchmarks against which we can judge active galaxies. In addition to luminosity, we will also consider the *total* energy output of these sources, which is perhaps the best measure of their overall activity. Because luminosity measures the amount of energy radiated per unit time, the total energy emitted by an object is just the product of an object's luminosity and its lifetime:

$$\textit{total energy} = \textit{luminosity} \times \textit{lifetime}.$$

We can use this equation to calculate the net emission of radiation for any object at all wavelengths.

To try to place in perspective the energies of normal and active galaxies, let's consider first the most violent of terrestrial events—the explosion of a one-megaton nuclear bomb, equivalent to 1 million tons of TNT. A one-megaton nuclear blast produces about 10^{28} erg/s. Of course, the blast is an instantaneous event. It lasts no more than a microsecond. Despite the large luminosity of a nuclear event at the moment of detonation, the total amount of energy released is puny by cosmic standards. The total energy released by a nuclear bomb is 10^{28} erg/s $\times$ 10^{-6} s $= 10^{22}$ erg.

Solar flares represent larger-scale energy production, although they too are nearly insignificant compared with active galaxies. The largest flare produces about 10^{30} erg/s and lasts a few minutes, making its total energy output about 10^{32} erg.

The entire Sun's luminosity is even larger, about 4×10^{33} erg/s—the equivalent, at any instant, of about 100 billion nuclear bombs. Because our Sun will burn for billions of years, the total amount of energy it will release is very large—approximately 10^{51} erg. In fact, we can establish a rule of thumb concerning luminosity: 10^{51} erg is about the total amount of energy released by *any* star during its main-sequence lifetime. The O- and B-type high-mass stars, which can have luminosities as high as 10^{38} erg/s, live for such short times that they still produce "only" about 10^{51} erg in total. Conversely, the low-mass M-type stars have only a fraction of the Sun's luminosity, but they live much longer, thereby yielding a similar total energy. The total energy released by a supernova explosion in the form of electromagnetic radiation and high-speed gas is also about 10^{51} erg.

An even greater source of energy is the central region of our Milky Way Galaxy, as well as the centers of other normal galaxies. These are regions of great star density, with many more stars per unit volume than in the outer spiral arms (look at the spiral galaxies pictured in Chapter 26 to confirm this). The central regions of normal galaxies have typical luminosities of about 10^{40} erg/s. Sustained over the lifetime of the galaxy, this amounts to a total energy release of about 10^{57} erg.

The entire Milky Way Galaxy has an even larger luminosity—hardly surprising, as it contains a huge number of stars. The accumulated galactic luminosity is a little less than 10^{44} erg/s. Our location within the Galaxy makes it hard for us to measure its total luminosity accurately, but the Milky Way has been studied carefully in some parts of the electromagnetic spectrum (radio and optical) and at least surveyed in all other parts (infrared, x-ray, ultraviolet, and gamma-ray). It is unlikely that additional large sources of energy lurk in the darkness.

The value of 10^{44} erg/s is generally considered a typical luminosity for a normal spiral or elliptical galaxy. However, we know of no dead galaxies, so we have little idea of galaxy lifetimes. Still, bearing in mind that our own Milky Way is 15 billion years old, we can reason as follows: A normal galaxy emitting at an average rate of 10^{44} erg/s for, say, several tens of billions of years would release a total energy of about 10^{62} erg. We can extend our earlier rule of thumb: 10^{51} erg makes a normal star, and 10^{62} erg—100 billion times more energy—makes a normal galaxy. The most luminous normal galaxies (the giant ellipticals) emit perhaps ten times more energy than this.

Table 27-1 summarizes the levels of activity of the various objects just discussed and includes also the types of active objects—Seyfert galaxies, radio galaxies, and quasars—to be studied in this chapter. Generally speaking, active galaxies start near the top of the galactic luminosity scale and extend several orders of magnitude (that is, powers of ten) beyond it.

TABLE 27-1 ***Energetics of Various Objects***

Object	Luminosity (erg/s)	Total Lifetime Energy Output (erg)
Nuclear bomb	10^{28}	10^{22}
Solar flare	10^{30}	10^{32}
Star	10^{33}	10^{51}
Supernova	10^{44}	10^{51}
Galactic center region	10^{40}	10^{57}
Milky Way Galaxy	10^{44}	10^{62}
Most energetic normal galaxy	10^{45}	10^{63}
Active galaxies		
Seyfert galaxy	10^{43}–10^{45}	Lifetimes currently unknown
Radio galaxy	10^{43}–10^{45}	
Quasar	10^{45}–10^{49}	

The last decade has seen a great leap forward in our understanding of these highly luminous objects, although substantial gaps in our knowledge remain. As mentioned earlier, some active galaxies lie close enough for astronomers to make detailed observations of them. In the next few sections we will discuss some of the key properties of these "neighboring" active galaxies and develop the leading current theory of their power source. If we can understand the nearby systems, then perhaps we can extend our knowledge to understand the truly remote objects residing near the limits of the observable universe.

SEYFERT GALAXIES

✓3 In 1943 Carl Seyfert, an American optical astronomer studying spiral galaxies from Mount Wilson Observatory, discovered the type of active galaxy that now bears his name. **Seyfert galaxies,** or simply *Seyferts,* are a class of astronomical objects whose properties lie between those of normal galaxies like the Milky Way and those of the most violent active galaxies known. The spectral lines of Seyfert galaxies are usually substantially redshifted, which implies that they lie far away. Most Seyferts seem to reside at large distances (hundreds of megaparsecs) from us, although a few are as close as 20 or 30 Mpc.

Figure 27.2 shows a series of optical photographs of a typical Seyfert galaxy. A short exposure shows just a fuzzy round patch of light. No more than a few parsecs in diameter, this faint blob is really the brilliant center of a much larger object. Moderate time exposures begin to show some evidence for a larger object—or at least faint spiral wisps of visible radiation, irregularly spread beyond the central blur of light. With long exposures, a complete spiral galaxy can be seen, its bright center now overexposed and showing little sign of anything unusual. Notice how different-length exposures highlight different features of an object. The range of brightness, from the center to the spiral wisps, is so great that it is very difficult to capture them on a single photographic image. Nowadays, computer-processed CCD (charge-coupled device) images can get around this practical problem, but the technique of showing different "exposures" of a single object is still widely used.

A casual glance at a long-exposure photograph of a Seyfert (such as Figure 27.2[c]) reveals nothing strange. Superficially Seyferts resemble normal spiral galaxies. However, closer study of Seyferts reveals some peculiar physical properties not found in normal spirals.

First, Seyfert spectral lines bear little or no resemblance to those produced by ordinary stars, although they do have many similarities to the spectral lines observed toward the center of our own Galaxy. Seyfert spectra contain strong

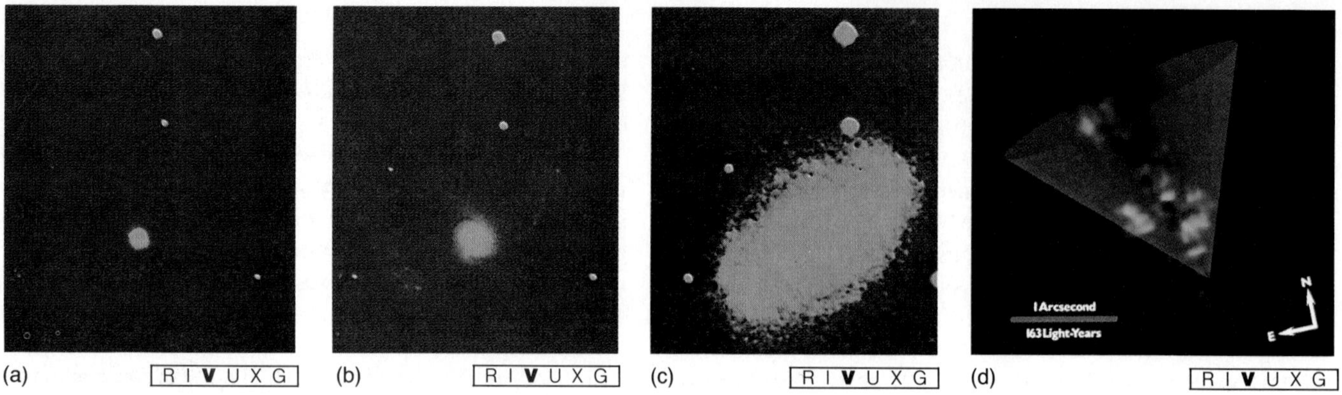

Figure 27.2 *Photographs of a Seyfert galaxy (NGC 4151) after (a) a short exposure, (b) a moderate exposure, and (c) a long exposure. (d) Recent image of Seyfert NGC 1068 made with the* Hubble *telescope shows a clutch of glowing blobs near the galaxy's core, perhaps illuminated by radiation (drawn in by an artist) arising from the accretion disk of a black hole. Only the bright central nucleus can be seen in (a), while in (c) the full normal-looking spiral galaxy is visible. (AURA; NASA).*

emission lines of highly ionized heavy elements, especially iron. The lines are very wide, indicating either that the galaxy's gases are tremendously hot (more than 10^8 K) or that they are rotating very rapidly (at about 1000 km/s) around some central object. The first possibility can be ruled out, since such a high temperature would cause all the gas to be ionized, so that no spectral lines would be produced. Thus, the broadening indicates rapid internal motion in the nucleus.

Second, maps of Seyfert energy emission show that nearly all of the radiation stems from a small central region known as the **galactic nucleus**—the bright region visible in short-exposure images like Figure 27.2(a). Astronomers suspect that the nucleus of a Seyfert may be quite similar in origin to the center of a normal galaxy, such as the Milky Way or Andromeda, but with one very important difference. The nucleus of a Seyfert is 10,000 times brighter than the center of our Galaxy. The brightest Seyfert nuclei are 10 times more energetic than the *entire* Milky Way Galaxy.

Third, Seyfert galaxies emit radiation in two broad ranges. They emit about the same amount of visible radiation as normal spiral galaxies. This emission stems from the stars in the Seyfert's galactic disk and spiral arms. But the nuclei of Seyferts emit most of their energy as low-frequency radio and infrared radiation. This radiation cannot be explained as coming from stars—it is nonstellar in origin.

Finally, extensive monitoring of Seyfert radiation over long periods of time has shown that the energy emission often varies over time. Figure 27.3 shows an example of luminosity variations for a typical Seyfert. These radiative changes are unlike anything found in the Milky Way or any other normal galaxy. A Seyfert's luminosity can double or halve within a fraction of a year.

These rather rapid fluctuations lead us to conclude that the source of energy emissions must be quite compact. As discussed in Chapter 24 (in the context of luminosity variations in neutron stars and black holes), for astronomers to be able to detect a coherent variation in brightness within a certain time interval, the source of radiation must be smaller in size than the distance traveled by light in that interval. Otherwise, the intensity variations would be blurred, not sharp, as observed. An object cannot "flicker" in less time than radiation takes to cross it. Because the rise and fall of a Seyfert's radiation usually occurs within a year, we can confidently conclude that the emitting region must be less than a light-year across.

High-resolution maps made with radio interferometers support this argument by proving that the variable, strongly emitting region resides within the very center of each Seyfert. These observations confirm that the nucleus is

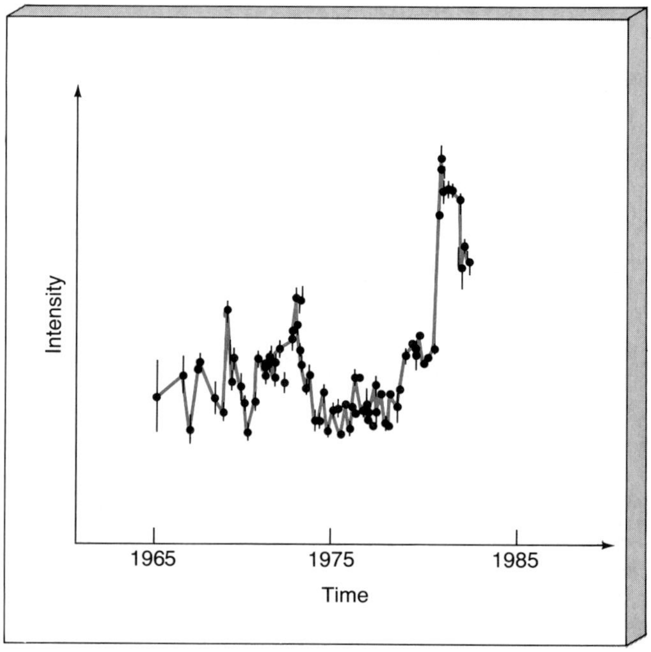

Figure 27.3 *The irregular variations of a particular Seyfert galaxy's luminosity over a period of two decades. Since this Seyfert, called 3C 84, emits most strongly in the radio part of the electromagnetic spectrum, these observations were made with large radio telescopes. The optical and x-ray luminosities vary as well.*

less than about a light-year across—an *extraordinarily* small region, considering that it emits more radiation than our entire Galaxy.

Seyfert galaxies are apparently experiencing huge explosions within their cores. Their time-variability and their large radio and infrared luminosities together strongly imply violent nonstellar activity. The *nature* of this activity may well resemble conditions in the center of our own Galaxy, but its *magnitude* is thousands of times greater than the comparatively mild events within our own Galaxy's center. Before going on to explore the nature of the energy source, let's continue our survey of active objects in the universe.

RADIO GALAXIES

Seyfert galaxies are not the only kind of active galaxy known. Radio observations have uncovered a wide variety of extremely energetic sources that also seem to be galactic in nature. The rather sudden realization in the 1960s that the universe is a lot more active than had been previously thought resulted largely from the Space Age surge of technology, especially radio astronomy. Theoreticians were almost totally unprepared for the discovery of these extraordinarily powerful objects. In this section we consider the bulk properties of the two types of active radio galaxies.

Core-Halo Radio Galaxies

The first type of radio galaxy is often called a **core-halo radio galaxy.** As illustrated in Figure 27.4, the energy from such an object comes mostly from an extremely small (less than 1 pc across) central nucleus, or core, with weaker emission coming from an extended halo surrounding it. The halo typically measures about 50 kpc across, about the size of the surrounding visible galaxy. Many of these objects can also be seen in visible light, although the associated optical galaxy, which is usually elliptical, is often quite faint. The radio luminosity from the core can be as great as 10^{44} erg/s, about the same as the emission from a Seyfert nucleus and comparable to the output from our entire Milky Way Galaxy at all wavelengths.

Figure 27.4 *Radio contour map of a typical core-halo radio galaxy. The radio emission from such a galaxy comes from a bright central nucleus, or core, surrounded by an extended, less intense halo. (Harvard-Smithsonian Center for Astrophysics)*

Figure 27.5 shows two optical photographs of a core-halo radio galaxy, along with contour maps of its radio and x-ray emission. This object is a giant elliptical galaxy known as M87—the eighty-seventh object in Messier's catalog. (We can be sure that this eighteenth-century Frenchman had no idea what he was really looking at. Nor perhaps do we!) M87 is roughly 20 Mpc distant, one of the closest active galaxies and a prominent member of the Virgo Cluster. Its nearness and interesting activity have made it one of the most intensely studied of all astronomical objects.

A long time exposure (Figure 27.5[a]) shows a large fuzzy ball of light—a fairly normal-looking type E1 elliptical galaxy, about 100 kpc across. A shorter exposure of M87 (Figure 27.5[b]) captures only the brightest regions of the galaxy and reveals a compact central region only a few hundred parsecs in diameter. Beyond the core, a long thin *jet* of matter ejected from M87's center appears. The jet is about 2 kpc long and is traveling outward at a velocity of nearly 25,000 km/s (almost a tenth of the speed of light). Computer enhancement shows that the jet is made up of a series of distinct "blobs," roughly evenly spaced along its length. This high-speed jet, which emits energy at the rate of almost 10^{42} erg/s, has been imaged in the radio and x-ray as well as in the optical (see Figures 27.5[c] and [d]). It has become one of the keys to our understanding of these energetic radio sources, as we will discuss later in this chapter.

Lobe Radio Galaxies

The other type of radio galaxies displays a somewhat different shape. Like the Seyfert and core-halo galaxies, this second class emits most of its radiation in the long-wavelength part of the spectrum, but unlike the others, very little of this radiation arises from a central nucleus. Most of the energy comes from giant **radio lobes**—roundish extensions of gas up to a megaparsec across, lying well beyond the center of the galaxy itself. For this reason, these objects are known as **lobe radio galaxies.** Their geometry suggests explosive events that expelled vast quantities of gas from some central source. The lobes contain matter too tenuous to emit optically, but their radio luminosity can range from 10^{43} to 10^{45} erg/s, between 1/10 and 10 times the total energy emitted by the entire Milky Way Galaxy. Several of these strange objects are relatively nearby, so we can study them at close range. One such system, known as Centaurus A, is shown in Figure 27.6. It lies about 4 Mpc from Earth.

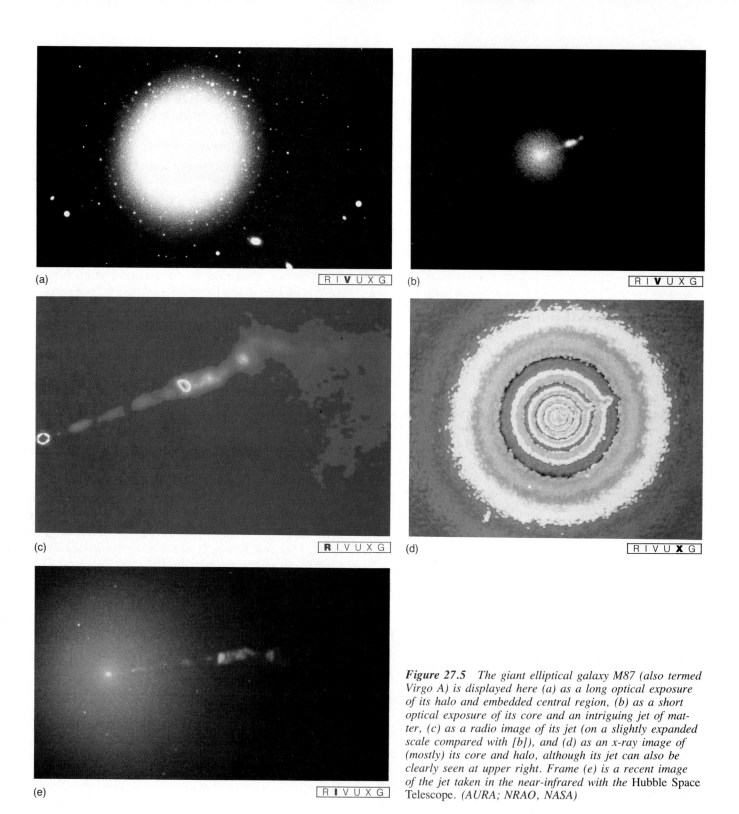

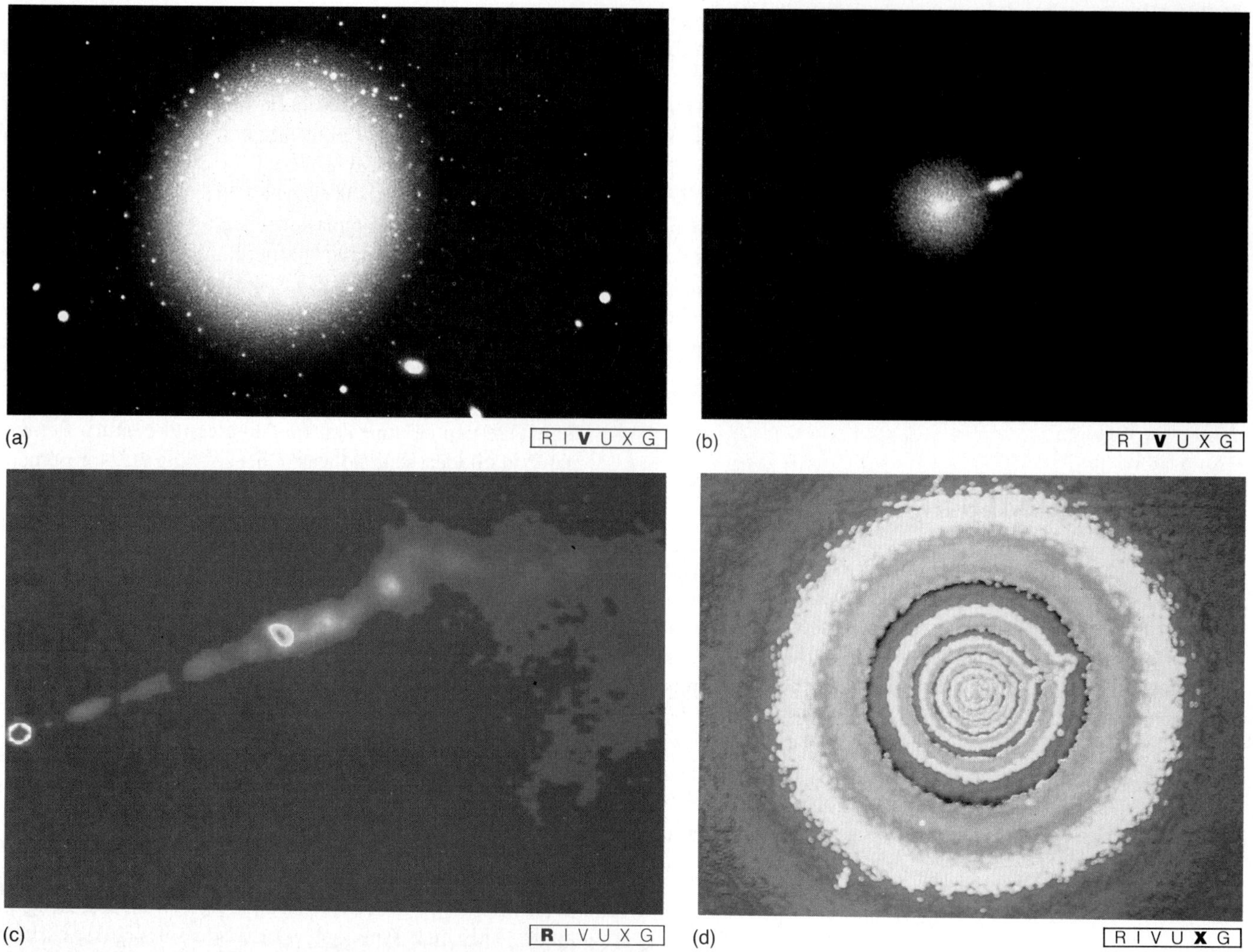

Figure 27.5 *The giant elliptical galaxy M87 (also termed Virgo A) is displayed here (a) as a long optical exposure of its halo and embedded central region, (b) as a short optical exposure of its core and an intriguing jet of matter, (c) as a radio image of its jet (on a slightly expanded scale compared with [b]), and (d) as an x-ray image of (mostly) its core and halo, although its jet can also be clearly seen at upper right. Frame (e) is a recent image of the jet taken in the near-infrared with the* Hubble Space Telescope. *(AURA; NRAO, NASA)*

Figure 27.7 is an optical image of Centaurus A, with a contour representation of Figure 27.6 superimposed to show the relation between the optical and radio emission. In visible light, Centaurus A is a rather peculiar-looking object, apparently an E2 galaxy bisected by an irregular band of obscuring dust. Numerical simulations indicate that this system is probably the result of a merger between an elliptical and a smaller spiral galaxy about 500 million years ago. The radio lobes are roughly symmetrically placed with respect to the center of the visible galaxy, and the line joining them is roughly perpendicular to the dust lane. The elliptical galaxy itself is very large—some 500 kpc in diameter. Relatively little radio emission is observed from the location of the optical image, however. Most of the radio radiation arises from the giant lobes well beyond the optical galaxy.

The alignment of the radio lobes with the center of the optical galaxy suggests that they were ejected by some violent event in the nucleus. This argument is strengthened by the presence of an additional pair of secondary lobes, smaller than the main lobes (about 15 kpc in length) and closer to the visible galaxy, shown in Figures 27.6 and 27.7. Both pairs of lobes share the same high degree of linear alignment, in general agreement with the idea that periodic explosions eject matter from the central galaxy. Still higher-resolution studies reveal the presence of a roughly 1-kpc-long jet in the center of Centaurus A, aligned with the larger lobes.

Further evidence in favor of the interpretation of radio lobes as material ejected from the center of a galaxy is provided by another object, Cygnus A, shown as an optical image in Figure 27.8(a) and as high-resolution radio map in Figure 27.8(b). The filamentary structure evident in the

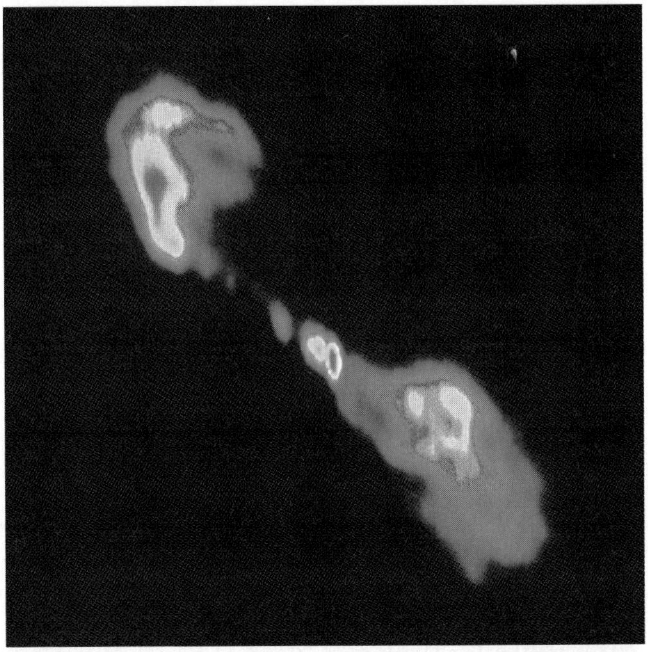

R I V U X G

Figure 27.6 Lobe radio galaxies, such as Centaurus A shown here, have giant radio-emitting regions extending a million parsecs or more beyond the central galaxy. The lobes are optically completely invisible and can be observed only with radio telescopes. (NRAO)

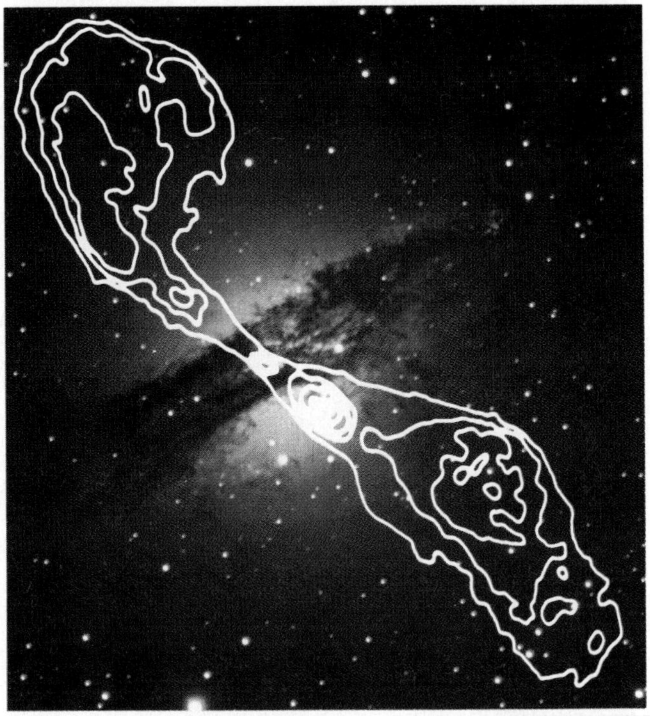

R I V U X G

Figure 27.7 An optical photograph of Centaurus A, one of the most massive and peculiar galaxies known and believed to be the result of a galaxy collision some 500 million years ago. Contours mark the radio emission shown in Figure 27.6. This entire image is of a "small" region no larger than the smallest blue blob amidst the lobes of the previous figure. (NRAO)

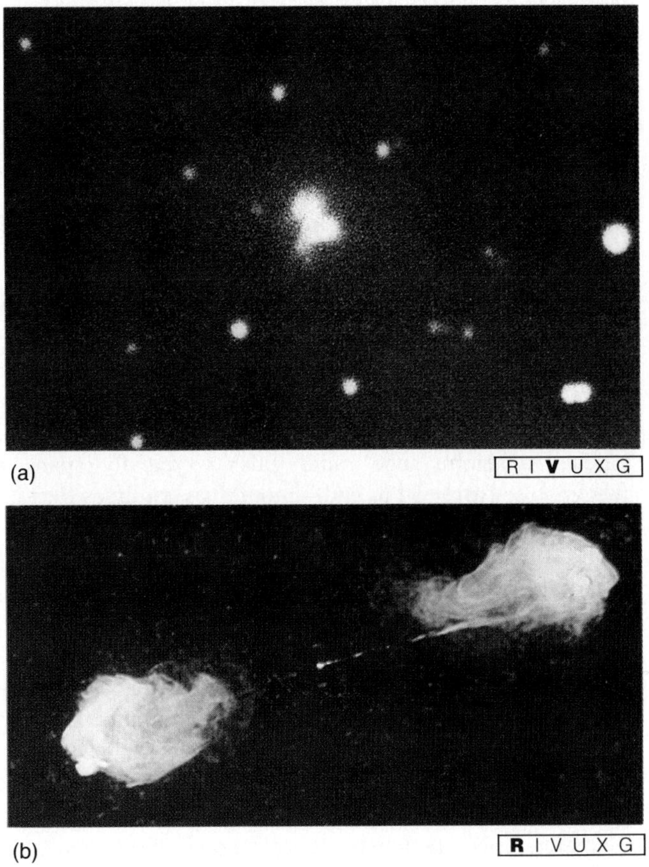

(a) R I V U X G

(b) R I V U X G

Figure 27.8 (a) Cygnus A also appears to be two galaxies in collision, although it is not completely clear that that is what is really happening. (b) On a much larger scale, this cosmic object displays radio-emitting lobes on either side of the optical image. To put these images into proper perspective, the optical galaxy in (a) is about the size of the small dot at the center of the "radiograph" in (b). Notice the thin line of radio-emitting material joining the right lobe to the central galaxy. (NOAO; NRAO)

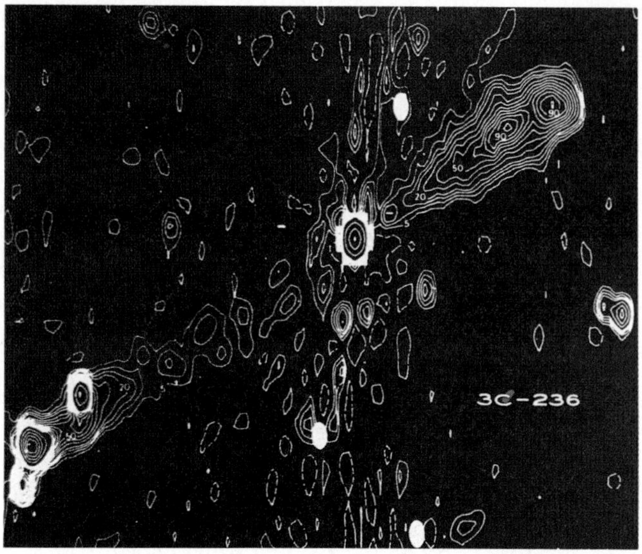

Figure 27.9 *Radio map of 3C 236, currently the largest known radio galaxy—and the largest known "galactic" object—in the universe. This object is about 600 Mpc away, and its lobes extend well over 3 Mpc. (NASA)*

radio lobes and the thin, radio-emitting line joining the right lobe to the center of the visible galaxy (the dot at the center of the radio image) strongly suggest that we are seeing two oppositely directed jets of material running into the intergalactic medium. The question then becomes, What is powering these jets? We will answer this question shortly.

The radio lobes of all lobe radio galaxies are enormous—often three or four times the size of a typical normal galaxy. The lobes vary in size and shape from source to source but maintain their precise alignment in nearly all cases. An entire lobe radio galaxy typically spans a megaparsec from end to end—more than ten times the size of the Milky Way.

Some objects have even larger lobes than Centaurus A or Cygnus A. Figure 27.9 shows a radio map of a gargantuan lobe radio galaxy. This one has the name 3C 236 (because it was the 236th object listed in the *Third Cambridge Catalog* of radio sources, an early compilation of observations made by radio astronomers during the late 1950s and early 1960s). 3C 236 is currently the largest known "galactic" object in the universe. Its lobes are more than 3 Mpc across. Remember, our Local Group is "only" 1 Mpc in diameter. This *one* active galaxy is about 100 times larger than our Milky Way Galaxy and three times larger than the entire Local Group of galaxies.

Very large lobe radio galaxies like 3C 236 are rare, but hundreds of "smaller" such objects, a mere half-megaparsec across, are known. Some lie relatively nearby—within 100 Mpc or so—although most are farther away, as indicated by the red shifts of spectral lines emitted by the optical objects between the lobes. In some systems, known as **head-tail radio galaxies,** the lobes seem to form a "tail" behind the main galaxy. For example, the lobes of radio galaxy NGC 1265 (shown in Figure 27.10) appear to be "swept back" by some onrushing wind. In fact, this is the most likely explanation for this galaxy's appearance. If NGC 1265 were at rest, it would be just another double-lobe source, perhaps looking quite similar to Centaurus A. However, the galaxy is traveling through the intergalactic medium of its parent galaxy cluster (known as the Perseus Cluster), and the outflowing matter forming the lobes tends to be left behind as the galaxy moves.

Radio galaxies share many characteristics with Seyfert galaxies. They emit comparably large amounts of energy, and there is good evidence that the energy source is a compact region at the center of an otherwise relatively normal-looking galaxy. In the lobe radio galaxies, that energy is fired out from the nucleus in the form of jets of

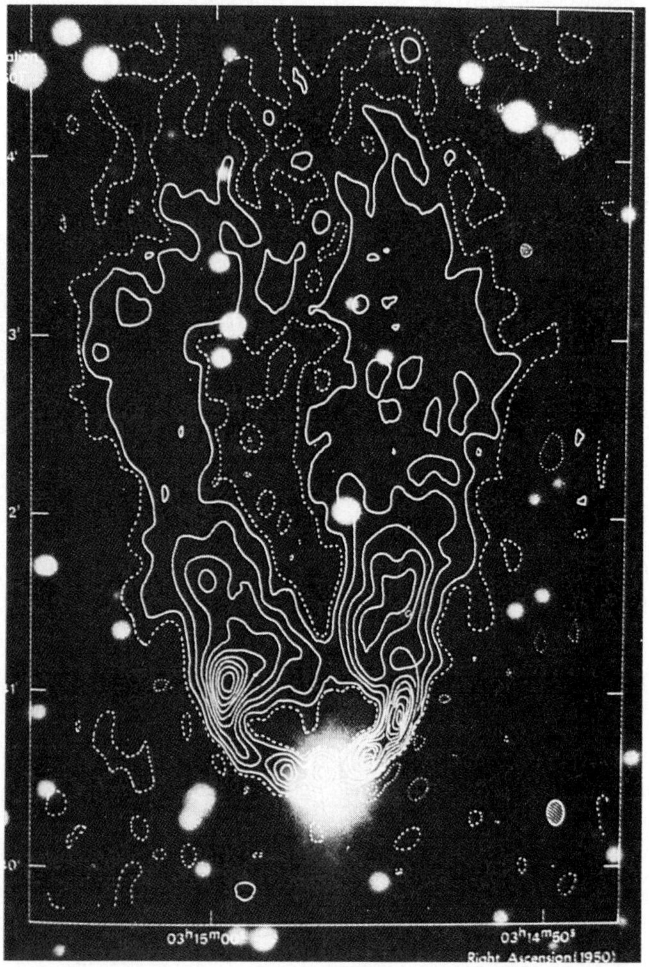

Figure 27.10 *Radio contour of the "head-tail" galaxy, superposed on the optical galaxy NGC 1265. Astronomers reason that this object is moving rapidly through space, trailing a "tail" behind as it goes. (Palomar Observatory/Westerbork Radio Observatory)*

matter and is ultimately *emitted* from far beyond the galaxy itself. But the central compact nucleus is still regarded as the place where the energy is actually *produced*. Before going on to probe even more distant, and more violent, objects, let us now consider the current view of the "engine" that powers all this activity.

THE CENTRAL ENGINE OF AN ACTIVE GALAXY

✓4 The behavior of active galaxies is contrary to that expected from vast collections of stars. The lobe radio galaxies in particular, with their huge energy emission from far beyond the optical galaxy, are among the most powerful objects in the universe. Can we explain this enormous nonstellar energy output in terms of known physics? Remarkably, the answer is yes. The present consensus among astronomers is that, despite the great differences in appearance, these objects—Seyferts and radio galaxies—may share a common energy-generation mechanism. The energy can be reprocessed into many different forms before it is finally emitted into intergalactic space, but the engine is probably the same in each case. Furthermore, the mechanism we will describe extends naturally to explain the even more energetic *quasars*, which we will discuss later in this chapter.

As a class, active galaxies (including quasars) show some or all of the following properties.

1. They have high luminosities, generally greater than the 10^{44} erg/s characterizing a fairly bright normal galaxy.
2. Their energy emission is nonstellar—it cannot be explained as the accumulation of even trillions of stars.
3. Their energy output can be highly variable, implying that it is emitted from a small central nucleus much less than a parsec across.
4. They often exhibit jets and other signs of explosive activity.
5. Their optical spectra may show broad emission lines, indicative of rapid internal motion within the energy-producing region.

The principal questions then are: How can such vast quantities of energy arise from these relatively small regions of space? Why is so much of the energy radiated at low frequencies, especially in the radio and infrared? And what is the origin of the extended radio-emitting lobes and jets? Let us first consider how the energy is produced.

Energy Requirements

To develop a feeling for the enormous emissions of active galaxies, consider for a moment an object with a luminosity of 10^{45} erg/s. In and of itself, this energy output is not inconceivably large. The brightest giant ellipticals are comparably powerful. Thus, some 10^{12} stars—a few normal galaxies' worth of material—could *equivalently* power a typical active galaxy. The difficulty arises when we consider that in an active galaxy this energy production is packed into an object much less than a parsec in diameter!

It is difficult to imagine how several Milky Way Galaxies could be squashed into a space no larger than a parsec. Even if we could somehow squeeze that much mass into such a volume, it would immediately collapse to form a huge black hole, and none of the light it produced could escape to the outside! The energy output of an active galaxy simply cannot be explained as the accumulated energy of many stars. We must think of some other energy source.

Energy Production

The twin requirements of large energy generation and small physical size bring to mind our discussion of x-ray sources in Chapter 24. The presence of the jet in M87 and the ejection of matter to form radio lobes in Centaurus A and Cygnus A strengthen the connection (see especially the discussion of SS433 in Chapter 24). Recall that the best current explanation for those "small-scale" phenomena involves the accretion of material onto a compact object—a neutron star or a black hole. Large amounts of energy are produced as the matter spirals down onto the central object, and high-speed jets may well be a common by-product of the process. In Chapter 25, we suggested that a similar mechanism, now involving a *supermassive black hole*—one with a mass of around a million suns—may also be responsible for the energetic radio and infrared emission observed at the center of our own Galaxy.

Illustrated in Figure 27.11, the leading model for the central engine of active galaxies is essentially a scaled-up version of the same accretion process, now involving black holes with masses between a few million and a billion times the mass of the Sun. In this case, however, the origin of the accreted gas is not a binary companion, as in stellar x-ray sources, but entire stars or clouds of interstellar gas that come too close to the hole's strong gravity and are torn apart by its tidal field.

Many of the observed characteristics of active galactic nuclei—for example, their small size, rapid motion, and large luminosity—can be explained by this energy-production mechanism. The small size of the energy-emitting region is a direct consequence of the compact nature of the central black hole. Even a billion-solar-mass hole has a radius of only 3×10^9 km, or 10^{-4} pc—about 20 A.U.—which fits our requirement that the source be less than a parsec across. The broadening of the spectral lines observed in the nuclei of Seyferts (and quasars) results from the rapid orbital motion of the gas in the hole's intense gravity. Large luminosities can be produced because the accretion process is remarkably efficient at converting the mass of the incom-

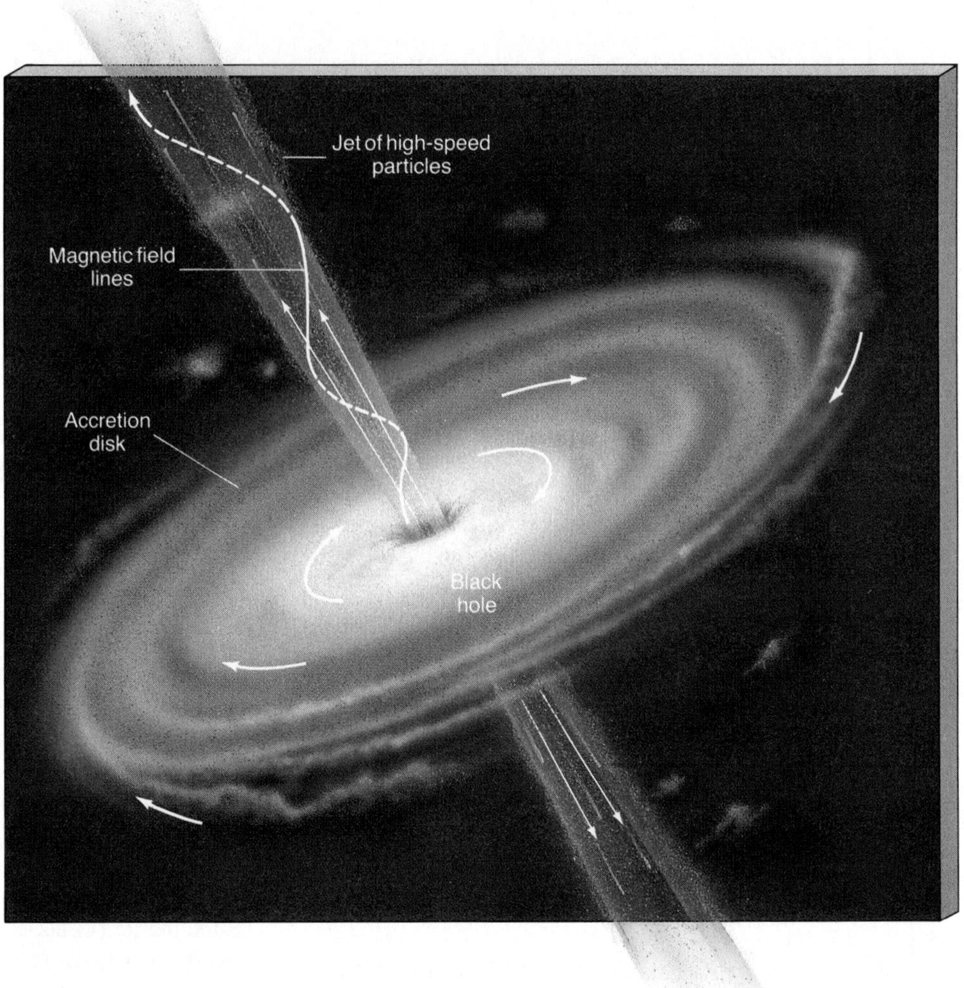

Figure 27.11 *The leading theory for the energy source in active galactic nuclei (and quasars) holds that these objects are powered by material accreting onto a supermassive black hole. As matter spirals toward the hole, it heats up, producing large amounts of energy. At the same time, high-speed beams of gas may be ejected perpendicular to the accretion disk, forming the jets and lobes seen in many active objects. Magnetic fields generated in the disk are carried by the jets out to the radio lobes, where they play a crucial role in producing the observed radiation.*

ing material into energy. As the infalling gas spirals around the hole, it is heated to high temperatures through friction and emits large amounts of energy. Detailed calculations indicate that as much as 10 or 20 percent of the total mass-energy of the infalling matter can be radiated away before it crosses the hole's event horizon and is lost forever.

The total mass-energy of a star like the Sun—the mass times the speed of light squared—is about 2×10^{54} erg. Assuming that 10 percent of that is actually converted into radiation, we find that the 10^{45} erg/s luminosity of our radio galaxy can be explained by the ingestion of about 1 star per decade by a 10^8–10^9-$M_\odot$ black hole. Thus, accretion into a black hole can in fact explain the energy output of active galaxies. Less luminous active galaxies require correspondingly less fuel—for example, a 10^{43} erg/s Seyfert would devour only one star every thousand years.

Energy Emission

Having come up with a means of supplying the energy needs of an active galaxy, we now turn to the way in which the energy finally gets emitted. The most extreme example of the separation between the source of the radiation and the place where it is emitted is the case of a lobe radio galaxy. If the energy is produced in the hot, compact regions surrounding a supermassive black hole, how does it come to be emitted from the distant lobes, primarily in the form of radio radiation?

An important clue regarding the emission of the radiation results from detailed studies of the jets in M87 (Figure 27.5[b] and [c]) and other active objects. The optical emission from the jet is *polarized,* as is much of the radio emission from the entire core-halo object. Recall from Chapter 20 that radiation is said to be polarized when its waves all vibrate in the same plane. The important point here is that only a few physical processes can produce polarized radiation. Starlight, for example, is unpolarized—the light waves making up sunlight and starlight are randomly oriented—as is the thermal radiation emitted from any other object, hot or cold.

In Chapter 20, we saw how starlight can become polarized by interacting with the dust particles in interstellar space. But when we look toward a distant galaxy, perpendicular to the plane of the Milky Way, we encounter little interstellar dust within the thin galactic disk along our line of sight, and there is no known dust or other polarizing matter in the space between the galaxy clusters. In any case, not all of M87's light is polarized—only its jet and radio emission, but not its starlight. We conclude that the polarization of the radiation from M87's jet is not caused by material intervening between us and M87. The polarization, then, must be part of the radiation-emission process itself.

To explain the luminosity and polarization of the active galaxies, we need to find an emission process that (1) is most intense at low frequencies and (2) is polarized. Only one such process is known: **synchrotron emission,** where high-speed charged particles (in this case, electrons) emit radiation as they are accelerated in a strong magnetic field. (On earth a synchrotron is a machine that accelerates particles to high speeds while confining them magnetically.) Synchrotron emission is an example of nonthermal radiation. There is no link between synchrotron emission and the temperature of the radiating object, so the radiation is not described by a Planck curve. Instead, its intensity increases with decreasing frequency, as shown in Figure 27.12. This is just what is needed to explain the range of radiation recorded from active galaxies. (Compare Figure 27.12 with Figure 27.1.)

As shown in Figure 27.13, whenever fast-moving electrons enter a magnetic field, they tend to spiral around the field lines. We have encountered this idea several times previously, in a variety of different contexts (as noted in the figure caption). As the electrons whirl around, they emit radiation. The faster the electrons, or the stronger the magnetic field, the greater the amount of energy radiated. Detailed calculations of the process predict the frequency distribution sketched in Figure 27.12. They also indicate that if the magnetic field is fairly orderly (as in Figure 27.13), the emitted radiation will indeed be polarized. These predictions have been confirmed in the laboratory. We can say with some confidence that we understand the emission mechanism leading to the nonthermal, polarized radiation observed in M87 and other active objects.

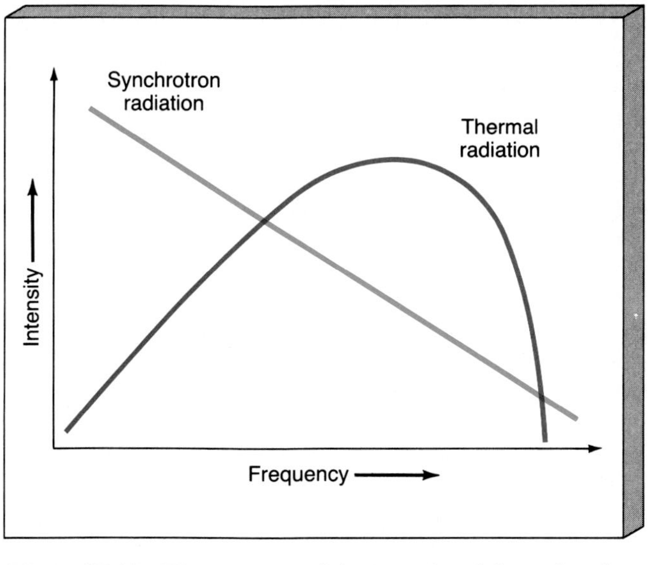

Figure 27.12 *The variation of the intensity of thermal and synchrotron (nonthermal) radiation with frequency. Thermal radiation, described by a Planck curve, is peaked at some frequency that depends on the temperature of the source. Nonthermal synchrotron radiation, by contrast, is most intense at low frequencies. It depends in no way on the temperature of the emitting object. Compare this figure with Figure 27.1.*

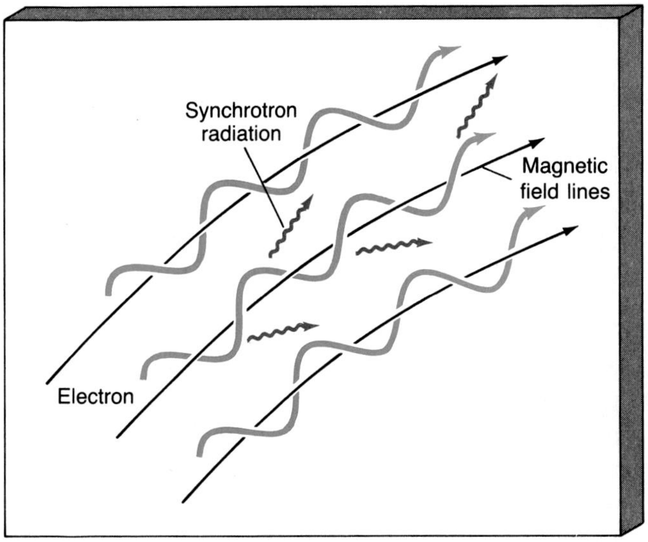

Figure 27.13 *Charged particles, especially fast electrons (blue lines), emit polarized synchrotron radiation (red lines) while spiraling in a magnetic field (black lines). This process is not confined to active galaxies. It occurs, on smaller scales, when charged particles interact with magnetism in Earth's Van Allen belts (see Chapter 7), when charged matter arches above sunspots on the Sun (see Chapter 18), in the vicinity of neutron stars (see Chapter 24), and when cosmic rays traverse our normal Galaxy (see Chapter 26).*

On the basis of its spectrum, the emission from the giant radio lobes of other radio galaxies is also believed to be synchrotron radiation, although it is generally not strongly polarized. The lack of polarization in this case is explained by assuming that the magnetic field lines have become twisted and tangled by the turbulent flow of the gas within the lobes, so there is no preferred orientation for the resulting radiation.

A General Scenario

We now have all of the basic ingredients necessary to understand the processes leading to the production and emission of radiation in active galaxies. As we mentioned earlier, many of the details are still only partly understood, but at least we can construct a general outline, or scenario.

At the center of an active galaxy lies a supermassive black hole. Stars and clouds of interstellar gas that pass too close to the hole are swept up by its strong gravitational field, torn apart, and dragged into the hole through a rotating accretion disk. As the material spirals in, it heats up, ultimately radiating away up to 20 percent of its total mass energy by the time it reaches the event horizon. This is the source of the radiation from Seyfert nuclei and the cores of radio galaxies. Instabilities in the accretion disk can cause fluctuations in the energy released, leading to the observed variability. The rotation of the disk accounts for the broadened spectral lines often observed. In the case of some Seyferts, it is necessary to assume further that the energy is "reprocessed"—absorbed and reemitted—by gas and dust surrounding the nucleus to account for the details of the observed radiation spectrum, but the basic power source is still the central black hole.

Some of the accreting matter never makes it to the black hole. Instead, it is blasted out into space in the form of high-speed jets of particles. The details of the way in which these jets form are uncertain—although several plausible competing theories exist—but there is a consensus among theorists that jets are a very common feature of accretion flows. The jets also contain strong magnetic fields, possibly generated by the gas motion within the disk itself, which accompany the gas as it leaves the galaxy. The interaction between the particles and the field emits the polarized synchrotron radiation by which we see the jet.

Eventually, the jet is slowed and stopped by the intergalactic medium, the flow becomes turbulent, and the field grows tangled. The result is a giant radio lobe, like those pictured in Figures 27.6 and 27.8. (Alternatively, the result may possibly be a so-called BL Lac object, as discussed in Interlude 27-1.) The existence of the inner lobes of Centaurus A and the blobs in M87's jet imply that jet formation may be an intermittent process, at least in some cases.

There is some evidence to suggest that much, if not all, of the activity observed in nearby active galaxies could have been sparked by recent interaction with a neighbor. Many nearby active galaxies appear to have been "caught in the act" of interacting with another galaxy, suggesting that the fuel supply can sometimes be turned on by the effect of a companion. Just as tidal forces may trigger star formation in starburst galaxies (mentioned in the previous chapter), they may also divert gas and stars into the galactic nucleus, triggering an active outburst that may last for millions or billions of years.

QUASI-STELLAR OBJECTS

The Discovery of Quasars

√5 In the early days of radio astronomy, many radio sources were detected for which no corresponding visible object was known. By 1960, several hundred such sources were listed in the *Third Cambridge Catalog,* and astronomers were scanning the skies in search of optical counter-

INTERLUDE 27-1 *BL Lac Objects*

In 1929, an object thought to be a variable star was discovered in the constellation Lacerta. Astronomers gave it the two-letter code BL, so it became known as BL Lacertae, or BL Lac for short. Not until the 1970s, when it became clear that BL Lac was a strong radio source, did anyone question its classification as a star. As more objects like BL Lac were found, astronomers began to realize that they had stumbled across a whole new class of extragalactic sources. *BL Lac objects* seem starlike in a telescope and vary greatly in luminosity. They are also powerful compact radio sources, sometimes colloquially referred to as "blazars."

Astronomers originally placed BL Lac objects into a special class because they displayed no spectral lines. Their red shifts could not be determined and their distances were unknown. In recent years, extremely faint spectral lines strongly suggest that BL Lac objects reside at great distances, making them nearly as luminous as quasars (discussed below in the text). Further careful observations have shown that blazars reside at the centers of relatively normal elliptical galaxies. We consider these galaxies active because of the tremendous strength of their radio emission.

The weak spectral lines may be the key needed to unravel the nature of the BL Lac objects. Evidently, the thermal emission from any stars they contain, which would provide spectral lines, is swamped by their nonthermal synchrotron radiation, which does not contain spectral features. Other types of active galaxies show a mixture of thermal and nonthermal radiation, so it is more difficult to study the nonthermal radiation alone. BL Lac objects, then, may offer astronomers a chance to study the "bare machine" (which generates the nonthermal radiation) powering these active galaxies.

In some ways, the BL Lac objects seem to represent a "link" or transition phase between radio galaxies and quasars. The luminosities of the compact radio sources inside BL Lac objects span the whole range from the relatively weak cores observed in radio galaxies to the stronger ones found in quasars. Perhaps there exists an evolutionary sequence of some sort, where the ancient quasars use up their fuel, turning into the weaker yet more erratic BL Lac objects, which subsequently become the rather inactive cores of radio galaxies. An alternative possibility is that BL Lacs *are* just radio galaxies, but radio galaxies from which the jet happens to be pointing nearly straight at us. We would see the galaxy *through* the nearer radio lobe. Only further observations and better statistics will tell us for sure which (if either) of these two possibilities is correct.

parts. Their job was made difficult both by the low resolution of the radio observations (which meant that the observers did not know exactly where to look) and by the faintness of many of these objects at visible wavelengths. In 1960, Thomas Matthews and Allan Sandage detected what appeared to be a faint blue star (Figure 27.14) at the location of the object 3C 48 (the 48th object on the Cambridge list) and obtained its spectrum. Containing many unknown broad emission lines, the unusual spectrum defied interpretation. 3C 48 remained a unique curiosity until 1962, when another similar-looking, and similarly mysterious, faint blue object with "odd" spectral lines was discovered and identified with the radio source 3C 273.

The following year, Maarten Schmidt, of the California Institute of Technology, made a breakthrough. He realized that the strongest unknown lines in 3C 273's spectrum were simply familiar hydrogen lines, redshifted by a very unfamiliar amount—in fact, about 16 percent—indicating a recessional velocity of about 48,000 km/s. Once Schmidt's discovery became known, the spectrum of 3C 48 was quickly found to have a similar explanation. Its 37 percent red shift implied that it is receding from Earth at almost one-third the speed of light. These huge velocities mean that neither of the two objects are members of our Galaxy. Applying the Hubble law (with our adopted value of the Hubble constant $H_0 = 75$ km/s/Mpc), we obtain distances of 640 Mpc and 1300 Mpc for 3C 273 and 3C 48, respectively. Clearly not stars, these objects became known as *quasi-stellar radio sources,* or **quasars** for short. We now know that not all such highly redshifted, starlike objects are strong

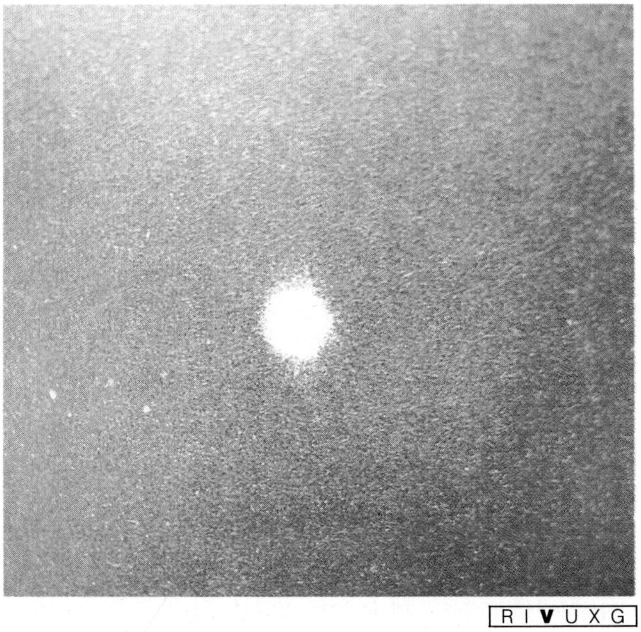

Figure 27.14 *Optical photograph of 3C 48, the first quasar discovered. Its starlike appearance shows no obvious structure and gives little outward indication of this object's enormous luminosity. 3C 48 has a red shift of 0.367, placing it at a distance of about 1300 Mpc. (Palomar Observatory)*

radio sources, so the term **quasi-stellar object** (or **QSO**) is more common today. However, the name *quasar* persists, and we will continue to use it here.

Relativistic Red Shifts and Look-back Time

When speaking of very distant objects, astronomers tend to talk about their red shifts, rather than their distances. Of course, because of the Hubble law, red shift and distance are equivalent, but the red shift has a rather more direct connection with the expansion of the universe, as we will see in the next chapter. The red shift of any object is, by definition, the fractional increase in its wavelength resulting from its recessional velocity. Using the formula for the Doppler shift presented in Chapter 4, we find that the red shift of a source of radiation is given by

$$\textit{redshift} = \frac{\textit{observed wavelength} - \textit{true wavelength}}{\textit{true wavelength}}$$

$$= \frac{v}{c},$$

where v is the source's recessional velocity and $c =$ 300,000 km/s (the speed of light). For example, a galaxy at a distance of 200 Mpc, with a recessional velocity (by Hubble's law) of 75 km/s/Mpc × 200 Mpc = 15,000 km/s, has a red shift of 15,000/300,000 = 0.05. An object with a red shift of 0.1 has $v = 0.1 \times c = 30{,}000$ km/s and hence a distance of 400 Mpc. To distinguish recessional red shift from red shifts caused by motion *within* an object—for example, galaxy orbits within a cluster, or explosive events in a galactic nucleus—the red shift resulting from the Hubble flow is often called the **cosmological red shift.** Objects so far away that they exhibit a large cosmological red shift are said to be at **cosmological distances**—distances comparable with the scale of the universe itself.

Unfortunately, the above equation does not take into account the effects of relativity. As we saw in Chapter 24, the rules of everyday physics have to be modified when velocities begin to approach the speed of light, and the formula for the Doppler shift is no exception. In particular, while our formula is valid for velocities much less than the speed of light, when $v \approx c$ the red shift is not 1, as the above equation suggests, but is in fact *infinite*. In other words, radiation received from an object moving away from us at nearly the speed of light would be redshifted to almost infinite wavelength. Thus, do not be alarmed to find below that many quasars have red shifts greater than 1. This does not mean that they are receding faster than light! It simply means that their recessional velocities are *relativistic*—comparable to the speed of light—and the above formula is inapplicable.

The actual formula relating the red shift of a galaxy to its recessional velocity is quite complicated, as it involves some assumptions about the overall expansion of the universe (some of which we will discuss in Chapter 28). In

TABLE 27-2 ***Red Shift and Recessional Velocity***

Red Shift	v/c	Look-back Time (billions of years)	Distance (Mpc)
0.00	0.000	0.00	0
0.10	0.095	1.24	380
0.25	0.220	2.87	880
0.50	0.385	5.02	1540
0.75	0.508	6.62	2031
1.00	0.600	7.82	2399
1.50	0.724	9.44	2896
2.00	0.800	10.4	3190
3.00	0.882	11.5	3528
4.00	0.923	12.0	3681
5.00	0.946	12.3	3773
∞	1.000	13.0	3988

addition (again as we will see in the next chapter), it is actually incorrect to think of the cosmological red shift as a Doppler shift, so we will not reproduce the formula here. However, Table 27-2 presents a brief ''conversion chart'' between red shift and velocity, usable even for $v \approx c$.

Another way that astronomers refer to distant objects is by their **look-back time,** which is simply how long ago they emitted the radiation we see today. Look-back time is another measure of distance. The light we receive tonight from a galaxy at a distance of 100 million light-years must have been emitted 100 million years ago. The notion of look-back time is valid because looking out into space is equivalent to looking back in time. It is often convenient to work with look-back time because measurements in the range of millions to billions of years are also appropriate in discussing the evolution of stars and galaxies. The look-back time, calculated simply by dividing the distance obtained from Hubble's law by the speed of light, is also listed in Table 27-2. As usual, a Hubble constant of 75 km/s/Mpc is assumed. Notice that, according to the table, the recessional velocity equals the speed of light—and the red shift becomes infinite—for objects that emitted their radiation about 13 billion years ago. This fact has deep implications for our notions about the age of the universe, as we will see in the next chapter.

Quasar Red Shifts and Luminosities

The most striking characteristic of the several hundred quasars now known is that their spectra all show large red shifts, ranging from 0.06 up to the current maximum of 4.9. Thus, *all* quasars lie at large distances from us—the closest is 240 Mpc away, the farthest resides at a distance of nearly 3800 Mpc (according to Table 27-2). The majority of quasars lie more than 1000 Mpc from Earth. As we saw above, these large red shifts mean that the visible spectra of quasars are often quite unfamiliar to most astronomers—ultraviolet lines are often shifted into the optical portion of the spectrum. Figure 27.15 shows the spectrum of 3C273, one of the most distant quasars known. Some prominent emission lines, and the extent of their red shift, are marked on the diagram.

Despite the unimpressive optical appearance of quasars, the large distances implied by their red shifts mean that these faint ''stars'' are in fact the brightest known objects in the universe! 3C 273, for example, has a luminosity of about 10^{47} erg/s. 3C 48 has a luminosity about a factor of 3 less. More generally, quasars range in luminosity from around 10^{45} erg/s—about the same as the brightest radio galaxies—up to nearly 10^{49} erg/s. A value of 10^{47} erg/s, comparable to 20 trillion Suns or 1000 Milky Way Galaxies, is fairly typical. Thus quasars outshine the brightest normal and active galaxies by about a factor of 1000.

Quasar Spectra

Like Seyferts, quasars have broad emission lines in their optical spectra, indicating internal velocities of 1000–

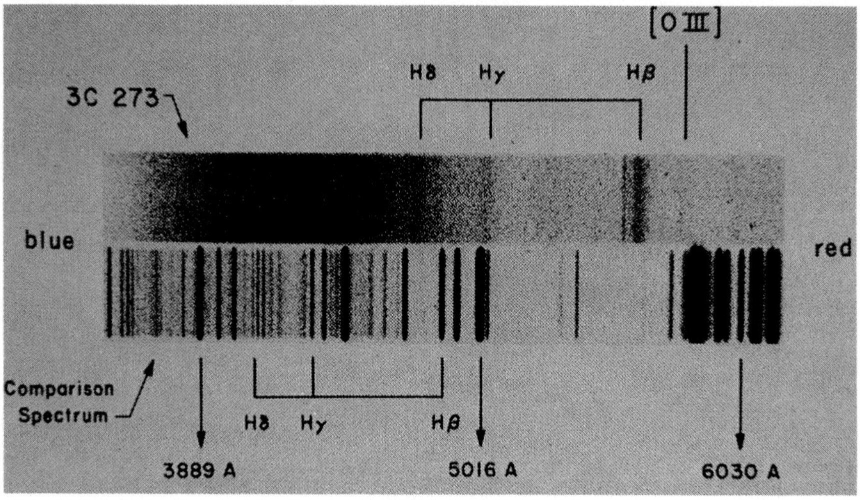

Figure 27.15 Optical spectrum of the distant quasar 3C273. Note the width of the marked spectral lines. (Palomar Observatory)

R I **V** U X G

1500 km/s. In addition, the spectra contain much narrower absorption lines, which are thought to form in regions of cooler gas lying between the energy-emitting region and Earth. These regions are still part of the quasar, but they are relatively far away from the source of the quasar's intense energy. These broad emission lines and narrower absorption lines all have the same red shift, which is how we know they all come from the same object.

Interestingly, many quasars also show additional absorption lines that are redshifted by *less* than the quasar emission lines. For example, the quasar known as PHL 938 has an emission red shift of 1.955, placing it at a distance of some 3200 Mpc. But it also shows three sets of absorption lines, with red shifts of 1.949, 1.945, and 0.613, respectively. The first two sets may well come from high-speed gas within the quasar itself (the velocity differences are only a few hundred kilometers per second), but the third is interpreted as arising from intervening gas that is much closer to us (only about 1800 Mpc away), which explains why it has a smaller red shift than the quasar itself. The most likely possibility is that this gas is part of an otherwise invisible galaxy lying along the line of sight. Quasar spectra, then, afford astronomers a means of probing previously undetected parts of the universe.

Quasar Appearance and Variability

Of the hundreds of quasars now known, few show more than a ball of luminous fuzz in visible-light images. Figure 27.16 is an optical photograph of 3C 273. Notice the jet of luminous matter, reminiscent of the jet in M87, extending nearly 3 kpc from the quasar itself. Quasars have been observed in the radio, infrared, optical, ultraviolet, and x-ray parts of the electromagnetic spectrum, and some have even been found to emit gamma-rays. However, most quasars emit most of their energy in the infrared. Often, as shown in Figure 27.17, quasar radio radiation arises from regions lying beyond the bright central core, much like the core-halo and lobe radio galaxies studied earlier. In other cases, the radio emission is confined to the central optical image. Again like the other active galaxies, the radiation is nonthermal and often polarized.

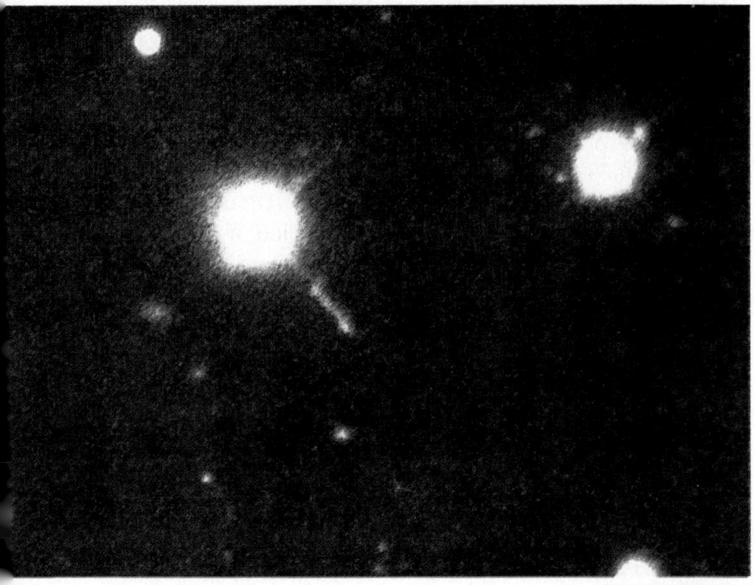

Figure 27.16 *The bright quasar 3C 273 displays a luminous jet of matter, but the main body of the quasar is starlike in appearance. The jet extends for about 100,000 light years. (Palomar Observatory)*

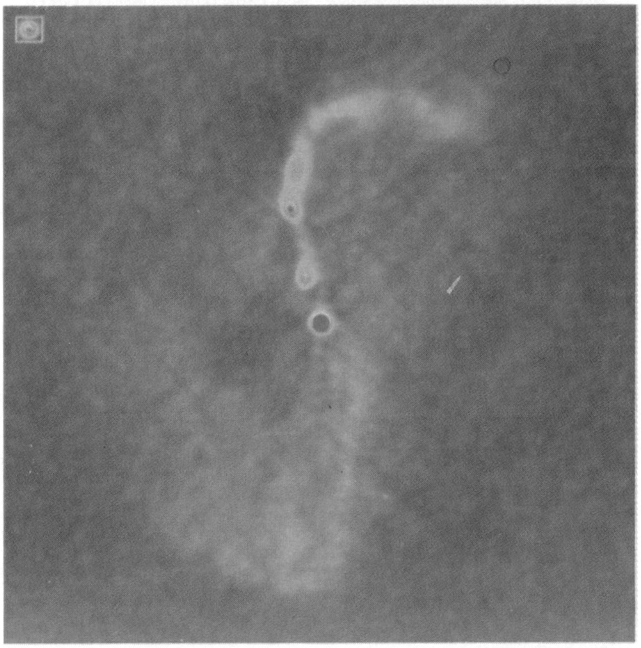

Figure 27.17 *Radio image of the quasar 2300-189 showing radio jets feeding faint radio lobes. The bright (red) central object is the quasar, some 400 Mpc away. Notice the extended radio lobes and jetlike structures arising in the central regions. (NRAO)*

The long-term variability of some quasars has been established through the study of archival astronomical data. As we noted in Interlude 19-2, astronomers throughout the past century have routinely photographed the sky in order to monitor the properties of stars. Unbeknownst to the original observers, many of the specks of light captured on their photographic patrols were not stellar, but quasi-stellar in nature. By consulting the comprehensive collection of photographic plates (extending back almost to the birth of photography) that has resulted from these routine surveys, modern astronomers can often piece together the "light history" of a quasar covering the past 100 or so years.

Figure 27.18 illustrates how the light history for one quasar reveals evidence for variations in its optical radiation. In part (a), a 1937 photograph clearly shows a starlike object (marked by an arrow). Its large red shift and other emissive peculiarities have since identified it as a quasar, labeled 3C 279. In part (b), taken in 1976, the quasar has nearly disappeared. Part (c) shows the light history of 3C

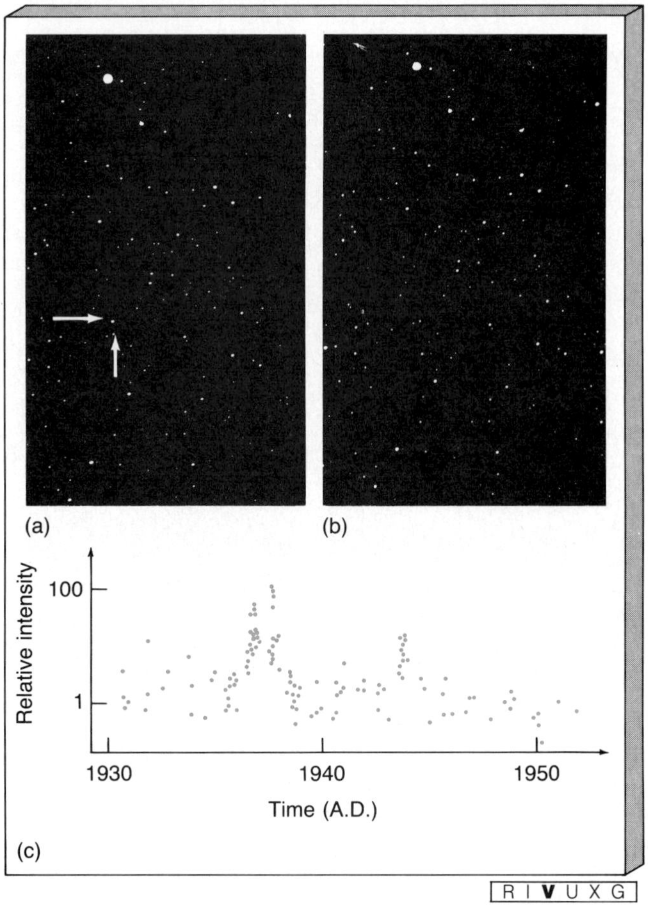

Figure 27.18 *(a) Quasar 3C 279 (at the intersection of the arrows), whose luminosity in 1937 made it the most intrinsically brilliant object known in the universe. (b) The same quasar in 1976, when its luminosity was much diminished. Part (c) shows this quasar's optical variations since 1930. Its 1937 outburst later gained 3C 279 a (temporary) place in the* Guinness Book of Records *for the greatest absolute brightness of any known cosmic object. (Harvard-Smithsonian Center for Astrophysics)*

279, based on photographs taken since 1930. 3C 279's great distance and measured apparent brightness imply that, in 1937, this faint speck of light was *intrinsically* one of the most luminous objects ever observed in the universe. At that time, its luminosity exceeded 10^{48} erg/s. Two years later, its brightness had dropped by almost a factor of 250, making it ''merely'' 10 times brighter than the brightest radio galaxies.

Quasar variability is a very common phenomenon. Quasars have been observed to vary in brightness over periods of months, weeks, days—in some cases, even hours—in many parts of the electromagnetic spectrum. The same argument as we used before for active galaxies leads to the conclusion that the region generating the energy in the quasar must be very *small*—not too much larger than our solar system in at least some cases. Interlude 27-2 discusses one curious aspect of quasar variability that may have much to tell us about the details of the energy source. Let us now turn to a more general description of the engine driving these enormously powerful objects.

Quasar Energy Generation and Lifetimes

Quasars exhibit all of the properties we earlier described for active galaxies—large luminosities, polarized nonthermal emission, jets, lobes, and rapid variability implying small size. In many respects, a quasar looks like a scaled-up version of a Seyfert or radio active galaxy, so it should come as no surprise that the best current explanation of the quasar engine is basically a scaled-up version of the mechanism powering lower-luminosity active galaxies—accretion onto a supermassive black hole residing at the galactic core.

We have seen that a 10^8 or 10^9 solar-mass black hole can emit enough energy to power even the brightest (10^{45} erg/s) radio galaxy by swallowing stars and gas at the relatively modest rate of 1 star every 10 years. To power a 10^{47} erg/s quasar, which is 100 times brighter, the hole simply consumes 100 times more fuel—10 stars per year. The ''reprocessing'' mechanisms that convert the quasar's power into the radiation we actually detect—namely, the ejection of matter in jets and lobes and the reemission of radiation by surrounding gas and dust—probably operate in much the same manner as the mechanisms we described earlier for Seyferts and radio galaxies. The most likely explanation for quasars' large luminosities is simply that there was more fuel available at very early times, perhaps left over from the formation of the galaxies in which the quasars reside. At the distances of most quasars, the galaxies themselves cannot be seen. Only their intensely bright nuclei are visible from Earth.

In this picture, the brightest known quasars devour about 1000 solar masses of material every year. A simple calculation indicates that if they kept up this rate of energy production for the 10^{10}-year (or so) age of the universe, a total of 10^{13} stars would have to be destroyed. Unless the galaxies housing quasars are much larger than any other galaxy we know of, most of the quasar's parent galaxy would be completely consumed, and the universe should contain many 10^{13}-$M_\odot$ black holes—''burned-out'' quasars. We have no evidence for the existence of any such objects. One way around this possible problem is to suppose that a quasar spends only a fairly short period of time in this highly luminous phase—perhaps a few tens of millions of years. There is theoretical evidence to suggest that black holes tend to eat out ''cavities'' at the centers of their host galaxies, effectively cutting off their fuel supply through their own greed. Alternatively, as with active galaxies nearby, the high luminosities may be the result of interactions between galaxies in the early universe. The fact that quasars have been observed in some distant galaxy clusters argues in favor of this latter view. For a radically different view of quasars, however, see Interlude 27-3.

Quasar "Mirages"

In 1979, astronomers were surprised to discover what appeared to be a binary quasar—two quasars with exactly the

INTERLUDE 27-2 *Faster-Than-Light Velocities?*

The compact sources of emission within radio galaxies and quasars vary not only in intensity but also in structure. Some quasars recently mapped with very-long-baseline interferometers have displayed dramatic changes in structure, often on time scales as short as months. For example, the accompanying illustrations show three radio images of the core of the well-studied "nearby" quasar 3C 273, made several years apart. The interior of this quasar is dominated by two large blobs of gas, which move over the course of time. Knowing the distance to 3C 273 (about 640 Mpc) and measuring the angle through which the blobs moved in the course of 3 years (about 2 milli-arc seconds), astronomers have calculated the blobs' velocities. Astonishingly, the result is nearly 10 times the speed of light!

The notion that the speed of light is the highest attainable velocity is central to modern physics. Scores of predictions made assuming this fact to be true have been verified to high accuracy since Einstein first published his Theory of Relativity early in the twentieth century (see Interlude 24-1). Astronomers almost universally agree that some reinterpretation of these apparent *superluminal* (that is, faster than light) quasar motions is needed.

One alternative assumes that quasars are not at cosmological distances from us (see Interlude 27-3). At comparatively local distances, the quasar blobs would not physically move too far within the quasar, and the problem of explaining how the individual blobs move faster than light does not exist. But if the quasars are local and not distant, the observed red shifts of their spectral lines cannot be a distance indicator, and we would be forced to find some other explanation for the large quasar red shifts. As discussed in more detail in Interlude 27-3, few astronomers are prepared to make that assertion.

Several alternative solutions have been proposed to account for the apparent superluminal motion of the blobs without requiring that the quasars are local or that the speeds are truly faster than light. The most straightforward model suggests that the observed changes in quasar structure are not caused by actual motions of the interior blobs at all, but rather by variations of the radio intensity of *stationary* blobs. In other words, the interiors of quasars may resemble the blinking lights that sometimes appear on movie theater marquees. These marquee bulbs blink on and off, suggesting motion, but they are of course stationary. Likewise, blinking radio sources within the quasar may suggest motion where none really exists. Because the motion is not real, there is no law of physics requiring it to remain within the speed of light. A variation on this model explains the motions as a different kind of illusion—a projection effect, produced by blobs moving almost precisely along our line of sight at slightly less than the speed of light. Calculations of how these blobs would look from Earth indicate that they could in fact appear to be moving faster than light.

None of the alternative models is simple. They all require peculiar geometries, and no one model is agreed upon by all researchers. We still lack a complete explanation of the puzzling phenomenon of superluminal velocities. Still, the existing models demonstrate that although the interiors of quasars are exceedingly complex and not terribly well understood, there is no compelling need to discard the laws of physics in order to explain them.

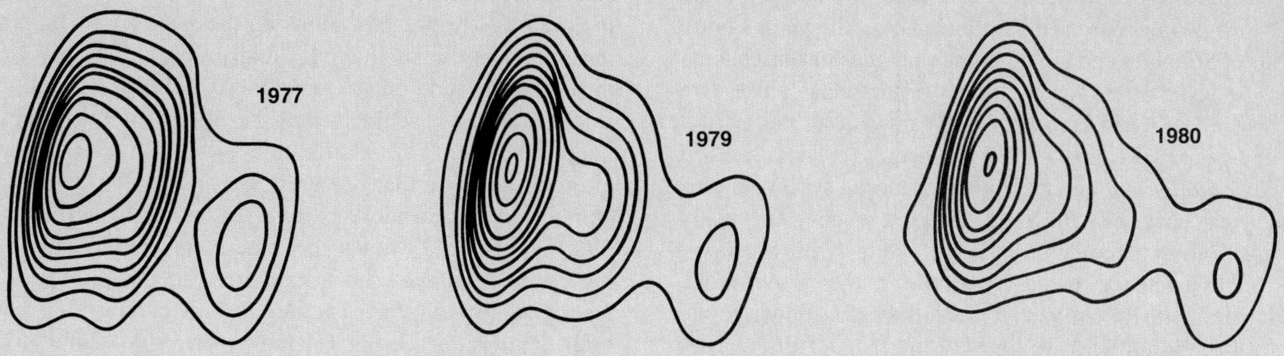

same red shift and very similar spectra, separated by only a few arc seconds on the sky. Remarkable as the discovery of such a binary would have been, the truth about this pair of quasars turned out to be even more amazing. Closer study of the quasars' radio emission revealed that they were *not* in fact two distinct objects. Instead, they were two separate images of the *same* quasar! Optical and radio views of this so-called *twin quasar* are shown in Figure 27.19.

What could produce such a "doubling" of a quasar image? The answer was provided long ago by Albert Einstein's General Theory of Relativity (see Interlude 24-1), which predicts that a beam of light will be deflected by the

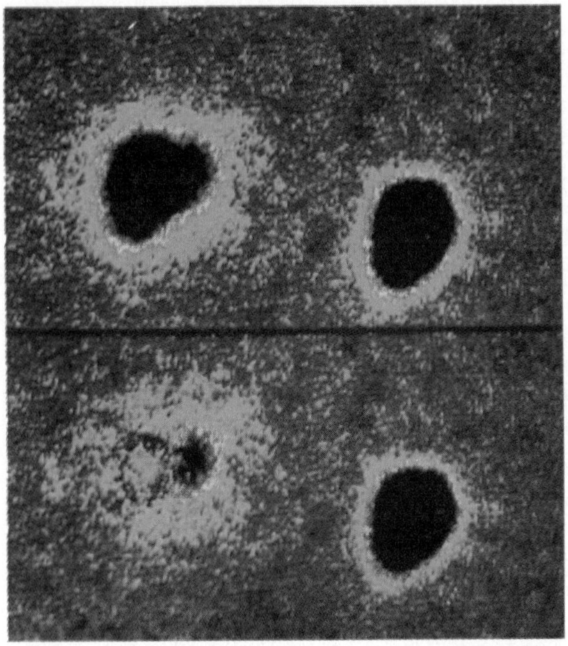

(a) R I V U X G

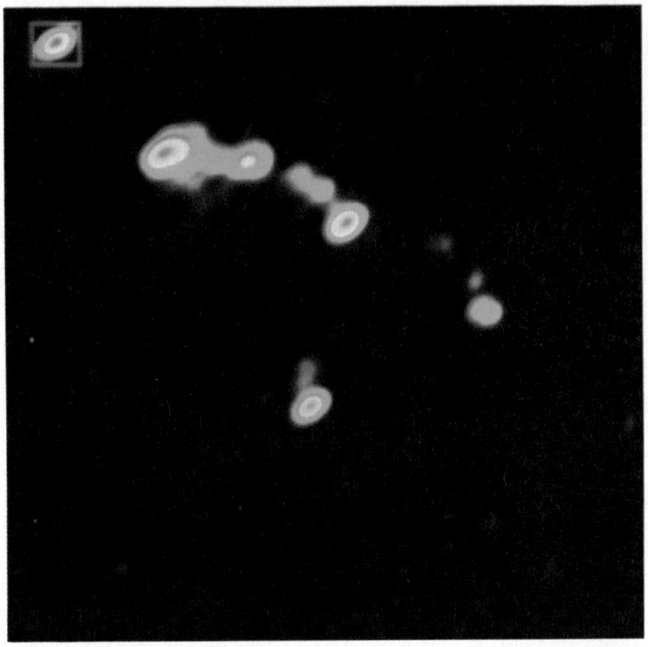

(b) R I V U X G

Figure 27.19 *This "twin" quasar (designated Q0957+561 and located about 7 billion light-years away) is not two separate objects at all. Instead, the two blobs shown here are images of the same object, created by a gravitational lens. The lens is barely visible in the optical image (a), but a high-resolution radio map (b) of the same region clearly shows both images and the lensing galaxy, a trillion-solar-mass giant elliptical. Notice the quasar jet, which is lensed along with the quasar itself in the upper A image but not in the bottom B image. Red is most intense radio emission, blue is faintest. (NOAO, NRAO)*

gravitational field of any massive object. We have already seen examples of this deflection in Chapter 24, in the bending of starlight by the Sun and the capture of light by a black hole. Given a massive enough body, such as a large galaxy, and the right geometrical alignment, it is possible for the light from a distant object (in this case, the quasar) to be split into two or more separate images, as shown in Figure 27.20. This phenomenon is known as **gravitational lensing.** About a dozen or so likely gravitational lenses are known, all of them involving a distant quasar and a foreground galaxy or galaxy cluster. Figure 27.21 shows a *Hubble Space Telescope* image of another lensed system. In this case, *four* images of the same quasar can be seen, neatly arranged around the central image of the lensing galaxy.

The existence of these multiple images provides astronomers with a number of useful observational tools. First, the lensing tends to amplify the light of the quasar, making it easier to observe. Second, because the light rays forming the two images usually follow paths of different lengths, there is often a *time delay,* ranging from several days to several years, between the two images. This delay provides advance notice of explosive events, such as sudden flare-ups in the quasar's brightness—that is, if one image flares up, astronomers know the other will too, and so they have a second opportunity to study the event. The time delay also permits astronomers to determine the *distance* to the lensing galaxy by carefully timing the measurements. If enough lenses are found, this method may provide a reliable alternative means of measuring the Hubble constant that is *independent* of any of the techniques discussed in Chapter 26. Third, so-called **microlensing**—lensing by individual stars in the foreground galaxy—can cause large fluctuations in a quasar's brightness, allowing the stellar content of the lensing galaxy to be studied. Finally, by studying the lensing of background quasars and galaxies by foreground galaxy clusters, astronomers may be able to obtain a better understanding of the distribution of dark matter in those clusters, an issue that has great bearing on the large-scale structure of the cosmos, as we will see in Chapters 28 and 29. Figure 27.22 shows the images of some faint, blue background galaxies bent into arcs by the gravity of a nearby galaxy cluster. The degree of bending allows the total mass of the cluster (including the dark matter) to be measured.

ACTIVE GALAXY EVOLUTION

In the previous chapter, we addressed the issue of evolutionary change among normal galaxies. Let us now briefly consider the possibility of evolutionary links among active

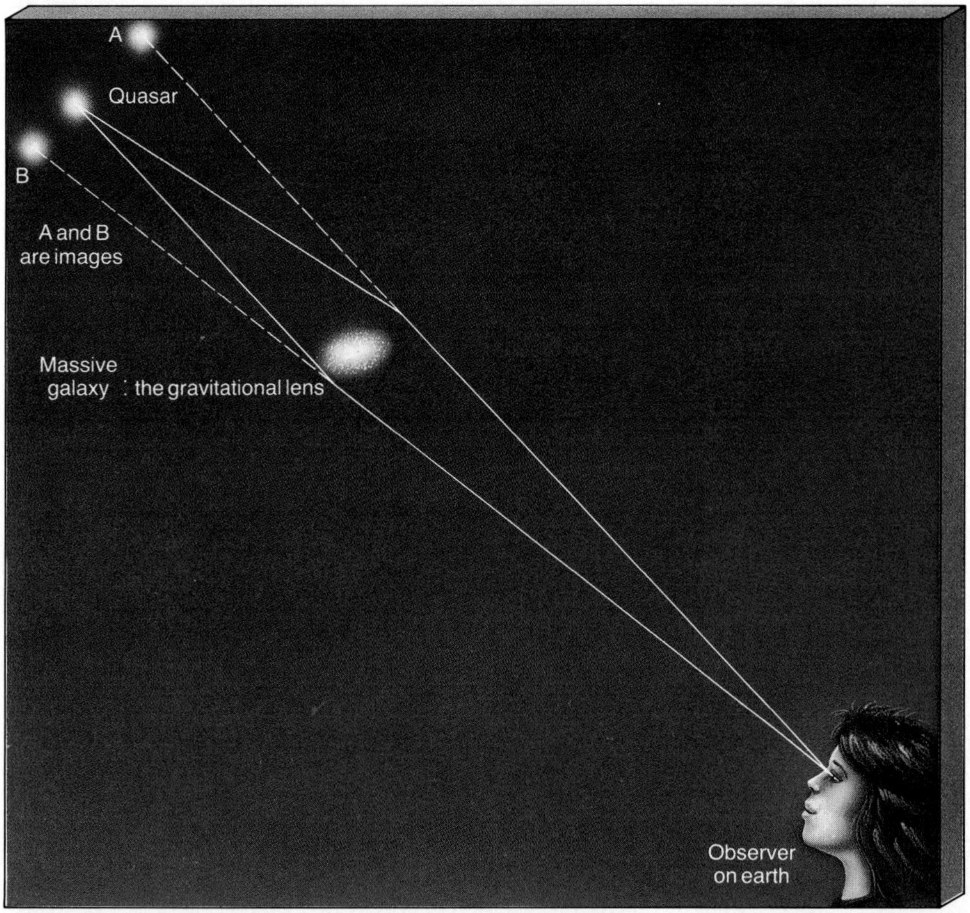

Figure 27.20 *When light from a distant object passes close to a galaxy or cluster of galaxies along the line of sight, the image of the background object (here, the quasar) can sometimes be split into two or more separate images (here, A and B). The foreground object is known as a gravitational lens.*

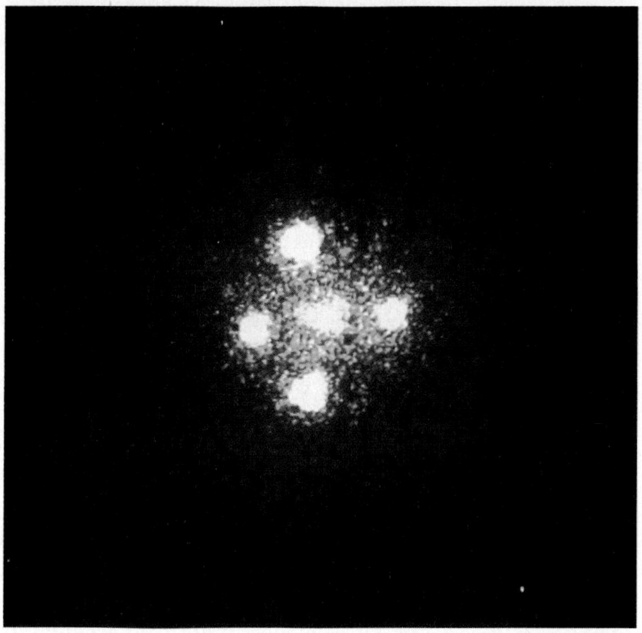

R I **V** U X G

Figure 27.21 *The "Einstein Cross," a multiply imaged quasar. In this* Hubble Space Telescope *image, four separate images of the same quasar have been produced by the galaxy at the center. (NASA)*

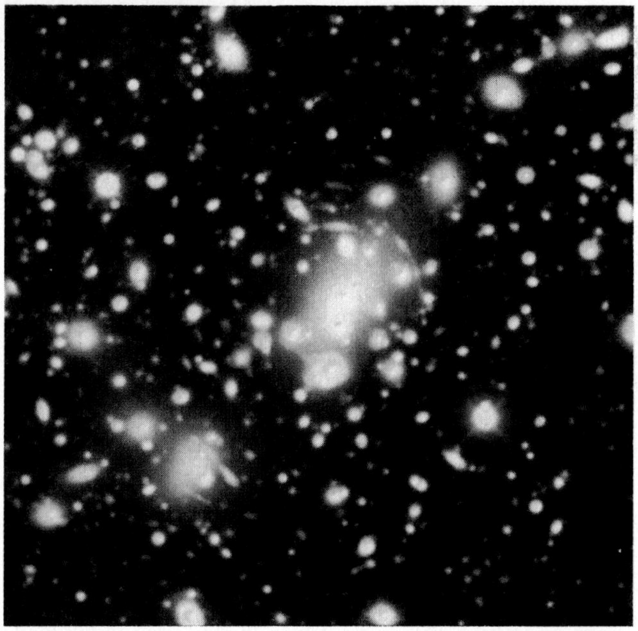

R I **V** U X G

Figure 27.22 *The blue arcs are images of young, blue background galaxies, bent into the observed shapes by the gravitational field of a foreground galaxy cluster, which deflects their light and distorts their appearance. By measuring the extent of this distortion, astronomers can estimate the mass of the intervening cluster. (G. Bernstein and T. Tyson)*

INTERLUDE 27-3 *Could Quasars Be Local?*

When quasars were first discovered, their large luminosities and small sizes troubled many astronomers. In the 1960s and 1970s, no known mechanism could account for the generation of 1000 Milky Ways' worth of energy within a region comparable in size to the solar system. The idea that supermassive black holes may be quite common in the cores of galaxies was unknown, and, as we have seen, other attempted explanations encounter serious difficulties.

In response to these problems, some astronomers, notably the respected observer Halton Arp and the equally reputable theorist Geoffrey Burbidge, sought an alternative explanation for quasars. Instead of believing that these objects were at cosmological distances and so very luminous, these researchers argued that perhaps there was an alternative, noncosmological explanation for the great red shift. Quasars could be relatively nearby and hence much less bright. For example, if the distance to 3C 273 were 100 times closer—only 6.4 Mpc, not 640 Mpc, away—the inverse-square law says that the luminosity required to account for its observed apparent brightness would be reduced by a factor of 10,000. In that case, 3C 273 would radiate ''only'' 10^{43} erg/s (1/10 the energy output of our Galaxy). This amount is still a lot of energy, but it is perhaps more easily explainable in terms of ''familiar'' stellar events: the formation of high-mass stars, supernovae, and so on.

Arp has reported many examples (two appear on the next page) of instances where galaxies and quasars are found close together on the sky but have greatly different, conflicting red shifts. He argues that there are simply too many of these ''coincidences,'' where a foreground galaxy lies in nearly the same direction as a supposed background quasar, for the distant-quasar hypothesis to be correct. Instead, he claims, the quasars are *physically* close to the galaxies, and the red shifts have some other, noncosmological, explanation. In the first image on the next page, a quasar appears to be connected to a galaxy with a very different red shift by a faint ''bridge'' of gas, reinforcing the view that the two really are close together in space. Arp and coworkers have gone further, citing similar examples where neighboring *galaxies* have conflicting red shifts (as shown in the second image), thus calling into question the cosmological interpretation of the red shift in all cases.

Most astronomers, however, would argue that the claims of conflicting red shifts are not statistically significant. In other words, given the numbers of known galaxies and quasars, accidental superpositions on the sky should be quite commonplace, and the observed quasar-galaxy and galaxy-galaxy alignments are quite consistent with pure chance. The apparent bridges may simply be photographic or image-processing defects. Usually, a lot of computer enhancement is needed to bring the images out, and there is ample opportunity for ''features'' to appear where none really exist. Finally, Hubble's law *is* well-established for galaxies within a few hundred megaparsecs, and some quasars have been found in galaxy clusters, *sharing* the red shift of their neighbors. Thus, at least some quasar red shifts are known to be cosmological, so the claimed violation of Hubble's law must be quite selective, if it exists at all.

If quasars are actually local, no convincing explanation has ever been advanced for their red shifts. If quasars are simply moving at high velocities, then we are led to the new ''energy problem'' of explaining how they were accelerated to such enormous speeds. More important, why are they *all* redshifted? If quasars were fired out of galaxies, we

galaxies and between normal and active galaxies. We emphasize that this section is really mostly speculation. Although the consensus is that galaxies began to form at a red shift of about 5—the red shift of the oldest quasars—and that quasars were an early stage of galaxy evolution, the details of the connections among different types of active and normal galaxies are still very uncertain.

Most quasars are very distant, indicating that they were more common in the past than they are today. At the same time, ''normal looking'' galaxies seem to be less common at high red shift. These two pieces of evidence suggest to many astronomers that, when galaxies first formed, they probably looked like quasars. This opinion is strengthened by the fact that the same black-hole energy-generation mechanism can account for the luminosity of quasars, active galaxies, and the central regions of normal galaxies like our own. Large black holes do not simply vanish, at least in the 10–20 billion years that the universe has existed (see Interlude 24-3). Thus, the presence of supermassive black holes in the centers of many, if not all, normal galaxies is consistent with the idea that they started off as quasars, then ''wound down'' to become the relatively quiescent objects we see today.

In this picture, the gradual reduction in violence from a quasar to a Seyfert galaxy to a normal spiral, for example, occurs primarily because the fuel supply is reduced as the galaxy evolves. A similar sequence might connect quasars to BL Lac objects to radio galaxies to normal ellipticals. These possible evolutionary connections among the active and normal galaxies are shown in Figure 27.23. Adjacent objects along this sequence are nearly indistinguishable from one another. For example, weak quasars share some

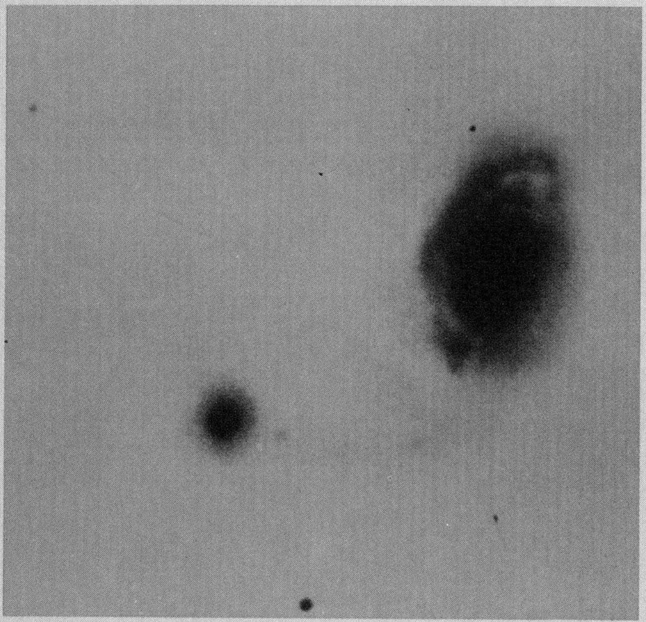

(*Halton Arp*) R I V U X G

would expect some to be moving toward us, and hence blueshifted. If they all came from our own Galaxy, then the Milky Way is special—a major violation of the principle of mediocrity and current scientific dogma—or the quasars must be so close (and thus quite faint) that we should have observed the motion of some of them across the sky. Arp's own supposed correlations between other galaxies and quasars would also argue against a Milky Way origin. In fact, no matter where we put the quasars, their red shifts pose problems. An alternative attempt to explain the red shift as a gravitational red shift suffered by light as it climbs out from the vicinity of a black hole (see Chapter 24) also fails because it cannot account for the observed widths of quasar spectral lines.

There is no clear observational evidence for conflicting quasar red shifts and no satisfactory mechanism of producing the red shifts by non-cosmological means. Furthermore, as we have seen, there is no "quasar luminosity problem" any more. Quasar luminosities can be explained by the same mechanism that powers active galaxies, with only a modest increase in fuel consumption and certainly without posing a serious challenge to the laws of physics. Consequently, the overwhelming majority of astronomers hold that quasar red shifts are cosmological in origin and that quasars really are the most distant objects known in the universe. Of course, there are many instances in the history of astronomy where the majority has later been proved totally wrong!

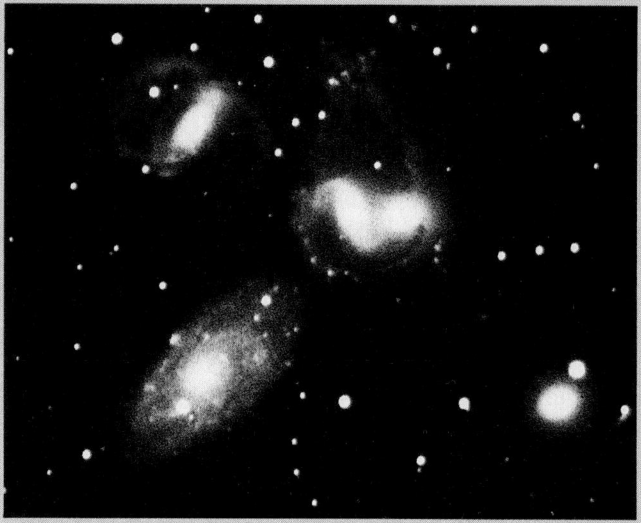

(*Halton Arp*) R I V U X G

characteristics with some very active galaxies, and the feeblest active galaxies often resemble the most explosive normal galaxies.

If we accept this appealing (but still unproven) view, we can construct the following possible scenario for the evolution of galaxies in the universe: Galaxies formed at a red shift of around 5. The early round of massive star formation that may have expelled galactic gas and helped determine a galaxy's Hubble type—spiral or elliptical—could also have given rise to many large, stellar-mass black holes, which sank to the center of the still-forming galaxy and merged into a supermassive black hole there. Alternatively, the supermassive hole may have formed directly by gravitational collapse of the dense central regions of the protogalaxy. Whatever the cause, large black holes appeared at the centers of many galaxies at a time when there was still plenty of fuel available to power them, resulting in many highly luminous quasars. The brightest quasars—the ones we see from Earth—were those with the greatest fuel supply. The young galaxies themselves were so faint compared with their bright quasar cores that we simply cannot see them.

As the galaxy developed and the black hole used up its fuel, the luminosity of the central nucleus diminished. While still active, it no longer completely overwhelmed the emission from the surrounding stars. The result was an active galaxy—a radio galaxy or a Seyfert—still emitting a lot of energy, but now with a noticeable "stellar" component in its spectrum. The central activity continued to decline. Eventually, only the surrounding galaxy remained visible—a normal galaxy, like the majority of those we now see around us. Today, the black holes that generated so

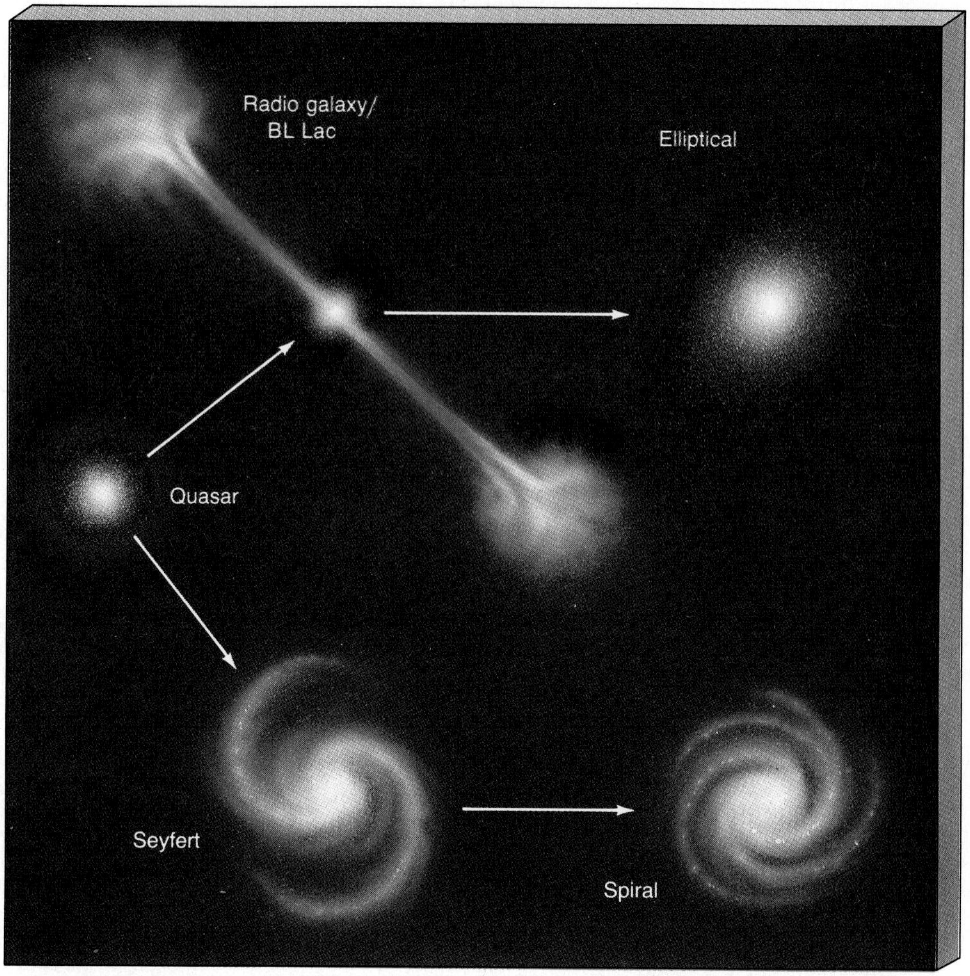

Figure 27.23 *A possible evolutionary sequence for galaxy activity, beginning with the highly luminous quasars, decreasing in violence through the radio and Seyfert galaxies, and ending with normal spirals and ellipticals. The central black holes that powered the early activity are still there at later times; they simply run out of fuel as time goes on.*

much youthful energy lie dormant in galactic cores, producing only a relative trickle of radiation. Occasionally, two nearby normal galaxies may interact with one another, causing a flood of new fuel to be directed toward the central black hole of one or both. The engine starts up for a while, giving rise to the nearby active galaxies we observe.

Should this picture be correct, then many normal galaxies, including perhaps our own Milky Way Galaxy, were once brilliant quasars. Perhaps some alien astronomer, thousands of megaparsecs away, is at this very moment observing our Galaxy—and seeing it as it was billions of years ago—and is commenting on its enormous luminosity and polarized jets and wondering what exotic physical process could possibly account for its violent activity!

When first discovered, active galaxies and quasars seemed to present astronomers with insurmountable problems. For a time, their dual properties of enormous energy output and small size appeared incompatible with the known laws of physics and threatened to overturn our modern view of the universe. Yet the problems were eventually solved, and the laws of physics remain intact. Far from jeopardizing our knowledge of the cosmos, these violent phenomena have become part of the thread of understanding that binds our own Galaxy to the earliest epochs of the universe we live in.

CHAPTER SUMMARY

Active galaxies emit colossal amounts of energy, mostly in the invisible part of the electromagnetic spectrum. Their energy production starts near the top of the luminosity scale for normal galaxies and extends several orders of magnitude beyond it. While some active galaxies lie relatively close to the Milky Way Galaxy, the most energetic objects are found at the largest red shifts, far from Earth. Many of these objects emit thousands of times more energy than the entire Milky Way.

Seyfert galaxies have properties lying between those of normal galaxies and those of the brightest active galaxies known. Rapid fluctuations in their luminosities indicate that in many cases their energy source must be less than a light-year across. Seyferts look like normal spiral galaxies except for their highly luminous galactic nuclei. Bright radio galaxies, however, tend to be elliptical. Most of the energy from a core-halo radio galaxy comes from an extremely small central nucleus, or core. Weaker emission comes from an extended halo surrounding the core. In lobe radio galaxies, most of the energy comes from roundish extensions of gas up to a megaparsec across that lie well beyond the visible galaxy. Their geometry suggests that violent events expel material from a central source.

The brightest giant elliptical galaxies are comparable in total luminosity to active galaxies, but active galaxies produce their energy in a very much smaller volume. Also, the energy output of an active galaxy is observed to be nonstellar in character—it cannot be explained simply as the accumulated energy of its component stars. The emission of active galaxies is most intense at low frequencies and is polarized. Synchrotron emission, where high-speed charged particles emit radiation as they are accelerated in a strong magnetic field, is the most likely explanation for these observations. The energy-emitting region may be far from the energy-production site. It is believed that the central powerhouse emits jets of energetic particles that can travel millions of parsecs before finally radiating their energy into space. The leading model for the central engines of active galaxies involves supermassive black holes, with up to a billion times more mass than our own Sun. In order to explain the luminosity of a typical active galaxy, we must assume that a central black hole disrupts and devours one star every few years.

The black-hole model can explain many of the features observed in active galaxies. There is also evidence to suggest that much of the activity observed in nearby active galaxies has been sparked by recent interactions with galactic neighbors. Thus, even after a supermassive black hole has used up most of its available fuel, the fuel supply may be replenished occasionally by the influence of a companion whose gravitational field can divert stars and gas into the central engine.

Quasars are the most distant and the most energetic objects yet observed in the universe. The several hundred known quasars all have spectra with large red shifts, implying great distances from Earth and therefore extremely high luminosities. Many quasars have red shifts greater than 1. This does not mean that they are moving faster than light, but rather that their recessional velocities are close to the speed of light. Astronomers often refer to distant objects by their look-back time, which is simply how long ago they emitted the radiation we see today.

Most quasars look like little more than balls of luminous fuzz in visible light. As with active galaxies, their radiation is nonthermal and often polarized. In addition, quasars have been observed to vary in brightness over periods of months, weeks, days, and even hours, so their energy must come from a very small region of space. The leading explanation of their luminosities is a scaled-up version of the energy-generation mechanism for active galaxies—accretion onto a supermassive black hole. If this model is correct, then quasars' central black holes must consume a lot of mass—as much as a thousand Suns per year. This may mean that quasars spend only a fairly short time in a highly luminous phase—perhaps a few tens of millions of years early in their lives, long ago.

Given a massive enough body, such as a large galaxy, and the right geometrical alignment, it is possible for the light from a distant quasar to be split into two or more separate images. This phenomenon is known as gravitational lensing. About a dozen gravitational lenses are known. These systems allow astronomers to study both the quasars themselves and the lensing galaxies (or galaxy clusters) in greater detail.

KEY WORDS

active galaxy
core-halo radio galaxy
cosmological distance
cosmological red shift
galactic nucleus
gravitational lensing
head-tail radio galaxy
lobe radio galaxy
look-back time
microlensing
nonthermal radiation
quasar
quasi-stellar object
radio lobe
Seyfert galaxy
synchrotron emission

REVIEW QUESTIONS

1. Name two basic differences between normal galaxies and active galaxies.
2. Contrast the energy emission of the most energetic normal galaxies with that of active galaxies.
3. Briefly describe some of the basic properties of Seyfert galaxies.
4. Give two reasons why the primary energy source of Seyferts is thought to be located in their centers.

5. What distinguishes a core-halo radio galaxy from a lobe radio galaxy?
6. What is the evidence that the radio lobes of some active galaxies consist of material ejected from the galaxy's center?
7. What conditions are thought to create a head-tail radio galaxy?
8. Briefly describe the leading model for the central engine of an active galaxy.
9. At what rate would a 10^8 or 10^9-$M_\odot$ black hole need to ingest stars in order to account for the brightest active galaxies (those with observed luminosities of about 10^{45} erg/s, say)?
10. What is synchrotron emission? How is this process linked to active galaxies?
11. What are BL Lac objects, and what do astronomers hope to learn from them?
12. What is look-back time, and why is it a helpful concept in discussions of active galaxies and quasars?
13. What might the large red shifts of quasar spectra reveal about the early universe?
14. How are the spectra of distant quasars used to probe the space between us and them?
15. What evidence suggests that quasars may spend only a few tens of millions of years in a highly luminous phase?
16. What are gravitational lenses?
17. Centaurus A—from one radio lobe to the other—spans about 1 Mpc. It lies at a distance of 4 Mpc from Earth. What is the angular size of Centaurus A? Compare this with the angular size of the Moon.
18. How much energy would an active galaxy generate if it consumed one Earth mass of material every day?
19. A certain quasar has a recessional velocity of 60,000 km/s and the same apparent brightness as the Sun would have if it were placed at a distance of 1 kpc. Assuming a Hubble constant of 75 km/s/Mpc, calculate the luminosity of the quasar.
20. A quasar has emission lines with a red shift of 0.15 and absorption lines with a red shift of 0.1. How far from the quasar is the object responsible for the absorption lines? (Assume H_0 = 75 km/s/Mpc.)

DISCUSSION QUESTIONS

1. Supermassive black holes may power the central engines of active galaxies. Galaxies may also experience a surge in energy production when they interact with their neighbors. Can you imagine other possible mechanisms for producing the tremendous energy of active galaxies? How might astronomers begin to test those possibilities?
2. Do you believe the argument put forth by respected astronomers like Halton Arp and Geoffrey Burbridge that the spectra of quasars do not necessarily imply cosmological distances? Why do you suppose they persist in their ideas, even though most other astronomers do not believe them?

PROJECTS

1. The core-halo radio galaxy M87 has been seen by observers with optical telescopes since the eighteenth century. This giant elliptical galaxy is one of the brightest members of the Virgo Cluster of galaxies, a wonderful assembly well known to observers who sweep overhead with small telescopes on May evenings. M87 is located nearly on a line about halfway between Vindemiatrix, an inconspicuous star in the constellation Virgo, and Denebola, a somewhat brighter star marking the tail of Leo the Lion. Telescopic observers typically see it as a hazy round ball, brighter toward the middle, with a star-like center. Sketch what you see. Compare your drawing with the images of M87 shown in this chapter.

SUGGESTED READINGS

Burns, J. O. ''Chasing the Monster's Tail.'' *Astronomy* (August 1990). Good article discussing new views of cosmic jets.

Courvoisier, T. J.-L. ''The Quasar 3C 273.'' *Scientific American* (June 1991). One of the most luminous objects in the universe and the nucleus of an active galaxy.

Mercury (January–February 1988). Special issue commemorating the twenty-fifth anniversary of the discovery of the red shift and bizarre properties of quasars.

Preston, R. *First Light*. The Atlantic Monthly Press, 1987. Insider's look at research at Palomar Observatory. Good section about early work on quasars.

Sheldon, E. ''Faster Than Light?'' *Sky and Telescope* (January 1990). A discussion of superluminal quasars.

Wilkes, B. ''The Emerging Picture of Quasars.'' *Astronomy* (December 1991). What many of today's astronomers believe about quasars and their relation to active galaxies.

CHAPTER

28

Cosmology

The Big Bang and the Fate of the Universe

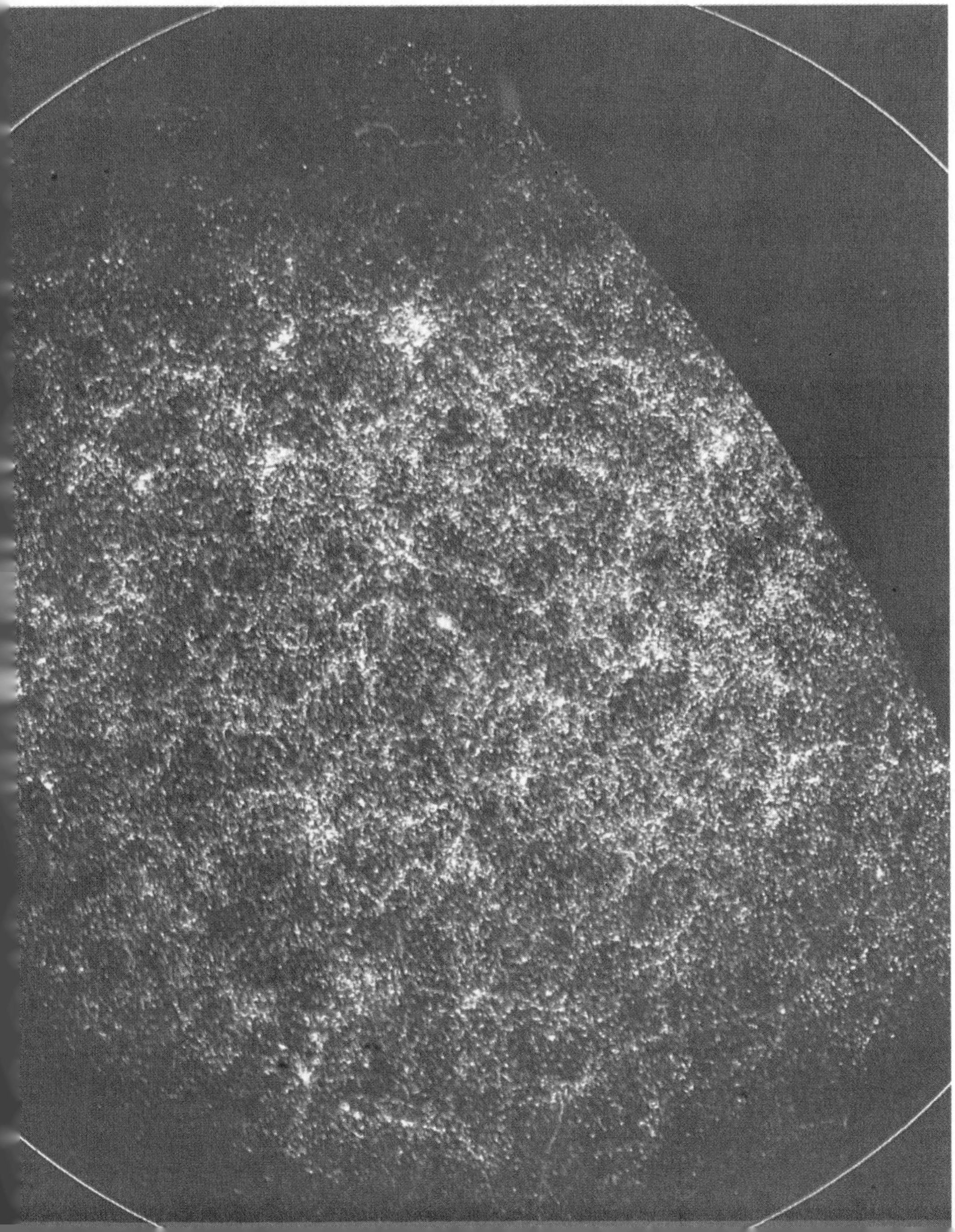

This computer-generated map of over a million galaxies is based on information contained in a catalog compiled in the 1970s. This is not the most extensive survey of the sky, but it is one of the broadest. The number of galaxies within an area on the map is represented by the brightest of that area—that is, the brighter the area, the more galaxies found there. The dark curved edge is that part of the sky obscured by the disk of our own Galaxy. Clearly, the distribution of galaxies across the sky is not uniform. Some regions are relatively crowded; others are relatively empty. The densest knots are galaxy clusters. This map, and others like it made during the 1970s and 1980s, provided astronomers with strong evidence that galaxy distribution is uneven and forced them to try to understand why this is so. (Courtesy Princeton University)

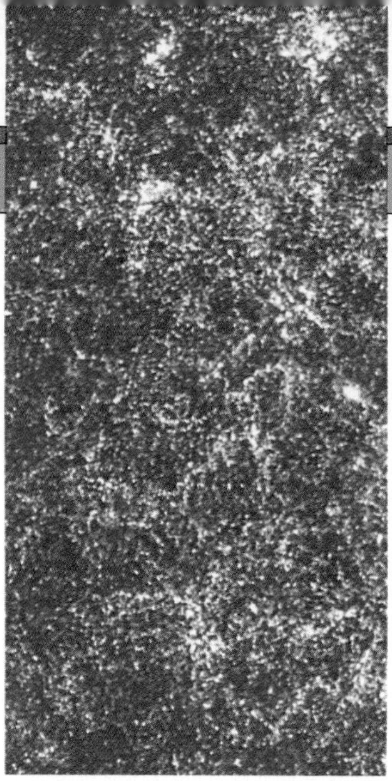

LEARNING GOALS

1. to recognize that we do not reside at any special place in the universe
2. to know the approximate age of the universe and how it is determined
3. to see how the future of the universe and the overall geometry of space are intertwined
4. to understand the leading evolutionary models of the universe
5. to study some of the methods used by astronomers to test their theories of the universe and to see what conclusions they have reached
6. to realize that the universe is filled with radiation left over from the Big Bang

We have now reached the limits of the observable universe. Along the way, we have studied the properties of planets, stars, and galaxies, all the way out to the distant quasars. Our field of view has widened, slowly at first, then at a rapid pace, to the point where it now extends for billions of parsecs into space and billions of years back in time. Now at last we are in a position to try to understand the central features of the biggest picture of all. In short, we are ready to tackle the entire universe.

In the next two chapters, our attention will be focused on the very largest scales of the cosmos. Even individual galaxies and galaxy clusters will be treated as insignificant points of mass and light. Only their bulk properties—the combined mass and gravity of millions of galaxies at once—will concern us here. We have asked and answered many questions about the universe on smaller scales; now we strive to understand the cosmos as a whole. What is the size of the universe? What is its structure, its shape? How long has the universe been around? How long will it last? What was its origin, and what will be its fate? Is the universe a one-time event, or does it recur and renew itself, in a grand cycle of birth, death, and rebirth?

Many cultures have asked these questions, in one form or another, and have developed their own cosmologies—theories of the nature, origin, and destiny of the universe—to answer them. In this and the next chapter, we will see how modern scientific cosmology addresses these important issues and what it has to tell us about the universe we inhabit.

THE UNIVERSE ON THE LARGEST SCALES

The End of Structure

✓1 The universe shows structure on every scale we have examined so far. Subatomic particles form nuclei and atoms. Atoms form planets and stars. Stars form star clusters and galaxies. Galaxies form galaxy clusters, superclusters, and even larger structures—voids and sheets that stretch across the sky. From the quarks in a proton to the galaxies in the Great Wall, we can trace a hierarchy of "clustering" of matter from the very smallest to the very largest scales. A fairly obvious and natural question is: Does the clustering ever end? Asked another way, Is there some scale on which the universe can be regarded as more or less smooth and featureless? Perhaps surprisingly, given the trend we have just described, most astronomers think the answer is *yes*.

We saw in Chapter 26 how surveys of the universe have revealed the existence of structures as large as 200 Mpc (megaparsecs) across. Although 200 Mpc is certainly an enormous size, it is still quite small relative to the 4000 Mpc distance of the remotest quasars. The most extensive surveys have studied only about one 1/10,000 of the total volume of the observable universe. As Margaret Geller (one of the researchers leading the CfA Redshift Survey) points out, extending those results to the entire visible universe is like using a map of Rhode Island to draw conclusions on the appearance of the entire surface of the Earth. The structure we see locally is not necessarily a reliable indicator of the appearance of the universe on very large scales.

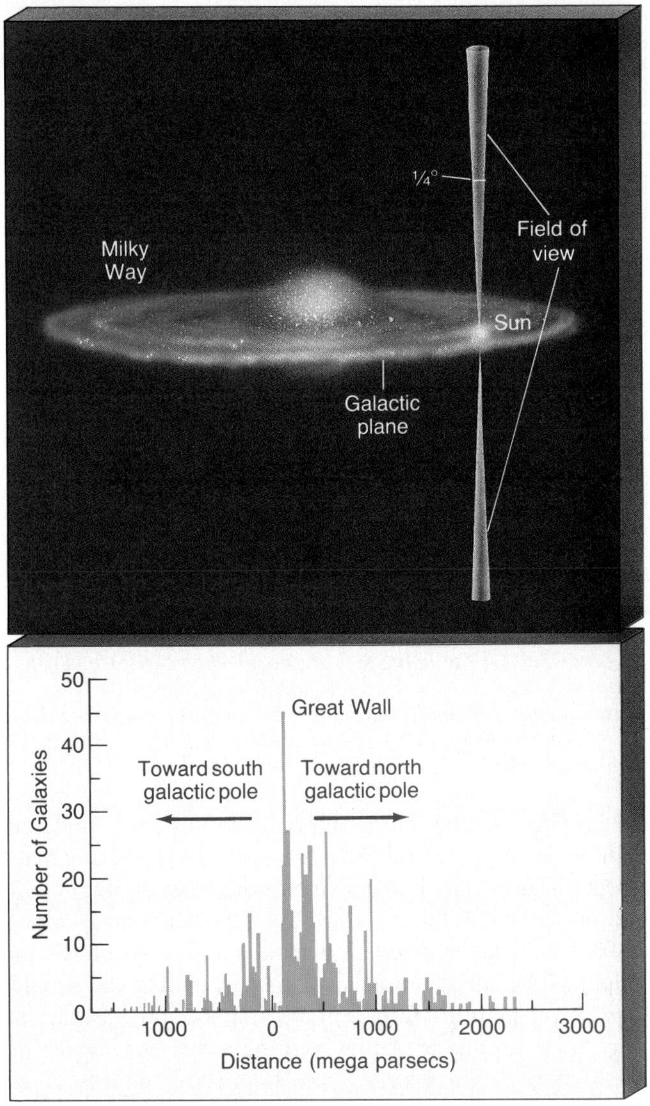

Figure 28.1 *The results of a deep "pencil-beam" survey of two small portions of the sky in opposite directions from Earth. The graph shows the number of galaxies at different distances from us, out to a maximum distance of about 2000 Mpc. The distinctive "picket fence" appearance seems to show voids and sheets of galaxies on scales of a 100 or 200 megaparsecs but provides no indication of any larger structure.*

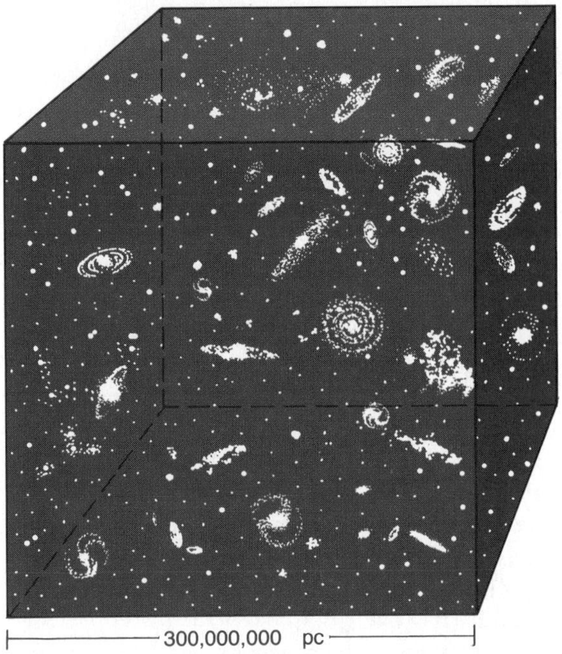

Figure 28.2 *Diagram of galaxies contained within an enormous cube, 300 Mpc on a side. Cosmologists believe that no matter where we placed this cube in the universe its contents would look similar.*

Do we have any larger studies of the universe? Yes and no. There are no surveys that cover larger volumes of space, although some are planned, but deeper surveys (that is, extending to greater distances) of small regions of the sky *do* exist. Figure 28.1 shows the result of such a "pencil-beam" survey, carried out by researchers in the U.S. and the U.K. It shows the distribution of galaxies with distance in two opposite directions out of the plane of our Galaxy, out to a distance of about 2000 Mpc, 10 times farther than the Harvard study. Although the numbers of galaxies fall off at large distances—mainly because of the simple fact that distant galaxies are very hard to see—a distinctive "on-off" pattern of galaxies, looking a little like a picket fence, can be seen. It is thought that the gaps between the "pickets" are voids like those seen closer to home and that the pickets themselves are places where the imaginary beam intersects sheets like the Great Wall.

The details of this and similar surveys remain somewhat controversial, but they do seem to indicate that the largest structures are only 100–200 Mpc across—no voids or clumps of galaxies much larger than that are evident in the data. However, absence of evidence is not evidence of absence, and there is no guarantee that larger structures will never be found. For that reason, you should bear in mind as you read this chapter that our discussion of the evolution of the universe is based on the assumption that there is no structure on the very largest scales. If this should turn out to be incorrect—for example, if a structure a few thousand megaparsecs across were to be discovered tomorrow—then much of the discussion that follows will be on very shaky ground indeed!

On the basis of a combination of sketchy data, theoretical insight, and not a little philosophical preference, **cosmologists**—astronomers who study the large-scale structure and dynamics of the universe—assume that the universe is roughly **homogeneous** on scales greater than a few hundred megaparsecs. In other words, if we imagine taking a huge cube—300 Mpc on a side, say—and placing it anywhere in the universe, as illustrated in Figure 28.2, the number of galaxies it enclosed would be pretty much the same—around 100,000, excluding the faint dwarf ellipticals and irregulars—no matter *where* in the universe we put it. Some of the galaxies would be clustered and clumped into fairly large structures, and some would not, but the total number would not vary much as the cube was moved from place to place. In short, the universe looks *smooth* on the largest scales.

The Cosmological Principle

Cosmic homogeneity is the first major assumption that astronomers make when studying the large-scale structure of the universe. As we have seen, this homogeneity is suggested, but not proven, by observations. We will see in Chapter 29 that there are also some plausible theoretical arguments in its favor. The second assumption made by cosmologists, also supported by observational evidence and theoretical reasoning, is that the universe is **isotropic**—that is, it looks the same in any direction. Isotropy is on much firmer observational ground than homogeneity. Apart from regions of the sky that are obscured by our Galaxy, the universe *does* look much the same in all directions, at any wavelength, provided we look far enough. In other words, any deep pencil-beam survey of the sky should count about the same number of galaxies as were found in the U.S./U.K. study cited above, regardless of which patch of the sky is chosen.

The assumptions of homogeneity and isotropy form the foundation of modern **cosmology**—the study of the structure and evolution of the entire universe. Together, these twin pillars of cosmology are known as the **cosmological principle.** No one really knows if this principle is absolutely correct. All that we can say is that, so far, it seems consistent with observations. From this point on in our discussion of the universe, we will simply assume that it holds.

The cosmological principle has very far-reaching implications. For example, it implies that there can be no *edge* to the universe, because that would violate the assumption of homogeneity. Furthermore, it implies that there is no *center,* because that would mean that the universe would not be the same in all directions from any noncentral point, a violation of the assumption of isotropy. Thus, this single principle strongly limits what the overall geometry of the universe can be. The cosmological principle is the ultimate expression of the principle of mediocrity. It states that not only are we not central to the universe, but *no one* can be central, because the universe *has* no center!

THE BIG BANG

Olbers's Paradox

✓2 It may not have occurred to you, but every time you go outside at night and notice that the sky is dark, you are making a profound cosmological observation. Here's why.

According to the cosmological principle, the universe is homogeneous and isotropic. Let's assume that it is also infinite in extent and unchanging in time—precisely the view of the universe that prevailed until the early part of the twentieth century. On average, then, the universe is uniformly populated with galaxies filled with stars. In that case, when you look up at the night sky, your line of sight *must* eventually encounter a star, as illustrated in Figure 28.3. The star may lie at an enormous distance, in some remote galaxy, but the laws of probability dictate that, sooner or later, any line drawn outward from the Earth will run into a bright stellar surface. This has a dramatic implication. No matter where you look, the sky should be as bright as the surface of a star—the entire night sky should be as brilliant as the surface of the Sun! The obvious difference between this prediction and the actual appearance of the night sky is known as **Olbers's paradox,** after the nineteenth-century German astronomer, Heinrich Olbers, who popularized the idea.

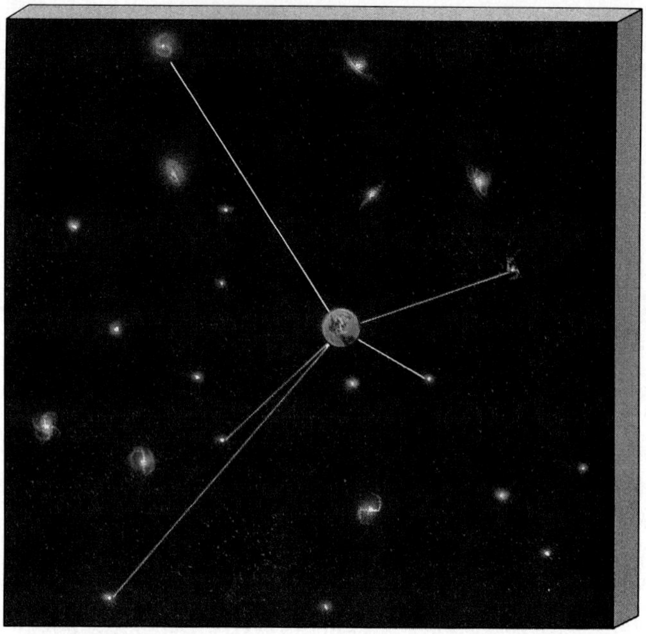

Figure 28.3 *If the universe were homogeneous, isotropic, infinite in extent, and unchanging, any line of sight from the Earth should eventually run into a star, and the entire night sky should be bright. This obvious contradiction of the facts is known as Olbers's paradox.*

What is the resolution to this paradox? We have accepted the cosmological principle, so we believe that the universe is homegeneous and isotropic. We must conclude, then, that one (or both) of the other two assumptions—that the universe is infinite in extent and unchanging in time—is false. Either the universe is finite in extent, or it evolves in time. As we will see in a moment, the answer involves a little of each and is intimately tied to the behavior of the universe on the largest scales.

Hubble's Law and the Age of the Universe

We saw in Chapter 26 that all the galaxies in the universe are rushing away from us in a manner described by Hubble's law:

$$\textit{recession velocity} = H_0 \times \textit{distance},$$

where $H_0 \approx 75$ km/s/Mpc is the Hubble constant. We used this relation in Chapters 26 and 27 as a convenient means of determining the distances to galaxies and quasars. Even though it appears to place us at the center of the expansion,

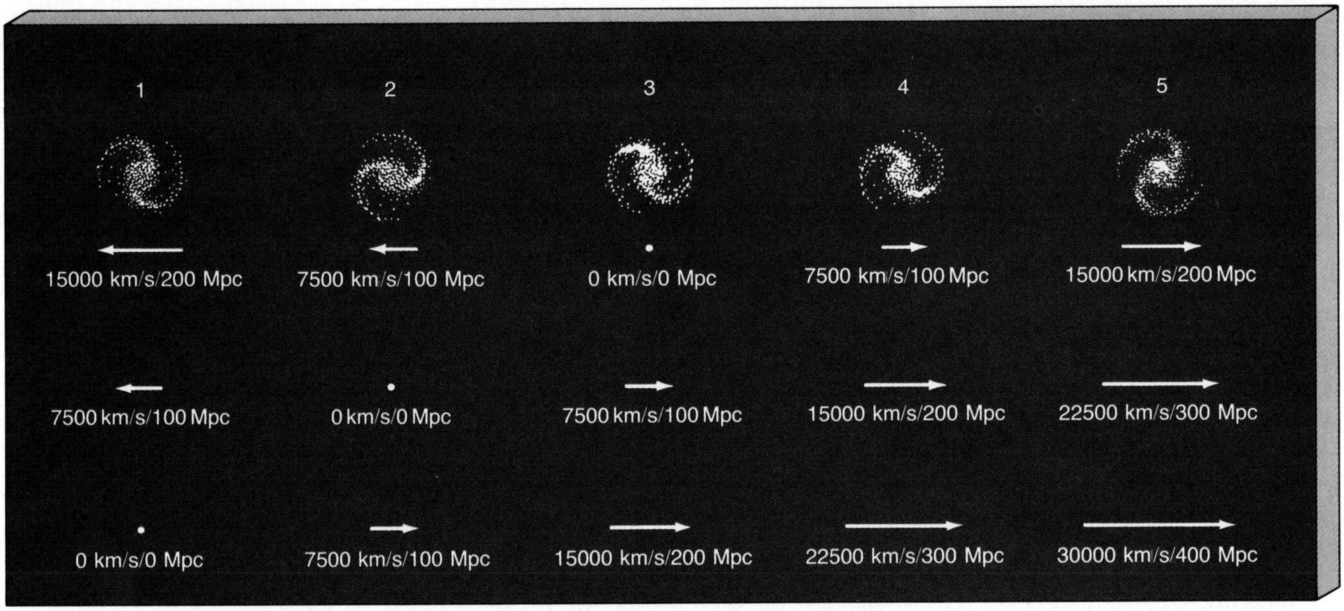

Figure 28.4 *The Hubble law is the same no matter which galaxy makes the measurements. The top set of numbers are the distances and recessional velocities as seen by an observer on galaxy 3. The bottom two sets are from the points of view of observers on galaxies 2 and 1, respectively. In all cases, the same Hubble law holds.*

Hubble's law does *not* violate the cosmological principle. This can be understood from Figure 28.4, which shows observers on five separate galaxies measuring the motion of their neighbors. For simplicity, the galaxies are taken to be equally spaced, 100 Mpc apart, and they are separating in accordance with Hubble's law, with $H_0 = 75$ km/s/Mpc. The first pair of numbers beneath each galaxy represents its distance and recessional velocity as measured by the observer on galaxy 3.

Now let's consider how the expansion looks from the point of view of the observer on galaxy 2. Galaxy 4, for example, is moving with velocity 7500 km/s to the right relative to galaxy 3, and galaxy 3 in turn is moving at 7500 km/s to the right as seen by observer 2. Therefore, galaxy 4 is moving at a velocity of 15,000 km/s to the right as seen by the observer on galaxy 2. The distances and velocities that would be measured by observer 2 are noted in the second row. Similarly, the measurements made by observer 1 are noted in the third row. The important point is this: *each* observer sees an overall expansion described by Hubble's law, and the constant of proportionality—the Hubble constant—is the same in all cases. In other words, each observer measures the same expansion rate, in accordance with the cosmological principle. Far from singling out any one observer as central, Hubble's law is in fact the *only* expansion law consistent with the cosmological principle.

Hubble's law has a major implication for the history of the universe. To see what it is, let us ask the following question: Assuming that all velocities have remained constant in time, how long has it taken for any given galaxy to reach its present distance from us? The answer follows from Hubble's law: the time taken is simply the distance traveled divided by the velocity, so

$$\begin{aligned} time &= distance/velocity \\ &= distance/(H_0 \times distance) \quad \text{[by Hubble's law]} \\ &= \frac{1}{H_0}. \end{aligned}$$

For a Hubble constant of 75 km/s/Mpc, this time turns out to be about 13 billion years. Notice that the time is *independent* of the distance—galaxies twice as far away move twice as fast, so the time taken to cross the intervening distance is the same in all cases. Hubble's law therefore implies that about 13 billion years ago *all* the galaxies in the universe were right on top of one another. In fact, it is believed that *everything* in the universe—matter and radiation alike—was confined to a single point at that instant. Then the point exploded, flying apart at high speeds. The present locations and velocities of the galaxies are a direct consequence of that primordial blast. This gargantuan explosion, involving everything in the universe, is known as the **Big Bang.**

The Big Bang marked the beginning of the universe. By measuring Hubble's constant, we can determine the *age* of the universe to be $1/H_0 \approx 13$ billion years. The range of possible error in this age is considerable because Hubble's constant is not known precisely. As noted in Chapter 26, its value is uncertain by about a factor of 2 because of uncertainties in the distances to the farthest galaxies. Most astronomers would agree that H_0 lies between 50 and 100 km/s/Mpc, so we conclude that the universe is somewhere between 10 and 20 billion years old.

This then is the explanation of why the sky is dark at night—the universe has a finite age. Olbers's paradox is resolved by the evolution of the universe itself. Whether the universe is actually finite or infinite in extent is irrelevant, at least as far as the appearance of the night sky is concerned. The important point is that we can see only a *finite* part of it—the region lying within 13 billion light years of us. What lies beyond is unknown; its light has not yet had time to reach us.

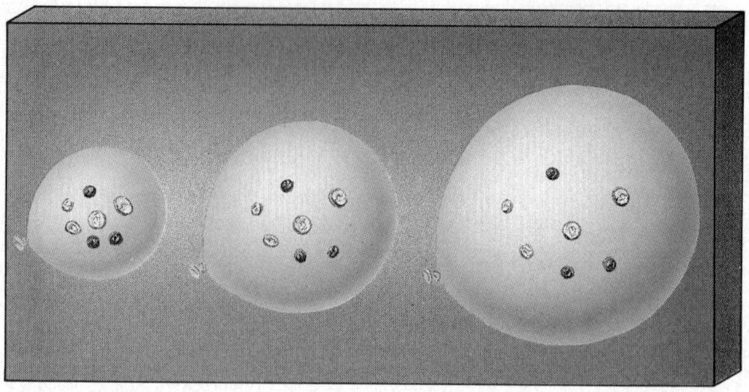

Figure 28.5 *The coins taped to the surface of a spherical balloon recede from one another as the balloon inflates (left to right). Similarly, galaxies recede from one another as the universe expands. As the coins recede, the distance between any two of them increases, and the rate of increase of this distance is proportional to the distance between them. Thus, our balloon expands according to Hubble's law.*

THE EXPANDING UNIVERSE

Where Was the Big Bang?

Let us pause for a moment to take stock of what we have learned so far and to shift our view of the Hubble expansion in a very important way. The cosmological principle indicates that the universe is the same everywhere, yet we have just seen that the observed recession of the galaxies described by Hubble's law implies that all of the galaxies exploded from a point some time in the past. Wasn't that point, then, different from the rest of the universe? And doesn't that violate the assumption of homogeneity within the cosmological principle? We know *when* the Big Bang occurred. Is there any way of telling *where*?

The answer to these questions requires us to make a great leap in our view of the universe. If we were to imagine that the Big Bang were simply an enormous explosion that spewed matter out into space, ultimately to form the galaxies we see, then the above reasoning would be quite correct. However, the Big Bang was *not* an explosion in an otherwise featureless, empty universe. The only way that we can have the Hubble expansion *and* retain the cosmological principle is to realize that the Big Bang involved the *entire* universe—not just the matter and radiation within it, but the universe *itself*. In other words, the galaxies are not flying apart into the rest of the universe. The universe itself is expanding. Like raisins in a loaf of raisin bread that move apart as the bread expands in an oven, the galaxies are just along for the ride.

Let us reconsider some of our earlier statements in light of this new perspective. We now recognize that Hubble's law actually describes the expansion of the universe. Although galaxies have some small-scale, individual random motions, on average they are not moving with respect to the fabric of space—any such overall motion would pick out a "special" direction space and violate the assumption of isotropy. On the contrary, the portion of the galaxies' motion that makes up the Hubble flow is really an expansion of space itself. The expanding universe remains homogeneous at all times. There is no "empty space" beyond the galaxies into which they rush. At the time of the Big Bang, the galaxies did not reside at a point located at some well-defined place within the universe. The entire universe was a point. That point was in no way different from the rest of the universe; that point *was* the universe. Therefore, there was no one point where the Big Bang "happened"—because the Big Bang involved the entire universe, it happened *everywhere* at once.

To illustrate these ideas, imagine an ordinary balloon with coins taped to its surface, as shown in Figure 28.5. The coins represent galaxies, and the two-dimensional surface of the balloon represents the "fabric" of our three-dimensional universe. The cosmological principle applies to the balloon because every point on the balloon looks pretty much the same as every other. The frames in the figure make up a movie—a hypothetical film strip of the expanding universe. Imagine yourself as a resident of one of the three coin "galaxies" in the left-most frame, and note your position relative to your neighbors. As the balloon inflates (as the universe expands), the other galaxies recede. Notice, incidentally, that the coins themselves do *not* expand along with the balloon, any more than people, planets, stars, or galaxies—all of which are held together by their own internal forces—expand along with the universe.

Regardless of which galaxy you choose to consider, you would note that all the other galaxies are receding. To appreciate this, imagine yourself a resident on a different coin "galaxy." Now again notice the change in the positions of the galaxies while viewing the frames from left to right. The galaxies recede for any observer in the universe. Nothing is special or peculiar about the fact that all the galaxies are receding from you. Such is the cosmological principle: No observer anywhere in the universe has a privileged position. There is no center to the expansion and no position that can be identified as the location from which the universal expansion began.

Now imagine letting the balloon deflate. This would correspond to running the universe backward from the present time to the Big Bang. *All* the galaxies (coins) would arrive at the same place at the same time—at the instant the balloon reached zero size. But there is no one point on the balloon that could be said to be *the* place where that occurred. The entire balloon expanded from a point, just as the Big Bang encompassed the entire universe and expanded from a point.

This analogy has its shortcomings. The main difficulty here is that we see the balloon, which in our illustration we imagined as two-dimensional, expanding into the third dimension of space. This might suggest that the three-dimensional universe is expanding "into" some fourth spatial dimension. It is not, so far as we know. At the very least, if higher spatial dimensions are involved, they are completely *irrelevant* to our theory of the universe.

Relativity Again

These concepts are fairly difficult to grasp. The notion of the entire universe shrinking to a point—with *nothing,* not even space and time, outside—takes some getting used to. Nevertheless, that is the picture of the universe that lies at the heart of modern cosmology. The description of the universe itself (not just its contents) as a dynamic, evolving object is far beyond the capabilities of Newtonian mechanics, which we discussed in Chapter 3 and which we have used almost everywhere throughout this book. Instead, the more powerful techniques of Einstein's *General Relativity,* with its built-in notions of warped space and dynamical spacetime, are needed.

We encountered General Relativity in Chapter 24 (see especially Interlude 24-1), when we discussed the strange properties of black holes. We can loosely summarize its description of the universe by saying that the presence of matter or energy causes a curvature of space (more correctly, space*time*) and that the curved trajetories of freely falling particles within warped space are what Newton thought of as orbits under the influence of gravity. The amount of curvature depends on the amount of matter present, and the orbits of particles in turn depend on the curvature. Put less formally, in the words of Princeton relativist John Archibald Wheeler, "Space tells matter how to move, and matter tells space how to curve." In the case of a homogeneous universe, the curvature of space must be uniform.

Must we keep this difficult notion of a warped, expanding, homogeneous, universe constantly in the forefront of our thoughts if we are to comprehend the evolution of the cosmos? Perhaps surprisingly, given that relativity is the only theory that properly describes the large-scale behavior of the universe, the answer is no. While General Relativity predicts some curious consequences for the overall *geometry* of space on large scales, as we will see in a moment, the *dynamics* of the universe can be understood using simple concepts that would have been thoroughly familiar to Newton.

Consider two neighboring points A and B in the expanding universe, as pictured in Figure 28.6. From the perspective of point A, every other point, including B, is rushing away from it, in accordance with Hubble's law. How does the distance between A and B change in time? Does it just keep growing, or does its rate of increase eventually slow down and stop? From a Newtonian viewpoint, we might expect that the overall gravitational pull of the universe would tend to slow the expansion, just as Earth's gravity tends to slow the upward motion of an object projected from the surface. But we have just argued that Newtonian physics is incapable of describing the motion of the universe in this way. However, a remarkable prediction of General Relativity is that the Newtonian picture gives the right result! The motion of point B relative to point A is *exactly* the same as if A were at the center of a planet and B were a projectile fired vertically upward from its surface. Thus, we *can* discuss the expansion of the universe in simple Newtonian terms, although we need relativity theory to justify our doing so.

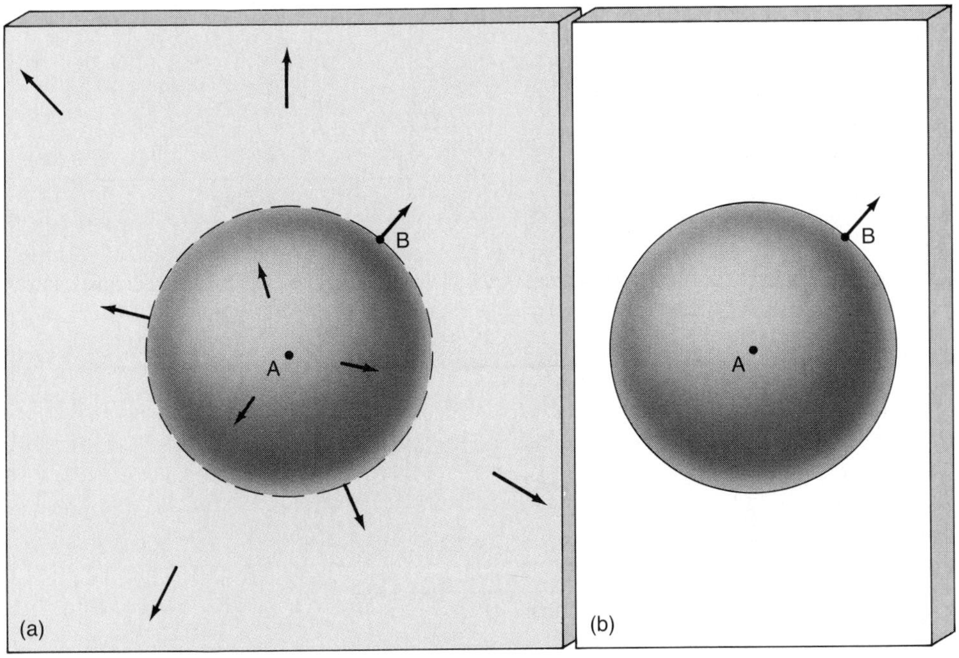

Figure 28.6 *The relative motion of any two points in the expanding universe can be addressed as a problem in Newtonian mechanics, although Einstein's theory of General Relativity is needed to explain why it is correct to do so. If all of the mass in the dark-shaded sphere centered on A and passing through B—part (a) in the figure—were concentrated at A and the rest of the universe ignored—as in part (b)—then the Newtonian calculation of B's motion relative to A would give the right answer.*

Three Futures

Now that we know we can confidently apply our familiar Newtonian concepts of motion under gravity to the behavior of the universe, we can ask what the possible outcomes of that motion might be. Let us first consider explicitly the case of a rocket ship launched from the surface of a planet. As we saw back in Chapter 3, there are three possible results depending on the speed of the ship. First, if the launch velocity is high enough or the mass of the planet is low enough, the speed of the ship will exceed the planet's *escape velocity,* and the ship will never return to the surface. The speed will diminish because of the planet's gravitational pull, but it will never reach zero. The spacecraft leaves the planet on an unbound, *hyperbolic* trajectory, as illustrated in Figure 28.7(a).

If the ship's launch speed is lower than the escape velocity, it reaches a maximum distance from the planet, then falls back to the surface. Its bound orbit is an *ellipse,* as shown in Figure 28.7(b). Intermediate between these two types of motion is the special case where the ship's launch speed is exactly equal to the escape velocity. In this situation, the ship never returns to the planet, but its velocity gets smaller and smaller as the distance from the planet increases, eventually reaching zero at an infinite distance away. This ''marginally bound'' orbit traces a geometric figure known as a *parabola*.

For spacecraft launched from the same planet, the deciding factor between these three possibilities is simply the launch speed. Alternatively, if both the spacecraft velocity and the planet size were fixed, the deciding factor would be the *mass* of the planet. A less massive planet will allow the ship to escape; the stronger gravity of a more massive planet will pull it back.

Similar reasoning applies to the expansion of the universe. Reconsider Figure 28.6. Imagine now that A and B are galaxies at some known distance from one another, their present relative velocity given by Hubble's law. The same three possibilities exist for these galaxies as for our escaping spacecraft—the distance between them can increase for-

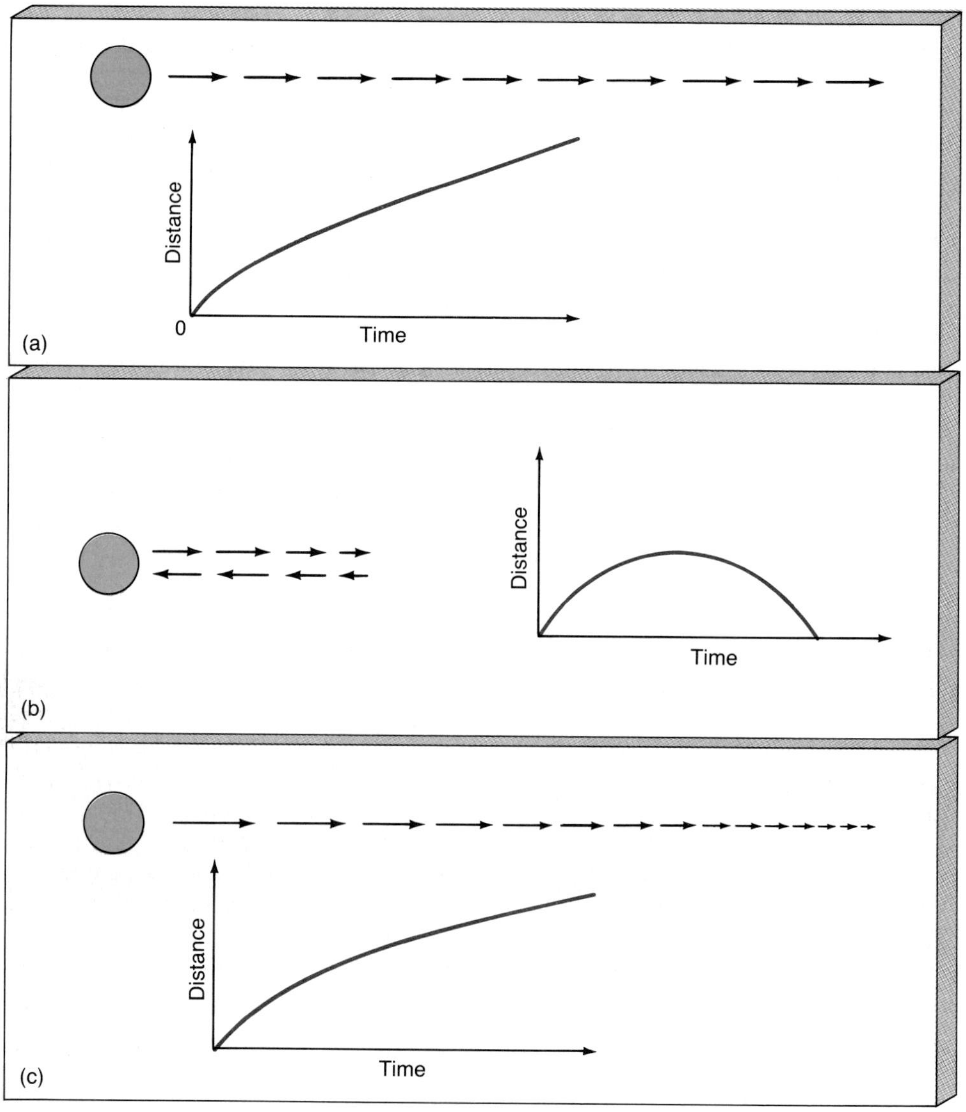

Figure 28.7 *(a) A spacecraft leaving a planet with a velocity greater than the escape velocity leaves on a hyperbolic trajectory (top). The lower graph shows the distance between the ship and the planet as a function of time. (b) If the velocity is less than the escape velocity, the ship eventually drops back to the planet. Its orbit is an ellipse; its distance from the planet first rises, then falls again. (c) In the intermediate case, where the ship leaves with precisely the escape velocity, the orbit is a parabola.*

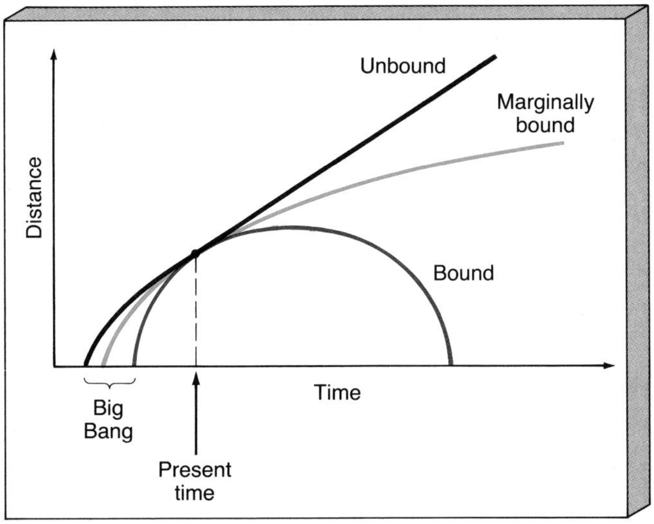

Figure 28.8 *Distance between two galaxies as a function of time in each of the three possible universes discussed in the text: (a) unbound, (b) bound, (c) marginally bound. The point where the three curves touch represents the present time.*

ever, increase for a while and then start to decrease, or increase forever, but at an ever-diminishing rate. The cosmological principle says that whatever the outcome, it must be the same for any two galaxies. In other words, the present overall expansion of the universe can do one of three things: It can continue forever, stop and turn around into a contraction, or continue forever, but more and more slowly. Figure 28.8 shows these three alternatives. The three curves are drawn so that they all pass through the same point at the present time. All three are possible descriptions of the present size and expansion rate of the universe.

What determines which of these alternatives will actually occur? The answer is the *density* of the universe. In all cases gravity decelerates the expansion over time. The more matter there is—the denser the universe—the more "pull" there is against the expansion, just as the more dense a planet (or the more matter the planet has) the less likely it is that the rocket ship can escape. In a high-density universe, there is enough mass to stop the expansion and cause a recollapse—the universe is bound. A low-density universe, conversely, is unbound and will expand forever. The intermediate density that corresponds to a marginally bound universe is usually referred to as the **critical density.** Its value depends on the Hubble constant, as more density is required to bind a more rapidly expanding universe. (In fact, the value depends on the *square* of H_0, so doubling the expansion rate of the universe would quadruple the critical density.) For $H_0 = 75$ km/s/Mpc, the present critical density turns out to be about 10^{-29} g/cm^3. That's an extraordinarily low density—about 6 hydrogen atoms per cubic *meter,* a volume the size of a typical household closet.

In more "cosmological" terms, the critical density corresponds to about one Milky Way Galaxy (excluding the dark matter) per cubic megaparsec. By comparing this density with the density of matter we can observe or infer in the universe, we should in principle be able to distinguish among the three possible futures for our universe. However, as we will see in a moment, this is not as straightforward as it might seem.

THE GEOMETRY OF SPACE

✓3 We have reverted to the familiar notion of gravity, away from the more valid concept of warped spacetime, because speaking in terms of gravity makes our discussion of the evolution of the universe much easier to understand. However, General Relativity makes some predictions that do *not* have a simple description in Newtonian terms. Foremost among these is the fact that space is *curved* and that the curvature is closely tied to the matter within it. Relativity asserts that mass alters the nature of spacetime. Matter effectively shapes or "warps" the geometry of space. The more matter, the greater the distortion. The degree of distortion—the curvature—must be the same everywhere (because of the cosmological principle), so there are only three distinct possibilities for the large-scale geometry of the universe, and they correspond precisely to the three possible futures we have just described. (For more information on the different types of geometry involved, see Interlude 28-1.)

If the average density of the cosmos is above the critical value, space is curved so much that it bends back on itself and closes off, making the universe *finite* in size. Such a universe is known as a **closed universe.** It is difficult to visualize a three-dimensional volume uniformly arching back on itself in this way, but the two-dimensional version is well known: It is just the surface of a sphere, like the balloon we discussed earlier. Figure 28.5, then, is the two-dimensional likeness of a three-dimensional closed universe. Like the surface of a sphere, a closed universe has no boundary, yet it is finite in extent.[1] One remarkable property of a closed universe is illustrated in Figure 28.9. Just as a traveler on the surface of a sphere can keep moving forward in a straight line and eventually return to her starting point, a flashlight beam shone in some direction in space might eventually traverse the entire universe and return from the opposite direction!

The surface of a sphere curves, loosely speaking, "in the same direction" no matter which way we move from a given point. A sphere is said to have *positive curvature*. However, if the average density of the universe is below the critical value, the surface curves like a saddle. It has *negative curvature*. Most people have seen a saddle—it curves

[1] Notice that, for the sphere analogy to work, we must imagine ourselves as two-dimensional "flatlanders" who cannot visualize or experience in any way the third dimension perpendicular to the sphere's surface. Flatlanders are confined to the sphere's surface, just as we are confined to the three-dimensional volume of our universe.

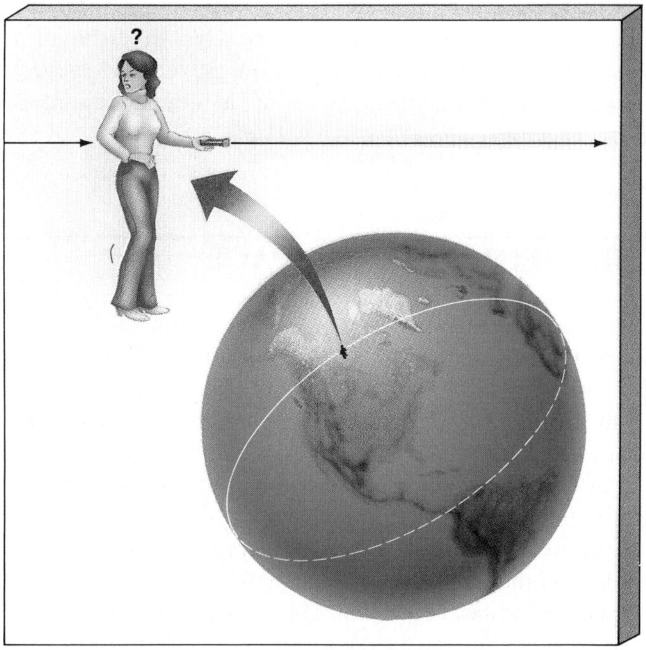

Figure 28.9 *In a closed universe, a beam of light launched in one direction might return someday from the opposite direction after circling the universe just as motion in a "straight line" upon Earth's surface will eventually encircle the globe.*

"up" in one direction and "down" in another, but no one has ever seen a uniformly negatively curved surface, for the simple reason that it cannot be constructed in three-dimensional space! It is just "too big" to fit. Interlude 28-1 provides a closer look at the possibility. A low-density, saddle-curved universe is infinite in extent and is usually called an **open universe.**

The third case, when the density is precisely equal to the critical density, is the easiest to visualize. This universe, called a **critical universe,** has no curvature. It is said to be "flat," and it is infinite in extent. In this case, and *only* in this case, the geometry of space on large scales is precisely the familiar Euclidean geometry taught in high schools. Apart from its overall expansion, this is basically the universe that Newton knew.

Euclidean geometry—the geometry of flat space—is familiar to most of us because it is a good description of space in the vicinity of the Earth. It is the geometry of everyday experience. Does this mean that the universe is flat, which would in turn mean that it has exactly the critical density? The answer is no. Just as a flat street map is a good representation of a city, even though we know the Earth is really a sphere, Euclidean geometry is a good description of space within the solar system, or even the Galaxy, because the curvature of the universe is negligible on scales smaller than about 1000 Mpc. Only on the very largest scales would the geometrical effects we have just discussed become evident.

THE EVOLUTION OF THE UNIVERSE

✓4 Now that we have seen the three possible futures for the universe, and the large-scale geometry they imply, let us discuss each of them in turn. As we saw in Figure 28.8, any of the three *could* describe the current state of the expanding universe. At the present time, astronomers do not know for certain which one of them represents reality. In a moment, we will see which is presently best supported by cosmological observations.

The Big Bang

In all three cases, the universe began in a state of incredibly high density and temperature—the Big Bang. The conditions that prevailed during the first few seconds of the universe's existence were crucial in determining its later development. They will be the topic of the next chapter.

The Big Bang grew from a singularity—an epoch where the present laws of physics say the universe had zero size and infinite temperature and density. As we saw in Chapter 24, where we discussed the singularities at the center of black holes, these predictions should not be taken too literally. Their presence signals that, under extreme conditions, the theory—in this case, General Relativity—making the predictions breaks down. At the present moment, no theory exists to let us penetrate the singularity at the start of the universe. We have no means of describing these earliest of times, so we have no way of answering the question, What came *before* the Big Bang? Indeed, given the laws of physics as we currently know them, the question itself may be meaningless. The Big Bang represented the beginning of the entire universe—mass, energy, space, *and* time came into being at that instant. With time not in existence, the notion of "before" does not exist. Consequently, some cosmologists maintain that asking what happened before the Big Bang is like asking what lies north of the North Pole! Other cosmologists disagree, arguing that one day the proper theory will explain the singularity and we can answer the question of what came before.

We cannot extend our theory of the universe all the way back to the Big Bang itself, but we can come fairly close. Theorists estimate that the "known" physics of today is probably adequate to describe the universe since about 10^{-43} s after the Big Bang. For the purposes of this chapter, at least, 10^{-43} is close enough to zero that we need not worry about this very early, and completely unknown, epoch.

An Open Universe

Our first model universe is one that emerged from the Big Bang with insufficient matter for gravity to counteract the expansion. Its density always has been, and always will be, too small to cause the universe to recontract. The low density also means that the spatial curvature is negative and that the universe is infinite in extent. As illustrated in Figure 28.10, the open universe expands forever.

Should this model be correct, the galaxies will continue to recede forever. Their radiation will weaken with increasing distance. In time, an observer on Earth will see no galaxies in the sky beyond the Local Group (which, as we have seen, is not itself expanding), even with the most powerful telescope. The rest of the observable universe will appear dark, the distant galaxies too faint to be seen. Eventually, the Milky Way and the Local Group too will peter out as their hydrogen supply is consumed. An open universe ultimately experiences a "cold death." All the radiation, matter, and life in such a universe are eventually destined to freeze.

A Closed Universe

A quite different fate awaits the universe if its density is above the critical value. Like the open case just considered, this model expands with time starting with the Big Bang. Unlike the open case, however, this finite, closed universe contains enough matter to halt its own expansion. The recession of the galaxies will eventually stop at some time in the future. Astronomers everywhere—on any planet within any galaxy—would then announce that the radiation received from nearby galaxies was no longer redshifted. (The light from *distant* galaxies would still be redshifted, however, because we would see them as they were in the past, at a time when the universe was still expanding.) The bulk motion of the universe, and of the galaxies within, will be stilled—at least momentarily.

The expansion might stop, but the inward pull of gravity will not. Gravity is relentless. The universe will begin to contract. Astronomers everywhere would announce that nearby galaxies had begun to show blue shifts. In many ways, the contraction would be a mirror image of the expansion that preceded it. As illustrated in Figure 28.11, the universe would recollapse to a point, requiring just as much time to fall back as it took to rise. The closed universe is not only finite in space; it is also finite in *time*.

This expansion-contraction scenario has many fascinating (and dire) implications. Following the Big Bang, the density of the universe thins to a rather small value by the time the expansion stops. Thereafter, the density rises again, returning to its earlier huge values as matter collapses onto itself. Toward the end of the contraction phase, galaxies will collide frequently as the available space diminishes. Just as compressing the air in a bicycle pump or rubbing our hands generates heat through friction, these collisions will generate heat as well. The entire universe will grow progressively denser and hotter as the end of the contraction is neared. Near total collapse, the temperature of the entire

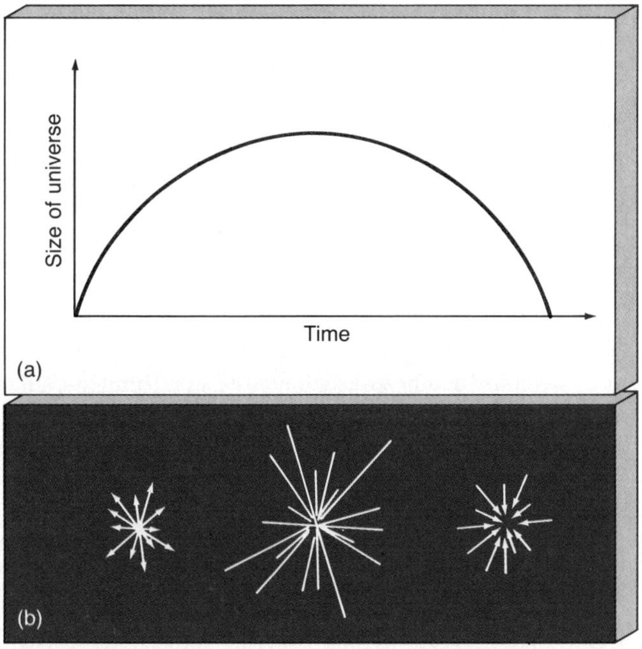

Figure 28.11 *A closed universe (a) has a beginning, an end, and a finite lifetime. The lower frames (b) illustrate its evolution, from explosion to maximum size to recollapse.*

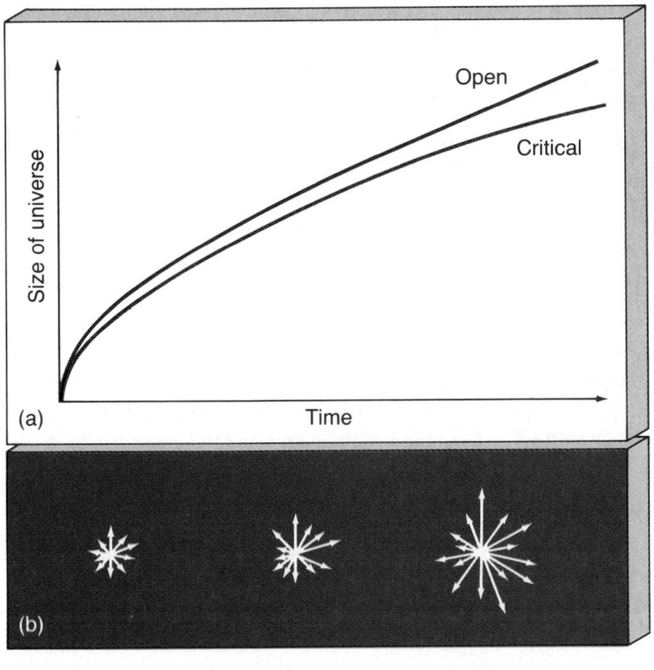

Figure 28.10 *A low-density universe (a) expands forever from its explosive beginning. The lower frames (b) illustrate the continuing expansion of the universe in this case. The top curve represents an open universe, with density less than the critical value. The lower curve is a critical universe, with density exactly equal to the critical value.*

INTERLUDE 28-1 *Curved Space*

Euclidean geometry is the geometry of flat space—the geometry taught in high schools everywhere. Set forth by one of the most famous of the ancient Greek mathematicians, Euclid, who lived around 300 B.C., it is the geometry of everyday experience. Houses are usually built with flat floors. Writing tablets and blackboards are also flat. We work easily with flat, straight objects, because the straight line is the shortest distance between any two points.

In constructing houses or any other straight-walled buildings on the surface of the Earth, the other basic axioms of Euclid's geometry would also apply: Parallel lines never meet even when extended to infinity; the angles of any triangle always sum to 180 degrees; the circumference of a circle equals π times its diameter. If these rules were not obeyed, walls and roof would never meet to form a house!

In reality, though, the geometry of Earth's surface is not really flat. It is curved. We live on the surface of a sphere, and on that *surface,* Euclidean geometry breaks down. Instead, the rules for the surface of a sphere are those of *Riemannian geometry,* named after the nineteenth-century German mathematician Georg Friedrich Riemann. For example, there are no parallel lines (or curves) on a sphere's surface; any lines drawn on the surface and around the full circumference will eventually intersect. The sum of a triangle's angles, when drawn on the surface of a sphere, exceeds 180 degrees—in the 90°-90°-90° triangle shown in the figure at right, the sum is actually 270°. And the circumference of a circle is less than π times its diameter.

We see that the curved surface of a sphere, governed by the spherical geometry of Riemann, differs greatly from the flat-space geometry of Euclid. These two geometries are approximately the same only if we confine ourselves to a small patch on the surface. If the patch is small enough compared with the sphere's radius, the surface looks "flat" nearby, and Eucludean geometry is approximately valid. This is why we can draw a usable map of our home, our city, even our state, on a flat sheet of paper, but an accurate map of the entire Earth must be drawn on a globe.

When we work with larger parts of the Earth, we must abandon Euclidean geometry. World navigators are fully aware of this. Aircraft do not fly along what you might regard as a straight-line path from one point to another. Instead they follow a "great circle," which, on the curved surface of a sphere, is the shortest distance between two points. For example, a flight from Los Angeles to London does not proceed directly across the U.S. and the Atlantic Ocean as you might expect from looking at a flat map. Instead, it goes far to the north, over Canada and Greenland, above the Arctic Circle, finally coming in over Scotland for a landing at London. This is the great circle—the shortest path—between the two cities.

The "positively curved" space of Riemann is not the only possible departure from flat space. Another is the "negatively curved" space first studied by Nikolaiivanovich Lobachevsky, a nineteenth-century Russian mathematician. In this geometry, there is an *infinite* number of lines through any given point parallel to another line, the sum of a triangle's angles is *less* than 180 degrees (see figure), and the circumference of a circle is *greater* than π times its diameter. Instead of the surface of a flat plane or a curved sphere, this type of space is described by the *surface* of a curved saddle. It is a hard geometry to visualize!

Most of the local realm of the *three*-dimensional universe (including the solar system, the neighboring stars, and even our Milky Way Galaxy) is correctly described by Euclidean geometry.

universe will have become greater than that of a typical star. Everything everywhere will have become so bright that the stars themselves will cease to shine for want of contrasting darkness. The universe will shrink toward a superdense, superhot singularity, much like the one from which it originated. Some astronomers call this event the "Big Crunch." In contrast to the open universe, which ends as a frozen cinder, this closed universe will experience a "heat death." Its contents are destined to fry.

Cosmologists are uncertain of the fate of a closed universe upon reaching the singularity. We cannot penetrate forward in time beyond the singularity at the Big Crunch any more than we can probe backward past the Big Bang. However, some cosmologists speculate that, with both density and temperature increasing as the contraction nears completion, the pressure might somehow be sufficient to overcome gravity, pushing the universe back out into another cycle of expansion. As depicted in Figure 28.12, the universe might not simply end—it might "bounce." A hypothetical universe having many—perhaps infinitely many—cycles of expansion and contraction is often termed an **oscillating universe.** Bear in mind, however, that any discussion of the universe outside of the current cycle is pure speculation.

Should the oscillating model be valid, we need not trouble ourselves with the philosophical concept of "existence" before the beginning of time. Such a universe always was and always will be.

A Critical Universe

The third possibility is the intermediate case—the critical universe, which is infinite in extent, and spatially flat. In this case, the accumulated matter is just sufficient eventually to halt the expansion—but only after an infinitely long time. Like the open universe, this model will expand for-

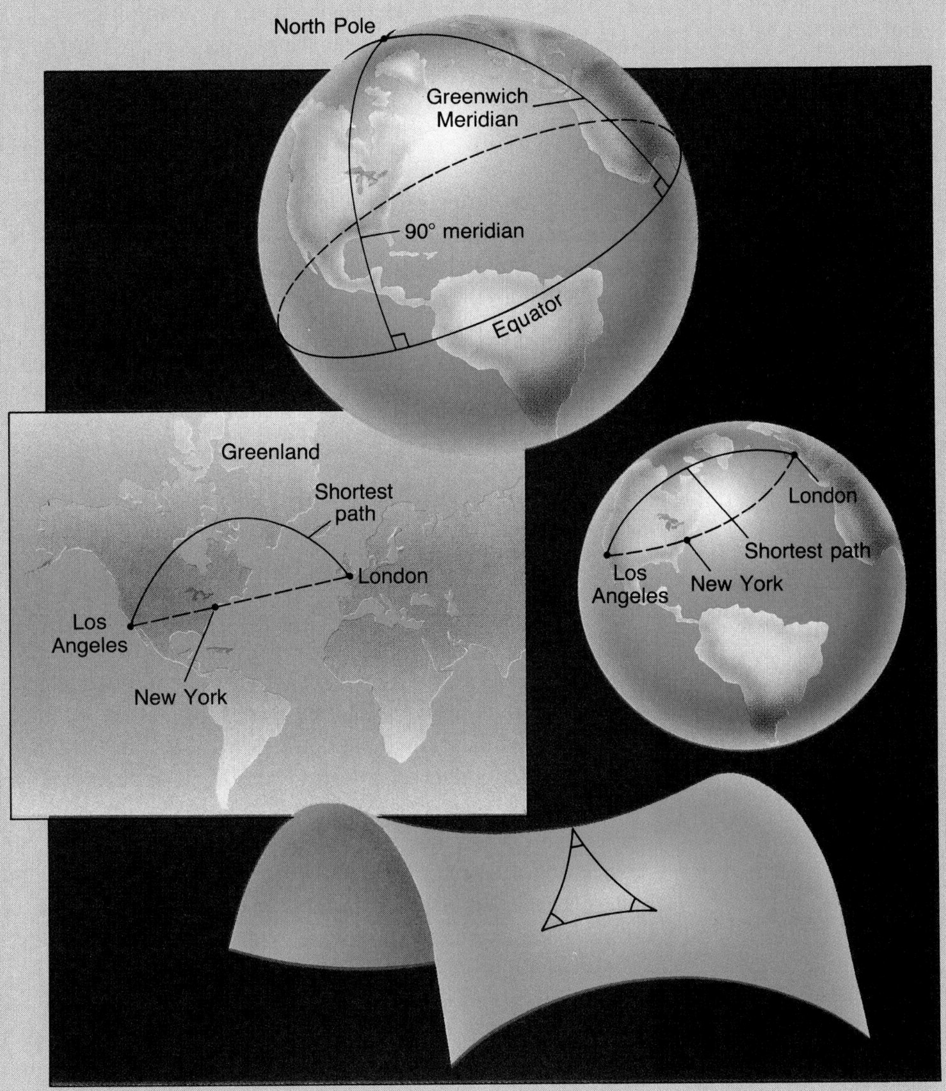

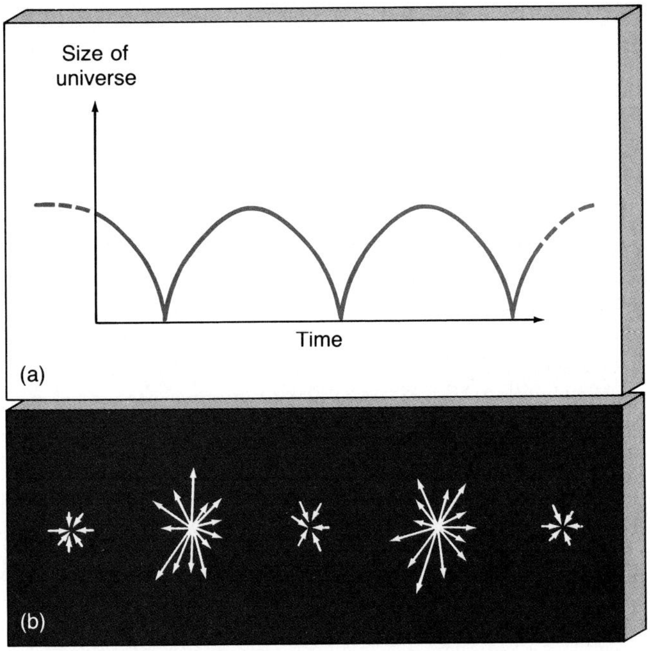

ever, its qualitative appearance resembling Figure 28.10(b).

One might wonder if there is any good reason to study this third special case at all. After all, it requires that the density of the universe be *exactly* equal to the critical value. A little more and the universe is closed, a little less and it is open. Can such a remarkable coincidence really occur? Why bother to consider it? As we will see in the next chapter, many astronomers now believe that conditions in the early universe may have conspired to produce a precisely critical expansion rate. If they are correct, the critical universe surely didn't arise by chance—it may be a consequence of the laws of physics.

Figure 28.12 *An oscillating universe has neither a beginning nor an end. Each expansion-contraction phase ends in a ''bounce'' which becomes the ''Big Bang'' of the next expansion. There is currently no information on whether or not this can actually occur.*

The Age of the Universe

When we estimated the age of the universe from the Hubble constant, we made the assumption that the expansion speeds of the galaxies were *constant* in the past. However, as we have now found, this is not the case. The effects of gravity have slowed the universe's expansion over time, so, regardless of which evolutionary model turns out to be correct, the universe expanded *faster* in the past. The assumption of a constant expansion rate leads to an *overestimate* of the universe's age. The universe is less than 13 billion years old, the age we had calculated earlier. How much younger the universe is depends on how much deceleration has occurred. Figure 28.13 illustrates this point. It is similar to Figure 28.8, except that we have added the line that corresponds to constant expansion at the present rate—which illustrates a universe completely *empty* of matter and so experiencing no deceleration. Because all the models we have discussed lie below this line, we can see graphically that the true age of the universe is indeed less than $1/H_0$.

In the special case of critical density, the expansion of the universe is easy to calculate, and it can be shown that the age of the universe is ⅔ of the above value, or 8.7 billion years. For an open universe, the deceleration is less than in the critical case, so the age of the universe lies between 8.7 and 13 billion years. In the closed case, the deceleration is greater than critical, so the universe must be less than 8.7 billion years old. If we knew the density of the universe, not only would we know its future, we would also know its precise age.

Impressive though it is that we can pin down the age of the universe to within a factor of two or so, these numbers indicate a potentially serious problem. Unless the universe is of *very* low density, and hence close to 13 billion years old, the age that we obtain from cosmology is less than the 12–15 billion year range implied by studies of globular clusters in our own Galaxy (see Chapters 19 and 22). Because the Galaxy cannot be older than the universe and because the density of the universe appears to be at least relatively close to the critical value (as we will see in a moment), we are forced to conclude that there may be a glaring contradiction between these two major areas of astronomy! If the Hubble constant turns out to be closer to 50 km/s/Mpc, and if the upper limit on the ages is correct, then the two age estimates might be reconciled. However, if H_0 turns out to be 100 km/s/Mpc, the discrepancy may become a serious embarrassment to astronomers. All people concerned agree that there is a problem, but no astronomer is yet prepared to admit that either estimate could be seriously in error!

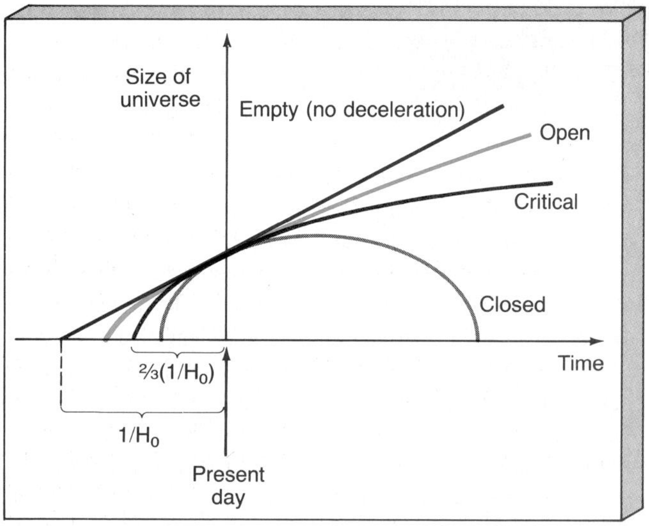

Figure 28.13 *As the density of the universe increases, its deceleration increases, too. The universe contains some matter, so whatever the model, its trajectory on this graph will lie below the line for the constant-velocity, empty universe. Thus, the age of the universe is always less than 1 over the Hubble constant. The true age is less for larger values of the present-day density.*

The Cosmological Red Shift

The theory of relativity also shows us the correct interpretation of the cosmological red shift. Up to now, we have viewed the red shift of galaxies as a Doppler shift, a consequence of their motion relative to us. However, we have just argued that the galaxies are *not* in fact moving with respect to the universe, in which case the Doppler interpretation is incorrect. The true explanation is that, as a photon moves through space, its wavelength is influenced by the expansion of the universe. In a sense, we can think of the photon as being attached to the expanding fabric of space, so its wavelength expands along with the universe, as illustrated in Figure 28.14. In other words, the red shift is a consequence of the changing size of the universe and is *not* related to velocity at all. The red shift of a photon measures the amount by which the universe has expanded since the photon was emitted: The light we see from a quasar with a red shift of 5 was emitted when the universe was one-fifth its present size. While it is standard practice in astronomy to refer to the cosmological red shift in terms of recessional velocity, bear in mind that, strictly speaking, this is not the right thing to do.

The Cosmological Constant

Let us relate an episode in the history of relativity that proves that even the greatest scientists are fallible. The first scientist to apply General Relativity to the universe was (not surprisingly) the theory's inventor, Albert Einstein. When he derived and solved the equations describing the behavior of the universe, he discovered that they predicted a universe that evolved in time. But in 1917, neither Einstein nor anyone else knew about the Hubble expansion, which would not be discovered for another 10 years. At the time, Einstein, like most scientists, believed that the universe was *static*—that is, unchanging and everlasting. The discovery

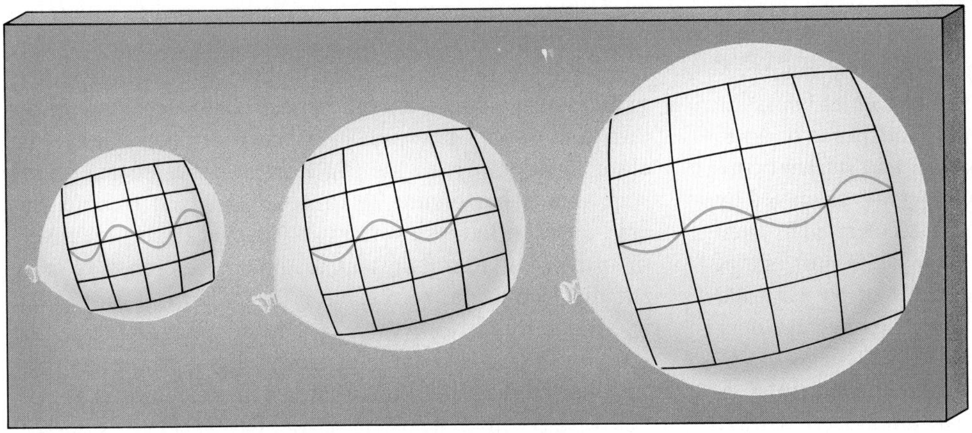

Figure 28.14 *As the universe expands, photons of radiation are stretched in wavelength, giving rise to the cosmological red shift.*

that there was no static solution to his equations seemed to Einstein to be a near-fatal flaw in his new theory.

To bring his theory into line with his beliefs, Einstein tinkered with his equations, introducing a ''fudge factor'' now known as the **cosmological constant.** This factor allowed his model universe to remain static. Instead of predicting an evolving universe, which would have been one of relativity's greatest triumphs, Einstein yielded to a preconceived notion of the way the universe ''should be.'' Later, when the expansion of the universe was discovered and Einstein's equations—without the fudge factor—were found to describe it perfectly, he declared that the cosmological constant was the biggest mistake of his scientific career.

COSMOLOGICAL TESTS

✓5 Is there any way for us to determine which of the above three models actually describes our universe (that is, apart from just waiting to find out)? Will the universe end as a small, dense point much like that from which it began? Or will it expand forever? And if the expansion ceases, when will that happen? Fortunately, we live at a time when astronomers are subjecting these questions to observational tests. Our observations and experiments, together with the theories underlying them, seek direct answers to these most basic questions of cosmology.

Galaxy Counting

The most straightforward way to distinguish between the open and closed models is to estimate the average density of matter in the universe, because density is what differentiates a closed universe from an open one. As noted above, for our ''standard'' Hubble constant of 75 km/s/Mpc, the critical density that separates these two possibile universes is about 10^{-29} g/cm^3. Cosmologists conventionally denote the ratio of the actual density to the critical value by the symbol Ω_0 (''omega-nought''). In terms of this quantity, then, a critical universe has $\Omega_0 = 1$. A universe with Ω_0 less than 1 is open and will expand forever; one with Ω_0 greater than 1 is closed and will recollapse.

How can we determine the average density of the universe? At face value, it would seem simple—just measure the average mass of the galaxies residing within any parcel of space, estimate the volume of that space, and calculate the total mass density. When astronomers do this, they usually find a little less than 10^{-31} g/cm^3 in the form of luminous matter. Largely independent of whether the chosen region contains only a few galaxies or a rich galaxy cluster, the resulting density is about the same, within a factor of 2 or 3. Galaxy counts thus yield $\Omega_0 \approx 0.01$, implying that the universe is open. If this measure is correct, then the universe will expand forever.

But there is an important additional consideration here. As we noted in Chapters 25 and 26, most of the matter in the universe is *dark*—it exists in the form of invisible material that has been detected only through its gravitational effect in galaxies and galaxy clusters. We currently do not know what the dark matter is, but we *do* know that it is there. Galaxies may contain as much as 10 times more dark matter than luminous material, and the figure for galaxy clusters is even higher—perhaps as much as 95 percent of the total mass in clusters is invisible. Even though we cannot see it, dark matter contributes to the average density of the universe and plays a part in opposing the expansion.

Unfortunately, the distribution of dark matter is not very well known. We can infer its presence in galaxies and galaxy clusters, but we are largely ignorant of its extent in superclusters, voids, or other larger structures. However, there are some indications that it may account for an even greater fraction of the mass on large scales than it does in galaxy clusters. For example, observations of gravitational lensing by galaxy clusters (see Chapter 27) suggest that dark matter may be more extensive than is indicated by the Virial method (described in Chapter 26). Furthermore, observations of the overall motion of galaxies in the local supercluster reveal the presence of a huge accumulation of mass known as the **Great Attractor,** with a total mass of about 10^{17} solar masses and a size of 100–150 Mpc.

The existence of this large concentration of matter is

inferred from careful studies of the velocities of galaxies within about 20 Mpc of the Milky Way. The observations seem to indicate that everything in our cosmic neighborhood is falling toward some large mass concentration in the direction of the constellation Centaurus, at speeds of up to 1000 km/s. Galaxy studies in the optical and infrared confirm its existence. The mass of this atttractor is estimated by calculating how large it must be to have accelerated us from rest to our present speed in the time available—the age of the universe. If the current best estimates of the size and mass of this gargantuan object are correct, its average density may be quite close to the critical value.

On the scale of galaxies, as we mentioned, there is about 10 times more dark matter than luminous material. Including this extra mass, our earlier estimate of Ω_0 increases by a factor of 10, to 0.1. On the scale of galaxy clusters, there may be 20 or 30 times more mass, and possibly even more, increasing our estimate further, to 0.2 or 0.3. On the scale of superclusters and voids, it is quite conceivable that invisible matter may account for as much as 99 percent of the total mass in the universe. In that case, the vast "voids" are not empty at all—they are huge seas of invisible matter, and the visible galaxies are merely insignificant "islands" of brightness within them. If the total amount of dark matter lurking in the darkness beyond the galaxy clusters does outweigh the luminous mass by a factor of nearly 100, Ω_0 might even exceed 1, and the universe could be closed. This is why it is so important to search for reservoirs of invisible matter beyond the galaxies. The measured value of Ω_0 has steadily increased over the past 20 years as larger and larger regions of the universe have been surveyed.

The Deceleration of the Universe

Determining the mass density of the universe is an example of a *local* measurement that provides an estimate of Ω_0. As we have seen, however, the result we obtain depends on just how local our measurement is—the larger the scale, the larger the result we obtain. In an attempt to get around this problem, astronomers have devised alternative methods that rely instead on so-called *global* measurements, which cover a much larger portion of the observable universe. The idea is that global tests should indicate the universe's *overall* density, not just its value in our vicinity.

One global method attempts to measure the deceleration of the universe directly by observing faraway galaxies. As we saw earlier, the greater the average density of the universe, the greater its deceleration should be, so a measurement of the change in the expansion rate translates directly into an estimate of the density. This test tries to answer the question, At what rate is matter everywhere causing the universal expansion to slow down? Put another way, how fast is the universe decelerating?

Astronomers cannot expect to measure the cosmic slowdown by watching the motion of any one galaxy. But by thinking more broadly, there may be another way to do it, at least in principle. Because the expansion is slowing, objects at great distances—that is, objects that emitted their radiation long ago—should appear to be receding *faster* than the Hubble law predicts. The older the galaxy, the closer in time it is to the Big Bang, so the greater its red shift should be. Figure 28.15 is an idealized Hubble diagram, similar to those we saw in Chapter 26. If the universal expansion were constant in time, recessional velocity and distance would be related by a straight line—the solid line in the figure. However, in a decelerating universe, the velocity is greater at large distances. Furthermore, as indicated in the diagram, this difference in velocities is greater for a denser universe, where gravity has been more effective at slowing the initial expansion.

What do the data indicate? Is there any evidence for faster recessional velocities among the more distant galaxies? The answer is yes. The most distant galaxies do indeed have substantially greater recessional velocities than the Hubble law would predict. Unfortunately, however, the Hubble diagrams constructed for galaxies at large distances are inconclusive in their implications. Distant galaxies are very faint, and accurate observations of their properties are difficult. As shown in Figure 28.16, the uncertainties in their measured luminosities mean that we simply cannot distinguish among the three possibilities. Furthermore, the observations do not take into account any evolutionary changes that might affect a galaxy's brightness over the

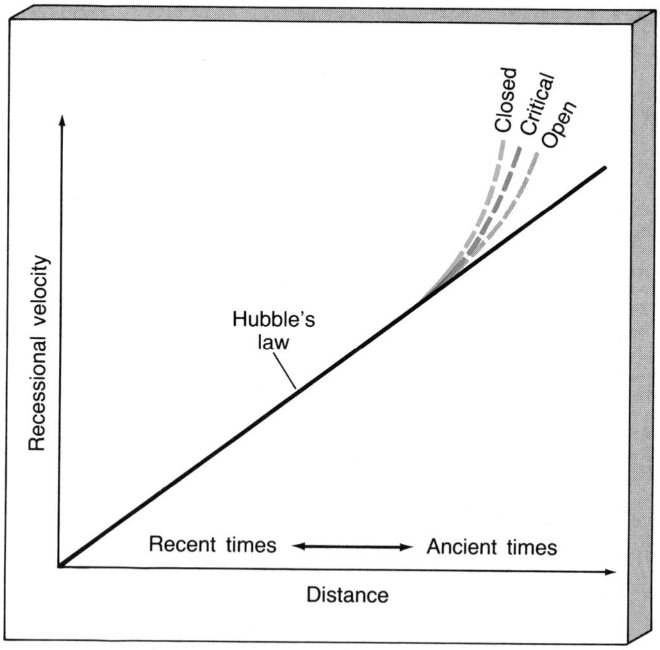

Figure 28.15 An idealized Hubble diagram, showing how we might detect evidence for a deceleration of the universe by observing a departure from the usual Hubble relationship (solid line). The dashed curves show the expected departure from the solid line for the three evolving models of the universe. (The departures of the dashed curves from the solid line are exaggerated for clarity.)

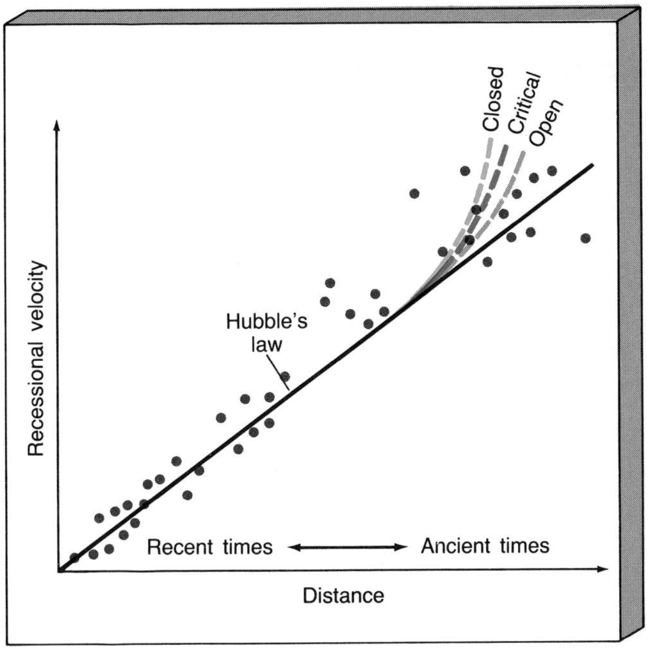

Figure 28.16 *A Hubble diagram for some distant galaxies. Uncertainties in the measured luminosities of distant galaxies affect estimates of their distances, so this technique is too imprecise to distinguish among the three possible models of the universe.*

course of time, making its luminosity, and hence its estimated distance, even more uncertain.

To the extent that there is any agreement among the various methods, the current data seem to favor an open universe destined to expand forever, although there is a large uncertainty in the value of Ω_0. In the next chapter, we will see a strong theoretical argument implying that the density of the universe should actually *equal* the critical value, so $\Omega_0 = 1$. While some astronomers remain unconvinced that the observations support this latter conclusion, most would at least agree that Ω_0 most likely lies between 0.1 and 1.

THE COSMIC MICROWAVE BACKGROUND

✓6 Looking out into space is equivalent to looking back into time. We have repeated that phrase many times now, in one form or another. We must always keep it in mind, especially when discussing cosmology. Just how far back in time can we probe? Is there any way to study the universe beyond the most distant quasar? How close can we come to perceiving directly the edge of time, the very origin of the universe?

The answer was discovered by accident in 1964, during an experiment designed to improve America's telephone system. As part of a project to identify and eliminate unwanted interference in satellite communications, Arno Penzias and Robert Wilson, two scientists at Bell Telephone Laboratories in New Jersey, were carrying out a study of the radio emission of the Milky Way at microwave wavelengths, using the horn-shaped antenna shown in Figure 28.17. In their data, they noticed a bothersome background "hiss" that just would not go away—a little like the background static on an AM radio station. Regardless of where and when they pointed their antenna, the hiss persisted. Never diminishing or intensifying, the weak signal was detectable at any time of the day, any day of the year, apparently filling all of space. Penzias and Wilson found that the hiss was equally intense in all directions in the sky—that is, *isotropic*—to very high accuracy.

What is the source of this radio noise? And why does it appear to come uniformly from all directions, unchanging in time? Unaware that they had detected a signal of great cosmological significance, Penzias and Wilson sought many different origins for the excess emission, including atmospheric storms, ground interference, equipment short circuits—even pigeon droppings inside the antenna. Eventually, after conversations with colleagues at Bell Labs and theorists at nearby Princeton University, the two experimentalists realized that the origin of the mysterious static was nothing less than the fiery creation of the universe itself. This discovery won Penzias and Wilson the 1978 Nobel Prize.

The radio hiss that Penzias and Wilson detected is now known as the **cosmic microwave background.** A team of researchers led by theorists Robert Dicke and P. J. E. Peebles at Princeton had predicted its existence and general properties a year before its discovery. In fact, as early as the 1940s, physicist George Gamow had realized that, in addition to being extremely dense, the early universe must also have been very *hot*. Shortly after the Big Bang, the universe

Figure 28.17 *This "sugarscoop" antenna, originally built to communicate with Earth-orbiting satellites, was used in discovering the 3-K cosmic background radiation. Pictured, left to right, are Robert Wilson and Arno Penzias, who used the antenna to make the discovery. (Bell Labs)*

was filled with extremely high-energy thermal radiation—gamma-rays of very short wavelength. Dicke and Peebles extended these ideas. They reasoned that as the universe expanded and cooled, the frequency of this primordial radiation would have been redshifted from gamma-ray, to x-ray, to ultraviolet, eventually all the way into the radio range of the electromagnetic spectrum. Figure 28.18 shows the theoretically expected change in the cosmic Planck curve as the universe expanded and cooled. By the present time, Dicke and Peebles argued, this redshifted "fossil remnant" of the primeval fireball should have a Planck curve with a very low temperature—no more than a few tens of kelvins—peaking in the microwave part of the spectrum. The Princeton group was in the process of constructing a microwave antenna to search for this radiation when Penzias and Wilson announced their discovery.

The Princeton researchers confirmed the existence of the microwave background and estimated its temperature at about 3K. However, proof that its frequency distribution really was described by a Planck curve was much harder to obtain. Wavelengths corresponding to the peak of the spectrum cannot be observed from the ground—the early radio measurements were made only in the low-frequency "tail"

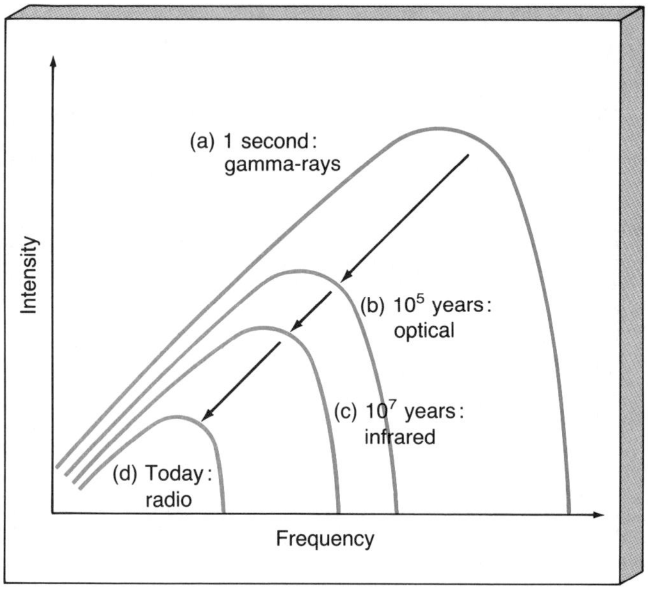

Figure 28.18 *Theoretically derived Planck curves for the entire universe (a) 1 second after the Big Bang, (b) 100,000 years after the Big Bang, (c) 10 million years after the Big Bang, and (d) at present, approximately 13 billion years after the Big Bang.*

INTERLUDE 28-2 *The Steady-State Universe*

All the cosmological models we have studied specify evolution as their central theme: The universe changes with time. These models derive from Einstein's theory of relativity, and they are favored in one form or another by the overwhelming majority of cosmologists. However, other models of the universe have been proposed from time to time. Most of them do not follow directly from relativity; some do not even call for change with time. One of the most prominent and long-lived of these alternative theories was the *steady-state universe*.

The discovery in the 1920s that the universe is expanding came as a blow to the traditionalists, who preferred the idea of a static, unchanging cosmos. Even Einstein introduced the fictitious "cosmological constant" into his equations purely to avoid the possibility of an evolving universe. The steady-state model was developed in the 1940s and 1950s by Hermann Bondi, Thomas Gold, and Fred Hoyle as an alternative to evolutionary models. For a time, it had many adherents. Motivated as much by philosophy as by science, it was an attempt to salvage as much of the old view as possible, given the new reality embodied in Hubble's law. It asserted that the universe appears the same for all observers (the cosmological principle), but it went one step further, maintaining that the universe has appeared the same *throughout all time*. This assumption is often called the *perfect* cosmological principle: To any observer at any time, the physical state of the universe is the same—the average density of the universe remains eternally constant. In this way, the steady-state model sought to avoid the thorny questions of the beginning of the universe and what happened before then. In it, the universe had no beginning and no end.

Steady-state cosmologists conceded that the universe was expanding, but because the idea of an initial explosion was unacceptable to them, they were forced to assume that an unknown repulsive force pushes the galaxies apart. Even so, the perfect cosmological principle demanded that the bulk properties of the universe—the average density of matter and the average distance between galaxy clusters—remain constant. Accordingly, to offset the dilution of the density due to the galaxies' recession, the steady-state model required the appearance of new matter in the universe. As illustrated below, the steady-staters proposed that this *new* matter was created from *nothing*. The emergence of new matter (center frame) would keep the average distance between galaxies in the future (right) the same as in the past (left), preserving forever the average density of matter in the universe.

The major problem with the steady-state model was its failure to specify how the additional matter could be created. Proponents of the theory argued that *very* little new matter would be needed to offset the natural thinning of the universe as the galaxies sped apart. The creation of a single hydrogen atom in a volume equivalent to the Houston As-of the distribution—and the equipment necessary to make precise measurements near the peak (from high-flying balloons or spacecraft) proved hard to design and build. As a result, years of painstaking observations left the shape of the distribution only poorly determined. All this changed in 1989 when the *Cosmic Background Explorer* satellite—*COBE* for short—measured the intensity of the microwave background at wavelengths straddling the peak, from half a millimeter up to about 10 cm. The results are shown in Figure 28.19. The solid line is the Planck curve that best fits the *COBE* data. The near-perfect fit corresponds to a universal temperature of about 2.7K. We will see in the next chapter that *COBE* is continuing to make spectacular contributions to the field of observational cosmology.

A striking aspect of the cosmic microwave background is its high degree of *isotropy*. Its intensity is virtually constant from one direction in the sky to another. This isotropy provides strong support for the assumption of the cosmological principle. It also provides us with a novel means of measuring Earth's ''true'' velocity through space, without reference to any neighboring galaxies or galaxy clusters. If we were at rest with respect to the universal expansion (like the coin taped to the surface of the expand-

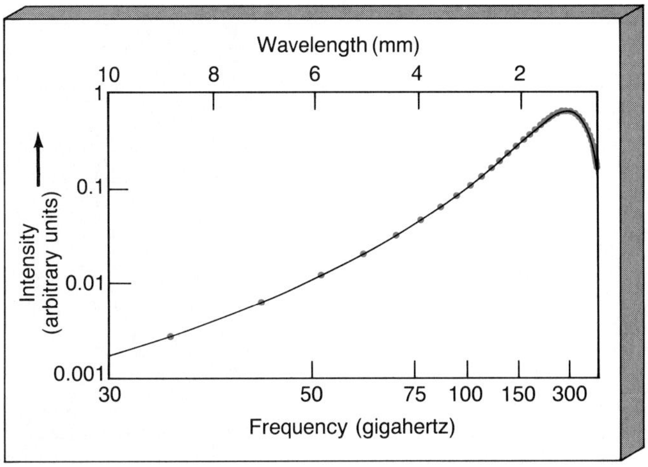

Figure 28.19 *The intensity of the cosmic background radiation, as measured by the* COBE *satellite, agrees very well with that expected from theory. The curve is the best fit to the data, corresponding to a temperature of 2.735K. The experimental errors in this remarkably accurate observation are smaller than the boxes representing the data points. When these results were presented at the 1990 meeting of the American Astronomical Society in Washington, D.C., they drew a standing ovation.*

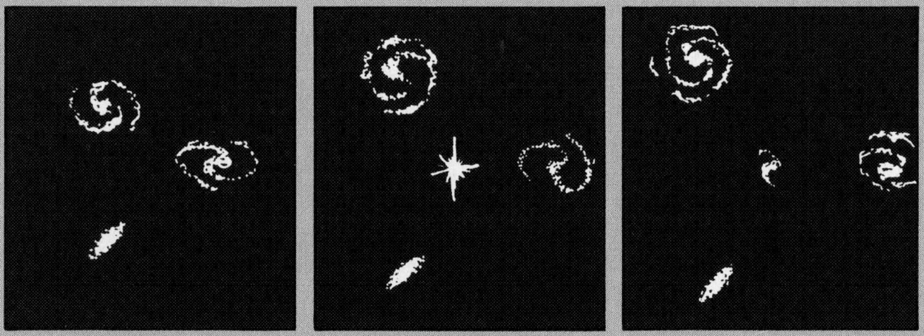

trodome every few years would do. The sudden appearance of such a minute quantity of matter, inside or outside a galaxy, would be quite impossible to detect, so it is not actually ruled out by either observation or experiment. Still, the sudden appearance of new matter from nothing, however little the mass involved, violates one of the most cherished concepts of modern science—the conservation of mass and energy. Matter may be created from energy, but it is very hard to understand how matter can be created spontaneously from nothing at all.

Few scientists are prepared to throw out most of the laws of physics just to avoid a philosophical difficulty. For this reason alone, the steady-state model is fatally flawed in the eyes of most astronomers. The steady-state model can be ruled out for at least two other good reasons. First, the spread of galaxies, especially the quasars, is *not* uniform throughout space. The distant quasars far outnumber those nearby. If we had lived 10 billion years ago, when quasars were the dominant cosmic objects, our view of the universe would have been much different, which is a clear violation of the perfect cosmological principle. Second, the steady-state universe offers no satisfactory explanation of the cosmic microwave background. In this theory, one simply has to imagine that the background radiation ''just is,'' with no particular reason for its existence.

This combination of objections far outweighs any benefits the steady-state theory might bring to cosmology. After a two-decade run as a cosmological contender, the steady-state universe was abandoned because it is inconsistent with reality. This episode illustrates how personal preferences can drive good scientists to abandon the scientific method, adopting theories that have no basis in fact. In the end, however, the scientific method prevails. In science, at least, theories unsupported by data simply do not last.

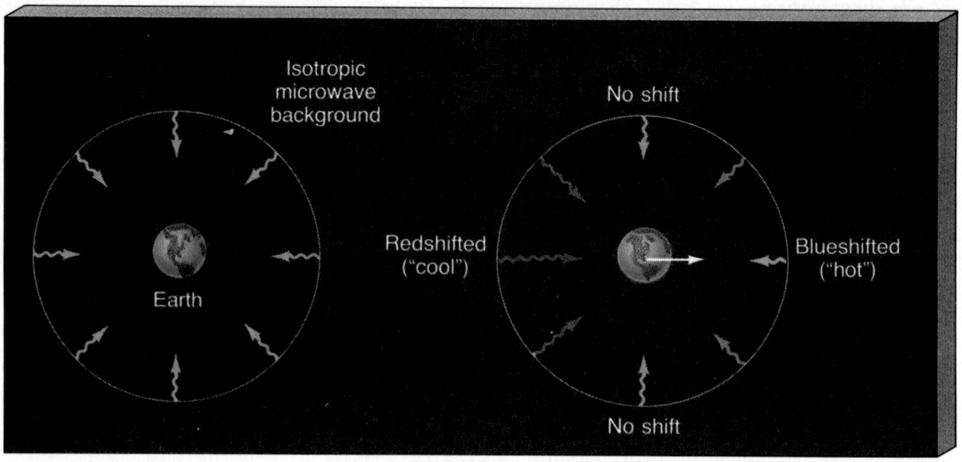

Figure 28.20 *To an observer at rest with respect to the expanding universe, the microwave background appears isotropic. A moving observer measures "hot" blueshifted radiation in one direction (the direction of motion) and "cool" redshifted radiation in the opposite direction.*

ing balloon in Figure 28.5), we would see the microwave background as almost perfectly isotropic (actually, to about 1 part in 10^5), as illustrated in Figure 28.20 . However, if we were moving with respect to that frame of reference, as in Figure 28.20 , the radiation from in front of us would be slightly blueshifted by our motion, while that from behind would be redshifted.

To a moving observer, the microwave background should appear a little hotter than average in front and slightly cooler behind. Figure 28.21 shows a *COBE* map of the microwave background temperature over the entire sky. The blue regions are hotter than average, by about 0.007K; the red regions cooler. The data indicate that the Earth's velocity is about 390 km/s in the general direction of the constellation Leo. Even though the principle of relativity says that there is no preferred frame of reference, as the laws of physics look the same to all observers, there *is* nevertheless a way to determine our absolute velocity with respect to the universe!

When we observe the microwave background, we are looking almost all the way to the very beginning of the universe. The photons that we receive as these radio waves today have not interacted with matter since the universe was a mere 100,000 years old, when, according to our models, it was less than 1⁄1000 of its present size. To probe further, back to the Big Bang itself, requires us to enter the world of nuclear and particle physics. Strange as it may seem, the sciences of the very large and the very small come together in the study of the early universe. The Big Bang was the biggest and the most powerful particle accelerator of all! In the next chapter, we will see how studies of conditions in the primeval fireball may aid us in understanding the present-day structure and future evolution of the universe we live in.

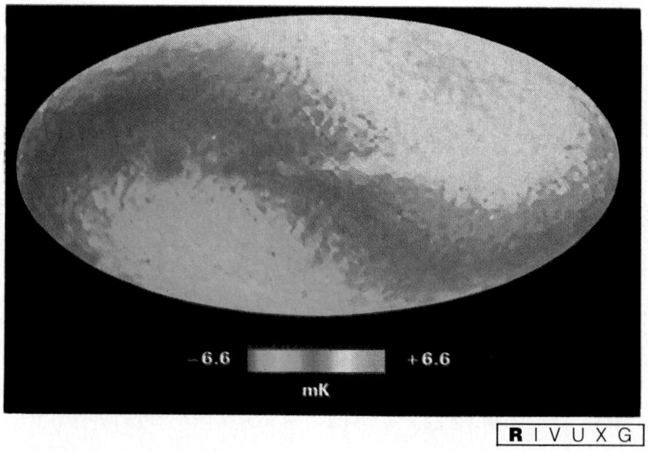

Figure 28.21 A COBE *map of the microwave sky reveals that the microwave background appears a little hotter in the direction of the constellation Leo and a little cooler in the opposite direction. The temperature difference is about 0.007K, corresponding to a velocity of 390 km/s. (NASA)*

CHAPTER SUMMARY

Cosmology is the study of the entire universe. In order to simplify their task, cosmologists usually assume that the universe is homogeneous on scales greater than a few hundred megaparsecs. A second assumption—supported by both observational evidence and theoretical reasoning—is that the universe is also isotropic. Together, the assumptions of homogeneity and isotropy are known as the cosmological principle.

Hubble's law implies that about 13 billion years ago all the galaxies in the universe were right on top of one another. In fact, cosmologists believe that everything in the universe—matter and radiation alike—was confined to a single point in that instant. The explosion of this point is known as the Big Bang. However, the galaxies are not simply flying apart into the rest of an otherwise empty universe. Space itself is expanding.

The dynamics of the universe can be understood using simple Newtonian concepts (although Einstein's theory of general relativity is needed to explain why these simple ideas work so well). A sufficiently dense universe will stop expanding, and then recollapse, as its gravity eventually overwhelms and reverses the outward motion. A low-density universe will expand forever—its gravity is simply too weak to stop the expansion. The dividing line between these two possibilities is a marginally bound universe, which is said to have critical density.

If the average density of the cosmos is above the critical value, space is curved so much that it bends back on itself, making the universe finite in size. Such a universe is said to be closed. A low-density universe is infinite in extent and has a ''saddle-shaped'' geometry. It is usually called an open universe. When the density exactly equals the critical value, the universe has no curvature—it is flat.

Astronomers do not know for certain whether our expanding universe is closed, open, or of critical density. In all three cases, the cosmos began in a state of incredibly high density and temperature—the Big Bang. In an open universe, the galaxies will recede forever, eventually growing dark as stars consume their fuel supplies. A closed universe will one day shrink toward a superdense, superhot singularity, much like the one from which it originated. The universe might conceivably rebound from this ''Big Crunch'' into another cycle of expansion. In the critical universe, the accumulated matter is just sufficient to halt the expansion, but only after an infinitely long time.

Attempts to measure the density of the universe have produced inconclusive results. Counts of visible galaxies suggest that the universe is open, but if the total amount of dark matter lurking in galaxy clusters were to outweigh the luminous mass by a factor of 100 or more, as could well be the case, the universe would be closed. A measurement of the change in the expansion rate of the universe translates directly into an estimate of the universe's density. The most distant galaxies do have substantially greater recessional velocities than the Hubble law would predict, but these galaxies are very faint, and accurate observations of their properties are difficult. Many astronomers now believe that conditions in the early universe may have conspired to produce a precisely critical expansion rate.

The universe expanded faster in the past. The assumption of a constant expansion rate leads to an overestimate of the universe's age. If H_0 = 75 km/s/Mpc, the universe can be no more than 13 billion years old.

It is standard practice in astronomy to refer to the cosmological red shift in terms of recessional velocity. It is more correct, however, to think that a photon is attached to the expanding fabric of space, so that its wavelength expands along with the universe.

In the 1960s, two scientists at Bell Telephone Laboratories discovered the cosmic microwave background, thought to be a ''fossil remnant'' of the primeval fireball from which our universe arose. The radiation from the fireball has been redshifted by the expansion of the universe all the way from the gamma-ray to the radio portion of the electromagnetic spectrum. The temperature of the microwave background is only a few kelvins and varies very little from one part of the sky to another. Its existence lends strong support to the Big Bang theory.

KEY WORDS

Big Bang
closed universe
cosmic microwave background
cosmological constant
cosmological principle
cosmologist
cosmology
critical density
critical universe
Great Attractor
homogeneous
isotropic
Olbers's paradox
open universe
oscillating universe

REVIEW QUESTIONS

1. What is the cosmological principle?
2. Assuming the cosmological principle is correct, how can we locate the center of the universe?
3. What is Olbers's paradox?
4. Explain how an accurate measure of Hubble's constant can lead to an estimate of the age of the universe.
5. We appear to be at the center of the Hubble flow. Why doesn't this violate the assumption of homogeneity within the cosmological principle?
6. Why isn't it correct to say that the expansion of the universe involves galaxies flying outward in space?
7. What is a closed universe, and what is its fate?
8. What is an open universe, and what is its fate?
9. What is a critical universe, and what is its fate?
10. What is the cosmological constant?
11. What do counts of visible galaxies reveal about the fate of the universe? Why are galaxy counts inconclusive?
12. Is there enough dark matter to close the universe?

13. Explain how a measurement of the change in the expansion rate of the universe translates directly into an estimate of the density of the universe.
14. What is the cosmic microwave background, and why is it so significant?
15. What is the perfect cosmological principle, and why do most astronomers reject it?
16. According to the Big Bang theory as described in this chapter, what is the maximum possible age of the universe if $H_0 = 50$ km/s/Mpc? 75 km/s/Mpc? 100 km/s/Mpc?
17. For our adopted value of the Hubble constant of 75 km/s/Mpc, the critical density is 10^{-29} g/cm^3. (a) How much mass does that correspond to within a volume of 1 cubic astronomical unit? (b) How large a cube would be required to enclose an amount of material equal to the mass of the Earth?
18. How far could light have traveled in the 13 billion years since the Big Bang?

DISCUSSION QUESTIONS

1. Many cultures throughout history have developed their own cosmologies. Do you think the modern scientific cosmology is more likely to endure than any other? Why or why not?
2. Which universe do you prefer philosophically: open, closed, critical, or steady-state? Why?
3. Estimates of the age of the universe based on the Hubble constant are glaringly different from estimates of the ages of globular clusters in our own Galaxy. Why do you think these two techniques have led to different conclusions? How do you think astronomers should proceed in resolving the controversy?

PROJECT

1. Go to a clear, dark country location on a night when the Moon is down. Spend an hour or more reclining comfortably, gazing up at the glittering stars. Think about the universe as revealed by modern astronomy: an expanding universe of billions of galaxies that are invisible to the eye but visible to the large telescopes and sophisticated detectors of astronomers. Can you imagine the universe of galaxies located beyond the visible stars? Can you imagine a time when those galaxies—and all that you see around you—were compressed into a single point?

SUGGESTED READINGS

Davies, P. "Everyone's Guide to Cosmology." *Sky and Telescope* (March 1991). Good basic article on cosmology by a prolific author.

Finkbeiner, A. "A Universe in Our Own Image." *Sky and Telescope* (August 1984). Is the presence of conscious beings such as ourselves crucial to the observed structure of the entire universe?

Lightman, A. *Ancient Light*. Cambridge: Harvard University Press, 1991. Big Bang cosmology presented on a popular level.

Odenwald, S. "Einstein's Fudge Factor." *Sky and Telescope* (April 1991). After 75 years, Einstein's "greatest blunder" keeps turning up in cosmologists' discussions about the origin, evolution, and fate of the universe.

Overbye, D. *Lonely Hearts of the Cosmos*. New York: Harper Perennial, 1991. Personal narrative on research at the front lines of modern cosmology.

Silk, J. "Probing the Primeval Fireball." *Sky and Telescope* (June 1990). Early article about *COBE*'s exploration of the cosmic microwave background.

CHAPTER

29

The Early Universe

Toward the Beginning of Time

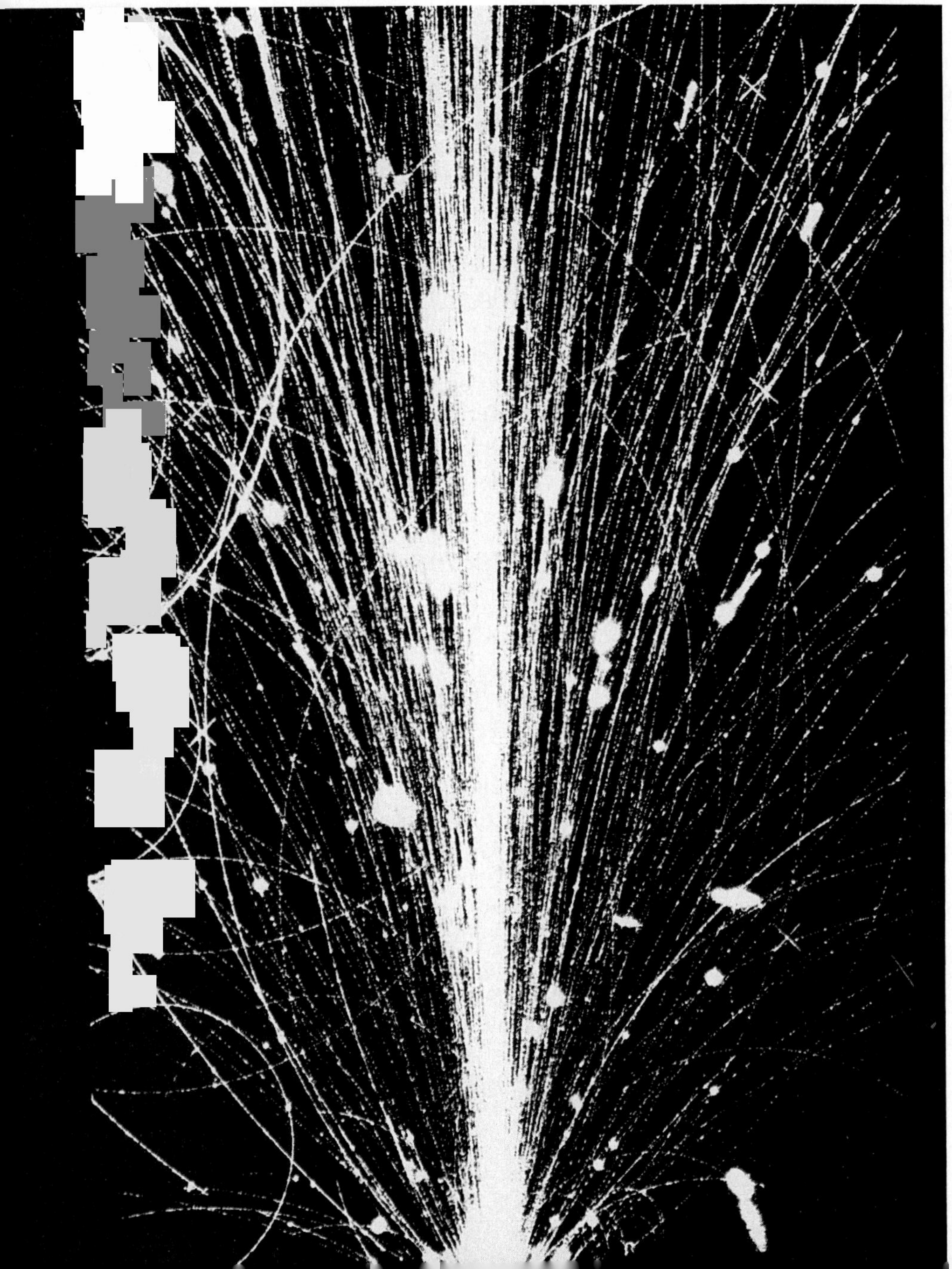

This picture shows the aftermath of a collision between very energetic particles in an accelerator. The violence of the collision produces a huge number of short-lived elementary particles that fly away in all directions at almost the speed of light. The straight lines are due to uncharged particles. The curved lines are due to charged particles whose paths are bent by the strong magnetic field that fills the detector area. The amount and direction of curvature allow particle physicists to probe the internal workings of subatomic particles on extremely small scales. The conditions produced in collisions like these are as close as we can come to the universe at epochs shortly after the Big Bang. (Courtesy Fermi Laboratory)

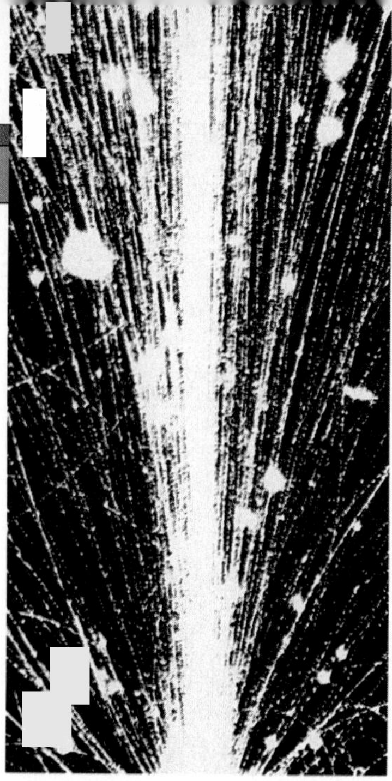

LEARNING GOALS

1 to know that the newborn universe was extremely hot and dense
2 to understand how matter emerged from the primeval fireball
3 to realize how and when the simplest nuclei and atoms formed
4 to recognize that the universe may have undergone a brief early phase of violent and rapid expansion
5 to see some of the efforts currently underway to unify the known forces in the universe
6 to understand the formation of large-scale structure in the cosmos

What was it like at the start of the universe? What happened at the origin of time? Can we say anything about the origin itself? What conditions existed during the first few moments of the universe? And how did those conditions change? How and when did our Galaxy form? These are surely basic questions, but they are hard questions. Until the twentieth century, they lay squarely in the domain of religion or philosophy. Now, after more than 10,000 years of civilization, science may be ready to provide some insight regarding the ultimate origin of all things.

The answers we obtain must still be considered tentative. Times long past are times long gone, and such ancient history cannot be observed directly. Nevertheless, astronomers have constructed models—mathematical sketches based on a large body of data on the shape and structure of our universe—that paint a compelling picture of the universe as it was long ago.

In studying the earliest moments of our universe, we enter a truly alien domain. As we move backward in time toward the Big Bang, our customary landmarks slip away one by one. Atoms vanish, then nuclei, then even the elementary particles themselves. In the beginning, the universe consisted of pure energy, at unimaginably high temperatures. As it expanded and cooled, the ancient energy changed into the particles that make up everything we see around us today. Modern physics has now arrived at the point where it can reach back almost to the instant of the Big Bang itself, allowing scientists to unravel some of the mysteries of our own beginning.

BACK TO THE BIG BANG

✓1 To appreciate the earliest epochs of the universe, we must think about times long, long ago. We must strive to imagine what it was like long before the Earth and the Sun originated—before any star existed. But before we can study some of the key events in the history of the cosmos, we need to construct a general picture of the physical conditions that prevailed during those early times and develop some important concepts and terminology to describe them.

Matter and Radiation

On the very largest scales, we can regard the universe as a roughly homogeneous mixture of matter and radiation. As we saw in Chapter 28, the overall density of matter is not known with certainty, but it is thought to be quite close to the critical density of 10^{-29} g/cm^3, above which the universe will eventually recollapse, and below which it will expand forever. The universe is apparently open, but barely so. The nature of the matter that makes up the universe is itself also partially understood. Some of it consists of the familiar building blocks of nuclei and atoms—protons, neutrons, and electrons—but much, if not most, of it may be in the form of dark matter, whose composition is still hotly debated by astronomers.

Most of the radiation in the universe is in the form of the cosmic microwave background, the low-temperature (3K) radiation field that fills all space. Surprisingly, although this background radiation is very weak, it still contains more energy than has been emitted by all the stars and galaxies that have ever existed in the universe! The reason for this is that stars and galaxies, though very intense sources of radiation, occupy only a small fraction of space. When their energy is averaged out over the volume of the entire universe, it falls short of the energy of the microwave background by at least a factor of 10. For our current purposes, then, we can ignore most of the first 27 chapters of this book and regard the cosmic microwave background as the only significant form of radiation in the universe!

In order to compare matter and radiation, we must first convert them to a "common currency"—either mass or energy. Let's choose to compare their masses. We can express the energy in the microwave background as an equivalent density by first calculating the number of photons in any cubic centimeter of space, then converting the total energy of these photons into a mass using the relation $E = mc^2$. When we do this, we arrive at an equivalent density for the microwave background of about 5×10^{-34} g/cm^3. Thus, *at the present moment* the density of matter (about 10^{-29} g/cm^3) in the universe far exceeds the density of radiation. In cosmological terminology, we say that we live in a **matter-dominated universe.**

Was the universe always matter-dominated? To answer this question, we must ask how the densities of both matter and radiation change as the universe expands. Both decrease, as the expansion dilutes the numbers of atoms and photons alike. But the radiation is also diminished in energy by the cosmological red shift, so its density falls *faster* than that of matter as the universe grows (see Figure 29.1). Conversely, as we look back in time, closer and closer to the Big Bang, the density of the radiation increases faster than that of matter. Accordingly, even though today the radiation density is much less than the matter density, there must have been a time in the past when they were equal. Before that time, radiation was the main constituent of the cosmos. The universe is said to have been **radiation-dominated** then. Given our best estimates of the present densities, the crossover point—the time at which the densities of matter and radiation were equal—occurred a few thousand years after the Big Bang, when the universe was about 20,000 times smaller than it is today. The temperature of the background radiation at that time was about 60,000K, so that it peaked well into the ultraviolet portion of the spectrum. The matter density was already very low—only 10^{-16} g/cm^3.

Our discussion of the universe naturally breaks into two main eras. For the first few thousand years after the Big Bang, the universe was small and dense and dominated by the effects of radiation. We call this the **Radiation Era.** Some matter existed during this time, but it was a mere contaminant in the blinding radiation of the primeval Big Bang fireball. Afterwards, matter came to dominate, in the **Matter Era.** Atoms, molecules, and galaxies formed as the universe cooled and thinned toward the state we see today.

Conditions during the Matter Era are relatively familiar. We have spent much of this book discussing the behavior of low-density gas and radiation at temperatures below 60,000K (the temperature at the crossover point). By contrast, the early phases of the Radiation Era are characterized by temperatures and densities far greater than anything we have encountered thus far, even in the hearts of supernovae. In order to fathom the early universe, we must delve a little more deeply into the behavior of matter and radiation at high temperatures.

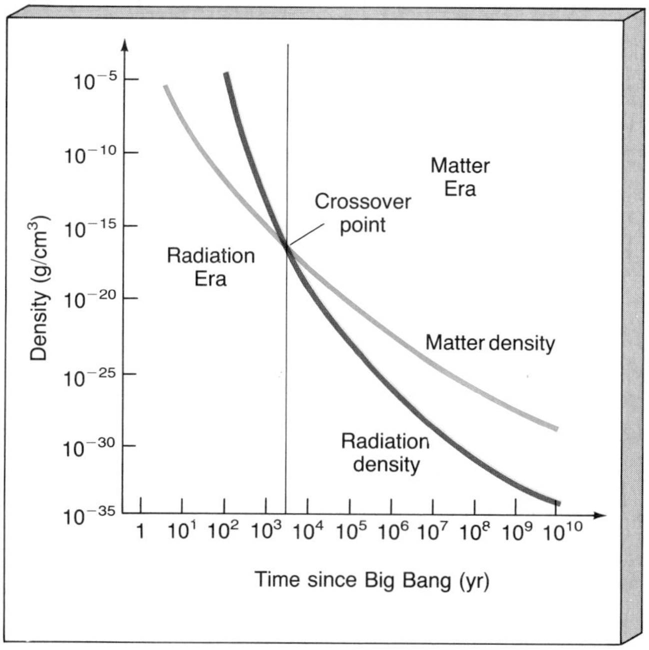

Figure 29.1 *As the universe expanded, the number of both matter particles and photons per unit volume decreased. However, the photons were also reduced in energy by the cosmological red shift, reducing their equivalent mass, and hence their density, still further. As a result, the density of radiation fell faster than the density of matter as the universe grew. Tracing the curves back from the densities we observe today, we see that radiation must have dominated matter at early times—that is, at times before the crossover point. The curves divide the universe into the Radiation Era and the Matter Era.*

Thermal Equilibrium in the Early Universe

✓2 The key to understanding events in these early times lies in a process known as **pair production,** in which two photons of electromagnetic radiation can give rise to a pair of particles—a particle and its antiparticle—as shown in Figure 29.2(a) for the case of electrons and positrons (indicated by the minus and the plus symbols in the figure). This process is just the opposite of particle-antiparticle *annihilation,* which we have already encountered in earlier chapters, where a particle and its antiparticle destroy each other in a burst of gamma-rays, as depicted in Figure 29.2(b). The point is this: Provided the total energy of the photons is greater than the combined mass-energy of the particle-antiparticle pair, pair production can and must occur. Fur-

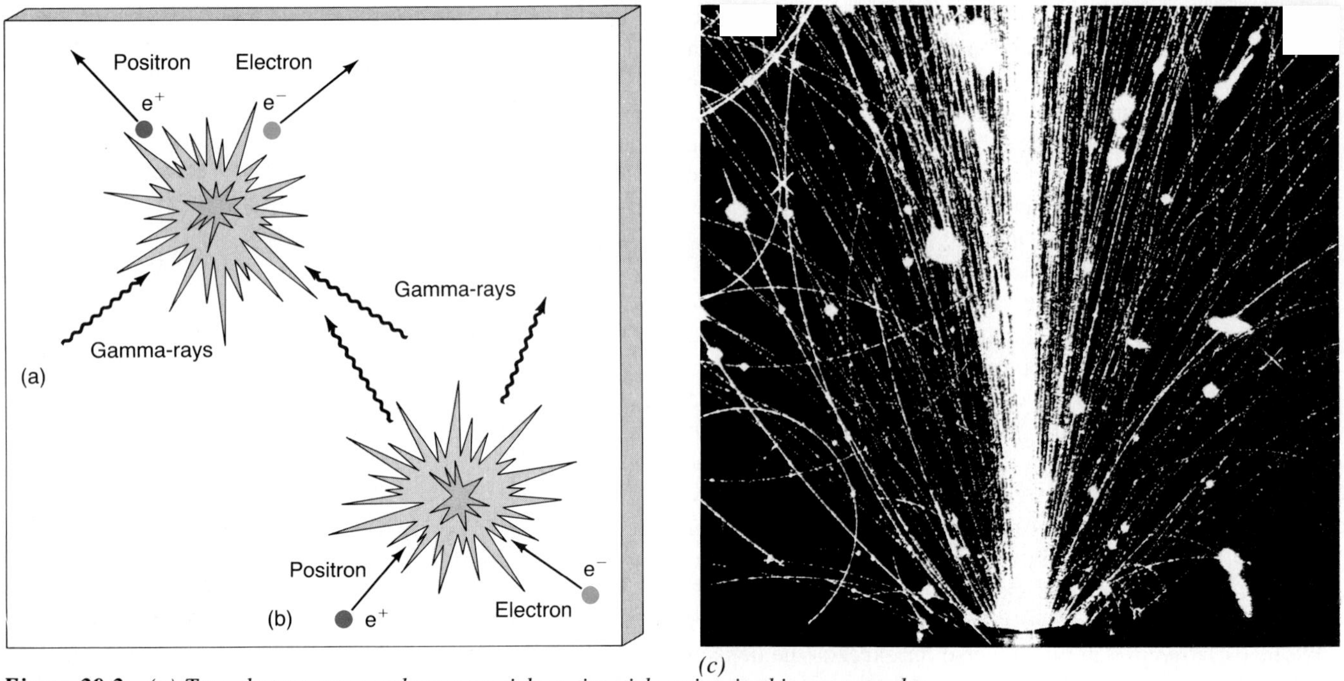

Figure 29.2 *(a) Two photons can produce a particle-antiparticle pair—in this case an electron and a positron—if their total energy exceeds the mass-energy of the particles produced. (b) The reverse process is particle-antiparticle annihilation, where our electron and positron destroy each other, vanishing in a flash of gamma-rays. (c) A photograph of the results of a particle-accelerator experiment, showing how radiation can lead to particle creation. (Fermilab)*

thermore, the process happens *spontaneously*. Any two high-energy photons can suddenly give rise to any particle-antiparticle pair, subject only to the law of energy conservation.

Recall that we encountered a similar notion in Interlude 24-3, where we saw how a particle and its antiparticle could be produced spontaneously from the vacuum. The creation of those so-called *virtual pairs* violated the law of conservation of mass and energy, but the rules of quantum mechanics permit this to occur as long as the violation is corrected quickly enough. The creation of a particle-antiparticle pair from two high-energy photons, however, violates no rules. Once created, the newborn particles can travel freely through space, at least until they are annihilated by a subsequent encounter with another antiparticle.

Whether or not pairs of particles are produced in this way depends on the temperature of the radiation. Higher temperatures mean more energetic photons and a greater chance that pair production will occur. As an example, let us consider the production of the familiar electron, along with its antiparticle, the positron. At "low" temperatures—less than a billion or so kelvins—photons simply do not have enough energy for pair production to occur. However, as the temperature rises, the average photon energy rises as well. Above about 10^{10} K, *most* photons have enough energy to form an electron or a positron, and pair production is commonplace. Where once there had been just radiation, space now seethes with electrons and positrons too, constantly created from the radiation and annihilating each other to form photons again. Figure 29.3 illustrates how this change takes place as the temperature rises.

The critical temperature at which particles of a given type can be spontaneously produced from radiation by the mechanism of pair production is called the **threshold temperature** for that type of particle. The threshold temperature increases as the mass of the particles increases. For electrons, the threshold temperature is about 6×10^9 K. For protons, which are nearly 2000 times more massive, it is about 10^{13} K.

Now let's consider what happens as the universe expands and cools. Above the threshold temperature for any given particle—let's assume a particle called X—X particles and anti-X particles existed in great numbers in the universe. They were formed by pair production at exactly the same rate as they annihilated each other. We say that the X particles were in **thermal equilibrium** with the radiation. As the universe expanded and cooled, its temperature dropped below the threshold for X and anti-X particles to form. The particles that had been created annihilated each other, and no more X and anti-X particles were created. The energy released by the annihilation was not enough to raise the temperature of the universe back above the threshold for forming more X and anti-X particles. In this scenario, all matter—the X and other particles—eventually became radiation. Only if the universe had an excess of matter over antimatter—more X particles than anti-X particles—or more antimatter over matter—more anti-X particles than X particles—could any particle have escaped annihilation.

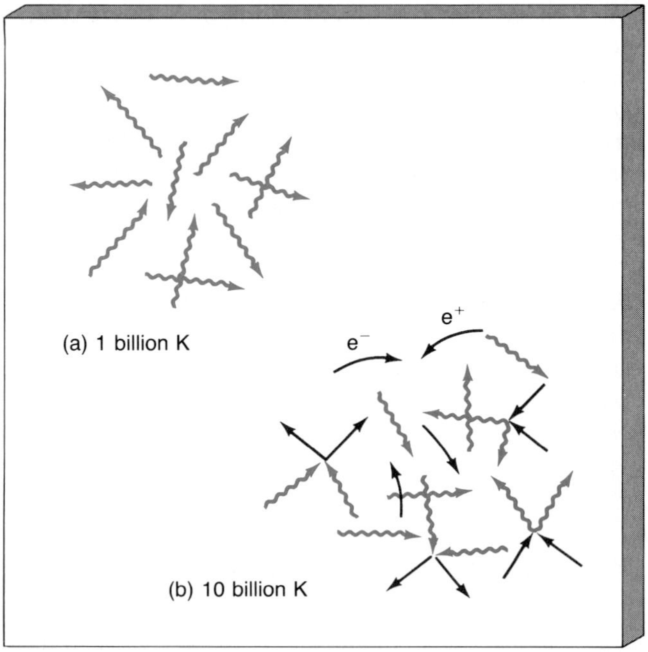

Figure 29.3 *(a) Below a temperature of about 1 billion K, photons have too little energy for pair production to occur. (b) At 10 billion K, most photons have enough energy to create particle-antiparticle (electron-positron) pairs, so these particles exist in great numbers, in equilibrium with the radiation.*

Because we are here to ponder the early universe and we are made of matter, we know that some matter did survive these early moments. The existence of matter means that, for some reason, there were unequal amounts of matter and antimatter early on. The particles that survived are said to have **frozen out** of the radiation. Once the temperature of the universe dropped below the threshold for production of a particular particle, no more such particles arose. The small residue of particles that outnumbered their antiparticles survived—there was nothing to annihilate them—and their numbers have remained constant ever since. That residue eventually came to dominate the universe and form the galaxies and large-scale structure we see today.

The Radiation Era

The universe began with an explosion from an incredibly hot and dense state. Precisely what state, we cannot say. And why it exploded, we really don't know. *Why* the universe began suddenly expanding some 10 or 15 billion years ago is a most difficult query. To understand why the universe began expanding, or even more fundamentally, why the universe exists at all, is currently beyond science—there are simply no relevant data.

Although still ignorant of the moment of creation itself (that is, precisely zero time), theorists believe that the physical conditions in the universe can be understood in terms of present-day physics back to an extraordinarily short time after the Big Bang—10^{-43} s, in fact. The period of time from the beginning to 10^{-43} s is usually referred to as the *Planck epoch,* after Max Planck, one of the creators of quantum mechanics. (This is the same Max Planck who gave his name to the thermal radiation curves we have studied throughout this book.)

Why can't theorists push our knowledge back to the Big Bang itself? The answer is simply that to do this we need a theory—a theory of *quantum gravity*—that unifies gravity with the other forces of nature (electromagnetism, the strong force, and the weak force, as described in Interlude 5-1 and later in this chapter). No such theory exists, so we simply cannot talk meaningfully about the universe during the Planck epoch.

Immediately after the Planck epoch, the universe was filled with radiation and a vast array of subatomic particles, created by the mechanism of pair production. The temperature exceeded 10^{28} K, and theory indicates that the energies of the particles were so high that the strong, weak, and electromagnetic forces were all indistinguishable from one another. These three forces are said to have been **unified** at this early time—apart from gravity, there was only one other fundamental force of nature. The present-day theories that describe this force are collectively known as **Grand Unified Theories,** or GUTs for short. Accordingly, we refer to this period of time as the *GUT epoch.*

Calculations of the temperature of the radiation-dominated universe, based on general relativity and incorporating the ideas described in this section, indicate that the temperature of the universe at any given time after the Big Bang is given approximately by

$$\textit{temperature (in kelvins)} = \frac{10^{10}}{\sqrt{\textit{time (in seconds)}}}.$$

This equation tells us that the temperature of the universe 0.0001 s (say) after the Big Bang was 1 trillion kelvins—10^{12} K. After 1 s, the temperature was 10^{10} K. At 100 s, the temperature was 10^9 K, and so on. By using this equation, we can see that the temperature of the universe was equal to the threshold temperature for proton-antiproton production (10^{13} K) about 1 μs (10^{-6} s) after the Big Bang and equaled the electron-positron threshold value of 6×10^9 K after about 3 s.[1]

Theory indicates that at temperatures below 10^{28} K the strong nuclear force becomes distinguishable from the electroweak force, and the forces of nature are no longer unified. When the universe had cooled to this temperature, about 10^{-35} s after the Big Bang, the GUT epoch ended. In effect, the strong force froze out of the expanding universe at that point. Our next major subdivision of the Radiation Era covers the period when all "heavy" elementary particles—that is, all the way down in mass to protons, neutrons, and their constituent quarks—were in thermal equi-

[1]The equation becomes less and less reliable as we move back in time toward the Big Bang, however. It could easily be in error by a factor of 10 or more when applied to the GUT epoch.

librium with the radiation. This stage is sometimes known as the *hadron epoch,* after the general term for particles that interact through the strong force.

As the universe expanded rapidly, it cooled and thinned. At a temperature of about 10^{15} K, 10^{-10} s after the Big Bang, the weak and the electromagnetic components of the electroweak force began to display their separate characters. By about 0.1 milliseconds (10^{-4} s) after the Big Bang, the temperature had dropped well below the 10^{13} K threshold for the creation of protons and neutrons, and the hadron epoch ended. The main constituents of the universe were now lighter particles—muons, electrons, neutrinos, and their antiparticles—all still in thermal equilibrium with the radiation. Compared with the numbers of these lighter particles, only very few protons and neutrons remained at this stage because most had been annihilated. Electrons, muons, and neutrinos are collectively known as *leptons,* after the Greek word meaning light (that is, not heavy). Accordingly, this period in the history of the universe is known as the *lepton epoch.*

The lepton epoch ended when the universe was about 100 s old, and the temperature became too low for electron-positron pair production to occur. The density of the universe by this time was about 10 times that of water, and the temperature was about 1 billion kelvins. During this epoch, the thinning universe became transparent to neutrinos, and these ghostly particles have been streaming freely through space ever since. Most have not interacted with any other particle since the universe was a few seconds old.

The final significant event in the Radiation Era occurred when protons and neutrons began to fuse into heavier nuclei. During the period, which we will call the *nuclear epoch,* the temperature was a few hundred million kelvins, and fusion occurred very rapidly, forming deuterium, helium, and some lithium in quick succession before conditions became too cool for further reactions to occur. By the time the universe was about 15 minutes old, much of the helium we observe today had been formed.

The Matter Era

As time elapsed, change continued. Our next major epoch extends in time from a few thousand years (the end of the Radiation Era) to about 1 million years after the Big Bang. Because atoms first formed then, we will call this time interval the *atomic epoch.* As the primeval fireball diminished in intensity, a crucial change occurred—perhaps the most important change in the history of the universe. At the beginning of the atomic epoch, radiation still overwhelmed matter. As soon as elementary particles tried to combine, radiation broke them apart, preventing the formation of even simple atoms or molecules. However, as the universe expanded and cooled, the early dominance of radiation eventually ended. Once formed, atoms remained intact.

The last two epochs together bring us to the current age of the universe. During these late stages, change happened at a much more sedate pace. By the time the universe was about a billion years old, galaxies and large-scale structure had formed. For the first time, the visible universe departed from homogeneity on macroscopic scales. The largely uniform universe of the Radiation Era became a universe containing large agglomerations of matter. We call the period from 10^6 to 10^9 years after the Big Bang the *galactic epoch.* At its end, quasars were shining brightly in the otherwise dark sky, and the first stars were burning and exploding, helping to determine the future shape of their parent galaxies. Since then, stars, planets, and life have appeared in the universe. This final *stellar epoch* has been the subject of the previous 28 chapters.

Epochs in the Evolution of the Universe

Let us pause to recap in broad terms the history of the cosmos, starting at the Big Bang. Table 29-1 and Figure 29.4 summarize the physical conditions that existed during the eight major stages in the development of the universe just discussed. Specified in this table and figure are the time, density, and temperature spans and a brief description of the main physical events that dominated the universe during each epoch. The numbers in this table and figure result from pushing the known laws of physics as far back in time as we can. In the next few sections, we will expand on some of the epochs in greater detail, but let's not lose sight of the big picture and the place of each epoch in it.

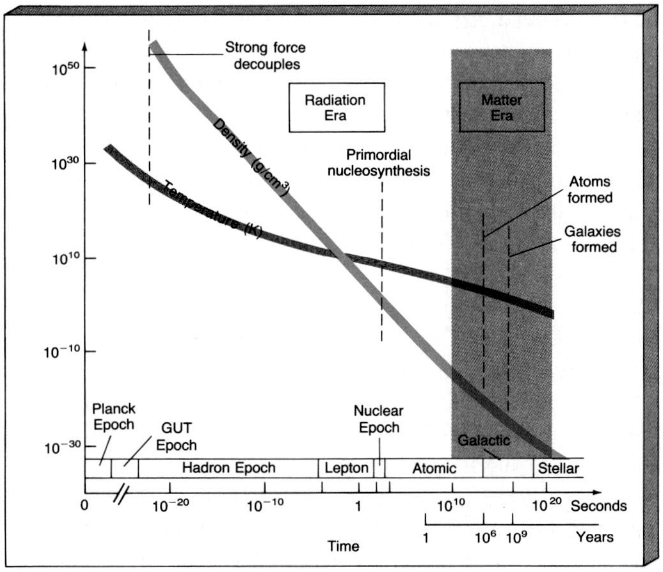

Figure 29.4 *The average temperature and average density throughout the history of the universe. The epochs listed in Table 29-1 are indicated. (Numerical values for these plots are taken from Table 29-1.) Some key events in the history of the universe are also marked.*

TABLE 29-1 ***Major Epochs in the History of the Universe***

Era	Epoch	Time (from Big Bang)	Density (g/cm^3)	Temperature (K)	Main Events
Radiation Era		0 s	∞	∞	
	Planck				Unknown physics; quantum gravity.
		10^{-43} s	10^{92}	10^{32}	
	GUT				Strong, weak, and electromagnetic forces unified.
		10^{-35} s	10^{72}	10^{27}	
	Hadron				Heavy and light particles all in thermal equilibrium.
		10^{-4} s	10^{13}	10^{12}	
	Lepton				Only light particles still in thermal equilibrium. Neutrinos decouple.
		10^{2} s	10^{1}	10^{9}	
	Nuclear				Deuterium and helium formed by fusion of protons and neutrons.
		10^{3} s	10^{-1}	3×10^{8}	
Matter Era		10^{3} yr ($\approx 3 \times 10^{10}$ s)	10^{-13}	6×10^{4}	
	Atomic				Matter begins to dominate. Atoms form.
		10^{6} yr ($\approx 3 \times 10^{13}$ s)	10^{-16}	4.5×10^{3}	
	Galactic				Galaxies and larger scale structures form.
		10^{9} yr ($\approx 3 \times 10^{16}$ s)	3×10^{-28}	10^{1}	
	Stellar				All galaxies have formed. Stars continue to form.
		$>10^{10}$ yr ($\approx 3 \times 10^{17}$ s)	10^{-29}	3	

PRIMORDIAL NUCLEOSYNTHESIS

✓3 Having outlined the main events in the evolution of the cosmos, we now consider in more detail some of the key episodes that have shaped the universe we see around us today: the formation of nuclei and atoms and the processes leading to the formation of large-scale structure. We will find that our studies take us far back into the past—almost to the beginning of the universe itself.

Helium Formation in the Early Universe

Let us start by completing our story of the creation of the elements, begun in Chapters 22 and 23 but never finished. We saw in those earlier chapters that the theory of stellar nucleosynthesis accounts very well for the observed abundances of heavy elements in the universe. However, there are some conflicts between theory and observations when it comes to the abundances of the light elements helium, lithium, beryllium, and boron. Simply put, there appears to be more of those elements than can be explained by nuclear fusion in stars over the lifetime of the Galaxy. What's more, astronomers find that, no matter where they look and no matter how low a star's abundance of heavy elements, there seems to be a minimum amount of helium—between 20 and 25 percent by mass—in all stars. The most obvious explanation is that this base level of helium is *primordial*—that is, formed during the early, hot epochs of the universe.

Could a large amount of helium have been created in the early universe? The answer is yes. The possibility of **primordial nucleosynthesis**—the production of elements heavier than hydrogen by nuclear fusion shortly after the Big Bang—was first realized in the 1940s, when physicist George Gamow pointed out that the hot, dense conditions in the early universe might have provided all of the ingredients necessary for the formation of helium. Later calculations, especially those performed in the 1970s, when the physical state of the cosmos at early times was much better understood, demonstrated that Gamow's idea was essentially correct.

During the nuclear epoch, the average temperature of the universe exceeded by a wide margin the 10 million K needed to fuse hydrogen into helium through the proton-proton chain (the Sun's energy source, as we discussed in Chapter 18). Was helium created within the primordial fireball in basically the same way that it now forms within stars? The answer is no. Helium did form, but the proton-proton chain was *not* the main route. There was an easier way, involving fusion of protons and neutrons instead.

By the end of the hadron epoch, 10^{-4} s after the Big Bang, protons and neutrons had frozen out of the cosmic fireball—the temperature of the universe had fallen below the threshold necessary to create them in quantity. Detailed calculations indicate that by the time the temperature had dropped to about a billion kelvins (100 s after the Big Bang) there was about 1 neutron for every 5 protons in the universe. The stage was set for a nuclear fusion to occur.

Protons and neutrons can combine to produce deuterium[2] nuclei:

$$^{1}\text{H (proton)} + \text{neutron} \longrightarrow {}^{2}\text{H (deuterium)} + \text{energy}.$$

Although the above reaction must have occurred very frequently during the lepton epoch, the temperature was still so high then that the deuterium nuclei were broken apart by the radiation as fast as they formed. For the same reason, the proton-proton chain could not operate because the deuterium created in its initial reaction:

$$^{1}\text{H} + {}^{1}\text{H} \longrightarrow {}^{2}\text{H} + \text{positron} + \text{neutrino}$$

could not remain intact. (Also, it turns out that this reaction is much slower than the proton-neutron reaction. It plays an important role in the Sun only because there are no neutrons around to make the other reaction possible.) Although temperatures and densities before the nuclear epoch were certainly high enough for fusion to occur, the process could not get under way because deuterium was destroyed as fast as it appeared. The universe had to wait until it became cool enough for the deuterium to survive. This waiting period is sometimes called the **deuterium bottleneck.**

When the temperature of the universe fell below about 900 million K, roughly 2 minutes after the Big Bang, deuterium was at last able to form and endure. When that occurred, numerous other reactions quickly converted deuterium into heavier elements:

$$\begin{aligned} {}^{2}\text{H} + {}^{1}\text{H} &\longrightarrow {}^{3}\text{He} + \text{energy} \\ {}^{3}\text{He} + \text{neutron} &\longrightarrow {}^{4}\text{He} + \text{energy} \\ {}^{2}\text{H} + {}^{2}\text{H} &\longrightarrow {}^{4}\text{He} + \text{energy} \end{aligned}$$

along with many others. The result was that, once the universe passed the deuterium bottleneck, fusion occurred rapidly, and large amounts of helium were formed. In just a few minutes, most of the free neutrons were consumed, leaving a universe whose matter content was primarily hydrogen and helium.

Heavier Elements?

We might imagine that the fusion chain would have continued to create heavier and heavier elements, just as in the cores of stars. However, this did not occur. In stars, the density and the temperature both *increase* slowly with time, allowing more and more massive nuclei to form, but in the early universe the opposite was true. The temperature and density were both *decreasing* rapidly, making conditions less and less favorable for fusion as time went on. Even before the supply of neutrons was completely used up, the nuclear reactions had effectively ceased. Reactions between helium nuclei and protons may have formed trace amounts of lithium (the next element beyond helium; see Interlude 29-1) by this time, but for all practical purposes the expansion of the universe caused the fusion process to stop at helium. The remaining neutrons soon decayed into protons:

$$\text{neutron} \longrightarrow \text{electron} + \text{proton} + \text{neutrino}$$

and the brief epoch of primordial nucleosynthesis was over about 15 minutes after it began.

By the end of the nuclear epoch, some 1000 s after the Big Bang, the temperature of the universe was about 300 million K, and the cosmic elemental abundance was set. Careful calculations indicate that by the end of the nuclear epoch about 1 helium nucleus had formed for every 12 protons remaining. Because a helium nucleus is four times more massive than a proton, helium accounted for about one quarter of the total mass of matter in the universe:

$$\frac{1 \text{ helium nucleus}}{12 \text{ protons} + 1 \text{ helium nucleus}} = \frac{4 \text{ mass units}}{12 \text{ mass units} + 4 \text{ mass units}} = \frac{4}{16} = \frac{1}{4}.$$

The remaining 75 percent of the matter in the universe was hydrogen. It would be almost a billion years before nucleosynthesis in stars would change these figures. Bear in mind, though, that the matter was still just an insignificant "contaminant" in the radiation-dominated universe at this early stage. Radiation outweighed matter by about a factor of 5000 at the time the helium formed. The existence of helium is important in determining the structure and appearance of stars today, but its creation was irrelevant to the evolution of the universe at the time.

Helium, Deuterium, and the Density of the Universe

Because helium-4 is produced in stars, it is difficult to disentangle the contributions made by primordial nucleosynthesis and by later stellar hydrogen burning. However, we can say that all cosmic objects should contain *at least* 25 percent helium by mass because fusion in stars is known to have created additional helium since then. The figure for the Sun, for example, is actually about 28 percent helium by mass.

Our best hope of determining the amount of helium formed in the early universe is to study the oldest stars known, but those stars are of low mass and quite cool, making the helium lines in their spectra very weak and hard to measure accurately. Our earlier figure of 20–25 percent helium is based on these difficult observations. Despite this uncertainty, the theoretical calculations of primordial helium formation are generally consistent with the observed abundance of helium in the universe today.

Deuterium was also created in the early universe as part of the helium-formation process. Although most deute-

[2] Recall from Chapter 18 that deuterium is simply a heavy isotopic form of hydrogen. Its nucleus contains one more neutron than normal hydrogen, but no additional protons.

rium was quickly burned into helium as soon as it formed, a small amount was left over when the nuclear reactions ceased. The more matter there was, the more particles there were to react with the deuterium, and the less deuterium remained when helium production stopped. But the total number of protons and neutrons present in the universe during the nuclear epoch was the *same* as it is today (or, at least, it is very nearly so—see Interlude 29-2). Protons and neutrons can combine into heavier nuclei and can interchange identities with one another, but they cannot be created or destroyed. Therefore, the amount of deuterium formed by primordial nucleosynthesis can be directly related to the *present-day* density of matter in the universe. The denser the universe is today, the *less* deuterium must have been produced at early times.

This connection between deuterium production and cosmic density leads us to ask the following question. Can we turn the whole problem around and use the *observed* abundance of deuterium in the universe to provide us with an estimate of the density of the universe? In other words, can we use primordial nucleosynthesis to measure that elusive, but all-important, quantity, Ω_0, which determines the ultimate fate of the cosmos? The answer is *yes*.

Unlike helium, deuterium is not likely to be produced in stars. In fact, deuterium is *destroyed* in stars, so whatever deuterium we can detect is an *underestimate* of the amount produced in the early universe. Observations of deuterium—especially those made by orbiting satellites that can capture deuterium's strongest spectral feature, which is emitted in the ultraviolet part of the spectrum—indicate a present-day abundance of about 2 deuterium nuclei for every 100,000 protons. A comparison of this number with theoretical calculations of primordial nucleosynthesis implies a present-day density of at most 10^{-30} g/cm^3—about 1/10 the critical density. Furthermore, the exact amount of deuterium produced depends very sharply on the density—a little less than 10^{-30} g/cm^3 and far too many deuterium nuclei would have been formed, a little more and far too few would now exist. Thus, measurements of the cosmic deuterium abundance provide us with a remarkably reliable estimate of Ω_0. (In fact, the amount of helium produced also depends a little on cosmic density, but the variation is so slight that it does not provide a very accurate estimate of Ω_0.)

But before we jump to the conclusion that $\Omega_0 = 0.1$—and therefore that the universe is open and will expand forever—we must make a very important qualification. Primordial nucleosynthesis as just described depends *only* on the presence of protons and neutrons in the early universe. Thus, measurements of the abundance of helium and deuterium tell us only about the density of so-called **baryonic** matter—matter made up of protons and neutrons—in the cosmos. Atoms, people, planets, and stars are all made of predominantly baryonic matter. Most of the mass in an atom, for example, is in its nucleus, which is composed of protons and neutrons. Its orbiting electrons, which are nonbaryonic, make up only a tiny fraction of the atom's total mass. Our above arguments thus imply only that the present-day density of baryonic matter is at most 1/10 the critical value.

In Chapter 26, we concluded on the basis of studies of the motions of galaxies in clusters and superclusters that $\Omega_0 = 0.2$ or 0.3 and possibly much more. If this value turns out to be correct, and if the density of matter in the form of protons and neutrons is at most 1/10 of the critical value, then we are forced to admit that not only is most of the matter in the universe dark, but the dark matter *cannot* be entirely

INTERLUDE 29-1 *Lithium, Beryllium, and Boron*

As best we can tell, the lightweight nuclei lithium, beryllium, and boron cannot be created by *any* of the processes described so far in this book. As we saw in Chapter 23 (see Table 23-2), all three of these elements are markedly underabundant compared with neighboring elements of similar mass. Lithium has the additional problem of not being able to withstand the heat of normal stars; it breaks apart at temperatures greater than a few million kelvins.

It seems unlikely that nuclei of these light elements were created in stars, although we can't be absolutely certain because the cores of stars cannot be observed directly. Yet it seems equally unlikely that they could have been created in any appreciable amount in the earliest moments of the universe. The temperature and density decreased too rapidly following the Big Bang.

Fortunately, several other processes are known that could conceivably form lithium, beryllium, and boron. Perhaps the most important of these processes is called *spallation,* a word meaning "to break up or reduce by chipping." This mechanism occurs not inside hot stars but in the much cooler interstellar space. As fast-moving galactic cosmic rays (see Chapter 25) bombard some of the carbon group of elements present in interstellar space, small amounts of the lithium group can be produced. Since the cosmic-ray particles have energies many times higher than that needed to break nuclear bonds, such collisions tend to split apart carbon, nitrogen, and oxygen nuclei, leaving a residue of lithium, beryllium, and boron.

The spallation process is very inefficient—one of the slowest in nature. Over the course of 100,000 years, a mere gram of beryllium is produced within each cubic astronomical unit of interstellar space. Along with their intolerance of high temperatures, this inefficiency explains the rarity of the lightweight elements.

baryonic in nature. Most of the matter in the universe is apparently in the form of elusive nonbaryonic particles—the WIMPS discussed in Chapters 18 and 25, whose nature we do not fully understand and whose existence has yet to be conclusively proven in laboratory experiments.

THE FORMATION OF ATOMS

At the start of the atomic epoch, the universe consisted of electromagnetic radiation, protons, helium nuclei, electrons, neutrinos, and (perhaps) dark matter particles of unknown nature. For the sake of brevity, we will refer to matter in the universe that is composed of the familiar protons, electrons, and neutrons simply as *normal* matter. We will also adopt the convention that the term *dark matter* refers only to nonbaryonic dark matter—that is, to WIMPs but not to MACHOs, in the language of Chapter 26. If neutrinos turn out to be massless, they play only a minor role in our discussion, and we can neglect them here. If they do have some mass, they fall into the category of dark matter. Accordingly, we can say that at this time the universe consisted of radiation, normal matter, and dark matter.

As the universe expanded, the temperature fell, and the radiation accounted for a smaller and smaller fraction of the total energy. A few thousand years after the Big Bang, radiation ceased to be the dominant component of the universe. The Matter Era had begun. From that point on, photons in the universe were reduced to the status of a **radiation background.**

At the start of the Matter Era, the temperature of the universe was several tens of thousands of kelvins—still too hot for atoms of hydrogen to exist, although some helium ions (missing one electron) may already have formed. The temperature at this time was comparable to that in the atmosphere of an O-type star. During the next few hundred thousand years, a major change occurred. The universe expanded by another factor of 10, the temperature dropped to a few thousand kelvins, and all of the normal matter in the universe combined to form neutral *atoms*. By the time the temperature had fallen to 4500K, the universe consisted of atoms, photons, and dark matter (whose weak interactions with normal matter meant that it played no part in the atom-formation process). The period during which atoms formed is sometimes called the **recombination** epoch, although we should stress that atoms had never existed previously, so the term *re*combination is a misnomer.

A better term for the process of atom formation is **decoupling,** for it was during this period that the radiation background parted company with normal matter. At early times, when matter was ionized, the universe was filled with large numbers of free electrons, which interacted easily with all radiation. As a result, a photon could not travel far before encountering an electron and scattering off it. In effect, the universe was opaque to radiation (rather like the deep interior of a star like the Sun). Matter and radiation were ''tied'' to one another by their frequent interactions. When the electrons combined with nuclei to form atoms, however, only certain wavelengths of radiation—the ones corresponding to the spectral lines of hydrogen and helium—could interact with matter. Radiation of other wavelengths could travel virtually forever without being absorbed. The universe became *transparent*. From that time on, the radiation background passed unhindered through space. As the universe expanded, the radiation simply cooled, eventually to become the microwave background we see today.

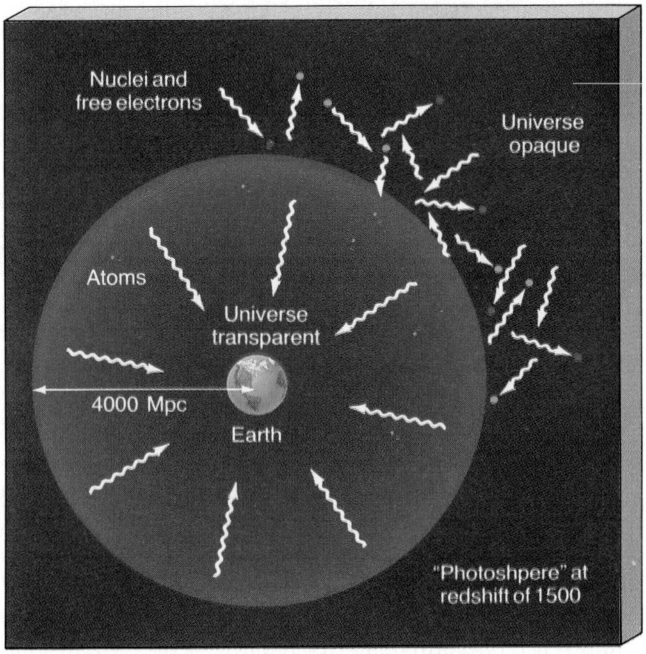

Figure 29.5 *When atoms formed, the universe became virtually transparent to radiation. Thus, observations of the cosmic background radiation allow us to study conditions in the universe around a time at a red shift of 1500, when the temperature dropped below about 4500K.*

The microwave photons now detected on Earth have been traveling through the universe since they decoupled. Their last interaction with matter occurred when the universe was a few hundred thousand years old and still opaque to radiation. As illustrated in Figure 29.5, the epoch of atom formation created a kind of ''photosphere'' in the universe, completely surrounding Earth at a distance of 4000 Mpc—the distance the photons have traveled in the 13 billion years since decoupling. On our side of the photosphere, the universe is transparent; on the far side, it was opaque. By observing the microwave background, we are therefore studying conditions in the universe around the time of decoupling, in much the same way as the study of sunlight tells us about the surface layers of the Sun. In terms of red shifts, if we suppose that atom formation was complete by the time the radiation temperature had dropped to 4500K, we can say that the last interaction between the background radiation and matter occurred at a red shift of about 1500.

COSMIC INFLATION

The Horizon and Flatness Problems

✓4 In the late 1970s, cosmologists trying to piece together the evolution of the universe were confronted with two nagging problems that had no easy explanation within the standard Big Bang model—basically, the sequence of events that we have just described. The resolution of these difficulties has caused cosmologists to completely rethink their view of the very early universe.

The first problem is known as the **horizon problem,** and it concerns the remarkable *isotropy* of the cosmic microwave background. Recall that the temperature of this radiation is virtually constant, at about 2.7K, in all directions. Imagine observing the microwave background in two opposite directions on the sky, as illustrated in Figure 29.6. As we have just seen, the radiation last interacted with mat-

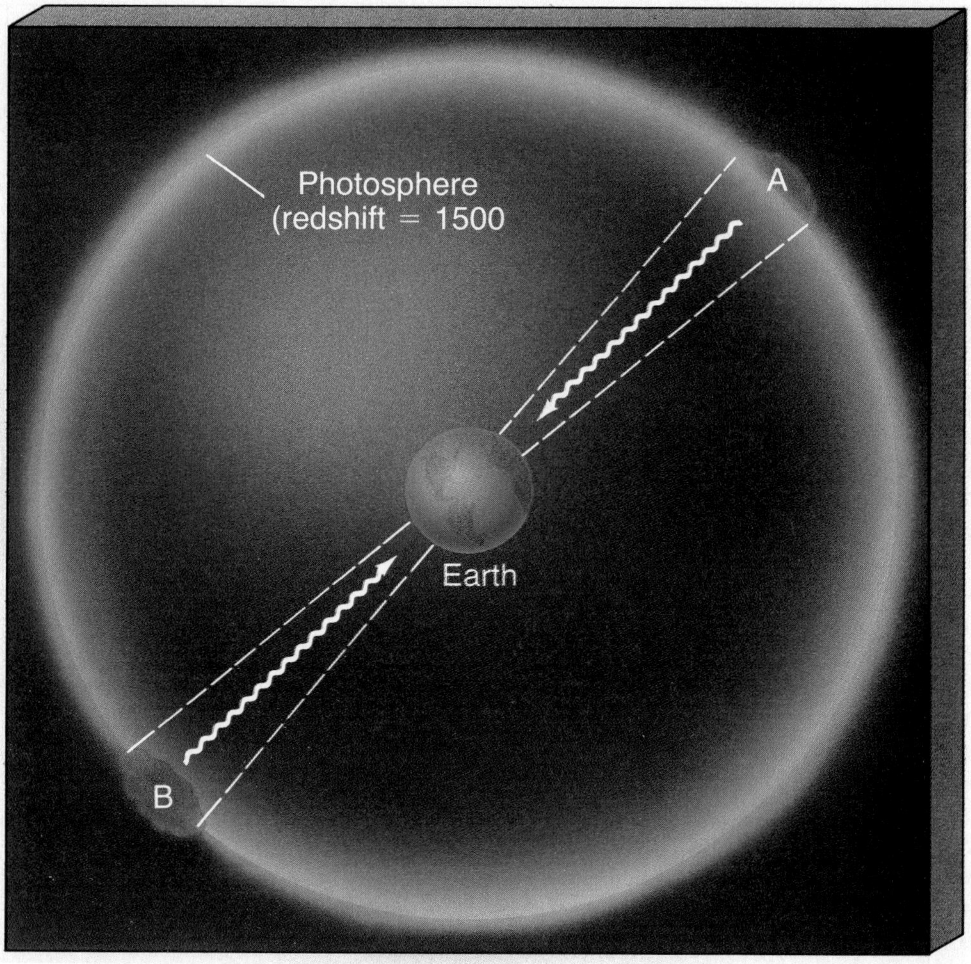

Figure 29.6 *The horizon problem. The isotropy of the microwave background indicates that regions A and B in the universe were very similar to one another when the radiation we observe left them, but there has not been enough time since the Big Bang for them ever to have "communicated" with one another. How do they "know" that they should look the same?*

ter in the universe around a red shift of 1500. In observing these two distant regions of the universe, marked A and B on the figure, we are studying regions separated by several thousand megaparsecs. The fact that the background radiation is known to be isotropic to high accuracy means that regions A and B had very similar densities and temperatures at the time the radiation we see left them, just as required by the cosmological principle. The problem is, within the Big Bang theory as just described, there is no particular reason *why* these regions should be so similar to one another.

Just as ripples on the surface of a pond tend to spread out, merge with one another, and die away, density and temperature fluctuations in the early universe would gradually have been smoothed out as neighboring regions interacted. But these interactions did not occur instantaneously, any more than ripples spread instantaneously across water. Energy is carried from place to place by photons, gravitational radiation, and other wave motions, and the speed at which it moves cannot exceed the speed of light. Calculations show that there simply has not been enough time since the Big Bang for light to have traveled from region A to region B—the regions are said to be beyond each other's *horizon*—so there has not yet been time for any initial differences between them to have been smoothed away. In other words, we cannot explain their similarity by saying that they have interacted with each other since the universe formed. Therefore, how do these two regions "know" that they are supposed to look the same? Unless we are prepared to stipulate that the universe started off perfectly homogeneous—something theorists are unwilling to do, because all models of the early universe predict fluctuations at some level—then there is no good reason for regions A and B to look alike.

The second problem with the standard Big Bang model is called the **flatness problem.** Its description begins with the observation that, whatever the exact value of Ω_0, it appears to be quite close to 1. The density of the universe is fairly near the critical value needed for the expansion to barely continue forever. In terms of spacetime curvature, we can say that the universe is remarkably close to being flat. We say "remarkably" here because again there is no good reason *why* the universe should have formed with a density very close to critical. Why not a millionth or a million times that value? Furthermore, as can be seen in Figure 29.7, a universe that starts off close to, but not exactly on, the critical curve soon deviates greatly from it, so if the universe is close to critical now, it must have been *extremely* close to critical in the past. For example, if $\Omega_0 = 0.1$ today, the departure from the critical density at the time of nucleosynthesis would have been only 1 part in 10^{15}.

The standard Big Bang theory simply provides no good reason why the universe should be so nearly isotropic and flat. These observations constitute "problems" because cosmologists want to be able to explain the present condition of the universe, not just accept it "as is." They would prefer to resolve the horizon and flatness problems in terms of physical processes that could have taken a universe with no special properties and caused it to evolve into the cosmos we now see. It turns out that there is a single physical process that provides a resolution to both problems.

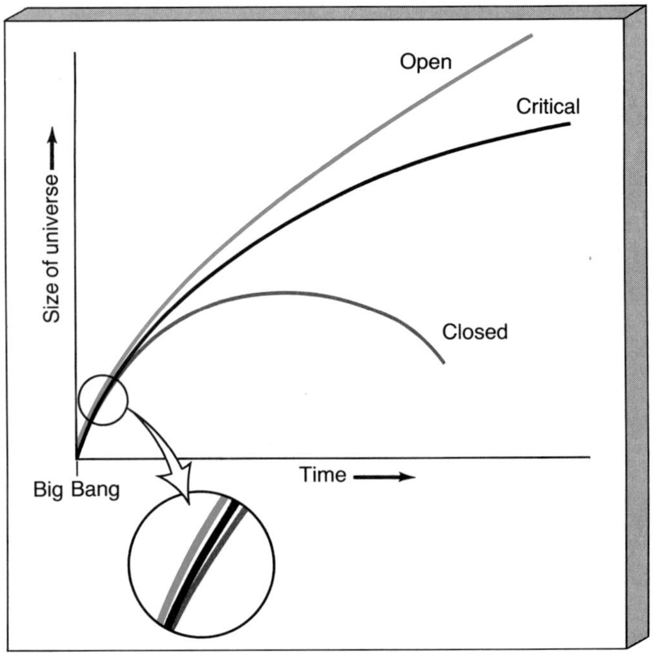

Figure 29.7 *The flatness problem. If the universe deviates even slightly from the critical case, that deviation will grow rapidly in time. For the universe to be as close to critical as it is today, it must have differed from the critical density in the past by only a tiny amount.*

Fundamental Forces Again

✓5 In order to understand the events in the very early universe that "solve" the horizon and flatness problems, we must pause to reconsider the four basic forces of nature, a topic that we discussed briefly in Interlude 5-1. In order of increasing strength, the forces are (1) gravity, (2) the weak force, (3) electromagnetism, and (4) the strong force. In terrestrial laboratories the four have very different properties from one another. Gravity and electromagnetism are long-range, inverse-square forces. The strong and weak forces have very short ranges—10^{-13} and 10^{-15} cm, respectively. The forces do not all affect the same particles. Gravity affects everything. The electromagnetic force affects only charged particles. The strong force operates between quarks and particles made from them (such as protons and neutrons), but it does not affect electrons and neutrinos. The weak force shows up in certain nuclear reactions and radioactive decays. The strong force is 137 times stronger that the electromagnetic force, 100,000 times stronger than the weak force, and 10^{39} times stronger than gravity. The attempt to unify these four forces—to describe them in terms of a single, all-encompassing "superforce"—is one of the great intellectual adventures in modern physics.

On the face of it, one might not imagine that there could be *any* deep underlying connection between four forces as dissimilar as those we have just described, yet there is growing evidence that at least three of them are really just different aspects of a single basic phenomenon. In the 1970s, theorists Sheldon Glashow, Abdus Salam, and Steven Weinberg succeeded in explaining the electromagnetic and weak forces in terms of a single **electroweak** force. Around the same time, Glashow and Howard Georgi proposed a method by which the strong and electroweak forces could also be merged—the first of several such schemes that would be developed by theorists over the next decade. As we mentioned earlier, theories that unite these three forces are called Grand Unified Theories or, less formally, just GUTs (and notice that the term is plural—no one GUT has yet been definitively proven to be "the" correct description of nature). Gravity, however, has not yet been incorporated into a single SuperGUT, in which all four fundamental forces are united.

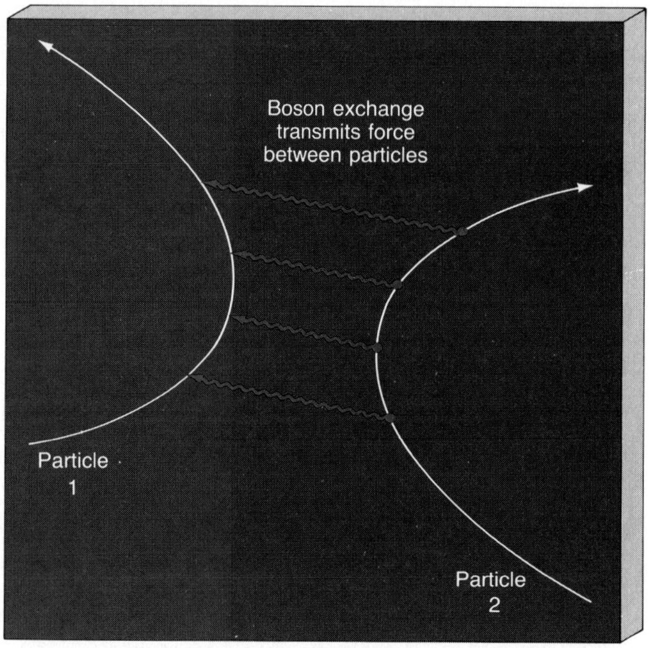

Figure 29.8 *Forces between elementary particles are transmitted through the exchange of particles called bosons. As two particles interact, they exchange bosons, a little like playing catch with a submicroscopic ball.*

Grand Unification

An important concept in quantum physics is the idea that forces between elementary particles are exerted, or *mediated*, by the exchange of another type of particle, called a **boson.** We might imagine the two particles as playing a rapid game of catch, using a boson as a ball, as illustrated in Figure 29.8. As the ball is thrown back and forth, the force is transmitted. In ordinary electromagnetism, the boson involved is just the photon—a bundle of electromagnetic energy that always travels at the speed of light. The strong force is mediated by particles known as **gluons.** The electroweak theory includes a total of four bosons: the massless photon and three other massive particles, called (for historical reasons) W^+, W^-, and Z^0, all of which have been observed in laboratory experiments. Most of the particles we have encountered so far in this book—electrons, protons, neutrons, neutrinos—play "catch" with at least some of these "balls."

At "low" temperatures—below about 10^{15} K, a range that includes almost everything we know about on Earth and in the stars—the four electroweak bosons split into two families. The photon expresses the electromagnetic force, while the other three bosons carry the weak force. While the photon is massless, the W and Z particles each have about 100 times the mass of a proton. The temperature of 10^{15} K—a million billion kelvins—is the threshold temperature for the production of W and Z particles from photons. Theory indicates that, above this temperature, the four bosons become indistinguishable from one another, and the weak and electromagnetic forces are unified. In particle physics parlance, there is said to be a **symmetry** in the electroweak force at high temperatures that is *broken* at low temperatures, allowing the separate characters of the weak and electromagnetic forces to become apparent.

In Grand Unified Theories, the strong and electroweak forces are connected by a new, very massive particle (actually, a family of particles) known generically as an **X-boson.** This particle mediates a very short-range "hyperweak" force that allows quarks (the building blocks of protons, neutrons, and other particles that interact through the strong force) to turn into leptons (electrons, neutrinos, and other particles that feel the electroweak force), and vice versa. However, because the X-boson is so massive—at least 10^{15} times the mass of the proton, and possibly much more—the symmetry that it represents between the strong and electroweak forces becomes evident only at extremely high temperatures. The threshold temperature for the production of X–anti-X pairs is at least 10^{28} K. Above this **grand unification temperature,** the strong and electroweak forces are indistinguishable from one another. As noted in Table 29-1, the temperature of the universe fell below this value about 10^{-35} s after the Big Bang. Unlike the electroweak theory, GUTs have not yet been experimentally verified, at least in part because of the extremely high energies that must be reached in order to observe their predictions. Interlude 29-2 discusses some ways of testing the GUTs without having to achieve temperatures in excess of 10^{28} K in the laboratory and suggests how GUTs may help resolve the important question of why matter exists at all.

Freeze-Out

Having considered a little of the particle physics necessary to understand the very early universe, we now return to our

INTERLUDE 29-2 *Testing the GUTs*

Scientists can never prove a theory correct. They can, however, prove it incorrect. By experimenting, observing, and testing their theories, scientists gradually rule out poor ideas. By this process of elimination, a more refined approximation to "the truth" is obtained.

How can we test Grand Unified Theories? We must do so indirectly, for we will never be able to recreate in the laboratory the energy of creation itself. The world's most powerful particle accelerators can just barely produce—and only for the briefest of instants—the temperatures of 10^{15} K needed to test the electroweak theory. GUTs become operative only at temperatures 10 *trillion* times greater than that. To accelerate elementary particles to such huge energies, we would need a machine that spans the distance between Earth and the Alpha Centauri star system a parsec away. Furthermore, the operation of such a truly cosmic device for even a microsecond would require a stupendous amount of energy, equal to several times the total power consumption of the United States for an entire year!

So what indirect tests can we apply? Some (but not all) GUTs predict that the proton is not the immortal building block we once thought—it can *decay*. Using the GUTs, we can estimate the average life expectancy of a proton. It turns out to be an extremely long time—at least 10^{31} years, or a billion trillion times the age of the universe. This extremely long lifetime means that there is a very small probability of decay in any given time span. Nevertheless, any proton is theoretically in danger of decaying at any moment. With enough protons, we might even hope to see a decay occur. Using the above figure, we can predict that, on average, a proton should decay every few centuries in each ton of water, an abundant source of protons. Experiments have been performed to detect such events in huge quantities of water stored in tanks in deep underground mines, much like that housing the neutrino telescope discussed in Chapter 18. There the water can be insulated from unwanted effects caused by exotic particles reaching Earth from outer space. To date, however, no research group has confirmed the prediction that protons do indeed decay. The present experimental limits imply that the proton lifetime must be greater than 10^{34} years, which is enough to rule out many of the simplest GUTs that theorists have proposed.

One important aspect of GUTs is that they may be able to account for the observed excess of matter over antimatter in the universe. As mentioned earlier, when the temperature of the universe dropped below the point at which proton-antiproton pairs could form—at the end of the hadron epoch—some neutrons and protons were left over. We know that because the material universe we see around us, ourselves included, is made of them. Some theorists wonder why nature is not more symmetrical. Why wasn't there *exact* symmetry between matter and antimatter in the early universe? In other words, why are we here at all?

Some GUTs predict that the X-bosons are unstable. As a result, although most of the X and anti-X particles that formed in great numbers during the GUT epoch did annihilate each other, some of them instead decayed into lighter, more stable particles, such as protons. It so happens that the decay of the X-bosons may have created slightly more protons than antiprotons, so that a slight excess of protons existed in the universe after the GUT epoch. During the hadron epoch, protons and antiprotons were created and destroyed, but the slight asymmetry was maintained—pair production and annihilation always increases or decreases the number of particles and antiparticles by the same amount. According to these theories, the present excess of matter over antimatter originated in the GUT epoch and has persisted ever since. It remains to be seen, however, whether these GUTs can account for the amount of matter actually observed.

study of the GUT epoch. We have already seen how particles "froze out" of the universe as its temperature dropped below the threshold temperature necessary for their creation through pair production. Now that we know that the forces of nature are mediated by particles, we can understand how the fundamental forces froze out too.

The freezing out of the electroweak force at a temperature of 10^{15} K, about 10^{-10} s after the Big Bang, had little overall effect on the cosmic expansion. Electrons and neutrinos stayed in thermal equilibrium with the rest of the universe until the end of the lepton epoch, after which the neutrinos decoupled, forming a "neutrino background" with a present-day temperature of about 1.9K (which we could use to probe the early universe if only we had a low-energy neutrino telescope). By contrast, the freeze-out of the strong force, 10^{-35} s after the Big Bang, produced one of the strangest events in the history of the cosmos.

In the early 1980s, cosmologist Alan Guth, at Stanford University, uncovered a remarkable prediction of many GUTs. He calculated that, as the universe cooled below 10^{28} K and the strong force decoupled, the universe must have briefly entered a very odd state, which he called the **false vacuum.** For a short while, empty space acquired an enormous *pressure,* which temporarily overcame the pull of gravity and accelerated the expansion of the universe at an enormous rate. What's more, the pressure remained constant as the cosmos expanded, and the acceleration grew more and more rapid with time—the size of the universe

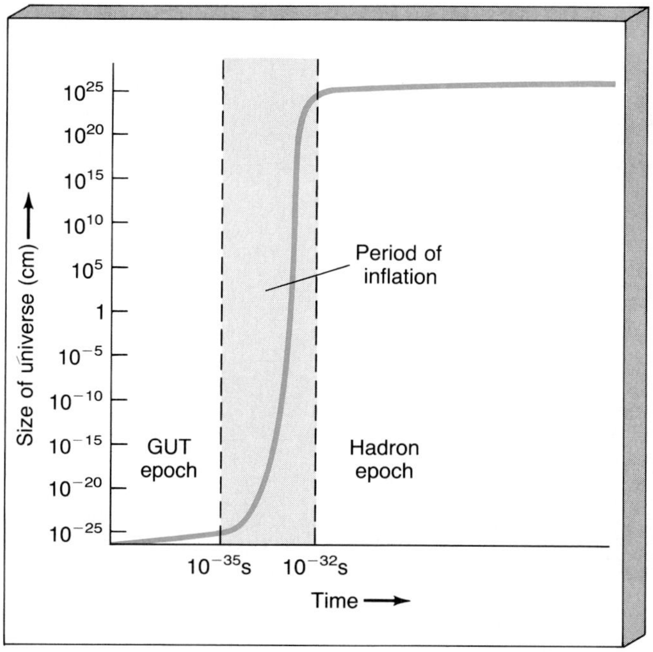

Figure 29.9 *During the period of inflation at the end of the GUT epoch, the universe expanded enormously in a very short time. Afterward, it resumed its earlier "normal" expansion, except that the size of the cosmos was about 10^{50} times bigger than it was before.*

doubled every 10^{-34} s or so! This period of unchecked cosmic expansion, illustrated in Figure 29.9, is generally known as the epoch of **inflation.** During this epoch, the universe was in an unstable, high-energy condition. Eventually, as in any quantum system, the universe made a transition down to the lower-energy "true vacuum" state. Regions of normal space began to appear within the false vacuum and rapidly spread to include the entire cosmos. With the return of the true vacuum, inflation stopped. The false vacuum persisted for a mere 10^{-32} s, but during that time the universe swelled in size by about a factor of about 10^{50}.

When the inflationary phase ended, the grand unified force was gone forever (with the possible exception of its role in such events as proton decay—see Interlude 29-2). In its place were the more familiar electroweak and strong forces that operate around us in the low-temperature universe of today. With the normal vacuum restored, the universe once again resumed its (relatively) leisurely expansion, slowly decelerated by the effect of gravity. However, a number of important changes had occurred that would have far-reaching ramifications for the evolution of the cosmos.

Implications for the Universe

The inflationary epoch provides a natural solution for the horizon and flatness problems described earlier. The horizon problem is solved because inflation took regions of the universe that had already had time to communicate with one another—and so had established similar physical properties—and then dragged them far apart, well out of communications range of one another. Regions A and B in Figure 29.6 have been out of contact since 10^{-32} s after creation, but they *were* in contact before then. As illustrated in Figure 29.10, their properties are the same today because they were the same long ago, before inflation separated them.

Figure 29.10(a) shows a small piece of the universe just before the onset of inflation. The point that will one day become the site of the Milky Way Galaxy is at the center of the shaded region, which represents the portion of space "visible" to that point at that time. That entire region is more or less homogeneous because there has been enough time since the Big Bang for light to have traveled from its center to its edge. The points A and B of Figure 29.6 are also marked. They lie well within the homogeneous patch, so they have very similar properties. The actual size of the shaded region is about 10^{-24} cm—only a trillionth the size of a proton. Immediately after inflation, as shown in Figure 29.10(b), the homogeneous region has expanded by 50 orders of magnitude, to a diameter of about 10^{26} cm, or some 30 Mpc—larger than the largest supercluster. By contrast, the visible portion of the universe, indicated by the dashed line, has grown only by a factor of a thousand, so it is still microscopic in size. In effect, the universe expanded much faster than the speed of light during the inflationary epoch, so what was once well within the horizon now lies far beyond it. In particular, points A and B are no longer visible, either to us or to each other, at this time.

Since then, the universe has expanded by a further factor of 10^{27}, so the size of the homogeneous region of space surrounding us is now about 10^{53} cm (10^{28} Mpc)—10 trillion trillion times greater than the distance to the most distant quasar. As shown in Figure 29.10(c), the horizon has expanded faster than the universe itself, so that points A and B are just now becoming visible again. As the portion of the universe now observable from Earth grows in time, it remains homogeneous because our cosmic field of view is simply reexpanding into a region of the universe that was within our horizon long ago. We will have to wait a very long time—at least 10^{35} years—before the edge of the homogeneous patch surrounding us comes back into view.

To see how inflation solves the flatness problem, let's return once more to the balloon analogy of Chapter 28. Imagine that you are a 1-mm-long ant sitting on the surface of the balloon as it expands, as illustrated in Figure 29.11. When the balloon is just a few centimeters across, you can easily perceive the surface to be curved—its circumference is only a few times your own size. When the balloon expands to, say, a few meters in diameter, the curvature of the surface will be less pronounced, but still perceptible. However, by the time the balloon has expanded to a few *kilo*meters across, an "ant-sized" patch of the surface will look quite flat, just as the surface of the Earth looks flat to us. Now imagine that the balloon expands 100 trillion trillion

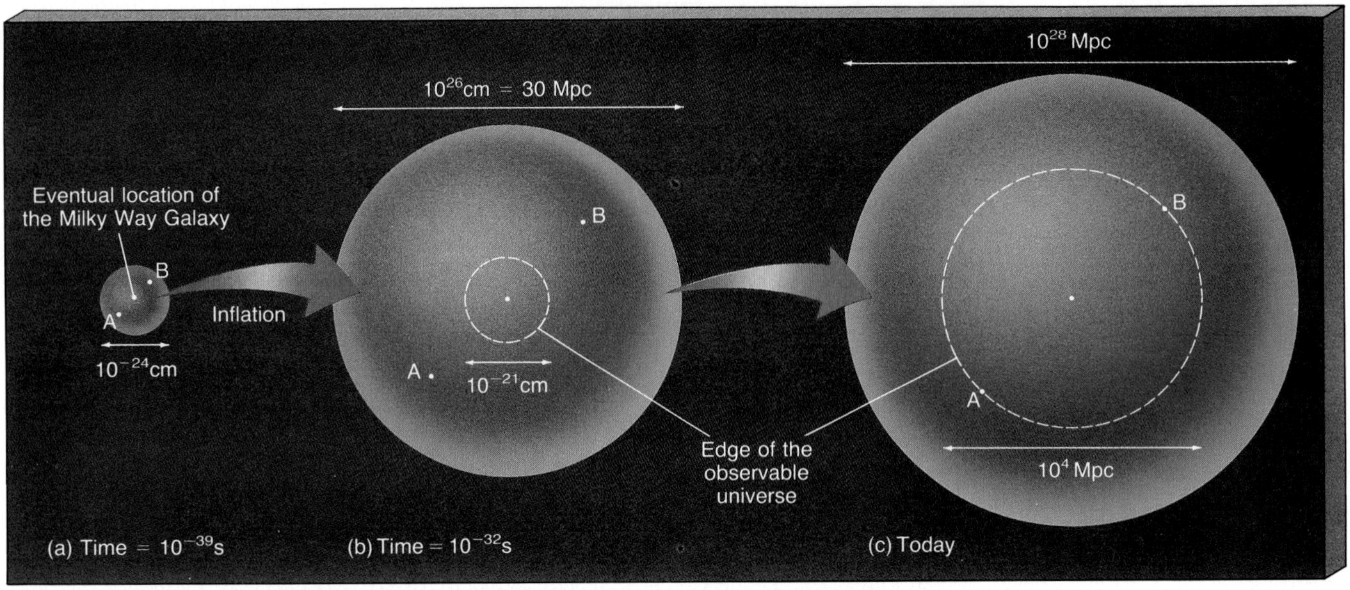

Figure 29.10 *Inflation solves the horizon problem by taking a small region of the very early universe, whose parts had already had time to communicate with one another and which had already become homogeneous, and expanding it to enormous size. In (a), points A and B are well within the (shaded) homogeneous region of the universe centered on the eventual site of the Milky Way Galaxy. In (b), after inflation, A and B are far outside the horizon (indicated by the dashed line), so they are no longer visible from our location. Subsequently, the horizon expands faster than the universe, so that today (c) A and B are just reentering our field of view. They have similar properties now because they had similar properties before the inflationary epoch.*

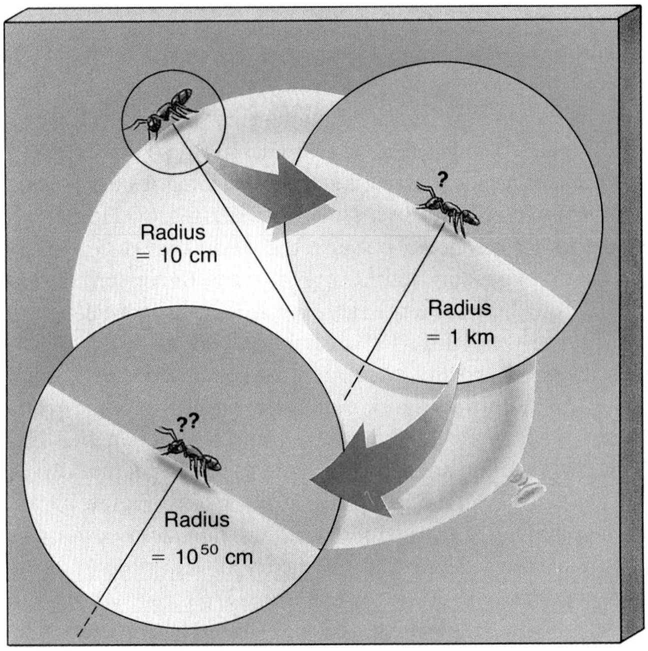

Figure 29.11 *Inflation solves the flatness problem by taking a curved surface, here represented by the surface of the expanding balloon, and expanding it enormously in size. To an ant on the surface, the balloon looks perfectly flat when the expansion is over.*

trillion trillion times, as the universe did during the period of inflation. Your local patch of the surface would now be indistinguishable from a perfectly flat plane.

Exactly the same argument applies to the universe. Any curvature the universe may have before inflation has been expanded so much that space is now perfectly flat, at least on the scale of the observable universe (and, as we just saw, on much larger scales too). But notice that this resolution to the flatness problem—the universe looks nearly flat because it is *precisely* flat—has a very important consequence. Because the universe is flat, the density of matter is *exactly* critical: $\Omega_0 = 1$. That means that there must be a lot of invisible matter in the universe beyond the clusters and the superclusters, filling the huge voids on the largest scales. And because we just saw that primordial nucleosynthesis implies that the density of "normal" baryonic matter is at most 1/10 the critical value, the rest of the mass—90 percent of all the matter in the universe—must be in the form of nonbaryonic dark matter.

When the idea of inflation was first suggested, many, if not most, astronomers were very skeptical. Some still are. They point out that there is *no* direct observational evidence for a cosmic density as high as the critical value, and in that contention they are correct. However, as we saw in Chapters 26 and 28, there is ample evidence for dark matter in the universe with as much as 20 or 30 percent of the critical density, and our estimates of Ω_0 increase as the scale under consideration increases. Indeed, observations on very large scales seem to indicate a value close to 1. Furthermore, inflation is a prediction of a group of theories—the GUTs—that are becoming more and more firmly established as the standard description of matter at high energies. If they are indeed correct, then inflation *must* have occurred. Finally, inflation provides a neat solution to two serious difficulties with the standard noninflationary Big Bang model. We will see in the next section another example of how an inflationary universe made up mostly of nonbaryonic dark matter actually makes it *easier* to understand how galaxies formed. For all these reasons, inflation has become entrenched (although as an unwelcome guest in some cases) in most cosmologists' models of the universe.

Toward Creation

Can science probe any closer to creation itself? As we mentioned earlier, our efforts to penetrate all the way back into the Planck epoch are currently hampered because physicists are unsure how to incorporate the force of gravity into GUTs. No one has yet invented a SuperGUT that merges gravity with the grand unified force. Our current knowledge of the behavior of gravity in microscopic domains suggests that the huge energies required for this unification could have prevailed only at times earlier than 10^{-43} s, when the temperature exceeded 10^{32} K. Before that time, many physicists believe, the four known basic forces were one—a single, truly fundamental force operating at energies characterizing the earliest part of the Planck epoch. Only later would the more familiar four forces begin to manifest themselves as separate entities.

An important idea that has arisen from the realization that strong and electroweak forces can be unified is known as **supersymmetry.** It takes the symmetries between particles that "feel" forces—the ball throwers—and the symmetries between particles that mediate the forces—the balls—one stage further, and places *all* particles—quarks, leptons, photons, X-bosons, whatever—on an equal footing. One important result of this theory is a prediction that all particles should have so-called *supersymmetric partners*—extra particles that must exist in order for the theory to remain self-consistent. Although none of these new particles has ever been detected, the sheer elegance of supersymmetry has convinced many physicists of its essential correctness. Experiments are planned that may soon provide evidence for supersymmetric partners of some of the known particles. If they exist, these particles would have been produced in abundance in the Big Bang and should still be around today. They are also expected to be very massive—at least a thousand times heavier than a proton. These so-called **supersymmetric relics** are the current leading candidates for the dark matter in the universe.

Supersymmetry also provides a possible avenue for introducing gravity into the unified theory, thus penetrating the Planck epoch of the early universe. In the resulting theory—known as **supergravity**—the gravitational force is transmitted by a boson called a **graviton.** This is a very different view of gravity from the geometric picture embodied in Einstein's general relativity, however, and merging the two into a consistent theory of quantum gravity has proved very difficult. Alternative theories start from the geometric view and attempt to explain the basic forces of nature in terms of additional curved dimensions of spacetime. They, too, encounter serious problems. At the present time, none of these theories has succeeded in making any definite statement about conditions in the very early universe.

Even in the absence of a working theory, many researchers have a "gut feeling" that once we have in hand the proper description of quantum gravity, our understanding might automatically include a natural description of creation itself. It is conceivable that the primal energy originally emerged from literally *nothing*. Even in a perfect vacuum—a region of space containing neither matter nor energy—virtual particle-antiparticle pairs are constantly appearing and disappearing within a time span too short to observe, causing natural **quantum fluctuations** to occur, even in empty space. Incredible though it may seem, we might be living in a sort of "self-creating universe" that erupted into existence spontaneously from just such a random quantum fluctuation! This sort of "statistical" creation of the primal cosmic energy from absolutely nothing has been dubbed "the ultimate free lunch."

THE FORMATION OF LARGE-SCALE STRUCTURE

The Growth of Perturbations

✓6 During the galactic epoch, gravity began pulling matter together into enormous clumps. Just as stars form from small perturbations in a larger cloud of matter, galaxies, galaxy clusters, and larger structures are also believed to have formed from small inhomogeneities in the matter of the expanding universe. Given the conditions found in the early universe, cosmologists calculate that regions of higher-than-average density containing more than about a million times the mass of the Sun could have begun to collapse during the atomic and galactic epochs listed in Table 29-1. As illustrated in Figure 29.12, natural instabilities in the early universe could have started the process of growing structures that ranged in mass from globular clusters to galactic nuclei to galaxies to galaxy clusters—all the way up to superclusters, voids, and beyond. In Chapter 26, we saw a little of how pregalactic fragments may have interacted and merged to form the galaxies we see today. Here, we will concern ourselves mostly with the formation of structure on much larger scales.

Because the collapsing matter had to "fight" the gen-

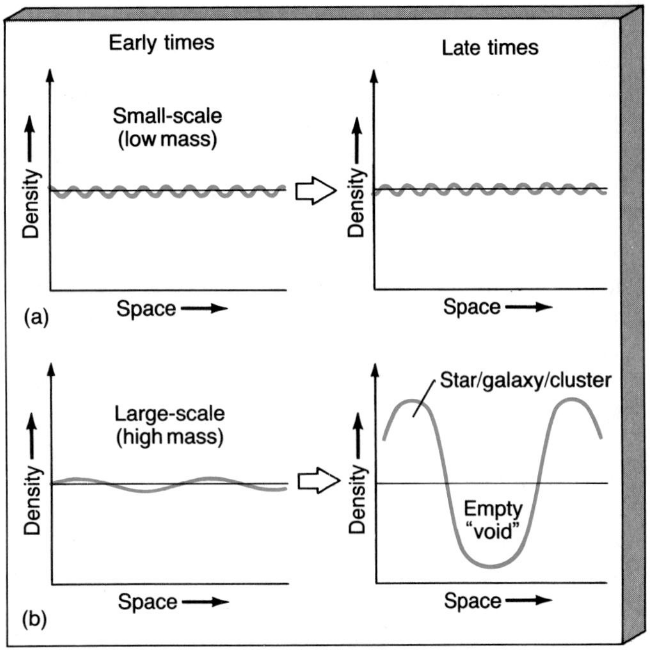

Figure 29.12 Growth of density inhomogeneities in the universe. At any instant, there is a minimum spatial scale (known as the Jeans length) below which density fluctuations do not grow. (a) Regions of above-average density that are smaller than the Jeans length tend to oscillate, like sound waves in air, and their density does not increase with time. The amount of mass in the overdense regions is simply not great enough for gravity to overcome gas pressure. (b) For fluctuations on spatial scales larger than the Jeans length, collapse will occur, just as a massive enough interstellar gas cloud will start to collapse to form a star. These denser-than-average regions will tend to become even denser, leaving empty ''voids'' between them.

eral expansion of the universe, it took time for these small inhomogeneities to grow into the structures we see today, and it was in the calculation of how much time they took, and how much was available, that the theory of galaxy formation first ran into serious problems. If we imagine an expanding universe filled with normal matter—hydrogen and helium—and radiation, and if we ask how rapidly a slightly overdense region would collapse, we find that the simplest idea of primordial gas clouds contracting and coalescing to form the luminous galaxies we see today just doesn't work. It is ruled out by observations. Here's why.

Normal Matter

Any theoretical attempt to make galaxies out of normal (baryonic) matter is constrained by three observations. First, we know that some quasars had already formed by the time the universe was one-fifth its present size—that is, at a red shift of 5. (Recall from Chapter 28 that the cosmological red shift we observe is just a measure of the size of the universe at the time radiation was emitted. Most cosmologists find it a convenient measure of time in the universe, and we will use it in that manner here. Large red shifts mean longer ago, closer in time to the Big Bang.) Furthermore, many theorists believe that, in order to produce the densest galactic bulges we see today, the formation process must have already been well-established as long ago as a red shift of 10 or 20. Thus, our first constraint is that the primordial density fluctuations must have grown into the first protogalactic fragments by a red shift of about 20—when the universe was about 100 million years old.

This time constraint might not present a problem if the galaxy-growing process could begin early enough. But there is a limit on how long ago it could have started. Calculations show that before the epoch of decoupling, at a red shift of 1500 or so, the background radiation would have prevented any galaxy-, cluster-, or even supercluster-sized clumps of matter from collapsing. The clumps would have had to wait until after decoupling before they could grow.

Even with this shortening of the time available, structure could still form if the initial amplitude of the fluctuations was large enough. In order for some inhomogeneities to grow into galaxies by a red shift of 20, the density at the epoch of decoupling would have had to vary by at least a few percent from one place to another. But since the radiation background was ''tied'' to the matter up until that time, any density fluctuations would have led to temperature variations in the cosmic background radiation—denser regions would have been a little hotter than less dense parts of the universe. The level of isotropy observed in the microwave background indicates that any density fluctuations must have been much less than 1 percent, effectively killing this whole theory of galaxy formation.

Dark Matter

By the early 1980s, cosmologists had come to realize that galaxies could not have grown from density fluctuations in the gaseous baryonic matter of the early universe. There just wasn't enough time, given the small initial size of the fluctuations implied by the cosmic microwave background. However, the existence of dark matter, and the growing evidence that most of the universe is *non*baryonic in nature, provide an alternative explanation. Because dark matter interacts only weakly with normal matter, it decoupled from the rest of the universe long ago—before the time of primordial nucleosynthesis, in fact—and its density fluctuations have been growing ever since the end of the Radiation Era, when matter first began to ''control'' the universe. Furthermore, because the dark matter is not directly tied to the radiation background, its variations could have been quite large at the time of radiation decoupling without having a large effect on the microwave background. In short, the dark matter could clump to form large-scale structure in the universe without running into the problems just described for the luminous material.

In this view (see Figure 29.13), the dark matter determines the overall distribution of mass in the universe. Then, at late times, gas is drawn by gravity into the regions of highest density, eventually forming the galaxies we actually

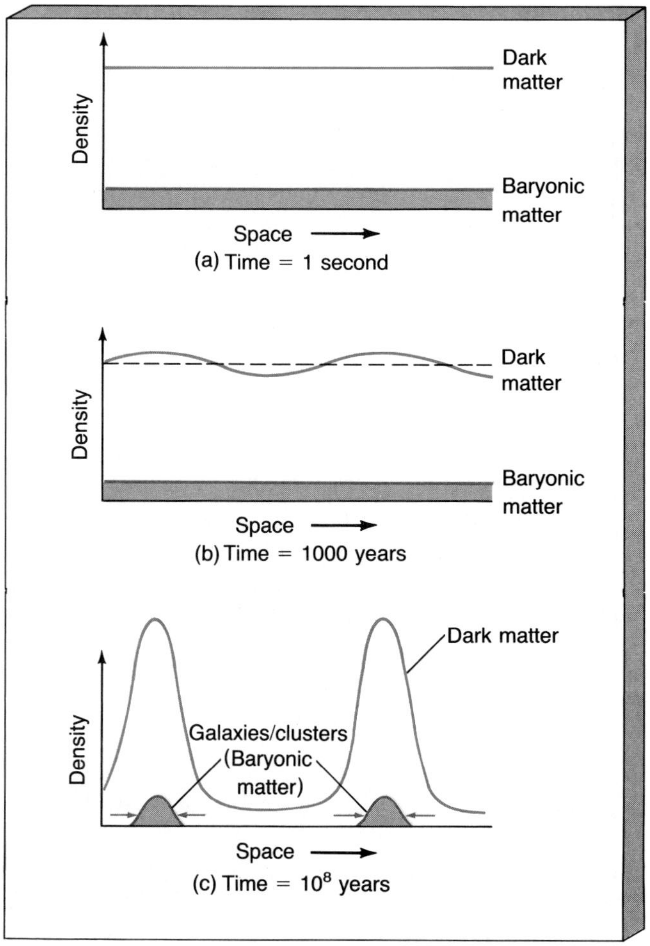

Figure 29.13 *The formation of structure in the cosmos. The universe started out (a) as a mixture of (mostly) dark and "normal" baryonic matter. The dark matter began to clump quite early on (b), eventually forming large structures (c) into which the baryonic matter flowed, ultimately to form the galaxies we see today.*

see. This picture explains why most dark matter lies outside the visible galaxies. The luminous material is strongly concentrated near the density peaks, dominating the dark matter there, but the rest of the universe is dark and essentially devoid of normal matter. This underscores the idea presented earlier (in Chapter 26) that, like foam on the crest of an ocean wave, the universe we can actually see is only a tiny part of the total.

Given that the nature of the dark matter is still unknown, theorists have considerable freedom in choosing its properties when they attempt to simulate the formation of structure in the universe. Cosmologists distinguish between two basic types of dark matter on the basis of its temperature at the time when galaxies began to form. These types are known as **hot dark matter** and **cold dark matter,** respectively, and they lead to quite different kinds of structure in the present-day universe. By performing computer simulations of model universes dominated by hot and by cold dark matter and comparing the results with observations of the real universe, cosmologists try to determine which, if either, of the two alternatives can account for the large-scale structure we see around us.

Hot dark matter consists of lightweight particles—much less massive than the electron. If neutrinos turn out to have a small mass, as many researchers suspect, they would be leading candidates for hot dark matter particles. Simulations of a universe filled with hot dark matter indicate that large structures, such as superclusters and voids, form fairly naturally, but the computer models cannot account for the existence of structure on smaller scales. Small amounts of hot material tend to disperse, not to clump together. Attempts to produce galaxies and clusters by other means after the formation of larger objects have been only partly successful, so most cosmologists have concluded that models based purely on hot dark matter are unable to explain the observed structure of the universe.

Cold dark matter consists of very heavy particles. As mentioned earlier, massive supersymmetric relics of the GUT era are presently the leading cold dark matter candidates. Computer simulations modeling the universe with these particles as the dark matter easily produce small-scale structure. With the understanding that galaxies form preferentially in the densest regions and with some fine-tuning, these models can also be made to produce large-scale structure comparable to what is actually observed. Figure 29.14 shows a few frames from a recent supercomputer simulation of a cold dark matter universe. Compare these images with the real observations of nearby structure shown in Figures 26.23 and 26.24. Although calculations like this cannot *prove* that cold dark matter models are the correct description of the universe, the similarities between the models and reality are certainly very striking.

The Microwave Background

One of the arguments supporting dark matter models of the universe is the fact that they are consistent with the high degree of isotropy seen in the microwave background. Because the dark matter does not interact with photons, its density fluctuations do not cause comparably large (and easily observable) temperature variations in the radiation. However, the radiation *is* influenced slightly by the gravity of the growing dark clumps. The radiation experiences a slight gravitational red shift that varies from place to place depending on the density of the invisible matter. As a result, dark matter models predict that there *should* be ripples (temperature fluctuations) in the microwave background, but ripples that are very small. Dark matter density fluctuations causing temperature variations of as little as a few parts per million could still have grown into the structure we see around us today.

Until the late 1980s, these ripples were too small to be accurately measured, but cosmologists were confident that they were there. As observations improved, especially with the launch of the *COBE* satellite in 1989 (see Chapter 28), the anticipated temperature fluctuations remained undetected. Many theorists began to worry that there was

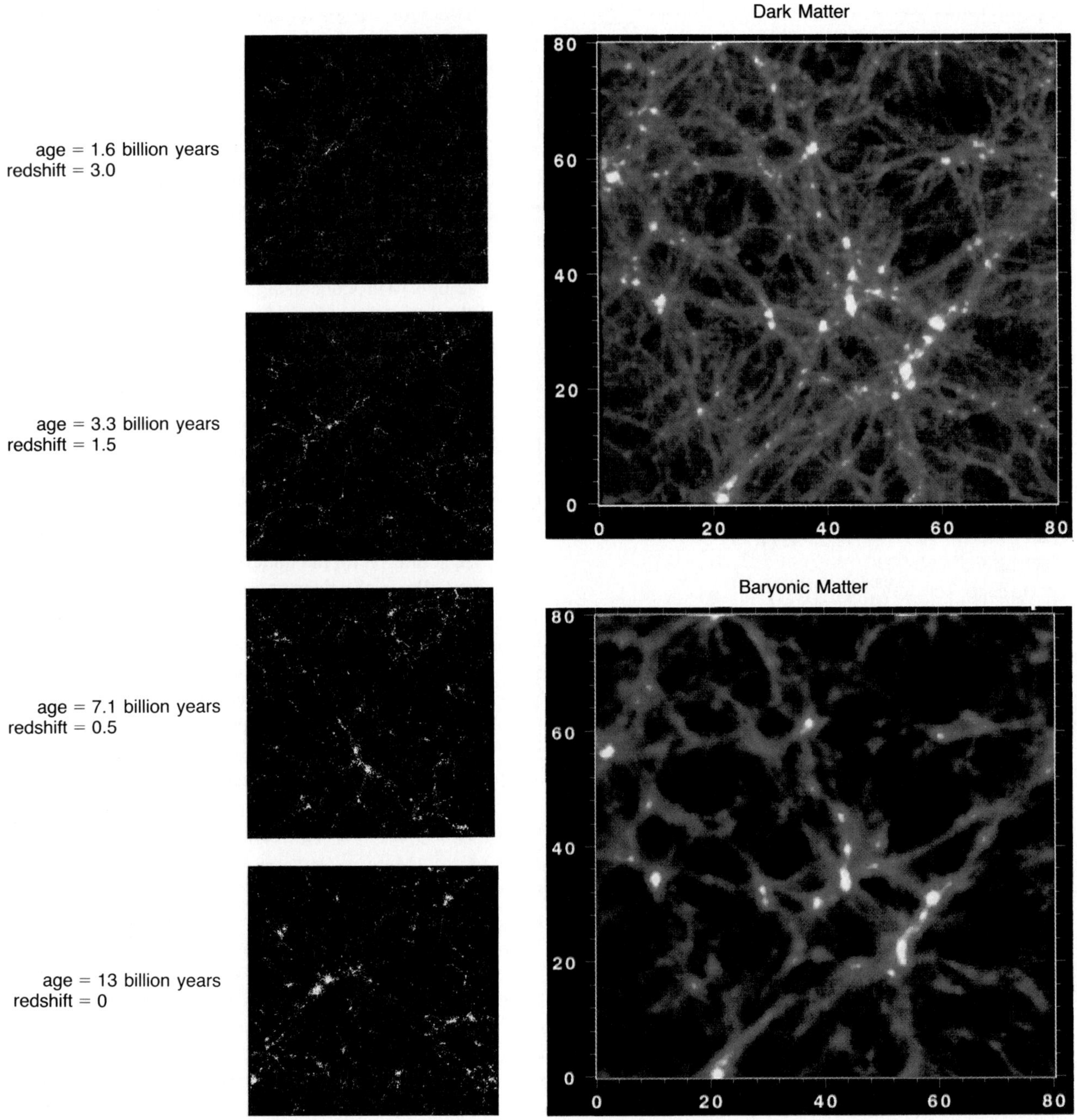

Figure 29.14 *Some frames from a supercomputer simulation (performed by Renyue Cen and Jeremiah Ostriker of Princeton University) of the formation of structure in a universe dominated by cold dark matter. The four small frames on the left are an evolutionary sequence showing how galaxies form as the universe ages (the overall expansion of the cosmos has been removed from all images). From top to bottom, the four frames depict the structure of the same portion of the universe at times 1.6, 3.3, 7.1, and 13 billion years after the Big Bang. The final view represents the present day, assuming a Hubble constant of 75 km/s/Mpc. The yellow points mark regions of space where galaxies have formed. The two larger images show in more detail the distribution of dark matter (top) and baryonic matter (bottom) in the same 80 by 80 Mpc (by 10 Mpc deep) patch of the universe depicted in the smaller frames. The densest regions are shown brightest. Notice that, while both types of matter form sheets and filaments on scales similar to those actually observed (compare Figures 26.23 and 26.24), the baryons are more strongly clumped than the dark matter. The large voids, while of generally low density, contain primarily non-baryonic matter.*

something seriously wrong with this theory of structure formation too. By 1990, the observational limits on the fluctuations were approaching 1 part in 100,000—still above the bare minimum needed for the cold dark matter theory to survive, but getting uncomfortably close to the point where the theory would have had to be abandoned.

In 1992, cosmologists breathed a huge collective sigh of relief when the *COBE* team, led by George Smoot of the Lawrence Berkeley Laboratory, announced that the expected fluctuations had been found. The temperature variations are tiny—only 17 *millionths* of a kelvin from place to place in the sky—but they are there. The *COBE* results are displayed as a temperature map of the microwave sky in Figure 29.15. The temperature variation due to the Earth's motion (see Figure 28.20) has been subtracted out, as has the radio emission from the Milky Way, and temperature *deviations* from the average are displayed.

Although the fluctuations in the microwave background are just big enough to be consistent with some cold dark matter models, they are not consistent with the models (like that shown in Figure 29.14) that provide the best agreement with actual observations of the present-day universe. The ripples seen by *COBE,* taken in conjunction with standard cold dark matter, imply *too little* structure on large scales—too few superclusters, voids, Great Walls, and so on. Thus the *COBE* results appear to confirm a central prediction of dark-matter theory, but at the same time they seriously undermine the leading candidate for cosmic dark matter! At the time of writing, the status of cold dark matter models as the proper description of invisible matter in the universe is quite uncertain. One thing does seem clear, though. If the *COBE* results hold up—as they are checked and rechecked by collaborators and competitors alike—they may one day come to rank alongside the discovery of the microwave background itself in terms of their importance to the field of cosmology.

In the late 1980s, many astronomers believed that cold dark matter models could successfully account for all of the large-scale structure found in the universe today. However, as both the quality and the detail of observations have continued to improve, it has become increasingly apparent that, like hot dark matter, cold dark matter may have serious, possibly even fatal, problems. If neither hot nor cold dark matter can provide all the answers, cosmologists may have to turn to more complex cosmological models, perhaps incorporating both types of dark matter simultaneously, or including new physical processes in their simulations. This is all in the future, though. The final word on dark matter in the universe has yet to be written.

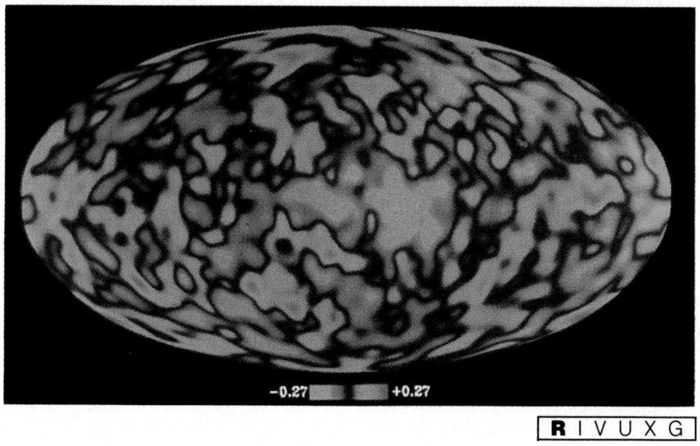

Figure 29.15 COBE *map of temperature fluctuations in the cosmic microwave background. Hotter-than-average regions are shown in red, cooler-than-average regions in blue. The total range in temperature is less than 20 millionths of a kelvin. (NASA)*

Quantum Fluctuations and the Universe Today

Finally, let us note that cold dark matter does not *necessarily* imply inflation, nor does inflation necessarily imply the existence of cold dark matter. A noninflated universe with $\Omega_0 = 0.5$, say, and cold dark matter could still have formed the structure we now see. Inflation and primordial nucleosynthesis together imply nonbaryonic dark matter, but it is not necessarily cold (although galaxy formation is hard to explain if none of it is cold). When inflation and cold dark matter are taken together, however, there is an added bonus. Inflation provides a possible explanation for the original density fluctuations that eventually grew into today's large-scale structure. They could have been microscopic *quantum fluctuations* that existed naturally in the universe during the GUT epoch, then grew to macroscopic size during the inflationary epoch. The fluctuations predicted by this idea agree quite well with the requirements of several cold dark matter models, although the *COBE* results have cast some doubt on their validity, as we have just seen.

We mentioned earlier that the entire universe might conceivably have grown out of a quantum fluctuation at the start of the Planck epoch. We also saw in Interlude 29-2 that the existence of matter may be a consequence of an asymmetric decay of particles during the GUT epoch. Now we see that the present structure of the universe may also be a consequence of a chance fluctuation at a very early time. If this view is correct—and some cosmologists believe that it is—then our Galaxy, the Sun, the Earth, even life itself, are direct consequences of a series of random events that occurred during an unimaginably short period of time some 10–15 billion years ago. Of course, these ideas are very speculative. In the strict sense, they are not really science at all, as they violate one of the central tenets of the scientific method—they are practically impossible to test experimentally. Still, whether or not you find them philosophically acceptable or intellectually pleasing, they do illustrate just how far the scope of physics and astronomy has expanded in the twentieth century. Astronomy is a subject that links the very big and the very small. Nowhere is that more evident than in our study of the most important question of all—the origin of the universe.

CHAPTER SUMMARY

For a few thousand years after the Big Bang, the universe was small, dense, and dominated radiation. As the universe slowly thinned and cooled, matter began to dominate.

High-energy photons can give rise spontaneously to particle-antiparticle pairs. The higher the temperature, the greater the chance that pair production will occur. Above the threshold temperature for pair production, in thermal equilibrium, new pairs are produced at exactly the same rate as particles and antiparticles annihilate one another. In the early universe, matter and radiation were linked together by this process. The existence of matter today means that there must have been unequal amounts of matter and antimatter early on.

Theorists believe that the physical state of the universe can be understood in terms of present-day physics back to about 10^{-43} s after the Big Bang. Before that, the four fundamental forces of nature—gravity, electromagnetism, the strong force, and the weak force—were all indistinguishable. There is presently no theory that can describe these extreme conditions. As the universe expanded and its temperature dropped, the forces became distinct from one another. First gravity, then the strong force, and then the weak and electromagnetic forces separated out.

All stars contain at least 20–25 percent helium by mass. This base level of helium is thought to have formed during the early, hot epochs of the universe. Elements heavier than helium did not form in the early universe, but in stars. The cores of stars become denser and hotter with time, but both the temperature and density of the early universe decreased rapidly as the universe expanded, so primordial nucleosynthesis stopped at helium. Deuterium—created in trace amounts along with primordial helium—can be directly related to the present-day density of matter in the universe, so measurements of the cosmic deuterium abundance provide a reliable estimate of Ω_0.

About a hundred thousand years after the beginning of the Matter Era, the universe became transparent to radiation. The first atoms appeared when the universe was approximately 1 million years old. By the time the universe was a billion years old, galaxies and large-scale structure had begun to form.

In the 1970s, theorists succeeded in explaining the electromagnetic and weak forces in terms of a single electroweak force. Electroweak theory has since been confirmed experimentally. In Grand Unified Theories (GUTs), also developed in the 1970s, the strong and electroweak forces are themselves connected by the influence of a very massive particle known as an X-boson. The GUTs have not yet been verified by experiment. Theorists have so far been unsuccessful in their attempts to incorporate gravity into any Grand Unified Theory.

Cosmologists wonder how regions of the universe that have not had time to "communicate" with one another are able to "know" they are supposed to look the same. This is called the horizon problem. Cosmologists also wonder why the density of the universe seems to be so near the critical value. This is called the flatness problem. Standard Big Bang theory does not provide the answers.

Some GUTs indicate that the universe may have undergone a brief phase of rapid expansion, during which the size of the universe increased by about 50 orders of magnitude in about 10^{-32} s. This idea is known as cosmic inflation. Inflation solves the horizon problem by taking a patch of the universe small enough to have become homogeneous before the period of rapid expansion began (at the end of the GUT epoch), then expanding it to enormous size—much larger than the presently observable universe. It solves the flatness problem by suggesting that any curvature the universe had before inflation has been expanded so much that space is now completely flat, for all practical purposes. If this is so, the density of matter is exactly critical.

Natural instabilities in the early universe could have started the process of growing structures that ranged in size up to superclusters, voids, and beyond. Theoretical attempts to make galaxies out of baryonic matter are hampered by the need for the first galactic fragments to have formed by the time the universe was 100 million years old.

The existence of dark matter, and the growing evidence that most of the universe is nonbaryonic in nature, provide an understandable mechanism for the formation of galaxies from small-scale density fluctuations in the early universe. Cosmologists distinguish between hot dark matter and cold dark matter, depending on its temperature at the end of the Radiation Era.

Dark matter models predict that there should be barely detectable ripples in the cosmic microwave background. In 1992, the *COBE* satellite discovered the expected fluctuations. However, the scale of the fluctuations is not in accord with the cold dark matter models whose predictions agree best with the universe we actually see.

KEY WORDS

baryonic matter
boson
cold dark matter
decoupling
deuterium bottleneck
electroweak force
false vacuum
flatness problem
freeze out
gluon
grand unification temperature

Grand Unified Theories
graviton
horizon problem
hot dark matter
inflation
matter-dominated universe
Matter Era
nonbaryonic matter
pair production
primordial nucleosynthesis
quantum fluctuation
radiation background
radiation-dominated universe
Radiation Era
recombination
supergravity
supersymmetric relic
supersymmetry
symmetry
thermal equilibrium
threshold temperature
unification of fundamental forces
X-boson

REVIEW QUESTIONS

1. For how long was the universe dominated by radiation? How hot was the universe when the dominance of radiation ended?
2. What conditions are needed in order for pair production to occur?
3. Why is our knowledge of the Planck epoch so limited?
4. When and how did the first atoms form?
5. Describe the universe at the end of the galactic epoch. When was this?
6. Why do all stars, regardless of their abundance of heavy elements, seem to contain at least one-quarter helium by mass?
7. Why didn't heavier and heavier elements form in the early universe, as they do in stars?
8. How do measurements of the cosmic deuterium abundance provide a reliable estimate of Ω_0?
9. When and how did the universe become transparent to radiation?
10. What are GUTs?
11. Describe how some GUTs may be able to account for the observed excess of matter over antimatter in the universe.
12. What is cosmic inflation?
13. How does cosmic inflation solve the horizon problem? The flatness problem?
14. What is "the ultimate free lunch"?
15. What was the significance of new information provided in 1992 by the *COBE* satellite?
16. Given that the threshold temperature for the production of electron-positron pairs is about 6×10^9 K and that a proton is 1800 times more massive than an electron, what is the threshold temperature for proton-antiproton pairs?
17. Using the formula given in the text, calculate (a) the (approximate) temperature of the universe 1 second, 1 minute, 1 hour, 1 day, and 1 year after the Big Bang, and (b) the age of the universe when its temperature was 1 trillion, 1 billion, and 1 million Kelvins.

DISCUSSION QUESTIONS

1. The modern picture of the early universe has at its base the cosmological principle and the twin assumptions of homogeny and isotropy. Do you think that a cosmology based on unproven assumptions can come close to representing the "truth"? Tell why you do or don't think so.
2. How do you feel about the possibility that the entire universe might have grown out of a quantum fluctuation at the start of the Planck epoch?

SUGGESTED READINGS

Alpher, R. A., and R. Herman. "Reflections on Early Work on 'Big Bang' Cosmology." *Physics Today* (August 1988). A historical exposition on the physics of the Big Bang cosmological model.

Cornell, J., ed. *Bubbles, Voids and Bumps in Time: The New Cosmology*. Cambridge: Cambridge University Press, 1989. Based on a series of popular lectures by various experts, this book includes a chapter called "Starting the Universe: The Big Bang and Cosmic Inflation."

Davies, P. *Superforce*. New York: Simon and Schuster, 1984. The search for a Grand Unified Theory of nature.

Flam, F. "Giving the Galaxies a History." *Science* (February 28, 1992). Astronomers discover one of the first galaxies.

Hawking, S. *A Brief History of Time*. New York: Bantam Books, 1988. On the best-seller list for many weeks, this book explores many basic ideas—and a few obscure ones—related to the physics of the early universe.

CHAPTER

30

Life on Earth and Life Beyond

Are We Alone?

This fanciful scene, obviously painted by an imaginative artist, depicts an alien planet that has been kicked out of its home galaxy. This has caused the civilization that once thrived there to die off, leaving behind only its Stonehenge-style ruins as a mute testimony to its existence. As the planet races away, its former galaxy provides a spectacular display in their darkening skies. (Courtesy D. Berry)

LEARNING GOALS

1 to understand the mix of proven facts and testable ideas that make up the scenario of cosmic evolution

2 to evaluate the chances of finding life in the solar system

3 to appreciate the various probabilities used to estimate the number of advanced civilizations now residing in the Galaxy

4 to understand some of the techniques we might use to search for extraterrestrials and to communicate with them

What are we? In essence, we are a combination of chemical elements produced eons ago inside the fiery cores of massive stars—the same elements that contribute to Earth's rocky continents, its atmosphere and its oceans. How did all these elements combine to form life? Next to the origin of the universe itself that question is one of the most fundamental we can ask. To try to answer it, to trace the specific steps that led to our origin, we must combine a wide variety of subjects—physics, chemistry, astronomy, geology, biology, and anthropology, among others—to examine the changes that have taken place in the universe since its beginning. The story that unfolds is one of continuous evolution of matter and energy, ultimately leading to the appearance of life on Earth.

The matter around us is not unique. But are we *unique? Is life on our planet the only example of life in the universe? These are tough questions, for the subject of extraterrestrial life is one on which there are no data. We can say only one thing for sure: Earth is the only place in the universe where we know for certain that life exists. In this chapter we take a look at how humans evolved on Earth, then consider whether those evolutionary steps might have happened elsewhere. Having done that, we will assess the likelihood of our having galactic neighbors and consider how we might learn about them if they exist.*

COSMIC EVOLUTION

The Arrow of Time

✓1 In order to trace the steps that have led to the origin of life and civilization, modern science must combine a wide variety of subjects. As illustrated in the "arrow of time" sketched in Figure 30.1, we can identify seven major phases

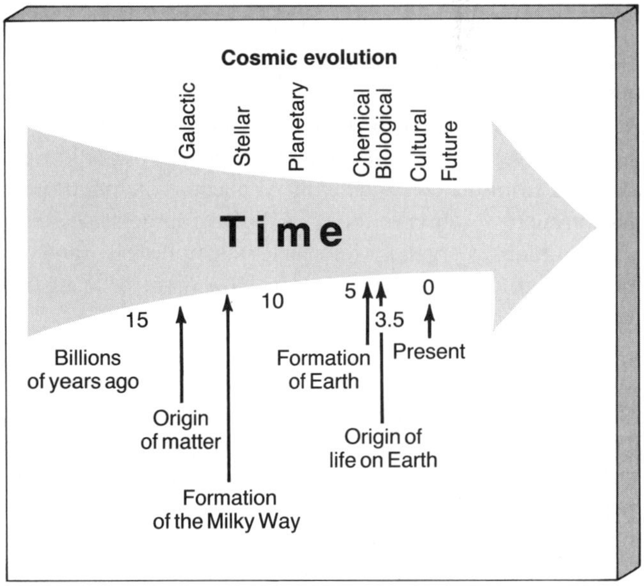

Figure 30.1 *Some highlights of cosmic history are noted along this arrow of time, from the beginning of the universe to the present. Noted along the top of the arrow are the major phases of cosmic evolution.*

in the history of the universe: *galactic*, *stellar*, *planetary*, *chemical*, *biological*, *cultural*, and *future* evolution. Together, these evolutionary stages make up the grand sweep of **cosmic evolution.**

We began this book by talking about the history of astronomy and the people that helped develop it into a vital science. When we reached astronomy as we know it today, we studied Earth, then expanded our perspective to include the entire solar system. We then widened our vision to study stars and galaxies, and in the last chapters we moved back in time to consider the origin of the universe and what went on as time itself came into being. In this final chapter we reverse direction. We move from the Big Bang to galaxies to stars to planetary systems to life today on planet Earth. Along the way we consider whether life may have evolved elsewhere in the universe and how we might get in touch with it if it has.

R I **V** U X G

Figure 30.2 *Clusters of galaxies top the hierarchy of material structures in the universe. Nearly every speck of light in this photograph arises from a full-fledged galaxy—members of the (Centaurus) Cluster, formed eons ago in the early universe. (NOAO)*

Galactic, Stellar, and Planetary Evolution

Let us start by summarizing briefly the evolutionary highlights of the preceding chapters. Modern cosmology holds that the universe began in a cataclysmic explosion approximately 15 billion years ago. At the very beginning, the temperature of the fireball, the Big Bang, was unimaginably hot. As the universe expanded, it cooled, allowing matter, nuclei, atoms, galaxies, and stars to form.

In those first few moments of the universe, radiation completely overwhelmed matter. As the temperature fell, the fundamental forces and elementary particles we know today came into being, although the intense radiation prevented the particles from assembling into anything more complex. After a few minutes, the temperature had dropped to the point where helium could form. Most of the helium that now exists in the universe appeared at that time. A few thousand years after the Big Bang, the universe had cooled sufficiently for matter to begin to dominate radiation. After a million years, the first atoms formed, and radiation decoupled from matter, eventually to become the microwave background we observe today. All of the large-scale structures—galaxies, galaxy clusters, superclusters, and so on—we now see in the universe have formed since that time.

Matter was probably distributed fairly uniformly at first. But if left alone, matter tends to clump. Small density fluctuations become denser and more compact in time, as they contract under the influence of their own gravity and pull additional matter onto themselves. In the early universe, clumps of many sizes, from small galaxies to superclusters, appeared and evolved, in time blooming into the galaxies and larger objects we see. The densest galaxies had already begun to form by the time the universe was about $\frac{1}{10}$ its current size, 100 million years after the Big Bang; most galaxy formation was well under way by the time the universe was a billion years old. Larger-scale structures, such as galaxy clusters and superclusters, are still forming today. As illustrated in Figures 30.2 and 30.3, we see galaxies in great abundance virtually everywhere in the universe.

Within galaxies, interstellar gas also clumps together and undergoes further contraction. The microscopic spaces between individual gas particles diminish, increasing the frequency of atomic collisions and generating heat. As a result, a contracting interstellar cloud heats up, until a small, dense region at its center exceeds a temperature of about 10 million K, and nuclear burning begins. This burning is what provides the steady solar energy that sustains life

R I **V** U X G

Figure 30.3 *A typical, relatively nearby galaxy—such as this spiral galaxy in Pegasus—contains hundreds of billions of individual stars held together in a swarm by the pull of gravity and distributed over at least 30,000 pc. Our Milky Way Galaxy, too large and complex for us to observe accurately from within, probably resembles this one. If so, our Sun would be seen as a rather undistinguished star in the galactic suburbs far from the Milky Way's center. (NOAO)*

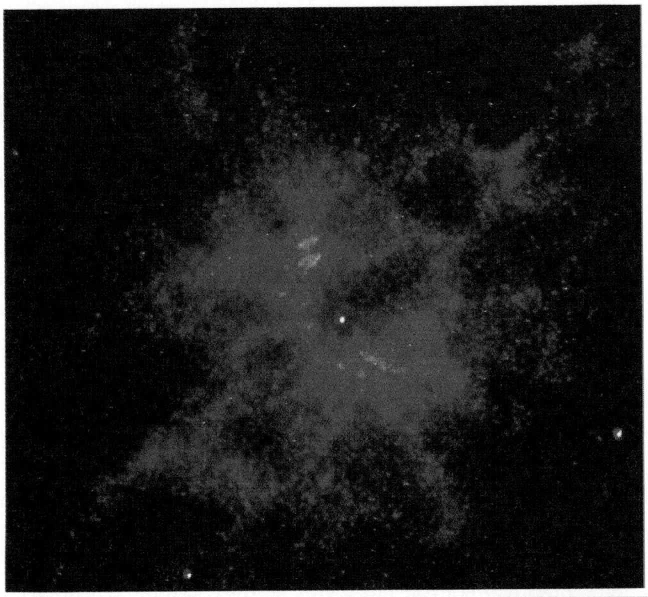

R I **V** U X G

Figure 30.4 *Patches of gas aglow in deep space, called emission nebulae, are sites of recent star formation. This one is called NGC 2440 and contains one of the hottest stars known (bright dot at center)—a torrid 200,000K. The region was imaged by the* Hubble Space Telescope. *(NASA)*

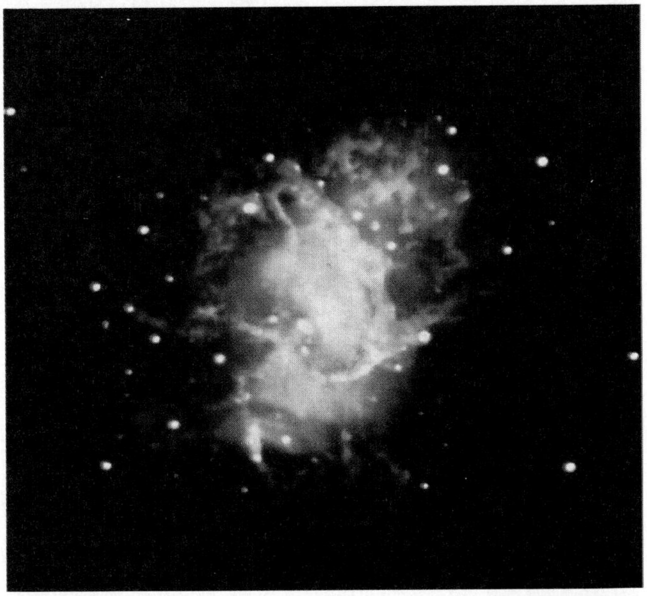

R I **V** U X G

Figure 30.5 *A clear and unmistakable image of a supernova explosion expelling heavy elements into the interstellar medium. The titanic death rattle of this expired star, now known as the Crab Nebula, was observed and recorded by Chinese astronomers in the eleventh century. (NOAO)*

on our planet. As illustrated in Figure 30.4, we see numerous regions in our Galaxy where star formation has just occurred. The entire process takes a few tens of millions of years to complete—several million human generations, yet still less than 1 percent of a typical star's lifetime. It amounts to a steady metamorphosis, an evolution, a gradual change of a cold, tenuous, flimsy pocket of gas into a hot, dense star.

The lifetime of a star depends on its mass. Stars much more massive than the Sun can achieve very high temperatures in their cores, temperatures capable of producing many of the heavier elements. After nuclear burning ceases, the whole stellar core collapses in a matter of seconds, then suddenly rebounds like a coiled spring, in a supernova explosion. The heavy elements and much of the star's original mass fly off into the interstellar medium at high speeds. Evidence of this violent process exists at many places in the sky, most notably in the Crab Nebula, shown in Figure 30.5.

The upshot is that the interstellar medium is regularly buffeted and enriched by exploding stars. Figure 30.6 depicts how the heavy elements they eject may become part of later-generation stars, planets, and other things, including living organisms. Because observations show that our Sun already contains minute amounts of heavy elements despite its relative youth and cool interior, we regard our Sun as a second- or later-generation star. Our Sun and its planets condensed about 4.5 billion years ago from a cool cloud of interstellar matter already enriched with heavy elements.

Astronomers generally believe that the birth of planets is a natural outgrowth of the star-formation process. Although they do not agree on all of the specifics, they do concur on the broad outline of the condensation theory. As

Figure 30.6 *In a continuous and ongoing process, many stars form from cool, dense regions of the interstellar medium, after which they explode much of their matter back into the galaxy. They fertilize the interstellar medium with heavy elements, which form later generations of stars, planets, and life. In this fanciful artist's rendition, we can see several planets, including Earth, emerging from the aftermath of a supernova detonation somewhere in the suburbs of a spiral galaxy. (D. Berry)*

Figure 30.7 *An artist's conception of a contracting, swirling interstellar cloud forming a central star and planetary system. Computer models suggest that a few million centuries were needed to generate nine protoplanets and scores of protomoons, as well as a big protosun in their midst.*

sketched in Figure 30.7, the huge ball of interstellar gas and dust that would form our Sun contracted and became flattened by rotation, after which it condensed and fragmented into boulder-sized chunks of solid matter orbiting at various distances from the central protosun. Far from the young Sun, where the temperature was low, conditions favored the production of low-density ices and gases that later accumulated into the giant gassy Jovian planets and their moons. Closer to the Sun, the solar nebula was too warm to form ice, but instead formed denser, rocky fragments that collided, stuck, and became the smaller terrestrial worlds.

Chemical Evolution

What information do we have concerning the earliest stages of planet Earth? Unfortunately, not much. Geological hints about the first billion years or so were steadily eroded by violent surface activity as volcanos erupted and meteorites bombarded our planet. The early Earth was barren, with shallow, lifeless seas washing upon grassless, treeless continents. Little of the light gas that made up the solar nebula—hydrogen and helium—remained in Earth's atmosphere. It escaped into space. Continued outgassing from the Earth's interior through volcanos, fissures, and geysers produced a secondary atmosphere rich in hydrogen, nitrogen, and carbon compounds and poor in free oxygen. This stage is sometimes referred to as the "Big Burp." As the Earth cooled, ammonia, methane, carbon dioxide, and water formed. The stage was set for the appearance of life.

In the absence of oxygen, no ozone layer could form. The Earth's surface was unprotected from solar ultraviolet radiation. This radiation energy synthesized the ammonia, methane, carbon dioxide, and water into more complex products known as **amino acids** and **nucleotide bases**—**organic molecules** (carbon-based) that are the building blocks of life as we know it. Amino acids build proteins, which direct metabolism (the daily income of energy and output of waste). Sequences of nucleotide bases form genes, which direct the passage of hereditary characteristics from one generation to the next in reproduction. In all living creatures on Earth—from bacteria to amoebas to humans—genes mastermind life, while proteins help maintain it.

The basic idea of **chemical evolution,** the notion that complex molecules could have evolved naturally from simpler ingredients found on the primitive Earth, has been around since the 1920s. The first experimental verification of the theory was provided in 1953 by University of Chicago scientists Harold Urey and Stanley Miller, using laboratory equipment somewhat similar to that shown in Figure 30.8. Urey and Miller took a mixture of the materials thought to be present on Earth long ago—a "primordial soup" of water, methane, carbon dioxide, and ammonia—and energized it by passing an electrical discharge ("lightning") through the gas. After a few days, they analyzed

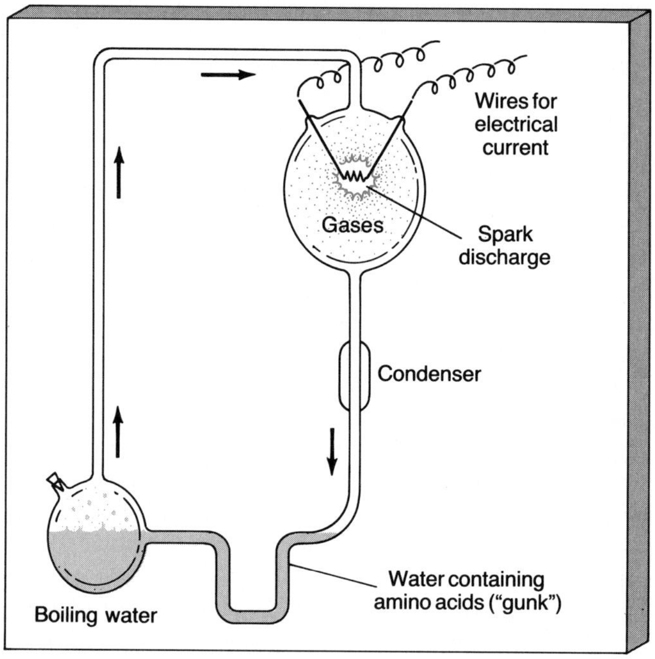

Figure 30.8 *This chemical apparatus is designed to synthesize complex biochemical molecules by energetically irradiating a mixture of simple chemicals. In the diagrammed procedure, the gases are placed in the upper bulb to simulate the primordial Earth atmosphere and then energized by spark-discharge electrodes. After about a week of recycled "cooking," the gases have formed amino acids and other prebiological molecules in the trap at the bottom, which simulates the primordial oceans into which heavy molecules would have fallen from the atmosphere.*

their mixture and found that it contained many of the same amino acids found today in all living things on Earth. About a decade later, biochemist Cyril Ponnamperuma succeeded in constructing nucleotide bases in a similar manner. In fact, we now know that the chemical mixtures used in those early simulations probably did not resemble Earth's primitive atmosphere particularly closely. Nevertheless, the importance of those experiments in demonstrating conclusively that biological molecules can be synthesized by strictly *non*biological means, using raw materials that were available on the early Earth, is undiminished.

Laboratory simulations of the primordial ocean and atmosphere have since shown that a wide variety of energy sources can produce amino acids and nucleotide bases. Not just solar ultraviolet radiation, but also lightning, volcanic heat, natural radioactivity, and the atmospheric shock waves produced by incoming meteorites can synthesize these molecules. The **Urey-Miller experiment** has been repeated in many different forms since the 1950s, with more realistic mixtures of gases, and always with the same basic outcome. Nothing has ever crawled out of a test tube at the end of an experiment, but the results clearly indicate that the first crucial steps from simplicity to complexity along the road to life could have occurred naturally on the early Earth.

Some scientists argue that Earth's primitive atmosphere may not have been a particularly suitable environment for the production of complex molecules. Instead, those scientists say, there may not have been sufficient energy available to power the chemical reactions, and the early atmosphere may not have contained enough raw material for the reactions to have become important in any case. They suggest that much, if not all, of the organic material that eventually combined to form living cells was produced in *interstellar space* and subsequently arrived on Earth in the form of interplanetary dust and small meteors that did not burn up during their descent through the atmosphere. We have seen that interstellar molecular clouds contain some very complex molecules (Chapter 20) and that large amounts of organic material were detected on comet Halley by space probes when Halley last visited the inner solar system (Chapter 16), so the idea that organic matter is constantly raining down on Earth from space is quite plausible. Whether or not this was the *primary* means by which complex molecules first appeared in Earth's oceans remains to be seen. For now, the issue is unresolved.

More advanced experiments, in which amino acids are united under the influence of heat, have fashioned proteinlike blobs that behave to some extent like true biological cells. Such near-protein material resists dissolution in water (so it would remain intact when it fell from the primitive atmosphere into the ocean) and tends to cluster into small droplets, sometimes called microspheres—a little like oil globules floating on the surface of water. Figure 30.9 shows some of these proteinlike microspheres. The walls of these laboratory-made droplets permit the inward passage of small molecules, which then combine within the droplet to construct more complex molecules too large to pass back out through the walls. As the droplets "grow," they tend to break up into an increased number of smaller droplets. Could this process of breaking up and forming new things be a primitive form of reproduction?

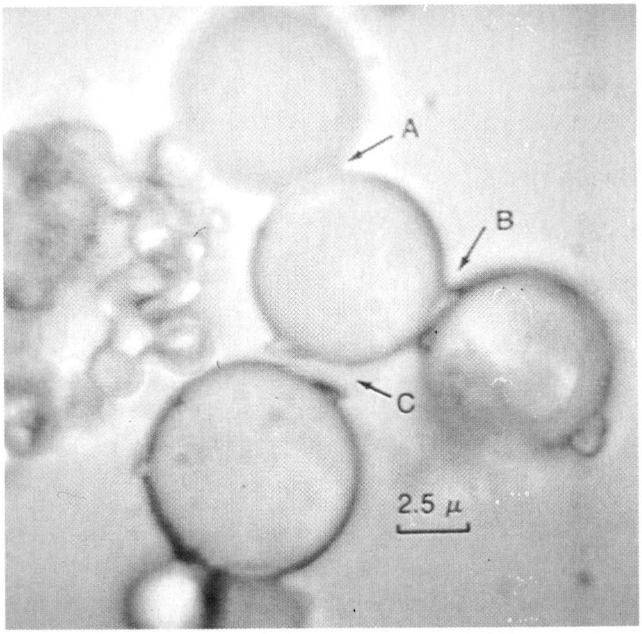

Figure 30.9 *These carbon-rich, proteinlike droplets display the clustering of as many as a billion amino-acid molecules in a liquid. Droplets can "grow," and parts of droplets can separate from the "parent" to become new individual droplets. The scale of 2.5 microns noted here is 1/4000 of a centimeter. (S. Fox)*

Can we consider these proteinlike microspheres to be alive? Almost certainly not. Most biochemists would say that the microspheres are not life itself but contain all the basic ingredients needed to form life. The microspheres lack the hereditary molecule DNA and a well-defined nucleus associated with many living cells. However, as illustrated in Figure 30.10, they have certain similarities to ancient cells found in the fossil record. Thus no actual living cells have yet been created "from scratch" in any laboratory, but many biochemists feel that the chain of events leading from simple nonbiological molecules almost to the point of life itself has been amply demonstrated over the past four decades. (For another example of an organism that seems to lie close to the boundary between life and nonlife, consult Interlude 30-1.)

Biological Evolution

How do we know anything about the early episodes of life on Earth? The key is the fossil record, which chronicles how life became widespread and diversified over the course

INTERLUDE 30-1 *The Virus*

The central idea of chemical evolution is that life evolved from nonlife. But aside from insight based on biochemical knowledge and laboratory simulations of some key events on primordial Earth, do we have any direct evidence that life could have developed from nonliving molecules? The answer is yes. The smallest and simplest entity that sometimes appears to be alive is a virus. We say "sometimes" because viruses seem to have the attributes of both nonliving molecules and living cells. *Virus* is the Latin word for "poison," an appropriate name since viruses are often a cause of disease. Although they come in many sizes and shapes, all viruses are smaller than the size of a typical modern cell. Some are made of only a few thousand atoms. In terms of size, then, viruses seem to bridge the gap between cells that are living and molecules that are not.

Viruses contain some proteins and genetic information, but not much else—none of the material by which living organisms normally grow and reproduce. How, then, can a virus be considered alive? When alone, it cannot; a virus is absolutely lifeless when isolated from living organisms. But when inside a living system, a virus has all the properties of life. Viruses come alive by transferring their genetic material into living cells. The genes of a virus seize control of a cell and establish themselves as the new master of chemical activity. Viruses grow and reproduce copies of themselves by using the material of the invaded cell, often robbing the cell of its usual function. Rapidly and wildly, some viruses multiply, spreading the disease and, if unchecked, eventually killing the invaded organism. In a sense, then, viruses exist within the gray area between the living and the nonliving.

of time. The study of fossil remains shows the initial appearance about 3.5 billion years ago of simple one-celled organisms such as blue-green algae. These were followed about 2 billion years ago by more complex one-celled creatures, like the amoeba. Multicellular organisms such as sponges did not appear until about 1 billion years ago, after which there flourished a wide variety of increasingly complex organisms—insects, reptiles, mammals, and humans. The fossil record also contains abundant evidence that many species did not adapt successfully to changing environments and so perished. In fact, despite the present existence of some 2 million living species, biologists estimate that more than 99 percent of all organisms that have ever lived on Earth are now extinct.

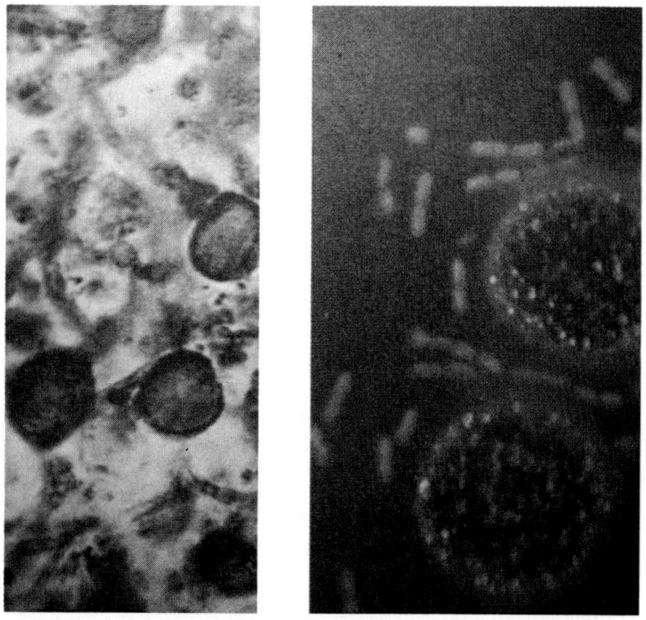

Figure 30.10 *The photograph on the left, taken through a microscope, shows a fossilized organism found in sediments radioactively dated as 2 billion years old. This primitive system possesses concentric spheres or walls connected by smaller spheroids. The roundish fossils here measure about a thousandth of a centimeter. The photograph on the right, also taken through a microscope and on approximately the same scale, displays modern blue-green algae. (E. Barghoorn)*

The fossil record leaves no doubt that biological organisms have changed over time—all scientists accept the reality of **biological evolution.** As conditions on the Earth have shifted and the Earth's surface itself has evolved, those organisms that could best take advantage of their new surroundings succeeded at the expense of those organisms that could not make the necessary adjustments. What led to changes? Chance. An organism that happened to have a certain useful characteristic—for example, the ability to run faster, climb higher, or even hide more easily—would find itself with the upper hand in a particular environment. This organism was, therefore, more likely to reproduce successfully, and its advantageous characteristic would then be more likely to be passed on to the next generation. The evolution of the rich variety of life on our planet—including human beings—occurred as chance mutations led to change in organisms over millions of years.

To put biological evolution into historical perspective, let's imagine the entire lifetime of Earth to be 45 years rather than 4.5 billion years. We have no reliable record of the first decade of our planet's existence. Life originated at least 35 years ago, when Earth was only about 10 years old. Our planet's middle age is largely a mystery, although we can be sure that life continued to evolve and that generations of mountain chains and oceanic trenches came and went.

Not until about 6 years ago did abundant life flourish throughout Earth's oceans. Life came ashore about 4 years ago, and plants and animals mastered the land only about 2 years ago. Dinosaurs reached their peak about 1 year ago, only to die suddenly about 8 months ago (see Interlude 30-2). Humanlike apes changed into apelike humans only last week, and the latest ice ages occurred only a few days ago. *Homo sapiens*—our species—did not emerge until about 4 hours ago. Agriculture was invented within the last hour, and the Renaissance is only 3 minutes old!

Apparently, it takes time—lots and lots of time—to construct life, intelligence, and civilization.

Cultural Evolution

We have now seen how chemical and biological evolution can account, in general terms at least, for the emergence of life and its proliferation into many different species. What about the development of intelligence? Many anthropologists believe that, like any other highly advantageous trait, intelligence is strongly favored by natural selection.

Changing environments caused some of the species that had claimed the land to evolve into ratlike, insect-eating, tree-dwelling creatures by about 75 million years ago. In the process, paws became hands for gripping tree branches. Vision became increasingly important for judging accurately the distance to the next branch or to the insects and fruit that would provide the next meal. For better protection, body size also increased. The fossil record of about 40 million years ago shows that competition for survival apparently favored animals that became increasingly like the monkeys we see today (see Figure 30.11). The precise line of descent from ape to human remains debatable, but as more and more fossils are unearthed—mainly along the Great Rift Valley in East Africa—the picture is becoming clearer.

Figure 30.11 *Dexterous hands and keen vision gave some tree-dwellers a distinct advantage in the constant struggle for survival. Here a monkey is shown grasping a branch while simultaneously reaching for food. This type of monkey has not changed much in the past 40 million years and is thought to closely resemble the advanced tree dwellers of long ago that were the ancestors of modern humans.*

Eventually, some of these creatures—our distant ancestors—left the trees and in time began to walk erect on the ground. We do not know precisely why they left their comfortable forest existence. Perhaps one type of ape began to hog the insects and the fruit. In order to survive, another type of ape was forced to develop into a ground-dwelling animal. Whatever the reason, the move to the ground began the chain of events that led directly to us. In time, those displaced apes—that is, humanity—would come to dominate the planet (including the apes left behind in the trees, still hogging the insects and the fruit).

For all the successful physical changes, what led to the development of the human brain? Studies of the behavioral patterns of chimpanzees (see Figure 30.12) today demonstrate their ability to strip leaves from a small twig, insert the twig into a termite mound, pull it out carefully, and lick off the termites. This meal requires both physical and mental dexterity. It is quite likely that increased dependence on the hands among early prehumans began the increase in brain size.

The brain became more and more elaborate. Humans learned about fire, tools, and agriculture. The social cooperation that went with coordinated hunting efforts was another important competitive advantage that developed as brain size increased. Perhaps most important of all was the development of language. By communicating, individuals could signal each other while hunting food or seeking protection. In the opinion of many anthropologists, language was a key factor in the development of our brain—in fact, some anthropologists have gone so far as to suggest that human intelligence *is* human language. Now our ancestors

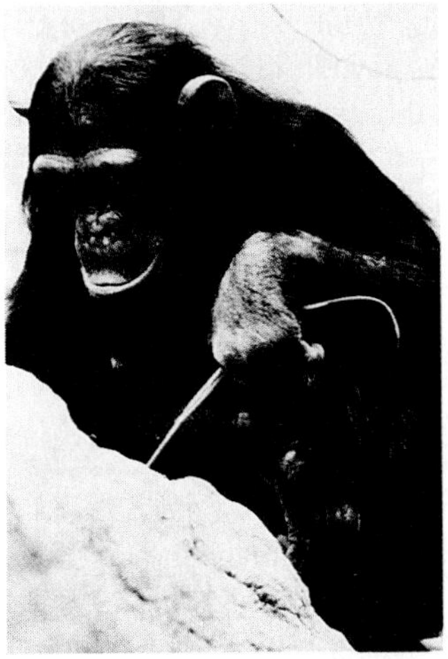

Figure 30.12 *Chimpanzees use sticks as tools to prod insects out of termite mounds, carefully extracting the stick and eating the insects. Notice also the chimp's relatively erect posture.*

INTERLUDE 30-2 *Where Did the Dinosaurs Go?*

The name *dinosaur* derives from the Greek words, *deinos* (terrible) and *sauros* (lizard). Dinosaurs were no ordinary reptiles. In their prime, roughly 100 million years ago, the dinosaurs were the all-powerful rulers of the Earth. Their fossilized remains have been uncovered on all the world's continents. Despite their dominance, according to the fossil record, these creatures vanished from Earth quite suddenly about 65 million years ago. What happened to them?

Until fairly recently, the prevailing view among paleontologists—scientists who study prehistoric life—was that dinosaurs were rather small-brained, cold-blooded creatures. In chilly climates, or even at night, the metabolisms of these huge reptiles would have become sluggish, making it difficult for them to move around and secure food. The suggestion was that they were poorly equipped to adapt to sudden changes in Earth's climate, so that they eventually died out.

However, a competing, and still controversial, view of dinosaurs has emerged. Recent fossil evidence suggests that many of these monsters might have had large, four-chambered hearts, like those of mammals and birds. They may also have been warm-blooded creatures and relatively fast-moving—not at all the dull-witted, slow-moving giants of earlier conception. Also, although the dinosaurs clearly had small brains compared with those of today's mammals, they were still relatively smart for their time. No species able to dominate the Earth for more than 100 million years could have been too dumb. By comparison, humans have thus far dominated for little more than 2 million years. If the dinosaurs didn't die out simply because of stupidity and inflexibility, then what happened to cause their sudden and complete disappearance?

Many explanations have been offered for the extinction of the dinosaurs. Devastating plagues, magnetic field reversals, increased tectonic activity, severe climate changes, and supernova explosions have all been proposed. In the 1980s, it was suggested that a huge extraterrestrial object collided with Earth 65 million years ago, and this is now (arguably) the leading explanation for the demise of the dinosaurs, although it is by no means universally accepted. According to this idea, a 10- to 15-km-wide asteroid or comet struck the Earth, releasing as much energy as 10 million or more of the largest hydrogen bombs humans have ever constructed and kicking huge quantities of dust (including the pulverized remnants of the asteroid itself) high into the atmosphere. The dust may have shrouded our planet for many years, virtually extinguishing the Sun's rays during this time. On the darkened surface, plants could not survive. The entire food chain was disrupted, and the dinosaurs, at the top of that chain, eventually died out.

Although we have no direct astronomical evidence to confirm or deny this idea, we can estimate the chances of a large asteroid or comet striking the Earth today, on the basis of observations of the number of objects presently on Earth-crossing orbits. The figure on the next page shows the likelihood of an impact as a function of the size of the impacting body. The horizontal scale indicates the energy released by the collision, measured in *megatons* of TNT—1 megaton is the explosive yield of a large nuclear warhead, the only common terrestrial measure of energy adequate to describe the violence of these occurrences. We see that 100 million megaton events, like the planetwide catastrophe that supposedly wiped out the dinosaurs, are very rare, occurring only once every 10 million years or so. However, smaller impacts, equivalent to "only" a few tens of kilotons of TNT (about one MX missile warhead), could happen every

could share ideas as well as food and shelter. Experience, stored in the brain as memory, could be passed down from generation to generation. A new kind of evolution had begun: **cultural evolution,** the changes in the ideas and behavior of society. Our more recent ancestors have created, within only the past 10,000 years or so, the entirety of human civilization.

LIFE IN THE SOLAR SYSTEM

√2 From the Big Bang, to the formation of galaxies, to the birth of the solar system, to the shift from the merely chemical to the biological, to the evolution of culture, the universe evolves from simplicity to complexity. The human brain is the most complex clump of matter we know of. But might there be other brains beyond our own planet? Are we alone in the universe? Or are we just one among countless other intelligent life forms in our Galaxy? How might we go about answering these all-important questions?

The general case in favor of extraterrestrial life is summed up in the so-called *assumptions of mediocrity*: (1) Because life on Earth depends on just a few basic molecules and (2) because the atoms that make up these molecules are common to all stars, then (3) if the laws of science we know apply to the entire universe, as we have supposed throughout this book, then, given sufficient time, life must have originated elsewhere in the cosmos. The opposing view maintains that intelligent life on Earth is the product of

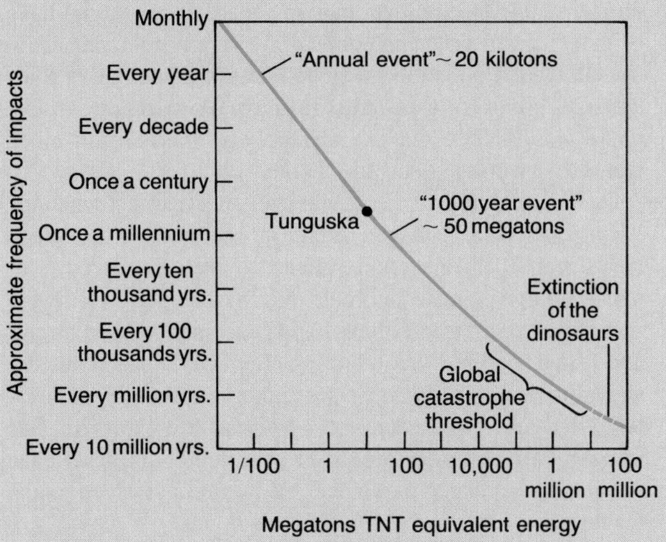

few years—we may be long overdue for one. The most recent large impact was the Tunguska explosion (mentioned in Chapter 16) in Siberia, in 1908, which packed a roughly 1-megaton punch.

The main geological evidence supporting this theory is a layer of clay enriched with the element iridium. This layer is found in 65-million-year-old rocky sediments all around our planet. Iridium on the Earth's surface is rare because most of it sank into our planet's interior. The iridium found in this one layer of clay is about 10 times more abundant than in other rocks, matching the amount of iridium found in meteorites (and, we assume, in asteroids and comets too).

This theory has its problems, however. The amount of iridium in this clay layer varies greatly from place to place across the globe, and there is no complete explanation of why that should be so. And if this body was so massive as to kick up enough dust to darken the entire planet, then where is its impact crater? One leading candidate has been found in the Yucatan Peninsula in Mexico, where possible evidence for a heavily eroded, but not completely obliterated, crater has been found. However, its identity as the site of the dinosaur-killing collision has not yet been conclusively proven. Perhaps, some scientists argue, the iridium layer was laid down by volcanos and had nothing to do with an extraterrestrial impact at all.

Another potential difficulty concerns the speed at which the dinosaurs disappeared. The fossil and geological records are rather imprecise in the ages they yield, and this translates into uncertainties in just how long the extinction process took. It seems to be no more than a million or so years, but that is still a very long time. If the dinosaurs vanished as a direct result of a collision and subsequent explosion, we might expect their disappearance to have been complete in a matter of decades, or maybe centuries. It is hard to see how the process could have spanned tens or hundreds of millennia. Greatly improved geological age–measurement techniques will be needed to settle this issue once and for all.

Whatever killed the dinosaurs, dramatic environmental change of some sort was almost surely responsible. It is important that we continue the search for the cause of their extinction, for there's no telling if and when that sudden change might strike again. As the dominant species on Earth, we are the ones who now stand to lose the most.

a series of extremely fortunate accidents—astronomical, geological, chemical, and biological events unlikely to have occurred anywhere else in the universe.

Life As We Know It

The first place to look for life beyond Earth is in our own solar system. Might we find life this close to us? Our search is hampered because we know about life only as it exists on Earth. Still, we can at least assess the likelihood of finding "life as we know it" in our cosmic neighborhood.

The Moon and Mercury lack a protective atmosphere and magnetic field and so are subjected to fierce bombardment by solar ultraviolet radiation, the solar wind, meteoroids, and cosmic rays. They also experience extremes in temperature. Simple molecules could not possibly survive in such hostile environments. The planet Venus, though, has far too much protective atmosphere! Its dense, dry, scorchingly hot atmospheric blanket effectively rules it out as a possible abode for life. The outermost planet, Pluto, and most of the moons of the outer planets are too cold. So far from the Sun, water could not possibly remain in the liquid state, so the building blocks of life, even if they exist there, would have no easy way to interact to form larger organic molecules.

The Jovian planets are all quite far from the Sun and have cloud tops well below the freezing point of water, but we should not be too quick to dismiss any possibility for life somewhere within these large gas balls. Greater warmth exists deep down in their clouds. Both the greenhouse effect (which traps solar radiation) and internal sources of energy (at least inside Jupiter, Saturn, and Neptune) provide some heat. Consequently, as we mentioned in Chapter 13, some researchers argue that intermediate levels of the Jovian atmospheres might be warm enough for the development and maintenance of life. However, most researchers believe that, despite the temperate conditions, life is unlikely. To survive and prosper, organisms would have to remain at a stable atmospheric level. If they floated too high in the atmosphere, they would freeze; too low, and they would fry. Freezing would not necessarily destroy large molecules, but it would surely prevent their further development. Too much heat will physically destroy any molecule.

Some of the moons of the Jovian planets may be better candidates than the planets themselves. For example, Saturn's moon Titan, with its atmosphere of methane, ammonia, and nitrogen and possibly a solid surface, is a potential site for life, although the results of the 1980 *Voyager 1* flyby suggest that its surface conditions are rather inhospitable for any kind of life familiar to us.

What about the cometary and meteoritic debris that orbits within our solar system? Comets contain many of the basic ingredients for life—for instance, ammonia, methane, and water vapor—and although comets are frozen, their icy matter warms while nearing the Sun. However, few heavy molecules have been observed in comet spectra. Still, a small fraction of the meteorites that survive the plunge to Earth's surface do contain organic compounds. These carbon-bearing meteorites, called *carbonaceous chondrites,* were noted briefly in Chapter 16.

Often it is hard to determine if the carbon-based molecules found in meteorites are really extraterrestrial in origin or if they are mere terrestrial contaminants picked up while plowing through Earth's atmosphere or accumulated while sitting for years on Earth's surface. Fortunately, we know of a few meteorites containing organic matter that could not possibly be due to terrestrial contamination. The so-called **Murchison meteorite,** shown in Figure 30.13, which fell near Murchison, Australia, in 1969, is a carbonaceous chondrite that has been studied extensively. Located soon after crashing to the ground, this meteorite contains many of the well-known amino acids normally found in living cells.

The moderately large molecules found in meteorites and in interstellar clouds are our only evidence that chemical evolution is occurring or has occurred elsewhere in the universe. Most researchers regard this organic matter as prebiotic—that is, matter that could eventually lead to life but that has not yet done so. Very few researchers think that the molecules in meteorites and interstellar clouds could be the decayed remains of already established life that exists some place else in the Galaxy.

The planet most likely to harbor life is Mars, so similar to Earth in physical properties and distance from the Sun. All things considered, Mars seems harsh by Earth standards. Liquid water is scarce, the atmosphere is thin, and the lack of magnetism and an ozone layer allows the solar high-energy particles and ultraviolet radiation to reach the surface unabated. Earth life would have a hard time surviving there. However, our views become somewhat tempered once we realize than any Martian life would have originated, evolved, and adapted in that environment. Continued adaptation could presumably have made native Martian life possible, especially in the past when the atmosphere was thicker and the surface was warmer and much wetter, as we discussed in Chapter 12.

The search for life on Mars began in earnest when the *Viking* spacecraft arrived there in the mid-1970s. Each lander carried a television camera to seek fossilized remnants of large plants or animals. No fossils of any kind were seen. Each robot (see Figure 30.14) scooped up Martian soil and tested for life by conducting chemical experiments designed to detect the waste gases and other products of metabolic activity. No evidence for Martian life emerged. The initial positive signals produced by some of the tests have all been adequately explained by nonbiological chemical reactions.

The consensus among biologists and chemists today is that Mars does not house any life similar to that on Earth. However, some scientists suspect that a different type of biology might be operating on the Martian surface. They suggest that Martian bugs capable of eating and digesting oxygen-rich compounds in the Martian soil could also explain the *Viking* results. In addition, microbial life as we know it might reside in more habitable regions on Mars, such as near the moist polar caps. After all, the two *Viking* spacecraft landed on the safest Martian terrain, not in the most interesting regions. Finally, no one has yet searched for fossils from an ancient Martian era. A solid verdict re-

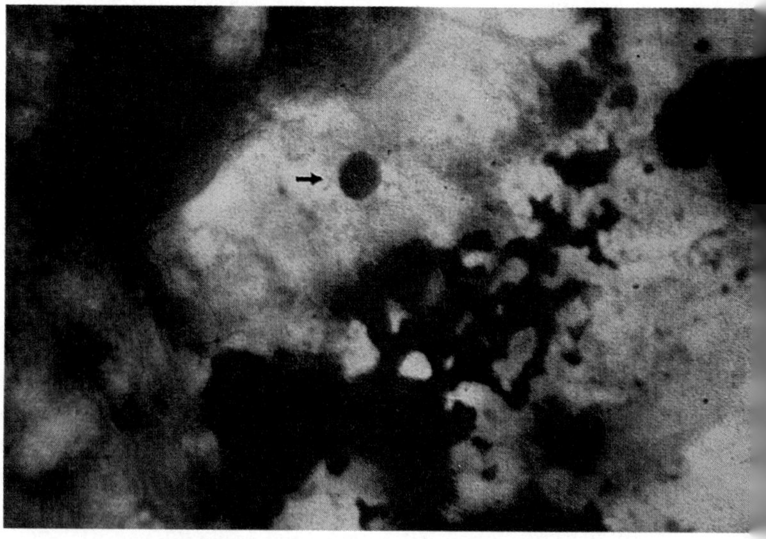

Figure 30.13 *The Murchison meteorite contains relatively large amounts of organic material, indicating that chemical evolution of some sort has occurred beyond our own planet. In this magnified view of a meteorite fragment, the arrow points to a microscopic sphere of organic matter. (Harvard-Smithsonian Center for Astrophysics)*

Figure 30.14 *A trench dug by the "arm" of one of the* Viking *robots can be seen at the right. Soil samples were scooped up and taken inside the robot, where instruments tested them for chemical composition and any signs of life. (NASA)*

garding life on Mars may not be reached until we have thoroughly explored our intriguing neighbor. The red planet may still hold some secrets.

Alternative Biochemistries

It seems that no environment in the solar system besides Earth is particularly well suited for sustaining life. Spacecraft that have landed on other solar system bodies or have flown by them have detected no evidence for extraterrestrial life. Yet, as we have seen, not all scientists completely dismiss the possibility of extraterrestrial life in our solar system. Some have even suggested that different types of biology may be at work out there, ones that we cannot recognize and that we do not know how to test for.

What other biologies might exist? We generally consider "life as we know it" to mean life based on carbon. Indeed, chemistry tells us that the carbon atom is an excellent atom from which to construct the more complex molecules needed for life. But carbon may not be the only possible building block. Life based on the abundant element silicon, however, appears at least chemically possible. Like carbon, silicon readily links with other elements, but a silicon bond is typically only about half as strong as a carbon bond. Silicon also forms chemical links with other silicon atoms, but the bonds are very much weaker than those connecting atoms of carbon. It seems most unlikely that silicon-based compounds could form the complex molecular chains that play such an important role in terrestrial biochemistry. Also, silicon-silicon bonds break apart in water, a medium in which carbon atoms remain tightly chained to each other. If life arose in water, then life is much more likely to be carbon-based than silicon-based.

Scientists generally believe that life on Earth originated in the oceans. A liquid medium better promotes the linking of atoms into life's complex molecules (certainly more so than a solid medium, in which movement is much more difficult, if not impossible). Several arguments favor water as the most reasonable liquid medium for life throughout the universe. The water molecule is made up of hydrogen and oxygen, both of which are plentiful throughout the cosmos. Also, water stays liquid over a very wide range of temperatures, and it bonds easily with many other molecules, making it more suitable for chemical and biological evolution than most other substances. But is there any real reason to believe that water is the *only* possible medium in which life could develop?

One possible alternative might be ammonia, which is made up of hydrogen and nitrogen. These atoms too are common in the universe, but ammonia must be colder than water to be liquid, suggesting that it might be a suitable medium on a cold planet. However, at lower temperatures, chemical reactions slow down. Life would evolve much more slowly within liquid ammonia than within liquid water. However, this doesn't necessarily mean that life could *not* evolve there, so we should perhaps consider ammonia as another possible life-giving medium, but one that would likely generate organisms with very different biochemistries.

Most biologists would argue that chemistry based on carbon and water is most likely to give rise to life. Carbon's flexible chemistry and water's wide liquid range are just what are needed for life to develop and thrive. Silicon and ammonia seem unlikely to fare as well as the basis for advanced life forms. Still, we must admit that we know next to nothing about noncarbon, nonwater biochemistries, for the very good reason that there are no examples of them to study experimentally.

PROSPECTS FOR INTELLIGENT LIFE IN THE GALAXY

[✓3] And so we look for life beyond our solar system, out into the Milky Way Galaxy and to other galaxies. At such distances, though, we cannot hope to detect actual life with our current equipment. The question before us becomes, How likely is it that life in any form—carbon-based, silicon-based, water-based, ammonia-based, or something we cannot even dream of—exists? The word *likely* in the last sentence speaks of probabilities. Let's look at some numbers to develop statistical estimates of the probability of life elsewhere in the universe.

The Green Bank Equation

An early version of this statistical problem was devised at a radio observatory in Green Bank, West Virginia, and so we refer to it as the **Green Bank Equation.** It is also known as the **Drake Equation,** after the U.S. astronomer who pioneered this analysis. As we will see, several of the terms in the equation are still largely a matter of opinion, so this formula is of little deterministic value, and it does nothing to constrain our personal prejudices. Its real value lies in the fact that it subdivides a large and very difficult question into smaller pieces that we can attempt to answer separately. It provides the framework within which the problem can be addressed and parcels out the responsibility for the final solution among many different scientific disciplines. The full equation appears below.

Figure 30.15 casts the Green Bank Equation into pictorial form. The figure illustrates how only a small fraction

the number of technological, intelligent civilizations now present in the Milky Way Galaxy = rate of star formation, averaged over the lifetime of the Galaxy × fraction of those stars having planetary systems × average number of planets within those planetary systems that are suitable for life × fraction of those habitable planets on which life actually arises × fraction of those life-bearing planets on which intelligence evolves × fraction of those intelligent-life planets that develop technological society × average lifetime of a technologically competent civilization.

(a)

(b)

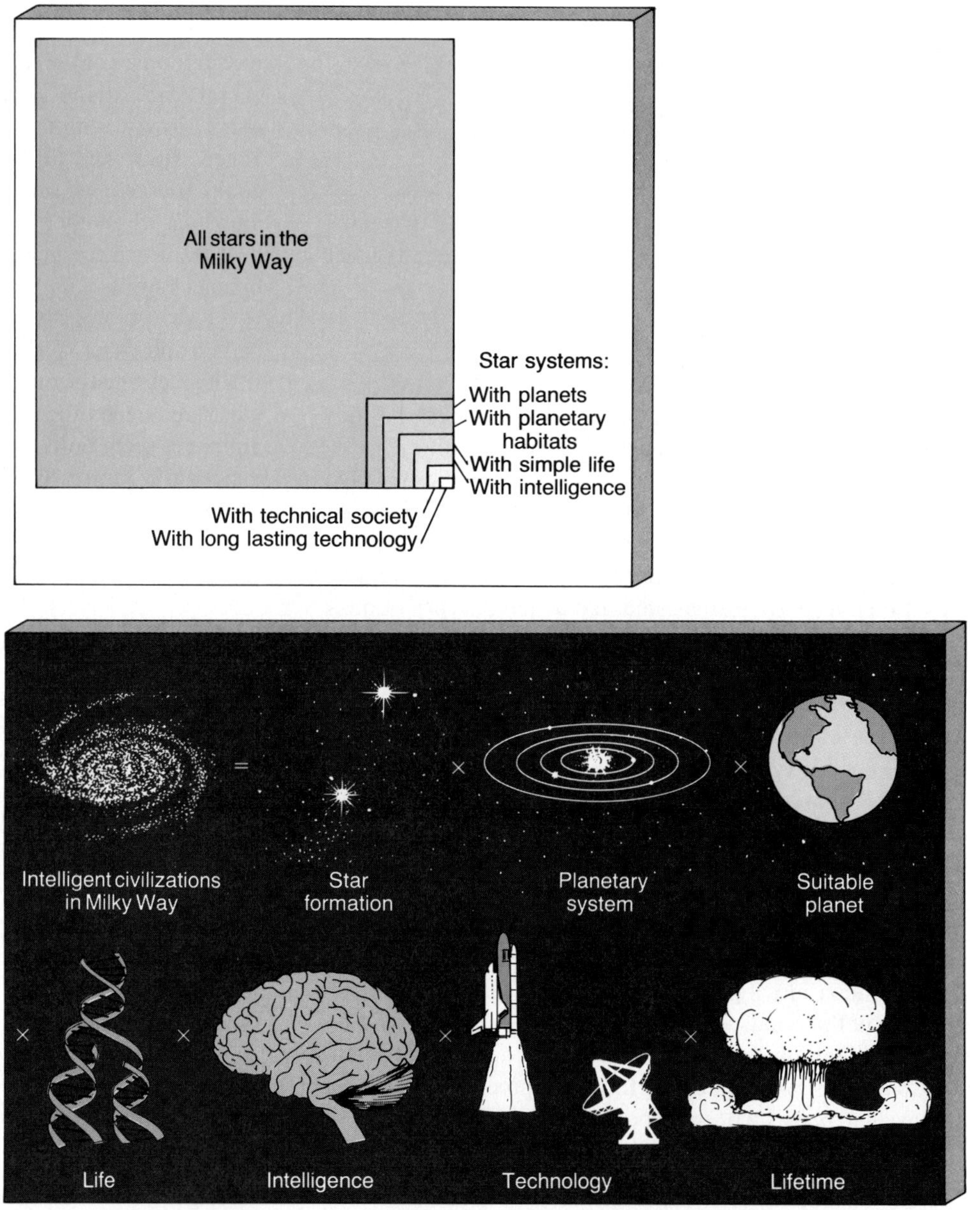

Figure 30.15 (a) *Of all the star systems in our Milky Way (represented by the largest box), progressively fewer and fewer have each of the qualities typical of a long-lasting technological society (represented by the smallest box at the lower right corner). (b) A pictorial illustration of the various terms in the Green Bank Equation.*

of star systems in the Milky Way are likely to generate the advanced qualities specified by the combination of terms on the right-hand side of the equation.

Evaluation of each of the terms in the Green Bank Equation requires familiarity with many fields of knowledge. We need a good deal of astronomy to estimate the rate of star formation and the fraction of stars having planets. More astronomy, along with some biology, is needed to specify the main properties of an ecologically suitable planet on which life might originate. We need insight into biology, chemistry, anthropology, and neurology (the study of the nervous system) to estimate the chances of any life at all, and then intelligent life, developing on any given planet. And finally, the lifetime of such an advanced civilization depends on a large number of additional factors, including history, politics, sociology, and psychology. Obviously we do not have all the information we need to be sure of these values. However, we can make some educated guesses in assigning values to these terms. Bear in mind, though, that if you ask two scientists for their best estimates of the values, you may well wind up with two quite different results. Let's now examine the terms in the equation one by one.

Rate of Star Formation

We can estimate the average number of stars forming each year in the Galaxy by noting that at least 100 billion stars now shine in the Milky Way. Dividing this number by the 10-billion-year lifetime of the Galaxy, we obtain a formation rate of 10 stars per year. This may be an overestimate because we think that fewer stars are forming now than formed at earlier epochs of the Galaxy, when more interstellar gas was available. However, we do know that many stars are forming today, and our estimate does not include stars that formed in the past and have since exploded, so our value of 10 stars per year seems reasonable when averaged over the lifetime of the Milky Way.

Fraction of Stars Having Planetary Systems

Many astronomers believe planet formation to be a natural result of the star-formation process. If the condensation theory is correct, and if there is nothing special about our Sun, as we have argued throughout this book, we would expect many stars to have at least one planet. Indeed, as we saw, increasingly sophisticated observations suggest the presence of disks around other young stars. Could these disks be proto-solar systems? The condensation theory suggests that the answer is yes.

Is there any direct evidence for planets circling other stars? At present, the answer is ambiguous. Certainly, there is no indication of Earth-like planets. The light reflected by any planet orbiting even the closest star would be too faint to detect with the very best equipment. This light would be lost in the glare of the parent star. Large orbiting telescopes may soon be able to detect Jupiter-sized planets circling the nearest stars. But even then these huge planets would barely be at the threshold of visibility, and the observations would not be able to provide direct evidence for Earth-like planets.

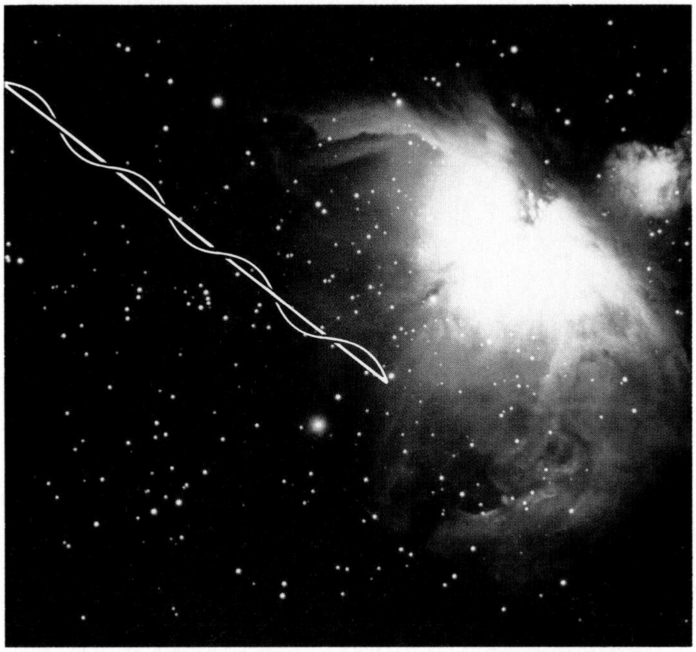

Figure 30.16 *Theoretically computed wobble in the path of a nearby, low-mass star having a Jupiter-sized object orbiting about it. The straight line is the path the star would take in the absence of any planets. The wavy curve is the resulting wobble in the star's motion caused by the to-and-fro gravitational pull of an unseen planet. For a dwarf star some 3 pc away, the expected deviation from a straight line would be a tiny 0.005 arc seconds. (NOAO)*

One possible indirect way of detecting a distant planet relies on the gravitational influence that all planets exert on their parent stars. A large planet orbiting a relatively low-mass star might produce a detectable change in that star's motion, even though the planet itself may be invisible to us. Figure 30.16 illustrates how this might look from Earth. The invisible planet pulls the star first one way and then the other during the course of its orbit. The result is a slight back-and-forth wobble in the visible star's path. Attempts to detect this wobble in Barnard's Star and other low-mass neighbors of the Sun have proven extremely frustrating, however. The amount of the back-and-forth shift is very small—only a few hundredths of an arcsecond for a Jupiter-sized object orbiting Barnard's Star at a distance of a few A.U.—and no unambiguous detection of the expected motion has yet been made for any star. Still, astronomers are confident that with steadily improving resolution they will one day detect a planet by this means.

Accepting the condensation theory and its consequences, and without being too conservative or naively optimistic, we assign a value near 1 to this term—that is, we believe that essentially all stars have planetary systems.

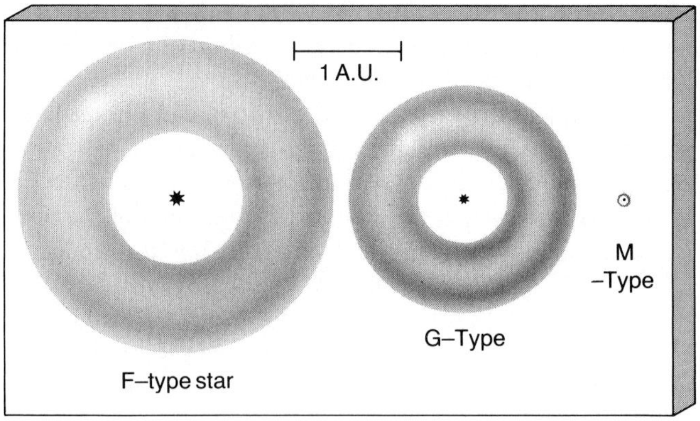

Figure 30.17 *The extent of the habitable zone is much larger around a hot star than around a cool one. For a star like the Sun (a G-type star), the zone extends from about 0.85 A.U. to 2.0 A.U. For an F-type star, the range is 1.2 to 2.8 A.U. For a faint M-type star only planets orbiting between about 0.02 and 0.06 A.U. would be habitable.*

Number of Habitable Planets per Planetary System

Temperature, more than any other single quantity, determines the feasibility of life on a given planet. The surface temperature of a planet depends on two things: the planet's distance from its parent star and the thickness of its atmosphere. Planets with a nearby parent star (but not too close) and some atmosphere (though not too thick) should be reasonably warm, like Earth or Mars. Planets far from the star and with no atmosphere, like Pluto, will surely be cold by our standards. And planets too close to the star and with a thick atmosphere, like Venus, will be very hot indeed.

Figure 30.17 illustrates that a three-dimensional zone of "comfortable" temperatures—often called a **habitable zone**—surrounds every star. It represents the range of radii within which an Earth-like planet would have a surface temperature between the freezing and boiling points of water. (Again, our Earth-based bias is evident.) The hotter the star, the larger this zone. For example, an A- or an F-type star has a rather large habitable zone. G-, K-, and M-type stars have increasingly smaller zones. O- and B-type stars are not considered here because they are not expected to last long enough for life to develop, even if they do have planets.

We can appreciate habitable zones by imagining regions surrounding various sources of heat on Earth. For example, skaters on a frozen lake know well the range of distance surrounding a bonfire (analogous to a star) where the warmth is comfortable. This range depends somewhat on the amount of clothing worn (analogous to a planet's atmosphere). Not much heat is felt far from a bonfire, but too much is felt when we come too close. Also, the larger the bonfire, the larger the zone of comfort and the more people who can benefit from the heat of the fire.

Three planets—Venus, Earth, and Mars—reside close to the habitable zone surrounding our Sun. Venus is too hot because of its thick atmosphere and proximity to the Sun. Mars is a little too cold because its atmosphere is too thin and it is too far from the Sun. It is interesting to note that if Venus had Mars's thin atmosphere and if Mars had Venus's thick atmosphere, both of these nearby planets might conceivably have surface conditions resembling those on Earth.

After taking inventory of how many stars of each type shine in our Galaxy and considering that stable planetary orbits are unlikely to occur in most binary-star systems (see Figure 30.18), we assign a value of 1/10 to this term. That is, we believe that 1/10 of all planetary systems that might exist in our Galaxy contain a potentially habitable planet. Single F-, G-, and K-type stars are the best candidates.

Fraction of Habitable Planets on Which Life Arises

How likely is it that life as we know it has arisen elsewhere? And how likely is it that life as we don't know it (say, silicon-based) might have arisen? We cannot answer these questions with any certainty. The number of possible combinations of atoms is incredibly large, so if the chemical reactions that led to the complex molecules like proteins or DNA that make up living organisms occurred completely at random, then it is extremely unlikely that those molecules could have formed at all, starting only with atoms or simple molecules. In that case, life is extraordinarily rare, and we are probably alone in the Galaxy, perhaps even in the entire observable universe.

However, laboratory experiments (like the Urey-Miller experiment described earlier) seem to suggest that certain chemical combinations are strongly favored over others. Of the billions upon billions of basic organic groupings that could possibly occur on Earth from the random combination of all sorts of simple atoms and molecules, only about 1500 actually do occur. Furthermore, these 1500 organic groups of terrestrial biology are made from only about 50 simple "building blocks." This suggests that molecules critical to life may not be assembled by pure chance. Apparently, additional forces are at work at the microscopic level. If that is the case, then a relatively small number of chemical "evolutionary tracks" are singled out, so that,

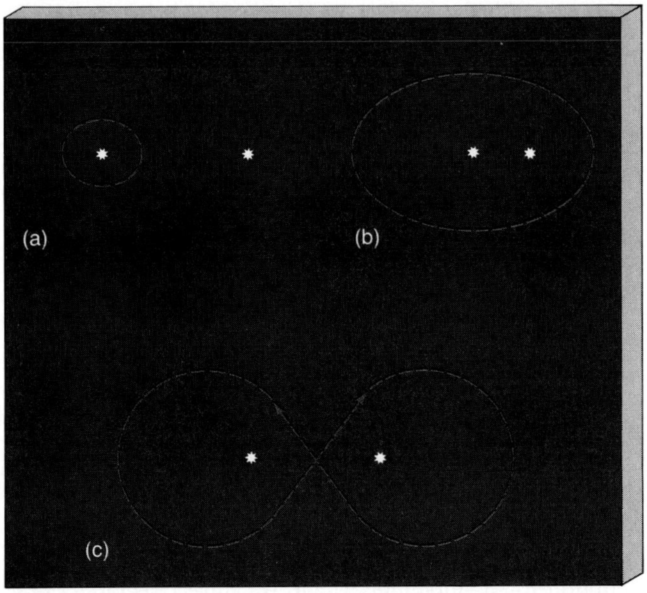

Figure 30.18 In binary-star systems, planets are restricted to only a few kinds of orbits that are gravitationally stable. For example, a planet is shown here (dashed curves) (a) closely orbiting one of the two stars and (b) circulating about both stars in an elliptical orbit. Another possible, but unstable path, (c) interweaves between the two stars in a "figure-8" pattern.

given the amount of time available, the formation of complex molecules—and hence, we assume, life—becomes much more likely.

To assign a value of 0 to this term in the equation is to believe that life arises randomly and rarely. To assign a value of 1 is to believe that life is inevitable, given the proper ingredients, a suitable environment, and a long enough period of time. No easy experiment can distinguish between these extreme alternatives, and there seems to be little or no middle ground. In the minds of many researchers, the discovery of life on Mars, Jupiter, Titan, or some other object in our solar system would convert the appearance of life from an unlikely miracle to a virtual certainty throughout the Galaxy. We will take the optimistic view, and adopt a value of 1.

Fraction of Life-bearing Planets on Which Intelligence Arises

As with the evolution of life, the appearance of a well-developed brain is a very unlikely event if only random chance is involved. However, biological evolution through natural selection is a mechanism that generates apparently highly improbable results by singling out and refining useful characteristics. Intelligence results from a series of adaptations. Organisms that profitably use those adaptations can develop more complex behavior, and complex behavior provides organisms with the *variety* of choices needed for more advanced development.

Members of one school of thought maintain that, given enough time, intelligence is inevitable. They argue that, assuming that natural selection is a universal phenomenon, at least one organism on a planet will always rise to the level of "intelligent life." If this view is correct, the fifth term in our Green Bank Equation equals or nearly equals 1. Other researchers argue that there is only one known case of intelligence, and that case is life on Earth. For 2.5 billion years—from the start of life about 3.5 billion years ago to the first appearance of multicellular organisms about 1 billion years ago—life did not advance beyond the one-celled stage. Life remained simple, and dumb, but it survived. If this view is correct, the fifth term in our equation is very small. In that case, we are faced with the (somewhat depressing) prospect that humans might be the smartest form of life anywhere in the Galaxy. As with the previous term, we will take the optimistic view and adopt a value of 1 here.

Fraction of Planets on Which Intelligent Life Develops Technology

To evaluate the sixth term of our equation, we need to estimate the probability that intelligent life eventually develops technological competence. Should the rise of technology be inevitable, this term is close to 1, given long enough periods of time. If it is not inevitable—if intelligent life can somehow "avoid" developing technology—then this term could be much less than 1. The latter possibility envisions a universe possibly teeming with intelligent civilizations, but very few among them ever becoming technologically competent. Perhaps only one managed it—ours.

Again, it is difficult to distinguish between these two views. We don't know how many prehistoric Earth cultures failed to develop technology, or rejected its use. We do know that the roots of our present civilization arose independently at several different places on Earth, including Mesopotamia, India, China, Egypt, Mexico, and Peru. Because so many of these ancient cultures originated at about the same time, it is tempting to conclude that the chances are good that some sort of technological society will inevitably develop, given some basic intelligence and enough time.

If technology is inevitable, why haven't other life forms on Earth also found it useful? Probably the competitive edge given to humans by intellectual and technological skills allowed them to rapidly dominate the planet. The fact that only one technological society exists on Earth does not imply that the sixth term in our Green Bank Equation must be very much less than 1. On the contrary, it is precisely because *some* species will probably always fill the niche of technological intelligence that we will take this term to be close to 1.

Average Lifetime of a Technological Civilization

The reliability of the estimate of each term in the Green Bank Equation declines markedly from left to right. For example, our knowledge of astronomy enables us to make a reasonably good stab at the first term, namely the rate of star formation in our Galaxy, but it is much harder to evaluate some of the later terms, such as the fraction of life-bearing planets that eventually develop intelligence. As for the term on the far right side of the equation, the longevity of technological civilizations is totally unknown. There is only one known example of such a civilization—that's us on planet Earth—and how long we will be around before a natural or humanmade catastrophe ends it all is impossible to tell.

One thing is certain: If the correct value for any one term in the equation is very small, then few technological civilizations now exist in the Galaxy. If the pessimistic view of the development of life or of intelligence is correct, then we are unique, and that is the end of our story. However, if both life and intelligence are inevitable consequences of chemical and biological evolution, as many scientists believe, and if intelligent life always becomes technological, then we can plug the higher, more optimistic values into the Green Bank Equation. In that case, the number of technological galactic civilizations is $10 \times 1 \times 1/10 \times 1 \times 1 \times 1 = 1 \times$ (the average lifetime of a civilization, in years). Thus, if civilizations typically survive for 1000 years, there should be 1000 of them currently in existence in the Galaxy. If they live for a million years, on average, we would expect there to be a million advanced civilizations in the Milky Way, and so on. Our own civilization has presently survived in its "technological" state for only approximately 100 years.

How long does an advanced civilization last? This term is the most difficult to estimate, for it requires that we speculate about the future evolution of our own culture. It is easy to be pessimistic about our chances of surviving even the next century, let alone the next millennium. Population growth, environmental pollution, depletion of natural resources, shortages of food and energy, the possibility of nuclear war—all threaten the existence of our civilization, and perhaps life on Earth itself. Still, the end of the Cold War in 1991 and the growing awareness among nations during the 1990s of the fragility of our planetary environment at least offer grounds for hope.

Are the difficulties we now face as a global civilization just a threshold beyond which we may be able to endure stably for many millennia, or will we simply encounter new problems as we go on, each of them fatal to our civilization if no solution is found? And as we solve each, what sort of society will result? Again, of course, we have no way of knowing. The only firm statement we can make is that predictions of the future are almost always wildly inaccurate (see Figure 30.19). We have only to look at previous generations' visions of life today, only a few decades in their future, to be immediately skeptical of any predictions looking centuries or millennia ahead.

Let us simply continue our optimistic assessment of the prospects for life and assume that, once their initial technological "teething problems" are past, civilizations enjoy a long stay on their parent planet, so they are likely to be plentiful in the Galaxy. In that case, how might we become aware of their existence? Is two-way communication with other planetary cultures possible—or, for that matter, even desirable?

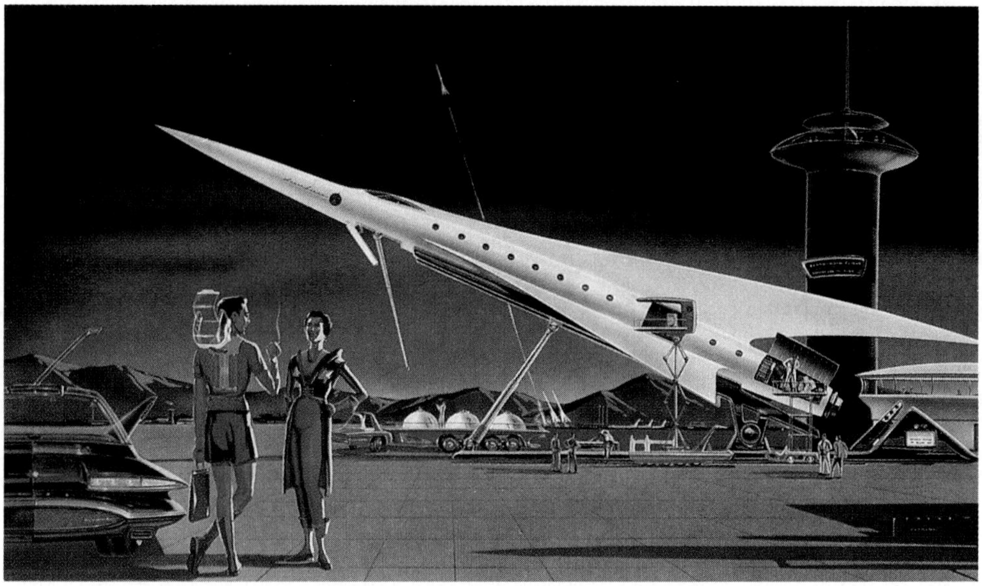

Figure 30.19 *A 1956 artist's conception of everyday life in the year 2000.*

THE SEARCH FOR EXTRATERRESTRIAL INTELLIGENCE

How Far to Our Neighbors?

✓4 For concreteness, let us imagine that the average lifetime of a technological civilization is 1 million years—only 1 percent of the reign of the dinosaurs, but 100 times longer than human civilization has survived thus far. Given the size and shape of our Galaxy, we can then estimate the average *distance* between these civilizations to be some 50 pc, or about 150 light-years. Thus, any two-way communication with our neighbor—using signals traveling at or below the speed of light—will take at least 300 years. Notice that, as the average lifetime decreases, the distance between neighbors goes up because civilizations become more thinly spread over the Galaxy. If the lifetime is less than about 3000 years, civilizations will die, on average, before any two-way conversation can be established. This then is the *minimum* lifetime for any dialog between civilizations to be possible.

Space Flight

One obvious way to search for extraterrestrial life would be to develop the capability to travel far outside our solar system. By involving many nations in the space programs begun by the United States and the former Soviet Union, our civilization might eventually develop the means to travel through interstellar space. However, such technology will not be achieved easily. It may not even be a practical possibility. The basic problems with interstellar space travel are the time and fuel requirements of the journey to even the nearest star. At a speed of 50 km/s, the speed of the current fastest unmanned space probes, the trip to our nearest neighbor, Alpha Centauri, would take about 25,000 years. The journey to the nearest civilization (assuming, as above, that a total of 1 million such cultures exist in the Galaxy) would take almost 1 million years. Interstellar travel at these speeds is clearly not feasible.

Could we remedy this by accelerating our ship to relativistic velocities—speeds close to that of light? (Despite the fervent hopes of science-fiction writers, the speed of light is still the upper limit on interstellar travel speeds.) This would certainly reduce the travel time, but the fuel requirements to boost a 10-ton spaceship to even 1⁄10 the speed of light are enormous—many times the mass of the ship itself, assuming that fusion is the energy source and that all the energy is converted into usable form. Schemes have been suggested to scoop up interstellar hydrogen as fuel, but these are far beyond our present technology.

The damaging effect of galactic radiation is an additional problem, as are the loneliness and boredom of many generations of humans having to spend their entire lives on board a spacecraft. As an alternative, interstellar robot probes could be programmed to orbit a particular star system and to listen for electromagnetic radiation emitted from a technological civilization. These probes would eliminate the dangers of human space travel, but they are still prohibitively expensive. We can hardly hope to use space flight as a practical means of making contact with extraterrestrials, either now or in the foreseeable future.

In a sense, our civilization has already launched several simple interstellar probes, although they lack the sophistication of the robots just mentioned and have no specific stellar destination. Figure 30.20 is a reproduction of a plaque mounted on board the *Pioneer 10* spacecraft launched in the mid-1970s and now well beyond the orbit of Pluto, with enough velocity to escape the solar system forever. Similar information was also included aboard the *Voyager* spacecraft launched in 1978.

Even if these spacecraft do accidentally encounter an alien star system housing an advanced civilization, these machines are incapable of reporting that news back to Earth. If the civilization on the other end intercepts the probe, it should be able to unravel most of its contents using the universal language of mathematics. The caption to Figure 30.20 notes how the aliens might discover from where and when the *Pioneer* and *Voyager* probes were originally launched. They would then know that we are here (or were when the probes were sent), although we would be unaware of their existence.

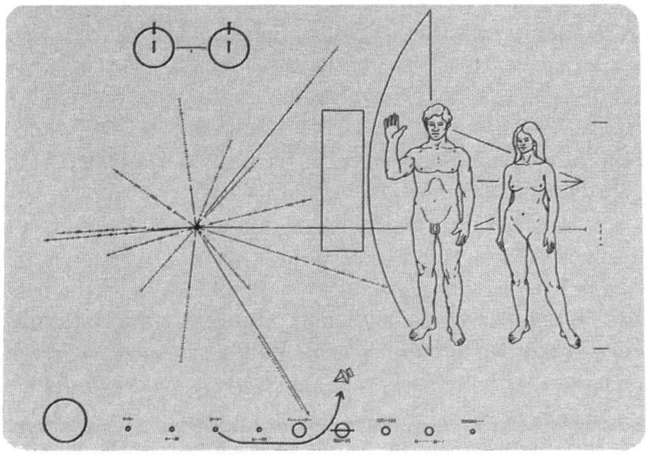

Figure 30.20 *A replica of a plaque mounted on board the* Pioneer 10 *spacecraft. The important features of the plaque include: a scale drawing of the spacecraft, a man, and a woman; a diagram of the hydrogen atom undergoing a change in energy (top left); a starburst pattern representing various pulsars and the frequencies of their radio waves that can be used to estimate when the craft was launched (middle left); a depiction of the solar system, showing that the spacecraft departed the third planet from the Sun and passed the fifth planet on its way into outer space (bottom). All the drawings have computer- (binary) coded markings from which actual sizes, distances, and times can be derived.*

Should We Make Contact?

Aside from these practical problems, some scientists have argued that it might not be a particularly good idea to signal extraterrestrials actively. Our recent emergence as a technological civilization implies that we must be one of the least advanced technological intelligences in the entire Galaxy. Any other civilization that we discover, or that discovers us, will almost surely be more advanced than us. Consequently, a healthy degree of caution is warranted. If extraterrestrials behave even remotely like human civilizations on Earth, then the most advanced aliens might naturally try to dominate all others. The behavior of the "advanced" European cultures toward the "primitive" races they encountered on their voyages of discovery in the seventeenth, eighteenth, and nineteenth centuries should serve as a clear warning of the possible undesirable consequences of contact. Of course, the aggressiveness of Earthlings may not apply to extraterrestrials, but given the history of the one intelligent species we know, the cautious approach may be in order.

Radio Communication

An alternative, and possibly safer, search technique is also much cheaper than interstellar travel, by human or by robot probe. With radio communication we would strive to make contact with extraterrestrials using only electromagnetic radiation, the fastest known means of transferring information from one place to another. Because light and other high-frequency radiation are heavily scattered while moving through dusty interstellar space, long-wavelength radio radiation seems to be the best choice. Radio telescopes on Earth would listen passively for radio signals emitted by other civilizations. We ourselves would not transmit radiation. Some preliminary searches of selected nearby stars are now under way, thus far without success.

In what direction should we aim our radio telescopes? The answer to this question, at least, is fairly easy. On the basis of our earlier reasoning, we should target all F-, G-, and K-type stars in our vicinity. But are extraterrestrials even broadcasting radio signals? If they are not, this search technique will obviously fail. If they are, how do we distinguish their artificially generated radio signals from signals naturally emitted by interstellar gas clouds? At what frequency should we tune our receivers? The answer depends on whether the signals are produced deliberately or are simply "waste radiation" escaping from a planet.

Consider how Earth would look at radio wavelengths to extraterrestrials. Figure 30.21 shows the pattern of radio signals we emit into space. From the viewpoint of a distant observer, the spinning Earth emits a bright flash of radio radiation every few hours. The flashes result from the periodic rising and setting of hundreds of FM radio stations and television transmitters (marked in Figure 30.21[a]), which broadcast mostly parallel to Earth's surface, so that each of them sends a great "sheet" of electromagnetic radiation into interstellar space, as illustrated in Figure 30.21(b). (The more common AM broadcasts are trapped below our ionosphere, so those signals never leave Earth.) Because the great majority of these transmitters are clustered in the eastern United States and western Europe, a distant observer would detect blasts of radiation from Earth as our planet rotates each day (Figure 30.21[c]). This radiation races out into space, and has been doing so since the invention of these technologies about six decades ago. In fact, Earth is now a more intense radio emitter than the Sun. Another civilization as advanced as ours might have constructed devices capable of detecting this radiation. If any sufficiently advanced (and sufficiently interested) civilization resides within about 60 light-years (20 pc) of Earth, then we have already broadcast our presence to them.

We currently have the engineering ability to build a radio system that could intercept the radio, radar, and television signals that leak into space from distant civilizations at our level of development. Figure 30.22 shows the major features of a gigantic array of radio telescopes capable of eavesdropping on Earth-like civilizations within 1000 light-years of us. Called Project Cyclops, this huge machine hasn't been built yet. Indeed, it may never be built—the price tag for such a device is at least $50 billion (in 1991 dollars).

The "Water Hole"

Now let us suppose that a civilization has decided to assist searchers by actively broadcasting its presence to the rest of the Galaxy. At what frequency should we listen for such an extraterrestrial beacon? The electromagnetic spectrum is enormous; the radio domain alone is vast. To hope to detect a signal at some unknown radio frequency is like searching for a needle in a haystack. Might there be frequencies more likely than others to carry alien transmissions?

Some basic arguments suggest that civilizations will probably communicate at a wavelength of nearly 20 cm. As we saw in Chapter 20, the basic building blocks of the universe, namely hydrogen atoms, naturally radiate at a wavelength of 21 cm. Also, one of the simplest molecules, hydroxyl (OH), radiates near 18 cm. Together, these two substances form water (H_2O). Arguing that water is likely to be the interaction medium for life anywhere, some researchers have proposed that the interval between 18 and 21 cm is the best wavelength range for civilizations to transmit or listen. Called the **water hole,** this radio interval might serve as an oasis where all advanced galactic civilizations would gather to conduct their electromagnetic business.

This water-hole frequency interval is only a guess, of course, but it is supported by other arguments as well. Figure 30.23 shows the water hole's location in the electromagnetic spectrum and plots the amount of natural emission from our Galaxy and from Earth's atmosphere. The 18 to 21 cm range lies within the quietest part of the spectrum, where the galactic "static" from stars and interstellar clouds hap-

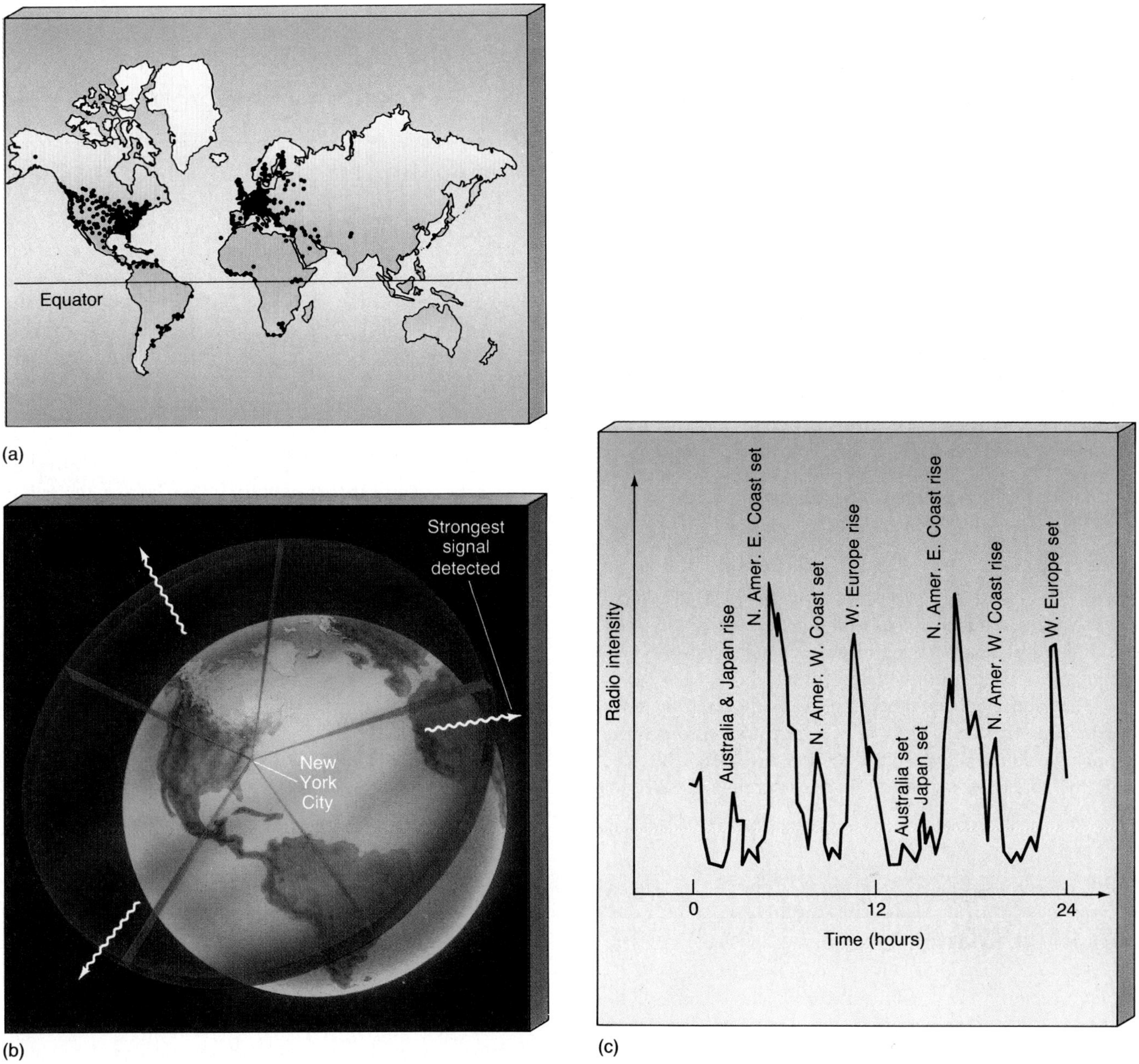

Figure 30.21 *Radio radiation now leaks from Earth into space because of the daily activities of our technological civilization. (a) Most of the radiation arises from FM radio stations and television transmitters (shown here as dots), most of them found in western Europe and North America. (b) These stations broadcast their energy parallel to Earth's surface, so they produce the strongest signal in any given direction when they happen to lie on Earth's horizon, as seen from that direction. The particular pattern (c) resulting from the sum of all Earth's television stations, as viewed from Barnard's Star, a few parsecs away.*

pens to be minimized. Furthermore, the atmospheres of typical planets are also expected to interfere least at these wavelengths. Thus the water hole seems like a good choice for the frequency of an interstellar beacon, although we cannot be sure of this reasoning until contact is actually achieved. Perhaps some other wavelength is better for reasons unknown to us at this time.

A few radio searches are now in progress at frequencies in and around the water hole. Thus far, nothing resembling an extraterrestrial signal has been detected. Proponents of continuing the search argue that we have good reason to suspect that extraterrestrials exist somewhere, given our development here on Earth. They admit that we have only a very small chance of making contact in the near

Figure 30.22 This is a design for a gigantic array of a thousand interconnected radio telescopes, called Project Cyclops. Such a device could be used to eavesdrop on civilizations similar to our own, provided they lie within about a thousand light-years of Earth. (NASA)

future, but they argue that now is the time to test the theory that advanced civilizations inhabit the Galaxy. To fail to try is to commit the cardinal sin of pre-Renaissance workers—thinking without experimentally testing.

The space surrounding all of us could be, right now, flooded with radio signals from extraterrestrial civilizations. If only we knew the proper direction and frequency, we might be able to make one of the most startling discoveries of all time. The result would likely provide whole new opportunities to study the cosmic evolution of energy, matter, and life throughout the universe.

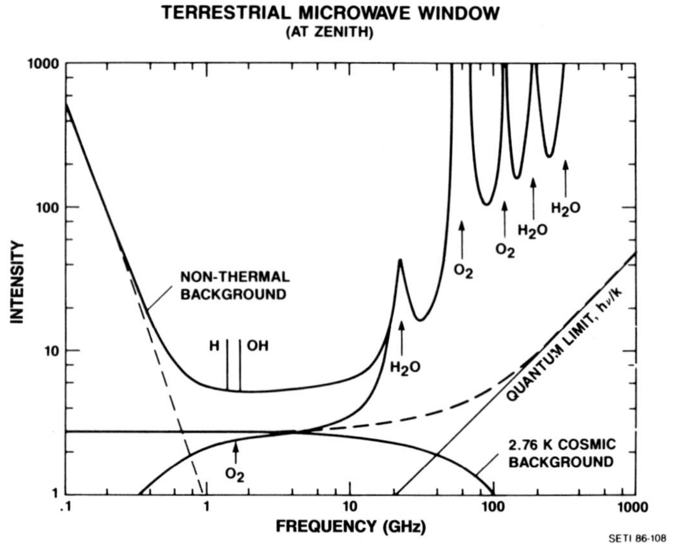

Figure 30.23 *The "water hole" is bounded by the natural emission frequencies of the hydrogen (H) atom (21-cm wavelength) and the hydroxyl (OH) molecule (18-cm wavelength). The top most solid curve sums the natural emissions of our Galaxy (leftside of diagram labeled "non-thermal background") and Earth's atmosphere (rightside of diagram denoted by various chemical symbols). This sum is minimized near the water-hole frequencies. Perhaps all intelligent civilizations conduct their interstellar communications within this quiet "electromagnetic oasis."*

CHAPTER SUMMARY

We can identify seven major phases in the history of the universe: galactic, stellar, planetary, chemical, biological, cultural, and future evolution. The natural processes that created the galaxies, stars, and planets may also have led to conditions ripe for life. Complex molecules could have evolved naturally from simpler ingredients found on the primitive Earth. It is also possible, but unproven, that some of the basic molecules needed to make proteins and nucleic acids might have been formed in interstellar space. No living cells have yet been created "from scratch" in any laboratory, but most biochemists feel that the chain of events leading from simple nonbiological molecules almost to the point of life itself has been amply demonstrated.

All scientists accept the reality of biological evolution. Organisms that can best take advantage of their new surroundings succeed at the expense of those organisms that cannot make the necessary adjustments. Time is crucial to the evolution of life, intelligence, and civilization. Intelligence is strongly favored by natural selection. We assume that intelligent life will give rise to civilization, and possibly to technology as well. Because the competitive edge given to humans by intellectual and technological skills allowed them to dominate this planet so rapidly, it is conceivable that other planets with intelligent life will eventually develop technology.

If there is life elsewhere in the solar system, the planet most likely to harbor it is Mars. Although no evidence for Martian life has been discovered, more exploration is needed before it can be conclusively ruled out. Biologists argue that chemistry based on carbon and water is most likely to give rise to life. Silicon and ammonia are possible, but perhaps far-fetched, alternatives.

It is possible to examine the statistical probabilities of life in the universe. The Green Bank Equation subdivides large and very difficult questions into small pieces that can be approached separately. Because at least 100 billion stars now shine in the Milky Way and because the Galaxy is thought to be about 10 billion years old, a reasonable star formation rate is about 10 new stars per year. The condensation theory of star formation suggests that essentially all stars have planetary systems, as planets are natural byproducts of star formation. A three-dimensional "habitable zone" surrounds every star. Habitable planets may have to lie within this zone. However, astronomers have no information about what fraction of those habitable planets may actually support life.

If technological civilizations typically live for only a short time, there are few such civilizations in the Galaxy at any given instant. If technological civilizations typically live a long time, then the Galaxy may be teeming with intelligent life. As the average lifetime of galactic civilizations decreases, the distance between neighbors increases. It is quite possible that civilizations may die before any two-way conversation can be established.

The basic problems with interstellar space travel are the time and fuel requirements of the journey to even the nearest star. Space flight is not a practical means of making contact with extraterrestrials, now or in the foreseeable future. Long-wavelength radio waves are the best choice we know for the transfer of information across galactic distances. Current radio systems could intercept radio, radar, and television signals leaking into space from distant civilizations. Hydrogen atoms radiate at a wavelength of 21 cm. Hydroxyl (OH) radiates near 18 cm. Together, these two substances form water (H_2O). If water is the interaction medium for life anywhere, the radio interval between 18 and 21 cm may be the best wavelength range for civilizations to transmit or listen. This radio interval is often called the water hole.

KEY WORDS

amino acid
biological evolution
carbonaceous chondrite
chemical evolution
cosmic evolution
cultural evolution
Drake Equation
Green Bank Equation
habitable zone
Murchison meteorite
nucleotide base
organic molecule
Urey-Miller experiment
water hole

REVIEW QUESTIONS

1. What is the Urey-Miller experiment?
2. How do we know anything at all about the early episodes of life on Earth?
3. What do many scientists think caused the extinction of the dinosaurs?
4. What is the role of language in cultural evolution?
5. Is extraterrestrial life thought to be common or rare?
6. Where—besides the planet Mars—might we find signs of life in our solar system?
7. What is the status of our search for life on Mars?
8. What is generally meant by "life as we know it"? What other forms of life might be possible?
9. What is the evidence for life beyond the solar system?
10. Describe a quantitative way in which scientists approach the question of life beyond the solar system.
11. How many of the terms in the Green Bank Equation are known with a fair degree of certainty?
12. What is the relationship between the average lifetime of galactic civilizations and the possibility of our someday communicating with them?
13. How long would it take our fastest current space probes to travel to the nearest star, Alpha Centauri? How long would it take to reach the nearest of a million hypothetical civilizations now populating the Galaxy?
14. How would Earth look at radio wavelengths to extraterrestrials?
15. What is the water hole?
16. Suppose that each of the "fraction" terms in the Green Bank Equation turns out to have a value of $1/10$, that stars form at an average rate of 20 per year, and that each star has exactly 1 habitable planet orbiting it. Estimate the present number of technological civilizations in the Milky Way Galaxy if the average lifetime of a civilization is (a) 100 years, (b) 10,000 years, (c) 1,000,000 years.
17. If the 4.5-billion-year age of the Earth were compressed to 45 years, as described in the text, what would be your age, in seconds? How long ago was the end of World War II? The Declaration of Independence? Columbus's discovery of the New World?
18. Assuming that there are 10,000 FM radio stations on Earth, each transmitting at a power level of 50 kilowatts (recall that 1 watt is 10^7 erg/s), calculate the total radio luminosity of the Earth in the FM band. Compare this with the 10^{13} erg/s radiated by the Sun in the same frequency range.

DISCUSSION QUESTIONS

1. Would our human culture would be altered profoundly—or only slightly—by the sure knowledge of extraterrestrial life forms? Tell why you think so.
2. What do you think of the assertion that advanced galactic civilizations might naturally try to dominate "primitive" civilizations such as ours? Do you think we should try to signal other life in the Galaxy?
3. For thousands of years, people lived out their lives in cosmological frameworks very different from our modern view. Suppose that in another few thousand years much of what you have learned in this book has been replaced by greater truths in a larger cosmological framework. What would this mean for our search for extraterrestrial intelligence? Does the search for life beyond Earth depend not only on alien intelligence having evolved a technology similar to ours, but also on its having evolved a similar cosmology?

SUGGESTED READINGS

Chaisson, E., *Cosmic Dawn:* The Origins of Matter and Life. New York: W.W. Norton, 1989. An introductory account of cosmic evolution.

Goldsmith, D., ed. *The Quest for Extraterrestrial Life*. Mill Valley, CA: University Science Books, 1980. A variety of fascinating readings.

Kutter, G. S. *The Universe and Life: Origins and Evolution*. Boston: Jones and Bartlett, 1987. A readable treatise on both the physical and biological aspects of cosmic evolution.

Miller, S. L., and C. Chyba. "Whence Came Life?" *Sky and Telescope* (June 1992). Two researchers present modern alternative views on the origin of life on Earth.

Naeye, R. *SETI at the Crossroads*. NASAs latest foray into the search for extraterrestrial intelligence.

Regis, E. *Extraterrestrials: Science and Alien Intelligence*. Cambridge: Cambridge University Press, 1985. Over a dozen scientists present their views on the subject.

Sagan, C. *The Cosmic Connection*. New York, Dell, 1973. Short, visionary book presenting life from a cosmic perspective.

Shklovskii, I. S., and C. Sagan. *Intelligent Life in the Universe*. New York: Dell, 1968. This classic book is a translation, extension, and revision of Shklovskii's earlier work called *Universe, Life and Mind*. It presents many of the basic questions concerning life in the universe.

Sullivan, W. T., S. Brown, and C. Wetherill. "Eavesdropping: The Radio Signature of the Earth." *Science* (January 27, 1978). Classic article describing how Earth's radio emissions would appear to extraterrestrial civilizations.

Appendix Tables

TABLE 1 ***Some Useful Constants and Physical Measurements****

1 astronomical unit
A.U. = 1.496×10^8 km (1.5×10^8 km)

1 light-year
ly = 9.46×10^{12} km (10^{13} km; 6 trillion miles)

1 parsec
pc = 3.09×10^{18} cm = 3.3 ly

speed of light
$c = 299{,}792 \pm 1$ km/s (3×10^5 km/s)

gravitational constant
$G = 6.67 \times 10^{-8}$ cm^3/g/s^2

Stefan-Boltzman constant
σ [Greek sigma] = 5.67×10^{-5} erg/s/cm^2/K^4

Planck's constant
$h = 6.63 \times 10^{-27}$ erg s

mass of the Earth
$M_\oplus = 5.97 \times 10^{27}$ g (6×10^{27} g; about 6000 billion billion tons)

radius of the Earth
$R_\oplus$ = 6378 km (6500 km)

mass of the Sun
$M_\odot = 1.99 \times 10^{33}$ g (2×10^{33} g)

radius of the Sun
$R_\odot = 6.96 \times 10^5$ km (7×10^5 km)

luminosity of the Sun
$L_\odot = 3.90 \times 10^{33}$ ergs/s (4×10^{33} erg/s)

effective temperature of the Sun
$T_\odot$ = 5778 K (5800 K)

Hubble constant
H_o = 75 km/s/Mpc

mass of the electron
$m_e = 9.11 \times 10^{-28}$ g

mass of the proton
$m_p = 1.67 \times 10^{-24}$ g

*The rounded-off values used in the text are shown above in parentheses.

TABLE 2 *Periodic Table of the Elements*

Atomic mass → 1.00797; Atomic number → 1 H ← Symbol; Hydrogen ← Name

Metals | Nonmetals

1.00797 1 H Hydrogen																1.00797 1 H Hydrogen	4.0026 2 He Helium
6.939 3 Li Lithium	9.012 4 Be Beryllium											10.811 5 B Boron	12.01 6 C Carbon	14.01 7 N Nitrogen	15.999 8 O Oxygen	18.998 9 F Fluorine	20.182 10 Ne Neon
22.99 11 Na Sodium	24.31 12 Mg Magnesium											26.98 13 Al Aluminum	28.09 14 Si Silicon	30.97 15 P Phosphorus	32.06 16 S Sulfur	35.45 17 Cl Chlorine	39.95 18 Ar Argon
39.10 19 K Potassium	40.08 20 Ca Calcium	44.96 21 Sc Scandium	47.90 22 Ti Titanium	50.94 23 V Vanadium	52.00 24 Cr Chromium	54.94 25 Mn Manganese	55.85 26 Fe Iron	58.93 27 Co Cobalt	58.71 28 Ni Nickel	63.54 29 Cu Copper	65.37 30 Zn Zinc	69.72 31 Ga Gallium	72.59 32 Ge Germanium	74.92 33 As Arsenic	78.96 34 Se Selenium	79.91 35 Br Bromine	83.80 36 Kr Krypton
85.47 37 Rb Rubidium	87.62 38 Sr Strontium	88.91 39 Y Yttrium	91.22 40 Zr Zirconium	92.91 41 Nb Niobium	95.94 42 Mo Molybdenum	(98) 43 Tc Technetium	101.07 44 Ru Ruthenium	102.9 45 Rh Rhodium	106.4 46 Pd Palladium	107.9 47 Ag Silver	112.4 48 Cd Cadmium	114.82 49 In Indium	118.69 50 Sn Tin	121.75 51 Sb Antimony	127.6 52 Te Tellurium	126.90 53 I Iodine	131.30 54 Xe Xenon
132.91 55 Cs Cesium	137.34 56 Ba Barium	Lanthanides 57–71	178.49 72 Hf Hafnium	180.95 73 Ta Tantalum	183.85 74 W Tungsten	186.2 75 Re Rhenium	190.2 76 Os Osmium	192.2 77 Ir Iridium	195.09 78 Pt Platinum	196.97 79 Au Gold	200.59 80 Hg Mercury	204.37 81 Tl Thallium	207.19 82 Pb Lead	208.98 83 Bi Bismuth	(210) 84 Po Polonium	(210) 85 At Astatine	(222) 86 Rn Radon
(223) 87 Fr Francium	(226) 88 Ra Radium	Actinides 89–103															

Lanthanides →	138.91 57 La Lanthanum	140.12 58 Ce Cerium	140.91 59 Pr Praseodymium	144.24 60 Nd Neodymium	(145) 61 Pm Promethium	150.4 62 Sm Samarium	151.96 63 Eu Europium	157.25 64 Gd Gadolinium	158.9 65 Tb Terbium	162.5 66 Dy Dysprosium	164.9 67 Ho Holmium	167.3 68 Er Erbium	168.9 69 Tm Thulium	173.0 70 Yb Ytterbium	175.0 71 Lu Lutetium
Actinides →	(227) 89 Ac Actinium	232.04 90 Th Thorium	(231) 91 Pa Protactinium	238.03 92 U Uranium	(237) 93 Np Neptunium	(242) 94 Pu Plutonium	(243) 95 Am Americium	(247) 96 Cm Curium	(247) 97 Bk Berkelium	(251) 98 Cf Californium	(254) 99 Es Einsteinium	(253) 100 Fm Fermium	(256) 101 Md Mendelevium	(253) 102 No Nobelium	(257) 103 Lr Lawrencium

†Atomic masses given in parentheses refer to the most stable isotope of an unstable element.

TABLE 3 *Planetary Data*

Planet	Semi-major Axis (A.U.)	Semi-major Axis (10^6 km)	Sidereal Period (tropical years)	Sidereal Period (days)	Synodic Period (days)	Mean Orbital Velocity (km/s)	Eccentricity e
Mercury	0.39	57.9	0.24	87.97	115.88	47.89	0.206
Venus	0.72	108.2	0.62	224.70	583.82	35.03	0.007
Earth	1.00	149.6	1.00	365.26		29.79	0.017
Mars	1.52	227.9	1.88	686.98	779.94	24.13	0.09
Jupiter	5.203	778.3	11.86		398.88	13.06	0.05
Saturn	9.54	1427.0	29.46		378.09	9.64	0.06
Uranus	19.19	2869.6	84.07		369.66	6.81	0.05
Neptune	30.06	4496.6	164.82		367.49	5.43	0.01
Pluto	39.53	5914	248.6		366.73	4.74	0.25

Planet	Inclination to the Ecliptic (degrees)	Perihelion (A.U.)	Perihelion (km)	Aphelion (A.U.)	Aphelion (10^6 km)	Equatorial Radius (km)	Equatorial Radius (Earth = 1)
Mercury	7.00	0.31	46.0	0.47	69.8	2,439	0.38
Venus	3.39	0.72	107.5	0.73	108.9	6,051	0.95
Earth	0.01	0.98	147.1	1.02	152.1	6,378	1.00
Mars	1.85	1.38	206.6	1.67	249.2	3,397	0.53
Jupiter	1.31	6.95	740.9	5.45	815.7	71,492	11.19
Saturn	2.49	9.01	1347	10.51	1507	60,268	9.46
Uranus	0.77	18.31	2738	20.07	3002	25,559	3.98
Neptune	1.77	29.76	4452	30.36	4542	24,764	3.81
Pluto	17.15	29.73	4447	49.33	7381	1,123	0.18

Planet	Greatest Angular Diameter as Seen from Earth (arc seconds)	Mass (g)	Mass (Earth = 1)	Mean Density (g/cm^3)	Sideral Rotation Period (days)*
Mercury	12.9	3.30×10^{26}	0.06	5.43	58.6
Venus	65.2	4.87×10^{27}	0.81	5.25	−243.0
Earth		5.97×10^{27}	1.00	5.52	0.9973
Mars	25.7	6.42×10^{26}	0.11	3.93	1.026
Jupiter	50.1	1.90×10^{30}	317.89	1.33	0.41
Saturn	20.8	5.69×10^{29}	95.18	0.71	0.43
Uranus	4.1	8.68×10^{28}	14.54	1.24	−0.65
Neptune	2.4	1.02×10^{29}	17.13	1.67	0.72
Pluto	0.11	1.36×10^{25}	0.003	1.89 to 2.14	−6.387

*A negative value indicates retrograde rotation

TABLE 3 ***(continued)***

Planet	Axial Tilt (degrees)	Oblateness	Surface Gravity (Earth = 1)	Escape Velocity (km/s)	Surface Magnetic Field (Earth = 1)
Mercury	7.0	0.0	0.38	4.3	0.011
Venus	177.4	0.0	0.91	10.4	<0.001
Earth	23.45	0.0034	1.00	11.2	1.0
Mars	23.98	0.0052	0.38	5.0	<0.002
Jupiter	3.08	0.065	2.53	60	13.89
Saturn	26.73	0.108	1.07	36	0.67
Uranus	97.92	0.030	0.92	21	0.74
Neptune	28.8	0.022	1.18	24	0.43
Pluto	118	0	0.08	1	?

Planet	Magnetic Axis Tilt (degrees relative to rotation axis)	Albedo	Surface Temperature[1] (K)	Surface Pressure[2] (Earth = 1)	Number of Moons
Mercury	<10	0.106	100 to 700		0
Venus		0.65	730 ± 5	90 ± 2	0
Earth	11.5	0.37	288 to 293	1.0	1
Mars		0.15	183 to 268	0.007 to 0.010	2
Jupiter	9.6	0.52	124 ± 0.3		16
Saturn	0.8	0.47	95.0 ± 0.4		20
Uranus	58.6	0.50	58 ± 2		15
Neptune	47.0	0.5	59.3 ± 1.0		8
Pluto		0.6	40 to 60		1

[1] Temperature is effective temperature for Jovian planets.

[2] Because the surfaces of the outer planets are not well defined, surface pressure there has no meaning.

TABLE 4 *The Twenty Brightest Stars*

Name	Star	Spectral Type[1]		Parallax (arc seconds)	Distance (pc)	Apparent Visual Magnitude[1]	
		A	B			A	B
Sirius	α CMa	A1V	wd[2]	0.37	2.7	−1.46	+8.7
Canopus	α Car	F01b-II		0.033	30	−0.72	
Rigel Kentaurus	α Cen	G2V	K0V	0.77	1.3	−0.01	+1.3
Arcturus	α Boo	K2IIIp		0.091	11	−0.06	
Vega	α Lyr	A0V		0.13	8.0	+0.04	
Capella	α Aur	GIII	M1V	0.071	14	+0.05	+10.2
Rigel	β Ori	B8 Ia	B9	—	250	+0.14	+6.6
Procyon	α CMi	F5IV-V	wd	0.29	3.5	+0.37	+10.7
Betelgeuse	α Ori	M2Iab		—	150	+0.41	
Achernar	α Eri	B5V		0.050	20	+0.51	
Hadar	β Cen	B1III	?	0.011	90	+0.63	+4
Altair	α Aql	A71V-V		0.20	5.1	+0.77	
Acrux	α Cru	B1IV	B3	0.008	120	+1.39	+1.9
Aldebaran	α Tau	K5III	M2V	0.063	16	+0.86	+13
Spica	α Vir	B1V		0.013	80	+0.91	
Antares	α Sco	MIIb	B4eV	0.008	120	+0.92	+5.1
Pollux	β Gem	K0III		0.083	12	+1.16	
Fomalhaut	α PsA	A3V		0.14	7.0	+1.19	+6.5
Deneb	α Cyg	A2Ia		—	430	+1.26	
Mimosa	β Cru	B0.5IV		—	150	+1.28	

Name	Luminosity (Sun = 1)		Absolute Visual Magnitude[1]		Proper Motion (arc seconds/yr)	Transverse Velocity (km/s)	Radial Velocity (km/s)
	A	B	A	B			
Sirius	23.5	0.003	+1.4	+11.6	1.33	17.0	−7.6[3]
Canopus	1510		−3.1		0.02	2.8	+20.5
Rigel Kentaurus	1.56	0.46	+4.4	+5.7	3.68	22.7	−24.6
Arcturus	115		−0.3		2.28	119	−5.2
Vega	55.0		+0.5		0.34	12.9	−13.9
Capella	166	0.01	−0.7	+9.5	0.44	29	+30.2[3]
Rigel	4.6×10^4	126	−6.8	−0.4	0.00	1.2	+20.7[3]
Procyon	7.7	0.0006	+2.6	+13.0	1.25	20.7	−3.2[3]
Betelgeuse	1.4×10^4		−5.5		0.03	21	+21.0[3]
Achernar	219		−1.0		0.10	9.5	+19
Hadar	3800	182	−4.1	−0.8	0.04	17	−12[3]
Altair	11.5		+2.2		0.66	16	−26.3
Acrux	3470	2190	−4.0	−3.5	0.04	24	−11.2
Aldebaran	105	0.0014	−0.2	+12	0.20	15	+54.1
Spica	2400		−3.6		0.05	19	+1.0[3]
Antares	5500	115	−4.5	−0.3	0.03	17	−3.2
Pollux	41.7		+0.8		0.62	35	+3.3
Fomalhaut	13.8	0.10	+2.0	+7.3	0.37	12	+6.5
Deneb	5.0×10^4		−6.9		0.003	6	−4.6[3]
Mimosa	6030		−4.6		0.05	36	

[1] A and B columns identify individual components of binary systems.

[2] "wd" stands for "white dwarf."

[3] Average value of variable velocity.

TABLE 5 *The Twenty Nearest Stars*

Name	Spectral Type[1] A	Spectral Type[1] B	Parallax (arc seconds)	Distance (pc)	Apparent Visual Magnitude[1] A	Apparent Visual Magnitude[1] B
Sun	G2V				−26.72	
Proxima Cen	M5e		0.772	1.30	+11.05	
Alpha Centauri	G2V	KOV	0.750	1.33	−0.01	+1.33
Barnard's Star	M5V		0.545	1.83	+9.54	
Wolf 359	M8V		0.421	2.38	+13.53	
BD + 36°2147	M2V		0.397	2.52	+7.50	
Luyten 726-8	M5.5V	M5.5V	0.387	2.58	+12.52	+13.02
Sirius	A1V	wd[2]	0.377	2.65	−1.46	+8.3
Ross 154	M4.5V		0.345	2.90	+10.45	
Ross 248	M6V		0.314	3.18	+12.29	
ϵ Eridani	K2V		0.303	3.30	+3.73	
Ross 128	M5V		0.298	3.36	+11.10	
61 Cygni	K5V	K7V	0.294	3.40	+5.22	+6.03
ϵ Indi	K5V		0.291	3.44	+4.68	
BD + 43°44	M1V	M6V	0.290	3.45	+8.08	+11.06
Luyten 789-6	M6V		0.290	3.45	+12.18	
Procyon	F5IV-V	wd	0.285	3.51	+0.37	+10.7
BD + 59°1915	M4V	M5V	0.282	3.55	+8.90	+9.69
CD − 36°15693	M2V		0.279	3.58	+7.35	
G51-15	MV		0.278	3.60	+14.81	

Name	Luminosity (Sun = 1) A	Luminosity (Sun = 1) B	Absolute Visual Magnitude[1] A	Absolute Visual Magnitude[1] B	Proper Motion (arc second/yr)	Transverse Velocity (km/s)	Radial Velocity (km/s)
Sun	1.0		+4.85				
Proxima Cen	0.00006		+15.5		3.86	23.8	−16
Alpha Centauri	1.6	0.45	+4.4	+5.7	3.68	23.2	−22
Barnard's Star	0.00045		+13.2		10.34	89.7	−108
Wolf 359	0.00002		+16.7		4.70	53.0	+13
BD + 36°2147	0.0055		+10.5		4.78	57.1	−84
Luyten 726-8	0.00006	0.00004	+15.5	+16.0	3.36	41.1	+30
Sirius	23.5	0.003	+1.4	+11.2	1.33	16.7	−8
Ross 154	0.00048		+13.3		0.72	9.9	−4
Ross 248	0.00011		+14.8		1.58	23.8	−81
ϵ Eridani	0.30		+6.1		0.98	15.3	+16
Ross 128	0.00036		+13.5		1.37	21.8	−13
61 Cygni	0.082	0.039	+7.6	+8.4	5.22	84.1	−64
ϵ Indi	0.14		+7.0		4.69	76.5	−40
BD + 43°44	0.0061	0.00039	+10.4	+13.4	2.89	47.3	+17
Luyten 789-6	0.00014		+14.6		3.26	53.3	−60
Procyon	7.65	0.00055	+2.6	+13.0	1.25	2.8	−3
BD + 59°1915	0.0030	0.0015	+11.2	+11.9	2.28	38.4	+5
CD − 36°15693	.013		+9.6		6.90	117	+10
G51-15	0.00001		+17.0		1.26	21.5	—

[1] A and B columns identify individual components of binary systems.

[2] "wd" stands for "white dwarf."

TABLE 6 ***The Local Group of Galaxies***

Galaxy	Hubble Type	Distance (kpc)	Apparent Visual Magnitude (m)	Absolute Visual Magnitude (M)	Luminosity (10^9 L☉)	Angular Diameter (degrees)	Diameter (kpc)	Radial Velocity (km/s)	Mass (Solar masses)
Our Galaxy	Sb	—	—	(−21)	20.7	—	30	—	2×10^{11}
Large Magellanic Cloud	Irr I	48	0.9	−17.7	0.99	11.9	10	+276	2.5×10^{10}
Small Magellanic Cloud	Irr I	56	2.5	−16.5	0.33	8.1	8	+168	
Ursa Minor system	E4 (dwarf)	70		(−9)	0.0003	0.81	1		
Sculptor system	E3 (dwarf)	83	8.0	−11.8	0.0043	1.51	2.2		(2 to 4×10^6)
Draco system	E2 (dwarf)	100		(−10)	0.0008	0.80	1.4		
Carina system	E3 (dwarf)	(170)		(−10)	0.0008	0.50	1.5		
Fornax system	E3 (dwarf)	250	8.4	−13.6	0.023	1.03	4.5	+39	(1.2 to 2×10^7)
Leo II system	E0 (dwarf)	230		−10.0	0.0008	0.40	1.6	+39	(1.1×10^6)
Leo I system	E4 (dwarf)	280	12.0	−10.4	0.0012	0.31	1.5		
NGC 6822	Irr I	460	8.9	−14.8	0.069	0.33	2.7	−32	
NGC 147	E6	570	9.73	−14.5	0.052	0.30	3		
NGC 185	E2	570	9.43	−14.8	0.069	0.23	2.3	−305	
NGC 205	E5	680	8.17	−16.5	0.33	0.42	5	−239	
NGC 221 (M32)	E3	680	8.16	−16.5	0.33	0.20	2.4	−214	
IC 1613	Irr I	680	9.61	−14.7	0.063	0.42	5	−238	
Andromeda galaxy (NGC 224; M31)	Sb	680	3.47	−21.2	24.9	3.35	40	−266	3×10^{11}
And I	E0 (dwarf)	(680)	(14)	(−11)	0.002	0.04	0.5		
And II	E0 (dwarf)	(680)	(14)	(−11)	0.002	0.06	0.7		
And III	E3 (dwarf)	(680)	(14)	(−11)	0.002	0.08	0.9		
NGC 598 (M33)	Sc	720	5.79	−18.9	3.0	1.35	17	−189	8×10^9

Glossary

A ring One of three Saturnian rings, visible from Earth. The A ring is farthest from the planet and is separated from the B ring by the Cassini division. (p. 289)

absolute brightness The apparent brightness a star would have if it were placed at a distance of 10 parsecs from Earth. (p. 395)

absolute zero Temperature at which all molecular and atomic motion is expected to cease (−273.15 celsius). (p. 74)

absorption line Dark line in an otherwise continuous bright spectrum, where light of one particular frequency has been removed. (p. 86)

acceleration The rate of change of velocity of a moving object. (p. 53)

accretion Gradual growth of bodies, such as planets, by the accumulation of other, smaller, bodies. (p. 356)

accretion disk Flat disk of matter spiraling down onto the surface of a star or black hole. Often, the matter originated on the surface of a companion star in a binary system. (p. 480)

active galaxy The most energetic galaxies, which can emit hundreds or thousands of times more energy per second than the Milky Way. (p. 567)

active optics Collection of techniques now being used to increase the resolution of ground-based telescopes. Minute modifications are made to the overall configuration of an instrument as its temperature and orientation change, to maintain the best possible focus at all times. (p. 120)

active prominence A solar prominence that changes its appearance rather rapidly in time, surging up from the photosphere and falling back on itself in a matter of hours. (p. 379)

active region Region of the photosphere of the Sun surrounding a sunspot group, which can erupt violently and unpredictably. During sunspot maximum, the number of active regions is also a maximum. (p. 379)

active Sun The unpredictable aspects of the Sun's behavior, such as sudden explosive outbursts of radiation in the form of prominences and flares. (p. 374)

adaptive optics Technique used to increase the resolution of a telescope by deforming the shape of the mirror's surface under computer control while a measurement is being taken, to undo the effects of atmospheric turbulence. (p. 120)

aesthenosphere Layer of the Earth's interior, just below the lithosphere, over which the surface plates slide. (p. 159)

albedo The fraction of incident sunlight reflected back into space by an object's surface. (p. 204)

alpha particle A helium nucleus. An alpha particle consists of two protons and two neutrons. (p. 460)

alpha process A two-step process occurring in the centers of stars which have evolved to the point of having silicon-28 in their cores. At this point, photodisintegration breaks nuclei apart into helium nuclei (alpha particles), which then are captured to form heavier elements. (p. 488)

ALSEP Collection of scientific instruments left behind by the Apollo program for the purpose of carrying out measurements of the lunar interior. (p. 186)

amino acid Organic molecules which are the basis for building proteins which direct metabolism in living creatures. (p. 640)

Amor asteroids Asteroids on orbits that cross the orbit of Mars, named after the first such asteroid discovered. (p. 331)

amplitude The maximum height above or below the zero point that a wave achieves. (p. 65)

angular momentum problem The fact that the Sun, which contains nearly all of the mass of the solar system, accounts for just 0.3 percent of the total angular momentum of the solar system. This is an aspect of the solar system that any acceptable formation theory must address. (p. 353)

angular resolution The ability of a telescope to distinguish between two adjacent objects in the sky. (p. 114)

annular eclipse Solar eclipse occurring at a time when the Moon is far enough away from the Earth that it fails to cover the disk of the Sun completely, leaving a ring of sunlight visible around its edge. (p. 16)

aphelion The point on the elliptical orbit of a planet about the Sun when the planet is at its greatest distance from the Sun. (p. 48)

apogee Greatest separation from the Earth of any object in orbit about it. (p. 170)

Apollo asteroids Asteroids on orbits that cross the orbit of Earth, named after the first such asteroid discovered. (p. 331)

Apollo program The US initiative begun in 1961 by President Kennedy with the stated goal of sending an astronaut to the Moon and returning him safely to Earth. (p. 185)

apparent brightness The brightness that a star appears to have, as measured by an observer on Earth. (p. 395)

arc degree Unit of angular measure. There are 360 arc degrees in one complete circle. (p. 9)

arc minute Unit of angular measure. There are 60 arc minutes in one arc degree. (p. 9)

arc second Unit of angular measure. There are 60 arc seconds in one arc minute. (p. 9)

association Small grouping of (typically 100 or less) stars, spanning up to a few tens of parsecs across, usually rich in very young stars. (p. 452)

asteroid One of thousands of very small members of the solar system orbiting the Sun between the orbits of Mars and Jupiter. Often referred to as "minor planets." (p. 191, p. 330)

asteroid belt A region of the solar system, lying between the orbits of Mars and Jupiter, in which the majority of the asteroids is found. (p. 191)

astronomical unit (A.U.) The average distance of the Earth from the Sun. Precise radar measurements yield a value for the A.U. of 149,603,500 km. (p. 49)

astronomy Branch of science dedicated to the study of everything in the universe that lies above Earth's atmosphere. (p. 5)

asymptotic giant branch Path on the HR diagram that corresponds to the changes that a star undergoes after helium burning in the core ceases. At this stage, the carbon core shrinks and drives the expansion of the envelope, and the star becomes a swollen red giant for a second time. (p. 463)

atmosphere The region above the surface of the Earth where air lies. (p. 141)

atom Building block of matter, composed of positively charged protons and neutral neutrons in the nucleus, surrounded by negatively charged neutrons. (p. 85)

aurora Event which occurs when atmospheric molecules are excited by incoming charged particles from the solar wind, then emit energy as they fall back to their ground states. (p. 151)

aurora australis Aurorae that occur in the southern hemisphere, also known as the southern lights. (p. 151)

aurora borealis Aurorae that occur in the northern hemisphere, also known as the northern lights. (p. 151)

autumnal equinox Date on which the Sun crosses the celestial equator moving southward, occurring on or near September 22. (p. 12)

B ring One of three Saturnian rings visible from Earth. The B ring lies just past the Cassini division, closer to the planet than the A ring, and is the brightest of the three. (p. 289)

barred-spiral galaxy Spiral galaxy in which a bar of material passes through the center of the galaxy, with the spiral arms beginning near the ends of the bar. (p. 547)

baryonic matter Matter that is composed primarily of baryons—protons and neutrons. "Normal" matter. (p. 621)

basalt Relatively dense rock, composed of an iron-magnesium-silicate mixture. Basalt is solidified lava. (p. 154)

baseline The distance between two observing locations, used for the purposes of triangulation measurements. The larger the baseline the better the resolution attainable. (p. 27)

belt Dark, low pressure region, where gas flows downward, in the atmosphere of a jovian planet. (p. 267)

Big Bang Event that cosmologists consider the beginning of the universe, in which all matter and radiation in the entire universe came into being. (p. 595)

binary asteroid Large asteroid that is orbited by a smaller satellite body, believed to form as the result of a close encounter or collision in the relatively congested asteroid belt. (p. 332)

binary-star system A system which consists of two stars in orbit about their common center of mass, held together by their mutual gravitational attraction. Most stars are found in binary-star systems. (p. 406)

biological evolution Steady changes in organisms in response to changes in the environment. (p. 642)

bipolar flow Two jets of matter expelled in opposite directions from a protostar. If the protostar is embedded in a disk of material, the protostellar wind encounters less resistance perpendicular to the disk than in the disk plane, so it flows outward along the poles. (p. 448)

black-body curve The characteristic way in which the intensity of radiation emitted by a hot object depends on frequency. The frequency at which the emitted intensity is highest is an indication of the temperature of the radiating object. Also referred to as the Planck curve. (p. 75)

black hole A region of space where the pull of gravity is so great that nothing—not even light—can escape. A possible outcome of the evolution of a very massive star. (p. 503)

blue giant Large, hot, bright star at the upper left end of the main sequence on the HR diagram. Its name comes from its color and size. (p. 402)

blue shift Motion-induced change in the observed wavelength from a source that is moving toward us. Relative approaching motion between the object and the observer causes the wavelength to appear shorter (and hence bluer) than if there were no motion at all. (p. 81)

blue supergiant The very largest of the large, hot, bright stars at the uppermost left end of the main sequence on the HR diagram. (p. 402)

Bohr model First theory of the hydrogen atom to explain the observed spectral lines. This model rests on three ideas: that there is a state of lowest energy for the electron, that there is a maximum energy, beyond which the electron is no longer bound to the nucleus, and that within these two energies the electron can only exist in certain energy levels. (p. 91)

boson Type of particle that mediates the forces felt between elementary particles. The force exerted on one particle by another is transmitted via the exchange of a boson. (p. 625)

brown dwarf Remnants of fragments of collapsing gas and dust that did not contain enough mass to initiate core nuclear fusion. Such objects are then frozen somewhere along the Kelvin-Helmholtz contraction phase, continually cooling into compact dark objects. Because of their small size and low temperature they are extremely difficult to detect observationally. (p. 442)

brown oval Feature of Jupiter's atmosphere that appears only at latitudes near 20 degrees N, this structure is a long-lived hole in the clouds that allows us to look down into Jupiter's lower atmosphere. (p. 269)

C ring One of three Saturnian rings visible from Earth. The C ring lies closest to the planet and is relatively thin compared with the A and B rings. (p. 289)

C-type asteroid Asteroid with relatively high carbon content; a carbonaceous asteroid. (p. 334)

caldera A volcanic crater often produced as the result of the collapse of the summit of a shield volcano. (p. 234)

carbon detonation supernova See type-I supernova. (p. 483)

carbonaceous asteroid A relatively dark asteroid. Its darkness is due to the presence of a large amount of carbon compared to other asteroids. (p. 334)

carbonaceous chondrite Believed to be the most primitive of all meteorites, these relatively dark meteorites, with high carbon content, contain chunks of rocky nebular material embedded in their interiors. (p. 344)

carbonaceous meteorite Dark meteorite, with relatively high carbon content believed to be among the most primitive of all meteorites. (p. 344)

Cassegrain telescope A type of reflecting telescope in which incoming light hits the primary mirror and is then reflected upward toward the prime focus. Here, a secondary mirror reflects the light back down through a small hole in the main mirror, into a detector or eyepiece. (p. 110)

Cassini Division A relatively empty gap in Saturn's ring system, discovered in 1675 by Giovanni Cassini. It is now known to contain a number of thin ringlets. (p. 289)

cataclysmic variable Star that undergoes sudden large changes in brightness. Novae and supernovae fall into this category. (p. 523)

catastrophic theory A theory that invokes statistically unlikely accidental events to account for observations. (p. 352)

celestial coordinates Pair of quantities, similar to latitude and longitude on Earth, used to pinpoint locations of objects on the celestial sphere. (p. 9)

celestial equator The projection of the Earth's equator onto the celestial sphere. (p. 9)

celestial sphere Imaginary sphere surrounding the Earth, to which all objects in the sky were once considered to be attached. (p. 8)

center of mass The ''average'' position in space of a collection of massive bodies, weighted by their masses. In an isolated system this point moves with constant velocity, according to Newtonian mechanics. (p. 55)

centripetal force The force, directed toward the center of curvature, required to divert straight-line motion into a curved path. (p. 54)

Cepheid variable Star whose luminosity varies in a characteristic way, with a rapid rise in brightness followed by a slow decline. The period of a Cepheid variable star is related to its luminosity, so a determination of this period can be used to obtain an estimate of the star's distance. (p. 523)

Chandrasekhar mass The maximum mass that a white dwarf star can have if electron degeneracy pressure is to prevent gravitational contraction. Once the Chandrasekhar mass is exceeded, gravity becomes stronger than electron degeneracy pressure, and the star collapses. This mass is about 1.4 solar masses. (p. 482)

chaotic rotation Unpredictable tumbling motion that non-spherical bodies in eccentric orbits, such as Saturn's satellite Hyperion, can exhibit. No amount of observation of an object rotating chaotically will ever show a well-defined period. (p. 303)

charge-coupled device (CCD) An electronic device used for data acquisition, composed of many tiny pixels each of which records a buildup of charge to measure the amount of light striking it. (p. 118)

chemical evolution The natural evolution of complex molecules from material that was present on the primitive Earth. (p. 640)

chondrule Rocky spheres embedded in the interiors of some carbonaceous meteorites. (p. 344)

chromatic aberration The tendency for a lens to focus red and blue light differently, causing images to become blurred. (p. 108)

chromosphere The Sun's lower atmosphere, lying just above the visible photosphere. (p. 365)

close-encounter theory A theory that explains the formation of the solar system by postulating a close encounter between the Sun and another star long ago. According to this theory, the near-collision tore off streamers of material from the two stars. This material later condensed to form planets. (p. 353)

closed universe Geometry that the universe as a whole would have if the density of matter is above the critical value. A closed universe is finite in extent, and has no edge, like the surface of a sphere. It has enough mass to stop the present expansion and eventually recollapse. (p. 599)

co-orbital satellites Pair of moons discovered just beyond Saturn's ring system which display a unique and curious motion. The two satellites are on nearly the same orbit. When they get close enough to begin to feel each other's gravity, they exchange orbits and continue on their way. (p. 302)

cocoon nebula Region of bright infrared emission. A cloud of dust, which surrounds a hot young star, absorbs ultraviolet radiation from the star, and reemits it in the infrared. (p. 447)

cold dark matter Class of dark-matter candidates made up of very heavy particles, such as supersymmetric relics. (p. 631)

collecting area The total area of a telescope that is capable of capturing incoming radiation. The larger the telescope, the greater its collecting area, and the fainter the objects it can detect. (p. 114)

collimator Instrument designed to detect x- or gamma-ray radiation, that rejects all photons except those moving nearly along the axis of the detector. (p. 130)

collision theory See close-encounter theory. (p. 353)

color index A convenient method of quantifying a star's color by comparing its apparent brightness as measured through different filters. Since the star's radiation is well described by a Plank spectrum, the ratio of its blue intensity (B) to its visual intensity (V) is a measure of the object's surface temperature. (p. 398)

color-magnitude diagram A way of plotting stellar properties, in which absolute magnitude is plotted against color index. (p. 401)

coma An effect occurring during the formation of an off-axis image in a telescope. Stars whose light enters the telescope at a large angle acquire comet-like tails on their images. (p. 112)

coma The brightest part of a comet, often referred to as the ''head.'' (p. 337)

comet A small body, composed mainly of ice and dust, in an elliptical orbit about the Sun. As it comes close to the Sun, some of its material is vaporized to form a gaseous head and extended tail. (p. 335)

common-envelope binary A binary star system in which both stars have expanded to fill their Roche lobes and the surfaces of the two stars merge. The binary system now consists of two nuclear burning stellar cores surrounded by a continuous common envelope. These systems are also referred to as contact binary systems. (p. 473)

comparative planetology Comparing and contrasting the properties of the nine planets to better understand the conditions under which the planets formed and developed. (p. 190)

condensation nuclei Dust grains in the interstellar medium which act as seeds around which other material can cluster. The presence of dust was very important in causing matter to clump during the formation of the solar system. (p. 355)

condensation theory Currently favored model of solar system formation which combines features of the old nebular theory with new information about interstellar dust grains, which acted as condensation nuclei. (p. 355)

conjunction The time at which a planet has the same celestial longitude as the Sun, or the time when two or more objects have the same celestial longitude. (p. 243)

conservation of mass and energy A fundamental law of modern physics which states that the sum of mass and energy must always remain constant in any physical process. In fusion reactions, the lost mass is converted into energy, primarily in the form of electromagnetic radiation. (p. 381)

constellation A grouping of stars in the night sky into a recognizable pattern. (p. 5)

constructive interference Alignment of two interfering waves such that their crests and troughs exactly coincide. The net effect is that the two waves reinforce each other, resulting in a wave of greater amplitude. (p. 66)

contact binary See common-envelope binary. (p. 473)

continuous spectrum Spectrum produced by a hot, solid body that emits radiation according to the Planck curve appropriate to its temperature. (p. 85)

convection Churning motion resulting from the constant upwelling of warm air and the concurrent downward flow of cooler air to take its place. (p. 144)

convection cells Regions of rising and falling air. (p. 146)

convection zone Region of the Sun's interior, lying just below the surface, where the material of the Sun is in constant convective motion. This region extends into the solar interior to a depth of about 200,000 km. (p. 365)

core hydrogen burning The energy burning stage for main sequence stars, in which helium is produced by hydrogen fusion in the central region of the star. A typical star spends up to 90% of its lifetime in hydrostatic equilibrium brought about by the balance between gravity and the energy generated by core hydrogen burning. (p. 457)

core The central region of the Earth, surrounded by the mantle. (p. 142) The central region of any planet or star. (p. 365)

core-halo radio galaxy Type of active galaxy in which most of the energy comes from a small central nucleus, with weaker emission coming from a surrounding halo. (p. 571)

corona The tenuous outer atmosphere of the Sun, which lies just above the chromosphere, and at great distances turns into the solar wind. (p. 365)

coronae Huge, roughly circular regions that are the largest volcanic structures found on Venus. (p. 234)

coronal hole Vast regions of the Sun's atmosphere where the density of matter is about 10 times lower than average. The gas there streams freely into space at high speeds, escaping the Sun competely. (p. 374)

cosmic distance scale Collection of indirect distance-measurement techniques that astronomers use to measure the scale of the universe. (p. 27)

cosmic evolution The collection of the seven major phases of the history of the universe, namely galactic, stellar, planetary, chemical, biological, cultural and future evolution. (p. 638)

cosmic microwave background The almost isotropic radio signal that is the remnant of the Big Bang explosion. (p. 607)

cosmological constant A term that Einstein added to his equations of General Relativity so that the universe would remain static. Without the addition of this term, his equations would have correctly predicted an evolving universe. (p. 605)

cosmological distance Any distance comparable with the scale of the universe itself. (p. 579)

cosmological principle Two assumptions which make up the basis of cosmology, namely that the universe is homogeneous and isotropic on large scales. (p. 594)

cosmological red shift The component of the redshift of an object which is due only to the Hubble flow of the universe. (p. 579)

cosmologist Astronomer who studies the large-scale structure and dynamics of the universe. (p. 593)

cosmology The study of the structure and evolution of the entire universe. (p. 31)

crater Bowl-shaped depressions on the surface of a planet or moon, resulting from collisions with interplanetary debris. (p. 174)

crater density The average number of craters per unit area of a solar-system object, usually per million square kilometers. (p. 178)

crescent Term used to describe a phase of the moon in which less than one half of the disk is visible. (p. 13)

critical density The cosmic density corresponding to the dividing line between a universe that recollapses and one that expands forever. (p. 599)

critical universe Geometry that the universe would have if the density of matter is exactly the critical density. The universe is infinite in extent, and has zero curvature. The expansion will continue for ever, but approach an expansion speed of zero. (p. 600)

crust Layer of the Earth which contains the solid continents and the seafloor. (p. 141)

cultural evolution Gradual adjustments in the attitudes and behavior of society. (p. 644)

current sheet The organization of charged particles around Jupiter into a flat sheet aligned with the planet's equator. (p. 272)

D ring Collection of very faint, thin rings, extending from the inner edge of the C ring down nearly to the cloud tops of Saturn. This region contains so few particles that it is completely invisible from Earth. (p. 295)

dark dust cloud A large cloud, often many parsecs across, which contains gas and dust in a ratio of about 10^{12} gas atoms for every dust particle. Typical densities are a few tens or hundreds of particles per cubic centimeter. (p. 423)

dark matter Term used to describe the mass in galaxies and clusters whose existence we infer from rotation curves and other techniques, but which has not been confirmed by observations at any electromagnetic wavelength. (p. 537)

declination Celestial coordinate used to measure latitude above or below the celestial equator on the celestial sphere. (p. 9)

decoupling Event in the early universe when atoms first formed, and after which photons could propagate freely through space. (p. 622)

deferent A construct of the geocentric model of the solar system which was needed to explain observed planetary motions. A deferent is a large circle encircling the Earth, on which an epicycle moves. (p. 32)

degree A unit of angular measure. There are 360 degrees in a complete circle. (p. 9)

density A measure of the compactness of the matter within an object, computed by dividing the mass by the volume of the object. (p. 140)

destructive interference Alignment of two interfering waves such that the crests of one wave align with the troughs of the other. The net effect is that the waves cancel each other out, so that no net motion remains. (p. 66)

detached binary A normal binary system, in which the two stars each lie well within their respective Roche lobes, and no mass is transferred between the two. (p. 473)

deuterium bottleneck Period of time in the early universe during which the formation of deuterium by fusion was possible, but the deuterium was destroyed as fast as it was created. (p. 620)

deuteron An isotope of hydrogen in which there is a neutron bound to the proton in the nucleus. Often called "heavy hydrogen" because of the extra mass of the neutron. (p. 381)

differential rotation The tendency for a gaseous sphere, such as a jovian planet or the Sun, to rotate more rapidly at the equator than at the poles. For a galaxy or other object, a condition where the angular velocity varies with distance from the object's center. (p. 262)

differentiated meteorite A meteorite in which the heavier material has fallen to the center, usually as a result of strong heating at some time in the past, indicating that the meteorite was either part of a larger body that underwent geological activity, or was partially melted during a collision. (p. 344)

differentiation Variation in the density and composition of a body, such as the Earth, with low density material on the surface and higher density material in the core. (p. 154)

diffraction The ability that waves have to bend around corners. The diffraction of light establishes its nature as a wave. (p. 65)

direct motion The usual eastward motion that planets have with respect to the background stars. (p. 32)

diurnal The daily progress of the Sun, Moon, planets and stars across the celestial sphere. (p. 10)

Doppler effect Any motion-induced change in the observed wavelength (or frequency) of a wave. (p. 79)

Drake Equation See Green Bank equation. (p. 648)

dust grain An interstellar dust particle, roughly 10^{-5} cm in size, comparable to the wavelength of visible light. (p. 417)

dust lane A lane of dark, obscuring dust in an emission nebula or galaxy. (p. 420)

dust tail The component of a comet's tail that is composed of dust particles. (p. 337)

dwarf Any star with radius comparable to, or smaller than, that of the Sun (including the Sun). (p. 394)

dwarf elliptical Smallest of the elliptical galaxies. These systems can be as small as a kiloparsec in size and contain less than one million stars. (p. 548)

dynamo theory Theory that explains planetary and stellar magnetic fields in terms of rotating, conducting material flowing in an object's interior. (p. 149)

E ring A faint ring, well outside the main ring system of Saturn, which was discovered by Voyager, and is believed to be associated with volcanism on the moon Enceladus. (p. 295)

earthquake A sudden dislocation of rocky material near the Earth's surface that causes the entire planet to vibrate. (p. 152)

eccentricity A measure of the flatness of an ellipse, given by the distance between the two foci divided by the length of the major axis. (p. 48)

eclipse Event during which one body passes in front of another, so that the light from the occulted body is blocked. (p. 15)

eclipse season Times of the year when the Moon lies in the same plane as the Earth and Sun, so that eclipses are possible. (p. 19)

eclipsing binary Rare binary-star system that is aligned in such a way that from Earth we observe one star pass in front of the other, eclipsing the other star. (p. 406)

ecliptic The apparent path of the Sun, relative to the stars on the celestial sphere, over the course of a year. (p. 11)

effective temperature The average temperature of a star's photosphere. This is the temperature that is measured from the Planck curve and Stefan's Law. (p. 366)

ejecta blanket The material thrown out by the impact explosion that created a crater. (p. 176)

electric field A field extending outward in all directions from a charged particle, such as a proton or an electron. The electric field determines the electric force exerted by the particle on all other charged particles in the universe; the strength of the electric field decreases with increasing distance from the charge according to an inverse-square law. (p. 67)

electromagnetic radiation Another term for light, electromagnetic radiation transfers energy and information from one place to another. (p. 68)

electromagnetic spectrum The complete range of electromagnetic radiation, from radio waves to gamma rays, including the visible spectrum. All types of electromagnetic radiation are basically the same phenomenon, differing only by wavelength, and all move at the speed of light. (p. 70)

electromagnetism The union of electricity and magnetism, which do not exist as independent quantities, but are two aspects of a single physical phenomenon. (p. 68)

electron An elementary particle, with a negative electric charge, which is one of the components of the atom. (p. 91)

electron degeneracy The incompressibility of electrons, due to the fact that they behave like hard spheres, and can't be squeezed beyond the point of contact. (p. 460)

electron degeneracy pressure The pressure produced by the resistance of electrons to compression once they are squeezed to the point of contact. (p. 460)

electroweak force The force which results when the electromagnetic and weak forces are unified. (p. 625)

element Matter made up of one particular atom. The number of protons in the nucleus of the atom determines which element it represents. (p. 94)

ellipse Geometric figure resembling an elongated circle. An ellipse is characterized by its degree of flatness, or eccentricity. An ellipse is the general bound orbit for objects moving under gravity. (p. 48)

elliptical galaxy Category of galaxy in which the stars are distributed in an elliptical shape on the sky, ranging from highly elongated to nearly circular in appearance. (p. 548)

elongation The angular distance on the sky between a planet and the Sun. (p. 203)

emission line Bright line in a specific location of the spectrum of radiating material, corresponding to emission of light at a certain frequency. A heated gas in a glass jar produces emission lines in its spectrum. (p. 86)

emission nebula A glowing cloud of hot interstellar gas. The gas glows as a result of a nearby young star which is ionizing the gas. Since this gas is mostly hydrogen, the emitted radiation falls predominantly in the red region of the spectrum, from the hydrogen alpha emission line. (p. 420)

emission spectrum The pattern of spectral emission lines, produced by an element. Each element has its own unique emission spectrum. (p. 86)

Encke Division A small gap in Saturn's A ring. (p. 289)

energy flux The energy emitted per unit area per unit time from a hot body. (p. 394)

epicycle A construct of the geocentric model of the solar system which was necessary to explain observed planetary motions. Each planet rides on a small epicycle whose center in turn rides on a larger circle (the deferent). (p. 32)

equinox One of two points on the celestial sphere where the ecliptic intersects the celestial equator. (p. 12)

erg The CGS unit for measuring energy. (p. 73)

erosion A slow wearing away of the surface of a planet or moon. On the Moon and other bodies without an atmosphere or surface water, erosion is caused primarily by interplanetary debris that collides with the surface. (p. 175)

escape velocity The velocity necessary for an object to escape the gravitational pull of an object. Anything that moves away from the object with more than the escape velocity will never return. (p. 56)

event horizon Imaginary spherical surface surrounding a collapsing star, with radius equal to the Schwarzschild radius, within which no event can be seen, heard, or known about by an outside observer. (p. 507)

evolutionary theory A theory which explains observations in a series of gradual steps, explainable in terms of well-established physical principles. (p. 352)

evolutionary track A graphical representation of a star's life, as a path on the HR diagram. (p. 437)

excited state State of an atom when one of its electrons is in a higher energy orbital than the ground state. Atoms can become excited by absorbing a photon of a specific energy, or by colliding with a nearby atom. (p. 93)

exosphere The outermost layers of the atmosphere, about 250 km from the Earth's surface. (p. 144)

extinction The dimming of starlight as it passes through the interstellar medium. (p. 417)

F ring Faint narrow outer ring of Saturn, discovered by Pioneer in 1979. The F ring lies just inside the Roche limit of Saturn, and was shown by Voyager to be made up of several ring strands apparently braided together. (pp. 294–295)

fall A meteorite discovered on the ground after having been seen shooting across the sky. (p. 343)

false vacuum Condition in the early universe caused by the enormous pressure created when the strong force decoupled. The false vacuum state gave rise to the epoch of inflation. (p. 626)

find A meteorite discovered on the ground which was not seen shooting across the sky. (p. 343)

flare Explosive event occurring in or near an active region on the Sun. (p. 379)

flare star Low mass main sequence star which undergoes violent surface activity and corresponding changes in luminosity. (p. 523)

flatness problem One of two conceptual problems with the Standard Big Bang model, which is that there is no natural way to explain why the density of the universe is so close to the critical density. (p. 624)

fluidized ejecta The ejecta blankets around some Martian craters, which apparently indicate that the ejected material was liquid at the time the crater formed. (p. 250)

focal length In a reflecting telescope, the distance from the mirror to the prime focus. (p. 107)

focus One of two special points within an ellipse, whose separation from each other indicate the eccentricity. In a bound orbit, objects move in ellipses about one focus. (p. 48)

forbidden line A spectral line seen in emission nebulae, but not seen in laboratory experiments, because collisions kick the electron in question into some other state before emission can occur. (p. 423)

force Action on an object that causes its momentum to change. The rate at which the momentum changes is numerically equal to the force. (p. 52)

fragmentation The breaking up of a large object into many smaller pieces (for example, as the result of high-speed collisions between planetesimals and protoplanets in the early solar system). (p. 357)

Fraunhofer lines The collection of over 600 absorption lines in the spectrum of the Sun, first catalogued by Joseph Fraunhofer in 1812. (p. 89)

freeze out Term given to the moment in the early universe when temperatures dropped below the threshold for production of a given type of particle. (p. 617)

frequency The number of wave crests passing any given point in a given period of time. (p. 65)

full moon Phase of the Moon in which it appears as a completely circular disk in the sky. (p. 13)

fusion Mechanism of energy generation in the core of the Sun, in which light nuclei are combined, or fused, into heavier ones, releasing energy in the process. (p. 381)

galactic bulge Thick distribution of warm gas and stars around the galactic center. (p. 530)

galactic cannibalism A galaxy merger in which a larger galaxy effectively consumes a smaller one. (p. 562)

galactic center The center of the Milky Way, or any other, galaxy. The point about which the disk of a spiral galaxy rotates. (p. 526)

galactic corona See outer halo. (p. 537)

galactic disk Flattened region of gas and dust that bisects the galactic halo in a spiral galaxy. This is the region of active star formation. (p. 526)

galactic halo Region of a galaxy extending far above and below the galactic disk, where globular clusters and other old stars reside. (p. 526)

galactic nucleus Small central region of a galaxy from which nearly all of the radiation is emitted in active galaxies. (p. 570)

galactic plane Plane in which the disk component of a spiral galaxy resides and rotates. (p. 526)

galaxy Gravitationally bound collection of a large number of stars. The Sun is a star in the Milky Way Galaxy. (p. 521)

galaxy cluster A collection of galaxies held together by their mutual gravitational attraction. (p. 551)

galaxy merger Result of the interaction of two galaxies, during which tidal forces exerted on each other cause the galaxies to spiral toward one another and coalesce. (p. 562)

Galilean moons The four brightest and largest moons of Jupiter (Io, Europa, Ganymede, Callisto), named after Galileo Galilei, the 17th century astronomer who first observed them. (p. 261)

gamma ray Region of the electromagnetic spectrum, far beyond the visible spectrum, corresponding to radiation of very high frequency, and very short wavelength. (p. 70)

gamma-ray burster Object that radiates tremendous amounts of energy in the form of gamma rays, believed to be due to the accretion of matter onto a neutron star from another star in a binary orbit. The accreting matter eventually reaches temperatures high enough to begin violent nuclear burning on the neutron star's surface. (p. 500)

gas exchange experiment A test for life on the Martian surface, performed by the Viking landers, in which a nutrient broth was introduced into the soil and measurements taken for any sign of metabolic activity. (p. 256)

geocentric A model of the solar system which holds that the Earth is at the center of the universe and all other bodies are in orbit around it. The earliest theories of the solar system were geocentric. (p. 32)

giant A star with a radius between 10 and 100 times that of the Sun. (p. 394)

giant elliptical Largest of the elliptical galaxies, which can span up to a few megaparsecs and contain trillions of stars. (p. 548)

gibbous Term used to describe any phase of the Moon where greater than one quarter and less than the full disk is visible. (p. 13)

globular cluster Tightly bound, roughly spherical collection of hundreds of thousands, and sometimes millions, of stars spanning about 50 parsecs. Globular clusters are distributed in the halo around the Milky Way and other galaxies. (p. 411)

gluon The boson that is responsible for mediating the strong force. (p. 625)

GONG Abbreviation for Global Oscillations Network Group, an ongoing project researching solar oscillations. Continuous observations of the Sun are made by observing from many different locations on Earth, and give precise information about the solar interior. (p. 367)

grand unification temperature The temperature at which X- and anti-X-bosons are spontaneously created, about 10^{28}K. (p. 625)

Grand Unified Theories Theories which describe the behavior of the single force that results from unification of the strong, weak, and electromagnetic forces in the early universe. (p. 617)

granite Relatively low density rock composed largely of aluminum and silicon. (p. 154)

granule The bright, topmost portion of a convective cell in the region of the Sun just below the photosphere. The brightness is due to the upwelling of hotter gas, and measures about 1000 km across. (p. 370)

gravitational constant The constant of proportionality in the law of Universal Gravitation, which gives the gravitational force between two objects as the ratio of the products of their masses to the square of the distance separating them. (p. 52)

gravitational field Field created by any object with mass, extending out from that object in all directions, which determines the influence of that massive object on all others. The strength of the gravitational field decreases as the square of the distance. (p. 53)

gravitational force The attractive effect that any massive object has on all other massive objects. The greater the mass of the object, the stronger its gravitational pull. (p. 53)

gravitational lensing The effect induced on the image of a distant object by a massive foreground object. Light from the distant object is bent into two or more separate images. (p. 584)

gravitational red shift A prediction of Einstein's general theory of relativity. Photons have to do work, and hence lose energy, to escape the gravitational field of a massive object. Because a photon's energy is proportional to its frequency, a photon that loses energy suffers a decrease in frequency, which corresponds to an increase, or redshift, in wavelength. (p. 512)

gravitational slingshot Mechanism for transferring energy from the orbit of a planet to a passing satellite. Some of the planet's momentum is transferred to the spacecraft as it passes by during a close approach. (p. 56)

graviton Particle which mediates the gravitational force, according to supergravity theory. (p. 629)

Great Attractor Huge accumulation of mass that has produced a measurable effect on the motion of galaxies in the local supercluster. (p. 605)

Great Dark Spot Prominent storm system in the atmosphere of Neptune, located near the equator of the planet and nearly the size of the Earth. (p. 313)

Great Red Spot A large, high-pressure, long-lived storm system visible in the atmosphere of Jupiter. (p. 265)

Great Wall Extended sheet of galaxies discovered in the CfA redshift survey, spanning at least 200 megaparsecs across. (p. 560)

Green Bank Equation Expression which gives an estimate of the probability that intelligence exists elsewhere in the galaxy, based on a number of supposedly necessary conditions for intelligent life to develop. (p. 648)

greenhouse effect The partial trapping of solar radiation by a planetary atmosphere, similar to the trapping of heat in a greenhouse. (p. 145)

ground state The lowest energy state that an electron can have within an atom. (p. 92)

habitable zone A three-dimensional zone around a star representing the radii within which an Earth-like planet would have a surface temperature between the freezing and boiling points of water. (p. 650)

Hawking radiation A process, first realized by Stephen Hawking, by which black holes slowly "evaporate." A black hole loses mass into space by Hawking radiation. (p. 511)

Hayashi track The final stages of stellar formation, during which nearly all of the protostar is convective. The temperature remains constant (about 3000K) while the protostar contracts and the luminosity decreases as the protostar nears the main sequence. (p. 439)

head-tail radio galaxy Type of active galaxy in which the lobes trail the galaxy as a result of motion of the galaxy through the intergalactic medium. (p. 574)

heliocentric A model of the solar system which is centered on the Sun, with the Earth in motion about the Sun. (p. 42)

helioseismology The study of conditions far below the Sun's surface through the analysis of internal "sound" waves that repeatedly cross the solar interior. (p. 367)

helium capture The formation of heavier elements by the capture of a helium nucleus. For example, carbon can form heavier elements by fusion with other carbon nuclei, but it is much more likely to occur by helium capture, which requires less energy. (p. 487)

helium flash An explosive event in the post-main-sequence evolution of a low mass star. When helium fusion begins in a degenerate core, the burning is explosive in nature. It continues until the energy released is enough to expand the core so that it is again nondegenerate. After this point the star can achieve stable equilibrium again. (p. 460)

helium precipitation Mechanism responsible for the low abundance of helium in Saturn's atmosphere. Helium condenses in the upper layers to form a mist, which rains down toward Saturn's interior, just as water vapor forms into rain in the atmosphere of Earth. (p. 288)

helium shell flash Event that occurs relatively late in a low mass star's post-main sequence evolution. Very high pressure in the helium burning shell, combined with the instability of the aging star, create temperature changes in the shell which drive shell flashes. (p. 464)

HI region Cloud of neutral atomic hydrogen gas. The Roman numeral I indicates that the gas is unionized. (p. 420)

high-energy telescope Telescope designed to detect radiation in x-ray and gamma ray. (p. 130)

highlands Regions on the surface of the Moon which are elevated several kilometers above the maria. (p. 174)

HII region Cloud of ionized hydrogen gas—an emission nebula. The Roman numeral II indicates that the gas is singly ionized. (p. 420)

Hirayama family A group of asteroids sharing the same orbit, believed to have resulted from the breakup of a large asteroid into a number of smaller ones. (p. 332)

homogeneity Assumed property of the universe such that the number of galaxies in an imaginary large cube of the universe is the same no matter where in the universe the cube is placed. (p. 593)

horizon problem One of two conceptual problems with the standard Big Bang model, which is that regions of the universe that are too far apart to have exchanged information in the age of the universe have very similar properties. (p. 623)

horizontal branch Region of the HR diagram where post-main sequence stars again reach hydrostatic equilibrium. At this point, the star is burning helium in its core, and hydrogen in a shell surrounding the core. (p. 460)

hot dark matter A class of candidates for the dark matter in the universe, composed of lightweight particles, such as neutrinos, much less massive than the electron. (p. 631)

hot longitudes Two diametrically opposite points on the surface of Mercury where, due to the spin-orbit resonance, temperatures on the surface reach their maximum values. (p. 212)

hour Unit used to measure right ascension on the celestial sphere. There are 24 hours in one complete rotation of the celestial sphere. (p. 10)

HR diagram A plot of luminosity (or absolute magnitude) against temperature (or spectral class, or color index) for a group of stars. (p. 400)

Hubble Classification scheme Method of classifying galaxies according to their appearance, developed by Edwin Hubble. (p. 545)

Hubble flow The universal recession of galaxies as described by Hubble's law. (p. 557)

Hubble's constant The constant of proportionality which gives the relation between recessional velocity and distance in Hubble's law. (p. 558)

Hubble's law Law that relates the observed velocity of recession of a galaxy to its distances from us. The velocity of recession of a galaxy is proportional to its distance. (p. 557)

hydrogen envelope An invisible region engulfing the coma of a comet, usually distorted by the solar wind, and extending across millions of kilometers of space. (p. 337)

hydrogen shell burning Fusion of hydrogen in a shell that is driven by contraction and heating of the helium core. Once hydrogen is depleted in the core of a star, hydrogen burning stops and the core contracts due to gravity, causing the temperature to rise, heating the surrounding layers of hydrogen in the star, and increasing the burning rate there. (p. 458)

hydrosphere Layer of the Earth which contains the liquid oceans and accounts for roughly 70 percent Earth's total surface area. (p. 141)

hyperbola Geometric shape of an unbound orbit. Objects under the influence of gravity, but not bound in elliptical orbits, describe hyperbolic trajectories. (p. 57)

image The optical representation of an object that is produced when light from the object is reflected or refracted by mirror or lens. (p. 112)

impact breccia Rock that consists of smaller fragments stuck together. Shock waves and high temperatures produced in meteoritic impacts redistribute rocks on the Moon's surface. (p. 178)

inertia The tendency of an object to continue motion at the same speed and in the same direction, unless acted upon by a force. (p. 52)

inferior conjunction The point in the orbit of an inferior planet when it is at the same longitude in the sky as the Sun, and is between the Sun and the Earth. (p. 219)

inflation Short period of unchecked cosmic expansion which was the result of a false vacuum state produced by the decoupling of the strong force. During inflation, the universe swelled in size by a factor of about 10^{50}. (p. 627)

infrared Region of the electromagnetic spectrum, just outside the visible spectrum, corresponding to light of a slightly longer wavelength than red light. (p. 70)

infrared telescope Telescope designed to detect infrared radiation. These telescopes are designed to be lightweight so that they can be carried above most of Earth's atmosphere by balloons, airplanes, or satellites. (p. 127)

inner core Innermost region of the Earth, surrounded by the outer core and mantle. (p. 142)

instability strip Region on the H-R diagram where stars undergo a pulsationally unstable phase of their evolution. (p. 524)

intensity A basic property of electromagnetic radiation that specifies the amount or strength of the radiation. (p. 74)

intercrater planes Regions on the surface of Mercury that do not show extensive cratering, but are relatively smooth. (p. 210)

interference The ability of two or more waves to interact in such a way that they either reinforce or cancel each other. (p. 66)

interferometer Collection of two or more radio telescopes working together as a team, observing the same object at the same time and at the same wavelength. The effective diameter of an interferometer is equal to the distance between its outermost dishes. (p. 125)

interferometry Technique in widespread use to dramatically improve the resolution of radio maps. More than one radio telescope observes the object at the same time and a computer analyses how the signals interfere with each other. (p. 124)

intergalactic space The virtual void that exists between individual galaxies and galaxy clusters in the universe. (p. 549)

interplanetary space The space between the objects in the solar system. (p. 191)

interstellar dust Microscopic dust grains that populate the space between stars, having their origins in the ejected matter of numerous long-dead stars. (p. 355)

interstellar matter Great clouds of gas and dust that obscure light from stars lying beyond them. This material is spread irregularly throughout space. (p. 416)

interstellar medium The matter between stars, composed of two components, gas and dust, intermixed throughout all of space. (p. 417)

inverse-square law The law that a field follows if the strength of the field decreases with the square of the distance. Fields that follow the inverse square law rapidly decrease in strength as the distance increases, but never quite reach zero. (p. 53)

invisible radiation Regions of the electromagnetic spectrum to which human eyes are completely insensitive. Invisible radiation includes radio waves, infrared and ultraviolet radiation, x-rays and gamma rays. (p. 70)

Io plasma torus Doughnut-shaped region of energetic ionized particles, emitted by the volcanos on Jupiter's moon Io, and swept up by Jupiter's magnetic field. (p. 275)

ion tail Thin stream of ionized gas that is pushed away from the head of a comet by the solar wind. It extends directly away from the Sun. Often referred to as a plasma tail. (p. 337)

ionized State of an atom that has had at least one of its electrons removed. (p. 92)

ionosphere Layer in Earth's atmosphere above about 100 km where the atmosphere is significantly ionized, and conducts electricity. (p. 144)

iron meteorite One of the major types of meteorites, composed mostly of nickel-iron, and of high density. (p. 344)

irregular galaxy A galaxy which does not fit into any of the other major categories in the Hubble classification scheme. (p. 548)

isotopes Nuclei containing the same number of protons but different numbers of neutrons. Most elements can exist in several isotopic forms. A common example of an isotope is deuterium, which differs from normal hydrogen by the presence of an extra neutron in the nucleus. (p. 382)

isotropy Assumed property of the universe such that the universe looks the same in every direction. (p. 593)

Jovian planet One of the four giant outer planets of the solar system, which resembles Jupiter in physical and chemical composition. (p. 195)

Kapteyn universe Incorrect model of the Milky Way Galaxy proposed around 1920, based on optical observations of stellar brightnesses and motions. (p. 528)

Kelvin scale A convenient temperature scale used by scientists. At 0 kelvins (0 K) all atomic and molecular motions stop. This temperature is absolute zero. (p. 74)

Kelvin-Helmholtz contraction phase The path followed on the HR diagram by a protostar as it contracts and heats. The release of gravitational energy gives rise to an increase in luminosity. (p. 439)

Kepler's Laws of Planetary Motion Three laws, based on precise observations of the motions of the planets by Tycho Brahe, which summarize the motions of the planets about the Sun. (p. 47)

Kirchhoff's Laws Three rules governing the formation of different types of spectra. (p. 90)

Kirkwood gaps Gaps in the spacings of semi-major axes of orbits of asteroids in the asteroid belt, produced by dynamical resonances with nearby planets, especially Jupiter. (p. 333)

labeled release experiment A test for life on the Martian surface, performed by the Viking landers, in which compounds containing radioactive carbon were introduced into the soil and measurements taken for signs that organisms had inhaled this carbon. (p. 256)

Lagrange point One of five special points in the plane of two massive bodies orbiting one another, where a third body of negligible mass can remain in equilibrium. (p. 303)

libration Apparent rocking of the side of the Moon facing Earth. The lunar rotation period is synchronized with the average orbital speed, but the orbit speed is not constant because the Moon's orbit is elliptical. (p. 173)

light curve The variation in brightness of a star with time. (p. 406)

lighthouse model The leading explanation for pulsars. A small region of the neutron star, near one of the magnetic poles, emits a steady stream of radiation which sweeps past Earth each time the star rotates. Thus the period of the pulses is just the star's rotation period. (p. 499)

light-year The distance that light, moving at a constant speed of 300,000 km/s, travels in one year. One light year is about 10 trillion kilometers. (p. 5)

limb The sharp edge of the Sun. (p. 370)

limb darkening Phenomenon arising from the temperature change in the solar interior that causes the center of the solar disk to appear brighter than the edge. Since the temperature increases with depth, photons coming from the center of the disk are more energetic than those coming from the limb. (p. 370)

line of nodes The line of intersection of the Moon's orbit with the ecliptic plane. (p. 19)

lithosphere Earth's crust and a small portion of the upper mantle that make up Earth's plates. This layer of the Earth undergoes tectonic activity. (p. 159)

lobe radio galaxy Type of active galaxy that emits most of its radiation from giant radio lobes far from the galactic center. (p. 571)

Local Group The galaxy cluster that includes the Milky Way Galaxy. (p. 551)

look-back time The time taken for the light emitted from a distant object to reach our detectors. Look-back time is a measure of distance. (p. 580)

luminosity One of the basic properties used to characterize stars, luminosity is defined as the total energy radiated by a star each second, at all wavelengths. (p. 367)

luminosity class A classification scheme which groups stars according to the width of their spectral lines. For a group of stars with the same temperature, luminosity class differentiates between supergiants, giants, main-sequence stars and subdwarfs. (p. 405)

lunar eclipse Celestial event during which the Moon passes through the shadow of the Earth, temporarily darkening its surface. (p. 15)

lunar phase The appearance of the moon at different points along its orbit. (p. 13)

MACHO Term used to describe stellar dark matter candidates residing in galactic halos, an acronym for MAssive Compact Halo Object. (p. 537)

Magellanic clouds Two small irregular galaxies that are gravitationally bound to the Milky Way Galaxy. (p. 549)

magnetic field Field which accompanies a changing electric field, and governs the influence of magnetized objects on one another. (p. 68)

magnetopause The boundary between the Earth's magnetosphere and the flow of high-energy particles in the solar wind. (p. 151)

magnetosphere A zone of charged particles trapped by the Earth's magnetic field, lying above the atmosphere. (p. 141)

main sequence A well-defined band on an HR diagram, on which most stars tend to fall, running from the top left of the diagram to the bottom right. (p. 401)

main-sequence turnoff Special point on an HR diagram for a cluster. If all the stars in a particular cluster are plotted, then the lower mass stars will trace out the main sequence up to the point where stars begin to evolve off the main sequence toward the red giant branch. The point where stars are just beginning to evolve off is the main-sequence turnoff point. (p. 470)

major axis The long axis of an ellipse, containing the two foci. (p. 48)

mantle Layer of the Earth just interior to the crust. (p. 142)

mare Relatively dark and smooth region on the surface of the Moon. (p. 174)

mass A measure of the total amount of matter contained within an object. (p. 140)

mass-luminosity relation The dependence of the luminosity of a main sequence star on its mass. The luminosity increases roughly like the mass raised to the third power. (p. 408)

mass-radius relation The dependence of the radius of a main sequence star on its mass. The radius rises roughly in proportion to the mass. (p. 408)

mass-transfer binary See semi-detached binary. (p. 473)

matter-dominated universe A universe in which the density of matter exceeds the density of radiation. The present-day universe is matter-dominated. (p. 615)

Matter Era The period of time during which the density of matter in the universe has exceeded the density of radiation. It includes the present epoch of the universe. (p. 615)

Maunder minimum Extended period between the years 1645 and 1715, during which time there was little or no solar activity, and the sunspot cycle apparently stopped. (p. 378)

mesosphere The layer of Earth's atmosphere from 50 to 90 km above the surface of the Earth. (p. 144)

meteor Bright streak in the sky, often referred to as a "shooting star," resulting from a small piece of interplanetary debris entering Earth's atmosphere and heating air molecules, which emit light as they return to their ground states. (p. 340)

meteor shower Event during which many meteors can be seen each hour, caused by the yearly passage of the Earth through the debris spread along the orbit of a comet. (p. 340)

meteorite Any part of a meteoroid that survives passage through the atmosphere and lands on the surface of Earth. (p. 340)

meteoroid Chunk of interplanetary debris prior to encountering the atmosphere of Earth. (p. 340)

meteoroid swarm Pebble-sized cometary fragments disloged from the main body, moving in nearly the same orbit as the parent comet. (p. 341)

microlensing Gravitational lensing arising from individual stars in a foreground galaxy. (p. 584)

micrometeoroid Relatively small chunks of interplanetary debris ranging from dust particle size to pebble-sized fragments. (p. 341)

millisecond pulsar A pulsar whose period indicates that the neutron star is rotating nearly 1000 times each second. The most likely explanation for these rapid rotators is that the neutron star has been spun up by drawing in matter from a companion star. (p. 502)

minute Unit used to measure right ascension. There are 60 minutes in one hour of right ascension. (p. 10)

modeling Answering questions about elusive objects, such as determining the distance to or composition of a planet, by constructing a mathematical description of the object. (p. 27)

molecular cloud Cold, dense interstellar cloud in which there is a high fraction of molecules. It is widely believed that the relatively high density of dust particles in these clouds plays an important role in the formation and protection of the molecules. (p. 428)

molecular cloud complex Collection of molecular clouds that spans as much as 50 parsecs and may contain enough material to make millions of solar-sized stars. (p. 429)

molecule A tightly bound collection of atoms held together by the electromagnetic fields of the atoms. Molecules, like atoms, emit and absorb photons at specific wavelengths. (p. 102)

momentum The product of a body's mass and its velocity, providing a measure of the inertia of the body. In the absence of a force, the momentum of an object stays constant. (p. 54)

moon A small body that is in orbit about a planet. (p. 190)

neap tide Weakest ocean tide which results when the line from the Sun to the Earth is perpendicular to the line from the Earth to the Moon. (p. 143)

nebula General term used for any fuzzy patch on the sky, either light or dark. (p. 420)

nebular theory One of the earliest models of solar system formation, dating back to Descartes, in which a large cloud of gas began to collapse under its own gravity to form the Sun and planets. (p. 352)

neutrino Virtually massless and chargeless particle that is one of the products of fusion reactions in the Sun. Neutrinos move very close to the speed of light, and interact with hardly anything. (p. 381)

neutrino oscillations Possible solution to the solar neutrino problem, in which the neutrino has a very tiny mass. In this case, the correct number of neutrinos can be produced in the solar core, but on their way to Earth, some can "oscillate," or transform into other particles, and go undetected. (p. 386)

neutron An elementary particle with roughly the same mass as a proton, but which is electrically neutral. Neutrons reside in the nuclei of atoms. (p. 94)

neutron degeneracy pressure Pressure that results when neutrons are pushed together to the point of contact. Neutrons can be considered to be hard spheres that resist compression when they are in contact, in much the same way that electrons create a degeneracy pressure in red giants and white dwarfs. (p. 481)

neutron star A dense ball of neutrons that remains at the core of a star after a supernova explosion has destroyed the rest of the star. Typical neutron stars are about 20 km across, and contain more mass than the Sun. (p. 497)

neutronization A process that occurs in the collapsing core of a high-mass star, prior to a supernova explosion. The collapse causes core densities to rise to the point where protons and electrons are crushed together to form neutrons and neutrinos. (p. 481)

new Moon Phase of the moon during which none of the disk is visible. (p. 13)

Newtonian mechanics The basic laws of motion, postulated by Newton, which are sufficient to explain and quantify virtually all of the complex dynamical behavior on Earth and elsewhere in the universe. (p. 51)

Newtonian telescope A reflecting telescope in which incoming light is intercepted before it reaches the prime focus and is deflected into an eyepiece at the side of the instrument. (p. 110)

nonthermal radiation Any type of radiation that is not described by a Planck curve, and therefore does not provide a simple relation to determine the temperature of the radiating object. (p. 272)

north celestial pole Point on the celestial sphere directly above the Earth's north pole. (p. 9)

nova A star that suddenly increases in brightness, often by a factor of as much as 10,000, then slowly fades back to its original luminosity. A nova is the result of an explosion on the surface of a white dwarf star, caused by matter falling onto its surface from the atmosphere of a binary companion. (p. 478)

nucleotide base An organic molecule which is the building block of genes that pass on hereditary characteristics from one generation to the next in living creatures. (p. 640)

nucleus Dense, central region of an atom, containing both protons and neutrons, and orbited by electrons. (p. 91)

nucleus The solid region of ice and dust that composes the central region of the head of the comet. (p. 337)

nutation The wobbling of the Earth's rotation axis caused by the Moon's gravitational pull. (p. 21)

oblateness The deviation of a body from a perfect sphere. Often the result of rotation, which produces a bulge at the planet's equator. (p. 263)

Olbers's paradox Thought experiment that concludes, if the universe is homogeneous, infinite and unchanging, then the entire sky should be as bright as the surface of the Sun. (p. 594)

Oort cloud Spherical halo of material surrounding the solar system, out to a distance of about 50,000 A.U. where most comets originate. (p. 336)

opacity A quantity that measures a material's ability to block electromagnetic radiation. Opacity is the opposite of transparency. (p. 70)

open cluster Loosely bound collection of tens to hundreds of stars, a few parsecs across, generally found in the plane of the Milky Way. (p. 411)

open universe Geometry that the universe would have if the density of matter were less than the critical value. In an open universe there is not enough matter to halt the expansion of the universe. (p. 600)

opposition The configuration of a planet when its elongation angle is 180 degrees, for example when the Earth lies between Mars and the Sun. (p. 242)

optical double A pair of stars that appear very close in the sky, even though they are unrelated, and are actually widely separated. (p. 407)

orbital One of a set of sharply defined energy states allowed to an electron that is bound to a particular atom. The electron cannot exist in an energy state other than these allowed orbitals. (p. 92)

organic molecule A carbon-based molecule, such as an amino acid or a nucleotide base—one of the building blocks of life as we know it. (p. 640)

oscillating universe A hypothetical universe that undergoes many cycles of expansion and recontraction. (p. 602)

outer core Interior region of the Earth, surrounded by the mantle, and itself surrounding the inner core. (p. 142)

outer halo Region of galaxy beyond the visible halo where dark matter is believed to reside. Also called the galactic corona. (p. 537)

outflow channel Surface feature on Mars, evidence that liquid water once existed there in great quantity, believed to be the relics of catastrophic flooding about 3 billion years ago. Outflow channels are found only in the equatorial regions of the planet. (p. 251)

ozone layer Layer of Earth's atmosphere at an altitude of 20 to 50 km where incoming ultraviolet solar radiation is absorbed by oxygen, ozone and nitrogen in the atmosphere. (p. 71)

P-wave Pressure waves that alternately expand and compress the Earth as they travel from the site of an earthquake. P-waves are the first waves to arrive at detectors after the earthquake. (p. 152)

pair production Process in which two photons of electromagnetic radiation give rise to a particle-antiparticle pair. (p. 615)

paleomagnetism The study of the ancient magnetic activity of the Earth, based on evidence found in rock samples. (p. 163)

parabola A special orbit, the transition between a bound ellipse and an unbound hyperbola, in which one object has just enough energy to escape the gravitational field of another. (p. 57)

parallactic angle Quantity that measures the parallax of a nearby object with respect to the background stars. (p. 29)

parallax The apparent motion of a relatively close object with respect to a more distant background as the location of the observer changes. (p. 29)

parsec The distance at which a star must lie in order that its measured parallax is exactly 1 arc second, equal to 206,000 A.U. (p. 390)

partial eclipse Celestial event during which only a part of the occulted body is blocked from view. (p. 15)

Pauli exclusion principle A rule of quantum mechanics which prohibits the electrons in the core of a star from being squeezed too closely together. Electrons can be thought of as tiny rigid spheres which are virtually incompressible if they are brought into contact. (p. 460)

penumbra Portion of the shadow cast by an eclipsing object in which the eclipse is seen as partial. (p. 16)

penumbra The outer region of a sunspot, surrounding the umbra, which is not as dark, and not as cool as the central region. (p. 374)

perigee Closest approach to the Earth of any object in orbit about it. (p. 170)

perihelion Closest approach to the Sun of any object in orbit about it. (p. 48)

period-luminosity relation A relation between the pulsation period of a Cepheid variable and its absolute brightness. Measurement of the pulsation period allows the distance of the star to be determined. (p. 525)

permafrost Layer of permanently frozen water ice, believed to lie just under the surface of Mars. (p. 250)

perturbation theory A mathematical method developed for understanding the orbital motions of planets in the solar system. The Sun is responsible for most of the motion, but the masses of the other planets cause small deviations, or perturbations, from purely "Keplerian" motion. (p. 308)

photodisintegration Reaction that happens in the highly evolved cores of massive stars. The core reaches such extreme temperatures that photons can break apart heavy elements into lighter nuclei, and eventually into protons and neutrons. Prior to a supernova explosion, photodisintegration in the collapsing core effectively undoes all of the effects of nuclear fusion that occurred during the previous 10 billion years. (p. 481)

photoelectric effect Experiment concerning the detection of electrons from a metal surface, whose speed off the surface was dependent on the frequency of light striking the surface. The theoretical explanation rests on viewing light as made up of photons, or individual "bullets" of energy. (p. 72)

photometer A device that measures the total amount of light received in all or part of the image. (p. 112)

photometry Branch of observational astronomy in which intensity measurements are made through each of a set of standard filters. (p. 398)

photon Individual packet of electromagnetic energy that makes up electromagnetic radiation. (p. 73)

photon sphere Imaginary sphere surrounding the center of a black hole, with radius equal to 1.5 times the Schwarzschild radius, where photons move in circular orbits, never escaping the black hole's gravity, but never crossing the event horizon. (p. 508)

photosphere The visible surface of the Sun, lying just above the uppermost layer of the Sun's interior, and just below the chromosphere. (p. 365)

pixel One of many tiny picture elements, organized into an array, making up a digital image. (p. 118)

Planck curve See black-body curve. (p. 75)

Planck's constant Constant of proportionality connecting the energy of a photon and the frequency of the electromagnetic radiation it represents. (p. 73)

planet One of nine major bodies that orbit the Sun, and are visible to us by reflected sunlight. (p. 190)

planetary nebula The ejected envelope of a red giant star, spread over a volume roughly the size of our solar system. (p. 464)

planetary ring system Material organized into thin, flat rings encircling a planet, such as Saturn. (p. 289)

planetesimal Term given to objects in the early solar system that had reached the size of small moons, at which point their gravitational fields were strong enough to begin to influence their neighbors. (p. 356)

plasma tail See ion tail.

plate tectonics The motions of regions of the Earth's lithosphere, which drift with respect to one another. The effect is also known as continental drift. (p. 157)

polarization The alignment of the electric fields of emitted photons, which are generally emitted with random orientations. (p. 418)

positron Atomic particle with properties identical to those of a negatively charged electron, except for its positive charge. The positron is the antiparticle of the electron. Positrons and electrons annihilate one another when they meet, producing pure energy in the form of gamma rays. (p. 381)

precession The slow change in the direction of the axis of a spinning object, caused by some external influence. (p. 20)

precession of the equinoxes The drift of the vernal equinox around the zodiac as the axis of the spinning Earth changes its orientation. (p. 21)

pressure broadening Smearing of an observed spectral line caused by electrons in an atom moving between orbitals while the atom is undergoing a collision. Common in dense gases. (p. 102)

primary atmosphere The chemical components that would have surrounded the Earth just after it formed. (p. 147)

prime focus The point in a reflecting telescope where the mirror focuses incoming light to a point. (p. 107)

primitive meteorite A meteorite showing no evidence for heating at any time in the past, so it retains the composition that it had during the formation of the solar system. All primitive meteorites are stones. (p. 344)

primordial nucleosynthesis The production of elements heavier than hydrogen by nuclear fusion in the high temperatures and densities which existed in the early universe. (p. 619)

Principle of Cosmic Censorship A proposition to separate the unexplained physics near a singularity from the rest of the well behaved universe. The principle states that nature always hides any singularity, such as a black hole, inside an event horizon, which insulates the rest of the universe from seeing the singularity. (p. 513)

progenitor A star that generates a supernova explosion. (p. 482)

prominence Loop or sheet of glowing gas ejected from an active region on the solar surface, which then moves through the inner parts of the corona under the influence of the Sun's magnetic field. (p. 379)

proper motion The angular movement of a star across the sky, as seen from Earth, measured in seconds of arc per year. This movement is a result of the star's actual motion through space. (p. 392)

proton An elementary particle, carrying a positive electric charge, a component of all atomic nuclei. The number of protons in the nucleus of an atom dictates what type of atom it is. (p. 91)

proton-proton chain The chain of fusion reactions, leading from hydrogen to helium, that powers low-mass stars. (p. 382)

protoplanet Clump of material, formed in the early stages of solar system formation, that was the forerunner of the planets we see today. (p. 352)

protostar Stage in star formation when the interior of a collapsing fragment of gas is hot and dense enough that it becomes opaque to its own radiation. The protostar is the dense region at the center of the fragment. (p. 438)

protoSun The central accumulation of material in the early stages of solar system formations, the forerunner of the present-day Sun. (p. 352)

Ptolemaic model Solar system model, developed by the second century astronomer Claudius Ptolemy, perhaps the best geocentric model to be proposed. It predicted with great accuracy the positions of the known planets, using more than 80 circles to model the system. (p. 33)

pulsar Object that emits radiation in the form of rapid pulses with a characteristic pulse period and duration. Charged particles, accelerated by the magnetic field of a rapidly rotating neutron star, flow along the magnetic field lines, producing radiation that beams outward as the star spins on its axis. (p. 497)

pulsar glitch Observed small drop in the pulse period of a pulsar. Such glitches are used to study the interiors of neutron stars. They occur when the thin crust that makes up the surface of the neutron star cracks and settles slightly atop the superfluid interior. (p. 499)

pulsating variable A star whose luminosity varies in a predictable, periodic way. (p. 523)

pyrolitic release experiment A test for life on the Martian surface, performed by the Viking landers, in which radioactively tagged carbon dioxide was introduced into a sample of Martian soil and atmosphere, and measurements taken for signs that something had absorbed the tagged gas. (p. 256)

quantum fluctuation Small deviation in the vacuum of empty space that arises due to the spontaneous creation of particle-antiparticle pairs. (p. 629)

quarter moon Lunar phase in which the moon appears as a half disk. (p. 13)

quasar Star-like radio source with an observed redshift that indicates extremely large distance. (p. 579)

quasi-stellar object (QSO) See quasar. (p. 579)

quiescent prominence Relatively stable prominence, which can persist for days or even weeks, hovering above the solar photosphere, suspended by the Sun's magnetic field. (p. 379)

quiet Sun The underlying predictable elements of the Sun's behavior, such as its average photospheric temperature, which do not change in time. (p. 374)

r-process Creation of heavy elements by neutron capture during supernova explosions. Free neutrons stream from the inner regions of the exploding star and collide with heavy elements surrounding the core. The neutron capture rate during supernovae explosions is so great that even unstable nuclei may capture another neutron before decaying. The r-process is responsible for the creation of the heaviest elements in the universe. (p. 489)

radar Acronym for RAdio Detection And Ranging. Radio waves are bounced off an object, and the echo indicates its distance. (p. 50)

radial motion Motion along a particular line of sight, which induces apparent changes in the wavelength (or frequency) of radiation received. (p. 79)

radial velocity Component of a star's space motion that is along our line of sight, and therefore does not contribute to the observed proper motion. (p. 393)

radiant The constellation from whose direction a meteor shower appears to come. (p. 340)

radiation A way in which energy is transferred from place to place in the form of a wave. Light is a form of electromagnetic radiation. (p. 64)

radiation background The remnant radiation field, at times after the Matter Era has begun. (p. 622)

radiation darkening The effect of chemical reactions that result when high-energy particles strike the icy surfaces of objects in the outer solar system. The reactions lead to a build-up of a dark layer of material. (p. 316)

radiation-dominated universe Early times in the universe, when the density of radiation in the cosmos exceeded the density of matter. (p. 615)

Radiation Era Early times in the history of the universe, when the cosmos was radiation dominated. (p. 615)

radiation zone Region of the Sun's interior where extremely high temperatures guarantee that the gas is completely ionized. Photons are only occasionally diverted by electrons, and travel through this region with relative ease. (p. 368)

radio Region of the electromagnetic spectrum corresponding to radiation of the longest wavelengths. (p. 70)

radio lobe Roundish region of radio-emitting gas, lying well beyond the center of an active galaxy. (p. 571)

radio telescope Large instrument designed to detect radiation from space in radio wavelengths. (p. 121)

radioactivity The release of energy by rare, heavy elements that emit energy when their nuclei decay into lighter nuclei. (p. 154)

radius-luminosity-temperature relation A mathematical proportionality, arising from Stefan's Law, which allows astronomers to indirectly determine the radius of a star once its luminosity and temperature are known. (p. 394)

recombination See decoupling. (p. 622)

recurrent nova A star that has been observed to "go nova" a number of times over the course of several decades. (p. 479)

red dwarf Small, cool faint star at the lower-right end of the main sequence on the HR diagram, whose color and size give them their name. (p. 402)

red giant A giant star whose surface temperature is relatively low, so that it glows with a red color. (p. 394)

red-giant branch The section of the evolutionary track of a star that corresponds to continued heating from rapid hydrogen shell burning, which drives a steady expansion and cooling of the outer envelope of the star. As the star gets larger in radius and its surface temperature cools, it becomes a red giant. (p. 459)

red giant region The upper right hand corner of the HR diagram, where the red giant stars are found. (p. 403)

red shift Motion-induced change in the wavelength of light emitted from a source moving away from us. The relative recessional motion causes the wave to have an observed wavelength longer (and hence redder) than it would if it were not moving. (p. 79)

red supergiant An extremely luminous red star. Often found on the asymptotic giant branch of the HR diagram. (p. 463)

reddening Dimming of starlight by interstellar matter, which tends to scatter higher-frequency (blue) components of the radiation more efficiently than the lower-frequency (red) components. (p. 417)

reflecting telescope A telescope which uses a carefully designed mirror to gather and focus light from a distant object. (p. 107)

refracting telescope A telescope which uses a lens to gather and focus light from a distant object. (p. 107)

regolith The layer of pulverized ejecta that covers the lunar landscape to an average depth of about 20 meters. (p. 179)

regression of the line of nodes Backward progression of the line of nodes of the Moon's orbit due to the Sun's gravitational pull. (p. 19)

Renaissance Historical period with its roots in early fifteenth century Italy, which marked a rebirth of artistic, philosophical, and scientific inquiry. (p. 41)

residual cap Portion of Martian polar ice caps that remains permanently frozen, undergoing no seasonal variations. (p. 252)

retrograde motion Backward, westward loop traced out by a planet with respect to the fixed stars. (p. 32)

revolution Orbital motion of one body about another, such as the Earth about the Sun. (p. 11)

right ascension Celestial coordinate used to measure longitude on the celestial sphere. The zero point is the position of the Sun on the vernal equinox. (p. 9)

rille A ditch on the surface of the Moon where molten lava flowed in the past. (p. 180)

ringlet Narrow region in Saturn's planetary ring system where the density of ring particles is high. Voyager discovered that the rings visible from Earth are actually composed of tens of thousands of ringlets. (p. 291)

Roche limit Often called the tidal stability limit, the Roche limit gives the distance from a planet at which the tidal force, due to the planet, between adjacent objects exceeds their mutual attraction. Objects within this limit are unlikely to accumulate into larger objects. The rings of Saturn occupy the region within Saturn's Roche limit. (p. 290)

Roche lobe An imaginary surface around a star. Each star in a binary system can be pictured as being surrounded by a tear-shaped zone of gravitational influence, called the Roche lobe. Any material within the Roche lobe of a star can be considered to be part of that star. During evolution, one member of the binary star can expand so that it overflows its own Roche lobe, and begins to transfer matter onto the other star. (p. 472)

rotation Spinning motion of a body about an axis. (p. 9)

rotation curve Plot of the orbital speed of disk material in a galaxy against its distance from the galactic center. Analysis of rotation curves of spiral galaxies indicates the existence of dark matter. (p. 536)

RR Lyrae star Variable star whose luminosity changes in a characteristic way. All RR Lyrae stars have more or less the same period. (p. 524)

runaway greenhouse effect A process in which the heating of a planet leads to an increase in its atmosphere's ability to retain heat and thus to further heating, quickly causing extreme changes in the temperature of the surface and the composition of the atmosphere. (p. 228)

runoff channel River-like surface feature on Mars, evidence that liquid water once existed there in great quantities. They are found in the southern highlands, and are thought to have been formed by water that flowed nearly 4 billion years ago. (p. 251)

s-process Creation of heavy elements in the cores of highly evolved stars by neutron capture. Neutrons are captured by nuclei until an unstable isotope is created. This nucleus then decays into a new stable nucleus, and the process continues until no heavier stable nuclei exist. In the cores of stars, the time between successive neutron captures is about a year. The "s" stands for "slow," meaning that the time between captures is long compared with the half-lives of the radioactive elements produced. (p. 489)

S-type asteroid A fairly reflective type of asteroid, containing rocky or silicate material. (p. 334)

S-wave Shear waves that travel from the site of an earthquake. S-waves cause side-to-side motion, and are detected after P-waves. (p. 152)

S0 galaxy Galaxy which shows evidence of a thin disk and a bulge, but which has no spiral arms and contains little or no gas. (p. 548)

SB0 galaxy An S0 type galaxy whose disk shows evidence of a bar. (p. 548)

scarp Surface feature on Mercury believed to be the result of cooling and shrinking of the crust forming a wrinkle on the face of the planet. (p. 210)

Schwarzschild radius The distance from the center of an object such that, if all the mass were compressed within that region, the escape velocity would equal the speed of light. Once a stellar remnant collapses within this radius, light cannot escape and the object is no longer visible. (p. 506)

scientific method The set of rules used to guide science, based on the idea that scientific "laws" be continually tested, and replaced if found inadequate. (p. 3)

seasonal cap Portion of Martian polar ice caps that is subject to seasonal variations, growing and shrinking once each Martian year. (p. 252)

seasons Changes in average temperature and length of day that result from the tilt of Earth's (or any planet's) axis with respect to the plane of its orbit. (p. 12)

second Unit used to measure right ascension. There are 60 seconds in one minute of right ascension. (p. 10)

secondary atmosphere The chemicals that composed the Earth's atmosphere after the planet's formation, once volcanic activity outgassed chemicals from the interior. (p. 147)

seeing A term used to describe the ease with which good telescopic observations can be made from Earth's surface, given the blurring effects of atmospheric turbulence. (p. 117)

seeing disk Roughly circular region on a detector over which a star's pointlike images is spread, due to atmospheric turbulence. (p. 117)

seismic wave A wave that travels outward from the site of an earthquake through the Earth. (p. 152)

self-propagating star formation A proposed explanation for the existence of spiral arms, in which the formation of an earlier group of stars provides the mechanism for the creation of a later group of stars. (p. 530)

semi-detached binary A binary system in which one star has evolved and its envelope has expanded so that it overflows its Roche lobe. Mass transfer begins as gas begins to flow onto the companion through the Lagrange point (where the Roche lobes meet). (p. 473)

semi-major axis One half of the major axis of an ellipse. The semi-major axis is the way in which the size of an ellipse is usually quantified. (p. 48)

Seyfert galaxy Type of active galaxy whose emission comes from a very small region in the galactic nucleus. (p. 569)

shepherd satellites Satellites whose gravitational effects on a ring preserve its shape, such as the two satellites of Saturn, Prometheus and Pandora, whose orbits lie on either side of the F ring. (p. 295)

shield volcano A volcano produced by repeated nonexplosive eruptions of fluid basalts, creating a gradually sloping, shield-shaped low dome. Often contains a caldera at its summit. (p. 234)

shock wave Wave of matter, which may be generated by a star, which pushes material outward into the surrounding molecular cloud. The material tends to pile up, forming a rapidly-expanding shell of dense gas. (p. 448)

sidereal day The time needed for a star on the celestial sphere to make one complete rotation in the sky. (p. 10)

sidereal month Time required for the Moon to complete one trip around the celestial sphere. (p. 15)

sidereal year The time required for the constellations to complete one cycle around the sky and return to their starting points, as seen from a given point on Earth. (p. 13)

singularity A point in the universe where the density of matter and the gravitational field are infinite, such as at the center of a black hole. (p. 504)

solar constant The amount of solar energy reaching the Earth per unit area per unit time, approximately 1.4×10^6 erg/cm^2/sec. (p. 366)

solar core The region at the center of the Sun, with a radius of nearly 200,000 km, where the powerful nuclear reactions occur that generate the Sun's energy output. (p. 365)

solar cycle The 22-year period that is needed for both the average number of spots and the Sun's magnetic polarity to repeat themselves. The Sun's polarity reverses on each new 11-year sunspot cycle. (p. 378)

solar day The period of time between the instant when the Sun is directly overhead (i.e. at noon) to the next time it is directly overhead. (p. 10)

solar disk The glowing surface of the Sun that we see with our eyes or through a heavily filtered telescope. (p. 365)

solar eclipse Celestial event during which the new Moon passes directly between the Earth and Sun, temporarily blocking the Sun's light. (p. 15)

solar interior The region of the Sun's interior between the solar core and the bottom of the convective zone. (p. 365)

solar maximum Approximately 4 years into a sunspot cycle, when the number of spots has increased markedly. At this point most spots are found between 15 and 20 degrees of the equator. (p. 378)

solar minimum The starting point of the sunspot cycle, during which only a few spots are seen. They are generally confined to narrow regions, one in each hemisphere, at about 25–30 degrees latitude. (p. 378)

solar nebula The swirling gas surrounding the early Sun during the epoch of solar system formation, also referred to as the primitive solar system. (p. 352)

solar neutrino problem The discrepancy between the theoretically predicted flux of neutrinos streaming from the Sun as a result of fusion reactions in the core and the flux which is actually observed. The observed number of neutrinos is only about half the predicted number. (p. 385)

solar system The Sun, and all the planets that orbit the Sun—Mercury, Venus, Earth, Mars, Jupiter, Saturn, Uranus, Neptune, and Pluto. (p. 31)

solar wind An outward flow of charged particles from the Sun. (p. 150)

south celestial pole Point on the celestial sphere directly above the Earth's south pole. (p. 9)

speckle interferometry Technique for resolving the disks of nearby, large stars. Many short-exposure images of the star, which are too brief to be affected by the turbulence of Earth's atmosphere, are used to form the composite image. (p. 394)

spectral class Classification scheme, based on the strength of its spectral lines, which is an indication of the temperature of a star. (p. 400)

spectrogram The resulting photograph or digitized computer image from a spectrograph measurement. (p. 85)

spectrograph A complex instrument which produces very detailed spectra of stars. Usually, a spectrograph records the spectrum on a photographic plate, or more recently, in electronic form on a computer. (p. 85)

spectroscope Instrument that is used to view a light source so that it is split into its component colors. (p. 85)

spectroscopic binary A binary-star system which from Earth appears as a single star, but whose spectral lines show back and forth Doppler shifts as two stars orbit one another. (p. 406)

spectroscopic parallax Method of determining distance to a star by measuring its temperature and then determining its absolute brightness by comparing with a standard HR diagram. The absolute and apparent brightnesses of the star give the star's distance from Earth. (p. 403)

spectroscopy The study of the way in which atoms absorb and emit electromagnetic radiation. Spectroscopy allows astronomers to determine the chemical composition of stars. (p. 85)

spin-orbit resonance State that a body is said to be in if its rotation period and its orbital period are related in a simple way. (p. 207)

spiral arm Distribution of material in a galaxy in a pinwheel-shaped design apparently emanating from near the galactic center. (p. 530)

spiral density wave A wave of matter formed in the plane of planetary rings, similar to ripples on the surface of a pond, which wrap around the rings forming spiral patterns similar to grooves in a record disk. Spiral density waves can lead to the appearance of ringlets. (p. 292) A proposed explanation for the existence of spiral arms, in which coiled waves of gas compression move through the galactic disk, triggering star formation. (p. 530)

spiral galaxy Galaxy composed of a flattened, star-forming disk component which may have spiral arms and a large central galactic bulge. (p. 545)

spring tides Highest ocean tides which result when the Sun, Earth and Moon are roughly lined up so that their gravitational effects reinforce each other. (p. 143)

standard candle Any object with an easily recognizable appearance and known luminosity, which can be used in estimating distances. Supernovae, which all have the same peak luminosity (depending on type), are good examples of standard candles and are used to determine distances to other galaxies. (p. 485)

Standard Solar Model A self-consistent picture of the Sun, developed by incorporating the important physical processes that are believed to be important in determining the Sun's internal structure, into a computer program. The results of the program are then compared with observations of the Sun, and modifications are made to the model. The Standard Solar Model is the result of this process, and enjoys widespread acceptance. (p. 367)

star A glowing ball of gas held together by its own gravity and powered by nuclear fusion at its center. (p. 365)

star cluster A grouping of anywhere from a dozen to a million stars which formed at the same time from the same cloud of interstellar gas. Stars in clusters are useful to aid our understanding of stellar evolution because they are all roughly the same age and chemical composition, and lie roughly at the same distance from Earth. (p. 410)

starburst galaxy Galaxy in which a violent event such as near collision has caused a sudden, intense burst of star formation in the recent past. (p. 564)

starquake A cracking and resettling of the crust of a neutron star, slightly reducing the radius of the star. As the radius decreases, the spin rate increases to conserve angular momentum, and a pulsar glitch is observed. (p. 499)

Stefan-Boltzmann constant Constant of proportionality used in Stefan's Law which gives the exact connection between the energy emitted per square centimeter per second for a body with a given temperature. (p. 77)

Stefan's Law Relation that gives the total energy emitted per square centimeter of its surface per second by an object of a given temperature. Stefan's Law shows that the energy emitted increases rapidly with an increase in temperature, proportional to the temperature raised to the fourth power. (p. 77)

stellar nucleosynthesis The formation of heavy elements by the fusion of lighter nuclei in the hearts of stars. Except for hydrogen and helium, all other elements in our universe result from stellar nucleosynthesis. (p. 485)

stellar occultation The dimming of starlight produced when a solar system object such as a planet, moon or ring, passes directly in front of a star. (p. 321)

stone meteorite A meteorite whose composition is comparable to that of terrestrial rocks. (p. 344)

stony-iron meteorite Meteorites which contain a mixture of stones and metals. (p. 344)

stratosphere The portion of Earth's atmosphere lying above the troposphere, extending up to an altitude of 40 to 50 km. (p. 144)

subduction zone Tectonic plate collision locations where one plate slides under another, ultimately being destroyed, such as in the deep trenches of the world's oceans. (p. 159)

subgiant branch The section of the evolutionary track of a star that corresponds to changes that occur just after hydrogen is depleted in the core, and core hydrogen burning ceases. Shell hydrogen burning heats the outer layers of the star, which causes a general expansion of the envelope of the star. (p. 459)

summer solstice Point on the ecliptic where the Sun is at its northernmost point above the celestial equator, occurring on or near June 21. (p. 11)

Sun-grazing comet A comet whose orbit takes it very close to the Sun, at which point it may be destroyed. (p. 339)

sunspot An Earth-sized dark blemish found on the surface of the Sun. The dark color of the sunspot indicates that it is a region of lower temperature than its surroundings. (p. 374)

sunspot cycle The fairly regular pattern that the number and distribution of sunspots follows, in which the average number of spots reaches a maximum every 11 or so years then falls off to almost zero. (p. 376)

supercluster Grouping of several clusters of galaxies into a larger, but not necessarily gravitationally bound, unit. (p. 553)

superconductor Any material that is able to conduct electricity without resistance. The interior of a neutron star may be made of superconducting material. (p. 499)

superfluid Any material that is able to flow without friction. The interior of a neutron star is thought to be a superfluid. (p. 499)

supergiant A star with a radius between 100 and 1000 times that of the Sun. (p. 394)

supergranulation Large-scale flow pattern on the surface of the Sun, consisting of cells measuring up to 30,000 km across, believed to be the imprint of large convective cells deep in the solar interior. (p. 367)

supergravity Theory that introduces gravity into the unified field theories with the addition of a boson called a graviton. (p. 629)

superior conjunction The point in the orbit of a planet when it is at the same longitude in the sky as the Sun, but is further from Earth than is the Sun. (p. 219)

supernova Explosive death of a star, caused by the sudden onset of nuclear burning (type I), or an enormously energetic shock wave (type II). One of the most energetic events of the universe, a supernova may temporarily outshine the rest of the galaxy in which it resides. (p. 482)

supernova remnant The scattered glowing remains from a supernova that occurred in the past. The Crab Nebula is one of the best-studied supernova remnants. (p. 483)

supersymmetric relic Theoretically postulated particles produced during the very early universe, which may exist today as WIMPS. (p. 629)

supersymmetry The theory that all known subatomic particles are actually members of a single family, and can be transformed from one into another.

symmetry A state that exists when forces are unified such that the bosons responsible for the original forces become indistinguishable. (p. 625)

synchronous orbit State of an object when its period of rotation is exactly equal to its average orbital period. The Moon is in a synchronous orbit, and so presents the same face toward Earth at all times. (p. 172)

synchrotron emission Type of nonthermal radiation caused by high-speed charged particles, such as electrons, emitting radiation as they are accelerated in a strong magnetic field. (p. 577)

synodic month Time required for the Moon to complete a full cycle of phases. (p. 15)

T Tauri star Protostars in the late stages of formation, following the Hayashi track on the HR diagram, which exhibit violent surface activity. T Tauri stars have been observed to brighten noticeably in a short period of time, consistent with the idea of rapid evolution during this final phase of stellar formation. (p. 440)

tectonic fracture Surface feature caused by large internal stresses in a planet warping and cracking the crust. Examples are found near the Martian Tharsis bulge, indicating crustal forces at work at least 2 billion years ago. (p. 251)

telescope Instrument used to capture as many photons as possible from a given region of the sky and concentrate them into a focused beam for analysis. (p. 107)

temperature A measure of the amount of heat in an object, and an indication of the speed at which the particles that make up the object move. (p. 74)

terminator Edge on the Moon separating day from night on the surface. (p. 174)

terrae Relatively light, higher-altitude regions on the surface of the Moon. (p. 174)

terrestrial planet One of the four innermost planets of the solar system, which resembles the Earth in general physical and chemical properties. (p. 195)

Theory of Relativity Einstein's theory, on which much of modern physics rests. Two essential facts of the theory are that nothing can travel faster than the speed of light, and that everything, including light, is affected by gravity. (p. 504)

thermal broadening The tendency of atoms in a gas cloud to spread out an observed spectral line. Atoms in the cloud have small thermal velocities, causing the radiation they absorb or emit to be Doppler shifted. The amount of broadening is an indication of the temperature of the gas. (p. 101)

thermal equilibrium Condition satisfied by two bodies when they are at the same temperature. (p. 616)

thermosphere The layer of Earth's atmosphere above 90 km from the surface. (p. 144)

thick disk Region of a spiral galaxy where an intermediate population of stars resides, younger than the halo stars, yet older than the disk stars. (p. 529)

threshold temperature Critical temperature at which particles of a given type can be spontaneously produced by pair production. (p. 616)

tidal bulge Elongation of the Earth caused by the difference between the gravitational force on the side nearest the Moon and the force on the side farthest from the Moon. The long axis of the tidal bulge points toward the Moon. (p. 142) More generally, the deformation of any body produced by the tidal effect of a nearby gravitating object.

tidal force The variation in one body's gravitational force from place to place across another body—for example, the variation of the Moon's gravity across the Earth. (p. 143)

tidal stability limit See Roche limit.

tides Rising and falling motion that bodies of water follow, exhibiting daily, monthly and yearly cycles. Ocean tides on Earth are caused by the competing gravitational pull of the Moon and Sun on different regions of the Earth. (p. 142)

time dilation A prediction of the theory of relativity, closely related to the gravitational redshift. To an outside observer, a clock lowered into a strong gravitational field will appear to run slow. (p. 513)

total eclipse Celestial event during which one body is completely blocked from view by another. (p. 15)

transit Event where a smaller, darker object passes in front of a larger, brighter one. (p. 205)

transition zone The region of rapid temperature increase that separates the Sun's chromosphere from the corona. (p. 373)

transverse motion Motion perpendicular to a particular line of sight, which does not result in a Doppler shift in radiation received. (p. 79)

transverse velocity Component of a star's space motion that is perpendicular to our line of sight, giving rise to the observed proper motion. (p. 393)

triangulation Method of determining distance based on the principles of geometry. A distant object is sighted from two well-separated locations. The distance between the two locations and the angle between the line joining them and the line to the distant object are all that are necessary to ascertain the object's distance. (p. 27)

triple-alpha process The generation of carbon-12 from the fusion of three helium-4 nuclei (alpha particles). Helium-burning stars occupy a region of the HR diagram known as the horizontal branch. (p. 460)

Trojan asteroid One of a group of asteroids locked into a 1:1 orbital resonance with the planet Jupiter, and which orbit at the same distance from the Sun as Jupiter, 60 degrees ahead and behind the planet. (p. 332)

tropical year The time needed for the Sun to make one complete trip around the celestial sphere on the ecliptic. (p. 11)

tropopause The top of the troposphere in Earth's atmosphere. (p. 141)

troposphere The portion of Earth's atmosphere from the surface to about 15 km. (p. 144)

Tully-Fisher relation A relation used to determine the absolute luminosity of a spiral galaxy. The rotational velocity, measured from the broadening of spectral lines, is related to the total mass, and hence the total luminosity. (p. 552)

turbulence Motion of a fluid that is seething and churning, creating eddies and vortices of varying sizes. Turbulence causes observed spectral lines to be broadened. (p. 101)

turnoff mass The mass of a star on a cluster HR diagram that corresponds to the point at which evolution off the main sequence has just begun. (p. 470)

21-centimeter line Radio spectral line emitted when an electron in the ground state of a hydrogen atom flips its spin to become parallel to the spin of the proton in the nucleus. (p. 426)

type–I supernova One possible explosive death of a star. A white dwarf in a binary system can accrete enough mass so that electron degeneracy pressure no longer supports the white dwarf. The star collapses and temperatures reach carbon fusion level. Carbon fusion begins in all parts of the white dwarf almost simultaneously and an explosion results. (p. 482)

type–II supernova One possible explosive death of a star, in which the highly evolved stellar core rapidly implodes and then explodes, destroying the surrounding star. (p. 482)

type I tail Component of material streaming from a comet which is approximately straight, and composed of ionized gases. Often referred to as an ion tail or plasma tail. (p. 337)

type II tail Broad diffuse component of a comet's tail, which curves gently and is composed of microscopic dust particles that reflect sunlight. Sometimes referred to as a dust tail. (p. 337)

UBV system A common method of photometry, in which a specific set of filters, one in the ultraviolet (U), one in the blue (B) and one in the visual (V) portion of the spectrum are used. (p. 398)

ultraviolet Region of the electromagnetic spectrum, just outside the range of the visible spectrum, corresponding to radiation of wavelengths slightly shorter than blue light. (p. 70)

ultraviolet telescope A telescope that is designed to collect radiation in the ultraviolet part of the spectrum. The Earth's atmosphere is partially opaque to these wavelengths, so ultraviolet telescopes are put on rockets, balloons or satellites to get high above most or all of the atmosphere. (p. 129)

umbra Central region of the shadow cast by an eclipsing body. (p. 16)

umbra The central region of a sunspot, which is its darker and cooler part. (p. 374)

unbound An orbit which does not stay in a specific region of space, but one in which an object escapes the gravitational field of another. Typical unbound orbits are hyperbolic. (p. 57)

unification of fundamental forces The joining together of the strong, weak, and electromagnetic forces at the extremely high energies which existed in the early universe. (p. 617)

universe The totality of all space, time, matter and energy. (p. 3)

Urey-Miller experiment Classic experiment in which a mixture of gases similar to the elements present on the primitive Earth is energized by an electrical discharge, mimicking lightning. In a few days' time, the result is the creation of amino acids. (p. 641)

Van Allen belts At least two doughnut-shaped regions of magnetically trapped charged particles high above the Earth's atmosphere. (p. 149)

variable star A star whose luminosity changes with time. (p. 523)

velocity of light The fastest possible speed, according to the currently known laws of physics. Electromagnetic radiation exists in the form of waves or photons moving at the speed of light. (p. 68)

vernal equinox Date on which the Sun crosses the celestial equator moving northward, occurring on or near March 21. (p. 12)

Virial method Technique used for estimating the total mass of a galaxy cluster from information about the speeds of the galaxies within the cluster. (p. 555)

visible spectrum The small range of the electromagnetic spectrum that human eyes perceive as light. The visible spectrum ranges from about 4000 to 7000 angstroms, corresponding to blue through red light. (p. 69–70)

visual binary A binary star system in which both members are resolvable from Earth. (p. 406)

void Large, relatively empty region of the universe around which superclusters of galaxies are organized. (p. 559)

volcano Upwelling of hot lava from below Earth's crust to the planet's surface. (p. 154)

wane Term used to describe section of the lunar cycle during which the total area of the visible moon is shrinking. (p. 13)

water hole The radio interval between 18 cm and 21 cm, the wavelengths at which hydroxyl (OH) and hydrogen (H) radiate, respectively, in which intelligent civilizations might send their communication signals. (p. 654)

wave A pattern that repeats itself cyclically in both time and space. Waves are characterized by the velocity with which they move, their frequency, and their wavelength. (p. 64)

wave crest The highest point of a wave. (p. 65)

wave period The amount of time required for a wave to repeat itself at a specific point in space. (p. 65)

wave trough The lowest point of a wave. (p. 65)

wavelength The length from one point on a wave to the point where it is repeated exactly in space, at a given time. (p. 65)

wax Term used to describe section of the lunar cycle during which the total area of the visible disk is increasing. (p. 13)

weight The force with which one body (such as a moon, a planet, or a star) attracts another. (p. xiv)

weird terrain A region on the surface of Mercury of oddly rippled features. This feature is thought to be the result of a strong impact which occurred on the other side of the planet, and sent seismic waves traveling around the planet, converging in the weird region. (p. 212)

white dwarf A dwarf star with a surface temperature that is hot, so that the object glows white. (p. 394)

white dwarf region The bottom left-hand corner of the HR diagram, where white dwarf stars are found. (p. 403)

white oval Light-colored region near the Great Red Spot in Jupiter's atmosphere. Like the red spot, such regions are apparently rotating storm systems. (p. 269)

Wien's Law Relation which gives the connection between the wavelength at which a black-body curve peaks and the temperature of the emitter. The temperature is inversely proportional to the peak wavelength, so the hotter the object, the bluer its radiation. (p. 76)

WIMP Abbreviation for Weakly Interacting Massive Particle, an "invented" particle that has mass but interacts hardly at all with matter or radiation, making it extremely difficult to detect with conventional means. (p. 386)

winter solstice Point on the ecliptic where the Sun is at its southernmost point below the celestial equator, occurring on or near December 21. (p. 12)

wispy terrain Surface feature on the trailing side of Saturn's moon Rhea, made up of prominent light-colored streaks. These wisps are thought to be associated with an event in the distant past during which water was released from the interior and condensed on the surface. (p. 299)

X boson The theoretically postulated boson that arises when the strong and electroweak forces are unified. (p. 625)

x-ray Region of the electromagnetic spectrum corresponding to radiation of high frequency and short wavelengths, far outside the visible spectrum. (p. 70)

x-ray burster X-ray source that radiates thousands of times more energy than our Sun, in short bursts that last only a few seconds. A neutron star in a binary system accretes matter onto its surface until temperatures reach the level needed for hydrogen fusion to occur. The result is a sudden period of rapid nuclear burning and release of energy. (p. 500)

zero-age main sequence The region on the HR diagram, as predicted by theoretical models, where stars are located at the onset of nuclear burning in their cores. (p. 442)

zodiac The twelve constellations through which the Sun moves as it follows its path on the ecliptic. (p. 11)

zonal flow Alternating regions of westward and eastward flow, roughly symmetrical about the equator of Jupiter, associated with the belts and zones in the planet's atmosphere. (p. 267)

zones Bright, high pressure regions, where gas flows upward, in the atmosphere of a jovian planet. (p. 267)

Photo Credits

Chapter 1 **CO** NASA **1.1** NASA **1.2** AURA **1.3** AURA **1.4** AURA **1.5** D. Berry **1.6b** Harvard-Smithsonian Center for Astrophysics **1.8** AURA **1.14** Lick Observatory **1.16** G. Schneider **1.17** G. Schneider **1.18a** G. Schneider **1.18b** NOAA

Chapter 2 **CO** C. Trost **2.1** G. Schneider **2.7** Boston Museum of Science **2.10** Harvard-Smithsonian Center for Astrophysics **2.12** © *English Heritage* **2.13** J. Cornell **2.14** J. Eddy

Chapter 3 **CO** A. Tannenbaum-Sygma **3.1** Erich Lessing/Art Resource **3.3** Art Resource **3.6** Erich Lessing/Art Resource **3.10** The Granger Collection

Chapter 4 **CO** A. Tannenbaum-Sygma **4.1** AURA **4.5** From the "Atlas of Optical Phenomena," Michel Cagnet, Maurice Francon, Jean Claude Thrierr. © Springer-Verlag OHG, Berlin 1962. Published by Prentice Hall, Inc. Englewood Cliffs, N.J. 07632 **4.13a** AURA **4.13b** NASA **4.13c** NASA **4.13d** NRAO **4.14a** Harvard-Smithsonian Center for Astrophysics **4.14b** Anglo-Australian Telescope Board **4.14c** AURA **4.14d** NASA

Chapter 5 **CO** Harvard College Observatory **5.3** Courtesy Bausch & Lomb **5.4** AURA **5.15** Harvard-Smithsonian Center for Astrophysics **5.21** Courtesy Bausch & Lomb

Chapter 6 **CO** D. Berry **6.4** Yerkes Observatory/University of Chicago **6.6b** Palomar Observatory **6.6c** California Institute of Technology **6.8** Palomar Observatory **6.9** Russia **6.10** AURA **6.11** Harvard-Smithsonian Center for Astrophysics **6.12** AURA **6.14** AURA **6.15** CARA **Int. 6-1** NASA **6.17** Harvard-Smithsonian Center for Astrophysics **6.18** NASA **6.19** European Southern Observatory **6.20a** NRAO **6.21** J. Cornell **6.22** Massachusetts Institute of Technology **6.23** AURA **6.25** NRAO **6.26a** NRAO **6.26b** AURA **6.27** University of Hawaii **6.28** Harvard-Smithsonian Center for Astrophysics **6.29** NASA **6.30** NASA **6.31** NASA **6.32** NASA **6.33** NASA **6.34** NASA **6.36a** AURA **6.36b** NASA **6.36c** NASA **6.37** AURA **6.38** NASA **6.39a** NRAO **6.39b** Lund Observatory **6.39c** NASA **6.39d** NASA **6.39e** NASA

Chapter 7 **CO** A. Tannenbaum-Sygma **Int. 7-1** NCAR **7.11** NCAR **7.17b** NCAR **7.17c** Bettmann Archive **7.19** NASA **7.21a** Lee Day/Black Star **7.21b** NASA **7.22b** Peter Menzel/Stock, Boston; NASA **7.25** Woods Hole Oceanography Institute

Chapter 8 **CO** D. Berry **Int. 8-1** NASA **8.3a** John Sanford/Scientific Photographic Library/Photo Researchers **8.3b/c** Palomar Observatory **8.3d** Lick Observatory **8.4** NASA **8.5** NASA **8.6** NASA **8.8** NASA **8.9** NASA **8.10** United States Geological Survey **8.11** NASA **8.12** NASA **8.13** NASA **8.14** NASA **8.15** NASA **8.17** W. Benz, W. L. Slattery, and A. G. W. Cameron **Int. 8-2** NASA **8.19** United States Geological Survey **8.20** NASA

Chapter 9 **CO** NASA **9.1** Art Resource **9.2** The Birr Scientific and Heritage Foundation **9.3** NASA **9.4** NASA **9.6** NASA **9.7** S. Westphal **9.9** NASA

Chapter 10 **CO** NASA **10.2** C. Trost **10.4** New Mexico State University **10.5** Yerkes Observatory/University of Chicago **10.6** Palomar Observatory **10.9** NASA **10.10** NASA **10.11** NASA **10.12** NASA **10.13** Palomar Observatory/NASA **10.14** NASA **10.15** NASA

Chapter 11 **CO** NASA **11.1** S. Westphal **11.6** Lick Observatory **11.7** NASA **11.11** NASA **11.13a** NAIC **11.13b** NASA **11.13c** Russia **11.13c** NASA **11.15a** Russia **11.15b/c** NASA **11.16** NASA **11.17** NASA **11.18** NASA **11.19** NASA **11.20** NASA **11.21** NASA **11.22** Russia **11.23** Russia

Chapter 12 **CO** NASA **12.2a** Lick Observatory **12.2b** Pic du Midi Observatory **12.2c/d** NASA **Int. 12-1** NASA **12.4a** NASA **12.5** NASA **12.6** NASA **12.7** NASA **12.8a** Lick Observatory **12.8b** NASA **12.9** NASA **12.10** NASA **12.11** NASA **12.12** NASA **12.13** NASA **12.14** NASA **12.15** NASA **12.17** NASA **12.18** NASA

Chapter 13 **CO** NASA **13.1a** Palomar Observatory **13.1b** NASA **13.2** NASA **13.6** NASA **13.9** NASA **13.10** NASA **13.11** NASA **13.13** NASA **13.17** NASA **13.18** NASA **13.19** NASA **13.20** NASA **13.22** NASA **13.23** NASA **13.24** NASA **13.25** NASA **13.26** NASA

Chapter 14 **CO** NASA **14.1** Lick Observatory **14.2** NASA **14.3** NASA **14.5** NASA **14.6** NASA **14.7** NASA **14.9a** Palomar Observatory **14.9b** NASA **14.10** NASA **14.11** NASA **14.12** NASA **14.13** NASA **14.14** NASA **14.16** NASA **14.17** NASA **14.18** NASA **14.20** NASA **14.21** NASA **14.23** NASA **14.24** NASA **14.25** NASA **14.26** NASA **14.27** NASA **14.28** NASA

Chapter 15 **CO** NASA/European Space Agency **15.1** Lick Observatory **15.2** NASA **15.3** Lick Observatory **15.4** NASA **15.5** NASA **15.8** NASA **15.9** NASA **15.12** NASA **15.13** NASA **15.14** NASA **15.15** NASA **15.16** NASA **15.17** NASA **15.18** NASA **15.19** NASA **15.21** NASA **15.22** NASA **15.23** NASA **15.25** Lick Observatory **15.26** NASA; Naval Research Laboratory **15.27** NASA

Chapter 16 **CO** D. Berry **16.5** NASA **16.8a** Palomar Observatory **16.8b** Lick Observatory **16.10a** Naval Research Laboratory **16.10b/c** NOAO **16.11a** Russia **16.11b** European Space Agency **16.13** NASA **16.15** Harvard-Smithsonian Center for Astrophysics **16.16** Brownley, University of Washington **16.17** Canadian Institute for Theoretical Astrophysics **16.19** NASA **16.20** Wood, Smithsonian Observatory

Chapter 17 **CO** D. Berry **17.4** Malin, Anglo-Australian Telescope Board **17.5** NASA **Insert** D. Berry

Chapter 18 **CO** IBM/Smithsonian Astrophysical Observatory **18.3** National Solar Observatory **18.6** Palomar Observatory **18.7** NOAO **18.9** G. Schneider **18.10** NOAO **18.11** Palomar Observatory **18.12** NOAO **18.14** NASA **18.15** Palomar Observatory **18.16** Palomar Observatory **18.17** NASA **18.23a/b** NOAO; NASA **18.24** Palomar Observatory **18.25** G. Schneider **18.28** Ray Davis, Brookhaven National Laboratory

Chapter 19 **CO** NASA **19.3** Harvard-Smithsonian Center for Astrophysics **19.5** NOAO **19.8** S. Westphal **19.15** Harvard-Smithsonian Center for Astrophysics **Int. 19-2** Harvard-Smithsonian Center for Astrophysics **19.18a** NOAO **19.19a** NOAO

Chapter 20 **CO** NASA **20.1** Palomar Observatory **20.5** Palomar Observatory **20.6** Harvard-Smithsonian Center for Astrophysics **20.7** Harvard-Smithsonian Center for Astrophysics **20.8** NOAO **Insert** Anglo-Australian Telescope Board **20.9** Palomar Observatory **20.10** Harvard-Smithsonian Center for Astrophysics **20.11** NASA **20.12** Harvard-Smithsonian Center for Astrophysics **20.13** Royal Observatory **Insert** Anglo-Australian Telescope Board **20.18** NOAO **Int. 20-1** NASA **20.19** NOAO **20.20** NASA

Chapter 21 **CO** D. Berry **Int. 21-1** NOAO **21.10** Anglo-Australian Telescope Board **Insert** Palomar Observatory **21.12** NOAO **21.13** NOAO **21.14a** Harvard-Smithsonian Center for Astrophysics **21.14b** NOAO **21.15** NASA **21.17** NRAO **Int. 21-2** Harvard-Smithsonian Center for Astrophysics **21.20** NASA **21.22** NOAO **Insert** NASA **21.23a** Lick Observatory **21.23b** NASA **21.24** NOAO **21.25** NOAO

Chapter 22 **CO** D. Berry **22.6b** Adapted from a diagram by F. Renzini and A. Fusi Pecci, University of Bologna. **22.10** NOAO **22.11b** Anglo-Australian Telescope Board **22.11c** NASA **22.14** Lick Observatory **22.17a** NOAO **22.18a** NASA **Insert** Harvard-Smithsonian Center for Astrophysics **22.19a** European Southern Observatory **22.19b** Adapted from data provided by J. Hesser, Dominion Astrophysical Observatory **Insert** NASA **Int. 22-1** NASA

Chapter 23 **CO** D. Berry **23.3** Palomar Observatory **23.6** Palomar Observatory **23.8** NOAO **23.8 L** ASP **23.8 R** NASA **23.9** Harvard-Smithsonian Center for Astrophysics **23.10** European Southern Observatory **Int. 23-1** Harvard-Smithsonian Center for Astrophysics **Int. 23-2** European Southern Observatory/NASA

Chapter 24 **CO** D. Berry **24.4 L** Lick Observatory **24.4 R** NASA **24.7** Harvard-Smithsonian Center for Astrophysics; NASA **24.8** NRAO **24.17** Harvard-Smithsonian Center for Astrophysics **24.18** NASA

Chapter 25 **CO** NASA/European Space Agency **25.1** NOAO **Inserts** Palomar Observatory **25.2** NASA **25.3a** Harvard-Smithsonian Center for Astrophysics **Int. 25-1** Harvard-Smithsonian Center for Astrophysics **25-10** NASA; NOAO **25.12a** L. Chaisson **25.12b** Palomar Observatory **25.17** European Southern Observatory **25.18** NASA **25.19** NRAO **25.20** MIT **25.21** L. Chaisson **Insert** NRAO

Chapter 26 **CO** D. Berry **26.1** Palomar Observatory **26.2** Smith, Harvard-Smithsonian Center for Astrophysics **26.4** Palomar Observatory **26.5** AURA; NASA **26.6** NOAO **26.7** Palomar Observatory **26.8** NOAO **26.9** NOAO; NASA; Palomar Observatory **26.10** NOAO **Ins.** H. Leavitt, Harvard-Smithsonian Center for Astrophysics **26.13** Royal Observatory **26.16** Palomar Observatory **26.18** Space Telescope Science Institute **26.19a** NASA **26.19b** AURA **26.20** Carnegie Institute of Washington **26.23** Harvard-Smithsonian Center for Astrophysics **26.27** Harvard-Smithsonian Center for Astrophysics **26.28** J. Barnes/L. Herquist **26.29** NOAO/NRAO

Chapter 27 **CO** NASA **27.2 L** AURA **27.2 R** NASA **27.4** Harvard-Smithsonian Center for Astrophysics **27.5a** AURA **27.5b** NASA **27.5c** NRAO **27.5d** AURA **27.5e** NASA **27.6** NRAO **27.7** NRAO **27.8a** NOAO **27.8b** NRAO **27.9** NASA **27.10** Palomar Observatory, Westerbork Radio Telescope **27.11** D. Berry **27.14** Palomar Observatory **27.15** Palomar Observatory **27.16** Palomar Observatory **27.17** NRAO **27.18** Harvard-Smithsonian Center for Astrophysics **27.19a** NOAO **27.19b** NRAO **27.21** NASA **Int. 27-3** H. Arp

Chapter 28 **CO** Princeton University **28.17** Bell Laboratories

Chapter 29 **CO** Fermi Laboratory **29.14** R. Cen, J. Ostriker; Princeton University **29.15** NASA

Chapter 30 **CO** D. Berry **30.2** NOAO **30.3** NOAO **30.4** NASA **30.5** NOAO **30.6** L. Chaisson **30.9** S. Fox **30.10** E. Barghoorn **30.13** Harvard-Smithsonian Center for Astrophysics **30.14** NASA **30.16** NOAO **30.20** NASA **30.22** NASA

Index

These star maps show the brighter stars and the prominent constellations as they appear on the dates and at the times indicated. To use these maps, face the south and hold the book overhead with top of the map toward the north and the right-hand edge toward the west. The brightest stars are indicated by the star symbol (☆) and the names are indicated. (Star maps courtesy of Robert Dixon, *Dynamic Astronomy,* 6th ed., Prentice Hall, 1992.)

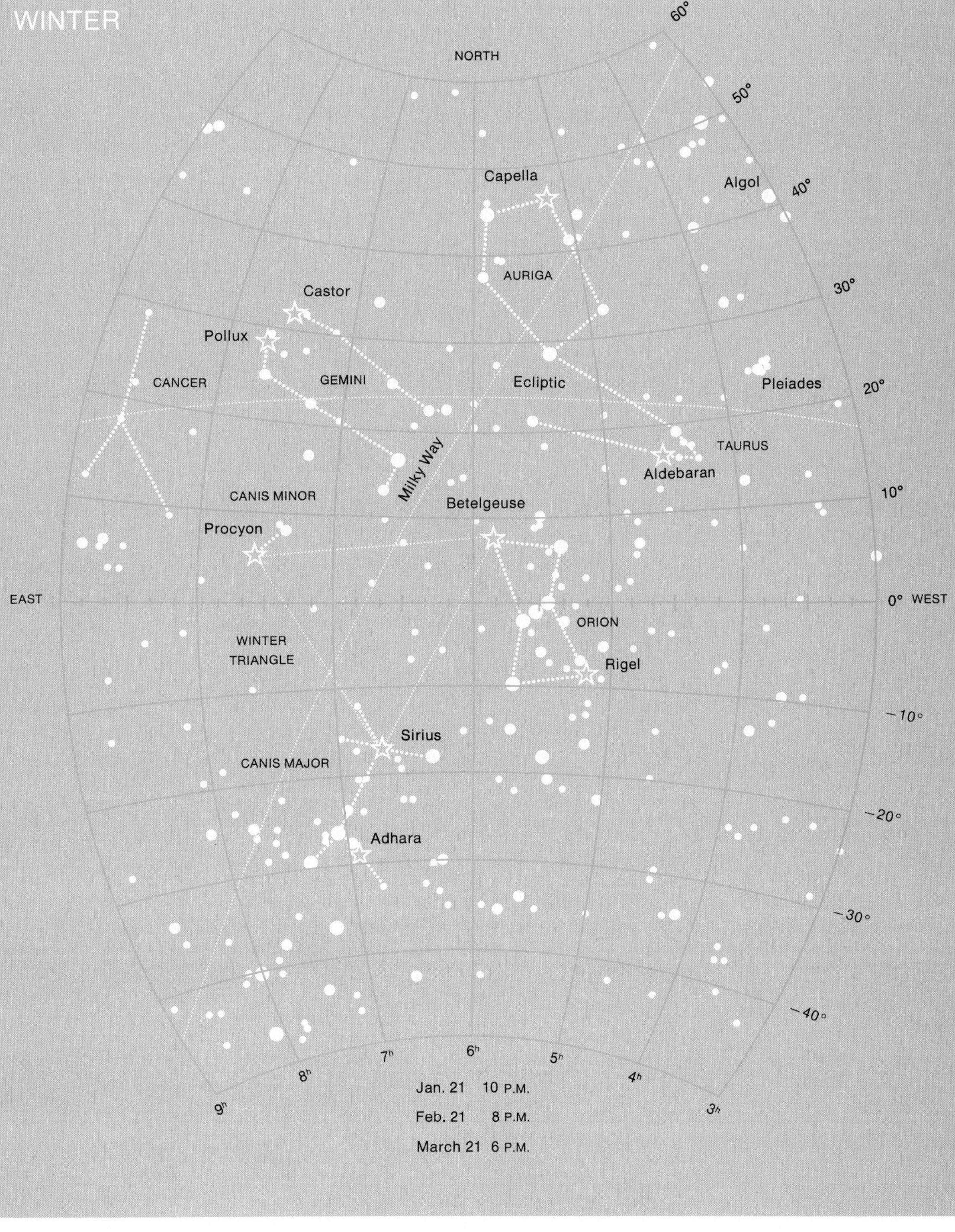

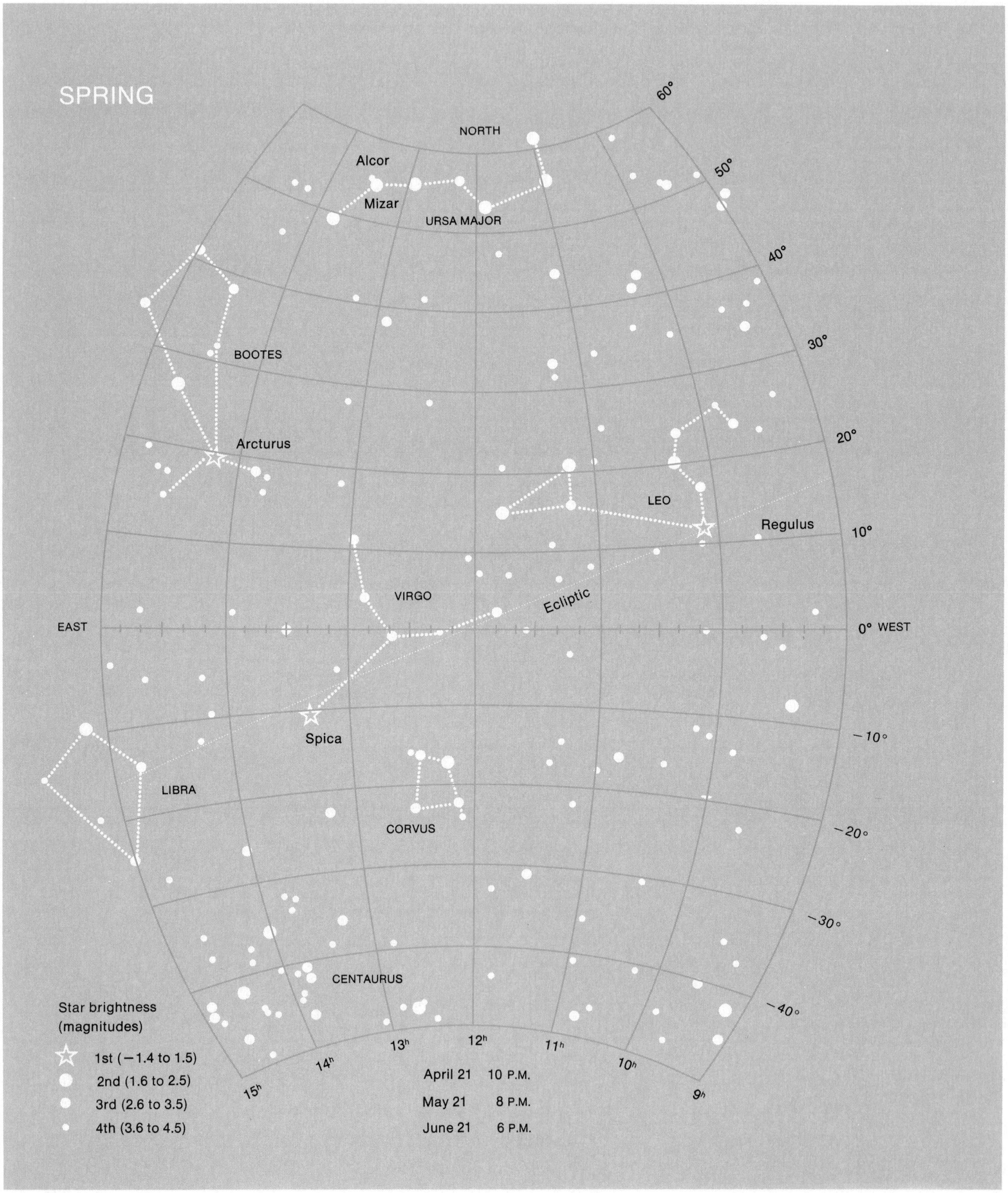
SPRING
NORTH
60°
50°
40°
30°
20°
10°
0° WEST
−10°
−20°
−30°
−40°
EAST
Alcor
Mizar
URSA MAJOR
BOOTES
Arcturus
LEO
Regulus
VIRGO
Ecliptic
Spica
LIBRA
CORVUS
CENTAURUS
15h
14h
13h
12h
11h
10h
9h
Star brightness
(magnitudes)
1st (−1.4 to 1.5)
2nd (1.6 to 2.5)
3rd (2.6 to 3.5)
4th (3.6 to 4.5)
April 21 10 P.M.
May 21 8 P.M.
June 21 6 P.M.

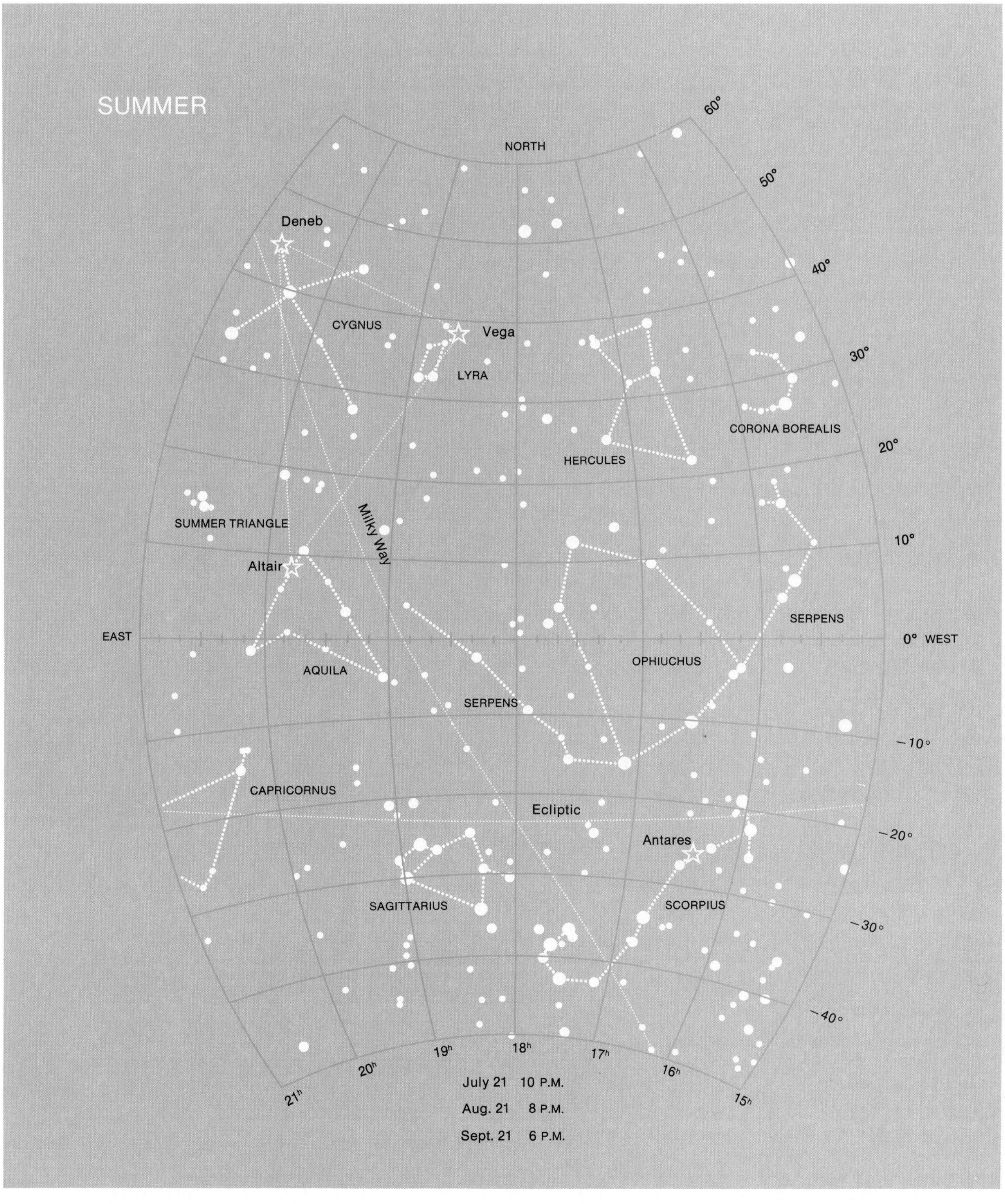
SUMMER
NORTH
60°
50°
40°
30°
20°
10°
0° WEST
EAST
−10°
−20°
−30°
−40°
Deneb
CYGNUS
Vega
LYRA
HERCULES
CORONA BOREALIS
SUMMER TRIANGLE
Milky Way
Altair
SERPENS
OPHIUCHUS
AQUILA
SERPENS
CAPRICORNUS
Ecliptic
Antares
SAGITTARIUS
SCORPIUS
21h
20h
19h
18h
17h
16h
15h
July 21 10 P.M.
Aug. 21 8 P.M.
Sept. 21 6 P.M.

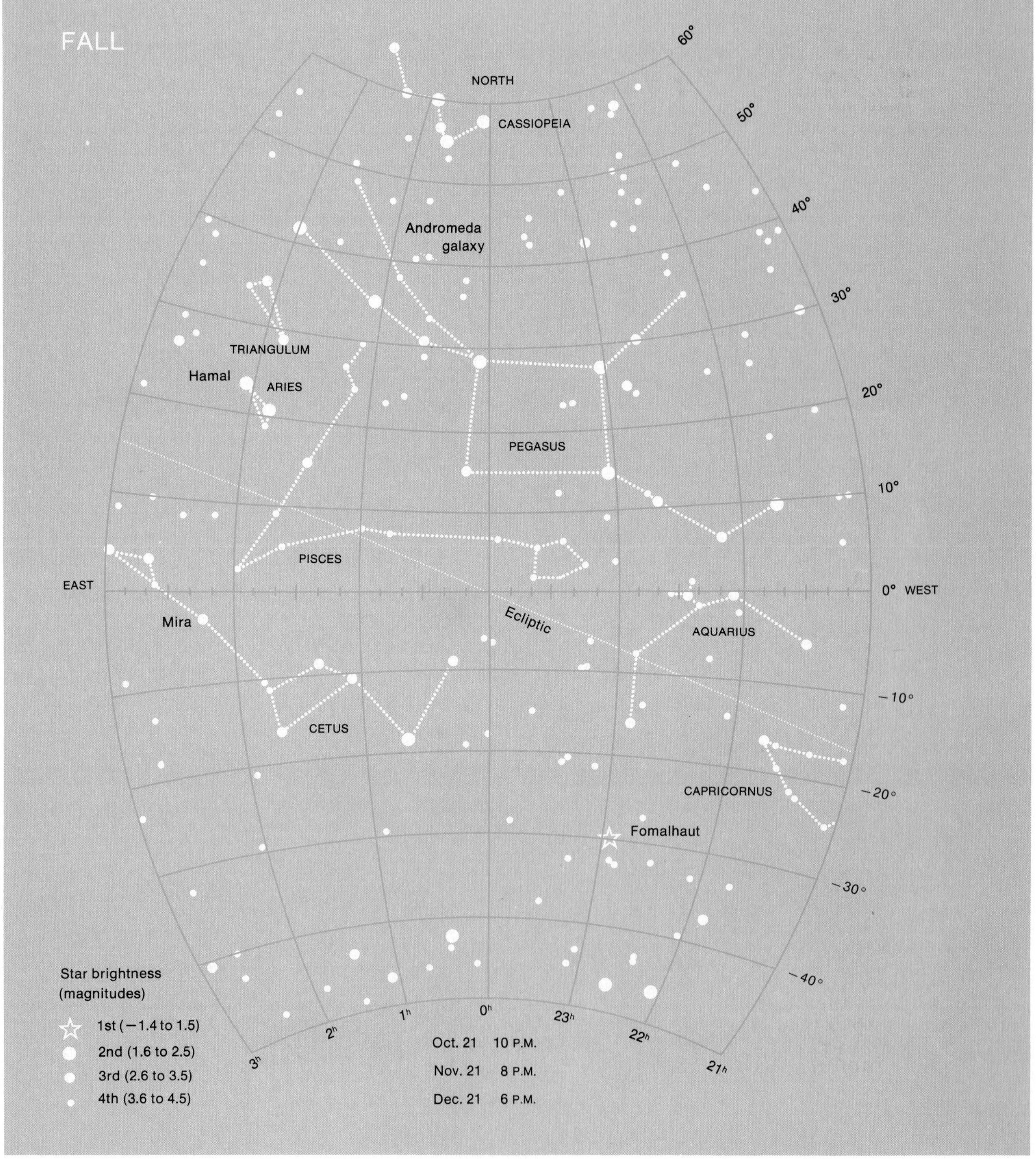
FALL
NORTH
CASSIOPEIA
Andromeda
galaxy
TRIANGULUM
Hamal
ARIES
PEGASUS
PISCES
EAST
0° WEST
Ecliptic
Mira
AQUARIUS
CETUS
CAPRICORNUS
Fomalhaut
60°
50°
40°
30°
20°
10°
−10°
−20°
−30°
−40°
3h
2h
1h
0h
23h
22h
21h
Star brightness
(magnitudes)
1st (−1.4 to 1.5)
2nd (1.6 to 2.5)
3rd (2.6 to 3.5)
4th (3.6 to 4.5)
Oct. 21 10 P.M.
Nov. 21 8 P.M.
Dec. 21 6 P.M.